Safety Symbols	Hazard	Examples	Precaution	Remedy
Disposal	Special disposal procedures need to be followed.	Certain chemicals, living organisms	Do not dispose of these materials in the sink or trash can.	Dispose of wastes as directed by your teacher.
Biological	Organisms or other biological materials that might be harmful to humans	Bacteria, fungi, blood, unpreserved tissues, plant materials	Avoid skin contact with these materials. Wear mask and gloves.	Notify your teacher if you suspect contact with material. Wash hands thoroughly.
Extreme Temperature	Objects that can burn skin by being too cold or too hot	Boiling liquids, hot plates, dry ice, liquid nitrogen	Use proper protection when handling.	Go to your teacher for first aid.
Sharp Object	Use of tools or glassware that can easily puncture or slice skin	Razor blades, pins, scalpels, pointed tools, dissecting probes, broken glass	Practice common-sense behavior and follow guidelines for use of the tool.	Go to your teacher for first aid.
Fume	Possible danger to respiratory tract from fumes	Ammonia, acetone, nail polish remover, heated sulfur, moth balls	Be sure there is good ventilation. Never smell fumes directly. Wear a mask.	Leave foul area and notify your teacher immediately.
Electrical	Possible danger from electrical shock or burn	Improper grounding, liquid spills, short circuits, exposed wires	Double-check setup with teacher. Check condition of wires and apparatus.	Do not attempt to fix electrical problems. Notify your teacher immediately.
Irritant	Substances that can irritate the skin or mucous membranes of the respiratory tract	Pollen, moth balls, steel wool, fiberglass, potassium permanganate	Wear dust mask and gloves. Practice extra care when handling these materials.	Go to your teacher for first aid.
Chemical	Chemicals that can react with and destroy tissue and other materials	Bleaches such as hydrogen peroxide; acids such as sulfuric acid, hydrochloric acid; bases such as ammonia, sodium hydroxide	Wear goggles, gloves, and an apron	Immediately flush the affected area with water and notify your teacher.
Toxic	Substance may be poisonous if touched, inhaled, or swallowed.	Mercury, many metal compounds, iodine, poinsettia plant parts	Follow your teacher's instructions.	Always wash hands thoroughly after use. Go to your teacher for first aid.
Flammable	Flammable chemicals may be ignited by open flame, spark, or exposed heat.	Alcohol, kerosene, potassium permanganate	Avoid open flames and heat when using flammable chemicals.	Notify your teacher immediately. Use fire safety equipment if applicable.
Open Flame	Open flame in use, may cause fire.	Hair, clothing, paper, synthetic materials	Tie back hair and loose clothing. Follow teacher's instruction on lighting and extinguishing flames.	Notify your teacher immediately. Use fire safety equipment if applicable.

Eye Safety
Proper eye protection should be worn at all times by anyone performing or observing science activities.

Clothing Protection
This symbol appears when substances could stain or burn clothing.

Radioactivity
This symbol appears when radioactive materials are used.

Handwashing
After the lab, wash hands with soap and water before removing goggles.

California Inspire Physics

411591852

McGraw Hill Education

Phenomenon: The Northern Lights

The Northern Lights, formally called the Aurora Borealis, are usually seen in the Artic region. While green is the most common color, red, blue, purple, yellow, and pink aurorae have also been observed. Aurorae also occur in the Antarctic region, where they are called the Southern Lights or the Aurora Australis.

Fun Fact

A comprehensive scientific understanding of the aurorae was not reached until the 20th century, but the phenomenon has been known for at least two millennia.

FRONT COVER: Frank Olsen, Norway/Moment/Getty Images
BACK COVER: Frank Olsen, Norway/Moment/Getty Images

mheducation.com/prek-12

Mc Graw Hill Education

Send all inquiries to:
McGraw-Hill Education
STEM Learning Solutions Center
8787 Orion Place
Columbus, OH 43240

ISBN: 978-0-07-674235-6
MHID: 0-07-674235-0

Printed in the United States of America.

3 4 5 6 7 8 9 LWI 23 22 21 20 19

STEM

McGraw-Hill is committed to providing instructional materials in Science, Technology, Engineering, and Mathematics (STEM) that give all students a solid foundation, one that prepares them for college and careers in the 21st century.

Welcome to

California Inspire Physics

Explore Our Phenomenal World

The Inspire High School Series brings phenomena to the forefront of learning to engage and inspire students to investigate key science concepts through their three-dimensional learning experience.

Start exploring now!

Inspire Curiosity • Inspire Investigation • Inspire Innovation

Owning Your Learning

1 Encounter the Phenomenon

Every day, you are surrounded by natural phenomena that make you wonder.

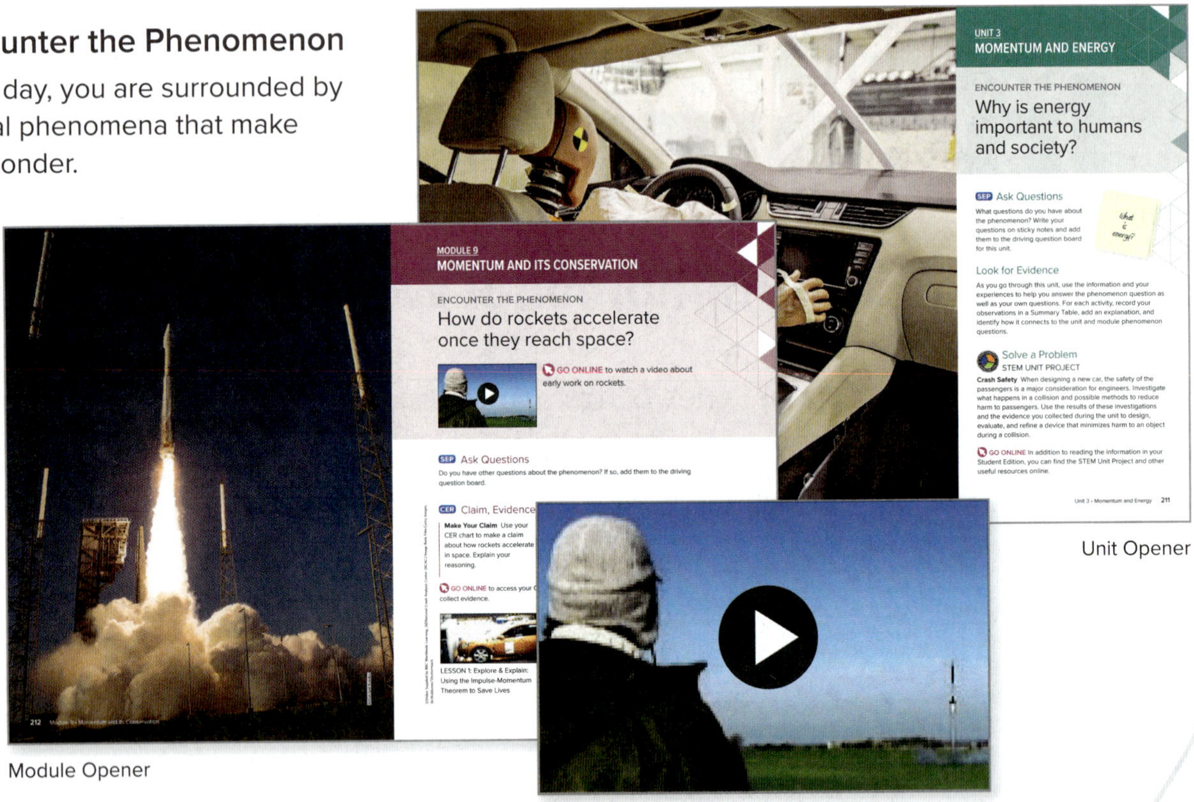

Unit Opener

Module Opener

Phenomenon Video

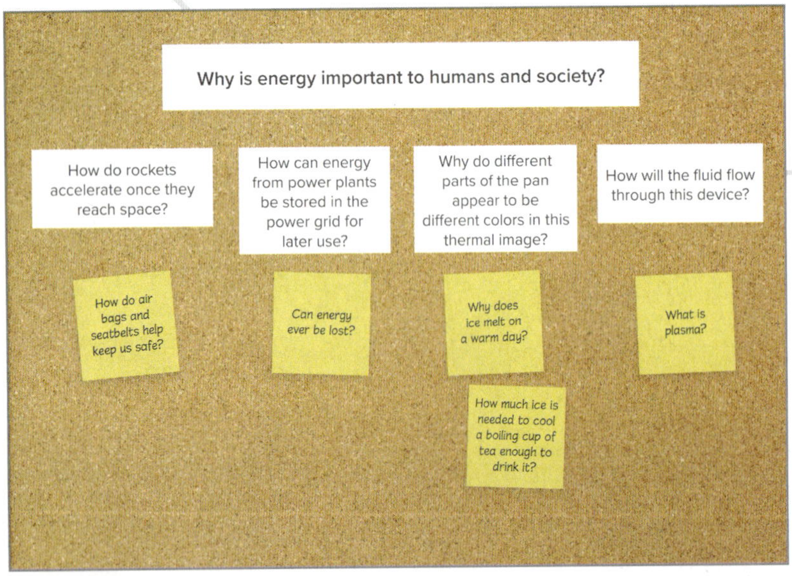

2 Ask Questions

At the beginning of each unit and module, make a list of the questions you have about the phenomenon. Talk about and share your questions with your classmates.

3 Claim, Evidence, Reasoning

As you investigate each phenomenon, you will write your claim, gather evidence by performing labs and completing reading assignments and Applying Practices, and explain your reasoning to answer the unit and module phenomena.

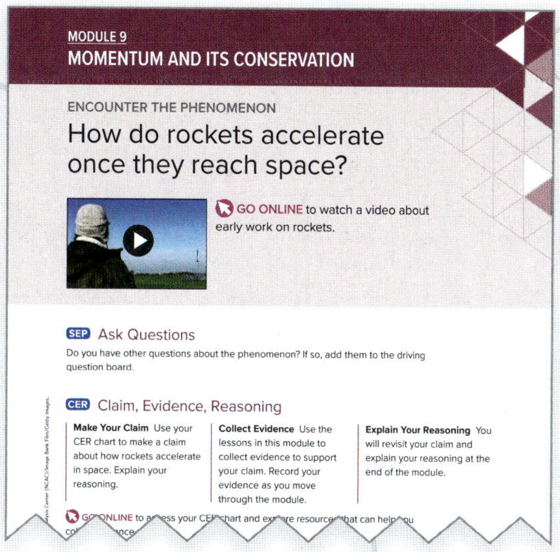
MODULE 9
MOMENTUM AND ITS CONSERVATION

ENCOUNTER THE PHENOMENON

How do rockets accelerate once they reach space?

GO ONLINE to watch a video about early work on rockets.

SEP Ask Questions

Do you have other questions about the phenomenon? If so, add them to the driving question board.

CER Claim, Evidence, Reasoning

Make Your Claim Use your CER chart to make a claim about how rockets accelerate in space. Explain your reasoning.	**Collect Evidence** Use the lessons in this module to collect evidence to support your claim. Record your evidence as you move through the module.	**Explain Your Reasoning** You will revisit your claim and explain your reasoning at the end of the module.

GO ONLINE to access your CER chart and explore resources that can help you

SUMMARY TABLE					
Activity Model	Observation Evidence	Explanation Reasoning	Connection to Phenom	Questions Answered	New Questions
Applying Practices: Coffee Cup Calorimetry	Over time, the metal cools and the water heats up slightly, until they are at the same temperature.	The calorimeter is an isolated system, so the thermal energy lost by the metal is equal to the thermal energy gained by the water.	Unit: Thermal energy and heat are part of everyday life. Module: Different materials heat at different rates because they have different specific heats.	How much ice is needed to cool a boiling cup of tea enough to drink it?	How can a calorimeter be used to measure the energy in food?

4 Summarize Your Work

When you collect evidence, you can record your data in a summary table and use the data to collaborate with others to answer the questions you had.

5 Apply Your Evidence and Reasoning

At the end of the unit, modules, and lessons, you can use all of the data you collected to help complete your STEM Unit Project.

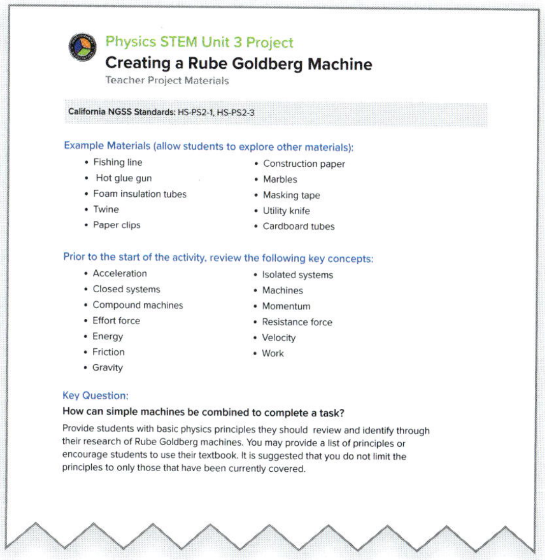
Physics STEM Unit 3 Project
Creating a Rube Goldberg Machine
Teacher Project Materials

California NGSS Standards: HS-PS2-1, HS-PS2-3

Example Materials (allow students to explore other materials):
- Fishing line
- Hot glue gun
- Foam insulation tubes
- Twine
- Paper clips
- Construction paper
- Marbles
- Masking tape
- Utility knife
- Cardboard tubes

Prior to the start of the activity, review the following key concepts:
- Acceleration
- Closed systems
- Compound machines
- Effort force
- Energy
- Friction
- Gravity
- Isolated systems
- Machines
- Momentum
- Resistance force
- Velocity
- Work

Key Question:
How can simple machines be combined to complete a task?
Provide students with basic physics principles they should review and identify through their research of Rube Goldberg machines. You may provide a list of principles or encourage students to use their textbook. It is suggested that you do not limit the principles to only those that have been currently covered.

ADVISORS AND CONSULTANTS

California Advisory Board

Each advisor provided valuable feedback and suggestions regarding effective science instruction. We thank them for their help.

Mercedes Alou-Hicks
Palm Desert High School
Desert Sands Unified School District
Palm Desert, CA

Lance Atkinson
Eisenhower High School
Rialto Unified School District
Rialto, CA

James Broderick
Highland High School
Antelope Valley Union High School District
Lancaster, CA

Jackie Johansen
Julian Charter School
Julian Union School District
Juliana, CA

Marc Jones
Heritage High School
Perris Union High School District
Menifee, CA

Shaun Ho
Monrovia High School
Monrovia Unified School District
Monrovia, CA

Raymond Pietersen
Elk Grove Unified School District
Elk Grove, CA

Gregory Tyler
Palm Desert High School
Desert Sands Unified School District
Palm Desert, CA

Joyce Wilkinson
Salinas Union High School District
Salinas, CA

Content Consultants

Content consultants each reviewed selected chapters of ***California Inspire Physics*** for content accuracy and clarity.

Dr. Solomon Bililign, PhD
Professor of Physics
North Carolina A&T State University
Greensboro, NC

Ruth Howes
Professor Emerita of Physics and Astronomy
Ball State University
Muncie, IN

Dr. Keith H. Jackson
Professor and Chair of Department of Physics
Morgan State University
Baltimore, MD

Dr. Kathleen Johnston, PhD
Professor of Physics
Louisiana Tech University
Ruston, LA

Dr. Monika Kress
Associate Professor
Department of Physics & Astronomy
San Jose State University
San Jose, CA

Jorge Lopez
Professor
University of Texas at El Paso
El Paso, TX

Dr. Ramon E. Lopez
Department of Physics
University of Texas at Arlington
Arlington, TX

Albert J. Osei
Professor of Physics
Oakwood University
Huntsville, AL

Toni Sauncy
Department Chair of Physics
Texas Lutheran University
Seguin, TX

Sally Seidel
Professor of Physics
University of New Mexico
Albuquerque, NM

Contributing Writers

Additional science writers added feature content, teacher materials, assessment, and laboratory investigations.

Molly Wetterschneider
Austin, TX

Steve Whitt
Columbus, OH

Safety Consultant

The safety consultant reviewed labs and lab materials for safety and implementation.

Kenneth Russell Roy, PhD
Director of Science and Safety
Glastonbury Public Schools
Glastonbury, CT

Authors

Paul W. Zitzewitz, PhD
Lead Author, *In Memorium*
Formerly Professor Emeritus of Physics and of Science Education
The University of Michigan-Dearborn
Dearborn, MI

David G. Haase, PhD
Professor Emeritus of Physics
North Carolina State University
Raleigh, NC

Kathleen A. Harper, PhD
Senior Lecturer, Engineering Education Innovation Center
The Ohio State University
Columbus, OH

KEY PARTNERS

Smithsonian

Smithsonian

Following the mission of its founder James Smithson for "an establishment for the increase and diffusion of knowledge," the Smithsonian Institution today is the world's largest museum, education, and research complex. To further their vision of shaping the future, a wealth of Smithsonian online resources are integrated within this program.

SPONGELAB
A GLOBAL STEM COMMUNITY
V5.4

SpongeLab Interactives

SpongeLab Interactives is a learning technology company that inspires learning and engagement by creating gamified environments that encourage students to interact with digital learning experiences.

Students participate in inquiry activities and problem-solving to explore a variety of topics using games, interactives, and video while teachers take advantage of formative, summative, or performance-based assessment information that is gathered through the learning management system.

PhET
INTERACTIVE SIMULATIONS
University of Colorado Boulder

PhET Interactive Simulations

The PhET Interactive Simulations project at the University of Colorado Boulder provides teachers and students with interactive science and math simulations. Based on extensive education research, PhET sims engage students through an intuitive, game-like environment where students learn through exploration and discovery.

UNIT 1

MECHANICS IN ONE DIMENSION

ENCOUNTER THE PHENOMENON

How can we model motion and forces?

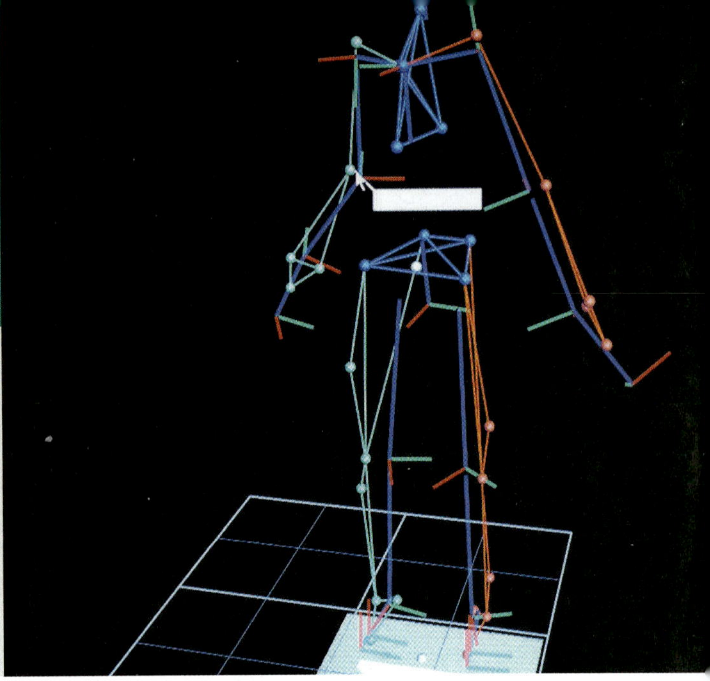

Science Photo Library/Alamy Stock Photo

INTRODUCTION TO PHYSICS

This module introduces the nature of science, what Physics is, and provides tools for the study of Physics.

MODULE 1: A PHYSICS TOOLKIT

MODULE 2: REPRESENTING MOTION

MODULE 3: ACCELERATED MOTION

MODULE 4: FORCES IN ONE DIMENSION

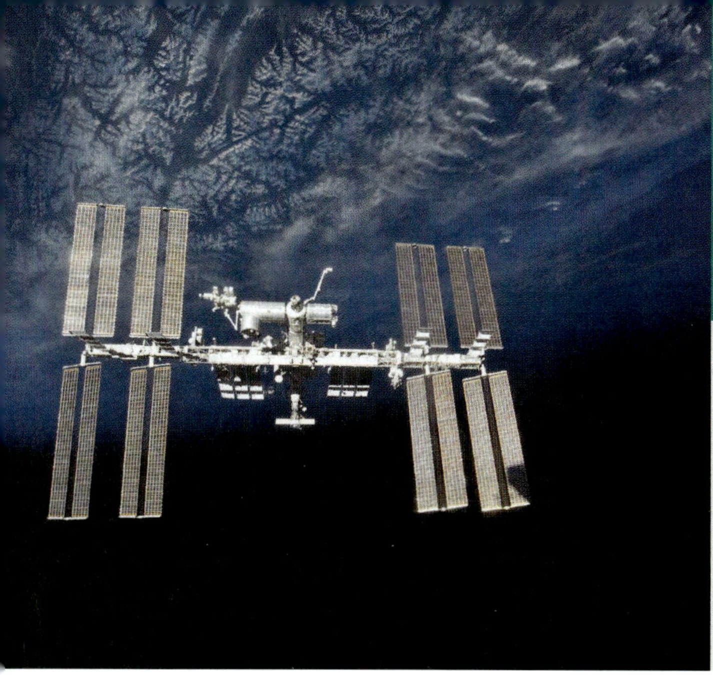

UNIT 2

MECHANICS IN TWO DIMENSIONS

ENCOUNTER THE PHENOMENON

How can forces cause so many different types of motion?

NASA

UNIT 3

MOMENTUM AND ENERGY

ENCOUNTER THE PHENOMENON

Why is energy important to humans and society?

MODULE 9: MOMENTUM AND ITS CONSERVATION

MODULE 10: ENERGY AND ITS CONSERVATION

MODULE 11: THERMAL ENERGY

MODULE 12: STATES OF MATTER

Martin Sterba/ZUMA Press/Newscom

Andrey VP/Shutterstock

UNIT 4

WAVES AND LIGHT

ENCOUNTER THE PHENOMENON

How do waves affect our everyday lives?

UNIT 5

ELECTRICITY AND MAGNETISM

ENCOUNTER THE PHENOMENON

What role do electricity and magnetism play in the technology we use every day?

MODULE 18: ELECTROSTATICS

MODULE 19: ELECTRIC CURRENT AND CIRCUITS

MODULE 20: MAGNETISM

MODULE 21: ELECTROMAGNETISM

UNIT 6

SUBATOMIC PHYSICS

ENCOUNTER THE PHENOMENON

What is the universe made of?

MODULE 22: QUANTUM THEORY AND THE ATOM

ENCOUNTER THE PHENOMENON

MODULE 23: SOLID-STATE ELECTRONICS

ENCOUNTER THE PHENOMENON

MODULE 24: NUCLEAR AND PARTICLE PHYSICS

ENCOUNTER THE PHENOMENON

FatCamera/E+/Getty Images

MODULE 1

A PHYSICS TOOLKIT

ENCOUNTER THE PHENOMENON

What tools and skills do physicists use?

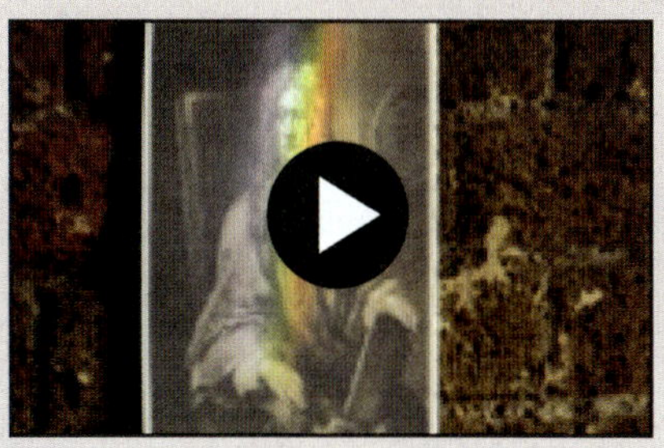

GO ONLINE to watch a video about the way Newton studied light.

SEP Ask Questions

Do you have other questions about the phenomenon? If so, add them to the driving question board.

CER Claim, Evidence, Reasoning

Make Your Claim Use your CER chart to make a claim the tools and skills physicists use. Explain your reasoning.

Collect Evidence Use the lessons in this module to collect evidence to support your claim. Record your evidence as you move through the module.

Explain Your Reasoning You will revisit your claim and explain your reasoning at the end of the module.

GO ONLINE to access your CER chart and explore resources that can help you collect evidence.

LESSON 1: Explore & Explain: What is physics?

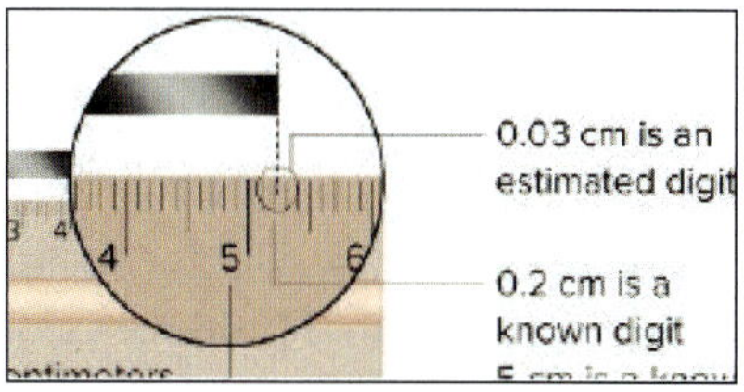

LESSON 2: Explore & Explain: Uncertainty in Data: Significant Figures

Smithsonian

Additional Resources

(t)Video Supplied by BBC Worldwide Learning, (b)Dusit/Shutterstock

LESSON 1
METHODS OF SCIENCE

FOCUS QUESTION

What do physicists do?

What is physics?

Science is not just a subject in school. It is a process based on inquiry that helps develop explanations about events in nature. **Physics** is a branch of science that involves the study of the physical world: energy, matter, and how they are related.

When you see the word *physics* you might picture a chalkboard full of formulas and mathematics: $E = mc^2$, $I = \frac{V}{R}$, $x = \left(\frac{1}{2}\right)at^2 + v_0t + x_0$. Maybe you picture scientists in white lab coats or well-known figures such as Marie Curie and Albert Einstein. Alternatively, you might think of the many modern technologies created with physics, such as weather satellites, laptop computers, and lasers.

Physicists investigate the motions of electrons and rockets, the energy in sound waves and electric circuits, and the structure of the proton and of the universe. The goal of this course is to help you better understand the physical world.

People who study physics go on to many different careers. Some become scientists at universities and colleges, at industries, or in research institutes. Others go into related fields, such as engineering, computer science, teaching, medicine, or astronomy, as shown in **Figure 1.** Still others use the problem-solving skills of physics to work in finance, construction, or other very different disciplines. In the last 50 years, research in the field of physics has led to many new technologies, including satellite-based communications and high-speed microscanners used to detect disease.

Figure 1 Physicists may choose from a variety of careers.

inga spence/Alamy Stock Photo

3D THINKING **DCI** Disciplinary Core Ideas **CCC** Crosscutting Concepts **SEP** Science & Engineering Practices

COLLECT EVIDENCE

Use your Science Journal to record the evidence you collect as you complete the readings and activities in this lesson.

INVESTIGATE

GO ONLINE to find these activities and more resources.

Quick Investigation: Measuring Change
Analyze data to find **patterns** that can be used for extrapolation.

Review the News
Obtain information from a current news story about current physics research. **Evaluate** your source and **communicate** your findings to your class.

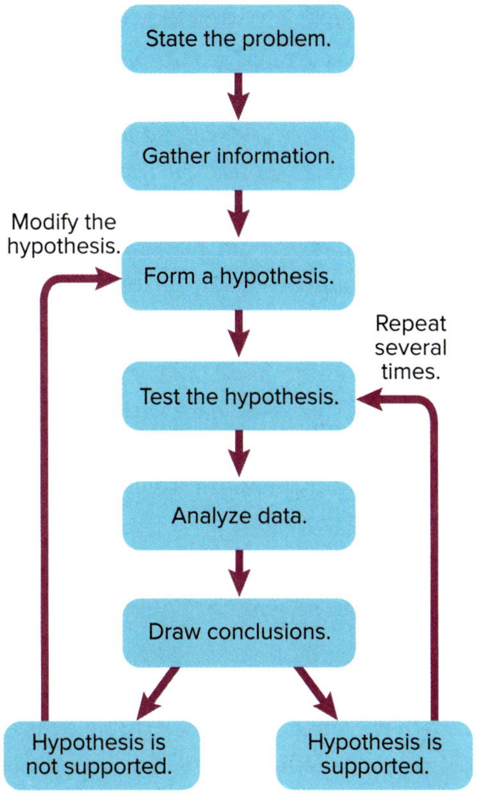

Figure 2 The series of procedures shown here is one way to use scientific methods to solve a problem.

Scientific Methods

Although physicists do not always follow a rigid set of steps, investigations often follow similar patterns called **scientific methods,** as shown in **Figure 2.** Depending on the particular investigation, a scientist might add new steps, repeat some steps, or skip steps altogether.

State the problem Investigations can begin when one observes an event in nature and wonders why or how it occurs. The question of "why" or "how" is the problem to be stated.

Sometimes a new question is posed during an investigation, leading to a new statement of a problem. In the 1940s, researcher Percy Spencer was trying to answer the question of how to mass-produce the magnetron tubes used in radar systems. When he stood in front of an operating magnetron, which produces microwaves, a candy bar in his pocket melted. The new question of how the magnetron was cooking food was then asked.

Research and gather information Before beginning an investigation, it is useful to research what is already known about the problem. Making and examining observations and interpretations from reliable sources helps fine-tune the question and form it into a hypothesis.

WORD ORIGINS

Science
comes from the Latin word *scientia*, which means *knowledge*

STEM CAREER Connection

Survey or Mapping Technician
Do you like being outside? Survey technicians spend most of their time outside in various weather conditions collecting geographic data such as elevation and contour. Mapping technicians use the data and sophisticated computer software to make maps of Earth's surface.

Form and test a hypothesis

A **hypothesis** is a possible explanation for a problem using what you know and have observed. A scientific hypothesis can be tested through experimentation and observation. Sometimes scientists must wait for new technologies before a hypothesis can be tested. For example, the first hypotheses about the existence of atoms were developed more than 2300 years ago, but the technologies to test these hypotheses were not available for many centuries.

Some hypotheses can be tested by making observations. Others can be tested by building a model and relating it to real-life situations. One common way to test a hypothesis is to perform an experiment. An experiment tests the effect of one thing on another, using a control. Sometimes it is not possible to perform experiments; in these cases, investigations become descriptive in nature. For example, physicists cannot conduct experiments in deep space. They can, however, collect and analyze valuable data to help us learn more about events occurring there.

Analyze the data

An important part of every investigation includes recording observations and organizing data into easy-to-read tables and graphs. Later in this module, you will study ways to display data. When you are making and recording observations, you should include all your results, even unexpected ones. Many important discoveries have been made from unexpected results. Scientific inferences are based on scientific observations. All possible scientific explanations must be considered. If the data are not organized in a logical manner, incorrect conclusions can be drawn.

When a scientist communicates and shares data, other scientists will examine those data and how the data were analyzed, and compare the data to the work of others. Scientists, such as the planetary scientist in **Figure 3,** share their data and analyses through reports, journals, and conferences.

Draw conclusions

Based on the analysis of the data, the next step is to decide whether the hypothesis is supported. For the hypothesis to be considered valid and widely accepted, the results of the experiment must be the same every time it is repeated. If the experiment does not support the hypothesis, the hypothesis must be reconsidered. Perhaps the hypothesis needs to be revised, or maybe the experimenter's procedure needs to be refined.

Figure 3 An important part of scientific methods is to share data and results with other scientists. This scientist is sharing predicted planetary orbital data with colleagues.

FREDERIC J BROWN/AFP/Getty Images

Peer review Before it is made public, science-based information is reviewed by scientists' peers—scientists who are in the same field of study. Peer review is a process by which the procedures and results of an experiment are evaluated by peer scientists of those who conducted the research. Reviewing other scientists' work is a responsibility that many scientists have.

Being objective One also should be careful to reduce bias in scientific investigations. Bias can occur when the scientist's expectations affect how the results are analyzed or the conclusions are made. This might cause a scientist to select a result from one trial over those from other trials. Bias might also be found if the advantages of a product being tested are used in a promotion and the drawbacks are not presented. Scientists can lessen bias by running as many trials as possible and by keeping accurate notes of each observation made.

Models

Sometimes, scientists cannot see everything they are testing. They might be observing an object that is too large or too small, a process that takes too much time to see completely, or a material that is hazardous. In these cases, scientists use models. A **model** is a representation of an idea, event, structure, or object that helps people better understand it.

Models in history Models have been used throughout history. In the early 1900s, British physicist J.J. Thomson created a model of the atom that consisted of electrons embedded in a ball of positive charge. Several years later, physicist Ernest Rutherford created a model of the atom based on new research. Later in the twentieth century, scientists discovered that the nucleus is not a solid ball but is made of protons and neutrons. The present-day model of the atom is a nucleus made of protons and neutrons surrounded by an electron cloud. All three of these models are shown in **Figure 4.** Scientists use models of atoms to represent their current understanding because of the small size of an atom.

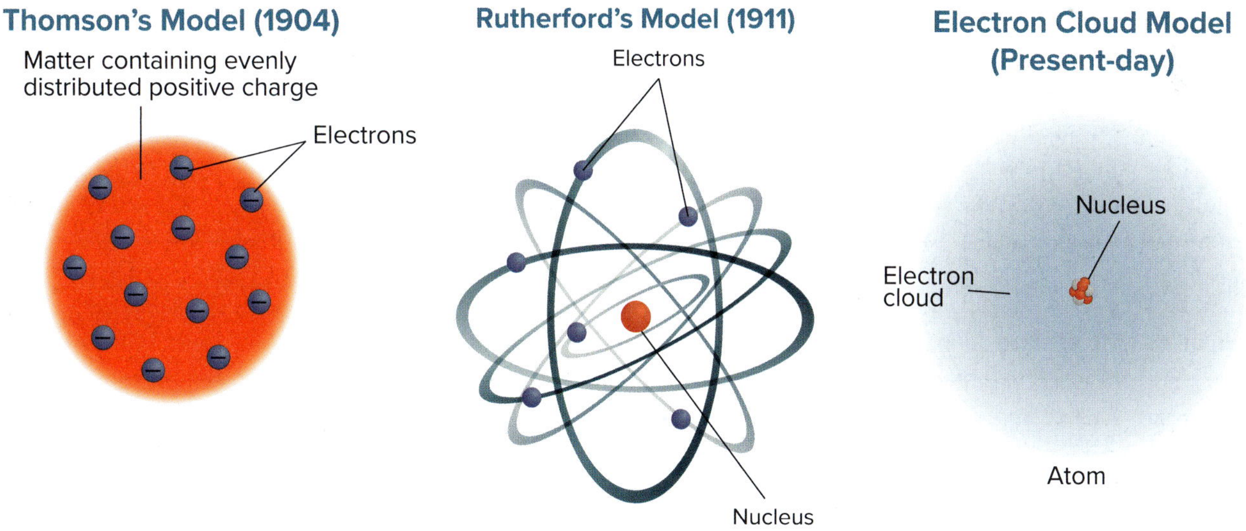

Figure 4 Throughout history, scientists have made models of the atom.
Infer *Why have models of the atom changed over the years?*

Get It?
Explain how the model shown at the beginning of this module is helpful.

High-tech models Both physical models and computers can be used in various ways to aid in the engineering design process. Computers are useful for a variety of purposes, such as running simulations to test different ways of solving a problem or to see which one is most efficient or economical; and in making a persuasive presentation to a client about how a given design will meet his or her needs. Computer software is designed to mimic the processes under study.

For instance, computer simulators, such as the one shown in **Figure 5,** help airplane pilots practice all aspects of flight without ever leaving the ground. In addition, computer simulations can simulate harsh weather conditions and other potentially dangerous in-flight challenges.

Figure 5 This is a flight simulator used to help train pilots. The image mimics what the pilot would see if flying a real plane.

Identify *other models around your classroom.*

Get It?

Discuss how computer simulations can help develop possible solutions to a problem.

Scientific Theories and Laws

A **scientific theory** is an explanation of things or events based on knowledge gained from many observations and investigations.

A theory is not a guess. If scientists repeat an investigation and the results always support the hypothesis, the hypothesis can be called a theory. Just because a scientific theory has data supporting it does not mean it will never change. As new information becomes available, theories can be refined or modified, as shown in **Figure 6** on the next page.

A **scientific law** is a statement about what happens in nature and seems to be true all the time. Laws tell you what will happen under certain conditions, but they don't explain why or how something happens. Gravity is an example of a scientific law. The law of gravity states that any one mass will attract another mass. To date, no experiments have been performed that disprove the law of gravity.

A theory can be used to explain a law, but theories do not become laws. For example, many theories have been proposed to explain how the law of gravity works. Even so, there are few accepted theories in science and even fewer laws.

CCC CROSSCUTTING CONCEPTS

Systems and System Models Models can be used to simulate systems. Choose a model not mentioned in this lesson. Remember that a model isn't necessarily a physical model, something you can build and touch, although it could be. Prepare a poster that shows how your model can help test a process or a procedure. What type of model will you use? What evidence supports your explanation?

Greek philosophers proposed that objects fall because they seek their natural places. The more massive the object, the faster it falls.

↓

Revision

Galileo showed that the speed at which an object falls depends on the amount of time for which that object has fallen and not on the object's mass.

↓

Revision

Newton provided an explanation for why objects fall. Newton proposed that objects fall because the object and Earth are attracted by a force. Newton also stated that there is a force of attraction between any two objects with mass.

↓

Revision

Einstein suggested that the force of attraction between two objects is due to mass causing the space around it to curve.

Figure 6 If experiments provide new insight and evidence about a theory, such as the theory describing the behavior of falling objects, the theory is modified accordingly.

The Limitations of Science

Science can help you explain many things about the world, but science cannot explain or solve everything. Scientists make guesses, but the guesses must be tested and verified. Questions about opinions, values, and emotions are not scientific because they cannot be tested. For example, some people may find a particular piece of art beautiful while others do not. Some people might think that certain foods, such as pizza, taste delicious while others do not. You might take a survey to gather opinions about such questions, but that would not prove that the opinions are true for everyone.

Check Your Progress

1. **Summarize** the steps you might use to carry out an investigation using scientific methods.
2. **Define** the term *hypothesis*. Identify three ways to test a hypothesis.
3. **Describe** why it is important for scientists to avoid bias.
4. **Explain** why scientists use models. Give an example of a scientific model not mentioned in this lesson, and explain how it is useful.
5. **Analyze** Your friend finds that 90 percent of students surveyed in the cafeteria like pizza. She says this scientifically proves that everyone likes pizza. How would you respond?
6. **Critical Thinking** An accepted value for free-fall acceleration is 9.8 m/s^2. In an experiment with pendulums, you calculate a value to be 9.4 m/s^2. Should the accepted value be tossed out because of your finding? Explain.

LEARNSMART® Go online to follow your personalized learning path to review, practice, and reinforce your understanding.

LESSON 2

MATHEMATICS AND PHYSICS

FOCUS QUESTION

How is math helpful to physicists?

Mathematics in Physics

Physicists often use the language of mathematics. In physics, equations are important tools for modeling observations and for making predictions. Equations are one way of representing relationships between measurements. Physicists rely on theories and experiments with numerical results to support their conclusions. For example, you can predict that if you drop a penny, it will fall, but can you predict how fast it will be going when it strikes the ground? Different models of falling objects give different answers to how the speed of the object changes as it falls or on what the speed depends. By measuring how an object falls, you can compare the experimental data with the results predicted by different models. This tests the models, allowing you to pick the best one or to develop a new model.

SI Units

To communicate results, it is helpful to use units that everyone understands. The worldwide scientific community uses an adaptation of the metric system for measurements. **Table 1** shows that the Système International d'Unités, or SI, uses seven base quantities. Other units, called derived units, are formed by combining the base units in various ways. Velocity is measured in meters per second (m/s). Often, derived units are given their own names. For example, electric charge is measured in ampere-seconds (A·s), which are also called coulombs (C).

SI is regulated by the International Bureau of Weights and Measures in Sèvres, France. In the past, this bureau and the National Institute of Science and Technology (NIST) in Gaithersburg, Maryland, kept physical samples on which the standards were based.

Table 1 SI Base Units

Base Quantity	Base Unit	Symbol
Length	meter	m
Mass	kilogram	kg
Time	second	s
Temperature	kelvin	K
Amount of a substance	mole	mol
Electric current	ampere	A
Luminous intensity	candela	cd

3D THINKING DCI Disciplinary Core Ideas CCC Crosscutting Concepts SEP Science & Engineering Practices

COLLECT EVIDENCE

Use your Science Journal to record the evidence you collect as you complete the readings and activities in this lesson.

INVESTIGATE

GO ONLINE to find these activities and more resources.

CCC **Identify Crosscutting Concepts**

Create a table of the crosscutting concepts and fill in examples you find as you read.

? **Revisit the Encounter the Phenomenon Question**

What information from this lesson can help you answer the Unit and Module questions?

However, the standards are being redefined in terms of phenomena. For example, the meter, as shown in **Figure 7,** is now defined as $\frac{1}{299,792,458}$, the distance that light travels in $\frac{1}{299,792,458}$ of a second.

The ease of switching between units is another convenient feature of SI. To convert between units, multiply or divide by the appropriate power of 10. Prefixes are used to change SI base units by powers of 10, as shown in **Table 2.** You often will encounter these prefixes in daily life, as in, for example, milligrams and centimeters.

Figure 7 You will use metersticks as your standard of measurement, but the official definition of a meter is $\frac{1}{299,792,458}$, the distance that light travels in $\frac{1}{299,792,458}$ of a second.

Describe *Why is it important to have standards for measurements?*

Dimensional Analysis

You often will need to manipulate a formula, or use a string of formulas, to solve a physics problem. One way to check whether you have set up a problem correctly is to write out the equation or set of equations you plan to use. Before doing calculations, check that the answer will be in the expected units. For example, if you are finding a car's speed and you see that your answer will have the units s/m or m/s^2, you have made an error in setting up the problem. This method of treating the units as algebraic quantities that can be canceled is called **dimensional analysis.** Knowing that your answer will be in the correct units is not a guarantee that your answer is right, but if you find that your answer has or will have the wrong units, you can be sure that you have made an error. Dimensional analysis also is used in choosing conversion factors. A conversion factor is a multiplier equal to 1.

Table 2 **Prefixes Used with SI Units**

Prefix	Symbol	Multiplier	Scientific Notation	Example
femto–	f	0.000000000000001	10^{-15}	femtosecond (fs)
pico–	p	0.000000000001	10^{-12}	picometer (pm)
nano–	n	0.000000001	10^{-9}	nanometer (nm)
micro–	μ	0.000001	10^{-6}	microgram (μg)
milli–	m	0.001	10^{-3}	milliamps (mA)
centi–	c	0.01	10^{-2}	centimeter (cm)
deci–	d	0.1	10^{-1}	deciliter (dL)
kilo–	k	1000	10^{3}	kilometer (km)
mega–	M	1,000,000	10^{6}	megagram (Mg)
giga–	G	1,000,000,000	10^{9}	gigameter (Gm)
tera–	T	1,000,000,000,000	10^{12}	terahertz (THz)

Get It?

Identify the prefix that would be used to express 2,000,000,000 bytes of computer memory.

For example, because 1 kg = 1000 g, you can construct the following conversion factors:

$$1 = \frac{1\,\text{kg}}{1000\,\text{g}} \quad \text{and} \quad 1 = \frac{1000\,\text{g}}{1\,\text{kg}}$$

Choose a conversion factor that will make the initial units cancel, leaving the answer in the desired units. For example, to convert a mass of 1.34 kg to grams, set up the conversion as shown below.

$$1.34\,\cancel{\text{kg}}\left(\frac{1000\,\text{g}}{1\,\cancel{\text{kg}}}\right) = 1340\,\text{g}$$

You also might need to do a series of conversions. To convert 43 km/h to m/s, do the following:

$$\left(\frac{43\,\cancel{\text{km}}}{1\,\cancel{\text{h}}}\right)\left(\frac{1000\,\text{m}}{1\,\cancel{\text{km}}}\right)\left(\frac{1\,\cancel{\text{h}}}{60\,\cancel{\text{min}}}\right)\left(\frac{1\,\cancel{\text{min}}}{60\,\text{s}}\right) = 12\,\text{m/s}$$

Significant Figures

Suppose you measure a pen and find that the end of the pen is just past 138 mm, as shown in **Figure 8.** You estimate that the pen is one-tenth of a millimeter past the last tick mark on the ruler and record the pen as being 138.1 mm long. This measurement has four valid digits: the first three digits are certain, and the last one is uncertain. The valid digits in a measurement are called **significant figures.** The last digit given for any measurement is the uncertain digit. All nonzero digits in a measurement are significant.

Are all zeros significant?

No. For example, in the measurement 0.0860 m, the first two zeros serve only to locate the decimal point and are not significant. The last zero, however, is the estimated digit and is significant. The measurement 172,000 m could have 3, 4, 5, or 6 significant figures. This ambiguity is one reason to use scientific notation. It is clear that the measurement 1.7200×10^5 m has five significant figures.

Arithmetic with significant figures

When you do any arithmetic operation, remember that the result never can be more precise than the least-precise measurement.

To add or subtract measurements, first do the operation, then round off the result to correspond to the least-precise value involved. For example, 3.86 m + 2.4 m = 6.3 m, not 6.26 m, because the least-precise measure is to one-tenth of a meter.

To multiply or divide measurements, first do the calculation, and then round to the same number of significant figures as the least-precise measurement. For example, $\frac{409.2\,\text{km}}{11.4\,\text{L}} = \frac{35.9\,\text{km}}{\text{L}}$, because the least-precise measurement has three significant figures. Some calculators display several additional digits, while others round at different points. Be sure to record your answers with the correct number of digits.

Solving Problems

Most practice problems in this course will be complex and require a strategy to solve. This textbook includes many example problems, each of which is solved using a three-step process. Example Problem 1 on the next page follows the steps to calculate a car's average speed using distance and time.

Figure 8 You recorded the length of this pen as 138.1 mm.
Infer *Why is the last digit uncertain?*

Richard Hutchings/Digital Light Source

EXAMPLE Problem 1

USING DISTANCE AND TIME TO FIND SPEED When a car travels 434 km in 4.5 h, what is the car's average speed?

1 ANALYZE AND SKETCH THE PROBLEM

The car's speed is unknown. The known values include the distance the car traveled and the time. Use the relationship among speed, distance, and time to solve for the car's speed.

Known	**Unknown**
distance = 434 km	speed = ?
time = 4.5 h	

2 SOLVE FOR THE UNKNOWN

distance = **speed** × time	State the relationship as an equation.
speed = $\frac{\text{distance}}{\text{time}}$	Solve the equation for speed.
speed = $\frac{434\text{ km}}{4.5\text{ h}}$	Substitute distance = 434 km and time = 4.5 h.
speed = 96.4 km/h	Calculate, and specify the units.

3 EVALUATE THE ANSWER

Check your answer by using it to calculate the distance the car traveled.

distance = speed × time = 96.4 km/~~h~~ × 4.5 ~~h~~ = 434 km

The calculated distance matches the distance stated in the problem. This means that the calculated average speed is correct.

THE PROBLEM

1. Read the problem carefully.
2. Be sure you understand what is being asked.

ANALYZE AND SKETCH THE PROBLEM

1. Read the problem again.
2. Identify what you are given, and list the known data. If needed, gather information from graphs, tables, or figures.
3. Identify and list the unknowns.
4. Determine whether you need a sketch to help solve the problem.
5. Plan the steps you will follow to find the answer.

SOLVE FOR THE UNKOWN

1. If the solution is mathematical, write the equation and isolate the unknown factor.
2. Substitute the known quantities into the equation.
3. Solve the equation.
4. Continue the solution process until you solve the problem.

EVALUATE THE ANSWER

1. Reread the problem. Is the answer reasonable?
2. Check your math. Are the units and significant figures correct?

Check Your Progress

7. **Modeling** Why are concepts in physics described with formulas?
8. **Significant Figures** Solve the following problems, using the correct number of significant figures each time.
   a. 10.8 g − 8.264 g
   b. 4.75 m − 0.4168 m
   c. 139 cm × 2.3 cm
   d. 13.78 g / 11.3 mL
   e. 1.6 km + 1.62 m + 1200 cm
9. **Dimensional Analysis** How many seconds are in a leap year?
10. **Solving Problems** Rewrite $F = Bqv$ to find v in terms of F, q, and B.
11. **Critical Thinking** Using values given in a problem and the equation for distance, distance = speed × time, you calculate a car's speed to be 290 km/h. Is this answer reasonable? Explain. Under what circumstances might this be a reasonable answer?

LEARNSMART® Go online to follow your personalized learning path to review, practice, and reinforce your understanding.

LESSON 3
MEASUREMENT

FOCUS QUESTION

Why is it important to make careful measurments?

What is measurement?

A **measurement** is a comparison between an unknown quantity and a standard. For example, if you measure the mass of a rolling cart used in an experiment, the unknown quantity is the mass of the cart and the standard is the gram, as defined by the balance or spring scale you use. If you do an experiment with a spring for which the length is unknown, the centimeter is the standard you might use for length. Measurements quantify our observations. For example, a person's blood pressure isn't just "pretty good"; it's $\frac{110}{60}$, the low end of the good range.

Comparing Results

As you learned in Lesson 1, scientists share their results. Before new data are fully accepted, other scientists examine the experiment, look for possible sources of error, and try to reproduce the results. Results often are reported with an uncertainty. A new measurement that is within the margin of uncertainty is in agreement with the old measurement.

For example, archaeologists use radiocarbon dating to determine the age of cave paintings, such as those from the Niaux cave in France, in **Figure 9**, and the Chauvet cave, also in France. Each radiocarbon date is reported with an uncertainty. Three radiocarbon ages from a panel in the Chauvet cave are 30,940 ± 610 years, 30,790 ± 600 years, and 30,230 ± 530 years. While none of the measurements matches, the uncertainties in all three overlap, and the measurements agree with each other.

Figure 9 These drawings are from the Niaux cave in France. Scientists estimate that the drawings were made about 17,000 years ago.

3D THINKING DCI Disciplinary Core Ideas CCC Crosscutting Concepts SEP Science & Engineering Practices

COLLECT EVIDENCE

Use your Science Journal to record the evidence you collect as you complete the readings and activities in this lesson.

INVESTIGATE

GO ONLINE to find these activities and more resources.

PhysicsLAB: Mass and Volume

Carry out an investigation to determine the relationship between mass, volume, and density.

Revisit the Encounter the Phenomenon Question

What information from this lesson can help you answer the Unit and Module questions?

Suppose three students performed an experiment several times starting with springs of the same length. With two washers on the spring, student 1 made repeated measurements, which ranged from 14.4 cm to 14.8 cm. The average of student 1's measurements was 14.6 cm, as shown in **Figure 10.** This result was reported as (14.6 ± 0.2) cm. Student 2 reported finding the spring's length to be (14.8 ± 0.3) cm. Student 3 reported a length of (14.0 ± 0.1) cm.

Could you conclude that the three measurements are in agreement? Is student 1's result reproducible? The ranges of the results of students 1 and 2 overlap between 14.5 cm and 14.8 cm. However, there is no overlap and, therefore, no agreement, between their results and the result of student 3.

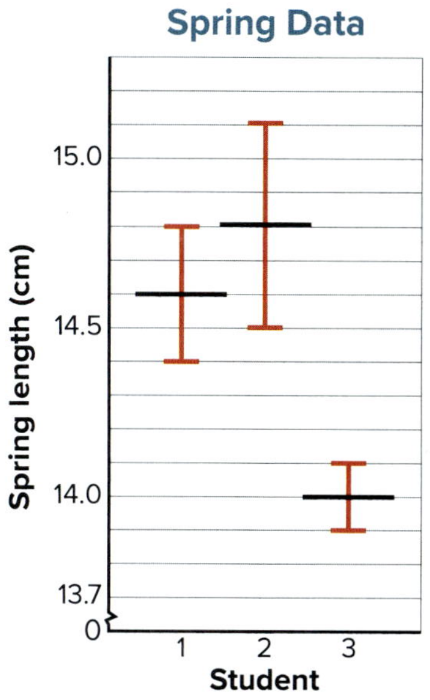

Figure 10 Three students took multiple measurements. The red bars show the uncertainty of each student's measurement.

Get It?

Explain Is student 3's result reproducible? Why or why not?

Precision Versus Accuracy

Both precision and accuracy are characteristics of measured values, as shown in **Figure 11.** How precise and accurate are the measurements of the three students above? The degree of exactness of a measurement is called its **precision.** In the example above, student 3's measurements are the most precise, within ± 0.1 cm. Both the measurements of student 1 and student 2 are less precise because they have a larger uncertainty (student 1 = ± 0.2 cm, student 2 = ± 0.3 cm).

Precision depends on the instrument and technique used to make the measurement. Generally, the device that has the finest division on its scale produces the most precise measurement. The precision of a measurement is one-half the smallest division of the instrument.

For example, suppose a graduated cylinder has divisions of 1 mL. You could measure an object to within 0.5 mL with this device. However, if the smallest division on a beaker is 50 mL, how precise would your measurements be compared to those taken with the graduated cylinder?

The significant figures in an answer show its precision. A measure of 67.100 g is precise to the nearest thousandth of a gram. Say you add 1.2 mL of acid to a beaker containing 2.4×10^2 mL of water—you cannot say you now have 2.412×10^2 mL of fluid because the volume of water was not measured to the nearest tenth of a milliliter, but to the nearest 10 mL.

Accuracy describes how well the results of a measurement agree with the "real" value; that is, the accepted value as measured by competent experimenters. **Figure 11** illustrates accuracy and precision.

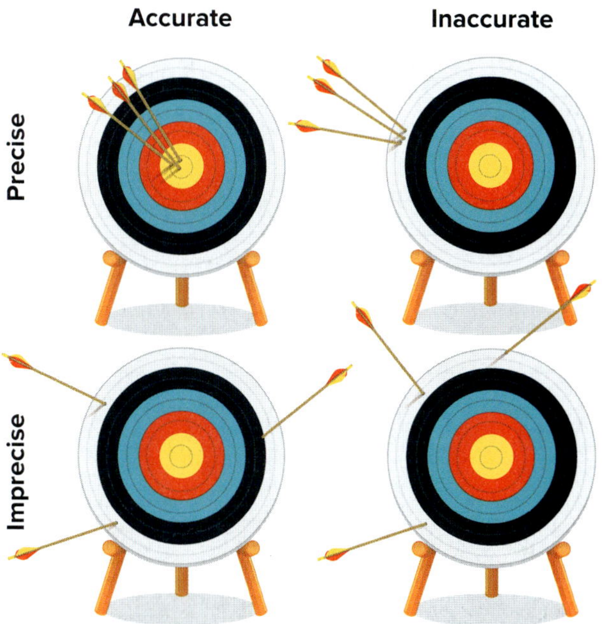

Figure 11 The yellow area in the center of each target represents an accepted value for a particular measurement. The arrows represent measurements taken by a scientist during an experiment.

If the length of the spring that the three students above measured had been 14.8 cm, then student 2 would have been most accurate and student 3 least accurate. What might have led someone to make inaccurate measurements? How could you check the accuracy of measurements?

A common method for checking the accuracy of an instrument is called the two-point calibration. First, does the instrument read zero when it should, as shown in **Figure 12?** Second, does it give the correct reading when it is measuring an accepted standard? Regular checks for accuracy are performed on critical measuring instruments, such as the radiation output of the machines used to treat cancer.

Get It?

Compare and contrast precision and accuracy.

Techniques of Good Measurement

To assure accuracy and precision, instruments also have to be used correctly. Measurements have to be made carefully if they are to be as precise as the instrument allows. One common source of error comes from the angle at which an instrument is read. Scales should be read with one's eye directly in front of the measure, as shown on the left of **Figure 13.** If the scale is read from an angle, as shown on the right of **Figure 13,** a different, less accurate, value will be obtained. The difference in the readings is caused by parallax, which is the apparent shift in the position of an object when it is viewed from different angles. To experiment with parallax, place your pen on a ruler and read the scale with your eye directly over the tip, then read the scale with your head shifted far to one side.

Figure 12 Accuracy is checked by zeroing an instrument before measuring.

Infer *Is this instrument accurate? Why or why not?*

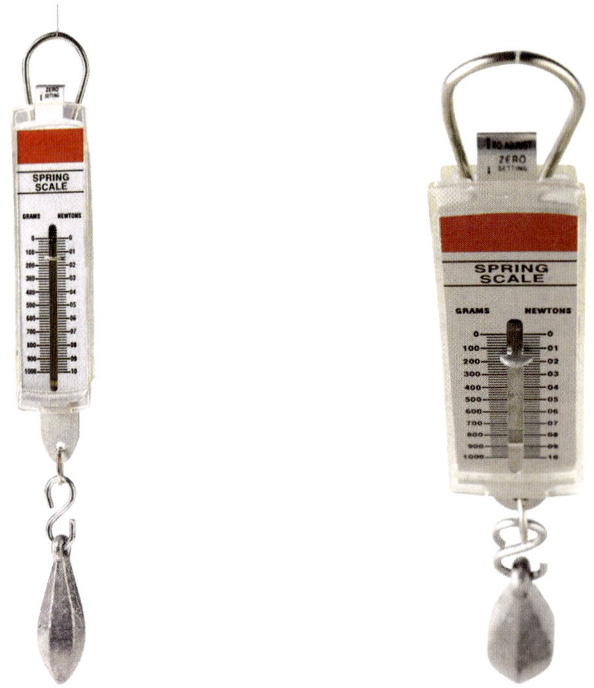

Figure 13 By positioning the scale head-on (left), your results will be more accurate than if you read your measurements at an angle (right).

Identify *How far did parallax shift the measurement on the right?*

GPS

The Global Positioning System, or GPS, offers an illustration of accuracy and precision in measurement. The GPS consists of 24 satellites with transmitters in orbit and several receivers on Earth. The satellites send signals with the time, measured by highly accurate atomic clocks.

The receiver uses the information from at least four satellites to determine latitude, longitude, and elevation. (The clocks in the receivers are not as accurate as those on the satellites.)

Receivers have different levels of precision. A device in an automobile might give your position to within a few meters. Devices used by geophysicists and geologists, as in **Figure 14,** can measure movements of millimeters in Earth's crust.

The GPS was developed by the United States Department of Defense. It uses atomic clocks, which were developed to test Einstein's theories of relativity and gravity. The GPS eventually was made available for civilian use.

GPS signals now are provided worldwide free of charge and are used in navigation on land, at sea, and in the air, for mapping and surveying, by telecommunications and satellite networks, and for scientific research into earthquakes and plate tectonics.

Figure 14 This geologist in the Mali Desert is using a highly accurate GPS receiver to record and analyze the movements of continental plates. His findings will help in the search for oil deposits.

Check Your Progress

12. **Precision and Accuracy** You find a micrometer (a tool used to measure objects to the nearest 0.001 mm) that has been bent. How does it compare to a new, high-quality meter-stick in its precision and accuracy?
13. **Accuracy** Some wooden rulers do not start with 0 at the edge, but have it set in a few millimeters. How could this improve the accuracy of the ruler?
14. **Parallax** Does parallax affect the precision of a measurement that you make? Explain.
15. **Uncertainty** Your friend tells you that his height is 182 cm. In your own words, explain the range of heights implied by this statement.
16. **Precision** A box has a length of 18.1 cm, a width of 19.2 cm, and is 20.3 cm tall.
    a. What is its volume?
    b. How precise is the measurement of length? Of volume?
    c. How tall is a stack of 12 of these boxes?
    d. How precise is the measurement of the height of one box? Of 12 boxes?
17. **Critical Thinking** Your friend states in a report that the average time required for a car to circle a 2.4-km track was 65.414 s. This was measured by timing 7 laps using a clock with a precision of 0.1 s. How much confidence do you have in the results of the report? Explain.

LEARNSMART® Go online to follow your personalized learning path to review, practice, and reinforce your understanding.

LESSON 4
GRAPHING DATA

FOCUS QUESTION
How do graphs help scientists analyze data?

Identifying Variables

When you perform an experiment, it is important to change only one factor at a time. For example, **Table 3** gives the length of a spring with different masses attached. Only the mass varies; if different masses were hung from different types of springs, you wouldn't know how much of the difference between two data pairs was due to the different masses and how much was due to the different springs.

Independent and dependent variables A variable is any factor that might affect the behavior of an experimental setup. The factor that is manipulated during an investigation is the **independent variable.** In the experiment that gave the data in **Table 3,** the mass was the independent variable. The factor that depends on the independent variable is the **dependent variable.** In this investigation, the amount the spring stretched depended on the mass, so the amount of stretch was the dependent variable.

Line of best fit A line graph shows how the dependent variable changes with the independent variable. The data from **Table 3** are graphed in **Figure 15** on the next page. The line in blue, drawn as close to all the data points as possible, is called a **line of best fit.** The line of best fit is a better model for predictions than any one point along the line. **Figure 15** gives detailed instructions on how to construct a graph, plot data, and sketch a line of best fit.

A well-designed graph allows patterns that are not immediately evident in a list of numbers to be seen quickly and simply. The graph in **Figure 15** shows that the length of the spring increases as the mass suspended from the spring increases.

Table 3 Length of a Spring for Different Masses

Mass Attached to Spring (g)	Length of Spring (cm)
0	13.7
5	14.1
10	14.5
15	14.9
20	15.3
25	15.7
30	16.0
35	16.4

3D THINKING DCI Disciplinary Core Ideas CCC Crosscutting Concepts SEP Science & Engineering Practices

COLLECT EVIDENCE

Use your Science Journal to record the evidence you collect as you complete the readings and activities in this lesson.

INVESTIGATE

GO ONLINE to find these activities and more resources.

Forensics Lab: It's in the Blood
Analize and interpret data to determine cause and effect at a crime scene.

Review the News
Obtain information from a current news story where graphs are used to present data. Evaluate your source and communicate your findings to your class.

Figure 15 Use the steps outlined here to plot line graphs from data tables.

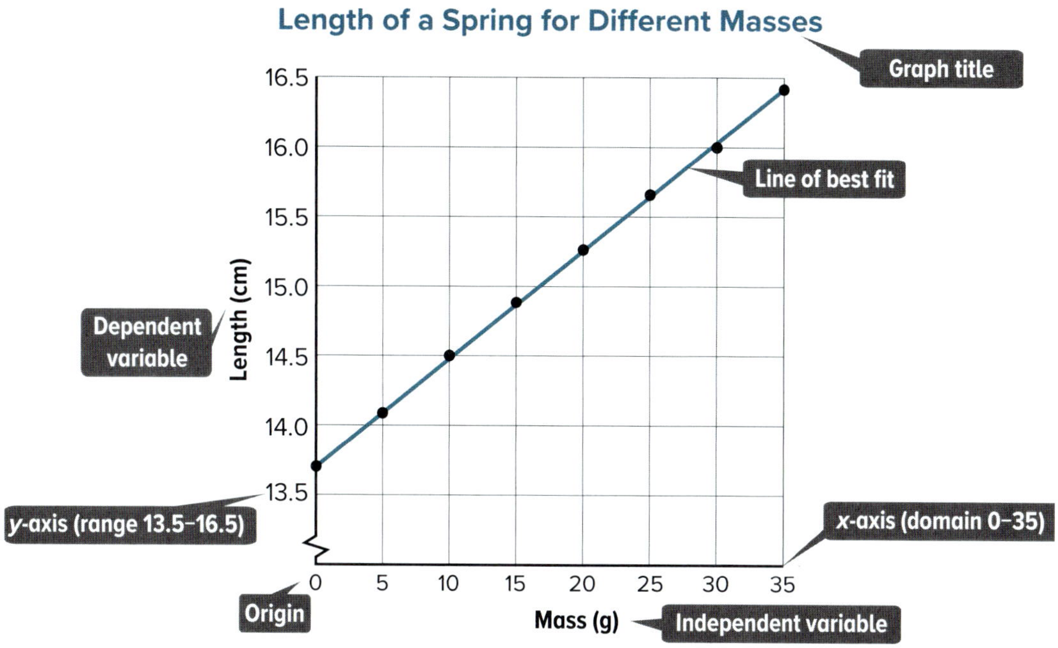

1. Identify the independent variable and the dependent variable in your data. In this example, the independent variable is mass (g) and the dependent variable is length (cm). The independent variable is plotted on the horizontal axis, the *x*-axis. The dependent variable is plotted on the vertical axis, the *y*-axis.
2. Determine the range of the independent variable to be plotted. In this case the range is 0–35.
3. Decide whether the origin (0, 0) is a valid data point.
4. Spread the data out as much as possible. Let each division on the graph paper stand for a convenient unit. This usually means units that are multiples of 2, 5, or 10.
5. Number and label the horizontal axis. The label should include the name of the variable and its units, for example, Mass (g).
6. Repeat steps 2–5 for the dependent variable.
7. Plot the data points on the graph.
8. Draw the best-fit straight line or smooth curve that passes through as many data points as possible. This is sometimes called *eyeballing*. Do not use a series of straight-line segments that connect the dots. The line that looks like the best fit to you may not be exactly the same as someone else's. There is a formal procedure, which many graphing calculators use, called the least-squares technique, that produces a unique best-fit line, but that is beyond the scope of this textbook.
9. Give the graph a title that clearly tells what the graph represents.

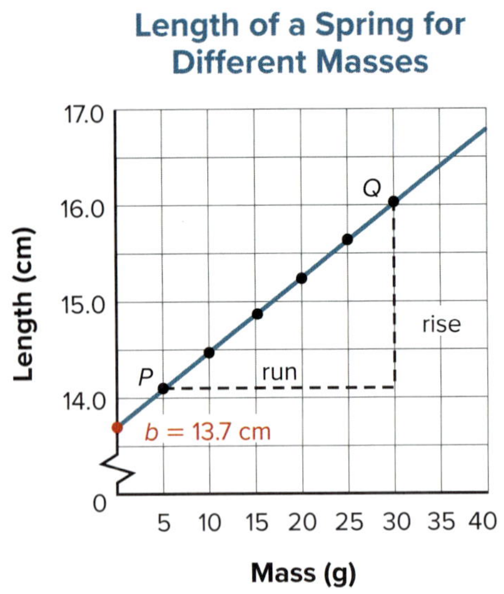

Figure 16 In a linear relationship, the dependent variable—in this case, length—varies linearly with the independent variable. The independent variable in this experiment is mass.

Describe *What happens to the length of the spring as mass decreases?*

Linear Relationships

Scatter plots of data take many different shapes, suggesting different relationships. Three of the most common relationships are linear relationships, quadratic relationships, and inverse relationships. You probably are familiar with them from math class.

When the line of best fit is a straight line, as in **Figure 15,** there is a linear relationship between the variables. In a **linear relationship,** the dependent variable varies linearly with the independent variable. The relationship can be written as the following equation.

Linear Relationship Between Two Variables

$$y = mx + b$$

Here, m is the slope of the line, or the ratio of the vertical change to the horizontal change, and b is the y-intercept, the point at which the line crosses the vertical axis. To find the slope, select two points, P and Q, far apart on the line–they may or may not be data points. The vertical change, or rise (Δy), is the difference between the vertical values of P and Q, as shown in **Figure 16.** The horizontal change, or run (Δx), is the difference between the horizontal values of P and Q.

Slope

The slope of a line is equal to the rise divided by the run, which also can be expressed as the vertical change divided by the horizontal change.

$$m = \frac{rise}{run} = \frac{\Delta y}{\Delta x}$$

In **Figure 16:** $m = \frac{(16.0 \text{ cm} - 14.1 \text{ cm})}{(30 \text{ g} - 5 \text{ g})} = 0.08 \text{ cm/g}$

If y gets smaller as x gets larger, then $\frac{\Delta y}{\Delta x}$ is negative, and the line slopes downward from left to right. The y-intercept, or the y-value when the value of x is zero, in this example is $b = 13.7$ cm (**Figure 16**). So, when no mass is suspended by the spring, it has a length of 13.7 cm. When $b = 0$, or $y = mx$, y is said to vary directly with x. In physics, the slope of the line and the y-intercept always contain information about the physical system that is described by the graph.

Nonlinear Relationships

Figure 17 graphs the distance a brass ball falls versus time. Note that the graph is not a straight line, meaning the relationship is not linear. There are many types of nonlinear relationships in science. Two of the most common are quadratic and inverse relationships.

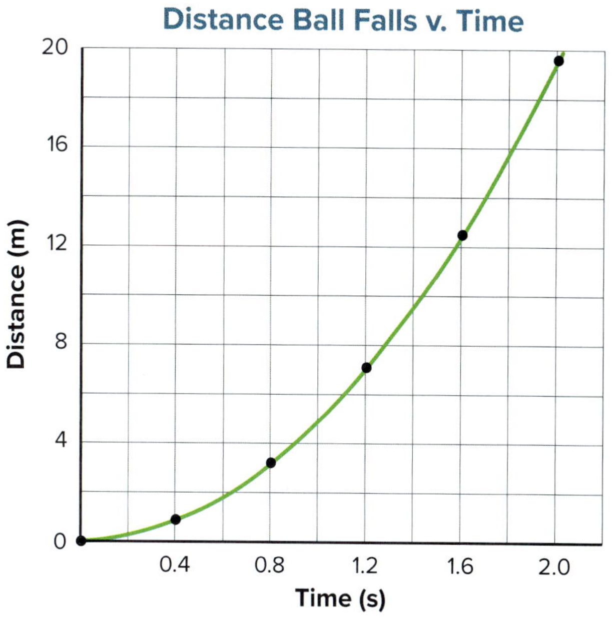

Figure 17 The quadratic, or parabolic, relationship shown here is an example of a nonlinear relationship.

Quadratic relationships

The graph in **Figure 17** is a quadratic relationship, represented by the equation below. A **quadratic relationship** exists when one variable depends on the square of another.

Quadratic Relationship Between Two Variables

$$y = ax^2 + bx + c$$

A computer program or graphing calculator can easily find the values of the constants a, b, and c in the above equation. In **Figure 17**, the equation is $d = 5t^2$. See the Math Skill Handbook in the back of this book or online for more on making and using line graphs.

Get It?

Explain how two variables are related to each other in a quadratic relationship.

PHYSICS Challenge

An object is suspended from spring 1, and the spring's elongation (the distance it stretches) is x_1. Then the same object is removed from the first spring and suspended from a second spring. The elongation of spring 2 is x_2. x_2 is greater than x_1.

1. On the same axes, sketch the graphs of the mass versus elongation for both springs.
2. Should the origin be included in the graph? Why or why not?
3. Which slope is steeper?
4. At a given mass, $x_2 = 1.6\ x_1$. If $x_2 = 5.3$ cm, what is x_1?

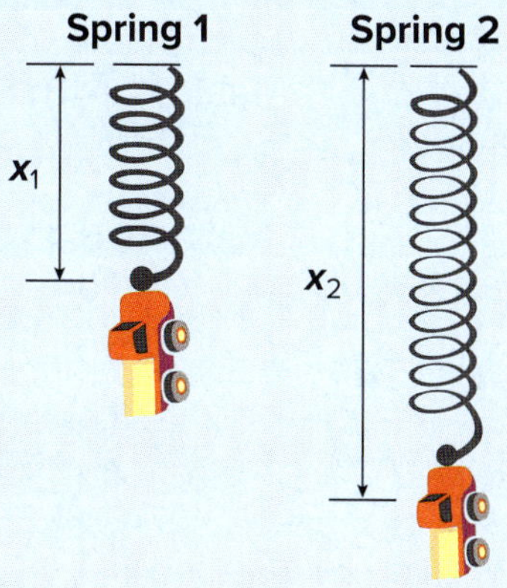

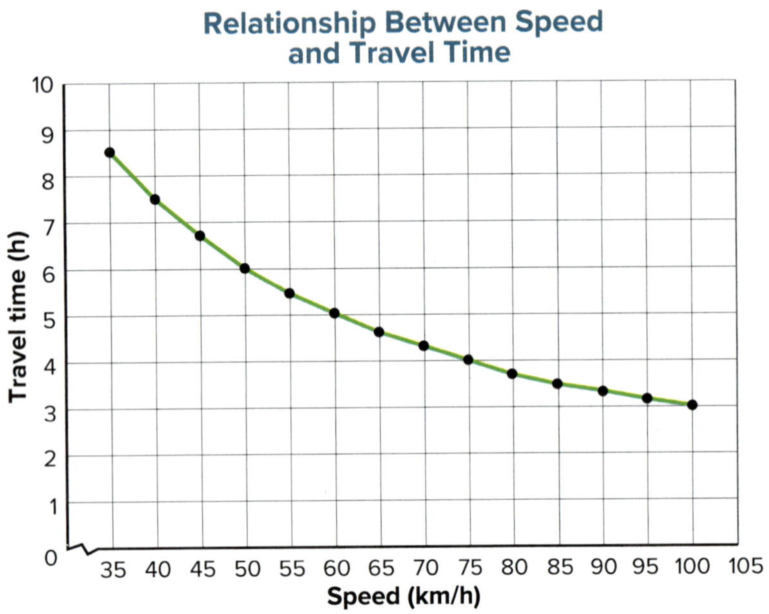

Figure 18 This graph shows the inverse relationship between speed and travel time.

Describe *How does travel time change as speed increases?*

Inverse relationships The graph in **Figure 18** shows how the time it takes to travel 300 km varies as a car's speed increases. This is an example of an inverse relationship, represented by the equation below. An **inverse relationship** is a hyperbolic relationship in which one variable depends on the inverse of the other variable.

Inverse Relationship Between Two Variables

$$y = \frac{a}{x}$$

The three relationships you have learned about are a sample of the relations you will most likely investigate in this course. Many other mathematical models are used. Important examples include sinusoids, used to model cyclical phenomena, and exponential growth and decay, used to study radioactivity. Combinations of different mathematical models represent even more complex phenomena.

Get It?

Explain how two variables are related to each other in an inverse relationship.

PRACTICE Problems

ADDITIONAL PRACTICE

18. Refer to the data listed in **Table 4.**

a. Plot mass versus volume, and draw the curve that best fits all points. Describe the curve.

b. What type of relationship exists between the mass of the gold nuggets and their volume?

c. What is the value of the slope of this graph? Include the proper units.

d. Write the equation showing mass as a function of volume for gold.

e. Write a word interpretation for the slope of the line.

Table 4 Mass of Pure Gold Nuggets

Volume (cm^3)	Mass (g)
1.0	19.4
2.0	38.6
3.0	58.1
4.0	77.4
5.0	96.5

Predicting Values

When scientists discover relationships like the ones shown in the graphs in this lesson, they use them to make predictions. For example, the equation for the linear graph in **Figure 16** is as follows:

$$y = (0.08 \text{ cm/g})x + 13.7 \text{ cm}$$

Relationships, either learned as formulas or developed from graphs, can be used to predict values you haven't measured directly. How far would the spring in **Table 3** stretch with 49 g of mass?

$$y = (0.08 \text{ cm/g})(49 \text{ g}) + 13.7 \text{ cm}$$
$$= 18 \text{ cm}$$

It is important to decide how far you can extrapolate from the data you have. For example, 90 g is a value far outside the ones measured and displayed in **Table 3,** and the spring might break rather than stretch that far.

Physicists use models to accurately predict how systems will behave: what circumstances might lead to a solar flare (an immense outburst of material from the Sun's surface into space) or how changes to a grandfather clock's pendulum will change its ability to keep accurate time. People in all walks of life use models in many ways. One example is shown in **Figure 19.** With the tools you have learned in this module, you can answer questions and produce models for the physics questions you will encounter in the rest of this textbook.

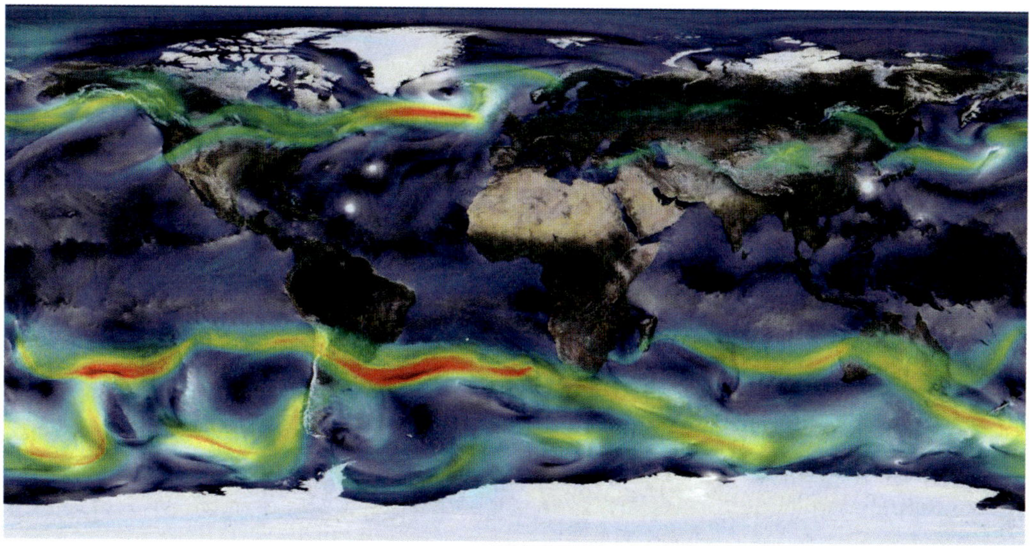

Figure 19 In order to create a realistic model, computer animators use mathematical models of the real world to help visualize global atmospheric conditions.

Check Your Progress

19. **Make a Graph** Graph the following data. Time is the independent variable.

Time (s)	0	5	10	15	20	25	30	35
Speed (m/s)	12	10	8	6	4	2	2	2

20. **Interpret a Graph** What would be the meaning of a nonzero *y*-intercept in a graph of total mass versus volume?

21. **Predict** Use the relationship illustrated in **Figure 16** to determine the mass required to stretch the spring 15 cm.

22. **Predict** Use the relationship shown in **Figure 18** to predict the travel time when speed is 110 km/h.

23. **Critical Thinking** Look again at the graph in **Figure 16.** In your own words, explain how the spring would be different if the line in the graph were shallower or had a smaller slope.

LEARNSMART® Go online to follow your personalized learning path to review, practice, and reinforce your understanding.

ENGINEERING & TECHNOLOGY

A Step in the Right Direction

Prosthetics, or artificial body parts, have been in use since ancient Greek times. Most prostheses of the past, however, were more decorative than functional. They were made mainly of metal and wood, and were attached to the body with straps or harnesses made of leather. Prosthetics have come a long way over the last few decades. Today's prosthetic limbs can transform people's lives.

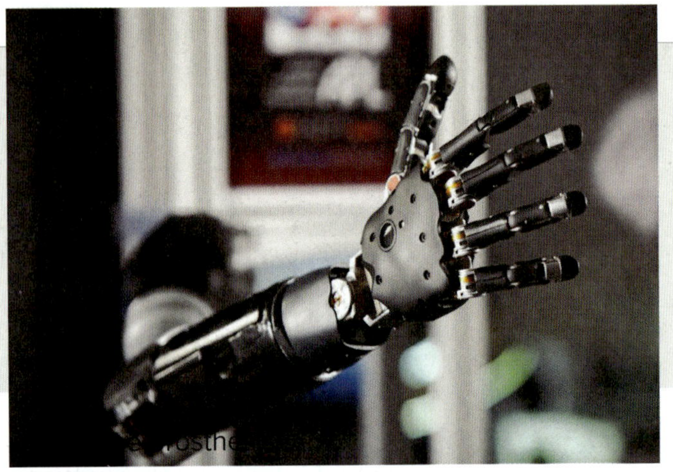

Some prosthetics can be attached directly to the patient's body, rather than being attached with slings or harnesses.

New materials, new look

The prostheses of today look and function more like biological limbs than ever before. The inner structure of a prosthesis, called a pylon, is now commonly made of carbon-fiber composites and new types of plastics. These materials make the prosthesis stronger, but also lighter, than metal or wood prostheses.

Electronic components give users better control over a prosthesis. For example, myoelectric prostheses use the electric signals generated by muscles to control a prosthesis. A person with a myoelectric prosthetic hand can move muscles in his or her arm to signal the hand to move or grip in different ways, with varying amounts of force.

Targeted muscle reinnervation (TMR) is another new method of giving people more control over cutting-edge prostheses, specifically arms. The nerves in a person's residual limb, which once controlled the amputated limb, are surgically "reassigned" to

Another game-changing advance in prosthetic technology is a change in the socket, the part of a prosthesis that connects to a person's residual limb. When prostheses are attached with slings or harnesses, the socket is sometimes uncomfortable. It can cause sores, blisters, and pain, and it can damage the tissues under the skin. To combat this problem, some new prostheses are mounted directly into the bone marrow of a person's residual limb. This method of attachment increases the person's comfort, as well as his or her control of the prosthesis.

Some prosthetic advances are still making the transition from prototypes to patient care. But scientists are determined to speedily move these new technologies into the everyday lives of patients.

COMMUNICATE TECHNICAL INFORMATION

Use print or online sources to research an advance in prosthetics that was not discussed in this feature. Share the results of your research with your class.

CALIFORNIA CONNECTION

Athletes with a variety of disabilities and impairments participate in the Paralympics Games. Visit the Web site of the U.S. Paralympics. Choose a California Paralympic athlete who uses a prosthesis and learn more about his or her life and sport. Do a presentation about that athlete for your class.

Food Collection/Alamy Stock Photo

STEM AT WORK

What do physicists study?

Physicist study a wide range of topics from the tiniest particles to stars, galaxies, and the universe itself.

Mechanics is the branch of physics that studies motion and forces.

Physicists often work with other branches of science. For example, geophysics is the study of Earth's physical properties and processes. Another area of collaboration is biomechanics, which is the study of the motion of living things. The researcher in the photo at the right is collecting data about the motion of an athlete.

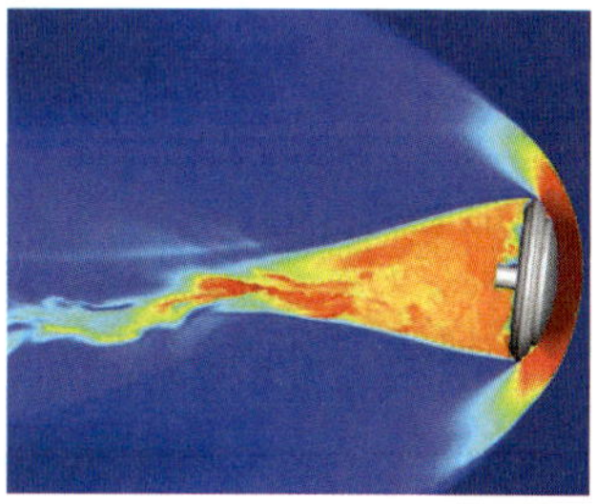

Thermodynamics is the study of the energy of a system's particles. It includes the study of temperature, heat, and the relationship between the characteristics of a system's particles and the macroscopic characteristics of the system. Because systems contain very large numbers of particles, physics often use computer models to study thermodynamics. The NASA model on the left allows scientists to predict how heat shields will perform in the atmosphere.

Electromagnetism is the study of electricity, magnetism, and the relationship between them. By understanding how currents produce magnetic fields, physicists and engineers can design and build electromagnets, like the one on the right.

Nuclear and Particle Physics Nuclear physicists study the atomic nucleus. Particle physicists will use this superconducting electromagnet in a particle accelerator to study even smaller structures—the elementary particles that make up universe.

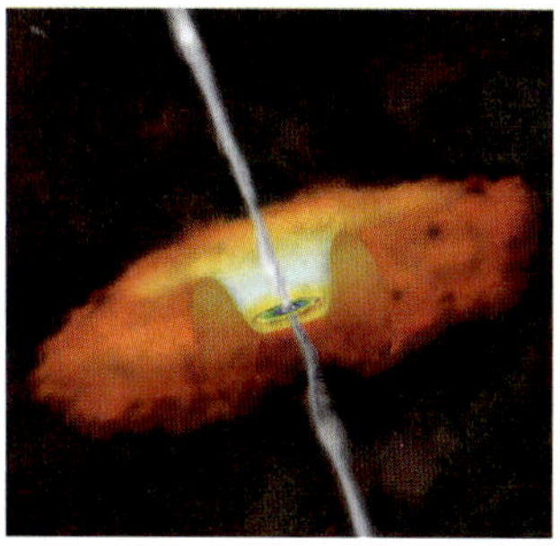

Optics is the study of the generation, transmission, and detection of visible, infrared, ultraviolet, and microwave radiation.

Astrophysics and Cosmology are, respectively, the study of the physical properties of celestial bodies such as the active galactic nucleus on the left and the development and overall structure of the universe.

OBTAIN AND COMMUNICATE INFORMATION

Choose one of the fields of physics that interests you. Research your chosen field, and identify a related career, what a person does in that career, and the types of education and training required. Develop a presentation in which you are recruiting candidates for a job opening in that career.

CALIFORNIA CONNECTION

Start your research by investigating information on Web sites by the California Department of Education's Career Zone and the American Institute of Physics.

MODULE 1
STUDY GUIDE

GO ONLINE to study with your Science Notebook.

Lesson 1 METHODS OF SCIENCE

- Scientific methods include making observations and asking questions about the natural world.
- Scientists use models to represent things that may be too small or too large, processes that take too much time to see completely, or a material that is hazardous.
- A scientific theory is an explanation of things or events based on knowledge gained from observations and investigations. A scientific law is a statement about what happens in nature, which seems to be true all the time.
- Science can't explain or solve everything. Questions about opinions and values cannot be tested.

- physics
- scientific methods
- hypothesis
- model
- scientific theory
- scientific law

Lesson 2 MATHEMATICS AND PHYSICS

- Using SI, an adaptation of the metric system, helps scientists around the world communicate more easily.
- Dimensional analysis is used to check that an answer is in the correct units.
- Significant figures are the valid digits in a measurement.

- dimensional analysis
- significant figures

Lesson 3 MEASUREMENT

- Measurements are reported with uncertainty because a new measurement that is within the margin of uncertainty confirms the old measurement.
- Precision is the degree of exactness with which a quantity is measured. Accuracy is the extent to which a measurement matches the true value.
- A common source of error that occurs when making a measurement is the angle at which an instrument is read. If the scale of an instrument is read at an angle, as opposed to at eye level, the measurement will be less accurate.

- measurement
- precision
- accuracy

Lesson 4 GRAPHING DATA

- Graphs contain information about the relationships between variables. Patterns that are not immediately evident in a list of numbers are seen more easily when the data are graphed.
- Common relationships shown in graphs are linear relationships, quadratic relationships, and inverse relationships. In a linear relationship, the dependent variable varies linearly with the independent variable. A quadratic relationship occurs when one variable depends on the square of another. In an inverse relationship, one variable depends on the inverse of the other variable.
- Scientists use models and relationships between variables to make predictions.

- independent variable
- dependent variable
- line of best fit
- linear relationship
- quadratic relationship
- inverse relationship

THREE-DIMENSIONAL THINKING

Module Wrap-Up

REVISIT THE PHENOMENON

What tools and skills do physicists use?

CER Claim, Evidence, Reasoning

Explain Your Reasoning Revisit the claim you made when you encountered the phenomenon. Summarize the evidence you gathered from your investigations and research and finalize your Summary Table. Does your evidence support your claim? If not, revise your claim. Explain why your evidence supports your claim.

GO FURTHER

SEP Data Analysis Lab

How is string length and square of the period of a pendulum related?

A group of students planned and carried out an investigation into the relationship between string length L and the time it took for one complete swing of a pendulum (its period) T.

Data and Observations Their measurements are given in the table.

String Length L (m)	Period T (s)
0.07	0.53
0.10	0.63
0.20	0.90
0.40	1.27
0.55	1.49
0.70	1.68
0.90	1.90

CER Analyze and Interpret Data

One of the students used the data to investigate the relationship between the string length L and the square of the period T^2.

1. **Calculate and record** the values of T^2.
2. **Construct a Model** Plot the values of L and T^2 and draw the best-fit curve. Write an equation that describes the curve.
3. **Claim** What is the relationship between the square of the period of a pendulum and the length of the string?
4. **Evidence and Reasoning** Use your graph as evidence and explain your reasoning to justify your claim.

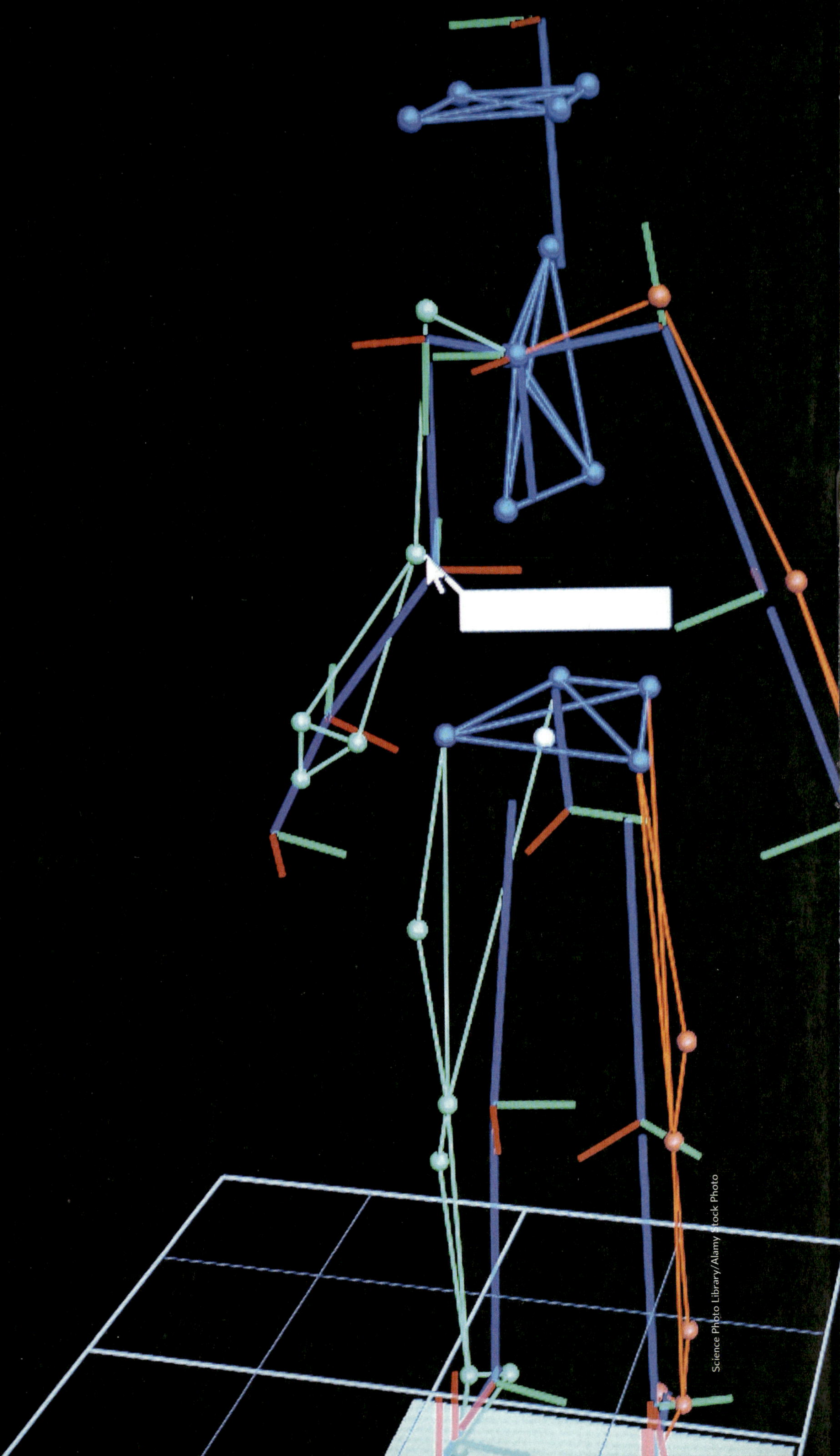

Science Photo Library/Alamy Stock Photo

UNIT 1

MECHANICS IN ONE DIMENSION

ENCOUNTER THE PHENOMENON

How can we model motion and forces?

SEP Ask Questions

What questions do you have about the phenomenon? Write your questions on sticky notes and add them to the driving question board for this unit.

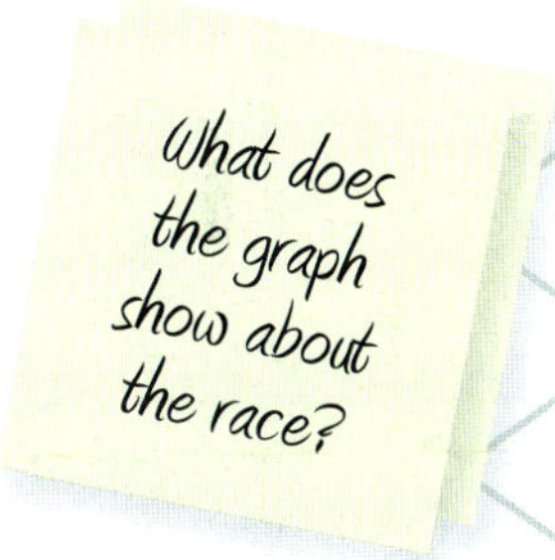

Look for Evidence

As you go through this unit, use the information and your experiences to help you answer the phenomenon question as well as your own questions. For each activity, record your observations in a Summary Table, add an explanation, and identify how it connects to the unit and module phenomenon questions.

Solve a Problem

STEM UNIT PROJECT

Build a Rocket A good understanding of motion and force is necessary for rocketry, whether you are a building models as a hobby or working with an organization like NASA to get to the Moon. Investigate the forces involved in a rocket launch. Use the results of these investigations and the evidence you collected during the unit to build and test a model rocket.

GO ONLINE In addition to reading the information in your Student Edition, you can find the STEM Unit Project and other useful resources online.

Aurora Photos/Alamy Stock Photo

MODULE 2

REPRESENTING MOTION

ENCOUNTER THE PHENOMENON

How does a GPS unit know where you are?

GO ONLINE to watch a video about how GPS works.

SEP Ask Questions

Do you have other questions about the phenomenon? If so, add them to the driving question board.

CER Claim, Evidence, Reasoning

Make Your Claim Use your CER chart to make a claim about how a GPS unit knows where you are. Explain your reasoning.

Collect Evidence Use the lessons in this module to collect evidence to support your claim. Record your evidence as you move through the module.

Explain Your Reasoning You will revisit your claim and explain your reasoning at the end of the module.

GO ONLINE to access your CER chart and explore resources that can help you collect evidence.

LESSON 2: Explore & Explain: Coordinates and Vectors

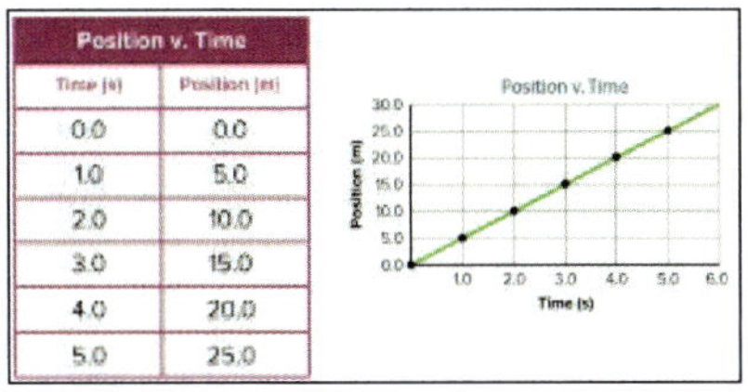

Position v. Time

Time (s)	Position (m)
0.0	0.0
1.0	5.0
2.0	10.0
3.0	15.0
4.0	20.0
5.0	25.0

LESSON 3: Explore & Explain: Velocity and Speed

Smithsonian

Additional Resources

LESSON 1

PICTURING MOTION

FOCUS QUESTION

How do you know that something is moving?

All Kinds of Motion

You have learned about scientific processes that will be useful in your study of physics. You will now begin to use these tools to analyze motion. In subsequent modules, you will apply these processes to many kinds of motion. You will use words, sketches, diagrams, graphs, and equations.

Changes in position What comes to your mind when you hear the word motion? A spinning ride at an amusement park? A baseball soaring over a fence for a home run? Motion is all around you—from fast trains and speedy skiers to slow breezes and lazy clouds. Objects move in many different ways, such as the straight-line path of a bowling ball in a bowling lane's gutter, the curved path of a car rounding a turn, the spiral of a falling kite, and swirls of water circling a drain. When an object is in motion, such as the train in **Figure 1,** its position changes.

Some types of motion are more complicated than others. When beginning a new area of study, it is generally a good idea to begin with the least complicated situation, learn as much as possible about it, and then gradually add more complexity to that simple model. In the case of motion, you will begin your study with movement along a straight line.

Get It?

Describe how the picure in **Figure 1** would be different if the train were sitting still.

Figure 1 The train appears blurry in the photograph because its position changed during the time the camera shutter was open.

Ingram Publishing

3D THINKING DCI Disciplinary Core Ideas CCC Crosscutting Concepts SEP Science & Engineering Practices

COLLECT EVIDENCE

Use your Science Journal to record the evidence you collect as you complete the readings and activities in this lesson.

INVESTIGATE

GO ONLINE to find these activities and more resources.

PhysicsLAB: Motion Diagrams
Use a model to identify patterns in motion.

Review the News
Obtain information from a current news story about motion capture technology. Evaluate your source and communicate your findings to your class.

Movement along a straight line In general, an object can move along many different kinds of paths, but straight-line motion follows a path directly between two points without turning left or right. For example, you might describe an object's motion as forward and backward, up and down, or north and south. In each of these cases, the object moves along a straight line.

Suppose you are reading this textbook at home. As you start to read, you glance over at your pet hamster and see that it is sitting in a corner of the cage. Sometime later you look over again, and you see that it now is sitting next to the food dish in the opposite corner of the cage. You can infer that your hamster has moved from one place to another in the time between your observations. What factors helped you make this inference about the hamster's movement?

The description of motion is a description of place and time. You must answer the questions of where an object is located and when it is at that position in order to describe its motion.

Get It?

Identify two factors you must know to describe the motion of an object along a straight line.

Motion Diagrams

Consider the following example of straight-line motion: a runner jogs along a straight path. One way of representing the runner's motion is to create a series of images showing the runner's position at equal time intervals. You can do this by photographing the runner in motion to obtain a sequence of pictures. Each photograph will show the runner at a point that is farther along the straight path.

Consecutive images Suppose you point a camera in a direction and a runner crosses the camera's field of view. Then you take a series of photographs of the runner at equal time intervals, without moving the camera. **Figure 2** shows what a series of consecutive images for a runner might look like. Notice that the runner is in a different position in each image, but everything in the background remains in the same position. This indicates that, relative to the camera and the ground, only the runner is in motion.

Figure 2 You can tell that the runner is in motion because her position changes relative to the tree and the ground.

Combining images Suppose that you layered the four images of the runner from **Figure 2** one on top of the other, as shown in **Figure 3**. A series of images showing the positions of a moving object at equal time intervals is called a **motion diagram.**

Particle Models

Keeping track of the runner's motion is easier if you disregard the movement of her arms and her legs and instead concentrate on a single point at the center of her body. In a **particle model,** you replace the object or objects with single points.

To use the particle model, the object's size must be much less than the distance it moves. In the photographic motion diagram, you could identify a central point at her waistline, and draw a dot to represent the position at different times. The bottom of **Figure 3** shows the particle model for the runner's motion.

Figure 3 Combining the images from **Figure 2** produces this motion diagram of the runner's movement. The series of dots at the bottom of the figure is a particle model that corresponds to the motion diagram.

Explain *how the particle model shows that the runner's speed is not changing.*

Get It?

Describe how you would model the motion of the hiker at the beginning of this module.

Check Your Progress

1. **Representing Motion** How does a motion diagram represent an object's motion?
2. **Bike Motion Diagram** Draw a particle model motion diagram for a bike rider moving at a constant pace along a straight path.
3. **Car Motion Diagram** Draw a particle motion diagram corresponding to the motion in **Figure 4** for a car coming to a stop at a stop sign. What point on the car did you use to represent the car?

Figure 4

4. **Bird Motion Diagram** Draw a particle model motion diagram corresponding to the motion diagram in **Figure 5.** What point on the bird did you choose to represent the bird?

Figure 5

5. **Critical Thinking** Draw particle model motion diagrams for two runners during a race in which the first runner crosses the finish line as the other runner is three-fourths of the way to the finish line.

LEARNSMART® Go online to follow your personalized learning path to review, practice, and reinforce your understanding.

LESSON 2

WHERE AND WHEN?

FOCUS QUESTION

What are some different ways of describing and representing motion?

Coordinate Systems

Is it possible to measure distance and time on a motion diagram? Before photographing a runner, you could place a long measuring tape on the ground to show where the runner is in each image. A stop-watch within the camera's view could show the time. But where should you place the end of the measuring tape? When should you start the stopwatch?

Position and distance It is useful to identify a system in which you have chosen where to place the zero point of the measuring tape and when to start the stopwatch. A **coordinate system** gives the location of the zero point of the variable you are studying and the direction in which the values of the variable increase, as shown in the diagram in **Figure 6.**

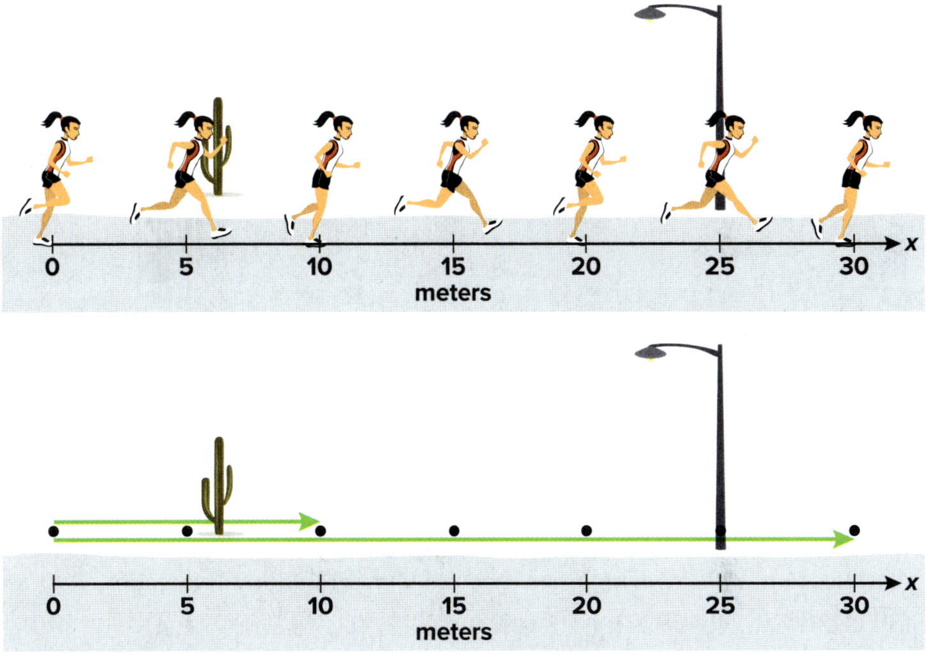

Figure 6 A simplified motion diagram uses dots to represent a moving object and arrows to indicate positions.

3D THINKING | DCI Disciplinary Core Ideas | CCC Crosscutting Concepts | SEP Science & Engineering Practices

COLLECT EVIDENCE

Use your Science Journal to record the evidence you collect as you complete the readings and activities in this lesson.

INVESTIGATE

GO ONLINE to find these activities and more resources.

Quick Investigation: Vector Models
Develop and use a model to visualize the result of adding vector quantities in one dimension.

Revisit the Encounter the Phenomenon Question
What information from this lesson can help you answer the Unit and Module questions?

The **origin** is the point at which all variables in a coordinate system have the value zero. In the example of the runner shown in **Figure 6,** the origin, which is the zero point of the measuring tape, could be 6 m to the left of the cactus. Because the motion is in a straight line, your measuring tape should lie along this line. The straight line is an axis of the coordinate system.

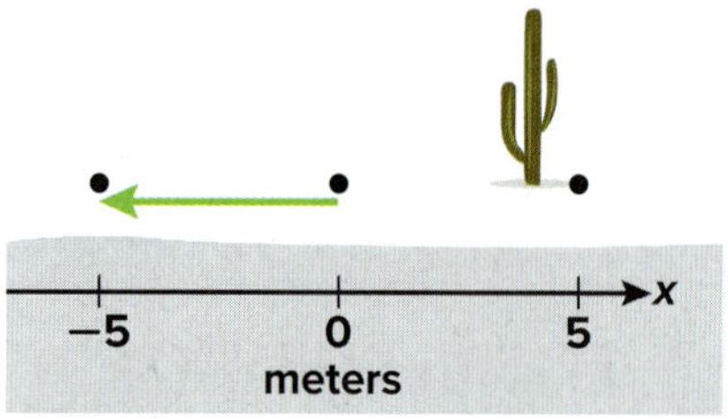

Figure 7 The green arrow indicates a negative position of −5 m, if the direction right of the origin is chosen as positive.

Infer *What position would the arrow indicate if you chose the direction left of the origin as positive?*

You can indicate how far the runner in **Figure 6** is from the origin at a certain time on the motion diagram by drawing an arrow from the origin to the point that represents the runner, shown at the bottom of **Figure 6**. This arrow represents the runner's **position,** the distance and direction from the origin to the object. In general, **distance** is the entire length of an object's path, even if the object moves in many directions. Because the motion in **Figure 6** is in one direction, the arrow lengths represent distance.

Negative position Is there such a thing as a negative position? Suppose you chose the coordinate system just described but this time placed the origin 4 m left of the cactus with the *x*-axis extending in a positive direction to the right. A position 9 m left of the cactus, or 5 m left of the origin, would be a negative position, as shown in **Figure 7.**

Get It?
Explain how positive and negative positions are determined.

Vectors and Scalars

As you might imagine, there are many kinds of measurements and numbers used to represent or describe motion. If you needed to describe how far you ran, you might say that you ran 1.6 km. If you needed to run to a specific location, you might say that you need to run 1.6 km north. Many quantities in physics have both size, also called **magnitude,** and direction. A quantity that has both magnitude and direction is called a **vector.** You can represent a vector with an arrow. The length of the arrow represents the magnitude of the vector, and the direction of the arrow represents the direction of the vector. A quantity that is just a number without any direction, such as distance, time, or temperature, is called a **scalar.** In this textbook, we will use boldface letters to represent vector quantities and regular letters to represent scalars.

Get It?
Describe the difference between a vector and a scalar.

SCIENCE USAGE V. COMMON USAGE

Magnitude

Science usage: a measure of size

When drawing vectors, the magnitude of a vector is proportional to that vector's length.

Common usage: great size or extent

The magnitude of the Grand Canyon is difficult to capture in photographs.

CCC CROSSCUTTING CONCEPTS

Systems and Systems Models Vectors represent a form of modeling. With a partner, create a vector drawing representing motion from one location to another. Add explanations to your drawing that explain how your motion was modeled.

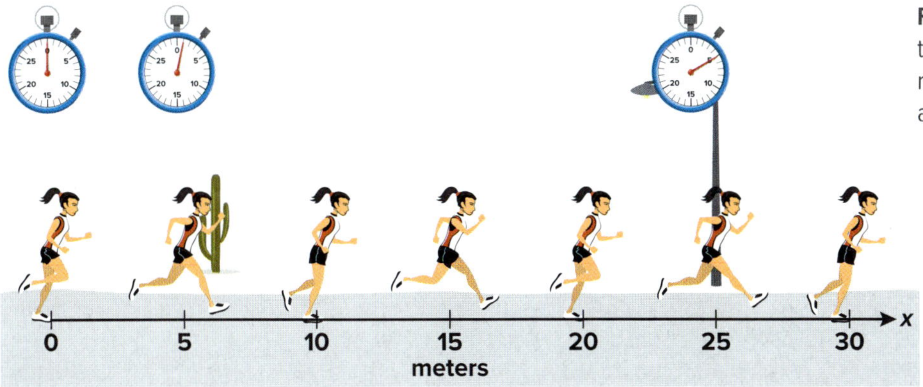

Figure 8 You can use the clocks in the figure to calculate the time interval (Δt) for the runner's movement from one position to another.

Time intervals are scalars.

When analyzing the runner's motion in **Figure 8,** you might want to know how long it took her to travel from the cactus to the lamppost. You can obtain this value by finding the difference between the stopwatch reading at the cactus and the stopwatch reading at the lamppost. **Figure 8** shows these stopwatch readings. The difference between two times is called a **time interval.**

A common symbol for a time interval is Δt, where the Greek letter delta (Δ) is used to represent a change in a quantity. Let t_i represent the initial (starting) time, when the runner was at the cactus. Let t_f represent the final (ending) time of the interval, when the runner was at the lamppost. We define a time interval mathematically as follows.

Time Interval

The time interval is equal to the change in time from the initial time to the final time.

$$\Delta t = t_f - t_i$$

The subscripts i and f represent the initial and final times, but they can be the initial and final times of any time interval you choose. In the example of the runner, the time it takes for her to go from the cactus to the lamppost is $t_f - t_i = 5.0\ \text{s} - 1.0\ \text{s} = 4.0\ \text{s}$. You could instead describe the time interval for the runner to go from the origin to the lamppost. In this case, the time interval would be $t_f - t_i = 5.0\ \text{s} - 0.0\ \text{s} = 5.0\ \text{s}$. The time interval is a scalar because it has no direction. Is the runner's position a scalar?

Positions and displacements are vectors.

You have already seen how a position can be described as negative or positive in order to indicate whether that position is to the left or the right of a coordinate system's origin. This suggests that position is a vector because position has direction—either right or left in this case.

STEM CAREER Connection

Geographic Information Systems (GIS) Manager

Would you like to work with computers, maps, and logistics? Then, a GIS manager might be a career for you. GIS managers work with teams that produce geographical information systems that are used for road traffic management, health care delivery, defense planning, market research, and community services such as garbage collection and utilities delivery.

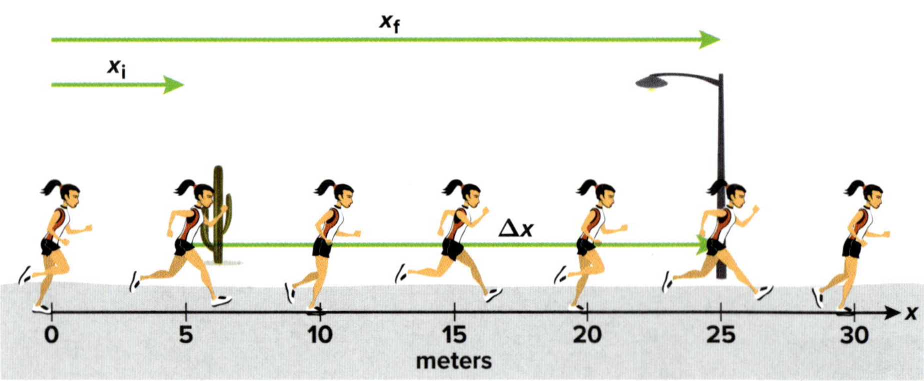

Figure 9 The vectors $\mathbf{x}_i$ and $\mathbf{x}_f$ represent positions. The vector $\Delta\mathbf{x}$ represents displacement from $\mathbf{x}_i$ to $\mathbf{x}_f$.

Describe *the displacement from the lamppost to the cactus.*

Figure 9 shows the position of the runner at both the cactus and the lamppost. Notice that you can draw an arrow from the origin to the location of the runner in each case. These arrows have magnitude and direction. In common speech, a position refers to a certain place, but in physics, the definition of a position is more precise. A position is a vector with the arrow's tail at the origin of a coordinate system and the arrow's tip at the place.

You can use the symbol $\mathbf{x}$ to represent position vectors mathematically. In **Figure 9**, the symbol $\mathbf{x}_i$ represents the position at the cactus, and the symbol $\mathbf{x}_f$ represents the position at the lamppost. The symbol $\Delta\mathbf{x}$ represents the change in position from the cactus to the lamppost. Because a change in position is described and analyzed so often in physics, it has a special name. In physics, a change in position is called a **displacement.** Because displacement has both magnitude and direction, it is a vector.

What was the runner's displacement when she ran from the cactus to the lamppost? By looking at **Figure 9**, you can see that this displacement is 20 m to the right. Notice also, that the displacement from the cactus to the lamppost ($\Delta\mathbf{x}$) equals the position at the lamppost ($\mathbf{x}_f$) minus the position at the cactus ($\mathbf{x}_i$). This is true in general; displacement equals final position minus initial position.

Displacement

Displacement is the change in position from initial position to final position.

$$\Delta\mathbf{x} = \mathbf{x}_f - \mathbf{x}_i$$

Remember that the initial and final positions are the start and the end of any interval you choose. Although position is a vector, sometimes the magnitude of a position is described without the boldface. In this case, a plus or minus sign might be used to indicate direction.

Vector addition and subtraction You will learn about many different types of vectors in physics, including velocity, acceleration, and momentum. Often, you will need to find the sum of two vectors or the difference between two vectors. A vector that represents the sum of two other vectors is called a **resultant.** **Figure 10** on the next page shows how to add and subtract vectors in one dimension. In a later module, you will learn how to add and subtract vectors in two dimensions.

Figure 10 You can use a diagram or an equation to combine vectors.
Analyze *What is the sum of a vector 12 m north and a vector 8 m north?*

COLOR CONVENTION
Displacement (x) ⟷ green

Example of Vector Addition

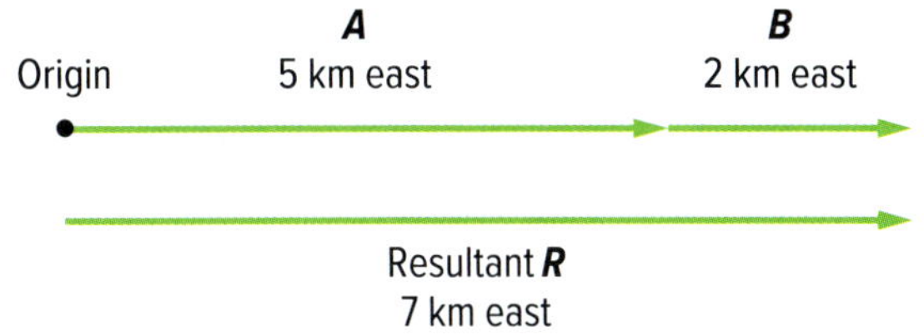

$$R = A + B = 5\text{ km} + 2\text{ km} = 7\text{ km}$$

$$\mathbf{R} = \mathbf{A} + \mathbf{B} = 7\text{ km east}$$

Examples of Vector Subtraction

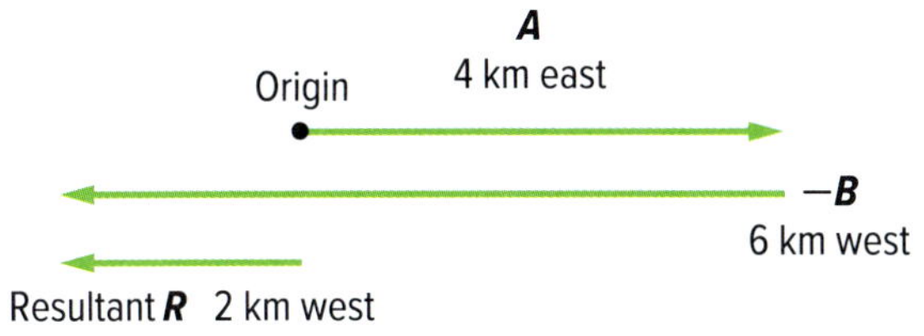

$$R = A - B = 4\text{ km} - 6\text{ km} = -2\text{ km}$$

$$\mathbf{R} = \mathbf{A} - \mathbf{B} = \mathbf{A} + (-\mathbf{B}) = 2\text{ km west}$$

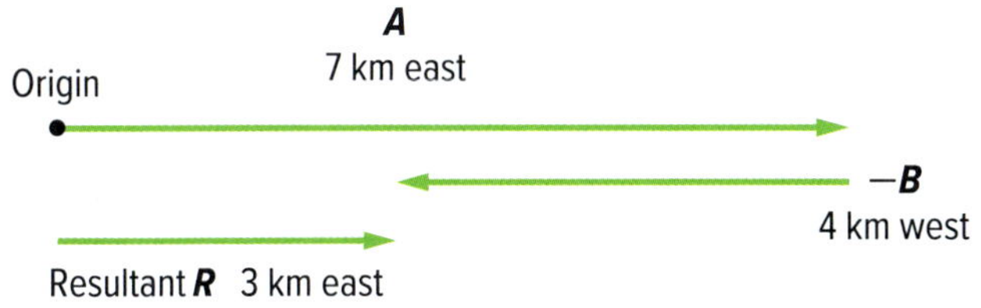

$$R = A - B = 7\text{ km} - 4\text{ km} = 3\text{ km}$$

$$\mathbf{R} = \mathbf{A} - \mathbf{B} = \mathbf{A} + (-\mathbf{B}) = 3\text{ km east}$$

Check Your Progress

6. **Coordinate System** Identify a coordinate system you could use to describe the motion of a girl swimming across a rectangular pool.

7. **Displacement** The motion diagram for a car traveling on an interstate highway is shown below. The starting and ending points are indicated.

   Start • • • • • • End

   Make a copy of the diagram. Draw a vector to represent the car's displacement from the starting time to the end of the third time interval.

8. **Position** Two students added a vector for a moving object's position at $t = 2$ s to a motion diagram. When they compared their diagrams, they found that their vectors did not point in the same direction. Explain.

9. **Displacement** The motion diagram for a boy walking to school is shown below.

   Home • • • • • • • • • • School

   Make a copy of this motion diagram, and draw vectors to represent the displacement between each pair of dots.

LEARNSMART® Go online to follow your personalized learning path to review, practice, and reinforce your understanding.

LESSON 3

POSITION-TIME GRAPHS

FOCUS QUESTION

What can you learn from position-time graphs?

Finding Positions

When analyzing complex motion, it often is useful to represent the motion in a variety of ways. A motion diagram contains information about an object's position at various times. Tables and graphs can also show this same information. Review the motion diagrams in **Figure 8** and **Figure 9**. You can use these diagrams to organize the times and corresponding positions of the runner, as in **Table 1.**

Plotting data The data listed in **Table 1** can be presented on a **position-time graph,** in which the time data is plotted on a horizontal axis and the position data is plotted on a vertical axis. The graph of the runner's motion is shown in **Figure 11**. To draw this graph, first plot the runner's positions. Then, draw a line that best fits the points.

Table 1 Position v. Time

Time (s)	Position (m)
0.0	0.0
1.0	5.0
2.0	10.0
3.0	15.0
4.0	20.0
5.0	25.0

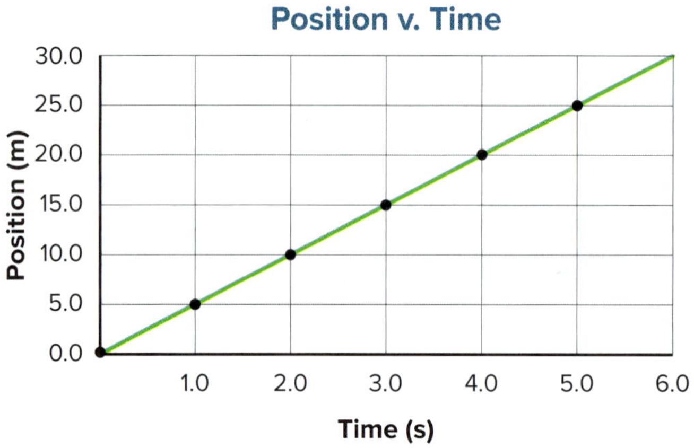

Figure 11 You can create a position-time graph by plotting the positions and times from the table. By drawing a best-fit line, you can estimate other times and positions.

3D THINKING DCI Disciplinary Core Ideas CCC Crosscutting Concepts SEP Science & Engineering Practices

COLLECT EVIDENCE

Use your Science Journal to record the evidence you collect as you complete the readings and activities in this lesson.

INVESTIGATE

GO ONLINE to find these activities and more resources.

CCC **Identify Crosscutting Concepts**

Create a table of the crosscutting concepts and fill in examples you find as you read.

Revisit the Encounter the Phenomenon Question

What information from this lesson can help you answer the Unit and Module questions?

Estimating time and position Notice that the graph is not a picture of the runner's path—the graphed line is sloped, but the runner's path was horizontal. Instead, the line represents the most likely positions of the runner at the times between the recorded data points. Even though there is no data point exactly when the runner was 12.0 m beyond her starting point or where she was at $t = 4.5$ s, you can use the graph to estimate the time or her position. The example problem on the next page shows how.

Instantaneous position How long did the runner spend at any location? Each position has been linked to a time, but how long did that time last? You could say "an instant," but how long is that? If an instant lasts for any finite amount of time, then the runner would have stayed at the same position during that time, and she would not have been moving. An instant is not a finite period of time, however. It lasts zero seconds. The symbol x represents the runner's **instantaneous position**—the position at a particular instant. Instantaneous position is usually simply called position.

Equivalent representations As shown in **Figure 12**, you now have several different ways to describe motion. You might describe motion using words, pictures (or pictorial representations), motion diagrams, data tables, or position-time graphs. All of these representations contain the same information about the runner's motion. However, depending on what you want to learn about an object's motion, some types of representations will be more useful than others.

Table 1 Position v. Time

Time (s)	Position (m)
0.0	0.0
1.0	5.0
2.0	10.0
3.0	15.0
4.0	20.0
5.0	25.0

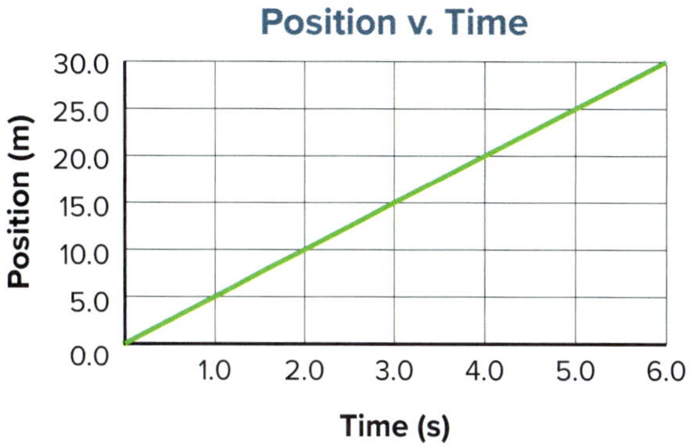

Figure 12 You can describe the runner's motion using the data table, the graph, and the motion diagram.

Identify *one benefit the table has over the graph.*

PHYSICS Challenge

POSITION-TIME GRAPHS Natana, Olivia, and Phil all enjoy exercising and often go to a path along the river for this purpose. Natana bicycles at a very consistent 40.25 km/h, Olivia runs south at a constant speed of 16.0 km/h, and Phil walks south at a brisk 6.5 km/h. Natana starts biking north at noon from the waterfalls. Olivia and Phil both start at 11:30 A.M. at the canoe dock, 20.0 km north of the falls.

1. Draw position-time graphs for each person.
2. At what time will the three exercise enthusiasts be located within the smallest distance interval from each other?
3. What is the length of that distance interval?

EXAMPLE Problem 1

ANALYZE A POSITION-TIME GRAPH When did the runner whose motion is described in **Figure 11** reach 12.0 m beyond the starting point? Where was she after 4.5 s?

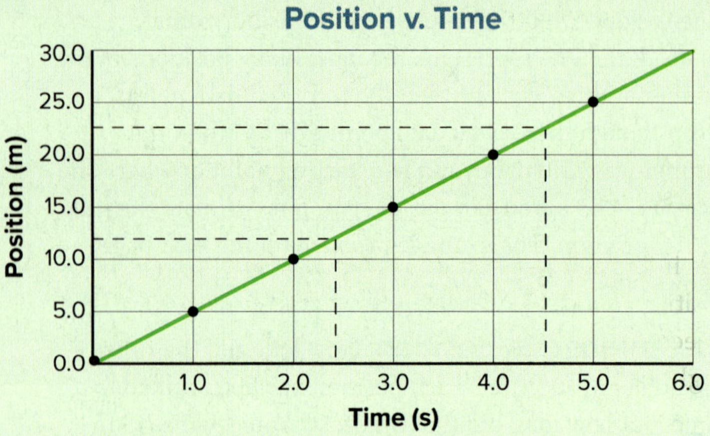

1 ANALYZE THE PROBLEM

Restate the questions.

Question 1: At what time was the magnitude of the runner's position (x) equal to 12.0 m?

Question 2: What was the runner's position at time $t = 4.5$ s?

2 SOLVE FOR THE UNKNOWN

Question 1

Examine the graph to find the intersection of the best-fit line with a horizontal line at the 12.0 m mark. Next, find where a vertical line from that point crosses the time axis. The value of t there is 2.4 s.

Question 2

Find the intersection of the graph with a vertical line at 4.5 s (halfway between 4.0 s and 5.0 s on this graph). Next, find where a horizontal line from that point crosses the position axis. The value of x is approximately 22.5 m.

PRACTICE Problems

ADDITIONAL PRACTICE

For problems 10–12, refer to ***Figure 13.***

10. The graph in **Figure 13** represents the motion of a car moving along a straight highway. Describe in words the car's motion.

11. Draw a particle model motion diagram that corresponds to the graph.

12. Answer the following questions about the car's motion. Assume that the positive x-direction is east of the origin and the negative x-direction is west of the origin.

a. At what time was the car's position 25.0 m east of the origin?

b. Where was the car at time $t = 1.0$ s?

c. What was the displacement of the car between times $t = 1.0$ s and $t = 3.0$ s?

13. The graph in **Figure 14** represents the motion of two pedestrians who are walking along a straight sidewalk in a city. Describe in words the motion of the pedestrians. Assume that the positive direction is east of the origin.

14. CHALLENGE Ari walked down the hall at school from the cafeteria to the band room, a distance of 100.0 m. A class of physics students recorded and graphed his position every 2.0 s, noting that he moved 2.6 m every 2.0 s. When was Ari at the following positions?

a. 25.0 m from the cafeteria

b. 25.0 m from the band room

c. Create a graph showing Ari's motion.

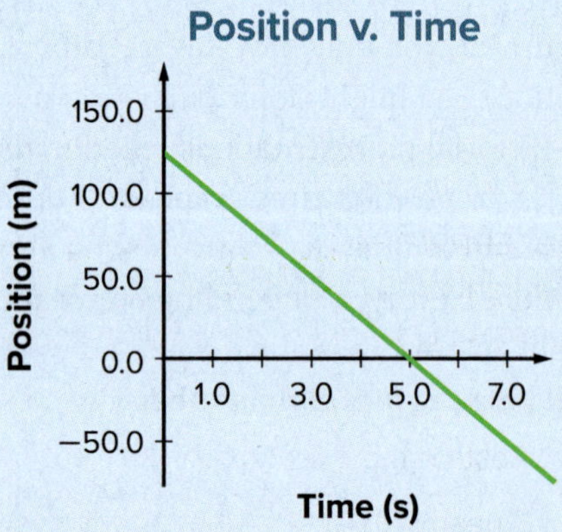

Figure 13

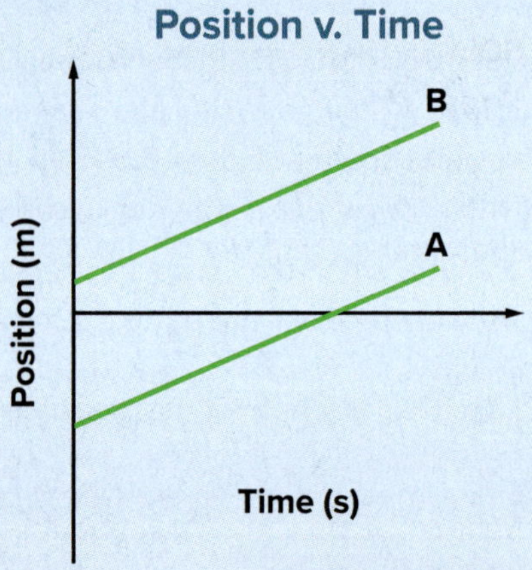

Figure 14

Multiple Objects on a Position-Time Graph

A position-time graph for two different runners is shown in Example Problem 2 below. Notice that runner A is ahead of runner B at time $t = 0$, but the motion of each runner is different. When and where does one runner pass the other? First, you should restate this question in physics terms: At what time are the two runners at the same position? What is their position at this time? You can evaluate these questions by identifying the point on the position-time graph at which the lines representing the two runners' motions intersect.

The intersection of two lines on a position-time graph tells you when objects have the same position, but does this mean that they will collide? Not necessarily. For example, if the two objects are runners and if they are in different lanes, they will not collide, even though they might be the same distance from the starting point.

Get It?

Explain what the intersection of two lines on a position-time graph means.

What else can you learn from a position-time graph? Notice in Example Problem 2 that the lines on the graph have different slopes. What does the slope of the line on a position-time graph tell you? In the next lesson, you will use the slope of a line on a position-time graph to determine the velocity of an object. When you study accelerated motion, you will draw other motion graphs and learn to interpret the areas under the plotted lines. In later studies, you will continue to refine your skills with creating and interpreting different types of motion graphs.

EXAMPLE Problem 2

INTERPRETING A GRAPH The graph to the right describes the motion of two runners moving along a straight path. The lines representing their motion are labeled A and B. When and where does runner B pass runner A?

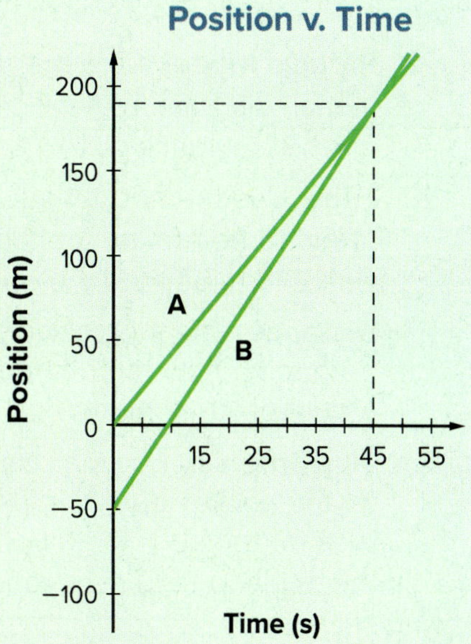

1 ANALYZE THE PROBLEM

Restate the questions.

Question 1: At what time are runner A and runner B at the same position?

Question 2: What is the position of runner A and runner B at this time?

2 SOLVE FOR THE UNKNOWN

Question 1

Examine the graph to find the intersection of the line representing the motion of runner A with the line representing the motion of runner B. These lines intersect at time 45 s.

Question 2

Examine the graph to determine the position when the lines representing the motion of the runners intersect. The position of both runners is about 190 m from the origin.

Runner B passes runner A about 190 m beyond the origin, 45 s after A has passed the origin.

PRACTICE Problems

ADDITIONAL PRACTICE

For problems 15–18, refer to the figure in Example Problem 2 on the previous page.

15. Where was runner A located at $t = 0$ s?

16. Which runner was ahead at $t = 48.0$ s?

17. When runner A was at 0.0 m, where was runner B?

18. How far apart were runners A and B at $t = 20.0$ s?

19. CHALLENGE Juanita goes for a walk to the north. Later her friend Heather starts to walk after her. Their motions are represented by the position-time graph in **Figure 15.**

a. How long had Juanita been walking when Heather started her walk?

b. Will Heather catch up to Juanita? How can you tell?

c. What was Juanita's position at $t = 0.2$ h?

d. At what time was Heather 5.0 km from the start?

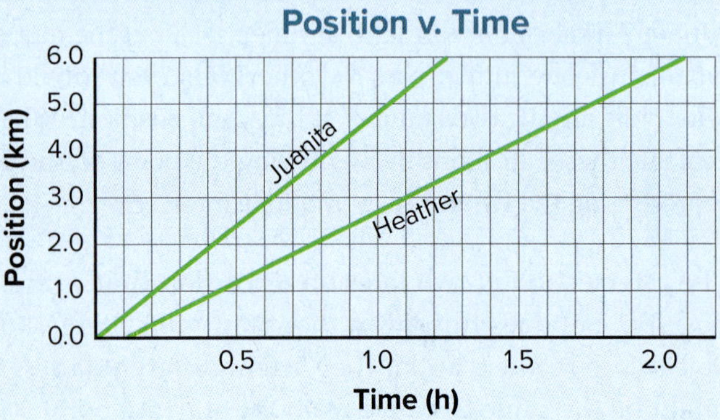

Figure 15

Check Your Progress

20. **Particle Diagram** Using the particle model motion diagram in **Figure 16** of a baby crawling across a kitchen floor, plot a position-time graph to represent the motion. The time interval between dots on the diagram is 1 s.

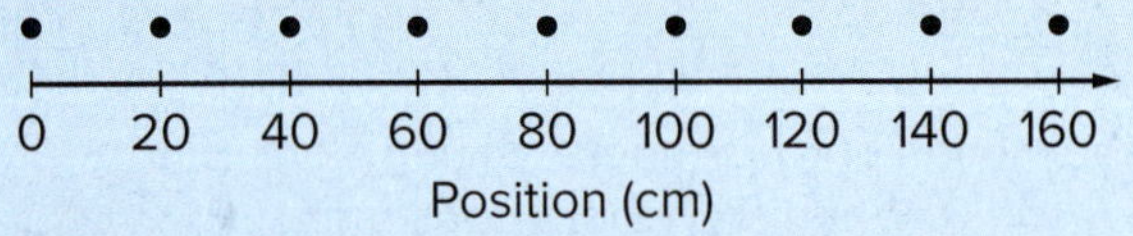

Figure 16

Figure 17

*For problems 21–24, refer to **Figure 17.***

21. **Particle Model** Create a particle model motion diagram from the position-time graph of a hockey puck gliding across the ice.

22. **Time** Use the hockey puck's position-time graph to determine the time when the puck was 10.0 m beyond the origin.

23. **Distance** Use the position-time graph to determine how far the hockey puck moved between 0.0 s and 5.0 s.

24. **Time Interval** Use the position-time graph for the hockey puck to determine the time it took for the puck to go from 40.0 m beyond the origin to 80.0 m beyond the origin.

25. **Critical Thinking** Look at the diagram and graph shown in **Figure 18**. Do they describe the same motion? Explain. The time intervals in the particle model diagram are 2 s.

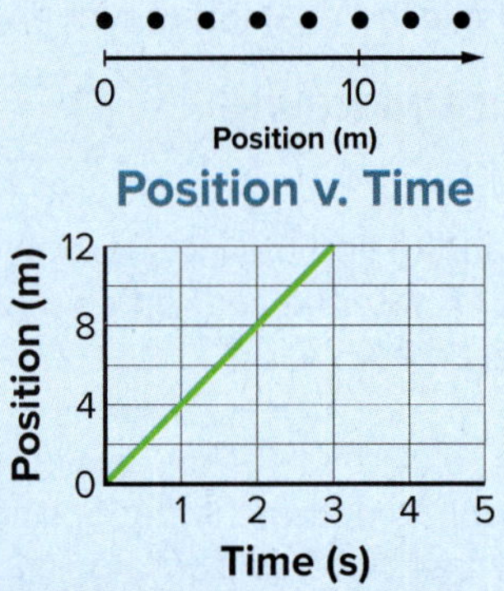

Figure 18

LEARNSMART® Go online to follow your personalized learning path to review, practice, and reinforce your understanding.

LESSON 4

HOW FAST?

FOCUS QUESITON

How do you describe how fast something is moving?

Velocity and Speed

Suppose you recorded the motion of two joggers on one diagram, as shown by the graph in **Figure 19.** The position of the jogger wearing red changes more than that of the jogger wearing blue. For a fixed time interval, the magnitude of the displacement (Δx) is greater for the jogger in red because she is moving faster. Now, suppose that each jogger travels 100 m. The time interval (Δt) for the 100 m would be smaller for the jogger in red than for the one in blue.

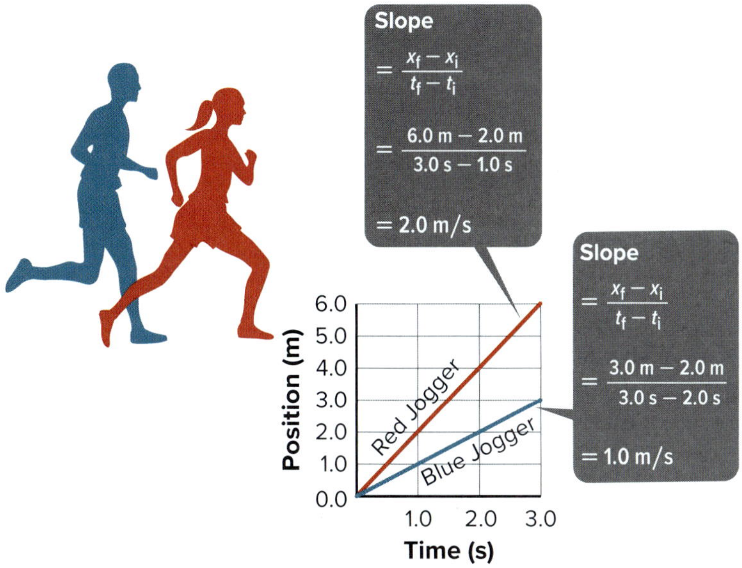

Figure 19 A greater slope shows that the red jogger traveled faster.

Analyze *How much farther did the red jogger travel than the blue jogger in the 3 s interval described by the graph?*

3D THINKING | DCI Disciplinary Core Ideas | CCC Crosscutting Concepts | SEP Science & Engineering Practices

COLLECT EVIDENCE

Use your Science Journal to record the evidence you collect as you complete the readings and activities in this lesson.

INVESTIGATE

GO ONLINE to find these activities and more resources.

PhysicsLAB: Constant Speed

Plan and carry out an investigation to relate distance, speed, and time interval.

Review the News

Obtain information from a current news story about speed or velocity. Evaluate your source and communicate your findings to your class.

Slope on a position-time graph Compare the lines representing the joggers in the graph in **Figure 19.** The slope of the red jogger's line is steeper, indicating a greater change in position during each time interval. Recall that you find the slope of a line by first choosing two points on the line. Next, you subtract the vertical coordinate (x in this case) of the first point from the vertical coordinate of the second point to obtain the rise of the line. After that, you subtract the horizontal coordinate (t in this case) of the first point from the horizontal coordinate of the second point to obtain the run. The rise divided by the run is the slope.

Get It?

Explain In the graph in **Figure 19**, the graph for each jogger starts at the coordinates (0, 0). Explain what this means in terms of position.

Average velocity Notice that the slope of the faster runner's line in **Figure 19** is a greater number. A greater slope indicates a faster speed. Also notice that the slope's units are meters per second. Looking at how the slope is calculated, you can see that slope is the change in the magnitude of the position divided by the time interval during which that change took place: $\frac{x_f - x_i}{t_f - t_i}$ or $\frac{\Delta x}{\Delta t}$. When Δx gets larger, the slope gets larger; when Δt gets larger, the slope gets smaller. This agrees with the interpretation given on the previous page of the speeds of the red and blue joggers. **Average velocity** is the ratio of an object's change in position to the time interval during which the change occurred. If the object is in uniform motion, so that its speed does not change, then its average velocity is the slope of its position-time graph.

Average velocity

Average velocity is defined as the change in position divided by the time during which the change occurred.

$$\bar{v} \equiv \frac{\Delta x}{\Delta t} = \frac{x_f - x_i}{t_f - t_i}$$

The symbol ≡ means that the left-hand side of the equation is defined by the right-hand side.

Interpreting slope The position-time graph's slope in **Figure 20** is −5.0 m/s. Notice that the slope of the graph indicates both magnitude and direction. By calculating the slope from the rise divided by the run between two points, you find that the object whose motion is represented by the graph has an average velocity of −5.0 m/s. The object started out at a positive position and moves toward the origin. After 4 s, it passes the origin and continues moving in the negative direction at a rate of 5.0 m/s.

Get It?

Explain the meaning of an upward and a downward slope of a position-time graph both above and below the x-axis.

Figure 20 The downward slope of this position-time graph shows that the motion is in the negative direction.

Analyze *What would the graph look like if the motion were at the same speed, but in the positive direction?*

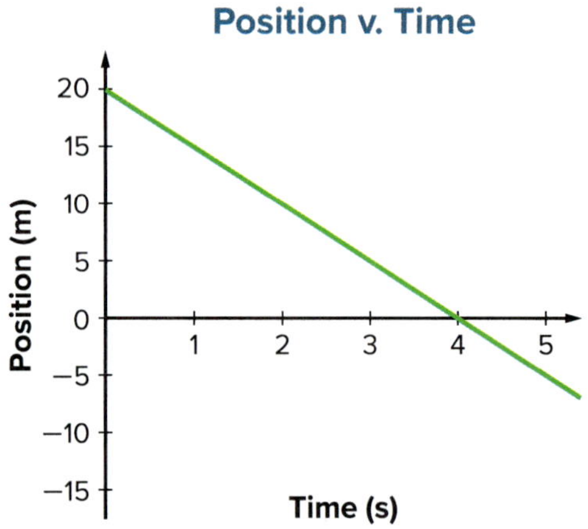

Average speed The slope's absolute value is the object's **average speed,** 5.0 m/s, which is the distance traveled divided by the time taken to travel that distance. For uniform motion, average speed is the absolute value of the slope of the object's position-time graph. The combination of an object's average speed ($\bar{v}$) and the direction in which it is moving is the average velocity ($\bar{v}$). Remember that if an object moves in the negative direction, its change in position is negative. This means that an object's displacement and velocity are both always in the same direction.

Instantaneous velocity Why do we call the quantity $\frac{\Delta x}{\Delta t}$ average velocity? Why don't we just call it velocity? A motion diagram shows the position of a moving object at the beginning and end of a time interval. It does not, however, indicate what happened within that time interval. During the time interval, the object's speed could have remained the same, increased, or decreased. The object may have stopped or even changed direction. You can find the average velocity for each time interval in the motion diagram, but you cannot find the speed and the direction of the object at any specific instant. The speed and the direction of an object at a particular instant is called the **instantaneous velocity.** In this textbook, the term velocity will refer to instantaneous velocity, represented by the symbol v.

Get It?

Explain how average velocity is different from velocity.

Average velocity on motion diagrams When an object moves between two points, its average velocity is in the same direction as its displacement. The two quantities are also proportional—when displacement is greater during a given time interval, so is average velocity. A motion diagram indicates the average velocity's direction and magnitude.

Imagine two cars driving down the road at different speeds. A video camera records the motion of the cars at the rate of one frame every second. Imagine that each car has a paintbrush attached to it that automatically descends and paints a red line on the ground for half a second every second. The faster car would paint a longer line on the ground. The vectors you draw on a motion diagram to represent the velocity are like the lines that the paintbrushes make on the ground below the cars. In this book, we use red to indicate velocity vectors on motion diagrams. **Figure 21** shows motion diagrams with velocity vectors for two cars. One is moving to the right, and the other is moving to the left.

Get It?

Identify what the lengths of velocity vectors mean.

Real-World Physics

SPEED RECORDS The world record for the men's 100-m dash is 9.58 s, established in 2009 by Usain Bolt. The world record for the women's 100-m dash is 10.49 s, established in 1988 by Florence Griffith-Joyner.

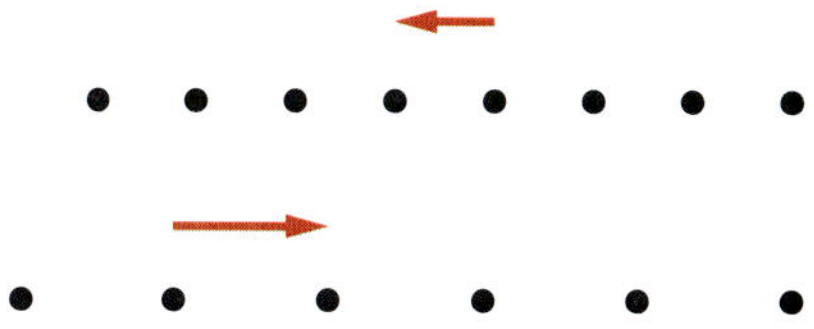

Figure 21 The length of each velocity vector is proportional to the magnitude of the velocity that it represents.

EXAMPLE Problem 3

AVERAGE VELOCITY The graph at the right describes the straight-line motion of a student riding her skateboard along a smooth, pedestrian-free sidewalk. What is her average velocity? What is her average speed?

1 ANALYZE AND SKETCH THE PROBLEM

Identify the graph's coordinate systems.

UNKNOWN

$\bar{v} = ?$ $\quad\quad \bar{v} = ?$

2 SOLVE FOR THE UNKNOWN

Find the average velocity using two points on the line.

$$\bar{v} = \frac{\Delta x}{\Delta t}$$

$$= \frac{x_f - x_i}{t_f - t_i}$$

$$= \frac{12.0\text{ m} - 0.0\text{ m}}{7.0\text{ s} - 0.0\text{ s}}$$ Substitute $x_f = 12.0$ m, $x_i = 0.0$ m, $t_f = 7.0$ s, $t_i = 0.0$ s.

$\bar{v} = 1.7$ m/s in the positive direction

The average speed ($\bar{v}$) is the absolute value of the average velocity, or 1.7 m/s.

3 EVALUATE THE ANSWER

Are the units correct? The units for both velocity and speed are meters per second.

Do the signs make sense? The positive sign for the velocity agrees with the coordinate system. No direction is associated with speed.

PRACTICE Problems

26. The graph in **Figure 22** describes the motion of a cruise ship drifting slowly through calm waters. The positive *x*-direction (along the vertical axis) is defined to be south.

a. What is the ship's average speed?

b. What is its average velocity?

27. Describe, in words, the cruise ship's motion in the previous problem.

28. What is the average velocity of an object that moves from 6.5 cm to 3.7 cm relative to the origin in 2.3 s?

29. The graph in **Figure 23** represents the motion of a bicycle.

a. What is the bicycle's average speed?

b. What is its average velocity?

30. Describe, in words, the bicycle's motion in the previous problem.

31. CHALLENGE When Marshall takes his pet dog for a walk, the dog walks at a very consistent pace of 0.55 m/s. Draw a motion diagram and a position-time graph to represent Marshall's dog walking the 19.8-m distance from in front of his house to the nearest stop sign.

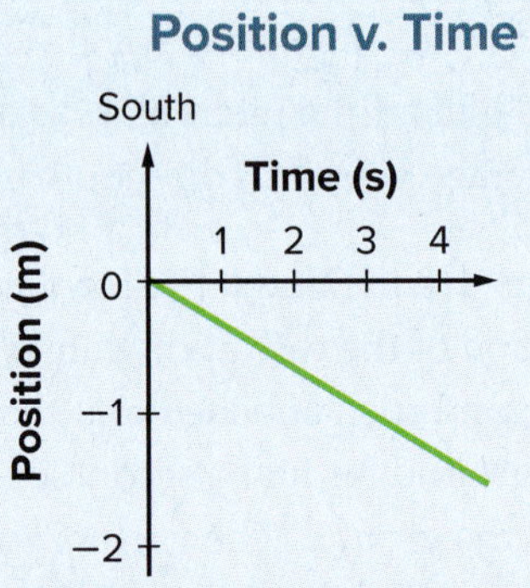

Figure 22

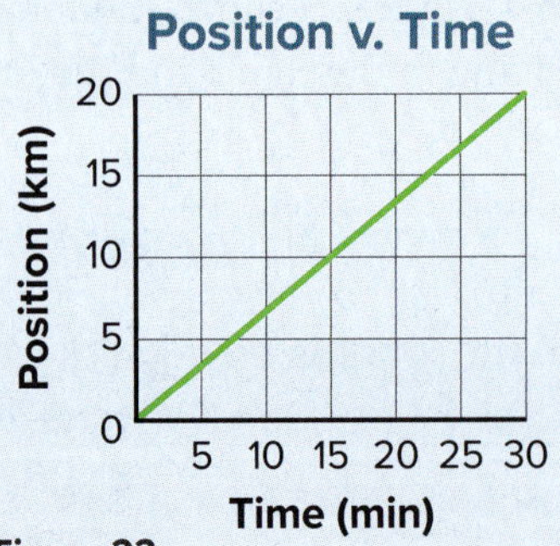

Figure 23

Equation of Motion

Often it is more efficient to use an equation, rather than a graph, to solve problems. Any time you graph a straight line, you can find an equation to describe it. Take another look at the graph in **Figure 20** for the object moving with a constant velocity of −5.0 m/s. Recall that you can represent any straight line with the equation $y = mx + b$, where y is the quantity plotted on the vertical axis, m is the line's slope, x is the quantity plotted on the horizontal axis, and b is the line's y-intercept.

For the graph in **Figure 20**, the quantity plotted on the vertical axis is position, represented by the variable $\boldsymbol{x}$. The line's slope is −5.0 m/s, which is the object's average velocity ($\bar{v}$). The quantity plotted on the horizontal axis is time (t). The y-intercept is 20.0 m. What does this 20.0 m represent? This shows that the object was at a position of 20.0 m when $t = 0.0$ s. This is called the initial position of the object and it is designated $\boldsymbol{x}_i$.

A summary is given below of how the general variables in the straight-line formula are changed to the specific variables you have been using to describe motion. The table also shows the numerical values for the average velocity and initial position. Consider the graph shown in **Figure 20.** The mathematical equation for the line graphed is as follows:

$$y = (-5.0\ \text{m/s})x + 20.0\ \text{m}$$

You can rewrite this equation, using $\boldsymbol{x}$ for position and t for time.

$$\boldsymbol{x} = (-5.0\ \text{m/s})t + 20.0\ \text{m}$$

It might be confusing to use y and x in math but use $\boldsymbol{x}$ and t in physics. You do this because there are many types of graphs in physics, including position v. time graphs, velocity v. time graphs, and force v. position graphs. For a position v. time graph, the math equation $y = mx + b$ can be rewritten as follows:

Position

An object's position is equal to the average velocity multiplied by time plus the initial position.

$$\boldsymbol{x} = \bar{\boldsymbol{v}}t + \boldsymbol{x}_i$$

This equation gives you another way to represent motion. Note that a graph of x v. t would be a straight line.

CONNECTING MATH to Physics

Lines and Graphs Symbols used in the point-slope equation of a line relate to symbols used for motion variables on a position-time graph.

General Variable	Specific Motion Variable	Value in Figure 20
y	$\boldsymbol{x}$	
m	$\bar{\boldsymbol{v}}$	−5.0 m/s
x	t	
b	$\boldsymbol{x}_i$	20.0 m

EXAMPLE Problem 4

POSITION The figure shows a motorcyclist traveling east along a straight road. After passing point ***B***, the cyclist continues to travel at an average velocity of 12 m/s east and arrives at point ***C*** 3.0 s later. What is the position of point ***C***?

1 ANALYZE THE PROBLEM

Choose a coordinate system with the origin at ***A***.

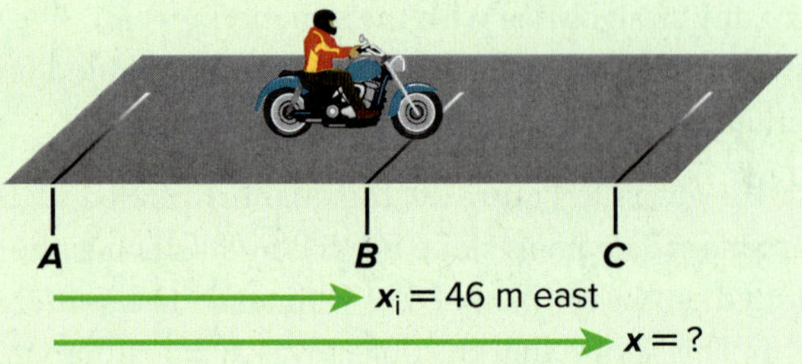

KNOWN	**UNKNOWN**
$\bar{\boldsymbol{v}}$ = 12 m/s east	$\boldsymbol{x}$ = ?
$\boldsymbol{x}_i$ = 46 m east	
t = 3.0 s	

2 SOLVE FOR THE UNKNOWN

$x = \bar{v}t + x_i$ — Use magnitudes for the calculations.

$= (12\text{ m/s})(3.0\text{ s}) + 46\text{ m}$ — Substitute $\bar{v}$ = 12 m/s, t = 3.0 s, and x_i = 46 m.

$= 82\text{ m}$

$\boldsymbol{x}$ = 82 m east

3 EVALUATE THE ANSWER

Are the units correct? Position is measured in meters.

Does the direction make sense? The motorcyclist is traveling east the entire time.

PRACTICE Problems

For problems 32–35, refer to ***Figure 24.***

32. The diagram at the right shows the path of a ship that sails at a constant velocity of 42 km/h east. What is the ship's position when it reaches point ***C***, relative to the starting point, ***A***, if it sails from point ***B*** to point ***C*** in exactly 1.5 h?

33. Another ship starts at the same time from point ***B***, but its average velocity is 58 km/h east. What is its position, relative to ***A***, after 1.5 h?

34. What would a ship's position be if that ship started at point ***B*** and traveled at an average velocity of 35 km/h west to point ***D*** in a time period of 1.2 h?

35. **CHALLENGE** Suppose two ships start from point ***B*** and travel west. One ship travels at an average velocity of 35 km/h for 2.2 h. Another ship travels at an average velocity of 26 km/h for 2.5 h. What is the final position of each ship?

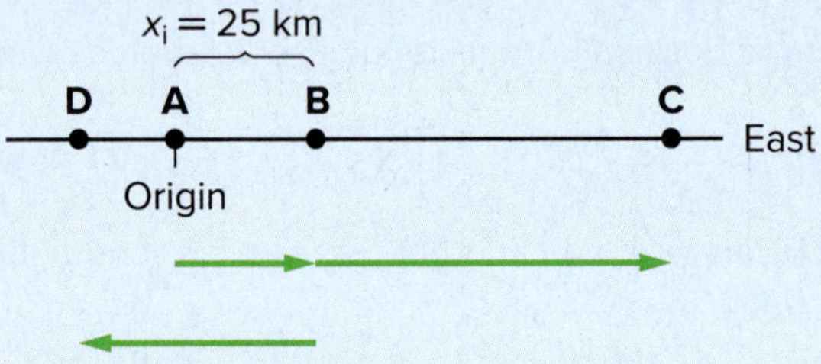

Figure 24

After your study of this lesson, you probably realize how closely linked physics is to mathematics. Algebraic thinking is used to examine scientific data and predict the effect of a change in one variable on another. For example, you can predict the change in average velocity of an object by examining the change in position over time. Using mathematics to model motion is a useful tool for physicists and physics students.

Check Your Progress

36. **Velocity and Position** How is an object's velocity related to its position?

For problems 37–39, refer to ***Figure 25.***

37. **Ranking Task** Rank the position-time graphs according to the average speed, from greatest average speed to least average speed. Specifically indicate any ties.

38. **Contrast Average Velocities** Describe differences in the average velocities shown on the graph for objects A and B. Describe differences in the average velocities shown on the graph for objects C and D.

39. **Ranking Task** Rank the graphs in **Figure 25** according to each object's initial position, from most positive position to most negative position. Specifically indicate any ties. Would your ranking be different if you ranked according to initial distance from the origin?

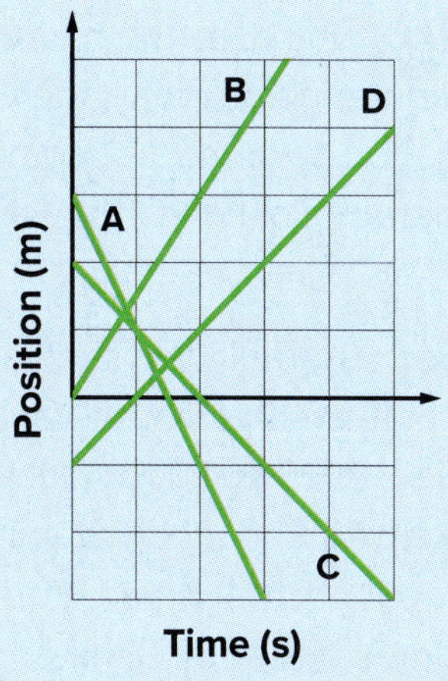

Figure 25

40. **Average Speed and Average Velocity** Explain how average speed and average velocity are related to each other for an object in uniform motion.

41. **Position** Two cars are traveling along a straight road, as shown in **Figure 26.** They pass each other at point B and then continue in opposite directions. The orange car travels for 0.25 h from point B to point C at a constant velocity of 32 km/h east. The blue car travels for 0.25 h from point B to point D at a constant velocity of 48 km/h west. How far has each car traveled from point B? What is the position of each car relative to the origin, point A?

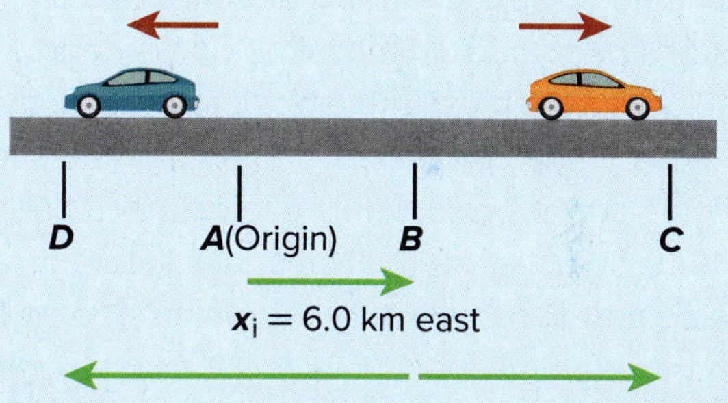

Figure 26

42. **Position** A car travels north along a straight highway at an average speed of 85 km/h. After driving 2.0 km, the car passes a gas station and continues along the highway. What is the car's position relative to the start of its trip 0.25 h after it passes the gas station?

43. **Critical Thinking** In solving a physics problem, why is it important to create pictorial and physical models before trying to solve an equation?

LEARNSMART® Go online to follow your personalized learning path to review, practice, and reinforce your understanding.

In the Nick of Time

Global Positioning System (GPS) and other navigation systems rely on satellites with extremely accurate atomic clocks to keep track of time. However, some navigation systems must operate in environments—such as inside buildings, in the depths of canyons, or deep under water—where radio signals from satellites cannot reach them. Navigation systems that rely on satellites are also susceptible to intentional interference due to hacking or jamming. To address this problem, scientists have developed tiny atomic clocks for navigation systems small enough to be carried by a person.

Chip-scale atomic clocks (CSACs) allow for astoundingly accurate and precise time-keeping in a very small package.

Chip-scale atomic clocks

Old-fashioned clocks count the seconds by the swings of a pendulum. Mechanical clocks—and even digital clocks, which use the electrical oscillations in power lines or the oscillations of quartz crystals to keep time—are not completely accurate and precise. So, in 1949, scientists at the National Institute of Standards and Technology (NIST) took a step toward solving this problem by developing the first atomic clock.

Atomic clocks count the seconds by oscillations between atomic energy states of various elements, including cesium, rubidium, and strontium. Over the years, advances in technology have led to the creation of atomic clocks that are more and more accurate and precise. The most current models can keep perfect time for the next 15 billion years.

In the early 2000s, NIST began working with the U.S. Department of Defense's Defense Advanced Research Projects Agency (DARPA) to create atomic clocks small enough to work in portable navigation systems. In 2004, they introduced the first chip-scale atomic clock (CSAC) which, as the name implies, is the size of a computer chip. Today, the CSAC is used in portable navigation systems, telecommunications systems, and other applications, both civilian and military.

Even CSACs have vulnerabilities, such as temperature fluctuations that cause variations in atomic frequencies. In 2016, DARPA kicked off the Atomic Clocks with Enhanced Stability (ACES) project with the goal of creating an atomic clock that performs 1,000 times better than the original CSAC.

USE A MODEL TO ILLUSTRATE

Use print or online sources to find a diagram of an atomic clock or a CSAC. Study the diagram and then use it to write a short text explaining how the device works.

CALIFORNIA CONNECTION

Use a search engine to find information about the Deep Space Atomic Clock, developed by NASA's Jet Propulsion Laboratory in Pasadena, California. Use a creative method to present your research.

MODULE 2

STUDY GUIDE

GO ONLINE to study with your Science Notebook.

Lesson 1 PICTURING MOTION

- A motion diagram shows the position of an object at successive equal time intervals.
- In a particle model motion diagram, an object's position at successive times is represented by a series of dots. The spacing between dots indicates whether the object is moving faster or slower.

- motion diagram
- particle model

Lesson 2 WHERE AND WHEN?

- A coordinate system gives the location of the zero point of the variable you are studying and the direction in which the values of the variable increase.
- A vector drawn from the origin of a coordinate system to an object indicates the object's position in that coordinate system. The directions chosen as positive and negative on the coordinate system determine whether the objects' positions are positive or negative in the coordinate system.
- A time interval is the difference between two times.

$$\Delta t = t_f - t_i$$

- Change in position is displacement, which has both magnitude and direction.

$$\Delta x = x_f - x_i$$

- On a motion diagram, the displacement vector's length represents how far the object was displaced. The vector points in the direction of the displacement, from x_i to x_f

- coordinate system
- origin
- position
- distance
- magnitude
- vector
- scalar
- time interval
- displacement
- resultant

Lesson 3 POSITION-TIME GRAPHS

- Position-time graphs provide information about the motion of objects. They also might indicate where and when two objects meet.
- The line on a position-time graph describes an object's position at each time.
- Motion can be described using words, motion diagrams, data tables, or graphs.

- position-time graph
- instantaneous position

Lesson 4 HOW FAST?

- An object's velocity tells how fast it is moving and in what direction it is moving.
- Speed is the magnitude of velocity.
- Slope on a position-time graph describes the average velocity of the object.

$$\bar{\boldsymbol{v}} \equiv \frac{\Delta x}{\Delta t} = \frac{x_f - x_i}{t_f - t_i}$$

- A simple equation relates an object's initial position (x_i), its constant average velocity ($\bar{\boldsymbol{v}}$), its position (x), and the time (t) since the object was at its initial position.

$$x = \bar{\boldsymbol{v}}t + x_i$$

- average velocity
- average speed
- instantaneous velocity

THREE-DIMENSIONAL THINKING

Module Wrap-Up

REVISIT THE PHENOMENON

How does a GPS unit know where you are?

CER Claim, Evidence, Reasoning

Explain Your Reasoning Revisit the claim you made when you encountered the phenomenon. Summarize the evidence you gathered from your investigations and research and finalize your Summary Table. Does your evidence support your claim? If not, revise your claim. Explain why your evidence supports your claim.

STEM UNIT PROJECT

Now that you've completed the module, revisit your STEM unit project. You will summarize your evidence and apply it to the project.

GO FURTHER

SEP Data Analysis Lab

How can you rank velocity from a graph?

Four cows were walking back to the barn at the end of the day. The position-time graph shows the motion of the four cows.

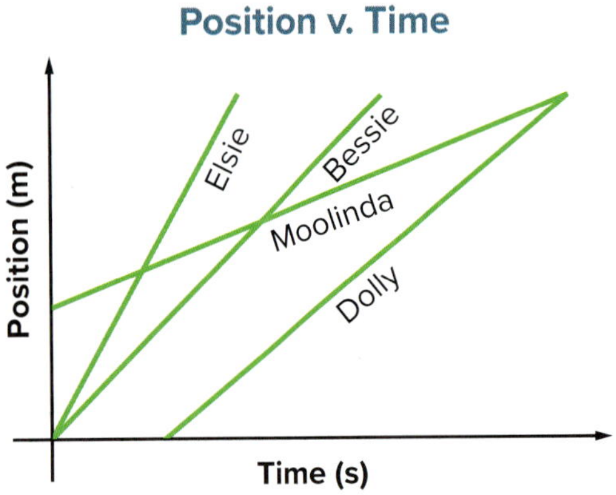

CER Analyze and Interpret Data

1. **Compare and contrast the average velocities of the cows.**
2. **Claim** Rank the cows according to their average velocities.
3. **Evidence and Reasoning** Explain the evidence you obtained from the position-time graph and your reasoning to justify your ranking.

Aurora Photos/Alamy Stock Photo

MODULE 3

ACCELERATED MOTION

Steve Whittington/Moment/Getty Images

ACCELERATED MOTION

ENCOUNTER THE PHENOMENON

Why do sudden changes in the direction or speed of jet planes affect pilots?

GO ONLINE to watch a video about the effects of acceleration on pilots.

SEP Ask Questions

Do you have other questions about the phenomenon? If so, add them to the driving question board.

CER Claim, Evidence, Reasoning

Make Your Claim Use your CER chart to make a claim about why sudden changes in the direction or speed of jet planes affect pilots. Explain your reasoning.

Collect Evidence Use the lessons in this module to collect evidence to support your claim. Record your evidence as you move through the module.

Explain Your Reasoning You will revisit your claim and explain your reasoning at the end of the module.

GO ONLINE to access your CER chart and explore resources that can help you collect evidence.

LESSON 1: Explore & Explain: Calculating Acceleration

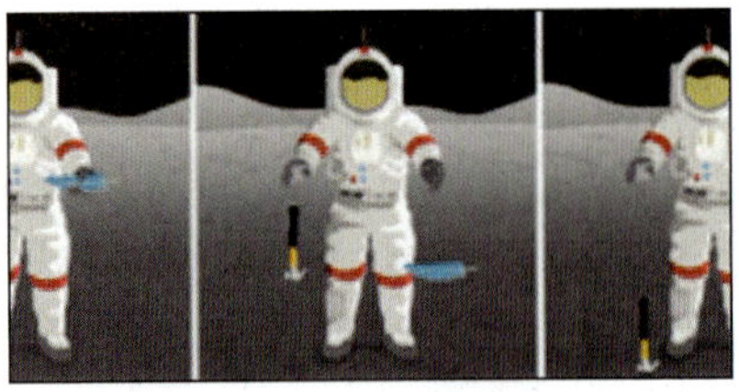

LESSON 3: Explore & Explain: Galileo's Discovery

Smithsonian

Additional Resources

(t)Video Supplied by BBC Worldwide Learning, (b)ImagineGolf/iStock/Getty Images

LESSON 1
ACCELERATION

FOCUS QUESTION

What are two ways velocity can change?

Nonuniform Motion Diagrams

An object in uniform motion moves along a straight line with an unchanging velocity, but few objects move this way all the time. More common is nonuniform motion, in which velocity is changing. In this module, you will study nonuniform motion along a straight line. Examples include balls rolling down hills, cars braking to a stop, and falling objects. In later modules you will analyze nonuniform motion that is not confined to a straight line, such as motion along a circular path and the motion of thrown objects, such as baseballs.

Describing nonuniform motion Uniform motion feels smooth. If you close your eyes, it feels as if you are not moving at all. In contrast, when you move around a curve or up and down a roller coaster hill, you feel pushed or pulled.

In the first diagram in **Figure 1**, the single image indicates that the person is motionless, like a runner waiting for the signal at the start of a race. In the other diagrams, the distances between successive positions change in different ways. In the second diagram, the distances are the same, which indicates that the jogger is in uniform motion at a constant velocity. In the third diagram, the distance increases because the jogger speeds up. In the fourth diagram, the distance decreases because the jogger slows down.

Motion Diagram

Figure 1 The distance the jogger moves in each time interval indicates the type of motion.

3D THINKING — DCI Disciplinary Core Ideas — CCC Crosscutting Concepts — SEP Science & Engineering Practices

COLLECT EVIDENCE

Use your Science Journal to record the evidence you collect as you complete the readings and activities in this lesson.

INVESTIGATE

GO ONLINE to find these activities and more resources.

Virtual Investigation: Accelerated Motion
Use a computer model to relate acceleration, change in velocity, and time.

Probeware Lab: Tossed Ball Motion
Carry out an investigation to determine how the position, velocity, and acceleration of a tossed ball change over time.

Particle model diagram What does a particle model motion diagram look like for an object with changing velocity? **Figure 2** shows particle model motion diagrams below the motion diagrams of the jogger when she is speeding up and slowing down. There are two major indicators of the change in velocity in this form of the motion diagram. The change in the spacing of the dots and the differences in the lengths of the velocity vectors indicate the changes in velocity. If an object speeds up, each subsequent velocity vector is longer, and the spacing between dots increases. If the object slows down, each vector is shorter than the previous one, and the spacing between dots decreases. Both types of motion diagrams indicate how an object's velocity is changing.

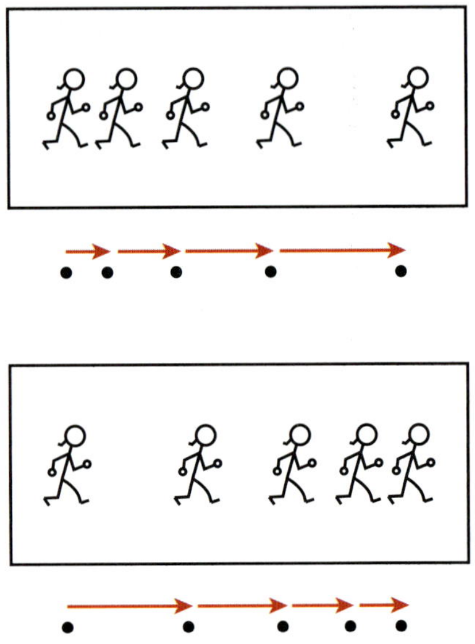

Figure 2 The change in length of the velocity vectors on these motion diagrams indicates whether the jogger is speeding up or slowing down.

Get It?

Analyze What do increasing and decreasing lengths of velocity vectors indicate on a motion diagram?

Displaying acceleration on a motion diagram For a motion diagram to give a full picture of an object's movement, it should contain information about the rate at which the object's velocity is changing. The rate at which an object's velocity changes is called the **acceleration** of the object. By including acceleration vectors on a motion diagram, you can indicate the rate of change for the velocity.

Figure 3 shows a particle motion diagram for an object with increasing velocity. Notice that the lengths of the red velocity vectors get longer from left to right along the diagram. The figure also describes how to use the diagram to draw an acceleration vector for the motion. The acceleration vector that describes the increasing velocity is shown in violet on the diagram.

Notice in the figure that if the object's acceleration is constant, you can determine the length and direction of an acceleration vector by subtracting two consecutive velocity vectors and dividing by the time interval. That is, first find the change in velocity, $\Delta v = v_f - v_i = v_f + (-v_i)$, where v_i and v_f refer to the velocities at the beginning and the end of the chosen time interval. Then divide by the time interval (Δt). The time interval between each dot in **Figure 3** is 1 s. You can draw the acceleration vector from the tail of the final velocity vector to the tip of the initial velocity vector.

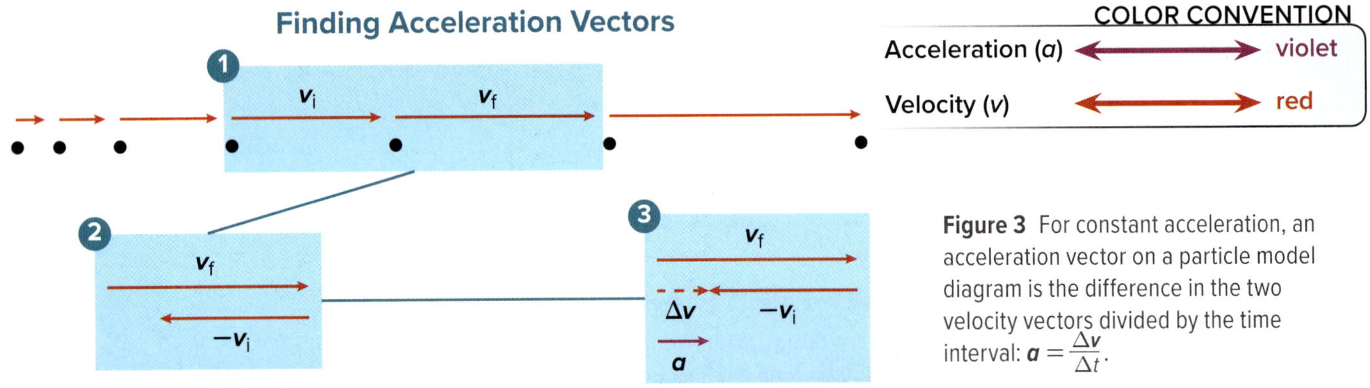

First, draw v_f. Below that, draw v_i with its tail aligned with the tip of v_f.

Next, draw the vector Δv from the tail of v_f to the tip of v_i. The acceleration vector a is the same as Δv divided by the time interval.

Figure 3 For constant acceleration, an acceleration vector on a particle model diagram is the difference in the two velocity vectors divided by the time interval: $a = \frac{\Delta v}{\Delta t}$.

Analyze *Can you draw an acceleration vector for two successive velocity vectors that are the same length and direction? Explain.*

Direction of Acceleration

Consider the four situations shown in **Figure 4** in which an object can accelerate by changing speed. The first motion diagram shows the car moving in the positive direction and speeding up. The second motion diagram shows the car moving in the positive direction and slowing down. The third shows the car speeding up in the negative direction, and the fourth shows the car slowing down as it moves in the negative direction. The figure also shows the velocity vectors for the second time interval of each diagram, along with the corresponding acceleration vectors. Note that Δt is equal to 1 s.

In the first and third situations, when the car is speeding up, the velocity and acceleration vectors point in the same direction. In the other two situations, in which the acceleration vector is in the opposite direction from the velocity vectors, the car is slowing down. In other words, when the car's acceleration is in the same direction as its velocity, the car's speed increases. When they are in opposite directions, the speed of the car decreases.

Both the direction of an object's velocity and its direction of acceleration are needed to determine whether it is speeding up or slowing down. An object has a positive acceleration when the acceleration vector points in the positive direction and a negative acceleration when the acceleration vector points in the negative direction. It is important to notice that the sign of acceleration alone does not indicate whether the object is speeding up or slowing down.

Get It?

Describe the motion of an object if its velocity and acceleration vectors have opposite signs.

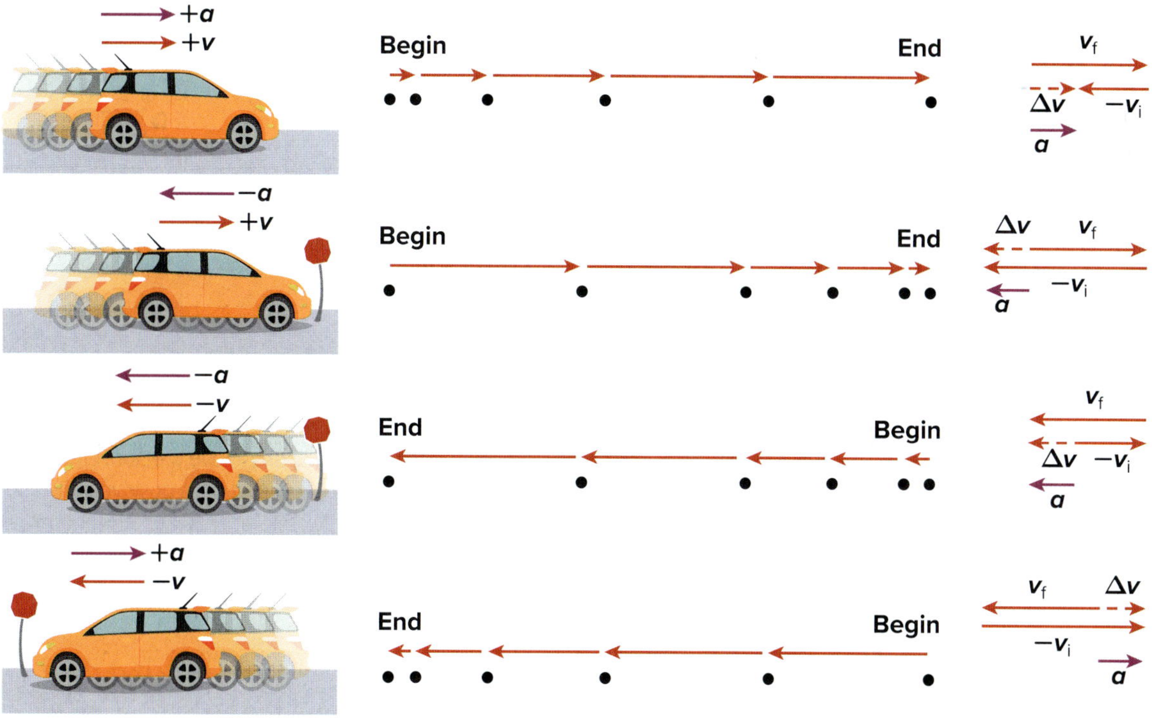

Figure 4 You need to know the direction of both the velocity and acceleration vectors in order to determine whether an object is speeding up or slowing down.

Velocity-Time Graphs

Just as it was useful to plot position versus time, it also is useful to plot velocity versus time to analyze an object's motion. On a **velocity-time graph,** or v-t graph, velocity is plotted on the vertical axis and time is plotted on the horizontal axis.

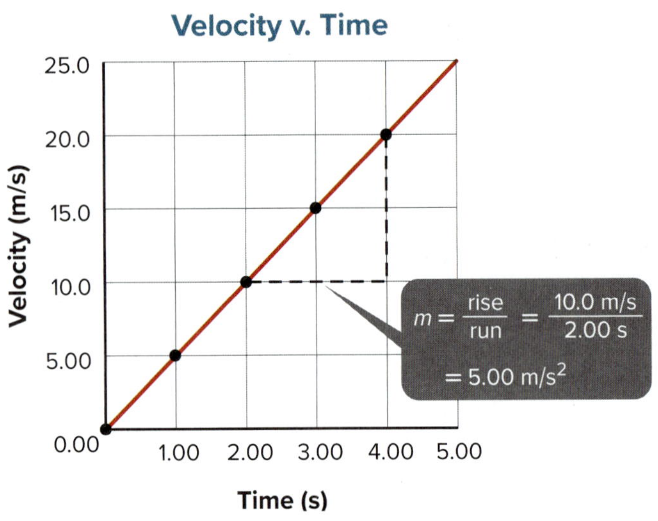

Figure 5 You can determine acceleration from a velocity-time graph by calculating the slope of the data. The slope is the rise divided by the run using any two points on the line.

Slope The velocity-time graph for a car that started at rest and sped up along a straight stretch of road is shown in **Figure 5.** The positive direction has been chosen to be the same as that of the car's motion. Notice that the graph is a straight line. This means the car sped up at a constant rate. The rate at which the car's velocity changed can be found by calculating the slope of the velocity-time graph

Get It?

Identify What can you conclude about the acceleration of an object if the graph of its motion is a straight line on a velocity-time graph?

The graph shows that the slope is 5.00 (m/s)/s, which is often written as 5.00 m/s^2. Consider the time interval between 4.00 s and 5.00 s. At 4.00 s, the car's velocity was 20.0 m/s in the positive direction. At 5.00 s, the car was traveling at 25.0 m/s in the same direction. Thus, in 1.00 s, the car's velocity increased by 5.0 m/s in the positive direction. When the velocity of an object changes at a constant rate, it has a constant acceleration.

Reading velocity-time graphs The motions of five runners are shown in **Figure 6.** Assume that the positive direction is east. The slopes of Graphs A and E are zero. Thus, the accelerations are zero. Both graphs show motion at a constant velocity–Graph A to the east and Graph E to the west. Graph B shows motion with a positive velocity eastward. Its slope indicates a constant, positive acceleration. You can infer that the speed increases because velocity and acceleration are positive. Graph C has a negative slope. It shows motion that begins with a positive velocity, slows down, and then stops. This means the acceleration and the velocity are in opposite directions. The point at which Graphs C and B cross shows that the runners' velocities are equal at that time. It does not, however, identify their positions.

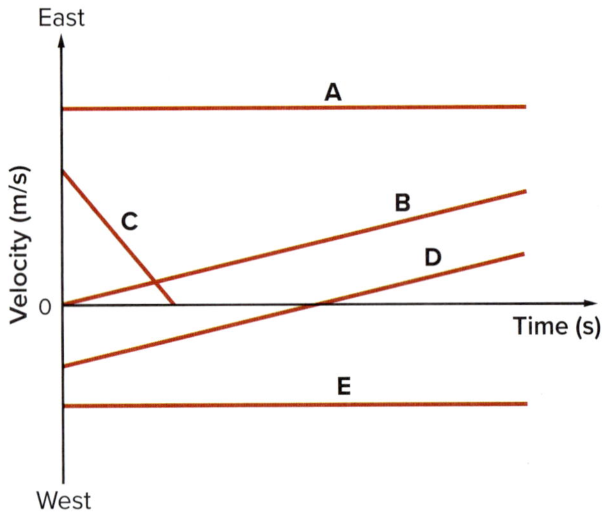

Figure 6 Because east is chosen as the positive direction on the graph, velocity is positive if the line is above the horizontal axis and negative if the line is below it. Acceleration is positive if the line is slanted upward on the graph. Acceleration is negative if the line is slanted downward on the graph. A horizontal line indicates constant velocity and zero acceleration.

Graph D indicates motion that starts out toward the west, slows down, for an instant has zero velocity, and then moves east with increasing speed. The slope of Graph D is positive. Because velocity and acceleration are initially in opposite directions, the speed decreases to zero at the time the graph crosses the *x*-axis. After that time, velocity and acceleration are in the same direction, and the speed increases.

Get It?

Describe the meaning of a line crossing the *x*-axis in a velocity-time graph.

Average and Instantaneous Acceleration

How does it feel differently if the car you ride in accelerates a little or a lot? As with velocity, the acceleration of most moving objects continually changes. If you want to describe an object's acceleration, it is often more convenient to describe the overall change in velocity during a certain time interval rather than describing the continual change.

The **average acceleration** of an object is its change in velocity during some measurable time interval divided by that time interval. Average acceleration is measured in meters per second per second (m/s/s), or simply meters per second squared (m/s^2). A car might accelerate quickly at times and more slowly at times. Just as average velocity depends only on the starting and ending displacement, average acceleration depends only on the starting and ending velocity during a time interval.

Figure 7 shows a graph of motion in which the acceleration is changing. The average acceleration during a certain time interval is determined just as it is in **Figure 5** for constant acceleration. Notice, however, that because the line is curved, the average acceleration in this graph varies depending on the time interval that you choose.

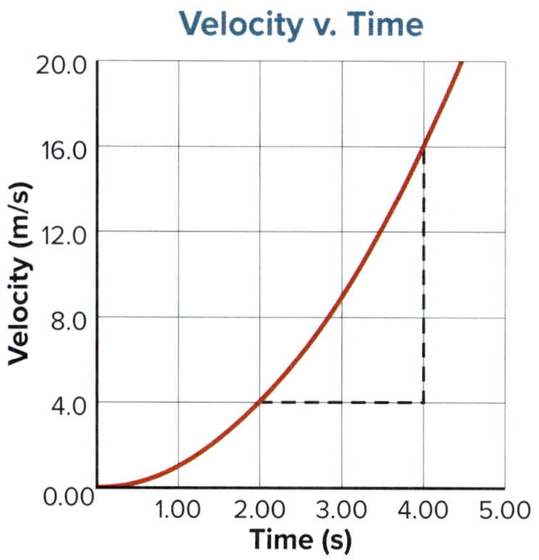

Figure 7 A curved line on a velocity-time graph shows that the acceleration is changing. The slope indicates the average acceleration during a time interval that you choose.

Calculate *How large is the average acceleration between 0.00 s and 2.00 s?*

The change in an object's velocity at an instant of time is called **instantaneous acceleration.** You can determine the instantaneous acceleration of an object by drawing a tangent line on the velocity-time graph at the point of time in which you are interested. A tangent line is a line that intersects the graph at one and only one point. The slope of this line is equal to the instantaneous acceleration. Most of the situations considered in this textbook assume an ideal case of constant acceleration. When the acceleration is the same at all points during a time interval, the average acceleration and the instantaneous accelerations are equal.

Get It?

Contrast How is instantaneous acceleration different from average acceleration?

Calculating Acceleration

How can you describe the acceleration of an object mathematically? Recall that the acceleration of an object is the slope of that object's velocity v. time graph. On a velocity v. time graph, slope equals $\frac{\Delta v}{\Delta t}$.

Average Acceleration

Average acceleration is defined as the change in velocity divided by the time it takes to make that change.

$$\bar{a} \equiv \frac{\Delta v}{\Delta t} = \frac{v_f - v_i}{t_f - t_i}$$

Suppose you run wind sprints back and forth across the gym. You run toward one wall, turn, then run toward the opposite wall. You first run at a speed of 4.0 m/s toward the wall. Then, 10.0 s later, your speed is 4.0 m/s as you run away from the wall. What is your average acceleration if the positive direction is toward the wall?

$$\bar{a} \equiv \frac{\Delta v}{\Delta t} = \frac{v_f - v_i}{t_f - t_i}$$

$$= \frac{-4.0 \text{ m/s} - 4.0 \text{ m/s}}{10.0 \text{ s}} = -0.80 \text{ m/s}^2$$

Get It?

Explain why the final velocity used in the equation is −4.0 m/s.

STEM CAREER Connection

Delivery Truck Driver

With more and more online sales, companies must have a way to get the products to your door. Delivery truck drivers pick up products at a distribution center and drive them to businesses and homes for delivery. Acceleration, velocity, and time are important for delivery truck drivers as they try to make their deliveries on a tight schedule.

EXAMPLE Problem 1

VELOCITY AND ACCELERATION How would you describe the sprinter's velocity and acceleration as shown on the graph?

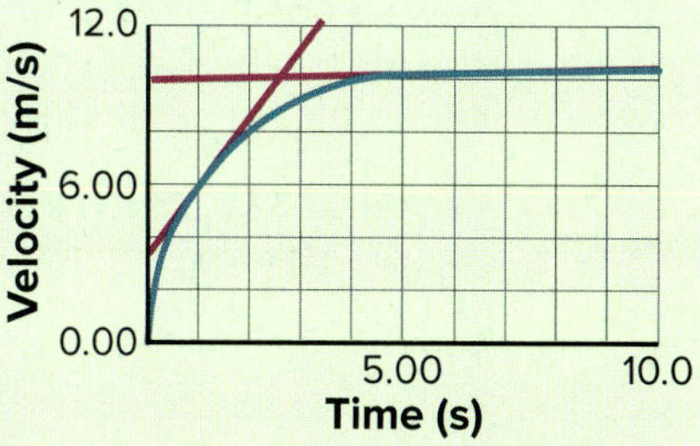

1 ANALYZE AND SKETCH THE PROBLEM

From the graph, note that the magnitude of the sprinter's velocity starts at zero, increases rapidly for the first few seconds, and then, after reaching about 10.0 m/s, remains almost constant.

KNOWN	**UNKNOWN**
v = varies	a = ?

2 SOLVE FOR THE UNKNOWN

Draw tangents to the curve at two points. Choose $t = 1.00$ s and $t = 5.00$ s. Solve for the magnitude of the instantaneous acceleration at 1.00 s:

$$a = \frac{\text{rise}}{\text{run}}$$ The slope of the line at 1.00 s is equal to the acceleration at that time.

$$= \frac{10.0\text{ m/s} - 6.00\text{ m/s}}{2.4\text{ s} - 1.00\text{ s}}$$

$$= 2.9\text{ m/s/s} = 2.9\text{ m/s}^2$$

Solve for the magnitude of the instantaneous acceleration at 5.00 s:

$$a = \frac{\text{rise}}{\text{run}}$$ The slope of the line at 5.00 s is equal to the acceleration at that time.

$$= \frac{10.3\text{ m/s} - 10.0\text{ m/s}}{10.0\text{ s} - 0.00\text{ s}}$$

$$= 0.030\text{ m/s/s} = 0.030\text{ m/s}^2$$

The acceleration is not constant because its magnitude changes from 2.9 m/s^2 at 1.00 s to 0.030 m/s^2 at 5.00 s.

The acceleration is in the direction chosen to be positive because both values are positive.

3 EVALUATE THE ANSWER

- **Are the units correct?** Acceleration is measured in m/s^2.

PRACTICE Problems

ADDITIONAL PRACTICE

1. The velocity-time graph in **Figure 8** describes Steven's motion as he walks along the midway at the state fair. Sketch the corresponding motion diagram. Include velocity vectors in your diagram.
2. Use the v-t graph of the toy train in **Figure 9** to answer these questions.
   a. When is the train's speed constant?
   b. During which time interval is the train's acceleration positive?
   c. When is the train's acceleration most negative?
3. Refer to **Figure 9** to find the average acceleration of the train during the following time intervals.
   a. 0.0 s to 5.0 s **b.** 15.0 s to 20.0 s **c.** 0.0 s to 40.0 s
4. **CHALLENGE** Plot a v-t graph representing the following motion: An elevator starts at rest from the ground floor of a three-story shopping mall. It accelerates upward for 2.0 s at a rate of 0.5 m/s^2, continues up at a constant velocity of 1.0 m/s for 12.0 s, and then slows down with a constant downward acceleration of 0.25 m/s^2 for 4.0 s as it reaches the third floor.

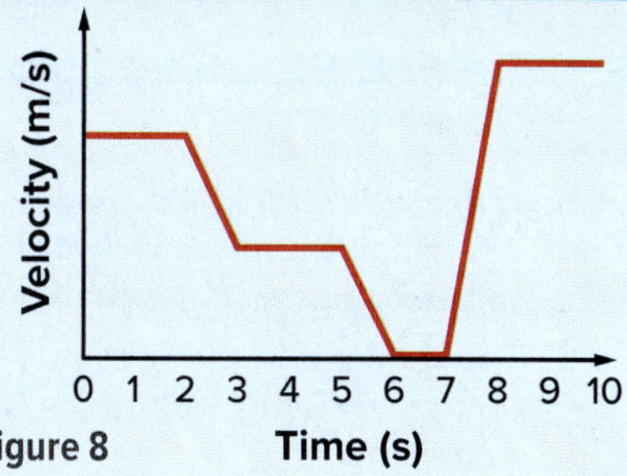

Figure 8

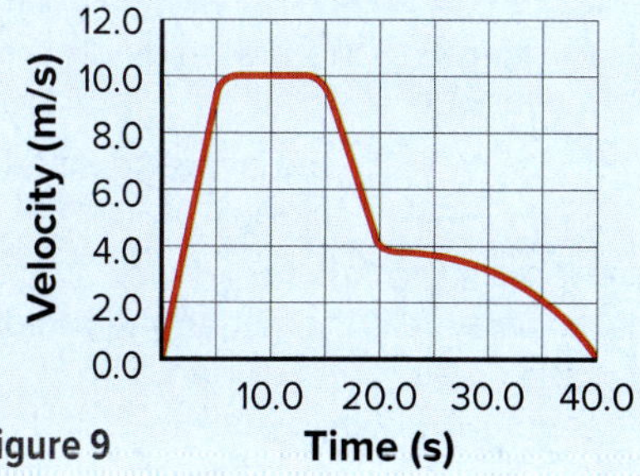

Figure 9

EXAMPLE Problem 2

ACCELERATION Describe a ball's motion as it rolls up a slanted driveway. It starts at 2.50 m/s, slows down for 5.00 s, stops for an instant, and then rolls back down. The positive direction is chosen to be up the driveway. The origin is where the motion begins. What are the sign and the magnitude of the ball's acceleration as it rolls up the driveway?

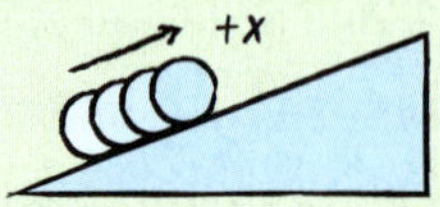

1 ANALYZE AND SKETCH THE PROBLEM

- Sketch the situation.
- Draw the coordinate system based on the motion diagram.

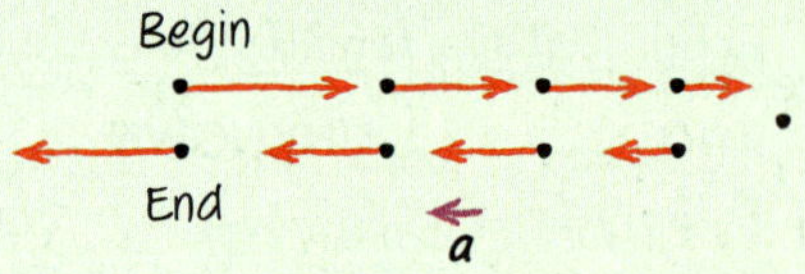

KNOWN	UNKNOWN
$v_i = +2.50$ m/s	$a = ?$
$v_f = 0.00$ m/s at $t = 5.00$ s	

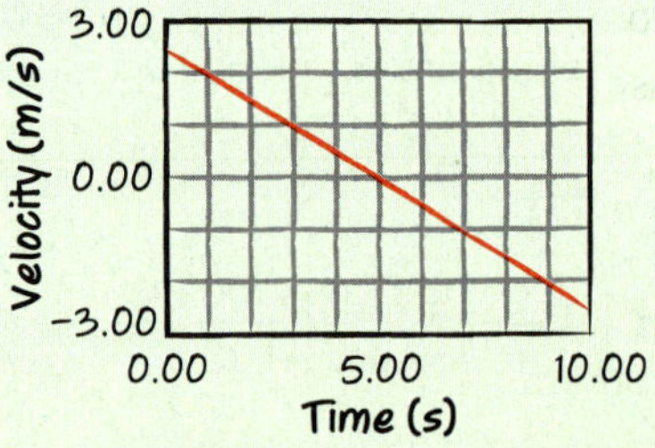

2 SOLVE FOR THE UNKNOWN

Find the acceleration from the slope of the graph.

Solve for the change in velocity and the time taken to make that change.

$\Delta v = v_f - v_i$

$= 0.00 \text{ m/s} - 2.50 \text{ m/s} = -2.50 \text{ m/s}$ Substitute $v_f = 0.00$ m/s at $t_f = 5.00$ s, $v_i = 2.50$ m/s at $t_i = 0.00$ s.

$\Delta t = t_f - t_i$

$= 5.00 \text{ s} - 0.00 \text{ s} = 5.00 \text{ s}$ Substitute $t_f = 5.00$ s, $t_i = 0.00$ s.

Solve for the acceleration.

$\bar{a} \equiv \frac{\Delta v}{\Delta t} = (-2.50 \text{ m/s}) / 5.00 \text{ s}$ Substitute $\Delta v = -2.50$ m/s, $\Delta t = 5.00$ s.

$= -0.500 \text{ m/s}^2$ or 0.500 m/s^2 down the driveway

3 EVALUATE THE ANSWER

- **Are the units correct?** Acceleration is measured in m/s^2.
- **Do the directions make sense?** As the ball slows down, the direction of acceleration is opposite that of velocity.

PRACTICE Problems

ADDITIONAL PRACTICE

5. A race car's forward velocity increases from 4.0 m/s to 36 m/s over a 4.0-s time interval. What is its average acceleration?
6. The race car in the previous problem slows from 36 m/s to 15 m/s over 3.0 s. What is its average acceleration?
7. A bus is moving west at 25 m/s when the driver steps on the brakes and brings the bus to a stop in 3.0 s.
   - **a.** What is the average acceleration of the bus while braking?
   - **b.** If the bus took twice as long to stop, how would the acceleration compare with what you found in part **a?**
8. A car is coasting backward downhill at a speed of 3.0 m/s when the driver gets the engine started. After 2.5 s, the car is moving uphill at 4.5 m/s. If uphill is chosen as the positive direction, what is the car's average acceleration?
9. Rohith has been jogging east toward the bus stop at 3.5 m/s when he looks at his watch and sees that he has plenty of time before the bus arrives. Over the next 10.0 s, he slows his pace to a leisurely 0.75 m/s. What was his average acceleration during this 10.0 s?
10. **CHALLENGE** If the rate of continental drift were to abruptly slow from 1.0 cm/y to 0.5 cm/y over the time interval of a year, what would be the average acceleration?

Acceleration with Constant Speed

Think again about running wind sprints across the gym. You first run at a speed of 4.0 m/s toward the wall. Then, 10.0 s later, you are running away from the wall and your speed is 4.0 m/s. Notice that your speed is the same as you move toward the wall of the gym and as you move away from it. In both cases, you are running at a speed of 4.0 m/s. How is it possible for you to be accelerating?

Acceleration can occur even when speed is constant. The average acceleration for the entire trip you make toward the wall of the gym and back again is $-0.80\ m/s^2$. The negative sign indicates that the direction of your acceleration is away from the wall because the positive direction was chosen as toward the wall. The velocity changes from positive to negative when the direction of motion changes. A change in velocity results in acceleration. Thus, acceleration can also be associated with a change in the direction of motion.

Get It?

Describe the evidence that the planes at the beginning of this module are accelerating even if they are traveling at a constant speed.

Check Your Progress

11. **Describing Motion** What are three ways an object can accelerate?

12. **Position-Time and Velocity-Time Graphs** Two joggers run at a constant velocity of 7.5 m/s east. **Figure 10** shows the positions of both joggers at time $t = 0$.
    a. What would be the difference(s) in the position-time graphs of their motion?
    b. What would be the difference(s) in their velocity-time graphs?

Figure 10

13. **Velocity-Time Graph** Sketch a velocity-time graph for a car that goes east at 25 m/s for 100 s, then west at 25 m/s for another 100 s.

14. **Average Velocity and Average Acceleration** A canoeist paddles upstream at a velocity of 2.0 m/s for 4.0 s and then floats downstream at 4.0 m/s for 4.0 s.
    a. What is the average velocity of the canoe during the 8.0-s time interval?
    b. What is the average acceleration of the canoe during the 8.0-s time interval?

15. **Critical Thinking** A police officer clocked a driver going 32 km/h over the speed limit just as the driver passed a slower car. When the officer stopped the car, the driver argued that the other driver should get a ticket as well. The driver said that the cars must have been going the same speed because they were observed next to each other. Is the driver correct? Explain with a sketch and a motion diagram.

LEARNSMART® Go online to follow your personalized learning path to review, practice, and reinforce your understanding.

LESSON 2

MOTION WITH CONSTANT ACCELERATION

FOCUS QUESTION

How are position, velocity, acceleration, and time related?

Position with Constant Acceleration

If an object experiences constant acceleration, its velocity changes at a constant rate. How does its position change? The positions at different times of a car with constant acceleration are graphed in **Figure 11.** The graph shows that the car's motion is not uniform. The displacements for equal time intervals on the graph get larger and larger. As a result, the slope of the line in **Figure 11** gets steeper as time goes on. For an object with constant acceleration, the position-time graph is a parabola.

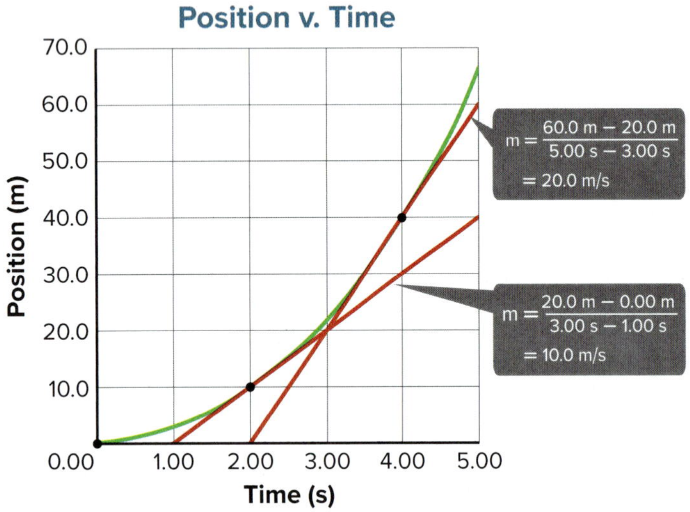

Figure 11 The slope of a position-time graph changes with time for an object with constant acceleration.

3D THINKING | DCI Disciplinary Core Ideas | CCC Crosscutting Concepts | SEP Science & Engineering Practices

COLLECT EVIDENCE

Use your Science Journal to record the evidence you collect as you complete the readings and activities in this lesson.

INVESTIGATE

GO ONLINE to find these activities and more resources.

Probeware Lab: Measuring Acceleration
Analyze data to determine how the displacement, velocity, and acceleration of an object change over time.

CCC Identify Crosscutting Concepts
Create a table of the crosscutting concepts and fill in examples you find as you read.

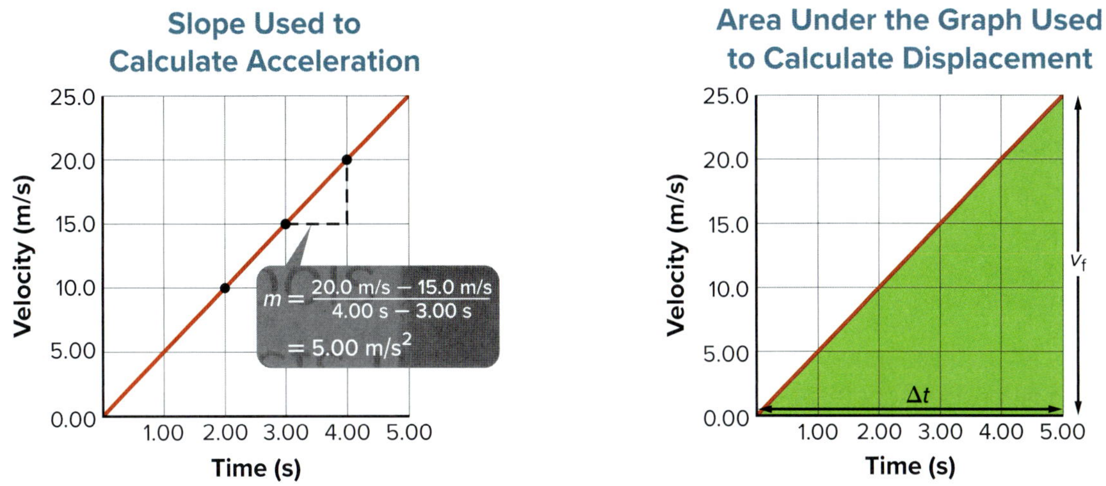

Figure 12 The slopes of the position-time graph in **Figure 11** are shown in these velocity-time graphs. The rise divided by the run gives the acceleration on the left. The area under the curve gives the displacement on the right.

Calculate *What is the slope of the velocity-time graph on the left between $t = 2.00$ s and $t = 5.00$ s?*

The slopes from the position-time graph in **Figure 11** have been used to create the velocity-time graph on the left in **Figure 12.** For an object with constant acceleration, the velocity-time graph is a straight line.

Get It?

Identify What is the shape of a position-time graph of an object traveling with constant acceleration?

A unique position-time graph cannot be created using a velocity-time graph because it does not contain information about position. It does, however, contain information about displacement. Recall that for an object moving at a constant velocity, the velocity is the displacement divided by the time interval. The displacement is then the product of the velocity and the time interval. On the right graph in **Figure 12**, v is the height of the plotted line above the horizontal axis, and Δt is the width of the shaded triangle. The area is $\left(\frac{1}{2}\right)v\Delta t$, or Δx. Thus, the area under the v-t graph equals the displacement.

Get It?

Identify What is the shape of a velocity-time graph of an object traveling with constant acceleration?

Velocity with Average Acceleration

You have read that the equation for average velocity can be algebraically rearranged to show the new position after a period of time, given the initial position and the average velocity. The definition of average acceleration can be manipulated similarly to show the new velocity after a period of time, given the initial velocity and the average acceleration.

If you know an object's average acceleration during a time interval, you can use it to determine how much the velocity changed during that time.

You can rewrite the definition of average acceleration $\left(\bar{a} \equiv \frac{\Delta v}{\Delta t}\right)$ as follows:

$$\Delta v = \bar{a}\Delta t$$

$$v_f - v_i = \bar{a}\Delta t$$

The equation for final velocity with average acceleration can be written:

Final Velocity with Average Acceleration

The final velocity is equal to the initial velocity plus the product of the average acceleration and the time interval.

$$v_f = v_i + \bar{a}\Delta t$$

In cases when the acceleration is constant, the average acceleration ($\bar{a}$) is the same as the instantaneous acceleration (a) at any point within the time interval. This equation can be rearranged to find the time at which an object with constant acceleration has a given velocity. You can also use it to calculate the initial velocity of an object when both a velocity and the time at which it occurred are given.

Real-World Physics

DRAG RACING A dragster driver tries to obtain maximum acceleration over a course. The fastest U.S. National Hot Rod Association time on record for the 402-m course is 3.771 s. The highest final speed on record is 145.3 m/s (324.98 mph).

PRACTICE Problems

ADDITIONAL PRACTICE

16. A golf ball rolls up a hill toward a miniature-golf hole. Assume the direction toward the hole is positive.

a. If the golf ball starts with a speed of 2.0 m/s and slows at a constant rate of 0.50 m/s^2, what is its velocity after 2.0 s?

b. What is the golf ball's velocity if the constant acceleration continues for 6.0 s?

c. Describe the motion of the golf ball in words and with a motion diagram.

17. A bus traveling 30.0 km/h east has a constant increase in speed of 1.5 m/s^2. What is its velocity 6.8 s later?

18. If a car accelerates from rest at a constant rate of 5.5 m/s^2 north, how long will it take for the car to reach a velocity of 28 m/s north?

19. CHALLENGE A car slows from 22 m/s to 3.0 m/s at a constant rate of 2.1 m/s^2. How many seconds are required before the car is traveling at a forward velocity of 3.0 m/s?

CCC CROSSCUTTING CONCEPTS

Scale, Proportion, and Quantity Using algebraic thinking, draw a conclusion about how changes in the initial velocity of an object affects the final position of the object. Make a table that provides evidence to support your conclusions.

EXAMPLE Problem 3

FINDING DISPLACEMENT FROM A VELOCITY-TIME GRAPH The velocity-time graph at the right shows the motion of an airplane. Find the displacement of the airplane for $\Delta t = 1.0$ s and for $\Delta t = 2.0$ s. Let the positive direction be forward.

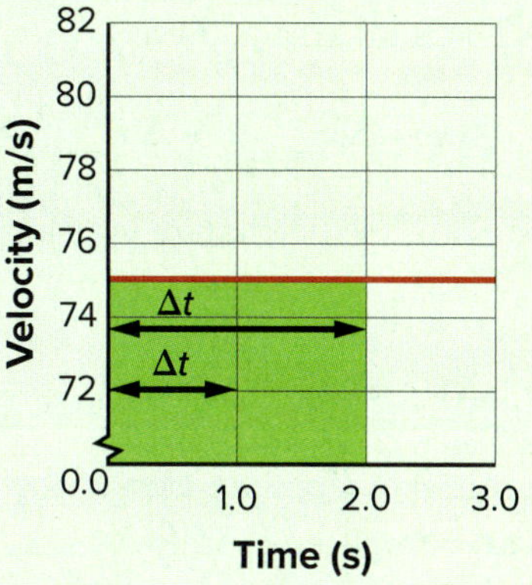

1 ANALYZE AND SKETCH THE PROBLEM

- The displacement is the area under the *v-t* graph.
- The time intervals begin at $t = 0.0$ s.

KNOWN	UNKNOWN
$v = +75$ m/s	$\Delta x = ?$
$\Delta t = 1.0$ s	
$\Delta t = 2.0$ s	

2 SOLVE FOR THE UNKNOWN

Use the relationship among displacement, velocity, and time interval to find Δx during $\Delta t = 1.0$ s.

$\Delta x = v\Delta t$

$= (+75 \text{ m/s})(1.0 \text{ s})$ — Substitute $v = +75$ m/s, $\Delta t = 1.0$ s.

$= +75$ m

Use the same relationship to find Δx during $\Delta t = 2.0$ s.

$\Delta x = v\Delta t$

$= (+75 \text{ m/s})(2.0 \text{ s})$ — Substitute $v = +75$ m/s, $\Delta t = 2.0$ s.

$= +150$ m

3 EVALUATE THE ANSWER

- **Are the units correct?** Displacement is measured in meters.
- **Do the signs make sense?** The positive sign agrees with the graph.
- **Is the magnitude realistic?** Moving a distance of about one football field per second is reasonable for an airplane.

PRACTICE Problems

ADDITIONAL PRACTICE

20. The graph in **Figure 13** describes the motion of two bicyclists, Akiko and Brian, who start from rest and travel north, increasing their speed with a constant acceleration. What was the total displacement of each bicyclist during the time shown for each?

Hint: Use the area of a triangle: area = $\left(\frac{1}{2}\right)$*(base)(height).*

21. The motion of two people, Carlos and Diana, moving south along a straight path is described by the graph in **Figure 14.** What is the total displacement of each person during the first 4.0-s interval shown on the graph?

22. CHALLENGE A car, just pulling onto a straight stretch of highway, has a constant acceleration from 0 m/s to 25 m/s west in 12 s.

a. Draw a *v-t* graph of the car's motion.

b. Use the graph to determine the car's displacement during the 12.0-s time interval.

c. Another car is traveling along the same stretch of highway. It travels the same distance in the same time as the first car, but its velocity is constant. Draw a *v-t* graph for this car's motion.

d. Explain how you knew this car's velocity.

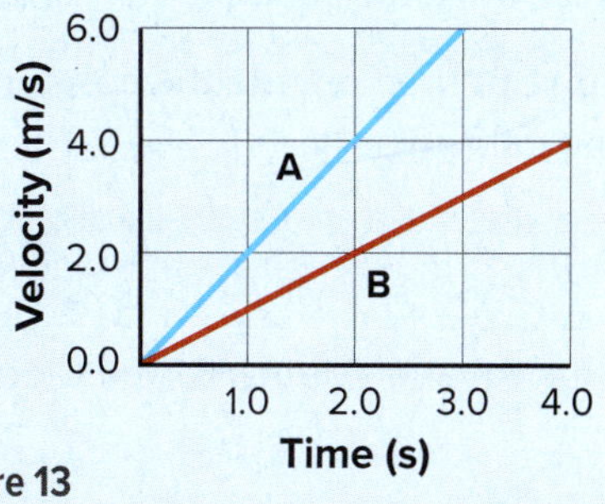

Figure 13

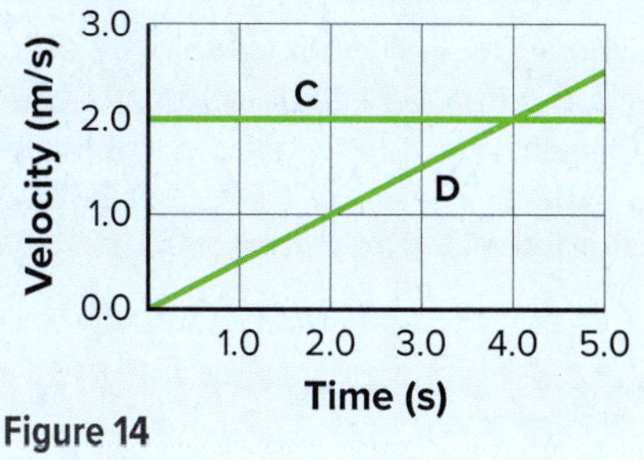

Figure 14

Motion with an initial nonzero velocity The graph in **Figure 15** describes constant acceleration that started with an initial velocity of v_i. To determine the displacement, you can divide the area under the graph into a rectangle and a triangle. The total area is then:

$$\Delta x = \Delta x_{\text{rectangle}} + \Delta x_{\text{triangle}} = v_i(\Delta t) + \left(\tfrac{1}{2}\right)\Delta v \Delta t$$

Substituting $a\Delta t$ for the change in velocity in the equation yields:

$$\Delta x = \Delta x_{\text{rectangle}} + \Delta x_{\text{triangle}} = v_i(\Delta t) + \left(\tfrac{1}{2}\right)a(\Delta t)^2$$

When the initial or final position of the object is known, the equation can be written as follows:

$$x_f - x_i = v_i(\Delta t) + \left(\tfrac{1}{2}\right)a(\Delta t)^2 \text{ or } x_f = x_i + v_i(\Delta t) + \left(\tfrac{1}{2}\right)a(\Delta t)^2$$

If the initial time is $t_i = 0$, the equation then becomes the following.

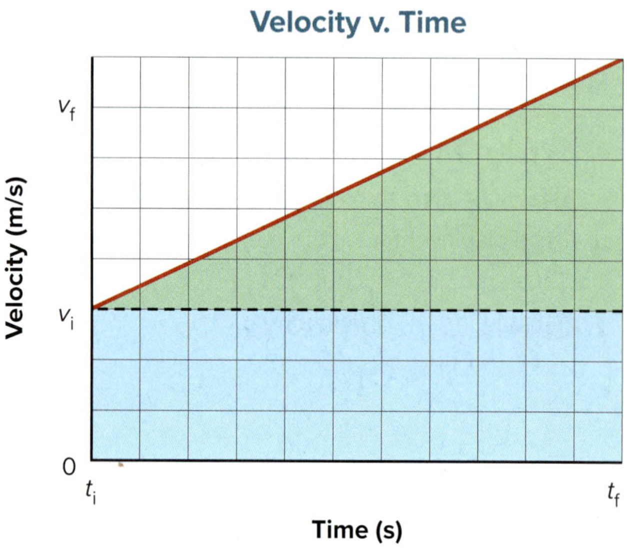

Figure 15 For motion with constant acceleration, if the initial velocity on a velocity-time graph is not zero, the area under the graph is the sum of a rectangular area and a triangular area.

Position with Average Acceleration

An object's final position is equal to the sum of its initial position, the product of the initial velocity and the final time, and half the product of the acceleration and the square of the final time.

$$x_f = x_i + v_i t_f + \left(\tfrac{1}{2}\right)a t_f^2$$

An Alternative Equation

Often, it is useful to relate position, velocity, and constant acceleration without including time.

Rearrange the equation $v_f = v_i + at_f$ to solve for time: $t_f = \frac{v_f - v_i}{a}$.

You can then rewrite the position with average acceleration equation by substituting t_f to obtain the following:

$$x_f = x_i + v_i\left(\frac{v_f - v_i}{a}\right) + \left(\tfrac{1}{2}\right)a\left(\frac{v_f - v_i}{a}\right)^2$$

This equation can be solved for the velocity (v_f) at any position (x_f).

Velocity with Constant Acceleration

The square of the final velocity equals the sum of the square of the initial velocity and twice the product of the acceleration and the displacement since the initial time.

$$v_f^2 = v_i^2 + 2a(x_f - x_i)$$

EXAMPLE Problem 4

DISPLACEMENT An automobile starts at rest and accelerates at 3.5 m/s^2 after a traffic light turns green. How far will it have gone when it is traveling at 25 m/s?

1 ANALYZE AND SKETCH THE PROBLEM

- Sketch the situation.
- Establish coordinate axes. Let the positive direction be to the right.
- Draw a motion diagram.

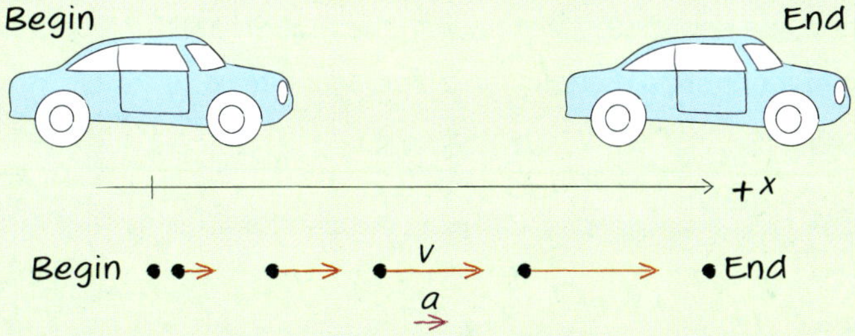

KNOWN	UNKNOWN
$x_i = 0.00$ m	$x_f = ?$
$v_i = 0.00$ m/s	
$v_f = +25$ m/s	
$\bar{a} = a = +3.5$ m/s^2	

2 SOLVE FOR THE UNKNOWN

Use the relationship among velocity, acceleration, and displacement to find x_f.

$$v_f^2 = v_i^2 + 2a(x_f - x_i)$$

$$x_f = x_i + \frac{v_f^2 - v_i^2}{2a}$$

$$= 0.00\text{ m} + \frac{(+25\text{ m/s})^2 - (0.00\text{ m/s})^2}{2(+3.5\text{ m/s}^2)}$$

Substitute $x_i = 0.00$ m, $v_f = +25$ m/s, $v_i = 0.00$ m/s, $a = +3.5$ m/s^2.

$$= +89\text{ m}$$

3 EVALUATE THE ANSWER

- **Are the units correct?** Position is measured in meters.
- **Do the signs make sense?** The positive sign agrees with both the pictorial and physical models.
- **Is the magnitude realistic?** The displacement is almost the length of a football field. The result is reasonable because 25 m/s (about 55 mph) is fast.

PRACTICE Problems

ADDITIONAL PRACTICE

23. A skateboarder is moving at a constant speed of 1.75 m/s when she starts up an incline that causes her to slow down with a constant acceleration of −0.20 m/s^2. How much time passes from when she begins to slow down until she begins to move back down the incline?

24. A race car travels on a straight racetrack with a forward velocity of 44 m/s and slows at a constant rate to a velocity of 22 m/s over 11 s. How far does it move during this time?

25. A car accelerates at a constant rate from 15 m/s to 25 m/s while it travels a distance of 125 m. How long does it take to achieve the final speed?

26. A bike rider pedals with constant acceleration to reach a velocity of 7.5 m/s north over a time of 4.5 s. During the period of acceleration, the bike's displacement is 19 m north. What was the initial velocity of the bike?

27. CHALLENGE The car in **Figure 16** travels west with a forward acceleration of 0.22 m/s^2. What was the car's velocity (v_i) at point x_i if it travels a distance of 350 m in 18.4 s?

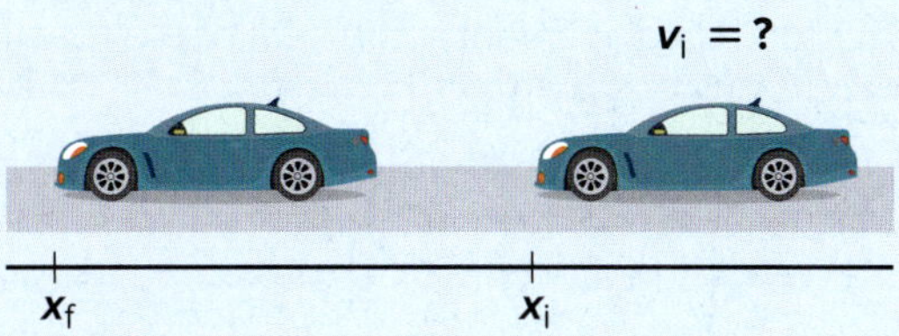

Figure 16

EXAMPLE Problem 5

TWO-PART MOTION You are driving a car, traveling at a constant velocity of 25 m/s along a straight road, when you see a child suddenly run onto the road. It takes 0.45 s for you to react and apply the brakes. As a result, the car slows with a steady acceleration of 8.5 m/s^2 in the direction opposite your motion and comes to a stop. What is the total displacement of the car before it stops?

1 ANALYZE AND SKETCH THE PROBLEM

- Sketch the situation.
- Choose a coordinate system with the motion of the car in the positive direction.
- Draw the motion diagram, and label $\boldsymbol{v}$ and $\boldsymbol{a}$.

KNOWN	UNKNOWN
$\boldsymbol{v}_{\text{reacting}} = +25$ m/s	$\boldsymbol{x}_{\text{reacting}} = ?$
$t_{\text{reacting}} = 0.45$ s	$\boldsymbol{x}_{\text{braking}} = ?$
$\boldsymbol{a}_{\text{braking}} = -8.5$ m/s^2	$\boldsymbol{x}_{\text{total}} = ?$
$\boldsymbol{v}_{\text{i, braking}} = +25$ m/s	
$\boldsymbol{v}_{\text{f, braking}} = 0.00$ m/s	

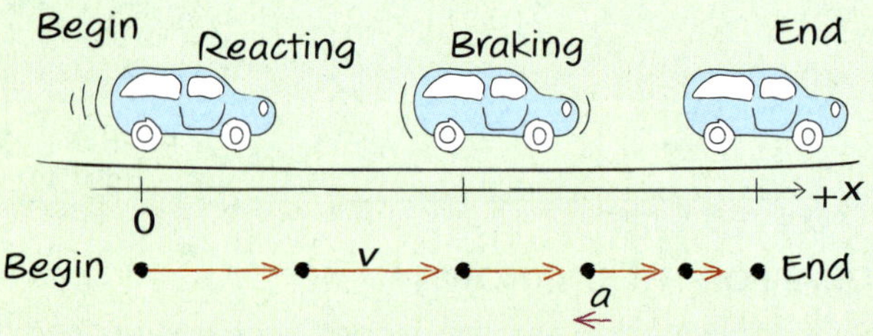

2 SOLVE FOR THE UNKNOWN

Reacting:

Use the relationship among displacement, velocity, and time interval to find the displacement of the car as it travels at a constant speed.

$\boldsymbol{x}_{\text{reacting}} = \boldsymbol{v}_{\text{reacting}} t_{\text{reacting}}$

$\boldsymbol{x}_{\text{reacting}} = (+25 \text{ m/s})(0.45 \text{ s})$ — Substitute $\boldsymbol{v}_{\text{reacting}} = +25$ m/s, $t_{\text{reacting}} = 0.45$ s.

$= +11$ m

Braking:

Use the relationship among velocity, acceleration, and displacement to find the displacement of the car while it is braking.

$v_{\text{f, braking}}{}^2 = v_{\text{reacting}}{}^2 + 2a_{\text{braking}}(x_{\text{braking}})$

Solve for x_{braking}.

$$x_{\text{braking}} = \frac{v_{\text{f, braking}}{}^2 - v_{\text{reacting}}{}^2}{2a_{\text{braking}}}$$

$$= \frac{(0.00 \text{ m/s})^2 - (+25 \text{ m/s})^2}{2(-8.5 \text{ m/s}^2)}$$

Substitute $v_{\text{f, braking}} = 0.00$ m/s, $v_{\text{reacting}} = +25$ m/s, $a_{\text{braking}} = -8.5$ m/s^2.

$= +37$ m

The total displacement is the sum of the reacting displacement and the braking displacement.

Solve for $\boldsymbol{x}_{\text{total}}$.

$\boldsymbol{x}_{\text{total}} = \boldsymbol{x}_{\text{reacting}} + \boldsymbol{x}_{\text{braking}}$

$= +11 \text{ m} + 37 \text{ m}$ — Substitute $\boldsymbol{x}_{\text{reacting}} = +11$ m, $\boldsymbol{x}_{\text{braking}} = +37$ m.

$= +48$ m

3 EVALUATE THE ANSWER

- **Are the units correct?** Displacement is measured in meters.
- **Do the signs make sense?** Both xreacting and xbraking are positive, as they should be.
- **Is the magnitude realistic?** The braking displacement is small because the magnitude of the acceleration is large.

PRACTICE Problems ADDITIONAL PRACTICE

28. A car with an initial velocity of 24.5 m/s east has an acceleration of 4.2 m/s^2 west. What is its displacement at the moment that its velocity is 18.3 m/s east?

29. A man runs along the path shown in **Figure 17.** From point A to point B, he runs at a forward velocity of 4.5 m/s for 15.0 min. From point B to point C, he runs up a hill. He slows down at a constant rate of 0.050 m/s^2 for 90.0 s and comes to a stop at point C. What was the total distance the man ran?

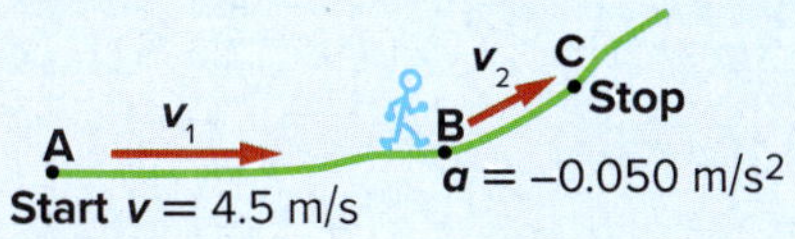

Figure 17

30. You start your bicycle ride at the top of a hill. You coast down the hill at a constant acceleration of 2.00 m/s^2. When you get to the bottom of the hill, you are moving at 18.0 m/s, and you pedal to maintain that speed. If you continue at this speed for 1.00 min, how far will you have gone from the time you left the hilltop?

31. Sunee is training for a 5.0-km race. She starts out her training run by moving at a constant pace of 4.3 m/s for 19 min. Then she accelerates at a constant rate until she crosses the finish line 19.4 s later. What is her acceleration during the last portion of the training run?

32. CHALLENGE Sekazi is learning to ride a bike without training wheels. His father pushes him with a constant acceleration of 0.50 m/s^2 east for 6.0 s. Sekazi then travels at 3.0 m/s east for another 6.0 s before falling. What is Sekazi's displacement? Solve this problem by constructing a velocity-time graph for Sekazi's motion and computing the area underneath the graphed line.

Check Your Progress

33. **Displacement** Given initial and final velocities and the constant acceleration of an object, what mathematical relationship would you use to find the displacement?

34. **Acceleration** A woman driving west along a straight road at a speed of 23 m/s sees a deer on the road ahead. She applies the brakes when she is 210 m from the deer. If the deer does not move and the car stops right before it hits the deer, what is the acceleration provided by the car's brakes?

35. **Distance** The airplane in **Figure 18** starts from rest and accelerates east at a constant 3.00 m/s^2 for 30.0 s before leaving the ground.
    a. What was the plane's displacement (Δx)?
    b. How fast was the airplane going when it took off?

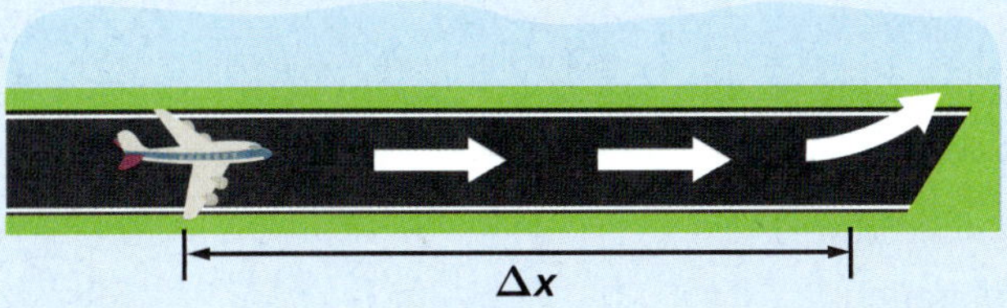

Figure 18

36. **Distance** An in-line skater accelerates from 0.0 m/s to 5.0 m/s in 4.5 s, then continues at this constant speed for another 4.5 s. What is the total distance traveled by the in-line skater?

37. **Final Velocity** A plane travels 5.0×10^2 m north while accelerating uniformly from rest at 5.0 m/s^2. What final velocity does it attain?

38. **Final Velocity** An airplane accelerated uniformly from rest at the rate of 5.0 m/s^2 south for 14 s. What final velocity did it attain?

39. **Graphs** A sprinter walks to the starting blocks at a constant speed, then waits. When the starting pistol sounds, she accelerates rapidly until she attains a constant velocity. She maintains this velocity until she crosses the finish line, and then she slows to a walk, taking more time to slow down than she did to speed up at the beginning of the race. Sketch a velocity-time and a position-time graph to represent her motion. Draw them one above the other using the same time scale. Indicate on your position-time graph where the starting blocks and finish line are.

LEARNSMART® Go online to follow your personalized learning path to review, practice, and reinforce your understanding.

LESSON 3

FREE FALL

FOCUS QUESTION

How does an object's speed change as it falls?

Galileo's Discovery

Do heavier objects accelerate more as they fall? If you release a paper and a book, the book hits the ground first. But if you put the paper flat on the book, they fall together. Collisions with particles of air have a greater effect on the paper. To understand falling objects, first consider the case in which air does not have an appreciable effect on motion. Recall that gravity is an attraction between objects. **Free fall** is the motion of an object when gravity is the only significant force acting on it.

About 400 years ago, Galileo Galilei discovered that, neglecting the effect of the air, all objects in free fall have the same acceleration. It doesn't matter what they are made of or how much they weigh. The acceleration of an object due only to the effect of gravity is known as **free-fall acceleration.** **Figure 19** depicts the results of a 1971 free-fall experiment on the Moon in which astronauts verified Galileo's results.

Figure 19 In 1971, astronaut David Scott dropped a hammer and a feather at the same time from the same height above the Moon's surface. The hammer's mass was greater, but both objects hit the ground at the same time because the Moon has gravity but no air.

3D THINKING DCI Disciplinary Core Ideas CCC Crosscutting Concepts SEP Science & Engineering Practices

COLLECT EVIDENCE

Use your Science Journal to record the evidence you collect as you complete the readings and activities in this lesson.

INVESTIGATE

GO ONLINE to find these activities and more resources.

PhysicsLAB: Free-Fall Acceleration
Calculate the acceleration due to gravity for a system in free-fall.

Revisit the Encounter the Phenomenon Question
What information from this lesson can help you answer the Unit and Module questions?

Near Earth's surface, free-fall acceleration is about 9.8 m/s^2 downward (which is equal to about 22 mph/s downward). Think about skydivers. Each second skydivers fall, their downward velocity increases by 9.8 m/s. When analyzing free fall, whether you treat the acceleration as positive or negative depends on the coordinate system you use. If you define upward as the positive direction, then the free-fall acceleration is negative. If you decide that downward is the positive direction, then free-fall acceleration is positive.

Get It?

Identify Why is it important to clearly define the coordinate system you want to use when analyzing objects in free fall?

Free-Fall Acceleration

Galileo's discovery explains why parachutists can form a ring in midair. Regardless of their masses, they fall with the same acceleration. To understand the acceleration that occurs during free fall, look at the multiflash photo of a dropped ball in **Figure 20.**

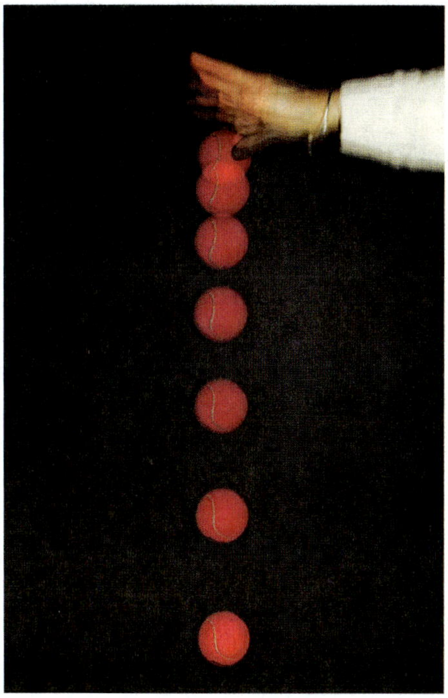

Figure 20 Because of free-fall acceleration, the speed of this falling ball increases 9.8 m/s each second.

When the photograph was taken, the time interval between flashes was set to 0.06 s. So, the same amount of time passes between one image of the ball and the next. Look at the distance that the ball travels during each time interval. The distance between each pair of images increases as the ball falls. So, the speed of the ball is increasing. If the upward direction is positive, then the velocity is becoming more and more negative.

Ball thrown upward Instead of a dropped ball, could this photo also illustrate a ball thrown upward? Suppose you throw a ball upward with a speed of 20.0 m/s. If you choose upward to be positive, then the ball starts at the bottom of the photo with a positive velocity. The acceleration is $a = -9.8\ \text{m/s}^2$. Because velocity and acceleration are in opposite directions, the speed of the ball decreases. If you think of the bottom of the photo as the start, this agrees with the multiflash photo.

Rising and falling motion After 1 s, the ball's velocity is reduced by 9.8 m/s, so it now is traveling at +10.2 m/s. After 2 s, the velocity is +0.4 m/s, and the ball still is moving upward. What happens during the next second? The ball's velocity is reduced by another 9.8 m/s and equals −9.4 m/s. The ball now is moving downward. After 4 s, the velocity is −19.2 m/s, meaning the ball is falling even faster.

Get It?

Analyze During which second does the rising ball stop and reverse direction? How can you tell?

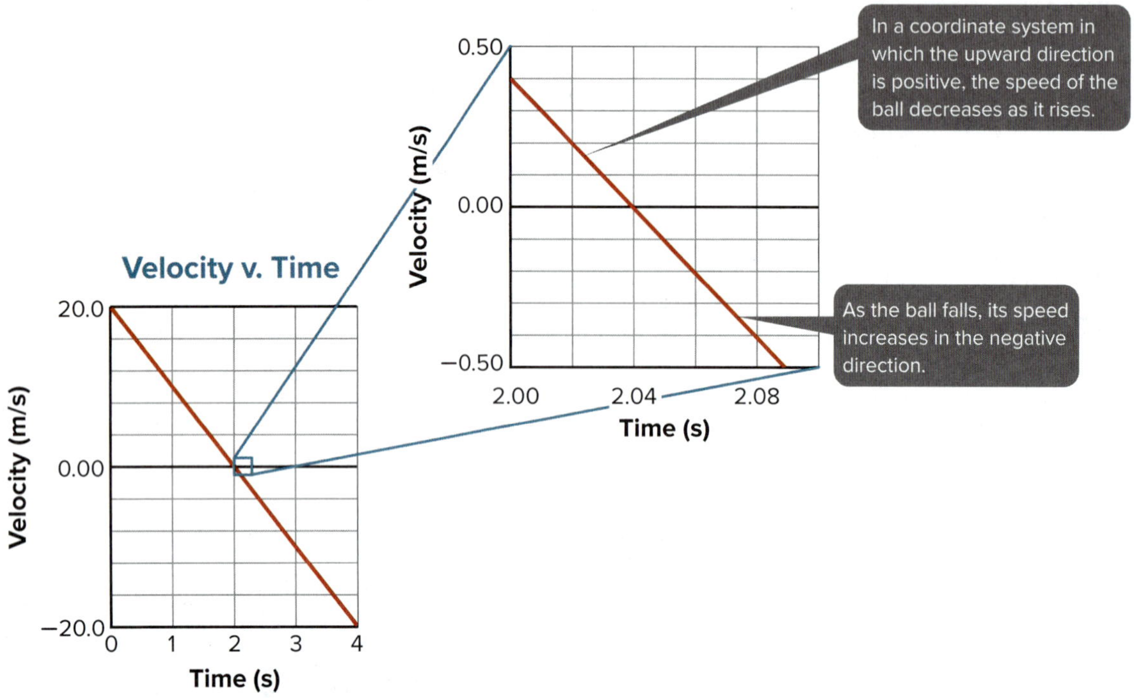

Figure 21 The velocity-time graph describes the change in the ball's speed as it rises and falls. The graph on the right gives a close-up view of the change in velocity at the top of the ball's trajectory.

Analyze *What would the graph look like if downward were chosen as the positive direction?*

Velocity-time graph The *v-t* graph for the ball as it goes up and down is shown in **Figure 21.** The straight line sloping downward does not mean that the speed is always decreasing. The speed decreases as the ball rises and increases as it falls. At around 2 s, the velocity changes smoothly from positive to negative. As the ball falls, its speed increases in the negative direction. The figure also shows a closer view of the *v-t* graph. At an instant of time, near 2.04 s, the velocity is zero.

Position-time graph Look at the position-time graphs in **Figure 22** on the next page. These graphs show how the ball's height changes as it rises and falls. If an object is moving with constant acceleration, its position-time graph forms a parabola. Because the ball is rising and falling, its graph is an inverted parabola. The shape of the graph shows the progression of time. It does not mean that the ball's path was in the shape of a parabola. The close-up graph on the right shows that at about 2.04 s, the ball reaches its maximum height.

Get It?

Describe If you throw a ball straight up, what would the shape of its position-time graph look like?

Maximum height Compare the close-up graphs in **Figure 21** and **Figure 22.** Just before the ball reaches its maximum height, its velocity is decreasing in the negative direction. At the instant of time when its height is maximum, its velocity is zero. Just after it reaches its maximum height, the ball's velocity is increasing in the negative direction.

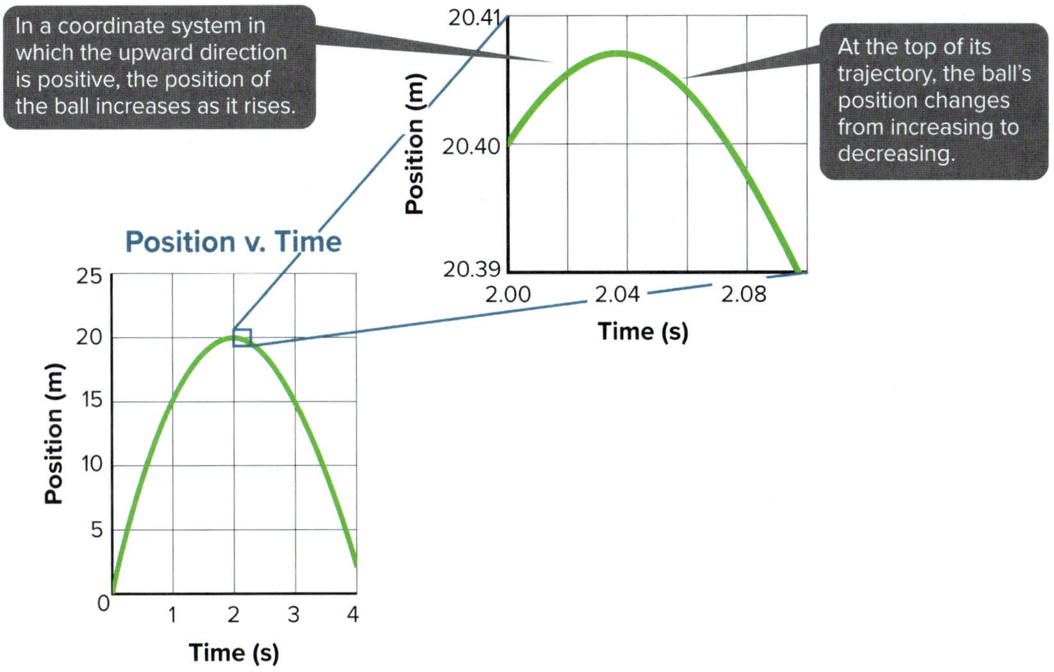

Figure 22 A position-time graph shows how the ball's position changes as it rises and falls. The graph at the right shows a close-up view of how the position changes at the top of the ball's trajectory.

Acceleration The slope of the line on the velocity-time graph in **Figure 21** is constant at $-9.8\ m/s^2$. This shows that the ball's free-fall acceleration is $9.8\ m/s^2$ in the downward direction the entire time the ball is rising and falling.

It may seem that the acceleration should be zero at the top of the trajectory, but this is not the case. At the top of the flight, the ball's velocity is 0 m/s. If its acceleration were also zero, the ball's velocity would not change and would remain at 0 m/s. The ball would not gain any downward velocity and would simply hover in the air. Have you ever seen that happen? Objects tossed in the air on Earth always fall, so you know the acceleration of an object at the top of its flight must not be zero. Further, because the object falls down, you know the acceleration must be downward.

Get It?

Analyze If you throw a ball straight up, what are its velocity and acceleration at the uppermost point of its path?

SCIENCE USAGE v. COMMON USAGE

Free fall

Science usage: motion of a body when air resistance is negligible and the acceleration can be considered due to gravity alone.
Acceleration during free fall is $9.8\ m/s^2$ downward.

Common usage: a rapid and continuing drop or decline
The stock market's free fall in 1929 marked the beginning of the Great Depression.

Figure 23 The people on this amusement-park ride experience free-fall acceleration.

Free-fall rides Amusement parks use the concept of acceleration to design rides that give the riders the sensation of free fall. These types of rides usually consist of three parts: the ride to the top, momentary suspension, and the fall downward. Motors provide the force needed to move the cars to the top of the ride. When the cars are in free fall, the most massive rider and the least massive rider will have the same acceleration.

Suppose the free-fall ride shown in **Figure 23** starts from the top at rest and is in free fall for 1.5 s. What would be its velocity at the end of 1.5 s? Choose a coordinate system with a positive axis upward and the origin at the initial position of the car. Because the car starts at rest, v_i would be equal to 0.0 m/s. To calculate the final velocity, use the equation for velocity with constant acceleration.

$$\begin{aligned} v_f &= v_i + \bar{a}t_f \\ &= 0.0\ \text{m/s} + (-9.8\ \text{m/s}^2)(1.5\ \text{s}) \\ &= -15\ \text{m/s} \end{aligned}$$

How far do people on the ride fall during this time? Use the equation for displacement when time and constant acceleration are known.

$$\begin{aligned} x_f &= x_i + v_it_f + \left(\tfrac{1}{2}\right)\bar{a}t_f^2 \\ &= 0.0\ \text{m} + (0.0\ \text{m/s})(1.5\text{s}) + \left(\tfrac{1}{2}\right)(-9.8\ \text{m/s}^2)(1.5\ \text{s})^2 \\ &= -11\ \text{m} \end{aligned}$$

PRACTICE Problems

ADDITIONAL PRACTICE

40. A construction worker accidentally drops a brick from a high scaffold.

a. What is the velocity of the brick after 4.0 s?

b. How far does the brick fall during this time?

41. Suppose for the previous problem you choose your coordinate system so that the opposite direction is positive.

a. What is the brick's velocity after 4.0 s?

b. How far does the brick fall during this time?

42. A student drops a ball from a window 3.5 m above the sidewalk. How fast is it moving when it hits the sidewalk?

43. A tennis ball is thrown straight up with an initial speed of 22.5 m/s. It is caught at the same distance above the ground.

a. How high does the ball rise?

b. How long does the ball remain in the air?

Hint: The time it takes the ball to rise equals the time it takes to fall.

44. You decide to flip a coin to determine whether to do your physics or English homework first. The coin is flipped straight up.

a. What are the velocity and acceleration of the coin at the top of its trajectory?

b. If the coin reaches a high point of 0.25 m above where you released it, what was its initial speed?

c. If you catch it at the same height as you released it, how much time was it in the air?

45. CHALLENGE A basketball player is holding a ball in her hands at a height of 1.5 m above the ground. She drops the ball, and it bounces several times. After the first bounce, the ball only returns to a height of 0.75 m. After the second bounce, the ball only returns to a height of 0.25 m.

a. Suppose downward is the positive direction. What would the shape of a velocity-time graph look like for the first two bounces?

b. What would be the shape of a position-time graph for the first two bounces?

Variations in Free Fall

When astronaut David Scott performed his free-fall experiment on the Moon, the hammer and the feather did not fall with an acceleration of magnitude 9.8 m/s^2. The value 9.8 m/s^2 is free-fall acceleration only near Earth's surface. The magnitude of free-fall acceleration on the Moon is approximately 1.6 m/s^2, which is about one-sixth its value on Earth.

When you study force and motion, you will learn about factors that affect the value of free-fall acceleration. One factor is the mass of the object, such as Earth or the Moon, that is responsible for the acceleration. Free-fall acceleration is not as great near the Moon as near Earth because the Moon has much less mass.

Free-fall acceleration also depends on the distance from the object responsible for it. The rings drawn around Earth in **Figure 24** show how free-fall acceleration decreases with distance from Earth. It is important to understand, however, that variations in free-fall acceleration at different locations on Earth's surface are very small, even with great variations in elevation. In New York City, for example, the magnitude of free-fall acceleration is about 9.81 m/s^2. In Denver, Colorado, it is about 9.79 m/s^2, despite a change in elevation of almost 1600 m greater. For calculations in this book, a value of 9.8 m/s^2 will be used for free-fall acceleration.

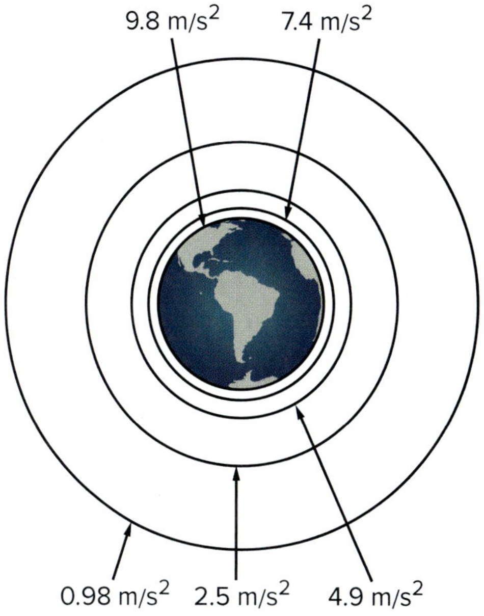

Figure 24 As the distance from Earth increases, the effect of free-fall acceleration decreases.

Analyze *According to the diagram, what is the magnitude of free-fall acceleration a distance above Earth's surface equal to Earth's radius?*

Check Your Progress

46. **Free Fall** Suppose you hold a book in one hand and a flat sheet of paper in your other hand. You drop them both, and they fall to the ground. Explain why the falling book is a good example of free fall, but the paper is not.

47. **Final Velocity** Your sister drops your house keys down to you from the second-floor window, as shown in **Figure 25.** What is the velocity of the keys when you catch them?

48. **Free-Fall Ride** Suppose a free-fall ride at an amusement park starts at rest and is in free fall. What is the velocity of the ride after 2.3 s? How far do people on the ride fall during the 2.3-s time period?

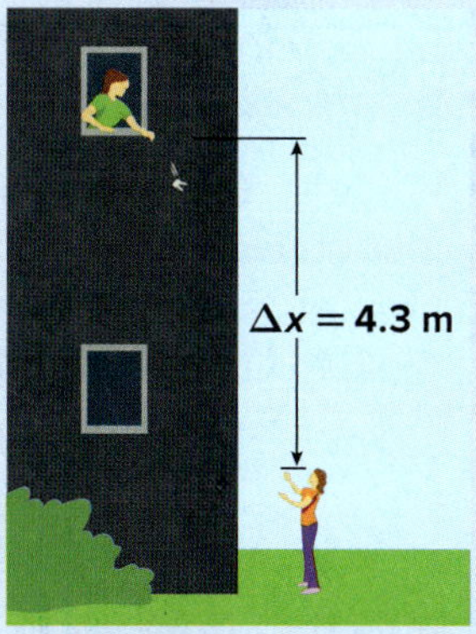

Figure 25

49. **Maximum Height and Flight Time** The free-fall acceleration on Mars is about one-third that on Earth. Suppose you throw a ball upward with the same velocity on Mars as on Earth.
    a. How would the ball's maximum height compare to that on Earth?
    b. How would its flight time compare?

50. **Velocity and Acceleration** Suppose you throw a ball straight up into the air. Describe the changes in the velocity of the ball. Describe the changes in the acceleration of the ball.

51. **Critical Thinking** A ball thrown vertically upward continues upward until it reaches a certain position, and then falls downward. The ball's velocity is instantaneously zero at that highest point. Is the ball accelerating at that point? Devise an experiment to prove or disprove your answer.

LEARNSMART® Go online to follow your personalized learning path to review, practice, and reinforce your understanding.

Designing Fun

Think about the last amusement park attraction that made your pulse quicken. Perhaps it was a heart-pounding roller coaster or a chilling haunted house. Did you ever wonder how theme park rides are created? The people who design the attractions at theme parks have many different titles—theme park designer, creative design engineer, industrial designer, building information modeling manager, and many others. No matter their title, they all concentrate on creating exciting attractions that draw visitors to the parks.

Roller coasters are just one of many types of attractions created by theme park designers.

Tools of the Trade

Most theme park designers have a background in physics, engineering, and mathematics. Some are also graphic designers or writers. They use a variety of tools to create the attractions that delight park visitors.

CAD (computer-aided design) drafting software, for example, allows designers to create 2D and 3D blueprints for attractions. Increasingly, designers are also employing BIM (building information modeling) systems, used to create 3D models that can be viewed at different levels of complexity. There are other modeling software programs that allow designers to work in 3D as well. Theme park designers also use computer graphics software to create graphics and animations, and they use computer programs to alter images.

Forces and Safety

Theme park designers work with acceleration and gravitational forces to produce thrills for riders. Roller coasters, for example, offer both changes in speed (such as slowly grinding up a hill and then barreling down it) and in direction (such as suddenly banking to the left or right). Attractions that drop riders from great heights and then slow their descent before they reach the ground make use of free-fall acceleration.

Designers must consider the attractions' effects on the human inner ear, heart, and other body systems. They want riders to be exhilarated, but they must also ensure that riders are safe. They must be mindful, for instance, that gravitational forces do not exceed the amount an average person can tolerate without suffering ill effects. Designers create elaborate systems of seatbelts and other restraints, headrests, and padding to protect riders from injury. They design improved braking systems to maximize safety.

DEVELOP AND USE MODELS TO ILLUSTRATE

With a partner, brainstorm, design, and build a model of a theme park ride that uses acceleration and gravitational forces. Use modeling software, if available.

CALIFORNIA CONNECTION

Use a search engine to learn more about a ride at a California theme park. Write a short text explaining how acceleration and gravitational forces impact riders.

MODULE 3
STUDY GUIDE

GO ONLINE to study with your Science Notebook.

Lesson 1 ACCELERATION

- acceleration
- velocity-time graph
- average acceleration
- instantaneous acceleration

- Acceleration is the rate at which an object's velocity changes.
- Velocity and acceleration are not the same thing. An object moving with constant velocity has zero acceleration. When the velocity and the acceleration of an object are in the same direction, the object speeds up; when they are in opposite directions, the object slows down.
- You can use a velocity-time graph to find the velocity and the acceleration of an object. The average acceleration of an object is the slope of its velocity-time graph.

$$\bar{a} \equiv \frac{\Delta v}{\Delta t} = \frac{v_f - v_i}{t_f - t_i}$$

Lesson 2 MOTION WITH CONSTANT ACCELERATION

- If an object is moving with constant acceleration, its position-time graph is a parabola, and its velocity-time graph is a straight line.
- The area under an object's velocity-time graph is its displacement.
- In motion with constant acceleration, position, velocity, acceleration, and time are related:

$$v_f = v_i + \bar{a}\Delta t$$
$$x_f = x_i + v_i t_f + \frac{1}{2}\bar{a}t_f^2$$
$$v_f^2 = v_i^2 + 2\bar{a}(x_f - x_i)$$

Lesson 3 FREE FALL

- free fall
- free-fall acceleration

- Free-fall acceleration on Earth is about 9.8 m/s^2 downward. The sign associated with free-fall acceleration in equations depends on the choice of the coordinate system.
- When an object is in free fall, gravity is the only force acting on it. Equations for motion with constant acceleration can be used to solve problems involving objects in free fall.

THREE-DIMENSIONAL THINKING

Module Wrap-Up

REVISIT THE PHENOMENON

Why do sudden changes in the direction or speed of jet planes affect pilots?

CER Claim, Evidence, Reasoning

Explain your Reasoning Revisit the claim you made when you encountered the phenomenon. Summarize the evidence you gathered from your investigations and research and finalize your Summary Table. Does your evidence support your claim? If not, revise your claim. Explain why your evidence supports your claim.

STEM UNIT PROJECT

Now that you've completed the module, revisit your STEM unit project. You will summarize your evidence and apply it to the project.

GO FURTHER

SEP Data Analysis Lab

How do free fall motion on Earth and Jupiter compare?

Suppose a ball could be thrown vertically upward with the same initial velocity on Earth and on Jupiter.

Data and Observations An object on the planet Jupiter has about three times the free-fall acceleration as on Earth. Neglect the effects of atmospheric resistance and assume gravity is the only force on the ball.

CER Analyze and Interpret Data

1. **Claim** How would the maximum height reached by the ball on Jupiter compare to the maximum height reached on Earth?
2. **Evidence and Reasoning** Use sketches and mathematics to explain your approach to this problem and your reasoning to justify your claim.
3. How would your claim change if the initial velocity of the ball on Jupiter were three times greater? Explain your reasoning.

Steve Whittington/Moment/Getty Images

MODULE 4
FORCES IN ONE DIMENSION

Oliver Furrer/Cultura/Getty Images

MODULE 4

FORCES IN ONE DIMENSION

ENCOUNTER THE PHENOMENON

How do wing suits help BASE jumpers control their velocity?

GO ONLINE to watch a video about da Vinci's parachute.

SEP Ask Questions

Do you have other questions about the phenomenon? If so, add them to the driving question board.

CER Claim, Evidence, Reasoning

Make Your Claim Use your CER chart to make a claim about how BASE jumpers control their velocity with wing suits. Explain your reasoning.

Collect Evidence Use the lessons in this module to collect evidence to support your claim. Record your evidence as you move through the module.

Explain Your Reasoning You will revisit your claim and explain your reasoning at the end of the module.

GO ONLINE to access your CER chart and explore resources that can help you collect evidence.

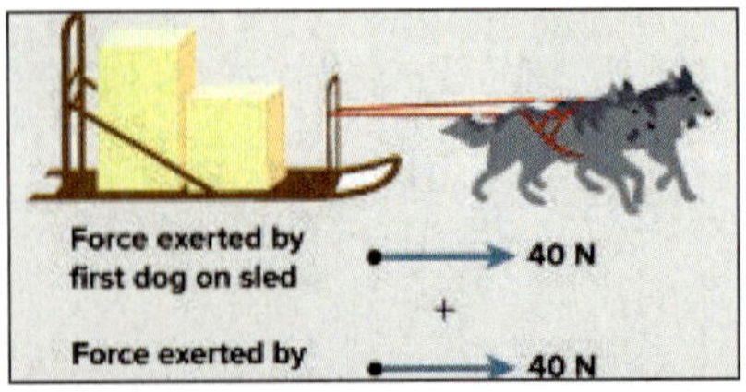

LESSON 1: Explore & Explain: Newton's Second Law

LESSON 2: Explore & Explain: Drag Force

Smithsonian

Additional Resources

(t)Video Supplied by BBC Worldwide Learning, (b)U.S. Army Photo by Sgt. Aaron Ricca, 115th Mobile Public Affairs Detachment

LESSON 1

FORCE AND MOTION

FOCUS QUESTION

What causes a change in motion?

Force

Consider a textbook resting on a table. To cause it to move, you could either push or pull on it. In physics, a push or a pull is called a **force.** If you push or pull harder on an object, you exert a greater force on the object. In other words, you increase the magnitude of the applied force. The direction in which the force is exerted also matters—if you push the resting book to the right, the book will start moving to the right. If you push the book to the left, it will start moving to the left. Because forces have both magnitude and direction, forces are vectors. The symbol $\mathbf{F}$ is vector notation that represents the size and direction of a force, while F represents only the magnitude. The magnitude of a force is measured in units called newtons (N).

Unbalanced forces change motion

Recall that motion diagrams describe the positions of an object at equal time intervals. For example, the motion diagram for the book in **Figure 1** shows the distance between the dots increasing. This means the speed of the book is increasing. At $t = 0$, it is at rest, but after 2 seconds it is moving at 1.5 m/s. This change in speed means it is accelerating. What is the cause of this acceleration? The book was at rest until you pushed it, so the cause of the acceleration is the force exerted by your hand. In fact, all accelerations are the result of an unbalanced force acting on an object.

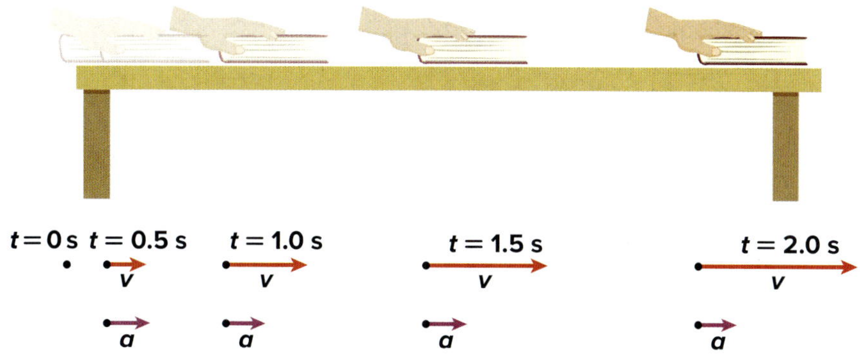

Figure 1 The hand pushing on the book exerts a force that causes the book to accelerate in the direction of the unbalanced force.

3D THINKING DCI Disciplinary Core Ideas CCC Crosscutting Concepts SEP Science & Engineering Practices

COLLECT EVIDENCE

Use your Science Journal to record the evidence you collect as you complete the readings and activities in this lesson.

INVESTIGATE

GO ONLINE to find these activities and more resources.

Applying Practices: Newton's Second Law

HS-PS2-1. Analyze data to support the claim that Newton's second law of motion describes the mathematical relationship among the net force on a macroscopic object, its mass, and its acceleration.

Systems and external world When considering how a force affects motion, it is important to identify the object or objects of interest, called the **system.** Everything around the system with which the system can interact is called the external world. In **Figure 2**, the book is the system. Your hand, Earth, string and the table are parts of the external world that interact with the book by pushing or pulling on it.

Contact forces Again, think about the different ways in which you could move a textbook. You could push or pull it by directly touching it, or you could tie a string around it and pull on the string. These are examples of contact forces. A contact force exists when an object from the external world touches a system, exerting a force on it. If you are holding this physics textbook right now, your hands are exerting a contact force on it. If you place the book on a table, you are no longer exerting a contact force on the book. The table, however, is exerting a contact force because the table and the book are in contact.

Field forces There are other ways in which the motion of the textbook can change. You could drop it, and as you learned in a previous chapter, it would accelerate as it falls to the ground. The gravitational force of Earth acting on the book causes this acceleration. This force affects the book whether or not Earth is actually touching it. Gravitational force is an example of a field force. Field forces are exerted without contact. Can you think of other kinds of field forces? If you have ever investigated magnets, you know that they exert forces without touching. You will investigate magnetism and other field forces in future chapters. For now, the only field force you need to consider is the gravitational force.

Agents Forces result from interactions; every contact and field force has a specific and identifiable cause, called the agent. You should be able to name the agent exerting each force as well as the system upon which the force is exerted. For example, when you push your textbook, your hand (the agent) exerts a force on the textbook (the system). If there are not both an agent and a system, a force does not exist. What about the gravitational force? The agent is the mass of Earth exerting a field force on the book. The labels on the forces in **Figure 2** are good examples of how to identify a force's agent and the system upon which the force acts.

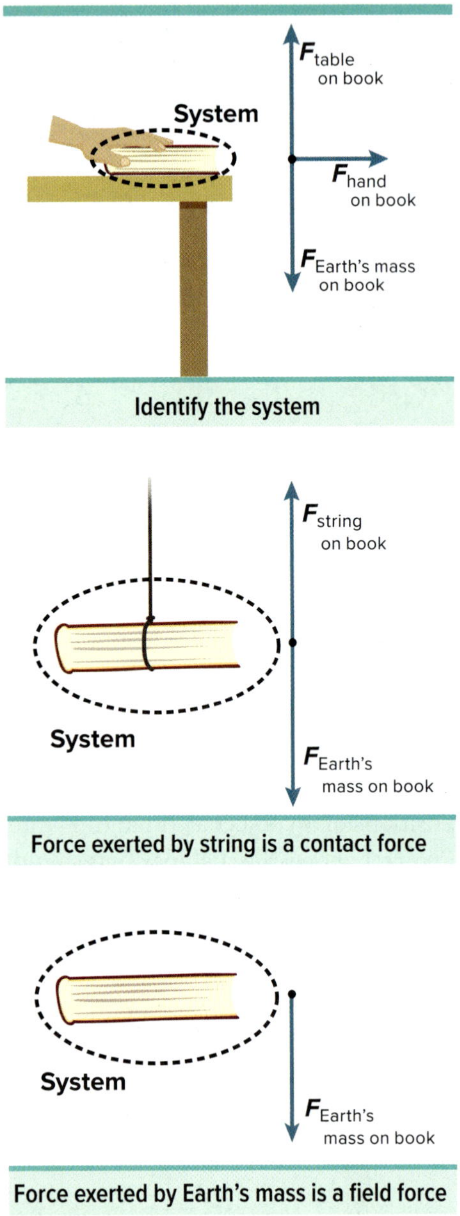

Figure 2 The book is the system in each of these situations.

Classify *each force in the first panel as either a contact force or a field force.*

Get It?

Explain how contact forces are different from field forces.

SCIENCE USAGE V. COMMON USAGE

Force

Science usage: a push or pull exerted on an object

The force of gravity exerted by the Sun on Earth pulls Earth into orbit around the Sun.

Common usage: to compel by physical, moral, or intellectual means

Emi forced her younger brother to wash the dishes.

STEM CAREER Connection

Astronaut

On Earth, we are used to accounting for the force exerted by Earth's gravitational field. To prepare for missions in space, astronauts must become familiar with the effects of forces on objects in a zero-gravity environment.

Figure 3 The drawings of the ball in the hand and the ball hanging from the string are both pictorial models. The free-body diagram for each situation is shown next to each pictorial model.

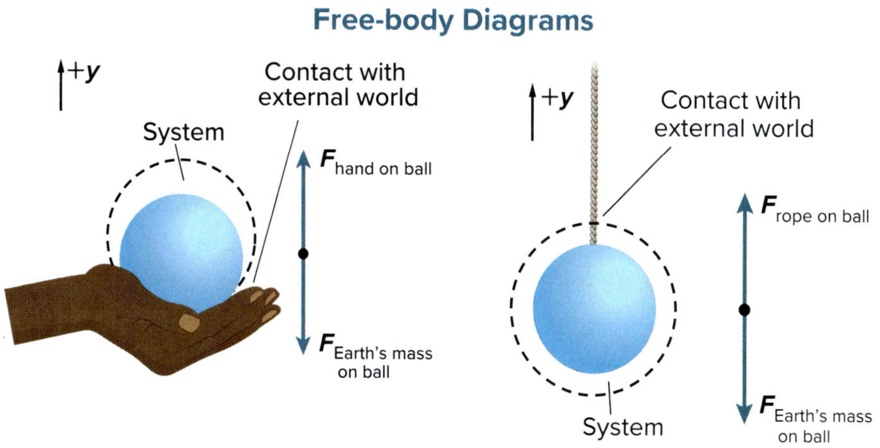

COLOR CONVENTION

Force (F) ⟷ blue

Free-body diagrams Just as pictorial representations and motion diagrams are useful in solving problems about motion, similar representations will help you analyze how forces affect motion. The first step is to sketch the situation, as shown in **Figure 3**. Circle the system, and identify every place where the system touches the external world. It is at these places that an agent exerts a contact force. Then identify any field forces on the system. This gives you the pictorial representation.

A **free-body diagram** is a physical representation that shows the forces acting on a system. Follow these guidelines when drawing a free-body diagram:

- The free-body diagram is drawn separately from the sketch of the problem situation.
- Apply the particle model, and represent the object with a dot.
- Represent each force with an arrow that points in the direction the force is applied. Always draw the force vectors pointing away from the particle, even when the force is a push.
- Make the length of each arrow proportional to the size of the force. Often you will draw these diagrams before you know the magnitudes of all the forces. In such cases, make your best estimate.
- Label each force. Use the symbol F with a subscript label to identify both the agent and the object on which the force is exerted.
- Choose a direction to be positive, and indicate this on the diagram.

Using free-body diagrams and motion diagrams Recall that all accelerations are the result of unbalanced forces. If a motion diagram shows that an object is accelerating, a free-body diagram of that object should have an unbalanced force in the same direction as the acceleration. Notice that the falling ball in **Figure 3** shows an unbalanced force downward, in the direction of the acceleration of the ball.

Get It?

Compare the direction of an object's acceleration with the direction of the unbalanced force exerted on the object.

CCC CROSSCUTTING CONCEPTS

Cause and Effect Study **Figure 3**. With a partner, plan a demonstration to show the specific cause and effect relationship between the motion of an object and the magnitudes of two forces acting on it in opposite directions. Vary the forces to show the different effects when the forces are unbalanced. Gather empirical evidence to show that the acceleration (or lack of acceleration) is caused directly by the imbalance (or balance) of the forces and not simply correlation.

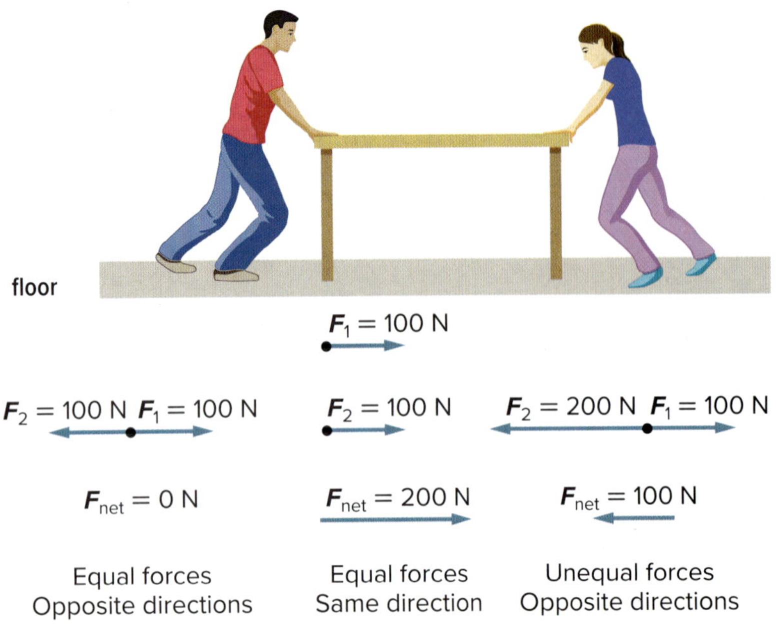

Figure 4 The net force acting on the table is the vector sum of all the forces acting on the table. This case only considers the horizontal forces acting on the table.

Combining Forces

What happens if you and a friend each push a table and exert 100 N of force on it? When you push together in the same direction, you give the table twice the acceleration that it would have if just one of you applied 100 N of force. When you push on the table in opposite directions with the same amount of force, as in **Figure 4,** there is no unbalanced force, so the table does not accelerate but remains at rest.

Net force The bottom portion of **Figure 4** shows free-body diagrams for these two situations. The third diagram in **Figure 4** shows the free-body diagram for a third situation in which your friend pushes on the table twice as hard as you in the opposite direction. Below each free-body diagram is a vector representing the resultant of the two forces. When the force vectors are in the same direction, they can be replaced by one vector with a length equal to their combined length. When the forces are in opposite directions, the resultant is the length of the difference between the two vectors. Another term for the vector sum of all the forces on an object is the **net force.**

You also can analyze the situation mathematically. Call the positive direction the direction in which you are pushing the table with a 100 N force. In the first case, your friend is pushing with a negative force of 100 N. Adding them together gives a total force of 0 N, which means there is no acceleration. In the second case, your friend's force is 100 N, so the total force is 200 N in the positive direction and the table accelerates in the positive direction. In the third case, your friend's force is −200 N, so the total force is −100 N and the table accelerates in the negative direction.

PRACTICE Problems

ADDITIONAL PRACTICE

For each of the following situations, specify the system and draw a motion diagram and a free-body diagram. Label all forces with their agents, and indicate the direction of the acceleration and of the net force. Draw vectors of appropriate lengths. Ignore air resistance unless otherwise indicated.

1. A skydiver falls downward through the air at constant velocity. (The air exerts an upward force on the person.)
2. You hold a softball in the palm of your hand and toss it up. Draw the diagrams while the ball is still touching your hand.
3. After the softball leaves your hand, it rises, slowing down.
4. After the softball reaches its maximum height, it falls down, speeding up.
5. **CHALLENGE** You catch the ball in your hand and bring it to rest.

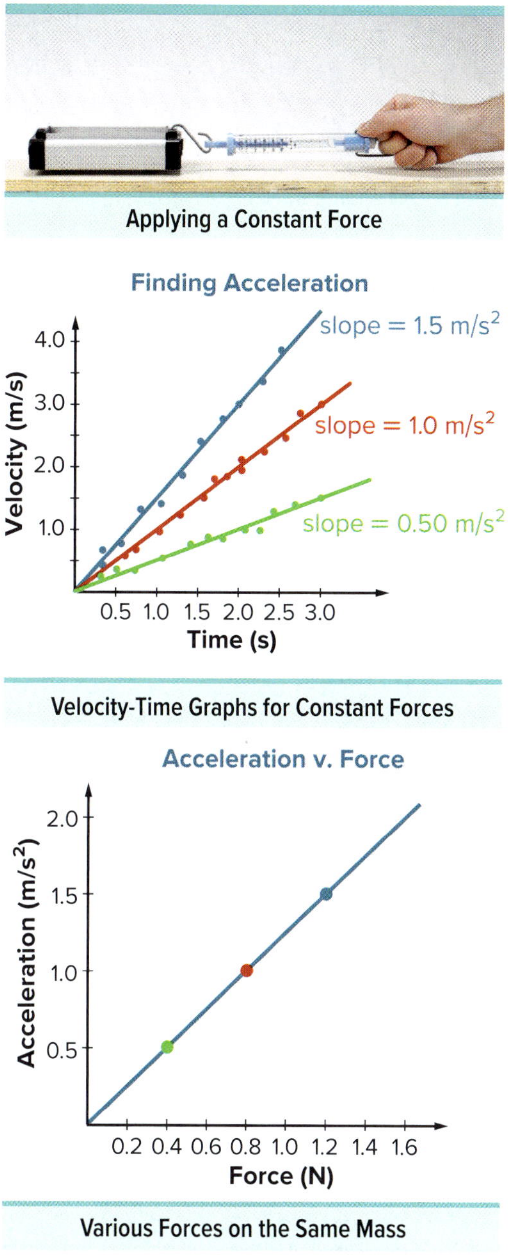

Figure 5 A spring scale exerts a constant unbalanced force on the cart. Repeating the investigation with different forces produces velocity-time graphs with different slopes.

Acceleration and Force

To explore how forces affect an object's motion, think about doing a series of investigations. Consider the simple situation shown in the top photo of **Figure 5** in which we exert one force horizontally on an object. Starting with the horizontal direction is helpful because gravity does not act horizontally. To reduce complications resulting from the object rubbing against the surface, the investigations should be done on a smooth surface, such as a well-polished table. We'll also use a cart with wheels that spin easily.

Apply constant force How can you exert a constant unbalanced force? One way is to use a device called a spring scale. Inside the scale is a spring that stretches proportionally to the magnitude of the applied force. The front of the scale is calibrated to read the force in newtons. If you pull on the scale so that the reading on the front stays constant, the applied force is constant. The top photo in **Figure 5** shows a spring scale pulling a low-resistance cart with a constant unbalanced force.

If you perform this investigation and measure the cart's velocity for a period of time, you could construct a graph like the green line shown in the velocity-time graphs for constant forces in the middle panel of **Figure 5**. The constant slope of the line in the velocity-time graph indicates the cart's velocity increases at a constant rate. The constant rate of change of velocity means the acceleration is constant. This constant acceleration is a result of the constant unbalanced force applied by the spring scale to the cart.

How does acceleration depend on force? Repeat the investigation with a larger constant force. Repeat it again with an even greater force. For each force, plot a velocity-time graph like the red and blue lines in the middle panel of **Figure 5.** Recall that the line's slope is the cart's acceleration. Calculate the slope of each line and plot the results for each force to make an acceleration-force graph, as shown in the bottom panel of **Figure 5**.

The graph indicates the relationship between force and acceleration is linear. Because the relationship is linear, you can apply the equation for a straight line:

$$y = kx + b$$

The y-intercept is 0, so the linear equation simplifies to $y = kx$. The y-variable is acceleration. The x-variable is force. Therefore, acceleration is equal to the slope of the line multiplied by the applied net force.

Get It?

Describe the relationship between applied net force and acceleration.

Interpreting slope What is the physical meaning of the slope of the acceleration-force graph? Does it describe something about the object that is accelerating? To see, change the object. Suppose that a second, identical cart is placed on top of the first, and then a third cart is added as in **Figure 6.** The spring scale would be pulling two carts and then three. A plot of the force versus acceleration for one, two, and three carts is shown in the graph in **Figure 6.**

The graph shows that if the same force is applied in each case, the acceleration of two carts is $\frac{1}{2}$ the acceleration of one cart, and the acceleration of three carts is $\frac{1}{3}$ the acceleration of one cart. This means that as the number of carts increases, the acceleration decreases. In other words, a greater force is needed to produce the same acceleration. The slopes of the lines in **Figure 6** depend upon the number of carts; that is, the slope depends on the total mass of the carts. In fact, the slope is the reciprocal of the mass (slope $= \frac{1}{\text{mass}}$). Using this value for slope, the mathematical equation $y = kx$ becomes the physics equation: $a = \frac{F_{net}}{m}$.

What information is contained in the equation $a = \frac{F_{net}}{m}$? It tells you that a net force applied to an object causes that object to experience a change in motion—the force causes the object to accelerate. It also tells you that for the same object, if you double the force, you will double the object's acceleration. Lastly, if you apply the same force to objects with different masses, the one with the most mass will have the smallest acceleration and the one with the least mass will have the greatest acceleration.

Changing the Mass of the System

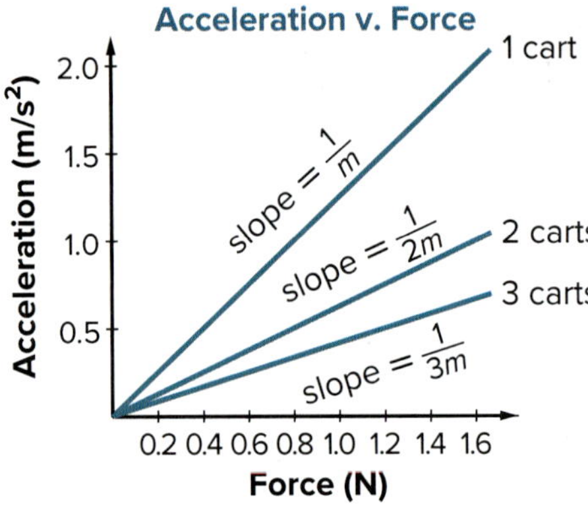

Same Force on Different Masses

Figure 6 Changing an object's mass affects that object's acceleration.

Compare *the acceleration of one cart to the acceleration of two carts for an applied force of 1 N.*

Get It?

Determine how the force exerted on an object must be changed to reduce the object's acceleration by half.

Recall that forces are measured in units called newtons. Because $F_{net} = ma$, a newton has the units of mass times the units of acceleration. So one newton is equal to one $kg{\cdot}m/s^2$. To get an approximate idea of the size of 1 N, think about the downward force you feel when you hold an apple in your hand. The force exerted by the apple on your hand is approximately one newton. **Table 1** shows the magnitudes of some other common forces.

Table 1 Common Forces

Description	*F*(N)
Force of gravity on a coin (nickel)	0.05
Force of gravity on a 0.45-kg bag of sugar	4.5
Force of gravity on a 70-kg person	686
Force exerted by road on an accelerating car	3000
Force of a rocket engine	5,000,000

Figure 7 The net force acting on an object is the vector sum of all the forces acting on that object.

Newton's Second Law

Figure 7 shows two dogs pulling a sled. Each dog pulls with a force of 40 N. From the cart and spring-scale investigations, you know that the sled accelerates as a result of the unbalanced force acting it. Would the acceleration change if instead of two dogs each exerting a 40-N force, there was one bigger, stronger dog exerting a single 80-N force on the sled? When considering forces and acceleration, it is important to find the sum of all forces, called the net force, acting on a system.

Newton's second law states that the acceleration of an object is proportional to the net force and inversely proportional to the mass of the object being accelerated. This law is based on observations of how forces affect masses and is represented by the following equation. Newton's second law accurately predicts changes in the motion of macroscopic objects.

Newton's Second Law

The acceleration of an object is equal to the sum of the forces acting on the object divided by the mass of the object.

$$a = \frac{F_{net}}{m}$$

Solving problems using Newton's second law One of the most important steps in correctly applying Newton's second law is determining the net force acting on the object. Often, more than one force acts on an object, so you must add the force vectors to determine the net force. Draw a free-body diagram showing the direction and relative strength of each force acting on the system. Then, add the force vectors to find the net force. Next, use Newton's second law to calculate the acceleration. Finally, if necessary, you can use what you know about accelerated motion to find the velocity or position of the object.

PRACTICE Problems

ADDITIONAL PRACTICE

6. Two horizontal forces, 225 N and 165 N, are exerted on a canoe. If these forces are applied in the same direction, find the net horizontal force on the dance.

7. If the same two forces as in the previous problem are exerted on the canoe in opposite directions, what is the net horizontal force on the canoe? Be sure to indicate the direction of the net force.

8. CHALLENGE Three confused sled dogs are trying to pull a sled across the Alaskan snow. Alutia pulls east with a force of 35 N, Seward also pulls east but with a force of 42 N, and big Kodiak pulls west with a force of 53 N. What is the net force on the sled? Explain how Newton's second law accurately predicts the sled's change in motion.

EXAMPLE Problem 1

FIGHTING OVER A PILLOW **Anudja is holding a pillow with a mass of 0.30 kg when Sarah decides that she wants it and tries to pull it away from Anudja. If Sarah pulls horizontally on the pillow with a force of 10.0 N and Anudja pulls with a horizontal force of 11.0 N, what is the horizontal acceleration of the pillow?**

1 ANALYZE AND SKETCH THE PROBLEM

- Sketch the situation.
- Identify the pillow as the system, and the direction in which Anudja pulls as positive.
- Draw the free-body diagram. Label the forces.

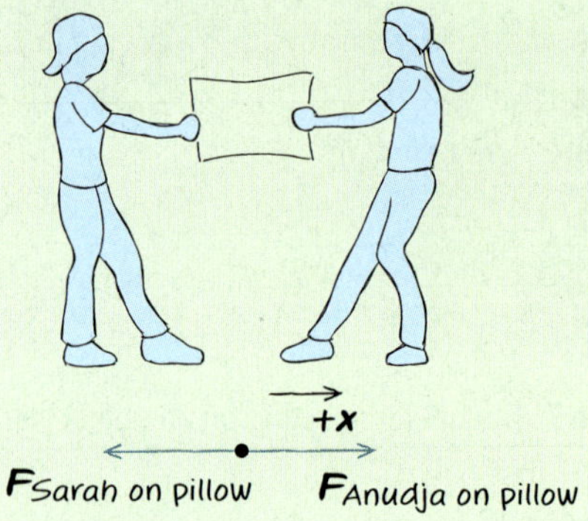

KNOWN	UNKNOWN
$m = 0.30\text{ kg}$	$a = ?$
$F_{\text{Anudja on pillow}} = 11.0\text{ N}$	
$F_{\text{Sarah on pillow}} = 10.0\text{ N}$	

2 SOLVE FOR THE UNKNOWN

$$F_{\text{net}} = F_{\text{Anudja on pillow}} + (-F_{\text{Sarah on pillow}})$$

Use Newton's second law.

$$a = \frac{F_{\text{net}}}{m}$$

$$= \frac{F_{\text{Anudja on pillow}} + (-F_{\text{Sarah on pillow}})}{m}$$

$$= \frac{11.0\text{ N} - 10.0\text{ N}}{0.30\text{ kg}}$$

Substitute $F_{\text{Anudja on pillow}} = 11.0\text{ N}$, $F_{\text{Sarah on pillow}} = 10.0\text{ N}$, $m = 0.30\text{ kg}$.

$$= 3.3\text{ m/s}^2$$

$\boldsymbol{a} = 3.3\text{ m/s}^2$ toward Anudja

3 EVALUATE THE ANSWER

- **Are the units correct?** m/s^2 is the correct unit for acceleration.
- **Does the sign make sense?** The acceleration is toward Anudja because Anudja is pulling toward herself with a greater force than Sarah is pulling in the opposite direction.
- **Is the magnitude realistic?** The net force is 1 N and the mass is 0.3 kg, so the acceleration is realistic.

PRACTICE Problems

ADDITIONAL PRACTICE

9. A spring scale is used to exert a net force of 2.7 N on a cart. If the cart's mass is 0.64 kg, what is the cart's acceleration?

10. Kamaria is learning how to ice skate. She wants her mother to pull her along so that she has an acceleration of 0.80 m/s^2. If Kamaria's mass is 27.2 kg, with what force does her mother need to pull her? (Neglect any resistance between the ice and Kamaria's skates.)

11. CHALLENGE Two horizontal forces are exerted on a large crate. The first force is 317 N to the right. The second force is 173 N to the left.

a. Draw a force diagram for the horizontal forces acting on the crate.

b. What is the net force acting on the crate?

c. The box is initially at rest. Five seconds later, its velocity is 6.5 m/s to the right. What is the crate's mass?

Newton's First Law

What is the motion of an object when the net force acting on it is zero? Newton's second law says that if $F_{net} = 0$, then acceleration equals zero. Recall that if acceleration equals zero, then velocity does not change. Thus a stationary object with no net force acting on it will remain at rest. What about a moving object, such as a ball rolling on a surface? How long will the ball continue to roll? The answer depends on the surface on which the ball is rolling. If the ball is rolled on a thick carpet that exerts a force on the ball, it will come to rest quickly. If it is rolled on a hard, smooth surface that exerts very little force, such as a bowling alley, the ball will roll for a long time with little change in velocity.

Galileo did many experiments and he concluded that if he could remove all forces opposing motion, horizontal motion would never stop. Galileo was the first to recognize that the general principles of motion could be found by extrapolating experimental results to an ideal case.

In the absence of a net force, the velocity of the moving ball and the lack of motion of the stationary object do not change. Newton recognized this and generalized Galileo's results into a single statement. Newton's statement, "an object that is at rest will remain at rest, and an object that is moving will continue to move in a straight line with constant speed, if and only if the net force acting on that object is zero," is called **Newton's first law.**

Inertia

Newton's first law is sometimes called the law of inertia because **inertia** is the tendency of an object to resist changes in velocity. The car and the red block in **Figure 8** demonstrate the law of inertia. In the left panel, both objects are moving to the right. In the right panel, the wooden box applies a force to the car, causing it to stop. The red block does not experience the force applied by the wooden box. It continues to move to the right with the same velocity as in the left panel.

Is inertia a force? No. Forces are results of interactions between two objects; they are not properties of single objects, so inertia cannot be a force. Remember that because velocity includes both the speed and direction of motion, a net force is required to change either the speed or the direction of motion. If the net force is zero, Newton's first law means the object will continue with the same speed and direction.

Figure 8 The car and the block approach the wooden box at the same speed. After the collision, the block continues on with the same horizontal speed.

Identify *the forces that will eventually cause the block to stop moving.*

Figure 9 An object is in equilibrium if its velocity isn't changing. In both cases pictured here, velocity isn't changing, so the net force must be zero.

Equilibrium According to Newton's first law, a net force causes the velocity of an object to change. If the net force on an object is zero, then the object is in **equilibrium.** An object is in equilibrium if it is moving at a constant velocity. Note that being at rest is simply of the state of constant velocity, $v = 0$. Newton's first law identifies a net force as something that disturbs a state of equilibrium. Thus, if there is no net force acting on the object, then the object does not experience a change in speed or direction and is in equilibrium. **Figure 9** indicates, at least in terms of net forces, there is no difference between lying on a sofa and falling at a constant velocity while skydiving–velocity isn't changing, so the net force is zero.

Keep in mind that the real world is full of forces that resist motion. Newton's ideal, friction-free world is not easy to obtain. If you analyze a situation and find that the result is different from your own experience, ask yourself if this s is because of the presence of frictional forces.

Check Your Progress

12. **Forces** Identify each of the following as either a, b, or c: mass, inertia, the push of a hand, friction, air resistance, spring force, gravity, and acceleration.
    a. contact force
    b. a field force
    c. not a force
13. **Free-Body Diagram** Draw a free-body diagram of a bag of sugar being lifted by your hand at an increasing speed. Specifically identify the system. Use subscripts to label all forces with their agents. Remember to make the arrows the correct lengths.
14. **Free-Body Diagram** Draw a free-body diagram of a water bucket being lifted by a rope at a decreasing speed. Specifically identify the system. Label all forces with their agents and make the arrows the correct lengths.
15. **Critical Thinking** A force of 1 N is the only horizontal force exerted on a block, and the horizontal acceleration of the block is measured. When the same horizontal force is the only force exerted on a second block, the horizontal acceleration is three times as large. What can you conclude about the masses of the two blocks?

LEARNSMART® Go online to follow your personalized learning path to review, practice, and reinforce your understanding.

LESSON 2
WEIGHT AND DRAG FORCE

FOCUS QUESTION

How does the drag force change after a sky diver deploys their parachute?

Weight

From Newton's second law, the fact that the ball in **Figure 10** is accelerating means there must be unbalanced forces acting on the ball. The only force acting on the ball is the gravitational force due to Earth's mass. An object's **weight** is the gravitational force experienced by that object. This gravitational force is a field force whose magnitude is directly proportional to the mass of the object experiencing the force. In equation form, the gravitational force, which equals weight, can be written $F_g = mg$. The mass of the object is m, and g, called the **gravitational field,** is a vector quantity that relates an object's mass to the gravitational force it experiences at a given location.

Near Earth's surface, g is 9.8 N/kg toward Earth's center. Objects near Earth's surface experience 9.8 N of force for every kilogram of mass.

Scales When you stand on a scale as shown in the right panel of **Figure 10**, the scale exerts an upward force on you. Because you are not accelerating, the net force acting on you must be zero. Therefore the magnitude of the force exerted by the scale ($F_{\text{scale on you}}$) pushing up must equal the magnitude of F_g pulling down on you. Inside the scale, springs provide the upward force necessary to make the net force equal zero. The scale is calibrated to convert the stretch of the springs to a weight. If you were on a different planet with a different g, the scale would exert a different force to keep you in equilibrium, and consequently, the scale's reading would be different. Because weight is a force, the proper unit used to measure weight is the newton.

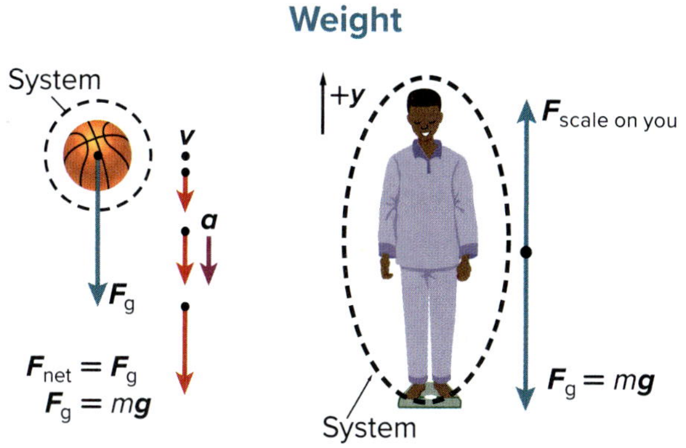

Figure 10 The gravitational force exerted by Earth's mass on an object equals the object's mass times the gravitational field, $F_g = mg$.

Identify *the forces acting on you when you are in equilibrium while standing on a scale.*

3D THINKING | DCI Disciplinary Core Ideas | CCC Crosscutting Concepts | SEP Science & Engineering Practices

COLLECT EVIDENCE

Use your Science Journal to record the evidence you collect as you complete the readings and activities in this lesson.

INVESTIGATE

GO ONLINE to find these activities and more resources.

Probeware Lab: Terminal Velocity
Analyze and interpret data to determine which factors affect the size of the drag force on a falling object.

Revisit the Encounter the Phenomenon Question
What information from this lesson can help you answer the Unit and Module questions?

EXAMPLE Problem 2

COMPARING WEIGHTS Kiran holds a brass cylinder in each hand. Cylinder A has a mass of 100.0 g and cylinder B has a mass of 300.0 g. What upward forces do his two hands exert to keep the cylinders at rest? If he then drops the two, with what acceleration do they fall? (Ignore air resistance.)

1 ANALYZE AND SKETCH THE PROBLEM

- Sketch the situation.
- Identify the two cylinders as the systems, and choose the upward direction as positive.
- Draw the free-body diagrams. Label the forces.

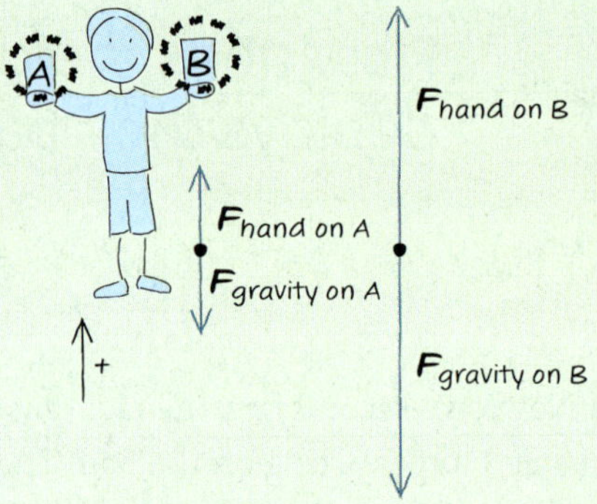

KNOWN	UNKNOWNS
$m_A = 0.1000$ kg	$F_{\text{Hand on A}} = ?$
$m_B = 0.3000$ kg	$F_{\text{Hand on B}} = ?$
$g = -9.8$ N/kg	$a_A = ?$ $a_B = ?$

2 SOLVE FOR THE UNKNOWNS

For cylinder A:

$$F_{\text{Net on A}} = F_{\text{Hand on A}} + F_{\text{Gravity on A}}$$
$$0 = F_{\text{Hand on A}} + F_{\text{Gravity on A}}$$
$$F_{\text{Hand on A}} = -F_{\text{Gravity on A}}$$
$$F_{\text{Hand on A}} = -m_A g$$
$$= -(0.1000 \text{ kg})(-9.8 \text{ N/kg})$$
$$= 0.98 \text{ N up}$$

For cylinder B:

$$F_{\text{Net on B}} = F_{\text{Hand on B}} + F_{\text{Gravity on B}}$$
$$= F_{\text{Hand on B}} + F_{\text{Gravity on B}}$$
$$F_{\text{Hand on B}} = -F_{\text{Gravity on B}}$$
$$F_{\text{Hand on B}} = m_B g$$
$$= -(0.3000 \text{ kg})(-9.8 \text{ N/Kg})$$
$$= 2.9 \text{ N up}$$

After the cylinders are dropped, the only force on each is the force of gravity. Use Newton's second law.

$$a_A = \frac{F_{\text{Net on A}}}{m_A} \qquad a_B = \frac{F_{\text{Net on B}}}{m_B}$$

$$a_A = \frac{m_A g}{m_A} = g \qquad a_B = \frac{m_B g}{m_B} = g$$

Substitute $F_{\text{Net on A}} = m_A g$ and $F_{\text{Net on B}} = m_B g$

$$= -9.8 \text{ m/s}^2 \qquad = -9.8 \text{ m/s}^2$$

Substitute $g = -9.8$ N/kg $= -9.8$ m/s^2.

3 EVALUATE THE ANSWER

- **Are the units correct?** N is the correct unit for force; m/s^2 is the correct unit for acceleration.
- **Does the sign make sense?** The direction of the fall is downward, the negative direction, and the object is speeding up, so the acceleration should be negative.
- **Is the magnitude realistic?** Forces are 1–5 N, typical of that exerted by objects that have a mass of one kg or less. The accelerations are both equal to free fall acceleration.

PRACTICE Problems

ADDITIONAL PRACTICE

16. You place a 4.0-kg watermelon on a spring scale that measures in newtons. What is the scale's reading?

17. You place a 22.50-kg television on a spring scale. If the scale reads 235.2 N, what is the gravitational field?

18. A 0.50-kg guinea pig is lifted up from the ground. What is the smallest force needed to lift it? Describe the particular motion resulting from this minimum force.

19. CHALLENGE A grocery sack can withstand a maximum of 230 N before it rips. Will a bag holding 15 kg of groceries that is lifted from the checkout counter at an acceleration of 7.0 m/s^2 hold?

Apparent weight What is weight? Because the weight force is defined as $F_g = mg$, F_g changes when g varies. On or near the surface of Earth, g is approximately constant, so an object's weight does not change appreciably as it moves around near Earth's surface. If a bathroom scale provides the only upward force on you, then it reads your weight. What would it read if you stood with one foot on the scale and one foot on the floor? What if a friend pushed down on your shoulders or lifted up on your elbows? Then there would be other contact forces on you, and the scale would not read your weight.

What happens if you stand on a scale in an elevator? As long as you are not accelerating, the scale will read your weight. What would the scale read if the elevator accelerated upward? **Figure 11** shows the pictorial and physical representations for this situation. You are the system, and upward is the positive direction. Because the acceleration of the system is upward, the net force must be upward. The upward force of the scale must be greater than the downward force of your weight. Therefore, the scale reading is greater than your weight.

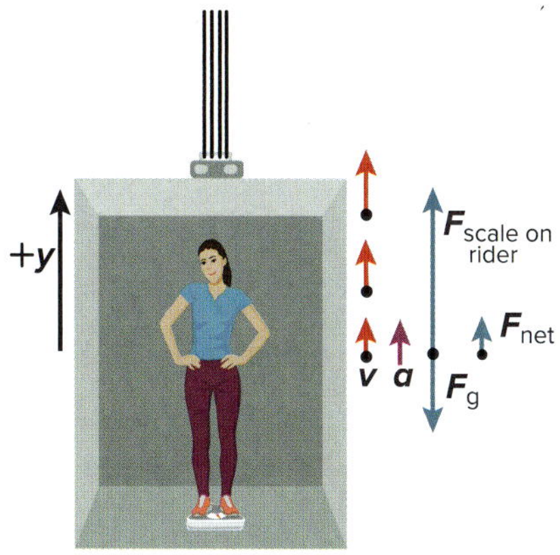

Figure 11 If you are accelerating upward, the net force acting on you must be upward. The scale must exert an upward force greater than the downward force of your weight.

Get It?

Describe the reading on the scale as the elevator accelerates upward from rest, reaches a constant speed, and then comes to a stop.

If you ride in an elevator accelerating upward, you feel as if you are heavier because the floor presses harder on your feet. On the other hand, if the acceleration is downward, then you feel lighter, and the scale reads less than your weight. The force exerted by the scale is an example of **apparent weight,** which is the support force exerted on an object.

Imagine that the cable holding the elevator breaks. What would the scale read then? The scale and you would both accelerate at $a = g$. According to this formula, the scale would read zero and your apparent weight would be zero. That is, you would be weightless. However, **weightlessness** does not mean that an object's weight is actually zero; rather, it means that there are no contact forces acting to support the object, and the object's *apparent weight* is zero. Similar to the falling elevator, astronauts experience weightlessness in orbit because they and their spacecraft are in free fall.

PROBLEM-SOLVING STRATEGY

Force and Motion

When solving force and motion problems, use the following strategies.

1. Read the problem carefully, and sketch a pictorial model.
2. Circle the system and choose a coordinate system.
3. Determine which quantities are known and which are unknown.
4. Create a physical model by drawing a motion diagram showing the direction of the acceleration.
5. Create a free-body diagram showing all the forces acting on the object.
6. Use Newton's laws to link acceleration and net force.
7. Rearrange the equation to solve for the unknown quantity.
8. Substitute known quantities with their units into the equation and solve.
9. Check your results to see whether they are reasonable.

EXAMPLE Problem 3

REAL AND APPARENT WEIGHT Your mass is 75.0 kg, and you are standing on a bathroom scale in an elevator. Starting from rest, the elevator accelerates upward at 2.00 m/s² for 2.00 s and then continues at a constant speed. Is the scale reading during acceleration greater than, equal to, or less than the scale reading when the elevator is at rest?

1 ANALYZE AND SKETCH THE PROBLEM

- Sketch the situation.
- Choose a coordinate system with the positive direction as upward.
- Draw the motion diagram. Label v and a.
- Draw the free-body diagram. The net force is in the same direction as the acceleration, so the upward force is greater than the downward force.

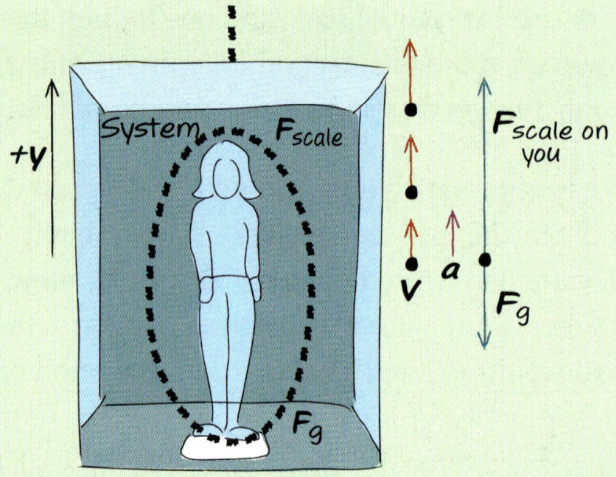

KNOWN	UNKNOWN
$m = 75.0$ kg	$F_{scale} = ?$
$a = 2.00$ m/s²	
$t = 2.00$ s	
$g = 9.8$ N/kg	

2 SOLVE FOR THE UNKNOWN

$F_{net} = ma$

$F_{net} = F_{scale} + (-F_g)$ — F_g is negative because it is in the negative direction defined by the coordinate system.

Solve for F_{scale}.

$F_{scale} = F_{net} + F_g$

Elevator at rest:

$F_{scale} = F_{net} + F_g$ — The elevator is not accelerating. Thus, $F_{net} = 0.00$ N.

$= F_g$ — Substitute $F_{net} = 0.00$ N.

$= mg$ — Substitute $F_g = mg$.

$= (75.0\ \text{kg})(9.8\ \text{N/kg})$ — Substitute $m = 75.0$ kg, $g = 9.8$ N/kg.

$= 735$ N

Elevator accelerating upward:

$F_{scale} = F_{net} + F_g$

$= ma + mg$ — Substitute $F_{net} = ma$, $F_g = mg$

$= (75.0\ \text{kg})(2.00\ \text{m/s}^2) + (75.0\ \text{kg})(9.8\ \text{N/kg})$ — Substitute $m = 75.0$ kg, $a = 2.00$ m/s², $g = 9.8$ N/kg

$= 885$ N

The scale reading when the elevator is accelerating (885 N) is larger than when it is at rest (735 N).

3 EVALUATE THE ANSWER

- **Are the units correct?** kg·m/s² is the force unit, N.
- **Does the sign make sense?** The positive sign agrees with the coordinate system.
- **Is the magnitude realistic?** $F_{Scale} = 885$ N is larger than it would be at rest when F_{scale} would be 735 N. The increase is 150 N, which is about 20 percent of the rest weight. The upward acceleration is about 20 percent of that due to gravity, so the magnitude is reasonable.

PRACTICE Problems

ADDITIONAL PRACTICE

20. On Earth, a scale shows that you weigh 585 N.
    a. What is your mass?
    b. What would the scale read on the Moon ($g = 1.60$ N/kg)?
21. **CHALLENGE** Use the results from Example Problem 3 to answer questions about a scale in an elevator on Earth. What force would be exerted by the scale on a person in the following situations?
    a. The elevator moves upward at constant speed.
    b. It slows at 2.0 m/s^2 while moving downward.
    c. It speeds up at 2.0 m/s^2 while moving downward.
    d. It moves downward at constant speed.
    e. In what direction is the net force as the elevator slows to a stop as it is moving down?

Drag Force

The particles in the air around an object exert forces on that object. In fact, air exerts huge forces, but in most cases, it exerts balanced forces on all sides, and therefore has no net effect. So far, you have neglected the force of air on an object moving through the air. In actuality, when an object moves through any fluid, such as air or water, the fluid exerts a force on the moving object in the direction opposite the object's motion. A **drag force** is the force exerted by a fluid on an object opposing motion through the fluid. This force is dependent on the motion of the object, the properties of the object, and the properties of the fluid that the object is moving through. For example, as the speed of the object increases, so does the magnitude of the drag force. The size and shape of the object also affect the drag force. The fluid's properties, such as its density and viscosity, also affect the drag force.

Get It?

Describe how the wingsuits shown in the photo at the beginning of the module affect the drag force experienced by the skydivers.

PHYSICS Challenge

A 415-kg container of food and water is dropped from an airplane at an altitude of 300 m. First, consider the situation ignoring air resistance. Then calculate the more realistic situation involving a drag force provided by a parachute.

1. If you ignore air resistance, how long will it take the container to fall 300 m to the ground?
2. Again, ignoring air resistance, what is the speed of the container just before it hits the ground?
3. The container is attached to a parachute designed to produce a drag force that allows the container to reach a constant downward velocity of 6 m/s. What is the magnitude of the drag force when the container is falling at a constant 6 m/s down?

Terminal velocity If you drop a tennis ball, as in **Figure 12,** it has very little velocity at the start and thus only a small drag force. The downward force of gravity is much stronger than the upward drag force, so there is a downward acceleration. As the ball's velocity increases, so does the drag force. Soon the drag force equals the force of gravity. When this happens, there is no net force, and so there is no acceleration. The constant velocity that is reached when the drag force equals the force of gravity is called the **terminal velocity.**

When light objects with large surface areas fall, the drag force has a substantial effect on their motion, and they quickly reach terminal velocity. Heavier, more compact objects are not affected as much by the drag force. For example, the terminal velocity of a table-tennis ball in air is 9 m/s, and that of a baseball is 42 m/s. Skydivers can increase or decrease their terminal velocity by changing their body orientation and shape. A horizontal, spreadeagle shape produces the slowest terminal velocity, about 60 m/s. After the parachute opens, the skydiver becomes part of a large object with a correspondingly large drag force and a terminal velocity of about 5 m/s.

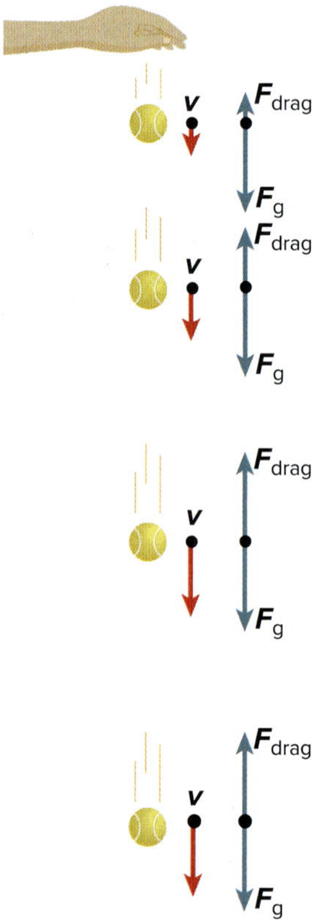

Figure 12 The drag force on an object increases as its velocity increases. When the drag force equals the gravitational force, the object is in equilibrium.

Check Your Progress

22. **Terminal Velocity** The skydiver in **Figure 13** falls at a constant speed in the spread-eagle position. Immediately after opening the parachute, is the skydiver accelerating? If so, in which direction? Explain your answer.

Figure 13

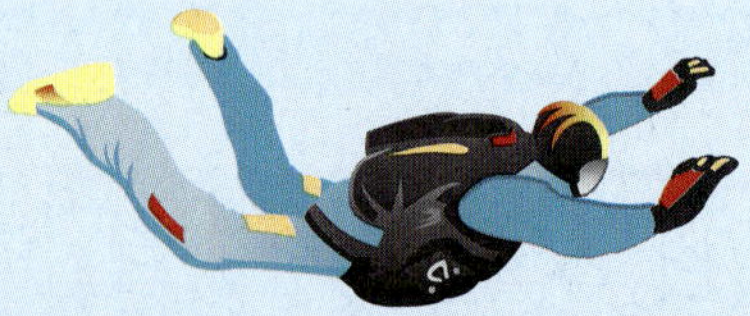

23. **Lunar Gravity** Compare the force holding a 10.0-kg rock on Earth and on the Moon. The gravitational field on the Moon is 1.6 N/kg.
24. **Motion of an Elevator** You are riding in an elevator holding a spring scale with a 1-kg mass suspended from it. You look at the scale and see that it reads 9.3 N. What does this tell you about the elevator's motion?
25. **Apparent Weight** You take a ride in a fast elevator to the top of a tall building and ride back down. Compare your apparent and real weights at each part of the journey. Sketch free-body diagrams to support your answers.
26. **Acceleration** Tecle, with a mass of 65.0 kg, is standing on an ice-skating rink. His friend applies a force of 9.0 N to him. What is Tecle's resulting acceleration?
27. **Critical Thinking** You have a job loading inventory onto trucks at a meat warehouse. Each truck has a weight limit of 10,000 N of cargo. You push each crate of meat along a low-resistance roller belt to a scale and weigh it before moving it onto the truck. One night, right after you weigh a 1000-N crate, the scale breaks. Describe a way in which you could apply Newton's laws to approximate the masses of the remaining crates.

LEARNSMART® Go online to follow your personalized learning path to review, practice, and reinforce your understanding.

LESSON 3
NEWTON'S THIRD LAW

FOCUS QUESTION

If you push on a wall, what force does the wall exert on you?

Interaction Pairs

Figure 14 illustrates the idea of forces as interaction pairs. There is a force from the woman on the dog's toy, and there is a force from the dog's toy on the woman. Forces always come in pairs similar to this example. Consider the woman (A) as one system and the toy (B) as another. What forces act on each of the two systems? Looking at the force diagrams in **Figure 14**, you can see that each system exerts a force on the other. The two forces, $\boldsymbol{F}_{\text{A on B}}$ and $\boldsymbol{F}_{\text{B on A}}$, are an example of an **interaction pair,** which is a set of two forces that are in opposite directions, have equal magnitudes, and act on different objects. Sometimes, an interaction pair is called an action-reaction pair. This might suggest that one causes the other; however, this is not true. For example, the force of the woman pulling on the toy doesn't cause the toy to pull on the woman. The two forces either exist together or not at all.

Definition of Newton's third law In **Figure 14**, the force exerted by the woman on the toy is equal in magnitude and opposite in direction to the force exerted by the toy on the woman. Such an interaction pair is an example of **Newton's third law,** which states that all forces come in pairs. The two forces in a pair act on different objects and are equal in strength and opposite in direction.

Figure 14 The force that the toy exerts on the woman and the force that the woman exerts on the toy are an interaction pair.

3D THINKING — DCI Disciplinary Core Ideas — CCC Crosscutting Concepts — SEP Science & Engineering Practices

COLLECT EVIDENCE

Use your Science Journal to record the evidence you collect as you complete the readings and activities in this lesson.

INVESTIGATE

GO ONLINE to find these activities and more resources.

PhysicsLAB: Newton's Third Law
Plan and carry out a investigation that applies Newton's laws of motion to different systems.

Review the News
Obtain information from a current news story about forces and motion. Evaluate your source and communicate your findings to your class.

Newton's Third Law
The force of A on B is equal in magnitude and opposite in direction of the force of B on A.

$$F_{\text{A on B}} = -F_{\text{B on A}}$$

Using Newton's third law Consider the situation of holding a book in your hand. You can draw one free-body diagram for you and one for the book. Are there any interaction pairs? When identifying interaction pairs, keep in mind that they always occur in two different free-body diagrams, and they always will have the symmetry of subscripts noted on the previous page. In this case, the interaction pair is $F_{\text{book on hand}}$ and $F_{\text{hand on book}}$.

The ball in **Figure 15** interacts with the table and with Earth. First, analyze the forces acting on one system, the ball. The table exerts an upward force on the ball, and the mass of Earth exerts a downward gravitational force on the ball. Even though these forces are in opposite directions, they are not an interaction pair because they act on the same object. Now consider the ball and the table together. In addition to the upward force exerted by the table on the ball, the ball exerts a downward force on the table. This is an interaction pair. Notice also that the ball has a weight. If the ball experiences a force due to Earth's mass, then there must be a force on Earth's mass due to the ball. In other words, they are an interaction pair.

$$F_{\text{Earth's mass on ball}} = -F_{\text{ball on Earth's mass}}$$

An unbalanced force on Earth would cause Earth to accelerate. But acceleration is inversely proportional to mass. Because Earth's mass is so huge in comparison to the masses of other objects that we normally consider, Earth's acceleration is so small that it can be neglected. In other words, Earth can be often treated as part of the external world rather than as a second system. The problem-solving strategies below summarize how to deal with interaction pairs.

Get It?
Explain why Earth's acceleration is usually very small compared to the acceleration of the object that Earth interacts with.

PROBLEM-SOLVING STRATEGY

Interaction Pairs

Use these strategies to solve problems in which there is an interaction between objects in two different systems.

1. Separate the system or systems from the external world.
2. Draw a pictorial model with coordinate systems for each system.
3. Draw a physical model that includes free-body diagrams for each system.
4. Connect interaction pairs by dashed lines.
5. To calculate your answer, use Newton's second law to relate the net force and acceleration for each system.
6. Use Newton's third law to equate the magnitudes of the interaction pairs and give the relative direction of each force.
7. Solve the problem and check the reasonableness of the answers' units, signs, and magnitudes.

Newton's Third Law

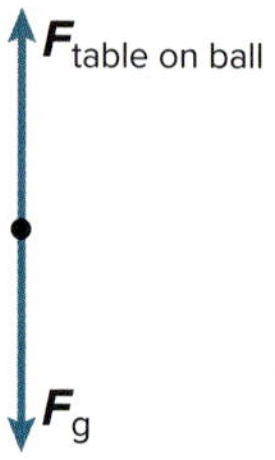

The two forces acting on the ball are $F_{\text{table on ball}}$ and $F_{\text{Earth's mass on ball}}$. These forces are not an interaction pair.

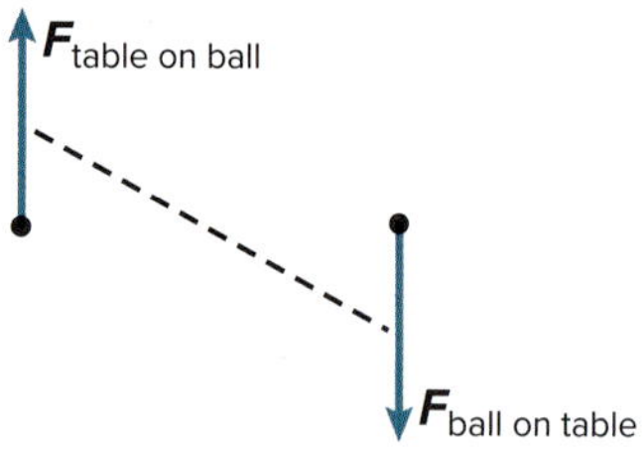

Force interaction pair between ball and table.

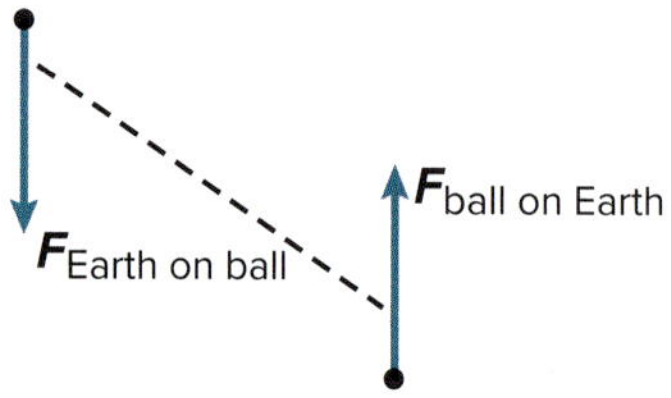

Force interaction pair between ball and Earth.

Figure 15 A ball resting on a table is part of two interaction pairs.

EXAMPLE Problem 4

EARTH'S ACCELERATION A softball has a mass of 0.18 kg. What is the gravitational force on Earth due to the ball, and what is Earth's resulting acceleration? Earth's mass is 6.0×10^{24} kg.

1 ANALYZE AND SKETCH THE PROBLEM

- Draw free-body diagrams for the two systems: the ball and Earth.
- Connect the interaction pair by a dashed line.

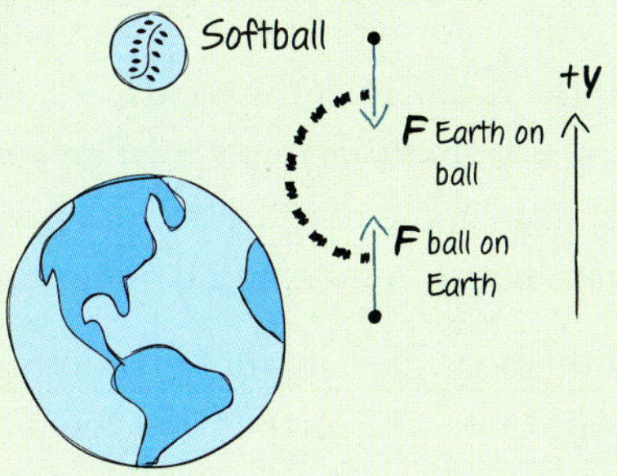

KNOWN

$m_{ball} = 0.18$ kg

$m_{Earth} = 6.0\times10^{24}$ kg $\quad g = 9.8$ N/kg

UNKNOWN

$F_{Earth\ on\ ball} = ?$

$a_{Earth} = ?$

2 SOLVE FOR THE UNKNOWN

Use Newton's second law to find the weight of the ball.

$$F_{Earth\ on\ ball} = m_{ball}g = (0.18\text{ kg})(-9.0\text{ N/kg}) = -1.8\text{ N}$$

Substitute $m_{ball} = 0.18$ kg.g $= -9.8$ N/kg.

Use Newton's third law to find $F_{ball\ on\ Earth}$.

$$F_{Ball\ on\ Earth} = -F_{Earth\ on\ ball} = -(-1.8\text{ N}) = +1.8\text{ N}$$

Substitute $F_{Earth\ on\ ball} = -1.8$ N.

Use Newton's second law to find a Earth.

$$a_{Earth} = \frac{F_{net}}{m_{Earth}}$$

$$= \frac{1.8\text{ N}}{6.0\times10^{24}\text{ kg}}$$

Substitute $F_{net} = 1.8$ N, $m_{Earth} = 6.0\times10^{24}$ kg.

$$= 2.9\times10^{-25}\text{ m/s}^2 \text{ toward the softball}$$

3 EVALUATE THE ANSWER

- **Are the units correct?** Force is in N and acceleration is in m/s^2.
- **Do the signs make sense?** Force and acceleration should be positive.
- **Is the magnitude realistic?** It makes sense that Earth's acceleration is small; Earth's mass is large.

PRACTICE Problems

ADDITIONAL PRACTICE

28. You lift a relatively light bowling ball with your hand, accelerating it upward. What are the forces on the ball? What forces does the ball exert? What objects are these forces exerted on?
29. A brick falls from a construction scaffold. Identify any forces acting on the brick. Also identify any forces the brick exerts and the objects on which these forces are exerted. (Air resistance may be ignored.)
30. A suitcase sits on a stationary airport luggage cart, as in **Figure 16.** Draw a free-body diagram for each object and specifically indicate any interaction pairs between the two.
31. **CHALLENGE** You toss a ball up in the air. Draw a free-body diagram for the ball after it has lost contact with your hand but while it is still moving upward. Identify any forces acting on the ball. Also identify any forces that the ball exerts and the objects on which these forces are exerted. Assume that air resistance is negligible.

Figure 16

Tension

Tension is simply a specific name for the force that a string or rope exerts. A simplification within this textbook is the assumption that all strings and ropes are massless. In **Figure 17**, the rope is about to break in the middle. If the rope breaks, the bucket will fall; before it breaks, there must be forces holding the rope together. The force that the top part of the rope exerts on the bottom part is $F_{\text{top on bottom}}$. Newton's third law states that this force must be part of an interaction pair. The other member of the pair is the force that the bottom part of the rope exerts on the top, $F_{\text{bottom on top}}$. These forces, equal in magnitude but opposite in direction, also are shown in **Figure 17**.

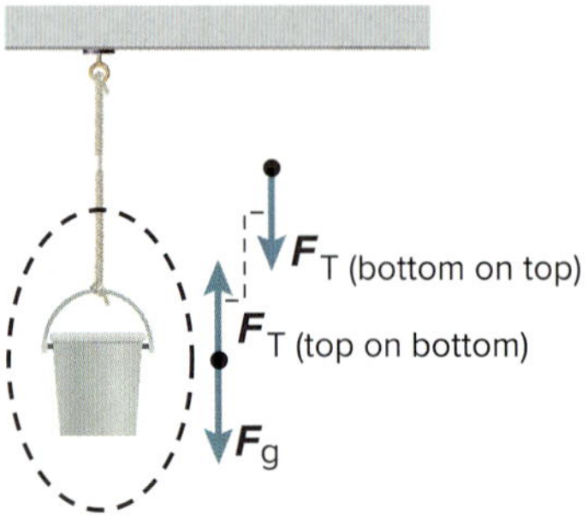

Figure 17 The tension in the rope is equal to the weight of all the objects hanging from it.

Think about this situation in another way. Before the rope breaks, the bucket is in equilibrium. This means that the force of its weight downward must be equal in magnitude but opposite in direction to the tension in the rope upward. Similarly, if you look at the point in the rope just above the bucket, it also is in equilibrium. Therefore, the tension of the rope below it pulling down must be equal to the tension of the rope above it pulling up. You can move up the rope, considering any point in the rope, and see that the tension forces at any point in the rope are pulling equally in both directions. Thus, the tension in the rope equals the weight of all objects below it.

Examine the tension forces shown in **Figure 18**. If team A is exerting a 500-N force and the rope does not accelerate, then team B also must be pulling with a force of 500 N. What is the tension in the rope? If each team pulls with 500 N of force, is the tension 1000 N? To decide, think of the rope as divided into two halves. The left side is not accelerating, so the net force on it is zero. Thus, $F_{\text{A on left side}} = F_{\text{right side on left side}} = 500$ N. Similarly, $F_{\text{B on right side}} = F_{\text{left side on right side}} = 500$ N. But the two tensions, $F_{\text{right side on left side}}$ and $F_{\text{left side on right side}}$, are an interaction pair, so they are equal and opposite. Thus, the tension in the rope equals the force with which each team pulls, or 500 N. To verify this, you could cut the rope in half and tie the ends to a spring scale. The scale would read 500 N.

Figure 18 The rope is not accelerating, so the tension in the rope equals the force with which each team pulls.

Rawpixel/iStock/Getty Images Plus/Getty Images

EXAMPLE Problem 5

LIFTING A BUCKET A 50.0-kg bucket is being lifted by a rope. The rope will not break if the tension is 525 N or less. The bucket started at rest, and after being lifted 3.0 m, it moves at 3.0 m/s. If the acceleration is constant, is the rope in danger of breaking?

1 ANALYZE AND SKETCH THE PROBLEM

- Draw the situation, and identify the forces on the system.
- Establish a coordinate system with the positive axis upward.
- Draw a motion diagram; include ***v*** and ***a***.
- Draw the free-body diagram, and label for forces.

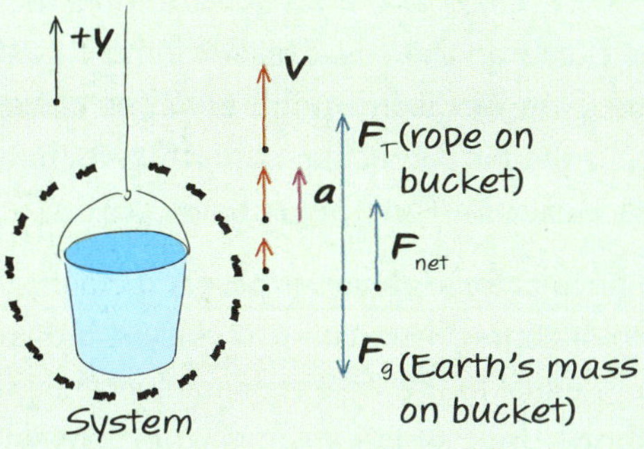

KNOWN		UNKNOWN
$m = 50.0$ kg	$v_t = 3.0$ m/s	$F_T = ?$
$v_i = 0.0$ m/s	$d = 3.0$ m	

2 SOLVE FOR THE UNKNOWN

F_{net} is the sum of the positive force of the rope pulling up (T_F) and the negative weight force ($-F_g$) pulling down as defined by the coordinate system.

$$F_{net} = F_T + (-F_g)$$

$$F_T = F_{net} + F_g = ma + mg$$ Substitue $F_{net} = ma$, $F_g = mg$

v_i, v_f, and d are known.

$$v_f^2 = v_i + 2ad$$

$$a = \frac{v_f^2 - v_i^2}{2d} = \frac{v_f^2}{2d}$$ Substitute $v_i = 0.0$ m/s

$$F_T = ma + mg$$ Substitute $a = v_f^2 / (2d)$.

$$= m\left(\frac{v_f^2}{2d}\right) + mg$$

$$= (50.0\text{ kg})\left(\frac{(3.0\text{ m/s})^2}{2(3.0\text{ m})}\right) + (50.0\text{ kg})(9.8\text{ N/kg})$$ Substitute $m = 50.0$ kg, $v_f = 3.0$ *m/s*, $d = 3.0$ *m*, g = 9.8 N/Kg.

$$= 560\text{ N}$$

The rope is in danger of breaking because the tension exceeds 525 N.

3 EVALUATE THE ANSWER

- **Are the units correct?** dimensional analysis verifies kg·m/s², which is N
- **Does the sign make sense?** The upward force should be positive.
- **Is the magnitude realistic?** The magnitude is a little larger than 490 N, which is the weight of the bucket. $F_g = mg = (50.0\text{ kg})(9.8\text{ N/kg}) = 490$ N

PRACTICE Problems

ADDITIONAL PRACTICE

32. Diego and Mika are trying to remove a tire from Diego's car. When they pull together in the same direction, Mika with a force of 23 N and Diego with a force of 31 N, they just barely get the tire to move off the wheel. What is the magnitude of the force between the tire and the wheel?

33. CHALLENGE You are loading equipment into a bucket that roofers will hoist to a rooftop. If the rope will not break as long as the tension does not exceed 450 N and you fill the bucket until it has a mass of 42 kg, what is the greatest acceleration the workers can give the bucket as they hoist it?

The Normal Force

Any time two objects are in contact, they exert a force on each other. Consider a box sitting on a table. There is a downward force on the box due to gravity. There also is an upward force that the table exerts on the box. This force must exist because the box is in equilibrium. The **normal force** is the perpendicular contact force that a surface exerts on another surface.

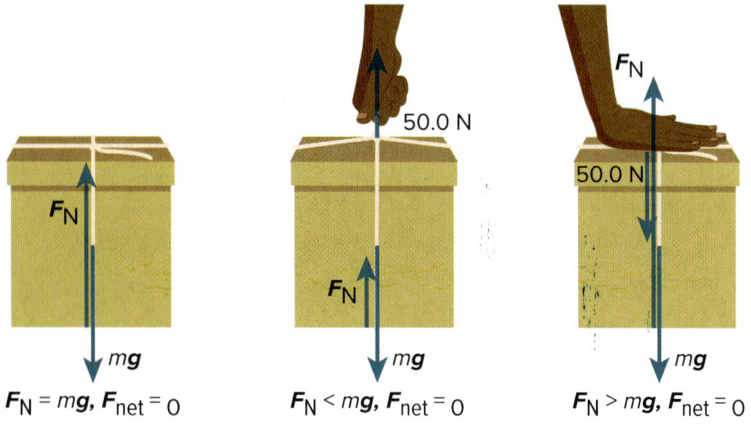

Figure 19 The normal force is not always equal to the object's weight.

The normal force always is perpendicular to the plane of contact between two objects, but is it always equal to the weight of an object? **Figure 19** shows three situations involving a box with the same weight. What if you tied a string to the box and pulled up on it a little bit, but not enough to accelerate the box, as shown in the middle panel in **Figure 19**? When you apply Newton's second law to the box and the forces acting on the box, you see $F_N + F_{\text{string on box}} - F_g = ma = 0$ N, which can be rearranged to show $F_N = F_g - F$ string on box.

You can see that in this case the normal force that the table exerts on the box is less than the box's weight (F_g). Similarly, if you pushed down on the box on the table as shown in the final panel in **Figure 19**, the normal force would be more than the box's weight. Finding the normal force will be important when you study friction in detail.

Check Your Progress

34. **Interaction Pair** Identify each force acting on the ball and its interaction pair in **Figure 20.**

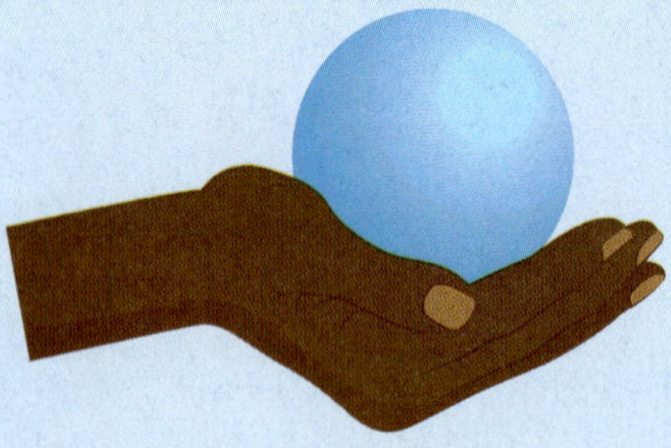

Figure 20

35. **Force** Imagine lowering the ball in **Figure 20** at increasing speed. Draw separate free-body diagrams for the forces acting on the ball and for each set of interaction pairs.

36. **Tension** A block hangs from the ceiling by a massless rope. A second block is attached to the first block and hangs below it on another piece of massless rope. If each of the two blocks has a mass of 5.0 kg, what is the tension in the rope?

37. **Tension** A block hangs from the ceiling by a massless rope. A 3.0-kg block is attached to the first block and hangs below it on another piece of massless rope. The tension in the top rope is 63.0 N. Find the tension in the bottom rope and the mass of the top block.

38. **Critical Thinking** A curtain prevents two tug-of-war teams from seeing each other. One team ties its end of the rope to a tree. If the other team pulls with a 500-N force, what is the tension in the rope? Explain.

LEARNSMART® Go online to follow your personalized learning path to review, practice, and reinforce your understanding.

Finding the Source of the Force

Racetrack Playa in Death Valley, CA has a flat surface dotted with rocks and boulders of varying sizes. The rocks occasionally move, leaving trails in the mud that show their path of motion. For a long time, no one knew what caused the rocks to move. This scientific mystery was finally solved during an investigation that gathered data using advanced technology.

For many years, scientists tried to determine the source of the forces that caused these rocks to move.

Wandering Stones

The rocks on Racetrack Playa have earned many nicknames over the years. They have been called wandering stones, sliding rocks, slithering stones, and sailing stones. The rocks vary greatly in size—some of the largest have a mass of more than 300 kg. The motion of the rocks had never been observed first-hand.

Hypotheses

Many scientists developed hypotheses to explain the motion of the rocks. Some thought that the force of gravity was pulling the rocks slowly down a very slight slope. Others thought that high winds occasionally pushed the rocks with enough force to cause their motion. However, evidence did not support these hypotheses. Gravity was ruled out by data showing the rocks were moving up a slight slope, not downhill. High winds were also ruled out, because data showed that more massive rocks often moved farther than smaller rocks.

Solving the Mystery

To solve the mystery of the rocks' movement, researchers embedded GPS in fifteen rocks to track their motion. They set up a weather station to keep track of the wind speed and direction. They also set up a camera to record video of the rocks' motion.

After analyzing the data, the researchers concluded that rock movement occurred when a shallow pond formed and froze on the Playa's surface. When the ice sheet on the top of the pond started to melt, very thin floating sheets of ice were blown by light breezes. The floating ice sheets pushed against rocks, making the rocks slide, similar to the way a sail makes a boat move across the water. Data providing evidence that thin ice and light breezes could generate enough force to cause the motion of the rocks was an unexpected answer to this scientific mystery.

DEVELOP A MODEL TO ILLUSTRATE

Work with a team to draw a free-body diagram of the forces acting on a sliding rock in Racetrack Playa. Indicate the direction of the acceleration and of the net force. Next to the free-body diagram, make an illustration that uses a circle to designate the system.

CALIFORNIA CONNECTION

Make a travel brochure about Racetrack Playa that would be useful for people from outside California who are curious about the sliding rocks. Include a map of California showing the location of Racetrack Playa in your brochure.

MODULE 4

STUDY GUIDE

GO ONLINE to study with your Science Notebook.

Lesson 1 FORCE AND MOTION

- A force is a push or a pull. Forces have both direction and magnitude. A force might be either a contact force or a field force.
- Newton's second law states that the acceleration of a system equals the net force acting on it divided by its mass.

$$a = \frac{F_{net}}{m}$$

- Newton's first law states that an object that is at rest will remain at rest and an object that is moving will continue to move in a straight line with constant speed, if and only if the net force acting on that object is zero. An object with zero net force acting on it is in equilibrium.

- force
- system
- free-body diagram
- net force
- Newton's second law
- Newton's first law
- inertia
- equilibrium

Lesson 2 WEIGHT AND DRAG FORCE

- The object's weight (F_g) depends on the object's mass and the gravitational field at the object's location.

$$F_g = mg$$

- An object's apparent weight is the magnitude of the support force exerted on it. An object with no apparent weight experiences weightlessness.
- A falling object reaches a constant velocity when the drag force is equal to the object's weight. The constant velocity is called the terminal velocity. The drag force on an object is determined by the object's weight, size, and shape as well as the fluid through which it moves.

- weight
- gravitational field
- apparent weight
- weightlessness
- drag force
- terminal velocity

Lesson 3 NEWTON'S THIRD LAW

- Newton's third law states that the two forces that make up an interaction pair of forces are equal in magnitude, but opposite in direction and act on different objects. In an interaction pair, $F_{\text{A on B}}$ does not cause $F_{\text{B on A}}$. The two forces either exist together or not at all.

$$F_{\text{A on B}} = -F_{\text{B on A}}$$

- The normal force is a support force resulting from the contact between two objects. It is always perpendicular to the plane of contact between the two objects.

- interaction pair
- Newton's third law
- tension
- normal force

THREE-DIMENSIONAL THINKING

Module Wrap-Up

REVISIT THE PHENOMENON

How do wing suits help BASE jumpers control their velocity?

CER Claim, Evidence, Reasoning

Explain Your Reasoning Revisit the claim you made when you encountered the phenomenon. Summarize the evidence you gathered from your investigations and research and finalize your Summary Table. Does your evidence support your claim? If not, revise your claim. Explain why your evidence supports your claim.

STEM UNIT PROJECT

Now that you've completed the module, revisit your STEM unit project. You will apply your evidence from this module and complete your project.

GO FURTHER

SEP Data Analysis Lab

How does weight change during a rocket launch?

A rocket is launched vertically. When the rocket reaches its maximum height, a parachute is deployed, and the rocket descends to the ground.

CER Analyze and Interpret Data

1. **Draw and label** a free-body diagram for the rocket during each of the following intervals:
   a. While the engine is firing
   b. After the engine shuts down but before the parachute is deployed
   c. The moment the parachute is deployed
2. **Claim** Imagine the rocket is full-sized and a person stands on a bathroom scale inside. Is the scale reading less than, equal to, or greater that the scale reading when the rocket is at rest for the following intervals?
   a. While the engine is firing
   b. After the engine shuts down but before the parachute is deployed
   c. The moment the parachute is deployed
3. **Evidence and Reasoning** Justify your answers by using your diagrams and explaining your reasoning.

Oliver Furrer/Cultura/Getty Images

NASA

UNIT 2

MECHANICS IN TWO DIMENSIONS

ENCOUNTER THE PHENOMENON

How can forces cause so many different types of motion?

SEP Ask Questions

What questions do you have about the phenomenon? Write your questions on sticky notes and add them to the driving question board for this unit.

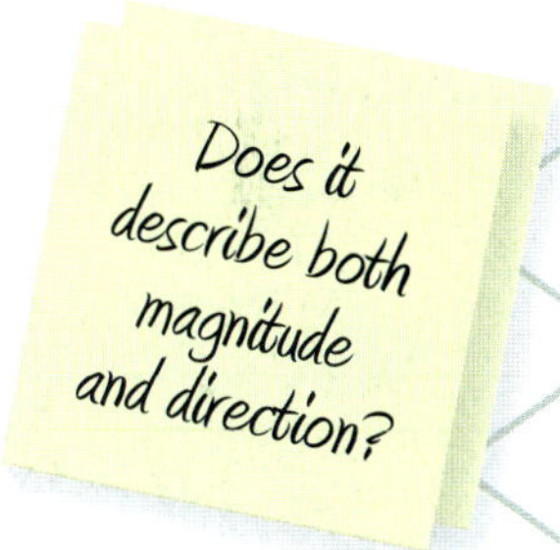

Look for Evidence

As you go through this unit, use the information and your experiences to help you answer the phenomenon question as well as your own questions. For each activity, record your observations in a Summary Table, add an explanation, and identify how it connects to the unit and module phenomenon questions.

Solve a Problem

STEM UNIT PROJECT

In Orbit From GPS satellites circling Earth to the planets traveling around the Sun, understanding orbital motion is key to understanding the dynamics of our solar system. Investigate orbital motion and gravity. Use the results of these investigations and the evidence you collected during the unit to predict the motion of orbiting objects in the solar system.

GO ONLINE In addition to reading the information in your Student Edition, you can find the STEM Unit Project and other useful resources online.

robert france/Alamy Stock Photo

MODULE 5

DISPLACEMENT AND FORCE IN TWO DIMENSIONS

ENCOUNTER THE PHENOMENON

Why is this specialized train washing the train tracks?

GO ONLINE to watch a video about the importance of cleaning train tracks.

SEP Ask Questions

Do you have other questions about the phenomenon? If so, add them to the driving question board.

CER Claim, Evidence, Reasoning

Make Your Claim Use your CER chart to make a claim about why the train is washing the tracks. Explain your reasoning.

Collect Evidence Use the lessons in this module to collect evidence to support your claim. Record your evidence as you move through the module.

Explain Your Reasoning You will revisit your claim and explain your reasoning at the end of the module.

GO ONLINE to access your CER chart and explore resources that can help you collect evidence.

LESSON 2: Explore & Explain: Kinetic and Static Friction

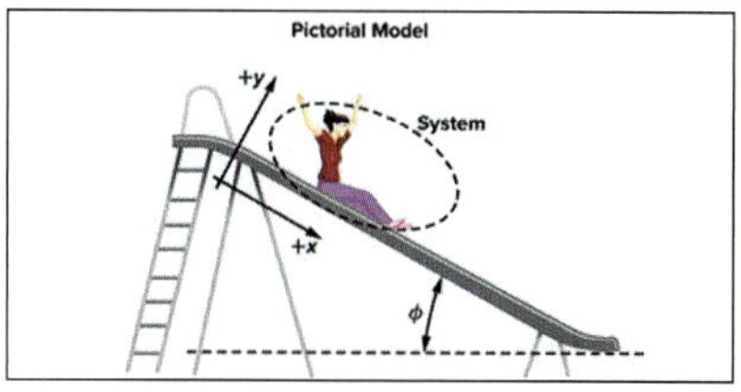

LESSON 3: Explore & Explain: Inclined Planes

Smithsonian

Additional Resources

LESSON 1
VECTORS

FOCUS QUESTION

How can you add forces in two dimensions?

Vectors in Two Dimensions

How do climbers cling to a rock wall? Often the climber has more than one support point. This means there are multiple forces acting on him. Because he grips crevices in the rock, the rock pulls back on him. Also the rope secures him to the rock, so there are two contact forces acting on him. Gravity pulls on him as well, so there are a total of three forces acting on the climber.

One aspect of this situation that is different from the ones you might have studied earlier is that the forces exerted by the rock face on the climber do not push or pull only in the horizontal or vertical direction. You know from previous study that you can pick your coordinate system and orient it in the way that is most useful to analyze a situation. But how can you set up a coordinate system for a net force when you are dealing with more than one dimension? And what happens when the forces are not at right angles to each other?

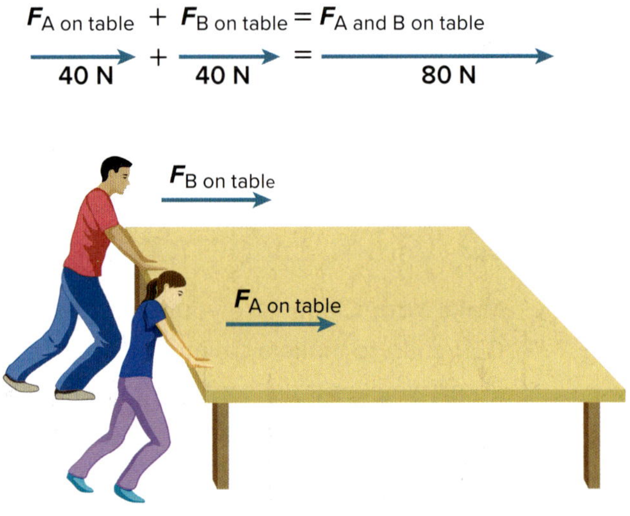

Figure 1 The sum of the two applied forces is 80 N to the right.

Vectors revisited

Let's begin by reviewing force vectors in one dimension. Consider the case in **Figure 1** in which you and a friend both push on a table together. Suppose that you each exert a 40-N force to the right. The sum of the forces is 80 N to the right, which is what you probably expected. But how do you find the sum of these force vectors?

In earlier discussions of adding vectors along a straight line, you learned that the resultant vector always points from the tail of the first vector to the tip of the final vector. Does this rule still apply when the vectors do not lie along a straight line?

3D THINKING DCI Disciplinary Core Ideas | CCC Crosscutting Concepts | SEP Science & Engineering Practices

COLLECT EVIDENCE

Use your Science Journal to record the evidence you collect as you complete the readings and activities in this lesson.

INVESTIGATE

GO ONLINE to find these activities and more resources.

Launch Lab: Adding Vectors
Carry out an investigation to determine the net force on a system when forces are exerted in two dimensions.

CCC Identify Crosscutting Concepts
Create a table of the crosscutting concepts and fill in examples you find as you read.

Adding vectors in two dimensions Even when vectors do not lie on a straight line, the resultant vector always points from the tail of the first vector to the tip of the final vector. You can use a protractor and a ruler both to draw the vectors at the correct angles and to measure the magnitude and the direction of the resultant vector.

Figure 2 illustrates how to add vectors graphically in two dimensions. Notice that when vector ***B*** is moved, its magnitude and direction are unchanged. This is always the case—you do not change a vector's length or direction when you move that vector.

Perpendicular vectors If you are adding together two vectors at right angles, such as vector ***A*** pointing east and vector ***B*** pointing north in **Figure 2,** you could use the Pythagorean theorem to find the magnitude (R) of the resultant vector.

$$R^2 = A^2 + B^2$$

Angles other than 90° If you are adding two vectors that are at an angle other than 90°, as in **Figure 3,** then you can use trigonometry, such as the law of sines or the law of cosines. It is best to use the law of sines when you are given two angle measurements and only one vector magnitude. The law of cosines is particularly useful when given two vectors and the angle between them.

Law of Sines: $\frac{R}{\sin \theta} = \frac{A}{\sin a} = \frac{B}{\sin b}$

Law of Cosines: $R^2 = A^2 + B^2 - 2AB \cos \theta$

What happens if you apply the law of cosines to a triangle in which $\theta = 90°$? Notice that the first three terms in the law of cosines are the same three terms found in the Pythagorean theorem. The final term in the law of cosines, $-2AB \cos \theta$, equals zero if $\theta = 90°$ because $\cos 90° = 0$.

$$R^2 = A^2 + B^2 - 2AB \cos 90°$$

$$R^2 = A^2 + B^2 - 2AB\,(0)$$

$$R^2 = A^2 + B^2$$

If $\theta = 90°$, the triangle is a right triangle and the law of cosines reduces to the Pythagorean theorem.

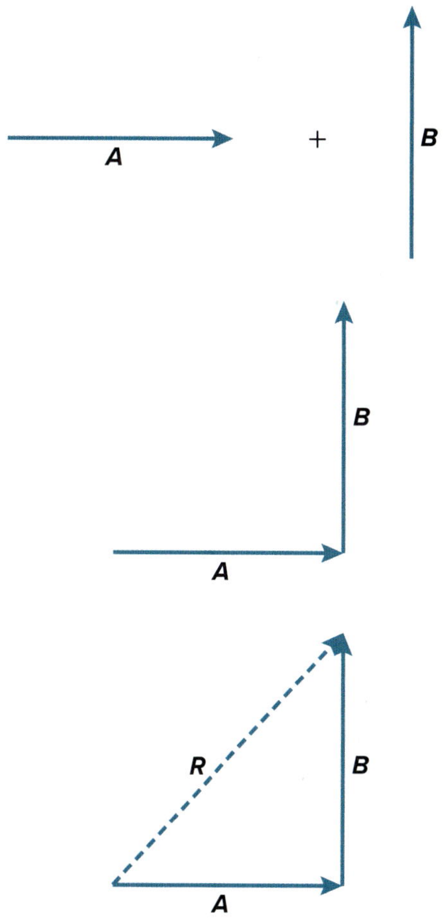

Figure 2 When adding vectors in two dimensions, the process is the same as adding vectors in one dimension: place vectors tip to tail, then connect the tail of the first vector to the tip of the second vector (or final vector if more than two) to find the resultant.

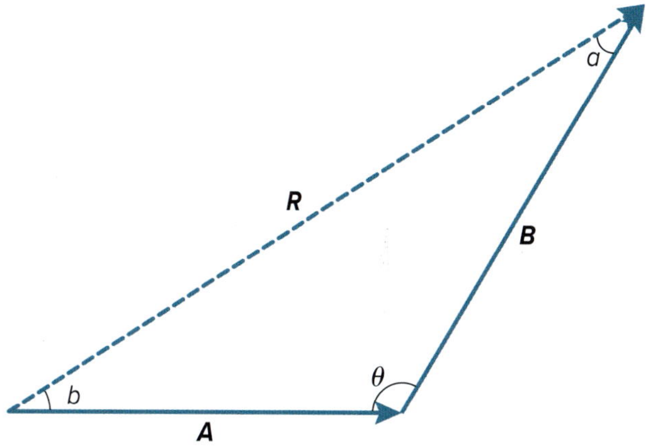

Figure 3 If the vectors are not at right angles, the Pythagorean theorem does not apply. Instead, use the law of cosines or the law of sines with the variables as shown here.

EXAMPLE Problem 1

FINDING THE MAGNITUDE OF THE SUM OF TWO VECTORS Find the magnitude of the sum of a 15-km displacement and a 25-km displacement when the angle θ between them is 90° and when the angle θ between them is 135°.

1 ANALYZE AND SKETCH THE PROBLEM

- Sketch the two displacement vectors, A and B, and the angle between them.

Known

$A = 25$ km $\quad \theta = 90°$ or $\theta_2 = 135°$

$B = 15$ km

Unknown

$R = ?$

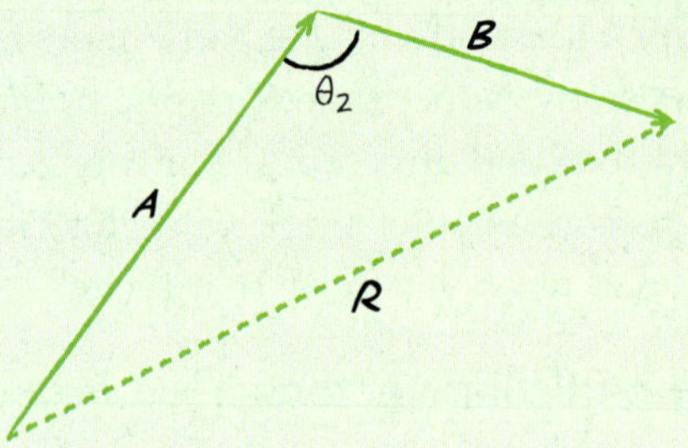

2 SOLVE FOR THE UNKNOWN

When θ is 90°, use the Pythagorean theorem to find the magnitude of the resultant vector.

$$R^2 = A^2 + B^2$$

$$R = \sqrt{A^2 + B^2}$$

$$= \sqrt{(25\text{ km})^2 + (15\text{ km})^2}$$ Substitute $A = 25$ km, $B = 15$ km

$$= 29\text{ km}$$

When θ does not equal 90°, use the law of cosines to find the magnitude of the resultant vector.

$$R^2 = A^2 + B^2 - 2AB(\cos\theta_2)$$

$$R = \sqrt{A^2 + B^2 - 2AB(\cos\theta_2)}$$

$$= \sqrt{(25\text{ km})^2 + (15\text{ km})^2 - 2(25\text{ km})(15\text{ km})(\cos 135°)}$$ Substitute $A = 25$ km, $B = 15$ km, $\theta_2 = 135°$

$$= 37\text{ km}$$

3 EVALUATE THE ANSWER

- **Are the units correct?** Each answer is a length measured in kilometers.
- **Do the signs make sense?** The sums are positive.
- **Are the magnitudes realistic?** From the sketch, you can see that the resultant should be longer than either vector.

PRACTICE Problems

ADDITIONAL PRACTICE

1. You and your family are out for a drive. You drive 125.0 km due west, then turn due south and drive for another 65.0 km. What is the magnitude of your displacement? Solve this problem both graphically and mathematically, and check your answers against each other.
2. On a fine, sunny day, you and your siblings decide to go for a nearby hike. You walk 4.5 km in one direction, then make a 45° turn to the right and walk another 6.4 km. What is the magnitude of your displacement?
3. After picking up school supplies, you and your caregiver walk from the door of the mall to the car. You first walk 250.0 m down a lane of cars, and then turn 90° to the right and walk an additional 60.0 m. How far is the car from the door to the mall? Solve this problem both graphically and mathematically, and check your answers against each other.
4. **CHALLENGE** An ant crawls on the sidewalk. It first moves south a distance of 5.0 cm. It then turns southwest and crawls 4.0 cm. What is the magnitude of the ant's displacement?

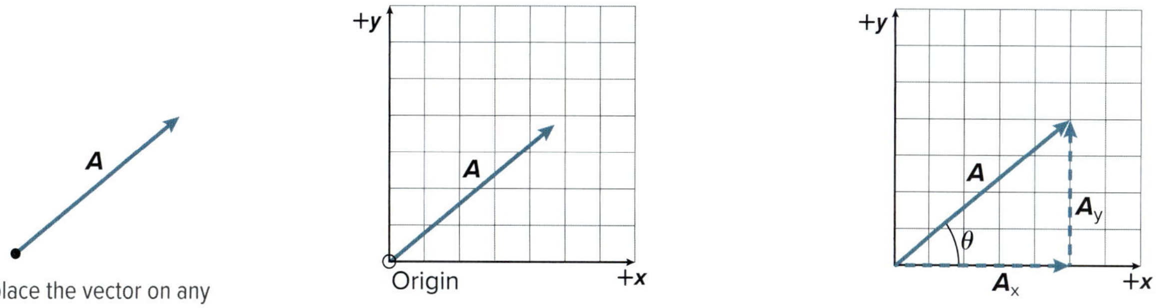

You may place the vector on any coordinate system as long as the vector's direction and magnitude remain unchanged.

The coordinate system can be oriented to make the problem easier to solve.

If the components of a vector are added together, they result in the original vector.

Figure 4 Vector ***A*** is placed on a coordinate system. Notice in the right panel that ***A***'s direction is measured counterclockwise from the positive *x*-axis. $\boldsymbol{A}_x$ is parallel to the *x*-axis, and $\boldsymbol{A}_y$ is parallel to the *y*-axis.

Describe *a vector whose y-component is zero.*

Vector Components

What is the direction of the vector shown on the left of **Figure 4?** To determine the exact direction, we need to choose a coordinate system. This coordinate system, such as the one in the center of **Figure 4,** is similar to laying down a grid drawn on transparent plastic on top of a vector problem. You choose where to put the center of the grid (the origin) and establish the directions in which the axes point. Notice that the *x*-axis is drawn through the origin with an arrow pointing in the positive direction. The positive *y*-axis is located 90° from the positive *x*-axis and crosses the *x*-axis at the origin.

How do you choose the direction of the *x*-axis? There is never a single correct answer, but some choices make the problem easier to solve than others. When the motion you are describing is confined to the surface of Earth, have the *x*-axis point east and the *y*-axis point north. If the motion is on an incline, place the positive *x*-axis in the direction of the motion parallel to the surface of the incline

Component vectors

Now you can describe a vector in a useful way. Vector ***A***, shown on the right in **Figure 4,** can be described as going 5 units in the positive *x*-direction ($\boldsymbol{A}_x$) and 4 units in the positive *y*-direction ($\boldsymbol{A}_y$). If you add $\boldsymbol{A}_x$ and $\boldsymbol{A}_y$, the resultant is the original vector, ***A***. A vector can be broken into its **components,** which are a vector parallel to the *x*-axis and another vector parallel to the *y*-axis: $\boldsymbol{A} = \boldsymbol{A}_x + \boldsymbol{A}_y$.

The process of breaking a vector into its components is sometimes called **vector resolution.** The original vector is the hypotenuse of a right triangle. This means that the magnitude of the original vector will always be greater than or equal to the magnitude of either component vector.

SCIENCE USAGE v. COMMON USAGE

Components

Science usage: projections of a vector parallel to the *x*- and *y*-axes
A vector is the algebraic sum of its components.

Common usage: constituent parts or ingredients
The computer's electronic components malfunctioned during the power surge.

STEM CAREER Connection

Civil Engineer

When in a moving car on a wet surface, do you notice the difference in friction between the tires and the wet road? One task for a civil engineer is to design roads and pavements to account for the friction between the tires and the road in different driving conditions.

Direction in coordinate systems In a two-dimensional force problem, you should carefully select your coordinate system because the direction of any vector will be specified relative to those coordinates. We define the direction of a vector as the angle that the vector makes with the x-axis, measured counterclockwise from the positive x-axis. In **Figure 5,** the angle θ represents the direction of the vector $\mathbf{A}$. All algebraic calculations involve only the components of vectors, not the vectors themselves.

In addition to measuring the lengths of the component vectors graphically, you can find the components by using trigonometry as shown below:

$$\cos\theta = \frac{\text{adjacent side}}{\text{hypotenuse}} = \frac{A_x}{A}; \text{ therefore, } A_x = A\cos\theta$$

$$\sin\theta = \frac{\text{opposite side}}{\text{hypotenuse}} = \frac{A_y}{A}; \text{ therefore, } A_y = A\sin\theta$$

When the angle that a vector makes with the x-axis is larger than 90°, the sign of one or more components is negative, as shown in the top coordinate system of **Figure 5.**

Get It?

Explain how you should measure the direction of a vector.

Coordinate System

Second quadrant $90° < \theta < 180°$ (II)	First quadrant $0° < \theta < 90°$ (I)
A_x is negative.	A_x is positive.
A_y is positive.	A_y is positive.
$\tan\theta$ is negative.	$\tan\theta$ is positive.

Third quadrant $180° < \theta < 270°$ (III)	Fourth quadrant $270° < \theta < 360°$ (IV)
A_x is negative.	A_x is positive.
A_y is negative.	A_y is negative.
$\tan\theta$ is positive.	$\tan\theta$ is negative.

+y, +x

Example:

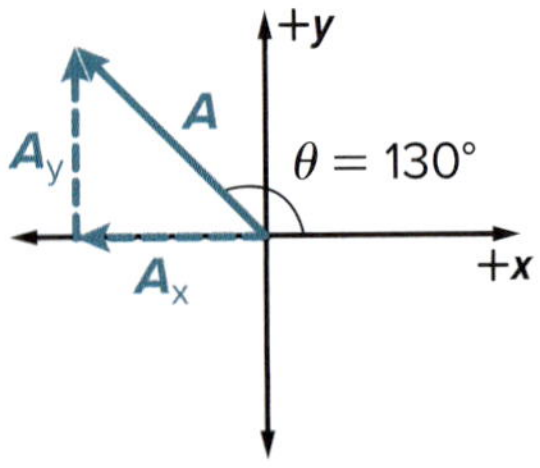

- **A** is in 2nd quadrant.
- Expect A_x to be negative:

$$A_x = A\cos\theta = (5.0\text{ N})\cos 130° = -3.2\text{ N}$$

- Expect A_y to be positive:

$$A_y = A\sin\theta = (5.0\text{ N})\sin 130° = 3.8\text{ N}$$

Figure 5 The coordinate plane is divided into four quadrants. A vector's components will be positive or negative depending on the vector's quadrant.

Classify *as positive or negative the components of a vector whose angle is 280°. In which quadrant does the vector lie?*

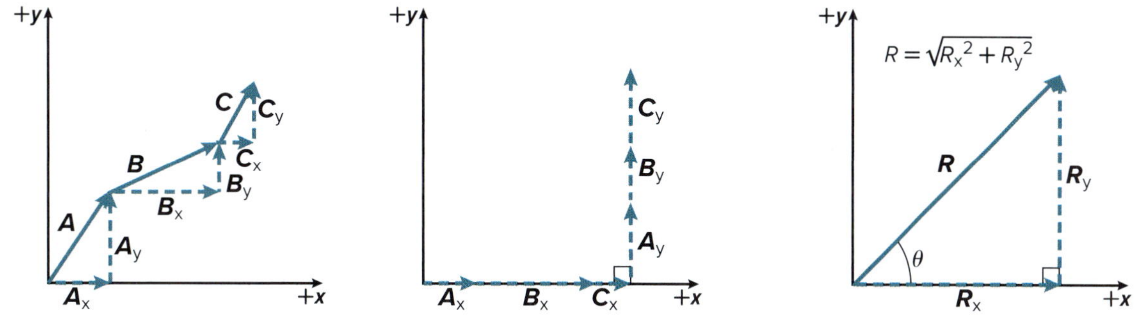

Add the vectors graphically by placing them tip to tail.

Add the *x*-components together and the *y*-components together.

The magnitude of ***R*** can be calculated using the Pythagorean theorem.

Figure 6 The vector sum of ***A***, ***B***, and ***C*** is the same as the vector sum of $\mathbf{R}_x$ and $\mathbf{R}_y$.

Algebraic Addition of Vectors

Figure 6 shows how resolving vectors into components makes adding vectors together easier. First, resolve each vector into its *x*- and *y*-components. Add the *x*-components to form the *x*-component of the resultant:

$$R_x = A_x + B_x + C_x.$$

Similarly, add the *y*-components to form the *y*-component of the resultant:

$$R_y = A_y + B_y + C_y.$$

Because $\mathbf{R}_x$ and $\mathbf{R}_y$ are at a right angle (90°), you can calculate the magnitude of the resultant vector using the Pythagorean theorem:

$$R^2 = R_x^{\,2} + R_y^{\,2}$$

Recall that the angle the vector makes with the *x*-axis is given by the following equation.

$$\theta = \tan^{-1}\left(\frac{R_y}{R_x}\right)$$

You can find the angle, or direction of the vector, using the $\tan^{-1}$ key on your calculator. When $\tan\theta > 0$, most calculators give the angle between 0° and 90°. When $\tan\theta < 0$, the angle will be between 0° and −90°.

PROBLEM-SOLVING STRATEGIES

Vector Addition

Use the following technique to solve problems for which you need to add or subtract vectors.

1. Choose a coordinate system.
2. Resolve the vectors into their *x*-components using $A_x = A\cos\theta$ and their *y*-components using $A_y = A\sin\theta$, where θ is the angle measured counterclockwise from the positive *x*-axis.
3. Add or subtract the component vectors in the *x*-direction.
4. Add or subtract the component vectors in the *y*-direction.
5. Use the Pythagorean theorem, $R = \sqrt{R_x^{\,2} + R_y^{\,2}}$, to find the magnitude of the resultant vector.
6. Use $\theta = \tan^{-1}\left(\frac{R_y}{R_x}\right)$ to find the angle of the resultant vector.

EXAMPLE Problem 2

FINDING YOUR WAY HOME You are on a hike. Your camp is 15.0 km away, in the direction 40.0° north of west. The only path through the woods leads directly north. If you follow the path 5.0 km before it opens into a field, how far, and in what direction, would you have to walk to reach your camp?

1 ANALYZE AND SKETCH THE PROBLEM

- Draw the resultant vector, ***R***, from your original location to your camp.
- Draw ***A***, the known vector, and draw ***B***, the unknown vector.

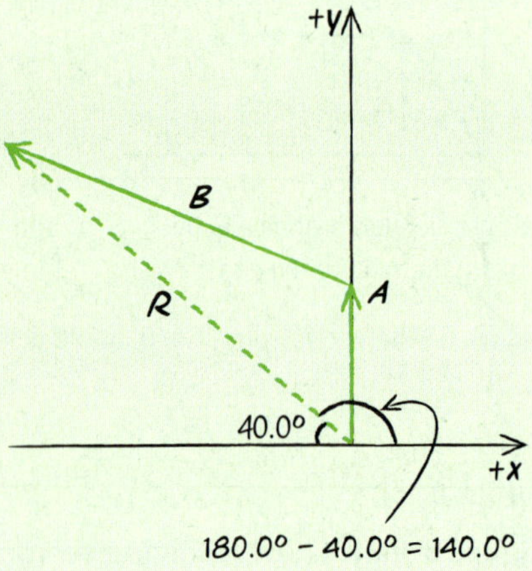

Known

A = 5.0 km, due north

R = 15.0 km, 40.0° north of west

$\theta = 140.0°$

Unknown

B = ?

2 SOLVE FOR THE UNKNOWN

Find the components of ***R***.

$$R_x = R\cos\theta$$
$$= (15.0\text{ km})\cos 140.0° \quad \text{Substitute } R = 15.0\text{ km}, \theta = 140.0°.$$
$$= 11.5\text{ km}$$

$$R_y = R\sin\theta$$
$$= (15.0\text{ km})\sin 140.0° \quad \text{Substitute } R = 15.0\text{ km}, \theta = 140.0°.$$
$$= 9.64\text{ km}$$

Because ***A*** is due north, $A_x = 0.0$ km and $A_y = 5.0$ km.
Use the components of ***R*** and ***A*** to find the components of ***B***.

$$B_x = R_x - A_x$$
$$= -11.5\text{ km} - 0.0\text{ km} \quad \text{Substitute } R_x = -11.5\text{ km}, A_x = 0.0\text{ km}.$$
$$= -11.5\text{ km} \quad \text{The negative sign means that this component points west.}$$

$$B_y = R_y - A_y$$
$$= 9.64\text{ km} - 5.0\text{ km} \quad \text{Substitute } R_y = 9.64\text{ km}, A_y = 5.0\text{ km}.$$
$$= 4.6\text{ km} \quad \text{This component points north.}$$

Use the components of vector ***B*** to find the magnitude of vector ***B***.

$$B = \sqrt{B_x^2 + B_y^2}$$
$$= \sqrt{(-11.5\text{ km})^2 + (4.6\text{ km})^2} \quad \text{Substitute } B_x = -11.5\text{ km}, B_y = 4.6\text{ km}.$$
$$= 12.4\text{ km}$$

Locate the tail of vector ***B*** at the origin of a coordinate system, and draw the components B_x and B_y. The vector ***B*** is in the second quadrant. Use the tangent to find the direction of vector ***B***.

$$\theta = \tan^{-1}\left(\frac{B_y}{B_x}\right)$$
$$\theta = \tan^{-1}\left(\frac{4.6\text{ km}}{-11.5\text{ km}}\right) \quad \text{Substitute } B_y = 4.6\text{ km}, B_x = -11.5\text{ km}.$$
$$= -22° \text{ or } 158° \quad \text{The tangent of an angle is negative in quadrants II and IV, so two answers are possible.}$$

Since ***B*** is in the second quadrant, θ, measured from the positive x-axis, must be 158°. This direction can also be given as 22° north of west. Thus, ***B*** = 12.4 km at 22° north of west.

3 EVALUATE THE ANSWER

- **Are the units correct?** Kilometers and degrees are correct.
- **Do the signs make sense?** They agree with the diagram.
- **Is the magnitude realistic?** The length of ***B*** should be longer than R_x because the angle between ***A*** and ***B*** is greater than 90°.

PRACTICE Problems

ADDITIONAL PRACTICE

Solve problems 5–10 algebraically. You may also solve some of them graphically to check your answers.

5. Sudhir walks 0.40 km in a direction 60.0° west of north, then goes 0.50 km due west. What is his displacement?

6. You first walk 8.0 km north from home then walk east until your displacement from home is 10.0 km. How far east did you walk?

7. In a coordinate system in which the positive *x*-axis is east, for what range of angles is the *x*-component positive? For what range is it negative?

8. Could a vector ever be shorter than one of its components? Could a vector ever be equal in length to one of its components? Explain.

9. Two ropes tied to a tree branch hold up a child's swing as shown in **Figure 7.** The tension in each rope is 2.28 N. What is the combined force (magnitude and direction) of the two ropes on the swing?

10. **CHALLENGE** Afua and Chrissy are going to sleep overnight in their tree house and are using some ropes to pull up a 3.20-kg box containing their pillows and blankets. The girls stand on different branches, as shown in **Figure 8,** and pull at the angles with the forces indicated. Find the *x*- and *y*-components of the initial net force on the box. *Hint: Draw a free-body diagram so you do not leave out a force.*

Figure 7

Figure 8

Check Your Progress

11. **Vectors** Use **Figure 9** for these questions.
 a. Find the components of vectors ***K***, ***L***, and ***M***.
 b. Find the sum of the three vectors.
 c. Subtract vector ***K*** from vector ***L***.

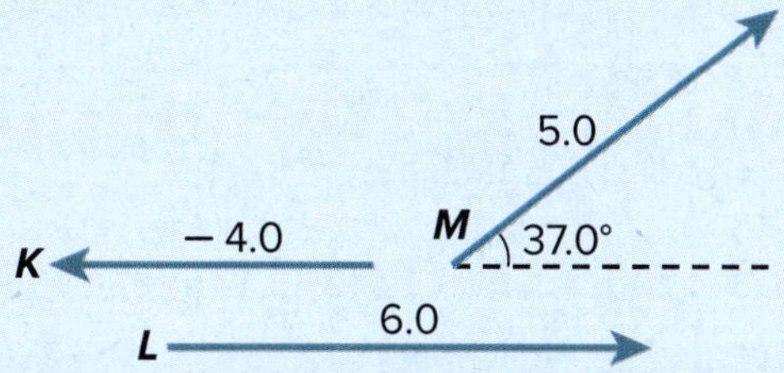

Figure 9

12. **Distance v. Displacement** Are distance and displacement always the same? Give an example that supports your conclusion.

13. **Commutative Operations** Vector addition is commutative because the order in which vectors are added does not matter.
 a. Use the vectors from **Figure 9** to show graphically that $\mathbf{M} + \mathbf{L} = \mathbf{L} + \mathbf{M}$.
 b. Are addition, subtraction, multiplication, and division commutative? Give an example of each operation to support your conclusion.

14. **Critical Thinking** You move a box through two unequal displacements. Could the displacements have directions such that the resultant displacement is zero? Suppose you now move the box through three displacements of unequal magnitude. Could the resultant displacement be zero? Support your conclusion with a diagram.

LEARNSMART® Go online to follow your personalized learning path to review, practice, and reinforce your understanding.

LESSON 2

FRICTION

FOCUS QUESTION

What force causes a train to stop?

Kinetic and Static Friction

You push a book across your desk. When you stop pushing, the book stops due to friction.

Types of friction When you pushed your book across the desk, the book experienced a type of friction that acts on moving objects—**kinetic friction** is exerted on one surface by another when the two surfaces rub against each other because one or both surfaces are moving.

Now imagine trying to push a couch across the floor, as shown on the left in **Figure 10.** You push on it with a small force, but it does not move. Newton's laws tell you that the net force on the couch must be zero. There must be a second horizontal force acting on the couch that opposes your force and is equal in size. This force is **static friction,** which is the force exerted on one surface by another when there is no motion between the two surfaces.

You push harder, and the couch still does not move. The static friction force is increasing in response to your applied force. When you push hard enough, the couch begins to move as in the right side of **Figure 10.** There is a limit to how large the static friction force can be. Once your force is greater than this maximum static friction, the couch moves and kinetic friction acts on it.

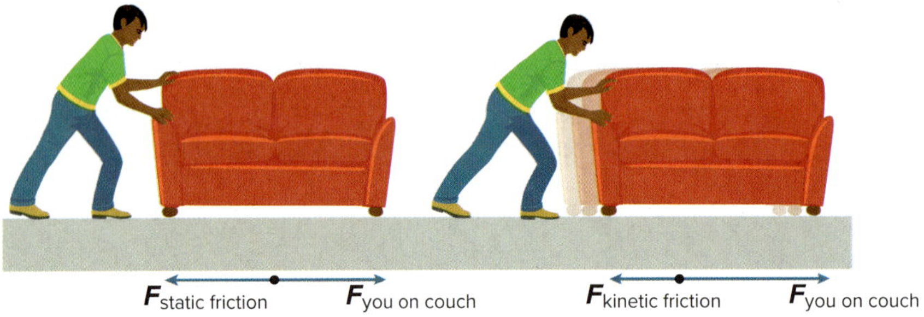

Figure 10 An applied force is balanced by static friction up to a maximum limit. When this limit is exceeded, the object begins to move.

Identify *the type of friction force acting on the couch when it begins to move.*

3D THINKING DCI Disciplinary Core Ideas CCC Crosscutting Concepts SEP Science & Engineering Practices

COLLECT EVIDENCE

Use your Science Journal to record the evidence you collect as you complete the readings and activities in this lesson.

INVESTIGATE

GO ONLINE to find these activities and more resources.

Forensics Lab: Hit-and-Run Driver
Analize data about friction forces to determine cause and effect at a crime scene.

Revisit the Encounter the Phenomenon Question
What information from this lesson can help you answer the Unit and Module questions?

Mathematical models for friction forces

The materials that the surfaces are made of affect the force of friction. For example, there is more friction between your shoes and concrete than there is between your socks and a polished wood floor. It might seem reasonable that the force of friction depends on either the surface area in contact or the speed at which the surfaces move past each other. Experiments have shown that this is not true. The normal force between the two objects does matter, however. The harder one object is pushed against the other, the greater the force of friction that results.

Get It?

Identify the two factors that affect friction forces.

Kinetic friction Imagine pulling a block along a surface at a constant velocity, as shown in **Figure 11.** Since the block is not accelerating, according to Newton's laws, the friction force must be equal and opposite to the force with which you pull. The spring scale in **Figure 11** indicates the force exerted on the block. You can then stack additional blocks on top of the first block to increase the normal force and repeat the measurement. **Table 1** shows the results of such an experiment.

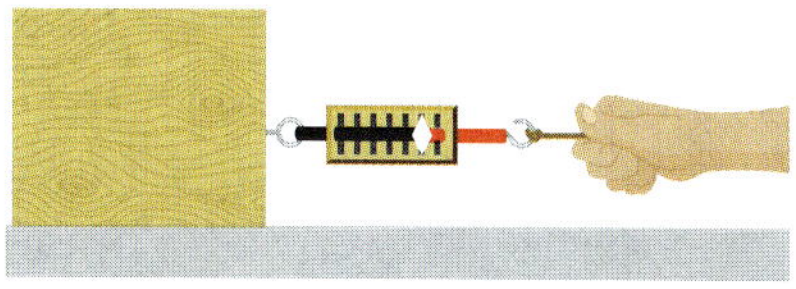

Figure 11 The spring scale applies a constant force on the block.

Plotting the data will yield a graph like the one in **Figure 12.** There is a linear relationship between the kinetic friction force and the normal force. The different lines on the graph correspond to dragging the block along different surfaces. Notice that the line for the sandpaper surface has a steeper slope than the line for the highly polished table. You would expect it to be much more difficult to pull the block along sandpaper than along a polished table, so the slope must be related to the magnitude of the resulting friction force. The slope of the line on a kinetic friction force v. normal force graph, designated μ_k, is called the **coefficient of kinetic friction** and relates the friction force to the normal force.

Kinetic Friction Force

The kinetic friction force equals the product of the coefficient of kinetic friction and the normal force.

$$F_{f,\,kinetic} = \mu_k F_N$$

Table 1 **Kinetic Friction v. Normal Force (sandpaper)**

Number of blocks	Normal force (N)	Kinetic friction (N)
1	0.98	0.53
2	1.96	0.95
3	2.94	1.4
4	3.92	1.8
5	4.90	2.3
6	5.88	3.1
7	6.86	3.3
8	7.84	4.0

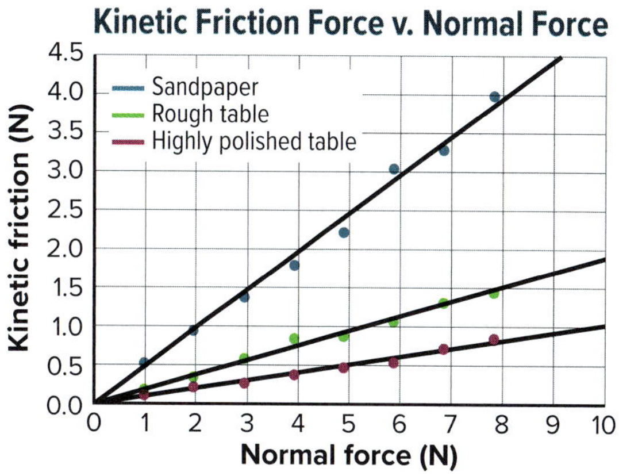

Figure 12 A plot of kinetic friction v. normal force for a block pulled along different surfaces shows a linear relationship between the two forces for each surface. The slope of the line is μ_k.

Compare *the coefficient of kinetic friction for the three surfaces shown on the graph.*

Get It?

Describe, in terms of friction, why the railway track maintenance mentioned at the beginning of this module should include wheel cleaning as well.

Static friction The maximum static friction force relates to the normal force in a similar way as the kinetic friction force. Remember that the force of static friction acts in response to a force trying to cause a stationary object to start moving. If there is no such force acting on an object, the force of static friction is zero. If there is a force trying to cause motion, the force of static friction will increase up to a maximum value before it is overcome and motion starts.

Static Friction Force

The static friction force is less than or equal to the product of the coefficient of static friction and the normal force.

$$F_{\text{f, static}} = \mu_s F_N$$

In the equation for the maximum static friction force, μ_s is the **coefficient of static friction** between the two surfaces. The maximum static friction force that must be overcome before motion can begin is $\mu_s F_N$. In the example of pushing the couch in **Figure 10,** the maximum force of static friction balances the force of the person pushing on the couch the instant before the couch begins to move.

Real-World Physics

ALL SURFACES, even surfaces that appear to be smooth, are rough at a microscopic level. If you look at a photograph of a graphite crystal magnified by a scanning tunneling microscope, the atomic level surface irregularities of the crystal are revealed. When two surfaces touch, the high points on each are in contact and temporarily bond. This is the origin of both static and kinetic friction. The details of this process are still unknown and are the subject of research in both physics and engineering.

Note that the equations for the kinetic and maximum static friction forces involve only the magnitudes of the forces. Also, the friction forces (F_f) are always perpendicular to the normal force (F_N).

Measuring coefficients of friction **Table 2** shows coefficients of friction between various surfaces. These coefficients are estimates for each combination of surfaces. Measurements of coefficients of friction are quite sensitive to the conditions of the surfaces. Surface impurities, such as dust or small amounts of oil from experimenters' hands, can significantly affect the results of the measurements.

Table 2 Typical Coefficients of Friction*

Surfaces	Coefficient of static friction (μ_s)	Coefficient of kinetic friction (μ_k)
Cast iron on cast iron	1.1	0.15
Glass on glass	0.94	0.4
Leather on oak	0.61	0.52
Nonstick coating on steel	0.04	0.04
Oak on oak	0.62	0.48
Steel on steel	0.78	0.42
Steel on steel (with castor oil)	0.15	0.08

*All measurements are for dry surfaces unless otherwise stated. All problems in this text assume dry surfaces.

CCC CROSSCUTTING CONCEPTS

Cause and Effect Systems can be designed to cause a desired effect. When designing and building vehicles such as cars and airplanes, manufacturers use steel for many of the components. Do you think that any of the steel components are in direct contact with each other in a vehicle? What would be the effect if they were? Explain your answer, using evidence from **Table 2.**

EXAMPLE Problem 3

BALANCED FRICTION FORCES You push a 25.0-kg wooden box across a wooden floor at a constant speed of 1.0 m/s. The coefficient of kinetic friction is 0.20. How large is the force that you exert on the box?

1 ANALYZE AND SKETCH THE PROBLEM

- Identify the forces, and establish a coordinate system.
- Draw a motion diagram indicating constant v and $a = 0$.
- Draw the free-body diagram.

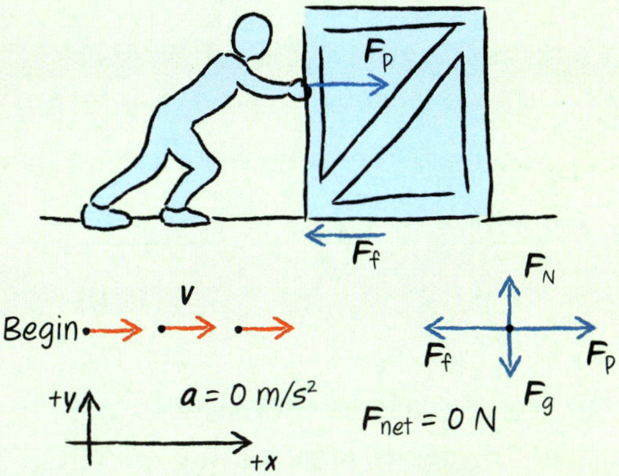

Known		Unknown
$m = 25.0$ kg	$v = 1.0$ m/s	$\mathbf{F}_{\text{person on box}} = ?$
$a = 0.0$ m/s^2	$\mu_k = 0.20$	

2 SOLVE FOR THE UNKNOWN

The normal force is in the y-direction, and the box does not accelerate in that direction.

$$\mathbf{F}_N = -\mathbf{F}_g$$

$$= -mg \qquad \text{Substitute } F_g = -mg$$

$$= -(25.0\text{ kg})(-9.8\text{ N/kg}) \qquad \text{Substitute } m = 25.0\text{ kg}, g = -9.8\text{ N/kg}$$

$$= +245\text{ N}$$

The pushing force is in the x-direction; v is constant; thus the box does not accelerate.

$$F_{\text{person on box}} = \mu_k F_N$$

$$= (0.20)(245\text{ N}) \qquad \text{Substitute } \mu_k = 0.20, F_N = 245\text{ N}$$

$$= 49\text{ N}$$

$$\mathbf{F}_{\text{person on box}} = 49\text{ N, to the right}$$

3 EVALUATE THE ANSWER

- **Are the units correct?** Force is measured in newtons.
- **Do the signs make sense?** The positive sign agrees with the sketch.
- **Is the magnitude realistic?** The pushing force is $\frac{1}{5}$ the weight of the box. This corresponds with $\mu_k = 0.20 = \frac{1}{5}$.

PRACTICE Problems

ADDITIONAL PRACTICE

15. Gwen exerts a 36-N horizontal force as she pulls a 52-N sled across a cement sidewalk at constant speed. What is the coefficient of kinetic friction between the sidewalk and the metal sled runners? Ignore air resistance.

16. Mr. Ames is dragging a box full of books from his office to his car. The box and books together have a combined weight of 134 N. If the coefficient of static friction between the pavement and the box is 0.55, how hard must Mr. Ames push horizontally on the box in order to start it moving?

17. Thomas sits on a small rug on a polished wooden floor. The coefficient of kinetic friction between the rug and the slippery wooden floor is only 0.12. If Thomas weighs 650 N, what horizontal force is needed to pull the rug and Thomas across the floor at a constant speed?

18. CHALLENGE You need to move a 105-kg sofa to a different location in the room. It takes a 403-N force to start the sofa moving. What is the coefficient of static friction between the sofa and the carpet?

EXAMPLE Problem 4

UNBALANCED FRICTION FORCES Imagine that the force you exert on the 25.0-kg box in Example Problem 3 is doubled.

a. What is the resulting acceleration of the box?

b. How far will you push the box if you push it for 3 s?

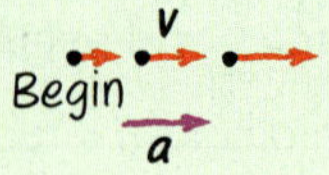

1 ANALYZE AND SKETCH THE PROBLEM

- Draw a motion diagram showing v and a.
- Draw the free-body diagram with a doubled $F_{\text{person on box}}$.

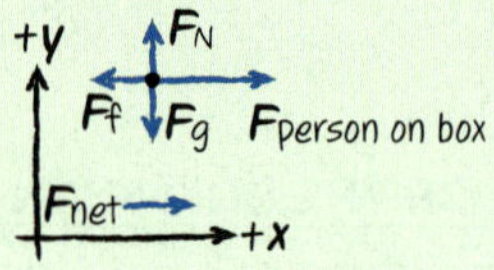

Known

$m = 25.0$ kg $\mu_k = 0.20$

$v = 1.0$ m/s $F_{\text{person on box}} = 2(49\text{ N}) = 98$ N

$t = 3.0$ s

Unknown

$a = ?$

2 SOLVE FOR THE UNKNOWN

a. The normal force is in the y-direction, and the box does not accelerate in that direction.

$$\boldsymbol{F}_N = -\boldsymbol{F}_g$$
$$= -mg$$

The box does accelerate in the x-direction. So the forces must be unequal.

$$F_{net} = F_{\text{person on box}} - F_f$$

$$ma = F_{\text{person on box}} - F_f$$ Substitute $F_{net} = ma$.

$$a = \frac{F_{\text{person on box}} - F_f}{m}$$

Find F_f and substitute it into the expression for a.

$$F_f = \mu_k F_N$$

$$= \mu_k mg$$ Substitute $F_N = mg$.

$$a = \frac{F_{\text{person on box}} - \mu_k mg}{m}$$ Substitute $F_f = \mu_k mg$.

$$= \frac{98\text{ N} - (0.20)(25.0\text{ kg})(9.8\text{ N/kg})}{25.0\text{ kg}}$$ Substitute $F_{\text{person on box}} = 98$ N, $\mu_k = 0.20$, $m = 25.0$ kg, $g = 9.8$ N/kg.

$$= 2.0\text{ m/s}^2$$

b. Use the relationship between distance, time, and constant acceleration.

$$x_f = x_i + v_i t_f + \left(\tfrac{1}{2}\right)a t_f^2$$

$$x_f = 0\text{ m} + v_i t + \left(\tfrac{1}{2}\right)a t^2$$ Substitute $x_i = 0$ m.

$$= (1.0\text{ m/s})(3.0\text{ s}) + \left(\tfrac{1}{2}\right)(2.0\text{ m/s}^2)(3.0\text{ s})^2$$ Substitute $v_i = 1.0$ m/s, $t = 3.0$ s, $a = 2.0$ m/s^2.

$$= 12\text{ m}$$

3 EVALUATE THE ANSWER

- **Are the units correct?** a is measured in m/s^2, and x_f is measured in meters.
- **Does the sign make sense?** In this coordinate system, the sign should be positive.
- **Is the magnitude realistic?** 12 m is about the length of two full-sized cars. It is realistic to push a box this distance in 3 s.

PRACTICE Problems

ADDITIONAL PRACTICE

19. You want to move a 41-kg bookcase to a different place in the living room. If you push with a force of 65 N and the bookcase accelerates at 0.12 m/s^2, what is the coefficient of kinetic friction between the bookcase and the carpet?

20. Consider the force pushing the box in Example Problem 4. How long would it take for the velocity of the box to double to 2.0 m/s?

21. Ke Min is driving at 23 m/s. He sees a tree branch lying across the road. He slams on the brakes when the branch is 60.0 m in front of him. If the coefficient of kinetic friction between the car's locked tires and the road is 0.41, will the car stop before hitting the branch? The car has a mass of 1200 kg.

22. CHALLENGE Isabel pushes a shuffleboard disk, accelerating it to a speed of 6.5 m/s before releasing it as indicated in **Figure 13.** If the coefficient of kinetic friction between the disk and the concrete court is 0.31, how far does the disk travel before it comes to a stop? Will Isabel's shot stop in the 10-point section of the board?

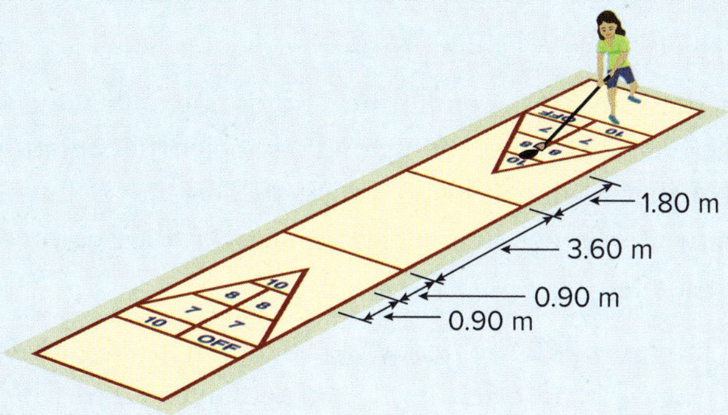

Figure 13

Check Your Progress

23. **Analyze** Compare static friction and kinetic friction. How are the frictional forces similar, and how do they differ?

24. **Friction** At a wedding reception, you notice a boy who looks like his mass is about 25 kg running across the dance floor, then sliding on his knees until he stops. If the coefficient of kinetic friction between the boy's pants and the floor is 0.15, what is the friction force acting on him as he slides?

25. **Velocity** Dinah is playing cards with her friends, and it is her turn to deal. A card with a mass of 2.3 g slides 0.35 m along the table before it stops. If the coefficient of kinetic friction between the card and the table is 0.24, what was the initial speed of the card as it left Dinah's hand?

26. **Force** The coefficient of static friction between a 40.0-kg picnic table and the ground is 0.43. Find the greatest horizontal force that could be exerted on the table without moving the table.

27. **Acceleration** You push a 13-kg table in the cafeteria with a horizontal force of 20 N, but the table does not move. You then push the table with a horizontal force of 25 N, and it accelerates at 0.26 m/s^2. What, if anything, can you conclude about the coefficients of static and kinetic friction?

28. **Critical Thinking** Rachel puts a dresser in the back of her pickup truck. When the truck accelerates forward, what force accelerates the dresser? Under what circumstances could the dresser slide? In which direction?

LEARNSMART® Go online to follow your personalized learning path to review, practice, and reinforce your understanding.

LESSON 3

FORCES IN TWO DIMENSIONS

FOCUS QUESTION

How can you analyze forces on a train as it climbs a hill?

Equilibrium Revisited

So far, you have considered motion along a horizontal surface. Now you will analyze situations in which the forces acting on an object are at angles other than 90°.

When the net force on an object is zero, the object is in equilibrium. According to Newton's laws, the object will not accelerate because there is no net force acting on it; an object in equilibrium moves with constant velocity. (Remember that staying at rest is a state of constant velocity.) Equilibrium can also occur if more than two forces act on an object. As long as the net force on the object is zero, the object is in equilibrium.

What is the net force acting on the ring in **Figure 14?** The free-body diagram in the figure shows the three forces acting on the ring. The ring is not accelerating, so the net force is zero. The free-body diagram, however, does not immediately indicate that the net force is zero. To find the net force, you must add all the vectors together. **Figure 15** on the next page shows the process of adding the force vectors to discover the net force.

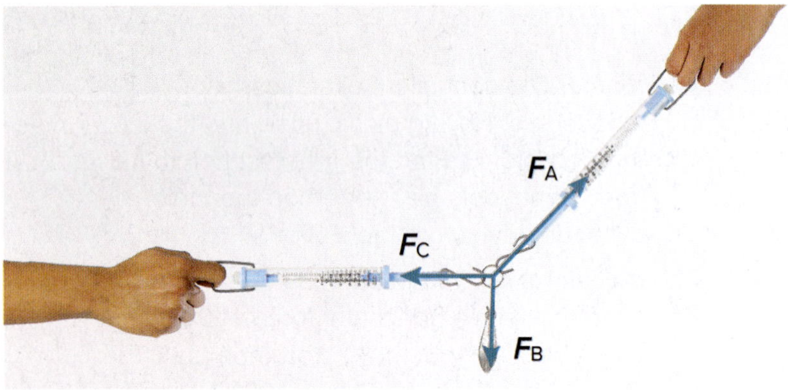

Richard Hutchings/Digital Light Source

Figure 14 The ring does not accelerate, so the net force acting on it must be zero.

Compare *the vertical component of the force pulling up and to the right to the weight of the mass hanging from the ring.*

3D THINKING DCI Disciplinary Core Ideas CCC Crosscutting Concepts SEP Science & Engineering Practices

COLLECT EVIDENCE

Use your Science Journal to record the evidence you collect as you complete the readings and activities in this lesson.

INVESTIGATE

GO ONLINE to find these activities and more resources.

PhysicsLAB: Sliding Down a Slope

Use mathematics to apply Newton's second law of motion to a system on an inclined plane.

Quick Investigation: Equilibrium

Carry out an investigation of the forces on a system in equilibrium

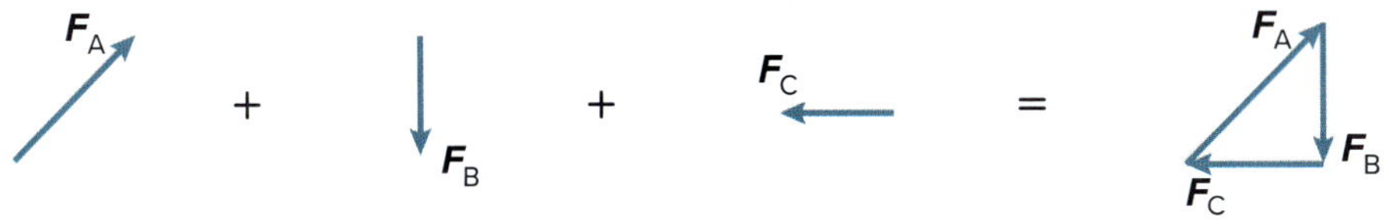

Figure 15 The forces acting on the ring sum to a zero net force.

Equilibrants **Figure 15** shows the addition of the three forces, F_A, F_B, and F_C, acting on the ring. Note that the three vectors form a closed triangle. The sum $F_A + F_B + F_C$ is zero. Now suppose two forces are exerted on an object and the sum is not zero, as shown in the free-body diagram in **Figure 16.** We want to find a third force that, when added to the other two, would add up to zero and put the object in equilibrium. First, add the forces already being exerted on the object. This single force that produces the same effect as the two or more individual forces together is called the resultant force. To put the object in equilibrium, add another force, called the **equilibrant,** that has the same magnitude as the resultant force but is in the opposite direction. Notice that in **Figure 15** the force F_C is the equilibrant of $F_A + F_B$. **Figure 16** illustrates the procedure for finding the equilibrant for two vectors.

Get It?

Identify the relationship between the equilibrant and the resultant vector.

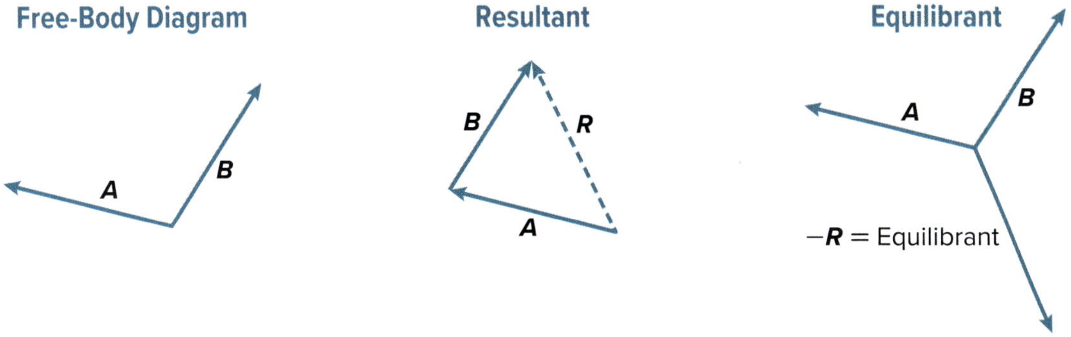

Figure 16 The equilibrant is the force required to put an object in equilibrium. This general procedure for finding the equilibrant works for any number of vectors.

PHYSICS Challenge

Find the equilibrant for the entire set of forces listed below, also shown in the figure.

F_1 = 61.0 N at 17.0° north of east

F_2 = 38.0 N at 64.0° north of east

F_3 = 54.0 N at 8.0° west of north

F_4 = 93.0 N at 53.0° west of north

F_5 = 65.0 N at 21.0° south of west

F_6 = 102.0 N at 15.0° west of south

F_7 = 26.0 N south

F_8 = 77.0 N at 22.0° east of south

F_9 = 51.0 N at 33.0° east of south

F_{10} = 82.0 N at 5.0° south of east

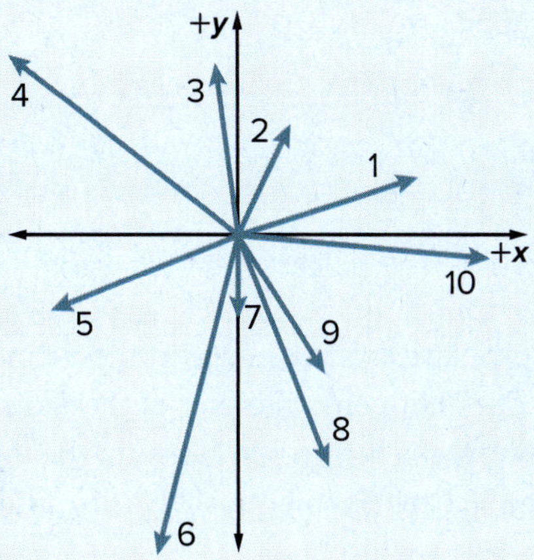

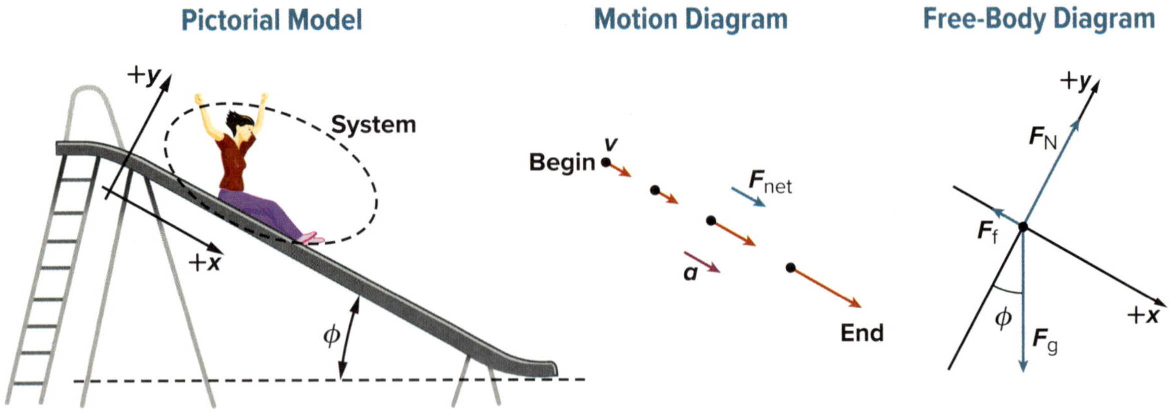

Figure 17 The girl's weight causes her to accelerate down the slide.

Describe *how the component of the girl's weight parallel to the incline changes when the angle between the slide and the horizontal increases.*

Inclined Planes

You have applied Newton's laws to a variety of situations but only to motions that were either horizontal or vertical. How would you apply Newton's laws in a situation such as the one in **Figure 17,** in which you want to find the net force on the girl?

First, identify the forces acting on the system. In this case, the girl is the system. The gravitational force on the girl is downward, toward the center of Earth. The normal force acts perpendicular to the slide, and the kinetic friction force acts parallel to the slide. You can see the resulting free-body diagram on the right in **Figure 17.** You know from experience that the girl's acceleration will be along the slide.

Choosing a coordinate system

Second, you should carefully choose your coordinate system. Because the girl's acceleration is parallel to the incline, one axis, the x-axis, should be in that direction. The y-axis is perpendicular to the x-axis and perpendicular to the incline's surface. With this coordinate system, you now have two forces, the normal force and the friction force, in the directions of the coordinate axes. The weight, however, has components in both the x- and y-directions. This means that when you place an object on an inclined plane, the magnitude of the normal force between the object and the plane usually will not equal the object's weight.

Get It?

Explain why you would choose the x-direction to be parallel to the slope of an inclined plane.

You will need to apply Newton's laws once in the x-direction and once in the y-direction. Because the weight does not point in either of these directions, you will need to break this vector into its x- and y-components before you can sum the forces in these two directions. Example Problem 5 and Example Problem 6 both show this procedure.

EXAMPLE Problem 5

COMPONENTS OF WEIGHT FOR AN OBJECT ON AN INCLINE A 562-N crate is resting on a plane inclined 30.0° above the horizontal. Find the components of the crate's weight that are parallel and perpendicular to the plane.

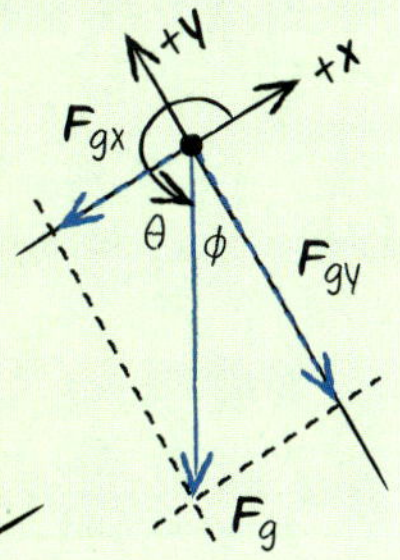

1 ANALYZE AND SKETCH THE PROBLEM

- Include a coordinate system with the positive *x*-axis pointing uphill.
- Draw the free-body diagram showing $\boldsymbol{F}_g$, the components F_{gx} and F_{gy}, and the angles θ and ϕ.

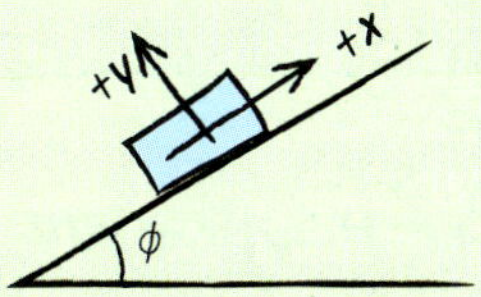

Known	Unknown
$\boldsymbol{F}_g$ = 562 N down	$F_{gx} = ?$
$\phi = 30.0°$	$F_{gy} = ?$
	$\theta = ?$

2 SOLVE FOR THE UNKNOWN

$\theta + \phi = 270°$ The angle from the positive *x*-axis to the negative *y*-axis is 270°.

$\theta = 270° - 30°$ Rearrange, substitute $\phi = 30°$, and solve for θ.

$= 240°$

$F_{gx} = F_g(\cos\theta)$

$= (562\text{ N})(\cos 240.0°)$ Substitute $F_g = 562$ N, $\theta = 240.0°$.

$= -281\text{ N}$

$F_{gy} = F_g(\sin\theta)$

$= (562\text{ N})(\sin 240.0°)$ Substitute $F_g = 562$ N, $\theta = 240.0°$.

$= -487\text{ N}$

3 EVALUATE THE ANSWER

- **Are the units correct?** Force is measured in newtons.
- **Do the signs make sense?** The components point in directions opposite to the positive axes.
- **Are the magnitudes realistic?** The values are less than F_g, as expected.

PRACTICE Problems

ADDITIONAL PRACTICE

29. An ant climbs at a steady speed up the side of its anthill, which is inclined 30.0° from the vertical. Sketch a free-body diagram for the ant.

30. Ryan and Becca are moving a folding table out of the sunlight. A cup of lemonade, with a mass of 0.44 kg, is on the table. Becca lifts her end of the table before Ryan does, and as a result, the table makes an angle of 15.0° with the horizontal. Find the components of the cup's weight that are parallel and perpendicular to the plane of the table.

31. Fernando, who has a mass of 43.0 kg, slides down the banister at his grandparents' house. If the banister makes an angle of 35.0° with the horizontal, what is the normal force between Fernando and the banister?

32. CHALLENGE A suitcase is on an inclined plane as shown in **Figure 18.** At what angle θ will the component of the suitcase's weight parallel to the plane be equal to half the component of its weight perpendicular to the plane?

Figure 18

EXAMPLE Problem 6

SLIDE Ichiko, who has a mass of 45 kg, is going down a slide sloped at 27°. The coefficient of kinetic friction is 0.23. How fast does she slide 1.0 s after starting from rest?

1 ANALYZE AND SKETCH THE PROBLEM

- Establish a coordinate system.
- Draw a free-body diagram that shows the forces acting on the girl as she travels down the slide.
- The three forces acting on the girl are the normal force, kinetic friction, and the girl's weight.

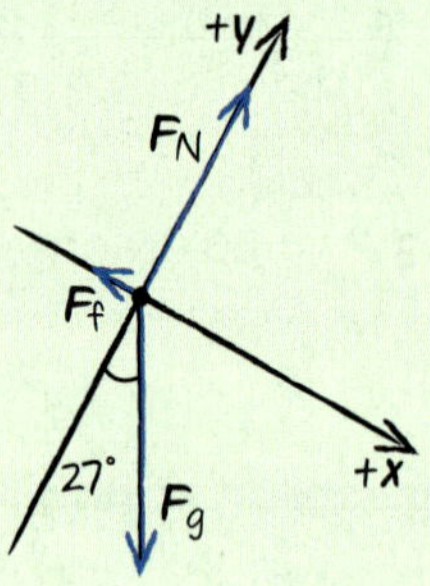

Known		Unknown
$m = 45$ kg	$\phi = 27°$	$a = ?$
$\mu_k = 0.23$	$v_i = 0.0$ m/s	$v_f = ?$
$t = 1.0$ s		$\theta = ?$

2 SOLVE FOR THE UNKNOWN

$\theta = 270° + \phi$ — The angle from the positive x-axis to the negative y-axis is 270°.

$= 270° + 27°$ — Substitute $\phi = 27°$ and solve for θ.

$= 297°$

y-direction:

$F_{net,y} = ma_y$ — There is no acceleration in the y-direction, so $a_y = 0.0$ m/s².

$= 0.0$ N

Add forces in the y-direction to find F_N.

$F_N + F_{gy} = F_{net,y}$

$F_N = -F_{gy}$ — Substitute $F_{net,y} = 0.0$ N, rearrange, and solve for F_N.

$= -mg(\sin\theta)$ — Substitute $F_{gy} = mg(\sin\theta)$.

x-direction:

Use the net force in the x-direction and Newton's second law to solve for a.

$F_{net,x} = F_{gx} - F_f$ — F_f is negative because it is in the negative x-direction.

$ma_x = mg(\cos\theta) - \mu_k F_N$ — Substitute $F_{net,x} = ma_x$, $F_{gx} = mg(\cos\theta)$, $F_f = \mu_k F_N$.

$ma = mg(\cos\theta) + \mu_k mg(\sin\theta)$ — Substitute $a_x = a$ (all acceleration is in the x-direction), $F_N = -mg(\sin\theta)$.

$a = g(\cos\theta + \mu_k \sin\theta)$

$= (9.8\text{ m/s}^2)(\cos 297° + (0.23)(\sin 297°))$ — Substitute $g = 9.8$ m/s², $\theta = 297°$, $\mu_k = 0.23$.

$= 2.4\text{ m/s}^2$

Because v_i, a, and t are all known, use the relationship between velocity, acceleration, and time.

$v_f = v_i + at$

$= 0.0\text{ m/s} + (2.4\text{ m/s}^2)(1.0\text{ s})$ — Substitute $v_i = 0.0$ m/s, $a = 2.4$ m/s², $t = 1.0$ s.

$= 2.4$ m/s

3 EVALUATE THE ANSWER

- **Are the units correct?** Performing dimensional analysis on the units verifies that v_f is in m/s and a is in m/s².
- **Do the signs make sense?** Because v_f and a are both in the +x-direction, the signs make sense.
- **Are the magnitudes realistic?** The velocity is similar to a person's running speed, which is realistic for a steep slide and a low coefficient of kinetic friction.

PRACTICE Problems

ADDITIONAL PRACTICE

33. Consider the crate on the incline in Example Problem 5. Calculate the magnitude of the acceleration. After 4.00 s, how fast will the crate be moving?

34. Jorge decides to try the slide discussed in Example Problem 6. Jorge's trip down the slide is quite different from Ichiko's. After giving himself a push to get started, Jorge slides at a constant speed. What is the coefficient of kinetic friction between Jorge's pants and the slide?

35. Stacie, who has a mass of 45 kg, starts down a slide that is inclined at an angle of 45° with the horizontal. If the coefficient of kinetic friction between Stacie's shorts and the slide is 0.25, what is her acceleration?

36. CHALLENGE You stack two physics books on top of each other as shown in **Figure 19.** You tilt the bottom book until the top book just begins to slide. You perform five trials and measure the angles given in **Table 3.**

a. What is the average ϕ measured during the five trials?

b. What is the coefficient of static friction between the covers of the two books? Use the average ϕ found in part **a.**

c. You measure the top book's acceleration down the incline to be 1.3 m/s^2. What is the coefficient of kinetic friction? Assume ϕ is the average value found in part **a.**

Figure 19

Table 3 Trial Number and Angle of Tilt

Trial	ϕ
1	21°
2	17°
3	21°
4	18°
5	19°

Check Your Progress

37. **Acceleration** A rope pulls a 63-kg water skier up a 14.0° incline with a tension of 512 N. The coefficient of kinetic friction between the skier and the ramp is 0.27. What are the magnitude and direction of the skier's acceleration?

38. **Forces** One way to get a car unstuck is to tie one end of a strong rope to the car and the other end to a tree, then pull the rope at its midpoint at right angles to the rope. Draw a free-body diagram. Explain how even a small force on the rope can exert a large force on the car.

39. **Mass** A large scoreboard is suspended from the ceiling of a sports arena by ten strong cables. Six of the cables make an angle of 8.0° with the verticals and the other four make an angle of 10.0°. If the tension in each cable is 1300 N, what is the scoreboard's mass?

40. **Equilibrium** You are hanging a painting using two lengths of wire. The wires will break if the force is too great. Should you hang the painting as shown on the left or the right of **Figure 20?** Explain.

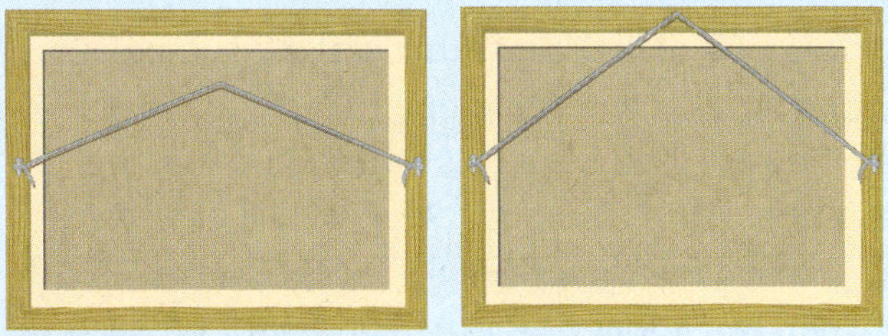

Figure 20

41. **Critical Thinking** Can the coefficient of friction ever have a value such that a child would be able to slide *up* a slide at a constant velocity? Explain. Assume that no one pushes or pulls on the child.

LEARNSMART® Go online to follow your personalized learning path to review, practice, and reinforce your understanding.

More or Less—Using Biomimetics to Control Friction

As solutions are designed to meet engineering challenges, the force of friction needs to be considered. In some applications, friction between surfaces needs to be maximized. In other applications, friction between surfaces needs to be minimized. Structures of living things have provided insight into how these goals can be accomplished.

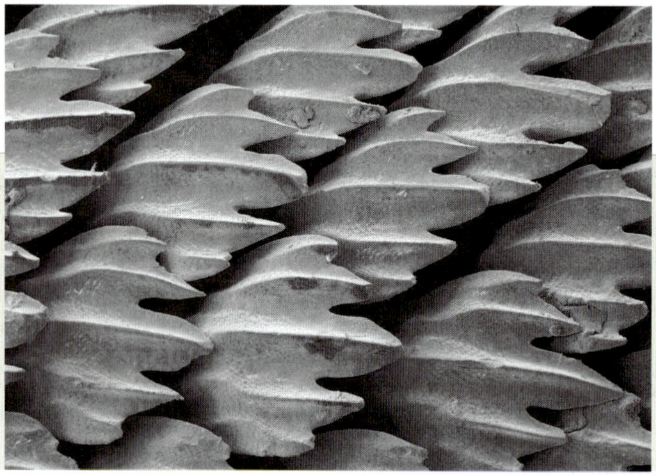

This SEM image shows the structure of shark skin. Scientists and engineers apply properties of this structure to reduce friction on everything from boats to airplanes.

Biomimetics

Biomimetics is the application of ideas from nature and living things to solve design problems. For example, burdock seeds stick to living things to move from place to place. Scientists imitated the structure of burdock seeds to develop hook-and-loop tape.

Using Biomimetics to Increase Friction

Clingfish are a species that live near the coast, so are subject to tides. These fish have a suction-cup-like disc with an astonishing ability to hold the fish in place in their ebbing and flowing environment.

Friction is one of the forces that make this clingfish structure so effective at sticking to surfaces, even when they are wet. Scientists have found that the outer edge of the disc is made up of hair-like structures so tiny they require a microscope to see. Recall that when two surfaces touch, high points on each surface temporarily bond. The tiny hair-like structures of the clingfish's disc vastly increase the number of points of contact between the fish and the surface to which they cling. Engineers can apply information about the structure of the clingfish disc when they need to maximize friction between two surfaces, such as in the development of adhesives.

Using Biomimetics to Decrease Friction

Friction can negatively affect the motion of vehicles and the function of machines, so engineers look for ways to reduce the friction between surfaces. Shark skin is made up of structures called denticles. Denticles have several properties that interest engineers, including the reduction of friction between the shark and the water it moves through. Applying this technology to vehicles and other machines has the potential to decrease friction, reducing fuel usage and energy consumption.

USE A MODEL TO DESCRIBE

Draw a labeled diagram to describe how friction can be increased or decreased using biomimetics. You might want to carry out additional research to gather details to add to your model.

CALIFORNIA CONNECTION

Research how biomimetics is being applied to solve a variety of engineering challenges at the Jet Propulsion Lab at UC Berkeley. Write a brief summary of what you learn.

STUDY GUIDE

GO ONLINE to study with your Science Notebook.

Lesson 1 VECTORS

- components
- vector resolution

- Vectors are added graphically by placing the tail of the second vector on the tip of the first vector. The resultant vector points from the tail of the first vector to the tip of the final vector.
- The components of a vector are projections of the component vectors onto axes. Vectors can be summed by separately adding the *x*- and *y*-components.

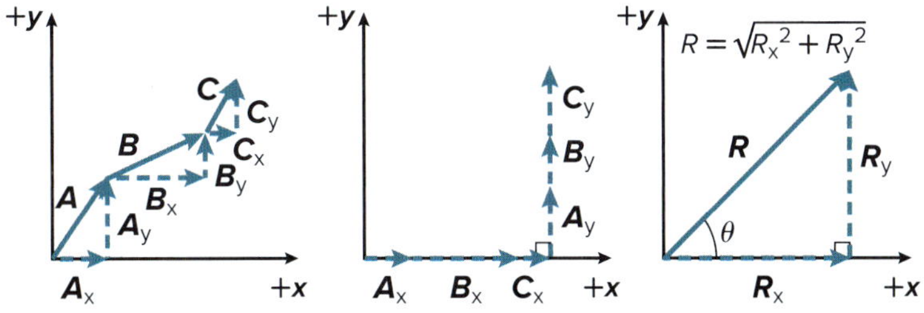

- When two vectors are at right angles, you can use the Pythagorean theorem to determine the magnitude of the resultant vector. The law of sines and the law of cosines can be used to find the resultant of any two vectors.

$$R^2 = A^2 + B^2$$

$$\text{Law of Sines: } \frac{R}{\sin \theta} = \frac{A}{\sin a} = \frac{B}{\sin b}$$

$$\text{Law of Cosines: } R^2 = A^2 + B^2 - 2AB \cos \theta$$

Lesson 2 FRICTION

- kinetic friction
- static friction
- coefficient of kinetic friction
- coefficient of static friction

- Friction is a force that acts parallel to the surfaces when two surfaces touch.
- The kinetic friction force equals the coefficient of kinetic friction times the normal force.

$$F_{\text{f, kinetic}} = \mu_k F_N$$

- The static friction force is less than or equal to the coefficient of static friction times the normal force.

$$F_{\text{f, static}} \leq \mu_s F_N$$

Lesson 3 FORCES IN TWO DIMENSIONS

- equilibrant

- The equilibrant is a force of the same magnitude but opposite direction as the resultant force, which is the sum of all the other forces acting on an object.
- Friction forces are parallel to an inclined plane but point up the plane if the motion of the object is down the plane. An object on an inclined plane has a component of the force of gravity parallel to the plane; this component can accelerate the object down the plane.

THREE-DIMENSIONAL THINKING

Module Wrap-Up

REVISIT THE PHENOMENON

Why is this specialized train washing the train tracks?

CER Claim, Evidence, Reasoning

Explain your Reasoning Revisit the claim you made when you encountered the phenomenon. Summarize the evidence you gathered from your investigations and research and finalize your Summary Table. Does your evidence support your claim? If not, revise your claim. Explain why your evidence supports your claim.

STEM UNIT PROJECT

Now that you've completed the module, revisit your STEM unit project. You will summarize your evidence and apply it to the project.

GO FURTHER

SEP Data Analysis Lab

Who goes down the slide fastest?

Margaret Mary, Doug, and Kako are at a local amusement park. They see an attraction called the Giant Slide, which is simply a very long and high inclined plane. Visitors at the amusement park climb a long flight of stairs to the top of the 27° inclined plane and are given canvas sacks. They sit on the sacks and slide down the plane 70 m long. At the time when the three friends walked past the slide, a 135-kg man and a 20-kg boy are each at the top preparing to slide down. "I wonder how much less time it will take the man to slide down than it will take the boy," said Margaret Mary. "I think the boy will take less time," said Doug. "You're both wrong," said Kako. "They will reach the bottom at the same time."

CER Analyze and Interpret Data

1. **Sketch the situation.**
2. **Claim** Who is correct?
3. **Evidence and Reasoning** Justify your claim.

MODULE 6
MOTION IN TWO DIMENSIONS

Thomas Barwick/Iconica/Getty Images

MODULE 6

MOTION IN TWO DIMENSIONS

ENCOUNTER THE PHENOMENON

Why do thrown basketballs travel in arcs?

GO ONLINE to watch a video about the path of a projectile.

SEP Ask Questions

Do you have other questions about the phenomenon? If so, add them to the driving question board.

CER Claim, Evidence, Reasoning

Make Your Claim Use your CER chart to make a claim about why thrown basketballs travel in arcs. Explain your reasoning.

Collect Evidence Use the lessons in this module to collect evidence to support your claim. Record your evidence as you move through the module.

Explain Your Reasoning You will revisit your claim and explain your reasoning at the end of the module.

GO ONLINE to access your CER chart and explore resources that can help you collect evidence.

LESSON 1: Explore & Explain: Angled Launches

LESSON 2: Explore & Explain: Centripetal Acceleration

Smithsonian

Additional Resources

(t)Video Supplied by BBC Worldwide Learning, (bl)David Madison/Digital Vision/Getty Images, (br)Exroy/Shutterstock

LESSON 1

PROJECTILE MOTION

FOCUS QUESTION

What forces affect a basketball's trajectory?

Path of a Projectile

A hopping frog, a tossed snowball, and an arrow shot from a bow all move along similar paths. Each path rises and then falls, always curving downward along a parabolic path. An object shot through the air is called a **projectile.** Its path through space is called its **trajectory.**

You can determine a projectile's trajectory if you know its initial velocity. **Figure 1** shows water that is launched as a projectile horizontally and at an angle. In both cases, gravity curves the path downward along a parabolic path.

You can draw a free-body diagram of a launched projectile and identify the forces acting on it. If you ignore air resistance, after an initial force launches a projectile, the only force on it as it moves through the air is gravity. Gravity causes the object to curve downward.

Figure 1 A projectile launched horizontally immediately curves downward, but if it is launched upward at an angle, it rises and then falls, always curving downward.

3D THINKING DCI Disciplinary Core Ideas CCC Crosscutting Concepts SEP Science & Engineering Practices

COLLECT EVIDENCE

Use your Science Journal to record the evidence you collect as you complete the readings and activities in this lesson.

INVESTIGATE

GO ONLINE to find these activities and more resources.

PhysicsLAB: On Target

Plan and carry out an investigation to determine what factors affect projectile motion.

Revisit the Encounter the Phenomenon Question

What information from this lesson can help you answer the Unit and Module questions?

Independence of Motion in Two Dimensions

Think about two softball players warming up for a game, tossing high fly balls back and forth. What does the path of the ball through the air look like? Because the ball is a projectile, it follows a parabolic path.

Imagine you are standing directly behind one of the players and you are watching the softball as it is being tossed. What would the motion of the ball look like? You would see it go up and back down, just like any object that is tossed straight up in the air.

If you were watching the softball from a hot-air balloon high above the field, what motion would you see? You would see the ball move from one player to the other at a constant speed, just like any object that is given an initial horizontal velocity, such as a hockey puck sliding across ice. The motion of projectiles is a combination of these two motions.

Get It?

Illustrate Draw sketches of a drone's-eye view of the person at the beginning of the module shooting a basketball. The sketches should show the view from directly behind the person, from directly above the court, and from the sideline.

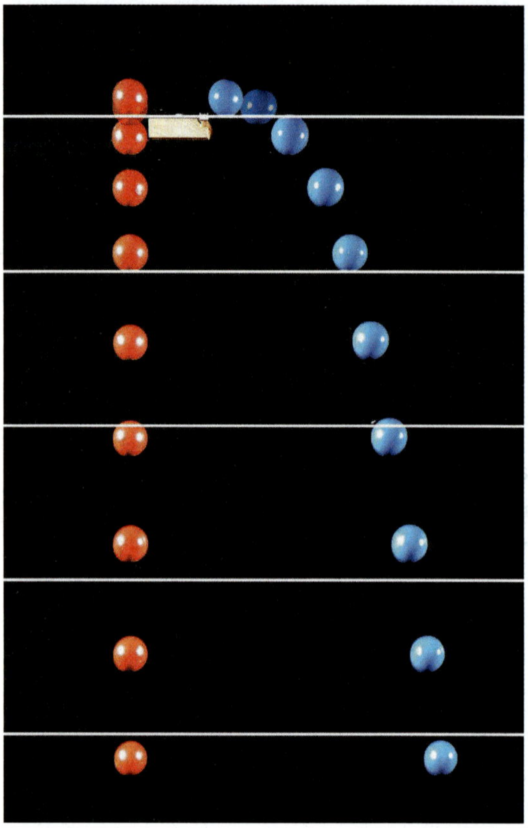

Figure 2 The ball on the left was dropped with no initial velocity. The ball on the right was given an initial horizontal velocity. The balls have the same vertical motion as they fall.

Identify *What is the vertical velocity of the balls after falling for 1 s?*

Why do projectiles behave in this way? After a softball leaves a player's hand, what forces are exerted on the ball? If you ignore air resistance, there are no contact forces on the ball. There is only the field force of gravity in the downward direction. How does this affect the ball's motion? Gravity causes the ball to have a downward acceleration.

Comparing motion diagrams The trajectories of two balls are shown in **Figure 2.** The red ball was dropped, and the blue ball was given an initial horizontal velocity of 2.0 m/s. What is similar about the two paths?

Look at their vertical positions. The horizontal lines indicate equal vertical distances. At each moment that a picture was taken, the heights of the two balls were the same. Because the change in vertical position was the same for both, their average vertical velocities during each interval were also the same. The increasingly large distance traveled vertically by the balls, from one time interval to the next, shows that they were accelerating downward due to the force of gravity.

Notice that the horizontal motion of the launched ball does not affect its vertical motion. A projectile launched horizontally has initial horizontal velocity, but it has no initial vertical velocity. Therefore, its vertical motion is like that of an object dropped from rest. Just like the red ball, the blue ball has a downward velocity that increases regularly because of the acceleration due to gravity.

Get It?

Explain why a dropped object has the same vertical velocity as an object launched horizontally.

Horizontally Launched Projectiles

Imagine a person standing near the edge of a cliff and kicking a pebble horizontally. Like all horizontally launched projectiles, the pebble will have an initial horizontal velocity and no initial vertical velocity. What will happen to the pebble as it falls from the cliff?

Separate motion diagrams Recall that the horizontal motion of a projectile does not affect its vertical motion. It is therefore easier to analyze the horizontal motion and the vertical motion separately. Separate motion diagrams for the *x*-components and *y*-components of a horizontally launched projectile, such as a pebble kicked off a cliff, are shown on the left in **Figure 3.**

Horizontal motion Notice the horizontal vectors in the diagram on the left. Each of the velocity vectors is the same length, which indicates that the object's horizontal velocity is not changing. The pebble is not accelerating horizontally. This constant velocity in the horizontal direction is exactly what should be expected, because after the initial kick, there is no horizontal force acting on the pebble. (In reality, the pebble's speed would decrease slightly because of air resistance, but remember that we are ignoring air resistance in this module.)

Get It?

Explain why the horizontal motion of a projectile is constant.

Vectors in Two Dimensions

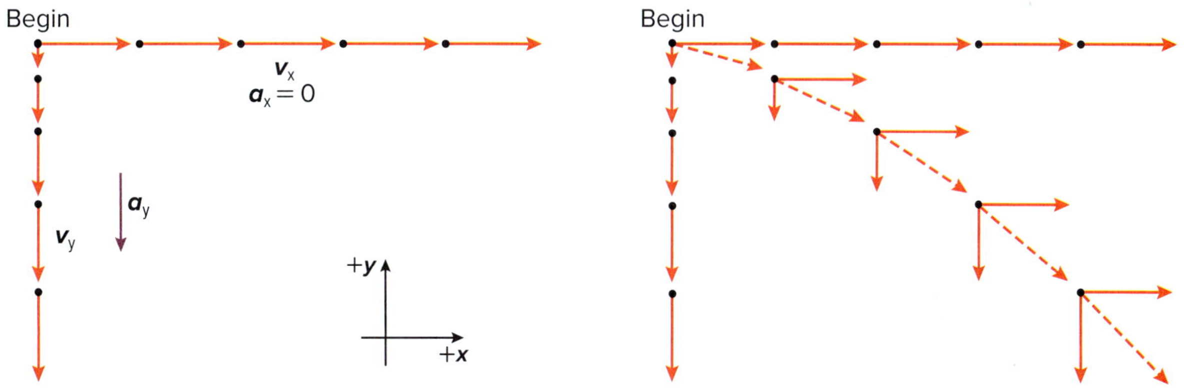

Figure 3 To describe the motion of a horizontally launched projectile, the *x*- and *y*-components can be treated independently. The resultant vectors of the projectile are tangent to a parabola.

Decide *What is the value of* $\mathbf{a}_y$*?*

Vertical motion Now look at the vertical velocity vectors in the diagram on the left. Each velocity vector has a slightly longer length than the one above it. The changing length shows that the object's vertical velocity is increasing and the object is accelerating downward. Again, this is what should be expected, because in this case the force of gravity is acting on the pebble.

Get It?

Describe why the vertical velocity vectors change length but the horizontal velocity vectors do not for a projectile.

Parabolic path When the x- and y-components of the object's motion are treated independently, each path is a straight line. The diagram on the right in **Figure 3** shows the actual parabolic path. The horizontal and vertical components at each moment are added to form the total velocity vector at that moment. You can see how the combination of constant horizontal velocity and uniform vertical acceleration produces a trajectory that has a parabolic shape.

PROBLEM-SOLVING STRATEGIES

Motion in Two Dimensions

When solving projectile problems, use the following strategies.

1. Draw a motion diagram with vectors for the projectile at its initial position and its final position. If the projectile is launched at an angle, also show its maximum height and the initial angle.
2. Consider vertical and horizontal motion independently. List known and unknown variables.
3. For horizontal motion, the acceleration is $a_x = 0.0\ \text{m/s}^2$. If the projectile is launched at an angle, its initial vertical velocity and its vertical velocity when it falls back to that same height have the same magnitude but different direction: $v_{yi} = -v_{yf}$.
4. For vertical motion, $a_y = -9.8\ \text{m/s}^2$ (if you choose up as positive). If the projectile is launched at an angle, its vertical velocity at its highest point is zero: $v_{y,\max} = 0.0$ m/s.
5. Choose the motion equations that will enable you to find the unknown variables. Apply them to vertical and horizontal motion separately. Remember that time is the same for horizontal and vertical motion. Solving for time in one of the dimensions identifies the time for the other dimension.
6. Sometimes it is useful to apply the motion equations to part of the projectile's path. You can choose any initial and final points to use in the equations.

Motion Equations

Horizontal (constant speed)

$x_f = vt_f + x_i$

Vertical (constant acceleration)

$v_f = v_i + at_f$

$x_f = x_i + v_it_f + \frac{1}{2}at_f^2$

$v_f^2 = v_i^2 + 2a(x_f - x_i)$

CCC CROSSCUTTING CONCEPTS

Patterns Gather empirical evidence to identify the pattern of motion that projectiles follow. Launch an object at various angles and record the motion it follows. Create a presentation that illustrates your findings.

EXAMPLE Problem 1

A SLIDING PLATE You are preparing breakfast and slide a plate on the countertop. Unfortunately, you slide it too fast, and it flies off the end of the countertop. If the countertop is 1.05 m above the floor and the plate leaves the top at 0.74 m/s, how long does it take to fall, and how far from the end of the counter does it land?

1 ANALYZE AND SKETCH THE PROBLEM

Draw horizontal and vertical motion diagrams. Choose the coordinate system so that the origin is at the top of the countertop. Choose the positive *x* direction in the direction of horizontal velocity and the positive *y* direction up.

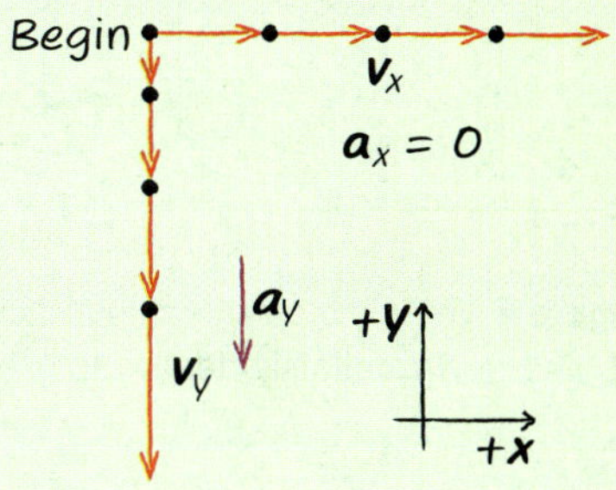

Known		Unknown
$x_i = y_i = 0$ m	$a_x = 0$ m/s^2	$t = ?$
$v_{xi} = 0.74$ m/s	$a_y = -9.8$ m/s^2	$x_f = ?$
$v_{yi} = 0$ m/s	$y_f = -1.05$ m	

2 SOLVE FOR THE UNKNOWN

Use the equation of motion in the *y* direction to find the time of fall.

$$y_f = y_i + \frac{1}{2}a_y t^2$$

$$t = \sqrt{\frac{2(y_f - y_i)}{a_y}}$$ Rearrange the equation to solve for time.

$$= \sqrt{\frac{2(-1.05\text{ m} - 0\text{ m})}{-9.8\text{ m/s}^2}} = 0.46\text{ s}$$ Substitute $y_f = -1.05$ m, $y_i = 0$ m, $a_y = -9.8$ m/s^2.

Use the equation of motion in the *x* direction to find where the plate hits the floor.

$\mathbf{x}_f = \mathbf{v}_x t =$ (0.74 m/s to the right)(0.46 s) = 0.34 m to the right of the counter

3 EVALUATE THE ANSWER

- **Are the units correct?** Time is measured in seconds. Position in measured in meters.
- **Do the signs make sense?** Both are positive. The position sign agrees with the coordinate choice.
- **Are the magnitudes realistic?** A fall of about a meter takes about 0.5 s. During this time, the horizontal displacement of the plate would be about 0.5 s × 0.74 m/s.

PRACTICE Problems

ADDITIONAL PRACTICE

1. You throw a stone horizontally at a speed of 5.0 m/s from the top of a cliff that is 78.4 m high.
   a. How long does it take the stone to reach the bottom of the cliff?
   b. How far from the base of the cliff does the stone hit the ground?
   c. What are the horizontal and vertical components of the stone's velocity just before it hits the ground?
2. Lucy and her friend are working at an assembly plant making wooden toy giraffes. At the end of the line, the giraffes go horizontally off the edge of a conveyor belt and fall into a box below. If the box is 0.60 m below the level of the conveyor belt and 0.40 m away from it, what must be the horizontal velocity of giraffes as they leave the conveyor belt?
3. **CHALLENGE** You are visiting a friend from elementary school who now lives in a small town. One local amusement is the ice-cream parlor, where Stan, the short-order cook, slides his completed ice-cream sundaes down the counter at a constant speed of 2.0 m/s to the servers. (The counter is kept very well polished for this purpose.) If the servers catch the sundaes 7.0 cm from the edge of the counter, how far do they fall from the edge of the counter to the point at which the servers catch them?

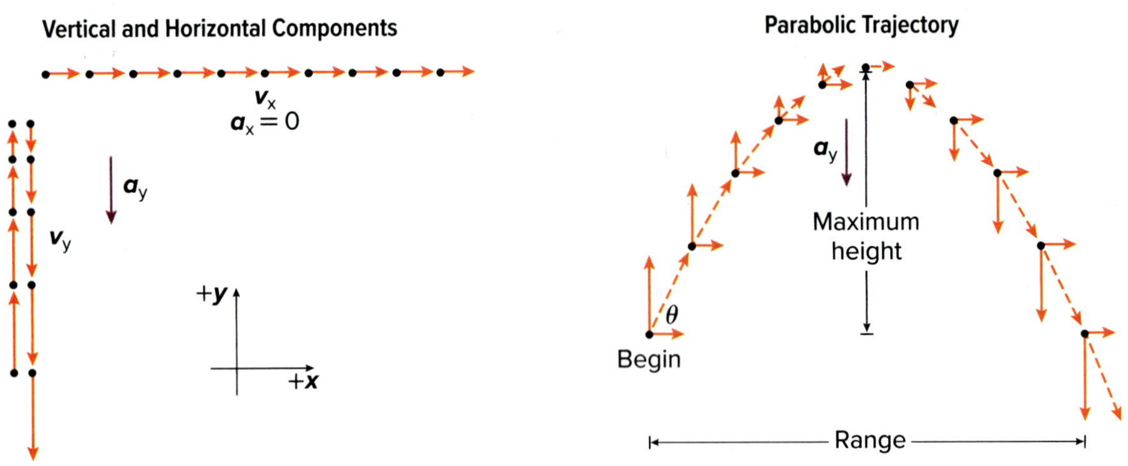

Figure 4 When a projectile is launched at an upward angle, its parabolic path is upward and then downward. The up-and-down motion is clearly represented in the vertical component of the vector diagram.

Angled Launches

When a projectile is launched at an angle, the initial velocity has a vertical component as well as a horizontal component. If the object is launched upward, like a ball tossed straight up in the air, it rises with slowing speed, reaches the top of its path where its speed is momentarily zero, and descends with increasing speed.

Separate motion diagrams

The left diagram of **Figure 4** shows the separate vertical- and horizontal-motion diagrams for the trajectory. In the coordinate system, the *x*-axis is horizontal and the *y*-axis is vertical. Note the symmetry. At each point in the vertical direction, the velocity of the object as it is moving upward has the same magnitude as when it is moving downward. The only difference is that the directions of the two velocities are opposite. When solving problems, it is sometimes useful to consider symmetry to determine unknown quantities.

Get It?

Analyze When a ball on the ground is kicked, the ball moves with an initial vertical velocity of 1.2 m/s up. When the ball hits the ground, what will its vertical velocity be? Explain your reasoning.

Parabolic path

The right diagram of **Figure 4** defines two quantities associated with the trajectory. One is the maximum height, which is the height of the projectile when the vertical velocity is zero and the projectile has only its horizontal-velocity component. The other quantity depicted is the range (*R*), which is the horizontal distance the projectile travels when the initial and final heights are the same. Not shown is the flight time, which is how much time the projectile is in the air. For football punts, flight time often is called hang time.

Get It?

Describe At what point of a projectile's trajectory is its vertical velocity zero?

EXAMPLE Problem 2

THE FLIGHT OF A BALL A ball is launched at 4.5 m/s at 66° above the horizontal. It starts and lands at the same distance from the ground. What are the maximum height above its launch level and the flight time of the ball?

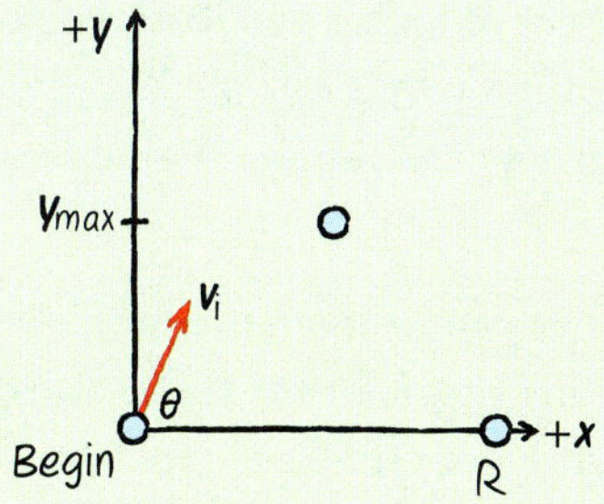

1 ANALYZE AND SKETCH THE PROBLEM

- Establish a coordinate system with the initial position of the ball at the origin.
- Show the positions of the ball at the beginning, at the maximum height, and at the end of the flight. Show the direction of $\boldsymbol{F}_{net}$.
- Draw a motion diagram showing $\boldsymbol{v}$ and $\boldsymbol{a}$.

Known			Unknown
$y_i = 0.0$ m	$\theta_i = 66°$	$v_{y,\,max} = 0.0$ m/s	$y_{max} = ?$
$v_i = 4.5$ m/s	$a_y = -9.8$ m/s^2		$t = ?$

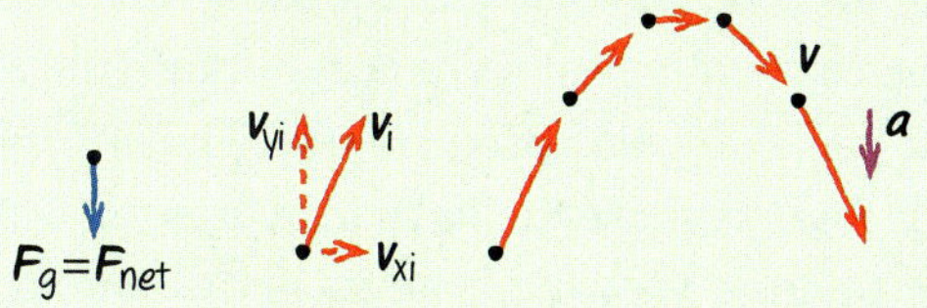

2 SOLVE FOR THE UNKNOWN

Find the y-component of v_i.

$$v_{yi} = v_i(\sin \theta_i)$$
$$= (4.5 \text{ m/s})(\sin 66°) = 4.1 \text{ m/s}$$

Substitute $v_i = 4.5$ m/s, $\theta_i = 66°$.

Use symmetry to find the y-component of v_f.

$$v_{yf} = -v_{yi} = -4.1 \text{ m/s}$$

Solve for the maximum height.

$$v_{y,\,max}{}^2 = v_{yi}{}^2 + 2a_y(y_{max} - y_i)$$
$$(0.0 \text{ m/s})^2 = v_{yi}{}^2 + 2a_y(y_{max} - 0.0 \text{ m})$$
$$y_{max} = -\frac{v_{yi}{}^2}{2a_y}$$
$$= -\frac{(4.1 \text{ m/s})^2}{2(-9.8 \text{ m/s}^2)} = 0.86 \text{ m}$$

Substitute $v_{yi} = -4.1$ m/s; $a_y = -9.8$ m/s^2.

Solve for the time to return to the launching height.

$$v_{yf} = v_{yi} + a_y t$$
$$t = \frac{v_{yf} - v_{yi}}{a_y}$$
$$= \frac{-4.1 \text{ m/s} - 4.1 \text{ m/s}}{-9.8 \text{ m/s}^2} = 0.84 \text{ s}$$

Substitute $v_{yf} = -4.1$ m/s; $v_{yi} = 4.1$ m/s; $a_y = -9.8$ m/s^2.

3 EVALUATE THE ANSWER

- **Are the magnitudes realistic?** For an object that rises less than 1 m, a time of less than 1 s is reasonable.

PRACTICE Problems

ADDITIONAL PRACTICE

4. A player kicks a football from ground level with an initial velocity of 27.0 m/s, 30.0° above the horizontal, as shown in **Figure 5.** Find each of the following. Assume that forces from the air on the ball are negligible.

a. the ball's hang time

b. the ball's maximum height

c. the horizontal distance the ball travels before hitting the ground

5. The player in the previous problem then kicks the ball with the same speed but at 60.0° from the horizontal. What is the ball's hang time, horizontal distance traveled, and maximum height?

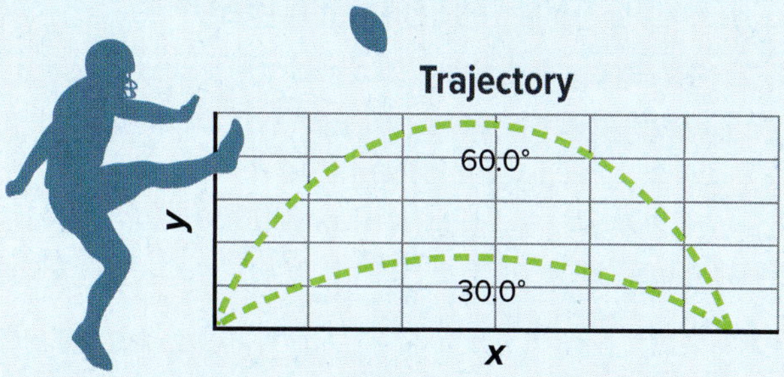

Figure 5

6. CHALLENGE A rock is thrown from a 50.0-m-high cliff with an initial velocity of 7.0 m/s at an angle of 53.0° above the horizontal. Find its velocity when it hits the ground below.

Forces from Air

The effect of forces due to air has been ignored so far in this chapter, but think about why a kite stays in the air or why a parachute helps a skydiver fall safely to the ground. Forces from the air can significantly change the motion of an object.

What happens if there is wind? Moving air can change the motion of a projectile. Consider the three cases shown in **Figure 6.** In the top photo, water is flowing from the hose pipe with almost no effect from air. In the middle photo, wind is blowing in the same direction as the water's initial movement. The path of the water changes because the air exerts a force on the water in the same direction as its motion. The horizontal distance the water travels increases because the force increases the water's horizontal speed. The direction of the wind changes in the bottom photo. The horizontal distance the water travels decreases because the air exerts a force in the direction opposite the water's motion.

What if the direction of wind is at an angle relative to a moving object? The horizontal component of the wind affects only the horizontal motion of an object. The vertical component of the wind affects only the vertical motion of the object. In the case of the water, for example, a strong updraft could decrease the downward speed of the water.

The effects shown in **Figure 6** occur because the air is moving enough to significantly change the motion of the water. Even air that is not moving, however, can have a significant effect on some moving objects. A piece of paper held horizontally and dropped, for example, falls slowly because of air resistance. The air resistance increases as the surface area of the object that faces the moving air increases.

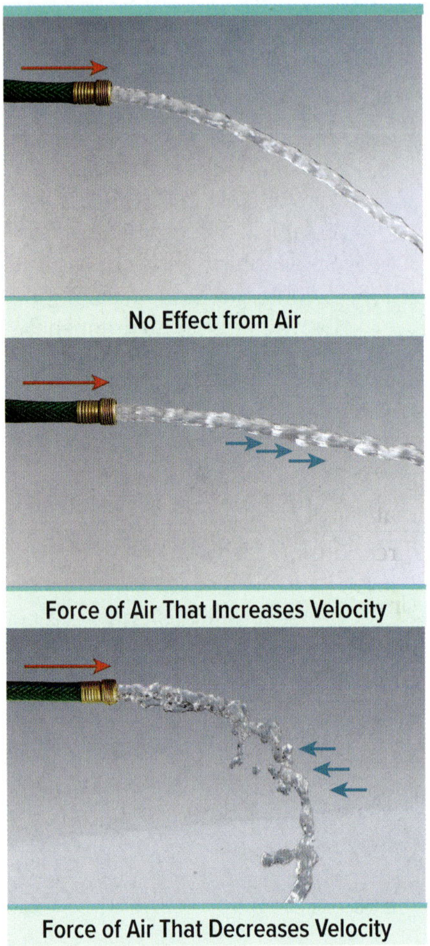

Figure 6 Forces from air can increase or decrease the velocity of a moving object.

Check Your Progress

7. **Initial Velocity** Two baseballs are pitched horizontally from the same height but at different speeds. The faster ball crosses home plate within the strike zone, but the slower ball is below the batter's knees. Why do the balls pass the batter at different heights?
8. **Free-Body Diagram** An ice cube slides without friction across a table at a constant velocity. It slides off the table and lands on the floor. Draw free-body and motion diagrams of the ice cube at two points on the table and at two points in the air.
9. **Projectile Motion** A ball is thrown out a window 28 m high at 15.0 m/s and 20.0° below the horizontal. How far does the ball move horizontally before it hits the ground?
10. **Projectile Motion** What is the maximum height of a softball thrown at 11.0 m/s and 50.0° above the horizontal?
11. **Critical Thinking** Suppose an object is thrown with the same initial velocity and direction on Earth and on the Moon, where the acceleration due to gravity is one-sixth its value on Earth. How will vertical velocity, time of flight, maximum height, and horizontal distance change?

LEARNSMART® Go online to follow your personalized learning path to review, practice, and reinforce your understanding.

LESSON 2
CIRCULAR MOTION

FOCUS QUESTION

What force causes an object to move in a circle?

Describing Circular Motion

Are objects moving in a circle at a constant speed, such as a fixed horse on a carousel, accelerating? At first, you might think they are not because their speeds do not change. But remember that acceleration is related to the change in velocity, not just the change in speed. Because their directions are changing, the objects must be accelerating.

Uniform circular motion is the movement of an object at a constant speed around a circle with a fixed radius. The position of an object in uniform circular motion, relative to the center of the circle, is given by the position vector $\boldsymbol{r}$. Remember that a position vector is a displacement vector with its tail at the origin. Two position vectors, $\boldsymbol{r}_1$ and $\boldsymbol{r}_2$, at the beginning and end of a time interval are shown on the left in **Figure 7.**

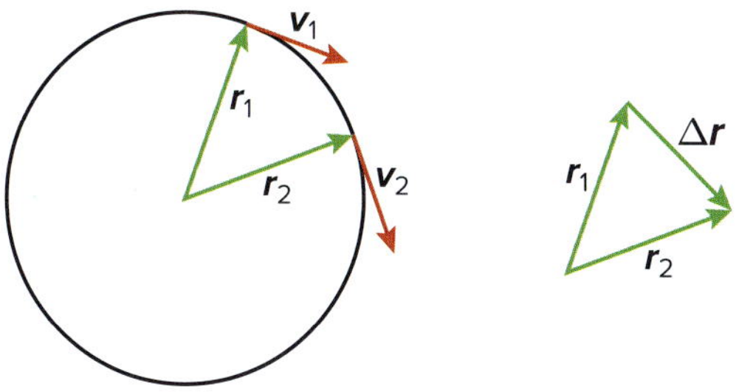

Figure 7 For an object in uniform circular motion, the velocity is tangent to the circle. It is in the same direction as the displacement.

Analyze *How can you tell from the diagram that the motion is uniform?*

3D THINKING | DCI Disciplinary Core Ideas | CCC Crosscutting Concepts | SEP Science & Engineering Practices

COLLECT EVIDENCE

Use your Science Journal to record the evidence you collect as you complete the readings and activities in this lesson.

INVESTIGATE

GO ONLINE to find these activities and more resources.

PhysicsLAB: Centripetal Force

Analyze data about the relationship between centripetal force and centripetal acceleration.

Review the News

Obtain information from a current news story about circular motion. Evaluate your source and communicate your findings to your class.

As the object moves around the circle, the length of the position vector does not change, but the direction does. The diagram also shows two instantaneous velocity vectors. Notice that each velocity vector is tangent to the circular path, at a right angle to the corresponding position vector. To determine the object's velocity, you first need to find its displacement vector over a time interval. You know that a moving object's average velocity is defined as $\frac{\Delta x}{\Delta t}$, so for an object in circular motion, $\bar{v} = \frac{\Delta \boldsymbol{r}}{\Delta t}$. The right side of **Figure 7** shows $\Delta \boldsymbol{r}$ drawn as the displacement from $\boldsymbol{r}_1$ to $\boldsymbol{r}_2$ during a time interval. The velocity for this time interval has the same direction as the displacement, but its length would be different because it is divided by Δt.

Velocity Vectors

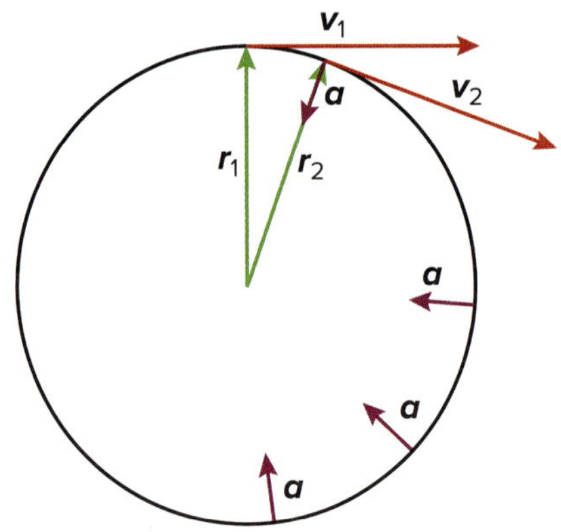

Centripetal Acceleration

A velocity vector of an object in uniform circular motion is tangent to the circle. What is the direction of the acceleration? **Figure 8** shows the velocity vectors v_1 and v_2 at the beginning and end of a time interval. The difference in the two vectors (Δv) is found by subtracting the vectors, as shown at the bottom of the figure. The average acceleration $\left(\bar{a} = \frac{\Delta v}{\Delta t}\right)$ for this time interval is in the same direction as Δv. For a very small time interval, Δv is so small that $\boldsymbol{a}$ points toward the center of the circle. As the object moves around the circle, the direction of the acceleration vector changes, but it always points toward the center of the circle. For this reason, the acceleration of an object in uniform circular motion is called center-seeking or **centripetal acceleration.**

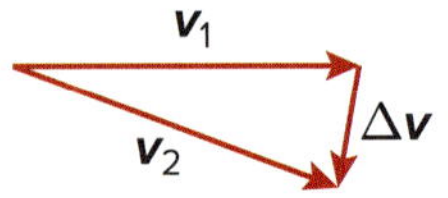

Figure 8 The acceleration of an object in uniform circular motion is the change in velocity divided by the time interval. The direction of centripetal acceleration is always toward the center of the circle.

Magnitude of acceleration

What is the magnitude of an object's centripetal acceleration? Look at the starting points of the velocity vectors in the top of **Figure 8.** Notice the triangle the position vectors at those points make with the center of the circle. An identical triangle is formed by the velocity vectors in the bottom of **Figure 8.** The angle between $\boldsymbol{r}_1$ and $\boldsymbol{r}_2$ is the same as that between v_1 and v_2. Therefore, similar triangles are formed by subtracting the two sets of vectors, and the ratios of the lengths of two corresponding sides are equal.

Thus, $\frac{\Delta r}{r} = \frac{\Delta v}{v}$. The equation is not changed if both sides are divided by Δt.

$$\frac{\Delta r}{r\Delta t} = \frac{\Delta v}{v\Delta t}$$

But, $v = \frac{\Delta r}{\Delta t}$ and $a = \frac{\Delta v}{\Delta t}$.

$$\left(\frac{1}{r}\right)\left(\frac{\Delta r}{\Delta t}\right) = \left(\frac{1}{v}\right)\left(\frac{\Delta v}{\Delta t}\right)$$

Substituting $v = \frac{\Delta r}{\Delta t}$ in the left-hand side and $a = \frac{\Delta v}{\Delta t}$ in the right-hand side gives the following equation:

$$\frac{v}{r} = \frac{a}{v}$$

Solve for acceleration, and use the symbol a_c for centripetal acceleration.

Centripetal Acceleration

Centripetal acceleration always points to the center of the circle. Its magnitude is equal to the square of the speed divided by the radius of motion.

$$a_c = \frac{v^2}{r}$$

Period of revolution One way to describe the speed of an object moving in a circle is to measure its period (T), the time needed for the object to make one complete revolution. During this time, the object travels a distance equal to the circumference of the circle ($2\pi r$).

The speed, then, is represented by $v = \frac{2\pi r}{T}$. If you substitute for v in the equation for centripetal acceleration, you obtain the following equation:

$$a_c = \frac{(2\pi r/T)^2}{r} = \frac{4\pi^2 r}{T^2}.$$

Centripetal force Because the acceleration of an object moving in a circle is always in the direction of the net force acting on it, there must be a net force toward the center of the circle. This force can be provided by any number of agents. For Earth circling the Sun, the force is the Sun's gravitational force on Earth. When a hammer thrower swings the hammer, as in **Figure 9**, the force is the tension in the chain attached to the massive ball. When an object moves in a circle, the net force toward the center of the circle is called the **centripetal force.** To accurately analyze centripetal acceleration situations, you must identify the agent of the force that causes the acceleration. Then you can apply Newton's second law for the component in the direction of the acceleration in the following way.

Newton's Second Law for Circular Motion

The net centripetal force on an object moving in a circle is equal to the object's mass times the centripetal acceleration.

$$F_{net} = ma_c$$

Figure 9 As the hammer thrower swings the ball around, tension in the chain is the force that causes the ball to have an inward acceleration.

Predict *Neglecting air resistance, how would the horizontal acceleration and velocity of the hammer change if the thrower released the chain?*

Direction of acceleration When solving problems, you have found it useful to choose a coordinate system with one axis in the direction of the acceleration. For circular motion, the direction of the acceleration is always toward the center of the circle. Rather than labeling this axis x or y, call it c, for centripetal acceleration. The other axis is in the direction of the velocity, tangent to the circle. It is labeled *tang* for tangential. You will apply Newton's second law in these directions, just as you did in the two-dimensional problems you have solved before. Remember that centripetal force is just another name for the net force in the centripetal direction. It is the sum of all the real forces, those for which you can identify agents that act along the centripetal axis.

In the case of the hammer thrower in **Figure 9**, in what direction does the hammer fly when the chain is released? Once the contact force of the chain is gone, there is no force accelerating the hammer toward the center of the circle, so the hammer flies off in the direction of its velocity, which is tangent to the circle. Remember, if you cannot identify the agent of a force, then it does not exist.

CCC CROSSCUTTING CONCEPTS

Systems and System Models Provide evidence to help overcome the misconception of centrifugal "force" that is held by many people by developing a model to demonstrate what occurs as a car makes a turn.

EXAMPLE Problem 3

UNIFORM CIRCULAR MOTION A 13-g rubber stopper is attached to a 0.93-m string. The stopper is swung in a horizontal circle, making one revolution in 1.18 s. Find the magnitude of the tension force exerted by the string on the stopper.

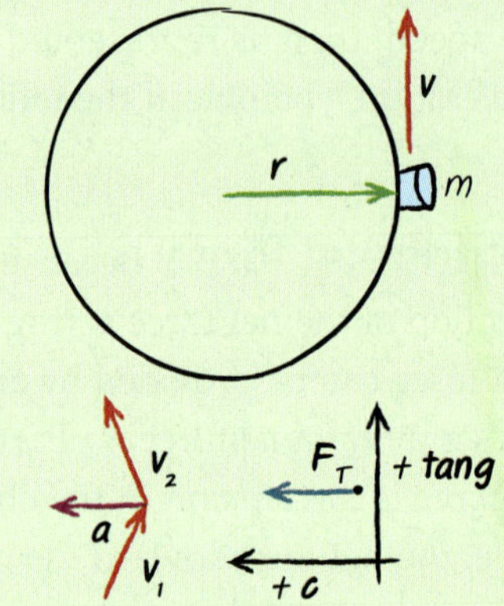

1 ANALYZE AND SKETCH THE PROBLEM

- Draw a free-body diagram for the swinging stopper.
- Include the radius and the direction of motion.
- Establish a coordinate system labeled *tang* and *c*. The directions of a_c and F_T are parallel to *c*.

Known	Unknowns
$m = 13$ g	$F_T = ?$
$r = 0.93$ m	
$T = 1.18$ s	

2 SOLVE FOR THE UNKNOWN

Find the magnitude of the centripetal acceleration.

$$a_c = \frac{4\pi^2 r}{T^2}$$

$$= \frac{4\pi^2(0.93\text{ m})}{(1.18\text{ s})^2}$$ Substitute $r = 0.93$ m, $T = 1.18$ s.

$$= 26\text{ m/s}^2$$

Use Newton's second law to find the magnitude of the tension in the string.

$$F_T = ma_c$$

$$= (0.013\text{ kg})(26\text{ m/s}^2)$$ Substitute $m = 0.013$ kg, $a_c = 26$ m/s^2.

$$= 0.34\text{ N}$$

3 EVALUATE THE ANSWER

- **Are the units correct?** Dimensional analysis verifies that a_c is in meters per second squared and F_T is in newtons.
- **Do the signs make sense?** The signs should all be positive.
- **Are the magnitudes realistic?** The force is almost three times the weight of the stopper, and the acceleration is almost three times that of gravity, which is reasonable for such a light object.

PRACTICE Problems

ADDITIONAL PRACTICE

12. A runner moving at a speed of 8.8 m/s rounds a bend with a radius of 25 m. What is the centripetal acceleration of the runner, and what agent exerts the centripetal force on the runner?

13. An airplane traveling at 201 m/s makes a turn. What is the smallest radius of the circular path (in kilometers) the pilot can make and keep the centripetal acceleration under 5.0 m/s^2?

14. A 45-kg merry-go-round worker stands on the ride's platform 6.3 m from the center, as shown in **Figure 10.** If her speed (v_{worker}) as she goes around the circle is 4.1 m/s, what is the force of friction (F_f) necessary to keep her from falling off the platform?

15. A 16-g ball at the end of a 1.4-m string is swung in a horizontal circle. It revolves once every 1.09 s. What is the magnitude of the string's tension?

16. CHALLENGE A car racing on a flat track travels at 22 m/s around a curve with a 56-m radius. Find the car's centripetal acceleration. What minimum coefficient of static friction between the tires and the road is necessary for the car to round the curve without slipping?

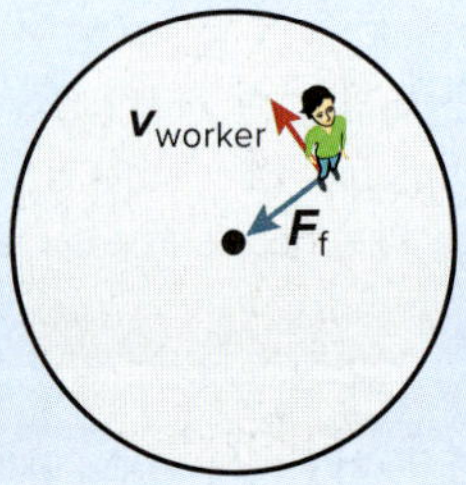

Figure 10

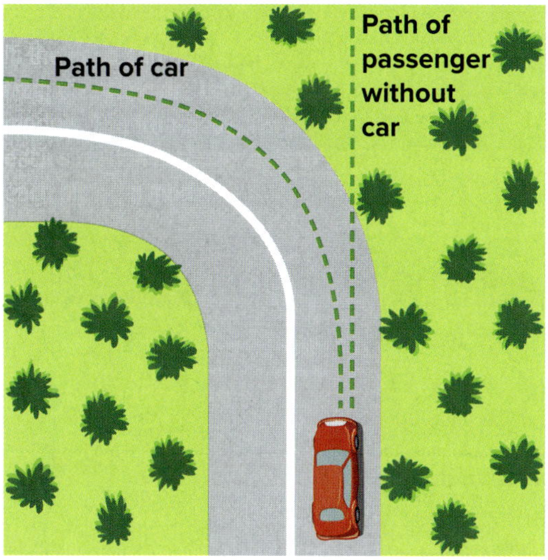

Figure 11 When a car moves around a curve, a rider feels a fictitious centrifugal force directed outward. In fact, the force on the person is the centripetal force, which is directed toward the center of the circle and is exerted by the seat on which the person is sitting.

Centrifugal "Force"

If a car you are in stops suddenly, you are thrown forward into your safety belt. Is there a forward force on you? No, because according to Newton's first law, you will continue moving with the same velocity unless there is a net force acting on you. The safety belt applies that force.

Figure 11 shows a car turning left as viewed from above. A passenger in the car would continue to move straight ahead if it were not for the force of the door acting in the direction of the acceleration. As the car goes around the curve, the car and the passenger are in circular motion, and the passenger experiences centripetal acceleration. Recall that centripetal acceleration is always directed toward the center of the circle. There is no outward force on the passenger. You feel as if you are being pushed only because you are accelerating relative to your surroundings. The so-called centrifugal, or outward, force is a fictitious, nonexistent force. There is no real force because there is no agent exerting a force.

Check Your Progress

17. **Circular Motion** A ball on a rope swung at a constant speed in a circle above your head is in uniform circular motion. In which direction does it accelerate? What force causes this?
18. **Circular Motion** What is the direction of the force that acts on clothes spinning in a top-load washing machine? What exerts the force?
19. **Centripetal Acceleration** A newspaper article states that when turning a corner, a driver must be careful to balance the centripetal and centrifugal forces to keep from skidding. Write a letter to the editor that describes physics errors in this article.
20. **Free-Body Diagram** You are sitting in the back seat of a car going around a curve to the right. Sketch motion and free-body diagrams to answer these questions:
    a. What is the direction of your acceleration?
    b. What is the direction of the net force on you?
    c. What exerts this force?
21. **Centripetal Acceleration** An object swings in a horizontal circle, supported by a 1.8-m string. It completes a revolution in 2.2 s. What is the object's centripetal acceleration?
22. **Centripetal Force** The 40.0-g stone in **Figure 12** is whirled horizontally at a speed of 2.2 m/s. What is the tension in the string?

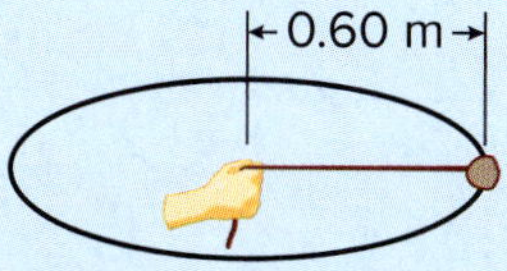

Figure 12

23. **Amusement-Park Ride** A ride park has people stand around a 4.0-m radius circle with their backs to a wall. The ride then spins them with a 1.7-s period of revolution. What are the centripetal acceleration and velocity of the riders?
24. **Critical Thinking** You are always in uniform circular motion due to Earth's rotation. What agent supplies the force? If you are on a scale, how does the motion affect your weight?

LEARNSMART® Go online to follow your personalized learning path to review, practice, and reinforce your understanding.

LESSON 3
RELATIVE VELOCITY

FOCUS QUESTION

Does your description of motion depend on your frame of reference?

Relative Motion in One Dimension

Suppose you are in a school bus that is traveling at a velocity of 8 m/s in a positive direction. You walk with a velocity of 1 m/s toward the front of the bus. If a friend is standing on the side of the road watching the bus go by, how fast would your friend say you are moving? If the bus is traveling at 8 m/s, its speed as measured by your friend in a coordinate system fixed to the road is 8 m/s. When you are standing still on the bus, your speed relative to the road is also 8 m/s, but your speed relative to the bus is zero. How can your speed be different?

Different reference frames In this example, your motion is viewed from different coordinate systems. A coordinate system from which motion is viewed is a **reference frame.** Walking at 1 m/s toward the front of the bus means your velocity is measured in the reference frame of the bus. Your velocity in the road's reference frame is different. You can rephrase the problem as follows: given the velocity of the bus relative to the road and your velocity relative to the bus, what is your velocity relative to the road? A vector representation of this problem is shown in **Figure 13.** If right is positive, your speed relative to the road is 9 m/s, the sum of 8 m/s and 1 m/s.

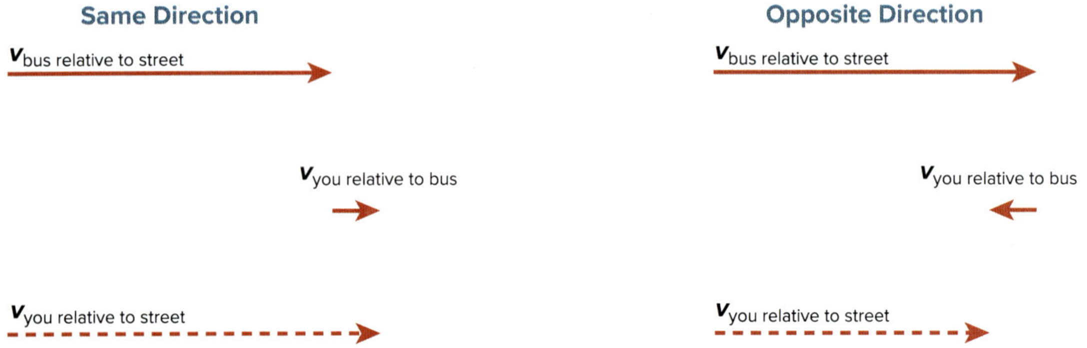

Figure 13 When an object moves in a moving reference frame, you add the velocities if they are in the same direction. You subtract one velocity from the other if they are in opposite directions.

Recall *What do the lengths of the velocity vectors indicate?*

3D THINKING DCI Disciplinary Core Ideas CCC Crosscutting Concepts SEP Science & Engineering Practices

COLLECT EVIDENCE

Use your Science Journal to record the evidence you collect as you complete the readings and activities in this lesson.

INVESTIGATE

GO ONLINE to find these activities and more resources.

PhysicsLAB: Moving Reference Frame
Carry out an investigation to examine the motion of a system in different reference frames.

CCC **Identify Crosscutting Concepts**
Create a table of the crosscutting concepts and fill in examples you find as you read.

Suppose that you now walk at the same speed toward the rear of the bus. What would be your velocity relative to the road? **Figure 13** shows that because the two velocities are in opposite directions, the resultant speed is 7 m/s, the difference between 8 m/s and 1 m/s. You can see that when the velocities are along the same line, simple addition or subtraction can be used to determine the relative velocity.

Get It?

Identify When would you use subtraction to determine the relative velocity?

PHYSICS Challenge

TENSION IN A ROPE Phillipe whirls a stone of mass *m* on a rope in a horizontal circle above his head such that the stone is at a height *h* above the ground. The circle has a radius of *r*, and the magnitude of the tension in the rope is *T*. Suddenly the rope breaks, and the stone falls to the ground. The stone travels a horizontal distance *x* from the time the rope breaks until it impacts the ground. Find a mathematical expression for *x* in terms of *T*, *r*, *m*, and *h*. Does your expression change if Phillipe is walking 0.50 m/s relative to the ground?

Combining velocity vectors Take a closer look at how the relative velocities in **Figure 13** were obtained. Can you find a mathematical rule to describe how velocities are combined when the motion is in a moving reference frame? For the situation in which you are walking in a bus, you can designate the velocity of the bus relative to the road as $\boldsymbol{v}_{b/r}$. You can designate your velocity relative to the bus as $\boldsymbol{v}_{y/b}$ and the velocity of you relative to the road as $\boldsymbol{v}_{y/r}$. To find the velocity of you relative to the road in both cases, you added the velocity vectors of you relative to the bus and the bus relative to the road. Mathematically, this is represented as $\boldsymbol{v}_{y/b} + \boldsymbol{v}_{b/r} = \boldsymbol{v}_{y/r}$. The more general form of this equation is as follows.

Relative Velocity

The relative velocity of object a to object c is the vector sum of object a's velocity relative to object b and object b's velocity relative to object c.

$$\boldsymbol{v}_{a/b} + \boldsymbol{v}_{b/c} = \boldsymbol{v}_{a/c}$$

Get It?

Explain A flea is running on the back of a dog that is running along a road. Write an equation to determine the velocity of the flea relative to the road.

STEM CAREER Connection

Commercial Pilot

A person wanting to be a commercial pilot must have an understanding of relative motion to get the plane or helicopter to its intended destination. Commercial pilots who drop fire-retardant gel to combat forest fires must also understand projectile motion to hit their targets.

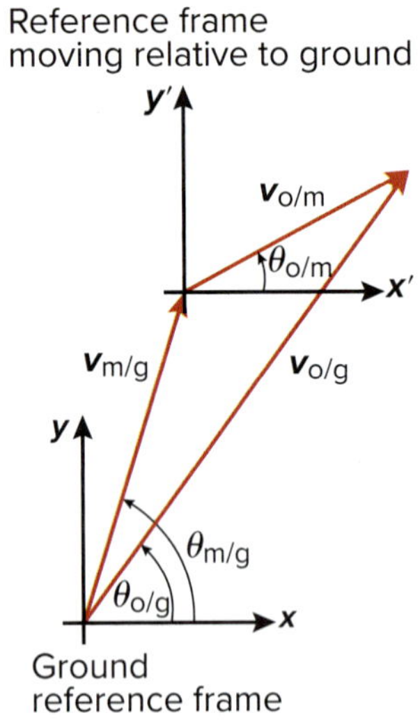

Figure 14 Vectors are placed tip-to-tail to find the relative velocity vector for two-dimensional motion. The subscript o/g refers to an object relative to ground, o/m refers to an object relative to a moving reference frame, and m/g refers to the moving frame relative to ground.

Analyze *How would the resultant vector change if the ground reference frame were considered to be the moving reference frame?*

Relative Motion in Two Dimensions

Adding relative velocities also applies to motion in two dimensions. As with one-dimensional motion, you first draw a vector diagram to describe the motion, and then you solve the problem mathematically.

Vector diagrams The method of drawing vector diagrams for relative motion in two dimensions is shown in **Figure 14.** The velocity vectors are drawn tip-to-tail. The reference frame from which you are viewing the motion, often called the ground reference frame, is considered to be at rest. One vector describes the velocity of the second reference frame relative to ground. The second vector describes the motion in that moving reference frame. The resultant shows the relative velocity, which is the velocity relative to the ground reference frame.

Get It?

Decide Can a moving car be a reference frame?

An example is the relative motion of an airplane. Airline pilots cannot expect to reach their destinations by simply aiming their planes along a compass direction. They must take into account the plane's speed relative to the air, which is given by their airspeed indicators, and their direction of flight relative to the air. They also must consider the velocity of the wind at the altitude they are flying relative to the ground. These two vectors must be combined to obtain the velocity of the airplane relative to the ground. The resultant vector tells the pilot how fast and in what direction the plane must travel relative to the ground to reach its destination. A similar situation occurs for boats that are traveling on water that is flowing.

Get It?

Describe How could a boat on a flowing river stay in one location relative to the shore?

Vector Resolution

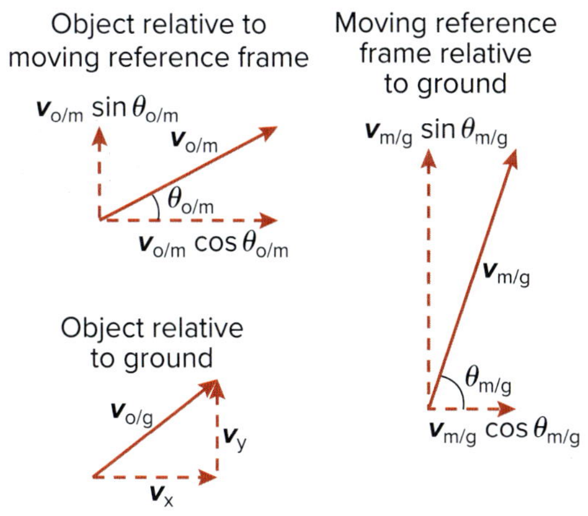

Equation

$$v_{o/g}{}^2 = v_x{}^2 + v_y{}^2$$

where $v_x = v_{o/m} \cos\theta_{o/m} + v_{m/g} \cos\theta_{m/g}$

$$v_y = v_{o/m} \sin\theta_{o/m} + v_{m/g} \sin\theta_{m/g}$$

Figure 15 To find the velocity of an object in a moving reference frame, resolve the vectors into *x*- and *y*-components.

Combining velocities

You can use the equations in **Figure 15** to solve problems for relative motion in two dimensions. The velocity of a reference frame moving relative to the ground is labeled $v_{m/g}$. The velocity of an object in the moving frame is labeled $v_{o/m}$. The relative velocity equation gives the object's velocity relative to the ground: $v_{o/g} = v_{o/m} + v_{m/g}$.

To determine the magnitude of the object's velocity relative to the ground ($v_{o/g}$), first resolve the velocity vectors of the object and the moving reference frame into *x*- and *y*-components. Then apply the Pythagorean theorem. The general equation is shown in **Figure 15**, but for many problems the equation is simpler because the vectors are along an axis. As shown in the example problem below, you can find the angle of the object's velocity relative to the ground by observing the vector diagram and applying a trigonometric relationship.

Get It?

Explain How are vectors used to describe relative motion in two dimensions?

EXAMPLE Problem 4

RELATIVE VELOCITY OF A MARBLE Ana and Sandra are riding on a ferry boat traveling east at 4.0 m/s. Sandra rolls a marble with a velocity of 0.75 m/s north, straight across the deck of the boat to Ana. What is the velocity of the marble relative to the water?

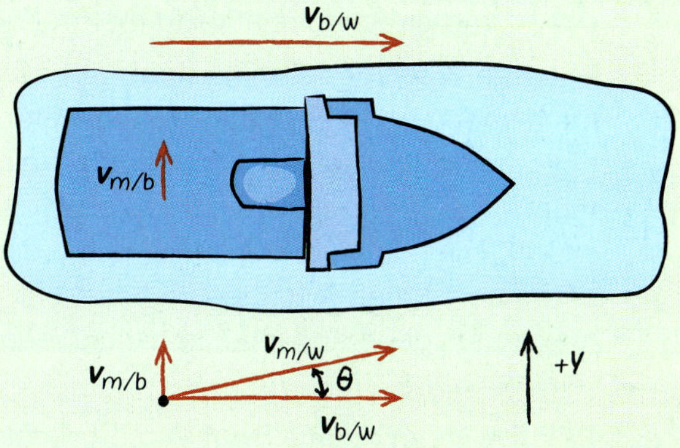

1 ANALYZE AND SKETCH THE PROBLEM

Establish a coordinate system. Draw vectors for the velocities.

Known		Unknown
$v_{b/w} = 4.0$ m/s	$v_{m/b} = 0.75$ m/s	$v_{m/w} = ?$

2 SOLVE FOR THE UNKNOWN

The velocities are perpendicular, so we can use the Pythagorean theorem.

$$v_{m/w}{}^2 = v_{b/w}{}^2 + v_{m/b}{}^2$$

$$v_{m/w} = \sqrt{v_{b/w}{}^2 + v_{m/b}{}^2}$$

$$= \sqrt{(4.0\text{ m/s})^2 + (0.75\text{ m/s})^2} = 4.1\text{ m/s}$$

Substitute $v_{b/w} = 4.0$ m/s, $v_{m/b} = 0.75$ m/s.

Find the angle of the marble's velocity.

$$\theta = \tan^{-1}\left(\frac{v_{m/b}}{v_{b/w}}\right)$$

$$= \tan^{-1}\left(\frac{0.75\text{ m/s}}{4.0\text{ m/s}}\right) = 11° \text{ north of east}$$

The marble travels 4.1 m/s at 11° north of east.

3 EVALUATE THE ANSWER

- **Are the units correct?** Dimensional analysis verifies units of meters per second for velocity.
- **Do the signs make sense?** The signs should all be positive.
- **Are the magnitudes realistic?** The resulting velocity is of the same order of magnitude as the velocities given in the problem and slightly larger than the larger of the two.

PRACTICE Problems ADDITIONAL PRACTICE

25. You are riding in a bus moving slowly through heavy traffic at 2.0 m/s. You hurry to the front of the bus at 4.0 m/s relative to the bus. What is your speed relative to the street?

26. Rafi is pulling a toy wagon through a neighborhood at a speed of 0.75 m/s. A caterpillar in the wagon is crawling toward the rear of the wagon at 2.0 cm/s. What is the caterpillar's velocity relative to the ground?

27. A boat moves directly upriver at 2.5 m/s relative to the water. Viewers on the shore see that the boat is moving at only 0.5 m/s relative to the shore. What is the speed of the river? Is it moving with or against the boat?

28. A boat is traveling east at a speed of 3.8 m/s. A person walks across the boat with a velocity of 1.3 m/s south.

a. What is the person's speed relative to the water?

b. In what direction, relative to the ground, does the person walk?

29. An airplane flies due north at 150 km/h relative to the air. There is a wind blowing at 75 km/h to the east relative to the ground. What is the plane's speed relative to the ground?

30. CHALLENGE The airplane in **Figure 16** flies at 200.0 km/h relative to the air. What is the velocity of the plane relative to the ground if it flies during the following wind conditions?

a. a 50.0-km/h tailwind

b. a 50.0-km/h headwind

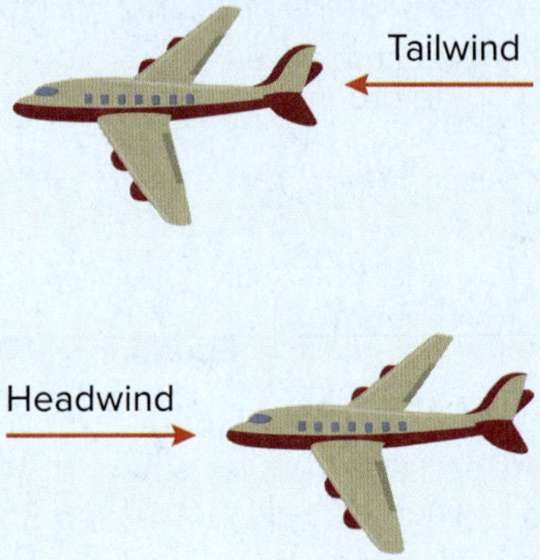

Figure 16

Check Your Progress

31. **Relative Velocity** A plane travels 285 km/h west relative to the air. A wind blows 25 km/h east relative to the ground. Find the plane's speed and direction relative to the ground.

32. **Relative Velocity** A fishing boat with a maximum speed of 3 m/s relative to the water is in a river that is flowing at 2 m/s. What is the maximum speed the boat can obtain relative to the shore? The minimum speed? Give the direction of the boat, relative to the river's current, for the maximum speed and the minimum speed relative to the shore.

33. **Relative Velocity of a Boat** A motorboat heads due west at 13 m/s relative to a river that flows due north at 5.0 m/s. What is the velocity (both magnitude and direction) of the motorboat relative to the shore?

34. **Boating** Martin is riding on a ferry boat that is traveling east at 3.8 m/s. He walks north across the deck of the boat at 0.62 m/s. What is Martin's velocity relative to the water?

35. **Relative Velocity** An airplane flies due south at 175 km/h relative to the air. There is a wind blowing at 85 km/h to the east relative to the ground. What are the plane's speed and direction relative to the ground?

36. **A Plane's Relative Velocity** An airplane flies due north at 235 km/h relative to the air. There is a wind blowing at 65 km/h to the northeast relative to the ground. What are the plane's speed and direction relative to the ground?

37. **Critical Thinking** You are piloting a boat across a fast-moving river. You want to reach a pier directly opposite your starting point. Describe how you would navigate the boat in terms of the components of your velocity relative to the water.

Fighting Fire with Forces

Smokejumpers are professional wildland firefighters who parachute into remote areas to control and extinguish wildfires. Their job is incredibly dangerous. They jump out of planes up to 450 meters above the ground into blazing fires. How do they land safely? Parachute jumping from a plane requires extensive training and an understanding of two-dimensional parabolic motion.

A smokejumper's leap from a plane is a classic example of the parabolic path of a projectile moving in two dimensions.

Smokejumpers and Parabolic Trajectory

Firefighters accepted into the smokejumper program must have years of previous experience fighting fires. Even then, applicants must go through a rigorous training program before they are ready to work as smokejumpers. They learn how to properly exit an airplane, deploy their parachute, and land safely on the ground. Smokejumpers also receive first aid training and practice climbing trees and digging fire lines.

When smokejumpers skydive out of airplanes to parachute down to the site of a fire, they are moving in two dimensions at the same time—along the x and y axes. They are subject to both the initial force that launches them and the force of gravity. Gravity causes their path to curve downward. Smokejumpers must plan their parabolic trajectories so that they land in a safe place.

A spotter flies with every smokejumper mission. Spotters take wind speed and direction into account as they help plan the trajectories of both the smokejumpers and packages of supplies. The packages—which contain food, water, first aid kit, and fire-fighting equipment—ensure that smokejumpers can work 48 hours or more without other outside assistance. Spotters also relay information about how and where the fire is moving.

Work on the Ground and in the Air

Smokejumpers on the ground build firebreaks and use water and fire retardant to contain fires. They build backfires, which burn in the direction of the existing fire, to eliminate possible fuel. Smokejumpers on planes dump water and fire retardant from above. Smokejumpers protect vast expanses of wildlands from destruction every year.

USE MATHEMATICAL REPRESENTATIONS

Write a problem about the parabolic trajectory of a smokejumper. Swap problems with a classmate. Solve each other's problem.

CALIFORNIA CONNECTION

Use an online search engine to find information on the role smokejumpers played in fighting the 2017 wildfires in California and other western states. Write a short paper on the results of your research and present it to the class.

MODULE 6
STUDY GUIDE

GO ONLINE to study with your Science Notebook.

Lesson 1 PROJECTILE MOTION

- projectile
- trajectory

- The vertical and horizontal motions of a projectile are independent. When there is no air resistance, the horizontal motion component does not experience an acceleration and has constant velocity; the vertical motion component of a projectile experiences a constant acceleration under these same conditions.
- The curved flight path a projectile follows is called a trajectory and is a parabola. The height, time of flight, initial velocity, and horizontal distance of this path are related by the equations of motion. The horizontal distance a projectile travels before returning to its initial height depends on the acceleration due to gravity and on both components of the initial velocity.

Lesson 2 CIRCULAR MOTION

- uniform circular motion
- centripetal acceleration
- centripetal force

- An object moving in a circle at a constant speed has an acceleration toward the center of the circle because the direction of its velocity is constantly changing.
- Acceleration toward the center of the circle is called centripetal acceleration. It depends directly on the square of the object's speed and inversely on the radius of the circle.

$$a_c = \frac{v^2}{r}$$

- A net force must be exerted by external agents toward the circle's center to cause centripetal acceleration.

$$\boldsymbol{F}_{net} = m\boldsymbol{a}_c$$

Lesson 3 RELATIVE VELOCITY

- reference frame

- A coordinate system from which you view motion is called a reference frame. Relative velocity is the velocity of an object observed in a different, moving reference frame.
- You can use vector addition to solve motion problems of an object in a moving reference frame.

THREE-DIMENSIONAL THINKING

Module Wrap-Up

REVISIT THE PHENOMENON

Why do thrown basketballs travel in arcs?

CER Claim, Evidence, Reasoning

Explain Your Reasoning Revisit the claim you made when you encountered the phenomenon. Summarize the evidence you gathered from your investigations and research and finalize your Summary Table. Does your evidence support your claim? If not, revise your claim. Explain why your evidence supports your claim.

STEM UNIT PROJECT

Now that you've completed the module, revisit your STEM unit project. You will summarize your evidence and apply it to the project.

GO FURTHER

SEP Data Analysis Lab

Does the baseball clear the wall?

A baseball player hits a belt-high (1.0 m) fastball down the left-field line. The player hits the ball with an initial velocity of 42.0 m/s at an angle of 26° above the horizontal. The left-field wall is 96.0 m from home plate at the foul pole and is 14 m high.

CER Analyze and Interpret Data

1. Write the equation for the height of the ball (y) as a function of its distance from home plate (x).
2. Use a computer or a graphing calculator to plot the path of the ball. Trace along the path to find how high above the ground the ball is when it is at the wall.
3. **Claim** Is the hit a home run?
4. **Evidence and Reasoning** Justify your claim.

NASA/Johns Hopkins University Applied Physics Laboratory/

GRAVITATION

ENCOUNTER THE PHENOMENON

How can gravity keep moons orbiting planets but also cause things to fall?

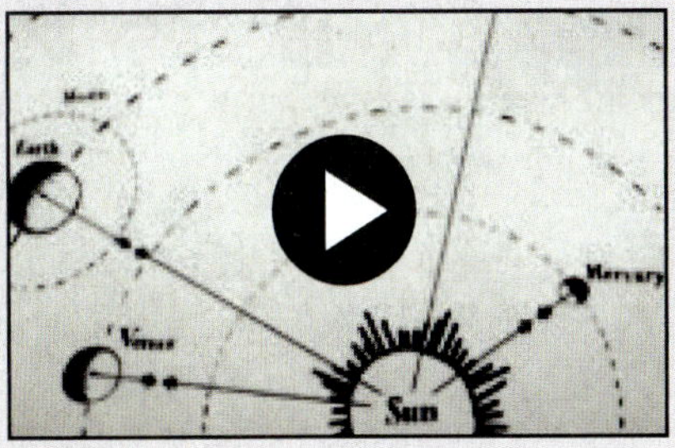

GO ONLINE to watch a video about Newton's work on gravity.

SEP Ask Questions

Do you have other questions about the phenomenon? If so, add them to the driving question board.

CER Claim, Evidence, Reasoning

Make Your Claim Use your CER chart to make a claim about how gravity both keeps moons in orbit and causes things to fall. Explain your reasoning.

Collect Evidence Use the lessons in this module to collect evidence to support your claim. Record your evidence as you move through the module.

Explain Your Reasoning You will revisit your claim and explain your reasoning at the end of the module.

GO ONLINE to access your CER chart and explore resources that can help you collect evidence.

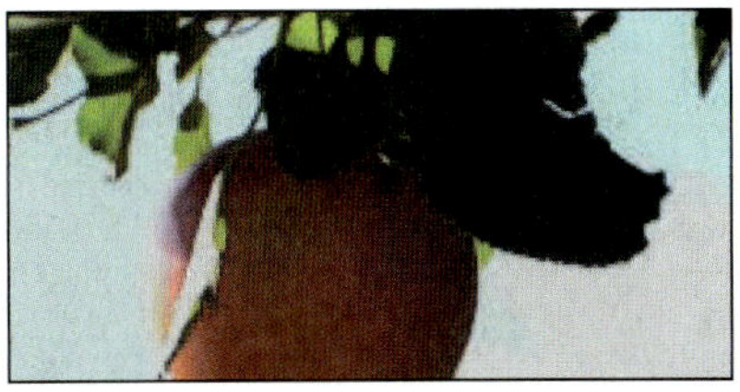

LESSON 1: Explore & Explain: Universal Gravitation

LESSON 2: Explore & Explain: Gravity and Free-Fall Acceleration

Smithsonian

Additional Resources

LESSON 1

PLANETARY MOTION AND GRAVITATION

FOCUS QUESTION

What role does gravity play in planetary motion?

Early Observations

In ancient times, the Sun, the Moon, the planets, and the stars were assumed to revolve around Earth. Nicholas Copernicus, a Polish astronomer, noticed that the movements of planets did not fully agree with the Earth-centered model. The results of his many years of work were published in 1543, when Copernicus was on his deathbed. His book showed that the motion of planets is much more easily understood by assuming that Earth and other planets revolve around the Sun. His model helped explain phenomena such as the inner planets Mercury and Venus always appearing near the Sun. Copernicus's view advanced our understanding of planetary motion.

Tycho Brahe was born a few years after Copernicus died. Tycho studied astronomy using instruments like those shown in **Figure 1.**

Figure 1 Instruments such as those shown on the left were used by Tycho Brahe to measure the positions of planets.

DeAgostini/Getty Images

3D THINKING DCI Disciplinary Core Ideas CCC Crosscutting Concepts SEP Science & Engineering Practices

COLLECT EVIDENCE

Use your Science Journal to record the evidence you collect as you complete the readings and activities in this lesson.

INVESTIGATE

GO ONLINE to find these activities and more resources.

PhysicsLAB: Modeling Orbits

Devlop a model of the orbits of various objects in the solar system to support Kepler's laws.

Virtual Investigation: Universal Gravitation

Carry out an investigation of the relationship between variables in Newton's law of universal gravitation.

Tycho Brahe realized that the charts of the time did not accurately predict astronomical events. He recognized that measurements were required from one location over a long period of time. Tycho was granted an estate on the Danish island of Hven and the funding to build an early research institute. Telescopes had not been invented, so to make measurements, Tycho used huge instruments that he designed and built in his own shop, like those shown in **Figure 1.** Tycho is credited with the most accurate measurements of the time.

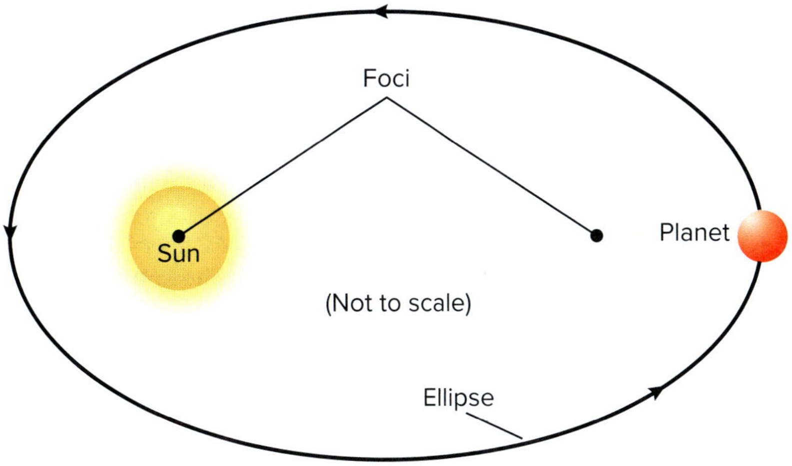

Figure 2 The orbit of each planet is an ellipse, with the Sun at one focus.

Kepler's Laws

In 1600 Tycho moved to Prague where Johannes Kepler, a 29-year-old German, became one of his assistants. Kepler analyzed Tycho's observations. After Tycho's death in 1601, Kepler continued to study Tycho's data and used geometry and mathematics to explain the motion of the planets. After seven years of careful analysis of Tycho's data on Mars, Kepler discovered the laws that describe the motion of every planet and satellite, natural or artificial. Here, the laws are presented in terms of planets.

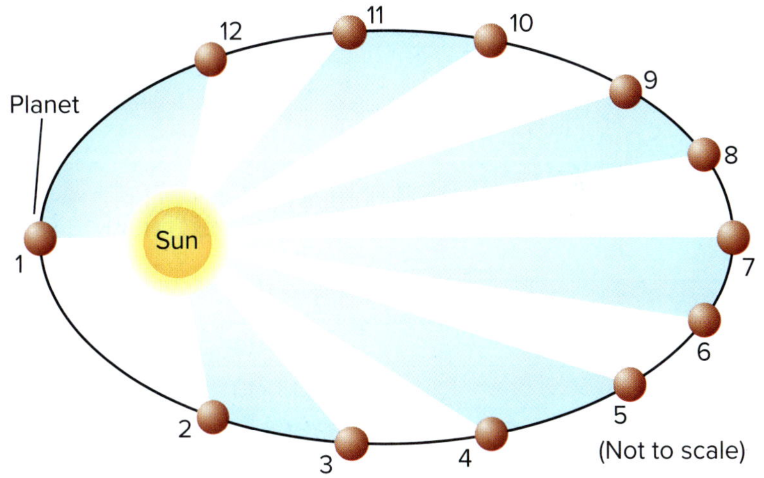

Figure 3 Kepler found that elliptical orbits sweep out equal areas in equal time periods.

Explain *why the equal time areas are shaped differently.*

Kepler's first law states that the paths of the planets are ellipses, with the Sun at one focus. An ellipse has two foci, as shown in **Figure 2**. Although exaggerated ellipses are used in the diagrams, Earth's actual orbit is very nearly circular. You would not be able to distinguish it from a circle visually.

Kepler found that orbits might change due to gravitational effects from, or collisions with, other objects in the solar system. He also found that the planets move faster when they are closer to the Sun and slower when they are farther away from the Sun. **Kepler's second law** states that an imaginary line from the Sun to a planet sweeps out equal areas in equal time intervals, as illustrated in **Figure 3.**

A period is the time it takes for one revolution of an orbiting body. Kepler also discovered a mathematical relationship between periods of planets and their mean distances away from the Sun.

Get It?

Describe the common feature that Kepler's first law found concerning the paths of orbiting objects around the Sun.

Kepler's third law states that the square of the ratio of the periods of any two planets revolving about the Sun is equal to the cube of the ratio of their average distances from the Sun. Thus, if the periods of the planets are T_A and T_B and their average distances from the Sun are r_A and r_B, Kepler's third law can be expressed as follows.

Kepler's Third Law

The square of the ratio of the period of planet A to the period of planet B is equal to the cube of the ratio of the distance between the centers of planet A and the Sun to the distance between the centers of planet B and the Sun.

$$\left(\frac{T_A}{T_B}\right)^2 = \left(\frac{r_A}{r_B}\right)^3$$

Note that Kepler's first two laws apply to each planet, moon, and satellite individually. The third law, however, relates the motion of two objects around a single body. For example, it can be used to compare the planets' distances from the Sun, shown in **Table 1,** to their periods around the Sun. It also can be used to compare distances and periods of the Moon and artificial satellites orbiting Earth.

Table 1 Solar System Data

Name	Average Radius (m)	Mass (kg)	Average Distance from the Sun (m)
Sun	6.96×10^{8}	1.99×10^{30}	—
Mercury	2.44×10^{6}	3.30×10^{23}	5.79×10^{10}
Venus	6.05×10^{6}	4.87×10^{24}	1.08×10^{11}
Earth	6.38×10^{6}	5.97×10^{24}	1.50×10^{11}
Mars	3.40×10^{6}	6.42×10^{23}	2.28×10^{11}
Jupiter	7.15×10^{7}	1.90×10^{27}	7.79×10^{11}
Saturn	6.03×10^{7}	5.68×10^{26}	1.43×10^{12}
Uranus	2.56×10^{7}	8.68×10^{25}	2.87×10^{12}
Neptune	2.48×10^{7}	1.02×10^{26}	4.50×10^{12}

Figure 4 Comet McNaught is a long-period comet, with a period of about 92,000 years. This photo was taken in 2007, when McNaught was highly visible.

Comet periods Comets are classified as long-period comets or short-period comets based on orbital periods. Long-period comets have orbital periods longer than 200 years, and short-period comets have orbital periods shorter than 200 years. Comet McNaught, also known as C/2006 P1, shown in **Figure 4,** with a period of about 92,000 years, is an example of a long-period comet. Comet Halley, with a period of 76 years, is an example of a short-period comet. Comets also obey Kepler's laws. Unlike planets, however, comets have highly elliptical orbits.

EXAMPLE Problem 1

CALLISTO'S DISTANCE FROM JUPITER Galileo measured the orbital radii of Jupiter's moons using the diameter of Jupiter as a unit of measure. He found that Io, the closest moon to Jupiter, has a period of 1.8 days and is 4.2 units from the center of Jupiter. Callisto, the fourth moon from Jupiter, has a period of 16.7 days. Using the same units that Galileo used, predict Callisto's distance from Jupiter.

1 ANALYZE AND SKETCH THE PROBLEM

- Sketch the orbits of Io and Callisto.
- Label the radii.

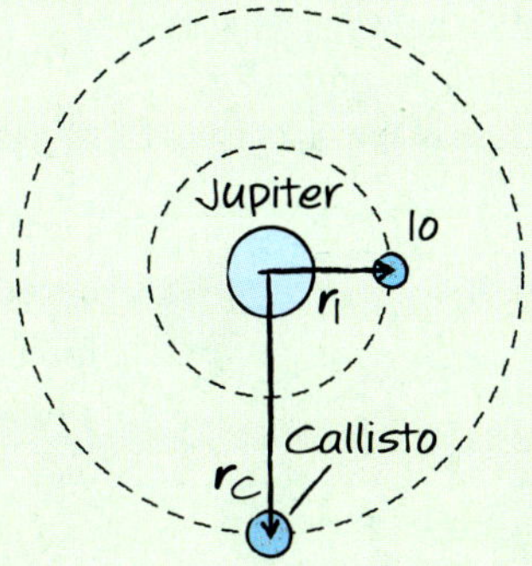

Known	Unknown
$T_C = 16.7$ days	$r_C = ?$
$T_I = 1.8$ days	
$r_I = 4.2$ units	

2 SOLVE FOR CALLISTO'S DISTANCE FROM JUPITER

Solve Kepler's third law for r_C.

$$\left(\frac{T_C}{T_I}\right)^2 = \left(\frac{r_C}{r_I}\right)^3$$

$$r_C^{\,3} = r_I^{\,3}\left(\frac{T_C}{T_I}\right)^2$$

$$r_C = \sqrt[3]{r_I^{\,3}\left(\frac{T_C}{T_I}\right)^2}$$

$$= \sqrt[3]{(4.2\text{ units})^3\left(\frac{16.7\text{ days}}{1.8\text{ days}}\right)^2}$$

Substitute $r_I = 4.2$ units, $T_C = 16.7$ days, $T_I = 1.8$ days.

$$= \sqrt[3]{6.4\times10^3\text{ units}^3}$$

$$= 19\text{ units}$$

3 EVALUATE THE ANSWER

- **Are the units correct?** r_C should be in Galileo's units, like r_I.
- **Is the magnitude realistic?** The period is larger, so the radius should be larger.

PRACTICE Problems

ADDITIONAL PRACTICE

1. If Ganymede, one of Jupiter's moons, has a period of 32 days, how many units is its orbital radius? Use the information given in Example Problem 1.
2. An asteroid revolves around the Sun with a mean orbital radius twice that of Earth's. Predict the period of the asteroid in Earth years.
3. Venus has a period of revolution of 225 Earth days. Find the distance between the Sun and Venus as a multiple of Earth's average distance from the Sun.
4. Uranus requires 84 years to circle the Sun. Find Uranus's average distance from the Sun as a multiple of Earth's average distance from the Sun.
5. From **Table 1** you can find that, on average, Mars is 1.52 times as far from the Sun as Earth is. Predict the time required for Mars to orbit the Sun in Earth days.
6. The Moon has a period of 27.3 days and a mean distance of 3.9×10^5 km from its center to the center of Earth.
   a. Use Kepler's laws to find the period of a satellite in orbit 6.70×10^3 km from the center of Earth.
   b. How far above Earth's surface is this satellite?
7. **CHALLENGE** Using the data in the previous problem for the period and radius of revolution of the Moon, predict what the mean distance from Earth's center would be for an artificial satellite that has a period of exactly 1.00 day.

Newton's Law of Universal Gravitation

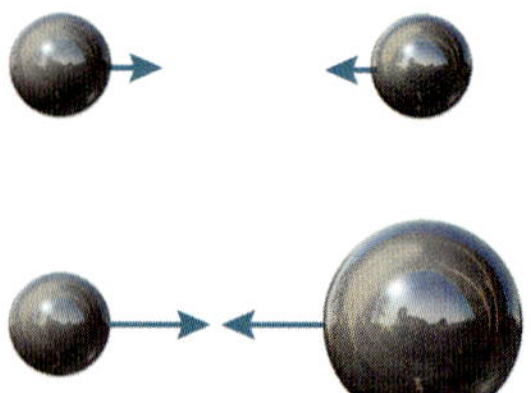

Gravitation between objects depends on the product of the masses of the objects.

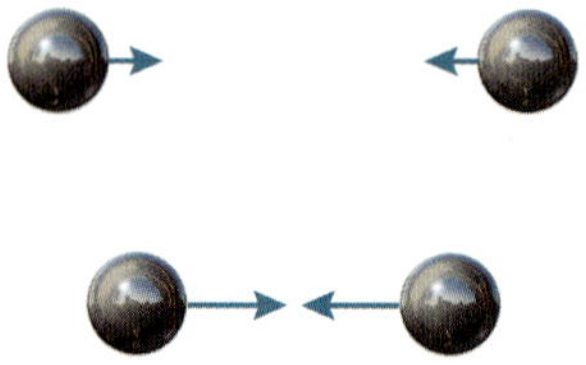

The gravitational force between objects is inversely proportional to the square of the distance between the objects.

Figure 5 Mass and distance affect the magnitude of the gravitational force between objects.

In 1666, Isaac Newton began his studies of planetary motion. It has been said that seeing an apple fall made Newton wonder if the force that caused the apple to fall might extend to the Moon, or even beyond. He found that the magnitude of the force (F_g) on a planet due to the Sun varies inversely with the square of the distance (r) between the centers of the planet and the Sun. That is, F_g is proportional to $\frac{1}{r^2}$. The force ($\mathbf{F}_g$) acts in the direction of the line connecting the centers of the two objects, as shown in **Figure 5.**

Newton found that both the apple's and the Moon's accelerations agree with the $\frac{1}{r^2}$ relationship. According to his own third law, the force Earth exerts on the apple is exactly the same as the force the apple exerts on Earth. Even though these forces are exactly the same, you can easily observe the effect of the force on the apple because it has much lower mass than Earth. The force of attraction between two objects must be proportional to the objects' masses and is known as the **gravitational force.**

Newton was confident that the same force of attraction would act between any two objects anywhere in the universe. He proposed the **law of universal gravitation,** which states that objects attract other objects with a force that is proportional to the product of their masses and inversely proportional to the square of the distance between them. Newton's law of universal gravitation, shown below, provides the mathematical models to describe and predict the effects of gravitational forces between distant objects.

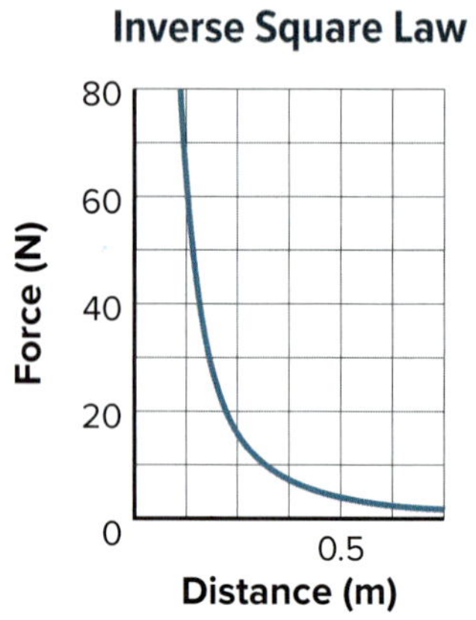

Figure 6 This is a graphical representation of the inverse square relationship.

Law of Universal Gravitation

The gravitational force is equal to the universal gravitational constant, times the mass of object 1, times the mass of object 2, divided by the distance between the centers of the objects, squared.

$$F_g = \frac{Gm_1m_2}{r^2}$$

According to Newton's equation, F is directly proportional to m_1 and m_2. If the mass of a planet near the Sun doubles, the force of attraction doubles. Use the Connecting Math to Physics feature on the next page to examine how changing one variable affects another. **Figure 6** illustrates the inverse square relationship graphically. The term G is the universal gravitational constant and will be discussed on the following pages.

CONNECTING MATH to Physics

Direct and Inverse Relationships Newton's law of universal gravitation has both direct and inverse relationships.

$F_g \propto m_1m_2$		$F_g \propto \frac{1}{r^2}$	
Change	Result	Change	Result
$(2m_1)m_2$	$2F_g$	$2r$	$\frac{1}{4}F_g$
$(3m_1)m_2$	$3F_g$	$3r$	$\frac{1}{9}F_g$
$(2m_1)(3m_2)$	$6F_g$	$\frac{1}{2}r$	$4F_g$
$\left(\frac{1}{2}\right)m_1m_2$	$\frac{1}{2}F_g$	$\frac{1}{3}r$	$9F_g$

Universal Gravitation and Kepler's Third Law

Newton stated the law of universal gravitation in terms that applied to the motion of planets about the Sun. This agreed with Kepler's third law and confirmed that Newton's law fit the best observations of the day.

Consider a planet orbiting the Sun, as shown in **Figure 7.** Newton's second law of motion, $F_{net} = ma$, can be written as $F_{net} = m_pa_c$, where F_{net} is the magnitude of the gravitational force, m_p is the mass of the planet, and a_c is the centripetal acceleration of the planet. For simplicity, assume circular orbits. Recall from your study of uniform circular motion that for a circular orbit $a_c = \frac{4\pi^2 r}{T^2}$. This means that $F_{net} = m_pa_c$ may now be written $F_{net} = \frac{m_p 4\pi^2 r}{T^2}$. In this equation, T is the time in seconds required for the planet to make one complete revolution about the Sun. If you set the right side of this equation equal to the right side of the law of universal gravitation, you arrive at the following result:

$$\frac{Gm_sm_p}{r^2} = \frac{m_p4\pi^2 r}{T^2}$$

$$T^2 = \left(\frac{4\pi^2}{Gm_s}\right)r^3$$

$$T = \sqrt{\left(\frac{4\pi^2}{Gm_s}\right)r^3}$$

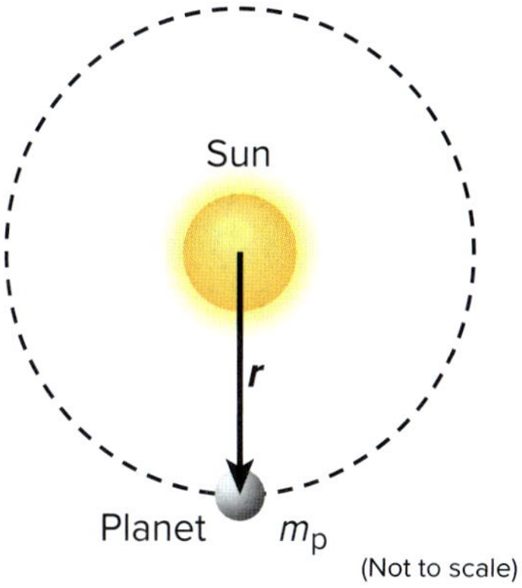

Figure 7 A planet with mass m_p and average distance from the Sun r orbits the Sun. The mass of the Sun is m_s.

Get It?

Summarize how the gravitational force can be used to explain the phenomena of a planet orbiting the Sun.

The period of a planet orbiting the Sun can be expressed as follows.

Period of a Planet Orbiting the Sun

The period of a planet orbiting the Sun is equal to 2π times the square root of the average distance from the Sun cubed, divided by the product of the universal gravitational constant and the mass of the Sun.

$$T = 2\pi\sqrt{\frac{r^3}{Gm_s}}$$

Squaring both sides makes it apparent that this equation is Kepler's third law of planetary motion: the square of the period is proportional to the cube of the distance that separates the masses. The factor $\frac{4\pi^2}{Gm_s}$ depends on the mass of the Sun and the universal gravitational constant. Newton found that this factor applied to elliptical orbits as well.

PHYSICS Challenge

Astronomers have detected three planets that orbit the star Upsilon Andromedae. Planet B has an average orbital radius of 0.0595 AU and a period of 4.6171 days. Planet C has an average orbital radius of 0.832 AU and a period of 241.33 days. Planet D has an average orbital radius of 2.53 AU and a period of 1278.1 days. (Distances are given in astronomical units (AU)—Earth's average distance from the Sun. The distance from Earth to the Sun is 1.00 AU.)

1. Do these planets obey Kepler's third law?
2. Find the mass of the star Upsilon Andromedae in units of the Sun's mass. *Hint: compare $\frac{r^3}{T^2}$ for these planets with that of Earth in the same units (AU and days).*

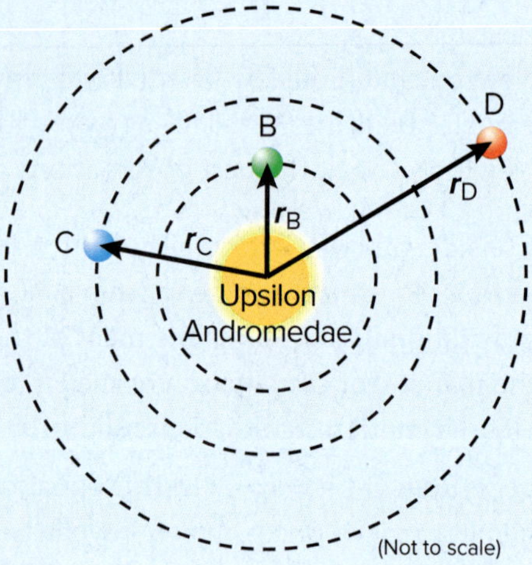

Measuring the Universal Gravitational Constant

How large is the constant G? As you know, the force of gravitational attraction between two objects on Earth is relatively small. The slightest attraction, even between two massive bowling balls, is almost impossible to detect. In fact, it took 100 years from the time of Newton's work for scientists to develop an apparatus that was sensitive enough to measure the force of gravitational attraction.

STEM CAREER Connection

Astronomer

Would you like to study the objects in the night sky or the origin of the universe? Astronomers study the universe, including stars, planets, and other objects. Astronomers work in academia, laboratories, and observatories.

Cavendish's apparatus In 1798, English scientist Henry Cavendish used equipment similar to the apparatus shown in **Figure 8** to measure the gravitational force between two objects. The apparatus has a horizontal rod with small lead spheres attached to each end. The rod is suspended at its midpoint so that it can rotate. Because the rod is suspended by a thin wire, the rod and spheres are very sensitive to horizontal forces.

To measure G, two large spheres are placed in a fixed position close to each of the two small spheres, as shown in **Figure 8.** The force of attraction between the large and small spheres causes the rod to rotate. When the force required to twist the wire equals the gravitational force between the spheres, the rod stops rotating. By measuring the angle through which the rod turns, the attractive force between the objects can be calculated.

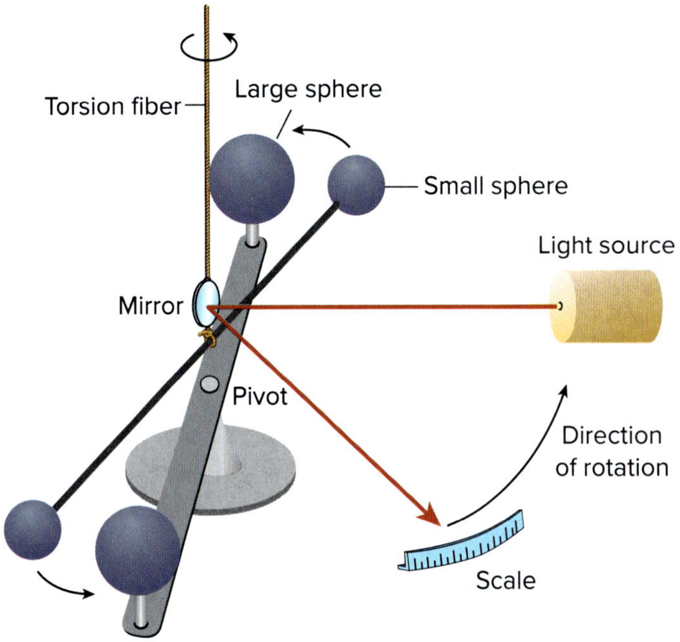

Figure 8 A Cavendish balance uses a light source and a mirror to measure the movement of the spheres.

The angle through which the rod turns is measured by using a beam of light that is reflected from the mirror. The distances between the sphere's centers and the force can both be measured. The masses of the spheres are known. By substituting the values for force, mass, and distance into Newton's law of universal gravitation, an experimental value for G is found: when m_1 and m_2 are measured in kilograms, r in meters, and F in newtons, $G = 6.67\times10^{-11}$ N·m^2/kg^2.

CCC CROSSCUTTING CONCEPTS

Patterns Create a presentation that explains how the patterns that Cavendish observed on his smaller scale experiment provide evidence to explain phenomena on a larger scale such as the interactions within the solar system.

The importance of G Cavendish's investigation often is called "weighing Earth" because it helped determine Earth's mass. Once the value of G is known, not only the mass of Earth, but also the mass of the Sun can be determined. In addition, the gravitational force between any two objects can be calculated by using Newton's law of universal gravitation. For example, the attractive gravitational force (F_g) between two bowling balls of mass 7.26 kg, with their centers separated by 0.30 m, can be calculated as follows:

$$F_g = \frac{(6.67\times10^{-11}\ \text{N·m}^2/\text{kg}^2)(7.26\ \text{kg})(7.26\ \text{kg})}{(0.30\ \text{m})^2} = 3.9\times10^{-8}\ \text{N}$$

You know that on Earth's surface, the weight of an object of mass m is a measure of Earth's gravitational attraction: $F_g = mg$. If Earth's mass is represented by m_E and Earth's radius is represented by r_E, the following is true: $F_g = \frac{Gm_Em}{r_E^{\,2}}$ and so $g = \frac{Gm_E}{r_E^{\,2}}$

This equation can be rearranged to solve for m_E. $m_E = \frac{gr_E^{\,2}}{G}$

Using $g = 9.8$ N/kg, $r_E = 6.38\times10^6$ m, and $G = 6.67\times10^{-11}$ N·m²/kg², the following result is obtained for Earth's mass: $m_E = \frac{(9.8\ \text{N/kg})(6.38\times10^6\ \text{m})^2}{6.67\times10^{-11}\ \text{N·m}^2/\text{kg}^2} = 5.98\times10^{24}\ \text{kg}$

Cavendish's investigation determined the value of G, confirmed Newton's prediction that a gravitational force exists between any two objects, and helped calculate the mass of Earth.

Check Your Progress

8. **Neptune's Orbital Period** Neptune orbits the Sun at an average distance given in **Figure 9,** which allows gases, such as methane, to condense and form an atmosphere. If the mass of the Sun is 1.99×10^{30} kg, calculate the period of Neptune's orbit.

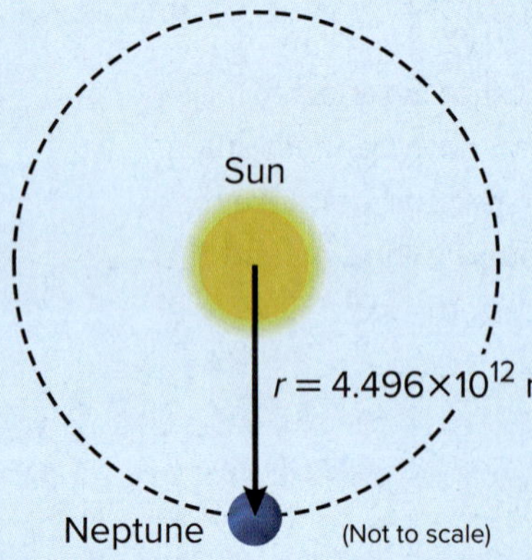

Figure 9

9. **Mathematical Representations** Predict the gravitational force between two 15-kg balls whose centers are 35 cm apart. What fraction is this of the weight of one ball?
10. **Gravity** If Earth began to shrink, but its mass remained the same, what would happen to the value of *g* on Earth's surface?
11. **The Value of G** Cavendish did his investigation using lead spheres. Would his value of G be the same or different if he used copper spheres of equal mass? Explain.
12. **Laws or Theories?** Kepler's three statements and Newton's equation for gravitational attraction are called laws. Were they ever theories? Will they ever become theories?
13. **Critical Thinking** Picking up a rock requires less effort on the Moon than on Earth. How will the Moon's gravitational force affect the path of the rock if it is thrown horizontally?

LEARNSMART® Go online to follow your personalized learning path to review, practice, and reinforce your understanding.

LESSON 2
USING THE LAW OF UNIVERSAL GRAVITATION

FOCUS QUESTION

How does gravity affect objects that aren't touching?

Orbits of Planets and Satellites

The planet Uranus was discovered in 1781 by William Herschel using a telescope that he built himself. By 1830 it was clear that the law of gravitation didn't correctly predict its orbit. Two astronomers proposed that Uranus was being attracted by the Sun and by an undiscovered planet. They calculated the orbit of such a planet in 1845, and, one year later, astronomers at the Berlin Observatory found the planet now called Neptune. How is it possible for planets, such as Neptune and Uranus, to remain in orbit around the Sun?

Sir Issac Newton used a drawing similar to the one shown in **Figure 10** to illustrate a thought experiment on the motion of satellites. Imagine a cannon, perched high atop a mountain, firing a cannonball horizontally with a given horizontal speed. The cannonball is a projectile, and its motion has both vertical and horizontal components. Like all projectiles under the influence of Earth's gravity, it would follow a parabolic trajectory and fall back to the ground.

If the cannonball's horizontal speed were increased, it would travel farther across the surface of Earth and still fall back to the ground. If a powerful cannon were used, the cannonball would travel all the way around Earth and keep going. It would fall toward Earth at the same rate that Earth's surface curves away. In other words, the curvature of the projectile would continue to just match the curvature of Earth so that the cannonball would never get any closer to or farther away from Earth's curved surface. The cannonball would, therefore, be in orbit.

Figure 10 Newton imagined a projectile launched parallel to Earth. If it has enough speed, it will fall toward Earth with a curvature that matches the curvature of Earth's surface.

3D THINKING — DCI Disciplinary Core Ideas — CCC Crosscutting Concepts — SEP Science & Engineering Practices

COLLECT EVIDENCE

Use your Science Journal to record the evidence you collect as you complete the readings and activities in this lesson.

INVESTIGATE

GO ONLINE to find these activities and more resources.

Applying Practices: Planetary Orbits

HS-ESS1-4. Use mathematical or computational representations to predict the motion of orbiting objects in the solar system.

Revisit the Encounter the Phenomenon Question

What information from this lesson can help you answer the Unit and Module questions?

Newton's thought experiment ignored air resistance. For the cannonball to be free of air resistance, the mountain on which the cannon is perched would have to be more than 150 km above Earth's surface. By way of comparison, the mountain would have to be much taller than the peak of Mount Everest, the world's tallest mountain, which is only 8.85 km in height. A cannonball launched from a mountain that is 150 km above Earth's surface would encounter little or no air resistance at an altitude of 150 km because the mountain would be above most of the atmosphere. Thus, a cannonball or any object or satellite at or above this altitude could orbit Earth with lttle or no air resistance.

A satellite's speed A satellite in an orbit that is always the same height above Earth moves in uniform circular motion. Recall that its centripetal acceleration is given by $a_c = \frac{v^2}{r}$. Newton's second law, $F_{net} = ma_c$, can thus be rewritten $F_{net} = \frac{mv^2}{r}$. If Earth's mass is m_E, then this expression combined with Newton's law of universal gravitation produces the following equation:

$$\frac{Gm_E m}{r^2} = \frac{mv^2}{r}$$

Solving for the speed of a satellite in circular orbit about Earth (v) yields the following.

Speed of a Satellite Orbiting Earth

The speed of a satellite orbiting Earth is equal to the square root of the universal gravitational constant times the mass of Earth, divided by the radius of the orbit.

$$v = \sqrt{\frac{Gm_E}{r}}$$

A satellite's orbital period A satellite's orbit around Earth is similar to a planet's orbit about the Sun. Recall that the period of a planet orbiting the Sun is expressed by the following equation:

$$T = 2\pi\sqrt{\frac{r^3}{Gm_S}}$$

Thus, the period for a satellite orbiting Earth is given by the following equation.

Period of a Satellite Orbiting Earth

The period for a satellite orbiting Earth is equal to 2π times the square root of the radius of the orbit cubed, divided by the product of the universal gravitational constant and the mass of Earth.

$$T = 2\pi\sqrt{\frac{r^3}{Gm_E}}$$

The equations for the speed and period of a satellite can be used for any object in orbit about another. The mass of the central body will replace m_E in the equations, and r will be the distance between the centers of the orbiting body and the central body. Orbital speed (v) and period (T) are independent of the mass of the satellite.

EXAMPLE Problem 2

ORBITAL SPEED AND PERIOD Assume that a satellite orbits Earth 225 km above its surface. Given that the mass of Earth is 5.97×10^{24} kg and the radius of Earth is 6.38×10^{6} m, what are the satellite's orbital speed and period?

1 ANALYZE AND SKETCH THE PROBLEM

Sketch the situation showing the height of the satellite's orbit.

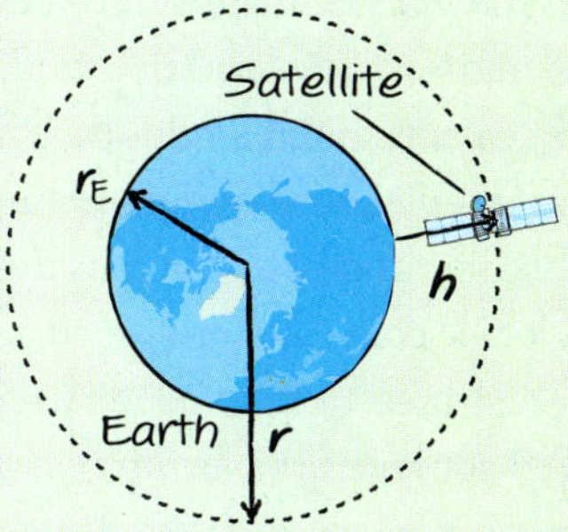

Known	**Unknown**
$h = 2.25\times10^{5}$ m	$v = ?$
$r_E = 6.38\times10^{6}$ m	$T = ?$
$m_E = 5.97\times10^{24}$ kg	
$G = 6.67\times10^{-11}$ N·m²/kg²	

2 SOLVE FOR ORBITAL SPEED AND PERIOD

Determine the orbital radius by adding the height of the satellite's orbit to Earth's radius.

$$r = h + r_E$$

$$= 2.25\times10^{5}\ \text{m} + 6.38\times10^{6}\ \text{m} = 6.60\times10^{6}\ \text{m}$$

Substitute $h = 2.25\times10^{5}$ m and $r_E = 6.38\times10^{6}$ m.

Solve for the speed.

$$v = \sqrt{\frac{Gm_E}{r}}$$

$$= \sqrt{\frac{(6.67\times10^{-11}\ \text{N·m}^2/\text{kg}^2)(5.97\times10^{24}\ \text{kg})}{6.60\times10^{6}\ \text{m}}}$$

Substitute $G = 6.67\times10^{-11}$ N·m²/kg², $m_E = 5.97\times10^{24}$ kg, and $r = 6.60\times10^{6}$ m.

$$= 7.77\times10^{3}\ \text{m/s}$$

Solve for the period.

$$T = 2\pi\sqrt{\frac{r^3}{Gm_E}}$$

$$= 2\pi\sqrt{\frac{(6.60\times10^{6}\ \text{m})^3}{(6.67\times10^{-11}\ \text{N·m/kg}^2)(5.97\times10^{24}\ \text{kg})}}$$

Substitute $r = 6.60\times10^{6}$ m, $G = 6.67\times10^{-11}$ N·m²/kg², and $m_E = 5.97\times10^{24}$ kg.

$$= 5.34\times10^{3}\ \text{s}$$

This is approximately 89 min, or 1.5 h.

3 EVALUATE THE ANSWER

- **Are the units correct?** The unit for speed is meters per second, and the unit for period is seconds.

PRACTICE Problems

ADDITIONAL PRACTICE

Assume a circular orbit for all calculations.

14. Suppose that the satellite in Example Problem 2 is moved to an orbit that is 24 km larger in radius than its previous orbit. What is its speed? Is this faster or slower than its previous speed? Explain.

15. Uranus has 27 known moons. One of these moons is Miranda, which orbits at a radius of 1.29×10^{8} m. Uranus has a mass of 8.68×10^{25} kg. Find the orbital speed of Miranda. How many Earth days does it take Miranda to complete one orbit?

16. Use Newton's thought experiment on the motion of satellites to calculate the speed that a satellite shot from a cannon must have to orbit Earth 150 km above its surface. How long, in seconds and minutes, would it take for the satellite to complete one orbit and return to the cannon?

17. **CHALLENGE** Use the data for Mercury in **Table 1** to find the speed of a satellite that is in orbit 260 km above Mercury's surface and the period of the satellite.

EARTH SCIENCE Connection

The *Landsat 8* satellite, shown in **Figure 11,** is one of eight artificial satellites that has provided images of Earth's continental surfaces. *Landsat* images have been used to create maps, study land use, and monitor resources and global changes. The *Landsat* system enables researchers to monitor small-scale processes, such as deforestation, on a global scale. Satellites, such as *Landsat 8,* are accelerated to the speeds necessary for them to achieve orbit by large rockets, such as shuttle-booster rockets. Because the acceleration of any mass must follow Newton's second law of motion, $F_{net} = ma$, more force is required to launch a more massive satellite into orbit. Thus, the mass of a satellite is limited by the capability of the rocket used to launch it.

Figure 11 The *Landsat 8* orbits Earth about 14 times per day.

Free-Fall Acceleration

The acceleration of objects due to Earth's gravity can be found by using Newton's law of universal gravitation and his second law of motion. For a free-falling object of mass m, the following is true:

$$F = \frac{Gm_E m}{r^2} = ma, \text{ so } a = \frac{Gm_E}{r^2}$$

If you set $a = g_E$ and $r = r_E$ on Earth's surface, the following equation can be written:

$$g = \frac{Gm_E}{r_E^{\,2}}, \text{ thus, } m_E = \frac{gr_E^{\,2}}{G}$$

You saw above that $a = \frac{Gm_E}{r^2}$ for a free-falling object. Substitution of the above expression for m_E yields the following:

$$a = G\frac{\left(\frac{gr_E^{\,2}}{G}\right)}{r^2}$$

$$a = g\left(\frac{r_E}{r}\right)^2$$

On the surface of Earth, $r = r_E$ and so $a = g$. But, as you move farther from Earth's center, r becomes larger than r_E, and the free-fall acceleration is reduced according to this inverse square relationship. What happens to your mass as you move farther and farther from Earth's center?

Real-World Physics

GEOSYNCHRONOUS ORBIT The GOES weather satellites orbit Earth once a day at an altitude of 35,785 km. The orbital speed of the satellite matches Earth's rate of rotation. Thus, to an observer on Earth, the satellite appears to remain above one spot. Dish antennas on Earth can be directed to one point in the sky and remain in a fixed position as the satellite orbits.

Weight and weightlessness You may have seen photos similar to **Figure 12** in which astronauts are on a spacecraft in an environment often called zero-g or weightlessness. The spacecraft orbits about 400 km above Earth's surface. At that distance, $g = 8.7$ N/kg, only slightly less than that on Earth's surface. Earth's gravitational force is certainly not zero in the shuttle. In fact, gravity causes the shuttle to orbit Earth. Why, then, do the astronauts appear to have no weight?

Remember that you sense weight when something, such as the floor or your chair, exerts a contact force on you. But if you, your chair, and the floor all are accelerating toward Earth together, then no contact forces are exerted on you. Thus, your apparent weight is zero and you experience apparent weightlessness. Similarly, the astronauts experience apparent weightlessness as the shuttle and everything in it falls freely toward Earth.

Figure 12 Astronauts in orbit around Earth are in free fall because their spacecraft and everything in it is accelerating toward Earth along with the astronauts. That is, the floor is constantly falling from beneath their feet.

Get It?

Calculate how much force is acting on a 75.0-kg astronaut that is 405 km above Earth's surface, assuming $g = 8.7$ N/kg.

The Gravitational Field

Recall from studying motion that many common forces are contact forces. Friction is exerted where two objects touch; for example, the floor and your chair or desk push on you when you are in contact with them. Gravity, however, is different. It acts on an apple falling from a tree and on the Moon in orbit. In other words, gravity acts over a distance. It acts between objects that are not touching or that are not close together. Newton was puzzled by this concept. He wondered how the Sun could exert a force on planet Earth, which is hundreds of millions of kilometers away.

Field concept The concept of a field was developed by Michael Faraday in the nineteenth century while he was studying the attraction between magnets. Fields, such as gravitational, electric, and magnetic, explain forces acting at a distance. These forces are between objects that transfer energy through space. How does this apply to gravitational forces?

Any object with mass is surrounded by a gravitational field, which exerts a force that is directly proportional to the mass of the object and inversely proportional to the square of the distance from the object's center. Another object experiences a force due to the interaction between its mass and the gravitational field (g) at its location. The direction of g and the gravitational force is toward the center of the object producing the field. Gravitational field strength is expressed by the following equation.

Gravitational Field

The gravitational field strength produced by an object is equal to the universal gravitational constant times the object's mass, divided by the square of the distance from the object's center.

$$g = \frac{Gm}{r^2}$$

Suppose the gravitational field is created by the Sun. Then a planet of mass m in the Sun's gravitational field has a force exerted on it that depends on its mass and the magnitude of the gravitational field at its location. That is, $\mathbf{F}_g = m\mathbf{g}$, toward the Sun. The force is caused by the interaction of the planet's mass with the gravitational field at its location, not with the Sun millions of kilometers away. To find the gravitational field caused by more than one object, calculate all gravitational fields and add them as vectors.

The gravitational field is measured by placing an object with a small mass (m) in the gravitational field and measuring the force ($\mathbf{F}_g$) on it. The gravitational field is calculated using $\mathbf{g} = \frac{\mathbf{F}_g}{m}$ The gravitational field is measured in units of newtons per kilogram (N/kg).

On Earth's surface, the strength of the gravitational field is 9.8 N/kg, and its direction is toward Earth's center. The field can be represented by a vector of length g pointing toward the center of the object producing the field. You can picture the gravitational field produced by Earth as a collection of vectors surrounding Earth and pointing toward it, as shown in **Figure 13.** The strength of Earth's gravitational field varies inversely with the square of the distance from Earth's center. Earth's gravitational field depends on Earth's mass but not on the mass of the object experiencing it.

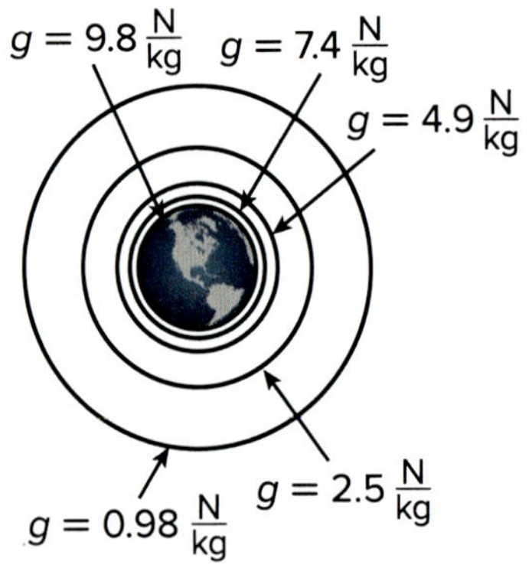

Figure 13 Earth's gravitational field can be represented by vectors pointing toward Earth's center. The decrease in **g**'s magnitude follows an inverse-square relationship as the distance from Earth's center increases.

Explain *why the value of* **g** *never reaches zero.*

Get It?

Explain the concept that is used to explain forces at a distance.

Two Kinds of Mass

You read that mass can be defined as the slope of a graph of force versus acceleration. That is, mass is equal to the net force exerted on an object divided by its acceleration. This kind of mass, related to the inertia of an object, is called inertial mass and is represented by the following equation.

Inertial Mass

Inertial mass is equal to the net force exerted on the object divided by the acceleration of the object.

$$m_{\text{inertial}} = \frac{F_{\text{net}}}{a}$$

STEM CAREER Connection

Theoretical Physicist

Do you enjoying using mathematics to solve problems? Theoretical physicists attempt to better understand the universe by attempting to model the workings of the universe using mathematics.

Inertial mass You know that it is much easier to push an empty cardboard box across the floor than it is to push one that is full of books. The full box has greater inertial mass than the empty one. The **inertial mass** of an object is a measure of the object's resistance to any type of force. Inertial mass of an object is measured by exerting a force on the object and measuring the object's acceleration. The more inertial mass an object has, the less acceleration it undergoes as a result of a net force exerted on it.

Gravitational mass Newton's law of universal gravitation, $F_g = \frac{Gm_1m_2}{r^2}$, also involves mass—but a different kind of mass. Mass as used in the law of universal gravitation is a quantity that measures an object's response to gravitational force and is called **gravitational mass.** It can be measured by using a simple balance, such as the one shown in **Figure 14.** If you measure the magnitude of the attractive force exerted on an object by another object of mass m, at a distance r, then you can define the gravitational mass in the following way.

Gravitational Mass

The gravitational mass of an object is equal to the distance between the centers of the objects squared, times the gravitational force, divided by the product of the universal gravitational constant, times the mass of the other object.

$$m_g = \frac{r^2F_g}{Gm}$$

How different are these two kinds of mass? Suppose you have a watermelon in the trunk of your car. If you accelerate the car forward, the watermelon will roll backward relative to the trunk. This is a result of its inertial mass—its resistance to acceleration. Now, suppose your car climbs a steep hill at a constant speed. The watermelon will again roll backward. But this time, it moves as a result of its gravitational mass. The watermelon is pulled downward toward Earth.

Newton made the claim that inertial mass and gravitational mass are equal in magnitude. This hypothesis is called the principle of equivalence. All investigations conducted so far have yielded data that support this principle. Most of the time we refer simply to the mass of an object. Albert Einstein also was intrigued by the principle of equivalence and made it a central point in his theory of gravity.

Figure 14 A simple balance is used to determine the gravitational mass of an object.

Einstein's Theory of Gravity

Newton's law of universal gravitation allows us to calculate the gravitational force that exists between two objects because of their masses. Newton was puzzled, however, as to how two objects could exert forces on each other if those two objects were millions of kilometers away from each other. Albert Einstein proposed that gravity is not a force but rather an effect of space itself. According to Einstein's explanation of gravity, mass changes the space around it. Mass causes space to be curved, and other bodies are accelerated because of the way they follow this curved space.

Curved space One way to picture how mass affects space is to model three-dimensional space as a large, two-dimensional sheet, as shown in the top part of **Figure 15.** The large blue ball on the sheet represents a massive object. The ball forms an indentation on the sheet. A small blue ball rolling across the sheet simulates the motion of an object in space. If the small ball moves near the sagging region of the sheet, it will be accelerated. In a similar way, Earth and the Sun are attracted to each other because of the way space is distorted by the two objects.

Planes traveling north The following is another analogy that might help you understand the curvature of space. Suppose you watch from space as two planes travel due north from the equator. At the equator, the planes are separated by 4000 km. As they approach the North Pole, the distance decreases to 1 km. To the pilots, their paths are straight lines, but because of Earth's curvature, they travel in a curve, as viewed far from Earth's surface, as in **Figure 15.**

Converging lines Consider a similar motion. Two apples are dropped to Earth, initially traveling in parallel paths, as in **Figure 15.** As they approach Earth, they are pulled toward Earth's center. Their paths converge. This convergence can be attributed to the curvature of space near Earth. Far from any massive object, such as a planet or star, space is flat, and parallel lines remain parallel. Then they begin to converge. In flat space, the parallel lines would remain parallel. In curved space, the lines converge.

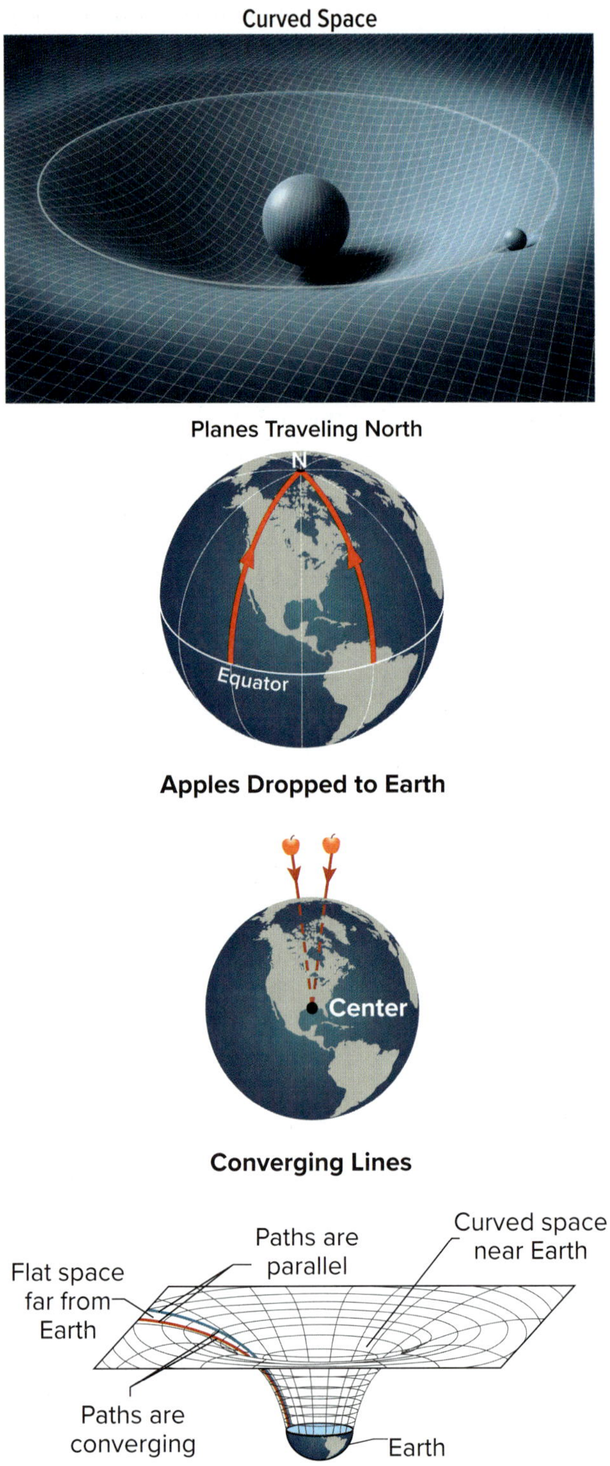

Figure 15 Visualizing how space is curved is difficult. Analogies can help you understand difficult concepts.

Einstein's theory or explanation, called the general theory of relativity, makes many predictions about how massive objects affect one another. In every test conducted to date, Einstein's theory has been shown to give the correct results.

posteriori/Shutterstock

Deflection of light Einstein's theory predicts that massive objects deflect and bend light. Light follows the curvature of space around the massive object and is deflected, as shown in **Figure 16.** In 1919, during an eclipse of the Sun, astronomers found that light from distant stars that passed near the Sun was deflected an amount that agreed with Einstein's predictions.

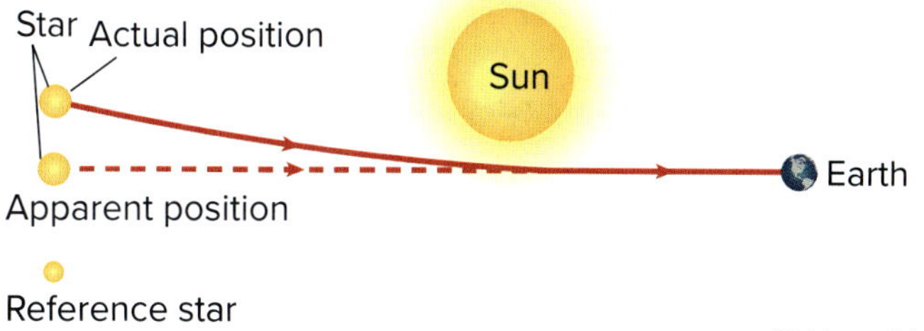

Figure 16 Light is bent around massive objects in space, altering the apparent position of the light's source.

Describe *how this effect contradicts your experience of light's behavior.*

Another result of general relativity is the effect of gravity on light from extremely massive objects. If an object is massive and dense enough, the light leaving it is totally bent back to the object. No light ever escapes the object. Objects such as these, called black holes, have been identified as a result of their effects on nearby stars. Black holes have been detected through the radiation produced when matter is pulled into them.

Check Your Progress

18. **Earth and Moon** The Moon is 3.9×10^5 km from Earth's center and Earth is 14.96×10^7 km from the Sun's center. The masses of Earth and the Sun are 5.97×10^{24} kg and 1.99×10^{30} kg, respectively. During a full moon, the Sun, Earth, and the Moon are in line with each other, as shown in **Figure 17.**
    a. Find the ratio of the gravitational fields due to Earth and the Sun at the center of the Moon.
    b. What is the net gravitational field due to the Sun and Earth at the center of the Moon?

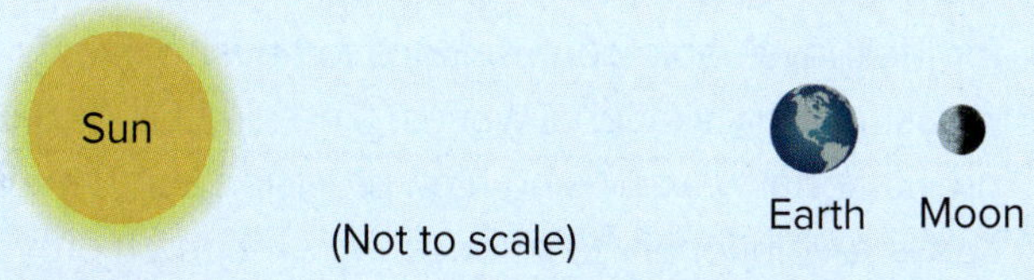

Figure 17

19. **Apparent Weightlessness** Chairs in an orbiting spacecraft are weightless. If you were on board such a spacecraft and you were barefoot, would you stub your toe if you kicked a chair? Explain.
20. **Gravitational Field** The mass of the Moon is 7.3×10^{22} kg and its radius is 1738 km. What is the strength of the gravitational field on the surface of the Moon?
21. **Orbital Period and Speed** Two satellites are in circular orbits about Earth. One is 150 km above the surface, the other is 160 km.
    a. Which satellite has the larger orbital period?
    b. Which has the greater speed?
22. **Theories and Laws** Why is Einstein's description of gravity called a theory, while Newton's is a law?
23. **Astronaut** What would be the strength of Earth's gravitational field at a point where an 80.0-kg astronaut would experience a 25.0 percent reduction in weight?
24. **A Satellite's Mass** When the first artificial satellite was launched into orbit by the former Soviet Union in 1957, U.S. president Dwight D. Eisenhower asked scientists to calculate the mass of the satellite. Would they have been able to make this calculation? Explain.
25. **Critical Thinking** It is easier to launch a satellite from Earth into an orbit that circles eastward than it is to launch one that circles westward. Explain.

LEARNSMART® Go online to follow your personalized learning path to review, practice, and reinforce your understanding.

Making Waves

In recent years, scientists have made groundbreaking observations of gravitational waves at Laser Interferometer Gravitational-Wave Observatories (LIGOs) in the United States and abroad. Before 2015, the waves, which are ripples in the space-time continuum traveling at the speed of light, had never before been detected on Earth.

This image from a computer simulation shows gravitational waves formed by the violent collision of two black holes.

Einstein and Gravitational Waves

In 1916, Albert Einstein predicted the existence of gravitational waves based on his theory of general relativity. His prediction was not confirmed until 2015, when two LIGOs—one in Louisiana and another in Washington—directly detected the waves for the first time. The waves were produced by the collision of two black holes that merged into one larger black hole about 1.3 billion years ago. As the black holes collided, some of their matter was converted into energy, as expressed in Einstein's famous equation, $E = mc^2$. The gravitational waves are the energy produced by the collision.

More Observatories, Higher Resolution Data

In 2017, the two LIGOs based in the United States—plus a similar facility in Italy, the Advanced Virgo Detector of the European Gravitational Observatory—detected the collision of two black holes that occurred about 1.8 billion years ago. Scientists were able to determine the general location of the collision more precisely because they triangulated the waves picked up by the three separate observatories.

That same year, the three observatories for the first time detected gravitational waves emitted by the collision of two neutron stars about 130 million years ago. Scientists were able to study both the gravitational waves and the entire electromagnetic spectrum emitted during the cosmic event. The stars continued to give off light for weeks after their collision.

A Window to the Cosmos

In 2015, three physicists were awarded the Nobel Prize for their invaluable contributions to advancements in measuring gravitational waves. The direct detection of gravitational waves using high precision instrumentation is revolutionary to science. This research provides a way to learn more about cataclysmic events in the universe, such as the collision of black holes. Scientists predict that the detection of gravitational waves will change the field of astrophysics.

USE A MODEL TO ILLUSTRATE

Work with a partner to research how a LIGO works. Draw or build a model to illustrate how a LIGO detects gravitational waves. Present the model to your class.

CALIFORNIA CONNECTION

The California Institute of Technology (Caltech) is one of the operators of LIGOs in the U.S. Visit Caltech's website to learn how the institute operates LIGOs. Write a report on your findings.

MODULE 7
STUDY GUIDE

GO ONLINE to study with your Science Notebook.

Lesson 1 PLANETARY MOTION AND GRAVITATION

- Kepler's first law states that planets move in elliptical orbits, with the Sun at one focus, and Kepler's second law states that an imaginary line from the Sun to a planet sweeps out equal areas in equal times. Kepler's third law states that the square of the ratio of the periods of any two planets is equal to the cube of the ratio of the distances between the centers of the planets and the center of the Sun.

$$\left(\frac{T_A}{T_B}\right)^2 = \left(\frac{r_A}{r_B}\right)^3$$

- Newton's law of universal gravitation can be used to rewrite Kepler's third law to relate the radius and period of a planet to the mass of the Sun. Newton's law of universal gravitation states that the gravitational force between any two objects is directly proportional to the product of their masses and inversely proportional to the square of the distance between their centers. The force is attractive and along a line connecting the centers of the masses.

$$F = \frac{Gm_1Gm_2}{r^2}$$

- Cavendish's investigation determined the value of G, confirmed Newton's prediction that a gravitational force exists between two objects, and helped calculate the mass of Earth.

- Kepler's first law
- Kepler's second law
- Kepler's third law
- gravitational force
- law of universal gravitation

Lesson 2 USING THE LAW OF UNIVERSAL GRAVITATION

- The speed and period of a satellite in circular orbit describe orbital motion. Orbital speed and period for any object in orbit around another are calculated with Newton's second law.
- Gravitational mass and inertial mass are two essentially different concepts. The gravitational and inertial masses of an object, however, are numerically equal.
- All objects have gravitational fields surrounding them. Any object within a gravitational field experiences a gravitational force exerted on it by the gravitational field. Einstein's general theory of relativity explains gravitational force as a property of space itself.

- inertial mass
- gravitational mass

THREE-DIMENSIONAL THINKING
Module Wrap-Up

REVISIT THE PHENOMENON

How can gravity keep moons orbiting planets but also cause things to fall?

CER Claim, Evidence, Reasoning

Explain Your Reasoning Revisit the claim you made when you encountered the phenomenon. Summarize the evidence you gathered from your investigations and research and finalize your Summary Table. Does your evidence support your claim? If not, revise your claim. Explain why your evidence supports your claim.

STEM UNIT PROJECT

Now that you've completed the module, revisit your STEM unit project. You will summarize your evidence and apply it to the project.

GO FURTHER

SEP Data Analysis Lab

How does Earth's orbit compare to Neptune's?

The mean distance of the planet Neptune from the Sun is 30.05 times the mean distance of Earth from the Sun.

CER Analyze and Interpret Data

1. Sketch the situation (not to scale).
2. **Claim, Evidence, Reasoning** Without doing any numerical calculations answer the following: Is the orbital speed of Earth less than, equal to, or greater than the orbital speed of Neptune? Explain your reasoning.
3. **Claim, Evidence, Reasoning** The radius of Neptune is 3.883 time the radius of Earth, and the mass of Neptune is 17.147 times the mass of Earth. Would it be easier to launch a satellite from the surface of Earth or Neptune? Explain your reasoning.

MODULE 8

ROTATIONAL MOTION

NASA Earth Observatory image by Joshua Stevens, using VIIRS data from LANCE/EOSDIS Rapid Response

MODULE 8

ROTATIONAL MOTION

ENCOUNTER THE PHENOMENON

Why do all tropical cyclones in the northern hemisphere rotate the same direction?

GO ONLINE to watch a video about the rotation of tropical cyclones.

SEP Ask Questions

Do you have other questions about the phenomenon? If so, add them to the driving question board.

CER Claim, Evidence, Reasoning

Make Your Claim Use your CER chart to make a claim about why tropical cyclones in the northern hemisphere rotate in the same direction. Explain your reasoning.

Collect Evidence Use the lessons in this module to collect evidence to support your claim. Record your evidence as you move through the module.

Explain Your Reasoning You will revisit your claim and explain your reasoning at the end of the module.

GO ONLINE to access your CER chart and explore resources that can help you collect evidence.

LESSON 2: Explore & Explain: Torque

LESSON 3: Explore & Explain: Rotating Frames of Reference

Smithsonian

Additional Resources

LESSON 1
DESCRIBING ROTATIONAL MOTION

FOCUS QUESTION

How would you describe the rotation of a hurricane?

Angular Displacement

You probably have observed a spinning object many times. How would you measure such an object's rotation? Find a circular object, such as a DVD. Mark one point on the edge of the DVD so that you can keep track of its position. Rotate the DVD to the left (counterclockwise), and as you do so, watch the location of the mark. When the mark returns to its original position, the DVD has made one complete revolution.

Measuring revolution How can you measure a fraction of one revolution? It can be measured in several different ways, but the two most used are degrees and radians. A degree is $\frac{1}{360}$ of a revolution and is the usual scale marking on a protractor. In mathematics and physics, the radian is related to the ratio of the circumference of a circle to its radius. In one revolution, a point on the edge of a wheel travels a distance equal to 2π times the radius of the wheel. For this reason, the **radian** is defined as $\frac{1}{(2\pi)}$ of a revolution. One complete revolution is an angle of 2π radians. The radian is abbreviated *rad*.

The Greek letter theta (θ) is used to represent the angle of revolution. **Figure 1** shows the angles in radians for several common fractions of a revolution. Note that counterclockwise rotation is designated as positive, while clockwise is negative. As an object rotates, the change in the angle is called the object's **angular displacement.**

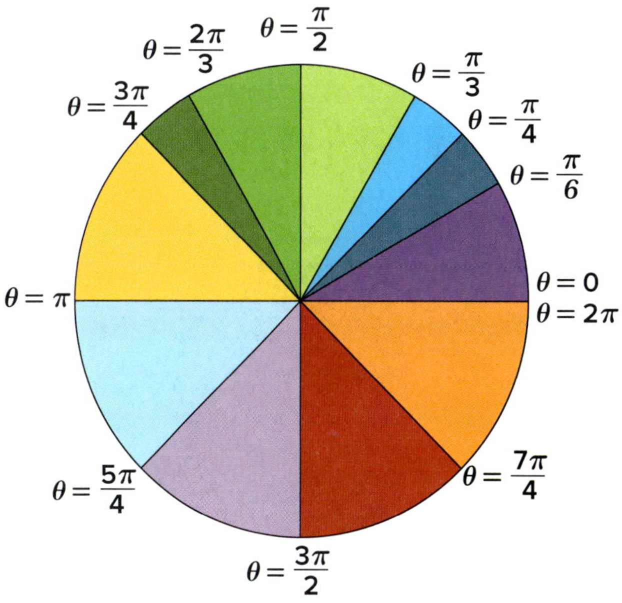

Figure 1 A fraction of a revolution can be measured in degrees or radians. Some common angles are shown below measured in radians. Each angle is measured in the counterclockwise direction from $\theta = 0$.

3D THINKING DCI Disciplinary Core Ideas CCC Crosscutting Concepts SEP Science & Engineering Practices

COLLECT EVIDENCE

Use your Science Journal to record the evidence you collect as you complete the readings and activities in this lesson.

INVESTIGATE

GO ONLINE to find these activities and more resources.

CCC **Identify Crosscutting Concepts**

Create a table of the crosscutting concepts and fill in examples you find as you read.

Review the News

Obtain information from a current news story about rotational motion. Evaluate your source and communicate your findings to your class.

Earth's revolution As you know, Earth turns one complete revolution, or 2π rad, in 24 h. In 12 h, it rotates through π rad. Through what angle does Earth rotate in 6 h? Because 6 h is one-fourth of a day, Earth rotates through an angle of $\frac{\pi}{2}$ rad during that period. Earth's rotation as seen from the North Pole is positive. Is it positive or negative when viewed from the South Pole?

Get It?

Identify the angle that Earth rotates in 48 h.

Measuring distance How far does a point on a rotating object move? You already found that a point on the edge of an object moves 2π times the radius in one revolution. In general, for rotation through an angle (θ), a point at a distance r from the center, as shown in **Figure 2**, moves a distance given by $x = r\theta$. If r is measured in meters, you might think that multiplying it by θ rad would result in x being measured in m·rad. However, this is not the case. Radians indicate the dimensionless ratio between x and r. Thus, x is measured in meters.

Measuring Distance

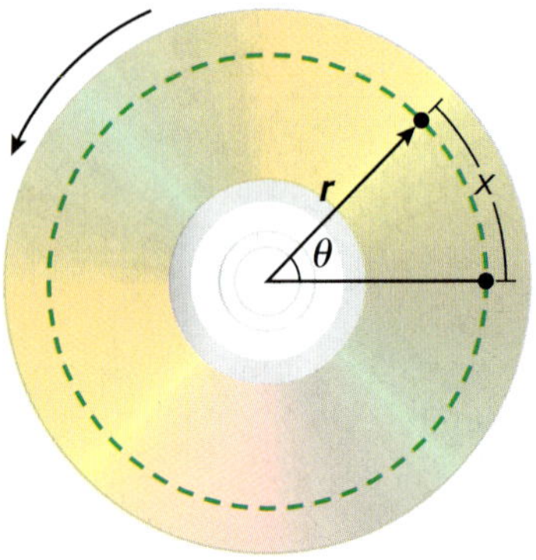

Figure 2 The dashed line shows the path of a point on a DVD as the DVD rotates counterclockwise about its center. The point is located a distance r from the center of the DVD and moves a distance x as it rotates.

Explain *what the variables r, x, and θ represent.*

Angular Velocity

How fast does a DVD spin? How can you determine its speed of rotation? Recall that velocity is displacement divided by the time taken to make the displacement. Likewise, the **angular velocity** of an object is angular displacement divided by the time taken to make the angular displacement. The angular velocity of an object is given by the following ratio, and is represented by the Greek letter omega (ω).

Average Angular Velocity of an Object

The angular velocity equals the angular displacement divided by the time required to make the rotation.

$$\omega = \frac{\Delta\theta}{\Delta t}$$

Recall that if the velocity changes over a time interval, the average velocity is not equal to the instantaneous velocity at any given instant. Similarly, the angular velocity calculated in this way is actually the average angular velocity over a time interval (Δt). Instantaneous angular velocity equals the slope of a graph of angular position versus time.

Earth's angular velocity

Angular velocity is measured in rad/s. You can calculate Earth's angular velocity as follows:

$$\omega_E = \frac{(2\pi \text{ rad})}{(24.0 \text{ h})(3600 \text{ s/h})} = 7.27\times10^{-5} \text{ (rad/s)}$$

In the same way that counterclockwise rotation produces positive angular displacement, it also results in positive angular velocity.

If an object's angular velocity is ω, then the linear velocity of a point a distance r from the axis of rotation is given by $v = r\omega$. The speed at which an object on Earth's equator moves as a result of Earth's rotation is given by $v = r\omega = (6.38\times10^6 \text{ m/rad})(7.27\times10^{-5} \text{ rad/s}) = 464 \text{ m/s}$. Earth is an example of a rotating, rigid body, as shown in **Figure 3**. Even though different points at different latitudes on Earth do not move the same distance in each revolution, all points rotate through the same angle. All parts of a rigid body rotate at the same rate. The Sun is not a rigid body. Different parts of the Sun rotate at different angular velocities. Most objects that we will consider in this chapter are rigid bodies.

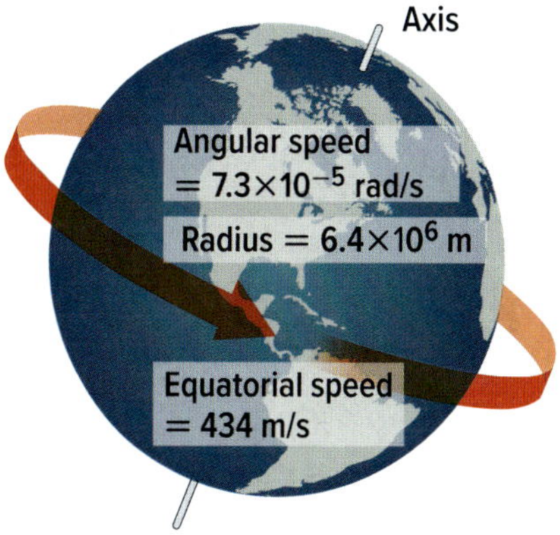

Figure 3 Earth is a rotating, rigid body and all parts rotate at the same rate.

Angular Acceleration

What if angular velocity is changing? For example, a car could accelerate from 0.0 m/s to 25 m/s in 15 s. In the same 15 s, the angular velocity of the car's 0.64 m diameter wheels would change from 0.0 rad/s to 78 rad/s. The wheels would undergo **angular acceleration,** which is the change in angular velocity divided by the time required to make the change. Angular acceleration (α) is represented by the following equation and is measured in rad/s^2.

Average Angular Acceleration of an Object

Angular acceleration is equal to the change in angular velocity divided by the time required to make that change.

$$\alpha = \frac{\Delta\omega}{\Delta t}$$

If the change in angular velocity is positive, then the angular acceleration also is positive. Angular acceleration defined in this way is also the average angular velocity over the time interval Δt. One way to find the instantaneous angular acceleration is to find the slope of a graph of angular velocity as a function of time. The linear acceleration of a point at a distance (r) from the axis of an object with angular acceleration (α) is given by $a = r\alpha$. Table 1 is a summary of linear and angular relationships discussed previously in this lesson.

Get It?

Compare the angular velocity and angular acceleration of a rotating body.

Table 1 **Linear and Angular Measures**

Quantity	Linear	Angular	Relationship
Displacement	x (m)	θ (rad)	$x = r\theta$
Velocity	v (m/s)	ω (rad/s)	$v = r\omega$
Acceleration	a (m/s^2)	α (rad/s^2)	$a = r\alpha$

PRACTICE Problems

ADDITIONAL PRACTICE

1. What is the angular displacement of each of the following hands of a clock in 1.00 h? State your answer in three significant digits.

- **a.** the second hand
- **b.** the minute hand
- **c.** the hour hand

2. A rotating toy above a crib makes one complete counterclockwise rotation in 1 min.

- **a.** What is its angular displacement in 3 min?
- **b.** What is the toy's angular velocity in rad/min?
- **c.** If the toy is turned off, does it have positive or negative angular acceleration? Explain.

3. If a truck has a linear acceleration of 1.85 m/s^2 and the wheels have an angular acceleration of 5.23 rad/s^2, what is the diameter of the wheels?

4. The truck in the previous problem is towing a trailer with wheels that have a diameter of 48 cm.

- **a.** How does the linear acceleration of the trailer compare with that of the truck?
- **b.** How do the angular accelerations of the wheels of the trailer and the wheels of the truck compare?

5. CHALLENGE You replace the tires on your car with tires of larger diameter. How will the angular velocity and number of revolutions change for trips at the same speed and over the same distance?

Angular frequency The number of complete revolutions made by an object in 1 s is called angular frequency. Angular frequency is defined as $f = \frac{\omega}{2\pi}$. One example of such a rotating object is a computer hard drive. Listen carefully when you start a computer. You often will hear the hard drive spinning. Hard drive frequencies are measured in revolutions per minute (RPM). Inexpensive hard drives rotate at 4800, 5400, and 7200 RPM. More advanced hard drives operate at 10,000 or 15,000 RPM. The faster the hard drive rotates, the quicker it can access or store information.

Check Your Progress

6. **Angular Velocity** The Moon ($r = 1.74\times10^6$ m) rotates once on its axis in 27.3 days.
   a. What is its period of rotation in seconds?
   b. What is the frequency of its rotation in rad/s?
   c. A rock sits on the surface at the Moon's equator. What is its linear speed due to the Moon's rotation?
   d. Compare this speed with that of a person at the equator due to Earth's rotation.

7. **Angular Displacement** A movie lasts 2 h. During that time, what is the angular displacement of each of the following?
   a. the hour hand
   b. the minute hand
   c. the second hand

8. **Angular Acceleration** In the spin cycle of a clothes washer, the drum turns at 635 rev/min. If the lid is opened, the motor is turned off. If the drum requires 8.0 s to slow to a stop, what is the angular acceleration of the drum?

9. **Angular Displacement** Do all parts of the minute hand on a watch such as the one shown in **Figure 4** have the same angular displacement? Do they move the same linear distance? Explain.

Figure 4

10. **Critical Thinking** A CD-ROM has a spiral track that starts 2.7 cm from the center of the disk and ends 5.5 cm from the center. The disk drive must turn the disk so that the linear velocity of the track is a constant 1.4 m/s. Find the following.
    a. the angular velocity of the disk (in rad/s and rev/min) for the start of the track
    b. the disk's angular velocity at the end of the track
    c. the disk's angular acceleration if the disk is played for 76 min

LEARNSMART® Go online to follow your personalized learning path to review, practice, and reinforce your understanding.

LESSON 2

ROTATIONAL DYNAMICS

FOCUS QUESTION

How does force affect rotation?

Force and Angular Velocity

How do you start the rotation of an object? That is, how do you increase its angular velocity? If you wrap a string around a toy top and pull hard, you can make the top spin rapidly. The force of the string is exerted at the outer edge of the top and at right angles to the line from the center of the top to the point where the string leaves the top's surface. You have learned that a force changes the velocity of a point object. In the case of a toy top, a force that is exerted in a very specific way changes the angular velocity of an extended object. An extended object is an object that has a definite shape and size.

How can you exert a force to open a door easily? To get the most effect from the least force, you exert the force as far from the axis of rotation as possible, as shown in **Figure 5.** In this case, the axis of rotation is an imaginary vertical line through the hinges. The doorknob is near the outer edge of the door. You exert the force on the doorknob at right angles to the door. Thus, the magnitude of the force, the distance from the axis to the point where the force is exerted, and the direction of the force determine the change in angular velocity.

Figure 5 When opening a door that is free to rotate about its hinges, apply the force farthest from the hinges, at an angle perpendicular to the door.

3D THINKING DCI Disciplinary Core Ideas CCC Crosscutting Concepts SEP Science & Engineering Practices

COLLECT EVIDENCE

Use your Science Journal to record the evidence you collect as you complete the readings and activities in this lesson.

INVESTIGATE

GO ONLINE to find these activities and more resources.

Forensics Lab: Leverage

Carry out an investigation to determine how a small force can produce a large torque.

PhysicsLAB: Torques

Calulate the torque on a system when the forces and lever arms are known.

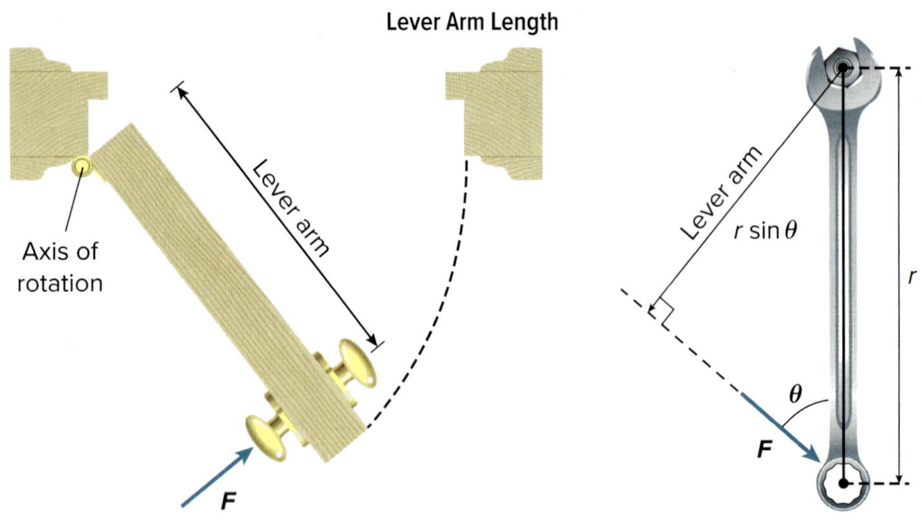

Figure 6 The lever arm is the perpendicular distance from the axis of rotation to the point where the force is exerted. For the door, the lever arm is along the width of the door, from the hinge to the point where the force is exerted. For the wrench, the lever arm is equal to $r \sin \theta$, when the angle (θ) between the force and the radius of rotation is not equal to 90°.

Explain *why the formula* r sin θ *is used to find the length of the lever arm.*

Lever arm For a given applied force, the change in angular velocity depends on the lever arm, which is the perpendicular distance from the axis of rotation to the point where the force is exerted. If the force is perpendicular to the radius of rotation, as it was with the toy top, then the **lever arm** is the distance from the axis (r). For the door example, the lever arm is the distance from the hinges to the point where you exert the force, as illustrated on the left in **Figure 6.** If a force is not exerted perpendicular to the radius, the length of the lever arm is reduced. You must use mathematics to find the length of the lever arm.

To find the lever arm, you must extend the line of the force until it forms a right angle with a line from the center of rotation, as shown in **Figure 6**. The distance between this intersection and the axis is the lever arm. Using trigonometry, the lever arm (L) can be calculated by the equation $L = r \sin \theta$. In this equation, r is the distance from the axis of rotation to the point where the force is exerted, and θ is the angle between the force and the radius from the axis of rotation to the point where the force is applied.

Get It?

Explain what each of the variables represents in the equation $L = r \sin \theta$.

Torque The term **torque** describes the combination of force and lever arm that can cause an object to rotate. The magnitude of a torque is the product of the force and the perpendicular lever arm. Because force is measured in newtons and distance is measured in meters, torque is measured in newton-meters (N·m). Torque is represented by the Greek letter tau (τ) and is represented by the equation shown below.

Torque

Torque is equal to the force F times the lever arm ($r \sin \theta$).

$$\tau = Fr \sin \theta$$

Get It?

Identify what each of the variables in the torque equation — τ, F, r, and θ— represents.

EXAMPLE Problem 1

LEVER ARM A bolt on a car engine must be tightened with a torque of 35 N·m. You use a 25-cm-long wrench and pull the end of the wrench at an angle of 60.0° to the handle of the wrench. How long is the lever arm, and how much force must you exert?

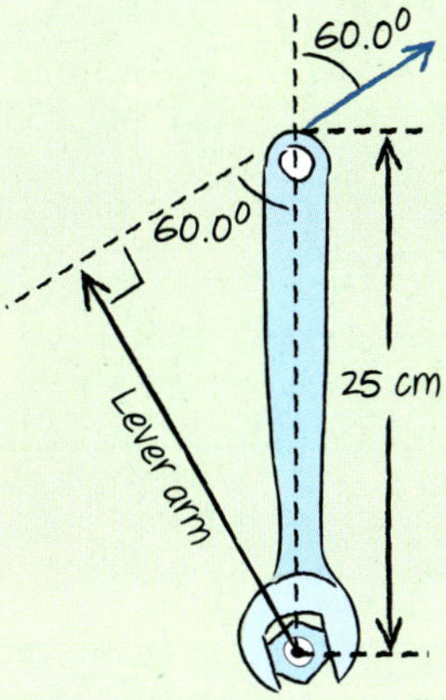

1 ANALYZE AND SKETCH THE PROBLEM

Sketch the situation. Find the lever arm by extending the force vector backward until a line that is perpendicular to it intersects the axis of rotation.

Known		Unknown
$r = 0.25$ m	$\tau = 35$ N·m	$L = ?$
$\theta = 60.0°$		$F = ?$

2 SOLVE FOR THE UNKNOWN

Solve for the length of the lever arm.

$$L = r \sin \theta$$

$$= (0.25 \text{ m})(\sin 60.0°)$$

Substitute $r = 0.25$ m and $\theta = 60.0°$ into the equation. Then, solve the equation.

$$= 0.22 \text{ m}$$

Solve for the force.

$$\tau = Fr \sin \theta$$

$$F = \frac{\tau}{r \sin \theta}$$

$$= \frac{35 \text{ N·m}}{(0.25 \text{ m})(\sin 60.0°)}$$

Substitute $\tau = 35$ N·m, $r = 0.25$ m, and $\theta = 60.0°$ into the equation.

$$= 1.6\times10^2 \text{ N}$$

Then, solve the equation. Remember to use significant digits.

3 EVALUATE THE ANSWER

- **Are the units correct?** Force is measured in newtons.
- **Does the sign make sense?** Only the magnitude of the force needed to rotate the wrench clockwise is calculated.

PRACTICE Problems

ADDITIONAL PRACTICE

11. Consider the wrench in Example Problem 1. What force is needed if it is applied to the wrench pointing perpendicular to the wrench?

12. If a torque of 55.0 N·m is required to turn a bolt and the largest force you can exert is 135 N, how long a lever arm must you use to turn the bolt?

13. You have a 0.234-m-long wrench. A job requires a torque of 32.4 N·m, and you can exert a force of 232 N.

a. What is the smallest angle, with respect to the handle of the wrench, at which you can pull on the wrench and get the job done?

b. A friend can exert 275 N. What is the smallest angle she can use to accomplish the job?

14. You stand on a bicycle pedal, as shown in **Figure 7.** Your mass is 65 kg. If the pedal makes an angle of 35° above the horizontal and the pedal is 18 cm from the center of the chain ring, how much torque would you exert?

Figure 7

15. CHALLENGE If the pedal in the previous problem is horizontal, how much torque would you exert? How much torque would you exert when it is vertical?

Richard Hutchings/Digital Light Source

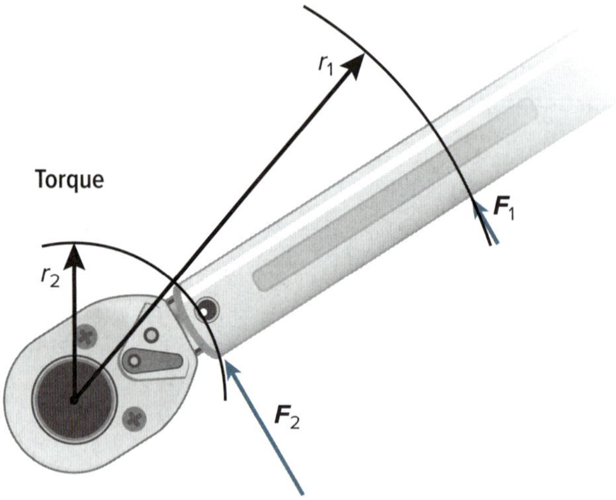

Figure 8 This worker uses a long wrench because it requires him to exert less force to tighten and loosen the nut. The wrench has a long lever arm, and less force is required if the force is applied farther from the axis of rotation (the center of the nut).

Finding Net Torque

Figure 8 shows a practical application of increasing torque to make a task easier. For another example of torque, try the following investigation. Collect two pencils, some coins, and some transparent tape. Tape two identical coins to the ends of the pencil and balance it on the second pencil, as shown in **Figure 9.** Each coin exerts a torque that is equal to the distance from the balance point to the center of the coin (r) times its weight (F_g), as follows:

$$\tau = rF_g$$

But the torques are equal and opposite in direction. Thus, the net torque is zero:

$$\tau_1 - \tau_2 = 0$$

or

$$r_1F_{g1} - r_2F_{g2} = 0$$

How can you make the pencil rotate? You could add a second coin on top of one of the two coins, thereby making the two forces different. You also could slide the balance point toward one end or the other of the pencil, thereby making the two lever arms of different length.

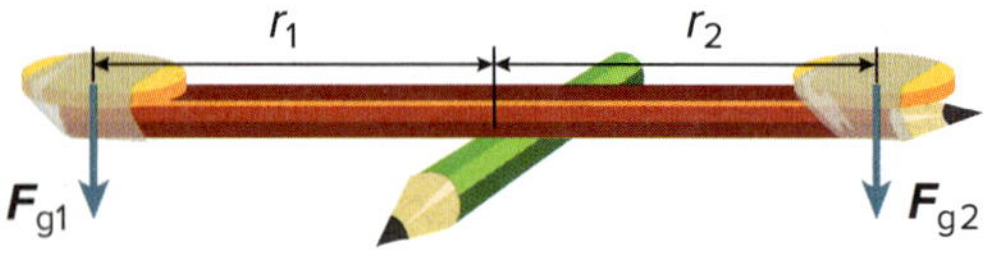

Figure 9 The torque exerted by the first coin ($F_{g1}r_1$) is equal and opposite in direction to the torque exerted by the second coin ($F_{g2}r_2$) when the pencil is balanced.

CCC CROSSCUTTING CONCEPTS

Systems and System Models Sketch models to describe the variations of the system in **Figure 9** described above. Use the models to predict how the different arrangements will affect the system. Test the arrangements and compare your predictions with your evidence.

STEM CAREER Connection

Race Car Mechanic

If you are competitive and love to hear the roar of a high-performance engine, race car mechanic might be the career for you. These specialists use their understanding of torque to help keep their team's car at the front of the pack.

EXAMPLE Problem 2

BALANCING TORQUES Kariann (56 kg) and Aysha (43 kg) want to balance on a 1.75-m-long seesaw. Where should they place the pivot point of the seesaw? Assume that the seesaw is massless.

1 ANALYZE AND SKETCH THE PROBLEM

- Sketch the situation.
- Draw and label the vectors.

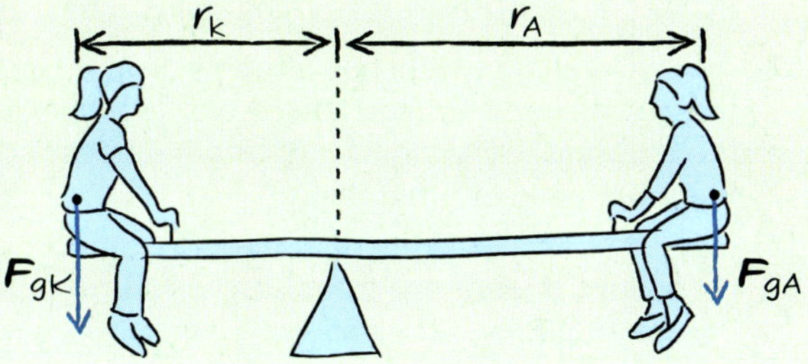

Known	**Unknown**
$m_K = 56$ kg	$r_K = ?$
$m_A = 43$ kg	$r_A = ?$
$r_K + r_A = 1.75$ m	

2 SOLVE FOR THE UNKNOWN

Find the two forces.

Kariann:

$$F_{gK} = m_K g$$
$$= (56\text{ kg})\,(9.8\text{ N/kg})$$
$$= 5.5\times10^2\text{ N}$$

Substitute the known values into the equation: $m_K = 56$ kg, $g = 9.8$ N/kg.

Aysha:

$$F_{gA} = m_A g$$
$$= (43\text{ kg})(9.8\text{ N/kg})$$
$$= 4.2\times10^2\text{ N}$$

Substitute the known values into the equation: $m_A = 43$ kg, $g = 9.8$ N/kg.

Define Kariann's distance in terms of the length of the seesaw and Aysha's distance.

$$r_K = 1.75\text{ m} - r_A$$

When there is no rotation, the sum of the torques is zero.

$$F_{gK}r_K = F_{gA}r_A$$
$$F_{gK}r_K - F_{g}A_{rA} = 0.0\text{ N·m}$$
$$F_{gK}(1.75\text{ m} - r_A) - F_{gA}r_A = 0.0\text{ N·m}$$

Substitute the relationship between Kariann's distance in terms of Aysha's distance in to the equation: $r_K = 1.75\ m - r_A$.

Solve for r_A.

$$F_{gK}(1.75\text{ m}) - F_{gK}(r_A) - F_{gA}r_A = 0.0\text{ N·m}$$
$$F_{gK}r_A + F_{gA}r_A = F_{gK}(1.75\text{ m})$$
$$(F_{gK} + F_{gA})r_A = F_{gK}\,(1.75\text{ m})$$
$$r_A = \frac{F_{gK}(1.75\text{ m})}{(F_{gK} + F_{gA})}$$
$$= \frac{(5.5\times10^2\text{ N})(1.75\text{ m})}{(5.5\times10^2\text{ N} + 4.2\times10^2\text{ N})}$$
$$= 0.99\text{ m}$$

Substitute $F_{gK} = 5.5\times10^2$ N and $F_{gA} = 4.2\times10^2$ N.

3 EVALUATE THE ANSWER

- **Are the units correct?** Distance is measured in meters.
- **Do the signs make sense?** Distances are positive.
- **Is the magnitude realistic?** Aysha is about 1 m from the center, so Kariann is about 0.75 m away from it. Because Kariann's weight is greater than Aysha's weight, the lever arm on Kariann's side should be shorter. Aysha is farther from the pivot, as expected.

PRACTICE Problems

ADDITIONAL PRACTICE

16. Ashok, whose mass is 43 kg, sits 1.8 m from a pivot at the center of a seesaw. Steve, whose mass is 52 kg, wants to seesaw with Ashok. How far from the center of the seesaw should Steve sit?

17. A bicycle-chain wheel has a radius of 7.70 cm. If the chain exerts a 35.0-N force on the wheel in the clockwise direction, what torque is needed to keep the wheel from turning?

18. Two stationary baskets of fruit hang from strings on pulleys of different diameters, as shown in **Figure 10.** What is the mass of basket A?

19. Suppose the radius of the larger pulley in problem 18 was increased to 6.0 cm. What is the mass of basket A now?

20. CHALLENGE A bicyclist, of mass 65.0 kg, stands on the pedal of a bicycle. The crank, which is 0.170 m long, makes a 45.0° angle with the vertical, as shown in **Figure 11.** The crank is attached to the chain wheel, which has a radius of 9.70 cm. What force must the chain exert to keep the wheel from turning?

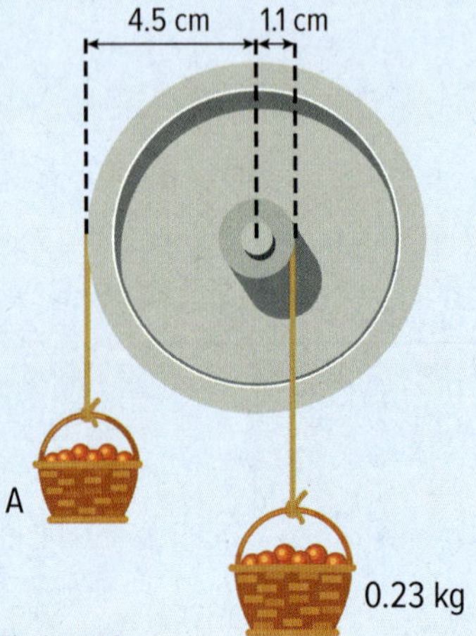

Figure 10

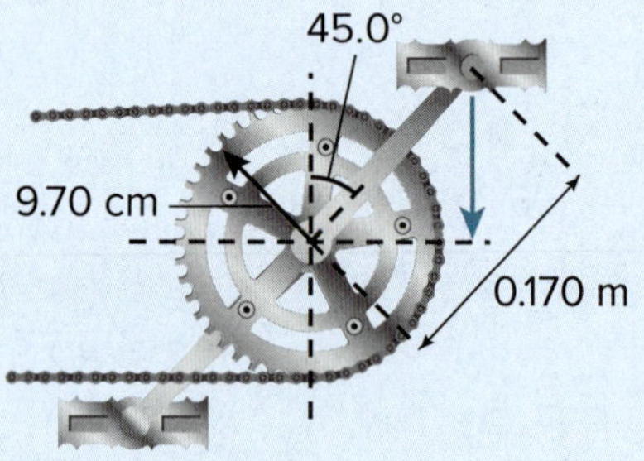

Figure 11

The Moment of Inertia

If you exert a force on a point mass, its acceleration will be inversely proportional to its mass. How does an extended object rotate when a torque is exerted on it? To observe firsthand, recover the pencil, the coins, and the transparent tape that you used earlier in this chapter. First, tape the coins at the ends of the pencil. Hold the pencil between your thumb and forefinger, and wiggle it back and forth. Take note of the forces that your thumb and forefinger exert. These forces create torques that change the angular velocity of the pencil and coins.

Now move the coins so that they are only 1 or 2 cm apart. Wiggle the pencil as before. Did you notice that the pencil is now easier to rotate? The torque that was required was much less this time. The mass of an object is not the only factor that determines how much torque is needed to change its angular velocity; the distribution or location of the mass also is important.

The resistance to rotation is called the **moment of inertia,** which is represented by the symbol I and has units of mass times the square of the distance. For a point object located at a distance (r) from the axis of rotation, the moment of inertia is given by the following equation.

Moment of Inertia of a Point Mass

The moment of inertia of a point mass is equal to the mass of the object times the square of the object's distance from the axis of rotation.

$$I = mr^2$$

Table 2 Moments of Inertia for Various Objects

Object	Location of Axis	Diagram	Moment of Inertia
Thin hoop of radius *r*	through central diameter	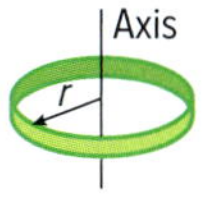	mr^2
Solid, uniform cylinder of radius *r*	through center	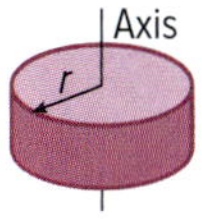	$\left(\frac{1}{2}\right)mr^2$
Uniform sphere of radius *r*	through center	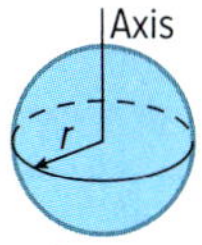	$\left(\frac{2}{5}\right)mr^2$
Long, uniform rod of length *l*	through center	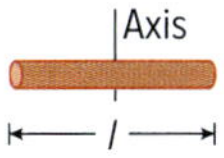	$\left(\frac{1}{12}\right)ml^2$
Long, uniform rod of length *l*	through end	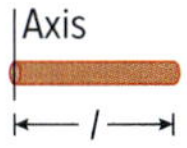	$\left(\frac{1}{3}\right)ml^2$
Thin, rectangular plate of length *l* and width *w*	through center	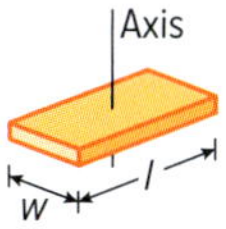	$\left(\frac{1}{12}\right)m\,(l^2+w^2)$

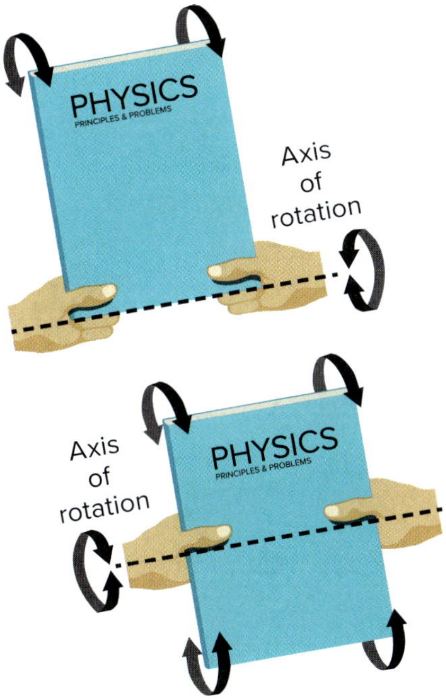

Figure 12 The moment of inertia of a book depends on the axis of rotation. The moment of inertia of the book on the top is larger than the moment of inertia of the book on the bottom because the average distance of the book's mass from the rotational axis is larger.

Identify *which book requires more torque to rotate it and why.*

Moment of inertia and mass

As you have seen, the moments of inertia for extended objects, such as the pencil and coins, depend on how far the masses are from the axis of rotation. A bicycle wheel, for example, has almost all of its mass in the rim and tire. Its moment of inertia about its axle is almost exactly equal to mr^2, where r is the radius of the wheel. For most objects, however, the mass is distributed closer to the axis so the moment of inertia is less than mr^2. For example, as shown in **Table 2,** for a solid cylinder of radius r, $I = \left(\frac{1}{2}\right) mr^2$, while for a solid sphere, $I = \left(\frac{2}{5}\right) mr^2$.

Get It?

Write the equation for the moment of inertia of a hoop.

Moment of inertia and rotational axis

The moment of inertia also depends on the location and direction of the rotational axis, as illustrated in **Figure 12.** To observe this effect, hold a book in the upright position by placing your hands at the bottom of the book. Feel the torque needed to rock the book toward you and then away from you. Now put your hands in the middle of the book and feel the torque needed to rock the book. Note that much less torque is needed when your hands are placed in the middle of the book because the average distance of the book's mass from the rotational axis is much less.

EXAMPLE Problem 3

MOMENT OF INERTIA A simplified model of a twirling baton is a thin rod with two round objects at each end. The length of the baton is 0.66 m, and the mass of each object is 0.30 kg. Find the moment of inertia of the baton as it is rotated about an axis at the midpoint between the round objects and perpendicular to the rod. What is the moment of inertia of the baton if the axis is moved to one end of the rod? Which is greater? Assume the mass of the rod is negligible.

1 ANALYZE AND SKETCH THE PROBLEM

Sketch the situation. Show the baton with the two different axes of rotation and the distances from the axes of rotation to the masses.

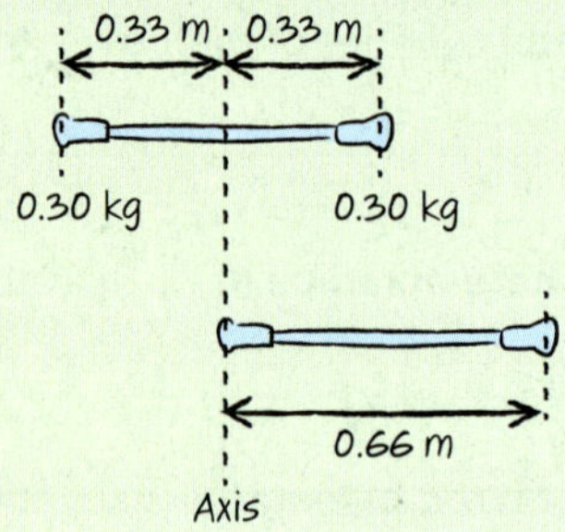

Known
$m = 0.30$ kg
$l = 0.66$ m

Unknown
$I = ?$

2 SOLVE FOR THE UNKNOWN

Calculate the moment of inertia of each mass separately.
Rotating about the center of the rod:

$$r = \left(\frac{1}{2}\right)l$$
$$= \left(\frac{1}{2}\right)(0.66\ \text{m})$$ Substitute the known value, $l = 0.66$ m, into the equation.
$$= 0.33\ \text{m}$$

$$I_{\text{single mass}} = mr^2$$
$$= (0.30\ \text{kg})(0.33\ \text{m})^2$$ Substitute $m = 0.30$ kg and $r = 0.33$ m into the equation.
$$= 0.033\ \text{kg·m}^2$$

Find the moment of inertia of the baton.

$$I = 2I_{\text{single mass}}$$
$$= 2(0.033\ \text{kg·m}^2)$$ Substitute $I_{\text{single mass}} = 0.033$ kg·m² into the equation.
$$= 0.066\ \text{kg·m}^2$$

Rotating about one end of the rod:

$$I_{\text{single mass}} = mr^2$$
$$= (0.30\ \text{kg})(0.66\ \text{m})^2$$ Substitute $m = 0.30$ kg and $r = 0.66$ m into the equation.
$$= 0.13\ \text{kg·m}^2$$

$$I_{\text{single mass}} = mr^2$$
$$= (0.30\ \text{kg})(0.0\ \text{m})^2$$
$$= 0 \text{ for the other mass.}$$

Find the moment of inertia of the baton.

$$I = I_{\text{single mass}} + 0$$
$$= 0.13\ \text{kg·m}^2$$

The moment of inertia is greater when the baton is swung around one end.

3 EVALUATE THE ANSWER

- **Are the units correct?** Moment of inertia is measured in kg·m².
- **Is the magnitude realistic?** Masses and distances are small, and so are the moments of inertia. Doubling the distance increases the moment of inertia by a factor of 4. Thus, doubling the distance increases the moment of inertia more than having only one mass decreases the moment of inertia.

PRACTICE Problems

ADDITIONAL PRACTICE

21. Two children of equal masses sit 0.3 m from the center of a seesaw. Assuming that their masses are much greater than that of the seesaw, by how much is the moment of inertia increased when they sit 0.6 m from the center? Ignore the moment of inertia for the seesaw.

22. Suppose there are two balls with equal diameters and masses. One is solid, and the other is hollow, with all its mass distributed at its surface. Are the moments of inertia of the balls equal? If not, which is greater?

23. Calculate the moments of inertia for each object below using the formulas in **Table 2.** Each object has a radius of 2.0 m and a mass of 1.0 kg.

a. a thin hoop

b. a solid, uniform cylinder

c. a solid, uniform sphere

24. CHALLENGE Figure 13 shows three equal-mass spheres on a rod of very small mass. Consider the moment of inertia of the system, first when it is rotated about sphere A and then when it is rotated about sphere C.

a. Are the moments of inertia the same or different? Explain. If the moments of inertia are different, in which case is the moment of inertia greater?

b. Each sphere has a mass of 0.10 kg. The distance between spheres A and C is 0.20 m. Find the moment of inertia in the following instances: rotation about sphere A, rotation about sphere C.

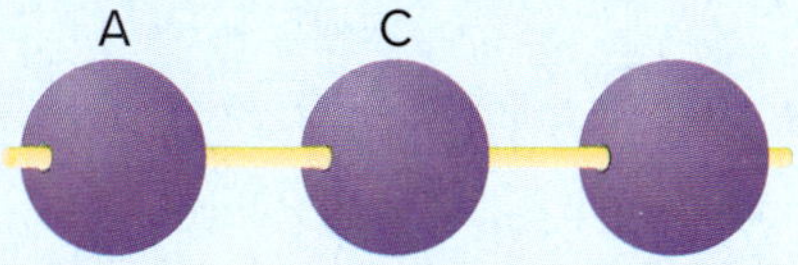

Figure 13

Newton's Second Law for Rotational Motion

Newton's second law for linear motion is expressed as $a = \frac{F_{net}}{m}$. If you rewrite this equation to represent rotational motion, acceleration is replaced by angular acceleration (α), force is replaced by net torque (τ_{net}), and mass is replaced by moment of inertia (I). Angular acceleration is directly proportional to the net torque and inversely proportional to the moment of inertia as stated in **Newton's second law for rotational motion.** This law is expressed by the following equation.

Newton's Second Law for Rotational Motion

The angular acceleration of an object about a particular axis equals the net torque on the object divided by the moment of inertia.

$$\alpha = \frac{\tau_{net}}{I}$$

If the torque on an object and the angular velocity of that object are in the same direction, then the angular velocity of the object increases. If the torque and angular velocity are in different directions, then the angular velocity decreases.

PHYSICS Challenge

Moments of Inertia Rank the objects shown in the diagram from least to greatest according to their moments of inertia about the indicated axes. All spheres have equal masses and all separations are the same. Assume that the rod's mass is negligible.

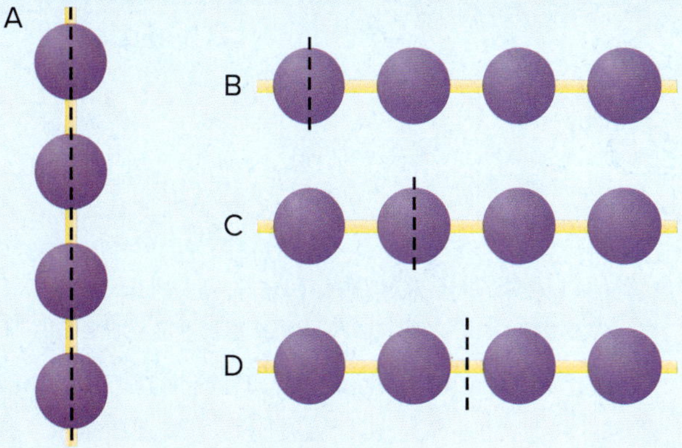

EXAMPLE Problem 4

TORQUE A solid steel wheel is free to rotate about a motionless central axis. It has a mass of 15 kg and a diameter of 0.44 m and starts at rest. You want to increase this wheel's rotation about its central axis to 8.0 rev/s in 15 s.

a. What torque must be applied to the wheel?

b. If you apply the torque by wrapping a strap around the outside of the wheel, how much force should you exert on the strap?

1 ANALYZE AND SKETCH THE PROBLEM

Sketch the situation. The torque must be applied in a counterclockwise direction; force must be exerted as shown.

Known	Unknown
$m = 15$ kg	$\alpha = ?$
$r = \left(\frac{1}{2}\right)(0.44\text{ m}) = 0.22$ m	$I = ?$
$\omega_i = 0.0$ rad/s	$\tau = ?$
$\omega_f = 2\pi(8.0\text{ rev/s})$	$F = ?$
$t = 15$ s	

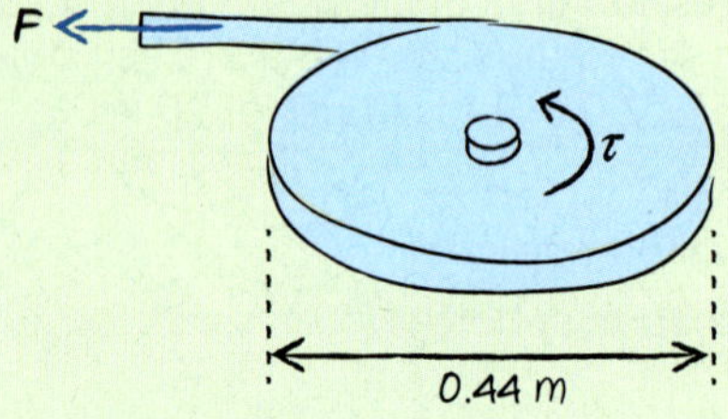

2 SOLVE FOR THE UNKNOWN

a. Solve for angular acceleration.

$$\alpha = \frac{\Delta\omega}{\Delta t}$$

$$= \frac{(16\pi\text{ rad/s} - (0.0\text{ rad/s}))}{15\text{ s}}$$

Substitute $\omega_f = 16\pi$ rad/s and $\omega_i = 0.0$ rad/s into the equation.

$$= 3.4\text{ rad/s}^2$$

Solve for the moment of inertia.

$$I = \left(\frac{1}{2}\right)mr^2$$

$$= \left(\frac{1}{2}\right)(15\text{ kg})(0.22\text{ m})^2$$

Substitute $m = 15$ kg and $r = 0.22$ m into the equation.

$$= 0.36\text{ kg·m}^2$$

Solve for torque.

$$\tau = I\alpha$$

$$= (0.36\text{ kg·m}^2)(3.4\text{ rad/s}^2)$$

Substitute $I = 0.36$ kg·m^2 and $\alpha = 3.4$ rad/s^2 into the equation.

$$= 1.2\text{ kg·m}^2\text{/s}^2$$

$$= 1.2\text{ N·m}$$

b. Solve for force.

$$\tau = Fr$$

$$F = \frac{\tau}{r}$$

$$= \frac{(1.2\text{ N·m})}{(0.22\text{ m})}$$

Substitute $\tau = 1.2$ N·m and $r = 0.22$ m into the equation.

$$= 5.5\text{ N}$$

3 EVALUATE THE ANSWER

- **Are the units correct?** Torque is measured in N·m and force is measured in N.
- **Is the magnitude realistic?** The size of the wheel makes it relatively easy to spin despite its mass.

PRACTICE Problems

ADDITIONAL PRACTICE

25. Consider the wheel in Example Problem 4. If the force on the strap were twice as great, what would be the angular frequency of the wheel after 15 s?

26. A solid wheel accelerates at 3.25 rad/s^2 when a force of 4.5 N exerts a torque on it. If the wheel is replaced by a wheel which has all of its mass on the rim, the moment of inertia is given by $I = mr^2$. What force should be exerted on the strap to give the same angular velocity?

27. A bicycle wheel on a repair bench can be accelerated either by pulling on the chain that is on the gear or by pulling on a string wrapped around the tire. The tire's radius is 0.38 m, while the radius of the gear is 0.14 m. What force would you need to pull on the string to produce the same acceleration you obtained with a force of 15 N on the chain?

28. The same bicycle wheel in the previous problem is used with a smaller gear whose radius is 0.11 m. If you obtained the needed acceleration with a force of 15 N on the chain, what force would you need to exert on the string?

29. A chain is wrapped around a pulley and pulled with a force of 16.0 N. The pulley has a radius of 0.20 m. The pulley's rotational speed increases from 0.0 to 17.0 rev/min in 5.00 s. What is the moment of inertia of the pulley?

30. CHALLENGE A disk with a moment of inertia of 0.26 kg·m^2 is attached to a smaller disk mounted on the same axle. The smaller disk has a diameter of 0.180 m and a mass of 2.5 kg. A strap is wrapped around the smaller disk, as shown in **Figure 14.** Find the force needed to give this system an angular acceleration of 2.57 rad/s^2.

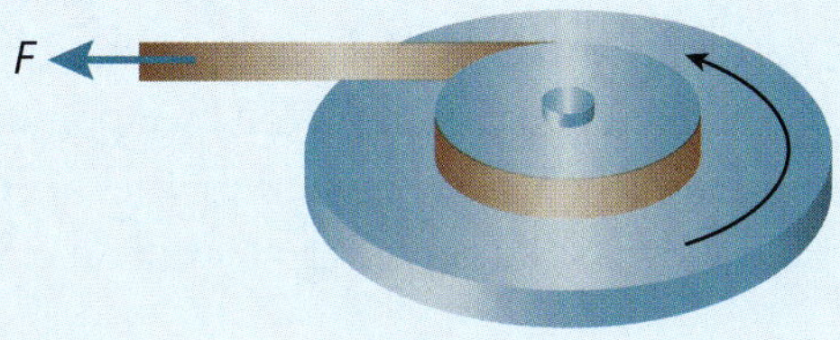

Figure 14

Check Your Progress

31. **Torque** Vijesh enters a revolving door that is not moving. Explain where and how Vijesh should push to produce a torque with the least amount of force.

32. **Lever Arm** You open a door by pushing at a right angle to the door. Your friend pushes at the same place, but at an angle of 55° from the perpendicular. If both you and your friend exert the same torque on the door, how do the forces you and your friend applied compare?

33. **Torque and Force** The solid wheel shown in **Figure 15** has a mass of 5.2 kg and a diameter of 0.55 m. It is at rest, and you need it to rotate at 12 rev/s in 35 s.

    a. What torque must be applied?

    b. If a nylon strap is wrapped around the outside of the wheel, how much force do you need to exert on the strap?

34. **Net Torque** Two people pull on ropes wrapped around the edge of a large wheel. The wheel has a mass of 12 kg and a diameter of 2.4 m. One person pulls in a clockwise direction with a 43-N force, and the other pulls counterclockwise with a 67-N force. What is the net torque?

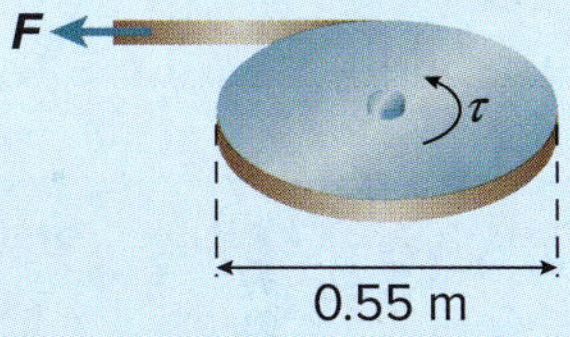

Figure 15

35. **Newton's Second Law for Rotational Motion** A rope is wrapped around a pulley and pulled with a force of 13.0 N. The pulley's radius is 0.150 m. The pulley's rotational speed increases from 0.0 to 14.0 rev/min in 4.50 s. What is the moment of inertia of the pulley?

36. **Critical Thinking** A ball on a low-friction, tilted surface will slide downhill without rotating. On a rough surface, the ball will roll. Explain why, using a free-body diagram.

LEARNSMART® Go online to follow your personalized learning path to review, practice, and reinforce your understanding.

LESSON 3
EQUILIBRIUM

FOCUS QUESTION

Why are some vehicles more likely to roll over than others?

The Center of Mass

How does a freely moving object rotate around its center of mass? A wrench may spin about its handle or end-over-end. Does any single point on the wrench follow a straight path? **Figure 16** shows the path of the wrench. You can see that there is a single point whose path traces a straight line, as if the wrench could be replaced by a point particle at that location. The white dot in the photo represents this point. The point on the object that moves in the same way that a point particle would move is the **center of mass** of an object.

Locating the center of mass How can you locate the center of mass of an object? First, suspend the object from any point. When the object stops swinging, the center of mass is somewhere along the vertical line drawn from the suspension point. Draw the line. Then, suspend the object from another point. Again, the center of mass must be directly below this point. Draw a second vertical line. The center of mass is at the point at which these two lines cross. The wrench and all other objects that are freely moving through space rotate about an axis that goes through their center of mass. Where would you think the center of mass of a human body is located?

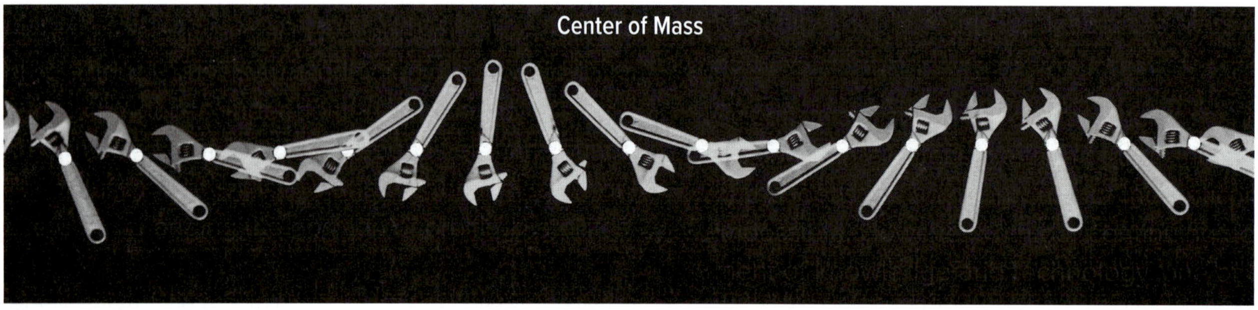

Figure 16 The path of the center of mass of a wrench is a straight line.

Berenice Abbott/Science Source

3D THINKING DCI Disciplinary Core Ideas CCC Crosscutting Concepts SEP Science & Engineering Practices

COLLECT EVIDENCE

Use your Science Journal to record the evidence you collect as you complete the readings and activities in this lesson.

INVESTIGATE

GO ONLINE to find these activities and more resources.

PhysicsLAB: Equilibrium
Use mathematics in the form of Newton's second law for both linear and rotational motion to determine if an object is in static equilibrium.

Revisit the Encounter the Phenomenon Question
What information from this lesson can help you answer the Unit and Module questions?

Figure 17 The upward motion of the ballet dancer's head is less than the upward motion of the center of mass. Thus, the head and torso move in a nearly horizontal path. This creates an illusion of floating.

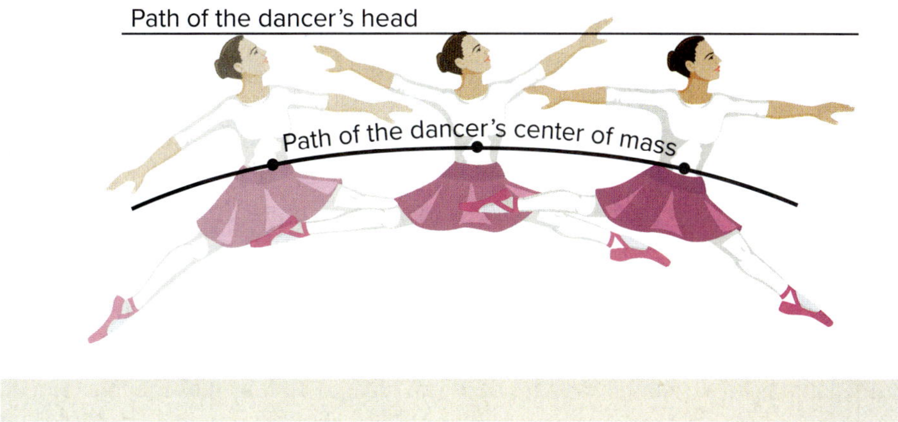

The human body's center of mass For a person who is standing with her arms hanging straight down, the center of mass is a few centimeters below the navel, midway between the front and back of the person's body. The center of mass is farther below the navel for women than men, which often results in better balance for women than men. Because the human body is flexible, however, its center of mass is not fixed. If you raise your hands above your head, your center of mass rises 6 to 10 cm. A ballet dancer, for example, can appear to be floating on air by changing her center of mass in a leap. By raising her arms and legs while in the air, as shown in **Figure 17,** the dancer moves her center of mass up. The path of the center of mass is a parabola, but the dancer's head stays at almost the same height for a surprisingly long time.

Real-World Physics

THE FOSBURY FLOP In high jumping, a technique called the Fosbury flop allows a high jumper to clear the bar when it is placed at the highest position. This is possible because the athlete's center of mass passes below the bar as he or she somersaults over the bar, with his or her back toward it.

Center of Mass and Stability

What factors determine whether a vehicle is stable or prone to roll over in an accident? To understand the problem, think about tipping over a box. A tall, narrow box standing on end tips more easily than a low, broad box. Why? To tip a box, as shown in **Figure 18,** you must rotate it about a corner. You pull at the top with a force (F) applying a torque (τ_F). The weight of the box, acting on the center of mass (F_g) applies an opposing torque (τ_w). When the center of mass is directly above the point of support, τ_w is zero. The only torque is the one applied by you. As the box rotates farther, its center of mass is no longer above its base of support, and both torques act in the same direction. At this point, the box tips over rapidly.

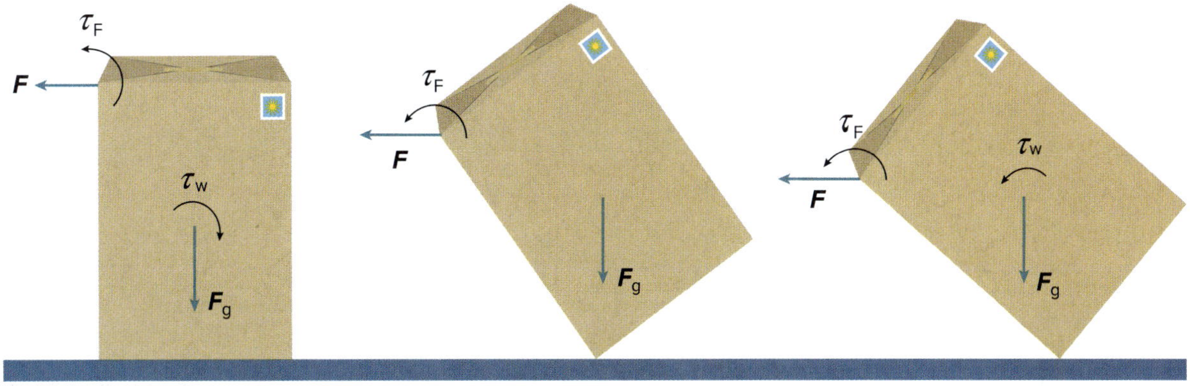

Figure 18 The curved arrows show the direction of the torque produced by the force exerted to tip over a box.

Stability An object is said to be stable if a large external force is required to tip it. The box in **Figure 18** is stable as long as the direction of the torque due to its weight (τ_w) tends to keep it upright. This occurs as long as the box's center of mass lies above its base. To tip the box over, you must rotate its center of mass around the axis of rotation until it is no longer above the base of the box. To rotate the box, you must lift its center of mass. The broader the base, the more stable the object is. Passengers on a city bus, for example, often stand with their feet spread apart to avoid toppling over as the bus starts and stops and weaves through traffic.

Why do vehicles roll over? **Figure 19** shows two vehicles about to roll over. Note that the one with the higher center of mass does not have to be tilted very far for its center of mass to be outside its base—its center of mass does not have to be raised as much as the other vehicle's center of mass. As demonstrated by the vehicles, the lower the location of an object's center of mass, the greater its stability.

You are stable when you stand flat on your feet. When you stand on tiptoe, however, your center of mass moves forward directly above the balls of your feet, and you have very little stability. In judo, aikido, and other martial arts, the fighter uses torque to rotate the opponent into an unstable position, where the opponent's center of mass does not lie above his or her feet. A small person can use torque, rather than force, to defend himself or herself against a stronger person.

In summary, if the center of mass is not located above the base of an object, it is unstable and will roll over without additional torque. If the center of mass is above the base of the object, it is stable. If the base of the object is very narrow and the center of mass is high, then the object might be stable, but the slightest force will cause it to tip over.

Get It?

Describe when an object is the most stable.

Conditions for Equilibrium

If your pen is at rest, what is needed to keep it at rest? You could either hold it up or place it on a desk or some other surface. An upward force must be exerted on the pen to balance the downward force of gravity. You must also hold the pen so that it will not rotate. An object is said to be in static equilibrium if both its velocity and angular velocity are zero or constant. Thus, for an object to be in static equilibrium, it must meet two conditions. First, it must be in translational equilibrium; that is, the net force exerted on the object must be zero. Second, it must be in rotational equilibrium; that is, the net torque exerted on the object must be zero.

Figure 19 Larger vehicles have a higher center of mass than smaller ones. The higher the center of mass, the smaller the tilt needed to cause the vehicle's center of mass to move outside its base and cause the vehicle to roll over.

EXAMPLE Problem 5

STATIC EQUILIBRIUM A 5.8-kg ladder, 1.80 m long, rests on two sawhorses. Sawhorse A is 0.60 m from one end of the ladder, and sawhorse B is 0.15 m from the other end of the ladder. What force does each sawhorse exert on the ladder?

1 ANALYZE AND SKETCH THE PROBLEM

- Sketch the situation.
- Choose the axis of rotation at the point where F_A acts on the ladder. Thus, the torque due to F_A is zero.

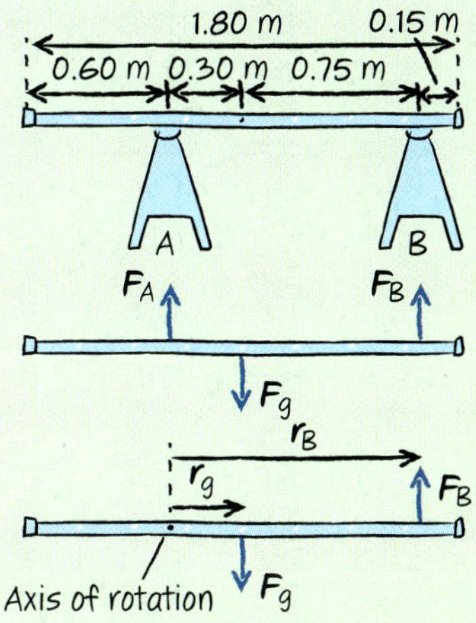

Known	Unknown
$m = 5.8$ kg	$F_A = ?$
$l = 1.80$ m	$F_B = ?$
$l_A = 0.60$ m	
$l_B = 0.15$ m	

2 SOLVE FOR THE UNKNOWN

For a ladder that has a constant density, the center of mass is at the center rung.

The net force is the sum of all forces on the ladder.

The ladder is in translational equilibrium, so the net force exerted on it is zero.

$$F_{net} = F_A + F_B + (-F_g)$$

$$0.0\text{ N} = F_A + F_B + (-F_g)$$ The ladder is in translational equilibrium, so the net force exerted on it is zero.

Solve for F_A.

$$F_A = F_g - F_B$$

Find the torques due to F_g and F_B.

$$\tau_g = -F_g r_g$$ τ_g is in the clockwise direction.

$$\tau_B = +F_B r_B$$ τ_B is in the counterclockwise direction.

The net torque is the sum of all torques on the object.

$$\tau_{net} = \tau_B + \tau_g$$

$$0.0\text{ N·m} = \tau_B + \tau_g$$ The ladder is in rotational equilibrium, so $\tau_{net} = 0.0$ N·m.

$$\tau_B = -\tau_g$$

$$F_B r_B = F_g r_g$$ Substitute $\tau_B = F_B r_B$ and $\tau_g = -F_g r_g$ into the equation.

Solve for F_B.

$$F_B = \frac{F_g r_g}{r_B}$$

$$= \frac{r_g mg}{r_B}$$ Substitute $F_g = mg$

Using the expression $F_A = F_g - F_B$, substitute in the expressions for F_B and F_g.

$$F_A = F_g - F_B$$

$$= F_g - \frac{r_g mg}{r_B}$$ Substitute $F_B = \frac{r_g mg}{r_B}$ into the equation.

$$= m_g - \frac{r_g mg}{r_B}$$ Substitute $F_g = mg$ into the equation.

$$= mg\left(1 - \frac{r_g}{r_B}\right)$$

EXAMPLE Problem 5 (continued)

Solve for r_g.

$$r_g = \left(\frac{1}{2}\right)l - l_A$$

$$= 0.90 \text{ m} - 0.60 \text{ m}$$

$$= 0.30 \text{ m}$$

For a ladder, which has a constant density, the center of mass is at the center rung. Substitute $\frac{l}{2} = 0.90$ m and $l_A = 0.60$ m into the equation.

Solve for r_B.

$$r_B = (0.90 \text{ m} - l_B) + (0.90 \text{ m} - l_A)$$

$$= (0.90 \text{ m} - 0.15 \text{ m}) + (0.90 \text{ m} - 0.60 \text{ m})$$

$$= 0.75 \text{ m} + 0.30 \text{ m}$$

$$= 1.05 \text{ m}$$

Substitute $l_B = 0.15$ m and $l_A = 0.60$ m into the equation.

Calculate F_B.

$$F_B = \frac{r_g mg}{r_B}$$

$$= \frac{(0.30 \text{ m})(5.8 \text{ kg})(9.8 \text{ N/kg})}{(1.05 \text{ m})}$$

$$= 16 \text{ N}$$

Substitute $r_g = 0.30$ m, $m = 5.8$ kg, $g = 9.8$ N/kg, and $r_B = 1.05$ m into the equation.

Calculate F_A.

$$F_A = mg\left(1 - \frac{r_g}{r_B}\right)$$

$$= (5.8 \text{ kg})(98 \text{ N/kg})\left(1 - \frac{(0.30 \text{ m})}{(1.05 \text{ m})}\right)$$

$$= 41 \text{ N}$$

Substitute $r_g = 0.30$ m, m = 5.8 kg, $g = 9.8$ N/kg, and $r_B = 1.05$ m into the equation.

3 EVALUATE THE ANSWER

- **Are the units correct? Forces are measured in newtons.**
- **Do the signs make sense? Both forces are upward.**
- **Is the magnitude realistic? The forces add up to the weight of the ladder, and the force exerted by the sawhorse closer to the center of mass is greater, which makes sense.**

PRACTICE Problems

ADDITIONAL PRACTICE

37. What would be the forces exerted by the two sawhorses if the ladder in Example Problem 5 had a mass of 11.4 kg?

38. A 7.3-kg ladder, 1.92 m long, rests on two sawhorses, as shown in **Figure 20.** Sawhorse A, on the left, is located 0.30 m from the end, and sawhorse B, on the right, is located 0.45 m from the other end. Choose the axis of rotation to be the center of mass of the ladder.

a. What are the torques acting on the ladder?

b. Write the equation for rotational equilibrium.

c. Solve the equation for F_A in terms of F_g.

d. How would the forces exerted by the two sawhorses change if A were moved very close to, but not directly under, the center of mass?

39. A 4.5-m-long wooden plank with a 24-kg mass has two supports. One is directly under the center of the board, and the other is at one end. What forces are exerted by the two supports?

40. CHALLENGE A 85-kg diver walks to the end of a diving board. The board, which is 3.5 m long with a mass of 14 kg, is supported at the center of mass of the board and at one end. What are the forces on the two supports?

Figure 20

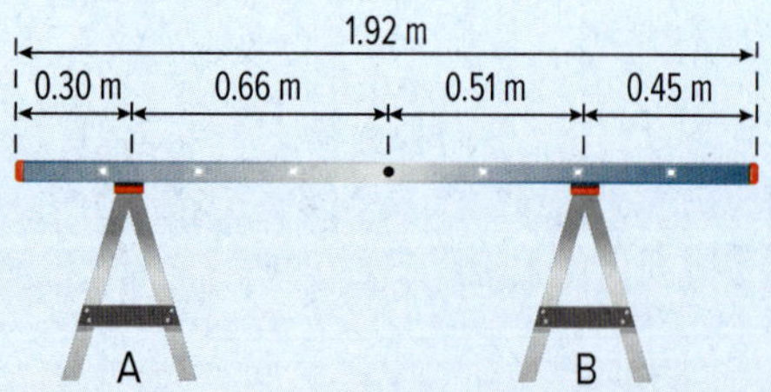

Rotating Frames of Reference

When you sit on a spinning amusement-park ride, it feels as if a strong force is pushing you to the outside. A pebble on the floor of the ride would accelerate outward without a horizontal force being exerted on it in the same direction. The pebble would not move in a straight line relative to the floor of the ride. In other words, Newton's laws of motion would not seem to apply. This is because in the ride your point of view, called your frame of reference, is rotating. Newton's laws are valid only in non-rotating or nonaccelerated frames of reference. Motion in a rotating reference frame is important to us because Earth rotates. The effects of the rotation of Earth are too small to be noticed in the classroom or lab, but they are significant influences on the motion of the atmosphere and, therefore, on climate and weather.

Centrifugal "Force" Suppose you fasten one end of a spring to the center of a rotating platform. An object lies on the platform and is attached to the other end of the spring. As the platform rotates, an observer on the platform sees the object stretch the spring. The observer might think that some force toward the outside of the platform is pulling on the object. This apparent force seems to pull on a moving object but does not exert a physical outward push on it. This apparent force, which seems to push an object outward, is observed only in rotating frames of reference and is called the **centrifugal "force."** It is not a real force because there is no physical outward push on the object.

As the platform rotates, an observer on the ground would see things differently. This observer sees the object moving in a circle, and it accelerates toward the center because of the force of the spring. As you know, this centripetal acceleration is given by $a_c = \frac{v^2}{r}$. It also can be written in terms of angular velocity, as $a_c = \omega^2 r$. Centripetal acceleration is proportional to the distance from the axis of rotation and depends on the square of the angular velocity. Thus, if you double the rotational frequency, the angular acceleration increases by a factor of four.

The Coriolis "Force" A second effect of rotation is shown in **Figure 21.** Suppose a person standing at the center of a rotating disk throws a ball toward the edge of the disk. Consider the horizontal motion of the ball as seen by two observers, and ignore the vertical motion of the ball as it falls. An observer standing outside the disk sees the ball travel in a straight line at a constant speed toward the edge of the disk. However, the other observer, who is stationed on the disk and rotating with it, sees the ball follow a curved path at a constant speed. The apparent force that seems to deflect a moving object from its path and is observed only in rotating frames of reference is called the **Coriolis "force."** Like the centrifugal "force," the Coriolis "force" is not a real force. It seems to exist because we observe a deflection in horizontal motion when we are in a rotating frame of reference.

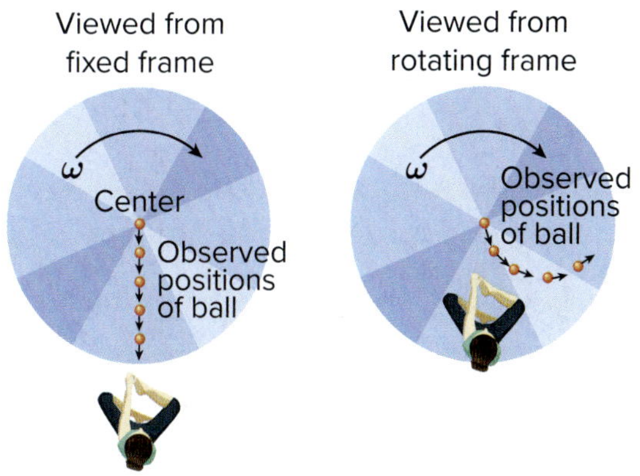

Figure 21 The Coriolis "force" is not a real force. It exists only in rotating reference frames.

Coriolis "force" due to Earth's rotation Suppose a cannon is fired from a point on the equator toward a target due north of it. If the projectile were fired directly northward, it would also have an eastward velocity component because of the rotation of Earth. This eastward speed is greater at the equator than at any other latitude. Thus, as the projectile moves northward, it also moves eastward faster than points on Earth below it do. The result is that the projectile lands east of the target as shown in **Figure 22.** While an observer in space would see Earth's rotation, an observer on Earth could claim that the projectile missed the target because of the Coriolis "force" on the rocket. Note that for objects moving toward the equator, the direction of the apparent force is westward. A projectile will land west of the target when fired due south.

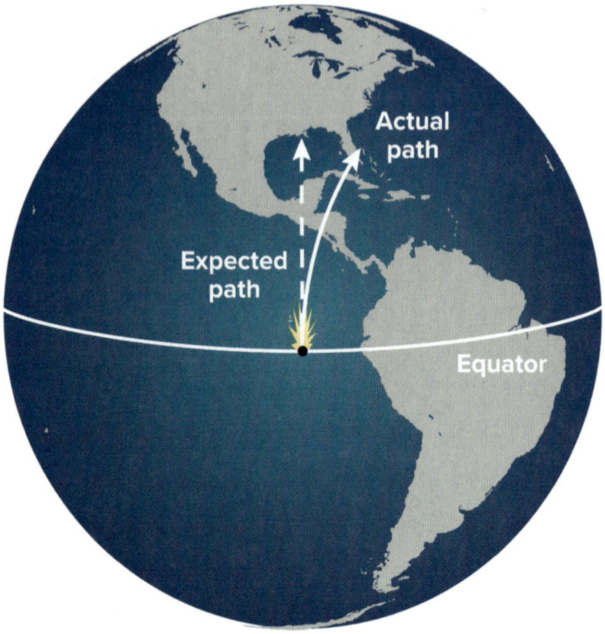

Figure 22 An observer on Earth sees the Coriolis "force" cause a projectile fired due north to deflect to the right of the intended target.

EARTH SCIENCE Connection The direction of winds around high- and low-pressure areas results from the Coriolis "force." Winds flow from areas of high to low pressure. Because of the Coriolis "force" in the northern hemisphere, winds from the south go to the east of low-pressure areas. Winds from the north end up west of low-pressure areas. Therefore, winds rotate counterclockwise around low-pressure areas in the northern hemisphere. In the southern hemisphere, however, winds rotate clockwise around low-pressure areas. This is why tropical cyclones, also called hurricanes, rotate clockwise in the southern hemisphere and counterclockwise in the northern hemisphere.

Get It?

Explain why hurricanes Katia, Irma, and Jose, mentioned at the beginning of this module, rotate counterclockwise.

Check Your Progress

41. **Equilibrium** Give an example of an object for each of the following conditions.
    a. rotational but not translational equilibrium
    b. translational but not rotational equilibrium
42. **Location of Center of Mass** Can the center of mass of an object be located in an area where the object has no mass? Explain.
43. **Stability of an Object** Why is a modified vehicle with its body raised high on risers less stable than a similar vehicle with its body at normal height?
44. **Center of Mass** Where is the center of mass on a roll of masking tape?
45. **Center of Mass** Describe how you would find the center of mass of this textbook.
46. **Rotating Frames of Reference** A penny is placed on a rotating old-fashioned record turntable. At the highest speed, the penny starts sliding outward. What are the forces acting on the penny?
47. **Critical Thinking** Predict the direction of the flow of surface ocean currents in the northern and southern hemispheres.

LEARNSMART® Go online to follow your personalized learning path to review, practice, and reinforce your understanding.

Quantum Jump

The sport of figure-skating is physics in motion. Some of the fundamental concepts of physics—force, motion, velocity, torque—are the key to success in this beautiful display of athleticism. The quadruple jump, one of most difficult in figure skating, is a stellar example of basic mechanics.

In this photo, Japanese figure-skater Yuzuru Hanyu completes a quadruple jump during the men's figure-skating short program of the 2014 Winter Olympics.

The "Quad" Revolution

Skaters complete a quadruple, "quad," jump when they leap into the air, rotate four times, and then land back on the ice. In 2017, 17-year-old American skater Nathan Chen became the first skater in history to land five quad jumps in a single program during competition.

The Physics of a Jump

When skaters set up for a jump, they first increase their velocity, and thus their linear momentum. Then, they push downward on the ice with the ball of one foot to launch themselves into the air. This torque generates angular momentum and puts them into a spin.

During a jump, skaters' legs are close together. This position increases their rotational speed. To spin even faster, they pull their arms in tight against their body. This action reduces their moment of inertia by changing the distribution of their weight, so their bodies rotate faster due to conservation of angular momentum. Skaters attempting a quadruple jump contract their bodies as much as possible in order to complete the four required rotations.

Skaters finish a jump by bringing their arms away from their bodies and extending one of their legs as they land on the other. Skaters coming down from a quad jump land with an enormous impact force—somewhere between eight and ten times their body weight. They do all this while keeping their balance and continuing with the rest of their routine.

Not every quadruple jump is the same. Depending on the type of jump, skaters launch themselves with their toe pick or the inside of their foot, and they launch from and land on different feet. The skater even switches direction while completing a quad jump called the Lutz. But no matter the type they choose, the quadruple jump is a marvel of motion.

ASK QUESTIONS TO CLARIFY

Brainstorm several questions about the physics of figure skating. Use print or online sources to find answers to your questions. Make a poster that lists your questions and answers. Display the poster in your classroom.

CALIFORNIA CONNECTION

Use an online search engine to research one of the many Olympic figure skaters from California. Choose one of the jumps the skater often uses (or used) and write an explanation of the mechanics involved in the jump.

STUDY GUIDE

GO ONLINE to study with your Science Notebook.

Lesson 1 DESCRIBING ROTATIONAL MOTION

- radian
- angular displacement
- angular velocity
- angular acceleration

- Angular displacement is the change in the angle (θ) as an object rotates. It is usually measured in degrees or radians.
- Average angular velocity is the object's angular displacement divided by the time taken to make the angular displacement. Average angular velocity is represented by the Greek letter omega (ω) and is determined by the following equation:

$$\omega = \frac{\Delta\theta}{\Delta t}$$

- Average angular acceleration is the change in angular velocity divided by the time required to make the change.

$$\alpha = \frac{\Delta\omega}{\Delta t}$$

Lesson 2 ROTATIONAL DYNAMICS

- lever arm
- torque
- moment of inertia
- Newton's second law for rotational motion

- Torque describes the combination of a force and a lever arm that can cause an object to rotate. Torque is represented by the Greek letter tau (τ) and is determined by the following equation:

$$\tau = Fr \sin\theta$$

- The moment of inertia is a point object's resistance to changes in angular velocity. The moment of inertia is represented by the letter I and for a point mass, it is represented by the following equation:

$$I = mr^2$$

- Newton's second law for rotational motion states that angular acceleration is directly proportional to the net torque and inversely proportional to the moment of inertia.

$$\alpha = \frac{\tau_{net}}{I}$$

Lesson 3 EQUILIBRIUM

- center of mass
- centrifugal "force"
- Coriolis "force"

- The center of mass of an object is the point on the object that moves in the same way that a point particle would move.
- An object is stable against rollover if its center of mass is above its base.
- An object in equilibrium has no net force exerted on it and there is no net torque acting on it.
- Centrifugal "force" and the Coriolis "force" are two apparent, but nonexistent, forces that seem to exist when an object is analyzed from a rotating frame of reference.

THREE-DIMENSIONAL THINKING

Module Wrap-Up

REVISIT THE PHENOMENON

Why do all tropical cyclones in the northern hemisphere rotate the same direction?

CER Claim, Evidence, Reasoning

Explain Your Reasoning Revisit the claim you made when you encountered the phenomenon. Summarize the evidence you gathered from your investigations and research and finalize your Summary Table. Does your evidence support your claim? If not, revise your claim. Explain why your evidence supports your claim.

STEM UNIT PROJECT

Now that you've completed the module, revisit your STEM unit project. You will apply your evidence from this module and complete your project.

GO FURTHER

SEP Data Analysis Lab

Is the system stable?

Madhu is installing a pivoted pole lamp in in front of her house, as shown in the diagram. The pole weighs 27 N, the lamp weighs 64 N, and the rope can support a tension of 60 N.

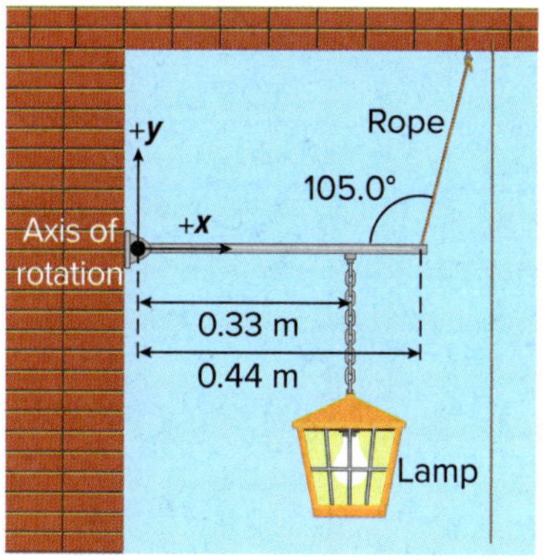

CER Analyze and Interpret Data

1. Determine the torque for each caused by the pole and by the lamp.
2. Determine the torque the rope must produce to keep the set-up in static equilibrium.
3. **Claim** Will Madhu's set up work or will the rope break?
4. **Evidence and Reasoning** Justify your claim.

Martin Sterba/ZUMA Press/Newscom

UNIT 3

MOMENTUM AND ENERGY

ENCOUNTER THE PHENOMENON

Why is energy important to humans and society?

SEP Ask Questions

What questions do you have about the phenomenon? Write your questions on sticky notes and add them to the driving question board for this unit.

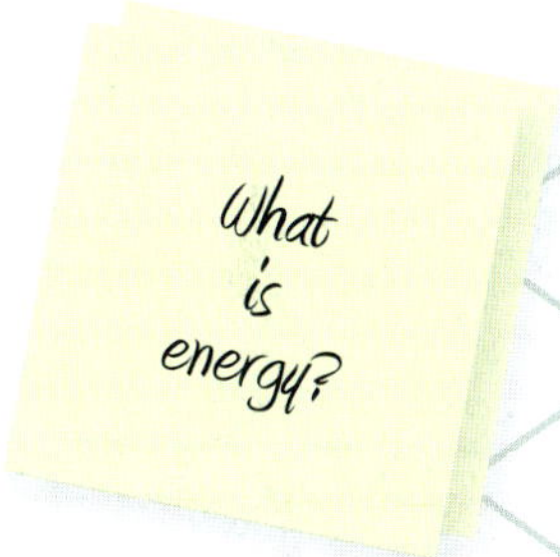

Look for Evidence

As you go through this unit, use the information and your experiences to help you answer the phenomenon question as well as your own questions. For each activity, record your observations in a Summary Table, add an explanation, and identify how it connects to the unit and module phenomenon questions.

Solve a Problem

STEM UNIT PROJECT

Crash Safety When designing a new car, the safety of the passengers is a major consideration for engineers. Investigate what happens in a collision and possible methods to reduce harm to passengers. Use the results of these investigations and the evidence you collected during the unit to design, evaluate, and refine a device that minimizes harm to an object during a collision.

GO ONLINE In addition to reading the information in your Student Edition, you can find the STEM Unit Project and other useful resources online.

NASA/Joel Kowsky

MODULE 9

MOMENTUM AND ITS CONSERVATION

ENCOUNTER THE PHENOMENON

How do rockets accelerate once they reach space?

GO ONLINE to watch a video about early work on rockets.

SEP Ask Questions

Do you have other questions about the phenomenon? If so, add them to the driving question board.

CER Claim, Evidence, Reasoning

Make Your Claim Use your CER chart to make a claim about how rockets accelerate in space. Explain your reasoning.

Collect Evidence Use the lessons in this module to collect evidence to support your claim. Record your evidence as you move through the module.

Explain Your Reasoning You will revisit your claim and explain your reasoning at the end of the module.

GO ONLINE to access your CER chart and explore resources that can help you collect evidence.

LESSON 1: Explore & Explain: Using the Impulse-Momentum Theorem to Save Lives

LESSON 2: Explore & Explain: Two Particle Collisions

Smithsonian

Additional Resources

(t)Video Supplied by BBC Worldwide Learning, (bl)National Crash Analysis Center (NCAC)/Image Bank Film/Getty Images, (br)Koldunov/Shutterstock

LESSON 1

IMPULSE AND MOMENTUM

FOCUS QUESTION

How do air bags work to save lives?

Impulse-Momentum Theorem

What happens when a baseball player hits a home run? According to Newton's second law of motion, the force from the bat changes the ball's velocity. This force changes over time, as shown in **Figure 1.** Just after contact, the ball is squeezed, and the force increases to a maximum more than 10,000 times the weight of the ball. The ball then recovers its shape and rebounds from the bat. The force on the ball rapidly returns to zero. This whole event takes place within about 3.0 ms. How can you calculate the change in velocity of the baseball?

Impulse

Newton's second law of motion ($\mathbf{F} = m\mathbf{a}$) can be rewritten in terms of the change in velocity divided by the time for that change.

$$\mathbf{F} = m\mathbf{a} = m\left(\frac{\Delta \mathbf{v}}{\Delta t}\right)$$

Multiplying both sides by the time interval (Δt) results in this equation.

$$\mathbf{F}\Delta t = m\Delta \mathbf{v} = m(\mathbf{v}_f - \mathbf{v}_i)$$

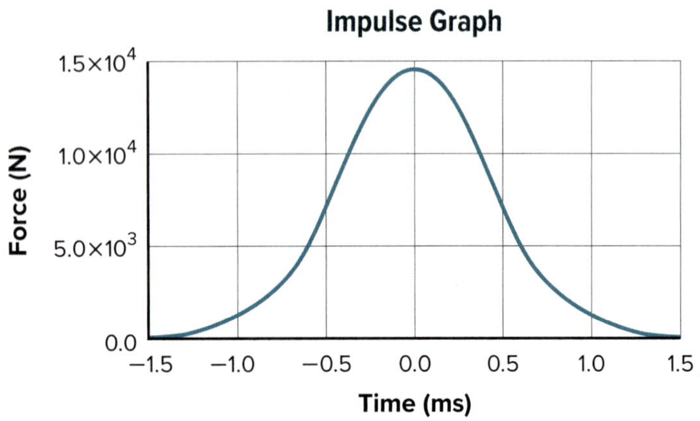

Figure 1 The force on a baseball increases then decreases during a collision.

The left side of the equation ($\mathbf{F}\Delta t$) is defined as the **impulse.** The impulse on an object is the product of the average force on an object and the time interval over which it acts. Impulse is measured in newton-seconds. If the force **F** varies with time, the magnitude of the impulse equals the area under the curve of a force-time graph, as in **Figure 1.**

3D THINKING DCI Disciplinary Core Ideas CCC Crosscutting Concepts SEP Science & Engineering Practices

COLLECT EVIDENCE

Use your Science Journal to record the evidence you collect as you complete the readings and activities in this lesson.

INVESTIGATE

GO ONLINE to find these activities and more resources.

Applying Practices: Egg Heads

HS-PS2-3. Apply science and engineering ideas to design, evaluate, and refine a device that minimizes the force on a macroscopic object during a collision.

PhysicsLAB: Sticky Collisions

Analyze data from a collision to determine how momentum is transferred within a system.

Momentum The right side of the equation ($m\Delta v$) involves the change in velocity, where $m\Delta v = mv_f - mv_i$. The product of the object's mass (m) and the object's velocity (v) is defined as the **momentum** of the object. Momentum is measured in kg·m/s. It is always defined for a particular frame of reference. From a stationary frame of reference, a moving object such as a car has momentum. To the driver of the car, the car has zero momentum because the car has zero velocity in its own frame of reference. When finding momentum, assume the frame of reference is stationary unless otherwise indicated.

Momentum

The momentum of an object equals the mass of the object times the object's velocity.

$$p = mv$$

Now, you can rewrite the impulse as $F\Delta t = m\Delta v = mv_f - mv_i = p_f - p_i$. Thus, the impulse on an object is equal to the change which is called the **impulse-momentum theorem.**

Impulse-Momentum Theorem

An impulse acting on an object is equal to the object's final momentum minus the object's initial momentum.

$$F\Delta t = p_f - p_i$$

If the force on an object is constant, the impulse is the product of the force multiplied by the time interval over which it acts. If force is not constant, you can find the impulse using an average force multiplied by the time interval over which it acts, or by finding the area under a force-time graph. Momentum and impulse are vectors because velocity and force are vectors.

Using the impulse-momentum theorem The impulse from a bat changes a baseball's momentum. In **Figure 1,** the area under the curve is about 15 N·s. The direction of the impulse is in the direction of the force. Therefore, the change in momentum of the ball also is 15 N·s, or 15 kg·m/s, in the direction of the force. Suppose a batter hits a fastball. Assume the positive direction is toward the pitcher. Before the collision, the ball, with a mass of 0.145 kg, has a velocity of −47 m/s. Therefore, the baseball's momentum is $p_i = (0.145\ \text{kg})(-47\ \text{m/s}) = -6.8\ \text{kg·m/s}$. The momentum of the ball after the bat hits it is found by solving $p_f = p_i + F\Delta t$.

$$p_f = p_i + F\Delta t = -6.8\ \text{kg·m/s} + 15\ \text{kg·m/s} = +8.2\ \text{kg·m/s}$$

Because $p_f = mv_f$, solving for v_f yields the ball's final velocity:

$$v_f = \frac{p_f}{m} = \frac{+8.2\ \text{kg·m/s}}{+0.145\ \text{kg}} = +57\ \text{m/s}$$

A speed of 52 m/s is fast enough to clear most outfields if the ball is hit in the right direction.

SCIENCE USAGE v. COMMON USAGE

Impulse

Science usage: the product of average net force and the time interval over which the force acts

The impulse on the rocket from its exhaust lifted the astronauts into orbit.

Common usage: a sudden tendency toward some action

After seeing the commercial, Amad had an impulse to eat a candy bar.

Momentum

Science usage: the product of an object's mass and its velocity

A moving truck usually has much more momentum than a moving bicycle.

Common usage: strength gained through the development of events

The team hoped to continue the momentum of the regular season into the play-offs.

EXAMPLE Problem 1

AVERAGE FORCE A 2200-kg vehicle traveling at 94 km/h (26 m/s) can be stopped in 21 s by gently applying the brakes. It can be stopped in 3.8 s if the driver slams on the brakes, or in 0.22 s if it hits a concrete wall. What is the impulse exerted on the vehicle? What average force is exerted on the vehicle in each of these stops?

1 ANALYZE AND SKETCH THE PROBLEM

- Sketch the system.
- Include a coordinate axis. Select the car's direction as positive.
- Draw a vector diagram for momentum and impulse.

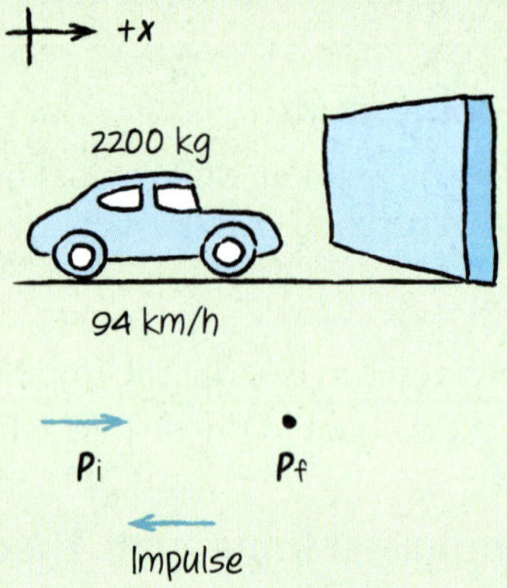

Known		Unknown
$m = 2200$ kg	$\Delta t_{\text{gentle braking}} = 21$ s	$F\Delta t = ?$
$\mathbf{v}_i = +26$ m/s	$\Delta t_{\text{hard braking}} = 3.8$ s	$\mathbf{F}_{\text{gentle braking}} = ?$
$\mathbf{v}_f = 0.0$ m/s	$\Delta t_{\text{hitting a wall}} = 0.22$ s	$\mathbf{F}_{\text{hard braking}} = ?$
		$\mathbf{F}_{\text{hitting a wall}} = ?$

2 SOLVE FOR THE UNKNOWN

Determine the initial momentum ($\mathbf{p}_i$).

$$\mathbf{p}_i = m\mathbf{v}_i$$

$$= (2200 \text{ kg})(+26 \text{ m/s})$$ Substitute $m = 2200$ kg, $\mathbf{v}_i = +26$ m/s.

$$= +5.7\times10^4 \text{ kg·m/s}$$

Determine the final momentum ($\mathbf{p}_f$).

$$\mathbf{p}_f = m\mathbf{v}_f$$

$$= (2200 \text{ kg})(0.0 \text{ m/s})$$ Substitute $m = 2200$ kg, $\mathbf{v}_f = 0.0$ m/s.

$$= 0.0 \text{ kg·m/s}$$

Apply the impulse-momentum theorem to determine the impulse and the force needed to stop the vehicle.

$$\mathbf{F}\Delta t = \mathbf{p}_f - \mathbf{p}_i$$

$$\mathbf{F}\Delta t = (0.0 \text{ kg·m/s}) - (5.7\times10^4 \text{ kg·m/s})$$ Substitute $\mathbf{p}_f = 0.0$ kg·m/s, $\mathbf{p}_i = 5.7\times10^4$ kg·m/s.

$$= -5.7\times10^4 \text{ kg·m/s}$$

$$\mathbf{F} = \frac{-5.7\times10^4 \text{ kg·m/s}}{\Delta t}$$

$$\mathbf{F}_{\text{gentle braking}} = \frac{-5.7\times10^4 \text{ kg·m/s}}{21 \text{ s}}$$ Substitute $\Delta t_{\text{gentle braking}} = 21$ s.

$$= -2.7\times10^3 \text{ N}$$

$$\mathbf{F}_{\text{hard braking}} = \frac{-5.7\times10^4 \text{ kg·m/s}}{3.8 \text{ s}}$$ Substitute $\Delta t_{\text{hard braking}} = 3.8$ s.

$$= -1.5\times10^4 \text{ N}$$

$$\mathbf{F}_{\text{hitting a wall}} = \frac{-5.7\times10^4 \text{ kg·m}}{0.22 \text{ s}}$$ Substitute $\Delta t_{\text{hitting a wall}} = 0.22$ s.

$$= -2.6\times10^5 \text{ N}$$

3 EVALUATE THE ANSWER

Are the units correct? Impulse is in kg·m/s. Force is measured in newtons.

Does the direction make sense? Force is exerted in the direction opposite to the velocity of the car, which means the force is negative.

Is the magnitude realistic? It is reasonable that the force needed to stop a car is thousands or hundreds of thousands of newtons. The impulse is the same for all stops. Thus, as the stopping time decreases by about a factor of 10, the force increases by about a factor of 10.

PRACTICE Problems

ADDITIONAL PRACTICE

1. A compact car, with mass 725 kg, is moving at 115 km/h toward the east. Sketch the moving car.
   a. Find the magnitude and direction of its momentum. Draw an arrow on your sketch showing the momentum.
   b. A second car, with a mass of 2175 kg, has the same momentum. What is its velocity?
2. The driver of the compact car in the previous problem suddenly applies the brakes hard for 2.0 s. As a result, an average force of 5.0×10^3 N is exerted on the car to slow it down.
   a. What is the change in momentum, or equivalently, what is the magnitude and direction of the impulse on the car?
   b. Complete the "before" and "after" sketches, and determine the momentum and the velocity of the car now.
3. A 7.0-kg object, moving at 2.0 m/s, receives two impulses (one after the other) along the direction of its motion. Both of these impulses are illustrated in **Figure 2.** Find the resulting speed and direction of motion of the object after each impulse.
4. The driver accelerates a 240.0-kg snowmobile, which results in a force being exerted that speeds up the snowmobile from 6.00 m/s to 28.0 m/s over a time interval of 60.0 s.
   a. Sketch the event, showing the initial and final situations.
   b. What is the snowmobile's change in momentum? What is the impulse on the snowmobile?
   c. What is the magnitude of the average force that is exerted on the snowmobile?
5. **CHALLENGE** Suppose a 60.0-kg person was in the vehicle that hit the concrete wall in Example Problem 1. The velocity of the person equals that of the car both before and after the crash, and the velocity changes in 0.20 s. Sketch the problem.
   a. What is the average force exerted on the person?
   b. Some people think they can stop their bodies from lurching forward in a vehicle that is suddenly braking by putting their hands on the dashboard. Find the mass of an object that has a weight equal to the force you just calculated. Could you lift such a mass? Are you strong enough to stop your body with your arms?

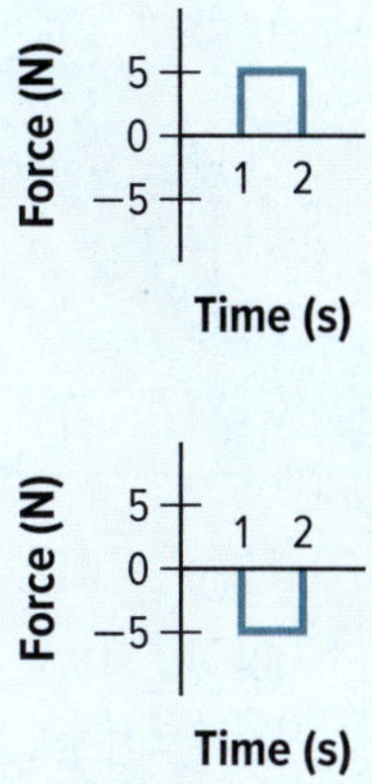

Figure 2

Using the impulse-momentum theorem to save lives The impulse-momentum theorem shows that a large impulse causes a large change in momentum. This large impulse could result either from a large force acting over a short period of time or from a smaller force acting over a longer period of time. A passenger in a car accident might experience a large impulse, but it is the force on the passenger that causes injuries.

What happens to the driver when a crash suddenly stops a car? An impulse, either from the dashboard or from an air bag, brings the driver's momentum to zero. According to the impulse-momentum equation, $F\Delta t = p_f - p_i$. The final momentum (p_f) is zero. The initial momentum (p_i) is the same with or without an air bag. Thus, the impulse ($F\Delta t$) also is the same. An air bag, such as the one shown in **Figure 3,** increases the time interval during which the force acts on the passenger. Therefore the required force is less. The air bag also spreads the force over a larger area of the person's body, thereby reducing the likelihood of injuries.

Figure 3 An air bag reduces injuries by making the force on a passenger less and by spreading that force over a larger area.

Figure 4 The ball's linear momentum is the product of its mass and its velocity. The ball's angular momentum as it rolls down the lane is the product of its moment of inertia and its angular velocity.

Angular Momentum

The impulse-momentum theorem is useful if the momentum of an object is linear, but how can you describe momentum that is angular, as it is for the rotating bowling ball in **Figure 4?** Recall Newton's law for rotational motion, $\tau = \frac{I\Delta\omega}{\Delta t}$. You can rearrange this relationship to produce $\tau\Delta t = I\Delta\omega$. The left side of this equation ($\tau\Delta t$) is the rotating object's angular impulse. You can rewrite the right side as $I\Delta\omega = I\omega_f - I\omega_i$. The product of a rotating object's moment of inertia and angular velocity is called **angular momentum,** which is represented by the symbol L. The following relationship describes an object's angular momentum.

Angular Momentum

The angular momentum of an object is defined as the product of the object's moment of inertia and the object's angular velocity.

$$L = I\omega$$

Angular momentum is measured in kg·m²/s. Just as an object's linear momentum changes when an impulse acts on it, the object's angular momentum changes when an angular impulse acts on it. The object's angular impulse is equal to the change in the object's angular momentum, as stated by the **angular impulse-angular momentum theorem.** This theorem can be represented by the following relationship.

Angular Impulse-Angular Momentum Theorem

The angular impulse on an object is equal to the object's final angular momentum minus the object's initial angular momentum.

$$\tau\Delta t = L_f - L_i$$

If the net force on an object is zero, its linear momentum is constant. If the net torque acting on an object is zero, its angular momentum is also constant, but the two situations are slightly different. Because an object's mass cannot change, if its momentum is constant, then its velocity is also constant. In the case of angular momentum, however, the object's angular velocity can change if its shape changes. This is because the moment of inertia depends on the object's mass and how it is distributed about the axis of rotation. Thus, an object's angular velocity can change even if no torques act on it.

ASTRONOMY Connection Astronomers have discovered many examples of two stars that orbit each other. Such stars are called a binary star system. The torque on the binary system is zero because the gravitational force acts only directly between the stars. Therefore, the binary system's angular momentum is constant. But sometimes the stars are so close that the strong gravitational force rips material from one star and deposits it on the other. This movement of matter changes the moment of inertia of the system. As a result, the angular velocity of the system changes, even though the angular momentum is unchanged.

Real-World Physics

PADDLE SPORTS A table-tennis player controls the flight of the ball by putting spin on the ball. The player hits the ball at an angle so that the paddle exerts both an impulse to change the momentum and an angular impulse to make the ball spin. The interaction with the air will cause a forward-spinning ball to drop rapidly, making it more difficult for the opponent to return.

Figure 5 Before she dives, the girl's center of mass is in front of her feet. During her dive, the path of her center of mass forms a parabola. She changes her moment of inertia by moving her arms and legs.

Analyze *How does extending her arms affect the diver's motion?*

Diving into a pool Consider the diver in **Figure 5.** How does she start rotating her body? She uses the diving board to apply an external force to her body. Then, she moves her center of mass in front of her feet and uses the board to give a final upward push to her feet. This provides a torque that acts over time (Δt) and increases the angular momentum of the diver. Before the diver reaches the water, she can change her angular velocity by changing her moment of inertia. She may go into a tuck position, grabbing her knees with her hands. By moving her mass closer to the axis of rotation, the diver decreases her moment of inertia and increases her angular velocity. When she nears the water, she stretches her body straight, thereby increasing the moment of inertia and reducing the angular velocity. As a result, she goes straight into the water.

Ice-skating An ice-skater uses a similar method to spin. To begin rotating on one foot, the ice-skater causes the ice to exert a force on her body by pushing a portion of one skate into the ice. If the skater in **Figure 6** pushes on the ice in one direction, the ice will exert a force on her in the opposite direction. The force results in a torque if the force is exerted some distance away from the pivot point and in a direction that is not toward it. The greatest torque for a given force will result if the push is perpendicular to the lever arm. The ice-skater then can control her angular velocity by changing her moment of inertia. Both arms and one leg can be extended from the body to slow the rotation or can be pulled in close to the axis of rotation to speed it up. To stop spinning, another torque must be exerted by using the free skate to create a way for the ice to exert the needed force.

Figure 6 When the skater pushes off the ice, the ice exerts a torque on the skater.

CCC CROSSCUTTING CONCEPTS

Cause and Effect Summarize the cause-and-effect relationships that ice-skaters use to spin. Describe how you could gather evidence to test the various ways the skater in **Figure 6** can control her angular velocity.

EXAMPLE Problem 2

ANGULAR IMPULSE AND CHANGE IN ANGULAR MOMENTUM At the conclusion of his competition routine, an ice-skater spins in one spot at a rate of 2.5 rotations per second. If you assume that his moment of inertia is 0.50 kg·m², what is his angular momentum? At the conclusion of the spin, the skater slowly and dramatically slows to a stop over 3.0 s by dragging the toe on one skate. What was the value and sign of the angular impulse on the skater? What average torque did the ice exert on the skater?

1 ANALYZE AND SKETCH THE PROBLEM

- Sketch the system.
- Draw the axis of rotation and the direction of the angular impulse.

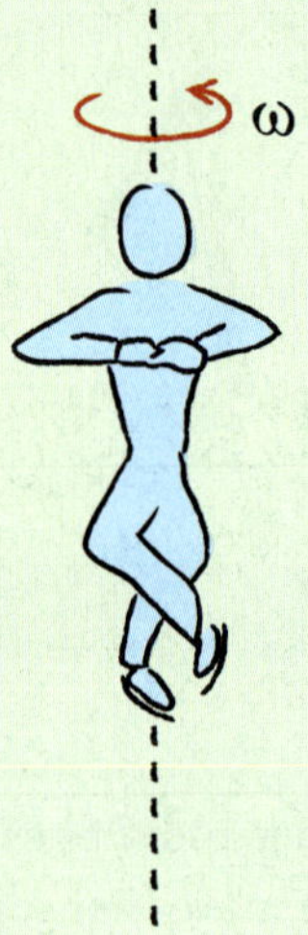

Known	Unknown
$\omega_i = 2.5$ rotations/s	$L_i = ?$
$\omega_f = 0$ rad/s	angular impulse = ?
$I = 0.50$ kg·m²	$\tau_{avg} = ?$
$\Delta t = 3.0$ s	
$L_f = 0.0$ kg·m²/s	

2 SOLVE FOR THE UNKNOWN

Determine the initial angular momentum.

$$L_i = I\omega_i$$

$$= (0.50\ \text{kg·m}^2)(2.5\ \text{rotations/s})(2\pi\ \text{rad/rotation})$$

Substitute $I = 0.50$ kg·m² and $\omega_i = 2.5$ rotations/s. Convert from rotations to radians.

$$= 7.9\ \text{kg·m}^2\text{/s}$$

Radians is a dimensionless unit and is removed.

Determine the angular impulse that the ice exerted on the skater.

$$\text{angular impulse} = \tau_{avg}\,\Delta t$$

$$= L_f - L_i$$

The angular impulse equals the change in angular momentum.

$$= 0.0\ \text{kg·m}^2\text{/s} - 7.9\ \text{kg·m}^2\text{/s}$$

Substitute $L_f = 0.0$ kg·m²/s, $L_i = 7.9$ kg·m²/s.

$$= -7.9\ \text{kg·m}^2\text{/s}$$

Determine the average torque that the ice exerted on the skater.

From the angular impulse, you can rearrange the quantities to find the average torque.

$$\tau_{avg} = \frac{L_f - L_i}{\Delta t}$$

Solve for the average torque.

$$= \frac{0.0\ \text{kg·m}^2\text{/s} - 7.9\ \text{kg·m}^2\text{/s}}{3.0\ \text{s}}$$

Substitute $L_f = 0.0$ kg·m²/s, $L_i = 7.9$ kg·m²/s, and $\Delta t = 3.0$ s.

$$= -2.6\ \text{N·m}$$

The unit kg·m²/s² is equivalent to N·m.

3 EVALUATE THE ANSWER

- **Are the units correct?** Yes, a kg·m²/s² is the equivalent of a N·m.
- **Do the signs make sense?** The angular impulse and the torque are each negative, indicating that the angular momentum decreases.
- **Are the magnitudes realistic?** Yes, the value of the torque is small, but the product of the torque and time equals the value of the angular momentum.

PRACTICE Problems

ADDITIONAL PRACTICE

6. A 0.25-m-diameter circular saw blade rotates at 5.0×10^3 rpm, as shown in **Figure 7.** The saw is turned off and takes several seconds for to slow to a complete stop. The moment of inertia of the blade is 8.0×10^{-3} kg·m². Friction in the axle provides an average torque of 2.3×10^{-1} N·m to slow the blade. How many seconds does it take for the blade to stop?

7. A baseball pitcher throws a 132 km/h (82 mph) curve ball that rotates about 6.0×10^2 rpm. What is the angular velocity of the ball? The pitcher's motion lasts about 0.15 s, and the moment of inertia of the ball is 8.0×10^{-5} kg·m². What average torque did the pitcher exert?

8. A bicycle is clamped upside down on a workbench for the bicycle repair woman to repair a front wheel axle. She gives the front wheel a spin with her hand, and the wheel rotates at 5.0 rev/s. What is the angular velocity of the wheel? If the moment of inertia of the wheel is 0.060 kg·m², what angular impulse did the repair woman give the wheel?

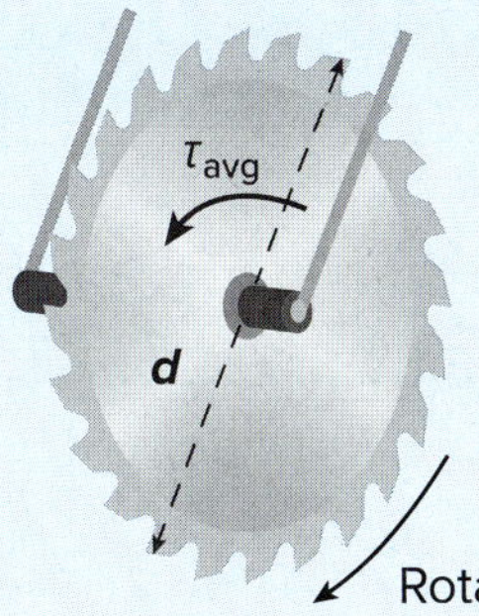

Figure 7

Check Your Progress

9. **Momentum** Which has more momentum, a truck that is parked or a falling raindrop? Explain.

10. **Momentum** Is the momentum of a car traveling south different from that of the same car when it travels north at the same speed? Draw the momentum vectors to support your answer.

11. **Impulse and Momentum** When you jump from a height to the ground, you bend at the knees as your feet hit the floor. Explain why you do this in terms of impulse and momentum.

12. **Impulse and Momentum** A 0.174-kg softball is pitched horizontally at 26.0 m/s. The ball moves in the opposite direction at 38.0 m/s after it is hit by the bat.
    a. Draw arrows showing the ball's momentum before and after the bat hits it.
    b. What is the change in momentum of the ball?
    c. What is the impulse delivered by the bat?
    d. If the bat and ball are in contact for 0.80 ms, what is the average force the bat exerts?

13. **Momentum** The speed of a basketball is the same just before and after the ball hits the floor. Are the impulse on and the change in momentum of the basketball zero when the ball hits the floor? If not, in which direction is the change in momentum? Draw the ball's momentum vectors before and after impact.

14. **Frame of Reference** Why is momentum defined for a particular frame of reference? Explain.

15. **Angular Momentum** The ice-skater shown in **Figure 8** spins with his arms outstretched. When he pulls his arms in and raises them above his head, he spins much faster than before. Did a torque act on the ice-skater? What caused his angular velocity to increase?

Figure 8

16. **Critical Thinking** An archer shoots arrows at a target. Some of the arrows stick in the target, while others bounce off. Assuming that the masses of the arrows and the velocities of the arrows are the same, which arrows produce a bigger impulse on the target? *Hint: Draw a diagram to show the momentum of the arrows before and after hitting the target for the two instances.*

LEARNSMART® Go online to follow your personalized learning path to review, practice, and reinforce your understanding.

LESSON 2

CONSERVATION OF MOMENTUM

FOCUS QUESTION

What is recoil and how is it useful?

Two-Particle Collisions

What will happen when the two balls shown in **Figure 9** collide? During the collision, each ball briefly exerts a force on the other. Despite the differences in sizes and velocities of the balls, the forces they exert on each other are equal and opposite, according to Newton's third law of motion. These forces are represented by the equation $\boldsymbol{F}_{\text{red on blue}} = -\boldsymbol{F}_{\text{blue on red}}$.

How do the impulses imparted by both balls compare? Because the time intervals over which the forces are exerted are the same, the impulses the balls exert on each other must also be equal in magnitude but opposite in direction: $(\boldsymbol{F}\Delta t)_{\text{red on blue}} = -(\boldsymbol{F}\Delta t)_{\text{blue on red}}$. Notice that no mention has been made of the masses of the balls. Even though the balls have different sizes and approach each other with different velocities, and even though they may have different masses, the forces and the impulses they exert on each other have equal strength but opposite directions.

Figure 9 The balls have different sizes, masses, and velocities, but if they interact the forces and impulses they exert on each other are equal and opposite.

Richard Hutchings/Digital Light Source

3D THINKING DCI Disciplinary Core Ideas CCC Crosscutting Concepts SEP Science & Engineering Practices

COLLECT EVIDENCE

Use your Science Journal to record the evidence you collect as you complete the readings and activities in this lesson.

INVESTIGATE

GO ONLINE to find these activities and more resources.

Applying Practices: Conservation of Momentum
HS-PS2-2. Use mathematical representations to support the claim that the total momentum of a system of objects is conserved when there is no net force on the system.

Revisit the Encounter the Phenomenon Question
What information from this lesson can help you answer the Unit and Module questions?

Momentum How did the momenta of the two balls the girl and the boy rolled toward each other in **Figure 9** change as a result of the collision? Compare the changes shown in **Figure 10.**

For ball C: $\boldsymbol{p}_{\text{Cf}} - \boldsymbol{p}_{\text{Ci}} = \boldsymbol{F}_{\text{D on C}}\Delta t$

For ball D: $\boldsymbol{p}_{\text{Df}} - \boldsymbol{p}_{\text{Di}} = \boldsymbol{F}_{\text{C on D}}\Delta t$

According to the impulse-momentum theorem, the change in momentum is equal to the impulse. Because the impulses are equal in magnitude but opposite in direction, we know that:

$$\boldsymbol{p}_{\text{Cf}} - \boldsymbol{p}_{\text{Ci}} = -(\boldsymbol{p}_{\text{Df}} - \boldsymbol{p}_{\text{Di}})$$

$$\boldsymbol{p}_{\text{Cf}} + \boldsymbol{p}_{\text{Df}} = \boldsymbol{p}_{\text{Ci}} + \boldsymbol{p}_{\text{Di}}$$

This equation states that the sum of the momenta of the balls in **Figure 10** is the same before and after the collision. That is, the momentum gained by ball D is equal to the momentum lost by ball C. If the system is defined as the two balls, the sum of the gain and loss in momentum is zero, and therefore, momentum is conserved for the system.

Momentum in a Closed, Isolated System

Under what conditions is the momentum of the system of two balls conserved? The first and most obvious condition is that no balls are lost and no balls are gained. Such a system, which does not gain or lose mass, is said to be a **closed system.**

The second condition required to conserve the momentum of a system is that the only forces that are involved are internal forces; that is, there are no unbalanced forces acting on the system from objects outside of it. When the net external force exerted on a closed system is zero, the resulting system is described as an **isolated system.** No system on Earth is absolutely isolated, however, because there will always be some interactions between a system and its surroundings. Often, these interactions are small enough to be ignored when solving physics problems.

Systems can contain any number of objects, and the objects that make up a system can stick together or come apart during a collision. But even under these conditions, momentum is conserved. The **law of conservation of momentum** states that the momentum of any closed, isolated system does not change. This law enables you to make a connection between conditions before and after an interaction without knowing any details of the interaction. If a system is not closed and isolated, and interacts with objects outside itself, the total momentum of the system can change. However, any such change to the system is balanced by changes in the momentum of objects outside the system.

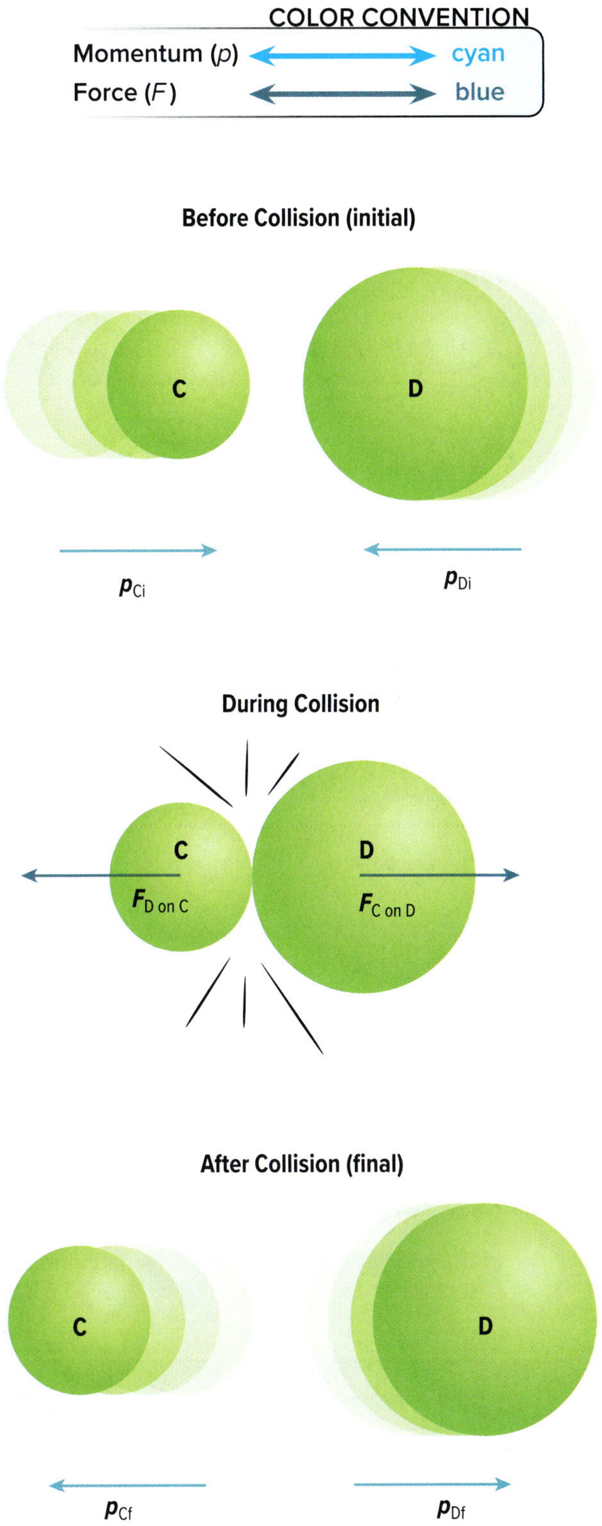

Figure 10 When two balls of different mass and velocity collide, the momentum of each ball is changed. If the system is isolated, however, the sum of their momenta before the collision equals the sum of their momenta after the collision.

Analyze *How can you tell from the diagram that the impulses exerted by the balls have the same magnitude?*

EXAMPLE Problem 3

SPEED A 1875-kg car going 23 m/s rear-ends a 1025-kg compact car going 17 m/s on ice in the same direction. The two cars stick together. How fast do the two attached cars move immediately after the collision?

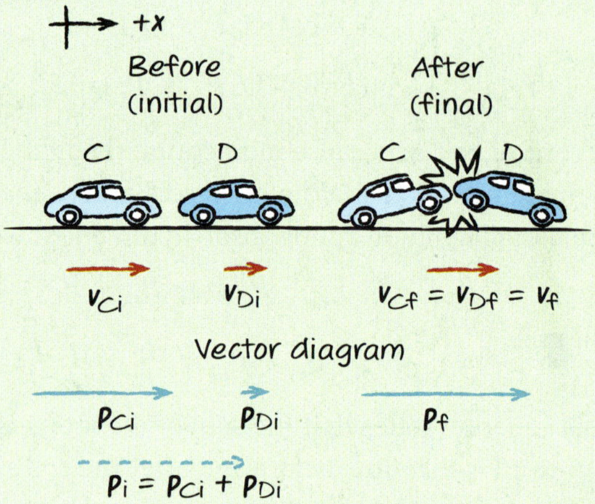

1 ANALYZE AND SKETCH THE PROBLEM

- Define the system.
- Establish a coordinate system.
- Sketch the situation showing the before and after states.
- Draw a vector diagram for the momentum

Known		**Unknown**
$m_C = 1875$ kg	$m_D = 1025$ kg	$v_f = ?$
$v_{Ci} = +23$ m/s	$v_{Di} = +17$ m/s	

2 SOLVE FOR THE UNKNOWN

The system is the two cars. Momentum is conserved because the ice makes the total external force on the cars nearly zero.

$$p_i = p_f$$
$$p_{Ci} + p_{Di} = p_{Cf} + p_{Df}$$
$$m_C v_{Ci} + m_D v_{Di} = m_C v_{Cf} + m_D v_{Df}$$

Because the two cars stick together, their velocities after the collision, denoted as v_f, are equal.

$$v_{Cf} = v_{Df} = v_f$$
$$m_C v_{Ci} + m_D v_{Di} = (m_C + m_D)v_f$$

Solve for v_f.

$$v_f = \frac{m_C v_{Ci} + m_D v_{Di}}{m_C + m_D}$$

$$= \frac{(1875\text{ kg})(+23\text{ ms}) + (1025\text{ kg})(+17\text{ ms})}{1875\text{ kg} + 1025\text{ kg}}$$

Substitute $m_C = 1875$ kg, $v_{Ci} = +23$ m/s, $m_D = 1025$ kg, $v_{Di} = +17$ m/s.

$$= +21\text{ m/s}$$

3 EVALUATE THE ANSWER

- **Are the units correct?** Velocity is measured in meters per second.
- **Does the direction make sense?** v_{Ci} and v_{Di} are in the positive direction; therefore, v_f should be positive.
- **Is the magnitude realistic?** The magnitude of v_f is between the initial speeds of the two cars, but closer to the speed of the more massive one, so it is reasonable.

PRACTICE Problems

ADDITIONAL PRACTICE

17. Two freight cars, each with a mass of 3.0×10^5 kg, collide and stick together. One was initially moving at 2.2 m/s, and the other was at rest. What is their final speed? Define the system as the two cars.

18. A 0.105-kg hockey puck moving at 24 m/s is caught and held by a 75-kg goalie at rest. With what speed does the goalie slide on the ice after catching the puck? Define the puck and the goalie as a system.

19. A 35.0-g bullet strikes a 2.5-kg steel ball that is at rest. After the collision, the bullet bounces backward at 5.0 m/s. How fast is the ball moving when the bullet bounces backward?

20. CHALLENGE A 0.50-kg ball traveling at 6.0 m/s collides head-on with a 1.00-kg ball moving in the opposite direction at 12.0 m/s. After colliding, the 0.50-kg ball bounces backward at 14 m/s. Find the other ball's speed and direction after the collision.

Figure 11 The boy exerts a force, causing the girl to move to the right. The boy's position relative to the ball shows that the equal but opposite force from the girl caused him to recoil to the left.

Infer *How can you tell from the motion that the boy's mass is greater than the girl's mass?*

Recoil

It is important to define a system carefully. To see why, consider the in-line skaters in **Figure 11.** Can you find their final velocities? Assume they are skating on a smooth surface with no external forces. They both start at rest, one behind the other. Skater C, the boy, gives skater D, the girl, a push. Now, both are moving in opposite directions. Because the push was an internal force, you can use the law of conservation of momentum to find the skaters' relative velocities. The total momentum of the system was zero before the push. Therefore, it must be zero after the push.

Before		After
$p_{Ci} + p_{Di}$	$=$	$p_{Cf} + p_{Df}$
0	$=$	$p_{Cf} + p_{Df}$
p_{Cf}	$=$	$-p_{Df}$
$m_C v_{Cf}$	$=$	$-m_D v_{Df}$

The coordinate system can be chosen so that the positive direction is to the right. The momenta of the skaters after the push are equal in magnitude but opposite in direction. The backward motion of skater C is an example of recoil. Are the skaters' velocities equal and opposite? The last equation can be written to solve for the velocity of skater C.

$$v_{Cf} = \left(\frac{-m_D}{m_C}\right) v_{Df}$$

The velocities depend on the skaters' relative masses. If skater C has a mass of 68.0 kg and skater D's mass is 45.4 kg, then the ratio of their velocities will be 68.0 : 45.4, or 1.50. The less massive skater moves at the greater velocity. Without more information about how hard skater C pushed skater D, however, you cannot find the velocity of each skater.

CCC CROSSCUTTING CONCEPTS

Systems and System Models Examine **Figure 11.** How does the red ball in the photographs provide evidence for analyzing the system, and why is it necessary to help define the system? Suggest another way this could have been achieved, and make a model of the system to illustrate.

STEM CAREER Connection

Computer Graphics Animator

To create realistic-looking computer graphics for movies and video games, animators must understand how objects move in the real world. Applying the laws of motion to animated interactions results in convincing imagery, be it of spaceship crashes or medieval battles.

Propulsion in Space

How does a rocket in space change velocity? The rocket carries both fuel and oxidizer. When the fuel and oxidizer combine in the rocket motor, the resulting hot gases leave the exhaust nozzle at high speed. The rocket and the chemicals form a closed system. The forces that expel the gases are internal forces, so the system is also an isolated system. Thus, objects in space can accelerate by using the law of conservation of momentum and Newton's third law of motion. A newer type of space propulsion uses the recoil that results from the force of ions. An ion thruster accelerates ions in an electric or magnetic field. As the ions are expelled in one direction, the conservation of momentum results in the movement of the spaceship in the opposite direction.

Get It?

Explain how the rocket shown at the beginning of the module can accelerate in space.

EXAMPLE Problem 4

SPEED An astronaut at rest in space fires a thruster pistol that expels a quick burst of 35 g of hot gas at 875 m/s. The combined mass of the astronaut and pistol is 84 kg. How fast and in what direction is the astronaut moving after firing the pistol?

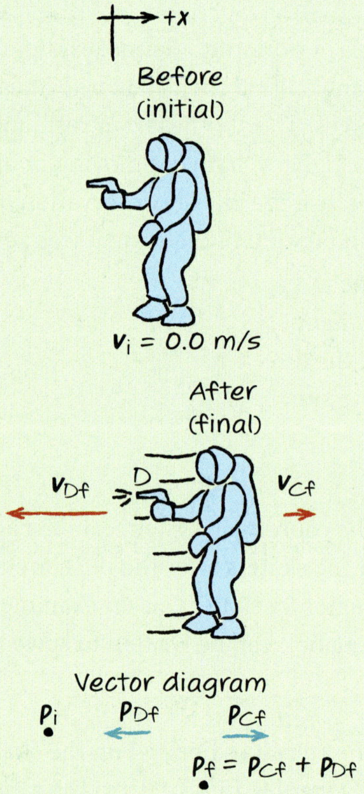

1 ANALYZE AND SKETCH THE PROBLEM

- Define the system and establish a coordinate axis.
- Sketch before and after conditions. Draw a vector diagram.

Known		Unknown
$m_C = 84$ kg	$v_{Ci} = v_{Di} = 0.0$ m/s	$v_{Cf} = ?$
$m_D = 0.035$ kg	$v_{Df} = -875$ m/s	

2 SOLVE FOR THE UNKNOWN

The system is the astronaut, the gun, and the gas.

$p_i = p_{Ci} + p_{Di} = 0.0\ \text{kg·m/s}$ — Before the pistol is fired, the system is at rest; initial momentum is zero.

Use the law of conservation of momentum to find p_{Cf}.

$$p_i = p_f$$
$$0.0\ \text{kg·m/s} = p_{Cf} + p_{Df}$$
$$p_{Cf} = -p_{Df}$$

The momentum of the astronaut is equal in magnitude, but opposite in direction, to the momentum of the gas leaving the pistol.

Solve for the final velocity of the astronaut (v_{Cf}).

$$m_C v_{Cf} = -m_D v_{Df}$$
$$v_{Cf} = \frac{-m_D v_{Df}}{m_C}$$
$$= \left(\frac{-(0.035\ \text{kg})(-875\ \text{m/s})}{84\ \text{kg}}\right)$$
$$= +0.36\ \text{m/s}$$

Substitute $m_D = 0.035$ kg, $v_{Df} = -875$ m/s, $m_C = 84$ kg.

3 EVALUATE THE ANSWER

- **Are the units correct?** The velocity is measured in meters per second.
- **Does the direction make sense?** The astronaut's direction is opposite that of the gas.
- **Is the magnitude realistic?** The astronaut's mass is much greater than that of the gas, so the velocity of the astronaut is much less than that of the expelled gas.

PRACTICE Problems

ADDITIONAL PRACTICE

21. A 4.00-kg model rocket is launched, expelling burned fuel with a mass of 50.0 g at a speed of 625 m/s. What is the velocity of the rocket after the fuel has burned? *Hint: Ignore the external forces of gravity and air resistance.*

22. A thread connects a 1.5-kg cart and a 4.5-kg cart. After the thread is burned, a compressed spring pushes the carts apart, giving the 1.5-kg cart a velocity of 27 cm/s to the left. What is the velocity of the 4.5-kg cart?

23. CHALLENGE Carmen and Judi row their canoe alongside a dock. They stop the canoe, but they do not secure it. The canoe can still move freely. Carmen, who has a mass of 80.0 kg, then steps out of the canoe onto the dock. As she leaves the canoe, Carmen moves forward at a speed of 4.0 m/s, causing the canoe, with Judi still in it, to move also. At what speed and in what direction do the canoe and Judi move if their combined mass is 115 kg?

Two-Dimensional Collisions

Until now you have considered momentum in only one dimension. The law of conservation of momentum holds for all closed systems with no external forces. It is valid regardless of the directions of the particles before or after they interact. But what happens in two or three dimensions? **Figure 12** shows a collision of two billiard balls. Consider the billiard balls to be the system. The original momentum of the moving ball is $\boldsymbol{p}_{\mathrm{Ci}}$, and the momentum of the stationary ball is zero. Therefore, the momentum of the system before the collision is $\boldsymbol{p}_{\mathrm{Ci}}$.

After the collision, both balls are moving and have momenta. Ignoring friction with the tabletop, the system is closed and isolated. Thus, the law of conservation of momentum can be used.

The initial momentum equals the sum of the final momenta, so $\boldsymbol{p}_{\mathrm{Ci}} = \boldsymbol{p}_{\mathrm{Cf}} + \boldsymbol{p}_{\mathrm{Df}}$. The sum of the components of the vectors before and after the collision must also be equal. If the *x*-axis is defined in the direction of the initial momentum, then the *y*-component of the initial momentum is zero. The sum of the final *y*-components also must be zero.

$$p_{\mathrm{Cf,\,y}} + p_{\mathrm{Df,\,y}} = 0$$

The *y*-components are equal in magnitude but are in the opposite direction and, thus, have opposite signs. The sums of the horizontal components before and after the collision also are equal.

$$p_{\mathrm{Ci}} = p_{\mathrm{Cf,\,x}} + p_{\mathrm{Df,\,x}}$$

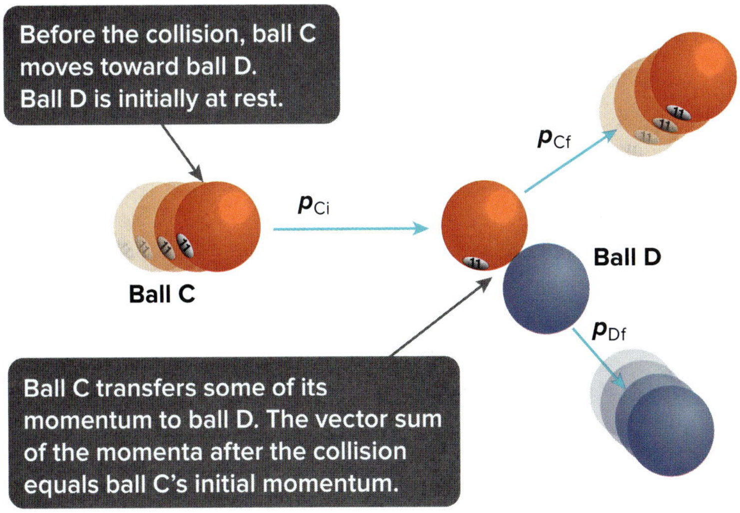

Figure 12 The momentum is conserved in a closed, isolated system, regardless of the velocities of the objects before a collision.

Analyze *The momentum of the system is not conserved if the balls are rolling on a high-friction carpet. Explain why, describing how the change in momentum of the system consisting only of the three balls is balanced by changes in momentum outside the system.*

EXAMPLE Problem 5

SPEED A 1325-kg car, C, moving north at 27.0 m/s, collides with a 2165-kg car, D, moving east at 11.0 m/s. The collision causes the two cars to stick together. In what direction and with what speed do they move after the collision?

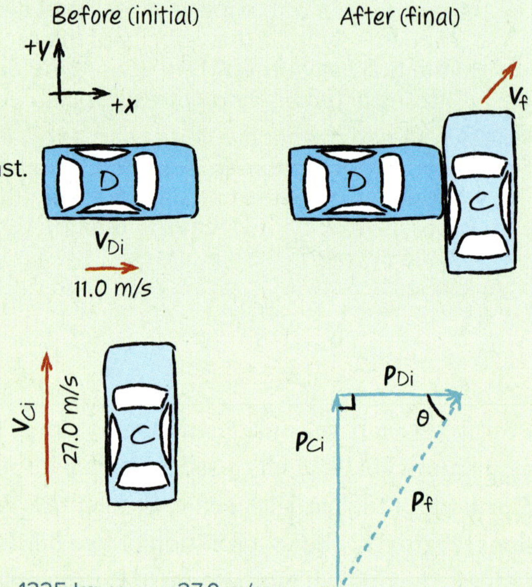

1 ANALYZE AND SKETCH THE PROBLEM

- Define the system.
- Sketch the before and after states.
- Establish the coordinate axis with the *y*-axis north and the *x*-axis east.
- Draw a momentum-vector diagram.

Known	Unknown
$m_C = 1325$ kg	$v_f = ?$
$m_D = 2165$ kg	$\theta = ?$
$v_{Ci,y} = 27.0$ m/s	
$v_{Di,x} = 11.0$ m/s	

2 SOLVE FOR THE UNKNOWN

Define the system as the two cars. Determine the magnitudes of the initial momenta of the cars and the momentum of the system.

$$p_{Ci} = m_C v_{Ci,y}$$
$$= (1325\text{ kg})(27.0\text{ m/s})$$ Substitute $m_C = 1325$ kg, $v_{Ci,y} = 27.0$ m/s.
$$= 3.58\times10^4\text{ kg·m/s (north)}$$

$$p_{Di} = m_D v_{Di,x}$$
$$= (2165\text{ kg})(11.0\text{ m/s})$$ Substitute $m_D = 2165$ kg, $v_{Di,x} = 11.0$ m/s.
$$= 2.38\times10^4\text{ kg·m/s (east)}$$

Use the law of conservation of momentum to find p_f.

$$p_{f,x} = p_{i,x} = 2.38\times10^4\text{ kg·m/s}$$ Substitute $p_{i,x} = p_{Di} = 2.38\times10^4$ kg·m/s.
$$p_{f,y} = p_{i,y} = 3.58\times10^4\text{ kg·m/s}$$ Substitute $p_{i,y} = p_{Ci} = 3.58\times10^4$ kg·m/s.

Use the diagram to set up equations for $p_{f,x}$ and $p_{f,y}$.

$$p_f = \sqrt{(p_{f,x})^2 + (p_{f,y})^2}$$
$$= \sqrt{(2.38\times10^4\text{ kg·m/s})^2 + (3.58\times10^4\text{ kg·m/s})^2}$$ Substitute $p_{f,x} = 2.38\times10^4$ kg·m/s, $p_{f,y} = 3.58\times10^4$ kg·m/s.
$$= 4.30\times10^4\text{ kg·m/s}$$

Solve for θ

$$\theta = \tan^{-1}\left(\frac{p_{f,y}}{p_{f,x}}\right)$$
$$= \tan^{-1}\left(\frac{3.58\times10^4\text{ kg·m/s}}{2.38\times10^4\text{ kg·m/s}}\right)$$ Substitute $p_{f,y} = 3.58\times10^4$ kg·m/s, $p_{f,x} = 2.38\times10^4$ kg·m/s.
$$= 56.4°$$

Determine the final speed.

$$V_f = \frac{p_f}{m_C + m_D}$$
$$= \frac{4.30\times10^4\text{ kg·m/s}}{1325\text{ kg} + 2165\text{ kg}}$$ Substitute $p_f = 4.30\times10^4$ kg·m/s, $m_C = 1325$ kg, $m_D = 2165$ kg.
$$= 12.3\text{ m/s}$$

3 EVALUATE THE ANSWER

- **Are the units correct?** The correct unit for speed is meters per second.
- **Does the direction make sense?** Answers are both positive and at the appropriate angles.
- **Is the magnitude realistic?** The cars stick together, so v_f must be smaller than v_{Ci}.

PRACTICE Problems

ADDITIONAL PRACTICE

24. A 925-kg car moving north at 20.1 m/s collides with a 1865-kg car moving west at 13.4 m/s. After the collision, the two cars are stuck together. In what direction and at what speed do they move after the collision? Define the system as the two cars.

25. A 1383-kg car moving south at 11.2 m/s is struck by a 1732-kg car moving east at 31.3 m/s. After the collision, the cars are stuck together. How fast and in what direction do they move immediately after the collision? Define the system as the two cars.

26. A 1345-kg car moving east at 15.7 m/s is struck by a 1923-kg car moving north. They stick together and move with a velocity of 14.5 m/s at $\theta = 63.5°$. Was the north-moving car exceeding the 20.1 m/s speed limit?

27. **CHALLENGE** A stationary billiard ball with mass 0.17 kg is struck by an identical ball moving 4.0 m/s. Afterwards, the second ball moves 60.0° to the left of its original direction. The stationary ball moves 30.0° to the right of the moving ball's original direction. What is the velocity of each ball after the collision?

Conservation of Angular Momentum

Like linear momentum, angular momentum can be conserved. The **law of conservation of angular momentum** states that if no net external torque acts on a closed system, then its angular momentum does not change, as represented by the following equation.

Law of Conservation of Angular Momentum

An isolated system's initial angular momentum is equal to its final angular momentum.

$$L_i = L_f$$

The spinning ice-skater in **Figure 13** demonstrates conservation of angular momentum. When he pulls in his arms, he spins faster. Without an external torque, his angular momentum does not change and $L = I\omega$ is constant. The increased angular velocity must be accompanied by a decreased moment of inertia. By pulling in his arms, the skater brings more mass closer to the axis of rotation, decreasing the radius of rotation and decreasing his moment of inertia. You can calculate changes in angular velocity using the law of conservation of angular momentum.

$$L_i = L_f$$

$$\text{thus, } I_i\omega_i = I_f\omega_f$$

$$\frac{\omega_f}{\omega_i} = \frac{I_i}{I_f}$$

Figure 13 When an ice-skater tucks his arms, his moment of inertia decreases. Because angular momentum is conserved, his angular velocity increases.

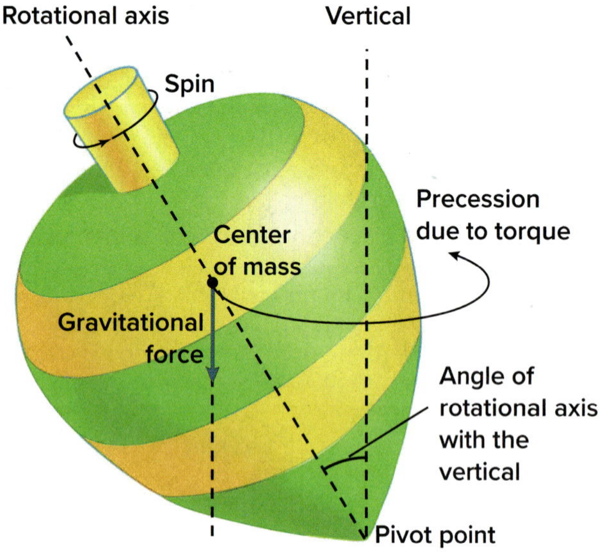

Figure 14 If you spin a top at a tilt, Earth's gravity exerts a torque on it, causing the upper end to process.

Decide *Why does the top precess only when it is tilted?*

Tops and Gyroscopes

Because of the conservation of angular momentum, the direction of rotation of a spinning object can be changed only by applying a torque. If you played with a top as a child, you may have spun it by twisting its axle between your fingers or by pulling the string wrapped around its axle. When a top is vertical, the force of gravity on the center of mass of the top points through the pivot point. As a result, there is no torque on the top, and the axis of its rotation does not change. If the top is tipped, as shown in **Figure 14,** a torque tends to rotate it downward. Rather than tipping over, however, the axis of the top slowly precesses, which means it revolves slightly away from the vertical line.

Earth's rotation is another example of precession of a spinning object. Earth is not a perfect sphere. Because it has an equatorial bulge, the gravitational pull of the Sun exerts a torque on it, causing it to precess. It takes about 26,000 years for Earth's rotational axis to go through one cycle of precession.

Get It?

Analyze Why does Earth precess as it rotates?

PHYSICS Challenge

Colliding Cars Your friend was driving her 1265-kg car north on Oak Street when she was hit by a 925-kg compact car going west on Maple Street. The cars stuck together and slid 23.1 m at 42° north of west. The speed limit on both streets is 22 m/s (50 mph). Define the two cars as a system. Assume that momentum was conserved during the collision and that acceleration was constant during the skid. The coefficient of kinetic friction between the tires and the pavement is 0.65.

1. Your friend claims she was not speeding but the driver of the other car was. How fast was your friend driving before the crash?
2. How fast was the other car moving before the crash? Can you support your friend's case in court?

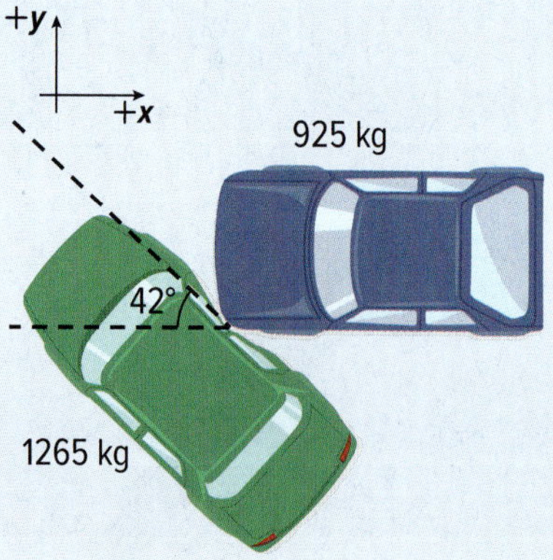

A gyroscope, such as the one shown in **Figure 15,** is a wheel or disk that spins rapidly around one axis while being free to rotate around one or two other axes. The direction of its large angular momentum can be changed only by applying an appropriate torque. Without such a torque, the direction of the axis of rotation does not change.

Gyroscopes are used in airplanes, submarines, and spacecraft to keep an unchanging reference direction. Gyroscopic compasses, unlike magnetic compasses, maintain direction even when they are not on a level surface.

A football quarterback uses the gyroscope effect to make an accurate forward pass. A spiral is a pass in which the football spins around its longer axis. If the quarterback throws the ball in the direction of its spin axis of rotation, the ball keeps its pointed end forward, thereby reducing air resistance. Thus, the ball can be thrown far and accurately. If its spin direction is slightly off, the ball wobbles. If the ball is not spun, it tumbles end over end. Spin also stabilizes the flight of a flying disk. A well-spun plastic disk can fly many meters through the air without wobbling. Some people are able to perform tricks with a yo-yo because the yo-yo's fast rotational speed keeps it rotating in one plane.

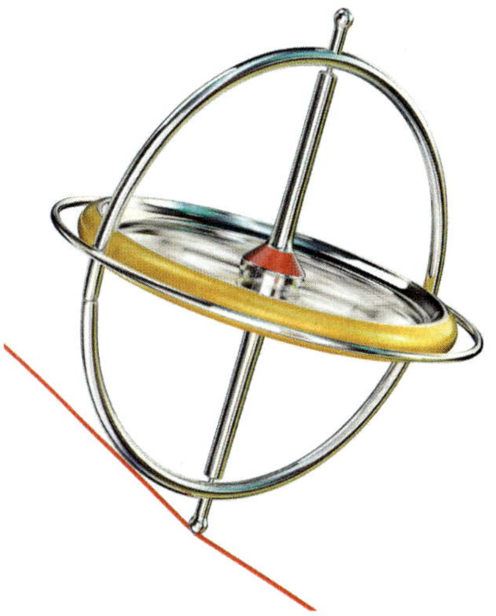

Figure 15 A gyroscope is useful for maintaining a fixed direction because its spin axis does not change orientation, even when the gyroscope is moved.

Check Your Progress

28. **Spin** The outer rim of a plastic disk is thick and heavy. Besides making it easier to catch, how does this affect the rotational properties of the plastic disk?

29. **Speed** A cart, weighing 24.5 N, is released from rest on a 1.00-m ramp, inclined at an angle of 30.0° as shown in **Figure 16.** The cart rolls down the incline and strikes a second cart weighing 36.8 N.
    a. Define the two carts as the system. Calculate the speed of the first cart at the bottom of the incline.
    b. If the two carts stick together, with what initial speed will they move along?

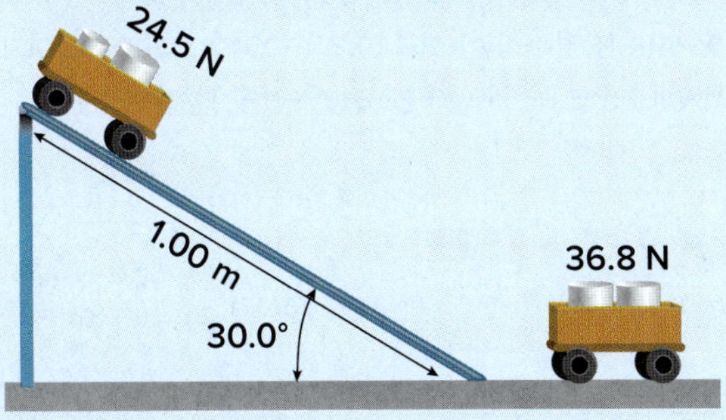

Figure 16

30. **Conservation of Momentum** During a tennis serve, the racket of a tennis player continues forward after it hits the ball. Is momentum conserved in the collision between the tennis racket and the ball? Explain, making sure you define the system.

31. **Momentum** A pole-vaulter runs toward the launch point with horizontal momentum. Where does the vertical momentum come from as the athlete vaults over the crossbar?

32. **Initial Momentum** During a soccer game, two players come from opposite directions and collide when trying to head the ball. The players come to rest in midair and fall to the ground. Describe their initial momenta.

33. **Critical Thinking** You catch a heavy ball while you are standing on a skateboard, and then you roll backward. If you were standing on the ground, however, you would be able to avoid moving while catching the ball.
    a. Identify the system you use in each case.
    b. Explain both situations using the law of conservation of momentum.

LEARNSMART® Go online to follow your personalized learning path to review, practice, and reinforce your understanding.

Pushing Beyond Our Solar System

Propulsion systems push or accelerate objects forward by producing thrust. Space propulsion systems push or accelerate spacecraft and satellites.

Most of today's spacecraft are propelled using liquid-fueled rockets.

Limitations of Rocket Science

Scientists are looking for new ways to move spacecraft and artificial satellites through space. And they are developing ideas for technology that might one day allow us to send spacecraft beyond our solar system.

Most of today's spacecraft are propelled using liquid-fueled rockets that work through chemical combustion. The technology was first developed 90 years ago. It is limited by the energy of the chemical bonds in the fuel. Scientists don't think this technology can be further developed and improved enough to thrust spacecraft to the outer reaches of our solar system and beyond.

Electrical energy has also been used for space propulsion. Thrust is created when the electrical energy produces super-heated plasma that is then fired through a nozzle at a speed faster than sound. The thrust, however, is low and the technology is only useful for changing the position of orbiting satellites.

Potential Future Technologies

One idea is to create thrust by catching photons using a "solar sail." Photons have momentum, and if enough are caught by a solar sail, there could be enough thrust to move spacecraft at very high velocities, without the need for fuel.

Plasma propulsion engines accelerate ions in plasma and generate thrust using electrical potential and magnetic currents. The idea has yet to be tested in space, but a company developing the technology has said it could send a spacecraft to Mars in 39 days.

Scientists are studying several other ideas, some less probable or feasible than others. Indeed, some of the ideas are seen only as not theoretically impossible. However, with our continuing development of knowledge and technology, one of these ideas might finally lead to a method of propulsion that will take us beyond our solar system.

DEVELOP A MODEL TO ILLUSTRATE

Choose a propulsion technology. Research how thrust is created. Develop a model to illustrate how the technology works.

CALIFORNIA CONNECTION

California's Jet Propulsion Laboratory (JPL) is a world leader in robotic exploration of the solar system. Find out more about one of its current missions. How does the spacecraft or satellite involved move through space? What is the purpose of the mission? Share your findings with your classmates.

MODULE 9
STUDY GUIDE

GO ONLINE to study with your Science Notebook.

Lesson 1 IMPULSE AND MOMENTUM

- The impulse on an object is the average net force exerted on the object multiplied by the time interval over which the force acts.

$$impulse = F\Delta t$$

- The momentum of an object is the product of its mass and velocity and is a vector quantity.

$$p = mv$$

When solving a momentum problem, first define the objects in the system and examine their momentum before and after the event. The impulse on an object is equal to the change in momentum of the object.

$$F\Delta t = p_f - p_i$$

- The angular momentum of a rotating object is the product of its moment of inertia and its angular velocity.

$$L = I\omega$$

- The angular impulse-angular momentum theorem states that the angular impulse on an object is equal to the change in the object's angular momentum.

$$\tau\Delta t = L_f - L_i$$

- impulse
- momentum
- impulse-momentum theorem
- angular momentum
- angular impulse–angular momentum theorem

Lesson 2 CONSERVATION OF MOMENTUM

- According to Newton's third law of motion and the law of conservation of momentum, the forces exerted by colliding objects on each other are equal in magnitude and opposite in direction.
- Momentum is conserved in a closed, isolated system.

$$p_f = p_i$$

- The law of conservation of momentum relates the momenta of objects before and after a collision. Use vector analysis to solve momentum-conservation problems in two dimensions. The law of conservation of angular momentum states that if there are no external torques acting on a system, then the angular momentum is conserved.

$$L_f = L_i$$

Because angular momentum is conserved, the direction of rotation of a spinning object can be changed only by applying a torque.

- closed system
- isolated system
- law of conservation of momentum
- law of conservation of angular momentum

THREE-DIMENSIONAL THINKING

Module Wrap-Up

REVISIT THE PHENOMENON

How do rockets accelerate once they reach space?

CER Claim, Evidence, Reasoning

Explain your Reasoning Revisit the claim you made when you encountered the phenomenon. Summarize the evidence you gathered from your investigations and research and finalize your Summary Table. Does your evidence support your claim? If not, revise your claim. Explain why your evidence supports your claim.

STEM UNIT PROJECT

Now that you've completed the module, revisit your STEM unit project. You will summarize your evidence and apply it to the project.

GO FURTHER

SEP Data Analysis Lab

How does velocity change in an inelastic collision?

A 92-kg fullback, running at a speed of 5.0 m/s, attempts to dive directly across the goal line for a touchdown. Just as he reaches the line, he is met head-on in midair by two 75-kg linebackers, both moving in the direction opposite the fullback. One is moving at 2.0 m/s, and the other at 4.0 m/s. They all become entangled as one mass.

CER Analyze and Interpret Data

1. Sketch the before and after situations.
2. What is the players' velocity after the collision?
3. **Claim** Does the fullback score a touchdown?
4. **Evidence and Reasoning** Justify your claim.

NASA/Joel Kowsky

MODULE 10

ENERGY AND ITS CONSERVATION

shan.shihan/Moment Open/Getty Images

MODULE 10

ENERGY AND ITS CONSERVATION

ENCOUNTER THE PHENOMENON

How can energy from power plants be stored in the power grid for later use?

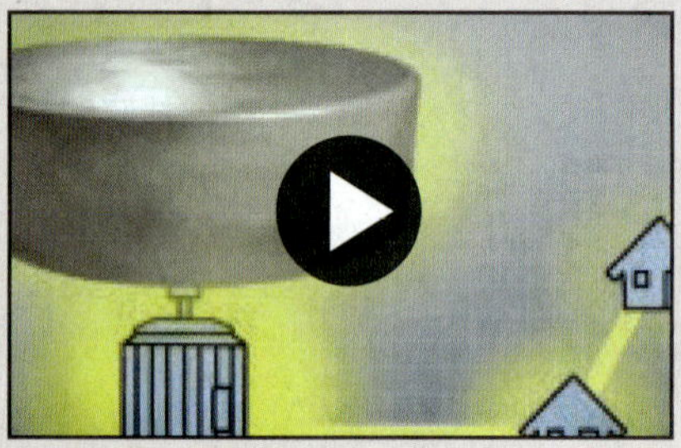

GO ONLINE to watch a video about energy can be stored for later use.

SEP Ask Questions

Do you have other questions about the phenomenon? If so, add them to the driving question board.

CER Claim, Evidence, Reasoning

Make Your Claim Use your CER chart to make a claim about how energy from power plants can be stored in the power grid for later use. Explain your reasoning.

Collect Evidence Use the lessons in this module to collect evidence to support your claim. Record your evidence as you move through the module.

Explain Your Reasoning You will revisit your claim and explain your reasoning at the end of the module.

GO ONLINE to access your CER chart and explore resources that can help you collect evidence.

LESSON 2: Explore & Explain: Gravitational Potential Energy

LESSON 3: Explore & Explain: Analyzing Collisions

Smithsonian

Additional Resources

LESSON 1
WORK AND ENERGY

FOCUS QUESTION
What is energy?

Work

Consider a force exerted on an object while the object moves a certain distance, such as the book bag in **Figure 1.** There is a net force, so the object is accelerated, $a = \frac{F}{m}$, and its velocity changes. Recall from your study of motion that acceleration, velocity, and distance are related by the equation $v_f^2 = v_i^2 + 2ad$. This can be rewritten as $2ad = v_f^2 - v_i^2$. Replace a with the term $\frac{F}{m}$ to get $2\left(\frac{F}{m}\right)d = v_f^2 - v_i^2$. Multiplying both sides by $\frac{m}{2}$ gives $Fd = \frac{1}{2}mv_f^2 - \frac{1}{2}mv_i^2$.

The left side of the equation describes an action that was done to the system by the external world. Recall that a system is the object or objects of interest and the external world is everything else. A force (F) was exerted on a system while the point of contact moved. When a force is applied through a displacement, **work** (W) is done on the system.

The SI unit of work is called a **joule** (J). One joule is equal to 1 N·m. One joule of work is done when a force of 1 N acts on a system over a displacement of 1 m. An apple weighs about 1 N, so it takes roughly 1 N of force to lift the apple at a constant velocity. Thus, when you lift an apple 1 m at a constant velocity, you do 1 J of work on it.

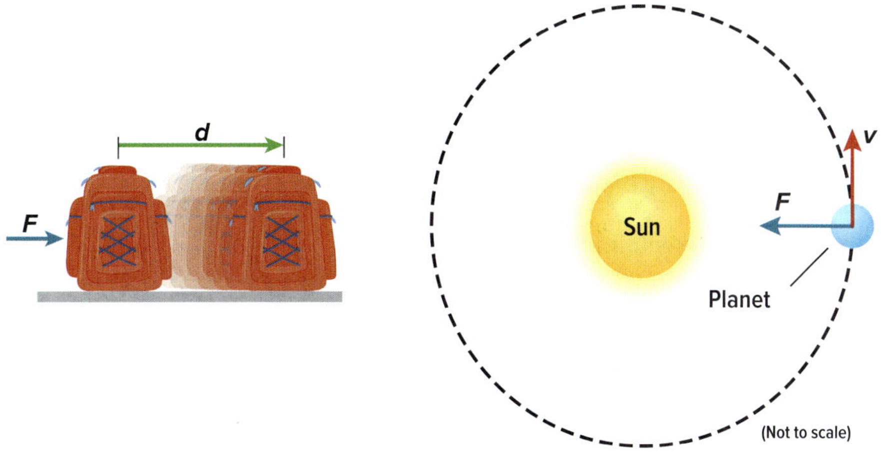

Figure 1 Work is done when a force is applied through a displacement.

Identify *another example of when a force does work on an object.*

3D THINKING | DCI Disciplinary Core Ideas | CCC Crosscutting Concepts | SEP Science & Engineering Practices

COLLECT EVIDENCE

Use your Science Journal to record the evidence you collect as you complete the readings and activities in this lesson.

INVESTIGATE

GO ONLINE to find these activities and more resources.

PhysicsLAB: Stair Climbing and Power
Analyze and interpret data to determine the relationships among force, power, and time.

Quick Investigation: Force Applied at an Angle
Carry out an investigation to determine how the angle at which you apply force affects the amount work done.

Work done by a constant force In the book bag example, F is a constant force exerted in the direction in which the object is moving. In this case, work (W) is the product of the force and the system's displacement. That is,

$$W = Fd.$$

What happens if the exerted force is perpendicular to the direction of motion? For example, for a planet in a circular orbit, the force is always perpendicular to the direction of motion, as shown in **Figure 1.** Recall that a perpendicular force only changes the direction. The speed of the planet doesn't change, so the right side of the equation, $\frac{1}{2}mv_f^2 - \frac{1}{2}mv_i^2$, is zero. Therefore, the work done is also zero.

Constant force exerted at an angle What work does a force exerted at an angle do? For example, what work does the person pushing the car in **Figure 2** do? Recall that any force can be replaced by its components. If you use the coordinate system shown in **Figure 2,** the 125-N force (F) exerted in the direction of the person's arm has two components.

The magnitude of the horizontal component (F_x) is related to the magnitude of the applied force (F) by $\cos 25.0° = \frac{F_x}{F}$. By solving for F_x, you obtain

$$F_x = F \cos 25.0° = (125\ \text{N})(\cos 25.0°) = 113\ \text{N}.$$

Using the same method, the vertical component is

$$F_y = -F \sin 25.0°$$

$$F_y = -(125\ \text{N})(\sin 25.0°) = -52.8\ \text{N}.$$

The negative sign shows that the force is downward. Because the displacement is in the x direction, only the x-component does work. The y-component does no work. The work you do when you exert a force on a system at an angle to the direction of motion is equal to the component of the force in the direction of the displacement multiplied by the displacement.

The magnitude of the component (F_x) force acting in the direction of displacement is found by multiplying the magnitude of force (F) by the cosine of the angle (θ) between the force and the direction of the displacement:

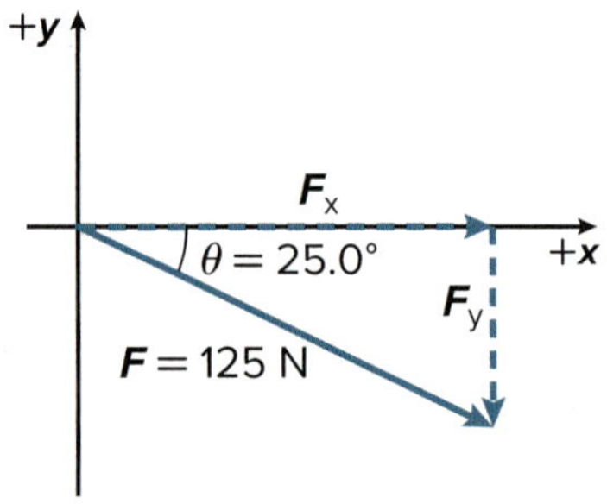

Figure 2 Only the horizontal component of the force that the man exerts on the car does work because the car's displacement is horizontal.

$F_x = F \cos \theta$. Thus, the work done is represented by the following equation.

Work

Work is equal to the product of the magnitude of the force and magnitude of displacement times the cosine of the angle between them.

$$W = Fd \cos \theta$$

Get It?

Determine the work you do when you exert a force of 3 N at an angle of 45° from the direction of motion for 1 m.

The equation above agrees with our expectations for constant forces exerted in the direction of displacement and for constant forces perpendicular to the displacement. In the book bag example, $\theta = 0°$ and $\cos 0° = 1$. Thus, $W = Fd(1) = Fd$, just as we found before. In the case of the orbiting planet, $\theta = 90°$ and $\cos 90° = 0$. Thus, $W = Fd(0) = 0$. This agrees with our previous conclusions.

Figure 3 The total work done on a system is the sum of the work done by each agent that exerts a force on the system.

Describe *the total work done on the box if your friend exerted a greater force than you did. Be sure to consider the direction of the displacement.*

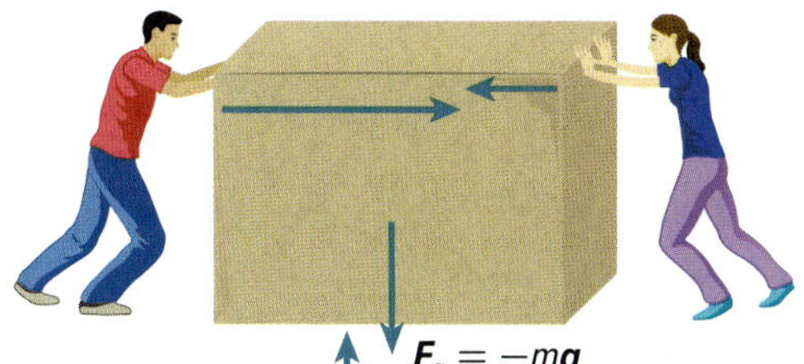

Work done by many forces

Suppose you are pushing a box on a frictionless surface while your friend is trying to prevent you from moving it, as shown in **Figure 3.** What forces are acting on the box? You are exerting a force to the right and your friend is exerting a force to the left. Earth's gravity exerts a downward force, and the ground exerts an upward normal force. How much work is done on the box?

When several forces are exerted on a system, calculate the work done by each force, and then add the results. For the box in **Figure 3,** the upward and downward forces (gravity and the normal force) are perpendicular ($\theta = 90^\circ$) to the direction of motion and do no work. For these forces, $\theta = 90^\circ$, which makes $\cos\theta = 0$, and thus, $W = 0$.

The force you exert ($F_{\text{on box by you}}$) is in the direction of the displacement, so the work is

$$W = F_{\text{on box by you}} d.$$

Your friend exerts a force (F_{friend}) in the direction opposite the displacement ($\theta = 180^\circ$). Because $\cos 180^\circ = -1$, your friend does negative work:

$$W = -F_{\text{on box by friend}} d$$

The total work done on the box is

$$W = F_{\text{on box by you}} d - F_{\text{on box by friend}} d.$$

Get It?

Explain why you do positive work on the box and your friend does negative work on the box.

It is also important to consider each distance separately. Say you push a box 1 m with a force of 3 N. You then pull the box back 1 m to the starting point with a force of 3 N. Is $W = 0$ because the total displacement is zero?

This would be true if the force was constant, but your force changed direction. Your push was in the box's direction of motion for the first part, so you did 3 J of work. In the second part, both the force you exerted and the direction of motion reversed. Your pull and the box's motion were in the same direction, and you did 3 J of work on the box. Therefore, you did a total of 6 J of work on the box.

Get It?

Describe another scenario in which you do work on a system, and explain how much work is done on the system.

SCIENCE USAGE v. COMMON USAGE

Work

Science usage: a force applied through a displacement
The student did work on the car, pushing it out of the mud.

Common usage: physical or mental effort
Studying physics can require a lot of work.

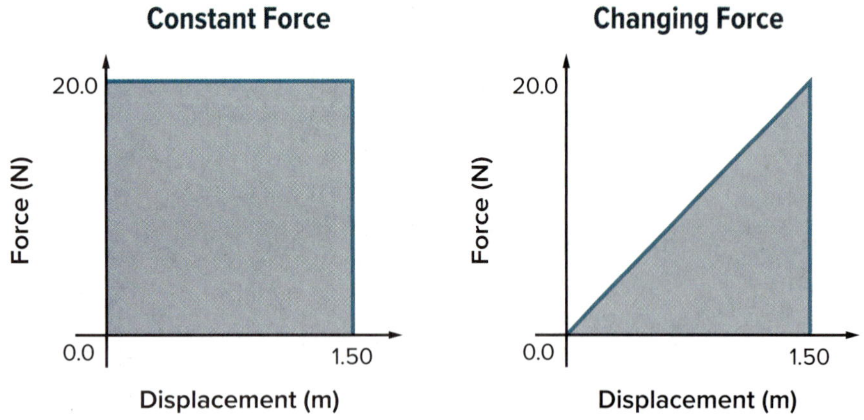

Figure 4 The area under a force-displacement graph is equal to the work.

Finding work done when forces change

In the last example, the force changed but we could determine the work done in each segment. But what if the force changes in a more complicated way?

A graph of force versus displacement lets you determine the work done by a force. This graphical method can be used to solve problems in which the force is changing. The left graph in **Figure 4** shows the work done by a constant force of 20.0 N that is exerted to lift an object 1.50 m. The work done by this force is represented by

$$W = Fd = (20.0\ \text{N})(1.50\ \text{m}) = 30.0\ \text{J}.$$

Note that the shaded area under the left graph is also equal to $(20.0\ \text{N})(1.50\ \text{m})$, or 30.0 J. The area under a force-displacement graph is equal to the work done by that force.

This is true even if the force changes. The right graph in **Figure 4** shows the force exerted by a spring that varies linearly from 0.0 to 20.0 N as it is compressed 1.50 m. The work done by the force that compressed the spring is the area under the graph, which is the area of a triangle, $\left(\frac{1}{2}\right)$(base)(altitude), or

$$W = \left(\tfrac{1}{2}\right)(20.0\ \text{N})(1.50\ \text{m}) = 15.0\ \text{J}.$$

Use the problem-solving strategy below when you solve problems related to work.

PROBLEM-SOLVING STRATEGY

Work

When solving problems related to work, use the following strategy:

1. Identify and sketch the system. Show any forces doing work on the system.
2. Establish a coordinate system. Draw the displacement vectors of the system and each force vector doing work on the system.
3. Find the angle (θ) between each force and displacement.
4. Calculate the work done by each force using $W = Fd \cos \theta$.
5. Calculate the net work done.

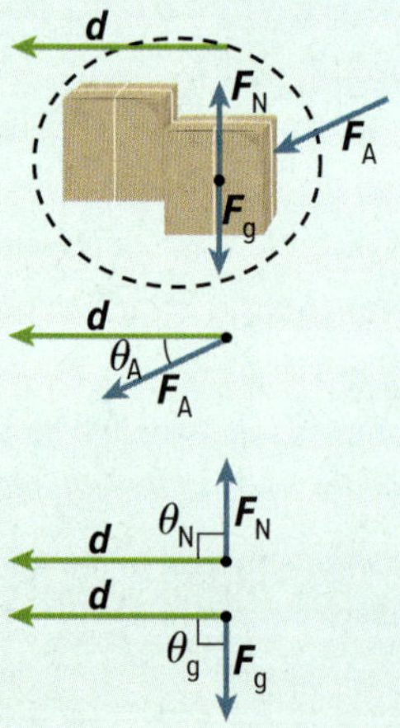

EXAMPLE Problem 1

WORK A hockey player uses a stick to apply a constant 4.50-N force forward to a 105-g puck sliding on ice over a displacement of 0.150 m forward. How much work does the stick do on the puck? Assume friction is negligible.

1 ANALYZE AND SKETCH THE PROBLEM

- Identify the system and the force doing work on it.
- Sketch the situation showing initial conditions.
- Establish a coordinate system with $+x$ to the right.
- Draw a vector diagram.

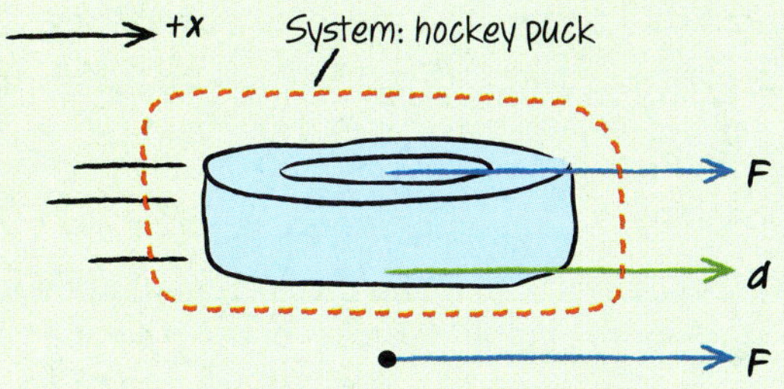

Player does work on hockey puck.

Known	Unknown
$m = 105$ g	$W = ?$
$F = 4.50$ N	
$d = 0.150$ m	
$\theta = 0°$	

2 SOLVE FOR THE UNKNOWN

Use the definition for work.

$$W = Fd\cos\theta$$

$$= (4.50\ \text{N})(0.150\ \text{m})(\cos\theta)$$ Substitute $F = 4.50$ N, $d = 0.150$ m, $\cos\theta = \cos 0° = 1$.

$$= 0.675\ \text{N·m}$$

$$= 0.675\ \text{J}$$ $1\ \text{J} = 1\ \text{N·m}$

3 EVALUATE THE ANSWER

- **Are the units correct?** Work is measured in joules.
- **Does the sign make sense?** The stick (external world) does work on the puck (the system), so the sign of work should be positive.

PRACTICE Problems

ADDITIONAL PRACTICE

1. Refer to Example Problem 1 to solve the following problem.
   a. If the hockey player exerted twice as much force (9.00 N) on the puck over the same distance, how would the amount of work the stick did on the puck be affected?
   b. If the player exerted a 9.00-N force, but the stick was in contact with the puck for only half the distance (0.075 m), how much work does the stick do on the puck?
2. Together, two students exert a force of 825 N in pushing a car a distance of 35 m.
   a. How much work do the students do on the car?
   b. If their force is doubled, how much work must they do on the car to push it the same distance?
3. A rock climber wears a 7.5-kg backpack while scaling a cliff. After 30.0 min, the climber is 8.2 m above the starting point.
   a. How much work does the climber do on the backpack?
   b. If the climber weighs 645 N, how much work does she do lifting herself and the backpack?
4. **CHALLENGE** Marisol pushes a 3.0-kg box 7.0 m across the floor with a force of 12 N. She then lifts the box to a shelf 1 m above the ground. How much work does Marisol do on the box?

EXAMPLE Problem 2

FORCE AND DISPLACEMENT AT AN ANGLE A sailor pulls a boat a distance of 30.0 m along a dock using a rope that makes a 25.0° angle with the horizontal. How much work does the rope do on the boat if its tension is 255 N?

1 ANALYZE AND SKETCH THE PROBLEM

- Identify the system and the force doing work on it.
- Establish coordinate axes.
- Sketch the situation showing the boat with initial conditions.
- Draw vectors showing the displacement, the force, and its component in the direction of the displacement.

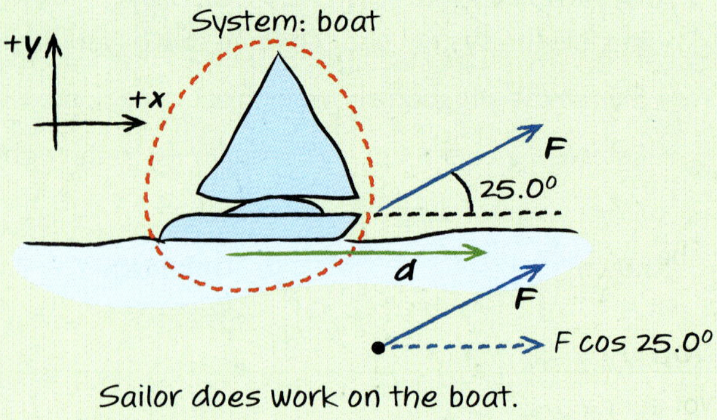

Known		Unknown
$F = 255$ N	$\theta = 25.0°$	$W = ?$
$d = 30.0$ m		

2 SOLVE FOR THE UNKNOWN

Use the definition of work.

$$W = Fd\cos\theta$$

$$= (255\text{ N})(30.0\text{ m})(\cos 25.0°) \quad \text{Substitute } F = 255\text{ N}, d = 30.0\text{ m}, \theta = 25.0°.$$

$$= 6.93\times10^3\text{ J}$$

3 EVALUATE THE ANSWER

- **Are the units correct?** Work is measured in joules.
- **Does the sign make sense?** The rope does work on the boat, which agrees with a positive sign for work.

PRACTICE Problems

ADDITIONAL PRACTICE

5. If the sailor in Example Problem 2 pulls with the same force through the same displacement but at an angle of 50.0°, how much work is done on the boat by the rope?

6. Two people lift a heavy box a distance of 15 m. They use ropes, each of which makes an angle of 15° with the vertical. Each person exerts a force of 225 N. Calculate the work done by the ropes.

7. An airplane passenger carries a 215-N suitcase up the stairs, a displacement of 4.20 m vertically and 4.60 m horizontally.

a. How much work does the passenger do on the suitcase?

b. The same passenger carries the same suitcase back down the same set of stairs. How much work does the passenger do on the suitcase to carry it down the stairs?

8. A rope is used to pull a metal box a distance of 15.0 m across the floor. The rope is held at an angle of 46.0° with the floor, and a force of 628 N is applied to the rope. How much work does the rope do on the box?

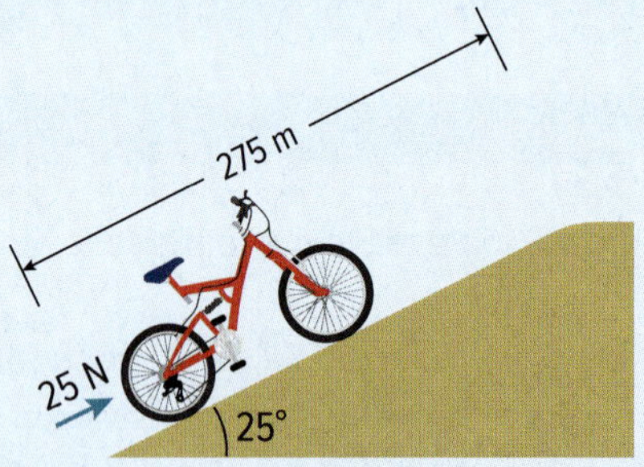

Figure 5

9. CHALLENGE A bicycle rider pushes a 13-kg bicycle up a steep hill. The incline is 25° and the hill is 275 m long, as shown in **Figure 5.** The rider pushes the bike parallel to the road with a force of 25 N.

a. How much work does the rider do on the bike?

b. How much work is done by the force of gravity on the bike?

Energy

Look again at the equation $W = \frac{1}{2}mv_f^2 - \frac{1}{2}mv_i^2$. What property of a system does $\frac{1}{2}mv^2$ describe? A massive, fast-moving vehicle can do damage to objects around it, and a baseball hit at high speed can rise high into the air. That is, a system with this property can produce a change in itself or the world around it. This ability of a system to produce a change in itself or the world around it is called **energy** and is represented by the symbol E.

The right side of the equation, $\frac{1}{2}mv_f^2 - \frac{1}{2}mv_i^2$, indicates a change in a specific kind of energy. That is, work causes a change in energy. This is the **work-energy theorem,** which states that when work is done on a system, the result is a change in the system's energy. The work-energy theorem can be represented by the following equation.

Work-Energy Theorem

Work done on a system is equal to the change in the system's energy.

$$W = \Delta E$$

Since work is measured in joules, energy must also be measured in joules. In fact, the unit gets its name from the nineteenth-century physicist James Prescott Joule, who established the relationship between work done and the change in energy. Recall that 1 joule equals 1 N·m and that 1 N equals 1 kg·m/s^2. Therefore, 1 joule equals 1 kg·m^2/s^2. These are the same units that $\frac{1}{2}mv^2$ has.

Through the process of doing work, energy can move between the external world and the system. The direction of energy transfer can be either way. If the external world does work on a system, then W is positive and the energy of the system increases. If a system does work on the external world, then W is negative and the energy of the system decreases. In summary, work is the transfer of energy that occurs when a force is applied through a displacement.

Changing kinetic energy So far, we have discussed the energy associated with a system's motion. For example, the bobsledders in **Figure 6** do work on their sled to get it moving at the beginning of a race. The energy associated with motion is called **kinetic energy** (KE).

In the case of the bobsled, work resulted in a change in the object's kinetic energy. There are, however, many other forms of energy.

Work can cause a change in these other forms, as well. Some of these forms, such as potential energy and thermal energy, will be explored later in this module and in subsequent modules.

Figure 6 The bobsledders do work on the bobsled when they push it. The result is a change in the bobsled's kinetic energy.

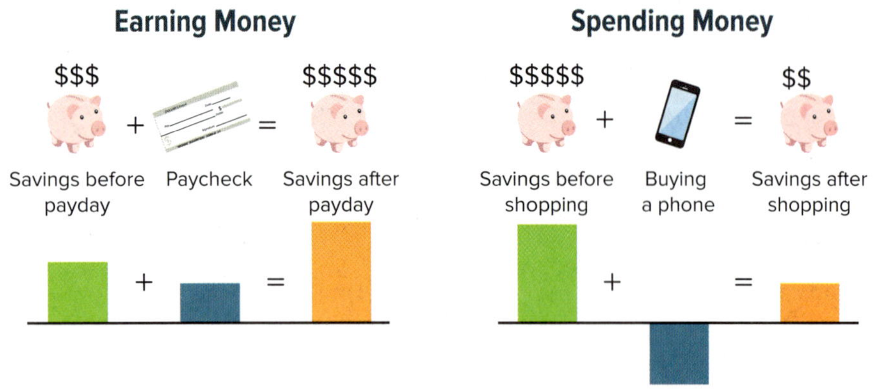

Figure 7 Energy is similar to a savings account. The amount can increase or decrease.

A Model of the Work-Energy Theorem

Keeping track of work and energy changes is like keeping track of your savings account. The amount of money you have changes when you earn more or spend it. Similarly, energy can be stored or it can be used. In each case, the behavior of a system is affected. We can use bar diagrams, such as those shown in **Figure 7,** to track money or energy.

Throwing and catching a ball

Suppose you exert a force to throw a stationary ball. Consequently, the ball gains kinetic energy, as shown on the left in **Figure 8.** In the energy bar diagram, the height of the bar represents the amount of work (in joules).

The kinetic energy after the work is done equals the initial kinetic energy plus the work done on the ball.

Before the moving ball is caught, it has kinetic energy. In catching it, you exert a force on the ball in the direction *opposite* to its motion. Therefore, you do negative work on it, causing it to stop. Now that the ball is not moving, it has no kinetic energy, as represented in the energy bar diagram on the right in **Figure 8.** Kinetic energy is always positive, so KE_{before} is positive. The work done on the ball is negative and KE_{after} is zero. Again,

$$KE_{after} = KE_{before} + W.$$

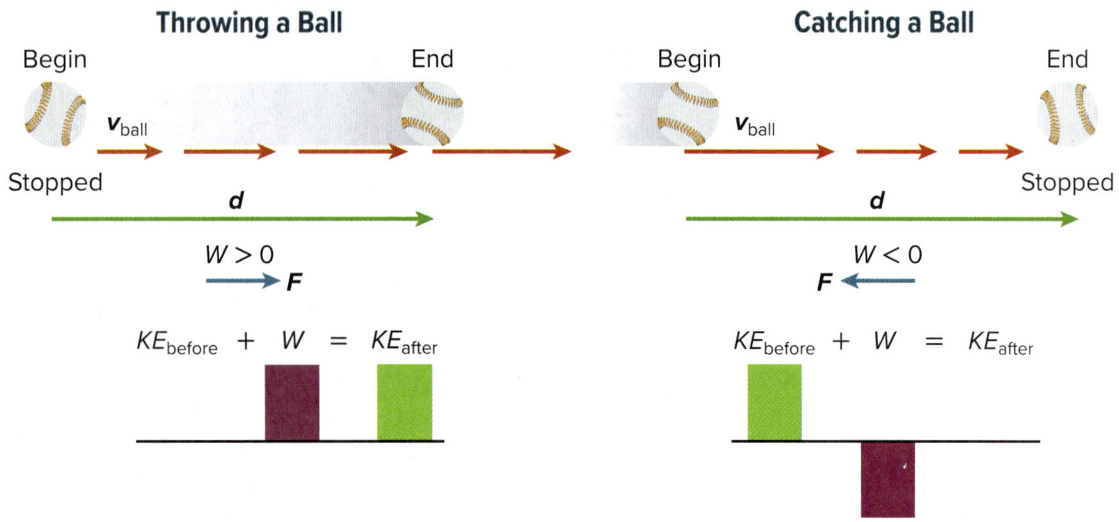

Figure 8 When you do work on an object, such as a ball, you change its energy.
Identify *two other events in which work causes a change in energy.*

PRACTICE Problems — ADDITIONAL PRACTICE

10. A catamaran with a mass of 5.44×10^3 kg is moving at 12 knots. How much work is required to increase the speed to 16 knots? (One knot = 0.51 m/s.)

11. A 52.0-kg skater moves at 2.5 m/s and stops over a distance of 24.0 m. Find the skater's initial kinetic energy. How much of her kinetic energy is transformed into other forms of energy by friction as she stops? How much work must she do to speed up to 2.5 m/s again?

12. An 875.0-kg car speeds up from 22.0 m/s to 44.0 m/s. What are the initial and final kinetic energies of the car? How much work is done on the car to increase its speed?

13. CHALLENGE A comet with a mass of 7.85×10^{11} kg strikes Earth at a speed of 25.0 km/s. Find the kinetic energy of the comet in joules, and compare the work that is done by Earth in stopping the comet to the 4.2×10^{15} J of energy that was released by the largest nuclear weapon ever exploded.

Power

Suppose you had a stack of books to move to a shelf. You could lift the stack at once, or you could move the books one at a time. Since the total force applied and the displacement are the same in both cases, the work is the same. However, the time needed is different. Recall that work causes a change in energy. The rate at which energy is transformed is **power.**

Power

Power is equal to the change in energy divided by the time required for the change.

$$P = \frac{\Delta E}{t}$$

When work causes the change in energy, power is equal to the work done divided by the time taken to do the work: $P = \frac{W}{t}$.

Consider the two forklifts in **Figure 9.** The left forklift raises its load in 5 seconds, and the right forklift raises the same load in 10 seconds. The left forklift is more powerful than the right forklift. Even though the same work is accomplished by both forklifts, the left forklift accomplishes the work in less time and thus develops more power.

Power is measured in watts (W). One **watt** is 1 J of energy transformed in 1 s. That is, 1 W = 1 J/s.

A watt is a relatively small unit of power. For example, a glass of water weighs about 2 N. If you lift it 0.5 m to your mouth at a constant speed, you do 1 J of work. If you lift it in 1 s, you are doing work at the rate of 1 W. Because a watt is such a small unit, power often is measured in kilowatts (kW). One kilowatt is equal to 1000 W.

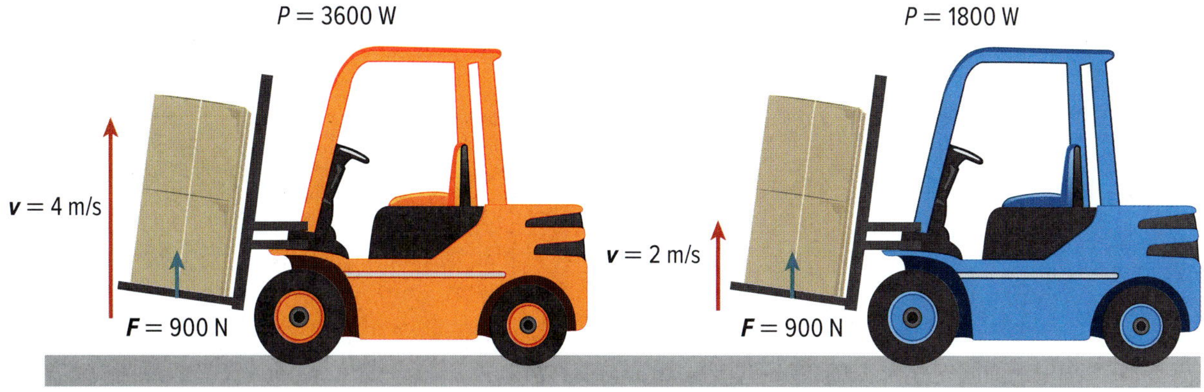

Figure 9 The forklift on the left develops more power than the forklift on the right because it lifts the load at a faster rate.

EXAMPLE Problem 3

POWER An electric motor lifts an elevator 9.00 m in 15.0 s by exerting an upward force of 1.20×10^4 N. What power does the motor produce in kW?

1 ANALYZE AND SKETCH THE PROBLEM

- Sketch the situation showing the system as the elevator with its initial conditions.
- Establish a coordinate system with up as positive.
- Draw a vector diagram for the force and displacement.

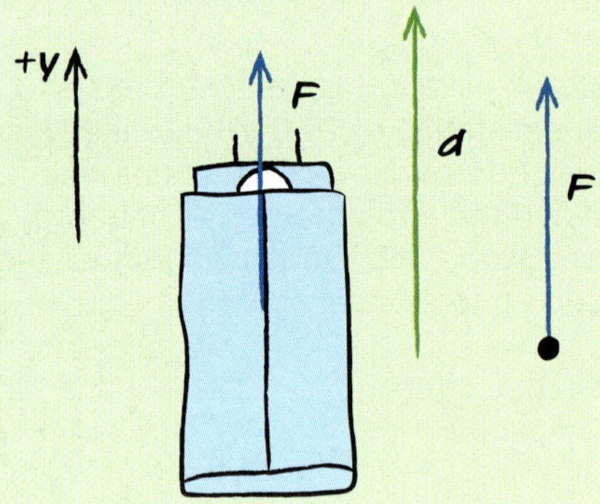

Known	Unknown
$d = 9.00$ m	$P = ?$
$t = 15.0$ s	
$F = 1.20\times10^4$ N	

2 SOLVE FOR THE UNKNOWN

Use the definition of power.

$$P = \frac{W}{t}$$

$$= \frac{Fd}{t}$$ Substitute $W = Fd\cos 0° = Fd$.

$$= \frac{(1.20\times10^4\ \text{N})(9.00\ \text{m})}{(15.0\ \text{s})}$$ Substitute $F = 1.20\times10^4$ N, $d = 9.00$ m, $t = 15.0$ s.

$$= 7.20\ \text{kW}$$

3 EVALUATE THE ANSWER

- **Are the units correct?** Power is measured in joules per second, or watts.
- **Does the sign make sense?** The positive sign agrees with the upward direction of the force.

PRACTICE Problems

ADDITIONAL PRACTICE

14. A cable attached to a motor lifts a 575-N box up a distance of 20.0 m. The box moves with a constant velocity and the job is done in 10.0 s. What power is developed by the motor in W and kW?

15. You push a wheelbarrow a distance of 60.0 m at a constant speed for 25.0 s by exerting a 145-N force horizontally.

a. What power do you develop?

b. If you move the wheelbarrow twice as fast, how much power is developed?

16. What power does a pump develop to lift 35 L of water per minute from a depth of 110 m? (One liter of water has a mass of 1.00 kg.)

17. An electric motor develops 65 kW of power as it lifts a loaded elevator 17.5 m in 35 s. How much force does the motor exert?

18. **CHALLENGE** A winch designed to be mounted on a truck, as shown in **Figure 10,** is advertised as being able to exert a 6.8×10^3-N force and to develop a power of 0.30 kW. How long would it take the truck and the winch to pull an object 15 m?

Figure 10

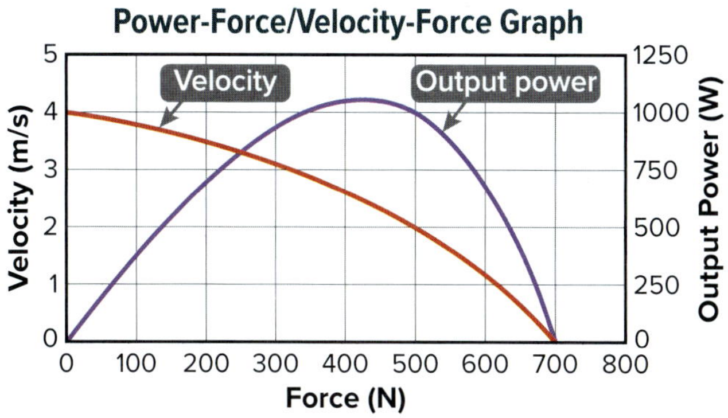

Figure 11 When riding a multispeed bicycle, the output power depends on the force you exert and your speed.

Power and speed When the force has a component (F_x) in the same direction as the displacement, $P = \frac{F_x d}{t}$. However, $v = \frac{d}{t}$ so power can be calculated using $P = F_x v$. From the equation $P = Fv$, you can see that either zero force or zero speed results in zero power. Our muscles cannot exert large forces nor can they move very fast. Some combination of moderate force and moderate speed will produce the largest amount of power when cycling, for example. **Figure 11** shows that for a certain cyclist, the maximum power output is over 1000 W when the force is about 400 N and speed is about 2.6 m/s.

Real-World Physics

A BICYCLIST in the Tour de France rides at an average speed of 8.94 m/s for more than 6 h a day. The power output of the racer is about 1 kW. One-fourth of that power goes into moving the bike against the resistance of the air, gears, and tires. Three-fourths of the power is used to cool the racer's body.

Check Your Progress

19. **Work-Energy Theorem** How can you apply the work-energy theorem to lifting a bowling ball from a storage rack to your shoulder?
20. **Work and Energy** If the work done on an object doubles its kinetic energy, does it double its speed? If not, by what ratio does it change the speed?
21. **Work** Murimi pushes a 20-kg mass 10 m across a floor with a horizontal force of 80 N. Calculate the amount of work done by Murimi on the mass.
22. **Work** Suppose you are pushing a stalled car. As the car gets going, you need less and less force to keep it going. For the first 15 m, your force decreases at a constant rate from 210.0 N to 40.0 N. How much work did you do on the car? Draw a force-displacement graph to represent the work done during this period.
23. **Work** A mover loads a 185-kg refrigerator into a moving van by pushing it at a constant speed up a 10.0-m, friction-free ramp at an angle of inclination of 11°. How much work is done by the mover on the refrigerator?
24. **Work** A 0.180-kg ball falls 2.5 m. How much work does the force of gravity do on the ball?
25. **Work and Power** Does the work required to lift a book to a shelf depend on how fast you raise it? Does the power required to lift it depend on how fast you raise it? Explain.
26. **Power** An elevator lifts a total mass of 1.1×10^3 kg a distance of 40.0 m in 12.5 s. How much power does the elevator deliver?
27. **Mass** A forklift raises a box 1.2 m and does 7.0 kJ of work on it. What is the mass of the box?
28. **Work** You and a friend carry identical boxes from the first floor of a building to a room on the second floor, farther down the hall. You carry the box first up the stairs, and then down the hall to the room. Your friend carries it down the hall on the first floor, then up a different stairwell to the second floor. How do the amounts of work done by the two of you on your boxes compare?
29. **Critical Thinking** Explain how to find the change in energy of a system if three agents exert forces on the system at once.

LEARNSMART® Go online to follow your personalized learning path to review, practice, and reinforce your understanding.

LESSON 2

THE MANY FORMS OF ENERGY

FOCUS QUESTION

What are some forms of energy?

Changing Forms of Energy

Recall that the work-energy theorem states that doing work on a system causes a change in the energy of that system. That is, work is the process that transfers energy between a system and the external world. When an agent performs work on a system, the system's energy increases. When the system does work on its surroundings, the system's energy decreases.

In some ways, energy is like ice cream—it comes in different varieties. You can have vanilla, chocolate, or peach ice cream. These are different varieties, but they are all ice cream. However, unlike ice cream, energy can be changed from one form to another. A system of objects can possess energy in a variety of forms.

Figure 12 Energy can manifest itself in a variety of ways, such as light and sound in a July 4 fireworks display.

At the macroscopic scale, energy manifests itself in multiple ways, such as in motion, sound, light, and thermal energy. For example, when you experience a fireworks display, such as the one shown in **Figure 12,** you hear and see the results of the energy released in the chemical reactions in the fireworks in the form of moving, colorful lights and loud pops. You may even feel the warmth given off during the explosion if you are close enough (but not too close).

In this module, you will learn how energy is transformed from one variety (or form) to another and how to keep track of the changes.

Get It?

Describe an example of an event in which more than one form of energy is evident.

LHB Photo/Alamy Stock Photo

3D THINKING DCI Disciplinary Core Ideas CCC Crosscutting Concepts SEP Science & Engineering Practices

COLLECT EVIDENCE

Use your Science Journal to record the evidence you collect as you complete the readings and activities in this lesson.

INVESTIGATE

GO ONLINE to find these activities and more resources.

Review the News
Obtain information from a current news story about energy or energy resources. Evaluate your source and communicate your findings to your class.

Revisit the Encounter the Phenomenon Question
What information from this lesson can help you answer the Unit and Module questions?

Translational and Rotational Kinetic Energy

In the examples we have considered so far, moving objects that were changing position had kinetic energy $\left(\frac{1}{2}mv^2\right)$ due to their motion. What about energy due to an object's changing position?

Translational kinetic energy

Energy due to changing position is called **translational kinetic energy** and can be represented by the following equation.

Translational Kinetic Energy

A system's translational kinetic energy is equal to one-half times the system's mass multiplied by the system's speed squared.

$$KE_{trans} = \frac{1}{2}mv^2$$

Translational kinetic energy is proportional to the object's mass. For example, a 7.26-kg bowling ball thrown through the air has more translational kinetic energy than a 0.148-kg baseball, like the one shown in **Figure 13,** moving with the same speed.

An object's translational kinetic energy is also proportional to the square of the object's speed. A car moving at 20 m/s has four times the translational kinetic energy of the same car moving at 10 m/s.

Figure 13 A baseball has less translational kinetic energy than a bowling ball moving with the same speed.

Get It?

Explain Using the equation for translational kinetic energy, show why a car moving at 20 m/s has four times the translational kinetic energy of the same car moving at 10 m/s.

SCIENCE USAGE v. COMMON USAGE

Energy

Science usage: the ability of a system to produce a change in itself or the world around it

The kinetic energy of the soccer ball decreased as it slowed down.

Common usage: the capacity of acting or being active

The young students had a lot of energy during recess.

Figure 14 The diving board does work on the diver. This work increases the diver's kinetic energy.

Rotational kinetic energy If you spin a toy top in one spot, you might say that it does not have kinetic energy because the top does not change its position. However, to make the top rotate, you had to do work on it. The top has **rotational kinetic energy,** which is energy due to rotational motion. Rotational kinetic energy can be calculated using $KE_{rot} = \frac{1}{2}I\omega^2$, where I is the object's moment of inertia and ω is the object's angular velocity.

In **Figure 14,** the diving board does work on a diver, transferring translational and rotational kinetic energy to the diver. Her center of mass moves as she leaps, so she has translational kinetic energy. She rotates about her center of mass, so she has rotational kinetic energy. When she slices into the water, she has mostly translational kinetic energy.

Potential Energy

Let's return to the money and energy model from Lesson 1. Money and energy can be in different forms. You can have one five-dollar bill, 500 pennies, or 20 quarters–in all forms, you still have five dollars. In the same way, energy can be kinetic energy, stored energy, or another form.

Consider some boulders on a mountain. They have been moved from Earth's center by geological processes against gravity. Thus, the Earth-boulders system has stored energy. Energy that is stored due to interactions between objects in a system is called **potential energy.** But not all potential energy is due to gravity. A spring-loaded toy is a system that has potential energy, but the energy is stored due to a compressed spring, not gravity.

CCC CROSSCUTTING CONCEPTS

Systems and System Models Mathematical models can be used to predict the behavior of a system, but models can be limited. Choose an example of a mathematical model used to make an energy calculation based on a realistic situation, such as ice-skating or pushing different materials across different surfaces. What evidence do you have that the model is limited? Explain your model as well as its limitations to a peer.

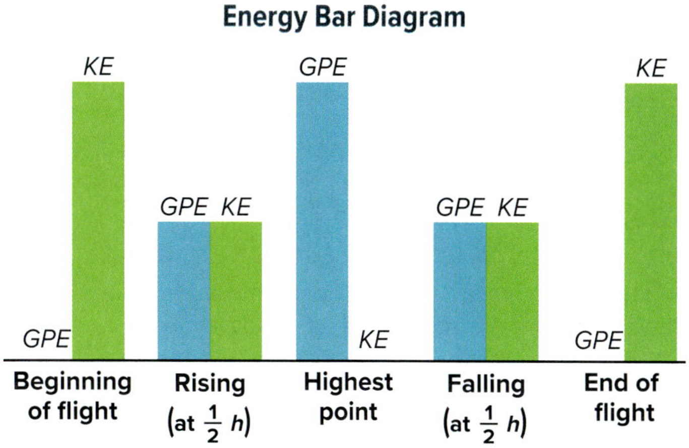

Figure 15 As the ball rises, its kinetic energy transforms into potential energy in the ball-Earth system. As it falls, the potential energy is transformed into kinetic energy. In this process, Earth also gains and loses very small amounts of kinetic energy. The energy bar diagrams show how the ball's energy changes.

Gravitational Potential Energy

Look at the balls being juggled in **Figure 15.** If you consider the system to be only one ball, then it has several external forces acting on it. The force of the juggler's hand does work, giving the ball its initial kinetic energy. After the ball leaves the juggler's hand, only the gravitational force acts on it, assuming that air resistance is negligible. How much work does gravity do on the ball as its height changes?

Work done by gravity Let h represent the ball's height measured from the juggler's hand. On the way up, its displacement is upward but the gravitational force on the ball (F_g) is downward, so the work done by gravity is negative: $W_g = -mgh$. On the way back down, the force and displacement are in the same direction, so the work done by gravity is positive: $W_g = mgh$. Thus, while the ball is moving upward, gravity does negative work, slowing the ball to a stop. On the way back down, gravity does positive work, increasing the ball's speed and thereby increasing its kinetic energy. The ball recovers all of its original kinetic energy when it returns to the same height as the juggler's hand, as shown in the energy bar diagrams in **Figure 15.** Notice in the diagrams that the ball has equal amounts of gravitational potential energy and kinetic energy at $\frac{1}{2}h$.

Get It?

Describe the energy transformations in the dam at the beginning of the module.

According to Newton's law of universal gravitation, there is a gravitational attraction between any pair of objects. This gravitational attraction between the objects is a force that performs work when one of the objects moves. Usually, we are most concerned with the gravitational potential energy between an object and Earth. For this particular example, we choose to discuss the object and Earth together as a system. If the object moves away from Earth, the system stores energy due to the gravitational force between the object and Earth. The stored energy due to gravity is called **gravitational potential energy,** or *GPE*. The height to which the object has risen is determined by using a **reference level,** the position where *GPE* is defined to be zero. In the example of the juggler, the reference level is the juggler's hand.

Calculating gravitational potential energy

For an object with mass m at height h above the reference level, the GPE of the object-Earth system is represented by the following equation.

Gravitational Potential Energy

The GPE of an object-Earth system is equal to the product of the object's mass, the gravitational field strength, and the object's distance above the reference level.

$$GPE = mgh$$

In this equation, g is the gravitational field strength. GPE, like kinetic energy, is measured in joules.

Energy transformations Consider the energy of a system consisting of Earth and a ball being juggled. The energy of the system is in two forms: kinetic energy and gravitational potential energy. At the beginning of the ball's flight, all the energy is in the form of kinetic energy, as shown on the left in **Figure 16.** As the ball rises, it slows down and the system's energy changes from kinetic energy to gravitational potential energy. At the highest point of the ball's flight, the velocity is zero and all the energy is in the form of gravitational potential energy.

Changes of energy in a system can be described in terms of energy flowing into, out of, and within that system. As the ball falls, gravitational potential energy changes back into kinetic energy, or flows within that system as it transforms back into kinetic energy. The sum of kinetic energy and gravitational potential energy is constant at all times because no work is done on the system by forces outside of the system.

Reference levels So far, we have set the reference level to be the juggler's hand. That is, the height of the ball is measured from where it left contact with the juggler's hand. Thus, at the juggler's hand, $h = 0$ m and $GPE = 0$ J. However, you can set the reference level at any height that is convenient for solving a given problem.

Suppose the reference level is set at the highest point of the ball's flight. Then, $h = 0$ m and the system's $GPE = 0$ J at that point, as illustrated on the right side of **Figure 16.** The system's potential energy is negative at the beginning of the ball's flight, zero at the highest point, and negative at the end of the ball's flight. If you were to calculate the total energy of the system represented on the left in **Figure 16,** it would be different from the total energy of the system represented on the right in **Figure 16.** This is because the reference levels are different in each case. However, the total energy of the system in each situation is constant at all times during the flight of the ball.

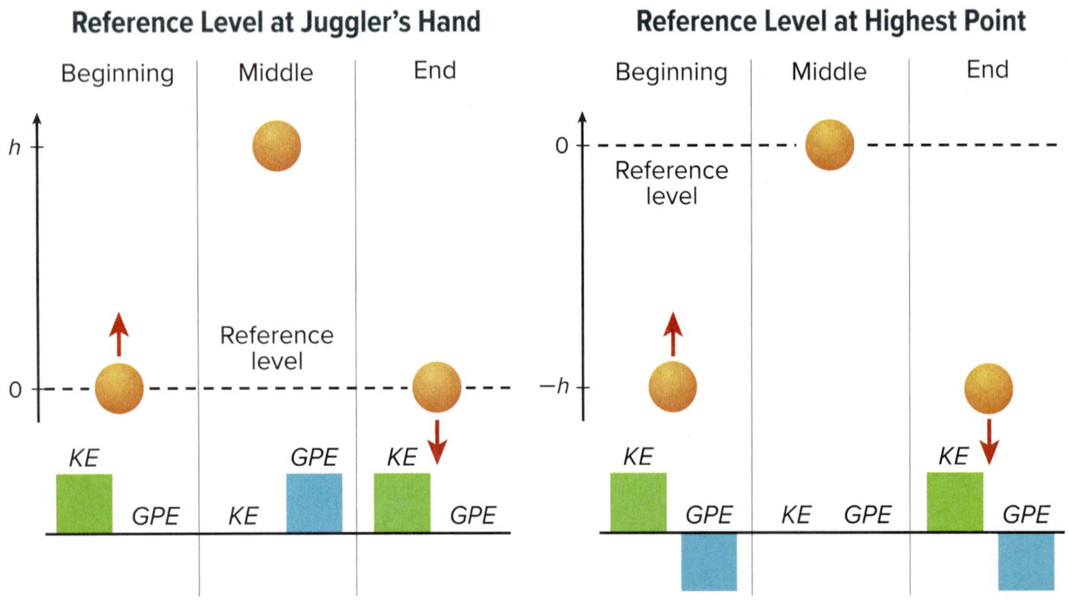

Figure 16 Where you put the reference level affects the system's GPE. Where you put the reference level does not affect how the system's GPE changes over time, however.

Analyze *Where else might you place the reference level?*

EXAMPLE Problem 4

GRAVITATIONAL POTENTIAL ENERGY You lift a 7.30-kg bowling ball from the storage rack and hold it up to your shoulder. The storage rack is 0.610 m above the floor and your shoulder is 1.12 m above the floor.

a. When the bowling ball is at your shoulder, what is the ball-Earth system's gravitational potential energy relative to the floor?

b. When the bowling ball is at your shoulder, what is the ball-Earth system's gravitational potential energy relative to the rack?

c. How much work was done by gravity as you lifted the ball from the rack to shoulder level?

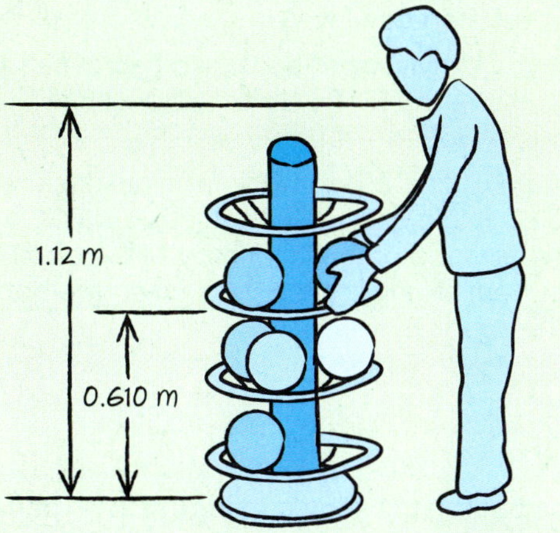

1 ANALYZE AND SKETCH THE PROBLEM

- Sketch the situation.
- Choose a reference level.
- Draw an energy bar diagram showing the gravitational potential energy with the floor as the reference level.

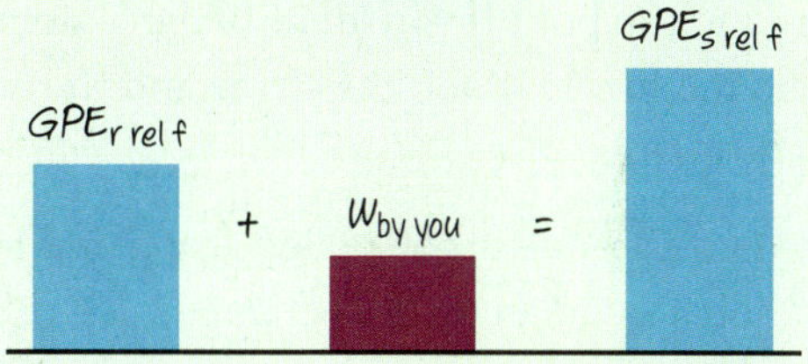

Known		**Unknown**
$m = 7.30$ kg	$g = 9.8$ N/kg	$GPE_{s\ rel\ f} = ?$
$h_r = 0.610$ m (rack relative to the floor)		$GPE_{s\ rel\ r} = ?$
$h_s = 1.12$ m (shoulder relative to the floor)		$W = ?$

2 SOLVE FOR THE UNKNOWN

a. Set the reference level to be at the floor.

Determine the gravitational potential energy of the system when the ball is at shoulder level.

$GPE_{s\ rel\ f} = mgh_s = (7.30\ \text{kg})(9.8\ \text{N/kg})(1.12\ \text{m})$ — Substitute $m = 7.30$ kg, $g = 9.8$ N/kg, $h_s = 1.12$ m

$= 8.0\times10^1\ \text{J}$

b. Set the reference level to be at the rack height.

Determine the height of your shoulder relative to the rack.

$h = h_s - h_r$

Determine the gravitational potential energy of the system when the ball is at shoulder level.

$GPE_{s\ rel\ r} = mgh = mg(h_s - h_r)$ — Substitute $h = h_s - h_r$

$= (7.30\ \text{kg})(9.8\ \text{N/kg})(1.12\ \text{m} - 0.610\ \text{m})$ — Substitute $m = 7.30$ kg, $g = 9.8$ N/kg, $h_s = 1.12$ m, $h_r = 0.610$ m

$= 36\ \text{J}$ — This also is equal to the work done by you as you lifted the ball.

c. The work done by gravity is the weight of the ball times the distance the ball was lifted.

$W = Fd = -(mg)h = -(mg)(h_s - h_r)$ — The weight opposes the motion of lifting, so the work is negative.

$= -(7.30\ \text{kg})(9.8\ \text{N/kg})(1.12\ \text{m} - 0.610\ \text{m})$ — Substitute $m = 7.30$ kg, $g = 9.8$ N/kg, $h_s = 1.12$ m, $h_r = 0.610$ m

$= -36\ \text{J}$

3 EVALUATE THE ANSWER

- **Are the units correct?** Both potential energy and work are measured in joules.
- **Does the answer make sense?** The system should have a greater *GPE* measured relative to the floor than relative to the rack because the ball's distance above the floor level is greater than the ball's distance above the rack.

PRACTICE Problems

ADDITIONAL PRACTICE

30. In Example Problem 4, what is the potential energy of the ball-Earth system when the bowling ball is on the floor? Use the rack as your reference level.
31. If you slowly lower a 20.0-kg bag of sand 1.20 m from the trunk of a car to the driveway, how much work do you do?
32. A boy lifts a 2.2-kg book from his desk, which is 0.80 m high, to a bookshelf that is 2.10 m high. What is the potential energy of the book-Earth system relative to the desk when the book is on the shelf?
33. You are walking around an old building and notice that it is falling apart. If a 1.8-kg brick falls to the ground from the building's chimney, which is 6.7 m high, what is the change in the potential energy of the brick-Earth system?
34. **CHALLENGE** A worker picks up a 10.1-kg box from the floor and sets it on a table that is 1.1 m high. He slides the box 5.0 m along the table and then lowers it back to the floor. What were the changes in the box-Earth system's energy, and how did the system's total energy change? (Ignore friction.)

Elastic Potential Energy

When the bow string shown in **Figure 17** is pulled, work is done on the bow string and energy is transferred to the bow. If you identify the system as the bow, the arrow, and Earth, the energy of the system increases. When the string and the arrow are released, the stored energy is changed into kinetic energy.

The stored energy due to the pulled string and bent bow is **elastic potential energy,** which is stored energy due to an object's change in shape. Systems that include springs, rubber bands, and trampolines have elastic potential energy.

Get It?

Define the term *elastic potential energy.*

You can also store elastic potential energy in a system when you bend an elastic object in that system. For example, when stiff metal or bamboo poles were used in pole-vaulting, the poles did not bend easily. Little work was done on the poles, and the poles did not store much potential energy. Since flexible fiberglass poles were introduced, however, record pole-vaulting heights have soared.

Figure 17 The archer-bow-arrow system has maximum elastic potential energy before the string is released, as shown on the left. When the arrow and string disengage, the elastic potential energy is completely transformed into kinetic energy, as shown on the right.

A pole-vaulter runs with a flexible pole and plants its end into a socket in the ground. When the pole-vaulter bends the pole, as shown in **Figure 18,** some of the pole-vaulter's kinetic energy is transformed to elastic potential energy. When the pole straightens, the elastic potential energy is transformed to gravitational potential energy and kinetic energy as the pole-vaulter is lifted as high as 6 m above the ground.

Figure 18 The pole-vaulter stores elastic potential energy in the pole-vaulter-track system. The elastic potential energy of this system then changes into kinetic energy and gravitational potential energy.

Other Forms of Energy

Think about all the forms and sources of energy you encounter every day. Gasoline provides energy to run your car. You eat food to obtain energy. Power plants harness the energy of wind, water, fossil fuels (**Figure 19**), and atoms and transform it into electrical energy.

Some of these forms of energy are best understood if we think of energy as a quantitative property of a system that depends on the motion and interactions of matter and radiation within that system. (Radiation, including light, is the transfer of energy by electromagnetic waves, which you will encounter in a later module.) For example, nuclear energy is the energy stored in the bonds inside the nucleus of an atom. Without understanding the structure and chemical subprocesses within an atom, we would not be able to develop mathematical models to predict the amount of energy that could be released.

Figure 19 Coal-burning plants harness energy and transform it into something more useful, such as electrical energy.

Get It?

Explain In your own words, explain why we need to understand the interactions in a process or subprocess in order to understand a system.

Chemical and nuclear energy

When you eat, your body uses the energy from the chemical bonds in your food as an energy source. The bonds inside an atom's nucleus also store energy. This energy is called nuclear energy and is released when the structure of an atom's nucleus changes.

CCC CROSSCUTTING CONCEPTS

Energy and Matter Changes of energy and matter in a system can be described in terms of energy and matter flow within that system. In a labeled poster, show the changes in energy flow in a pole-vaulter-track system from the moment the vaulter places the pole in the socket in the ground to the moment the vaulter lands. What evidence is there of a change in energy?

Thermal energy

Thermal energy, which is associated with temperature, is the sum of the kinetic energy and potential energy of the particles in a system. When you rub your hands together, you transform kinetic energy into thermal energy. Thermal energy can also be transferred, such as a stove like the one in **Figure 20** transferring thermal energy to a pan.

Figure 20 In your daily activities, you encounter many types of energy, such as thermal energy when cooking your food.

Electrical and radiant energy

Power plants transform various types of energy into electrical energy, which is energy associated with charged particles. Electrical energy can be transmitted through wires like those in **Figure 21** to your home, where appliances transform it into other forms of energy, such as radiant energy in lightbulbs.

Figure 21 Transmitting electrical energy is critical for our day-to-day needs.

Mass

Albert Einstein demonstrated that mass represents a form of energy, called rest energy (E_0): $E_0 = mc^2$. Mass is a form of energy that can be transformed into other forms of energy. Mass-energy equivalence is only apparent when large amounts of energy are involved, such as during nuclear explosions and particle physics experiments.

Check Your Progress

35. **Elastic Potential Energy** You compress the spring in a jumping toy, then release it. The toy then flies straight up. Draw bar graphs that show the forms of energy present in the following instances. Assume the system includes the spring toy and Earth.
    a. The toy is pushed down thereby compressing the spring.
    b. The spring expands and the toy jumps.
    c. The toy reaches the top of its flight.
    d. Describe how energy manifests itself in different ways in this system.
36. **Potential Energy** A 25.0-kg shell is shot from a cannon at Earth's surface. The reference level is Earth's surface.
    a. What is the shell-Earth system's gravitational potential energy when the shell's height is 425 m?
    b. What is the change in the system's potential energy when the shell falls to a height of 225 m?
    c. What motions and interactions does this system depend on?
37. **Rotational Kinetic Energy** On a playground, some children push a merry-go-round so that it turns twice as fast as it did before they pushed it. What are the relative changes in angular momentum and rotational kinetic energy of the merry-go-round?
38. **Critical Thinking** In the school lab, Karl uses an air hose to exert a constant horizontal force to move a puck a fixed distance on a frictionless air table.
    a. Explain what happens in terms of work and energy. Draw bar graphs.
    b. Karl uses a different puck with half the first one's mass. All other conditions remain the same. How do the kinetic energy and work done differ from the kinetic energy and work done in part a?
    c. Describe what happened in parts a and b in terms of impulse and momentum.

LEARNSMART® Go online to follow your personalized learning path to review, practice, and reinforce your understanding.

LESSON 3

CONSERVATION OF ENERGY

FOCUS QUESTION

Can energy be lost?

The Law of Conservation of Energy

The **law of conservation of energy** states that in a closed, isolated system, energy can neither be created nor destroyed. So, the total change of energy in any system is always equal to the total energy transferred into or out of the system. In other words, energy can be transported from one place to another, transferred between systems, and converted to different forms, such as thermal energy released to the surrounding environment from a running car. However, the system's total energy in all of its forms remains constant.

Within a system, energy is continually transferred from one object to another and between its various possible forms. Therefore, energy is considered a single quantity due to the law of conservation of energy.

Get It?

Explain whether the light from a battery-powered lamp can have more energy than the battery's energy.

Mechanical energy You often focus on the energy that comes from the motions of and interactions between objects. The sum of the *KE* and *PE* of the objects in a system is the system's **mechanical energy** (*ME*). The *KE* includes both the translational and rotational kinetic energies of the objects in the system. Potential energy includes the gravitational and elastic potential energies of the system. In any system, mechanical energy is represented as follows.

Mechanical Energy of a System

The mechanical energy of a system is equal to the sum of the *KE* and *PE* of the system's objects.

$$ME = KE + PE$$

3D THINKING — DCI Disciplinary Core Ideas — CCC Crosscutting Concepts — SEP Science & Engineering Practices

COLLECT EVIDENCE

Use your Science Journal to record the evidence you collect as you complete the readings and activities in this lesson.

INVESTIGATE

GO ONLINE to find these activities and more resources.

Applying Practices: Earth Power
HS-PS3-3. Design, build, and refine a device that works within given constraints to convert one form of energy into another form of energy.

Applying Practices: Modeling Changes in Energy
HS-PS3-1. Create a computational model to calculate the change in the energy of...

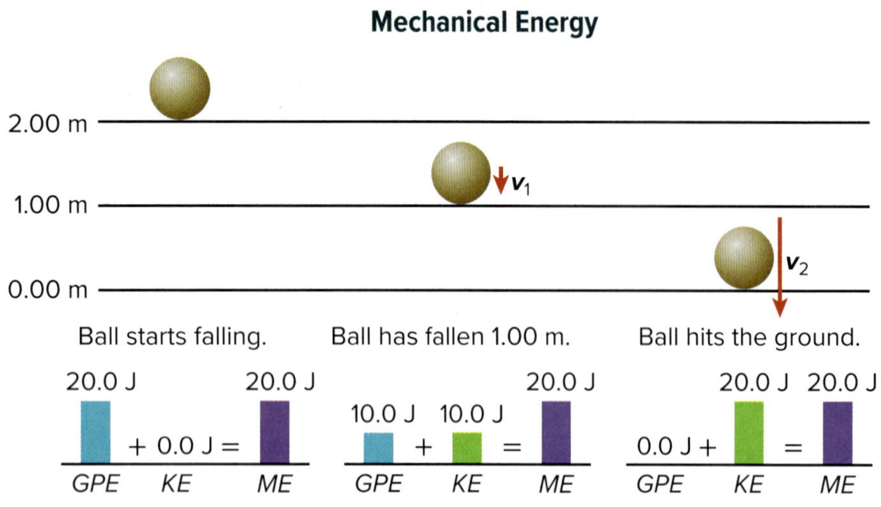

Figure 22 When a bowling ball is dropped, mechanical energy is conserved.

Predict *Suppose the ball started with an upward velocity. What would its energy graph look like?*

Imagine a system consisting of a 10.0-N bowling ball falling to Earth, as shown in **Figure 22.** For this case, the system's mechanical energy is 20.0 J.

Conservation of mechanical energy

Suppose that you let go of the bowling ball. As it falls, the *GPE* of the ball-Earth system is transformed into kinetic energy. When the ball is 1.00 m above Earth's surface, $GPE = mgh = F_g h = (10.0 \text{ N})(1.00 \text{ m}) = 10.0 \text{ J}$. What is the system's kinetic energy when the ball is at a height of 1.00 m? The system consisting of the ball and Earth is closed and isolated because no external forces are doing work on it. Therefore, the total mechanical energy of the system remains constant at 20.0 J.

$$ME = KE + PE$$

$$KE = ME - PE$$

$$KE = 20.0 \text{ J} - 10.0 \text{ J} = 10.0 \text{ J}$$

When the ball reaches ground level, the system's potential energy is zero, and the system's kinetic energy is 20.0 J. The availability of energy limits what can occur in this and any system; for example, at a height of 1.00 m, the *GPE* of this 10-N bowling ball cannot exceed 10.0 J.

Conservation of Mechanical Energy

When mechanical energy is conserved, the sum of the system's *KE* and *PE* before an event equals the sum of the system's *KE* and *PE* after the event.

$$KE_i + PE_i = KE_f + PE_f$$

CCC CROSSCUTTING CONCEPTS

Systems and System Models Models can be used to predict the behavior of a system, but these predictions have limited precision and liability. Develop an argument that explains why the bowling ball model discussed in the text is limited. What evidence from your life supports your argument? How would you present your case to your peers?

What happens if the ball does not fall straight down but rolls down a ramp, as shown in **Figure 23?** If friction does not affect the system, then the system of the ball and Earth remains closed and isolated. The ball still moves down a vertical distance of 2.00 m, so the decrease in potential energy is 20.0 J. Therefore, the increase in kinetic energy is 20.0 J.

As long as friction doesn't affect the system's energy, the path the ball takes does not matter. However, in the case with the ramp, the kinetic energy is split between translational kinetic energy (it is moving forward) and rotational kinetic energy (it is rolling). Three situations involving conservation of mechanical energy are shown in **Figure 24** on the next page.

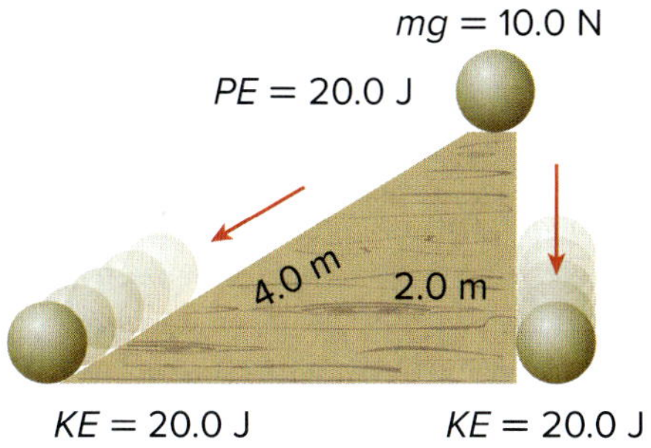

Figure 23 If friction does no work on the ball, the ball's final kinetic energy is equal to its initial gravitational potential energy regardless of which path it follows.

Get It?

Analyze how the ball's final kinetic energy in **Figure 23** would be different if friction transformed some of the system's energy.

Conservation and other forms of energy A swinging pendulum eventually stops, a bouncing ball comes to rest, and the heights of roller-coaster hills get lower and lower as the ride progresses. Where does the mechanical energy in such systems go? Objects moving through the air experience the forces of air resistance. In a roller coaster, there are also frictional forces between the wheels and the tracks. Each force transforms mechanical energy into other forms of energy.

When a ball bounces from a surface, it compresses. Most of the kinetic energy transforms into elastic potential energy. Some, but not all, of this elastic potential energy transforms back into kinetic energy during the bounce. But as in the cases of the pendulum and the roller coaster, some of the original mechanical energy in the system transforms into another form of energy within the system or transmits to energy outside the system, as in air resistance.

Get It?

Analyze If most of the kinetic energy of a bouncing ball transforms into elastic potential energy, what happens to the rest of the kinetic energy?

Mathematical expressions, which quantify how kinetic energy depends on mass and speed, allow the concept of conservation of energy to be used to predict and describe system behavior. The Example Problems in this lesson use the concept of conservation of energy to solve problems, including collisions.

STEM CAREER Connection

Solar Photovoltaic Installer

Alternative energy is an area of global importance today. Solar photovoltaic installers, or PV installers, install and maintain solar panels on homes and for industry. They may also assemble systems. Installers in this rapidly growing field work mostly outside, sometimes work inside as well. They also must travel to their job sites.

CCC CROSSCUTTING CONCEPTS

Energy and Matter Make a simple ramp with a book or a board. Roll a ball down the ramp. Explain how your ramp and ball provide evidence of the law of conservation of energy.

Figure 24 The conservation of mechanical energy is an important consideration in designing roller coasters, ski slopes, and the pendulums for grandfather clocks.

Roller Coasters The roller-coaster car is nearly at rest at the top of the first hill, and the total mechanical energy in the Earth-roller coaster car system is the system's gravitational potential energy at that point. If a hill farther along the track were higher than the first one, the roller-coaster car would not be able to climb the higher hill because the energy required to do so would be greater than the total mechanical energy of the system.

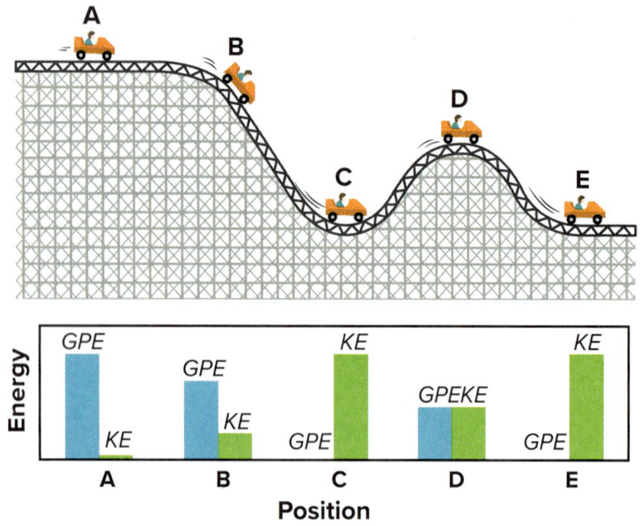

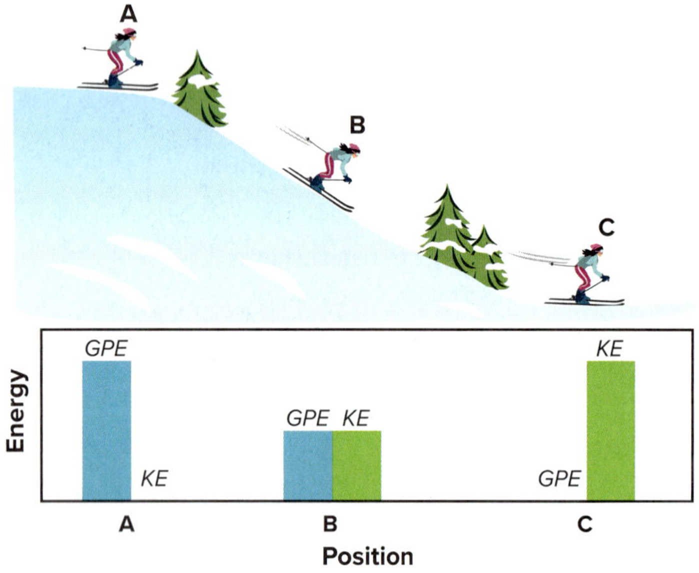

Skiing When you ski down a steep slope, you begin from rest at the top of the slope and the total mechanical energy is equal to the gravitational potential energy. Once you start skiing downhill, the gravitational potential energy is transformed to kinetic energy. As you ski down the slope, your speed increases as more potential energy is transformed to kinetic energy.

Pendulums The simple oscillation of a pendulum also demonstrates conservation of mechanical energy. The system is the pendulum bob and Earth. The reference level is chosen to be the height of the bob at the lowest point. When an external force pulls the bob to one side, the force does work that adds mechanical energy to the system. At the instant the bob is released, all the energy is in the form of potential energy, but as the bob swings downward, potential energy is transformed to kinetic energy. When the bob is at the lowest point, the gravitational potential energy is zero, and the kinetic energy is equal to the total mechanical energy.

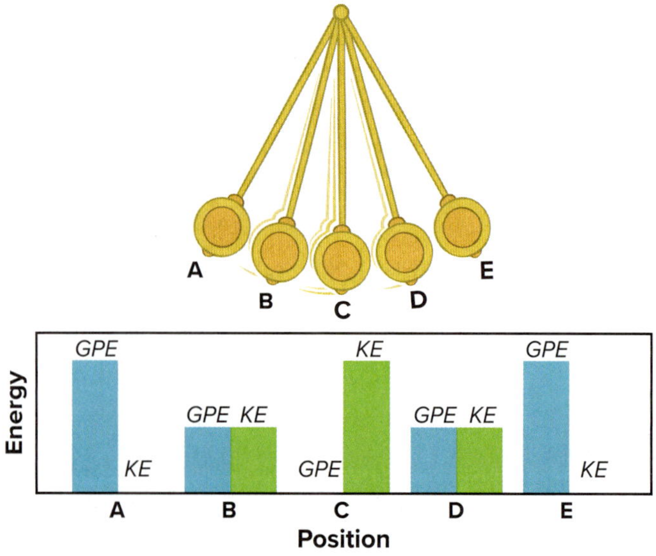

EXAMPLE Problem 5

CONSERVATION OF MECHANICAL ENERGY A 22.0-kg tree limb is 13.3 m above the ground. During a tropical storm, it falls on a roof that is 6.0 m above the ground.

a. Find the kinetic energy of the limb when it reaches the roof. Assume that the air does no work on the tree limb.

b. What is the limb's speed when it reaches the roof?

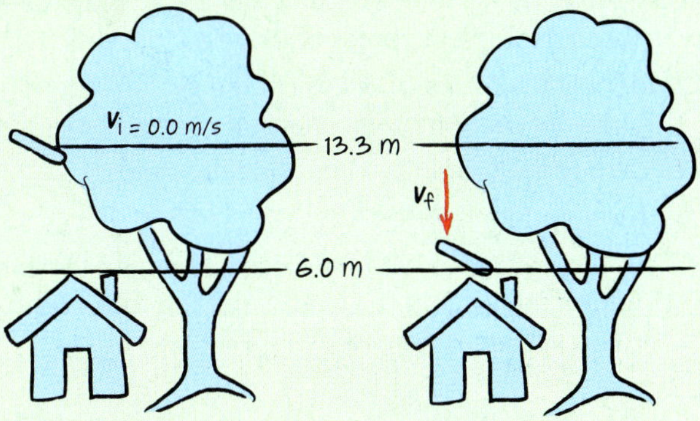

1 ANALYZE AND SKETCH THE PROBLEM

- Sketch the initial and final conditions.
- Choose a reference level.
- Draw an energy bar diagram.

Known		Unknown	
$m = 22.0$ kg	$g = 9.8$ N/kg	$GPE_i = ?$	$KE_f = ?$
$h_{limb} = 13.3$ m	$v_i = 0.0$ m/s	$GPE_f = ?$	$v_f = ?$
$h_{roof} = 6.0$ m	$KE_i = 0.0$ J		

2 SOLVE FOR THE UNKNOWN

a. Set the reference level as the height of the roof.

Find the initial height of the limb relative to the roof.

$h = h_{limb} - h_{roof} = 13.3 \text{ m} - 6.0 \text{ m} = 7.3 \text{ m}$ — Substitute $h_{limb} = 13.3$ m, $h_{roof} = 6.0$ m

Determine the initial gravitational potential energy of the limb-Earth system.

$GPE_i = mgh = (22.0 \text{ kg})(9.8 \text{ N/kg})(7.3 \text{ m})$ — Substitute $m = 22.0$ kg, $g = 9.8$ N/kg, $h = 7.3$ m

$= 1.6\times10^3 \text{ J}$

Identify the initial kinetic energy of the system.

$KE_i = 0.0 \text{ J}$ — The tree limb is initially at rest.

Identify the final potential energy of the system.

$GPE_f = 0.0 \text{ J}$ — $h = 0.0$ m at the roof.

Use the law of conservation of energy to find KE_f.

$KE_f + GPE_f = KE_i + GPE_i$

$KE_f = KE_i + GPE_i - GPE_f$

$= 0.0 \text{ J} + 1.6\times10^3 \text{ J} - 0.0 \text{ J}$ — Substitute $KE_i = 0.0$ J, $GPE_i = 1.6\times10^3$ J, and $GPE_f = 0.0$ J.

$= 1.6\times10^3 \text{ J}$

b. Determine the speed of the limb.

$KE_f = \frac{1}{2}mv_f^2$

$v_f = \sqrt{\frac{2KE_f}{m}} = \sqrt{\frac{2(1.6\times10^3 \text{ J})}{22.0 \text{ kg}}}$ — Substitute $KE_f = 1.6\times10^3$ J, $m = 22.0$ kg

$= 12 \text{ m/s}$

3 EVALUATE THE ANSWER

- **Are the units correct?** The magnitude of velocity is measured in m/s, and energy is measured in kg·m²/s² = J.
- **Do the signs make sense?** Kinetic energy and speed are both positive in this scenario.

PROBLEM-SOLVING STRATEGY

CONSERVATION OF ENERGY

Use the following strategy when solving problems about the conservation of energy.

1. Identify the system. Determine whether the system is closed. No objects enter or leave a closed system.
2. Identify the forms of energy in the system. Identify which forms are part of the mechanical energy of the system.
3. Identify the system's initial and final states.
4. Is the system isolated?
   a. If there are no external forces acting on the system, then the system is isolated and the total energy of the system is constant:
   $$E_{\text{initial}} = E_{\text{final}}$$
   b. If there are external forces, then the final energy is the sum of the initial energy and the work done on the system.

   Remember that work can be negative:
   $$E_{\text{final}} = E_{\text{initial}} + W$$
5. For an isolated system, identify the types of energy in the system before and after. If the only forms of energy are potential and kinetic, mechanical energy is conserved.
   $$KE_{\text{initial}} + PE_{\text{initial}} = KE_{\text{final}} + PE_{\text{final}}$$
   Decide on the reference level for gravitational potential energy. Draw energy bar diagrams showing initial and final energies like the diagram above.

Energy Bar Diagram

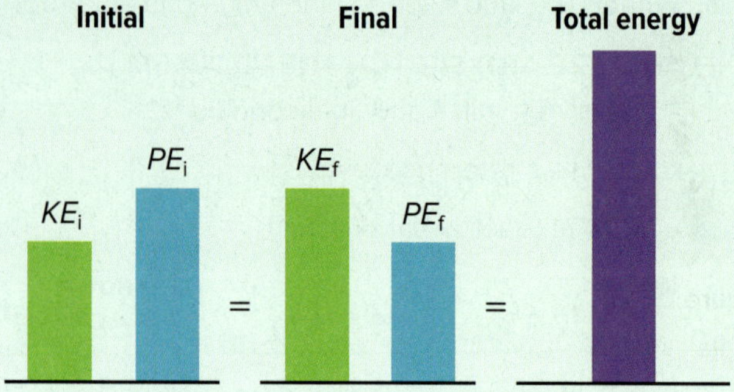

PRACTICE Problems

ADDITIONAL PRACTICE

39. A bike rider approaches a hill at a speed of 8.5 m/s. The combined mass of the bike and the rider is 85.0 kg. Choose a suitable system. Find the initial kinetic energy of the system. The rider coasts up the hill. Assuming friction is negligible, at what height will the bike come to rest?

40. Suppose that the bike rider in the previous problem pedaled up the hill and never came to a stop. In what system is energy conserved? From what form of energy did the bike gain mechanical energy?

41. A skier starts from rest at the top of a hill that is 45.0 m high, skis down a 30° incline into a valley, and continues up a hill that is 40.0 m high. The heights of both hills are measured from the valley floor. Assume that friction is negligible and ignore the effect of the ski poles.

   a. How fast is the skier moving at the bottom of the valley?

   b. What is the skier's speed at the top of the second hill?

   c. Do the angles of the hills affect your answers?

42. In a belly-flop diving contest, the winner is the diver who makes the biggest splash upon hitting the water. The size of the splash depends not only on the diver's style, but also on the amount of kinetic energy the diver has. Consider a contest in which each diver jumps from a 3.00-m platform. One diver has a mass of 136 kg and simply steps off the platform. Another diver has a mass of 100 kg and leaps upward from the platform. How high would the second diver have to leap to make a competitive splash?

43. CHALLENGE The spring in a pinball machine exerts an average force of 2 N on a 0.08-kg pinball over 5 cm. As a result, the ball has both translational and rotational kinetic energy. If the ball is a uniform sphere $\left(I = \frac{5}{2}mr^2\right)$, what is its linear speed after leaving the spring? (Ignore the table's tilt.)

Figure 25 In a car crash, the *KE* decreases.

Explain *What happened to the energy?*

Analyzing Collisions

A collision between any two objects, whether the objects include automobiles such as the ones shown in **Figure 25,** hockey players, or subatomic particles, is one of the most common situations analyzed in physics. The details of a collision can be complex during the collision itself, so the strategy is to find the motion of the objects just before and just after the collision. What conservation laws can be used to analyze such a system? If the system is closed and isolated, then momentum and energy are conserved. However, the potential or thermal energy in the system might decrease, remain the same, or increase. Therefore, you cannot predict whether kinetic energy is conserved. **Figure 26** on the next page shows four different kinds of collisions. Look at the calculations for kinetic energy for each of the cases. In all the cases except case 1, there is a change in the amount of kinetic energy.

In case 1, the momentum of the system before and after the collision is represented as

$$\begin{aligned} p_i &= p_{Ai} + p_{Bi} \\ &= (1.00\ \text{kg})(1.00\ \text{m/s}) + (1.00\ \text{kg})(0.00\ \text{m/s}) \\ &= 1.00\ \text{kg}\cdot\text{m/s} \end{aligned}$$

$$\begin{aligned} p_f = p_{Af} + p_{Bf} &= (1.00\ \text{kg})(0.00\ \text{m/s}) + (1.00\ \text{kg})(1.00\ \text{m/s}) \\ &= 1.00\ \text{kg}\cdot\text{m/s} \end{aligned}$$

Thus, in case 1, the momentum is conserved.

Get It?

Calculate the initial and final momenta for the remaining cases in **Figure 26.** Verify that momentum is conserved for each case.

Elastic and inelastic collisions Next, consider the kinetic energy of the system in each case. In case 1, the kinetic energy of the system before and after the collision was the same. A collision in which the kinetic energy does not change is called an **elastic collision.** Collisions between hard objects, such as those made of steel or hard plastic, often are called nearly elastic collisions.

Kinetic energy decreased in cases 2 and 3. Some of it may have been transformed to thermal energy. A collision in which kinetic energy decreases is called an **inelastic collision.** Objects made of soft, sticky materials, such as clay, act in this way. In case 2, the objects stuck together after colliding. A collision in which colliding objects stick together is called a perfectly inelastic collision.

In case 4, the kinetic energy of the system increased. If energy in the system is conserved, then one or more other forms of energy must have decreased. Perhaps potential energy from a compressed spring was released during the collision, adding kinetic energy to the system. This kind of collision is called a superelastic or explosive collision.

The four kinds of collisions can be represented using energy bar diagrams, also shown in **Figure 26.** The kinetic energies before and after the collisions can be calculated, but we do not always have detailed information about the other forms of energy involved.

In automobile collisions, kinetic energy can be transformed into thermal energy and sound, but they are difficult to measure. We infer their amounts from the amount of kinetic energy that is transformed.

Get It?

Compare and contrast elastic collisions and inelastic collisions.

Figure 26 In a closed, isolated system, momentum is conserved when objects collide. Whether kinetic energy is conserved depends on whether the collision is elastic, inelastic, or superelastic.

Case 1: Elastic Collision

$KE_f = KE_i$

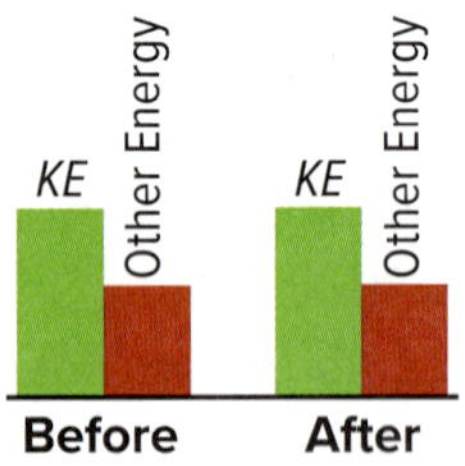

Before collision

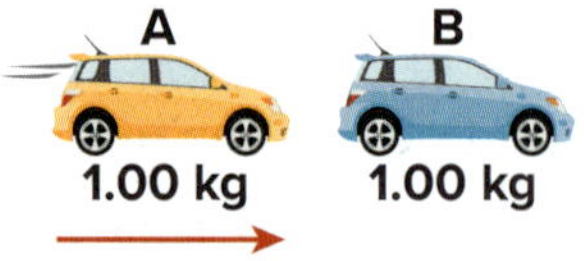

$v_{Ai} = 1.00$ m/s $v_{Bi} = 0.00$ m/s

$$KE_i = KE_{Ai} + KE_{Bi}$$
$$= \frac{1}{2}(1.00\text{ kg})(1.00\text{ m/s})^2 + \frac{1}{2}(1.00\text{ kg})(0.00\text{ m/s})^2$$
$$= 0.50\text{ J}$$

After collision

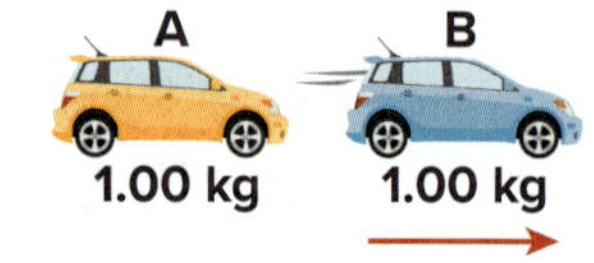

$v_{Af} = 0.00$ m/s $v_{Bf} = 1.00$ m/s

$$KE_f = KE_{Af} + KE_{Bf}$$
$$= \frac{1}{2}(1.00\text{ kg})(0.00\text{ m/s})^2 + \frac{1}{2}(1.00\text{ kg})(1.00\text{ m/s})^2$$
$$= 0.50\text{ J}$$

Case 2: Perfectly Inelastic Collision

$KE_f < KE_i$

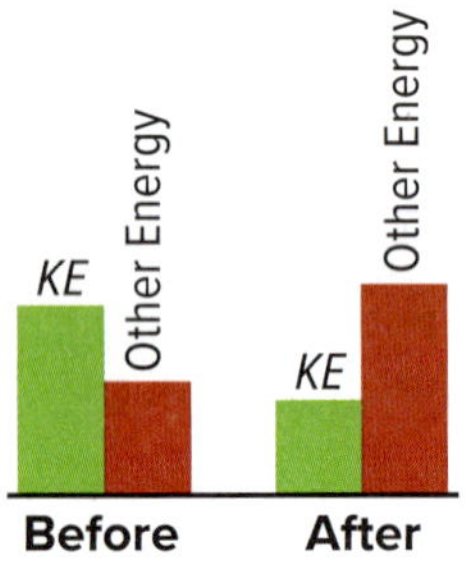

Before collision

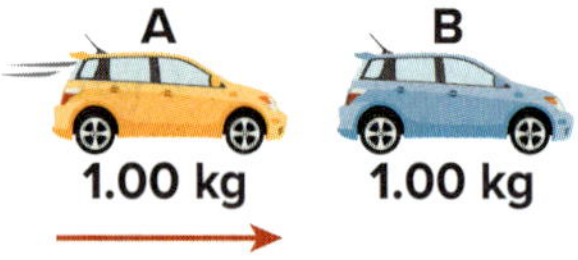

$v_{Ai} = 1.00$ m/s $v_{Bi} = 0.00$ m/s

$$KE_i = KE_{Ai} + KE_{Bi}$$
$$= \frac{1}{2}(1.00\text{ kg})(1.00\text{ m/s})^2 + \frac{1}{2}(1.00\text{ kg})(0.00\text{ m/s})^2$$
$$= 0.50\text{ J}$$

After collision

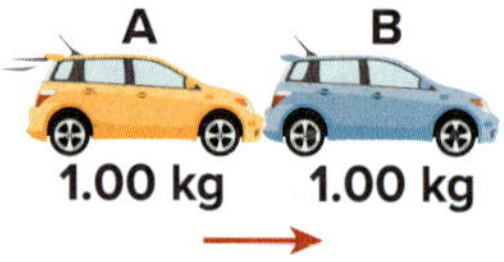

$v_{Af} = v_{Bf} = 0.50$ m/s

$$KE_f = KE_{Af} + KE_{Bf}$$
$$= \frac{1}{2}(1.00\text{ kg})(0.50\text{ m/s})^2 + \frac{1}{2}(1.00\text{ kg})(0.50\text{ m/s})^2$$
$$= 0.25\text{ J}$$

Case 3: Inelastic Collision

$KE_f < KE_i$

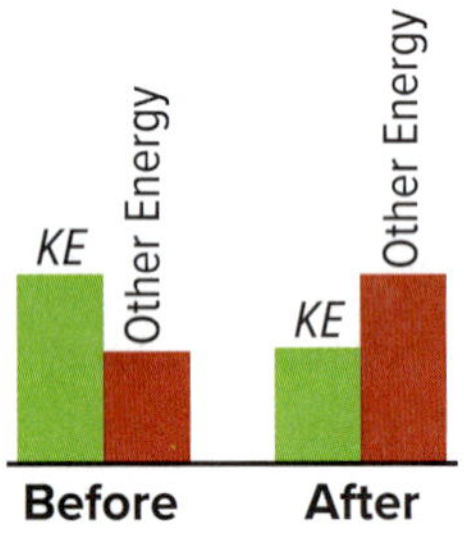

Before collision

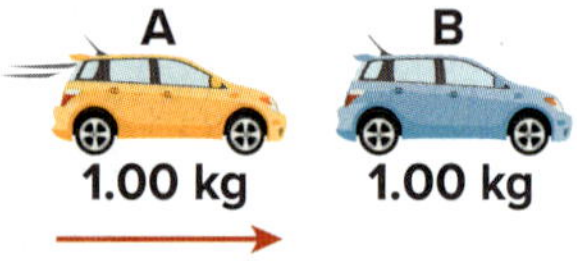

$v_{Ai} = 1.00$ m/s $v_{Bi} = 0.00$ m/s

$$KE_i = KE_{Ai} + KE_{Bi}$$
$$= \frac{1}{2}(1.00\text{ kg})(1.00\text{ m/s})^2 + \frac{1}{2}(1.00\text{ kg})(0.00\text{ m/s})^2$$
$$= 0.50\text{ J}$$

After collision

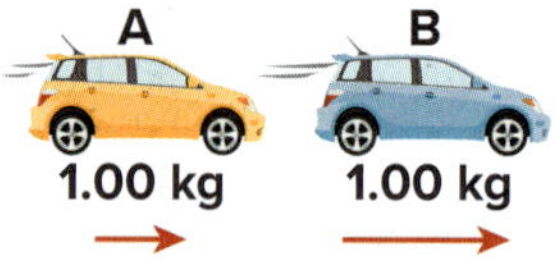

$v_{Af} = 0.25$ m/s $v_{Bf} = 0.75$ m/s

$$KE_f = KE_{Af} + KE_{Bf}$$
$$= \frac{1}{2}(1.00\text{ kg})(0.25\text{ m/s})^2 + \frac{1}{2}(1.00\text{ kg})(0.75\text{ m/s})^2$$
$$= 0.31\text{ J}$$

Case 4: Superelastic Collision

$KE_f > KE_i$

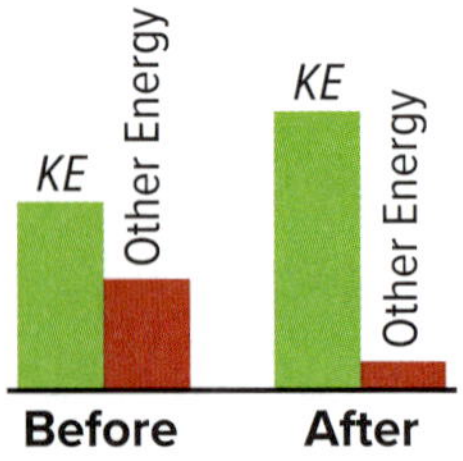

Before collision

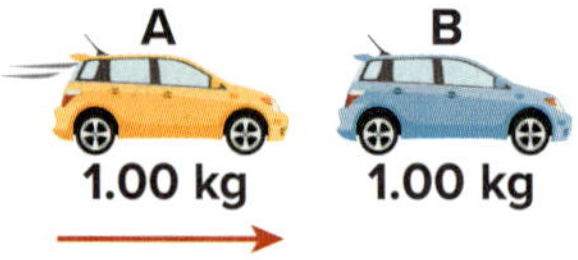

$v_{Ai} = 1.00$ m/s $v_{Bi} = 0.00$ m/s

$$KE_i = KE_{Ai} + KE_{Bi}$$
$$= \frac{1}{2}(1.00\text{ kg})(1.00\text{ m/s})^2 + \frac{1}{2}(1.00\text{ kg})(0.00\text{ m/s})^2$$
$$= 0.50\text{ J}$$

After collision

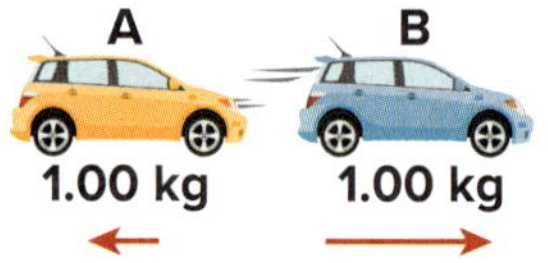

$v_{Af} = 0.20$ m/s $v_{Bf} = 1.20$ m/s

$$KE_f = KE_{Af} + KE_{Bf}$$
$$= \frac{1}{2}(1.00\text{ kg})(-0.20\text{ m/s})^2 + \frac{1}{2}(1.00\text{ kg})(1.20\text{ m/s})^2$$
$$= 0.74\text{ J}$$

EXAMPLE Problem 6

INELASTIC COLLISION In an accident on a slippery road, a compact car with a mass of 1150 kg moving at 15.0 m/s smashes into the rear end of a car with a mass of 1575 kg moving at 5.0 m/s in the same direction.

a. What is the final velocity if the wrecked cars lock together?

b. How much is the kinetic energy decreased in the collision?

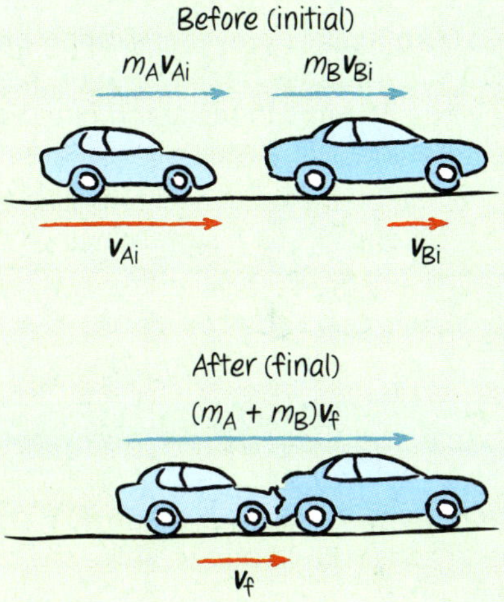

1 ANALYZE AND SKETCH THE PROBLEM

- Sketch the initial and final conditions.
- Sketch the momentum diagram.

Known	**Unknown**
$m_A = 1150$ kg	$v_f = ?$
$m_B = 1575$ kg	$\Delta KE = KE_f - KE_i = ?$
$v_{Ai} = 15.0$ m/s	
$v_{Bi} = 5.0$ m/s	
$v_{Af} = v_{Bf} = v_f$	

2 SOLVE FOR THE UNKNOWN

a. Use conservation of momentum to find the final velocity.

$$p_{Ai} + p_{Bi} = p_{Af} + p_{Bf}$$

$$m_A v_{Ai} + m_B v_{Bi} = (m_A + m_B)v_f$$

$$v_f = \frac{m_A v_{Ai} + m_B v_{Bi}}{m_A + m_B}$$

$$= \frac{(1150 \text{ kg})(15.0 \text{ m/s}) + (1575 \text{ kg})(5.0 \text{ m/s})}{(1150 \text{ kg} + 1575 \text{ kg})}$$

Substitute $m_A = 1150$ kg, $v_{Ai} = 15.0$ m/s, $m_B = 1575$ kg, $v_{Bi} = 5.0$ m/s

$= 9.2$ m/s, in the direction of the motion before the collision

b. To determine the change in kinetic energy of the system, KE_f and KE_i are needed.

$$KE_f = \frac{1}{2}mv_f^2 = \frac{1}{2}(m_A + m_B)v_f^2$$

Substitute $m = m_A + m_B$

$$= \frac{1}{2}(1150 \text{ kg} + 1575 \text{ kg})(9.2 \text{ m/s})^2$$

Substitute $m_A = 1150$ kg, $m_B = 1575$ kg, $v_f = 9.2$ m/s

$$= 1.2\times10^5 \text{ J}$$

$$KE_i = KE_{Ai} + KE_{Bi}$$

$$= \frac{1}{2}(m_A v_{Ai}^2) + \frac{1}{2}(m_B v_{Bi}^2)$$

Substitute $KE_{Ai} = \frac{1}{2}(m_A v_{Ai}^2)$, $KE_{Bi} = \frac{1}{2}(m_B v_{Bi}^2)$

$$= \frac{1}{2}(1150 \text{ kg})(15.0 \text{ m/s})^2 + \frac{1}{2}(1575 \text{ kg})(5.0 \text{ m/s})^2$$

Substitute $m_A = 1150$ kg, $m_B = 1575$ kg, $v_{Ai} = 15.0$ m/s, $v_{Bi} = 5.0$ m/s

$$= 1.5\times10^5 \text{ J}$$

Solve for the change in kinetic energy of the system.

$$\Delta KE = KE_f - KE_i$$

$$= 1.2\times10^5 \text{ J} - 1.5\times10^5 \text{ J} = -3\times10^4 \text{ J}$$

Substitute $KE_f = 1.2\times10^5$ J, $KE_i = 1.5\times10^5$ J

3 EVALUATE THE ANSWER

- **Are the units correct?** Velocity is measured in meters per second; energy is measured in joules.
- **Does the sign make sense?** The change in kinetic energy is negative, meaning that kinetic energy decreased as expected in an inelastic collision.

PRACTICE Problems

ADDITIONAL PRACTICE

44. An 8.00-g bullet is fired horizontally into a 9.00-kg block of wood on an air table and is embedded in it. After the collision, the block and bullet slide along the frictionless surface together with a speed of 10.0 cm/s. Find the initial speed of the bullet.

45. A 91.0-kg hockey player is skating on ice at 5.50 m/s. Another hockey player of equal mass, moving at 8.1 m/s in the same direction, hits him from behind. They slide off together.

a. What are the total mechanical energy and momentum of the system before the collision?

b. What is the velocity of the two hockey players after the collision?

c. How much was the system's kinetic energy decreased in the collision?

46. CHALLENGE A 0.73-kg magnetic target is suspended on a string. A 0.025-kg magnetic dart, shot horizontally, strikes the target head-on. The dart and the target together act like a pendulum and swing 12.0 cm above the initial level before instantaneously coming to rest.

a. Sketch the situation and choose a system.

b. Decide what is conserved in each step of the process and explain why.

c. What was the initial velocity of the dart?

PHYSICS Challenge

A bullet of mass *m*, moving at speed v_1, goes through a motionless wooden block and exits with speed v_2. After the collision, the block, which has mass m_B, is moving.

1. What is the final speed (v_B) of the block?

2. What was the change in the bullet's mechanical energy?

3. How much energy was lost to friction inside the block?

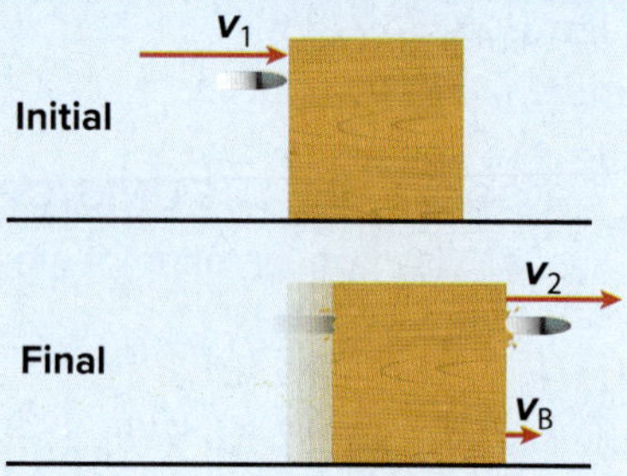

Check Your Progress

47. **Energy Diagrams** A child jumps on a trampoline. Draw energy bar diagrams to show the forms of energy in the following situations.

a. The child is at the highest point.

b. The child is at the lowest point.

48. **Energy** Explain why energy is considered a single quantity.

49. **Kinetic Energy** Suppose a glob of chewing gum and a small, rubber ball collide head-on in midair and then rebound apart. Would you expect kinetic energy to be conserved? If not, what happens to the energy?

50. **Potential Energy** A rubber ball drops from a height of 8.0 m onto a concrete floor and bounces repeatedly. Each time it hits the floor, the ball-Earth system loses $\frac{1}{5}$ of its *ME*. How many times will the ball bounce before it bounces back up to a height of only 4 m?

51. **Energy** In **Figure 27,** a child slides down a playground slide. At the bottom, she is moving at 3.0 m/s. How much energy was transformed by friction as she slid down the slide?

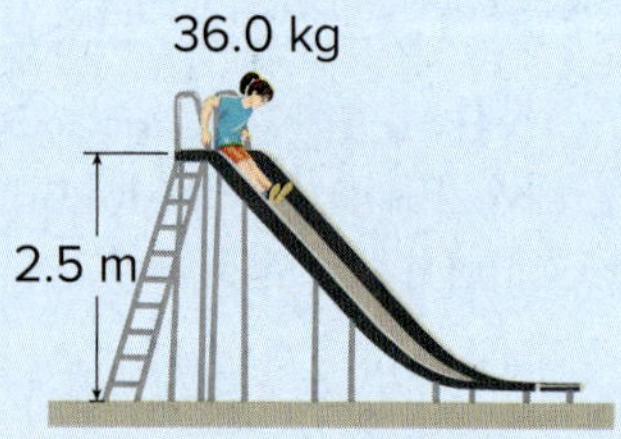

Figure 27

52. **Conservation of Energy** Your friend wants to solve the world's energy problems by inventing a device that will deliver ten times more energy than put into the device. Can this device work? Explain.

53. **Critical Thinking** A ball drops 20 m. When it has fallen 10 m, half of the energy is potential energy and half is kinetic energy. When the ball has fallen for half the amount of time it takes to fall, will more, less, or exactly half of the energy be potential energy?

LEARNSMART® Go online to follow your personalized learning path to review, practice, and reinforce your understanding.

LESSON 4
MACHINES

FOCUS QUESTION
How do machines make tasks easier?

Benefits of Machines

A **machine** is a device that makes tasks easier by changing either the magnitude or the direction of the applied force. When you lift the handle of the bottle opener in **Figure 28,** you transfer energy to the opener and do work on the opener. The opener lifts the cap, doing work on it. The work you do is called the input work (W_i). The work the machine does is called the output work (W_o).

The opener is not a source of energy. So, the cap cannot receive more energy than what you put into the opener. Thus, the output work can never be greater than the input work. The machine only helps transfer energy from you to the bottle cap.

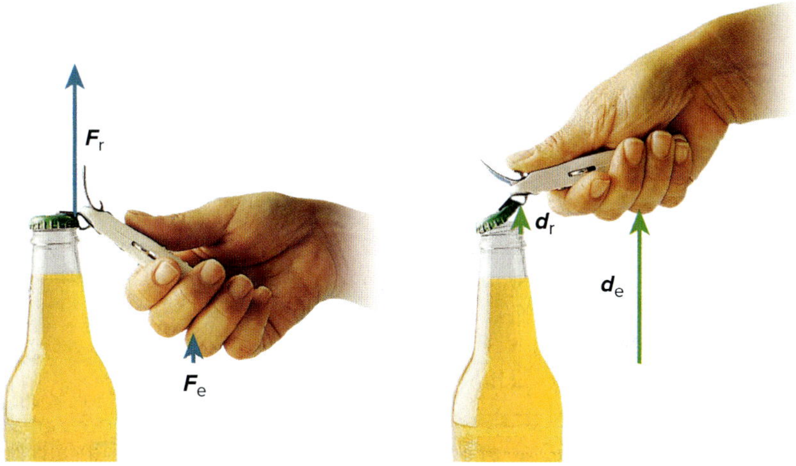

Figure 28 This bottle opener makes opening a bottle easier. However, it does not lessen the energy required to do so.

Mechanical advantage The force exerted by a user on a machine is called the **effort force** (F_e). The force exerted by the machine is called the **resistance force** (F_r). For the bottle opener in **Figure 28,** F_e is the upward force exerted by the person using the bottle opener and F_r is the upward force exerted by the bottle opener. The ratio of resistance force to effort force $\left(\frac{F_r}{F_e}\right)$ is called the **mechanical advantage** (MA) of the machine.

Mechanical Advantage

The mechanical advantage of a machine is equal to the resistance force divided by the effort force.

$$MA = \frac{F_r}{F_e}$$

3D THINKING | DCI Disciplinary Core Ideas | CCC Crosscutting Concepts | SEP Science & Engineering Practices

COLLECT EVIDENCE

Use your Science Journal to record the evidence you collect as you complete the readings and activities in this lesson.

INVESTIGATE

GO ONLINE to find these activities and more resources.

PhysicsLAB: Lifting with Pulleys
Carry out an investigation into how the arrangement of a pulley system affects the force needed to lift a weight.

Quick Investigation: Wheel and Axle
Carry out an investigation into the effort and resistance forces on a wheel-and-axle system.

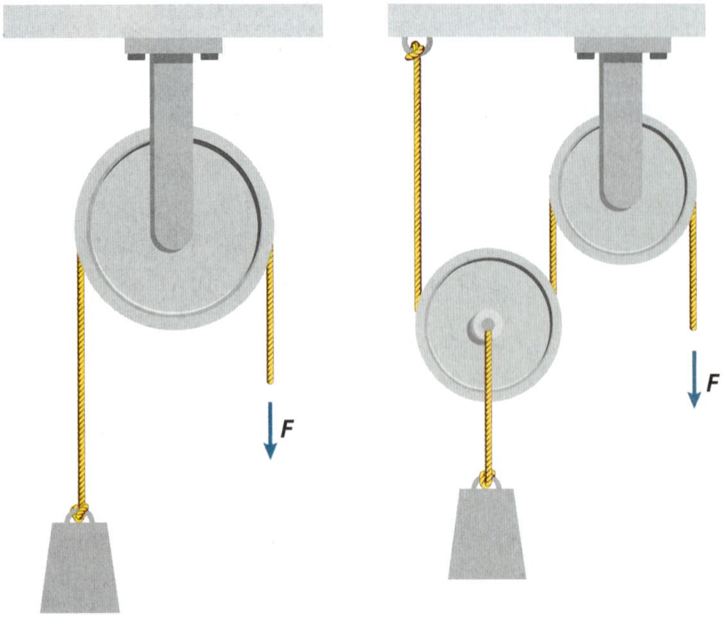

Figure 29 A fixed pulley has a mechanical advantage equal to 1. It is useful because it changes the direction of the force. A pulley system with a movable pulley has a mechanical advantage greater than 1, making the force applied to the object larger than the force originally exerted.

For a fixed pulley, such as the one shown on the left in **Figure 29,** the effort force (F_e) and the resistance force (F_r) are equal. Thus, the mechanical advantage is 1. What is the advantage of this machine? The fixed pulley is useful, not because the effort force is lessened, but because the direction of the effort force is changed. Many machines, such as the bottle opener shown in **Figure 28** and the pulley system shown on the right in **Figure 29,** have a mechanical advantage greater than 1. When the mechanical advantage is greater than 1, the machine increases the force applied by a person.

Get It?

Calculate A machine's *MA* is 3. If the input force is 2 N, what is the output force?

Ideal mechanical advantage A machine can increase force, but it cannot increase energy. An ideal machine transfers all the energy, so the output work equals the input work: $W_o = W_i$. The input work is the product of the effort force and the displacement the effort force acts through: $W_i = F_e d_e$. The output work is the product of the resistance force and the displacement the resistance force acts through: $W_o = F_r d_r$. Substituting these expressions into $W_o = W_i$ gives $F_r d_r = F_e d_e$. This equation can be rewritten $\frac{F_r}{F_e} = \frac{d_e}{d_r}$.

Recall that mechanical advantage is given by $MA = \frac{F_r}{F_e}$. Thus, for an ideal machine, **ideal mechanical advantage** (*IMA*) is equal to the displacement of the effort force divided by the displacement of the resistance force. *IMA* can be represented as follows.

Ideal Mechanical Advantage

The *IMA* of a machine is equal to the displacement of the effort force divided by the displacement of the resistance force.

$$IMA = \frac{d_e}{d_r}$$

Notice that you measure the displacements to calculate the ideal mechanical advantage, but you measure the forces exerted to find the actual mechanical advantage.

SCIENCE USAGE v. COMMON USAGE

Efficiency

Science usage: the ratio of output work to input work

The efficiency of the system of pulleys is 86 percent.

Common usage: production without waste

The new, high-efficiency washing machine uses less water and electricity than an older model of machine.

Efficiency In a real machine, not all of the input work is available as output work. Energy removed from the system through heat or sound means there is less output work from the machine. Consequently, the machine is less efficient at accomplishing the task. The **efficiency** of a machine (e) is defined as the ratio of output work to input work.

Efficiency

The efficiency of a machine (in %) is equal to the output work, divided by the input work, multiplied by 100.

$$e = \frac{W_o}{W_i} \times 100$$

An ideal machine has equal output and input work, $\frac{W_o}{W_i} = 1$, and its efficiency is 100 percent. All real machines have efficiencies that are less than 100 percent.

Efficiency can be expressed in terms of the mechanical advantage and ideal mechanical advantage. Efficiency, $e = \frac{W_o}{W_i}$, can be rewritten as follows:

$$e = \frac{W_o}{W_i} = \frac{F_r d_r}{F_e d_e}$$

Because $MA = \frac{F_r}{F_e}$ and $IMA = \frac{d_e}{d_r}$, the following expression can be written for efficiency.

Efficiency

The efficiency of a machine (in %) is equal to its mechanical advantage, divided by the ideal mechanical advantage, multiplied by 100.

$$e = \frac{MA}{IMA} \times 100$$

A machine's design determines its ideal mechanical advantage. An efficient machine has an *MA* almost equal to its *IMA*. A less-efficient machine has a small *MA* relative to its *IMA*. To obtain the same resistance force, a greater force must be exerted in a machine of lower efficiency than in a machine of higher efficiency.

Get It?

Determine A machine has an efficiency of 50 percent and an *MA* of 3. What is its *IMA?*

PHYSICS Challenge

In many rural areas, residents do not get their water supply from a central water tower or other common source. Many residents need to use an electric pump to get their water from a well. A certain electric pump pulls water at a rate of 0.25 m^3/s from a well that is 25 m deep. The water leaves the pump at a speed of 8.5 m/s.

1. What power is needed to lift the water to the surface?
2. What power is needed to increase the water's kinetic energy?
3. If the pump's efficiency is 80 percent, how much power must be delivered to the pump?

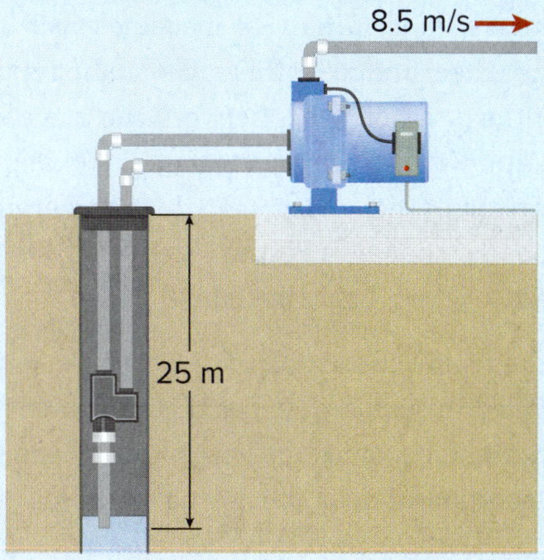

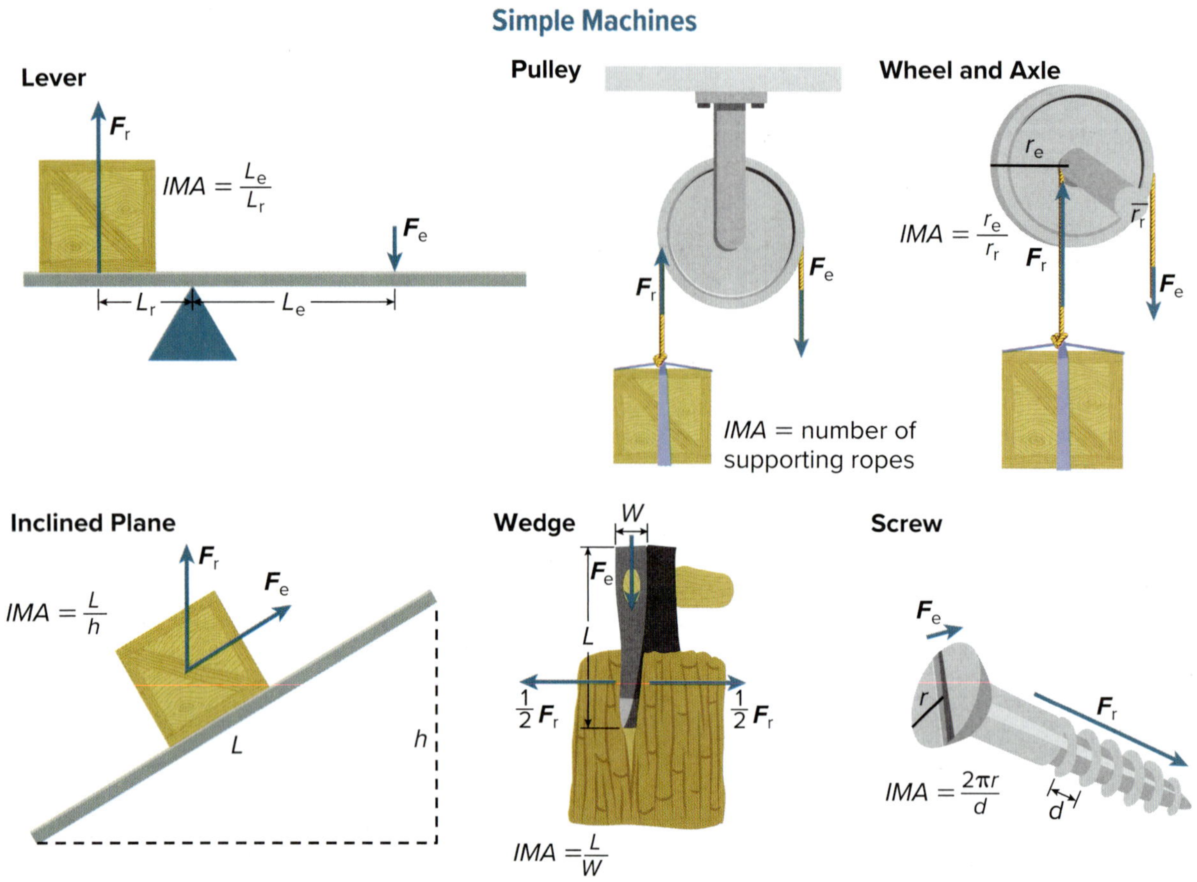

Figure 30 Each type of simple machine makes work easier by changing the direction or magnitude of the force. The *IMA* of each machine is related to the properties of the machine. For example, the *IMA* of a lever is equal to the length of the effort arm (L_e) divided by the length of the resistance arm (L_r). Similarly, the *IMA* of a wedge is equal to its length (L) divided by its width (W).
Identify *an everyday example of each simple machine.*

Compound Machines

Most machines are combinations of one or more of the six simple machines: the lever, pulley, wheel and axle, inclined plane, wedge, and screw, as shown in **Figure 30.**

The *IMA* of each machine in **Figure 30** is the ratio of the displacement of the effort force to the displacement of the resistance force. For machines such as the lever and the wheel and axle, this ratio can be replaced by the ratio of the displacements between the place where the force is applied and the pivot point. For example, a common version of the wheel and axle is a steering wheel, such as the one shown in **Figure 31.** The *IMA* is the ratio of the radii of the wheel and axle.

A machine consisting of two or more simple machines linked in such a way that the resistance force of one machine becomes the effort force of the second is called a **compound machine.** Some examples of compound machines are scissors (wedges and levers) and wheelbarrows (lever and wheel and axle).

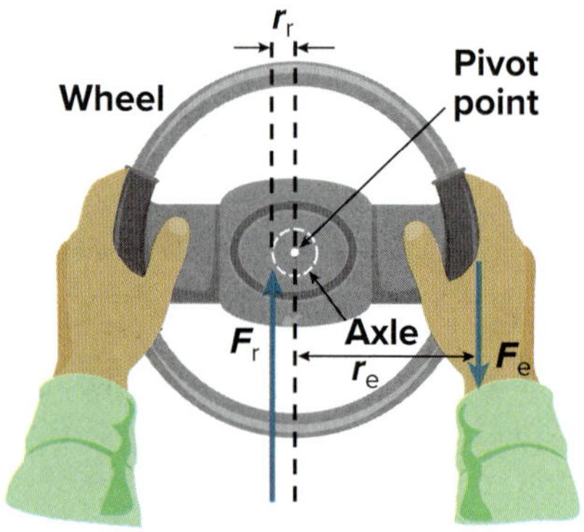

Figure 31 A steering wheel is an example of a wheel and axle. Its *IMA* is $\frac{r_e}{r_r}$.

Get It?

Compare and contrast simple machines and compound machines.

Figure 32 A bicycle is a compound machine. The rider exerts a force on the pedals, which exert a force on the chain. The chain then exerts a force on the rear gear.

Mechanical advantage and bicycles A bicycle, like the one shown in **Figure 32,** is a compound machine, consisting of two wheel-and-axle systems. The first is the pedal and front gear. Here, the effort force is the force that the rider exerts on the pedal, $F_{\text{rider on pedal}}$. The force of your foot is most effective when the force is exerted perpendicular to the arm of the pedal; that is, when the torque is largest. Therefore, we will assume that $F_{\text{rider on pedal}}$ is applied perpendicular to the pedal arm. The resistance is the force that the front gear exerts on the chain, $F_{\text{gear on chain}}$.

Get It?

Identify the effort force and the resistance force for the pedal and front gear.

The rear gear and the rear wheel act like another wheel and axle. The chain exerts an effort force on the rear gear, $F_{\text{chain on gear}}$. This force is equal to the force of the front gear on the chain. That is, $F_{\text{gear on chain}} = F_{\text{chain on gear}}$. The resistance force is the force that the wheel exerts on the road, $F_{\text{wheel on road}}$.

The mechanical advantage of a compound machine is the product of the mechanical advantages of the simple machines from which it is made. Therefore, the bicycle's mechanical advantage is given by the following equation:

$$MA_{\text{total}} = MA_{\text{pedal gear}} \times MA_{\text{rear wheel}}$$

$$MA_{\text{total}} = \left(\frac{F_{\text{gear on chain}}}{F_{\text{rider on pedal}}}\right)\left(\frac{F_{\text{wheel on road}}}{F_{\text{chain on gear}}}\right) = \frac{F_{\text{wheel on road}}}{F_{\text{rider on pedal}}}$$

Recall that $F_{\text{gear on chain}} = F_{\text{chain on gear}}$. Therefore, they cancel in the equation shown above.

Similarly, the *IMA* of the bicycle is $IMA = IMA_{\text{pedal gear}} \times IMA_{\text{rear wheel}}$. For each wheel and axle, the IMA is the ratio of the wheel radius to the axle radius.

For the pedal gear, $IMA = \dfrac{\text{pedal radius}}{\text{front gear radius}}$

For the rear wheel, $IMA = \dfrac{\text{rear gear radius}}{\text{wheel radius}}$

For the bicycle, then

$$IMA = \left(\frac{\text{pedal radius}}{\text{front gear radius}}\right)\left(\frac{\text{rear gear radius}}{\text{wheel radius)}}\right)$$
$$= \left(\frac{\text{rear gear radius}}{\text{front gear radius}}\right)\left(\frac{\text{pedal radius}}{\text{wheel radius}}\right)$$

EXAMPLE Problem 7

MECHANICAL ADVANTAGE You examine the rear wheel on your bicycle. It has a radius of 35.6 cm and has a gear with a radius of 4.00 cm. When the chain is pulled with a force of 155 N, the wheel rim moves 14.0 cm. The efficiency of this part of the bicycle is 95.0 percent.

a. What is the *IMA* of the wheel and gear?

b. What is the *MA* of the wheel and gear?

c. What is the resistance force?

d. How far was the chain pulled to move the rim 14.0 cm?

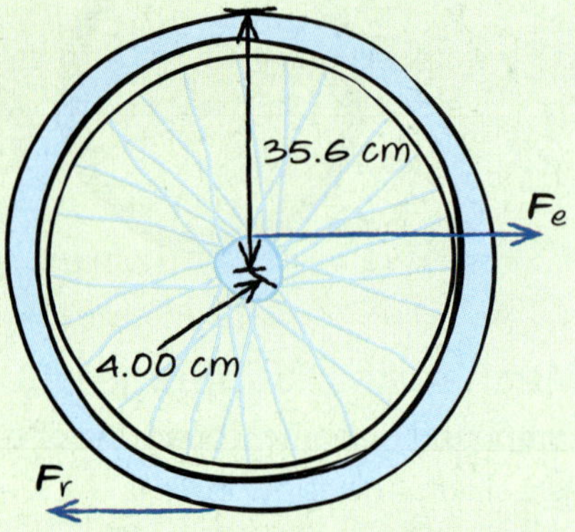

1 ANALYZE AND SKETCH THE PROBLEM

- Sketch the wheel and axle.
- Sketch the force vectors.

Known			Unknown	
$r_e = 4.00$ cm	$e = 95.0\%$	$F_e = 155$ N	$IMA = ?$	$F_r = ?$
$r_r = 35.6$ cm	$d_r = 14.0$ cm		$MA = ?$	$d_e = ?$

2 SOLVE FOR THE UNKNOWN

a. Solve for *IMA*.

$$IMA = \frac{r_e}{r_r}$$ For a wheel-and-axle machine, *IMA* is equal to the ratio of radii.

$$= \frac{4.00\text{ cm}}{35.6\text{ cm}} = 0.112$$ Substitute $r_e = 4.00$ cm, $r_r = 35.6$ cm.

b. Solve for *MA*.

$$e = \frac{MA}{IMA} \times 100$$

$$MA = \left(\frac{e}{100}\right) \times IMA$$

$$= \left(\frac{95.0}{100}\right) \times 0.112 = 0.106$$ Substitute $e = 95.0\%$, $IMA = 0.112$.

c. Solve for force.

$$MA = \frac{F_r}{F_e}$$

$$F_r = (MA)(F_e)$$

$$F_r = (0.106)(155\text{ N}) = 16.4\text{ N}$$ Substitute $MA = 0.106$, $F_e = 155$ N.

d. Solve for distance.

$$IMA = \frac{d_e}{d_r}$$

$$d_e = (IMA)(d_r)$$

$$= (0.112)(14.0\text{ cm}) = 1.57\text{ cm}$$ Substitute $IMA = 0.112$, $d_r = 14.0$ cm.

3 EVALUATE THE ANSWER

- **Are the units correct?** Force is measured in newtons and distance in centimeters.
- **Is the magnitude realistic?** *IMA* is low for a bicycle because a greater F_e is traded for a greater d_r. *MA* is always smaller than *IMA*. Because *MA* is low, F_r also will be low. The small distance the axle moves results in a large distance covered by the wheel. Thus, d_e should be very small.

PRACTICE Problems

ADDITIONAL PRACTICE

54. If the gear radius of the bicycle in Example Problem 7 is doubled while the force exerted on the chain and the distance the wheel rim moves remain the same, what quantities change, and by how much?

55. A sledgehammer is used to drive a wedge into a log to split it. When the wedge is driven 0.20 m into the log, the log is separated a distance of 5.0 cm. A force of 1.7×10^4 N is needed to split the log, and the sledgehammer exerts a force of 1.1×10^4 N.

a. What is the *IMA* of the wedge?

b. What is the *MA* of the wedge?

c. Calculate the efficiency of the wedge as a machine.

56. A worker uses a pulley to raise a 24.0-kg carton 16.5 m, as shown in **Figure 33.** A force of 129 N is exerted, and the rope is pulled 33.0 m.

a. What is the *MA* of the pulley?

b. What is the efficiency of the pulley?

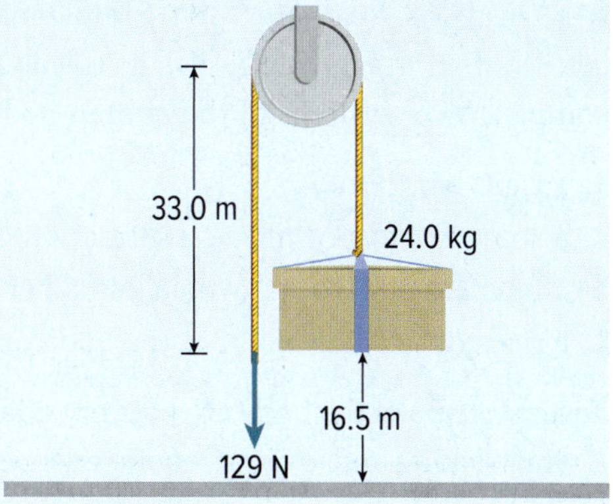

Figure 33

57. A winch has a crank with a 45-cm radius. A rope is wrapped around a drum with a 7.5-cm radius. One revolution of the crank turns the drum one revolution.

a. What is the ideal mechanical advantage of this machine?

b. If, due to friction, the machine is only 75 percent efficient, how much force would have to be exerted on the handle of the crank to exert 750 N of force on the rope?

58. CHALLENGE You exert a force of 225 N on a lever to raise a 1.25×10^3-N rock a distance of 13 cm. If the efficiency of the lever is 88.7 percent, how far did you move your end of the lever?

Multi-gear bicycle Shifting gears on a bicycle is a way of adjusting the ratio of gear radii to obtain the desired *IMA*. On a multi-gear bicycle, the rider can change the *IMA* of the machine by choosing the size of one or both gears. **Figure 34** shows a rear gear with five different gear sizes. When accelerating or climbing a hill, the rider increases *IMA* to increase the force that the wheel exerts on the road.

To do this, the rider makes the rear gear radius large compared to the front gear radius. For the same force exerted by the rider, a larger force is exerted by the wheel on the road—the rider rotates the pedals through more turns for each revolution of the wheel.

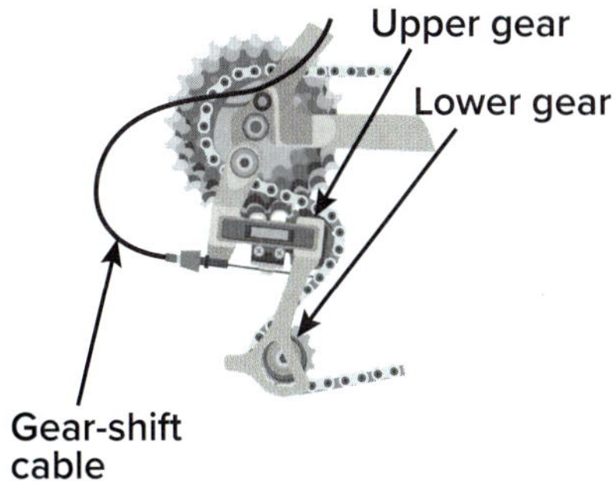

Figure 34 A rider can change the *IMA* of the bicycle by shifting gears.

However, less force is needed to ride the bicycle at high speed on a level road. The rider needs a small rear gear and a large front gear, resulting in a smaller *IMA*. Thus, for the same force exerted by the rider, a smaller force is exerted by the wheel on the road. In return, the rider does not have to move the pedals as far for each revolution of the wheel.

An automobile transmission works in the same way. To accelerate a car from rest, large forces are needed and the transmission increases the *IMA* by increasing the gear ratio.

At high speeds, however, the transmission reduces the gear ratio and the *IMA* because smaller forces are needed. Even though the speedometer shows a high speed, the tachometer indicates the engine's low angular speed.

Get It?

Explain why a car needs multiple gears.

The Human Walking Machine

BIOLOGY Connection Simple machines, in the form of levers, such as the lever system in **Figure 35,** give us the ability to walk, run, and do other activities. The complex lever systems of the human body have the following four basic parts:

1. a rigid bar (bone)
2. a source of force (muscle contraction)
3. a fulcrum or pivot (movable joints between bones)
4. a resistance (the weight of the body or an object being lifted or moved)

Lever systems of the body are not very efficient, and mechanical advantages are low. This is why walking and jogging require energy (burn calories) and help people lose weight.

When a person walks, the hip acts as a fulcrum and moves through the arc of a circle, centered on the foot. The center of mass of the body moves as a resistance around the fulcrum in the same arc. The length of the radius of the circle is the length of the lever formed by the bones of the leg. Athletes in walking races increase their velocity by swinging their hips upward to increase this radius.

Although tall people usually can walk faster than short people can, a tall person must apply a greater force to move the longer lever formed by the leg bones. Because of the inefficiency of their lever systems and the length of a 20- or 50-km walking race, very tall people rarely have the stamina to win.

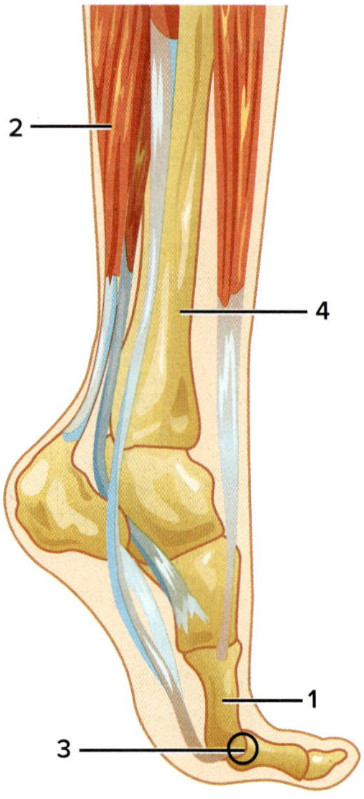

Figure 35 The human leg and foot function as a compound machine.

Check Your Progress

59. **Machines** Classify each tool below as a lever, a wheel and axle, an inclined plane, or a wedge. Describe how it changes the force to make the task easier.
    a. screwdriver
    b. pliers
    c. chisel
    d. nail puller

60. **IMA** A worker is testing a multiple pulley system to estimate the heaviest object that he could lift. The largest downward force he can exert is equal to his weight, 875 N. When the worker moves the rope 1.5 m, the object moves 0.25 m. What is the heaviest object that he could lift?

61. **Compound Machines** A winch has a crank on a 45-cm arm that turns a drum with a 7.5-cm radius through a set of gears. It takes three revolutions of the crank to rotate the drum through one revolution. What is the *IMA* of this compound machine?

62. **Efficiency** Suppose you increase the efficiency of a simple machine. Do the *MA* and *IMA* increase, decrease, or remain the same?

63. **Critical Thinking** The mechanical advantage of a multi-gear bicycle is changed by moving the chain to a suitable rear gear.
    a. To start out, you must accelerate the bicycle, so you want to have the bicycle exert the greatest possible force. Should you choose a small or large gear?
    b. As you reach your traveling speed, you want to rotate the pedals as few times as possible. Should you choose a small or large gear?
    c. Many bicycles also let you choose the size of the front gear. If you want even more force to accelerate while climbing a hill, would you move to a larger or smaller front gear? Explain.

LEARNSMART® Go online to follow your personalized learning path to review, practice, and reinforce your understanding.

Greenhouse Gas Emission Efficiency: Excrement to Energy

Raising cattle is big business. It is also a big environmental headache. The waste produced by cows pollutes Earth's air and water. Cows release as much as 25% of global methane (CH4) emissions, which is one of the main greenhouse gases contributing to climate change. Thanks to anaerobic digesters, methane gas from animal waste such as cow manure also has the potential to be harnessed to provide a source of renewable energy.

Anaerobic digesters on dairy farms help provide a source of renewable energy and eliminate harmful animal waste.

Anaerobic Digestion

There are over 90 million dairy and beef cattle in the United States. Each cow produces up to 40 kg of manure per day. How much manure is produced in the United States every year? A lot! In addition to contaminating runoff water with harmful bacteria and excess nitrogen, the methane gas emitted from decomposing manure and cow flatulence impacts Earth's atmosphere and climate change.

Anaerobic digestion is one process that is used to help mitigate this waste. Specific microorganisms are mixed with a watery manure slurry in a digestion tank. The conditions in the tank are kept oxygen-free and at a temperature optimal for anaerobic bacteria. The bacteria decompose the manure into smaller compounds, which are converted into biogas. Some impurities are filtered from the biogas as it rises in the fermenting tank. The result is a mixture of methane and carbon dioxide called *biogas,* which is fed into power plants through underground pipes. The gas is later burned in a power plant as a source of fuel in a process that generates heat and electricity.

The results of anaerobic digestion are overwhelmingly positive. The electricity and heat generated by biogas have been used to power farms and thousands of communities. And the remaining organic nutrient-rich materials left from production can be used to fertilize crops.

Some engineers are working on technology that can cheaply clean biogas to make it usable in converted car and airplane engines. Others are trying to make biogas feasible for home use. In the near future, manure may become one of the best renewable energy alternatives for our fast-depleting supply of fossil fuels, such as coal and oil.

OBTAIN, EVALUATE, AND COMMUNICATE INFORMATION

Using the internet or other appropriate resources, research other types of technology being used to generate biofuels. Write a brief summary of your findings.

CALIFORNIA CONNECTION

Several dairy farms in California use technology to generate biogas. Find a farm that is geographically close to where you live. Make a presentation that summarizes how the farm's technology affects local communities.

MODULE 10
STUDY GUIDE

GO ONLINE to study with your Science Notebook.

Lesson 1 WORK AND ENERGY

- Work is done when a force is applied through a displacement. Work is the product of the force exerted on a system and the component of the distance through which the system moves:

$$W = Fd \cos \theta$$

- Energy is the ability of a system to produce a change in itself or its environment. A moving object has kinetic energy.
- The work done on a system is equal to the change in energy of the system. This is called the work-energy theorem:

$$W = \Delta E$$

- Power is the rate at which energy is transformed: $P = \frac{\Delta E}{t}$.

$$P = \frac{\Delta E}{t} = \frac{W}{t}$$

- work
- joule
- energy
- work-energy theorem
- kinetic energy
- power
- watt

Lesson 2 THE MANY FORMS OF ENERGY

- The translational kinetic energy of an object is proportional to its mass and the square of its velocity. The rotational kinetic energy of an object is proportional to the object's moment of inertia and the square of its angular velocity.
- The gravitational potential energy of an object-Earth system is the product of mass, gravitational field, and height above a reference level: $GPE = mgh$.
- Elastic potential energy is energy stored due to stretching, compressing, or bending an object. Other forms of energy are chemical, nuclear, thermal, electrical, radiant, and mass energy.

- translational kinetic energy
- rotational kinetic energy
- potential energy
- gravitational potential energy
- reference level
- elastic potential energy
- thermal energy

Lesson 3 CONSERVATION OF ENERGY

- The total energy of a closed, isolated system is constant. Within the system, energy can change form, but the total amount of energy does not change. Thus, energy is conserved.
- Mechanical energy is the sum of kinetic and potential energy:

$$ME = KE + PE$$

- Momentum is conserved in collisions if the external force is zero. The kinetic energy may be unchanged or decreased by the collision, depending on whether the collision is elastic or inelastic.

- law of conservation of energy
- mechanical energy
- elastic collision
- inelastic collision

Lesson 4 MACHINES

- Machines do not change the work done. They can make a task easier by changing the magnitude or direction of the effort force.
- Mechanical advantage is the ratio of resistance force to effort force.

$$MA = \frac{F_r}{F_e}$$

- Ideal mechanical advantage is the ratio of the distances moved.

$$IMA = \frac{d_e}{d_r}$$

- Efficiency is the ratio of output work to input work, in percent.

$$e = \frac{W_o}{W_i} \times 100$$

- machine
- effort force
- resistance force
- mechanical advantage
- ideal mechanical advantage
- efficiency
- compound machine

THREE-DIMENSIONAL THINKING

Module Wrap-Up

REVISIT THE PHENOMENON

How can energy from power plants be stored in the power grid for later use?

CER Claim, Evidence, Reasoning

Explain Your Reasoning Revisit the claim you made when you encountered the phenomenon. Summarize the evidence you gathered from your investigations and research and finalize your Summary Table. Does your evidence support your claim? If not, revise your claim. Explain why your evidence supports your claim.

STEM UNIT PROJECT

Now that you've completed the module, revisit your STEM unit project. You will summarize your evidence and apply it to the project.

GO FURTHER

SEP Data Analysis Lab

How much kinetic energy does a falling object have?

Five clay pots are dropped from the same rooftop. The results are summarized in the table.

Clay Pot	Mass (kg)	Initial Velocity (m/s)
A	1.0	0
B	1.0	2.0; downward
C	1.0	2.0; upward
D	1.0	2.0; horizontally
E	2.0	0

CER Analyze and Interpret Data

1. **Claim** Rank the clay pots according to speed when they strike the ground, from least to greatest. Specifically indicate any ties.
2. **Evidence and Reasoning** Explain how you determined your ranking.

shan.shihan/Moment Open/Getty Images

Joseph Giacomin/Cultura RM/Alamy Stock Photo

MODULE 11

THERMAL ENERGY

ENCOUNTER THE PHENOMENON

Why do different parts of the pan appear to be different colors in this thermal image?

GO ONLINE to watch a video about how different materials heat at different rates.

SEP Ask Questions

Do you have other questions about the phenomenon? If so, add them to the driving question board.

CER Claim, Evidence, Reasoning

Make Your Claim Use your CER chart to make a claim about why different parts of the pan appear to be different colors in the thermal image. Explain your reasoning.

Collect Evidence Use the lessons in this module to collect evidence to support your claim. Record your evidence as you move through the module.

Explain Your Reasoning You will revisit your claim and explain your reasoning at the end of the module.

GO ONLINE to access your CER chart and explore resources that can help you collect evidence.

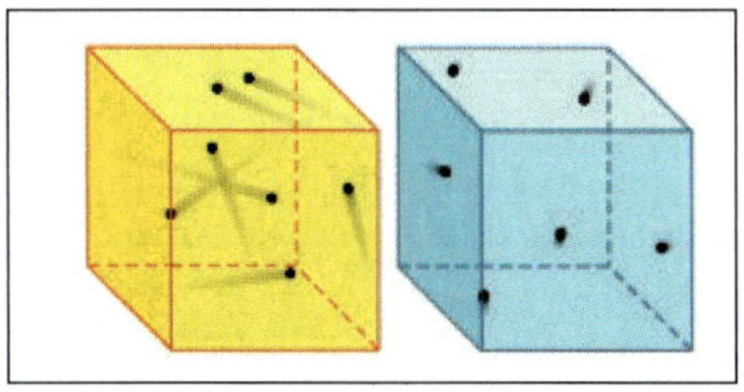

LESSON 1: Explore & Explain: Equilibrium and Thermometers

LESSON 2: Explore & Explain: The First Law of Thermodynamics

Smithsonian

Additional Resources

LESSON 1
TEMPERATURE, HEAT, AND THERMAL ENERGY

FOCUS QUESTION

How is thermal energy transferred when you cook something?

Thermal Energy

You have studied how objects collide and trade kinetic energies. Every material is made of microscopic particles. The many particles present in a gas have linear and rotational kinetic energies. The particles also might have potential energy due to their internal bonds and interactions with each other. As gas particles collide with each other and with the walls of the container, as shown in **Figure 1,** they transfer energy. There are numerous molecules that make up the gas, resulting in many collisions. The energies of the particles become randomly distributed.

Thus, it is convenient to discuss the total energy of the particles that compose the gas and the average energy per particle in the gas. Recall that the sum of the particles' energies is the object's thermal energy. The average kinetic energy per particle is related to the temperature of the gas. The relationship between the particles' random motions and the bulk property of the material is described by kinetic theory.

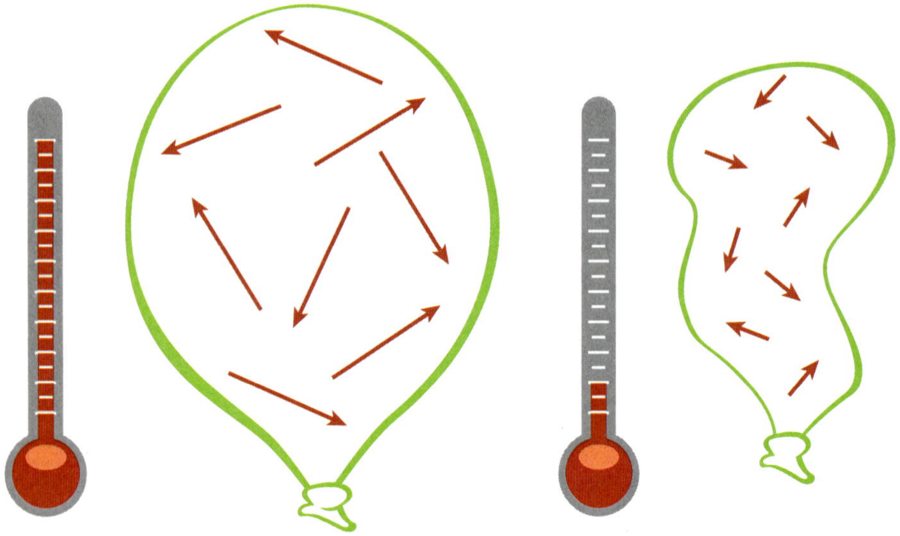

Figure 1 Gas particles collide with each other and the walls of the container. The temperature of an object is related to the average kinetic energy of its particles. The average kinetic energy of the particles that make up a hot object is greater than the average kinetic energy of the particles that make up a cold object.

Explain *what the arrows represent and why the arrows are longer in the balloon on the left than in the balloon on the right.*

3D THINKING — DCI Disciplinary Core Ideas — CCC Crosscutting Concepts — SEP Science & Engineering Practices

COLLECT EVIDENCE

Use your Science Journal to record the evidence you collect as you complete the readings and activities in this lesson.

INVESTIGATE

GO ONLINE to find these activities and more resources.

Applying Practices: Coffee Cup Calorimetry

HS-PS3-4. Plan and conduct an investigation to provide evidence that the transfer of thermal energy when two components of different temperature are combined within a closed system results in a more uniform energy distribution among the components in the system (second law of thermodynamics).

Hot objects and cold objects What makes an object hot? Consider a helium-filled balloon. The balloon is kept inflated by the repeated pounding from helium atoms on the balloon wall. Each of the approximately 10^{22} helium atoms in the balloon collides with the balloon wall, bounces back, and hits the balloon wall again somewhere else. The size and the temperature of the balloon are affected by the average kinetic energy of the helium atoms, as shown in **Figure 1.**

If you put a balloon in sunlight, energy absorbed from the sunlight makes each of the helium atoms move faster in random directions and bounce off the rubber walls of the balloon more often. Each atomic collision with the balloon wall puts a greater force on the balloon and stretches the rubber. Thus, the warmed balloon expands. On the other hand, if you refrigerate a balloon, you will find that it shrinks. It must do so because the particles are moving more slowly. The refrigeration has removed some of the atoms' thermal energy.

Thermal energy in solids The atoms or molecules in solids also have kinetic energy, but they are unable to move everywhere as gas atoms do. One way to illustrate the structure of a solid is to picture a number of atoms that are held in place next to each other by atomic forces that act like springs. The atoms cannot move freely, but they do bounce back and forth, with some bouncing more than others. Each atom has some kinetic energy and some potential energy. If a solid has N atoms, then the total thermal energy in the solid is equal to the average kinetic energy plus potential energy per atom times N.

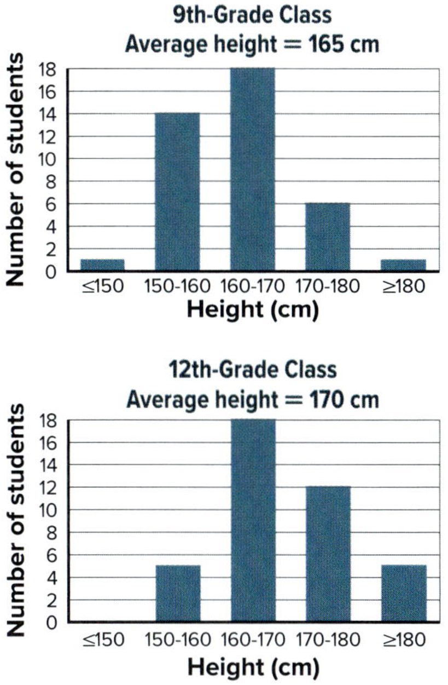

Figure 2 The average height in a 9th-grade class is less than the average height in a 12th-grade class. Similarly, the average kinetic energy of a hot object's particles is greater than the average kinetic energy of a cold object's particles.

Thermal Energy and Temperature

You have just seen that, on average, a particle in a hot object has more kinetic energy than a particle in a cold object. This does not mean that each of the particles that compose an object has the same amount of energy; they have a wide range of energies. The average kinetic energy of the particles that compose a hot object, however, is greater than the average kinetic energy of the particles that compose a cold object.

To understand this, consider the heights of students in a 9th-grade class and the heights of students in a 12th-grade class. The students' heights vary, as shown in **Figure 2,** but you can calculate the average height for each class. The average height of the 12th-grade class is greater than the average height of the 9th-grade class, even though some 9th-grade students might be taller than some 12th-grade students.

STEM CAREER Connection

HVAC Mechanic/Installer

If you like working with your hands and working with mechanical systems, a career as a heating, ventilation, air conditioning (HVAC) mechanic/installer might be for you. HVAC installers install heating and air conditioning systems in schools, homes, and businesses to improve the comfort and health of the occupants.

Figure 3 Two muffins at the same temperature can have different thermal energies.

Temperature depends only on the average kinetic energy of the particles in the object. It does not depend on the number of particles that compose the object. For example, consider the two muffins shown in **Figure 3.** They are at the same temperature, but the large muffin has ten times as many particles as the small muffin. Thus, the large muffin has ten times the thermal energy of the small muffin. The thermal energy of an object depends on both its temperature and the number of particles that make up that object.

Equilibrium and Thermometers

To measure your body temperature, you might place a thermometer in your mouth and wait for a moment before checking the reading. Measuring your temperature involves random collisions and energy transfers between the particles of the thermometer and the particles of your body. Your body is hotter than the thermometer. That is, the average kinetic energy of the particles that compose your body is greater than the average kinetic energy of the thermometer's particles. When the cool thermometer touches your skin, the particles of your skin collide with the particles of the thermometer. On average, the more energetic skin particles will transfer energy to the less energetic particles of the thermometer. **Thermal conduction** is the transfer of thermal energy that occurs when particles collide. As a result of these collisions, the thermal energy of the thermometer's particles increases and the thermal energy of your skin's particles decreases.

Thermal equilibrium

The thermometer's particles also transfer energy to your body's particles. As the thermometer's particles gain more energy, the amount of energy they give back to the skin increases. At some point, the rate of energy transfer from the thermometer to your body is equal to the rate of transfer in the other direction. At this point, your body and the thermometer have reached thermal equilibrium. **Thermal equilibrium** is the state in which the rates of thermal energy transfer between two objects are equal and the objects are at the same temperature. In other words, uncontrolled systems always evolve toward more stable states—that is, toward more uniform energy distribution (objects hotter than their surrounding environment cool down until thermal equilibrium is reached by the object and its surroundings). **Figure 4** on the next page shows two blocks reaching equilibrium.

Get It?

Describe an example in your everyday life in which an object and its surroundings move toward equilibrium, or more uniform energy distribution.

Reaching Thermal Equilibrium

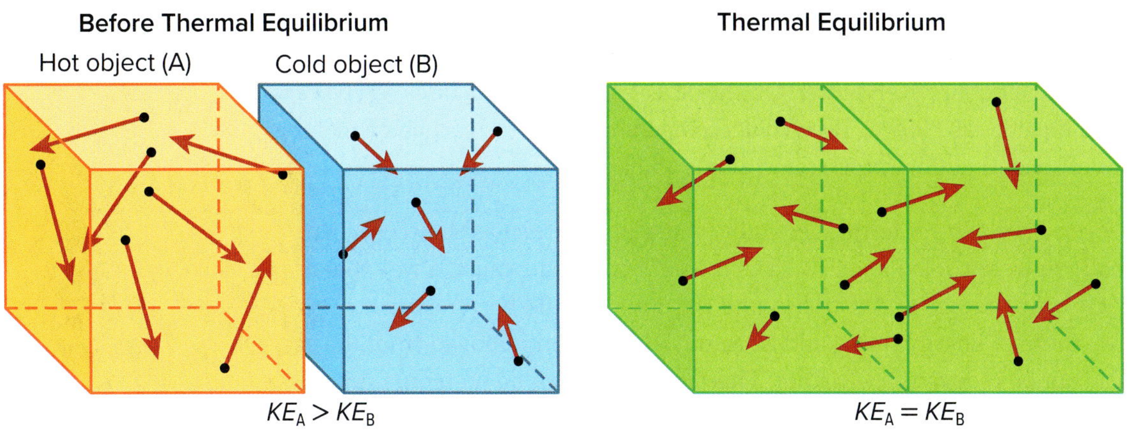

Figure 4 When a hot object and a cold object are in contact, there is a net transfer of thermal energy from the hot object to the cold object. When the two objects reach thermal equilibrium, the transfer of energy between the objects is equal, and the objects are at the same temperature.

Thermometers Every thermometer has some useful property that changes with temperature. Household thermometers often contain colored alcohol that expands when heated. The hotter the thermometer, the more the alcohol expands and the higher it rises in the tube. The liquid crystal thermometer in **Figure 5** uses a variety of long molecules that rearrange and cause a color change at specific temperatures. Medical thermometers and the thermometers that monitor automobile engines use very small, temperature-sensitive electronic circuits to take rapid temperature measurements.

Get It?

Describe in your own words how an alcohol thermometer works.

Figure 5 Liquid crystal thermometers change color with temperature.

Summarize *the process that occurs when this thermometer is placed on a person's forehead.*

Temperature limits You might say that a fire is hot and a freezer is cold. But the temperatures of everyday objects are only a small part of the wide range of temperatures present in the universe, as shown in **Figure 6.** Temperatures do not appear to have an upper limit. The interior of the Sun is at least 1.5×10^{7} °C. Supernova cores are even hotter. On the other hand, liquefied gases can be very cold. For example, helium liquefies at −269°C. Even colder temperatures can be reached by making use of special properties of solids, helium isotopes, and atoms and lasers.

Temperatures do, however, have a lower limit. Generally, materials contract as they cool. If an ideal atomic gas in a balloon were cooled to −273.15°C, it would contract in such a way that it occupied a volume that is only the size of the atoms, and the atoms would become motionless. At this temperature, all the thermal energy that could be removed has been removed from the gas, and the temperature cannot be reduced any further. Therefore, there can be no temperature less than −273.15°C, which is called absolute zero.

Get It?

Explain why the term *absolute zero* is appropriate for the coldest temperature possible.

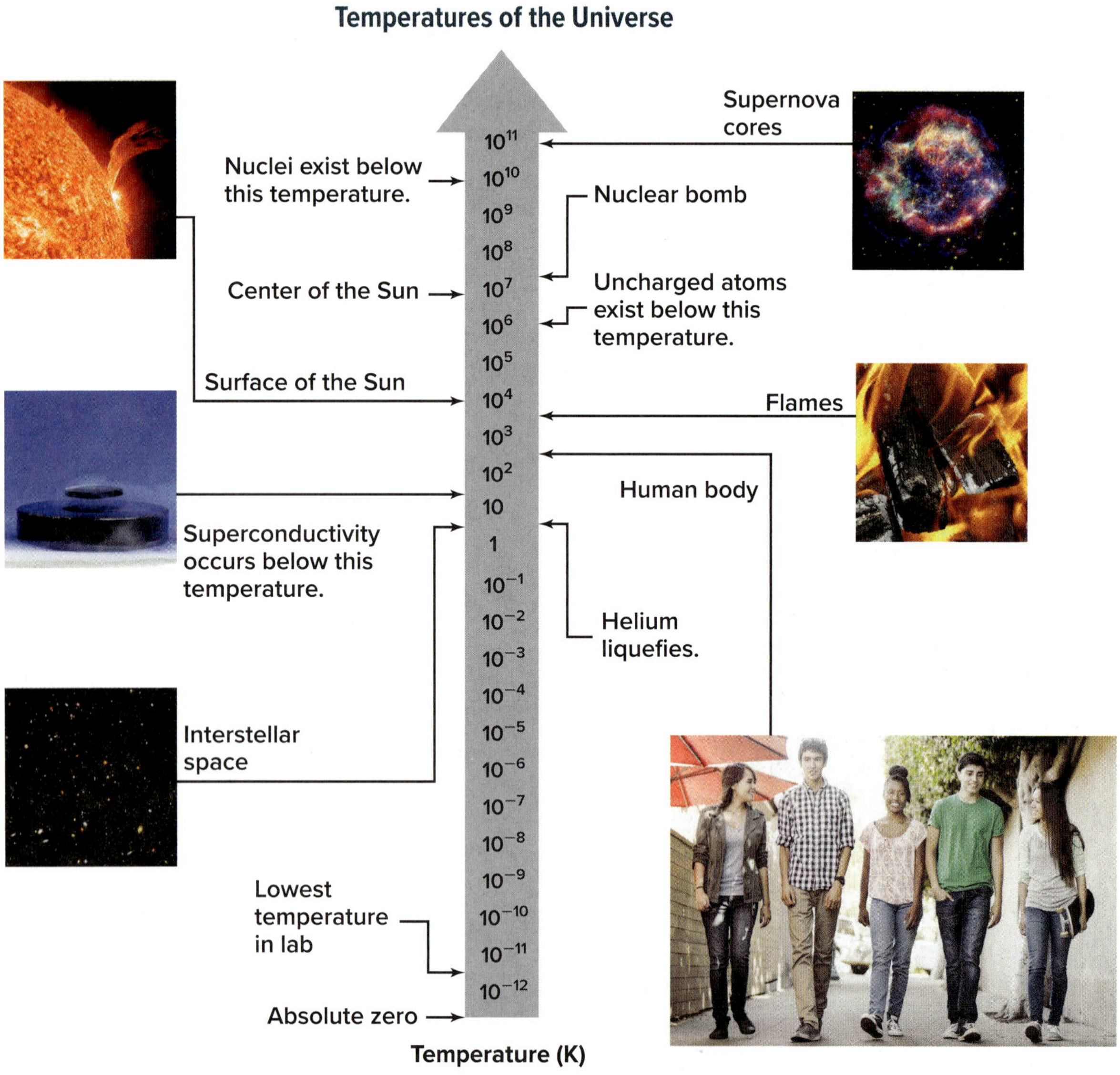

Figure 6 Temperatures in the universe range from just above absolute zero to more than 10^{10}K.

(l to r, t to b)NASA/GSFC/SDO; NASA/JPL-Caltech/STScI/CXC/SAO; Takeshi Takahara/Science Source; Nature-Elements/Alamy Stock Photo;

Temperature scales

In the United States, weather agencies report temperatures in degrees Fahrenheit. Scientists, however, use the Celsius and Kelvin scales. The Celsius scale is based on the properties of water and was devised in 1741 by Swedish physicist Anders Celsius. On the Celsius scale, the freezing point of pure water at sea level is defined to be 0°C. The boiling point of pure water at sea level is defined to be 100°C. The Celsius scale is useful for day-to-day measurements of temperature.

The Kelvin scale

On the Celsius scale, temperatures can be negative. Negative temperatures suggest a particle could have negative kinetic energy. Because temperature represents average kinetic energy of the object's particles, it makes more sense to use a temperature scale whose zero temperature is where the particles' kinetic energy is also zero. Therefore the zero point of the Kelvin scale is defined to be absolute zero. On the Kelvin scale, the freezing point of water (0°C) is about 273 K and the boiling point of water is about 373 K. Each interval on this scale, called a kelvin, is equal to 1°C. Thus, $T_C + 273 = T_K$. **Figure 7** compares the Fahrenheit, Celsius, and Kelvin scales.

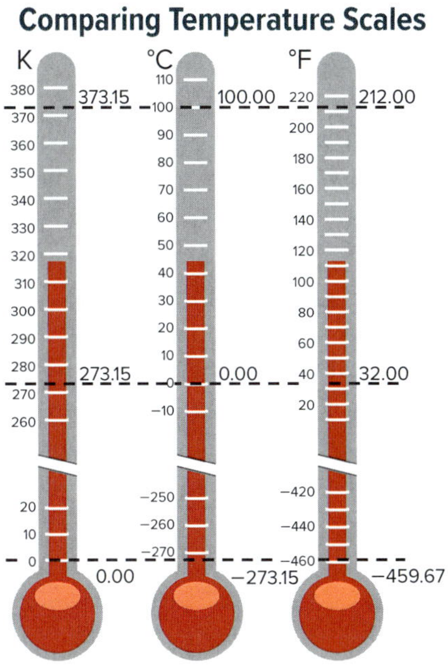

Figure 7 The Kelvin and Celsius scales are used by scientists. In the United States, the Fahrenheit scale is often used for weather reports and cooking.

Heat and Thermal Energy Transfer

When two objects come in contact with each other, they redistribute their thermal energies. **Heat** (Q) is the transfer of thermal energy, which occurs spontaneously from a hotter object to a cooler object. Thermal energy cannot be transferred from a colder object to a hotter object without work being done. Like work and energy, heat is measured in joules. In the thermometer example, thermal energy was transferred from the warm skin to the cold thermometer because of the collisions of particles. If thermal energy has been absorbed by an object, Q is positive. If thermal energy is transferred from an object, Q is negative.

Conduction, convection, and radiation

Figure 8 shows the three types of heat—conduction, convection, and radiation. If you place one end of a metal rod in a flame, the hot gas conducts heat to the rod. The other end of the rod also becomes warm because the particles that make up the rod conduct thermal energy to their neighbors.

Figure 8 Thermal energy can be transferred by conduction, convection, and radiation.
Identify *other common occurrences of conduction, convection, and radiation.*

Have you ever noticed the motion on the surface of a pot of water just about to boil? The water at the bottom of the pot is heated by conduction, expands, and floats to the top, while the colder, denser water at the top sinks to the bottom. The motion of the hot water rapidly carries heat from the bottom of the pot to the top surface of the water. Heating caused by the motion of fluid in a liquid or gas due to temperature differences is called **convection.** Atmospheric turbulence is caused by convection of gases in the atmosphere. Thunderstorms and hurricanes are excellent examples of large-scale atmospheric convection. Convection also contributes to ocean currents that move water and materials over large distances.

The third method of thermal transfer, unlike the first two, does not depend on the presence of matter. The Sun warms Earth from over 150 million km away via **radiation,** which is the transfer of energy by electromagnetic waves. These waves carry the energy from the hot Sun through the vacuum of space to the much cooler Earth.

Specific Heat

Some objects are easier to warm up than others. On a bright summer day, the Sun radiates thermal energy to the sand on a beach and to the ocean water. The sand on the beach becomes quite hot, while the ocean water stays relatively cool. When an object is heated, its thermal energy increases and its temperature can increase. The amount of the increase in temperature depends on the size of the object and its composition.

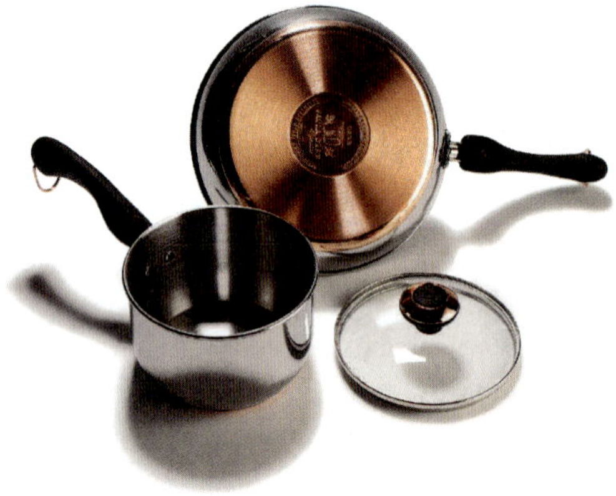

Figure 9 These pans are made of stainless steel and have copper bottoms and plastic handles.
Explain *how the selection of these materials is related to their specific heats.*

The **specific heat** of a material is the amount of energy that must be added to a unit mass of the material to raise its temperature by one temperature unit. In SI units, specific heat (C) is measured in J/(kg·K). **Table 1** provides values of specific heat for some common substances. For example, 897 J must be added to 1 kg of aluminum to raise its temperature by 1 K. The specific heat of aluminum is therefore 897 J/(kg·K). Materials with different specific heats are used for different purposes. Metals, such as those used to make the pans in **Figure 9,** have low specific heats and are good thermal conductors. Notice that liquid water has a high specific heat compared to other substances. Ice and water vapor also have relatively high specific heats. These high specific heats have had significant effects on our climate and our bodies.

Table 1 Specific Heat of Common Substances

Material	Specific Heat (J/kg·K))	Material	Specific Heat (J/kg·K))
Aluminum	897	Lead	130
Brass	376	Methanol	2450
Carbon	710	Silver	235
Copper	385	Water Vapor	2020
Glass	840	Water	4180
Ice	2060	Zinc	388
Iron	450		

Measuring heat When a substance is heated, the substance's temperature can change. The change of temperature (ΔT) depends on heat (Q), the mass of the substance (m), and the specific heat of the substance (C). By using the following equation, you can calculate the heat (Q) required to change the temperature of an object.

Heat

Heat is equal to the mass of an object times the specific heat of the object times the difference between the final and initial temperatures.

$$Q = mC\Delta T = mC\,(T_f - T_i)$$

For example, when the temperature of 10.0 kg of water is increased from 80 K to 85 K, the heat is $Q = (10.0\text{ kg})(4180\text{ J/(kg·K)})(85\text{ K} - 80\text{ K}) = 2.1\times10^5\text{ J}$. Remember that the temperature interval for the Kelvin scale is the same as that for the Celsius scale. For this reason, you can calculate ΔT on the Kelvin scale or on the Celsius scale.

EXAMPLE Problem 1

HEAT TRANSFER A 5.10-kg cast-iron skillet is heated on the stove from 295 K to 373 K. How much thermal energy had to be transferred to the iron?

1 ANALYZE AND SKETCH THE PROBLEM

Sketch the thermal energy transfer into the skillet from the stove top.

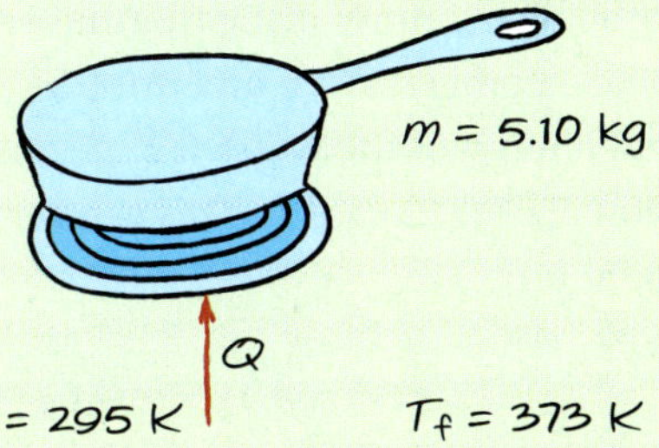

Known

$m = 5.10$ kg $C = 450$ J/(kg·K)

$T_i = 295$ K $T_f = 373$ K

Unknown

$Q = ?$

2 SOLVE FOR THE UNKNOWN

$Q = mC(T_f - T_i)$

$= (5.10\text{ kg})(450\text{ J/(kg·K)})(373\text{ K} - 295\text{ K})$ Substitute $m = 5.10$ kg, $C = 450$ J/(kg·K), $T_f = 373$ K, $T_i = 295$ K.

$= 1.8\times10^5$ J

3 EVALUATE THE ANSWER

- **Are the units correct?** Heat is measured in joules.
- **Do the signs make sense?** Temperature increased, so Q is positive.

PRACTICE Problems

ADDITIONAL PRACTICE

1. When you turn on the hot water to wash dishes, the water pipes heat up. How much thermal energy is absorbed by a copper water pipe with a mass of 2.3 kg when its temperature is raised from 20.0°C to 80.0°C?
2. Electrical power companies sell electrical energy by the kilowatt-hour, where 1 kWh = 3.6×10^6 J. Suppose that it costs $0.15 per kWh to run your electric water heater. How much does it cost to heat 75 kg of water from 15°C to 43°C to fill a bathtub?
3. **CHALLENGE** A car engine's cooling system contains 20.0 L of water (1 L of water has a mass of 1 kg).
   a. What is the change in the temperature of the water if 836.0 kJ of thermal energy is added?
   b. Suppose that it is winter, and the car's cooling system is filled with methanol. The density of methanol is 0.80 g/cm^3. What would be the increase in temperature of the methanol if it absorbed 836.0 kJ of thermal energy?
   c. Which coolant, water or methanol, would better remove thermal energy from a car's engine? Explain.

Measuring Specific Heat

A calorimeter, such as the simple one shown in **Figure 10,** is a device that measures changes in thermal energy. A calorimeter is carefully insulated so that thermal energy transfer to the external world is kept to a minimum. A measured mass of a substance that has been heated to a high temperature (T_A) is placed in the calorimeter. The calorimeter also contains a known mass of cold water at a measured temperature (T_B). Thermal energy is transferred from the warmer substance to the cooler water until they come to an equilibrium temperature (T_f). By measuring these three temperatures, the specific heat of the unknown substance can be calculated.

Calorimeter

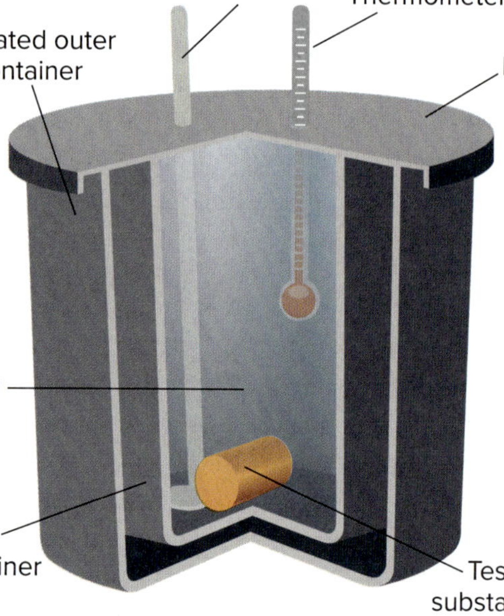

Figure 10 In a simple calorimeter, a hot test substance and a known volume of cold water are placed in an isolated system and allowed to come to thermal equilibrium. The ideal calorimeter has perfect insulation and does not transfer thermal energy to or from the outside. More sophisticated types of calorimeters are used to measure chemical reactions and the energy content of various foods.

Energy conservation

The operation of a calorimeter depends on the conservation of energy in an isolated, closed system composed of the water and the substance being measured. Energy can neither enter nor leave this system but can be transferred from one part of the system to another. Therefore, if the thermal energy of the test substance changes by an amount (ΔE_A) then the change in thermal energy of the water (ΔE_B) must be related by the equation $\Delta E_A + \Delta E_B = 0$. This can be rearranged to form the equation:

$$\Delta E_A = -\Delta E_B$$

The change in energy of the cold water is positive, and the change in energy of the hot test substance is negative. A positive change in energy indicates a rise in temperature, and a negative change in energy indicates a fall in temperature.

In an isolated, closed system, no work is done, so the change in thermal energy for each substance is equal to the heat and can be expressed by the following equation:

$$\Delta E = Q = mC\Delta T = mC(T_f - T_i)$$

Combining this equation with $\Delta E_A = -\Delta E_B$ gives:

$$m_A C_A(T_f - T_A) = -m_B C_B(T_f - T_B)$$

The final temperatures of the two substances are equal because they are in thermal equilibrium. Solving for the unknown specific heat (C_A) gives the equation:

$$C_A = \frac{-m_B C_B \Delta T_B}{m_A \Delta T_A}$$

Get It?

Evaluate the cup of liquid on the first page of this module. How is energy transferred and transformed?

EXAMPLE Problem 2

TRANSFERRING HEAT IN A CALORIMETER A calorimeter contains 0.50 kg of water at 15°C. A 0.10-kg block of an unknown substance at 62°C is placed in the water. The final temperature of the system is 16°C. What is the substance?

1 ANALYZE AND SKETCH THE PROBLEM

- Let the unknown be sample A and water be sample B.
- Sketch the transfer of thermal energy from the hotter unknown sample to the cooler water.

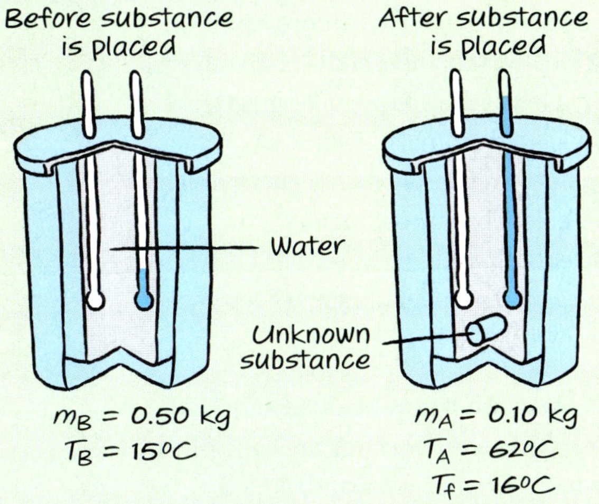

Known

$m_A = 0.10$ kg

$T_A = 62°C$

$m_B = 0.50$ kg

$C_B = 4180$ J/(kg·K)

$T_B = 15°C$

$T_f = 16°C$

Unknown

$C_A = ?$

2 SOLVE FOR THE UNKNOWN

Determine the final temperature using the following equation. Beware of the minus signs.

$$C_A = \frac{-m_B C_B \Delta T_B}{m_A \Delta T_A}$$

$$= \frac{-(0.50\text{ kg})(4180\text{ J/(kg·K)})(16°C - 15°C)}{(0.10\text{ kg})(16°C - 62°C)}$$

Substitute $m_A = 0.10$ kg, $T_A = 62°C$, $m_B = 0.50$ kg, $C_B = 4180$ J/(kg·K), $T_B = 15°C$, $T_f = 16°C$.

$$= 450\text{ J/(kg·K)}$$

According to **Table 1,** the specific heat of the unknown substance equals that of iron.

3 EVALUATE THE ANWER

- **Are the magnitudes realistic?** Specific heat is measured in J/(kg·K).
- **Is the magnitude realistic?** The answer is of the same magnitude as most metals listed in **Table 1.**

PRACTICE Problems

ADDITIONAL PRACTICE

4. A 1.00×10^2-g aluminum block at 100.0°C is placed in 1.00×10^2g of water at 10.0°C. The final temperature of the mixture is 26.0°C. What is the specific heat of the aluminum?
5. Three metal fishing weights, each with a mass of 1.00×10^2 g and at a temperature of 100.0°C, are placed in 1.00×10^2 g of water at 35.0°C. The final temperature of the mixture is 45.0°C. What is the specific heat of the metal in the weights?
6. A 2.00×10^2-g sample of water at 80.0°C is mixed with 2.00×10^2 g of water at 10.0°C in a calorimeter. What is the final temperature of the mixture?
7. A 1.50×10^2-g piece of glass at a temperature of 70.0°C is placed in a container with 1.00×10^2g of water initially at a temperature of 16.0°C. What is the equilibrium temperature of the water?
8. **CHALLENGE** A 4.00×10^2-g sample of water at 15.0°C is mixed with 4.00×10^2g of water at 85.0°C. After the system reaches thermal equilibrium, 4.00×10^2g of methanol at 15°C is added. Assume there is no thermal energy lost to the surroundings. What is the final temperature of the mixture?

Animals and Thermal Energy

PHYSICS Connection Animals can be divided into two groups based on how they control their body temperatures. Most, such as the spider in **Figure 11,** are cold-blooded animals. Their body temperatures depend on the environment. A cold-blooded animal regulates the transfer of thermal energy by its behavior, such as hiding under a rock to keep cool or sunning itself to keep warm.

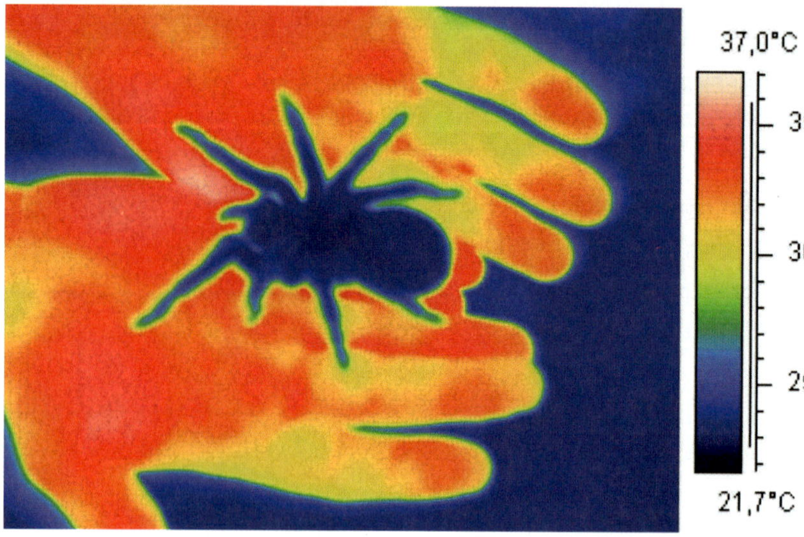

Figure 11 Cold-blooded animals depend on external sources of thermal energy to maintain their body temperatures. In contrast, warm-blooded animals maintain their body temperatures internally. In this thermal image, the spider is at the same temperature as the surrounding air, but the person's hands are significantly warmer than the surrounding air.

The others are warm-blooded animals whose body temperatures are controlled internally. That is, a warm-blooded animal's body temperature remains stable regardless of the temperature of the environment. For example, humans are warm-blooded and have body temperatures close to 37°C. To regulate its body temperature, a warm-blooded animal relies on bodily responses initiated by the brain, such as shivering and sweating, to counteract a rise or fall in body temperature.

Check Your Progress

9. **Conduction** The hard tile floor of a bathroom always feels cold to bare feet even though the rest of the room is warm. Is the floor colder than the rest of the room? Explain.
10. **Temperature** Make the following conversions:
    a. 5°C to kelvin
    b. 34 K to degrees Celsius
    c. 212°C to kelvin
    d. 316 K to degrees Celsius
11. **Units** Are the units the same for heat (Q) and specific heat (C)? Explain.
12. **Types of Energy** Describe the mechanical energy and the thermal energy of a bouncing basketball.
13. **Thermal Energy** Could the thermal energy of a bowl of hot water equal that of a bowl of cold water? Explain your answer.
14. **Cooling** On a dinner plate, a baked potato always stays hot longer than any other food. Why?
15. **Heat and Food** It takes much longer to bake a whole potato than potatoes that have been cut into pieces. Why?
16. **Cooking** Stovetop pans are made from metals such as copper, iron, and aluminum. Why are these materials used?
17. **Specific Heat** If you take a plastic spoon out of a cup of hot cocoa and put it in your mouth, you are not likely to burn your tongue. However, you could very easily burn your tongue if you put the hot cocoa in your mouth. Why?
18. **Critical Thinking** As water heats in a pot on a stove, it might produce some mist above its surface right before the water begins to roll. What is happening?

LEARNSMART® Go online to follow your personalized learning path to review, practice, and reinforce your understanding.

LESSON 2
CHANGES OF STATE AND THERMODYNAMICS

FOCUS QUESTION

Why doesn't a bowl of hot soup spontaneously freeze?

Change of State

The three most common states of matter on Earth are solid, liquid, and gas. As the temperature of a solid rises, that solid usually changes to a liquid. At even higher temperatures, it becomes a gas. If the gas cools, it returns to the liquid state. If the cooling continues, the liquid will return to the solid state. How can these changes be explained? Recall that when the thermal energy of a material changes, the motion of its particles also changes, as does the temperature.

Figure 12 diagrams the changes of state as thermal energy is added to 1.0 kg of water starting at 243 K (ice) and continuing until that water reaches 473 K (water vapor). Between points A and B, the ice is warmed to 273 K. At this point, the added thermal energy gives the water molecules enough energy to partially overcome the forces holding them together. The particles are still touching each other, but they have more freedom of movement. Eventually, the molecules become free enough to slide past each other.

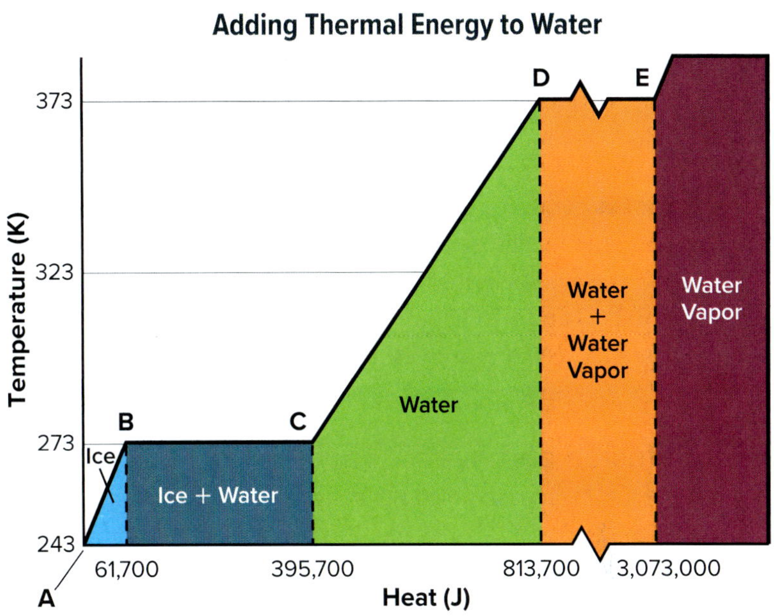

Figure 12 Thermal energy added to a substance can raise the temperature or cause a change in state. Note that the scale is broken between points D and E.

3D THINKING | DCI Disciplinary Core Ideas | CCC Crosscutting Concepts | SEP Science & Engineering Practices

COLLECT EVIDENCE

Use your Science Journal to record the evidence you collect as you complete the readings and activities in this lesson.

INVESTIGATE

GO ONLINE to find these activities and more resources.

Applying Practices: Modeling Energy at Different Scales
HS-PS3-2. Develop and use models to illustrate that energy at the macroscopic scale can be accounted for as either motions of particles or energy stored in fields.

Melting point At this point, water changes from a solid to a liquid, just as the snowman in **Figure 13** does. The temperature at which this change occurs is the melting point of a substance. When a substance melts, the addition of thermal energy allows the particles to move, rotate, and vibrate in ways not available in the solid. Each of these new motions might add new modes of kinetic or potential energy. The added thermal energy does not change the temperature of the material. This can be observed between points B and C in **Figure 12,** where the added thermal energy melts all the ice at a constant 273 K.

Figure 13 Thermal energy is transferred from the warmer air to the snowman, causing the snowman to melt.

Boiling point Once the ice is completely melted, adding more thermal energy increases the motion of the water molecules again. The temperature rises as shown between points C and D in **Figure 12.** As the temperature increases further, some of the particles that make up the liquid acquire enough energy to break free from the other particles.

At a specific temperature, known as the boiling point, adding more energy to a substance causes it to undergo another change of state. All the added thermal energy converts it from the liquid state to the gas state. As in melting, the temperature does not rise while a liquid boils, as shown between points D and E in **Figure 12.** After the water is entirely converted to gas, any added thermal energy again increases the motion of the molecules, and the temperature rises. After point E, the water vapor is heated to temperatures greater than 373 K.

Heat of fusion and heat of vaporization The amount of thermal energy needed to melt 1 kg of a substance is called the substance's **heat of fusion** (H_f). For ice, the heat of fusion is 3.34×10^5 J/kg. If 1 kg of ice at its melting point, 273 K, absorbs 3.34×10^5 J, the ice becomes 1 kg of water at the same temperature. The added energy causes a change in state but not in tempera-ture. The horizontal distance from point B to point C in **Figure 12** represents the heat of fusion.

The thermal energy needed to vaporize 1 kg of a liquid is called the **heat of vaporization** (H_v). Water's heat of vaporization is 2.26×10^6 J/kg. The horizontal distance from point D to point E in **Figure 12** represents the heat of vaporization. Every material has a characteristic heat of fusion (H_f) and heat of vaporization (H_v). Some values are shown in **Table 2.**

Table 2 Heats of Fusion and Vaporization of Common Substances

Material	Heat of Fusion H_f (J/kg)	Heat of Vaporization H_v (J/kg)
Copper	2.05×10^5	5.07×10^6
Mercury	1.15×10^4	2.72×10^5
Gold	6.30×10^4	1.64×10^6
Methanol	1.09×10^5	8.78×10^5
Iron	2.66×10^5	6.29×10^6
Silver	1.04×10^5	2.36×10^6
Lead	2.04×10^4	8.64×10^5
Water (ice)	3.34×10^5	2.26×10^6

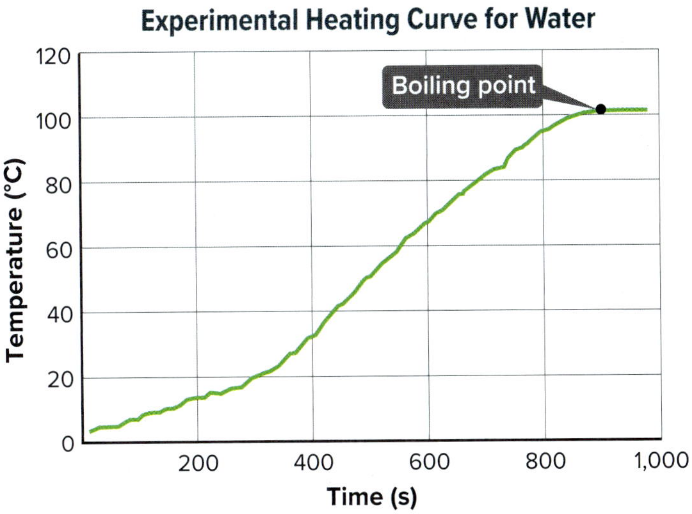

Figure 14 One way to measure the absorption of energy by a material is to add energy from a constant source of thermal energy and measure the change in temperature over time. The plot of temperature v. time is called a heating curve. For this figure, a beaker of cold water was placed on a hot plate. The resulting heating curve is graphed.

Explain *why thermal energy must be added at a constant rate to calculate the specific heat of water from this graph.*

Energy and changes of state

There is a definite slope to the graph in **Figure 14** between about 300 s and 800 s. Heat is added at a constant rate, so this slope is proportional to the reciprocal of the specific heat of water. The slope between points A and B in **Figure 12** is proportional to the reciprocal of the specific heat of ice, and the slope above point E is proportional to the reciprocal of the specific heat of water vapor. The slope for water is less than those of both ice and water vapor. This is because water has a greater specific heat than does ice or water vapor. The heat (Q) required to melt a solid of mass (m) is given by the following equation.

Heat Required to Melt a Solid

The heat required to melt a solid is equal to the mass of the solid times the heat of fusion of the solid.

$$Q = mH_f$$

Similarly, the heat (Q) required to vaporize a mass (m) of liquid is given by the following equation.

Heat Required to Vaporize a Liquid

The heat required to vaporize a liquid is equal to the mass of the liquid times the heat of vaporization of the liquid.

$$Q = mH_v$$

When a liquid freezes, an amount of thermal energy ($Q = -mH_f$) must be removed from the liquid to turn it into a solid. The negative sign indicates that the thermal energy is transferred from the sample to the external world. In the same way, when a vapor condenses to a liquid, an amount of thermal energy ($Q = -mH_v$) must be removed from the vapor.

Water absorbs significant amounts of thermal energy when it melts or evaporates. Every day you use the large heats of fusion and vaporization of water. Each gram of sweat that evaporates from your skin carries off about 2.3 kJ of thermal energy. This is one cooling process that many warm-blooded animals use to regulate their body temperatures. Similarly the melting of a 24-g cube of ice absorbs enough thermal energy, 8 kJ, to cool a glass of water by about 30°C.

EXAMPLE Problem 3

HEAT Suppose that you are camping in the mountains. You need to melt 1.50 kg of snow at 0.0°C and heat it to 70.0°C to make hot cocoa. How much heat will you need?

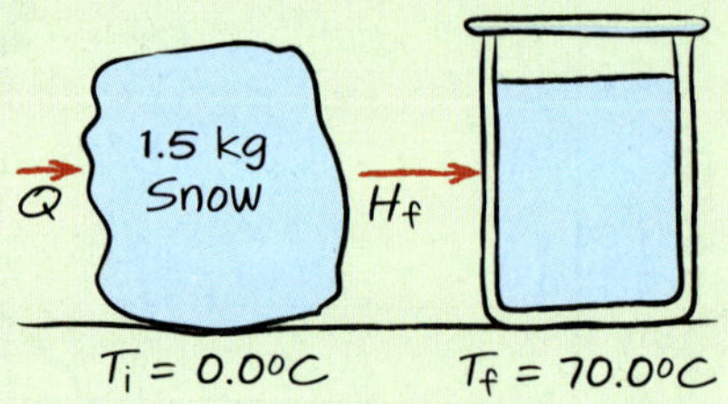

1 ANALYZE AND SKETCH THE PROBLEM

- Sketch the relationship between heat and water in its solid and liquid states.
- Sketch the transfer of heat as the temperature of the water increases.

Known

$m = 1.50$ kg

$T_i = 0.0°C$

$C = 4180$ J/(kg·K)

$H_f = 3.34\times10^5$ J/kg

$T_f = 70.0°C$

Unknown

$Q_{melt\ ice} = ?$

$Q_{heat\ liquid} = ?$

$Q_{total} = ?$

2 SOLVE FOR THE UNKNOWN

Calculate the heat needed to melt ice.

$$Q_{melt\ ice} = mH_f$$
$$= (1.50\text{ kg})(3.34\times10^5\text{ J/kg})$$
$$= 5.01\times10^5\text{ J} = 5.01\times10^2\text{ kJ}$$

Substitute $m = 1.50$ kg, $H_f = 3.34\times10^5$ J/kg.

Calculate the temperature change.

$$\Delta T = T_f - T_i$$
$$= 70.0°\text{C} - 0.0°\text{C} = 70.0°\text{C} = 70.0\text{K}$$

Substitute $T_f = 70.0°C$, $T_i = 0.0°C$. $\Delta°C = \Delta K$

Calculate the heat needed to raise the water temperature.

$$Q_{heat\ liquid} = mC\Delta T$$
$$= (1.50\text{ kg})(4180\text{ J/(kg·K)})(70.0\text{K})$$
$$= 4.39\times10^5\text{ J} = 4.39\times10^2\text{ kJ}$$

Substitute $m = 1.50$ kg, $C = 4180$ J/(kg·K), $\Delta T = 70.0$K.

Calculate the total amount of heat needed.

$$Q_{total} = Q_{melt\ ice} + Q_{heat\ liquid}$$
$$= 5.01\times10^2\text{ kJ} + 4.39\times10^2\text{ kJ}$$
$$= 9.40\times10^2\text{ kJ}$$

Substitute $Q_{melt\ ice} = 5.01\times10^2$ kJ, $Q_{heat\ liquid} = 4.39\times10^2$ kJ.

3 EVALUATE THE ANSWER

- **Are the units correct?** Energy units are in joules.
- **Does the sign make sense?** Q is positive when thermal energy is absorbed.
- **Is the magnitude realistic?** To check the magnitude, perform a quick estimation: $Q = (1.5\text{ kg})(300{,}000\text{ J/kg}) + (1.5\text{ kg})(4000\text{ J/(kg·K)})(70\text{ K}) = 9\times10^2\text{ kJ}$.

PRACTICE Problems

ADDITIONAL PRACTICE

19. How much thermal energy is absorbed by 1.00×10^2 g of ice at −20.0°C to become water at 0.0°C?

20. A 2.00×10^2-g sample of water at 60.0°C is heated to water vapor at 140.0°C. How much thermal energy is absorbed?

21. Use **Figure 15** to calculate the heat of fusion and heat of vaporization of water in J/kg.

22. A steel plant worker will change 100.0 kg of 25°C iron into molten iron (melting point = 1538°C). How much thermal energy must be added?

23. CHALLENGE How much thermal energy is needed to change 3.00×10^2 g of ice at −30.0°C to water vapor at 130.0°C?

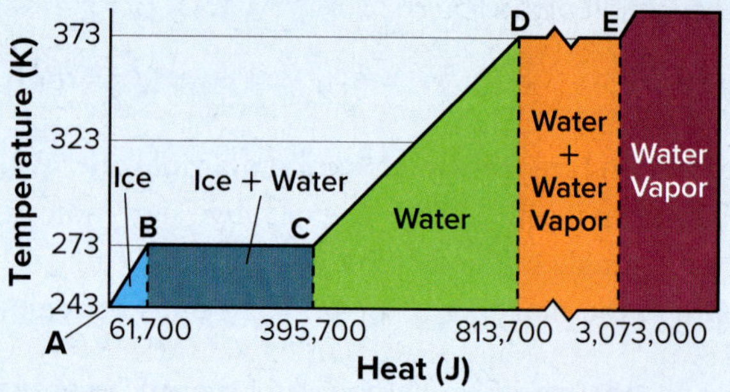

Figure 15

The First Law of Thermodynamics

The first steam engines were built in the eighteenth century and used to power trains and factories. Steam engines change thermal energy into mechanical energy. The invention of the steam engine contributed greatly to the Industrial Revolution and to the study of how heat is related to work. The study of how thermal energy is transformed into other forms of energy is called thermodynamics.

Before 1900, scientists did not realize that the concepts of thermodynamics were linked to the motions of particles in an object, and they considered thermodynamics to be a separate topic from mechanics. Today, engineers routinely apply the concepts of thermodynamics to develop higher performance refrigerators, automobile engines, aircraft engines, and numerous other kinds of machines.

The first law developed for thermodynamics was a statement about what thermal energy is and where it can go. As you know, you can raise the temperature of a glass of cold water by placing it on a hot plate or by stirring it. That is, you can increase the water's thermal energy by heating or by doing work on it. If we consider the system to be the water, the work the system does on you is equal to the negative of the work you do on the system. The **first law of thermodynamics** states that the change in thermal energy (ΔU) of an object is equal to the heat (Q) that is added to the object minus the work (W) done by the object. Note that ΔU, Q, and W are all measured in joules, the unit of energy.

The First Law of Thermodynamics

The change in thermal energy of an object is equal to the heat added to the object minus the work done by the object.

$$\Delta U = Q - W$$

The first law of thermodynamics is merely a restatement of the law of conservation of energy, which states that energy is neither created nor destroyed but can be transported from one place to another or transformed into another kind of energy. An example of transformed thermal energy in a system is a hand pump used to inflate a bicycle tire. As a person pumps, the air and the hand pump become warm. The mechanical energy in the moving piston is transformed into thermal energy of the gas. Similarly, other forms of energy, such as light, sound, and electrical energy, can be transformed into thermal energy. For example, a toaster transforms electrical energy into radiant energy when it toasts bread.

PRACTICE Problems

ADDITIONAL PRACTICE

24. A gas balloon absorbs 75 J of thermal energy. The balloon expands but stays at the same temperature. How much work did the balloon do in expanding?

25. A drill bores a small hole in a 0.40-kg block of aluminum and heats the aluminum by 5.0°C. How much work did the drill do in boring the hole?

26. How many times would you have to drop a 0.50-kg bag of lead shot from a height of 1.5 m to heat the shot by 1.0°C?

27. When you stir a cup of tea, you do about 0.050 J of work each time you circle the spoon in the cup. How many times would you have to stir the spoon to heat a 0.15-kg cup of tea by 2.0°C?

28. CHALLENGE An expansion valve does work on 100 g of water. The system is isolated, and all of the work is used to convert the 90°C water into water vapor at 110°C. How much work does the expansion valve do on the water?

Heat engines A **heat engine** is a device that is able to continuously convert thermal energy to mechanical energy. A heat engine requires a high-temperature source, a low-temperature receptacle, called a sink, and a way to convert the thermal energy into work, as shown in **Figure 16.**

Internal combustion engines An automobile's internal combustion engine, shown in **Figure 17,** is one example of a heat engine. In the engine, input heat (Q_H) is transferred from a high-temperature flame to a mixture of air and gas vapor in the cylinder. The hot air expands and pushes on a piston, thereby changing thermal energy into mechanical energy. The heated air is expelled, the piston returns to the top of the cylinder, and the cycle repeats. Car engines repeat this cycle many times each minute. The thermal energy from the flame is converted into mechanical energy, which propels the car.

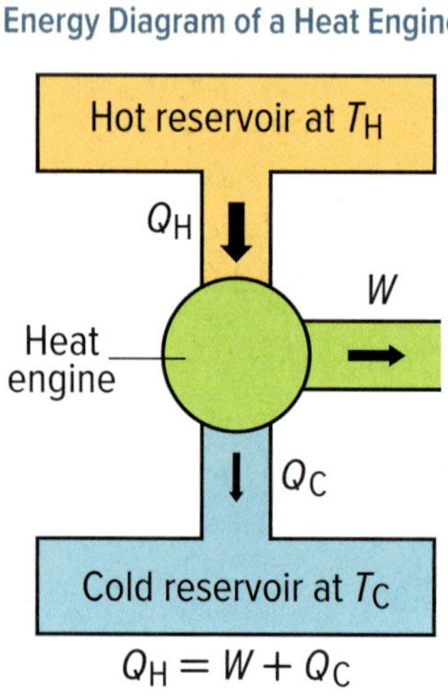

Figure 16 Heat engines transform thermal energy into mechanical energy and waste heat. This schematic shows the energy transfers and transformations.

Waste heat Not all thermal energy from the flame is converted into mechanical energy. When the engine is running, the gases and the engine parts become hot. The exhaust comes in contact with outside air and heats the air. In addition, thermal energy from the engine is transferred to a radiator. Outside air passes through the radiator and becomes warmer. All of this energy (Q_C) transferred out of the automobile engine is called waste heat. When the engine is working continuously, its internal energy does not change. That is, $\Delta U = 0 = Q - W$. The net heat going into the engine is $Q = Q_H - Q_C$. Thus, the work done by the engine is $W = Q_H - Q_C$. All heat engines generate waste heat, and no engine can convert all the energy into useful kinetic energy.

Figure 17 Internal combustion engines are one type of heat engine. They are used in automobiles.

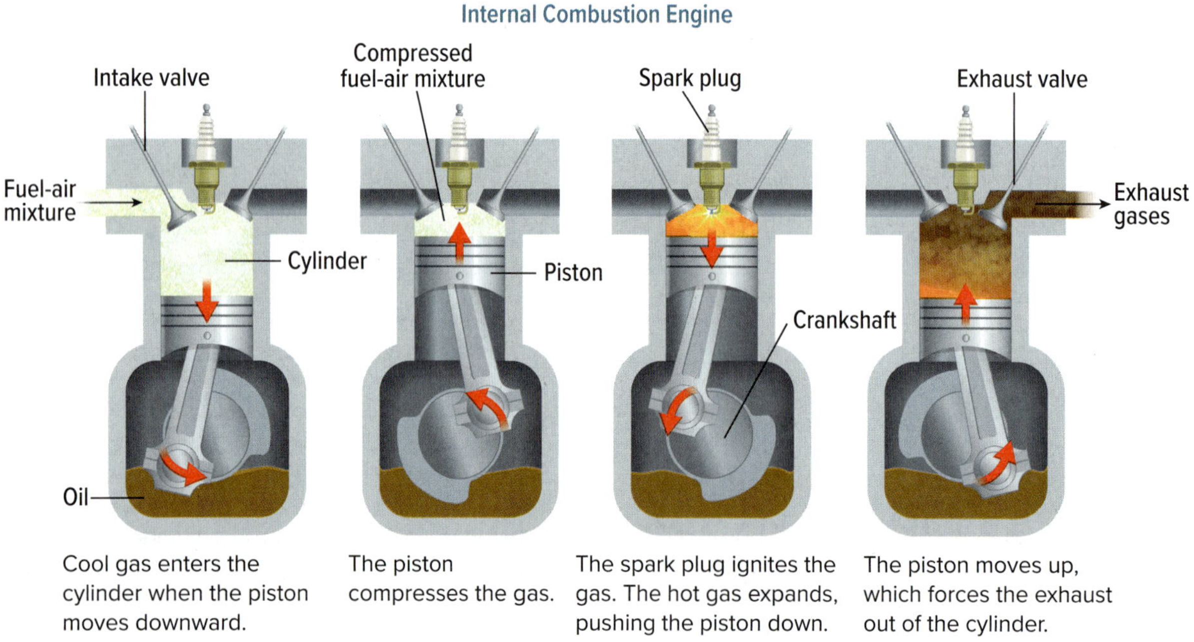

Efficiency Engineers and car salespeople often discuss the fuel efficiencies of automobile engines. They are referring to the amount of the input heat (Q_H) that is turned into useful work (W). The actual efficiency of an engine is given by the ratio W/Q_H. If all the input heat could be turned into useful work by the engine, the engine would have an efficiency of 100 percent. Because there is always waste heat (Q_C), even the most efficient engines fall short of 100 percent efficiency.

In fact, most heat engines are significantly less than 100 percent efficient. For example, even the most efficient automobile gasoline engines have an efficiency of less than 40 percent. A typical gasoline engine in an automobile has an efficiency that is closer to 20 percent. Energy is not destroyed, but a considerable amount of thermal energy transfers from a hot automobile engine to the surrounding environment.

Get It?

Describe an example of another instance where energy is converted to a less useful form.

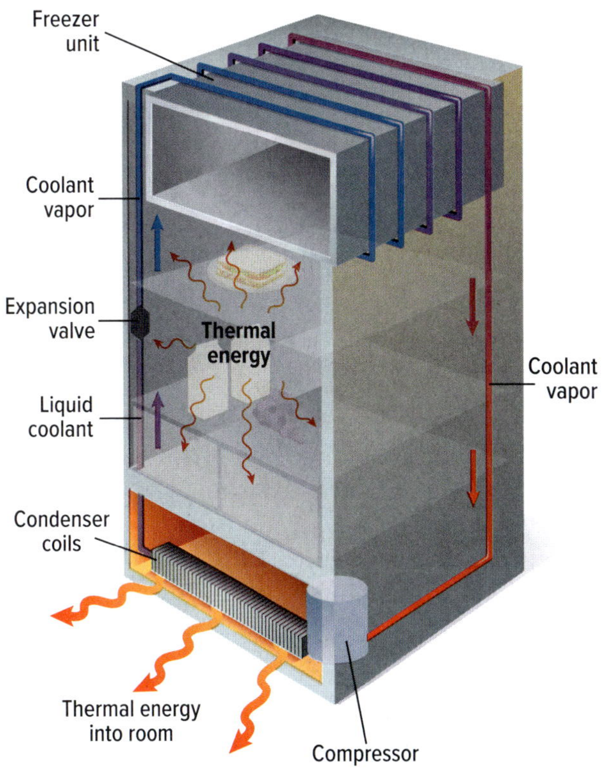

Figure 18 Liquid coolant is pumped into an expansion valve, where it absorbs energy from its surroundings and becomes a gas. The gas then heats up as it absorbs thermal energy from inside the refrigerator. A compressor does work on the gas to cool it to a liquid, and the cycle begins again.

Refrigerators Thermal energy is transferred from a warm object to a cold object spontaneously. But it is also possible to remove thermal energy from a colder object and add it to a warmer object if work is done. A refrigerator, such as the one in **Figure 18,** is a common example of a device that accomplishes this transfer. Electrical energy runs a motor that does work on a gas and compresses it.

The gas draws thermal energy from the interior of the refrigerator. It passes from the compressor through the condenser coils on the outside of the refrigerator. The gas cools into a liquid. Thermal energy is transferred into the air in the room. The liquid reenters the interior, vaporizes, and absorbs thermal energy from its surroundings. The gas returns to the compressor, and the process is repeated. The overall change in the thermal energy of the gas is zero. Thus, according to the first law of thermodynamics, the sum of the thermal energy removed from the refrigerator's contents and the work done by the motor is equal to the thermal energy expelled. These energy transfers and transformations are shown in **Figure 19.**

Energy Diagram of a Refrigerator

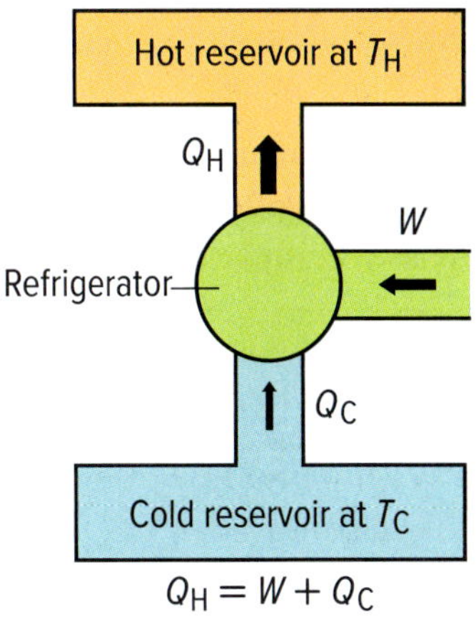

Figure 19 When work is done on the refrigerator, thermal energy is transferred from the cold reservoir to the hot reservoir.

Heat pumps A heat pump is a refrigerator that can be run in two directions. In summer, the pump removes thermal energy from a house and cools the house. In winter, thermal energy is removed from the cold outside air and transferred into the warmer house. In both cases, mechanical energy is required to transfer thermal energy from a colder object to a warmer one.

Second Law of Thermodynamics

You are not surprised when a metal spoon, heated at one end, soon becomes uniformly hot. Consider your reaction, however, if a spoon lying on a table suddenly, on its own, became red hot at one end and icy cold at the other. This reverse process would not violate the first law of thermodynamics—the thermal energy of the spoon would remain the same. Many processes that are consistent with the first law of thermodynamics have never been observed to occur spontaneously. However, there is more to modeling thermal events than making sure energy is conserved.

Energy spreads out Examine the melting ice pop and the cooling pizza in **Figure 20.** The first law of thermodynamics does not prohibit net thermal energy transfers from the cold ice pop to the air or from the air to the hot pizza. This does not occur, however, because of the second law of thermodynamics. When a hot object is placed in contact with cooler surroundings, the thermal energy in the hot object has the opportunity to disperse, or spread out more. Some of the thermal energy moves into the cold object, warming it and therefore cooling the originally hotter object. The **second law of thermodynamics** states that in an uncontrolled system whenever there is an opportunity for energy dispersal, the energy always spreads out toward more uniform energy distribution and to a more stable state.

Consider the cooling pizza. The particles in the pizza have a greater average kinetic energy than the particles in the air. Some of the pizza's original thermal energy disperses into the air. As a result, the pizza's temperature decreases and the air temperature increases a small amount. When the pizza and the air reach the same temperature, the average kinetic energy of the particles in the pizza and the air will be the same. That is, the energy spreads out among the particles. Similarly, if you leave the ice pop sitting on the counter, thermal energy from the air will be dispersed to the ice pop. The ice pop will heat up and melt, while the air will experience a small temperature decrease.

Figure 20 According to the second law of thermodynamics, thermal energy always spreads out. The red arrows represent thermal energy flow. The purple arrows represent time lapses. Thermal energy spontaneously flows from a warmer object to a colder object.

(t to b) Ingram Publishing/SuperStock; (2) Freer/Shutterstock; (3,4)Lucky Business/Shutterstock

Get It?

Explain how the pizza in **Figure 20** represents an uncontrolled system and how the thermal energy became more uniformly distributed.

CCC CROSSCUTTING CONCEPTS

Systems and System Models Using **Figure 20,** describe a system surrounding the pizza. Describe the system boundaries, inputs, and outputs in your defined system.

Entropy The measure of this dispersal of energy is known as **entropy** (S). A system in which the thermal energy is concentrated in one place is referred to as a system with low entropy. A system in which the thermal energy is spread throughout the system has high entropy.

Another way of stating the second law of thermodynamics is that natural processes go in a direction that maintains or increases the total entropy of the universe. That is, energy will naturally disperse unless some action is taken to localize it. Once a system is in a high-entropy state, it is highly unlikely that it will return to a lower entropy state on its own. Events that occur spontaneously, such as the melting ice pop or the cooling pizza, are events in which the entropy of the system increases. Processes that would decrease the entropy of a system do not tend to occur spontaneously but require work done by an external agent.

Get It?

State the second law of thermodynamics using the term *entropy*.

Entropy and heat engines How does entropy relate to heat engines? If heat engines completely converted thermal energy into mechanical energy with no waste heat, energy would still be conserved, and so the first law of thermodynamics would be obeyed. However, waste heat is always generated, dispersing thermal energy beyond the engine. In the nineteenth century, French engineer Sadi Carnot studied the ability of engines to convert thermal energy into mechanical energy. He developed a logical proof that even an ideal engine would generate some waste heat. Carnot's result was one of the first formal analyses leading to the development of the concept of entropy.

Changes in entropy Like energy, entropy is a property of a system. If thermal energy is added to a system, the entropy increases. If thermal energy is removed from a system, its entropy decreases. If a system does work on its surroundings without any transfer of thermal energy, the entropy does not change. For a reversible process, the change in entropy (ΔS) is expressed by the following equation. Entropy has units of J/K. The temperature, in kelvins, is constant.

Change in Entropy

For a reversible process, the change in entropy of a system is equal to the heat added to the system divided by the temperature of the system in kelvins.

$$\Delta S = \frac{Q}{T}$$

PHYSICS Challenge

Entropy has some interesting properties. Calculate the change in entropy for the following situations. Explain how and why these changes in entropy are different from each other. For these small temperature changes, you can use the original temperature to find the change in entropy.

1. Heating 1.0 kg of water from 273 K to 274 K.
2. Heating 1.0 kg of water from 353 K to 354 K.
3. Heating 1.0 kg of lead from 273 K to 274 K.
4. Completely melting 1.0 kg of ice at 273 K.

Entropy and the energy crisis The second law of thermodynamics and the increase in entropy are exemplifed by the energy crisis. The energy crisis refers to the continued use of limited resources such as petroleum. When you use a resource, you do not use up the energy in the resource. For example, when you drive a car, such as those in **Figure 21,** the gas ignites and the chemical energy contained in the molecules of the gas is converted into the kinetic energy that runs the car and the thermal energy that heats the engine. The chemical energy of the unburned gas has dispersed into many more objects contained within a much larger volume. While it is mathematically possible for all of the dispersed energy to be brought back together in one object, the probability of this happening is very near zero. For this reason, entropy often is used as a measure of the unavailability of useful energy. The energy in the warmed air in a home is not as available to do mechanical work or to transfer thermal energy to other objects as the energy in the original gas molecules was. The lack of usable energy is actually a surplus of entropy.

Figure 21 Burning gasoline uses up natural resources and increases entropy but does not use up energy. The energy is no longer in a useful form.

Check Your Progress

29. **Heat Engine** Describe the energy transformations and transfers made by a heat engine, and explain why operating a heat engine causes an increase in entropy.

30. **Heat of Vaporization** Old heating systems sent water vapor into radiators in each room of a house. In the radiators, the water vapor condensed to water. Analyze this process and explain how it heated a room.

31. **Heat of Fusion** How much thermal energy is needed to change 50.0 g of ice at −20.0°C to water at 10.0°C?

32. **Heat of Vaporization** How much energy is needed to heat 1.0 kg of mercury metal from 10.0°C to its boiling point (357°C) and vaporize it completely? For mercury, $C = 140$ J/kg·°C and $H_v = 3.06\times10^5$ J/kg.

33. **Mechanical Energy and Thermal Energy** A man uses a 320-kg hammer moving at 5.0 m/s to smash a 3.0-kg block of lead against a 450-kg rock. When he measured the temperature of the lead block, he found that it had increased by 5.0°C. Explain how this happened.

34. **Mechanical Energy and Thermal Energy** James Joule carefully measured the difference in temperature of water at the top and the bottom of a waterfall. Why did he expect a difference?

35. **Mechanical Energy and Thermal Energy** For the waterfall in **Figure 22,** calculate the temperature difference between the water at the top and the bottom of the fall. Assume that the potential energy of the water is all converted to thermal energy.

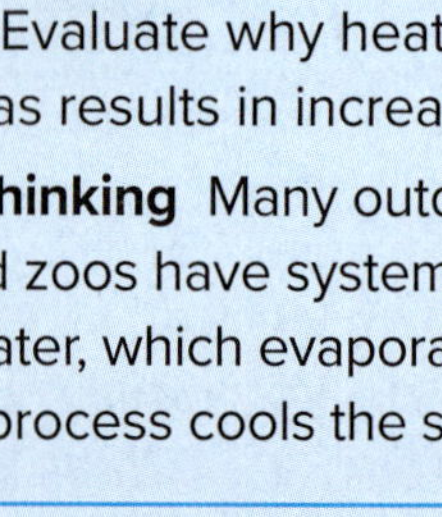

Figure 22

36. **Entropy** Evaluate why heating a home with natural gas results in increased entropy.

37. **Critical Thinking** Many outdoor amusement parks and zoos have systems that spray a fine mist of water, which evaporates quickly. Explain why this process cools the surrounding air.

LEARNSMART® Go online to follow your personalized learning path to review, practice, and reinforce your understanding.

Under Pressure

Diesel engines, like gasoline engines, are internal combustion engines in which the fuel is burned inside the engine. They both convert the chemical energy in fuel into the mechanical energy that moves the vehicle. However, diesel engines are more powerful and efficient than gasoline engines because of the way the engine burns fuel.

Diesel engines differ from gasoline engines in several crucial ways that increase their efficiency and power.

Diesel Versus Gasoline Engines

In gasoline engines, air and fuel are injected into the engine's cylinders. The piston rises up the cylinder, compressing the air and fuel and making them explosive. When the piston reaches the top of the cylinder, the car's spark plugs provide the electricity needed to make the air and fuel explode. The explosion pushes the piston back down the cylinder. The piston is connected to the crankshaft, so when it travels downward, it pushes the crankshaft, which turns the wheels of the vehicle. As the crankshaft turns, it pushes the piston up the cylinder again, and the cycle repeats.

In a diesel engine, the process is slightly different. First, air is injected into the cylinder. The piston then compresses the air with much more force than it is compressed in a gasoline engine. The air produces so much heat as it is compressed that, when fuel is sprayed into the cylinder, there is no need for a spark plug to ignite it with electricity. Instead, the fuel instantly explodes, driving the piston back down the cylinder and turning the crankshaft.

Diesel Fuel Versus Gasoline

Both gasoline and diesel fuel are made from crude oil. However, the fuels are made from different hydrocarbons that are separated during the refining process. Gasoline is lighter and less oily than diesel fuel because it contains fewer carbon and hydrogen atoms, but it also has a lower energy density than diesel fuel. Diesel fuel's high energy density contributes to its greater power and efficiency.

Historically, one downside of diesel fuel is that it contributes to air pollution by releasing large amounts of particulate matter and nitrogen compounds into the atmosphere. However, improved refining techniques and new technologies are helping to lessen pollution caused by diesel fuel.

DEVELOP AND USE MODELS TO ILLUSTRATE

Work with a partner to create a 2D or 3D model of the combustion cycle of a gasoline or diesel engine. Use sources for additional information. Display your model in your classroom.

CALIFORNIA CONNECTION

Visit the state's official government website to learn how diesel engines and diesel fuel are used in California. Write a short report that explains regulations on diesel, improvements in diesel technologies, or another topic related to diesel.

MODULE 11

STUDY GUIDE

GO ONLINE to study with your Science Notebook.

Lesson 1 TEMPERATURE, HEAT, AND THERMAL ENERGY

- Thermal energy is the sum of the kinetic and potential energies of an object's particles. An object's temperature is a measure of the average kinetic energy of its particles.
- When two objects are in thermal equilibrium, there is no net transfer of thermal energy between the objects, and the two objects are at the same temperature. A thermometer measures temperature by reaching thermal equilibrium with its surroundings. When an object's temperature is at absolute zero, the average kinetic energy of its particles is zero, and that object cannot transfer thermal energy.
- Heat is the transfer of thermal energy. Thermal energy is spontaneously transferred from a warm object to a cool object. Thermal energy is transferred by three processes: conduction, convection, and radiation.
- Substances heat differently, based on their specific heats. Specific heat (C) is the heat required to raise the temperature of 1 kg of a substance 1 K.

$$Q = mC\,\Delta T = mC(T_f - T_i)$$

- A calorimeter is a closed system used to measure changes in thermal energies. Specific heat is calculated by using measurements from a calorimeter.

- thermal conduction
- thermal equilibrium
- heat
- convection
- radiation
- specific heat

Lesson 2 CHANGES OF STATE AND THERMODYNAMICS

- Thermal energy transferred during a change of state does not change the temperature of a substance. The heat of fusion is the quantity of heat needed to change 1 kg of a substance from a solid state to a liquid state at its melting point.

$$Q = mH_f$$

- The heat of vaporization is the quantity of heat needed to change 1 kg of a substance from a liquid state to a gaseous state at its boiling point.

$$Q = mH_v$$

- The first law of thermodynamics states that the change in the thermal energy of an object is equal to the heat added to the object minus the work done by the object.

$$\Delta U = Q - W$$

- A heat engine converts thermal energy to mechanical energy. A heat pump and a refrigerator use mechanical energy to transfer thermal energy from a region of lower temperature to one of higher temperature.
- The second law of thermodynamics states that whenever there is an opportunity for energy dispersal, the energy always spreads out. Entropy (S) is a measure of the energy dispersal of a system. The second law of thermodynamics indicates that natural processes go in a direction that maintains or increases the total entropy of the universe. The change in entropy of an object is defined as the heat added to the object divided by the object's temperature.

- heat of fusion
- heat of vaporization
- first law of thermodynamics
- heat engine
- second law of thermodynamics
- entropy

THREE-DIMENSIONAL THINKING

Module Wrap-Up

REVISIT THE PHENOMENON

Why do different parts of the pan appear to be different colors in this thermal image?

CER Claim, Evidence, Reasoning

Explain Your Reasoning Revisit the claim you made when you encountered the phenomenon. Summarize the evidence you gathered from your investigations and research and finalize your Summary Table. Does your evidence support your claim? If not, revise your claim. Explain why your evidence supports your claim.

STEM UNIT PROJECT

Now that you've completed the module, revisit your STEM unit project. You will summarize your evidence and apply it to the project.

GO FURTHER

SEP Data Analysis Lab

How do materials differ in how fast they heat?

The following materials are each placed in identical containers holding equal amounts of room-temperature methanol.

Material	Mass (g)	Initial Temperature (°C)
Aluminum	50	30
Aluminum	60	40
Glass	50	30
Silver	50	30
Zinc	50	30

CER Analyze and Interpret Data

1. **Claim** Rank the materials according to the amount of thermal energy they transfer to the methanol, from least to greatest. Specifically indicate any ties.
2. **Evidence and Reasoning** Explain how you decided your rankings.

Joseph Giacomin/Cultura RM/Alamy Stock Photo

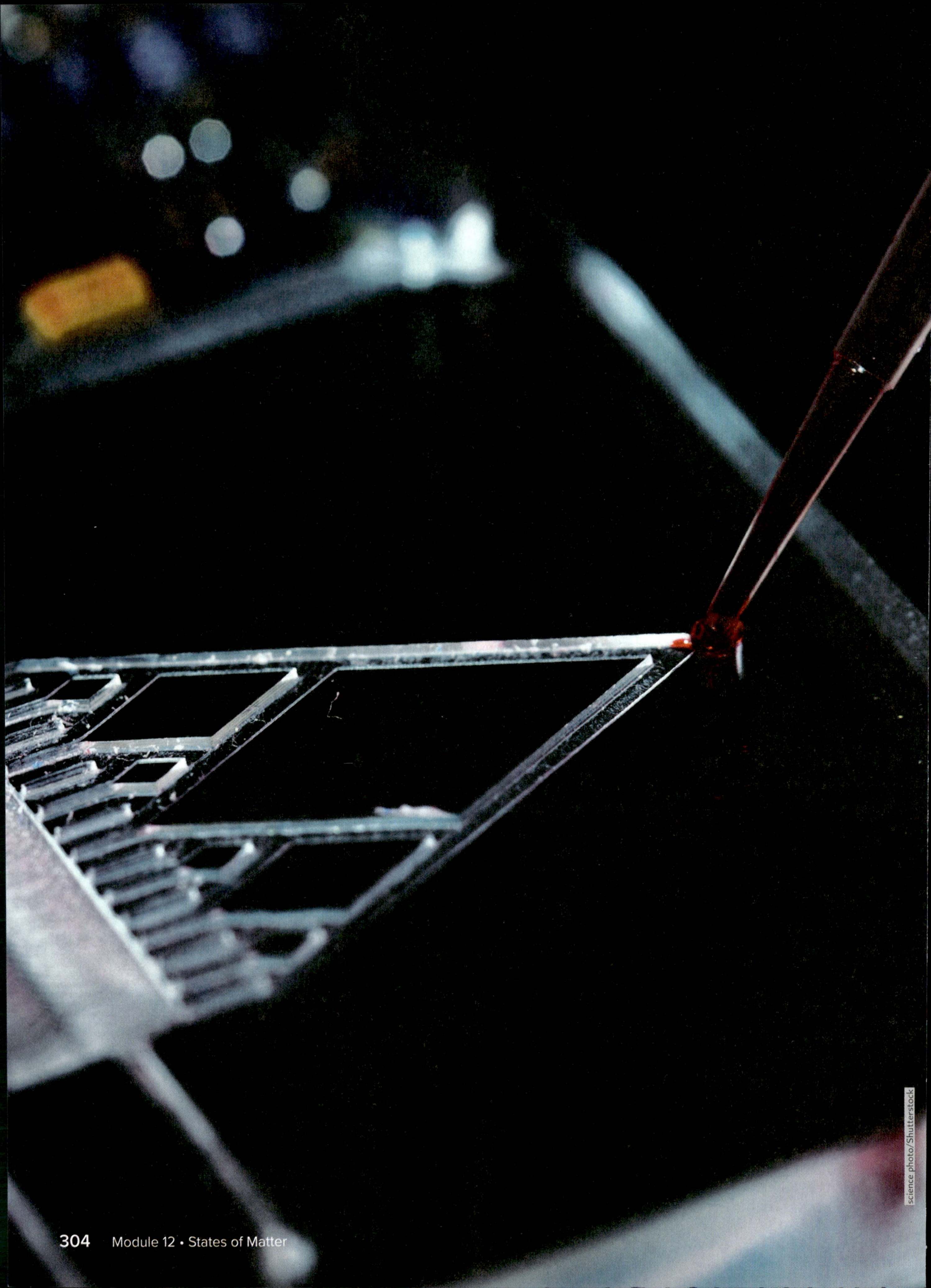

science photo/Shutterstock

MODULE 12
STATES OF MATTER

ENCOUNTER THE PHENOMENON

How will the fluid flow through this device?

GO ONLINE to watch a video about how air flows through a pipe.

SEP Ask Questions

Do you have other questions about the phenomenon? If so, add them to the driving question board.

CER Claim, Evidence, Reasoning

Make Your Claim Use your CER chart to make a claim about how the fluids will flow through the device. Explain your reasoning.

Collect Evidence Use the lessons in this module to collect evidence to support your claim. Record your evidence as you move through the module.

Explain Your Reasoning You will revisit your claim and explain your reasoning at the end of the module.

GO ONLINE to access your CER chart and explore resources that can help you collect evidence.

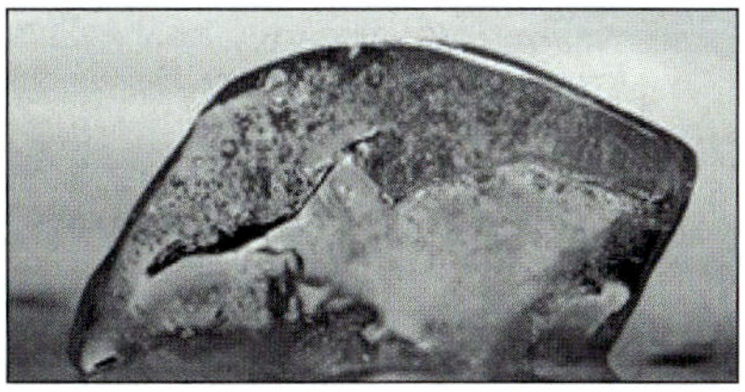

LESSON 1: Explore & Explain: Liquids and Gases

LESSON 4: Explore & Explain: Solid Bodies

Smithsonian

Additional Resources

LESSON 1

PROPERTIES OF FLUIDS

FOCUS QUESTION

What is a fluid and what are its properties?

Liquids and Gases

Water and air are common substances in the everyday lives of people. Both water and air flow, and unlike solids, neither one has a definite shape. Gases and liquids are two states of matter in which atoms and molecules have great freedom to move.

Fluids Like ice cubes, the chunks of ice in **Figure 1** have a certain mass and a certain shape that do not depend on the size or shape of the lake basin. When the ice melts, its mass remains the same, but its shape changes. The water flows to take the shape of the basin and forms a definite, flat, upper surface. As evaporation occurs, the liquid water changes into a gas in the form of water vapor. Like liquid water, water vapor flows and does not have any definite shape. Both liquids and gases are fluids. **Fluids** are materials that can flow and have no definite shape of their own.

Figure 1 The chunks of ice in this lake, which are solids, have definite shapes. However, the liquid water, a fluid, takes the shape of the lake basin.

Identify *What fluid is filling the space above the water?*

3D THINKING DCI Disciplinary Core Ideas CCC Crosscutting Concepts SEP Science & Engineering Practices

COLLECT EVIDENCE

Use your Science Journal to record the evidence you collect as you complete the readings and activities in this lesson.

INVESTIGATE

GO ONLINE to find these activities and more resources.

Quick Investigation: Pressure

Analyze data to determine how the pressure is affected by the force an object exerts over an area.

Review the News

Obtain information from a current news story about the properties or behavior of fluids. Evaluate your source and communicate your findings to your class.

Pressure

When considering fluids (as well as solids), it is often useful to think about pressure as well as force. You have probably heard people talk about water pressure and air pressure, and you might already have a sense that pressure and force are related. Pressure and force are not the same, however. **Pressure** is the perpendicular component of a force on a surface divided by the area of the surface. Since pressure is force exerted on a surface, anything that exerts pressure is capable of producing change and doing work. In **Figure 2,** both the astronaut and the legs of the landing module are exerting pressure on the Moon's surface.

Figure 2 The landing module and the astronaut both exert pressure on the surface of the Moon.

Calculate *If the lunar module weighed approximately 12,000 N and rested on four pads that were each 91 cm in diameter, what pressure did it exert on the Moon's surface? How could you estimate the pressure exerted by the astronaut?*

Pressure

Pressure equals the perpendicular component of the force divided by the surface area to which it is applied.

$$P = \frac{F}{A}$$

Pressure is a scalar. In the SI system, the unit of pressure is the **pascal** (Pa), which is 1 N/m^2. One pascal is a very small amount of pressure, about equal to the pressure that a flat dollar bill exerts on a tabletop. Thus the kilopascal (kPa), equal to 1000 Pa, is usually used. **Table 1** shows pressures in various locations.

Solids, liquids, and pressure Imagine you are standing on the surface of a frozen lake. The forces your feet exert on the ice are spread over the area of your shoes, resulting in pressure on the ice. Ice is a solid that is made up of vibrating water molecules, and the forces that hold the water molecules in place cause the ice to exert upward forces on your feet that equal your weight. If the ice melted, most of the bonds between the water molecules would be weakened. Although the molecules would continue to vibrate and remain close to each other, they also would slide past one another, and you would break through the surface. The moving water molecules would continue to exert forces on your body.

Table 1 Some Typical Pressures

Location	Pressure (kPa)
The center of Earth	4×10^{8}
The deepest ocean trench	1.1×10^{5}
Standard atmosphere	1.01325×10^{2}
Blood pressure	1.6×10^{1}
Air pressure on top of Mt. Everest	3×10^{1}
The best vacuum	1×10^{-10}

Gases and pressure The pressure exerted by a gas can be understood by applying the kinetic-molecular theory of gases, which explains the properties of an ideal gas. In this model, particles are treated as taking up no space and having no intermolecular attractive forces. In spite of the fact that particles of a real gas take up space and exert attractive forces, an ideal gas is an accurate model of a real gas under most conditions.

CCC CROSSCUTTING CONCEPTS

Structure and Function The molecular structure of a material at a given temperature and pressure allows you to predict its physical state and some of its properties. Make a model to provide evidence that the properties of fluids are based on their molecular structures.

According to the kinetic-molecular theory, the particles in a gas are in random motion at high speeds and colliding elastically with each other. When a gas particle hits a container's surface, it rebounds, which changes its momentum. The impulses exerted by these collisions result in gas pressure on the surface.

Atmospheric pressure At sea level, gases of the atmosphere exert a force in all directions of approximately 10 N, about the weight of a 1-kg object, on each square centimeter of surface area. This atmospheric pressure on your body is so well balanced by your body's outward forces that you seldom notice it. You probably become aware of this pressure only when your ears pop as the result of pressure changes, as when you ride an elevator in a tall building or fly in an airplane. Atmospheric pressure is about 10 N per 1 cm^2 (10^{-4} m^2), which is about 1.0×10^5 N/m^2, or 100 kPa. Other planets in our solar system also have atmospheres that exert pressure. For example, the pressure at the surface of Venus is about 92 times that at Earth's surface.

EXAMPLE Problem 1

CALCULATING PRESSURE A child weighs 364 N and sits on a three-legged stool, which weighs 41 N. The bottoms of the stool's legs touch the ground over a total area of 19.3 cm^2.

a. What is the average pressure that the child and the stool exert on the ground?

b. How does the pressure change when the child leans over so that only two legs of the stool touch the floor?

1 ANALYZE AND SKETCH THE PROBLEM

- Sketch the child and the stool, labeling the total force that they exert on the ground.
- List the variables, including the force that the child and the stool exert on the ground and the areas for parts **a** and **b.**

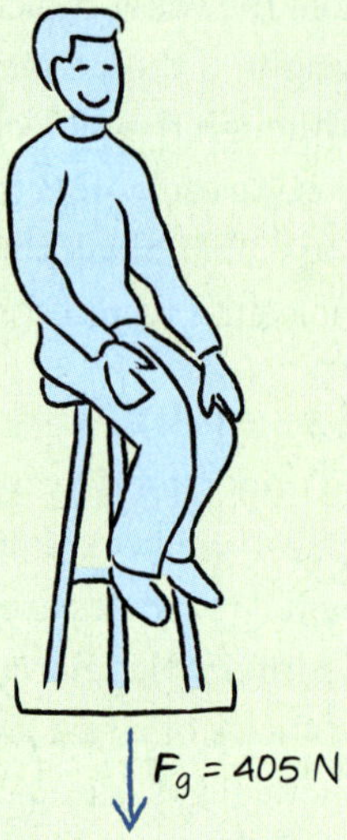

Known

$F_{g\ child} = 364$ N

$F_{g\ stool} = 41$ N

$F_{g\ total} = F_{g\ child} + F_{g\ stool}$

$= 364\ N + 41\ N$

$= 405$ N

$A_a = 19.3\ cm^2$

$A_b = \frac{2}{3} \times 19.3\ cm^2$

$= 12.9\ cm^2$

Unknown

$P_a = ?$

$P_b = ?$

2 SOLVE FOR THE UNKNOWN

Find each pressure.

$$P = \frac{F}{A}$$

a. $P_a = \left(\frac{405\ N}{19.3\ cm^2}\right)\left(\frac{(100\ cm)^2}{(1\ m)^2}\right)$ Substitute $F = F_{g\ total} = 405$ N, $A = A_a = 19.3\ cm^2$.

$= 2.10\times10^2$ kPa

b. $P_b = \left(\frac{405\ N}{12.9\ cm^2}\right)\left(\frac{(100\ cm)^2}{(1\ m)^2}\right)$ Substitute $F = F_{g\ total} = 405$ N, $A = A_b = 12.9\ cm^2$.

$= 3.14\times10^2$ kPa

3 EVALUATE THE ANSWER

- **Are the units correct?** The units for pressure should be Pa, and 1 N/m^2 = 1 Pa.

PRACTICE Problems

ADDITIONAL PRACTICE

1. The atmospheric pressure at sea level is about 1.0×10^5 Pa. What is the force at sea level that air exerts on the top of a desk that is 152 cm long and 76 cm wide?
2. A car tire makes contact with the ground on a rectangular area of 12 cm by 18 cm. If the car's mass is 925 kg, what pressure does the car exert on the ground as it rests on all four tires?
3. A lead brick, 5.0 cm × 10.0 cm × 20.0 cm, rests on the ground on its smallest face. Lead has a density of 11.8 g/cm^3. What pressure does the brick exert on the ground?
4. Suppose that during a storm, the atmospheric pressure suddenly drops by 15 percent outside. What net force would be exerted on a front door to a house that is 195 cm high and 91 cm wide? In what direction would this force be exerted?
5. **CHALLENGE** Large pieces of industrial equipment are placed on wide steel plates that spread the weight of the equipment over larger areas. If an engineer plans to install a 454-kg device on a floor that is rated to withstand additional pressure of 5.0×10^4 Pa, how large should the steel support plate be?

The Gas Laws

Think about a container of gas that is held at a constant temperature. If you reduced the volume, what would happen to the pressure of the gas? There would be more collisions between the particles and the container's walls, and so the pressure would increase. Similarly, if you increased the volume, there would be fewer collisions, decreasing the pressure. This inverse relationship was found by seventeenth-century chemist and physicist Robert Boyle. Because the product of inversely related variables is a constant, Boyle's law can be written $PV = \text{constant}$, or $P_1V_1 = P_2V_2$. The subscripts that you see in the gas laws help keep track of different variables, such as pressure and volume, as they change throughout a problem. The relationship between the pressure and the volume of a gas is critical to the scuba diver in **Figure 3.**

About 100 years after Boyle's work, Jacques Charles cooled a gas and found that the volume shrank by $\frac{1}{273}$ of its original volume for every degree cooled, which is a linear relationship. At the time, Charles could not cool gases to the extremely low temperatures achieved in modern laboratories. In order to see what lower limits might be possible, he extended, or extrapolated, the graph of his data to these temperatures. This extrapolation suggested that if the temperature were reduced to −273°C, a gas would have zero volume. The temperature at which a gas would have zero volume is now called absolute zero, which is represented by the zero of the Kelvin temperature scale.

Figure 3 The gas in the tank on a diver's back is at high pressure. The regulator in the diver's mouth reduces the pressure, making the pressure of the gas the diver breathes equal to that of the water pressure. Bubbles are emitted from the regulator as the diver exhales.

These experiments indicated that under constant pressure, the volume of a sample of gas varies directly with its Kelvin temperature, a result that is now called Charles's law. Charles's law can be written $\frac{V}{T} = \text{constant}$, or $\frac{V_1}{T_1} = \frac{V_2}{T_2}$.

Combined gas law

Combining Boyle's law and Charles's law relates the pressure, temperature, and volume of a fixed amount of ideal gas, which leads to the **combined gas law.**

Combined Gas Law

For a fixed amount of an ideal gas, the pressure times the volume, divided by the Kelvin temperature equals a constant.

$$\frac{P_1V_1}{T_1} = \frac{P_2V_2}{T_2} = \text{constant}$$

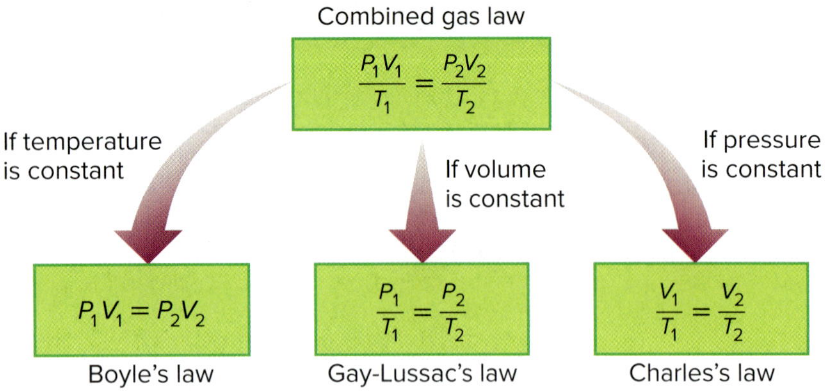

Figure 4 The combined gas law shows the relationship among pressure, temperature, and volume of a fixed amount of an ideal gas. Boyle's law and Charles's law can each be derived from the combined gas law under certain conditions.

Explain *What happens if you hold volume constant?*

As shown in **Figure 4,** the combined gas law reduces to Boyle's law under conditions of constant temperature and to Charles's law under conditions of constant pressure.

The ideal gas law What might the constant in the combined gas law depend on? Suppose the volume and temperature of an ideal gas are held constant while the number of particles (N) is increased. What happens to the pressure? The particles will have more collisions with the sides of the container, thus increasing the pressure. Removing particles decreases the number of collisions, and thus, decreases the pressure. Therefore, the constant in the combined gas law equation is proportional to N.

$$\frac{PV}{T} = \kappa N$$

The constant (κ) is called Boltzmann's constant, and its value is 1.38×10^{-23} Pa·m^3/K. In any practical application, the number of particles (N) is very large. Instead of using N, scientists often use a unit called a mole. One mole (abbreviated *mol*) is similar to one dozen, except that instead of representing 12 items, one mole represents 6.022×10^{23} particles. This number is called Avogadro's number, after Italian scientist Amedeo Avogadro.

Avogadro's number is numerically equal to the number of particles in a sample of matter whose mass equals the molar mass of the substance. You can use this relationship to convert between mass and n, the number of moles present. Using moles instead of the number of particles, however, changes Boltzmann's constant. This new constant is abbreviated R, and it has the value 8.31 Pa·m^3/(mol·K). Rearranging, you can write the ideal gas law in its most familiar form. The **ideal gas law** states that for an ideal gas, the pressure times the volume is equal to the number of moles multiplied by the constant R and the Kelvin temperature.

Ideal Gas Law

For an ideal gas, the pressure times the volume is equal to the number of moles times the constant R times the temperature.

$$PV = n\mathrm{R}T$$

Note that with the given value of R, volume must be expressed in cubic meters, temperature in kelvins, and pressure in pascals. In practice, the ideal gas law predicts the behavior of gases remarkably well, except under conditions of high pressures or low temperatures.

EXAMPLE Problem 2

GAS LAWS A 20.0-L sample of argon gas at 273 K is at atmospheric pressure (101.3 kPa). The temperature is lowered to 120 K, and the pressure is increased to 145 kPa.

a. What is the new volume of the argon sample?

b. Find the number of moles of argon atoms in the argon sample.

c. Find the mass of the argon sample. The molar mass (M) of argon is 39.9 g/mol.

1 ANALYZE AND SKETCH THE PROBLEM

- Sketch the situation. Indicate the conditions in the container of argon before and after the change in temperature and pressure.
- List the known and unknown variables.

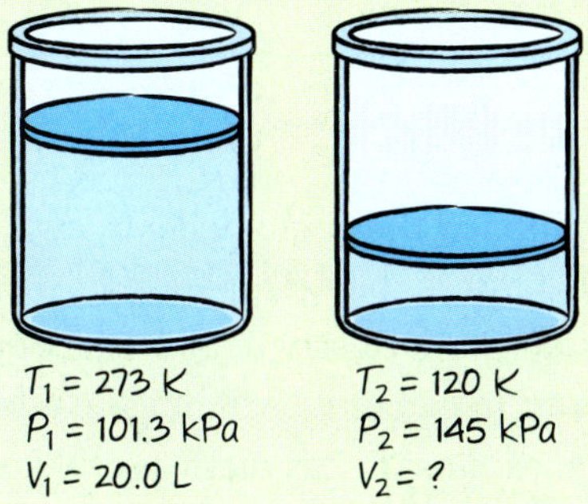

Known	Unknown
$V_1 = 20.0$ L	$V_2 = ?$
$P_1 = 101.3$ kPa	moles of argon = ?
$T_1 = 273$ K	mass of argon sample = ?
$P_2 = 145$ kPa	
$T_2 = 120$ K	
$R = 8.31\ \text{Pa}\cdot\text{m}^3/(\text{mol}\cdot\text{K})$	
$M_{\text{argon}} = 39.9$ g/mol	

2 SOLVE FOR THE UNKNOWN

a. Use the combined gas law and solve for V_2.

$$\frac{P_1V_1}{T_1} = \frac{P_2V_2}{T_2}$$

$$V_2 = \frac{P_1V_1T_2}{P_2T_1}$$

$$= \frac{(101.3\ \text{kPa})(20.0\ \text{L})(120\ \text{K})}{(145\ \text{kPa})(273\ \text{K})}$$

Substitute $P_1 = 101.3$ kPa, $P_2 = 145$ kPa, $V_1 = 20.0$ L, $T_1 = 273$ K, $T_2 = 120$ K.

$$= 6.1\ \text{L}$$

b. Use the ideal gas law, and solve for n.

$$PV = nRT$$

$$n = \frac{PV}{RT}$$

$$= \frac{(101.3\times10^3\ \text{Pa})(0.0200\ \text{m}^3)}{(8.31\ \text{Pa}\cdot\text{m}^3/(\text{mol}\cdot\text{K}))(273\ \text{K})}$$

Substitute $P = 101.3\times10^3$ Pa, $V = 0.0200\ \text{m}^3$, $R = 8.31\ \text{Pa}\cdot\text{m}^3/(\text{mol}\cdot\text{K})$, $T = 273$ K.

$$= 0.893\ \text{mol}$$

c. Use the molar mass to convert from moles of argon in the sample to mass of the sample.

$$m = Mn$$

$$m_{\text{argon sample}} = (39.9\ \text{g/mol})(0.893\ \text{mol})$$

Substitute $M = 39.9$ g/mol, $n = 0.893$ mol.

$$= 35.6\ \text{g}$$

3 EVALUATE THE ANSWER

- **Are the units correct?** The volume (V_2) is in liters, and the mass of the sample is in grams.
- **Is the magnitude realistic?** The change in volume is consistent with an increase in pressure and a decrease in temperature. The calculated mass of the argon sample is reasonable.

PRACTICE Problems

ADDITIONAL PRACTICE

6. A tank of helium gas used to inflate toy balloons is at a pressure of 15.5×10^6 Pa and a temperature of 293 K. The tank's volume is 0.020 m^3. How large a balloon would it fill at 1.00 atmosphere and 323 K?
7. What is the mass of the helium gas in the previous problem? The molar mass of helium gas is 4.00 g/mol.
8. A tank containing 200.0 L of hydrogen gas at 0.0°C is kept at 156 kPa. The temperature is raised to 95°C, and the volume is decreased to 175 L. What is the new pressure of the gas?
9. **CHALLENGE** The average molar mass of the components of air (mainly diatomic nitrogen gas and diatomic oxygen gas) is about 29 g/mol. What is the volume of 1.0 kg of air at atmospheric pressure and 20.0°C?

Thermal Expansion

As you applied the combined gas law, you discovered how gases expand as their temperatures increase. **Thermal expansion** is a property of all forms of matter that causes the matter to expand, becoming less dense, when heated. Thermal expansion has many useful applications, such as circulating air in a room.

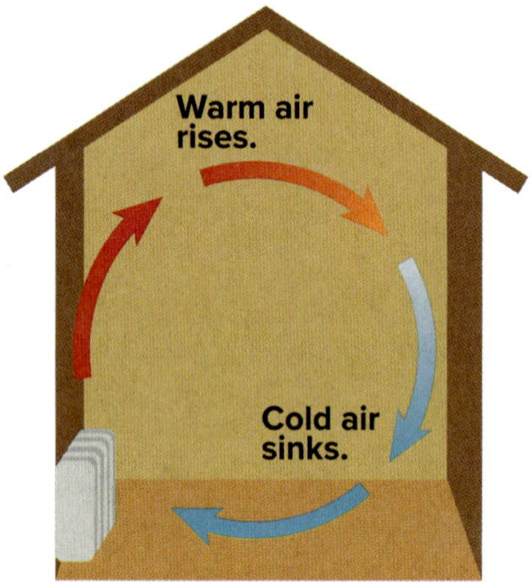

Figure 5 Convection currents occur as warmer, less dense air rises and cooler, denser air sinks.

Convection currents **Figure 5** shows that when the air near the radiator of a room is heated, it becomes less dense and, therefore, rises. Gravity pulls the denser, colder air near the ceiling down. The cold air is subsequently warmed by the radiator, and air continues to circulate. This circulation of air within a room is called a convection current. Convection currents also occur in a pot of hot, but not boiling, water on a stove. When the pot is heated from the bottom, the colder and denser water sinks to the bottom where it is warmed and then pushed up by the continuous flow of cooler water from the top.

This thermal expansion occurs in most fluids. A good model for all liquids does not exist, but it is useful to think of a liquid as a finely ground solid. Groups of two, three, or more particles move together as if they were tiny pieces of a solid. When a liquid is heated, particle motion causes these groups to expand in the same way that particles in a solid are pushed apart. The spaces between groups increase. As a result, the whole liquid expands. With an equal change in temperature, liquids expand considerably more than solids but not as much as gases.

Why ice floats Because matter expands as it is heated, you might predict that ice would be more dense than water, and therefore, it should sink. However, when water is heated from 0°C to 4°C, instead of expanding, it contracts as the forces between particles increase and the ice crystals collapse. These forces between water molecules are strong, and the crystals that make up ice have a very open structure. Even when ice melts, tiny crystals remain. These remaining crystals are melting, and the volume of the water decreases until the temperature reaches 4°C. However, once the temperature of water moves above 4°C, its volume increases because of greater molecular motion. The practical result is that water is most dense at 4°C, and ice floats. This unique property of water is very important to our lives and environment. If ice sank, lakes would freeze from the bottom each winter and many would never melt completely in the summer.

Plasma

If you heat a solid, it melts to form a liquid. Further heating results in a gas. What happens if you increase the temperature still further? Collisions between the particles become violent enough to tear the electrons off the atoms, thereby producing positively charged ions. The gaslike state of negatively charged electrons and positively charged ions is called **plasma.** Plasma is considered to be another state of matter.

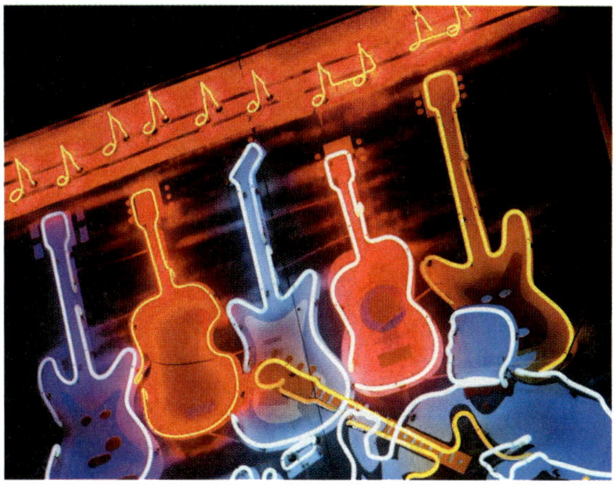

Figure 6 Plasma emits light as it conducts electricity. The color produced by glowing plasma depends on the gas inside the tube.

The plasma state may seem to be uncommon, but plasma is actually the most common state of matter in the universe. Stars consist mostly of plasma at extremely high temperatures. Much of the matter between stars and galaxies consists of energetic hydrogen that has no electrons. This hydrogen is in the plasma state. The primary difference between gas and plasma is that plasmas can conduct an electric current, whereas gases cannot. Lightning bolts are in the plasma state. Neon signs, such as the one shown in **Figure 6,** contain plasma. The fluorescent bulbs that probably light your school also contain plasma.

Check Your Progress

10. **Comparing States** Compare and contrast liquids, gases, and plasmas.
11. **Pressure and Force** Two boxes are each suspended by thin strings in midair. One is 20 cm × 20 cm × 20 cm. The other is 20 cm × 20 cm × 40 cm.
    a. How does the pressure of the air on the outside of the two boxes compare?
    b. How does the magnitude of the total force of the air on the two boxes compare?
12. **Meteorology** A weather balloon used by meteorologists is made of a flexible bag that allows the gas inside to freely expand. If a weather balloon containing 25.0 m^3 of helium gas is released from sea level, what is the volume of gas when the balloon reaches a height of 2100 m, where the pressure is 0.82×10^5 Pa? Assume the temperature is unchanged.
13. **Density and Temperature** Starting at 0°C, how will the density of water change as it is heated to 4°C? To 8°C?
14. **Gas Compression** In a certain internal-combustion engine, 0.0021 m^3 of air at atmospheric pressure and 303 K is rapidly compressed to a pressure of 20.1×10^5 Pa and a volume of 0.0003 m^3. What is the final temperature of the compressed gas?
15. **The Standard Molar Volume** What is the volume of 1.00 mol of a gas at atmospheric pressure and a temperature of 273 K?
16. **The Air in a Refrigerator** How many moles of air are in a refrigerator with a volume of 0.635 m^3 at a temperature of 2.00°C? If the average molar mass of air is 29 g/mol, what is the mass of the air in the refrigerator?
17. **Critical Thinking** Compared to the particles that make up carbon dioxide gas, the particles that make up helium gas are very small. What can you conclude about the number of particles in a 2.0-L sample of carbon dioxide gas compared to the number of particles in a 2.0-L sample of helium gas if both samples are at the same temperature and pressure?

LEARNSMART® Go online to follow your personalized learning path to review, practice, and reinforce your understanding.

LESSON 2
FORCES WITHIN LIQUIDS

FOCUS QUESTION
How do the forces between particles in a fluid affect its behavior?

Cohesive Forces

Figure 7 shows a water strider walking across the surface of a pond. This lightweight insect can do this because of surface tension—the tendency of the surface of a liquid to contract to the smallest possible area. Surface tension results from the cohesive forces among the particles of a liquid. **Cohesive forces** are the forces of attraction that like particles exert on one another. Notice that beneath the liquid's surface in **Figure 7,** each particle of the liquid is attracted equally in all directions by neighboring particles. As a result, no net force acts on any of the particles beneath the surface. At the surface, however, the particles are attracted downward and to the sides but not upward. There is a net downward force, which acts on the top layers and causes the surface layer to be slightly compressed. The surface layer acts like a tightly stretched sheet that is strong enough to support the weight of very light objects, such as the water strider.

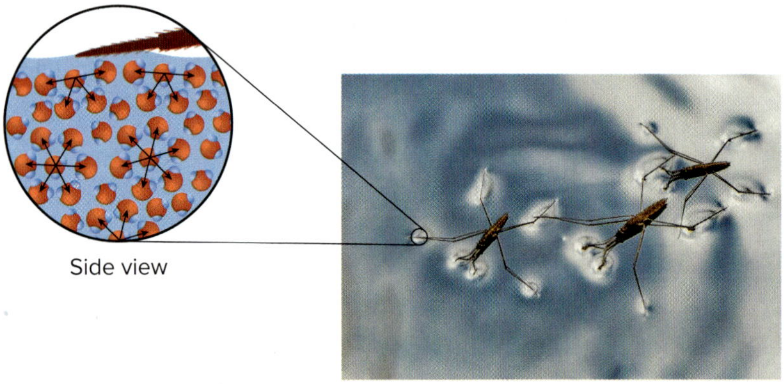

Jan Miko/Shutterstock.com

Figure 7 A water strider can walk on water because molecules at the surface experience a net downward force. Below the surface, each particle of liquid is equally attracted in all directions.

3D THINKING DCI Disciplinary Core Ideas CCC Crosscutting Concepts SEP Science & Engineering Practices

COLLECT EVIDENCE
Use your Science Journal to record the evidence you collect as you complete the readings and activities in this lesson.

INVESTIGATE
GO ONLINE to find these activities and more resources.

PhysicsLAB: Evaporative Cooling
Construct an explanation based on evidence to explain how cohesive forces affect the rate of evaporation of a liquid.

CCC **Identify Crosscutting Concepts**
Create a table of the crosscutting concepts and fill in examples you find as you read.

You might have seen beaded water droplets on a freshly washed and waxed car. Why do these spherical drops form? The force pulling the surface particles into a liquid causes the surface to become as small as possible, and the shape that has the least surface for a given volume is a sphere. The higher the surface tension of the liquid, the more resistant the liquid is to having its surface broken. For example, liquid mercury has much stronger cohesive forces than water does. Liquid mercury forms spherical drops, even when it is placed on a smooth surface. A drop of water flattens out on a smooth surface.

Viscosity In nonideal fluids, the cohesive forces and collisions between fluid molecules cause internal friction that slows the fluid flow and dissipates mechanical energy. The measure of this internal friction is called the viscosity of the liquid. Water is not very viscous, but motor oil is very viscous. As a result of its viscosity, motor oil flows slowly over the parts of an engine to coat the metal and reduce rubbing. Lava, molten rock that flows from a volcano or vent in Earth's surface, is one of the most viscous fluids. There are several types of lava, and the viscosity of each type varies with composition and temperature.

Adhesive Forces

Similar to cohesive forces, **adhesive forces** are attractive forces that act between particles of different substances. When a glass tube is placed in a beaker of water, the surface of the water climbs the outside of the tube, as shown in **Figure 8.** The adhesive forces between the particles that make up the glass and the water molecules are greater than the cohesive forces between the water molecules. In contrast, the cohesive forces between mercury atoms are greater than the adhesive forces between the mercury and the glass, so the liquid does not climb the tube. These forces also cause the mercury's surface to depress, as shown in **Figure 8.**

Real-World Physics

PLANTS The combination of adhesive and cohesive forces acting on water molecules in plant tissues results in a certain amount of tension. As water molecules evaporate from cells in leaves, this tension pulls adjacent water molecules further into the leaves. Cohesive forces keep water from separating into individual droplets, and water is pulled the entire length of a plant, in some cases as much as 115 m.

Figure 8 Due to adhesive forces, water climbs the outside wall of the glass tube. In the mercury, however, the forces of attraction between mercury atoms are stronger than any adhesive forces between the mercury and the glass. Therefore, the mercury is depressed by the tube.

If a glass tube with a small inner diameter is placed in water, the water rises inside the tube. This happens because the adhesive forces between glass and water molecules are stronger than the cohesive forces between water molecules. The water continues to rise until the weight of the water that is lifted balances the total adhesive force between the glass and water molecules. If the radius of the tube increases, the volume and the weight of the water will increase proportionally faster than the surface area of the tube. Thus, water is lifted higher in a narrow tube than in a wider one. This phenomenon is called capillary action. It causes molten wax to rise in a candle's wick and water to move up through the soil and into the roots of plants.

Get It?

Compare How do the adhesive forces between water and glass compare with the cohesive forces between water molecules?

Evaporation and Condensation

Why does a puddle of water disappear on a hot, dry day? As you have previously read, the particles in a liquid are moving at random speeds. If a fast-moving particle can break through the surface layer, it will escape from the liquid. Because there is a net downward cohesive force at the surface, only the most energetic particles escape. This escape of particles is called evaporation.

Evaporative cooling Evaporation has a cooling effect. On a hot day, your body perspires, and the evaporation of your sweat cools you. In a puddle of water, evaporation causes the remaining liquid to cool. Each time a particle with higher-than-average kinetic energy escapes from the water, the average kinetic energy of the remaining particles decreases. As you learned earlier, a decrease in average kinetic energy is a decrease in temperature. Rubbing alcohol has a noticeable cooling effect when it evaporates from a person's skin. Alcohol molecules evaporate easily because they have weak cohesive forces. A liquid that evaporates quickly is called a volatile liquid.

Have you ever wondered why humid days feel warmer than dry days at the same temperature? On a humid day, the water vapor content of the air is high. Because there are already many water molecules in the air, the water molecules in perspiration are less likely to evaporate from the skin. Evaporation is the body's primary cooling mechanism, so the body is not able to cool itself as effectively on a humid day.

Get It?

Explain why evaporation has a cooling effect.

CCC CROSSCUTTING CONCEPTS

Energy and Matter Prepare a skit that provides evidence to explain the changes between the liquid and gas states through a study of the flow of energy and matter.

Condensation Particles of liquid that have evaporated into the air can also return to the liquid phase if the kinetic energy or temperature decreases in a process called condensation.

What happens if you bring a cold glass into a hot, humid area? The outside of the glass soon becomes coated with condensed water. Water molecules moving randomly in the air surrounding the glass strike the cold surface, and if the water molecules lose enough energy, the cohesive forces between them become strong enough to prevent their escape.

Get It?

Define What happens during the process of condensation?

The air above any body of water, as shown in **Figure 9,** contains evaporated water vapor, which is water in the form of gas. If the temperature is reduced, the water vapor condenses around tiny dust particles in the air and produces droplets only 0.01 mm in diameter. A cloud of these droplets that forms at Earth's surface is called fog. Fog often forms when moist air is chilled by the cold ground. Fog also occurs briefly when a carbonated drink is opened. The sudden decrease in pressure causes the temperature of the gas in the container to drop, which condenses the water vapor dissolved in that gas.

Figure 9 Warm, moist, surface air rises. Clouds form when the air cools and the water vapor condenses.

Get It?

Identify What are two changes in conditions that can cause water vapor in the air to condense?

Check Your Progress

18. **Science Terms** The English language includes the term *adhesive tape* and the phrase *working as a cohesive group*. In these examples, are the words *adhesive* and *cohesive* being used in the same context as their meanings in physics? Explain your answer.
19. **Surface Tension** A paper clip, which has a density greater than that of water, can be made to stay on the surface of water. What procedures must you follow for this to happen? Explain.
20. **Floating** How can you tell that the paper clip in the previous problem was not floating?
21. **Adhesion and Cohesion** In terms of adhesion and cohesion, explain why alcohol clings to the surface of a glass rod but mercury does not.
22. **Evaporation and Cooling** In the past when a baby had a high fever, the doctor might have suggested gently sponging off the baby with a liquid that evaporates easily. Why would this help?
23. **Critical Thinking** On a hot, humid day, Beth took a glass of cold water outside. Soon, water coated the outside of the glass. Her friend, Sally, suggested that the water had leaked out through the glass. Design an experiment for Beth to show Sally where the water came from.

LEARNSMART® Go online to follow your personalized learning path to review, practice, and reinforce your understanding.

LESSON 3

FLUIDS AT REST AND IN MOTION

FOCUS QUESTION

What principles describe the behavior and motion of fluids?

Fluids at Rest

If you have ever dived deep into a swimming pool or a lake, you likely felt pressure on your ears. Blaise Pascal, a French physician, found that the pressure at a point in a fluid depends on its depth in the fluid and is unrelated to the shape of the fluid's container. He also noted that any change in pressure applied at any point on a confined fluid is transferred undiminished throughout the fluid, a fact that is now known as **Pascal's principle.** One application of Pascal's principle is using fluids in machines to multiply forces. In the hydraulic system shown in **Figure 10,** a fluid is confined to two connecting chambers. Each chamber has a piston that is free to move, and the pistons have different surface areas. Recall that if a force (F_1) is exerted on the first piston with a surface area of A_1, the pressure (P_1) exerted on the fluid is $P_1 = \frac{F_1}{A_1}$. The pressure exerted by the fluid on the second piston, with a surface area A_2, is $P_2 = \frac{F_2}{A_2}$.

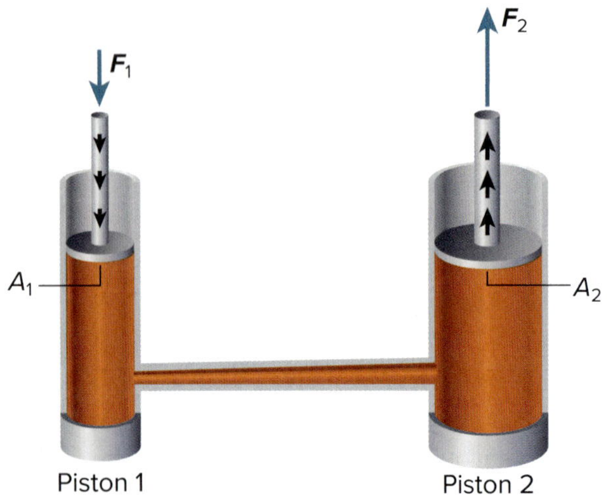

Figure 10 As F_1 exerts pressure on the smaller piston (piston 1), the pressure is transmitted throughout the fluid. As a result, a multiplied force (F_2) is exerted on the larger piston (piston 2).

Infer *How would F_2 change if F_1 increased? Explain why.*

3D THINKING — DCI Disciplinary Core Ideas — CCC Crosscutting Concepts — SEP Science & Engineering Practices

COLLECT EVIDENCE

Use your Science Journal to record the evidence you collect as you complete the readings and activities in this lesson.

INVESTIGATE

GO ONLINE to find these activities and more resources.

PhysicsLAB: Under Pressure

Carry out an investigation to determine the effect of force in a fluid.

Revisit the Encounter the Phenomenon Question

What information from this lesson can help you answer the Unit and Module questions?

PRACTICE Problems ADDITIONAL PRACTICE

24. Dentists' chairs are examples of hydraulic-lift systems. If a chair weighs 1600 N and rests on a piston with a cross-sectional area of 1440 cm^2, what force must be applied to the smaller piston, with a cross-sectional area of 72 cm^2, to lift the chair?

25. A mechanic exerts a force of 55 N on a 0.015 m^2 hydraulic piston to lift a small automobile. The piston the automobile sits on has an area of 2.4 m^2. What is the weight of the automobile?

26. CHALLENGE By multiplying a force, a hydraulic system serves the same purpose as a lever or a seesaw. If a 400-N child standing on one piston is balanced by a 1100-N adult standing on another piston, what is the ratio of the areas of their pistons?

According to Pascal's principle, pressure is transmitted without change throughout a fluid, so pressure P_2 is equal in value to P_1. You can determine the force exerted by the second piston by setting the pressures equal to each other and solving for F_2.

Force Exerted by a Hydraulic Lift

The force exerted by the second piston is equal to the force exerted by the first piston multiplied by the ratio of the area of the second piston to the area of the first piston.

$$F_2 = \frac{F_1 A_2}{A_1}$$

Swimming Under Pressure

When you are swimming, you feel the pressure of the water increase as you dive deeper. This pressure is a result of gravity; it is related to the weight of the water above you. The deeper you go, the more water there is above you, and the greater the pressure. The pressure of the water is equal to the weight (F_g) of the column of water above you divided by the column's cross-sectional area (A). Even though gravity pulls only in the downward direction, the fluid transmits the pressure in all directions: up, down, and to the sides. As before, the pressure of the water is $P = \frac{F_g}{A}$.

The weight of the column of fluid is $F_g = mg$, and the mass is equal to the density (ρ) of the fluid times its volume, $m = \rho V$. You also know that the volume of the fluid is the area of the base of the column times its height, $V = Ah$. Therefore, $F_g = \rho Ahg$. Substituting ρAhg for F_g gives $P = \frac{F_g}{A} = \frac{\rho Ahg}{A}$. Divide A from the numerator and denominator to arrive at the simplest form of the equation for the pressure exerted by a column of fluid on a submerged body.

Pressure of Fluid on a Body

The pressure a column of fluid exerts on a body is equal to the density of the fluid times the height of the column times the free-fall acceleration.

$$P = \rho hg$$

The pressure of a fluid on a body depends on the density of the fluid, its depth, and g. As shown in **Figure 11,** submersibles have explored the deepest ocean trenches and encountered pressures in excess of 1000 times standard air pressure.

BIOLOGY Connection Scientists use submersibles to learn more about deep ocean ecosystems. In 1977 the first hydrothermal vents were discovered as the crewed submersible ALVIN cruised over the Pacific Ocean floor. Hydrothermal vents form when superheated water flows up from cracks in the seafloor.

Figure 11 Submersibles are built to withstand the crushing pressure exerted by the water column.

Because these vents are located thousands of meters below the ocean surface, the fluid pressure can be over one hundred times the standard atmospheric pressure. Despite the high pressure and the fact that sunlight does not reach them, hydrothermal vents thrive with life. Giant tube worms harbor bacteria in their tissues. The bacteria use hydrogen sulfide from the vent water to produce sugar, which provides the energy that supports the entire ecosystem. Other organisms that live at hydrothermal vents include fish, mussels, shrimp, clams, and octopuses. Submersibles have been used to explore hydrothermal vents in the Atlantic, Indian, and Arctic Oceans.

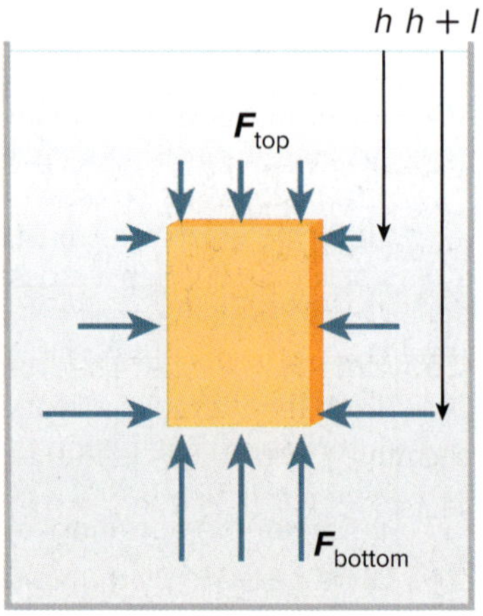

Figure 12 A fluid exerts a greater upward force on the bottom of an immersed object than the downward force on the top of the object. The net upward force is called the buoyant force.

Buoyancy What produces the upward force that allows you to swim? The increase in pressure with increasing depth creates an upward force called the **buoyant force.** By comparing the buoyant force on an object with its weight, you can predict whether the object will sink or float.

Suppose that a box is immersed in water. It has a height of l, and its top and bottom each have a surface area of A. Its volume, then, is $V = lA$. Water pressure exerts forces on all sides, as shown in **Figure 12.** Will the box sink or float? As you know, the pressure on the box depends on its depth (h). To find out whether the box will float in water, you will need to analyze the forces acting on it, which are its weight and the forces on each side due to the pressure of the fluid. Compare these two equations:

$$F_{top} = P_{top}A = \rho hgA$$

$$F_{bottom} = P_{bottom}A = \rho(l + h)gA$$

On the four vertical sides, the forces are equal in all directions, so there is no net horizontal force. The upward force on the bottom is larger than the downward force on the top, so there is a net upward force. The buoyant force can now be determined.

$$\begin{aligned} F_{buoyant} &= F_{bottom} - F_{top} \\ &= \rho(l + h)gA - \rho hgA \\ &= \rho lgA = \rho Vg \end{aligned}$$

These calculations show the net upward force to be proportional to the volume of the box. This volume equals the volume of the fluid displaced, or pushed out of the way, by the box. Therefore, the magnitude of the buoyant force (ρVg) equals the weight of the fluid displaced by the object.

Buoyant Force

The buoyant force on an object is equal to the weight of the fluid displaced by the object, which is equal to the density of the fluid in which the object is immersed multiplied by the object's volume and the free-fall acceleration.

$$F_{buoyant} = \rho_{fluid}Vg$$

Archimedes' principle The relationship between buoyant force and the weight of the fluid displaced by an object was discovered in the third century B.C. by Greek scientist and mathematician Archimedes. **Archimedes' principle** states that an object immersed in a fluid has an upward force on it that is equal to the weight of the fluid displaced by the object. The force does not depend on the weight of the object, only on the weight of the displaced fluid.

Sink or Float? If you want to know whether an object sinks or floats, you have to take into account all of the forces acting on the object. The buoyant force pushes up, but the weight of the object pulls it down. The difference between the buoyant force and the object's weight determines whether an object sinks or floats.

Suppose you submerge three objects in a tank of water ($\rho_{water} = 1.00\times10^3$ kg/m^3). Each object has a volume of 400 cm^3, or 4.00×10^{-4} m^3. The first object is a steel block with a mass of 3.60 kg. The second is a can of soda with a mass of 0.40 kg. The third is an ice block with a mass of 0.36 kg. How will each item move when it is immersed in water and released? Since the three objects have the same volume, they displace the same volume of water, and the upward force on all three objects is the same, as shown in **Figure 13.** This buoyant force can be calculated as follows.

$$F_{buoyant} = \rho_{water}Vg$$

$$= (1.00\times10^3 \text{ kg/m}^3)(4.00\times10^{-4} \text{ m}^3)(9.8 \text{ N/kg})$$

$$= 3.9 \text{ N}$$

Figure 13 All of the forces on an object must be accounted for to determine whether an object will sink or float.
Describe *the circumstances under which an object will float.*

Buoyant Force

Sinking The weight of the block of steel is 35 N, much greater than the buoyant force. There is a net downward force, so the block will sink to the bottom of the tank. The net downward force is less than the object's weight. All objects in a liquid, even those that sink, experience a net force that is less than the net force when the object is in air.

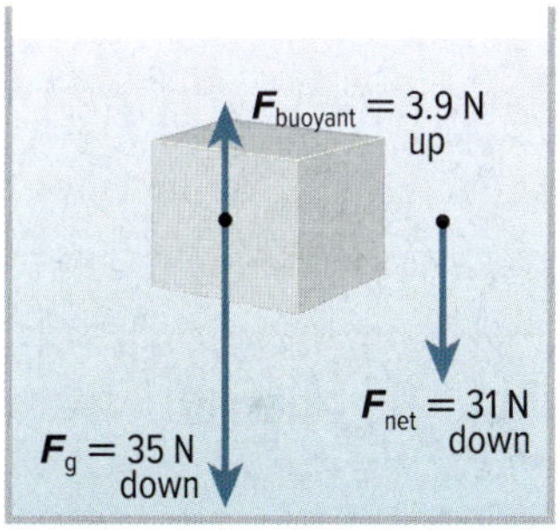

Neutral The weight of the soda can is 3.9 N, the same as the weight of the water displaced. There is, therefore, no net force, and the can will remain wherever it is placed in the water. It is said to have neutral buoyancy.

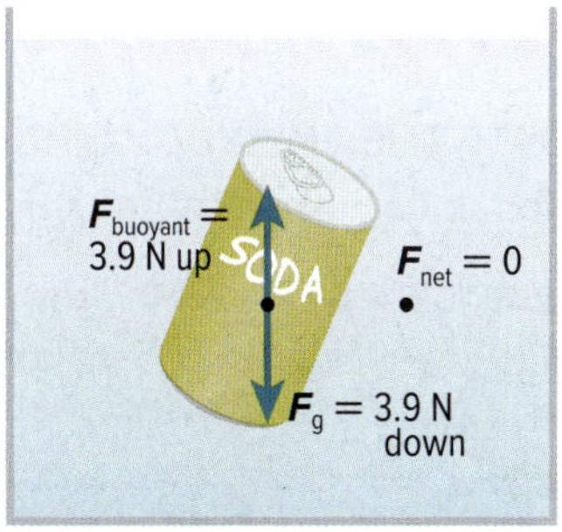

Floating The weight of the ice cube is 3.5 N, less than the buoyant force, so there is a net upward force, and the ice cube will rise. An object will float if its density is less than the density of the fluid in which it is placed.

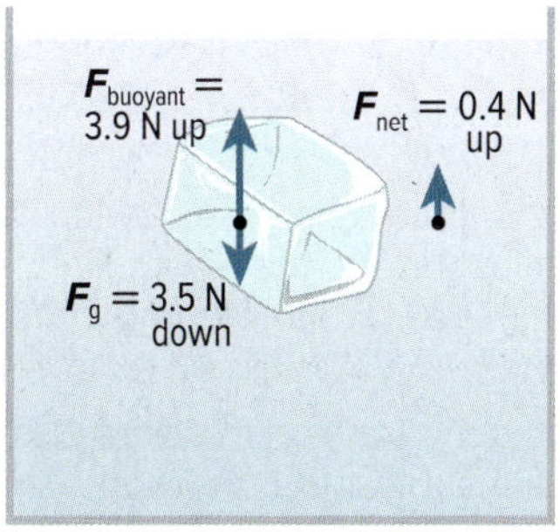

Note: Force vectors are not drawn to scale.

EXAMPLE Problem 3

ARCHIMEDES' PRINCIPLE A cubic decimeter (1.00×10^{-3} m^3) of a granite building block is submerged in water. The density of granite is 2.70×10^3 kg/m^3.

- What is the magnitude of the buoyant force acting on the block?
- What is the net force on the block?

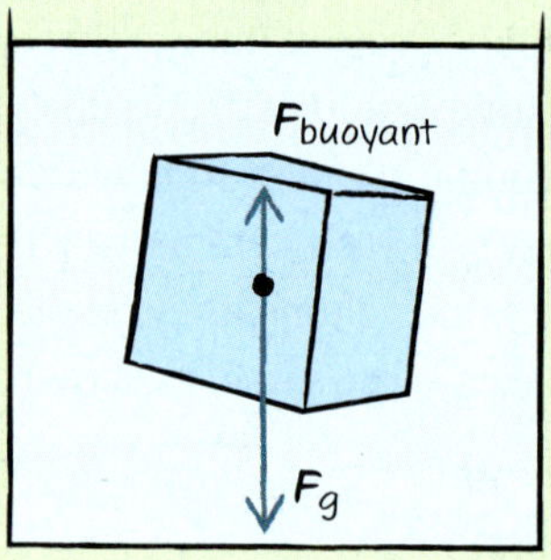

1 ANALYZE AND SKETCH THE PROBLEM

- Sketch the cubic decimeter of granite immersed in water.
- Show the upward buoyant force and the downward force due to gravity acting on the granite.

Known	**Unknown**
$V = 1.00\times10^{-3}$ m^3	$F_{buoyant} = ?$
$\rho_{granite} = 2.70\times10^3$ kg/m^3	$F_{net} = ?$
$\rho_{water} = 1.00\times10^3$ kg/m^3	

2 SOLVE FOR THE UNKNOWN

a. Calculate the buoyant force on the granite block.

$F_{buoyant} = \rho_{water}Vg$

$= (1.00\times10^3\text{ kg/m}^3)(1.00\times10^{-3}\text{ m}^3)(9.8\text{ N/kg})$ — Substitute $\rho_{water} = 1.00\times10^3$ kg/m^3, $V = 1.00\times10^{-3}$ m^3, $g = 9.8$ N/kg.

$= 9.8$ N

b. Calculate the granite's weight, and then find its net force.

$F_g = mg = \rho_{granite}Vg$

$= (2.70\times10^3\text{ kg/m}^3)(1.00\times10^{-3}\text{ m}^3)(9.8\text{ N/kg})$ — Substitute $\rho_{granite} = 2.70\times10^3$ kg/m^3, $V = 1.00\times10^{-3}$ m^3, $g = 9.8$ N/kg.

$= 26.5$ N

$F_{net} = F_g - F_{buoyant}$

$= 26.5\text{ N} - 9.8\text{ N}$ — Substitute $F_g = 26.5$ N, $F_{buoyant} = 9.8$ N.

$= 17$ N

3 EVALUATE THE ANSWER

- **Are the units correct?** The forces and the weight are in newtons, as expected.
- **Is the magnitude realistic?** The buoyant force is about one-third the weight of the granite, a sensible answer because the density of water is about one-third that of granite.

PRACTICE Problems

ADDITIONAL PRACTICE

27. Common brick is about 1.8 times denser than water. What is the net force on a 0.20 m^3 block of bricks under water?

28. A girl is floating in a freshwater lake with her head just above the water. If she weighs 610 N, what is the volume of the submerged part of her body?

29. What is the tension in a wire supporting a 1250-N camera submerged in water? The volume of the camera is 16.5×10^{-3} m^3.

30. Plastic foam is about 0.10 times as dense as water. What weight of bricks could you stack on a 1.0-m × 1.0-m × 0.10-m slab of foam so that the slab of foam floats in water and is barely submerged, leaving the bricks dry?

31. **CHALLENGE** Canoes often have plastic foam blocks mounted under the seats for flotation in case the canoe fills with water. What is the approximate minimum volume of foam needed for flotation for a 480-N canoe?

Ships How can steel ships float? If you make a small boat out of folded aluminum foil, it should float easily. Add a cargo of pennies, and it will ride lower in the water. Crumple the foil into a tight ball, and it will sink. The boat floats when it is hollow and large enough so that its average density is less than water's density. As you add cargo, the density increases, and more of the boat is submerged. The crumpled boat has a density greater than the water's and sinks. Archimedes' principle also applies to submarines and fishes. Submarines can pump water into or out of chambers to change the submarine's average density, causing it to sink or rise. Fishes that have swim bladders move upward in the water by filling the bladder with gas to displace more water and increase the buoyant force. The fish moves downward by contracting the swim bladder.

Get It?

Explain how the submersible at the beginning of the module can rise, sink, or stay suspended in the water.

Bernoulli's Principle

Study the flow of water from the hose in **Figure 14.** In the top photo, the water flows from the hose unobstructed. In the bottom photo, the hose opening has been narrowed by a person's thumb. Notice that the streams of water look different. The velocity of the water stream in the bottom photo is greater compared to the velocity of the stream in the top photo. What you can't see is that pressure exerted by the water in the bottom photo decreased. The relationship between the velocity and pressure exerted by a moving fluid is named for Swiss scientist Daniel Bernoulli. **Bernoulli's principle** states that as the velocity of a fluid increases, the pressure exerted by that fluid decreases. This principle is a statement of work and energy conservation as applied to fluids.

Another instance in which the velocity of water can change is in a stream. You might have seen the water in a stream speed up as it passed through narrowed sections of the stream bed. As the opening of the hose and the stream channel become wider or narrower, the velocity of the fluid changes to maintain the overall flow of water. The pressure of blood in our circulatory systems depends partly on Bernoulli's principle. Bernoulli's principle also helps explain how smoke is pulled up a fireplace chimney.

Figure 14 You can demonstrate Bernoulli's principle by narrowing the opening of the hose as water flows out. As the velocity of the water increases, the pressure it exerts decreases.

Consider a horizontal pipe completely filled with a smoothly flowing ideal fluid. If a certain mass of the fluid enters one end of the pipe, then an equal mass must come out the other end. What happens if the cross section becomes narrower, as shown in **Figure 15?** To keep the same mass of fluid moving through the narrow section in a fixed amount of time, the velocity of the fluid in the tube must increase. As the fluid's velocity increases, so does its kinetic energy. This means that net work has been done on the swifter fluid. This net work comes from the difference between the work that was done to move the mass of fluid into the pipe and the work that was done by the fluid pushing the same mass out of the pipe. The work is proportional to the force on the fluid, which, in turn, depends on the pressure. If the net work is positive, the pressure at the input end of the section, where the velocity is lower, must be larger than the pressure at the output end, where the velocity is higher.

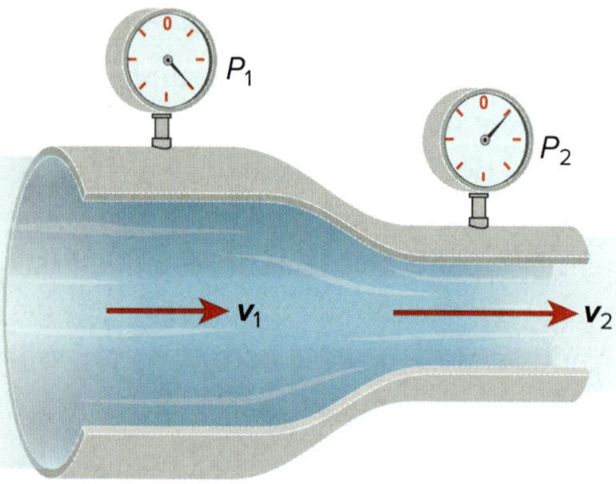

Figure 15 The fluid flowing through this pipe also demonstrates Bernoulli's principle. As the velocity of the fluid increases (v_2 is greater than v_1), the pressure it exerts decreases (P_2 is less than P_1).

Get It?

Describe the relationship between the velocity of a fluid and the pressure it exerts according to Bernoulli's principle.

Applications of Bernoulli's principle There are many common applications of Bernoulli's principle, such as paint sprayers and sprayers attached to garden hoses to apply fertilizers and pesticides to lawns and gardens. In a hose-end sprayer, a strawlike tube is sunk into the chemical solution in the sprayer. The sprayer is attached to a hose. A trigger on the sprayer allows water from the hose to flow at a high speed, producing an area of low pressure above the tube. The solution is then sucked up through the tube and into the stream of water.

A gasoline engine's carburetor, which is where air and gasoline are mixed, is another common application of Bernoulli's principle. Part of the carburetor is a tube with a constriction, as shown in the diagram in **Figure 16.** The pressure on the gasoline in the fuel supply is the same as the pressure in the thicker part of the tube. Air flowing through the narrow section of the tube, which is attached to the fuel supply, is at a lower pressure, so fuel is forced into the air flow. By regulating the flow of air in the tube, the amount of fuel mixed into the air can be varied. Carburetors are used in motorcycles, stock car race cars, and the motors of small gasoline-powered machines, such as lawn mowers.

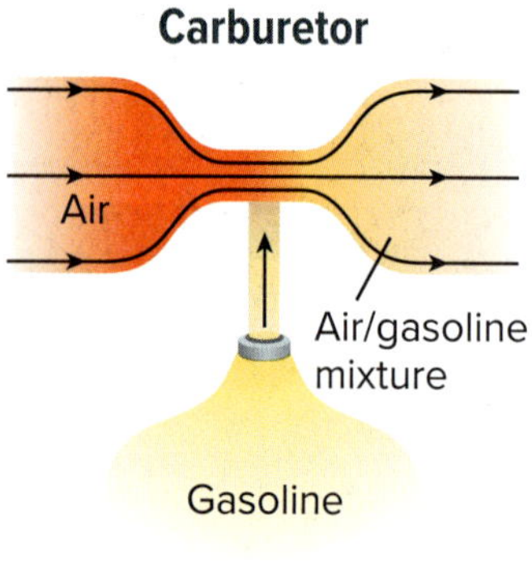

Figure 16 In a carburetor, low pressure in the narrow part of the tube draws fuel into the air flow.

SCIENCE USAGE v. COMMON USAGE

Streamlines

Science usage: lines representing the flow of fluids around objects

Simulations of the streamlines around the airplane indicated a design flaw.

Common usage: makes simpler or more efficient

The new computer system streamlines the registration process.

Streamlines Automobile and aircraft manufacturers spend a great deal of time and money testing new designs in wind tunnels to ensure the greatest efficiency of movement through air. The flow of fluids around objects is represented by **streamlines,** as shown in **Figure 17.** Objects require less energy to move through a smooth streamlined flow.

Streamlines can best be illustrated by a simple demonstration. Imagine carefully squeezing tiny drops of food coloring into a smoothly flowing fluid. If the colored lines that form stay thin and well defined, the flow is said to be streamlined. Notice that if the flow narrows, the streamlines move closer together. Closely spaced streamlines indicate greater velocity and, therefore, reduced pressure. If streamlines swirl and become diffused, the flow of the fluid is said to be turbulent. Bernoulli's principle does not apply to turbulent flow.

Figure 17 The streamlines show the air flowing above a car in a wind tunnel.

Check Your Progress

32. **Buoyancy** All soda cans contain the same volume of liquid, 354 mL, and displace the same volume of water. What might be a difference between a can that sinks and one that floats? *Hint: Place a full can of regular soda and a full can of diet soda in water.*

33. **Transmission of Pressure** A toy rocket launcher is designed so that a child stomps on a rubber cylinder, which increases the air pressure in a launching tube and pushes a foam rocket into the sky. If the child stomps with a force of 150 N on a 2.5×10^{-3}-m^2 area piston, what is the additional force transmitted to the 4.0×10^{-4}-m^2 launch tube?

34. **Floating in Air** A helium balloon rises because of the buoyant force of the air lifting it. The density of helium is 0.18 kg/m^3, and the density of air is 1.3 kg/m^3. How large a volume would a helium balloon need to lift the lead brick shown in **Figure 18?**

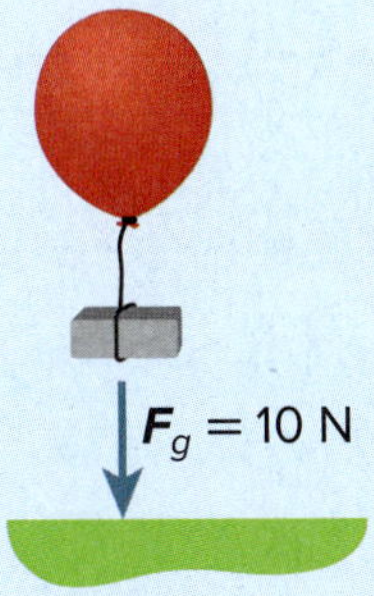

Figure 18

35. **Floating and Density** A fishing bobber made of cork floats with one-tenth of its volume below the water's surface. What is the density of cork?

36. **Pressure and Force** An automobile weighing 2.3×10^4 N is lifted by a hydraulic cylinder with an area of 0.15 m^2.
    a. What is the pressure in the hydraulic cylinder?
    b. The pressure in the lifting cylinder is produced by pushing on a 0.0082-m^2 cylinder. What force must be exerted on this small cylinder to lift the automobile?

37. **Volume Displacement** Which displaces more water when placed in a pool?
    a. a 1.0-kg block of aluminum or a 1.0-kg block of lead
    b. a 10-cm^3 block of aluminum or a 10-cm^3 block of lead

38. **Critical Thinking** A tornado passing over a house sometimes makes the house explode from the inside out. How might Bernoulli's principle explain this phenomenon? What could be done to reduce the danger of a door or window exploding outward?

LEARNSMART® Go online to follow your personalized learning path to review, practice, and reinforce your understanding.

LESSON 4
SOLIDS

FOCUS QUESTION

How are solids different from fluids?

Solid Bodies

Solids are stiff, can be cut, and retain their shapes. Liquids flow and do not retain their shapes. When the temperature of a liquid is lowered, the average kinetic energy of the particles decreases. As the particles slow down, the cohesive forces have more effect, and for many solids, the particles become frozen into a fixed pattern called a **crystal lattice,** shown in **Figure 19.** Although the cohesive forces hold the particles in place, the particles in a crystalline solid do not stop moving completely. Rather, they vibrate around their fixed positions. In other materials, such as butter and glass, the particles do not form a fixed crystalline pattern. A substance that has no regular crystal structure but does have a definite volume and shape is called an **amorphous solid.**

Pressure and freezing As a liquid becomes a solid, its particles usually fit more closely together than in the liquid state, making solids more dense than liquids. Also, an increase in the pressure on the surface of the liquid increases its freezing point. However, water is an exception. It is most dense at 4°C. Because water expands as it freezes, an increase in pressure forces the molecules closer together and opposes the freezing. Therefore, higher pressure lowers the freezing point of water very slightly.

Figure 19 As the temperature of water is lowered and it changes from a liquid to a solid, the particles are frozen in a pattern called a crystal lattice.

Don Paulson Photography/Purestock/SuperStock

Get It?

Compare For most substances, how do the densities of the solid and liquid compare?

3D THINKING DCI Disciplinary Core Ideas CCC Crosscutting Concepts SEP Science & Engineering Practices

COLLECT EVIDENCE

Use your Science Journal to record the evidence you collect as you complete the readings and activities in this lesson.

INVESTIGATE

GO ONLINE to find these activities and more resources.

Quick Investigation: Jumpers
Carry out an investigation into the expansion rates for different solids.

Review the News
Obtain information from a current news story about the properties or behavior of solids. Evaluate your source and communicate your findings to your class.

Elasticity of solids External forces applied to a solid object may twist, stretch, or bend it out of shape. The ability of a solid object to return to its original form when the external forces are removed is called the elasticity of the solid. If too much deformation occurs, the object will not return to its original shape because its elastic limit has been exceeded. Elasticity is a property of each substance and depends on the forces holding its particles together. Malleability and ductility are two properties that depend on the structure and elasticity of a substance. Because gold can be flattened and shaped into thin sheets, it is said to be malleable. Copper is a ductile metal because it can be pulled into thin strands of wire.

Get It?

Identify List one product or device that makes use of the property of malleability and one that makes use of the property of ductility.

Thermal Expansion of Solids

It is standard practice for engineers to design small gaps, called expansion joints, into concrete-and-steel highway bridges to allow for the expansion of parts in the heat of summer. Expansion joints are shown in **Figure 20.** Objects expand only a small amount when they are heated, but that small amount could be several centimeters in a 100-m-long bridge. If expansion gaps were not present, the bridge could buckle, or parts of it could break.

Some materials, such as the ovenproof glass used for laboratory experiments and cooking, are designed to have the least possible thermal expansion. Large telescope mirrors are made of a ceramic material that is designed to undergo essentially no thermal expansion.

To understand the expansion of heated solids, picture a solid as a collection of particles connected by springs that represent the attractive forces between the particles. When the particles get too close, the springs push them apart. When a solid is heated, the kinetic energy of the particles increases, and they vibrate rapidly and move farther apart, weakening the attractive forces between the particles. As a result, when the particles vibrate more violently with increased temperature, their average separation increases and the solid expands.

Figure 20 Expansion joints are included when bridges, roads, and train tracks are built.

Infer *If there were no expansion joints in this road, what might happen during the summer?*

STEM CAREER Connection

Industrial Designer

Have you bought a product because you liked its design? Did you ever have a product break and wonder what other material the manufacturer could have used? An industrial designer develops the products people use every day, such as toys, phones, and cars. They must consider both the aesthetics of the product and the durability of materials.

The change in length of a solid is proportional to the change in temperature, as shown in **Figure 21.** A solid will expand in length twice as much when its temperature is increased by 20°C than when it is increased by 10°C. The expansion also is proportional to its length. This means that a 2-m bar will expand twice as much as a 1-m bar with the same change in temperature. The length (L_2) of a solid at temperature T_2 can be found using the following relationship, where L_1 is the length at temperature T_1 and alpha (α) is the coefficient of linear expansion. The **coefficient of linear expansion** is equal to the change in length divided by the original length and the change in temperature.

$$L_2 = L_1 + \alpha L_1(T_2 - T_1)$$

With some algebra, you can solve for α.

$$L_2 - L_1 = \alpha L_1(T_2 - T_1)$$

$$\Delta L = \alpha L_1 \Delta T$$

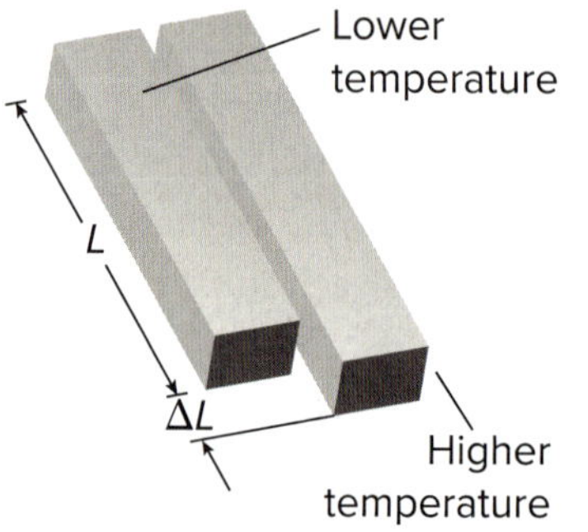

Figure 21 The change in length of a material is proportional to its original length and the change in temperature.

Coefficient of Linear Expansion

The coefficient of linear expansion is equal to the change in length divided by the product of the original length and the change in temperature.

$$\alpha = \frac{\Delta L}{L_1 \Delta T}$$

The unit for the coefficient of linear expansion is the reciprocal of degrees Celsius (which can be written as 1/°C or °C^{-1}). The **coefficient of volume expansion** is equal to the change in volume divided by the product of the original volume and the change in temperature. The coefficient of volume expansion (β) is generally about three times the coefficient of linear expansion.

Coefficient of Volume Expansion

The coefficient of volume expansion is equal to the change in volume divided by the product of the original volume and the change in temperature.

$$\beta = \frac{\Delta V}{V_1 \Delta T}$$

The unit for β is also 1/°C (°C^{-1}). The two coefficients of thermal expansion for a variety of materials are given in **Table 2.** (See reference table in Student Resources for additional materials.)

Table 2 **Coefficients of Thermal Expansion at 20°C**

Material	Coefficient of Linear Expansion, α(°C^{-1})	Coefficient of Volume Expansion, β(°C^{-1})
Solids		
Aluminum	23×10^{-6}	69×10^{-6}
Glass (soft)	9×10^{-6}	27×10^{-6}
Glass (ovenproof)	3×10^{-6}	9×10^{-6}
Concrete	12×10^{-6}	36×10^{-6}
Copper	17×10^{-6}	51×10^{-6}
Liquids		
Methanol	Not Applicable	1200×10^{-6}
Gasoline	Not Applicable	950×10^{-6}
Water	Not Applicable	210×10^{-6}

EXAMPLE Problem 4

LINEAR EXPANSION A metal bar is 1.60 m long at room temperature (21°C). The bar is put into an oven and heated to a temperature of 84°C. It is then measured and found to be 1.7 mm longer. What is the coefficient of linear expansion of this material?

1 ANALYZE AND SKETCH THE PROBLEM

- Sketch the bar, which is 1.7 mm longer at 84°C than at 21°C.
- Identify the initial length of the bar (L_1) and the change in length (ΔL).

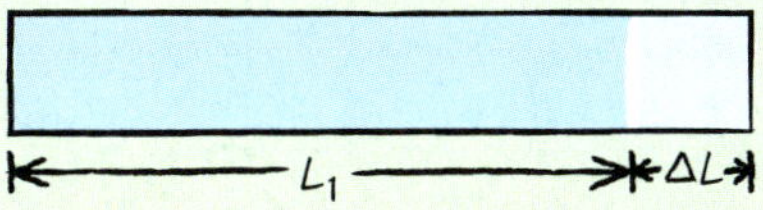

Known	Unknown
$L_1 = 1.60$ m	$\alpha = ?$
$\Delta L = 1.7\times10^{-3}$ m	
$T_1 = 21°C$	
$T_2 = 84°C$	

2 SOLVE FOR THE UNKNOWN

Calculate the coefficient of linear expansion using the relationship among known length, change in length, and change in temperature.

$$\alpha = \frac{\Delta L}{L_1 \Delta T}$$

$$= \frac{1.7\times10^{-3}\ \text{m}}{(1.60\ \text{m})(84°\text{C} - 21°\text{C})}$$

Substitute $\Delta L = 1.7\times10^{-3}$ m, $L_1 = 1.60$ m, $\Delta T = (T_2 - T_1) = 84°C - 21°C$

$$= 1.7\times10^{-5}\ °\text{C}^{-1}$$

3 EVALUATE THE ANSWER

- **Are the units correct?** The units are correctly expressed in $°C^{-1}$.
- **Is the magnitude realistic?** The magnitude of the coefficient is close to the accepted value for copper.

PRACTICE Problems

ADDITIONAL PRACTICE

39. A piece of aluminum house siding is 3.66 m long on a cold winter day of −28°C. How much longer is it on the hot summer day shown in **Figure 22?**

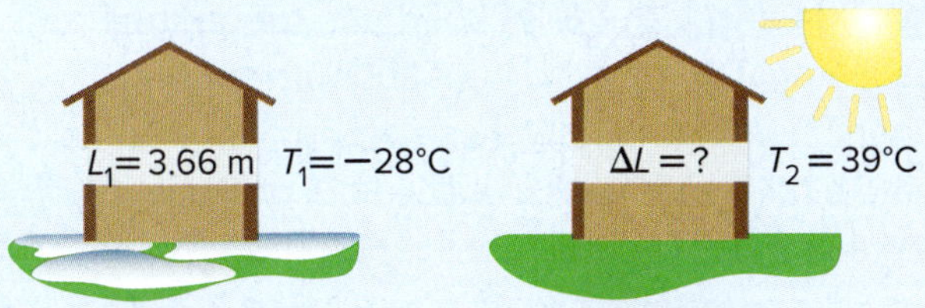

Figure 22

40. A piece of steel is 11.5 cm long at 22°C. It is heated to 1221°C, close to its melting temperature. How long is it?

41. A 400-mL glass beaker at room temperature is filled to the brim with cold water at 4.4°C. When the water warms up to 30.0°C, how much water will spill from the beaker?

42. A tank truck takes on a load of 45,725 L of gasoline in Houston, where the temperature is 28.0°C. The truck delivers its load in Minneapolis, where the temperature is −12.0°C.

a. How many liters of gasoline does the truck deliver?

b. What happened to the gasoline?

43. A hole with a diameter of 0.85 cm is drilled into a steel plate. At 30.0°C, the hole exactly accommodates an aluminum rod of the same diameter. What is the spacing between the plate and the rod when they are cooled to 0.0°C?

44. **CHALLENGE** A steel ruler is marked in millimeters so that the ruler is absolutely correct at 30.0°C. By what percentage would the ruler be incorrect at −30.0°C?

PHYSICS Challenge

You need to make a 1.00-m-long bar that expands with temperature in the same way as a 1.00-m-long bar of copper would. As shown in the figure at the right, your bar must be made from a bar of iron and a bar of aluminum attached end to end. How long should each of the bars be?

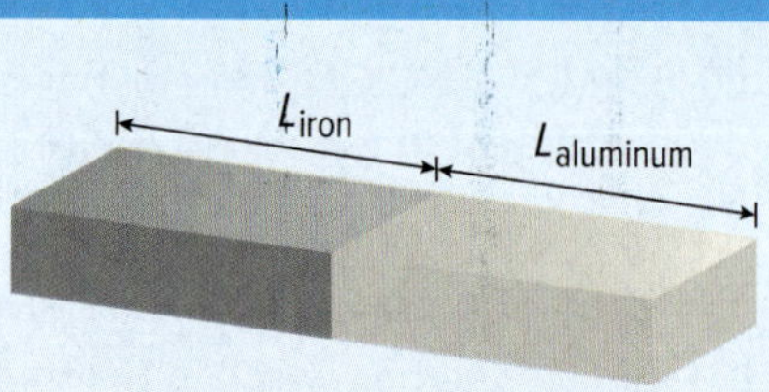

Applications of thermal expansion Engineers take the thermal expansion of materials into consideration as they design structures. You've already read about the expansion joints that are installed on concrete highways and bridges. The regular gaps between slabs of concrete in sidewalks also help keep sidewalks from buckling when the concrete expands during hot weather. Different materials expand at different rates, as indicated by the different coefficients of expansion given in **Table 2.** Engineers also consider different expansion rates when designing systems. Steel bars are often used to reinforce concrete. The steel and concrete must have the same expansion coefficient. Otherwise, the structure could crack on a hot day. Similarly, filling materials used to repair teeth must expand and contract at the same rate as tooth enamel.

Different rates of expansion have useful applications. For example, engineers have taken advantage of these differences to construct a useful device called a bimetallic strip, which is used in thermostats. A bimetallic strip consists of two strips of different metals welded or riveted together. Usually, one strip is brass, and the other is iron. When heated, brass expands more than iron does. Thus, when the bimetallic strip of brass and iron is heated, the brass part of the strip becomes longer than the iron part. As a result, the bimetallic strip bends with the brass on the outside of the curve. If the bimetallic strip is cooled, it bends in the opposite direction. The brass is then on the inside of the curve.

In a home thermostat, the bimetallic strip is installed so that it bends toward an electric contact as the room cools. When the room cools below the setting on the thermostat, the bimetallic strip bends enough to make electric contact with the switch, which turns on the heater. As the room warms, the bimetallic strip bends in the other direction. When the room's temperature reaches the setting on the thermostat, the electric circuit is broken, and the heater switches off.

Check Your Progress

45. **Thermal Expansion** On a hot day, you are installing an aluminum screen door in a concrete doorframe. You want the door to fit well on a cold winter day. Should you make the door fit tightly in the frame or leave extra room?
46. **Types of Solids** What is the difference between the structure of candle wax and that of ice?
47. **Thermal Expansion** Can you heat a piece of copper enough to double its length?
48. **States of Matter** Does **Table 2** provide a way to distinguish between solids and liquids?
49. **Solids and Liquids** A solid can be defined as a material that can be bent and will resist bending. Explain how these properties relate to the binding of atoms in a solid but do not apply to a liquid.
50. **Critical Thinking** The iron ring in **Figure 23** was made by cutting a small piece from a solid ring. If the ring in the figure is heated, will the gap become wider or narrower? Explain your answer.

Figure 23

A Solid That Flows

Solids, liquids, and gases are the three most familiar and abundant states of matter on Earth. Plasma, a fourth state of matter that is essentially an ionized gas, is the state of matter most common throughout the universe. However, other states of matter exist. Recently, two different research groups succeeded in making an unusual state of matter that had previously been only hypothetical: a supersolid.

Sodium, shown here as a solid, was one of the metals used to make supersolids.

How to Make a Supersolid

You can lower the temperature of liquid water to make ice, but it is far more difficult to make a supersolid. A supersolid is a state of matter that has particles arranged in a regular, repeating pattern, like a solid, but are also frictionless and can flow. Scientists have been trying for decades to produce one. In 2004, a group of scientists thought they had produced one using helium-4, but later testing debunked the discovery.

A Case of Multiple Discovery

In 2017, two independent teams of scientists, one from MIT in Cambridge, Massachusetts, and one from a research institute in Zurich, Switzerland, published results in a scientific journal showing they had produced supersolids. Although they used different elements and different methods, the fact that they both achieved similar results provides strong evidence that they succeeded.

The MIT team used sodium, cooling it to nearly absolute zero. When sodium becomes this cold, it forms another state of matter called a Bose-Einstein condensate. A Bose-Einstein condensate is a superfluid, which is a fluid with particles that have zero viscosity. The researchers then used an ultra-high-vacuum chamber and lasers to cause the sodium atoms to form a crystalline pattern.

The Swiss team also began with a Bose-Einstein condensate, but they used rubidium as a starting material instead of sodium. Their method for causing the particles of the superfluid to form a crystalline arrangement involved mirrors as well as lasers.

To provide evidence that the atoms had formed a supersolid, both teams used lasers to show that the material produced a diffraction pattern characteristic of a solid crystal.

No applications exist yet for the new state of matter, but the discovery may lead to insights that will improve existing superconductor technology. Further work by the research groups and others will aim to provide more evidence for and uncover further information about, the properties of the supersolid state.

COMMUNICATE SCIENTIFIC IDEAS

Two independent teams made supersolids nearly simultaneously. Research to find another example of multiple discovery. Share your findings by writing a blog post.

CALIFORNIA CONNECTION

A research group at USC in Los Angeles studies helium superfluids. Find out more about superfluids, and make a table to compare their properties to those of supersolids.

MODULE 12
STUDY GUIDE

GO ONLINE to study with your Science Notebook.

Lesson 1 PROPERTIES OF FLUIDS

- Matter in the fluid state flows and has no definite shape of its own.
- The combined gas law represents the relationships among the pressure, volume, and temperature of gases.

$$\frac{P_1V_1}{T_1} = \frac{P_2V_2}{T_2}$$

- The ideal gas law represents the relationships among pressure, volume, temperature, and the number of moles of a gas.
- Plasma is a gaslike state of negatively charged electrons and positively charged ions.

- fluids
- pressure
- pascal
- combined gas law
- ideal gas law
- thermal expansion
- plasma

Lesson 2 FORCES WITHIN LIQUIDS

- Surface tension is the tendency of the surface of a liquid to contract to the smallest possible area. Surface tension results from the attractive forces that like particles exert on one another.
- Capillary action occurs when a liquid rises in a thin tube because the adhesive forces between the tube and the liquid are stronger than the cohesive forces between the liquid's molecules.
- Clouds form when water vapor in the atmosphere cools and condenses, forming droplets around dust particles.

- cohesive forces
- adhesive forces

Lesson 3 FLUIDS AT REST AND IN MOTION

- Pascal's principle states that an applied pressure change is transmitted undiminished throughout a fluid.
- According to Archimedes' principle, the buoyant force equals the weight of the fluid displaced by an object.
- Bernoulli's principle states that the pressure exerted by a fluid decreases as its velocity increases.

- Pascal's principle
- buoyant force
- Archimedes' principle
- Bernoulli's principle
- streamlines

Lesson 4 SOLIDS

- A crystalline solid has a regular pattern of particles, and an amorphous solid has an irregular pattern of particles. Malleability and ductility depend on structure type.
- As the temperature of a solid changes, the kinetic energy of its particles changes accordingly. As the vibrations of the particles change, a solid expands as temperature increases and contracts as temperature decreases.
- Expansion rates of different materials must be considered when designing structures.

THREE-DIMENSIONAL THINKING

Module Wrap-Up

REVISIT THE PHENOMENON

How will the fluid flow through this device?

CER Claim, Evidence, Reasoning

Explain Your Reasoning Revisit the claim you made when you encountered the phenomenon. Summarize the evidence you gathered from your investigations and research and finalize your Summary Table. Does your evidence support your claim? If not, revise your claim. Explain why your evidence supports your claim.

STEM UNIT PROJECT

Now that you've completed the module, revisit your STEM unit project. You will apply your evidence from this module and complete your project.

GO FURTHER

SEP Data Analysis Lab

Sinking or floating?

Five objects with the following densities are placed in a tank of water. The diagram shows six possible positions of the objects.

Object	Density (g/cm^3)
A	0.85
B	0.95
C	1.05
D	1.15
E	1.25

CER Analyze and Interpret Data

1. **Claim** Select a position, from 1 to 6, for each of the five objects. Not all positions need to be selected.
2. **Evidence and Reasoning** Explain your selections.

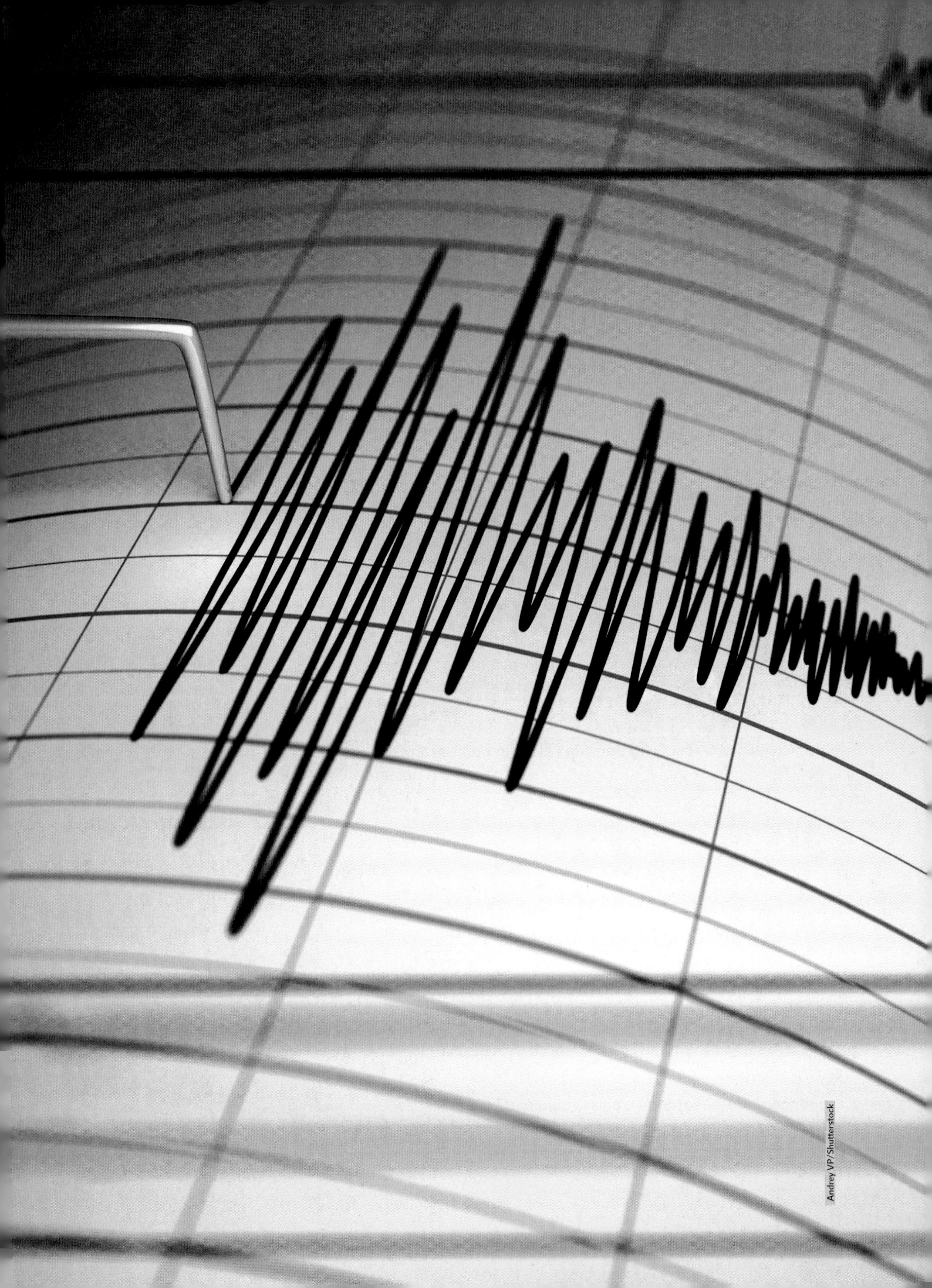

Andrey VP/Shutterstock

UNIT 4

WAVES AND LIGHT

ENCOUNTER THE PHENOMENON

How do waves affect our everyday lives?

SEP Ask Questions

What questions do you have about the phenomenon? Write your questions on sticky notes and add them to the driving question board for this unit.

Look for Evidence

As you go through this unit, use the information and your experiences to help you answer the phenomenon question as well as your own questions. For each activity, record your observations in a Summary Table, add an explanation, and identify how it connects to the unit and module phenomenon questions.

Solve a Problem

STEM UNIT PROJECT

What's inside Earth? Scientists cannot simply dig down thousands of kilometers to determine the composition of Earth's interior. Investigate how scientists use waves to understand Earth's structure. Use the results of these investigations and your knowledge from this unit to explain the evidence for the current understanding of Earth's interior.

GO ONLINE In addition to reading the information in your Student Edition, you can find the STEM Unit Project and other useful resources online.

Kayte Deioma/Alamy Stock Photo

MODULE 13
VIBRATIONS AND WAVES

ENCOUNTER THE PHENOMENON

How can this pendulum could save a building from earthquake damage?

GO ONLINE to watch a video about how dampers are used to prevent earthquake damage.

SEP Ask Questions

Do you have other questions about the phenomenon? If so, add them to the driving question board.

CER Claim, Evidence, Reasoning

Make Your Claim Use your CER chart to make a claim about how a pendulum could save a building from earthquake damage. Explain your reasoning.

Collect Evidence Use the lessons in this module to collect evidence to support your claim. Record your evidence as you move through the module.

Explain Your Reasoning You will revisit your claim and explain your reasoning at the end of the module.

GO ONLINE to access your CER chart and explore resources that can help you collect evidence.

LESSON 1: Explore & Explain: Pendulums

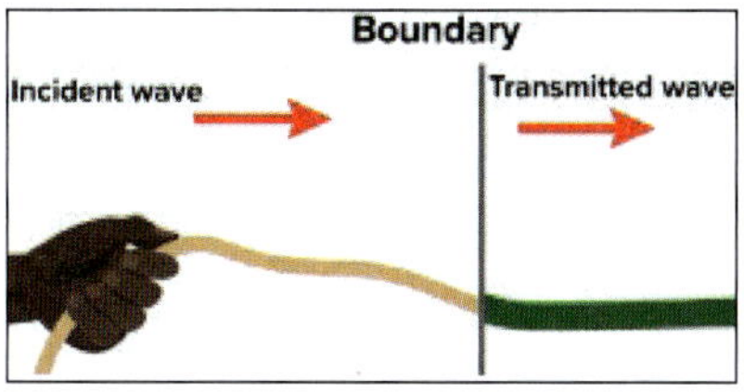

LESSON 2: Explore & Explain: Waves at Boundaries

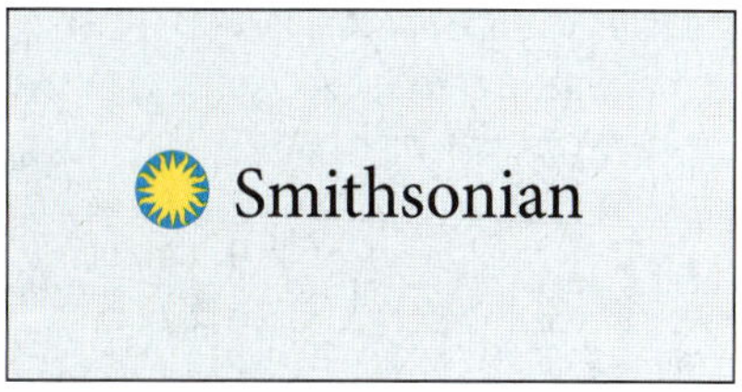

Additional Resources

LESSON 1
PERIODIC MOTION

FOCUS QUESTION

What are some types of repetitive motion?

Mass on a Spring

The bobbing of a mass on a spring or the swaying of a pendulum are examples of **periodic motion.** In each example, at one position the net force on the object is zero and the object is in equilibrium. When the object moves away from its equilibrium position, the net force on the system becomes nonzero. This net force acts to bring the object back toward equilibrium. The **period** (T) is the time needed for one full cycle of the motion. The **amplitude** of the motion is the maximum distance the object moves from the equilibrium position.

Simple harmonic motion In **Figure 1,** the force exerted by the spring is directly proportional to the distance it is stretched. When pulled down and released, the mass bobs up and down through equilibrium. Any system where the force acting to restore an object to its equilibrium position is directly proportional to the object's displacement shows **simple harmonic motion.**

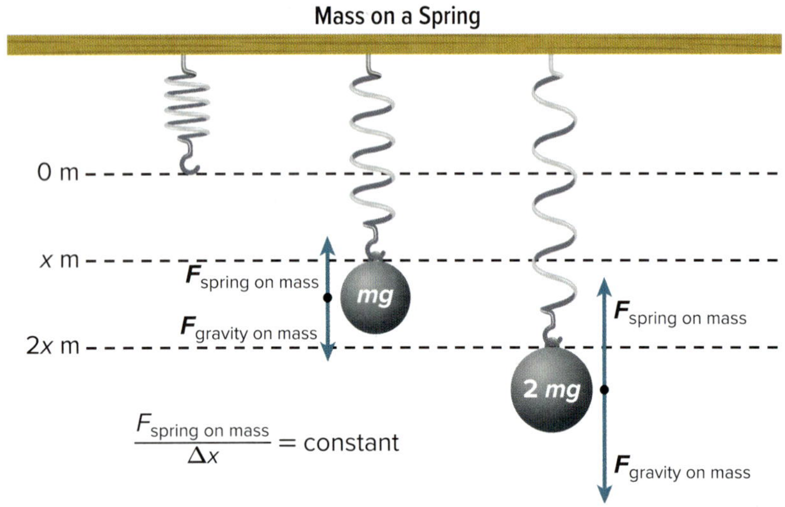

Figure 1 The force exerted on the mass by the spring is directly proportional to the mass's displacement.

Determine *the displacement if the mass is 0.5 mg.*

3D THINKING | DCI Disciplinary Core Ideas | CCC Crosscutting Concepts | SEP Science & Engineering Practices

COLLECT EVIDENCE

Use your Science Journal to record the evidence you collect as you complete the readings and activities in this lesson.

INVESTIGATE

GO ONLINE to find these activities and more resources.

PhysicsLAB: Pendulum Vibrations

Plan and carry out an investigation to determine what variables affect a pendulum's period and use the resulting data to calculate the magnitude of the gravitation field.

Revisit the Encounter the Phenomenon Question

What information from this lesson can help you answer the Unit and Module questions?

Hooke's Law

Table 1 **Force Magnitude-Stretch Distance in a Spring**

Stretch Distance (m)	Magnitude of Force Exerted by Spring (N)
0.0	0.0
0.030	1.9
0.060	3.7
0.090	6.3
0.12	7.8

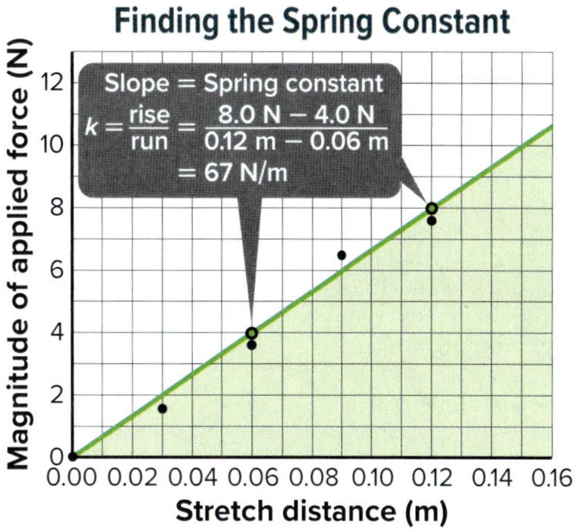

Figure 2 The spring constant can be determined from the slope of the force magnitude-stretch distance graph. The area under the curve is equal to the potential energy stored in the spring.

Hooke's Law

Table 1 shows the relationship between the magnitude of the force exerted by a spring and the distance the spring stretches. **Figure 2** is a graph of the data with the line of best fit. The linear relationship indicates that the magnitude of the force exerted by the spring is directly proportional to the amount the spring is stretched. A spring that exerts a force directly proportional to the distance stretched obeys **Hooke's law.**

Hooke's Law

The magnitude of the force exerted by a spring is equal to the spring constant times the distance the spring is stretched or compressed from its equilibrium position.

$$F = -kx$$

In this equation, k is the spring constant, which depends on the stiffness and other properties of the spring, and x is the distance the spring is stretched from its equilibrium position. Notice that k is the slope of the line in the magnitude of the force v. stretch distance graph. A steeper slope—a larger k—indicates that the spring is harder to stretch. The constant k has the same units as the slope, newtons/meter (N/m). The negative sign in Hooke's law indicates that the force is in the direction opposite the stretch or compression direction. The force exerted by the spring on the mass is always directed toward the spring's equilibrium position.

Hooke's law and real springs Not all springs obey Hooke's law. For example, rubber bands do not. Those that do obey Hooke's law are called elastic springs. Even for elastic springs, Hooke's law only applies over a limited range of distances. If a spring is stretched too far, it can become so deformed that the force is no longer proportional to the displacement.

Potential energy When you stretch a spring you transfer energy to the spring, giving it elastic potential energy. The work done by an applied force is equal to the area under a force v. distance graph like the one shown in **Figure 2.** This work is equal to the elastic potential energy stored in the spring. To calculate this stored energy, find the area of the triangle by multiplying one-half the base of the triangle, which is x, by the height of the triangle. According to Hooke's law, the height of the triangle—the magnitude of the force—is equal to kx.

Potential Energy in a Spring

The potential energy in a spring is equal to one-half times the product of the spring constant and the square of the displacement.

$$PE_{spring} = -\frac{1}{2}kx^2$$

This mathematical expression, which quantifies how the stored energy in a spring depends on its configuration, together with kinetic energy calculated from mass and speed data, allows the concept of the conservation of energy to be used to predict and describe system behavior. As shown in **Figure 3** on the next page, during horizontal simple harmonic motion the spring's elastic potential energy is converted to kinetic energy and then back to potential energy.

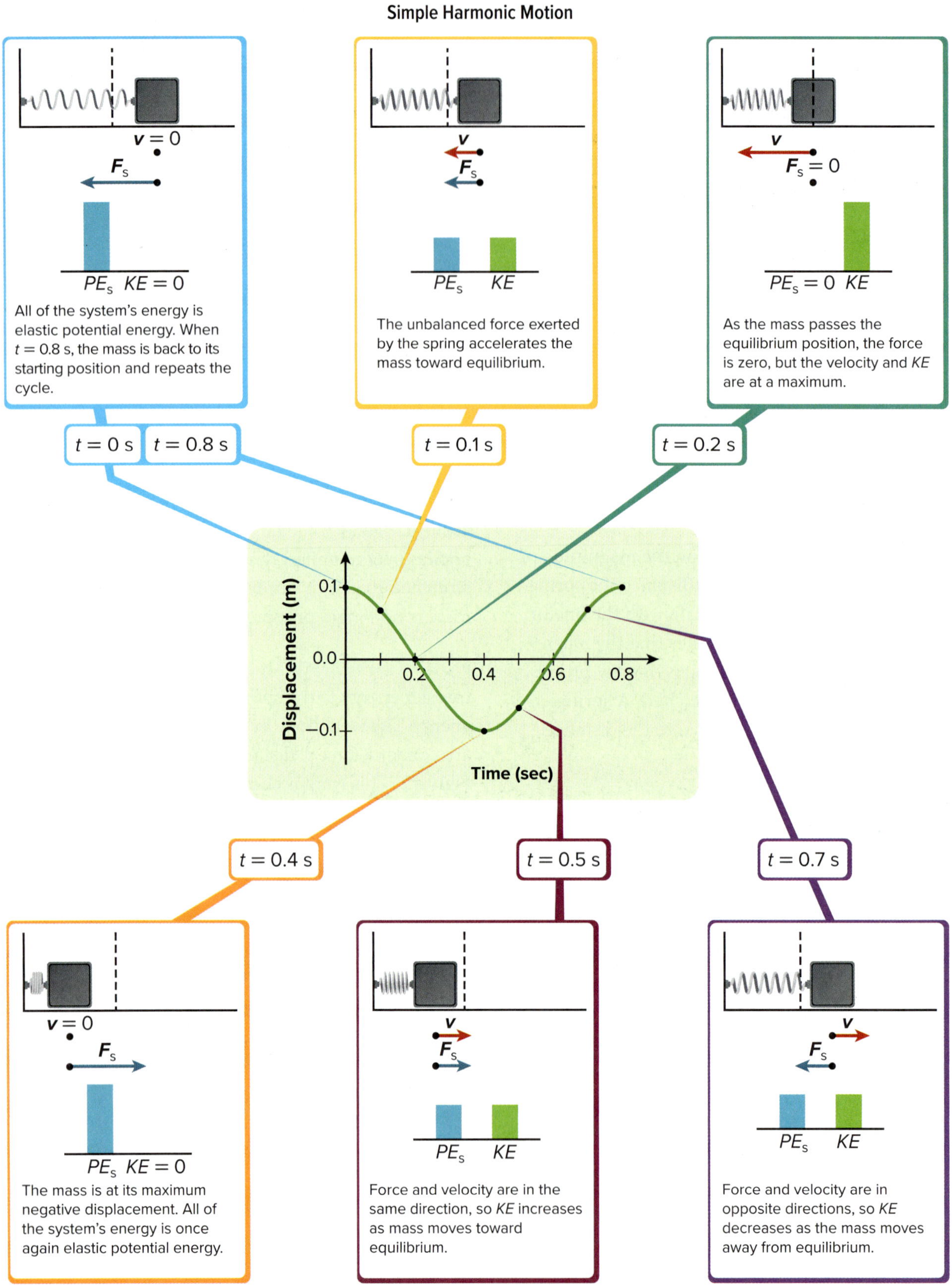

Figure 3 The total mechanical energy of the system is constant throughout the oscillation.

Explain *How can Hooke's law and the concept of conservation of energy be used to predict and describe the behavior of a system such as this?*

EXAMPLE Problem 1

THE SPRING CONSTANT AND THE ENERGY OF A SPRING A spring stretches by 18 cm when a bag of potatoes weighing 56 N is suspended from its end.

a. Determine the spring constant.

b. How much elastic potential energy does the stretched spring have?

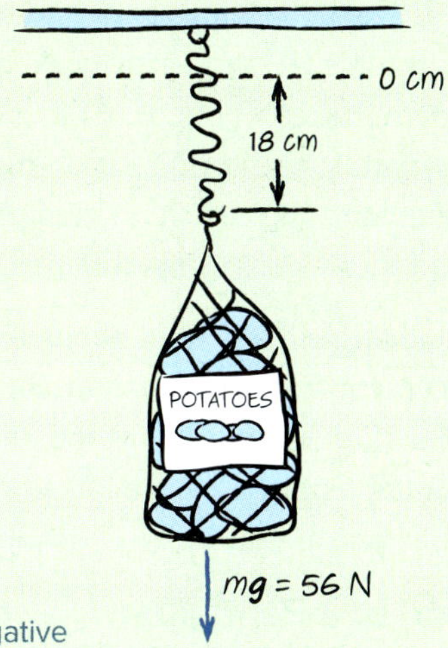

1 ANALYZE AND SKETCH THE PROBLEM

- Sketch the situation.
- Show and label the distance the spring has stretched and its equilibrium position.

Known	Unknown
$x = 18$ cm	$k = ?$
$F = -56$ N	$PE_s = ?$

2 SOLVE FOR THE UNKNOWN

a. Use Hooke's law and isolate k.

$$k = -\frac{F}{x}$$

$$= -\frac{-56\text{ N}}{0.18\text{ m}}$$ Substitute $F = -56$ N, $x = 0.18$ m. The force is negative because it is in the opposite direction of x.

$$= 310\text{ N/m}$$

b. $$PE_s = \frac{1}{2}kx^2$$

$$= \frac{1}{2}(310\text{ N/m})(0.18\text{ m})^2$$ Substitute $k = 310$ N/m, $x = 0.18$ m.

$$= 5.0\text{ J}$$

3 EVALUATE THE ANSWER

- **Are the magnitudes realistic?** N/m is the correct unit for the spring constant. $(\text{N/m})(\text{m}^2) = \text{N}\cdot\text{m} = \text{J}$, which is the correct unit for energy.
- **Is the magnitude realistic?** The average magnitude of the force the spring exerts is the average of 0 and 56 N. The work done is $W = Fx = (28\text{ N})(0.18\text{ m}) = 5.0$ J.

PRACTICE Problems

ADDITIONAL PRACTICE

1. What is the spring constant of a spring that stretches 12 cm when an object weighing 24 N is hung from it?
2. What is the elastic potential energy of a spring with $k = 144$ N/m that is compressed by 16.5 cm?
3. A spring has a spring constant of 56 N/m. How far will a block weighing 18 N stretch it?
4. **CHALLENGE** A spring has a spring constant of 256 N/m. How far must it be stretched to give it an elastic potential energy of 48 J?

PHYSICS Challenge

A car of mass m rests at the top of a hill of height h before rolling without friction into a crash barrier located at the bottom of the hill. The crash barrier contains a spring with a spring constant k, which is designed to bring the car to rest with minimum damage.

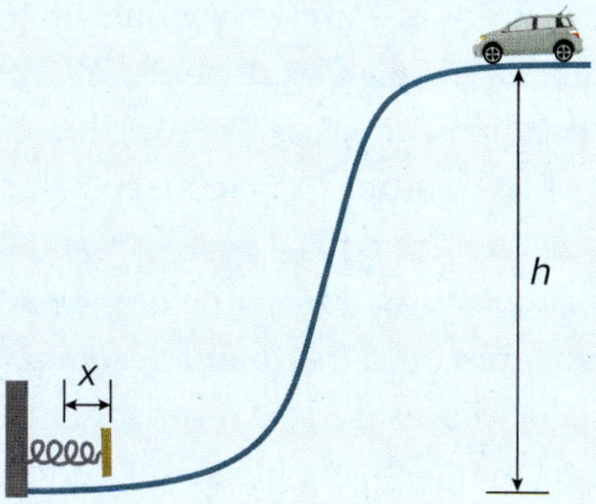

1. Determine, in terms of m, h, k, and g, the maximum distance (x) the spring will be compressed when the car hits it.
2. If the car rolls down a hill that is twice as high, by what factor will the spring compression increase?
3. What will happen after the car has been brought to rest?

Pendulums

Simple harmonic motion also occurs in the swing of a pendulum. A **simple pendulum** consists of a massive object, called the bob, suspended by a string or a light rod of length ℓ. The bob swings back and forth, as shown in **Figure 4.** The string or rod exerts a tension force (F_T), and gravity exerts a force (F_g) on the bob. Throughout the pendulum's path, the component of the gravitational force in the direction of the pendulum's circular path is a restoring force. At the left and right positions, the restoring force is at a maximum and the velocity is zero. At the equilibrium position, the restoring force is zero and the velocity is maximum.

For small angles (less than about 15°), the restoring force is proportional to the displacement from equilibrium. Similar to the motion of the mass on a spring discussed earlier, the motion of the pendulum is simple harmonic motion. The period of a pendulum is given by the following equation.

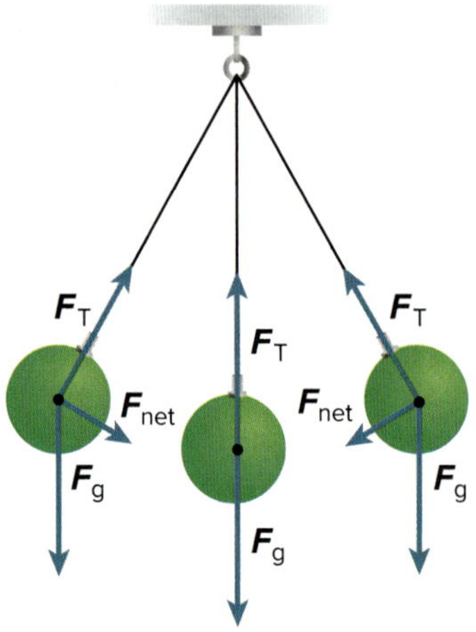

Figure 4 The pendulum's motion is an example of simple harmonic motion because the restoring force is directly proportional to the displacement from equilibrium.

Period of a Pendulum

The period of a pendulum is equal to 2π times the square root of the length of the pendulum divided by the gravitational field.

$$T = 2\pi\sqrt{\frac{\ell}{g}}$$

Notice that the period depends only on the length of the pendulum and the gravitational field, not on the mass of the bob or the amplitude of oscillation. One practical use of the pendulum is to measure g, which can vary slightly at different locations on Earth.

Get It?

Compare the period of a very massive pendulum, like the one shown at the beginning of the module, with the period of a pendulum with the same length but a tiny mass.

Resonance

To get a playground swing going, you can "pump" it by leaning back and pulling the chains at the same point in each swing. Another option is to have a friend give you repeated pushes at just the right times. **Resonance** occurs when forces are applied to a vibrating or oscillating object at time intervals equal to the period of oscillation. As a result, the amplitude of the vibration increases. Other familiar examples of resonance include rocking a car to free it from a snow bank and jumping rhythmically on a trampoline or a diving board to go higher.

Resonance in simple harmonic motion systems causes a larger and larger displacement as energy is added in small increments. As a child you may have been told to hold a seashell such as a conch up to your ear to "hear the sound of the ocean." The sound you hear when you hold a seashell or other similar-shaped object up to your ear actually comes from resonance. Sound waves resulting from background noise in the room interact with the seashell. Sounds with frequencies matching one of the natural frequencies at which the seashell vibrates result in resonance, and the sound becomes amplified and loud enough to hear. The large amplitude oscillations caused by resonance can also produce useful results. Resonance is used in musical instruments to amplify sounds and in clocks to increase accuracy.

EXAMPLE Problem 2

FINDING *g* USING A PENDULUM A pendulum with a length of 36.9 cm has a period of 1.22 s. What is the gravitational field at the pendulum's location?

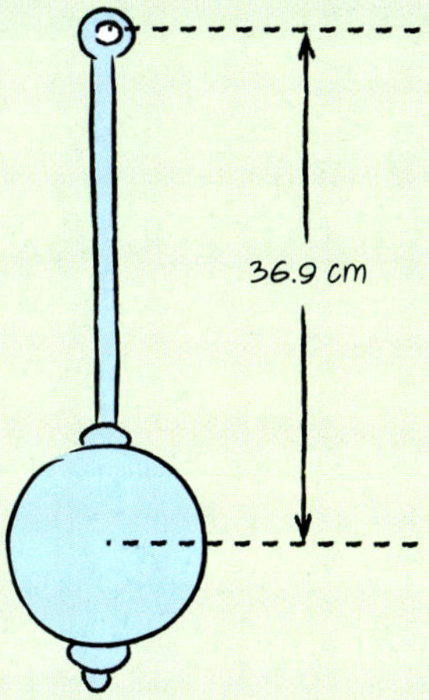

1 ANALYZE AND SKETCH THE PROBLEM

- Sketch the situation.
- Label the length of the pendulum.

Known	Unknown
$\ell = 36.9$ cm	$g = ?$
$T = 1.22$ s	

2 SOLVE FOR THE UNKNOWN

$$T = 2\pi\sqrt{\frac{\ell}{g}}$$ Solve for g.

$$g = (2\pi)^2\frac{\ell}{T^2}$$

$$= \frac{4\pi^2(0.369\text{ m})}{(1.22\text{ s})^2}$$ Substitute $l = 0.369$ m, $T = 1.22$ s.

$$= 9.78\text{ m/s}^2 = 9.78\text{ N/kg}$$

3 EVALUATE THE ANSWER

- **Are the magnitudes realistic?** N/kg is the correct unit for gravitational field.
- **Is the magnitude realistic?** The calculated value of g is quite close to the accepted value of g, 9.8 N/kg. This pendulum could be at a higher elevation above sea level.

PRACTICE Problems

ADDITIONAL PRACTICE

5. What is the period on Earth of a pendulum with a length of 1.0 m?

6. How long must a pendulum be on the Moon, where $g = 1.6$ N/kg, to have a period of 2.0 s?

7. **CHALLENGE** On a certain planet, the period of a 0.75-m-long pendulum is 1.8 s. What is g for this planet?

Check Your Progress

8. **Periodic Motion** Explain why a pendulum is an example of periodic motion.

9. **Energy of a Spring** The springs shown in **Figure 5** are identical. Contrast the potential energies of the bottom two springs.

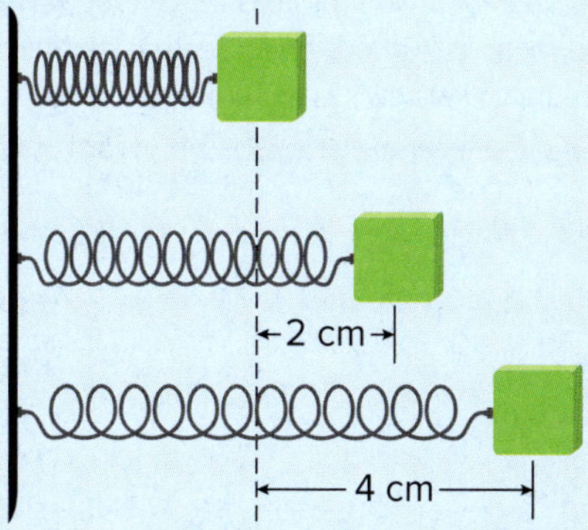

Figure 5

10. **Hooke's Law** Objects of various weights are hung from a rubber band that is suspended from a hook. The weights of the objects are plotted on a graph against the stretch of the rubber band. How can you tell from the graph whether the rubber band obeys Hooke's law?

11. **Pendulum** How must the length of a pendulum be changed to double its period? How must the length be changed to halve the period?

12. **Resonance** If a car's wheel is out of balance, the car will shake strongly at a specific speed but not at a higher or lower speed. Explain.

13. **Critical Thinking** How is uniform circular motion similar to simple harmonic motion? How are they different?

LEARNSMART® Go online to follow your personalized learning path to review, practice, and reinforce your understanding.

LESSON 2
WAVE PROPERTIES

FOCUS QUESTION

What are some common types of waves?

Mechanical Waves

A **wave** is a disturbance that carries energy through matter or space without transferring matter. You have learned how Newton's laws of motion and the law of conservation of energy govern the behavior of particles. These laws also govern the behavior of waves. Water waves, sound waves, and the waves that travel along a rope or a spring are mechanical waves. Mechanical waves pass through a physical medium, such as water, air, or a rope.

Transverse waves

A **wave pulse** is a single bump or disturbance that passes through a medium. In the left panel of **Figure 6,** the wave pulse disturbs the rope in the vertical direction, but the pulse travels horizontally. A wave that disturbs the particles in the medium perpendicular to the direction of the wave's travel is called a **transverse wave.** If the disturbances continue at a constant rate, a **periodic wave** is generated.

Longitudinal waves

In a coiled spring toy, you can produce another type of wave. If you squeeze together several turns of the coiled spring toy and then suddenly release them, pulses will move away in both directions. The result is called a **longitudinal wave** because the disturbance is parallel to the direction of the wave's travel. Sound waves are longitudinal waves in which the molecules are alternately compressed or decompressed along the path of the wave.

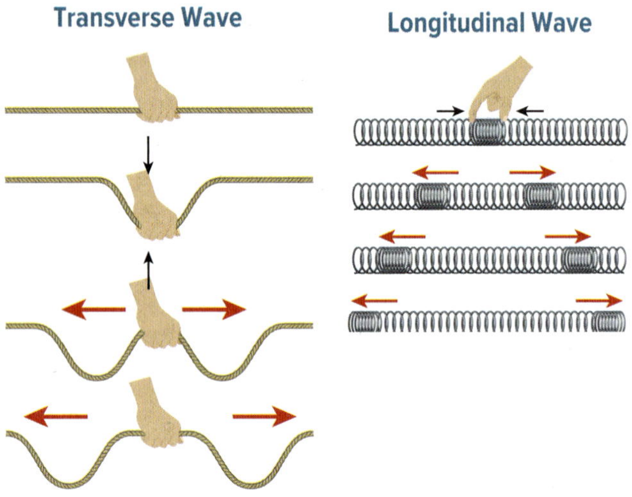

Figure 6 Shaking a rope up and down produces transverse wave pulses traveling in both directions. Squeezing and releasing the coils of a spring produces longitudinal wave pulses in both directions.

Explain *the difference between transverse and longitudinal waves.*

3D THINKING | DCI Disciplinary Core Ideas | CCC Crosscutting Concepts | SEP Science & Engineering Practices

COLLECT EVIDENCE

Use your Science Journal to record the evidence you collect as you complete the readings and activities in this lesson.

INVESTIGATE

GO ONLINE to find these activities and more resources.

Applying Practices: Wave Characteristics

HS-PS4-1. Use mathematical representations to support a claim regarding relationships among the frequency, wavelength, and speed of waves traveling in various media.

CCC **Identify Crosscutting Concepts**

Create a table of the crosscutting concepts and fill in examples you find as you read.

Figure 7 Surface waves in water cause movement both parallel and perpendicular to the direction of wave motion. When these waves interact with the shore, the regular, circular motion is disrupted and the waves break on the beach.

Surface waves

Waves that are deep in a lake or an ocean are longitudinal. In a **surface wave,** however, the medium's particles follow a circular path that is at times parallel to the direction of travel and at other times perpendicular to the direction of wave travel, as shown in **Figure 7.** Surface waves set particles in the medium, in this case water, moving in a circular pattern. At the top and bottom of the circular path, particles are moving parallel to the direction of the wave's travel. This is similar to a longitudinal wave. At the left and right sides of each circle, particles are moving up or down. This up-and-down motion is perpendicular to the wave's direction, similar to a transverse wave.

Real-World Physics

Tsunamis

On March 11, 2011, a wall of water estimated to be ten meters high hit areas on the east coast of Japan—tsunami! A tsunami is a series of ocean waves that can have wavelengths over 100 km, periods of one hour, and wave speeds of 500–1000 km/h.

Wave Properties

Many types of waves share a common set of wave properties. Some wave properties depend on how the wave is produced, whereas others depend on the medium through which the wave is passing.

Amplitude

How does the pulse generated by gently shaking a rope differ from the pulse produced by a violent shake? The difference is similar to the difference between a ripple in a pond and an ocean breaker—they have different amplitudes. You read earlier that the amplitude of periodic motion is the greatest distance from equilibrium. Similarly, as shown in **Figure 8,** a transverse wave's amplitude is the maximum distance of the wave from equilibrium. Since amplitude is a distance, it is always positive. You will learn more about measuring the amplitude of longitudinal waves when you study sound.

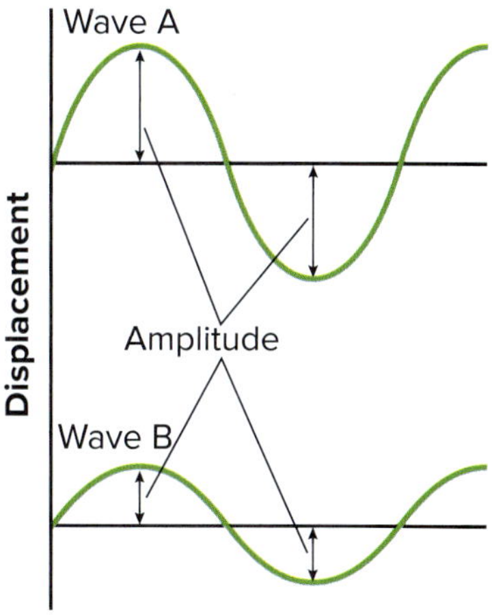

Figure 8 A wave's amplitude is measured from the equilibrium position to the highest or lowest point on the wave.

Energy of a wave

Waves, including water waves, are examples of the many ways that energy manifests at the macroscopic scale. The energy of a wave is related to its amplitude, and a wave's amplitude depends on how the wave is generated. More energy must be added to the system to generate a wave with a greater amplitude. For example, strong winds produce larger water waves than those formed by gentle breezes. Waves with greater amplitudes transfer more energy. Whereas a wave with a small amplitude might move sand on a beach a few centimeters, a giant wave can uproot and move a tree.

For waves that move at the same speed, the rate at which energy is transferred is proportional to the square of the amplitude. Thus, doubling the amplitude of a wave increases the amount of energy that wave transfers each second by a factor of four.

Wavelength Rather than focusing on one point on a wave, imagine taking a snapshot of the wave so you can see the whole wave at one instant in time. The top image in **Figure 9** shows each low point on a transverse wave, called a **trough,** and each high point on a transverse wave, called a **crest,** of a wave. The shortest distance between points where the wave pattern repeats itself is called the **wavelength.** Crests are spaced by one wavelength. Each trough also is one wavelength from the next. The Greek letter lambda (λ) represents wavelength.

Speed and velocity How fast does a wave travel? The speed of a wave pulse can be found in the same way the speed of a moving car is determined. First, measure the displacement of one of the wave's crests or compressions (Δd), then divide this by the time interval (Δt) to find the speed.

$$v = \frac{\Delta d}{\Delta t}$$

For most mechanical waves, transverse and longitudinal, except water surface waves, wave speed does not depend on amplitude, frequency, or wavelength. Speed depends only on the medium through which the waves are passing and on the type of wave. For example, sound waves have characteristic speeds in water and in air. The temperature of the medium also makes a difference. Sound waves travel faster in warm, dry air than in cool, dry air.

Get It?

Summarize how changing a wave's amplitude, frequency, or wavelength affects the wave's speed.

Phase Any two points on a wave that are one or more whole wavelengths apart are said to be in phase. Particles in the medium are in phase with one another when they have the same displacement from the equilibrium position and the same velocity. Particles with opposite displacements from the equilibrium position and opposite velocities are 180° out of phase. A crest and a trough, for example, are 180° out of phase with each other. Two particles in a wave medium can be anywhere from 0° to 360° out of phase with each other.

Period Although wave speed and amplitude can describe both wave pulses and periodic waves, period (T) applies only to periodic waves. You have learned that the period of simple harmonic motion, such as the motion of a simple pendulum, is the time it takes for the motion to complete one cycle. Such motion is usually the source, or cause, of a periodic wave. The period of a wave is equal to the period of the source. In **Figure 9** the period (T) equals 0.04 s, which is the time it takes the source to complete one cycle. The same time is taken by P, a point on the rope, to return to its initial position and velocity.

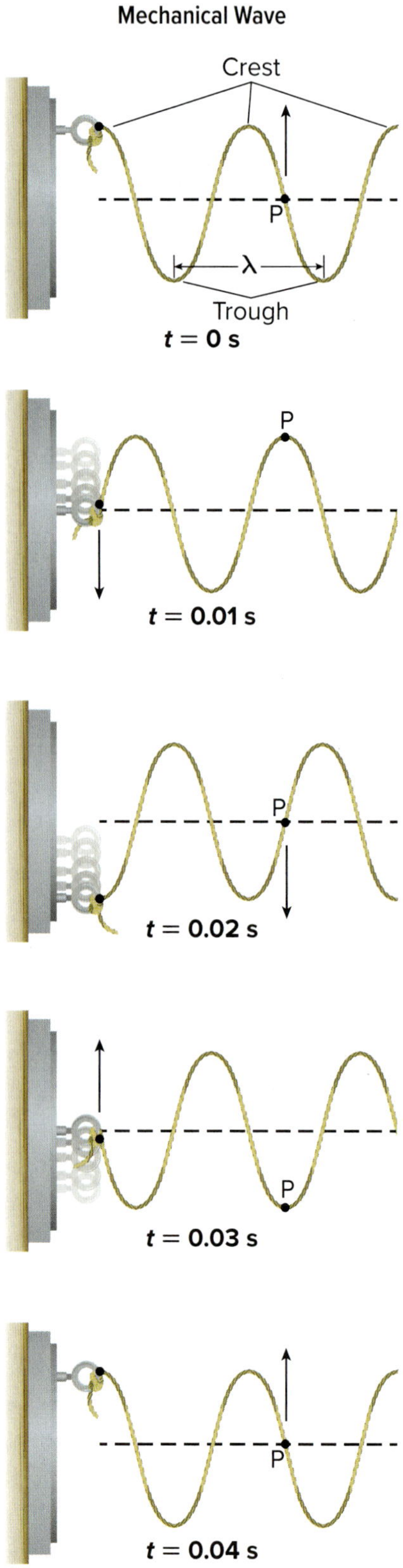

Figure 9 A mechanical oscillator moves the left end of the rope up and down, completing the cycle in 0.04 s.

Calculating frequency The **frequency** of a wave (f) is the number of complete oscillations a point on that wave makes each second. Frequency is measured in hertz (Hz). One hertz is one oscillation per second and is equal to 1/s or s^{-1}. The frequency and the period of a wave are related by the following equation.

Frequency of a Wave

The frequency of a wave is equal to the reciprocal of the period.

$$f = \frac{1}{T}$$

Both the period and the frequency of a wave depend only on the wave's source. They do not depend on the wave's speed or the medium.

Calculating wavelength You can directly measure a wave's wavelength by measuring the distance between adjacent crests or troughs. You can also calculate it because the wavelength and frequency of a wave are related to one another by the speed of travel of the wave, which depends on the type of wave and the medium through which it is passing. In the time interval of one period, a wave moves one wavelength. Therefore, the wavelength of a wave is the speed multiplied by the period, $\lambda = vT$. Using the relation that $f = \frac{1}{T}$, the wavelength equation is very often written in the following way.

Wavelength

The wavelength of a wave is equal to the velocity divided by the frequency.

$$\lambda = \frac{v}{f}$$

Graphing waves If you took a snapshot of a transverse wave on a coiled spring toy, it might look like one of the waves shown in **Figure 9.** This snapshot could be placed on a graph grid to show more information about the wave, as in the left panel of **Figure 10.** Measuring from peak to peak or trough to trough on such a snapshot provides the wavelength. Now consider recording the motion of a single particle, such as point P in **Figure 9.** That motion can be plotted on a displacement v. time graph, as in the graph on the right in **Figure 10.** Measuring from peak to peak or trough to trough in this graph provides the wave's period.

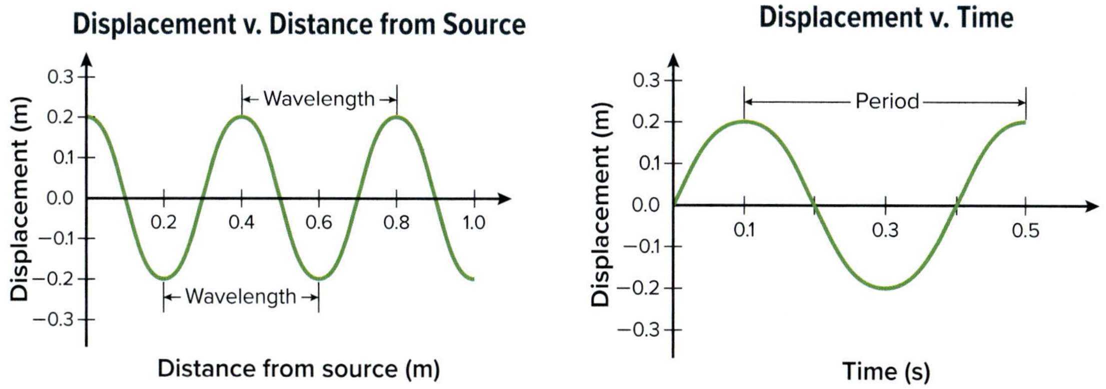

Figure 10 Graphing waves on different axes provides different kinds of information.

Determine *the period of the wave shown in the displacement v. time graph.*

EXAMPLE Problem 3

CHARACTERISTICS OF A WAVE A sound wave has a frequency of 192 Hz and travels the length of a football field, 91.4 m, in 0.271 s.

a. What is the speed of the wave?

b. What is the wavelength of the wave?

c. What is the period of the wave?

d. If the frequency were changed to 442 Hz, what would be the new wavelength and period?

1 ANALYZE AND SKETCH THE PROBLEM

- Draw a diagram of the wave.
- Draw a velocity vector for the wave.

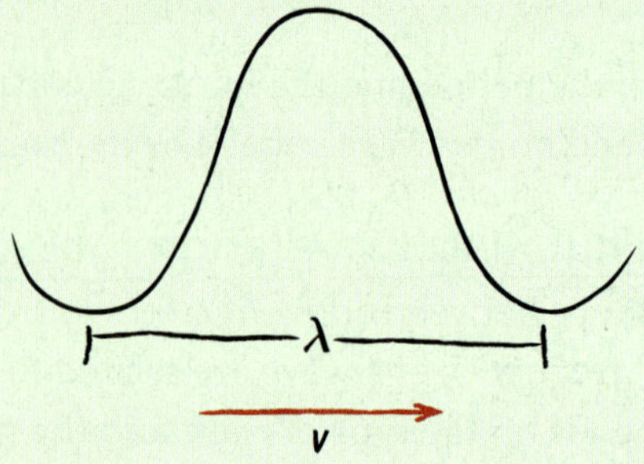

Known	Unknown
$f = 192\text{ Hz}$	$v = ?$
$\Delta d = 91.4\text{ m}$	$\lambda = ?$
$\Delta t = 0.271\text{ s}$	$T = ?$

2 SOLVE FOR THE UNKNOWN

a. Use the definition of velocity.

$$v = \frac{\Delta d}{\Delta t}$$

$$= \frac{91.4\text{ m}}{0.271\text{ s}} \quad \text{Substitute } \Delta d = 91.4\text{ m}, \Delta t = 0.271\text{ s}$$

$$= 337\text{ m/s}$$

b. Use the relationship between wave velocity, wavelength, and frequency.

$$\lambda = \frac{v}{f}$$

$$= \frac{337\text{ m/s}}{192\text{ Hz}} \quad \text{Substitute } v = 337\text{ m/s}, f = 192\text{ Hz.}$$

$$= 1.76\text{ m}$$

c. Use the relationship between period and frequency.

$$T = \frac{1}{f}$$

$$= \frac{1}{192\text{ Hz}} \quad \text{Substitute } f = 192\text{ Hz.}$$

$$= 0.00521\text{ s}$$

d. $\lambda = \frac{v}{f}$

$$= \frac{337\text{ m/s}}{442\text{ Hz}} \quad \text{Substitute } v = 337\text{ m/s}, f = 442\text{ Hz.}$$

$$= 0.762\text{ m}$$

$$T = \frac{1}{f}$$

$$= \frac{1}{442\text{ Hz}} \quad \text{Substitute } f = 442\text{ Hz.}$$

$$= 0.00226\text{ s}$$

3 EVALUATE THE ANSWER

- **Are the magnitudes realistic?** Hz has the unit s^{-1}, so $\left(\frac{m}{s}\right)/\text{Hz} = \left(\frac{m}{s}\right)\cdot s = m$, which is correct for λ.
- **Are the magnitudes realistic?** A typical sound wave travels at approximately 340 m/s in air, so 337 m/s is reasonable. The frequencies and periods are reasonable for sound waves. The frequency of 442 Hz is close to a 440-Hz A, which is A above middle-C on a piano.

PRACTICE Problems

ADDITIONAL PRACTICE

14. A sound wave produced by a clock chime is heard 515 m away 1.50 s later.

a. Based on these measurements, what is the speed of sound in air?

b. The sound wave has a frequency of 436 Hz. What is the period of the wave?

c. What is its wavelength?

15. How are the wavelength, frequency, and speed of a wave related? How do they depend on the medium through which the wave is passing and the type of wave?

16. What is the speed of a periodic wave disturbance that has a frequency of 3.50 Hz and a wavelength of 0.700 m?

17. How does increasing the wavelength by 50 percent affect the frequency of a wave on a rope?

18. The speed of a transverse wave in a string is 15.0 m/s. If a source produces a disturbance that has a frequency of 6.00 Hz, what is its wavelength?

19. Five wavelengths are generated every 0.100 s in a tank of water. What is the speed of the wave if the wavelength of the surface wave is 1.20 cm?

20. A periodic longitudinal wave that has a frequency of 20.0 Hz travels along a coiled spring toy. If the distance between successive compressions is 0.600 m, what is the speed of the wave?

21. How does the frequency of a wave change when the period of the wave is doubled?

22. Describe the change in the wavelength of a wave when the period is reduced by one-half.

23. If the speed of a wave increases to 1.5 times its original speed while the frequency remains constant, how does the wavelength change?

24. CHALLENGE A hiker shouts toward a vertical cliff as shown in **Figure 11.** The echo is heard 2.75 s later.

a. What is the speed of sound of the hiker's voice in air?

b. The wavelength of the sound is 0.750 m. What is its frequency?

c. What is the period of the wave?

Figure 11

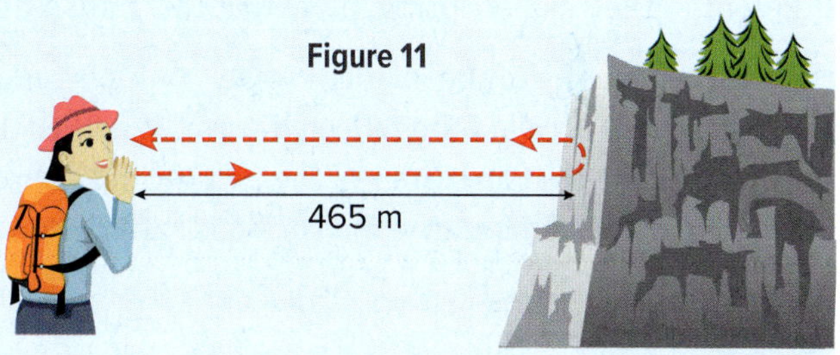

Check Your Progress

25. **Transverse Waves** Suppose you and your lab partner are asked to demonstrate that a transverse wave transports energy without transferring matter. How could you do it?

26. **Wave Characteristics** You are creating transverse waves on a rope by shaking your hand from side to side. Without changing the distance your hand moves, you begin to shake it faster and faster. What happens to the amplitude, wavelength, frequency, period, and velocity of the wave?

27. **Longitudinal Waves** Describe longitudinal waves. What types of mediums transmit longitudinal waves?

28. **Speeds in Different Mediums** If you pull on one end of a coiled spring toy, does the pulse reach the other end instantaneously? What happens if you pull on a rope? What happens if you hit the end of a metal rod? Compare the pulses that are traveling through each of these three materials.

29. **Critical Thinking** If a raindrop falls into a pool, it produces waves with small amplitudes. If a swimmer jumps into a pool, he or she produces waves with large amplitudes. Why doesn't the heavy rain in a thunderstorm produce large waves?

LEARNSMART® Go online to follow your personalized learning path to review, practice, and reinforce your understanding.

LESSON 3
WAVE BEHAVIOR

FOCUS QUESTION

What happens when two waves meet?

Waves at Boundaries

Recall from Lesson 2 that the speed of a mechanical wave depends only on the properties of the medium it passes through, not on the wave's amplitude or frequency. For example, for waves on a spring, the speed depends on the spring's tension and mass per unit length. Examine what happens when a wave travels from one medium to another. **Figure 12** shows a wave pulse traveling from a larger spring into a smaller one. The pulse that strikes the boundary is called the **incident wave.** One pulse from the larger spring continues in the smaller spring, but the speed is different in the smaller spring. Note that this transmitted wave pulse remains upward.

Some of the energy of the incident wave's pulse is reflected backward into the larger spring. This returning wave is called the reflected wave. Whether the **reflected wave** is upright or inverted depends on the characteristics of the two springs. For example, if the waves in the smaller spring have a greater speed because the spring is stiffer, then the reflected wave will be inverted.

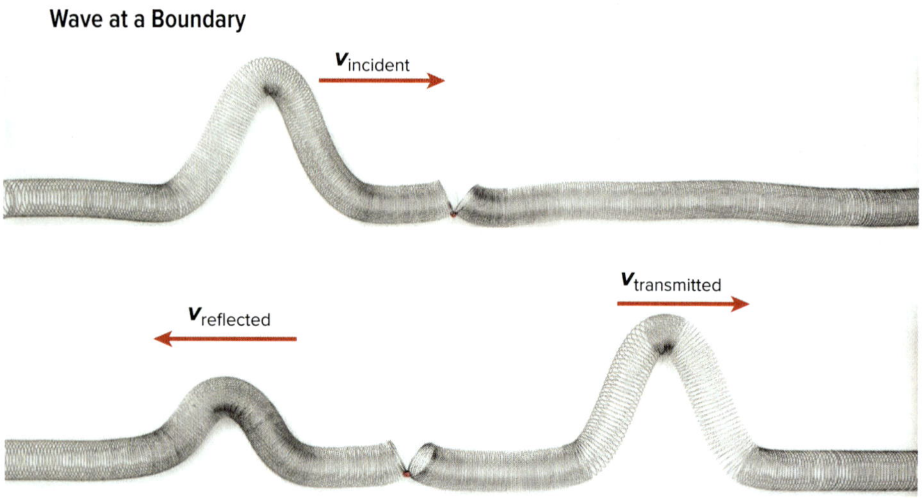

Figure 12 When the wave pulse meets the boundary between the two springs, a transmitted wave pulse and a reflected wave form.

Compare *the energy of the incident wave to the energy of the reflected wave.*

Matt Meadows/McGraw-Hill Education

3D THINKING — DCI Disciplinary Core Ideas — CCC Crosscutting Concepts — SEP Science & Engineering Practices

COLLECT EVIDENCE

Use your Science Journal to record the evidence you collect as you complete the readings and activities in this lesson.

INVESTIGATE

GO ONLINE to find these activities and more resources.

PhysicsLAB: Reflection and Refraction

Analyze data from wave patterns in a ripple tank to predict the behavior of waves in water.

Review the News

Obtain information from a current news story about mechanical waves, such as seismic or water waves. Evaluate your source and communicate your findings to your class.

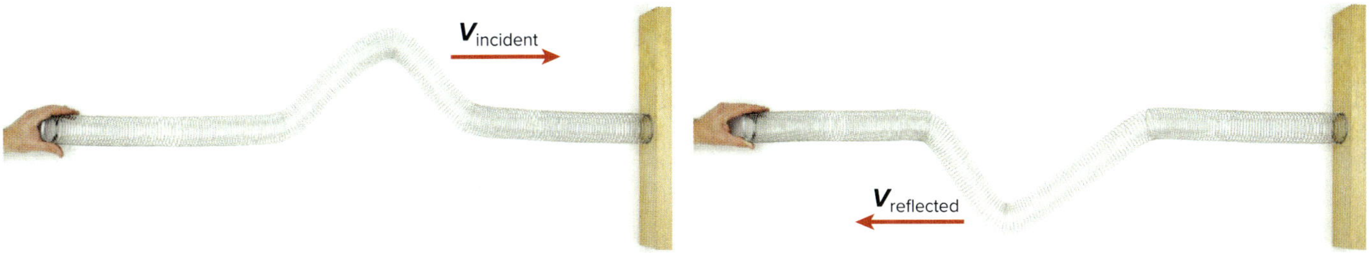

Figure 13 When a wave encounters a rigid boundary, the reflected wave is inverted. Note that the amplitude is not affected by the rigid boundary.

Rigid boundaries When a wave pulse hits a rigid boundary, the energy is reflected back, as shown in **Figure 13.** The wall is the boundary of a new medium through which the wave attempts to pass. Instead of passing through, the pulse is reflected from the wall with almost exactly the same amplitude as the pulse of the incident wave. Thus, almost all the wave's energy is reflected back. Very little energy is transmitted into the wall. Also note that the pulse is inverted.

Superposition of Waves

Suppose a pulse traveling along a spring meets a reflected pulse that is coming back from a boundary, as shown in **Figure 14.** In this case, two waves exist in the same place in the medium at the same time. Each wave affects the medium independently. The **principle of superposition** states that the displacement of a medium caused by two or more waves is the algebraic sum of the displacements caused by the individual waves. In other words, two or more waves can combine to form a new wave. If the waves move in the same medium, they can cancel or add or subtract to form a new wave of lesser or greater amplitude. They emerge unaffected by each other. The result of the superposition of two or more waves is called **interference.**

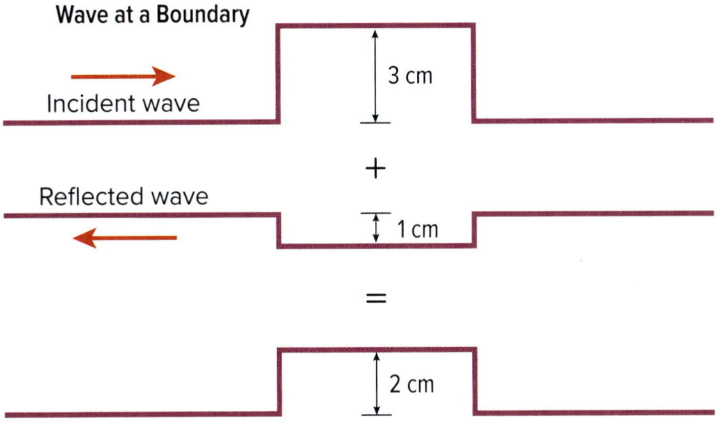

Figure 14 Waves add algebraically during superposition.

SCIENCE USAGE v. COMMON USAGE

Interference

Science usage: the result of superposition of two or more waves

The amplitude of the interference of several waves was much larger than the amplitude of the individual waves.

Common usage: the act of coming between in a way that hinders or impedes

Ehud was ejected from the game for an interference foul.

STEM CAREER Connection

Boat Designer

If you are drawn to the water, the career of boat designer might be for you. Whether designing ocean liners, freighters, yachts, or wakeboard boats, a boat designer must understand how a vessel will perform in various wave conditions.

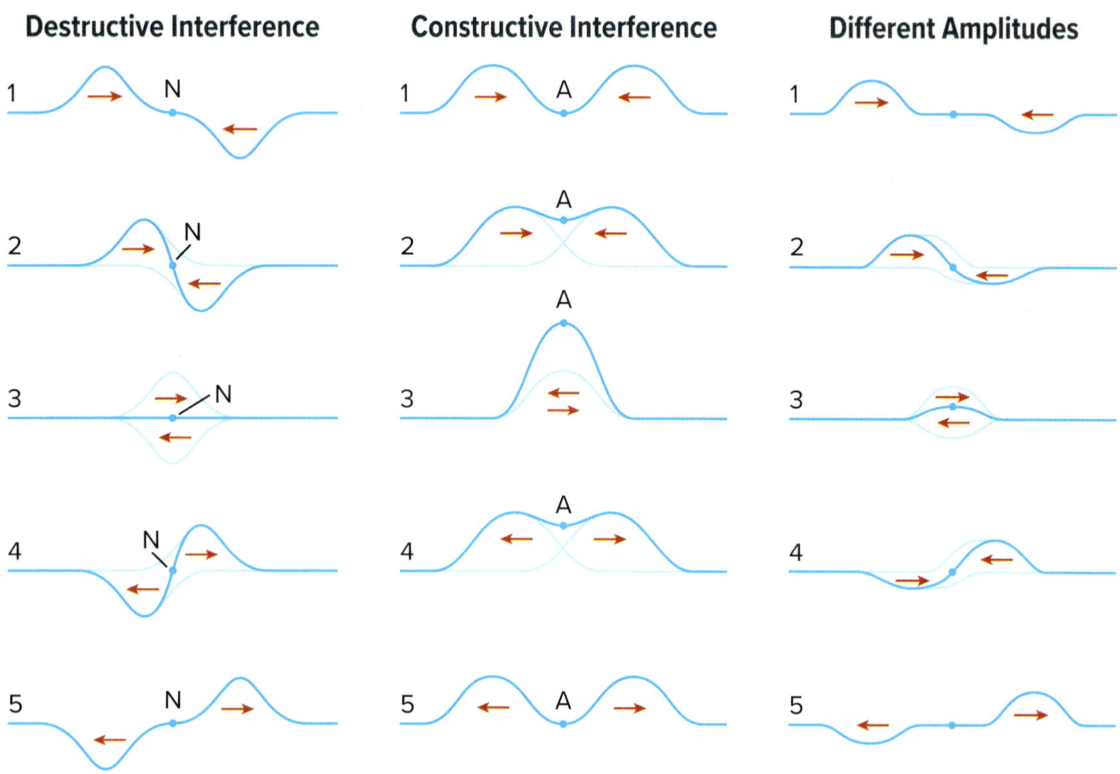

Figure 15 When waves add algebraically, the resulting combined waves can be quite different from the individual waves. **Summarize** *how waves behave during and after superposition.*

Wave interference Wave interference can be either constructive or destructive. The first panel in **Figure 15** shows the superposition of waves with equal but opposite displacements, causing destructive interference. When the pulses meet and are in the same location, the displacement is zero. Point N, which does not move at all, is called a **node.** The pulses travel horizontally and eventually emerge unaffected by each other.

Constructive interference occurs when wave displacements are in the same direction. The result is a wave that has an amplitude greater than those of the individual waves. A larger pulse appears at point A when the two waves meet. Point A has the largest displacement and is called the **antinode.** The two pulses pass through each other without changing their shapes or sizes. Even if the pulses have unequal amplitudes, the resultant pulse at the overlap is still the algebraic sum of the two pulses, as shown in the final panel of **Figure 15.**

Get It?

Compare the wave medium's displacement at a node and at an antinode.

Two reflections You can apply the concept of superimposed waves to the control of large-amplitude waves. Imagine attaching one end of a rope to a fixed point, such as a doorknob, a distance *L* away. When you vibrate the free end, the wave leaves your hand, travels along the rope toward the fixed end, is reflected and inverted at the fixed end, and returns to your hand. When it reaches your hand, the reflected wave is inverted and travels back down the rope. Thus, when the wave leaves your hand the second time, its displacement is in the same direction as it was when it left your hand the first time.

Standing waves

Suppose you adjust the motion of your hand so that the period of vibration equals the time needed for the wave to make one round-trip from your hand to the door and back. Then, the displacement given by your hand to the rope each time will add to the displacement of the reflected wave. As a result, the amplitude of oscillation of the rope will be much greater than the motion of your hand. This large-amplitude oscillation is an example of mechanical resonance.

The ends of the rope are nodes and an antinode is in the middle, as shown in the top photo in **Figure 16.** Thus, the wave appears to be standing still and is called a **standing wave.** Note, however, that the standing wave is the interference of waves traveling in opposite directions. If you double the frequency of vibration, you can produce one more node and one more antinode on the rope. Then it appears to vibrate in two segments. When you further increase the vibration frequency, it produces even more nodes and antinodes, as shown in the bottom photo in **Figure 16.**

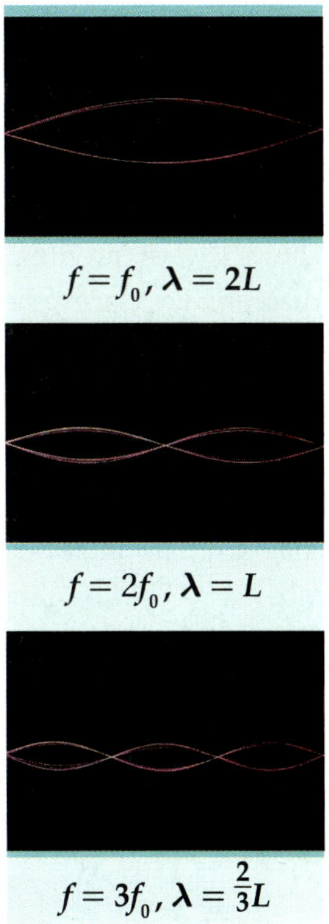

Figure 16 Interference produces standing waves only at certain frequencies.

Predict *the wavelength if the frequency is four times the lowest frequency.*

Waves in Two Dimensions

You have studied waves on a rope and on a spring reflecting from rigid supports. During some of these interactions, the amplitude of the waves is forced to be zero by destructive interference. These mechanical waves travel in only one dimension. Waves on the surface of water, however, travel in two dimensions, and sound waves and electromagnetic waves will later be shown to travel in three dimensions. How can two-dimensional waves be represented?

Picturing waves in two dimensions

When you throw a small stone into a calm pool of water, you see the circular crests and troughs of the resulting waves spreading out in all directions. You can sketch those waves by drawing circles to represent the wave crests. If you repeatedly dip your finger into water with a constant frequency, the resulting sketch would be a series of concentric circles, called wavefronts, centered on your finger. A **wavefront** is a line that represents the crest of a wave in two dimensions. Wavefronts can be used to show two-dimensional waves of any shape, including circular waves. The photo in **Figure 17** shows circular waves in water. The circles drawn on the diagram show the wavefronts that represent those water waves.

Figure 17 Waves spread out in a circular pattern from the oscillating source.

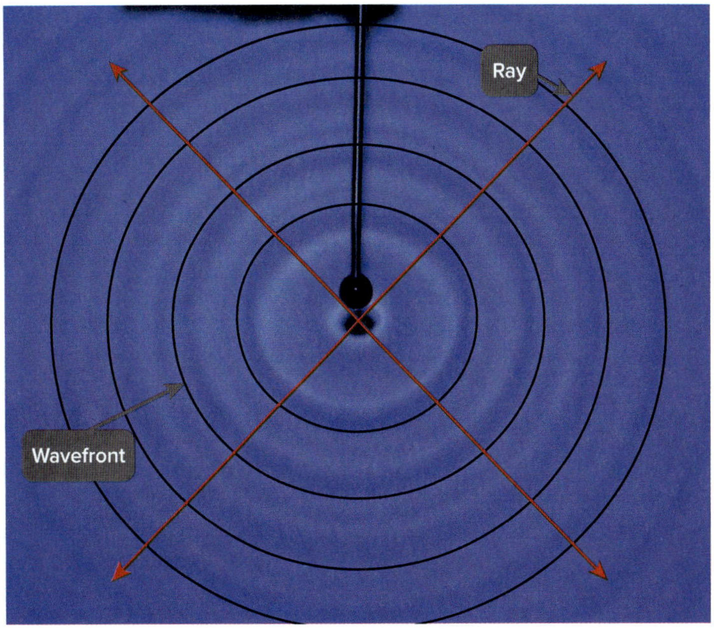

Whatever their shape, two-dimensional waves always travel in a direction that is perpendicular to their wavefronts. That direction can be represented by a **ray,** which is a line drawn at a right angle to the wavefront. When all you want to show is the direction in which a wave is traveling, it is convenient to draw rays instead of wavefronts. The red arrows in **Figure 17** are rays that show the water waves' direction of motion. One advantage of drawing wavefronts is when wavefronts are drawn to scale, they show the wave's wavelengths. In **Figure 17,** the wavelength equals the distance from one circle to the next.

Creating two-dimensional waves A ripple tank is a piece of laboratory equipment that is used to investigate the properties of two-dimensional waves. The main portion of the ripple tank shown in **Figure 18** is a shallow tank that contains a thin layer of water. A board attached to a mechanical oscillator produces waves with long, straight wavefronts. A lamp above the tank produces shadows below the tank that show the locations of the crests of the waves. The top photo in **Figure 18** shows a wave traveling through the ripple tank. The direction the wave travels is modeled by a ray diagram. For clarity, the wavefronts are not extended the entire length of the wave.

Reflection of two-dimensional waves The bottom row of pictures in **Figure 18** shows an incident ray encountering a rigid barrier placed at an angle to the ray's path. The orientation of the barrier is shown by a line, called the **normal,** drawn perpendicular to the barrier. The angle between the incident ray and the normal is called the angle of incidence and is labeled θ_i in the diagram. The angle between the normal and the reflected ray is called the angle of reflection and is labeled θ_r. The **law of reflection** states that the angle of incidence is equal to the angle of reflection. The law of reflection applies to many different kinds of waves, not just the waves in a ripple tank.

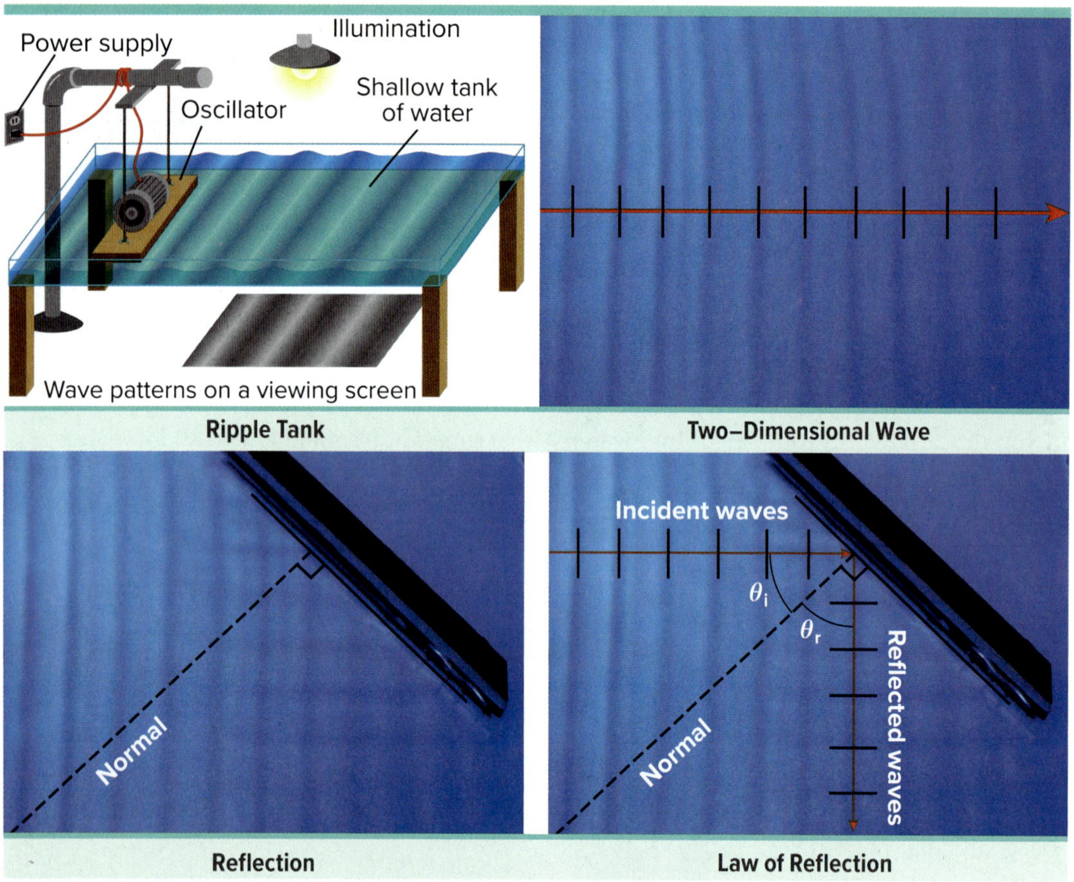

Figure 18 The ripple tank produces uniform waves that are useful for modeling wave behavior.

(t)Matt Meadows/McGraw-Hill Education; (b)McGraw-Hill Education

SCIENCE USAGE v. COMMON USAGE

Normal

Science usage: a line in a diagram that is drawn perpendicular to a surface
As measured from the normal, angles of incidence and reflection are equal.

Common usage: conforming to a type, standard, or regular pattern
Such cold temperatures in July are not normal.

CCC CROSSCUTTING CONCEPTS

Cause and Effect Write a procedure describing how you could use a ripple tank to gather empirical evidence to support the cause-and-effect relationship between the angle of a barrier and the angle of a reflected wave.

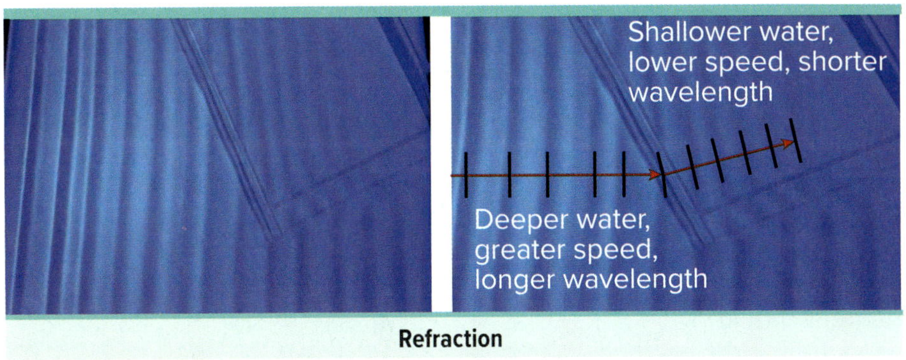

Figure 19 Waves in the ripple tank change direction as they enter shallower water.

Describe *how the wavelength changes as the wave travels into the shallow water.*

Refraction of waves in two dimensions A ripple tank can also model the behavior of waves as they travel from one medium into another. **Figure 19** shows a glass plate placed under the water in a ripple tank. The water above the plate is shallower than the water in the rest of the tank. As the waves travel from deep to shallow water, their speeds decrease and the direction of the waves changes. Such changes in speed are common when waves pass from one medium to another.

The waves in the shallow water are connected to the waves in the deep water. As a result, the frequency of the waves in the two mediums is the same. Based on the equation $\lambda = \frac{v}{f}$, the decrease in the speed of the waves means the wavelength is shorter in the shallower water. The change in the direction of waves at the boundary between two different mediums is known as **refraction. Figure 19** shows a wave front and ray model of refraction. Part of the wave will refract through the boundary, and part will be reflected from the boundary. Reflection and refraction occur for many different types of waves. Echoes are an example of reflection of sound waves by hard surfaces, such as the walls of a large gymnasium. Rainbows are the result of the reflection and refraction of light. As sunlight passes through a raindrop, reflection and refraction separate the light into its individual colors, producing a rainbow.

Check Your Progress

30. **Wave Characteristics** Which characteristics remain unchanged when a wave crosses a boundary into a different medium: frequency, amplitude, wavelength, velocity, direction?
31. **Superposition of Waves** Sketch two wave pulses whose interference produces a pulse with an amplitude greater than either of the individual waves.
32. **Refraction of Waves** In **Figure 19,** the wave changes direction as it passes from one medium to another. Can two-dimensional waves cross a boundary between two mediums without changing direction? Explain.
33. **Standing Waves** In a standing wave on a string fixed at both ends, how is the number of nodes related to the number of antinodes?
34. **Critical Thinking** As another way to understand wave reflection, cover the right-hand side of each drawing in the left panel in **Figure 15** with a piece of paper. The edge of the paper should be at point N, the node. Now, concentrate on the resultant wave, shown in darker blue. Note that it acts as a wave reflected from a boundary. Is the boundary a rigid wall? Repeat this exercise for the middle panel in **Figure 15.**

LEARNSMART® Go online to follow your personalized learning path to review, practice, and reinforce your understanding.

Harnessing the Motion of the Ocean

Ocean waves represent a huge, almost untapped potential source of energy. Estimates suggest that in the United States, wave power could produce one-quarter to one-third of the electricity used each year. Currently wave power technology lags behind wind and solar power, but around the world, engineers are working to change this and harness the power of ocean waves.

Wave power generators convert mechanical energy from ocean waves to electrical energy.

Wave Power Technologies

One wave power technology, sometimes called a point absorber, looks like a large floating buoy. The buoy has parts that are attached but are able to move independently. As waves rise and fall, the buoy's different parts move relative to each other. This motion drives an energy converter device such as a hydraulic pump, which transforms the mechanical energy of motion into electrical energy. Two different versions of these buoys are being tested off the coast of Hawaii.

Another technology that is designed for offshore use looks like a giant metal sea snake. Its segments can move independently but are attached to one another as it floats on the surface of the water. The whole device is anchored to the ocean floor. The motion of ocean waves causes the segments to move relative to one another. The mechanical energy of the segments is converted by a hydraulic pump or other device to electrical energy. The resulting electrical current is transported to shore via a cable.

A different style of generator, called an oscillating water column or a terminator, is usually designed for onshore use. Waves push water into a column-shaped chamber through an underwater opening. Air is trapped above. The captured water column moves up and down like a piston. Air is forced in and out of an opening at the top of the column. The movement of the air turns a turbine and this mechanical motion is converted to electricity.

Successful wave power devices need to do more than just convert mechanical energy from waves into electricity. They must also be extremely durable to last in harsh ocean conditions. Engineers must design devices that will withstand storms, constant impacts from waves, and corrosive seawater.

So far none of these technologies have been adopted for large-scale use. Engineers and scientists will continue to develop and test prototypes, working towards a goal of wave power devices that are effective, reliable, durable, and affordable.

DEVELOP AND USE MODEL TO ILLUSTRATE

Choose one type of wave power technology. Find out how it converts mechanical energy from waves to electricity. Develop and use a model to illustrate how the technology works.

CALIFORNIA CONNECTION

Would wave power be a good fit for coastal California? Write an email to the state government describing why the state should or should not invest in developing wave power technologies.

MODULE 13

STUDY GUIDE

GO ONLINE to study with your Science Notebook.

Lesson 1 PERIODIC MOTION

- Simple harmonic motion results when the restoring force on an object is directly proportional to the object's displacement from equilibrium.
- The elastic potential energy of a spring that obeys Hooke's law is expressed by the following equation:

$$PE_s = \frac{1}{2}kx^2$$

- The period of a pendulum depends on the pendulum's length and the gravitational field strength at the pendulum's location. The period can be found using the following equation

$$T = 2\pi\sqrt{\frac{\ell}{g}}$$

- periodic motion
- period
- amplitude
- simple harmonic motion
- Hooke's law
- simple pendulum
- resonance

Lesson 2 PHOTOSYNTHESIS

- Waves are disturbances that transfer energy without transferring matter.
- In transverse waves, the displacement of the medium is perpendicular to the direction the wave travels. In longitudinal waves, the displacement is parallel to the direction the wave travels.
- The velocity of a continuous wave is equal to the wave's frequency times its wavelength.

$$v = f\lambda$$

- wave
- wave pulse
- transverse wave
- periodic wave
- longitudinal wave
- surface wave
- trough
- crest
- wavelength
- frequency

Lesson 3 WAVE BEHAVIOR

- When two-dimensional waves are reflected from boundaries, the angles of incidence and reflection are equal. The change in direction of waves at the boundary between two different mediums is called refraction.
- Interference occurs when two or more waves travel through the same medium at the same time. The principle of superposition states that the displacement of a medium resulting from two or more waves is the algebraic sum of the displacements of the individual waves.

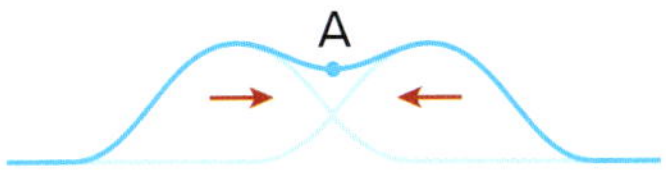

- incident wave
- reflected wave
- principle of superposition
- interference
- node
- antinode
- standing wave
- wavefront
- ray
- normal
- law of reflection
- refraction

THREE-DIMENSIONAL THINKING

Module Wrap-Up

REVISIT THE PHENOMENON

How can this pendulum could save a building from earthquake damage?

CER Claim, Evidence, Reasoning

Explain your Reasoning Revisit the claim you made when you encountered the phenomenon. Summarize the evidence you gathered from your investigations and research and finalize your Summary Table. Does your evidence support your claim? If not, revise your claim. Explain why your evidence supports your claim.

STEM UNIT PROJECT

Now that you've completed the module, revisit your STEM unit project. You will summarize your evidence and apply it to the project.

GO FURTHER

SEP Data Analysis Lab

How does the strength of gravity affect periodic motion?

An astronaut lands on an unknown planet and must determine the value of the gravitational acceleration *g*. He has the following instruments on hand.

Instrument A rests on the ground and shoots a ball vertically with a known speed. The astronaut can measure the time the ball takes to rise from its launch position and the time it takes to fall back to the launch position.

Instrument B is a simple unmarked weight suspended from a spring with constant $k = 22.5$ N/m.

Instrument C is a simple pendulum at the end of an arm that is 0.500 m long. The astronaut counts exactly 47 full swings in 1.00 min.

CER Analyze and Interpret Data

1. **Claim** Which instrument(s) can be used to determine *g*?
2. **Evidence and Reasoning** Explain how you made your decision.

SOUND

Mike Long/Alamy Stock Photo

MODULE 14

SOUND

ENCOUNTER THE PHENOMENON

Why does a fire truck siren pitch change as it passes you?

GO ONLINE to watch a video how pitch changes when a source of sound moves past an observer.

SEP Ask Questions

Do you have other questions about the phenomenon? If so, add them to the driving question board.

CER Claim, Evidence, Reasoning

Make Your Claim Use your CER chart to make a claim about why a fire truck's siren changes pitch as it passes you.

Collect Evidence Use the lessons in this module to collect evidence to support your claim. Record your evidence as you move through the module.

Explain Your Reasoning You will revisit your claim and explain your reasoning at the end of the module.

GO ONLINE to access your CER chart and explore resources that can help you collect evidence.

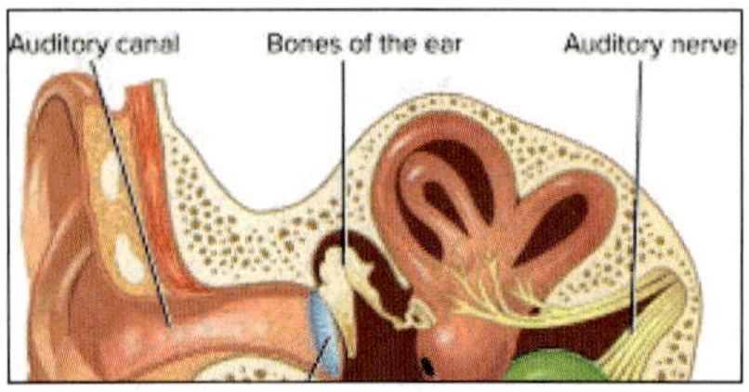

LESSON 1: Explore & Explain: Detecting Sound Waves

LESSON 2: Explore & Explain: Sources of Sound

Smithsonian

Additional Resources

LESSON 1

PROPERTIES AND DETECTION OF SOUND

FOCUS QUESTION

What factors affect the pitch of a sound?

Sound Waves

You already are familiar with several of the characteristics of sound, including volume, tone, and pitch, from your everyday experiences. Without thinking about it, you can use these, and other characteristics, to categorize many of the sounds that you hear.

Pressure variations Put your fingers against your throat as you hum or speak. Can you feel the vibrations? **Figure 1** shows a vibrating bell that can represent your vocal cords, a loudspeaker, or any other sound source. As it moves back and forth, the edge of the bell strikes the particles in the air. When the edge moves forward, air particles bounce off the bell with a greater velocity. When the edge moves backward, air particles bounce off the bell with a lower velocity.

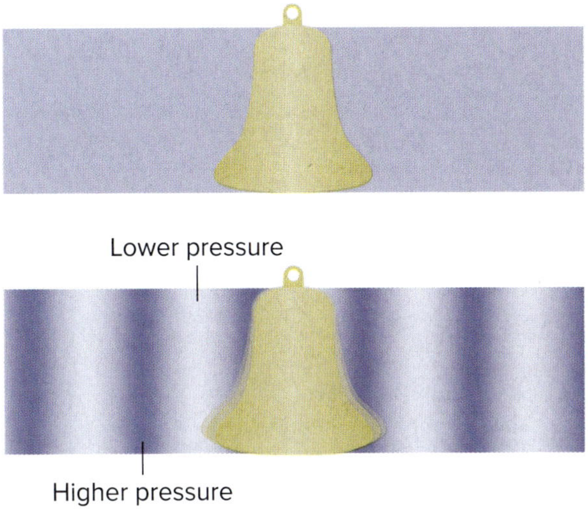

Figure 1 When the bell is at rest (top), the surrounding air is at average pressure. When the bell is struck (bottom), the vibrating edge creates regions of high and low pressure. Although the diagram shows the pressure regions moving in one direction, the waves move out in all directions.

3D THINKING — DCI Disciplinary Core Ideas — CCC Crosscutting Concepts — SEP Science & Engineering Practices

COLLECT EVIDENCE

Use your Science Journal to record the evidence you collect as you complete the readings and activities in this lesson.

INVESTIGATE

GO ONLINE to find these activities and more resources.

Virtual Investigation: Doppler Effect
Analyze data to determine how a moving source affects the frequency of the detected sound wave.

Revisit the Encounter the Phenomenon Question
What information from this lesson can help you answer the Unit and Module questions?

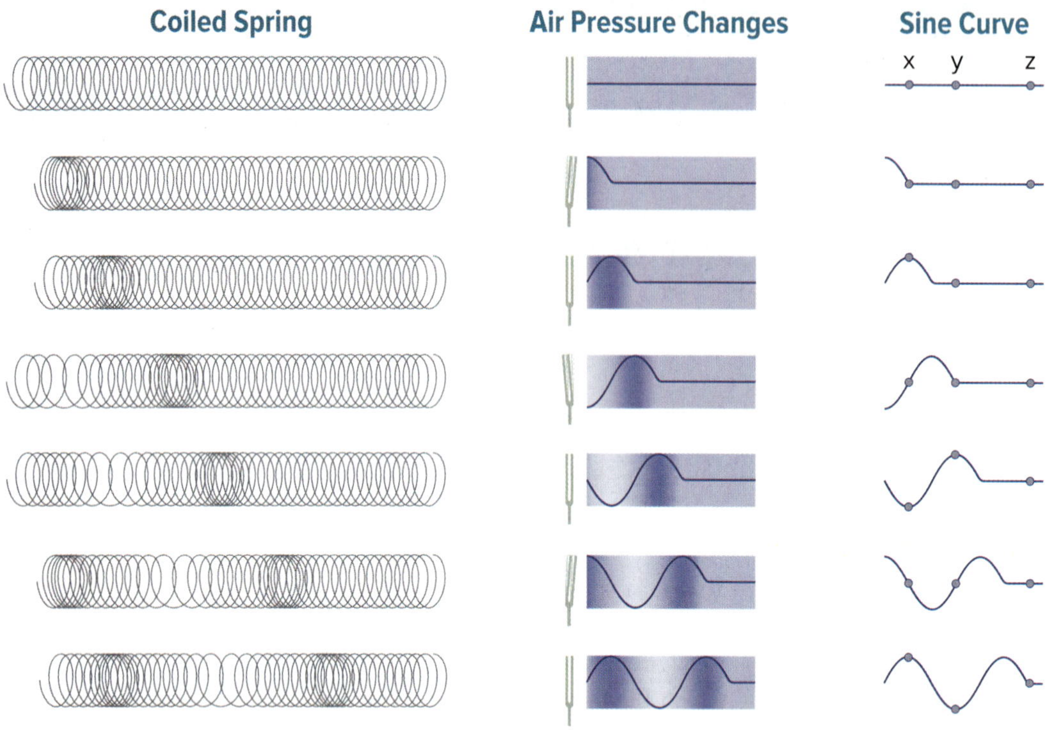

Figure 2 A coiled spring models the oscillations created by a sound wave. As the sound wave travels through the air, the air pressure rises and falls. The changes in the sine curves correspond to the changes in air pressure. Note that the positions of *x*, *y*, and *z* show that the wave, not the matter, travels forward.

The result of these velocity changes is that the forward motion of the bell produces a region where the air pressure is slightly higher than average, as shown in **Figure 1.** The backward motion produces slightly below-average pressure. Collisions of the air particles cause the pressure variations to move away from the bell in all directions. If you were to focus on one spot, you would see the value of the air pressure rise and fall. In this way, the pressure variations are transmitted through matter.

Describing sound

A pressure oscillation that is transmitted through matter is a **sound wave.** Sound waves travel through air because a vibrating source produces regular variations, or oscillations, in air pressure. The air particles collide, transmitting the pressure variations away from the source of the sound. The pressure of the air oscillates about the mean air pressure, as shown in **Figure 2.** The frequency of the wave is the number of oscillations in pressure each second. The wavelength is the distance between successive regions of high or low pressure. Because the motion of the particles in air is parallel to the direction of the wave's motion, sound is a longitudinal wave.

The wavelength and frequency of a wave are related to one another by the speed of travel of the wave. Just like any other wave, the speed of sound depends on the medium through which the sound passes. In air, the speed depends on the temperature, increasing by about 0.6 m/s for each 1°C increase in air temperature. At room temperature (20°C) and sea level, the speed of sound is 343 m/s. For the problems in this book, you may assume these conditions unless otherwise stated.

Get It?

Estimate the speed of sound through air at sea level if the temperature is 25°C.

In general, the speed of sound is greater in solids and liquids than in gases. **Table 1** lists the speeds of sound waves in various media. Sound cannot travel in a vacuum because there are no particles to collide.

Sound waves share the general properties of other waves. For example, they reflect off hard objects, such as the walls of a room. Reflected sound waves are called echoes. The time required for an echo to return to the source of the sound can be used to find the distance between the source and the reflective object. This principle is used by bats, by some cameras, and by ships that employ sonar. Two sound waves can interfere, causing dead spots at nodes where little sound can be heard. Recall that the frequency and wavelength of a wave are related to the speed of the wave by the equation $\lambda = \frac{v}{f}$.

Table 1 Speed of Sound in Various Media

Medium	Speed (m/s)
Air (0°C)	331
Air (20°C)	343
Helium (0°C)	965
Water (25°C)	1497
Seawater (25°C)	1535
Copper (20°C)	4760
Iron (20°C)	4994

Detection of Pressure Waves

Sound detectors transform sound energy—the kinetic energy of the vibrating particles of the transmitting medium—into another form of energy. A common detector is a microphone, which transforms sound energy into electrical energy. A microphone consists of a thin disk that vibrates in response to sound waves and produces an electrical signal.

The human ear

As shown in **Figure 3,** the human ear is a sound detector that receives pressure waves and converts them to electrical impulses. The tympanic membrane, also called the eardrum, vibrates when sound waves enter the auditory canal. Three tiny bones in the middle ear then transfer these vibrations to fluid in the cochlea. Tiny hairs lining the spiral-shaped cochlea detect certain frequencies in the vibrating fluid. These hairs stimulate nerve cells, which send impulses to the brain and produce the sensation of sound. The ear detects sound waves over a wide range of frequencies and is sensitive to an enormous range of amplitudes. In addition, human hearing can distinguish many different qualities of sound. Knowledge of both physics and biology is required to understand the complexities of the ear. The interpretation of sounds by the brain is even more complex, and is not totally understood.

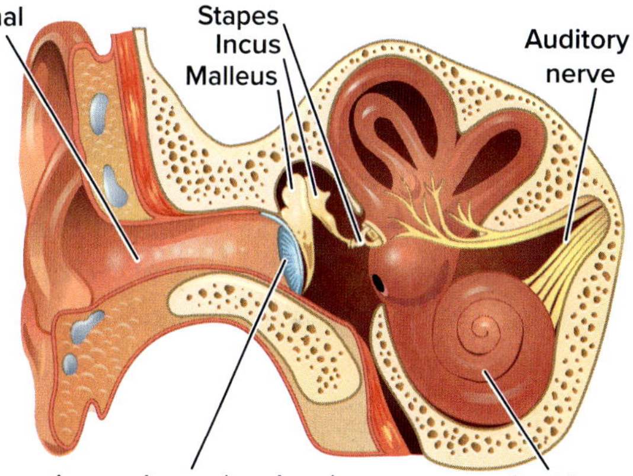

Figure 3 The human ear is a sense organ that translates sound vibrations from the external environment into nerve impulses that are sent to the brain for interpretation. The eardrum vibrates when sound waves enter the auditory canal. The bones in the middle ear—the malleus, the incus, and the stapes—move as a result of the vibrations. The vibrations are then transmitted to the inner ear, where they trigger nerve impulses to the brain. (Note: Diagram is not to scale.)

STEM CAREER Connection

Hearing Aid Specialist

Would you enjoy helping people improve their quality of life? A hearing aid specialist administers hearing tests and helps select and fit hearing aids for people.

Perceiving Sound

How humans perceive sound depends partly on the physical characteristics of sound waves, such as frequency and amplitude.

Pitch Marin Mersenne and Galileo first determined that the pitch we hear depends on the frequency of vibration. **Pitch** is the highness or lowness of a sound, and it can be given a name on the musical scale. For instance, the note known as middle C has a frequency of 262 Hz. The highest note on a piano has a frequency of 4186 Hz. The human ear is not equally sensitive to all frequencies. Most people cannot hear sounds with frequencies below 20 Hz or above 16,000 Hz. Many animals, such as dogs, cats, elephants, and bats, are capable of hearing frequencies that humans cannot hear.

Get It?

Identify What characteristic of waves is pitch most closely linked to?

Loudness Frequency and wavelength are two physical characteristics of sound waves. Another physical characteristic of sound waves is amplitude. Amplitude is the measure of the variation in pressure in a wave. The **loudness** of a sound is the intensity of the sound as perceived by the ear and interpreted by the brain. This intensity depends primarily on the amplitude of the pressure wave.

The human ear is extremely sensitive to variations in the intensity of sound waves. Recall that 1 atmosphere of pressure equals 1.01×10^{5} Pa. The ear can detect pressure-wave amplitudes of less than one-billionth of an atmosphere, or 2×10^{-5} Pa. At the other end of the audible range, pressure variations of approximately 20 Pa or greater cause pain. It is important to remember that the ear detects pressure variations only at certain frequencies. Driving over a mountain pass changes the pressure on your ears by thousands of pascals, but this change does not take place at audible frequencies.

Because humans can detect a wide range of intensities, it is convenient to measure these intensities on a logarithmic scale called the **sound level.** The most common unit of measurement for sound level is the **decibel** (dB). The sound level depends on the ratio of the intensity of a given sound wave to that of the most faintly heard sound. This faintest sound is measured at 0 dB. A sound that is ten times more intense registers 20 dB. A sound that is another ten times more intense is 40 dB.

Get It?

Compare How much more intense is a sound that registers 80 dB than one of 40 dB?

SCIENCE USAGE v. COMMON USAGE

Pitch

Science usage: the highness or lowness of a sound, which depends on the frequency of vibration

The flute and the tuba produce very different pitches.

Common usage: the delivery of a ball by a pitcher to a batter

Sharon hit the pitch over the fence for a home run.

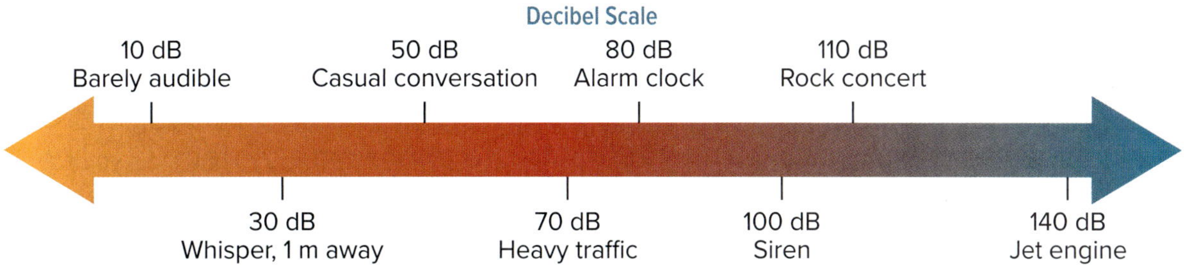

Figure 4 This decibel scale shows the sound level for a variety of sounds.

Infer *About how many times louder does an alarm clock sound than heavy traffic?*

Most people perceive a 10-dB increase in sound level as about twice as loud as the original level. **Figure 4** shows the sound level for a variety of sounds. In addition to intensity, pressure variations and the power of sound waves can be described by decibel scales.

The ear can lose its sensitivity, especially to high frequencies, after exposure to loud sounds in the form of noise or music. The longer a person is exposed to loud sounds, the greater the effect. A person can recover from short-term exposure in a period of hours, but the effects of long-term exposure can last for days or weeks. Long exposure to 100-dB or greater sound levels can produce permanent damage. Hearing loss also can result from loud music being transmitted to stereo headphones from personal music devices. In some cases, the listeners are unaware of just how high the sound levels really are. Cotton earplugs reduce the sound level only by about 10 dB. Special ear inserts can provide a 25-dB reduction. Specifically designed earmuffs and inserts, as shown in **Figure 5,** can reduce the sound level by up to 45 dB.

Figure 5 Hearing loss can occur with continuous exposure to loud sounds. Workers in many occupations, such as construction, wear ear protection. The jackhammer this worker is operating has a sound level of 130 dB.

The Doppler Effect

Have you ever noticed that the pitch of a fast car changed as the vehicle sped past you? The pitch was higher when the vehicle was moving toward you, then it dropped to a lower pitch as the vehicle moved away. The change in frequency of sound caused by the movement of either the source, the detector, or both is called the **Doppler effect.** The Doppler effect is illustrated in **Figure 6.**

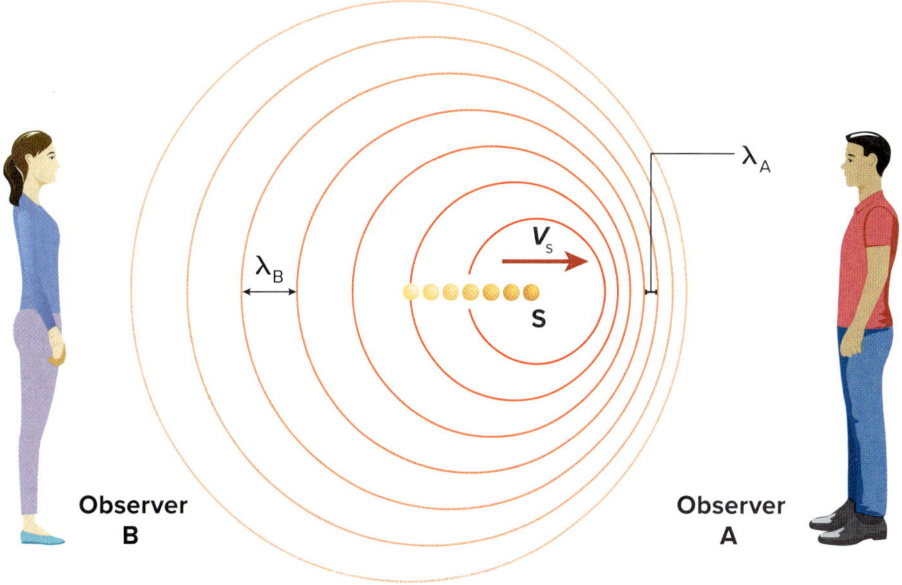

Figure 6 As a sound producing source moves toward observer A, the wavelength is shortened to λ_A. As the source moves away from observer B, the wavelength is lengthened to λ_B.

Describe *What is the relative difference in the frequency of the detected sound for each observer?*

The sound source (S) is moving to the right with a speed of v_s. The waves that the source emits spread in circles centered on the source at the time it produced the waves. As the source moves toward the sound detector, observer A in **Figure 6,** more waves are crowded into the space between them. The wavelength is shortened to λ_A. Because the speed of sound is not changed, more crests reach the ear each second, which means that the frequency of the detected sound increases. When the source is moving away from the detector, observer B in **Figure 6,** the wavelength is lengthened to λ_B, fewer crests reach the ear each second, and the detected frequency is lower.

A Doppler shift also occurs if the detector is moving and the source is stationary. As the detector approaches a stationary source, it encounters more wave crests each second than if it were still, and a higher frequency is detected. If the detector recedes from the source, fewer crests reach it each second, resulting in a lower detected frequency.

Get It?

Compare the wavelength and frequency heard by an observer in front of the moving fire engine at the beginning of the module with the wavelength and frequency heard by an observer behind the fire engine.

For any combination of moving source and moving observer, the frequency that the observer hears can be found using the relationship below.

Doppler Effect

The frequency perceived by a detector is equal to the velocity of the detector relative to the velocity of the wave, divided by the velocity of the source relative to the velocity of the wave, multiplied by the wave's frequency.

$$f_d = f_s \frac{v - v_d}{v - v_s}$$

In the Doppler effect equation, v is the velocity of the sound wave, v_s is the velocity of the sound's source, and v_d is the velocity of the observer of interest, who is detecting the sound. The subscript d is used instead of the letter o to avoid confusion with the number zero. The same subscripts are used to denote the corresponding frequencies.

Defining the coordinate system As you solve problems using the above equation, be sure to define the coordinate system so that the positive direction is from the source to the detector. The sound waves will be approaching the detector from the source, so the velocity of sound is always positive.

Try drawing diagrams to confirm that the term $\frac{v - v_d}{v - v_s}$ behaves as you would predict based on what you have learned about the Doppler effect. Notice that for a source moving toward the detector (positive direction, which results in a smaller denominator compared to a stationary source) and for a detector moving toward the source (negative direction and increased numerator compared to a stationary detector), the detected frequency (f_d) increases.

Similarly, if the source moves away from the detector or if the detector moves away from the source, then f_d decreases. Read the Connecting Math to Physics feature below to see how the Doppler effect equation reduces when the source or observer is stationary.

CONNECTING MATH to Physics

Reducing Equations When an element in an equation is equal to zero, the equation might reduce to a form that is easier to use.

Stationary detector, source in motion: $v_d = 0$	Stationary source, detector in motion: $v_s = 0$
$f_d = f_s \frac{v - v_d}{v - v_s}$	$f_d = f_s \frac{v - v_d}{v - v_s}$
$= f_s \frac{v}{v - v_s}$	$= f_s \frac{v - v_d}{v}$
$= f_s \frac{\frac{v}{v}}{\frac{v}{v} - \frac{v_s}{v}}$	$= f_s\left(\frac{v}{v} - \frac{v_d}{v}\right)$
$= f_s \frac{1}{1 - \frac{v_s}{v}}$	$= f_s\left(1 - \frac{v_d}{v}\right)$

EXAMPLE Problem 1

THE DOPPLER EFFECT A guitar player sounds C above middle C (523 Hz) while traveling in a convertible at 24.6 m/s. If the car is coming toward you, what frequency would you hear? Assume that the temperature is 20°C.

1 ANALYZE AND SKETCH THE PROBLEM

- Sketch the situation.
- Establish a coordinate axis. Make sure that the positive direction is from the source to the detector.
- Show the velocities of the source and detector.

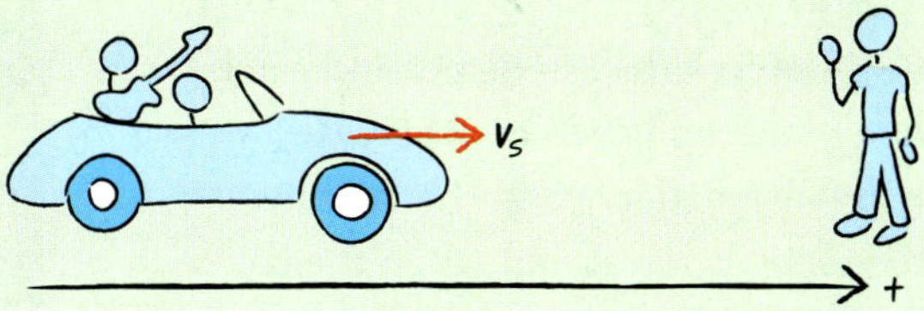

Known	Unknown
$v = +343$ m/s	$f_d = ?$
$v_s = +24.6$ m/s	
$v_d = 0$ m/s	
$f_s = 523$ Hz	

2 SOLVE FOR THE UNKNOWN

Use $f_d = f_s \frac{v - v_d}{v - v_s}$ with $v_d = 0$ m/s.

$$f_d = f_s \frac{1}{1 - \frac{v_s}{v}}$$

$$= 523\text{ Hz}\left(\frac{1}{1 - \frac{24.6\text{ m/s}}{343\text{ m/s}}}\right)$$ Substitute $v = +343$ m/s, $v_s = +24.6$ m/s, and $f_s = 523$ Hz.

$$= 564\text{ Hz}$$

3 EVALUATE THE ANSWER

- **Are the units correct?** Frequency is measured in hertz.
- **Is the magnitude realistic?** The source is moving toward you, so the frequency should be increased.

PRACTICE Problems

ADDITIONAL PRACTICE

1. Repeat Example Problem 1, but with the car moving away from you. What frequency would you hear?
2. You are in an automobile, like the one in **Figure 7,** traveling toward a pole-mounted warning siren. If the siren's frequency is 365 Hz, what frequency do you hear? Use 343 m/s as the speed of sound.

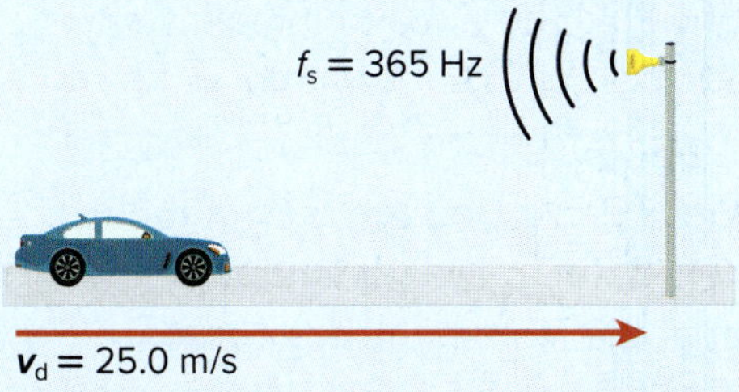

Figure 7

3. You are in an automobile traveling at 55 mph (24.6 m/s). A second automobile is moving toward you at the same speed. Its horn is sounding at 475 Hz. What frequency do you hear? Use 343 m/s as the speed of sound.
4. A submarine is moving toward another submarine at 9.20 m/s. It emits a 3.50-MHz ultrasound. What frequency would the second sub, at rest, detect? The speed of sound in water at the depth the submarines are moving is 1482 m/s.
5. **CHALLENGE** A trumpet plays middle C (262 Hz). How fast would it have to be moving to raise the pitch to C sharp (277 Hz)? Use 343 m/s as the speed of sound.

Applications of the Doppler effect The Doppler effect occurs in all wave motion, both mechanical and electromagnetic. It has many applications. Radar detectors use the Doppler effect to measure the speed of baseballs and automobiles. Astronomers observe light from distant galaxies and use the Doppler effect to measure their speeds. Physicians can detect the speed of the moving heart wall in a fetus by means of the Doppler effect in ultrasound.

BIOLOGY Connection Bats use the Doppler effect to detect and catch flying insects. When an insect is flying faster than a bat, the reflected frequency is lower, but when the bat is catching up to the insect, as in **Figure 8,** the reflected frequency is higher. Not only do bats use sound waves to navigate and locate their prey, but they often must do so in the presence of other bats. This means they must discriminate their own calls and reflections against a background of many other sounds of many frequencies. Scientists continue to study bats and their amazing abilities to use sound waves.

Figure 8 Bats use the Doppler effect to locate and catch flying insects, such as the moth shown here. As the bat catches up to the moth, the frequency of reflected sound waves increases.

Check Your Progress

6. **Wave Characteristics** What physical characteristic of a sound wave should be changed to alter the pitch? The loudness?
7. **Graph** The eardrum moves back and forth in response to the pressure variations of a sound wave. Sketch a graph of the displacement of the eardrum versus time for two cycles of a 1.0-kHz tone and of a 2.0-kHz tone.
8. **Effect of Medium** List two characteristics of sound that are affected by the medium through which the sound passes and two characteristics that are not affected.
9. **Decibel Scale** How many times greater is the sound pressure level of a typical rock concert (110 dB) than a normal conversation (50 dB)?
10. **Early Detection** In the nineteenth century, people put their ears to a railroad track to get an early warning of an approaching train. Why did this work?
11. **Bats** A bat emits short pulses of high-frequency sound and detects the echoes.
    a. In what way would the echoes from large and small insects compare if they were the same distance from the bat?
    b. In what way would the echo from an insect flying toward the bat differ from that of an insect flying away from the bat?
12. **Critical Thinking** Can a trooper using a radar detector at the side of the road determine the speed of a car at the instant the car passes the trooper? Explain.

LEARNSMART® Go online to follow your personalized learning path to review, practice, and reinforce your understanding.

LESSON 2
THE PHYSICS OF MUSIC

FOCUS QUESTION

How is pitch controlled in a musical instrument?

Sources of Sound

In the middle of the nineteenth century, German physicist Hermann Helmholtz studied sound production in musical instruments and the human voice. In the twentieth century, scientists and engineers developed electronic equipment that permits a detailed study of sound and the creation of electronic instruments and recording devices so we can listen to music anywhere.

Recall that sound is produced by a vibrating object. The vibrations of the object create particle motions that cause pressure oscillations in the air. In brass instruments, such as the trumpet, the tuba, and the bugle, the lips of the performer vibrate, as shown in **Figure 9.** Reed instruments, such as the clarinet and the saxophone, have a thin wooden strip called a reed that vibrates as a result of air blown across it, as shown in **Figure 9.** In flutes and organ pipes, air is forced across an opening in a pipe. Air moving past the opening sets the column of air in the instrument into vibration.

Figure 9 The sound produced by an instrument is partly determined by the structure of the mouthpiece.

3D THINKING | DCI Disciplinary Core Ideas | CCC Crosscutting Concepts | SEP Science & Engineering Practices

COLLECT EVIDENCE

Use your Science Journal to record the evidence you collect as you complete the readings and activities in this lesson.

INVESTIGATE

GO ONLINE to find these activities and more resources.

PhysicsLAB: Speed of Sound
Calculate the speed of sound using the relationship among frequency, wavelength, and speed.

Review the News
Obtain information from a current news story about sound waves and their applications. Evaluate your source and communicate your findings to your class.

Stringed instruments, such as the piano, the guitar, and the violin, work by setting wires or strings into vibration. In the piano, the wires are struck; for the guitar, they are plucked; and for the violin, the friction of the bow causes the strings to vibrate. The strings are attached to a sounding board that vibrates with the strings. The vibrations of the sounding board cause the pressure oscillations in the air that we hear as sound. Electric guitars use electronic devices to detect and amplify the vibrations of the guitar strings.

A loudspeaker has a cone that is made to vibrate by electrical currents. The surface of the cone creates the sound waves that travel to your ear and allow you to hear music. Musical instruments such as gongs, cymbals, and drums are other examples of vibrating surfaces that are sources of sound.

The human voice is produced by vibrations of the vocal cords, which are two membranes located in the throat. Air from the lungs rushing through the throat starts the vocal cords vibrating. The frequency of vibration is controlled by the muscular tension placed on the vocal cords. The more tension on the vocal cords, the more rapidly they vibrate, resulting in a higher pitch sound. If the vocal cords are more relaxed, they vibrate more slowly and produce lower-pitched sounds.

Get It?

Describe How does a vocalist sing higher pitched notes?

Resonance in Air Columns

If you have ever used just the mouthpiece of a brass or wind instrument, you know that while the vibration of your lips or the reed alone makes a sound, it is difficult to control the pitch. The long tube that makes up the instrument must be attached if music is to result. When the instrument is played, the air within this tube vibrates at the same frequency, or in resonance, with a particular vibration of the lips or reed. Remember that resonance increases the amplitude of a vibration by repeatedly applying a small external force to the vibrating air particles at the natural frequency of the air column. The length of the air column determines the frequencies of the vibrating air that will resonate. For wind and brass instruments, such as flutes, trumpets, and trombones, changing the length of the column of vibrating air varies the pitch of the instrument. The mouthpiece simply creates a mixture of different frequencies, and the resonating air column acts on a particular set of frequencies to amplify a single note, turning noise into music. A tuning fork above a hollow tube can provide resonance in an air column, as shown in **Figure 10.**

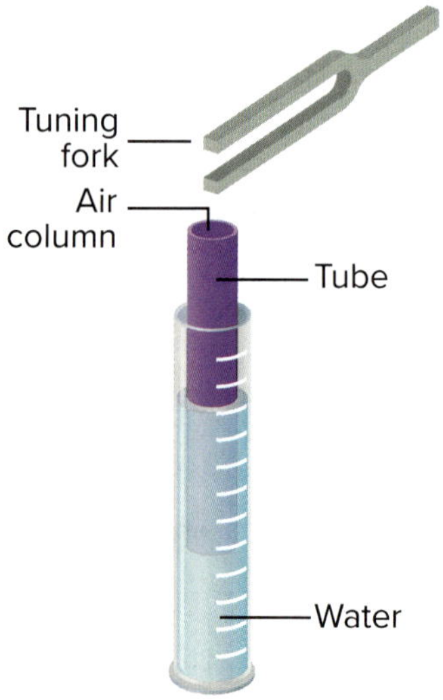

Figure 10 As the tube is raised or lowered, the length of the air column changes, which causes the sound's volume to change.

CCC CROSSCUTTING CONCEPTS

Cause and Effect Moving the tube in **Figure 10** up and down causes the volume of the sound to change. Develop a mathematical representation to relate the frequency, wavelength, speed of sound, and length of air column in the tube to explain this effect.

The tube is placed in water so that the bottom end of the tube is below the water surface. A resonating tube with one end closed to air is called a **closed-pipe resonator.** The length of the air column is changed by adjusting the height of the top of the tube above the water. If the tuning fork is struck with a rubber hammer, the sound alternately becomes louder and softer as the length of the air column is varied by moving the tube up and down in the water. The sound is loud when the air column is in resonance with the tuning fork because the resonating air column intensifies the sound of the tuning fork.

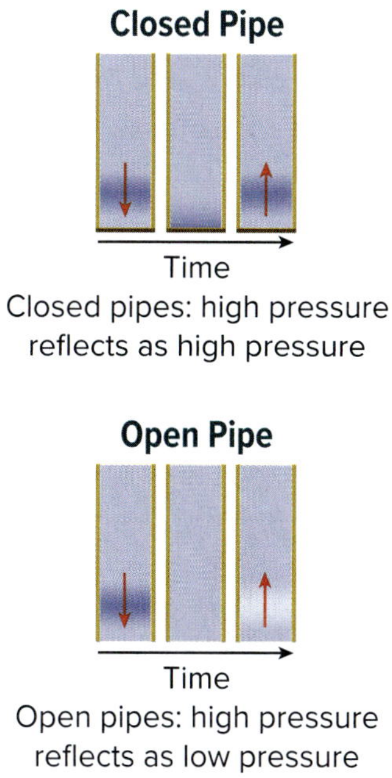

Figure 11 In closed pipes, the sound wave reflects off the closed end. High-pressure waves reflect as high pressure. In open pipes, the sound wave reflects off an open end. High-pressure waves are reflected as low pressure.

Standing pressure wave How does resonance occur? The vibrating tuning fork produces a sound wave. This wave of alternate high- and low-pressure variations moves down the air column. When the wave hits the water surface, it is reflected back up to the tuning fork, as shown in **Figure 11.** If the reflected high-pressure wave reaches the tuning fork at the same moment that the fork produces another high-pressure wave, then the emitted and returning waves reinforce each other. This reinforcement of waves creates a standing wave, and resonance occurs.

An **open-pipe resonator** is a resonating tube with both ends open that will resonate with a sound source. In this case, the sound wave does not reflect off a closed end, but rather off an open end. If the high-pressure part of the wave strikes the open end, the rebounding wave will be low-pressure at that point, as shown in **Figure 11.**

Resonance lengths **Figure 12** shows a standing sound wave in a pipe represented by a sine wave. Sine waves can represent either the air pressure or the displacement of the air particles. Recall that standing waves have nodes and antinodes. A node is the stationary point where two equal wave pulses meet and are in the same location. An antinode is the place of largest displacement when two wave pulses meet.

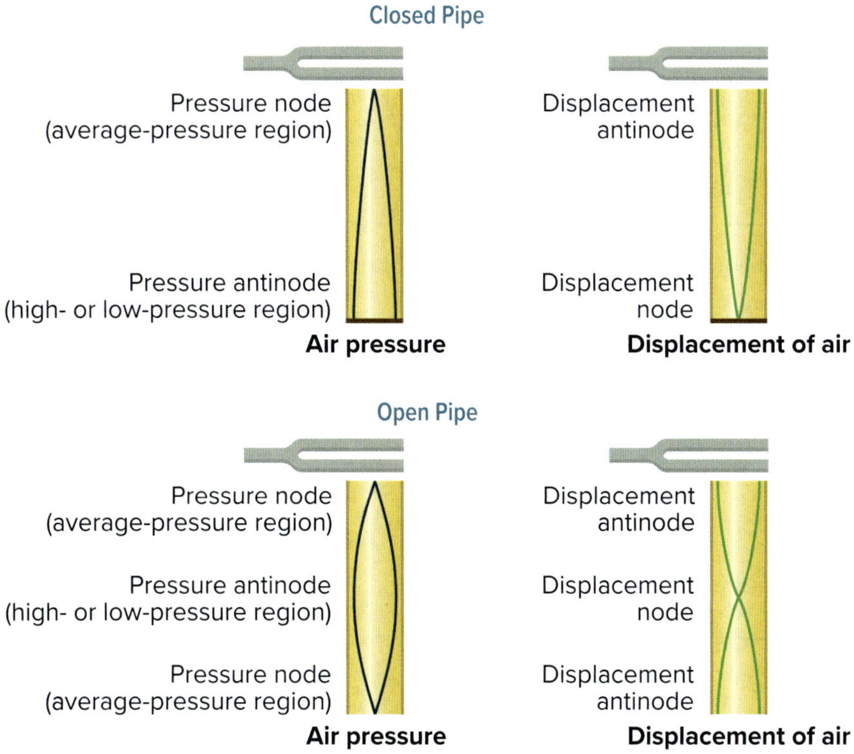

Figure 12 Standing waves in pipes can be represented by sine waves.

Identify *Which are the areas of mean atmospheric pressure in the air pressure graphs?*

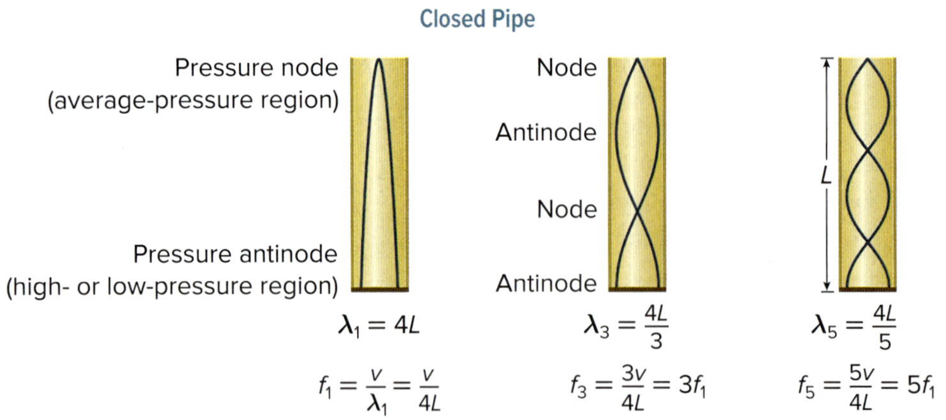

Figure 13 A closed pipe resonates when its length is an odd number of quarter wavelengths.

In the pressure graphs, the nodes are regions of mean atmospheric pressure. At the antinodes, the pressure oscillates between its maximum and minimum values. In the case of the displacement graph, the antinodes are regions of high displacement and the nodes are regions of low displacement. In both cases, two adjacent antinodes (or two nodes) are separated by one-half wavelength.

Get It?

Explain the difference between a node and an antinode on a displacement graph.

Resonance frequencies in a closed pipe If a closed end must act as a node, and an open end must act as an antinode, what is the shortest column of air that will resonate in a closed pipe? **Figure 13** shows that it must be one-fourth of a wavelength. As the frequency is increased, additional resonance lengths are found at half-wavelength intervals. Thus, columns of length $\frac{\lambda}{4}$, $\frac{3\lambda}{4}$, $\frac{5\lambda}{4}$, $\frac{7\lambda}{4}$, and so on will all be in resonance with a tuning fork that produces sound of wavelength λ.

In practice, the first resonance length is slightly longer than one-fourth of a wavelength. This is because the pressure variations do not drop to zero exactly at the open end of the pipe. Actually, the node is approximately 0.4 pipe diameters beyond the end. Additional resonance lengths, however, are spaced by exactly one-half of a wavelength. Measurements of the spacing between resonances can be used to find the velocity of sound in air, as in Example Problem 2.

Resonance frequencies in an open pipe The shortest column of air that can have nodes at both ends is one-half of a wavelength long, as shown in **Figure 14.** As the frequency is increased, additional resonance lengths are found at half-wavelength intervals. Thus, columns of length $\frac{\lambda}{2}$, λ, $\frac{3\lambda}{2}$, 2λ, and so on will be in resonance with a tuning fork.

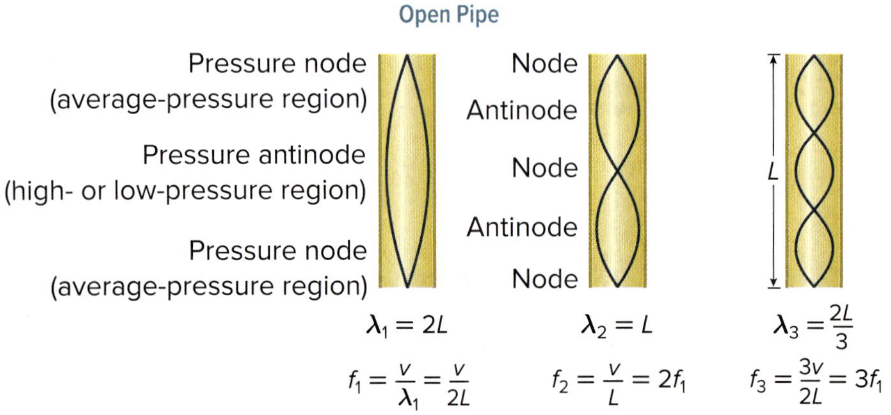

Figure 14 An open pipe resonates when its length is an even number of quarter wavelengths.

Explain *How does the length at which an open pipe resonates differ from the length at which a closed pipe resonates?*

Figure 15 A flute is an example of an open-pipe resonator. The hanging pipes of a marimba and seashells are examples of closed-pipe resonators.

If open and closed pipes of the same length are used as resonators, the wavelength of the resonant sound for the open pipe will be half as long as that for the closed pipe. Therefore, the frequency will be twice as high for the open pipe as for the closed pipe. For both pipes, resonance lengths are spaced by half-wavelength intervals.

Get It?

Predict A tuning fork plays a sound that has a wavelength of 0.78 m. A pipe that is 0.39 m long resonates with the tuning fork. Is the pipe open or closed? Explain your reasoning.

Hearing resonance

Musical instruments use resonance to increase the loudness of particular notes. Open-pipe resonators include flutes, shown in **Figure 15.** Clarinets and the hanging pipes under marimbas and xylophones are examples of closed-pipe resonators. If you shout into a long tunnel, the booming sound you hear is the tunnel acting as a resonator. The seashell in **Figure 15** also acts as a closed-pipe resonator.

Real-World Physics

HEARING AND FREQUENCY The human auditory canal acts as a closed-pipe resonator that increases the ear's sensitivity for frequencies between 2000 and 5000 Hz, but the full range of frequencies that people hear extends from 20 to 20,000 Hz. A dog's hearing extends to frequencies as high as 45,000 Hz, and a cat's extends to frequencies as high as 100,000 Hz.

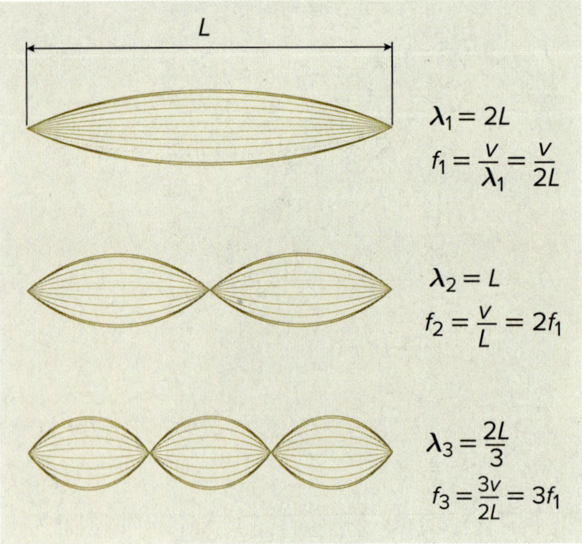

Figure 16 A string resonates with standing waves when its length is a whole number of half wavelengths.

Resonance on Strings

Although plucking, bowing, or striking strings produces variation in waveforms, waveforms on vibrating strings have many characteristics in common with standing waves on springs and ropes. A string on an instrument is clamped at both ends, and therefore, the string must have a node at each end when it vibrates. In **Figure 16,** you can see that the first mode of vibration has an antinode at the center and is one-half a wavelength long. The next resonance occurs when one wavelength fits on the string, and additional standing waves arise when the string length is $\frac{3\lambda}{2}$, 2λ, $\frac{5\lambda}{2}$, and so on. As with an open pipe, the resonant frequencies are whole-number multiples of the lowest frequency.

Recall that the speed of travel of a wave depends on the type of wave and the medium through which it is passing. For a string, the speed depends on the tension of the string, as well as its mass per unit length. The tighter the string, the faster the wave moves along it, and therefore, the higher the frequency of its standing waves. This makes it possible to tune a stringed instrument by changing the tension of its strings. Because strings are so small in cross-sectional area, they move very little air when they vibrate. This makes it necessary to attach them to a sounding board, which transfers their vibrations to the air and produces a stronger sound wave. Unlike the strings themselves, the sounding board should not resonate at any single frequency. Its purpose is to convey the vibrations of all the strings to the air, and therefore it should vibrate well at all frequencies produced by the instrument. Because of the complicated interactions among the strings, the sounding board, and the air, the design and construction of stringed instruments are complex processes, considered by many to be as much an art as a science.

Get It?

Describe the relationship between the tension of a string and the speed of a wave as it travels along the string.

EXAMPLE Problem 2

FINDING THE SPEED OF SOUND USING RESONANCE When a tuning fork with a frequency of 392 Hz is used with a closed-pipe resonator, the loudest sound is heard when the column is 21.0 cm and 65.3 cm long. What is the speed of sound in this case? Is the temperature warmer or cooler than normal room temperature, which is 20°C? Explain your answer.

1 ANALYZE AND SKETCH THE PROBLEM

- Sketch the closed-pipe resonator.
- Mark the resonance lengths.

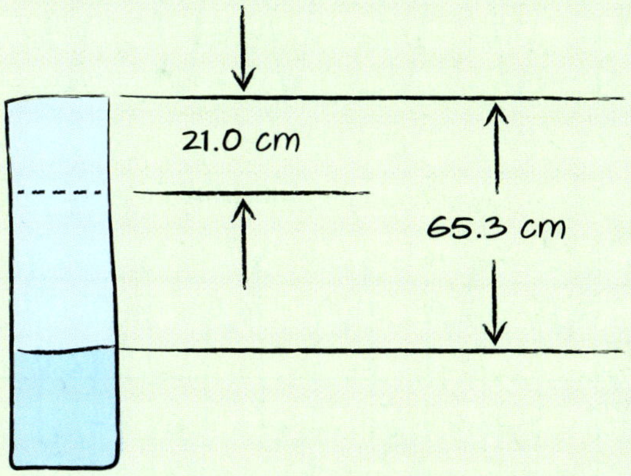

Known	Unknown
$f = 392$ Hz	$v = ?$
$L_A = 21.0$ cm	
$L_B = 65.3$ cm	

2 SOLVE FOR THE UNKNOWN

Solve for the length of the wave using the length-wavelength relationship for a closed pipe.

$L_B - L_A = \frac{1}{2}\lambda$

$\lambda = 2(L_B - L_A)$ — Rearrange the equation for λ.

$= 2(0.653 \text{ m} - 0.210 \text{ m})$ — Substitute $L_B = 0.653$ m, $L_A = 0.210$ m.

$= 0.886$ m

Use $\lambda = \frac{v}{f}$

$v = f\lambda$ — Rearrange the equation for v.

$= (392 \text{ Hz})(0.886 \text{ m})$ — Substitute $f = 392$ Hz, $\lambda = 0.886$ m.

$= 347$ m/s

The speed is slightly greater than the speed of sound at 20°C, indicating that the temperature is slightly higher than normal room temperature.

3 EVALUATE THE ANSWER

- **Are the units correct?** (Hz)(m) = $\left(\frac{1}{s}\right)$(m) = m/s. The answer's units are correct.
- **Is the magnitude realistic?** The speed is slightly greater than 343 m/s, which is the speed of sound at 20°C.

PRACTICE Problems

ADDITIONAL PRACTICE

13. A 440-Hz tuning fork is used with a resonating column to determine the velocity of sound in helium gas. If the spacing between resonances is 110 cm, what is the velocity of sound in helium gas?

14. The frequency of a tuning fork is unknown. A student uses an air column at 27°C and finds resonances spaced by 20.2 cm. What is the frequency of the tuning fork? Use the speed calculated in Example Problem 2 for the speed of sound in air at 27°C.

15. A 440-Hz tuning fork is held above a closed pipe. Find the spacing between the resonances when the air temperature is 20°C.

16. **CHALLENGE** A bugle can be thought of as an open pipe. If a bugle were straightened out, it would be 2.65-m long.

a. If the speed of sound is 343 m/s, find the lowest frequency that is resonant for a bugle (ignoring end corrections).

b. Find the next two resonant frequencies for the bugle.

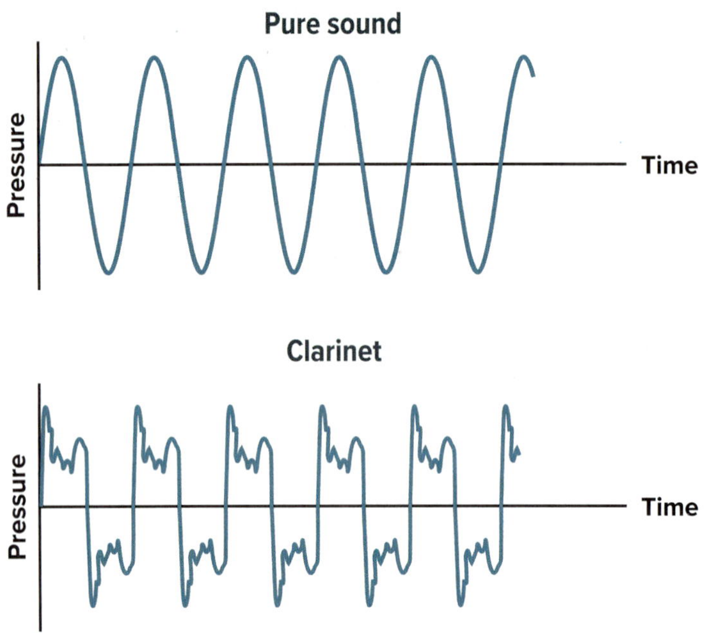

Figure 17 The pure sound produced by a tuning fork is represented by a simple sine wave. The more complex sound produced by a clarinet is represented in the bottom graph.

Sound Quality

A tuning fork produces a soft and uniform sound. This is because its tines vibrate like simple harmonic oscillators and produce the simple sine wave shown in the top graph in **Figure 17.** Sounds made by the human voice and musical instruments are much more complex, like the wave in the bottom graph in **Figure 17.** Both waves have the same frequency, or pitch, but they sound very different.

The complex sound wave is actually a blend of several different frequencies. The shape of the wave depends on the relative amplitudes of these frequencies. Different sources provide different combinations of frequencies. In musical terms, the difference between the waves from different instruments is called timbre (TAM bur), tone quality, or color.

The sound spectrum: fundamental and harmonics

The complex sound wave in **Figure 17** was made by a clarinet. Why does the clarinet produce such a sound wave? The air column in a clarinet acts as a closed pipe. Look back at **Figure 13,** which shows three resonant frequencies for a closed pipe. The clarinet acts as a closed pipe, so for a clarinet of length L the lowest frequency (f_1) that will be resonant is $\frac{v}{4L}$.

For a musical instrument, the lowest frequency of sound that resonates is called the **fundamental.** A closed pipe also will resonate at $3f_1$, $5f_1$, and so on. These higher frequencies, which are whole-number multiples of the fundamental frequency, are called **harmonics.** It is the addition of these harmonics that gives a clarinet its distinctive timbre.

Get It?

Explain the relationship between the fundamental and the harmonics of a musical instrument.

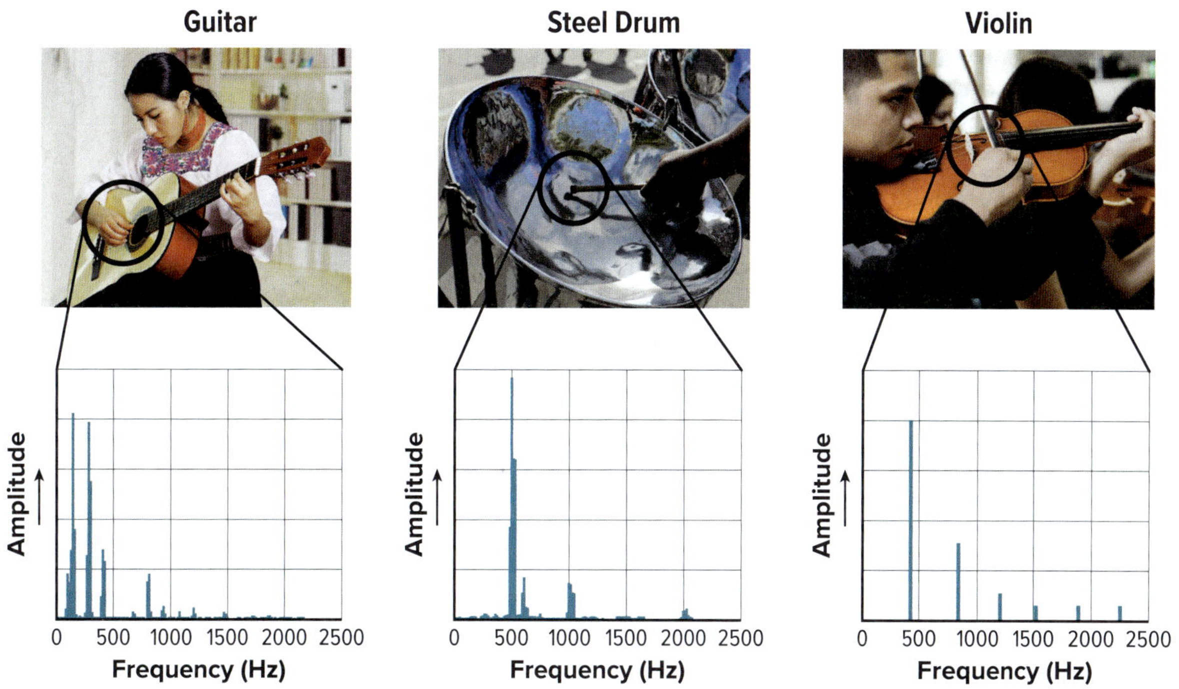

Figure 18 A guitar, a steel drum, and a violin produce characteristic sound spectra. Each spectrum is unique, as is the timbre of the instrument.

Some instruments, such as a flute, act as open-pipe resonators. Their fundamental frequency, which is also the first harmonic, is $f_1 = \frac{v}{2L}$ with subsequent harmonics at $2f_1$, $3f_1$, $4f_1$, and so on. Different combinations of these harmonics give each instrument its own unique timbre. Each harmonic on the instrument can have a different amplitude as well. A graph of the amplitude of a wave versus its frequency is called a sound spectrum. The spectra of three instruments are shown in **Figure 18.**

Consonance and dissonance When two different pitches are played at the same time, the resulting sound can be either pleasant or jarring. In musical terms, several pitches played together are called a chord. An unpleasant set of pitches is called **dissonance.** If the combination of pitches is pleasant, the sounds are said to be in **consonance.** What sounds pleasing varies between cultures, but most Western music is based upon the observations of Pythagoras of ancient Greece. He noted that pleasing sounds resulted when strings had lengths in small, whole-number ratios, such as 1:2, 2:3, or 3:4. This means their pitches (frequencies) will also have small, whole-number ratios.

PHYSICS Challenge

1. Determine the tension, F_T, in a violin string of mass m and length L that will play the fundamental note at the same frequency as a closed pipe also of length L. Express your answer in terms of m, L, and the speed of sound in air, v. The equation for the speed of a wave on a string is $v_{string} = \sqrt{\frac{F_T}{\mu}}$, where F_T is the tension in the string and μ is the mass per unit length of the string.
2. What is the tension in a string of mass 1.0 g and 40.0 cm long that plays the same note as a closed pipe of the same length?

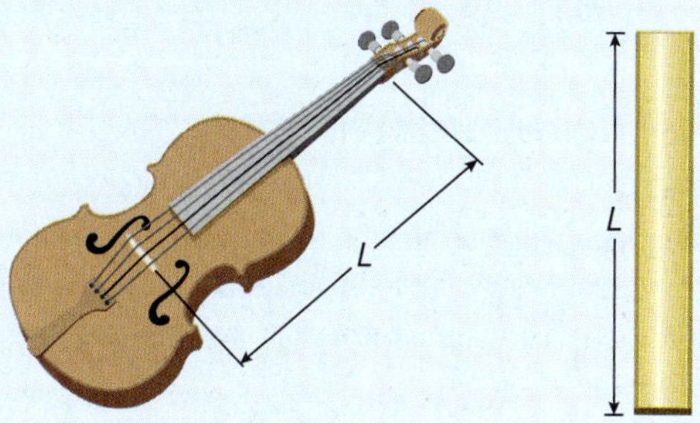

Musical intervals Two notes with frequencies related by the ratio 1:2 are said to differ by an octave. For example, if a note has a frequency of 440 Hz, a note that is one octave higher has a frequency of 880 Hz. The fundamental and its harmonics are related by octaves; the first harmonic is one octave higher than the fundamental, the second is two octaves higher, and so on. It is the ratio of two frequencies, not the size of the interval between them, that determines the musical interval.

In other musical intervals, two pitches may be close together. For example, the ratio of frequencies for a "major third" is 4:5. An example is the notes C and E. The note C has a frequency of 262 Hz, so E has a $\left(\frac{5}{4}\right)(262\text{ Hz}) = 327\text{ Hz}$. In the same way, notes in a "fourth" (C and F) have a frequency ratio of 3:4, and those in a "fifth" (C and G) have a ratio of 2:3. More than two notes sounded together also can produce consonance. The three notes called do, mi, and sol make a major chord. For at least 2500 years, western music has recognized this as the sweetest of the three-note chords; it has the frequency ratio of 4:5:6.

Beats

You have seen that consonance is defined in terms of the ratio of frequencies. When the ratio becomes nearly 1:1, the frequencies become very close. Two frequencies that are nearly identical interfere to produce oscillating high and low sound levels called a **beat.** This phenomenon is illustrated in **Figure 19.** The frequency of a beat is the magnitude of difference between the frequencies of the two waves, $f_{\text{beat}} = |f_A - f_B|$. When the difference is less than 7 Hz, the ear detects this as a pulsation of loudness. Musical instruments often are tuned by sounding one against another and adjusting the frequency of one until the beat disappears.

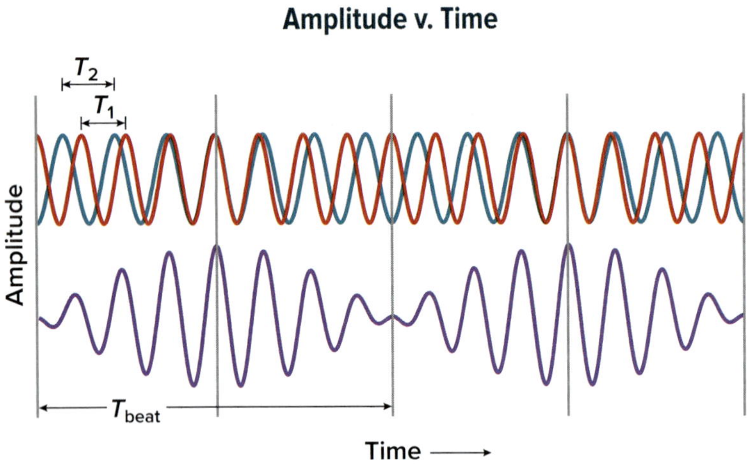

Figure 19 Beats occur as a result of the superposition of two sound waves of slightly different frequencies.

SCIENCE USAGE v. COMMON USAGE

Beat

Science usage: oscillation of wave amplitude that results from the superposition of two sound waves with almost identical frequencies

When the piano tuner no longer heard beats, she knew the piano was tuned properly.

Common usage: to strike repeatedly

Richard beat the drums while John played the guitar.

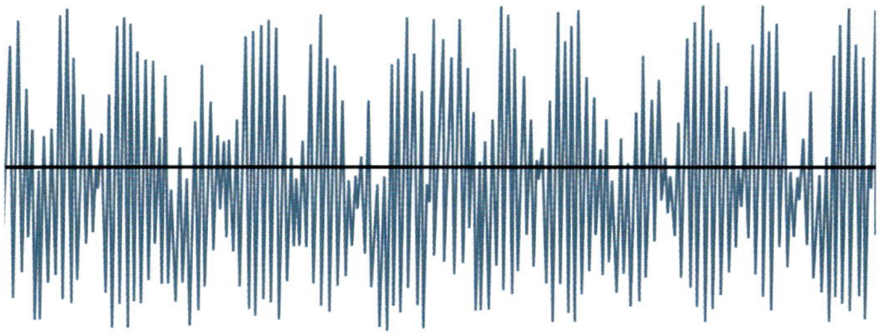

Figure 20 A noise wave consists of many different frequencies, all with about the same amplitude.

Sound Reproduction and Noise

When you listen to a live band or hear your school band practicing, you are hearing music produced directly by a human voice or musical instruments. You may want to hear live music every time you choose to listen to music. Most of the time, however, you likely listen to music that has been recorded and is played via electronic systems. To reproduce the sound faithfully, the system must accommodate all frequencies equally. A good stereo system keeps the amplitudes of all frequencies between 20 and 20,000 Hz to within a range of 3 dB.

A telephone system, on the other hand, needs only to transmit the information in spoken language. Frequencies between 300 and 3000 Hz are sufficient. Reducing the number of frequencies present helps reduce the noise. A noise wave is shown in **Figure 20.** Many frequencies are present with approximately the same amplitude. While noise is not helpful in a telephone system, some people claim that listening to white noise has a calming effect. For this reason, some dentists use noise to help their patients relax.

Check Your Progress

17. **Sound Sources** What is the vibrating object that produces sounds in each of the following?
    a. a human voice c. a tuba
    b. a clarinet d. a violin
18. **Resonance in Air Columns** Why is the tube from which a tuba is made much longer than that of a cornet?
19. **Resonance in Open Tubes** How must the length of an open tube compare to the wavelength of the sound to produce the strongest resonance?
20. **Resonance on Strings** A violin sounds a note of F sharp, with a pitch of 370 Hz. What are the frequencies of the next three harmonics produced with this note?
21. **Resonance in Closed Pipes** One closed organ pipe has a length of 2.40 m.
    a. What is the frequency of the note played?
    b. When a second pipe is played at the same time, a 1.40-Hz beat note is heard. By how much is the second pipe too long?
22. **Timbre** Why do various instruments sound different even when they play the same note?
23. **Beats** A tuning fork produces three beats per second with a second, 392-Hz tuning fork. What is the frequency of the first tuning fork?
24. **Critical Thinking** Strike a tuning fork with a rubber hammer and hold it at arm's length. Then press its handle against a desk, a door, a filing cabinet, and other objects. What do you hear? Why?

LEARNSMART® Go online to follow your personalized learning path to review, practice, and reinforce your understanding.

Out of Sight—Sonar Detection

Animals such as bats, whales, dolphins, and some birds use echolocation, also called bio-sonar, to detect objects when it is difficult for them to see (for example, at night or in cloudy waters). The animals emit high-frequency clicking noises. The sound waves from these noises bounce off nearby objects as echoes. The animals detect the echoes with their ears and process the information to determine the spatial imagery of their surroundings, including the location of prey.

Daniel Kish is blind. He uses echolocation to navigate around in his environment.

Human Echolocation

Some people who are blind also use echolocation to orient to their environment and move around safely without the help of a guide dog or cane. These people produce rapid clicking noises with their mouths and listen for the echoes as they reflect back from different surfaces and objects nearby. This acoustic information received from their ears is interpreted in the same part of the brain where people who have sight process visual information about their surroundings—in essence, they "see" by sound.

Californian Daniel Kish developed retinal cancer as an infant and soon became blind. Without the guidance of anyone, he instinctively used clicking noises at a young age as a compass to navigate in his environment—before the concept of human echolocation existed. Daniel is known as the "real-life batman" and can even mountain bike and sketch his surroundings using echolocation.

Daniel received his Master's degree in developmental psychology and developed the first step-by-step methods for teaching echolocation. He now trains visually impaired people to use echolocation as a valuable skill to move around in day-to-day life. Scientists have also found that it is possible to train people who are not blind to use echolocation.

Developing Echolocation Technologies

New assistive devices that are based on echolocation are being developed for people who are visually impaired. These include vibrating clothing and a wrist band that detects echoes bouncing off objects up to 14 feet away. Echolocation research may also be useful in the development of technology that uses artificial sonar, such as self-driving cars.

COMMUNICATE TECHNICAL INFORMATION

Research and choose a technology that uses echolocation. Summarize how the technology works in a short paragraph.

CALIFORNIA CONNECTION

Find out more about Daniel Kish and the work he does teaching other people how to echolocate. Share your findings with a partner.

MODULE 14
STUDY GUIDE

GO ONLINE to study with your Science Notebook.

Lesson 1 PROPERTIES AND DETECTION OF SOUND

- Sound is a pressure variation transmitted through matter as a longitudinal wave. A sound wave has frequency, wavelength, speed, and amplitude. Sound waves reflect and interfere.
- Sound detectors convert the energy carried by a sound wave into another form of energy. The human ear is a highly efficient and sensitive detector of sound waves. The frequency of a sound wave is heard as its pitch. The loudness of sound as perceived by the ear and brain depends mainly on its amplitude. The pressure amplitude of a sound wave can be measured in decibels (dB).

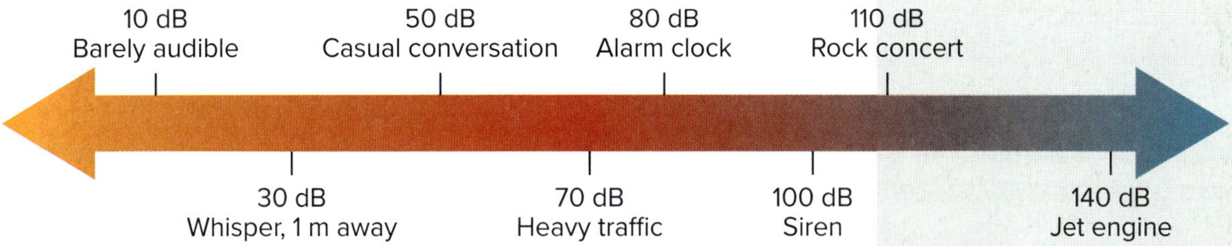

- The Doppler effect is the change in frequency of sound caused by the motion of either the source or the detector.
- The Doppler effect is used in radar detectors, in medical ultrasound machines, and by astronomers. Bats also use the Doppler effect to detect and catch flying insects.

- sound wave
- pitch
- loudness
- sound level
- decibel
- Doppler effect

Lesson 2 THE PHYSICS OF MUSIC

- Sound is produced by a vibrating object in a medium.
- An air column can resonate with a sound source, thereby increasing the amplitude of its resonant frequency. A closed pipe resonates when its length is $\frac{\lambda}{4}$, $\frac{3\lambda}{4}$, $\frac{5\lambda}{4}$, and so on. Its resonant frequencies are odd-numbered multiples of the fundamental. An open pipe resonates when its length is $\frac{\lambda}{2}$, $\frac{2\lambda}{2}$, $\frac{3\lambda}{2}$, and so on. Its resonant frequencies are whole-number multiples of the fundamental.
- A clamped string has a node at each end and resonates when its length is $\frac{\lambda}{2}$, $\frac{2\lambda}{2}$, $\frac{3\lambda}{2}$, and so on, just as with an open pipe. The string's resonant frequencies are also whole-number multiples of the fundamental.
- The frequencies and intensities of the complex waves produced by a musical instrument determine the timbre that is characteristic of that instrument.
- Two waves with almost the same frequency interfere to produce beats.

- closed-pipe resonator
- open-pipe resonator
- fundamental
- harmonics
- dissonance
- consonance
- beat

THREE-DIMENSIONAL THINKING

Module Wrap-Up

REVISIT THE PHENOMENON

Why does a fire truck siren pitch change as it passes you?

CER Claim, Evidence, Reasoning

Explain your Reasoning Revisit the claim you made when you encountered the phenomenon. Summarize the evidence you gathered from your investigations and research and finalize your Summary Table. Does your evidence support your claim? If not, revise your claim. Explain why your evidence supports your claim.

STEM UNIT PROJECT

Now that you've completed the module, revisit your STEM unit project. You will summarize your evidence and apply it to the project.

GO FURTHER

SEP Data Analysis Lab

Can you test the Doppler effect?

In 1845, Dutch astronomer Christoph Buys Ballot tested the Doppler effect. A trumpet player sounded an A (440 Hz) while riding on a flatcar pulled by a train. Your project team wants to repeat the experiment. Rather than using a train and listening for beats, you plan to have a trumpet played in a rapidly moving car and have the car move fast enough so that the moving trumpet sounds one major third above a stationary trumpet.

CER Analyze and Interpret Data

1. **Claim** Should you try your experiment?
2. **Evidence and Reasoning** Justify your claim.

Mike Long/Alamy Stock Photo

MODULE 15

FUNDAMENTALS OF LIGHT

NASA, ESA, J. Hester and A. Loll (Arizona State University)

MODULE 15

FUNDAMENTALS OF LIGHT

ENCOUNTER THE PHENOMENON

What does the light from a distant star or supernova tell us about it?

GO ONLINE to watch a video about measuring the speed of light.

SEP Ask Questions

Do you have other questions about the phenomenon? If so, add them to the driving question board.

CER Claim, Evidence, Reasoning

Make Your Claim Use your CER chart to make a claim about what light from a distant star or supernova tells us about it. Explain your reasoning.

Collect Evidence Use the lessons in this module to collect evidence to support your claim. Record your evidence as you move through the module.

Explain Your Reasoning You will revisit your claim and explain your reasoning at the end of the module.

GO ONLINE to access your CER chart and explore resources that can help you collect evidence.

LESSON 1: Explore & Explain: How Light Travels

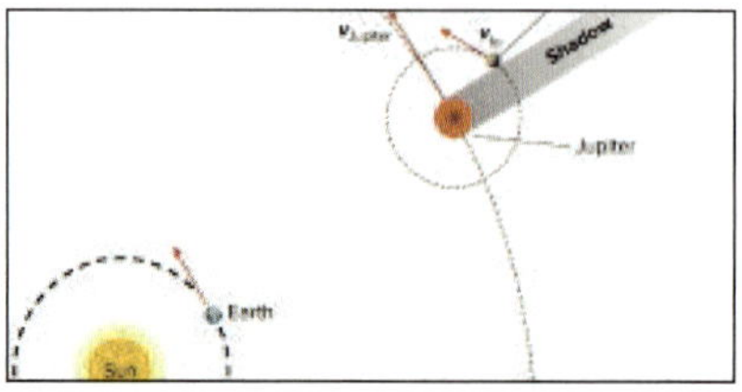

LESSON 2: Explore & Explain: Speed of Light

Smithsonian

Additional Resources

(t)Video Supplied by BBC Worldwide Learning; (b)Dwight Nadig/E+/Getty Images

LESSON 1

ILLUMINATION

FOCUS QUESTION

How does distance affect how bright a star appears?

Light

Light is electromagnetic radiation, like radio and microwaves. It can be modeled as a wave of changing electric and magnetic fields or as particles called photons. The wave model is useful for explaining many features of electromagnetic radiation, and the particle model explains other features. In this module, you will learn about the wave properties of light. You will learn about photons in a later module.

Light's path How does your body receive information? Many people respond to this question with the five senses, starting with sight and hearing. The sense of sight depends on light from your surroundings reaching your eyes. Did you ever wonder how light travels? Think of how a narrow beam of light, such as that of a flashlight or sunlight streaming through a small window, is made visible by dust particles in the air. You see the path of the light as a straight line. When your body blocks sunlight, you see your outline in a shadow, a result of light's straight path. **Figure 1** depicts light's straight path.

Figure 1 Light rays traveling in straight lines are evident in many situations.

Explain *how this photo demonstrates the wave properties of light.*

3D THINKING DCI Disciplinary Core Ideas CCC Crosscutting Concepts SEP Science & Engineering Practices

COLLECT EVIDENCE

Use your Science Journal to record the evidence you collect as you complete the readings and activities in this lesson.

INVESTIGATE

GO ONLINE to find these activities and more resources.

Quick Investigation: The Speed of Light

Use the relationship among the frequency, wavelength, and speed of a wave to calculate the speed of light.

Probeware Lab: Light Intensity and Distance

Analyze data to determine how distance from the source affects the intensity of light.

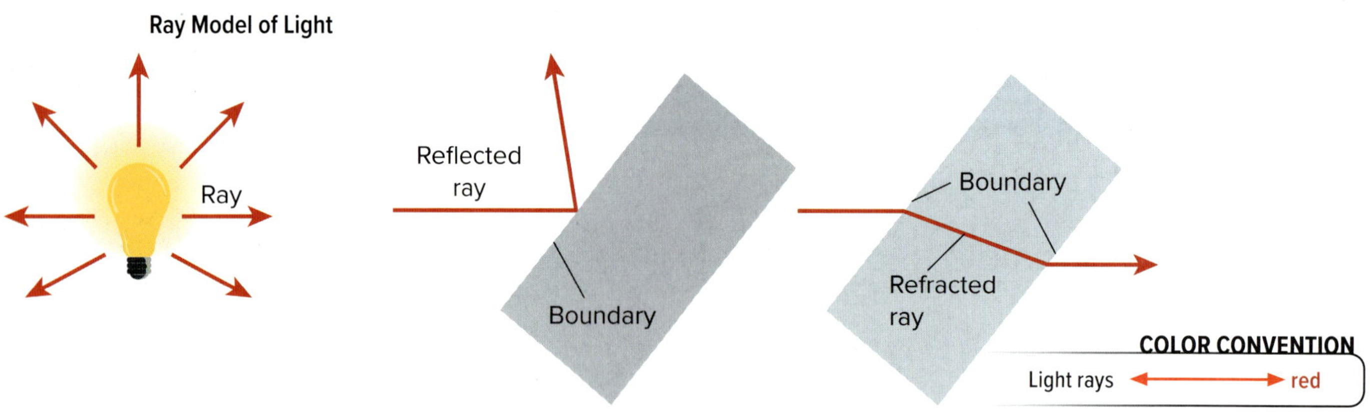

Figure 2 Light's straight-line path is demonstrated by the ray model. Light rays change direction when they are reflected or refracted by matter. In either case, light rays continue in a straight path.

Ray Model of Light

Imagine being in an empty, dark room when a small lightbulb is turned on in the center of the room. You can see around the room, and you can look at the bulb and see it. This must mean the bulb sends light in all directions. You could visualize the light coming from the bulb as an infinite number of arrows traveling straight away from the bulb in all directions. Each arrow represents a ray of light, which travels in a straight path until it reaches a boundary, as shown in **Figure 2.**

After interacting at a boundary, the ray still moves in a straight line, but its direction is changed. These basic principles—that light travels in straight lines and that its direction can be changed by encountering a boundary—constitute the **ray model of light.** The study of light interacting with matter is called ray optics or geometric optics.

Sources of light

What is the difference between sunlight and moonlight? For one, sunlight is much brighter. Another important fundamental difference is that the Sun produces and emits its own light, while the Moon is only visible because it reflects the Sun's light. Everything you see fits into one of these two categories. Objects such as the Sun and other stars that emit their own light are **luminous sources,** while those that you see due to light reflecting from them are illuminated sources, like the Moon and planets.

Luminous sources include natural sources such as flames and fireflies and human-made devices such as television screens, computer monitors, lasers, and tiny, light-emitting diodes. In **Figure 3,** the luminous source is fluorescent bulbs. They produce light from electrical energy. The other objects in the room are illuminated when the light from the bulbs is reflected off of them. In a room with no light, it would be impossible to see anything, because there is no light reflecting off of objects into your eyes.

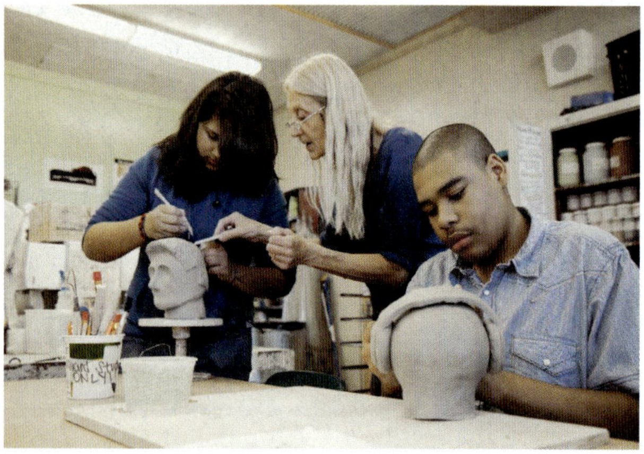

Figure 3 Objects in the room are visible because of reflected light.

Recognize *What, if anything, would be visible if the room had no luminous source? Explain.*

Figure 4 Light is transmitted through the transparent and translucent candle holders. The candle cannot be seen clearly in the opaque holder because the light is absorbed.

Light and matter Objects can absorb, reflect, or transmit light. Objects that reflect and absorb light but do not transmit it are **opaque.** Many common objects—such as books, people, and backpacks—are opaque. Mediums that transmit and reflect light but do not allow objects to be seen clearly through them are **translucent** mediums. The frosted glass in **Figure 4** is a translucent medium. A **transparent** medium, such as air or glass, transmits most of the light that reaches it. **Figure 4** illustrates objects that are opaque, translucent, and transparent. Transparent mediums transmit light, but they also often reflect some light. For example, you can see the glass of the transparent candle holder. This is possible because light is reflected off the glass.

All three types of objects also absorb some light. There are various factors that determine how much light will be absorbed, but opaque objects usually absorb a greater portion of light than translucent or transparent objects.

Get It?

Explain why it is possible to see a fish through a glass fishbowl and also see the glass of the bowl.

Quantity of Light

If you were to have a flashlight shone at you from across the room, what factors would determine how bright that light would appear to you? Three main factors determine the brightness: the quantity of light the flashlight produces, the distance between the lightbulb and your eye, and the angle at which the light rays hit your eye. In this lesson, you will read about the first two of these factors.

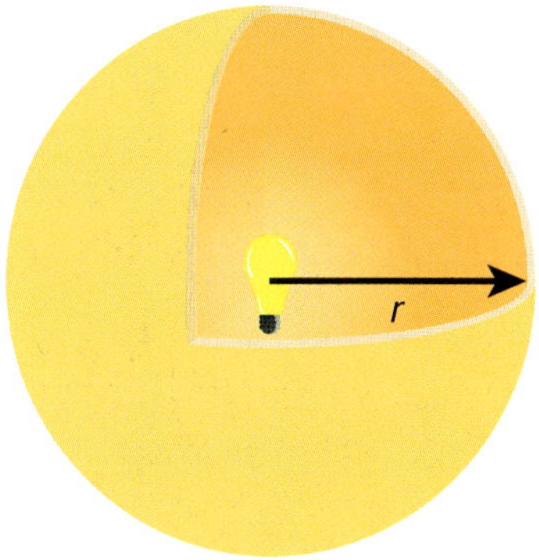

Figure 5 Luminous flux is the rate at which light rays are emitted from a luminous source.

Luminous flux With the ray model of light, a source that is brighter produces more light rays than a less bright source. Imagine again a single lightbulb sending rays in nearly all directions. How could you capture all the light it emits? You would need to construct a surface that completely encloses the bulb, as in **Figure 5.** The rate at which the bulb, a luminous source, produces light energy is called the **luminous flux** (*P*) and is measured in lumens (lm). The total amount of light that strikes the surface in a given unit of time depends only on the luminous flux of the source.

Illuminance Once you know the quantity of light being emitted by a luminous source, you can determine the amount of illumination the source provides to an object, such as a book. The luminous flux falling on a given surface area at any instant is called **illuminance** (E). It is measured in lux (lx), which is equivalent to lumens per square meter (lm/m^2). In this module, we assume, for simplification, that all light sources are point sources.

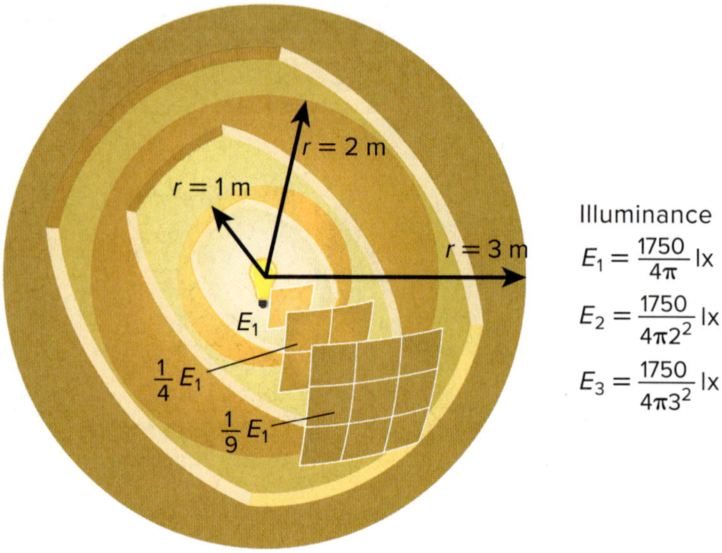

Figure 6 Illuminance (E) is the quantity of light that strikes a surface. As the distance from the luminous source (r) increases, E decreases. E depends on the inverse of r squared.

Consider the setup shown in **Figure 6.** The luminous flux of the source is 1750 lm (typical of a 26-W compact fluorescent bulb). What is the illuminance of the sphere's inside surface at $r = 1$ m? Because all the bulb's luminous flux strikes the surface, divide the luminous flux by the surface area of the sphere, $4\pi r^2$. The surface area is $4\pi(1.00\text{ m})^2 = 4\pi\text{ m}^2$, so the illuminance is $\frac{1750\text{ lm}}{4\pi\text{ m}^2} = 139$ lx. This mathematical relationship means that at a distance of 1.00 m from the bulb, 139 lm strikes each square meter.

Get It?

Define *illuminance*, and state the units that are used for illuminance.

Inverse-square relationship What if the sphere surrounding the lamp were larger? If the sphere's radius were 2.00 m, the luminous flux still would total 1750 lm because it only depends on the bulb. With a radius of 2.00 m, however, the area of the sphere would now be equal to $4\pi(2.00\text{ m})^2 = 16.0\pi\text{ m}^2$. The new area is four times larger than that of the 1.00-m sphere, as shown in **Figure 6.** The illuminance of the inside of the 2.00-m sphere is $\frac{1750\text{ lm}}{(160\pi\text{ m}^2)} = 34.8$ lx, so 34.8 lm strikes each square meter.

The illuminance on the inside surface of the 2.00-m sphere (E_2) is one-fourth the illuminance on the inside of the 1.00-m sphere. In the same way, the inside of a sphere with a 3.00-m radius has an illuminance only one-ninth $\left[\left(\frac{1}{3}\right)^2\right]$ as large as that of the 1.00-m sphere. **Figure 6** shows that the illuminance produced by a point source is proportional to $\frac{1}{r^2}$: an inverse-square relationship. In the case of the 3.00-m radius, only 15.5 lm strike each square meter inside the sphere. As the light rays spread out in straight lines in all directions from a point source, the number of light rays that illuminate a unit of area decreases as the square of the distance from the point source.

CCC CROSSCUTTING CONCEPTS

Systems and System Models Write a paragraph explaining how mathematical models were used in the study of light. Include information about the system and system boundaries that scientists used.

CONNECTING MATH to Physics

Direct and Inverse Relationships The illuminance provided by a source of light has both a direct and an inverse relationship.

Math	Physics
$y = \frac{x}{az^2}$	$E = \frac{P}{4\pi r^2}$
If z is constant, then y is directly proportional to x. • When x increases, y increases. • When x decreases, y decreases.	If r is constant, then E is directly proportional to P. • When P increases, E increases. • When P decreases, E decreases.
If x is constant, then y is inversely proportional to z^2. • When z^2 increases, y decreases. • When z^2 decreases, y increases.	If P is constant, then E is inversely proportional to r^2. • When r^2 increases, E increases. • When r^2 decreases, E decreases.

Luminous intensity Some luminous sources are specified in candelas (cd). A candela is not a measure of luminous flux but of luminous intensity. The luminous intensity of a point source is the luminous flux that falls on 1 m^2 of the inside of a 1-m-radius sphere, so luminous intensity is luminous flux divided by 4π. A bulb with 1750 lm of flux has an intensity of $\frac{1750\ \text{lm}}{4\pi} = 139$ cd.

In **Figure 7,** the lightbulb is twice as far away from the screen as the candle. For the bulb to provide the same illuminance on its side of the screen as the candle does on the candle side of the screen, the bulb would have to be four times brighter than the candle. The lightbulb's luminous intensity, therefore, would have to be four times the candle's luminous intensity. If both sources in **Figure 7** had the same luminous intensity, the source at $2r$ would only provide one-quarter the illuminance to the screen. This is consistent with the inverse-square relationship we just developed.

Get It?

Describe what luminous intensity is a measure of and what its relationship is to illuminance.

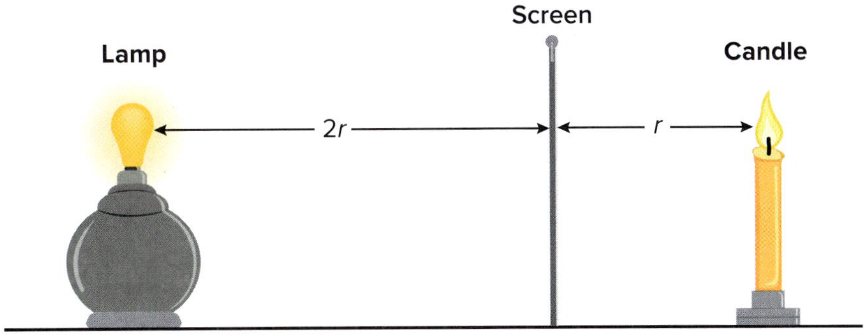

Figure 7 For the lightbulb and the candle to provide the same illuminance to the screen, the luminous intensity of the lightbulb is four times that of the candle.

Surface Illumination

Think again about the scenario in which a flashlight is shining at you from across the room. If the bulb has a small luminous intensity, the light will not be very bright. To increase the brightness, you could use a brighter bulb, thereby increasing the luminous flux, or you could move so that your eyes are closer to the light, decreasing the distance between the light source and your eyes. Following the simplification that we are treating all light sources as point sources, the illuminance and distance will follow the inverse-square relationship. In this case, and in all the cases we will deal with in this book, the illuminance caused by a point light source is represented by the following equation.

Real-World Physics

Illuminated Minds

When deciding how to achieve the correct illuminance on students' desktops in a classroom, architects must consider the luminous flux of the lights as well as the distance of the lights above the desktops. In addition, the efficiencies of the light sources are an important economic factor.

Point-Source Illuminance

If an object is illuminated by a point source of light, then the illuminance at the object is equal to the luminous flux of the light source divided by the surface area of the sphere whose radius is equal to the distance the object is from the light source.

$$E = \frac{P}{4\pi r^2}$$

Remember that the luminous flux of the light source is spreading out in all directions, so only some fraction of the luminous flux is available to illuminate the object. Use of this equation is valid only if the light from the luminous source strikes perpendicular to the surface it is illuminating. It is also only valid if the luminous source is small enough or far enough away to be considered a point source. Thus, the equation does not give accurate values of illuminance for long fluorescent lamps or lightbulbs that are close to the surfaces they illuminate.

Engineers who design lighting systems must understand how the light will be used. If an even illumination is needed to prevent dark areas, the common practice is to evenly space normal lights over the area to be illuminated, as was most likely done with the lights in your classroom. Because such light sources do not produce truly uniform light, however, engineers also design special light sources that control the spread of the light, such that they produce even illuminations over large surface areas. For safety reasons, this is extremely important for automobile headlights, as in **Figure 8.** Automobile engineers must consider these factors when designing headlights.

Figure 8 Dark areas can occur if headlights on cars are not set at the correct angles to adequately illuminate the road.

EXAMPLE Problem 1

ILLUMINATION OF A SURFACE What is the illuminance on your desktop if it is lit by a 1750-lm lamp that is 2.50 m above your desk?

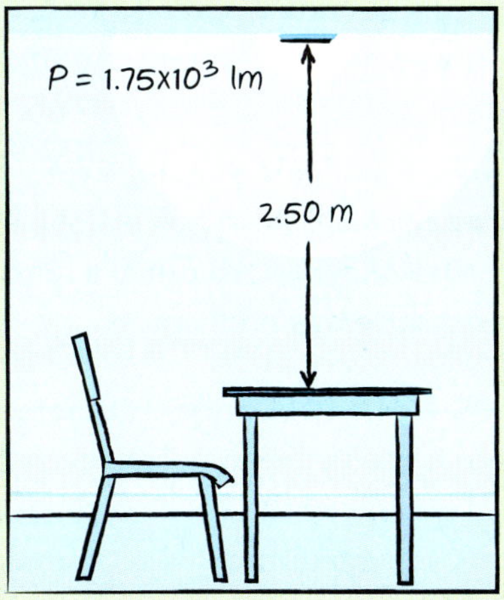

1 ANALYZE AND SKETCH THE PROBLEM

- Assume the lightbulb is the point source.
- Diagram the position of the bulb and the desktop. Label *P* and *r*.

Known	Unknown
$P = 1.75 \times 10^3$ lm	$E = ?$
$r = 2.50$ m	

2 SOLVE FOR THE UNKNOWN

The surface is perpendicular to the direction in which the light ray is traveling, so you can use the point-source illuminance equation.

$$E = \frac{P}{4\pi r^2}$$

$$= \frac{1.75 \times 10^3 \text{ lm}}{4\pi(2.50 \text{ m})^2}$$ Substitute $P = 1.75 \times 10^3$ lm, $r = 2.50$ m

$$= 22.3 \text{ lm/m}^2$$

$$= 22.3 \text{ lx}$$

3 EVALUATE THE ANSWER

- **Are the units correct?** The units of luminance are $\text{lm/m}^2 = \text{lx}$, which the answer agrees with.
- **Do the signs make sense?** All quantities are positive, as they should be.
- **Is the magnitude realistic?** Illuminance from an 1800-lm lamp at a distance of 2 m is about 20 lx.

PRACTICE Problems

ADDITIONAL PRACTICE

1. A lamp is moved from 30 cm to 90 cm above the pages of a book. Compare the illumination on the book before and after the lamp is moved.
2. Draw a graph of the illuminance produced by a lamp with a luminous flux of 2275 lm at distances from 0.50 m and 5.0 m.
3. A 64-cd point source of light is 3.0 m away from a painting. What is the illumination on the painting in lux?
4. A screen is placed between two lamps so that they illuminate the screen equally, as shown in **Figure 9.** The first lamp emits a luminous flux of 1445 lm and is 2.5 m from the screen. What is the distance of the second lamp from the screen if the luminous flux is 2375 lm?

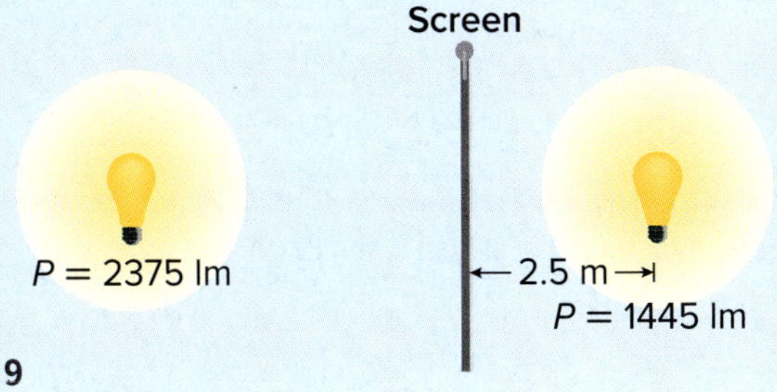

Figure 9

5. What is the illumination on a surface that is 3.0 m below a 150-W incandescent lamp that emits a luminous flux of 2275 lm?
6. A public school law requires a minimum illuminance of 160 lx at the surface of each student's desk. An architect's specifications call for classroom lights to be located 2.0 m above the desks. What is the minimum luminous flux that the lights must produce?
7. **CHALLENGE** Your local public library is planning to remodel the computer lab. The contractors have purchased fluorescent lamps with a rated luminous flux of 1750 lm. The desired illumination on the keyboard surfaces is 175 lx. Assume a single lamp illuminates each keyboard. What distance above the surface should the lights be placed to achieve the desired illumination? If the contractors had also already purchased fixtures to hold the lights that when installed would be 1.5 m above the keyboard surface, would the desired illuminance be achieved? If not, would the illuminance be greater or less than desired? What change in the lamp's luminous flux would be required to achieve the desired illuminance?

The Speed of Light

Arguments that light must travel at a finite speed have existed for more than 2400 years. By the seventeenth century, several scientists had performed experiments that supported the view that light travels at a finite speed, but that this speed is much faster than the speed of sound.

Actually measuring the speed of light was not an easy task in the seventeenth century. As you know from studying motion, if you can measure the time light takes to travel a certain distance, you can calculate the speed of light. However, the time that it takes light to travel between objects on Earth is much shorter than a human's reaction time. How could a seventeenth-century scientist solve this problem?

Clues from Io Danish astronomer Ole Roemer was the first to measure the time it took for light to travel between two points with any success. Between 1668 and 1674, Roemer made 70 measurements of the 1.8-day orbital period of Io, one of Jupiter's moons. He recorded the times when Io emerged from Jupiter's shadow, as shown in **Figure 10.** He made his measurements as part of a project to improve maps by calculating the longitude of locations on Earth. This is an early example of the needs of technology driving scientific advances.

After making many measurements, Roemer was able to predict when the next eclipse of Io would occur. He compared his predictions with the actual measured times and found that Io's observed orbital period increased on average by about 13 s per orbit when Earth was moving away from Jupiter and decreased on average by about 13 s per orbit when Earth was approaching Jupiter. Roemer believed that Jupiter's moons were just as regular in their orbits as Earth's moon; thus, he wondered what might cause this discrepancy in the measurement of Io's orbital period. He considered another variable within the system, the movement and position of Earth relative to Jupiter.

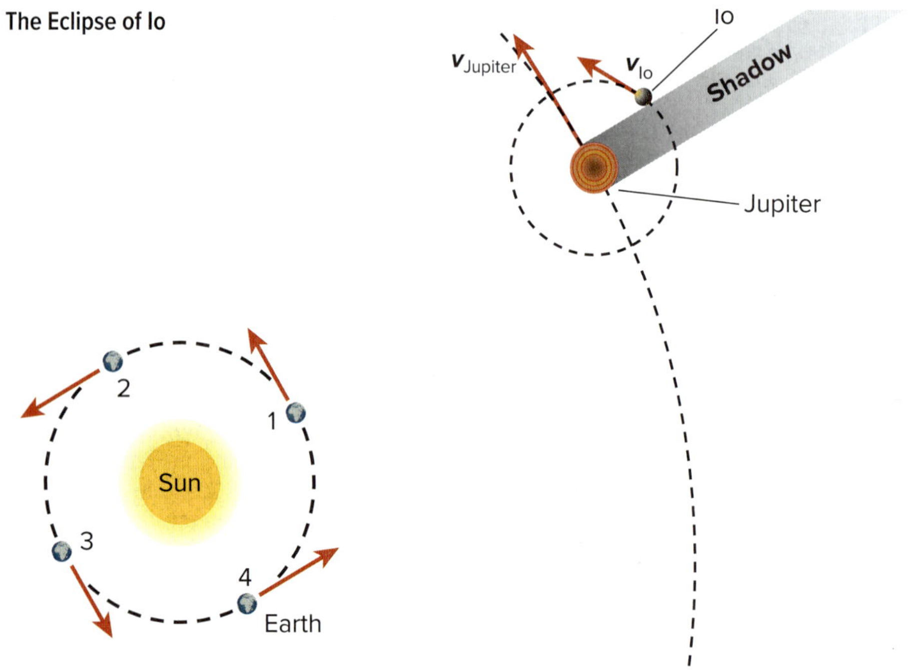

Figure 10 As Earth approaches Jupiter, the light reflected from Io takes less time to reach Earth than when Earth moves away from Jupiter. (Illustration is not to scale.)

Measuring the speed of light Roemer concluded that as Earth moved away from Jupiter, the light from each new appearance of Io took longer to reach Earth because it traveled farther. As Earth moved toward Jupiter, Io's orbital period seemed to decrease. During the 182.5 days it took for Earth to travel from position 1 to position 3, shown in **Figure 10,** there were (182.5 days)(1 Io eclipse/1.8 days) = 1.0×10^2 Io eclipses. Thus, for light to travel the diameter of Earth's orbit, he calculated that it takes (1.0×10^2 eclipses)(13 s/eclipse) = 1.3×10^3 s, or 22 min.

Using the presently known value of the diameter of Earth's orbit (2.93×10^{11} m), Roemer's value of 22 min gives a value for the speed of light of 2.9×10^{11} m/((22 min)(60 s/min)) = 2.2×10^8 m/s. Today, the speed of light is known to be closer to 3.0×10^8 m/s, so light takes 16.5 min, not 22 min, to cross Earth's orbit. Nevertheless, Roemer had successfully proved that light travels at a finite speed.

Michelson's measurements Although many measurements of the speed of light have been made, the most notable were performed by American physicist Albert A. Michelson. Between 1880 and the 1920s, he developed Earth-based techniques to measure the speed of light. In 1926 Michelson measured the time required for light to make a round trip between two California mountains 35 km apart. Michelson's best result was $(2.99796 \pm 0.00004)\times10^8$ m/s. For this work, he became the first American to receive a Nobel Prize in science.

The speed of light in a vacuum has its own special symbol, c. The International Committee on Weights and Measurements has measured and defined the speed of light in a vacuum to be c = 299,792,458 m/s. For many calculations, the value c = 3.00×10^8 m/s is precise enough. At this speed, light travels 9.46×10^{12} km in a year. This distance is called a light-year.

Check Your Progress

8. **Light** What evidence have you observed that light travels in a straight line?
9. **Light Properties** Why might you choose a window shade that is translucent? Opaque?
10. **Illuminance** Does one lightbulb provide more or less illuminance than two identical light-bulbs at twice the distance? Explain.
11. **Luminous Intensity** Two lamps illuminate a screen equally from distances shown in **Figure 11.** If Lamp A is rated 75 cd, what is Lamp B rated?

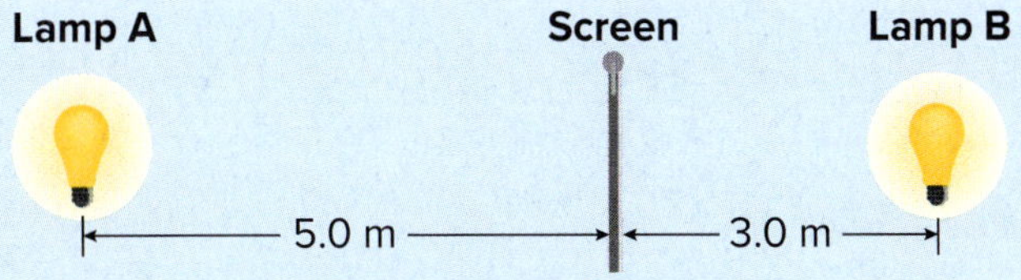

Figure 11

12. **Distance of a Light Source** A lightbulb illuminating your computer keyboard provides only half the illuminance that it should. If it is currently 1.0 m away, how far should it be to provide the correct illuminance?
13. **Light and Sound Travel** How far does light travel in the time it takes sound to travel 1 cm in air at 20°C?
14. **Distance of Light Travel** The distance to the Moon can be found with the help of mirrors left on the Moon by astronauts. A pulse of light is sent to the Moon and returns to Earth in 2.562 s. Using the defined value for the speed of light to the same precision, calculate the distance from Earth to the Moon.
15. **Critical Thinking** The correct time taken for light to cross Earth's orbit is 16.5 min, and the diameter of Earth's orbit is 2.98×10^{11} m. Calculate the speed of light using Roemer's method. Does this method appear to be accurate? Why or why not?

LEARNSMART® Go online to follow your personalized learning path to review, practice, and reinforce your understanding.

LESSON 2
THE WAVE NATURE OF LIGHT

FOCUS QUESTION

How do scientists use the Doppler shift to determine how stars and galaxies are moving?

Diffraction and the Wave Model

In 1665 Italian scientist Francesco Maria Grimaldi observed that the edges of shadows are not perfectly sharp. He introduced a narrow beam of light into a dark room and held a rod in front of the light such that it cast a shadow on a white surface. The shadow cast by the rod was wider than the shadow should have been if light traveled in a straight line past the edges of the rod. Grimaldi also noted that the shadow was bordered by colored bands. He determined that both of these observations could be explained if light bent slightly. He called the bending of light as it passes the edge of a barrier **diffraction.**

Huygens' principle In 1678 Dutch scientist Christiaan Huygens used a wave model to explain diffraction. According to Huygens' principle, all the points of a wavefront of light can be thought of as new sources of smaller waves. These smaller waves, or wavelets, expand in every direction and are in step with one another. A flat, or plane, wavefront of light consists of an infinite number of point sources in a line. **Figure 12** illustrates Huygen's principle.

Huygens' Principle

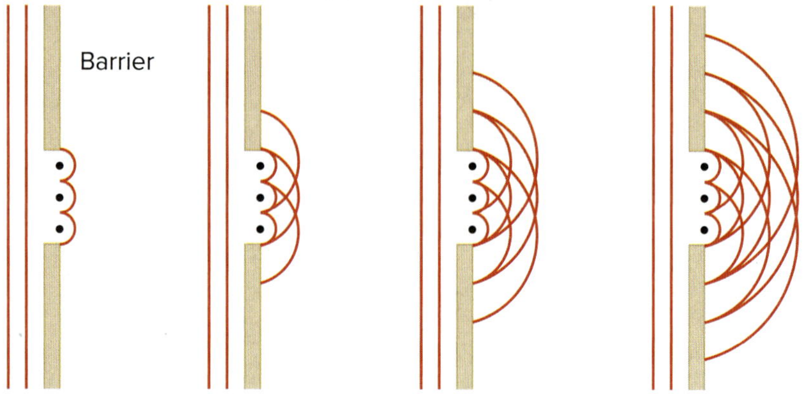

Figure 12 Huygens' wavelets combine to form a straight wavefront, except at the edges of the wave. The wavelets spread out in a circular manner when a barrier creates an edge.

3D THINKING DCI Disciplinary Core Ideas | CCC Crosscutting Concepts | SEP Science & Engineering Practices

COLLECT EVIDENCE

Use your Science Journal to record the evidence you collect as you complete the readings and activities in this lesson.

INVESTIGATE

GO ONLINE to find these activities and more resources.

PhysicsLAB: Polarization
Carry out an investigation to determine the effect of polarizting filters on light.

Revisit the Encounter the Phenomenon Question
What information from this lesson can help you answer the Unit and Module questions?

As a wavefront passes by an edge, the edge cuts the wavefront such that each circular wavelet generated by each Huygens' point will propagate as a circular wave in the region where the original wavefront was bent. This wave model explained the diffraction Grimaldi.

Color

In 1666 Newton performed experiments on the colors produced when a narrow beam of sunlight passed through a glass prism, as shown in **Figure 13.** Newton called the ordered arrangement of colors a spectrum. Using his later-disproved corpuscle (or particle) model of light, he thought that particles of light interacted with some unevenness in the glass to produce the spectrum.

To test this assumption, Newton allowed the spectrum from one prism to fall on a second prism. If the spectrum was caused by irregularities in the glass, he reasoned that the second prism would increase the spread in colors. Instead, the second prism reversed the spreading of colors and recombined them to form white light. After more experiments, Newton concluded that white light is composed of colors and that a property of the glass other than unevenness caused the light to separate into colors.

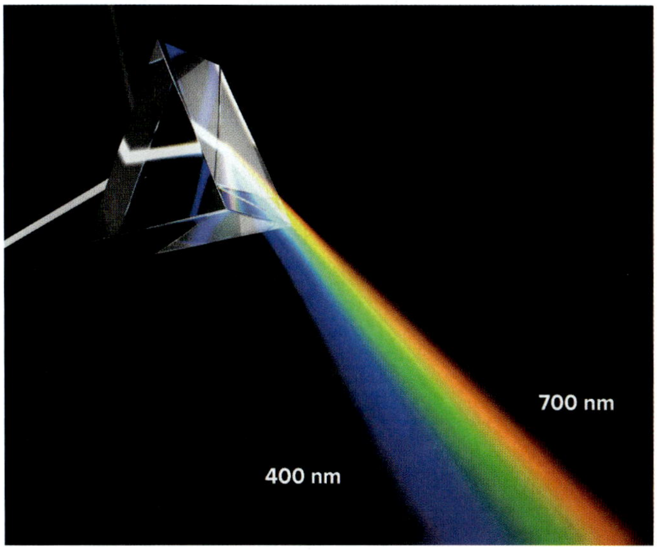

Figure 13 White light is separated into bands of color by a prism. Each color has a different wavelength.

Different wavelengths

Can the wave model of light explain Newton's observations? For light to be a wave, it must have wavelength and frequency. The work of Grimaldi, Huygens, Newton, and others suggested that the color of light is related to wavelength. Visible light falls within the range of wavelengths from about 400 nm (4.00×10^{-7} m) to 700 nm (7.00×10^{-7} m), as shown in **Figure 13.** The longest visible wavelengths are seen as red light and the shortest as violet.

As white light crosses the boundary from air into glass and back into air in **Figure 13,** its wave nature causes each different color of light to be bent at a different angle. The shorter the wavelength, the more the light is bent. This unequal bending of the different colors causes the white light to be spread into a spectrum.

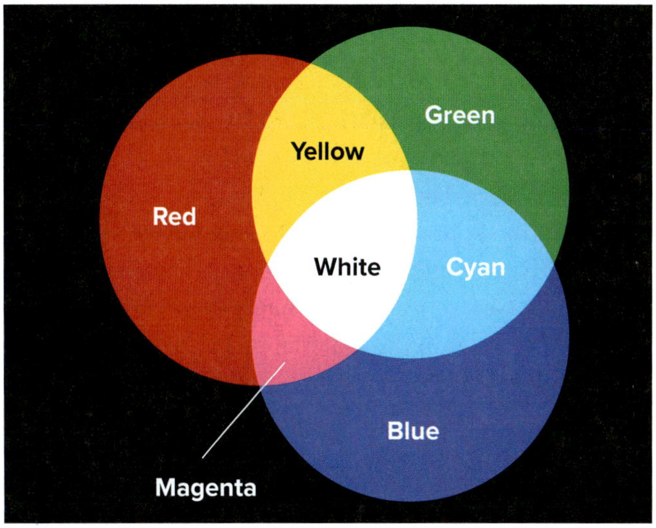

Figure 14 Red, green, and blue light, the primary colors, combine in pairs to produce yellow, cyan, or magenta light. The region where these three colors overlap on the screen appears white.

Color by addition of light

White light can be formed from colored light in a variety of ways. For example, when the correct intensities of red, green, and blue light are projected onto a white screen, as in **Figure 14,** white light is formed. This is called the additive color process, which is used in many television screens. A television screen uses three colors—red, green, and blue. Combinations of these produce the colors you see.

The correct intensities of red, green, and blue light appear on a screen as white light when combined. Because of this phenomenon, they are called the **primary colors** of light. The primary colors can be mixed in pairs to form three additional colors, as shown in **Figure 14.** Red and green light together produce yellow light, blue and green light produce cyan, and red and blue light produce magenta. The colors yellow, cyan, and magenta are called secondary colors. A **secondary color** is a combination of two primary colors. Note that these are slightly different from the primary and secondary colors you might have learned in art class; the reasons for this will be explained later in this module.

As shown in **Figure 14,** yellow light can be made from red light and green light. If yellow light and blue light are projected onto a white screen with the correct intensities, the surface will appear to be white. **Complementary colors** are two colors of light that can be combined to produce white light. Thus, yellow is a complementary color of blue, and vice versa, because the two colors of light combine to make white light. In the same way, cyan and red are complementary colors. Magenta and green are the other pair of complementary colors. A practical application of this is that yellowish laundry can be whitened with a bluing agent added to detergent.

Figure 15 The colors of objects we see are determined by which wavelengths of light are absorbed and which are reflected.

Explain *why the die that is yellow in the white light appears red in red light.*

Color by subtraction of light

As you learned in the first lesson of this module, objects can reflect and transmit light. They also can absorb light. The color of an object depends on the wavelengths present in the light that illuminates the object. The color also depends on which wavelengths are absorbed by the object and which wavelengths are reflected. The natural existence or artificial placement of dyes in the material of an object, or pigments on its surface, gives the object color.

Dyes You are probably familiar with dyes that are used to color cloth. Dyes can be made from plant or insect extracts. For example, purple dye can be extracted from the berries of a black mulberry tree. The saffron crocus is a source of yellow dye. One type of red dye is extracted from an insect called a cochineal. A dye is a molecule that absorbs certain wavelengths of light and transmits or reflects others. When light is absorbed, its energy is transferred to the object that it strikes and is transformed into other forms of energy. A red shirt is red because the dyes in it reflect mostly red light to our eyes. When white light falls on the red object shown in **Figure 15,** the dye molecules in the object absorb most of the blue and green light and reflect mostly red light. When only blue light falls on the red object, very little light is reflected and the object appears to be almost black.

Get It?

Distinguish the difference between color by subtraction and color by addition.

Pigments The difference between a dye and a pigment is that pigments usually are made of crushed minerals rather than plant or insect extracts. For example, hematite produces a red pigment, and blue pigment can be obtained from azurite. Pigment particles can be seen with a microscope. A pigment that absorbs only one primary color and reflects two from white light is called a **primary pigment.** Yellow pigment absorbs blue light and reflects red and green light. Yellow, cyan, and magenta are the colors of primary pigments.

A pigment that absorbs two primary colors and reflects one color is called a **secondary pigment.** The colors of secondary pigments are red (which absorbs green and blue light), green (which absorbs red and blue light), and blue (which absorbs red and green light). Note that the primary pigment colors are these secondary colors of light. In the same way, the secondary pigment colors are light's primary colors.

Figure 16 Magenta, cyan, and yellow are the primary pigments. Secondary pigments, red, green, and blue, are produced from mixing the primary pigments in pairs.

The primary and secondary pigments are shown in **Figure 16.** When the primary pigments yellow and cyan are mixed, the yellow absorbs blue light and the cyan absorbs red light. **Figure 16** shows yellow and cyan combining to make green pigment. When yellow pigment is mixed with the secondary pigment blue, which absorbs green and red light, all the primary colors are absorbed, and the result is black. Yellow and blue are complementary pigments. Cyan and red, as well as magenta and green, are also complementary pigments.

CHEMISTRY Connection A color printer uses yellow, magenta, and cyan dots of pigment to make a color image on paper. Often, pigments that are used are finely ground compounds, such as titanium(IV) oxide (white), chromium(III) oxide (green), and cadmium sulfide (yellow). Pigments mix to form suspensions rather than solutions. Their chemical form is not changed in a mixture, so they still absorb and reflect the same wavelengths.

BIOLOGY Connection You can now begin to understand the colors that you see in **Figure 17.** The plants on the mountain look green because of the chlorophyll in them. One type of chlorophyll absorbs mostly red light and the other absorbs mostly blue light, but they both reflect green light. The energy in the red and blue light that is absorbed is used by the plants during photosynthesis to make food.

Figure 17 Chlorophyll in green leaves reflects mostly green light, giving the leaves their color.

Explain *why the plants are various shades of green.*

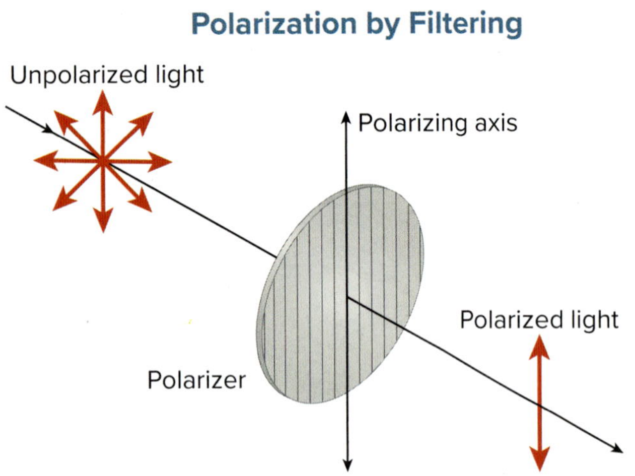

Figure 18 Nonpolarized light rays vibrate randomly in every direction perpendicular to the direction they travel. A polarizing medium blocks light that is not parallel to the polarizing axis.

Polarization of Light

Have you ever looked at light reflected off a road through polarizing sunglasses? If you rotate the glasses, the road first appears to be dark, then light, and then dark again. Light from a lamp, however, changes very little as the glasses are rotated. Why is there a difference? Normal lamplight is not polarized. However, the light that is coming from the road is reflected and has become polarized. **Polarization** is the production of light with a specific pattern of oscillation.

Recall that light behaves as a transverse wave. For waves on a rope, the oscillating medium is the rope. For light waves, the oscillating medium is the electric field. When this electric field oscillates in random directions, the light is nonpolarized. How can you filter nonpolarized light so that whatever passes through the filter is polarized light?

Get It?
Draw a diagram, with text explanations, showing nonpolarized light.

Polarization by filtering The lines in the polarizer in **Figure 18** represent a polarizing axis. The light with the portion of the electric field that oscillates parallel to these lines passes through. The light with the portion of the electric field that oscillates perpendicular to these lines is absorbed. If a polarizer is placed in a beam of nonpolarized light, only the components of the waves in the same direction as the polarizing axis can pass through. As a result, half of the total light passes through, reducing the intensity of the light by half.

Get It?
Draw a second diagram, with text explanations, showing nonpolarized light passing through a filter as polarized light.

CHEMISTRY Connection Polarizing mediums contain long molecules in which electrons can oscillate, or move back and forth, all in the same direction. As light travels past the molecules, the electrons absorb light waves that oscillate in the same direction as the electrons. This allows light waves vibrating in one direction to pass, while the waves vibrating in the other direction are absorbed. The direction of a polarizing medium perpendicular to the long molecules is called the polarizing axis. Only waves oscillating parallel to that axis can pass through.

Polarization by reflection When you look through a polarizing filter at the light reflected by a sheet of glass and rotate the filter, you will see the light brighten and dim. The light is partially polarized parallel to the plane of the glass when it is reflected. Polarized reflected light causes glare. Polarizing sunglasses reduce glare from the polarized light reflected off roads. Photographers can use polarizing filters over camera lenses to block reflected light. This result is shown in **Figure 19.**

Figure 19 Glare is light that has been polarized by reflection. Photographers use polarizing filters to reduce glare.

Malus's law Suppose you produce polarized light with a polarizing filter. What would happen if you place a second polarizing filter in the path of the polarized light? If the polarizing axis of the second filter is parallel to that of the first, the light will pass through. If the polarizing axis of the second filter is perpendicular to that of the first, no light will pass through, as shown in **Figure 20.**

If the light intensity after the first polarizing filter is I_1 and the intensity after the second filter is I_2, how can you control I_2? I_2 depends only on I_1 and the angle between the axes of the filters, θ. If θ is 0, I_2 equals I_1; if θ is 90°, all of the light is blocked, resulting in I_2 being 0. This indicates that the intensity might depend on the cosine of θ. The actual relationship is a that of a cosine squared. The law that explains the reduction of light intensity as light passes through a second polarizing filter is **Malus's law.**

Malus's Law

The intensity of light coming out of a second polarizing filter is equal to the intensity of polarized light coming out of a first polarizing filter multiplied by the cosine, squared, of the angle between the polarizing axes of the two filters.

$$I_2 = I_1 \cos^2 \theta$$

Figure 20 Polarizing filters with their axes parallel will allow the light with the same orientation to pass. With the polarizing axes at a 45° angle, the filters allow some light to pass. If the axes of the filters are perpendicular, the second filter will block the light that has passed through the first filter.

PHYSICS Challenge

You place an analyzer filter between the two cross-polarized filters, such that its polarizing axis is not parallel to either of the two filters, as shown in the figure to the right.

1. You observe that some light passes through filter 2, though no light passed through filter 2 before you inserted the analyzer filter. Why does this happen?
2. The analyzer filter is placed at an angle of θ relative to the polarizing axis of filter 1. Derive an equation for the intensity of light coming out of filter 2 compared to the intensity of light coming out of filter 1.

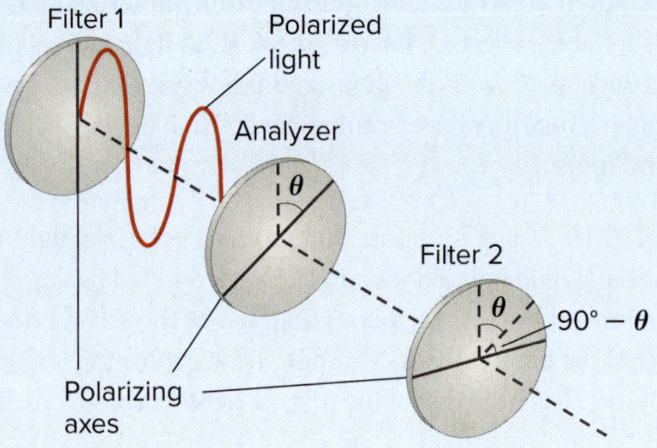

Speed, Wavelength, and Frequency of Light

As you have learned, the source of a wave determines that wave's frequency (f), and the medium and the frequency together determine the wavelength (λ) of a wave. Because light has wave properties, the same mathematical models used to describe waves in general can be used to describe light. For light of a given frequency traveling through a vacuum, wavelength is a function of the speed of light (c), which can be written as $\lambda_0 = c/f$. The development of the laser in the 1960s provided new ways to measure the speed of light. The frequency of light can be measured with extreme precision using lasers and the time standard of atomic clocks. Measurements of wavelengths of light, however, are much less precise.

All colors of light travel at c in a vacuum, though the wavelengths are different. Since $\lambda_0 = c/f$, once the frequency of a light wave in a vacuum is measured, the wavelength can be determined.

Relative motion and light What happens if a light source travels toward you or you move toward the source? You have learned that the frequency of a sound heard by a listener changes if either the source or the listener of the sound is moving. The same is true for light. However, when you consider the velocities of a sound source and the observer, you are really considering each one's velocity relative to the medium through which the sound travels. This is not the case for light.

Get It?

Explain how the wavelength, frequency, and speed of light are mathematically related.

The Doppler effect The nature of light waves is such that they are not vibrations of the particles of a medium. The Doppler effect for light can involve only the relative velocity between the source and the observer. Remember that the only factors in the Doppler effect are the velocity components along the axis between the source and the observer, as shown in **Figure 21** on the next page.

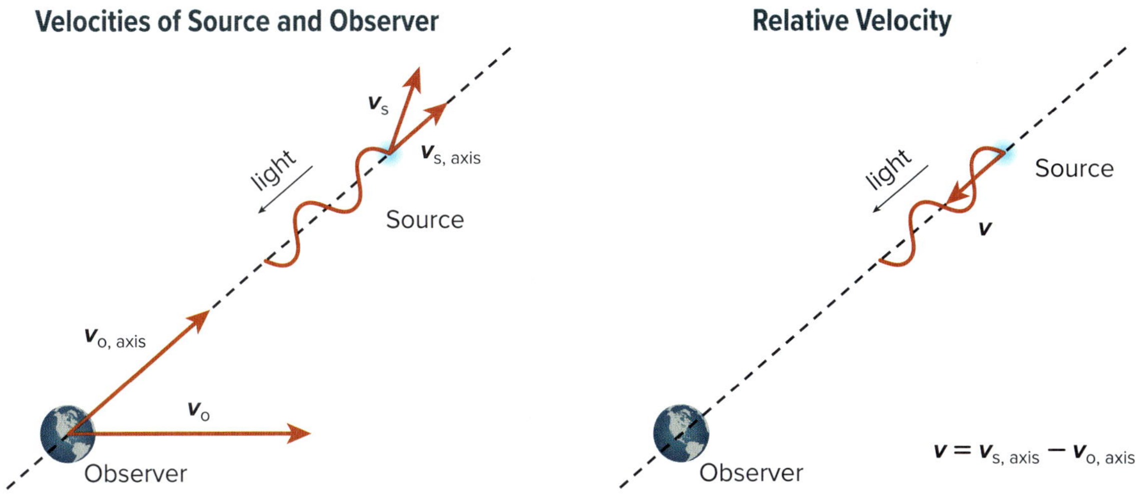

Figure 21 The Doppler effect describes how light frequency changes if an observer and a light source are moving toward or away from each other.

Doppler effect for light problems can be simplified by considering axial relative speeds that are much less than the speed of light ($v << c$). This simplification is used to develop the equation for the observed light frequency (f_{obs}), shown below.

Observed Light Frequency

The observed frequency of light from a source is equal to the actual frequency of the light generated by the source, times the quantity 1 plus the relative speed along the axis, divided by the speed of light, between the source and the observer if they are moving toward each other, or 1 minus the relative speed, divided by the speed of light, if they are moving away from each other.

$$F_{obs} = f\left(1 \pm \frac{v}{c}\right)$$

Applications Most applications of the Doppler effect for light are in astronomy, where phenomena are discussed more in terms of wavelength. Using the relationship $\lambda = c/f$ and the $v << c$ simplification, the following equation describes the Doppler shift ($\Delta\lambda$), the difference between the observed and the actual wavelengths.

Doppler Shift

The difference between the observed wavelength of light and the actual wavelength of light generated by a source is equal to the actual wavelength of light generated by the source, times the relative speed of the source and observer, divided by the speed of light.

$$(\lambda_{obs} - \lambda) = \Delta\lambda = \pm\left(\frac{v}{c}\right)\lambda$$

A positive change in wavelength occurs when the relative velocity of the source is away from the observer. In this case, the observed wavelength is longer than the original wavelength. The light appears closer to the red end of the spectrum than it normally would. We say this light is red shifted. A negative change in wavelength occurs when the relative velocity of the source is in a direction toward the observer. In this case, the observed wavelength is shorter than the original wavelength. This is known as a blue shift.

Because the speed of light is constant, when the wavelength is red shifted, the observed frequency is lower than the original due to the inverse relationship between the two variables. When light is blue shifted, the observed frequency is higher.

ASTRONOMY Connection Astronomers can determine how objects, such as galaxies, are moving relative to Earth by observing the Doppler shift of their light. This is done by observing the spectrum of light coming from stars in the galaxy using a spectrometer, as shown in **Figure 22.** The same elements that are present in the stars of galaxies emit light of specific wavelengths in labs on Earth. By comparing wavelength, astronomers can learn the velocities of objects toward or away from Earth.

Figure 22 The bottom hydrogen emission spectrum is red shifted compared to the laboratory spectrum, indicating the light source is moving away from Earth.

PRACTICE Problems

ADDITIONAL PRACTICE

16. Oxygen can be made to produce light with a wavelength of 513 nm. What is the frequency of this light?

17. A hydrogen atom in a galaxy moving with a speed of 6.55×10^6 m/s away from Earth emits light with a frequency of 6.16×10^{14} Hz. What frequency of light from that hydrogen atom would be observed by an astronomer on Earth?

18. A hydrogen atom in a galaxy moving with a speed of 6.55×10^6 m/s away from Earth emits light with a wavelength of 486 nm. What wavelength would be observed on Earth from that hydrogen atom?

19. CHALLENGE An astronomer is looking at the spectrum of a galaxy and finds that it has an oxygen spectral line of 525 nm, while the laboratory value is measured at 513 nm. Calculate how fast the galaxy would be moving toward or away from Earth and how you know.

STEM CAREER Connection

Photographer

Do you like capturing important events with a camera? Photographers must use light to compose a perfect picture or to achieve a desired effect. Understanding the fundamentals of light is an important part of being a good photographer.

In 1929, Edwin Hubble analyzed the light from many galaxies like the one shown in **Figure 23.** He observed that the light produced by familiar elements were at longer wavelengths than he had expected them to be. The light was shifted toward the red end of the spectrum. No matter what area of the sky he observed, almost all the galaxies were sending red shifted light to Earth. What do you think caused the spectral lines to be red shifted?

Hubble concluded that galaxies are moving away from Earth and suggested that the universe is expanding. Additional studies since then have supported this conclusion. As galaxies move, they sometimes collide and merge. The results of the collision depend on the size and speed of the galaxies.

Figure 23 Edwin Hubble observed that galaxies were sending red shifted light to Earth, indicating that they were moving away from Earth and the universe was expanding.

You have learned that some characteristics of light can be explained with a simple ray model of light, whereas others require a wave model of light. You can use both of these models to study how light interacts with mirrors and lenses. There are some aspects of light that can be understood only through the use of the wave model of light.

Check Your Progress

20. **Doppler Effect** Describe the relative motions of objects when light is red shifted and when light is blue shifted. Answer using the term *Doppler effect.*

21. **Addition of Light Colors** What color of light must be combined with blue light to obtain white light?

22. **Light and Pigment Interaction** What color will a yellow banana appear to be when illuminated by each of the following?
    a. white light
    b. green and red light
    c. blue light

23. **Pigment Colors** What are the secondary pigment colors, and why do they give objects the appearance of those colors?

24. **Combination of Pigments** What primary pigment colors must be mixed to produce red? Explain your answer in terms of color subtraction for pigment colors.

25. **Polarization** Describe a simple experiment you could do to determine whether sunglasses in a store are polarizing.

26. **Polarizing Sunglasses** Use **Figure 24** to determine the direction the polarizing axis of polarizing sunglasses should be oriented to reduce glare from the surface of a road: vertically or horizontally? Explain.

Figure 24

27. **Red Light** The speed of red light is slower in air and water than in a vacuum. The frequency, however, does not change when red light enters water. Does the wavelength change? If so, how?

28. **Critical Thinking** Astronomers have determined that our galaxy, the Milky Way, is moving toward Andromeda, a neighboring galaxy. Explain how they determined this. Can you think of a possible reason why the Milky Way is moving toward Andromeda?

LEARNSMART® Go online to follow your personalized learning path to review, practice, and reinforce your understanding.

Super-Efficient Solar Cells

Solar panels on homes and businesses are an increasingly popular alternative source of energy in today's society. This technology converts energy from the Sun into electricity people can use for the needs of everyday life. Solar energy is an inexhaustible source of clean energy. However, the current technology used to convert solar energy to electricity has limitations. Scientists working to optimize this technology have recently made a breakthrough that could greatly increase its efficiency.

Solar panels convert light waves into electricity.

What is a photovoltaic cell?

A solar panel is made up of many photovoltaic (PV) cells, which convert light waves into electrical current. Incoming light waves strike the semiconductor, usually silicon, and transfer energy to electrons. The electrons flow in one direction through the material, creating an electric current.

Limitations of Current Technology

Energy emitted by the Sun encompasses the entire electromagnetic spectrum. The silicon used in PV cells can absorb only a portion of these waves—from red to violet on the visible light spectrum. The remainder of the Sun's energy cannot be utilized by existing PV cells. As a result, PV cells are very inefficient. For example, most residential solar panels convert about 10 to 20 percent of the incoming energy to electricity, so solar panels need to be very large, especially in areas with cold and cloudy climates.

Hot Solar Cells

Scientists have developed a new solar thermophotovoltaic (STPV) cell for converting solar energy to electricity. These STPV cells, called "hot solar cells," couple conventional PV cells with a light concentrator to utilize a greater percentage of the incoming energy from the Sun. The absorbing layer of carbon nanotubes turns energy from the Sun into thermal energy—with temperatures up to 1000°C! The emitting layer of nanophotonic crystals converts this thermal energy back into light that is specifically in the range of wavelengths that can be absorbed by the traditional PV cell. Initial results show that STPV cells create up to twice the amount of electrical energy compared to basic PV cells. This increased energy production can reduce the dependence on nonrenewable energy sources such as fossil fuels.

EVALUATE DESIGN SOLUTIONS

Using the Internet, gather information about the new STPV cell technology and evaluate this design solution. With a partner, discuss how the new technology might impact the cost, safety, reliability, or aesthetics of using solar energy to produce electricity. Summarize your discussion in a written paragraph.

CALIFORNIA CONNECTION

California leads all other states in the production of electricity using solar energy. Research the history of legislation related to the use of renewable energy in California, and discuss what you learn with a small group.

MODULE 15
STUDY GUIDE

GO ONLINE to study with your Science Notebook.

Lesson 1 ILLUMINATION

- Light can be modeled as a ray that travels in a straight path until it encounters a boundary. Mediums can be characterized as being transparent, translucent, or opaque, depending on how light interacts with them.
- The luminous flux of a light source is the rate at which light is emitted. It is measured in lumens (lm). Illuminance is the luminous flux per unit area. Illuminance is measured in lux (lx), or lumens per square meter (lm/m^2). For a point source, illuminance follows an inverse-square relationship with distance and a direct relationship with luminous flux.

$$E = \frac{P}{4\pi r^2}$$

- Early measurements of the speed of light involved measurement of the time it takes for light to reach Earth from Jupiter's moon Io. Michelson used a land-based technique that involved the distance between two mountains and a set of rotating mirrors. In a vacuum, light has a constant speed of $c = 3.00\times10^8$ m/s.

- ray model of light
- luminous sources
- opaque
- translucent
- transparent
- luminous flux
- illuminance

Lesson 2 THE WAVE NATURE OF LIGHT

- In the wave model of light, all the points in a wavefront can be thought of as sources of smaller waves. As light travels past an edge, the wavefront is cut and each new wavelet generates a new circular wave.
- Visible light can have wavelengths between 400 and 700 nm. White light is a combination of the spectrum of colors, each color having a different wavelength. Combining the primary colors—red, blue, and green—forms white light. Combinations of two primary colors form the secondary colors, yellow, cyan, and magenta.

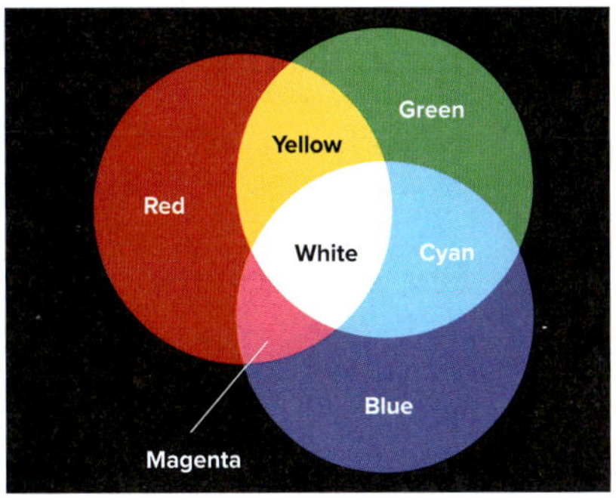

The primary pigments, cyan, magenta, and yellow, are used in combinations of two to produce the secondary pigments, red, blue, and green.

- Polarized light consists of waves whose electric fields oscillate with a specific pattern. Often, the oscillation is in a single plane. Light can be polarized with a polarizing filter or by reflection. Light waves traveling through a vacuum can be characterized in terms of frequency, wavelength, and the speed of light. Light waves are Doppler shifted based on the relative speed of the observer and light source along the axis of the observer and the light source.

- diffraction
- primary colors
- secondary color
- complementary colors
- primary pigment
- secondary pigment
- polarization
- Malus's law

THREE-DIMENSIONAL THINKING

Module Wrap-Up

REVISIT THE PHENOMENON

What does the light from a distant star or supernova tell us about it?

CER Claim, Evidence, Reasoning

Explain your Reasoning Revisit the claim you made when you encountered the phenomenon. Summarize the evidence you gathered from your investigations and research and finalize your Summary Table. Does your evidence support your claim? If not, revise your claim. Explain why your evidence supports your claim.

STEM UNIT PROJECT

Now that you've completed the module, revisit your STEM unit project. You will summarize your evidence and apply it to the project.

GO FURTHER

SEP Data Analysis Lab

Can a traffic light appear to change color?

Suppose you are a traffic officer and you stop a driver for going through a red light. Further suppose the driver draws you a picture and explains that the light looked green because of the Doppler effect when he drove through it. The wavelength for red light is 645 nm and it is 545 nm for a green light.

CER Analyze and Interpret Data

1. **Claim** Would you have given the driver a ticket for running the red light or accepted his explanation?
2. **Evidence and Reasoning** How would you explain your decision to the driver?

MODULE 16

REFLECTION AND REFRACTION

Cenerі/E+/Getty Images

MODULE 16
REFLECTION AND REFRACTION

ENCOUNTER THE PHENOMENON

How does light transmit information through a communication network?

GO ONLINE to watch a video about how communication networks use light to transmit information.

SEP Ask Questions

Do you have other questions about the phenomenon? If so, add them to the driving question board.

CER Claim, Evidence, Reasoning

Make Your Claim Use your CER chart to make a claim about how light is used to transmit information in a communication network. Explain your reasoning.

Collect Evidence Use the lessons in this module to collect evidence to support your claim. Record your evidence as you move through the module.

Explain Your Reasoning You will revisit your claim and explain your reasoning at the end of the module.

GO ONLINE to access your CER chart and explore resources that can help you collect evidence.

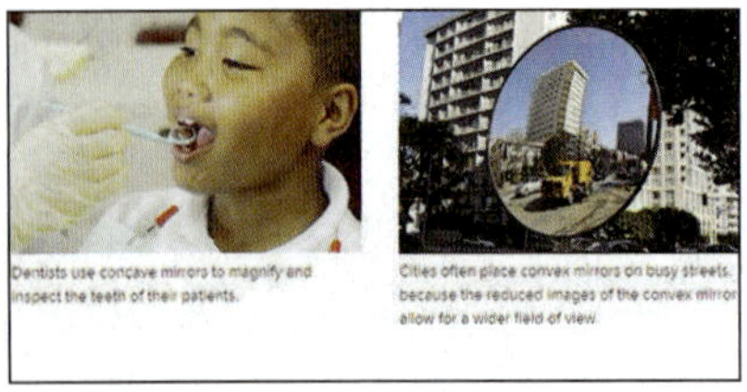

LESSON 2: Explore & Explain: Images in Curved Mirrors

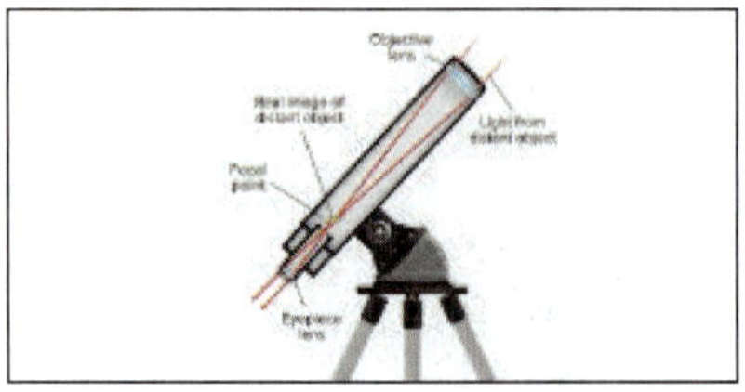

LESSON 4: Explore & Explain: Applications of Lenses

Additional Resources

LESSON 1
REFLECTION OF LIGHT

FOCUS QUESTION

Why does light reflected from a mirror make an image while light reflected from a piece of paper does not?

Reflected Images

When you look at the surface of a body of water, you don't always see a clear reflection as in **Figure 1.** Sometimes, the wind causes ripples in the water, or passing boats produce waves. Disturbances on the water's surface prevent the light from reflecting in a manner such that a clear image is visible.

Almost 4000 years ago, Egyptians understood that the type of reflection you see from a still pond requires a smooth surface. They used polished metal mirrors to view their images. Artisans in sixteenth-century Venice created mirrors by coating the back of a flat piece of glass with a thin sheet of metal. Sharp, well-defined, reflected images were not possible until 1857, however, when Jean Foucault, a French scientist, developed a method of coating glass with silver.

Today, we don't only use mirrors to view our reflections. Mirrors are key components in lasers, telescopes, and other precise optical systems. We use ever-increasing precision to make modern mirrors. Today's mirrors are made by evaporating aluminum or silver onto highly polished glass. The same basic physics principles, however, govern reflection from the smooth surface of a pond, from a sixteenth-century mirror, and from a tiny mirror inside a laser.

Figure 1 Disturbances on the surface of a pond or lake produce a distorted reflected image.

3D THINKING — DCI Disciplinary Core Ideas — CCC Crosscutting Concepts — SEP Science & Engineering Practices

COLLECT EVIDENCE

Use your Science Journal to record the evidence you collect as you complete the readings and activities in this lesson.

INVESTIGATE

GO ONLINE to find these activities and more resources.

Forensics Lab: A Little Time to Reflect
Carry out an investigation into the relationship between the angles of incident and reflection of light on a plane mirror.

CCC **Identify Crosscutting Concepts**
Create a table of the crosscutting concepts and fill in examples you find as you read.

The Law of Reflection

Figure 2 shows a light ray striking a reflecting surface. The normal is an imaginary line that is perpendicular to a surface at the location where light strikes the surface. The incident ray, the reflected ray, and the normal are in the same plane, which is perpendicular to the surface. Although light travels in three dimensions, the reflection of light is planar (two-dimensional). Planar relationships are known as the law of reflection.

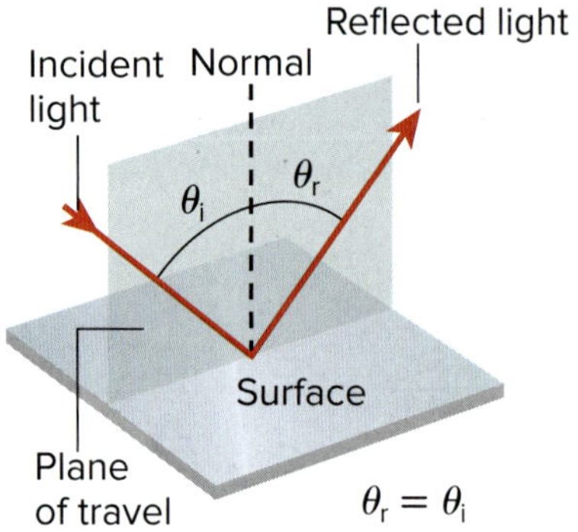

Figure 2 Reflecting light waves have an angle of reflection in the same plane as and equal to the angle of incidence.

COLOR CONVENTION

light ray ⟷ red

Law of Reflection

The angle that a reflected ray makes as measured from the normal to a reflective surface equals the angle that the incident ray makes as measured from the same normal.

$$\theta_r = \theta_i$$

Wave model We can describe this law in terms of the wave model of light. **Figure 3** shows a wavefront of light approaching a surface. The wavefront is perpendicular to the light ray. As each point along the wavefront reaches the surface, it reflects off that surface. Because all points are traveling at the same speed, they all travel the same total distance in the same time. Thus, the wavefront as a whole leaves the surface at an angle equal to its incident angle. Note that the wavelength of the light does not affect this process. The surface reflects red, green, and blue light all in the same direction.

Smooth surfaces Consider the light rays shown on the left in **Figure 4** on the next page. All the rays reflect off the surface parallel to one another. This occurs only if the reflecting surface is not rough on the scale of the wavelength of the light. We consider such a surface to be smooth. A smooth surface, such as a mirror, produces **specular reflection,** in which parallel light rays reflect in parallel.

Rough surfaces What happens when light strikes a sheet of paper? A sheet of paper might appear smooth, but on the scale of the wavelength of light, paper is actually quite rough. The right side of **Figure 4** shows light rays reflecting off a sheet of paper. All of the light rays are parallel before they strike the surface, but the reflected rays are not parallel, as shown. This scattering of light off a rough surface is called **diffuse reflection.**

The law of reflection applies to smooth and rough surfaces. For a rough surface, the angle that each incident ray makes with the normal equals the angle that its reflected ray makes with the normal. On a microscopic scale, however, the normals to the surface locations are not parallel.

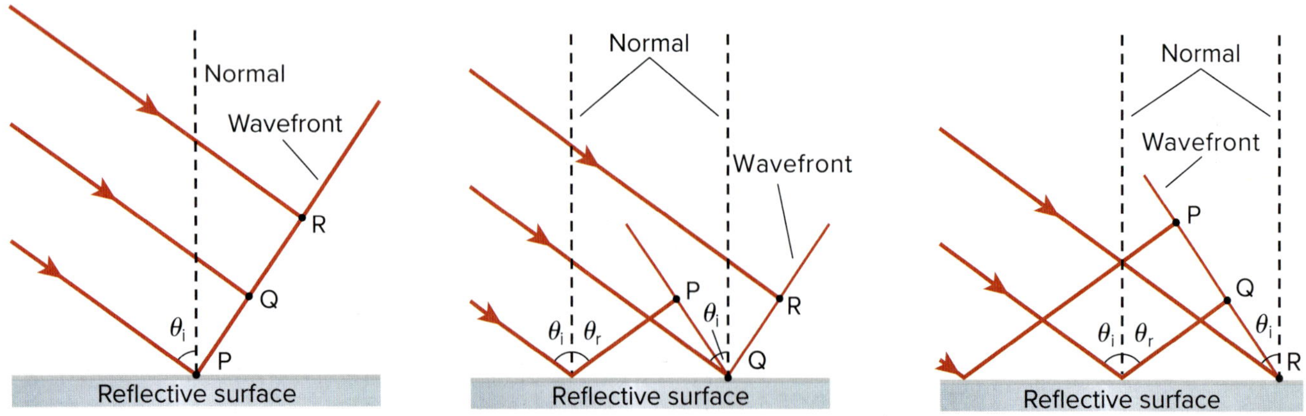

Figure 3 All parts of a wavefront reflect from a surface at the same angle. The angle of incidence equals the angle of reflection.

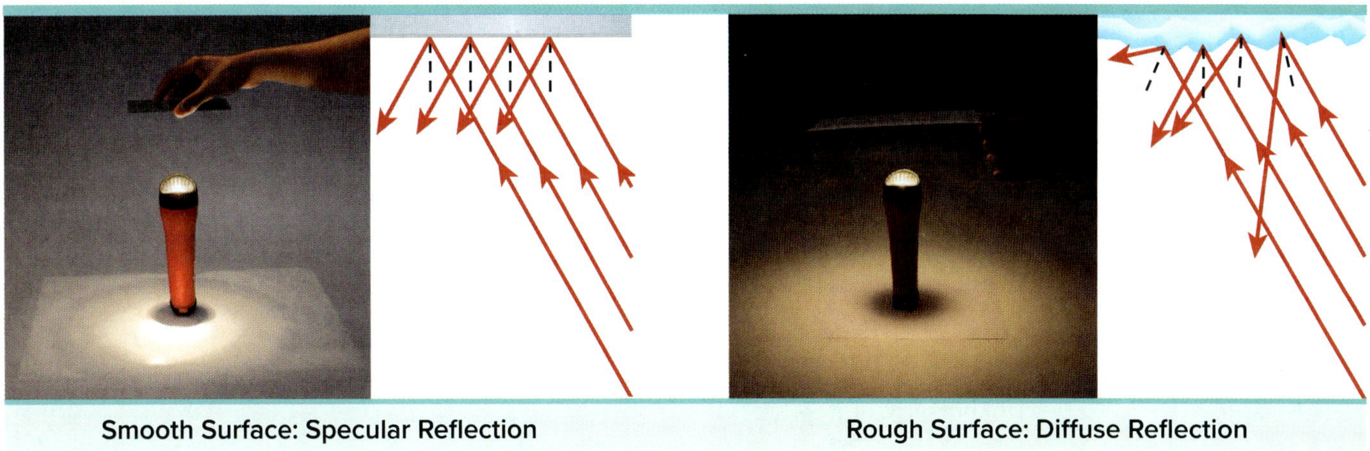

Figure 4 Notice that the image of the lightbulb in the flashlight is reflected on the table by the smooth mirror. The surface of the paper reflects a featureless area of light. Diffuse reflection enables you to read this page from various angles.

Thus, the rough surface prevents the reflected rays from being parallel, so the reflected rays are scattered in different directions. With specular reflection, as with a mirror, you can see your face. But no matter how much light reflects off a wall or a sheet of paper, as shown in **Figure 4,** you will never be able to use them as mirrors.

Objects and Plane-Mirror Images

A **plane mirror,** such as a mirror we use to look into, is a flat, smooth surface from which light is reflected by specular reflection.

When studying illumination, an **object** is either a luminous source of light rays, such as a lightbulb, or an illuminated source of light rays, such as a person. For most of the light sources that you will study in this module, light spreads out from that source in all directions. A mirrored surface can reflect these light rays so an image is visible, as shown in **Figure 5.**

In **Figure 6** light reflects diffusely from all parts of the bird (the object). The girl sees the light that reflects into her eye. Because her brain processes this information as if the rays travel in a straight path, it seems to the girl as if the light follows the dashed lines. The light seems to have come from a point behind the mirror. Just as the rays diverge from the object, they also diverge from the image.

The combination of light rays reflected from the bird in **Figure 6** forms the **image** of the bird. It is a **virtual image,** a type of image formed by diverging light rays. A virtual image is always on the opposite side of the mirror from the object. The image is virtual because there are no light rays at the image location. Plane mirrors produce only virtual images.

Figure 5 Reflection from the woman to the mirror is diffuse reflection. Reflection from the mirror is specular reflection.

Describe *how the woman is illuminated.*

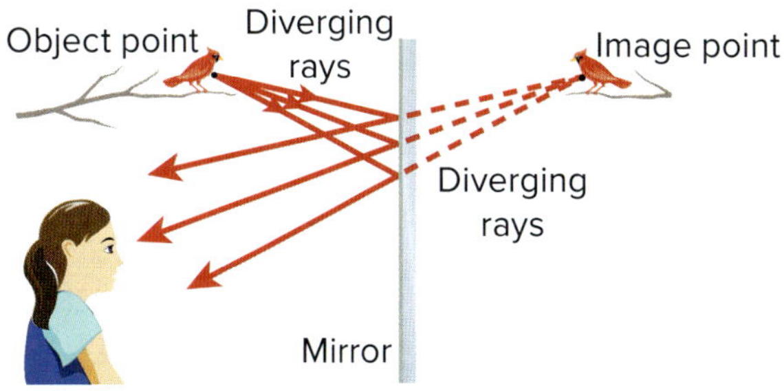

Figure 6 Rays that reflect from the bird disperse in many directions. Only a few that travel toward the mirror are shown. The image is located where multiple light rays from a point on an object seem to converge.

EXAMPLE Problem 1

CHANGING THE ANGLE OF INCIDENCE A light ray strikes a plane mirror at an angle of 52.0° to the normal. The mirror then rotates 35.0° around the point where the ray strikes the mirror so that the angle of incidence of the light ray decreases. The axis of rotation is perpendicular to the plane of the incident and the reflected rays. What is the angle between the initial and final reflected ray?

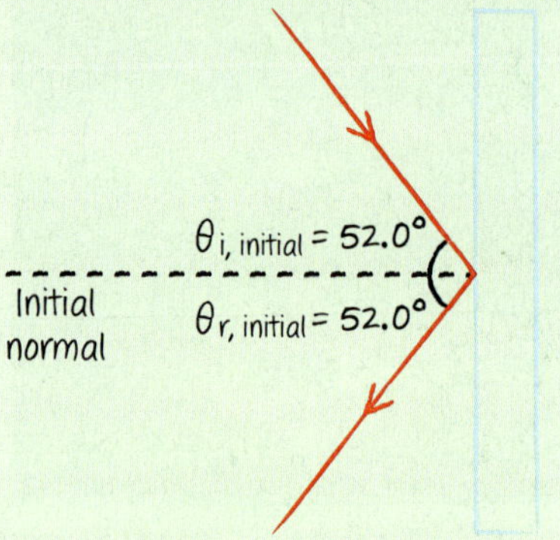

1 ANALYZE AND SKETCH THE PROBLEM

- Sketch the situation before the rotation of the mirror.
- Draw another sketch with the angle of rotation applied to the mirror.
- Draw a third sketch of the reflected rays.

Known	Unknown
$\theta_{i,\ initial} = 52.0°$	$\Delta\theta_r = ?$
$\Delta\theta_{mirror} = 35.0°$	

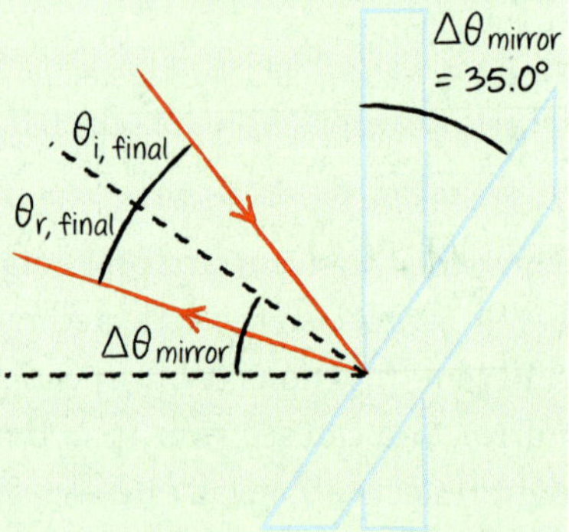

2 SOLVE FOR THE ANGLE DIFFERENCE

For the angle of incidence to reduce, rotate clockwise.

$\theta_{i,\ final} = \theta_{i,\ initial} - \Delta\theta_{mirror}$

$= 52.0° - 35.0°$ Substitute $\theta_{i,\ initial} = 52.0°$, $\Delta\theta_{mirror} = 35.0°$.

$= 17.0°$ clockwise from the new normal

Apply the law of reflection.

$\theta_{r,\ final} = \theta_{i,\ final}$

$= 17.0°$ counterclockwise from the new normal Substitute $\theta_{i,\ final} = 17.0°$.

Use the sketches to help determine the angle through which the reflected ray has rotated.

$\Delta\theta_r = \theta_{r,\ initial} + \Delta\theta_{mirror} - \theta_{r,\ final} = 52.0° + 35.0° - 17.0°$

$= 70.0°$ clockwise from the original angle

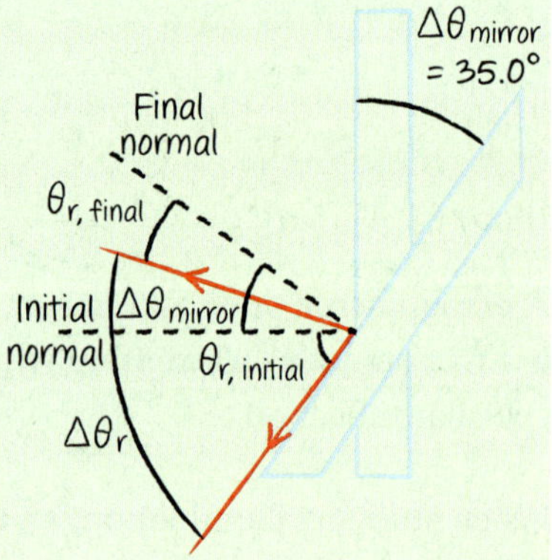

3 EVALUATE THE ANSWER

- **Is the magnitude realistic?** Comparing the final sketch with the initial sketch shows that the angle the light ray makes with the normal decreases as the mirror rotates clockwise. It makes sense, then, that the reflected ray also rotates clockwise and rotates through an angle twice as large as that of the mirror rotation.

PRACTICE Problems

ADDITIONAL PRACTICE

1. Explain why the reflection of light off ground glass changes from diffuse to specular if you spill water on it.
2. What is the angle of incidence of a light ray reflected off a plane mirror at an angle of 35° to the normal?
3. Suppose the angle of incidence of a light ray is 42°.
   a. What is the angle of reflection?
   b. What is the angle the incident ray makes with the mirror?
   c. What is the angle between the incident ray and the reflected ray?
4. Light from a laser strikes a plane mirror at an angle of 38° to the normal. If the angle of incidence increases by 13°, what is the new angle of reflection?
5. You position two plane mirrors at right angles to each other. A light ray strikes one mirror at an angle of 60° to the normal and reflects toward the second mirror. What is its angle of reflection off the second mirror?
6. **CHALLENGE** You are asked to design a retro-reflector using two mirrors that will reflect a laser beam by 180° independent of the incident direction of the beam. What should be the angle between the two mirrors?

Properties of Plane-Mirror Images

Looking at yourself in a mirror, you can see that your image in the mirror appears to be the same distance behind the mirror as you are in front of the mirror. How could you test this? Place a ruler between you and the mirror. Where does the image touch the ruler? You see that your image is vertically oriented as you are, and it matches your size. This is where the expression *mirror image* originates.

Image position and height The geometric model in **Figure 7** shows why the distances are the same. Two rays from point O at the tip of the candle strike the mirror at points P_1 and P_2. The mirror reflects both rays according to the law of reflection.

We can extend the reflected rays behind the mirror as sight lines (the dashed lines in the figure). These sight lines converge at point I, which is the image of point O. Ray 1, which strikes the mirror at an angle of incidence of 0°, reflects back on itself, so the sight line is at 90° to the mirror, just as ray 1 is. Ray 2 also reflects at an angle with respect to the normal equal to the angle of incidence, so the sight line is at the same angle to the mirror as ray 2.

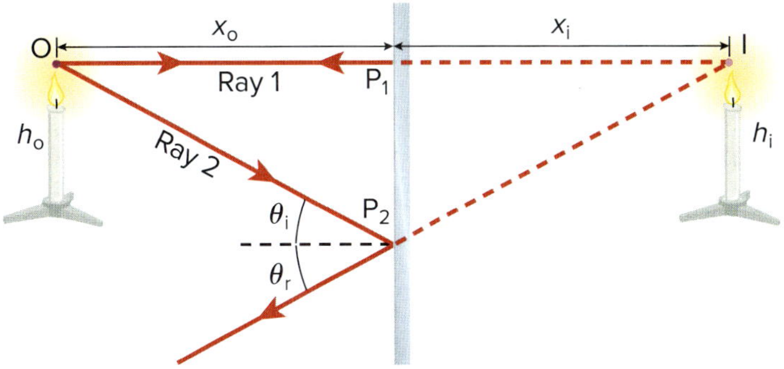

Figure 7 Reflected light rays from the candle (two rays are shown) strike the mirror. Some of those rays reach the viewer's eye. Sight lines (dashed lines) are drawn from where the rays reflect from the mirror to where they converge. The image is located where the sight lines converge.

Explain *why* $x_i = -x_o$.

This geometric model reveals that line segments OP_1 and IP_1 are corresponding sides of two congruent triangles, ΔOP_1P_2 and ΔIP_1P_2. The object's position with respect to the mirror (x_o) has a magnitude equal to the length of $\overline{OP_1}$. The apparent position of the image with respect to the mirror, which is called the image position (x_i), has a magnitude equal to the length of $\overline{IP_1}$. If image position is negative, indicating that the image is virtual, the following is true.

Plane-Mirror Image Position

With a plane mirror, the image position is equal to the negative of the object position. The negative sign indicates that the image is behind the mirror (and therefore virtual).

$$x_i = -x_o$$

Get It?

Summarize the properties of a plane-mirror image.

STEM CAREER Connection

Set Designer

Have you ever attended a shool play or other production and noticed eerie light effects? You could use your understanding of the properties of light to create sets for movies, theaters, or concerts. Set designers must understand how to use different properties of light, such as reflection, refraction, color, intensity, and brightness, to create mood and special effects.

You can draw more light rays from the object to the mirror to determine the size of the image. The sight lines of two rays originating from the bottom of the candle in **Figure 7** will converge at the bottom of the image. From the law of reflection and congruent-triangle geometry, we find the following is true of the object height (h_o) and image height (h_i) and any other dimension of the object and image.

Plane-Mirror Image Height

With a plane mirror, image height is equal to object height.

$$h_i = h_o$$

Image orientation A plane mirror produces an image with the same orientation as the object. If you stand upright, a plane mirror produces an image of you standing upright. If you do a headstand, the mirror shows you doing a headstand. However, the mirror gives a front-to-back reversal. You are looking face-to-face at the image of yourself.

Follow the sight lines in **Figure 8.** The image is behind the mirror such that the boy appears to be facing himself. When the boy extends his right hand, it appears as though the left hand of the image is extended, but from the boy's perspective, the hand that is extended is on his right-hand side. This happens because a plane mirror does not really reverse left and right. The mirror in **Figure 8** only reverses the boy's image such that it is facing in the opposite direction as the boy, or, in other words, it produces a front-to-back reversal.

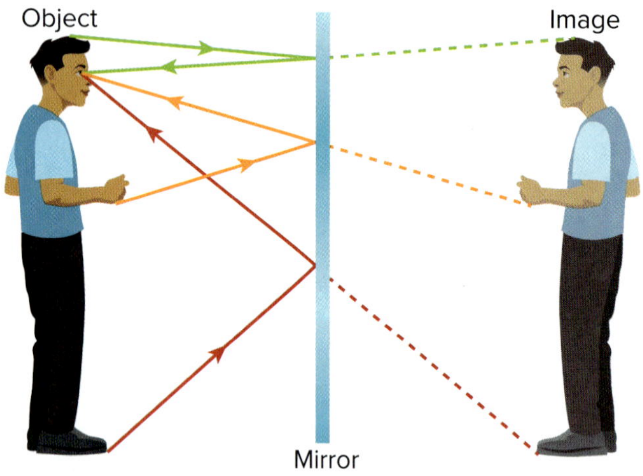

Figure 8 Viewed in a mirror, your height and distance from the mirror appear the same. There is, however, a difference. You are facing the opposite direction.

Check Your Progress

7. **Reflection** A light ray strikes a flat, smooth, reflecting surface at an angle of 80° to the normal. What angle does the reflected ray make with the surface of the mirror?
8. **Image Properties** A dog looks at its image, as shown in **Figure 9.** What are the position, height, and type of image?

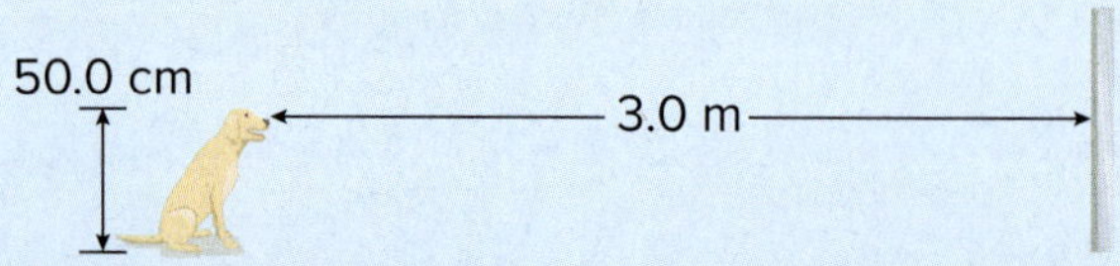

Figure 9

9. **Law of Reflection** Explain how the law of reflection applies to diffuse reflection.
10. **Image Diagram** A car is following another car along a straight road. The first car has a rear window tilted at 45° to the horizontal. Draw a ray diagram showing the position of the Sun that would cause sunlight to reflect into the eyes of the driver of the second car.
11. **Critical Thinking** Explain how diffuse reflection of light off an object enables you to see that object from any angle.

LEARNSMART® Go online to follow your personalized learning path to review, practice, and reinforce your understanding.

LESSON 2
CURVED MIRRORS

FOCUS QUESTION

What are some advantages and disadvantages of curved mirrors compared to plane mirrors?

Properties of Curved Spherical Mirrors

Move a metal spoon toward and away from your face. Look at your image on both sides of the spoon. Your image may appear larger or smaller, or it may even be inverted. The properties of curved mirrors, such as a metal spoon, and the images that they form depend on the shape of the mirror and the object's position.

The way that a spherical mirror reflects light depends on its curve. **Figure 10** shows two spherical mirrors and some important points and distances for understanding how they form images. A spherical mirror is shaped as if it were a section of a hollow sphere. The mirror (M) has the same geometric center (C) as a sphere of radius *r*. The distance *r* is also called the radius of curvature.

The straight line that includes line segment CM is called the **principal axis,** or line perpendicular to the mirror's surface that divides the mirror in half.

Focal point Light rays from the Sun to Earth are almost parallel because the Sun is so far away. When you point the principal axis of a concave mirror toward the Sun, all the rays reflect through a point. You can locate this point by moving a sheet of paper toward and away from the mirror until the smallest and sharpest spot of sunlight is focused on the paper. This spot is the mirror's **focal point,** which is the point at which incident light rays that are parallel to the principal axis converge after reflecting from the mirror. This is point F.

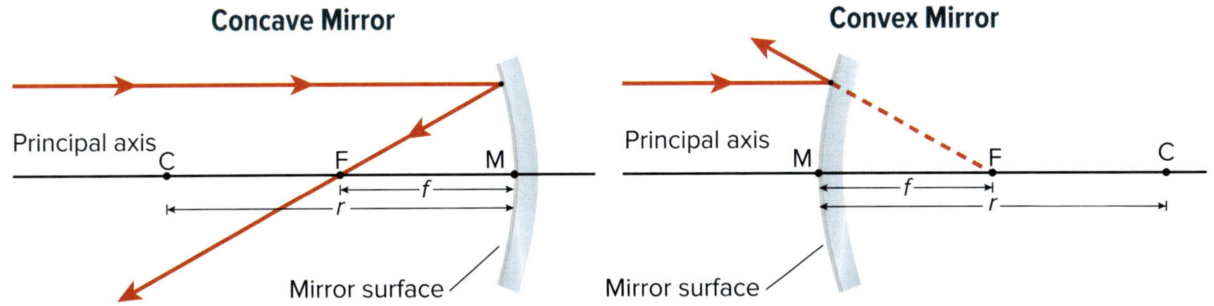

Figure 10 For concave and convex mirrors, the distance from M to F (the focal point) is half the distance from M to C.

3D THINKING DCI Disciplinary Core Ideas CCC Crosscutting Concepts SEP Science & Engineering Practices

COLLECT EVIDENCE

Use your Science Journal to record the evidence you collect as you complete the readings and activities in this lesson.

INVESTIGATE

GO ONLINE to find these activities and more resources.

Virtual Investigation: Ray Tracing for Mirrors
Use a computer model to investigate how the properties of a curved mirror affect how light is reflected.

PhysicsLAB: Concave Mirror Images
Construct an explanation for why different mirrors produce different images when reflecting light.

For the mirror on the left in **Figure 10**, a ray parallel to the principal axis is reflected and crosses the principal axis at point F. Note that the distance from M to F is half the distance from M to C. The **focal length** (f) is the distance between the mirror and the focal point and can be expressed as $f = \frac{r}{2}$.

Concave mirrors The inside surface of a shiny metal spoon acts as a concave mirror. A **concave mirror,** such as the one on the left in **Figure 10,** has an inwardly curving reflective surface, the edges of which curve toward the observer. Incident light rays parallel to the principal axis are reflected and intersect the principal axis at point F. The focal length is positive for a concave mirror because it is a point of converging rays. Rays reflected from concave mirrors are parallel, as shown in **Figure 11.** Concave mirrors are placed in car headlights. It is a concave mirror that allows the light from a spotlight to be focused into a spot.

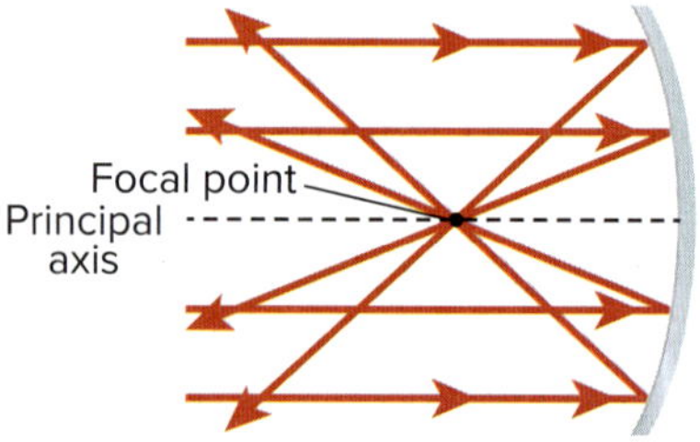

Light rays parallel to the principal axis reflect off a concave mirror and converge at the focal point.

A lightbulb is placed at the focal point of a concave mirror. The parallel reflected rays are evident in this flashlight beam.

Figure 11 Rays traced from the focal point to a concave mirror reflect off the surface as parallel rays.

Convex mirrors A spoon's outer surface acts as a **convex mirror,** an outwardly curving reflective surface with edges that curve away from the observer, as shown on the right side of **Figure 10.** Properties of a spherical convex mirror are also shown in **Figure 10.** The focal point (F) and the geometric center of the mirror (C) are behind the mirror. Store security mirrors and many side-view mirrors on cars are convex mirrors.

Magnification Another property of a spherical mirror is **magnification** (m), which is the ratio of the image height to the object height.

The sign of the magnification tells you whether the image is upright or inverted. For virtual images such as those formed by convex mirrors, x_i is negative. This means m is positive. Note that virtual images are always upright, so the height is always positive.

Magnification

The magnification of an object by a spherical mirror, defined as the image height divided by the object height, is equal to the negative of the image position, divided by the object position.

$$m \equiv \frac{h_i}{h_o} = -\frac{x_i}{x_o}$$

Get It?

Explain What does it mean if $m < 1$?

Nicholas Eveleigh/Stockbyte

Ray Diagrams for Curved Mirrors

Ray diagrams give you a visual representation of images formed by a curved mirror, depending on where the object is placed in relation to the focal point.

Figure 12 shows rays reflected from a concave mirror converging at the point I, where an image is located. The converging light rays form a **real image** that is inverted and larger than the object. You can see the image floating in space if you place your eye so that the rays that form the image fall on your eye, as in **Figure 12.** You must face a direction that allows the light rays coming from the image to enter your eyes, however. You cannot look at the image from behind. If you were to place a piece of paper or a movie screen at this point, the image would appear on the screen, as shown on the right in **Figure 12.** You cannot project virtual images on a screen, since they are not formed by converging light rays.

In **Figure 12,** if the object is beyond point C, the absolute value of the magnification for the real image is less than 1. This means that the image is smaller than the object. If the object is placed between point C and point F, the absolute value of the magnification is greater than 1, which means the image is larger than the object.

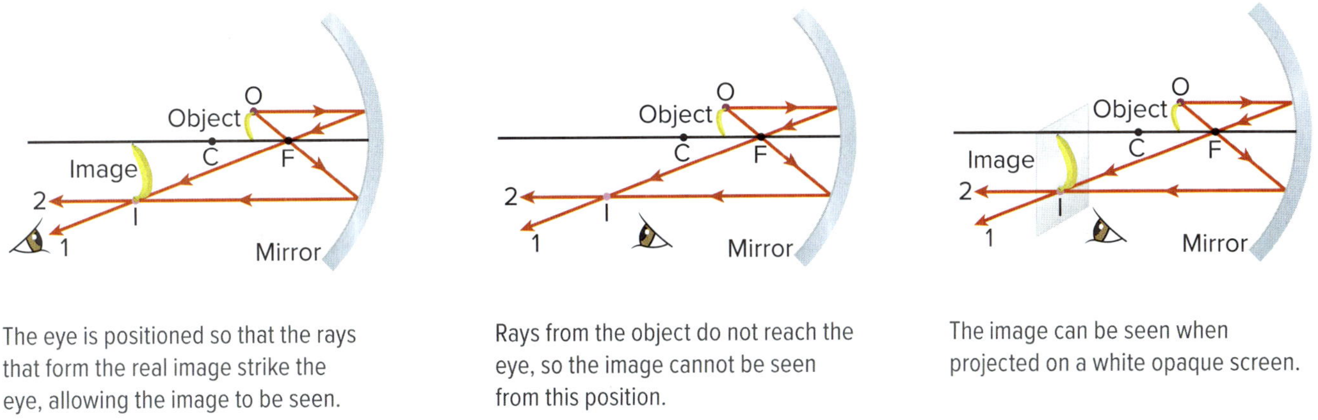

Figure 12 Ray diagrams can be used to locate an image reflected from a curved mirror.

Real Images with Concave Mirrors

You can often simplify ray diagrams by using simple, one-dimensional objects, such as the arrow shown in **Figure 13** on the next page. A spherical concave mirror produces an inverted real image if the object position (x_o) is greater than twice the focal length (f). The object is then beyond the center of curvature (C). If the object is placed between the center of curvature and the focal point, as shown on the right in **Figure 13,** the image is again real and inverted. The size of the image is now greater than the size of the object.

CCC CROSSCUTTING CONCEPTS

Cause and Effect Systems can be designed to cause a desired effect. You want to design a mirror that can be used for shaving, applying makeup, or both, so you want your mirror to produce a large image. Would you use a plane mirror, a convex mirror, or a concave mirror? What evidence will you use to support your choice? Make a simple sketch of your design.

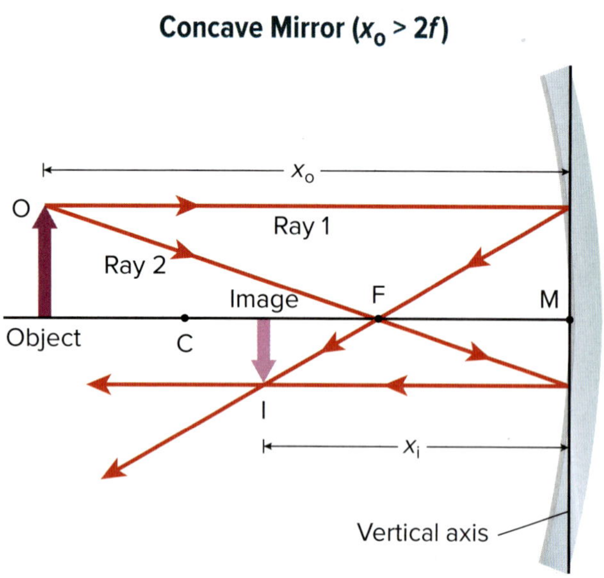

The image is reduced for an object located a distance greater than $2f$ from the mirror.

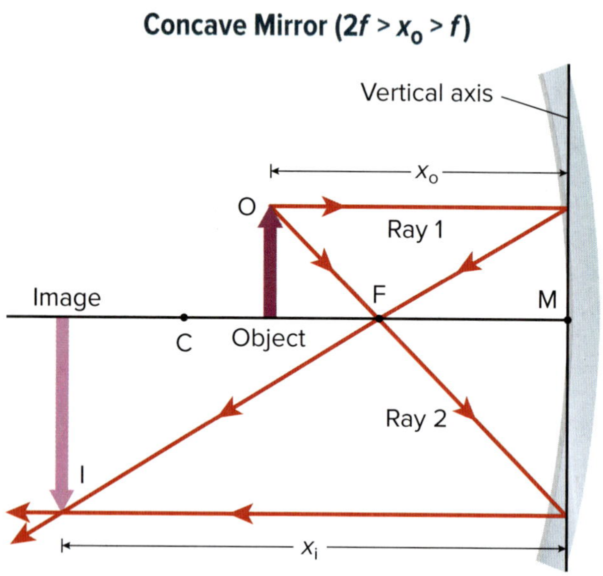

The image is enlarged for an object between f and $2f$.

Figure 13 The type of image depends on the object's distance from the mirror. A real, inverted image is formed in both situations shown here. Remember that f, the focal length, is the distance from M to F.

Diagram *how other points on the image are formed.*

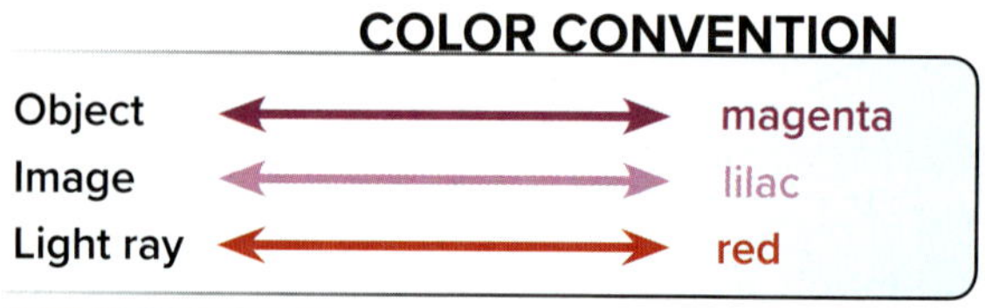

PROBLEM-SOLVING STRATEGIES

Using Ray Diagrams to Locate Images Produced by Spherical Mirrors

Use the following strategies for curved spherical-mirror problems. Refer to **Figure 13,** as well as **Figure 14** and **Figure 15** on the next page.

1. Using lined or graph paper, draw the principal axis of the mirror as a horizontal line from the left side to the right side of your paper, leaving six blank lines above and six blank lines below.
2. Place points and labels on the principal axis for the object, C, and F as follows:
   a. If the mirror is concave and the object is beyond C away from the mirror, place the mirror at the right side of the page, the object at the left side of the page, and make C and F to scale.
   b. If the mirror is concave and the object is between C and F, place the mirror at the right side of the page, C at the center of the paper, F halfway between the mirror and C, and the object to scale.
   c. For any other situation, place the mirror in the center of the page, the object or F (whichever is the greater distance from the mirror) at the left side of the page, and the other to scale.
3. To represent the mirror, draw a vertical line at the mirror point that extends for 12 lines. This is the mirror's principal plane.
4. Draw the object as an arrow, and label its top O. For concave mirrors, objects inside of C should not be higher than three lines. For all other situations, the objects should be six lines high. The scale for the height of the object will be different from the scale along the principal axis.
5. Draw ray 1, the parallel ray. It is parallel to the principal axis, reflects off the principal plane, and passes through F.
6. Draw ray 2, the focus ray. It passes through F, reflects off the principal plane, and is reflected parallel to the principal axis.
7. If reflected rays 1 and 2 diverge from each other, then extend the sight lines behind the mirror as dashed lines.
8. The image is located where rays 1 and 2 or their sight lines cross after reflection. For either case, label the point of intersection I. Draw the image as a perpendicular arrow from the principal axis to I.

Concave Mirror ($f > x_o > 0$)

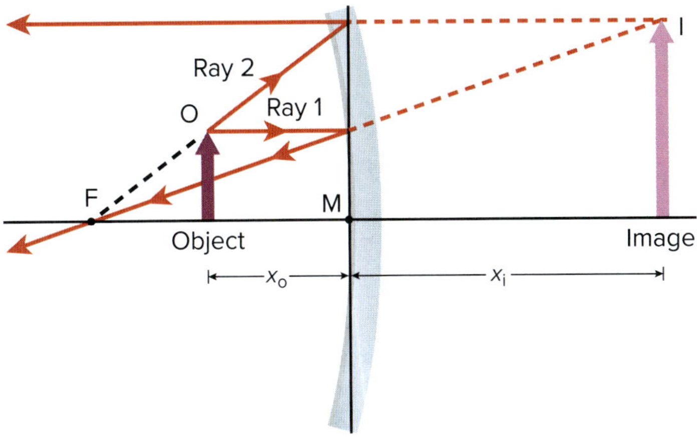

Figure 14 A virtual, upright, enlarged image is formed when an object, such as the block tower on the right, is placed between the focal point and the surface of a concave mirror.

Consider *the cause of the appearance of the other images in the mirror.*

Virtual Images with Concave Mirrors

You have seen that as an object approaches the focal point (F) of a concave mirror, the image moves farther away from the mirror. If the object is at the focal point, all reflected rays are parallel. They never meet, therefore, and the image is said to be at infinity, so the image could never be seen.

What happens if you move the object even closer to the mirror? The image is upright and behind the mirror. A concave mirror produces a virtual image if the object is located between the mirror and the focal point. This situation is shown in the ray diagram in **Figure 14.**

You draw two rays to locate the image of a point on an object. As before, you draw ray 1 parallel to the principal axis and then reflected through the focal point. Ray 2 is drawn as a line from the point on the object to the mirror, along a line defined by the focal point and the point on the object. At the mirror, ray 2 is reflected parallel to the principal axis. Note in **Figure 14** that ray 1 and ray 2 diverge as they leave the mirror, so there cannot be a real image. However, sight lines extended behind the mirror converge, showing that a virtual image forms behind the mirror. From the position of the point of convergence, you can see that the image is upright and larger than the object.

Get It?

Explain how you can determine whether an image is virtual.

Virtual Images with Convex Mirrors

Rays reflected from a convex mirror always diverge. Thus, convex mirrors form only virtual images. The ray diagram in **Figure 15** shows how a spherical convex mirror forms an image. The figure shows two rays, but remember that there is an infinite number of rays reflecting from the object. Ray 1 approaches the mirror parallel to the principal axis. Ray 1 reflects off the mirror following the line between F and the point where ray 1 strikes the principal plane.

Convex Mirror

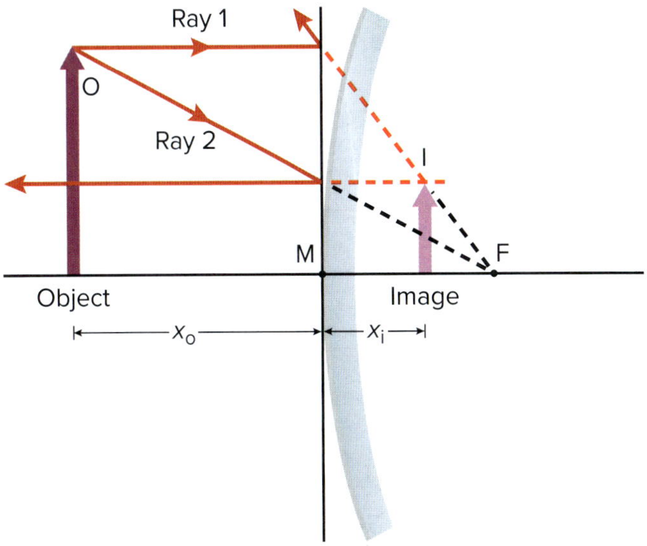

Figure 15 Convex mirrors form images that are virtual, upright, and smaller than the object.

A sight line extends from the point where ray 1 strikes the principle plane through F. Ray 2 approaches the mirror on a path that, if extended behind the mirror, would pass through F. The reflected part of ray 2 and its sight line are parallel to the principal axis. The two reflected rays diverge. However, the sight lines intersect behind the mirror. The point where they intersect is the location of the image. An image produced by a convex mirror is a virtual image that is upright and smaller than the object.

Field of view It might seem that convex mirrors would have little use because the images they form are smaller than the objects. This property of convex mirrors, however, does have practical uses. By forming smaller images, convex mirrors enlarge the area that an observer sees, which is called the field of view. Also, the image is visible from a large range of angles; thus, the field of view is visible from a wide perspective.

Because of the increased field of view, convex mirrors are used in cars as passenger-side rearview mirrors, as shown in **Figure 16.** However, because the reduced image size makes the objects appear farther away than they really are, there is often a warning on passenger-side rearview mirrors of cars stating that objects may be closer than they appear.

Figure 16 Images from convex mirrors are smaller than the object. This increases the field of view and decreases the driver's blind spot.

Explain *why a warning stating that objects in a convex side-view mirror are closer than they appear may be useful.*

Defects in Concave Mirrors

In diagramming rays, you used a vertical line to represent the mirror. In reality, light rays reflect off the mirror itself, as shown in **Figure 17.** Only parallel rays that are close to the principal axis reflect through the focal point. Other rays converge at points closer to the mirror. This defect, called **spherical aberration,** occurs because the light rays do not all converge at the focal point, which makes an image look fuzzy.

A mirror ground to the shape of a parabola, as in **Figure 17,** suffers no spherical aberration. Large, parabolic glass mirrors are expensive to manufacture; so, many telescopes use spherical mirrors and smaller, specially designed secondary mirrors or lenses to correct for spherical aberration.

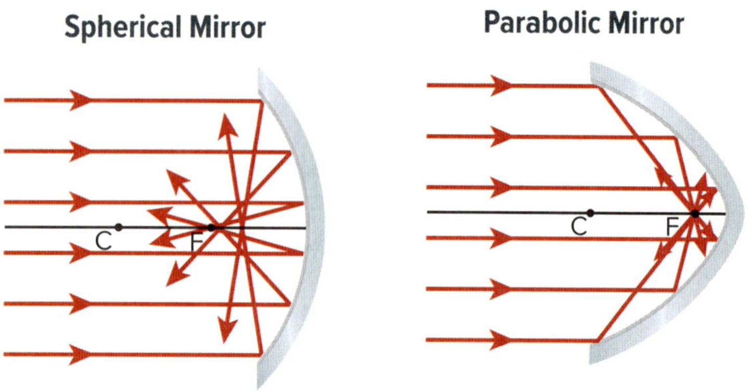

Figure 17 Spherical aberration occurs for spherical mirrors but does not occur for parabolic mirrors.

Get It?

Explain why a spherical mirror produces spherical aberration but a mirror in the shape of a parabola does not produce spherical aberration.

Calculating Image Position

We can use the spherical mirror model to develop a simple equation for spherical mirrors. You must use the paraxial ray approximation, which states that only rays that are close and almost parallel to the principal axis are used to form an image. Using this with the law of reflection leads to the mirror equation, relating the focal length (f), object position (x_o), and image position (x_i) of a spherical mirror.

Mirror Equation

The reciprocal of the focal length of a spherical mirror is equal to the sum of the reciprocals of the image position and the object position.

$$\frac{1}{f} = \frac{1}{x_i} + \frac{1}{x_o}$$

Real-World Physics

IN 1990, NASA launched the *Hubble Space Telescope* into orbit around Earth. Scientists expected *Hubble* to provide clear images without atmospheric distortions. Soon after it was deployed, however, scientists found that *Hubble* had a spherical aberration. In 1993, astronauts installed corrective optics called COSTAR on *Hubble*, enabling *Hubble* to produce clear images. For over twenty-five years, *Hubble* provided the world with stunning images and enabled scientists to make important discoveries about the universe.

Negative values When virtual images are formed, the image position (x_i) has a negative value, indicating that it is located behind the mirror. For concave mirrors, a virtual image only forms when the object is between a concave mirror and the focal point. The focal point is in front of the mirror and the focal length has a positive value. For convex mirrors, the focal point is always behind the mirror, and the focal length has a negative value.

Remember that the mirror equation is only approximately correct. In reality, light coming from an object toward a mirror is diverging, so not all of the light is close to or parallel to the axis. When the mirror diameter is large relative to the radius of curvature to minimize spherical aberration, this equation predicts image properties more precisely.

CONNECTING MATH to Physics

Adding and Subtracting Fractions When using the mirror equation, you first use math to move the fraction that contains the quantity you are seeking to the left-hand side of the equation and everything else to the right. Then you combine the two fractions on the right-hand side by using a common denominator that results from multiplying the denominators.

Math	Physics
$\frac{1}{x} = \frac{1}{y} + \frac{1}{z}$	$\frac{1}{f} = \frac{1}{x_i} + \frac{1}{x_o}$
$\frac{1}{y} = \frac{1}{x} - \frac{1}{z}$	$\frac{1}{x_i} = \frac{1}{f} - \frac{1}{x_o}$
$\frac{1}{y} = \left(\frac{1}{x}\right)\left(\frac{z}{z}\right) - \left(\frac{1}{z}\right)\left(\frac{x}{x}\right)$	$\frac{1}{x_i} = \left(\frac{1}{f}\right)\left(\frac{x_o}{x_o}\right) - \left(\frac{1}{x_o}\right)\left(\frac{f}{f}\right)$
$\frac{1}{y} = \frac{z-x}{xz}$	$\frac{1}{x_i} = \frac{x_o - f}{fx_o}$
$y = \frac{xz}{z-x}$	$x_i = \frac{fx_o}{x_o - f}$

Using this approach, the following relationships can be derived for image position, object position, and focal length:

$$x_i = \frac{fx_o}{x_o - f} \qquad x_o = \frac{fx_i}{x_i - f} \qquad f = \frac{x_i x_o}{x_i + x_o}$$

EXAMPLE Problem 2

REAL IMAGE FORMATION BY A CONCAVE MIRROR A concave mirror has a radius of curvature of 20.0 cm. You place a 2.0-cm-tall object 30.0 cm from the mirror. What are the image position and image height?

1 ANALYZE AND SKETCH THE PROBLEM

- Draw a diagram with the object and the mirror.
- Draw two principal rays to locate the image in the diagram.

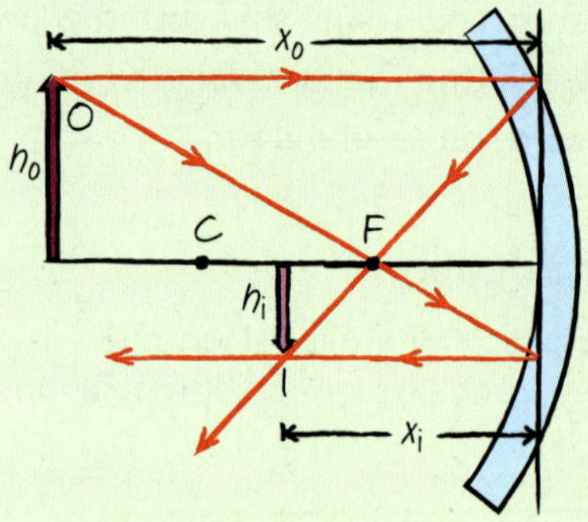

Known	Unknown
$h_o = 2.0$ cm	$x_i = ?$
$x_o = 30.0$ cm	$h_i = ?$
$r = 20.0$ cm	

2 SOLVE FOR THE UNKNOWN

Focal length is half the radius of curvature.

$$f = \frac{r}{2}$$

$$= \frac{20.0\text{ cm}}{2} \qquad \text{Substitute } r = 20.0\text{ cm.}$$

$$= 10.0\text{ cm}$$

Use the relationship between focal length and object position to solve for image position.

$$\frac{1}{f} = \frac{1}{x_i} + \frac{1}{x_o}$$

$$x_i = \frac{fx_o}{x_o - f}$$

$$= \frac{(10.0\text{ cm})(30.0\text{ cm})}{30.0\text{ cm} - 10.0\text{ cm}} \qquad \text{Substitute } f = 10.0\text{ cm, } x_o = 30.0\text{ cm.}$$

$$= 15.0\text{ cm (real image, in front of the mirror)}$$

Use the relationship between object height and object and image position to solve for image height.

$$m \equiv \frac{h_i}{h_o} = \frac{-x_i}{x_o}$$

$$h_i = \frac{-x_i h_o}{x_o}$$

$$= -\frac{(15.0\text{ cm})(2.0\text{ cm})}{30.0\text{ cm}} \qquad \text{Substitute } x_i = 15.0\text{ cm, } h_o = 2.0\text{ cm, } x_o = 30.0\text{ cm.}$$

$$= -1.0\text{ cm (inverted, smaller image)}$$

3 EVALUATE THE ANSWER

- **Are the units correct?** All positions and heights are in centimeters.
- **Do the signs make sense?** Positive position and negative height agree with the drawing.

PRACTICE Problems

ADDITIONAL PRACTICE

12. Use a ray diagram drawn to scale to solve Example Problem 2.

13. You place an object 36.0 cm in front of a concave mirror with a 16.0-cm focal length. Determine the image position.

14. You place a 3.0-cm-tall object 20.0 cm from a 16.0-cm-radius concave mirror. Determine the image position and image height.

15. A concave mirror has a 7.0-cm focal length. You place a 2.4-cm-tall object 16.0 cm from the mirror. Determine the image height.

16. CHALLENGE You place an object near a concave mirror with a 10.0-cm focal length. The image is 3.0 cm tall, inverted, and 16.0 cm from the mirror. What are the object position and object height?

EXAMPLE Problem 3

IMAGE IN A SECURITY MIRROR A convex security mirror in a warehouse has a −0.50-m focal length. A 2.0-m-tall forklift is 5.0 m from the mirror. What are the forklift's image position and image height?

1 ANALYZE AND SKETCH THE PROBLEM

- Draw a diagram with the mirror and the object.
- Draw two principal rays to locate the image in the diagram.

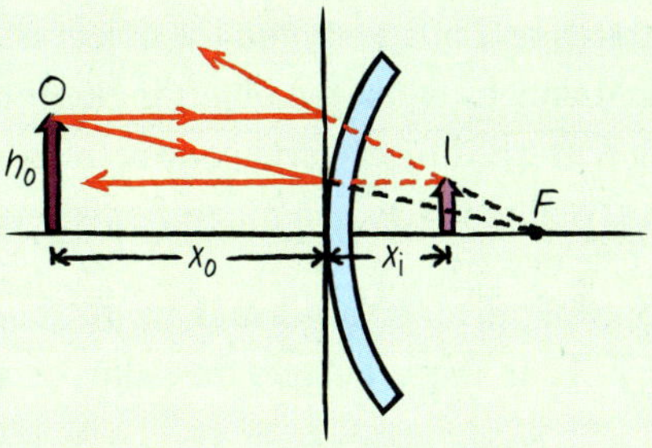

Known	Unknown
$h_o = 2.0$ m	$x_i = ?$
$x_o = 5.0$ m	$h_i = ?$
$f = -0.50$ m	

2 SOLVE FOR THE UNKNOWN

Use the relationship between focal length and object position to solve for image position.

$$\frac{1}{f} = \frac{1}{x_i} + \frac{1}{x_o}$$

$$x_i = \frac{fx_o}{x_o - f}$$

$$= \frac{(-0.50\text{ m})(5.0\text{ m})}{5.0\text{ m} - (-0.50\text{ m})}$$ Substitute $f = -0.50$ m, $x_o = 5.0$ m.

$= -0.45$ m (virtual image, behind the mirror)

Use the relationship between object height and object and image position to solve for image height.

$$m \equiv \frac{h_i}{h_o} = \frac{-x_i}{x_o}$$

$$h_i = \frac{-x_i h_o}{x_o}$$

$$= \frac{-(-0.45\text{ m})(2.0\text{ m})}{5.0\text{ m}}$$ Substitute $x_i = -0.45$ m, $h_o = 2.0$ m, $x_o = 5.0$ m.

$= 0.18$ m (upright, smaller image)

3 EVALUATE THE ANSWER

- **Are the units correct?** All positions and heights are in meters.
- **Do the signs make sense?** A negative position indicates a virtual image; a positive height indicates an image that is upright. These agree with the diagram.

PRACTICE Problems

ADDITIONAL PRACTICE

17. You place an object 20.0 cm in front of a convex mirror with a −15.0-cm focal length. Find the image position using both a scale diagram and the mirror equation.

18. A convex mirror has a focal length of −13.0 cm. You place a 6.0-cm diameter lightbulb 60.0 cm from that mirror. What are the lightbulb's image position and diameter?

19. A 7.6-cm-diameter ball is located 22.0 cm from a convex mirror with a radius of curvature of 60.0 cm. What are the ball's image position and diameter?

20. A 1.8-m-tall girl stands 2.4 m from a store's security mirror. Her image appears to be 0.36 m tall.

a. What is the image's distance?

b. What is the focal length of the mirror?

21. CHALLENGE A convex mirror is needed to produce an image that is three-fourths the size of an object and located 24 cm behind the mirror.

a. What is the object's distance?

b. What focal length should be specified?

Mirror Comparison

Table 1 summarizes the properties of single-mirror systems with objects that are located on the principal axis of the mirror.

Notice that the single plane mirror and convex mirror produce only virtual images. A concave mirror produces real images when the object is father than the focal distance. A concave mirror produces virtual images when the object is closer than the focal distance. Plane mirrors give reflections on scale with the objects, and convex mirrors provide reduced images, expanding the field of view. A concave mirror acts as a magnifier when an object is within the focal length of the mirror.

A concave mirror enlarges and inverts the image when the object is between the focal length and the radius of curvature. Beyond the radius of curvature, a concave mirror produces a reduced, inverted image.

Table 1 Single-Mirror System Properties

Mirror Type	f	x_o	x_i	m	Image
Plane	∞	$x_o > 0$	$\lvert x_i \rvert = x_o$ (negative)	same size	virtual
Concave	+	$x_o > r$	$r > x_i > f$	reduced, inverted	real
		$r > x_o > f$	$x_i > r$	enlarged, inverted	real
		$f > x_o > 0$	$\lvert x_i \rvert > x_o$ (negative)	enlarged	virtual
Convex	−	$x_o > 0$	$\lvert f \rvert > \lvert x_i \rvert > 0$ (negative)	reduced	virtual

Check Your Progress

22. **Image Properties** If you know the focal length of a concave mirror, where should you place an object so that its image is upright and larger compared to the object? Will this produce a real or virtual image?
23. **Magnification** You place an object 20.0 cm in front of a concave mirror with a focal length of 9.0 cm. What is the magnification of the image?
24. **Object Position** The placement of an object in front of a concave mirror with a focal length of 12.0 cm forms a real image that is 22.3 cm from the mirror. What is the object's position?
25. **Image Position and Height** You place a 3.0-cm-tall object 22.0 cm in front of a concave mirror that has a focal length of 12.0 cm. Find the image position and height by drawing a ray diagram to scale. Verify your answer using the mirror and magnification equations.
26. **Ray Diagram** You place a 4.0-cm-tall object 14.0 cm from a convex mirror with a focal length of −12.0 cm. In a scale ray diagram show the image position and height. Verify your answer using the mirror and the magnification equations.
27. **Radius of Curvature** You place a 6.0-cm-tall object 16.4 cm from a convex mirror. If the image of the object is 2.8 cm tall, what is the mirror's radius of curvature?
28. **Focal Length** A convex mirror is used to produce an image that is two-thirds the size of an object and located 12 cm behind the mirror. What is the focal length of the mirror?
29. **Critical Thinking** Would spherical aberration be less for a mirror whose height, compared to its radius of curvature, is small or large? Explain your answer.

LEARNSMART® Go online to follow your personalized learning path to review, practice, and reinforce your understanding.

LESSON 3

REFRACTION OF LIGHT

FOCUS QUESTION

What happens to light when it enters a new medium?

Light and Boundaries

Recall that when light crosses a boundary between two mediums, it bends. You might be familiar with this phenomenon already. Have you ever looked down through the water of a swimming pool and thought the water looked shallower than it was? This strange appearance is due to the bending of light at the boundaries between two mediums, or refraction.

Refraction Observe the light rays in **Figure 18.** Identical rays of light start in air and pass into three different mediums: water, glass, and diamond. The light rays hit the surface of each medium at the same angle and bend, or refract, as they cross the boundaries.

What difference do you notice between the three mediums shown? The light rays bend more when traveling from air to diamond than from air to water or air to glass. The amount of refraction depends on properties of the mediums that the light rays are traveling from and into.

What do you think the relationship is between the angle of the light as it crosses the boundary between mediums and refraction?

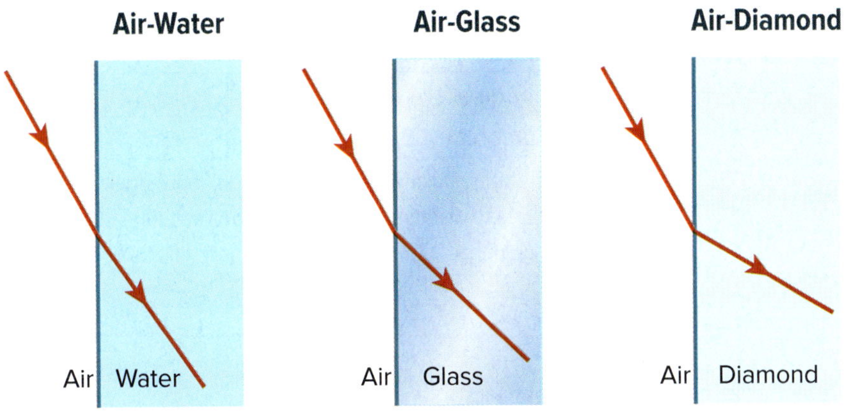

Figure 18 Light refracts as it crosses a boundary. The amount of refraction depends on the properties of the mediums. (Angles are not to scale.)

3D THINKING | DCI Disciplinary Core Ideas | CCC Crosscutting Concepts | SEP Science & Engineering Practices

COLLECT EVIDENCE

Use your Science Journal to record the evidence you collect as you complete the readings and activities in this lesson.

INVESTIGATE

GO ONLINE to find these activities and more resources.

Forensics Lab: A Whole Spectrum of Possibilities

Analyze data about how the index of refraction being dependent on wavelength causes the dipersion of light.

Revisit the Encounter the Phenomenon Question

What information from this lesson can help you answer the Unit and Module questions?

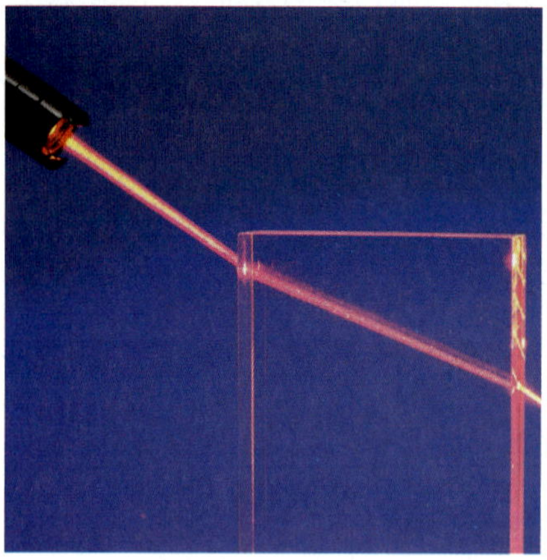

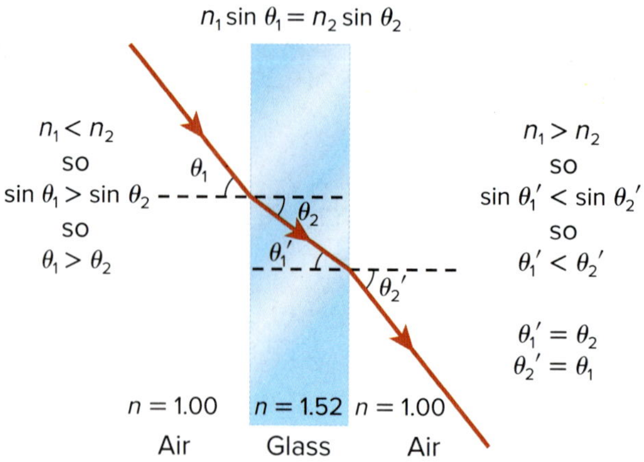

Figure 19 When light travels from air through glass and back to air, it refracts toward and then away from the normal.

Snell's Law of Refraction

René Descartes and Willebrord Snell first studied refraction in the seventeenth century by shining a narrow beam of light onto a transparent medium, such as the glass shown in **Figure 19.** They measured two angles—the angle of incidence and the angle of refraction. The angle of incidence (θ_1) is the angle at which the light ray strikes the surface. The angle of refraction (θ_2) is the angle at which the transmitted light leaves the surface. Both angles are measured with respect to the normal.

Index of refraction

Snell found that when light passes from air into a transparent medium, the sines of the angles are related. This **index of refraction** (n) determines the angle of refraction of light as it crosses the boundary between two mediums. Properties of the mediums light is traveling through determine indices of refraction. Values of n for several mediums are listed in **Table 2.**

Table 2 Indices of Refraction for Yellow Light ($\lambda = 589$ nm in a vacuum)

Medium	n
Vacuum	1.00
Air	1.0003*
Water	1.33
Ethanol	1.36
Float glass	1.52
Quartz	1.54
Flint glass	1.62
Diamond	2.42

*The value given for air contains additional significant figures to distinguish it from that for a vacuum. Use a value of $n = 1.00$ for air in problems in this text.

Snell's Law of Refraction

The product of the index of refraction of the first medium and the sine of the angle of incidence is equal to the product of the index of refraction of the second medium and the sine of the angle of refraction.

$$n_1 \sin \theta_1 = n_2 \sin \theta_2$$

We can use **Figure 19** and **Table 2** to show how Snell's law applies when light travels through a piece of glass with parallel surfaces, such as a windowpane. The light is refracted when it enters the glass and again when it leaves the glass. When light goes from air into glass, it travels from a medium with a lower n of 1.00 to a medium with a higher n of 1.52. The light bends toward the normal.

Traveling from glass to air, light moves from a medium with a higher n (1.52) to one with a lower n (1.00). The light is bent away from the normal. The relative values of n determine whether the light will bend toward or away from the normal. Note the direction of the ray when it leaves the glass. It is the same as before it struck the glass, $\theta_1 = \theta_2'$. This is because the boundaries are between the same two mediums.

EXAMPLE Problem 4

ANGLE OF REFRACTION A beam of light in air hits a sheet of float glass at an angle of 30.0°. What is the angle of refraction of the light ray?

1 ANALYZE AND SKETCH THE PROBLEM

- Make a sketch of the air and float glass boundary.
- Draw the light ray and label θ_1, θ_2, n_1, and n_2.

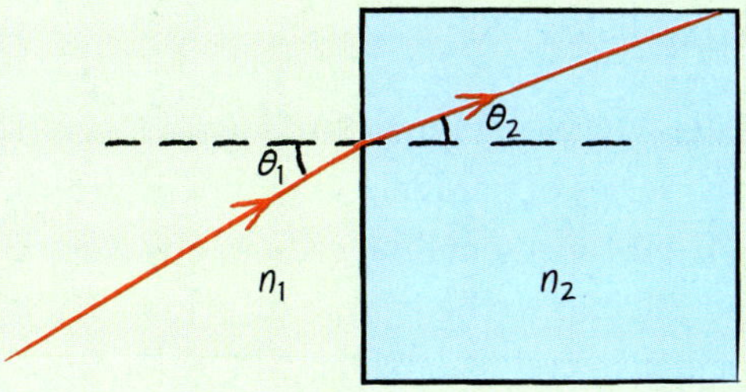

Known	Unknown
$\theta_1 = 30.0°$	$\theta_2 = ?$
$n_1 = 1.00$	
$n_2 = 1.52$	

2 SOLVE FOR THE ANGLE OF REFRACTION

Use Snell's law to solve for the sine of the angle of refraction.

$$n_1 \sin \theta_1 = n_2 \sin \theta_2$$

$$\sin \theta_2 = \left(\frac{n_1}{n_2}\right) \sin \theta_1$$

$$\theta_2 = \sin^{-1}\left(\left(\frac{n_1}{n_2}\right) \sin \theta_1\right)$$

$$= \sin^{-1}\left(\left(\frac{1.00}{1.52}\right) \sin 30.0°\right)$$ Substitute $n_1 = 1.00$, $n_2 = 1.52$, $\theta_1 = 30.0°$.

$$= 19.2°$$

3 EVALUATE THE ANSWER

- **Does the answer make sense?** The light beam travels into a medium with a higher n. Therefore, θ_2 should be less than θ_1.

PRACTICE Problems

ADDITIONAL PRACTICE

30. A laser beam in air enters ethanol at an angle of incidence of 37.0°. What is the angle of refraction?

31. As light travels from air into water, the angle of refraction is 25.0° to the normal. Find the angle of incidence.

32. Light in air enters a diamond facet at 45.0°. What is the angle of refraction?

33. A block of unknown material is submerged in water. Light in the water enters the block at an angle of incidence of 31°. The angle of refraction of the light in the block is 27°. What is the index of refraction of the material of the block?

34. CHALLENGE Light travels from air into another medium. The angle of incidence is 45.0° and the angle of refraction is 27.7°. What is the other medium?

The Meaning of the Index of Refraction

Looking at light as a wave yields an interesting connection between the propagation of light through a medium and its index of refraction. Snell's work was done prior to the development of the wave model of light; after the development of the wave model, it was understood that light interacts with atoms in such a way that it moves more slowly through a medium than it does through a vacuum.

Remember that for light traveling through a vacuum, $\lambda_0 = \frac{c}{f}$. This can be written more generally as $\lambda = \frac{v}{f}$, where v is the speed of light in any medium and λ is the wavelength in that medium. Frequency (f) is the number of oscillations a wave makes per second.

Frequency does not change at the boundary, so the wavelength decreases when light slows down as it crosses a boundary. And because the speed of light in any medium is slower than the speed of light in a vacuum, the wavelength in any medium is shorter than it is in a vacuum.

Wave model **Figure 20** shows a beam of light made up of a series of parallel, straight wavefronts. Each wavefront represents the crest of a wave and is perpendicular to the beam's direction. The beam's angle of incidence is θ_1. Consider ΔPQR. Because the wavefronts are perpendicular to the beam's direction, $\angle$PQR is a right angle. $\angle$QRP is equal to θ_1. Therefore, sin θ_1 is equal to $\overline{PQ}$ divided by $\overline{PR}$.

$$\sin \theta_1 = \frac{\overline{PQ}}{\overline{PR}}$$

The angle of refraction (θ_2) can be related in a similar way to ΔPSR. In this case,

$$\sin \theta_2 = \frac{\overline{RS}}{\overline{PR}}$$

By taking the ratio of the sines of the two angles, $\overline{PR}$ is canceled, leaving the following equation.

$$\frac{\sin \theta_2}{\sin \theta_1} = \frac{\overline{RS}}{\overline{PQ}}$$

In **Figure 20,** $\overline{PQ}$ is the length of three wavelengths of light in air, which can be written as $\overline{PQ} = 3\lambda_1$. In a similar way, $\overline{RS} = 3\lambda_2$. Substituting these values into the previous equation and canceling the common factor of 3 yields an equation relating the angles of incidence and refraction to the wavelength of light in each medium.

$$\frac{\sin \theta_2}{\sin \theta_1} = \frac{3\lambda_2}{3\lambda_1} = \frac{\lambda_2}{\lambda_1}$$

Using $\lambda = \frac{v}{f}$ and canceling the common factor f, because frequency stays constant, the equation can be rewritten:

$$\frac{\sin \theta_2}{\sin \theta_1} = \frac{v_2}{v_1}$$

Snell's law also can be written as a ratio of the sines of the angles of incidence and refraction.

$$\frac{\sin \theta_2}{\sin \theta_1} = \frac{n_1}{n_2}$$

Using the transitive property of equality, the previous two equations lead to the following equation:

$$\frac{n_1}{n_2} = \frac{v_2}{v_1}$$

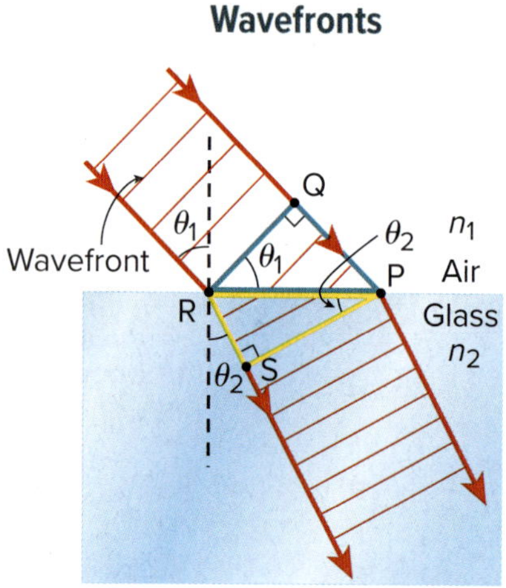

Figure 20 Each wavefront passes the boundary from air to glass at an angle. Part of the wavefront slows, and the ray bends. Since the wave slows and the frequency stays constant, for $\lambda = \frac{v}{f}$ to be true, the wavelength must decrease.

Infer *Which medium has a higher index of refraction?*

In a vacuum, $n = 1.00$ and $v = c$. If either medium is a vacuum, then the equation is simplified to an equation that relates the index of refraction to the speed of light in a medium.

Index of Refraction

The index of refraction of a medium is equal to the speed of light in a vacuum divided by the speed of light in the medium.

$$n = \frac{c}{v}$$

This definition of the index of refraction can be used to find the wavelength of light in a medium. In a medium with an index of refraction n, the speed of light is given by $v = \frac{c}{n}$. The wavelength of the light in a vacuum is $\lambda_0 = \frac{c}{f}$. Solve for frequency, and substitute $f = \frac{c}{\lambda_0}$ and $v = \frac{c}{n}$ into $\lambda = \frac{v}{f}$.

$$\lambda = \frac{(c/n)}{(c/\lambda_0)} = \frac{\lambda_0}{n}$$

Note that the wavelength of light in a medium is smaller than its wavelength in a vacuum.

Get It?

Describe the relationship between the index of refraction and the speed of light in a medium.

Total Internal Reflection

Recall that when light strikes a transparent boundary, some of the light is transmitted and some is reflected. As light travels from a medium of higher index of refraction to a medium of lower index of refraction, the angle of refraction is larger than the angle of incidence, as shown in **Figure 21.**

This leads to an interesting phenomenon. As the angle of incidence increases, the angle of refraction increases. At a certain angle of incidence known as the **critical angle** (θ_c), the refracted light ray lies along the boundary of the two mediums.

Total internal reflection occurs when light traveling from a region of a higher index of refraction to a region of a lower index of refraction strikes the boundary at an angle greater than θ_c such that all the light reflects back into the region of the higher index of refraction. This is shown in the third diagram of **Figure 21.** Using Snell's law and substituting $\theta_1 = \theta_c$, and $\sin \theta_2 = \sin 90.0° = 1$:

$$\sin \theta_c = \frac{n_2}{n_1}$$

Total internal reflection causes some curious effects. Suppose you are looking up at the surface from under water in a pool. You might see an upside-down reflection of another nearby object that is also under water. You might see a reflection of the bottom of the pool itself. The surface of the water acts like a mirror, reflecting the image back into the water.

Real-World Physics

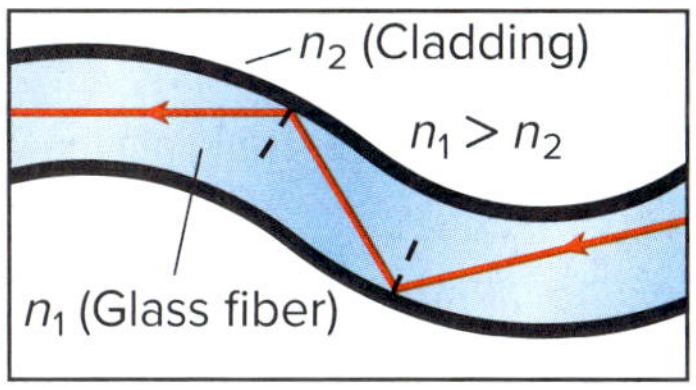

TOTAL INTERNAL REFLECTION is used in communication via optical fibers. The light traveling through the transparent fiber always hits the internal boundary of the optical fiber at an angle greater than the critical angle. All the light is reflected, and none of the light is transmitted through the boundary. Light pulses in fiber optic cables carry larger amounts of information over longer distances than other forms of communication. The outer material of each fiber is called cladding.

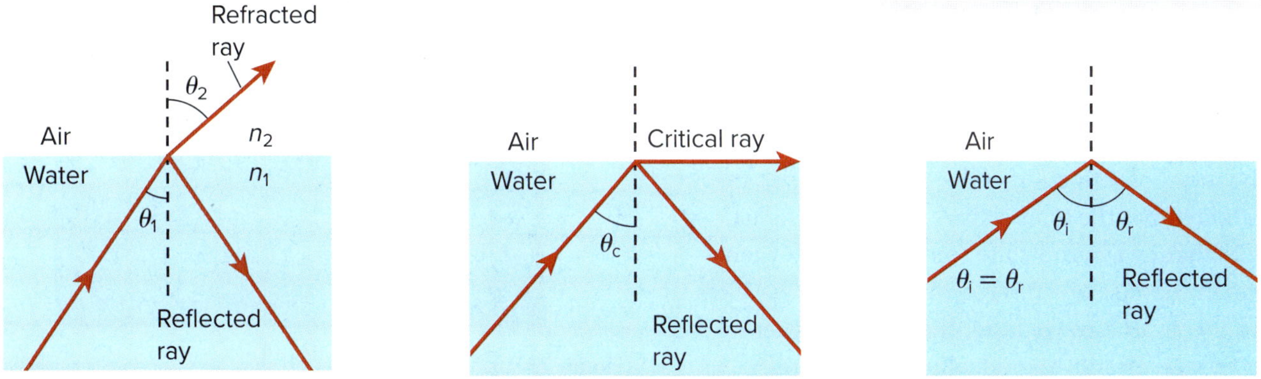

At an angle of incidence less than the critical angle, light is partially refracted and partially reflected.

A ray refracted along the boundary of the medium forms the critical angle.

An angle of incidence greater than the critical angle results in total internal reflection, which follows the law of reflection.

Figure 21 Refraction and reflection of light traveling between mediums depend on the angle of incidence θ_1.

Get It?

Compare and contrast data transmission using the fiber optics cable at the beginning of this module and data transmission by satellite.

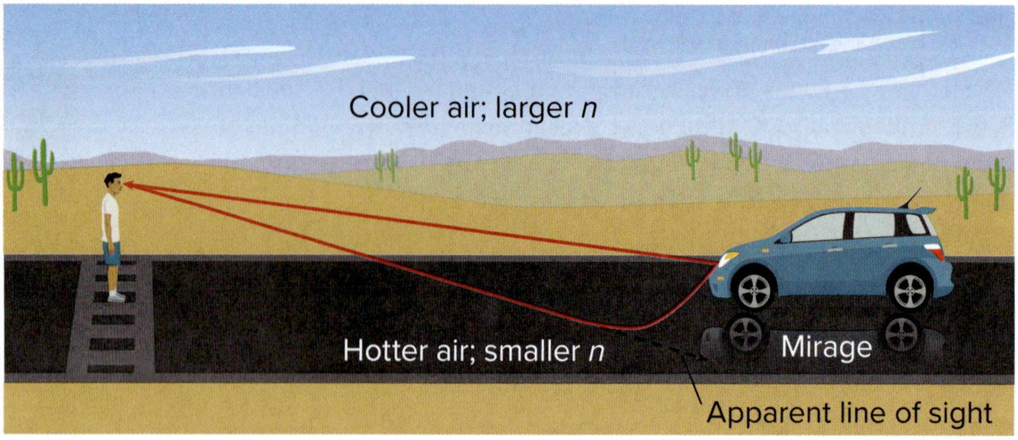

Figure 22 Light waves reflected from the vehicle on the road are refracted when the air near the surface of the road is hotter than the air above it.

Apply *Are the light waves traveling faster near the surface of the road or near the top of the vehicle?*

Mirages

On a hot summer day, you sometimes can see what appears to be the reflection of an oncoming vehicle in a pool of water as shown on the left in **Figure 22.** The pool, however, disappears as you approach it. This mirage is the result of the Sun heating the road. The speed of light and, therefore, the index of refraction, for a gaseous medium can change slightly with temperature. The hot road heats the air above it and produces a thermal layering of air, causing light traveling toward the road from the car to gradually bend upward. This makes the light appear to be coming from a reflection in a pool.

The diagram in **Figure 22** shows how this occurs. As light from a distant object travels downward toward the road, the index of refraction of the air decreases as the air gets hotter, but the temperature change is gradual. Recall that light wavefronts are comprised of Huygens' wavelets. In the case of a mirage, the Huygens' wavelets closer to the ground travel faster than those higher up, causing the wavefronts to gradually turn upward.

Dispersion of Light

If you look carefully at the spectrum formed by light passing through the prism in **Figure 23,** you will notice that violet light is refracted more than red light. This occurs because the speed of violet light through glass is less than the speed of red light through glass. Violet light has a higher frequency than red light, which causes it to interact differently with the atoms of the glass. This results in glass having a slightly higher index of refraction for violet light than it has for red light. This separation of white light into a spectrum of colors is called **dispersion.**

The speed of light in a medium is determined by interactions between the light and the atoms that make up the medium. The speed of light and the index of refraction for a solid or liquid medium vary for different wavelengths of light.

Figure 23 Dispersion through a prism occurs because the index of refraction varies with the wavelength of light.

Rainbows A rainbow is a spectrum formed when sunlight is dispersed by water droplets in the atmosphere. Sunlight that falls on a water droplet is refracted. Since each color has a different wavelength, it is refracted at a slightly different angle, as shown in **Figure 24** on the next page, resulting in dispersion. At the back surface of the droplet, some of the light undergoes internal reflection. On the way out of the droplet, the light once again is refracted and dispersed.

Although each droplet produces a complete spectrum, an observer positioned between the Sun and the rain will see only a certain wavelength of light from each droplet. The wavelength observed depends on the relative positions of the Sun, the droplet, and the observer, as shown in **Figure 24.**

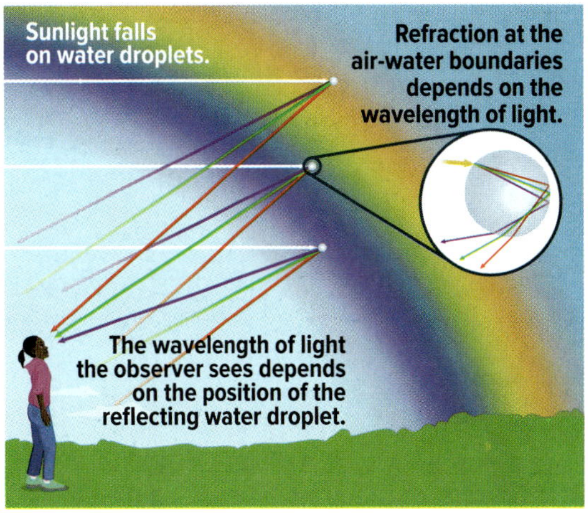

Figure 24 Light from the Sun is dispersed by water droplets to form rainbows. Because there are many droplets at various positions in the sky, a complete spectrum is visible.

Consider *whether it is possible for a rainbow to be close enough to touch.*

Second-order rainbow Sometimes, you can see a faint second-order rainbow like the one shown in **Figure 25.** The second rainbow is outside of the first. It is also fainter and has the order of the colors reversed. Light rays that are reflected twice inside water droplets produce this effect.

Figure 25 Light reflected within raindrops can cause the appearance of a secondary rainbow.

Check Your Progress

35. **Index of Refraction** When a ray of light enters a certain liquid from water, it is bent toward the normal, but when it enters the same liquid from float glass, it is bent away from the normal. What can you conclude about the liquid's index of refraction?

36. **Index of Refraction** A ray of light in air has an angle of incidence of 30.0° on a block of unknown material and an angle of refraction of 20.0°, as shown in **Figure 26.** What is the index of refraction of the material?

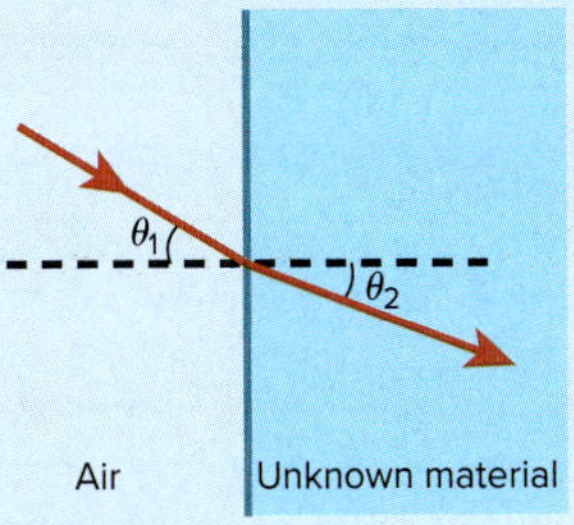

Figure 26

37. **Angle of Refraction** A beam of light passes from water into polyethylene ($n = 1.50$). If $\theta_1 = 57.5°$, what is the angle of refraction in the polyethylene?

38. **Speed of Light** What is the speed of light in chloroform ($n = 1.51$)?

39. **Critical Angle** Is there a critical angle for light traveling from glass to water? From water to glass? Explain your answer.

40. **Total Internal Reflection** Would you use quartz or float glass for the cladding of an optical fiber? Explain your answer.

41. **Speed of Light** Could an index of refraction ever be less than 1? What would this imply about the speed of light in that medium?

42. **Critical Thinking** In what direction would you have to be looking to see a rainbow on a late rainy afternoon? Explain.

LEARNSMART® Go online to follow your personalized learning path to review, practice, and reinforce your understanding.

LESSON 4

CONVEX AND CONCAVE LENSES

FOCUS QUESTION

How are systems of lenses used to make optical devices?

Types of Lenses

Refraction of light can be quite useful. In 1303, French physician Bernard of Gordon wrote of using lenses to correct eyesight. Around 1610, Galileo discovered the four major moons of Jupiter using a telescope made with two lenses. Since that time, lenses have been used in instruments such as microscopes and cameras.

A **lens** is a piece of transparent material that focuses light and forms an image. Each of a lens's two faces might be either curved or flat. A lens that is thicker at the center than at the edges is called a **convex lens,** as shown on the left in **Figure 27.** When it is surrounded by material with a lower index of refraction, such as air, it refracts parallel light rays so that they all pass through a common point, called the focal point, after going through the lens. For this reason, a convex lens is often called a converging lens.

A lens that is thinner in the middle than at the edges is called a **concave lens,** as shown on the right in **Figure 27.** When surrounded by material with a lower index of refraction, rays passing through it spread out, so it is often called a diverging lens.

You can determine the paths of rays passing through lenses using Snell's law. To simplify such problems, we assume that all refraction occurs at the principal plane, which passes through the center of the lens. This approximation, called the thin lens model, applies to all the lenses discussed in this module.

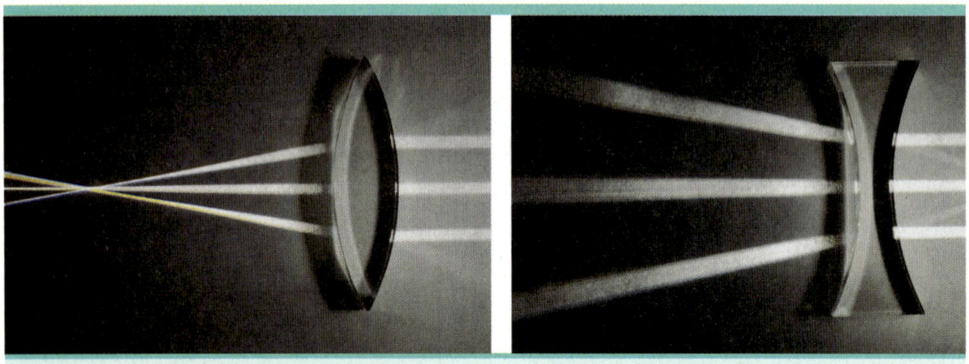

Convex Lens

Concave Lens

Figure 27 Convex lenses refract light so that the rays converge after passing through. The light passing through a concave lens does not meet at the focal point.

3D THINKING DCI Disciplinary Core Ideas CCC Crosscutting Concepts SEP Science & Engineering Practices

COLLECT EVIDENCE

Use your Science Journal to record the evidence you collect as you complete the readings and activities in this lesson.

INVESTIGATE

GO ONLINE to find these activities and more resources.

Virtual Investigation: Ray Tracing for Lenses
Use a computer model to investigate the properties of a lens affect how light is transmitted.

Review the News
Obtain information from a current news story about technologies that use light. Evaluate your source and communicate your findings to your class.

Convex Lenses

To determine an image's location, we can use ray diagrams. In a ray diagram, as with mirrors you represent a few important rays to find out how a lens affects the light that passes through it. You can use any two rays to locate the image; with experience, you will learn that some rays are easier to draw and use than others.

In ray diagrams for convex lenses, x_o is the distance of the object from the lens and x_i is the distance of the image from the lens. In all ray diagrams in this module, the thin lens model is used. In this model, light refracts at the center of the lens rather than at the boundaries between air and the surface of the lens.

$x_o \geq 2f$ In **Figure 28,** rays are traced from an object located far from the lens. Ray 1 is parallel to the principal axis. It refracts and passes through the focal point (F) on the other side of the lens. Ray 2 passes through F on its way to the lens. After refraction, its path is parallel to the principal axis.

The two rays intersect at a point beyond F and locate the image. Rays selected from other points on the object converge at corresponding points to form the complete image. Note that this is a real image that is inverted and smaller compared to the object.

$x_o = 2f$ If the object is placed at twice the focal length from the lens at the point 2F, the diagram is similar. The image is real and is found at 2F, but it is no longer reduced in size. Because of symmetry, the image and the object have the same size. Thus, you can conclude that if an object is more than twice the focal length from the lens, the image is smaller than the object.

$2f > x_o > f$ You can use **Figure 29** to locate the image of an object that is between F and 2F viewed through a convex lens. This is similar to the ray diagram for the object located at a distance greater than twice the focal length with the image and object interchanged. The direction of the rays would be reversed. In this case, the image is also real and inverted.

For an object placed between F and 2F, the image is enlarged. When an object is placed at the focal point of a convex lens, ray diagrams cannot be drawn. The refracted rays will emerge in a parallel beam and no image will be seen.

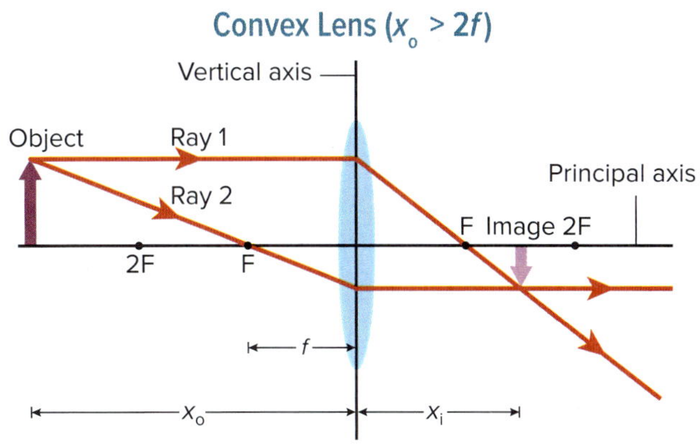

Figure 28 An object placed at a distance greater than twice the focal length from the lens will produce an image that is real, reduced in size, and inverted.

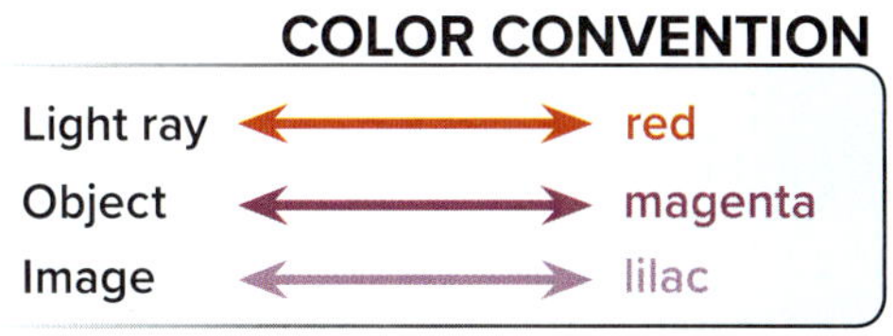

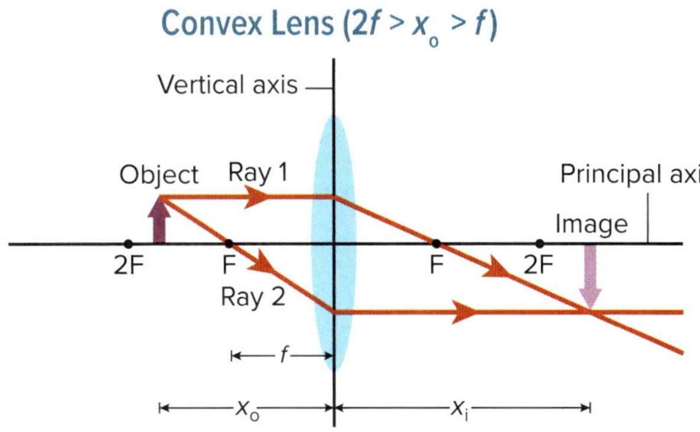

Figure 29 An object placed at a distance less than twice the focal length but greater than one focal length from the lens will produce an image that is real, enlarged, and inverted.

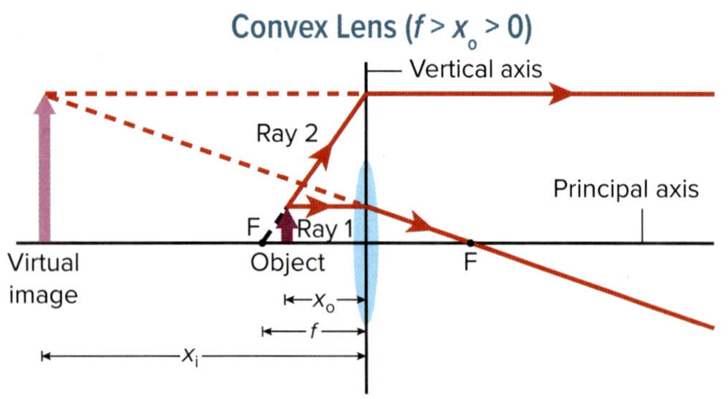

Figure 30 An object placed at a distance less than the focal length from the lens will produce an image that is virtual and enlarged.

$f > x_o > 0$ **Figure 30** shows how a convex lens forms a virtual image. The object in this situation is located between F and the lens. Ray 1, as usual, approaches the lens parallel to the principal axis and is refracted through the focal point, F. Ray 2 travels from the tip of the object in the direction it would have if it had started at F on the object side of the lens. The dashed line from F to the object shows you how to draw ray 2. Ray 2 leaves the lens parallel to the principal axis. Ray 1 and ray 2 diverge as they leave the lens.

The reflection appears to an observer to come from a spot on the same side of the lens as the object. This is a virtual image that is upright and larger compared to the object. No real image is possible. Drawing sight lines for the two rays back to their apparent intersection locates the virtual image. Note that the actual image is formed by light that passes through the lens, but you can still determine the location of the image by drawing rays that do not have to pass through the lens.

Get It?

Classify an image as virtual or real based on the side of the lens it is on.

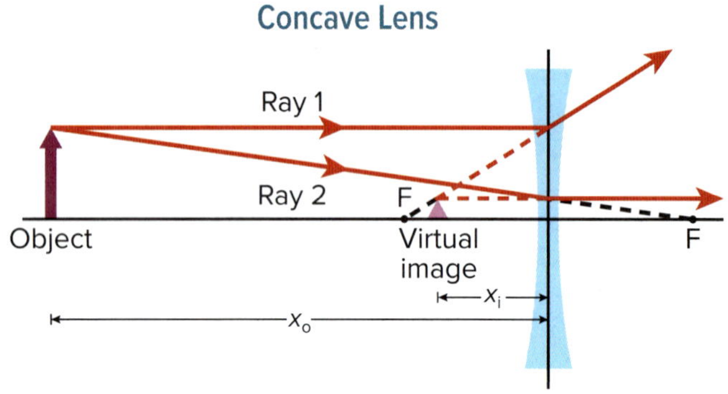

Figure 31 An object placed any distance from a concave lens will always produce an image that is virtual and reduced.

Concave Lenses

A concave lens causes all rays to diverge. **Figure 31** shows how a concave lens forms a virtual image. Ray 1 approaches the lens parallel to the principal axis. It leaves the lens along a line that extends back through the focal point on the object side of the lens. The focal point of a concave lens is found on the same side of the lens as the incoming light. Ray 2 approaches the lens as if it is going to pass through the focal point on the opposite side but leaves the lens parallel to the principal axis.

The sight lines of rays 1 and 2 intersect on the same side of the lens as the object. The image is located at the point from where the two rays appear to intersect, creating a virtual image. The image is upright and smaller than the object. This is true no matter how far from the lens the object is located. The focal length for a diverging lens is negative.

Get It?

Describe why a concave lens always produces a virtual image.

Table 3 Properties of a Single Spherical Lens System

Lens Type	f	x_o	x_i	m	Image
Convex	+	$x_o > 2f$	$2f > x_i > f$	reduced, inverted	real
		$2f > x_o > f$	$x_i > 2f$	enlarged, inverted	real
		$f > x_o > 0$	$\lvert x_i \rvert > x_o$ (negative)	enlarged	virtual
Concave	−	$x_o > 0$	$\lvert f \rvert > \lvert x_i \rvert > 0$ (negative)	reduced	virtual

Lens Equations

Lenses can be constructed with a variety of shapes, but you will only consider thin lenses, with faces having spherical curvatures. Based on this model of thin spherical lenses, an equation has been developed that looks exactly like the one for spherical mirrors. For lenses, virtual images are always on the same side of the lens as the object, which means the image position is negative. Note that a concave lens produces only virtual images, whereas a convex lens can produce real images or virtual images.

Thin lens equation

The **thin lens equation** relates the focal length of a spherical thin lens, the object position, and the image position.

Thin Lens Equation

The inverse of the focal length of a spherical lens is equal to the sum of the inverses of the image position and the object position.

$$\frac{1}{f} = \frac{1}{x_i} + \frac{1}{x_o}$$

When solving problems for concave lenses using the thin lens equation, remember that the sign convention for focal length is different from that of convex lenses. This is because a concave lens is a divergent lens. If the focal point for a concave lens is 24 cm from the lens, you should use the value $f = -24$ cm in the thin lens equation. All images for a concave lens are virtual. Thus, if an image distance is given as 20 cm from the lens, then you should use $x_i = -20$ cm. The object position always will be positive.

Magnification

A property of spherical thin lenses that measures how much larger or smaller the image is than the object is magnification. The magnification equation for spherical mirrors also can be used for spherical thin lenses. It is used to determine the height and the orientation of the image formed by a spherical thin lens.

Magnification

The magnification of an object by a spherical lens, defined as the image height divided by the object height, is equal to the negative of the image position divided by the object position.

$$m \equiv \frac{h_i}{h_o} = -\frac{x_i}{x_o}$$

Magnification gives information about the size and orientation of the image relative to the object. When the absolute value of a magnification is between zero and one, the image is smaller than the object. Magnifications with absolute values greater than one represent images that are larger than the objects. A negative magnification means the image is inverted compared to the object.

Using the equations for lenses

It is important that you use the proper sign conventions when using these equations. **Table 3** shows a comparison of the image position, magnification (m), and type of image formed by single convex and concave lenses when an object is placed at various object positions (x_o) relative to the lens. For convex lenses, the object position relative to the focal point influences the image type.

Notice how this table is similar to **Table 1** in Lesson 1, the table for mirrors. As with mirrors, the distance from the principal plane of a lens to its focal point is the focal length (f). The focal length depends on the shape of the lens and the index of refraction of the lens material. Focal lengths, image positions, and image heights can be negative.

EXAMPLE Problem 5

AN IMAGE FORMED BY A CONVEX LENS An object is placed 32.0 cm from a convex lens that has a focal length of 8.0 cm.

a. Where is the image?

b. If the object is 3.0 cm high, how tall is the image?

c. What is the orientation of the image?

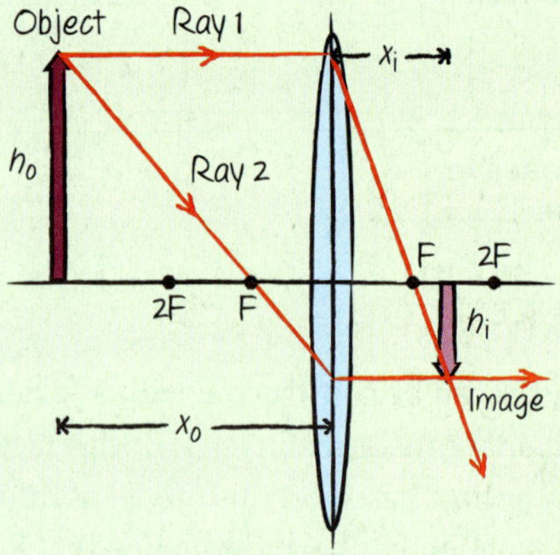

1 ANALYZE AND SKETCH THE PROBLEM

- Sketch the situation, locating the object and the lens.
- Draw the two principal rays.

Known	Unknown
$x_o = 32.0$ cm	$x_i = ?$
$h_o = 3.0$ cm	$h_i = ?$
$f = 8.0$ cm	

2 SOLVE FOR THE IMAGE POSITION AND HEIGHT

a. Use the thin lens equation to determine x_i.

$$\frac{1}{f} = \frac{1}{x_i} + \frac{1}{x_o}$$

$$x_i = \frac{fx_o}{x_o - f}$$

$$= \frac{(8.0\text{ cm})(32.0\text{ cm})}{32.0\text{ cm} - 8.0\text{ cm}}$$ Substitute $f = 8.0$ cm, $x_o = 32.0$ cm.

$= 11$ cm (11 cm away from the lens on the side opposite the object)

b. Use the magnification equation to solve for image height.

$$m \equiv \frac{h_i}{h_o} = -\frac{x_i}{x_o}$$

$$h_i = -\frac{x_i h_o}{x_o}$$

$$= -\frac{(11\text{ cm})(3.0\text{ cm})}{32.0\text{ cm}}$$ Substitute $x_i = 11$ cm, $h_o = 3.0$ cm, $x_o = 32.0$ cm.

$= -1.0$ cm (1.0 cm tall)

c. The negative sign for the height in part **b** means the image is inverted.

3 EVALUATE THE ANSWER

- **Are the units correct?** All are in centimeters.
- **Do the signs make sense?** Image position is positive (real image), and image height is negative (inverted compared to the object), which make sense for a convex lens.

PRACTICE Problems

ADDITIONAL PRACTICE

43. A 2.25-cm-tall object is 8.5 cm to the left of a convex lens whose focal length is 5.5 cm. Find the image position and height.

44. An object near a convex lens produces a 1.8-cm-tall real image that is 10.4 cm from the lens and inverted. If the focal length of the lens is 6.8 cm, what are the object position and height?

45. An object is placed to the left of a convex lens with a 25-mm focal length so that its image is the same size as the object. What are the image and object positions?

46. Calculate the image position and height of a 2.0-cm-tall object located 25 cm from a convex lens with a focal length of 5.0 cm. What is the orientation of the image?

47. Use a scale ray diagram to find the image position of an object that is 30 cm to the left of a convex lens with a 10-cm focal length.

48. **CHALLENGE** A magnifier with a focal length of 30 cm is used to view a 1-cm-tall object. Use a ray diagram to determine the location and size of the image when the magnifier is positioned 10 cm from the object.

Defects of Spherical Lenses

Throughout this section, you have studied lenses that produce perfect images at specific positions. In reality, spherical lenses, just like spherical mirrors, have intrinsic defects that cause problems with the focus and color of images. Spherical lenses exhibit an aberration associated with their spherical design, just as mirrors do. In addition, the dispersion of light through a spherical lens causes an aberration that mirrors do not exhibit.

Spherical aberration

The model you have used for drawing rays through spherical lenses suggests that all parallel rays focus at the same position. However, this is only an approximation. In reality, parallel rays that pass through the edges of a spherical lens focus at positions different from those of parallel rays that pass through the center. This inability of a spherical lens to focus all parallel rays to a single point is called spherical aberration. The effects are shown in **Figure 32.** In reality, most lenses have a slightly different shape to address this, but the spherical approximation works well enough for our purposes. In high-precision instruments, many lenses, often five or more, are used to form sharp, well-defined images.

Figure 32 Spherical aberration from the camera lens blurs the edges of this image.

Chromatic aberration

Lenses have a second defect that mirrors do not have. Because the index of refraction of a medium depends on wavelength, different wavelengths of light are refracted at slightly different angles, as you can see on the right in **Figure 33.** Light that passes through a lens is slightly dispersed, especially near the edges, causing an effect called **chromatic aberration.** This is seen as an apparent ring of color around an object viewed through a lens, as shown on the left in **Figure 33.**

Chromatic aberration is always present when a single lens is used. However, this defect can be greatly reduced by an **achromatic lens,** which is a system of two or more lenses, such as a convex lens with a concave lens, that have different indices of refraction. A simple lens and an achromatic lens are shown in **Figure 33.** Both lenses disperse light, but the dispersion caused by the achromatic lens is less than the dispersion caused by the simple lens. The index of refraction of the achromatic lens is chosen so that the light will still converge at the desired location.

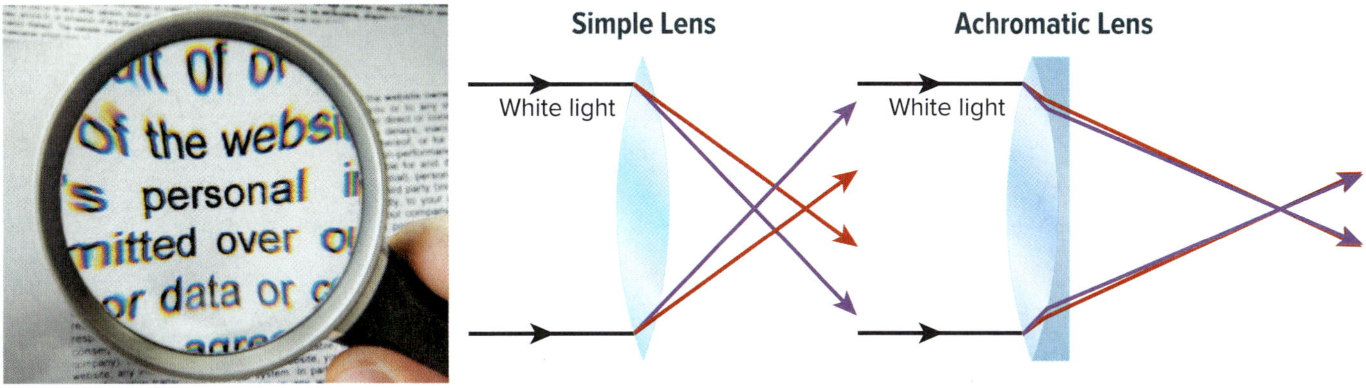

Figure 33 Simple lenses, such as the one shown in the photo on the left, exhibit a rainbow fringing effect called chromatic aberration. Achromatic lenses reduce chromatic aberration.

Explain *why the index of refraction is important for achromatic lenses.*

Lenses in Eyes

The concepts that you have learned for the refraction of light through lenses apply to almost every optical device, including the eye. The eye is a remarkable optical device. As shown in **Figure 34,** the eye is a fluid-filled, almost spherical vessel. Light that is emitted from or reflected off an object travels into the eye through the cornea and pupil. The light then passes through the lens and focuses onto the retina, which is at the back of the eye. Specialized cells on the retina absorb this light and send information about the image along the optic nerve to the brain.

BIOLOGY Connection Because of its name, you might assume that the lens of an eye is responsible for focusing light onto the retina. In fact, light entering the eye is primarily focused by the cornea, because the air-cornea boundary has the greater difference in indices of refraction. The lens is responsible for the fine focus that allows you to clearly see both distant and nearby objects.

Get It?
Describe the roles of the cornea and the lens in your eye.

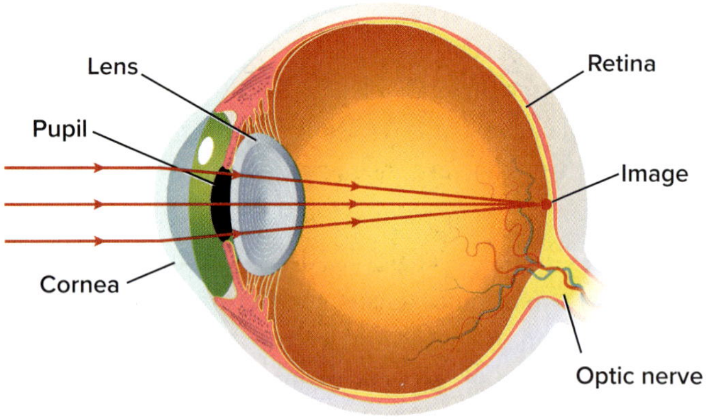

Figure 34 The cornea and the lens of your eye refract light that is reflected off every object you see.
Summarize *why most of the refraction occurs at the air-cornea boundary rather than at the lens.*

The eye lens uses a process called accommodation. Muscles surrounding the lens can contract or relax, thereby changing the shape of the lens. This, in turn, changes the focal length of the eye. For a healthy eye, when the muscles are relaxed, the image of distant objects is focused on the retina. When the muscles contract, the focal length is shortened, and this allows images of closer objects to be focused on the retina.

Nearsightedness

The eyes of many people focus images either in front of the retina or behind it. External lenses, such as eyeglasses and contact lenses, adjust the focal length and move images to the retina. **Figure 35** on the next page shows the condition of **nearsightedness,** also called myopia.

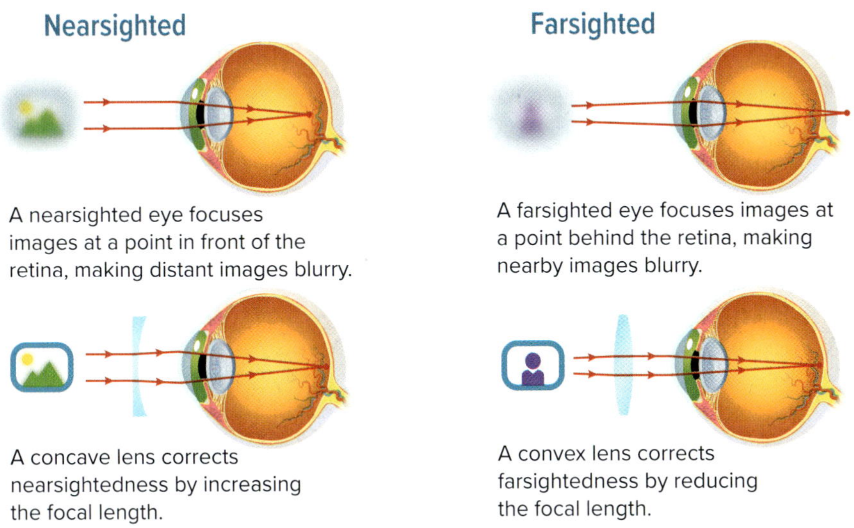

Figure 35 Objects at a distance are blurred for a nearsighted person. For a farsighted person, objects nearby are blurred.

Explain *how bifocal lenses might be made.*

With myopia, the focal length of the eye is too short to focus light on the retina. Images are formed in front of the retina. As shown in **Figure 35,** concave lenses correct this by diverging light, thereby increasing images' distances from the lens, and forming images on the retina.

Farsightedness You also can see in **Figure 35** that **farsightedness,** also called hyperopia, is the condition in which the focal length of the eye is too long. Images are therefore not formed on the retina.

A similar result is caused by the increasing rigidity of the lenses in the eyes of people who are more than about 45 years old. Their muscles cannot shorten the focal length enough to focus images of close objects on the retina.

For either condition, convex lenses produce virtual images farther from the eye than the associated objects, as shown in **Figure 35.** The images then become the objects for the eye lens and can be focused on the retina, thereby correcting the condition.

Real-World Physics

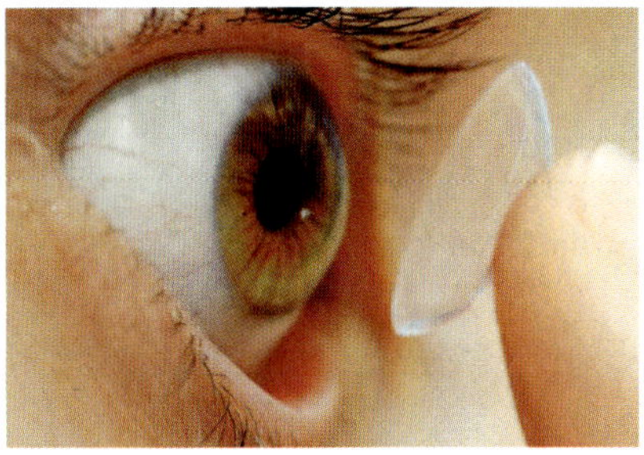

CONTACT LENSES produce the same results as eyeglasses. These small, thin lenses are placed directly on the corneas. A thin layer of tears between the cornea and lens keeps the lens in place. Most of the refraction occurs at the air-lens boundary, where the difference in indices of refraction is greatest.

PHYSICS Challenge

As light enters the eye, it first encounters the air-cornea boundary. Consider a ray of light that strikes the interface between the air and a person's cornea at an angle of 30.0° to the normal. The index of refraction of the cornea is approximately 1.4.

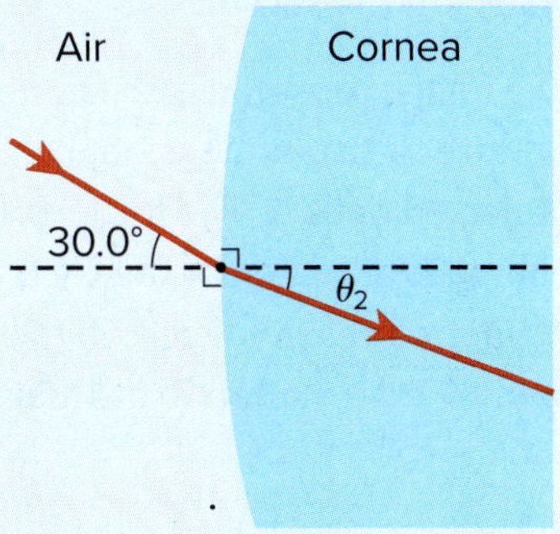

1. Use Snell's law to calculate the angle of refraction.
2. What would the angle of refraction be if the person were swimming underwater with his or her eyes open?
3. Is the refraction greater in air or in water? Does this mean objects underwater seem closer or more distant than they would in air?
4. If you want the angle of refraction for the light ray in water to be the same as it is for air, what should the new angle of incidence be?

Refracting Telescopes

An astronomical refracting telescope uses lenses to magnify distant objects. **Figure 36** shows the optical system for a Keplerian telescope. Light from stars and other astronomical objects comes from so far away that the incoming rays can be considered parallel. These rays enter the objective convex lens, which focuses them as a real image at the focal point of the objective lens. The image is inverted compared to the object. This image then becomes the object for the convex lens of the eyepiece.

Notice that the eyepiece lens is positioned so that the focal point of the objective lens is between the eyepiece lens and its focal point. This means a virtual image is produced that is upright and larger than the first image. However, because the first image was already inverted, the final image is still inverted. For viewing astronomical objects, an image that is inverted is acceptable.

In a telescope, the convex lens of the eyepiece is almost always an achromatic lens. Recall that an achromatic lens is a combination of lenses that functions as one lens. The combination of lenses greatly reduces the peripheral colors that can form on images due to chromatic aberration.

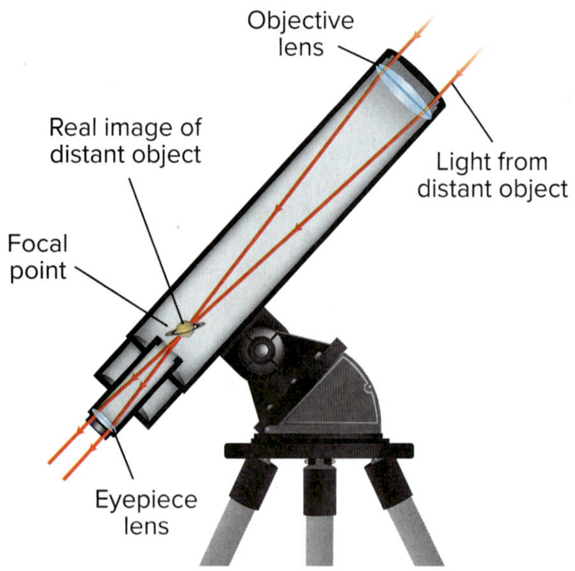

Figure 36 Light from distant objects is gathered by the objective lens and the eyepiece of a refracting telescope.

Evaluate *Why is an inverted image acceptable for viewing astronomical objects?*

Get It?

Explain how a refracting telescope works.

Cameras

Figure 37 shows the optical system used in a single-lens reflex camera. As light enters the camera through the aperture, it passes through an achromatic lens. This lens system refracts the light much like a single convex lens would, forming an image that is inverted on the reflex mirror. The image is reflected upward to a prism that inverts and redirects the light to the viewfinder.

When the person holding the camera takes a photograph, he or she presses the shutter-release button, which briefly raises the mirror, as shown in **Figure 37.** The light, instead of being diverted upward to the prism, then travels along a straight path to form an image on the sensor.

The image sensor, which is often a charge-coupled device (CCD), captures a two-dimensional image that corresponds to the image projected onto it. The image information is collected on the photoactive region of the CCD as electric charge proportional to the light intensity. The CCD then processes the charge and transfers the information to a storage device.

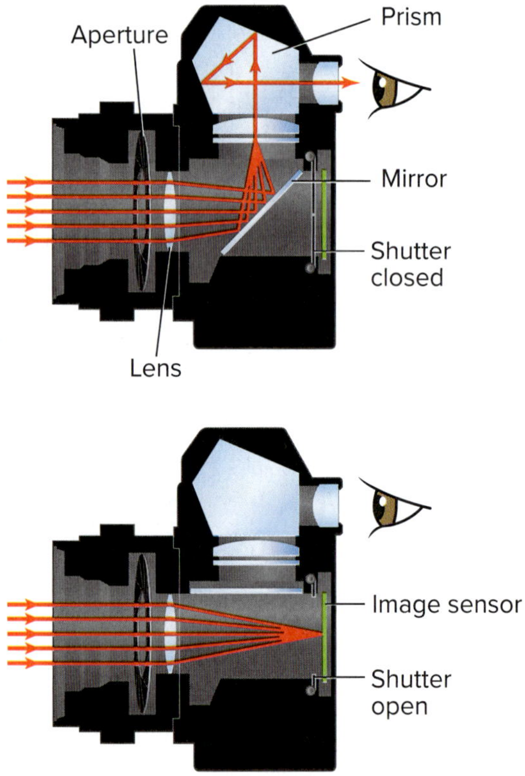

Figure 37 The lens of a camera can be adjusted to focus an image on the image sensor. When the shutter is closed, a mirror diverts the image to the viewer's eye.

Infer *Why would you want the shutter to be open longer when taking a photo in dim light?*

Check Your Progress

49. **Lenses** Magnifying lenses normally are used to produce images that are larger than the related objects, but they also can produce images that are smaller than the related objects. Explain.

50. **Types of Lenses** The cross sections of four different thin lenses are shown in **Figure 38.**
    a. Which of these lenses, if any, are convex, or converging, lenses?
    b. Which of these lenses, if any, are concave, or diverging, lenses?

Figure 38

51. **Image Position and Height** An object is placed 1.5 m from a convex lens with a focal length of 1.0 m. Use the thin lens equation to determine the distance of the image from the lens. If the object height is 2.0 m, what is the image height? Is the image real or virtual? Is the image inverted or upright?

52. **Thin Lens Approximation** What is a thin lens approximation and why is it used?

53. **Chromatic Aberration** All simple lenses have chromatic aberration. Infer why you do not see this effect when you look through a microscope, which has two convex lenses.

54. **Type of Image** Use the ray diagram in **Figure 39** to determine whether the image for object 1 will be reduced or enlarged, inverted or upright, and real or virtual. Do the same for object 2.

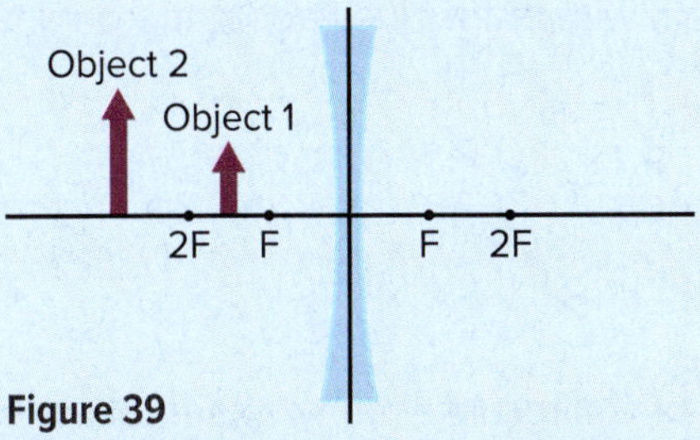

Figure 39

55. **Image Position and Height** A 6.0-cm-tall object is placed 5.0 cm from a convex lens with a focal length of 4.0 cm. Draw a ray diagram to determine the image location and size. Confirm your results using the thin lens equation and the magnification equation.

56. **Diverging Lens** A 6.5-cm-tall salt shaker is viewed through a diverging lens with a focal length of 5.0 cm.
    a. If the shaker is 6.0 cm from the lens, what is the image distance from the lens? Is the image virtual or real?
    b. What is the magnification? Is the image smaller or larger than the object?
    c. The salt shaker is now 4 cm from the lens. What is the distance of the image from the lens? The magnification? Is the image now smaller or larger than the object?

57. **Eyeglass Lenses** Which type of lens, convex or concave, should a nearsighted person use? Which type should a farsighted person use? See **Figure 40.** Explain.

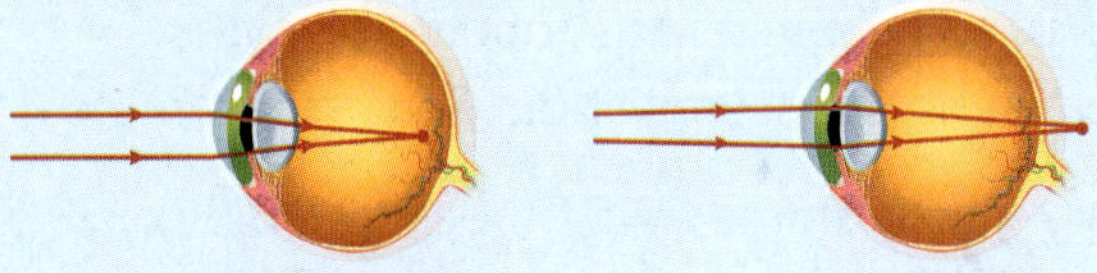

Figure 40

58. **Critical Thinking** An air lens constructed of two watch glasses is placed in a tank of water. Copy **Figure 41** and draw the effect of this lens on parallel light rays incident on the lens.

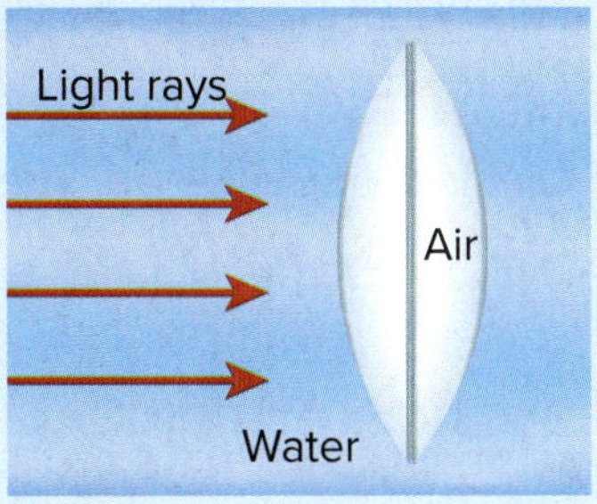

Figure 41

LEARNSMART® Go online to follow your personalized learning path to review, practice, and reinforce your understanding.

Scientists Hope the James Webb Space Telescope Will Help Them See the Light

Expected to launch in 2020, the James Webb Space Telescope (JWST) has been designed to produce images of deep space objects. Its ability to do so relies heavily on how its state of the art mirror captures and reflects light.

The ability of the JWST to produce images of deep space objects is largely due to the construction of its primary mirror.

Building a Better Telescope

The way in which backyard reflecting telescopes produce images is fairly straightforward. A concave primary mirror captures light from an object. This light is then reflected by another mirror to produce an image at a focal point. The larger the primary mirror, the more light it can capture. Many backyard reflecting telescopes used by amateur astronomers have a primary mirror that measures about 25 cm in diameter.

The primary mirror on the JWST measures 6.5 m in diameter. Its diameter not only dwarfs that of the average backyard telescope, but it's also much larger than that of two existing space telescopes. The diameter of the *Hubble Space Telescope*'s primary mirror, which has been in use since 1990, is 2.4 m. The telescope that has orbited Earth in the *Herschel Space Observatory* since 2009 has a primary mirror with a diameter of 3.5 m.

Scientists think that the JWST will be able to see some of the oldest objects in the universe. Some of the universe's oldest galaxies are over 13 billion light years away. At such a distance, the light that reaches Earth is faint. The JWST will be able to capture more of this light. The JWST will collect data in the infrared to see processes and objects not seen in visible light.

Unlike the *Hubble* and *Herschel* telescopes, the JWST will not be in Earth's orbit. It will be set about 1.5 million kilometers from Earth. Its primary mirror is too massive to launch in one piece, so it is constructed out of 18 smaller, six-sided mirrors made out of lightweight gold-plated beryllium. The components will fold up tightly for launch and unfold once in orbit. Each individual mirror can also be controlled to ensure that the pieces work together as a single large mirror and that light is reflected from the mirror accurately to produce an image that's not distorted.

Scientists are hopeful that the images from the JWST will provide them with more clues about the origins of the universe.

EVALUATE DESIGN SOLUTIONS

NASA officials have stated that the JWST has been designed to succeed, but not replace, other existing space telescopes. Compare the JWST to another telescope, giving an example of a space object that can be visualized using each.

CALIFORNIA CONNECTION

Research one of California's many observatories. Develop a presentation that both summarizes how the observatory's telescope works and discusses what types of objects it makes visible.

MODULE 16
STUDY GUIDE

GO ONLINE to study with your Science Notebook.

Lesson 1 REFLECTION OF LIGHT

- The law of reflection says that the angle of incidence equals the angle of reflection, and applies to specular and diffuse reflection.

$$\theta_i = \theta_r$$

- Plane mirrors form virtual images, which are the same size, orientation, and distance from the mirror as the object.

$$x_i = -x_o\,; \quad h_i = h_o$$

- specular reflection
- diffuse reflection
- plane mirror
- object
- image
- virtual image

Lesson 2 CURVED MIRRORS

- A spherical concave mirror is shaped as if it were a section of a hollow sphere with the same geometric center (C) and radius of curvature (r) as a sphere of radius r. The focal point (F) is the point where rays parallel to the principal axis converge after reflection.
- You can locate the image created by a curved mirror by drawing ray diagrams. The type of image formed by a concave mirror depends on the object's position. A convex mirror always forms a virtual image that is upright and smaller compared to the object.
- The magnification of a mirror image is given by equations relating either the positions or the heights of the image and the object.

$$m \equiv \frac{h_i}{h_o} = -\frac{x_i}{x_o}$$

- The mirror equation gives the relationship between image position, object position, and focal length of a spherical mirror.

$$\frac{1}{f} = \frac{1}{x_i} + \frac{1}{x_o}$$

- principal axis
- focal point
- focal length
- concave mirror
- convex mirror
- magnification
- real image
- spherical aberration

Lesson 3 REFRACTION OF LIGHT

- Light refracts when it travels across a boundary from one medium with an index of refraction (n_1) into a medium with a different index of refraction (n_2) according to Snell's law.

$$n_1 \sin \theta_1 = n_2 \sin \theta_2$$

- The ratio of the speed of light in a vacuum (c) to the speed of light in a medium (v) is the index of refraction (n) of the medium.
- When light traveling through a medium hits a boundary with a medium of a smaller index of refraction, if the angle of incidence exceeds the critical angle (θ_c), the light will be reflected back into the original medium by total internal reflection.
- Refraction causes optical effects such as mirages and rainbows.

- index of refraction
- critical angle
- total internal reflection
- dispersion

Lesson 4 CONVEX AND CONCAVE LENSES

- The image formed by a convex lens depends on the position of the object. The image formed by a concave lens is always virtual.
- The thin lens equation provides the relationship between focal length (f), object position (x_o), and image position (x_i).

$$\frac{1}{f} = \frac{1}{x_i} + \frac{1}{x_o}$$

- Concave and convex lenses correct eye conditions such as nearsightedness and farsightedness.

- lens
- convex lens
- concave lens
- thin lens equation
- chromatic aberration
- achromatic lens
- nearsightedness
- farsightedness

THREE-DIMENSIONAL THINKING

Module Wrap-Up

REVISIT THE PHENOMENON

How does light transmit information through a communication network?

CER Claim, Evidence, Reasoning

Explain your Reasoning Revisit the claim you made when you encountered the phenomenon. Summarize the evidence you gathered from your investigations and research and finalize your Summary Table. Does your evidence support your claim? If not, revise your claim. Explain why your evidence supports your claim.

STEM UNIT PROJECT

Now that you've completed the module, revisit your STEM unit project. You will summarize your evidence and apply it to the project.

GO FURTHER

SEP Data Analysis Lab

Which material should you use for fiber optic cables?

You are planning a colony on a very cold, icy planet. The engineer in charge of communications suggests that instead of hauling glass fiber optic cables from Earth, the expedition should make them from ice once they arrive on the planet.

CER Analyze and Interpret Data

1. Find the critical angle for light traveling from ice ($n = 1.31$) to air.
2. **Claim** Would fiber optic cables made of ice or those made of glass do a better job of keeping light inside the cable?
3. **Evidence and Reasoning** Use your calculations and what you know about refraction to support and explain your claims.

MODULE 17

INTERFERENCE AND DIFFRACTION

Anand Varma/National Geographic Magazines/Getty Images

MODULE 17

INTERFERENCE AND DIFFRACTION

ENCOUNTER THE PHENOMENON

What makes this hummingbird's feathers appear shiny and shimmery?

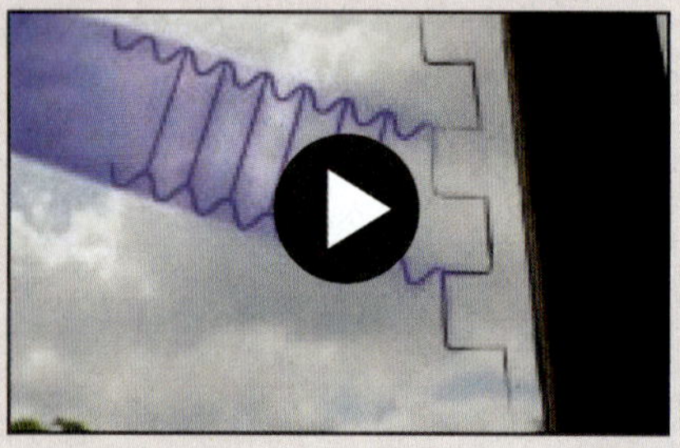

GO ONLINE to watch a video about iridescence in nature.

SEP Ask Questions

Do you have other questions about the phenomenon? If so, add them to the driving question board.

CER Claim, Evidence, Reasoning

Make Your Claim Use your CER chart to make a claim about why the hummingbird's feathers are shiny and shimmery. Explain your reasoning.

Collect Evidence Use the lessons in this module to collect evidence to support your claim. Record your evidence as you move through the module.

Explain Your Reasoning You will revisit your claim and explain your reasoning at the end of the module.

GO ONLINE to access your CER chart and explore resources that can help you collect evidence.

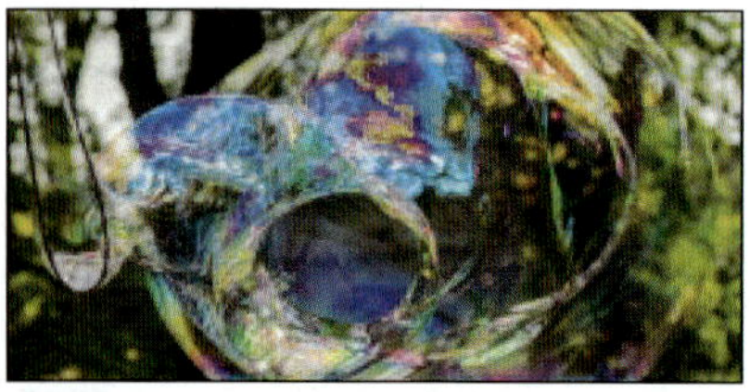

LESSON 1: Explore & Explain: Thin-Film Interference

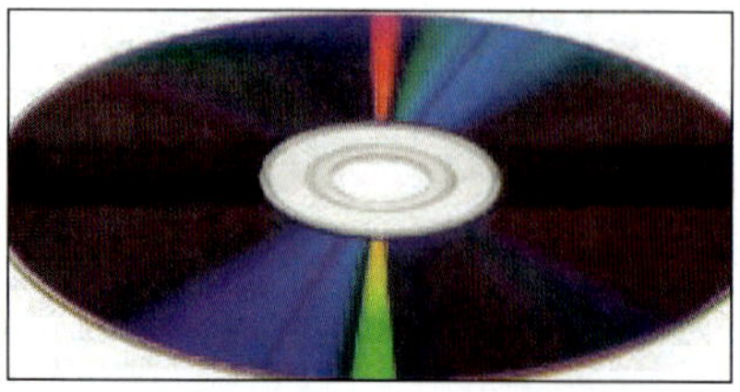

LESSON 2: Explore & Explain: Diffraction Gratings

Smithsonian

Additional Resources

(t)Video Supplied by BBC Worldwide Learning; (bl)Snova/Shutterstock; (br)RTimages/Alamy

LESSON 1
INTERFERENCE

FOCUS QUESTION

How do bubbles produce a rainbow effect?

Incoherent and Coherent Light

As you know, light has properties of a wave. Light diffracts as it passes an edge. When studying mirrors and lenses, you learned that reflection and refraction can be explained when light is modeled as a wave. When scientists discovered that light could be made to interfere, which results from the superposition of waves, they concluded that light has wave properties.

Incoherent light When you look at objects that are illuminated by a white light source such as a lightbulb, you are seeing **incoherent light,** which is light whose waves are not in phase. The effect of incoherence in waves can be seen in the example of heavy rain falling on still water. The surface of the water is choppy and does not have a regular pattern of waves, as shown in **Figure 1.** Because light waves have such a high frequency, incoherent light does not appear choppy to you. Instead, as light from an incoherent white light source illuminates an object, you see the combination of the incoherent light waves as an even, white light.

Coherent light Light made up of waves of the same wavelength that are in phase with each other is **coherent light.** A regular wavefront, which is made of coherent light, can be created by a single point source, as shown in **Figure 1.**

Figure 1 Choppy, irregular wave patterns model incoherent light. Regular wave patterns model coherent light.

3D THINKING DCI Disciplinary Core Ideas CCC Crosscutting Concepts SEP Science & Engineering Practices

COLLECT EVIDENCE

Use your Science Journal to record the evidence you collect as you complete the readings and activities in this lesson.

INVESTIGATE

GO ONLINE to find these activities and more resources.

Quick Investigation: Soap Film
Carry out an investigation to discover the patterns produced when light shines on a soap film.

PhysicsLAB: Double-Slit Interference
Calculate the wavelength of light using a doublt-slit interference pattern.

A regular wavefront also can be created by multiple point sources when all point sources are in phase. This type of coherent light is produced by a laser.

Figure 2 Thomas Young (1773–1829) is famous for his contributions in many different subject areas. Along with his role in establishing the wave nature of light, he is known for his work deciphering Egyptian hieroglyphics.

Interference of Coherent Light

Between 1801 and 1803, English physician Thomas Young, **Figure 2,** performed a number of investigations establishing the wave properties of light. In a crucial investigation attributed to Young, light from a small source that was passed through two closely spaced slits produced an interference pattern.

Young selected light from a tiny region of a source and made it coherent by passing it through a single, narrow slit. The light was then passed through two closely spaced narrow slits in a barrier. The overlapping light from the two slits fell on an observing screen. The overlap created a pattern of bright and dark bands called **interference fringes.** Young explained that the bands resulted from constructive and destructive interference of light waves from the two slits in the barrier.

Consider **monochromatic light,** which is light of only one wavelength. In a double-slit interference investigation that uses monochromatic light, constructive interference produces a bright central band of the given color on the screen, as well as other bright bands of near-equal spacing and near-equal width on either side, as shown in **Figure 3.** The intensity of the bright bands decreases the farther the band is from the central band, as you can see. Between the bright bands are dark areas where destructive interference occurs. The positions of the constructive and destructive interference bands depend on the light's wavelength.

When white light is used in a double-slit investigation, however, interference causes the appearance of colored spectra, as shown on the right in **Figure 3.** The various bands of color from the visible spectrum overlap on the screen. All these colors have constructive interference, and the central band is white. Because the positions of the other bright bands of constructive interference depend on wavelength, each color's band is at a different position, resulting in spectra of color.

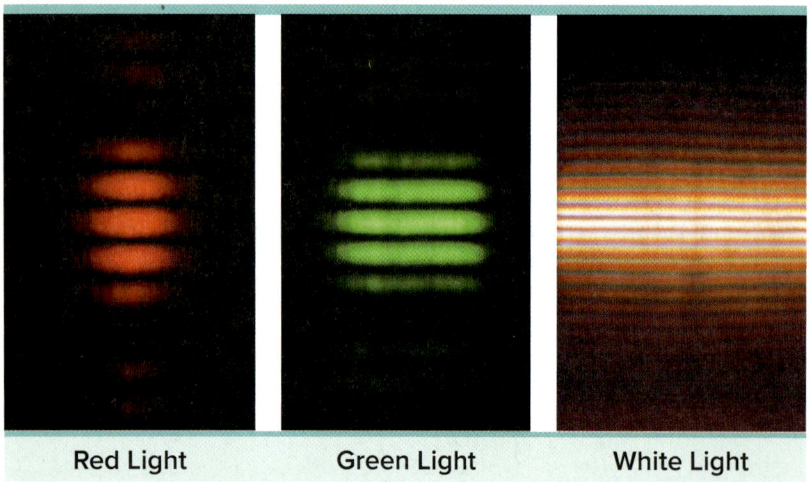

Figure 3 Double-slit interference patterns show a bright central band with a pattern of dark and bright bands on either side.

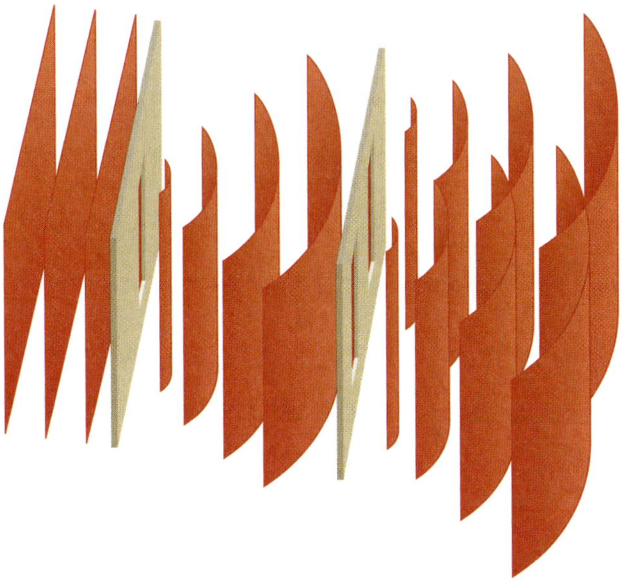

Figure 4 Nearly cylindrical wavefronts are generated as light passes through the slits.

Consider *why the width of the slits is tens to hundreds of light wavelengths.*

Generation of coherent light Light from a monochromatic source produces incoherent light. Placing a light barrier with a narrow slit in front of the monochromatic light produces coherent light. Because the width of the slit is very small, only light from a tiny region of the source passes through the slit. Diffraction by the slit produces nearly cylindrical wavefronts, as shown in **Figure 4.**

The second barrier has two very small slits. Because a cylinder is symmetrical, the two portions of the wavefront arriving at the second barrier are in phase. The two slits at the second barrier produce nearly cylindrical wavefronts. These two wavefronts can then interfere, as shown in **Figure 4.** Depending on their phase relationship, the two waves undergo constructive or destructive interference, as shown in **Figure 5.** If the interference is constructive when the light hits a screen, you will see a bright band. If it is destructive, you will see a dark band.

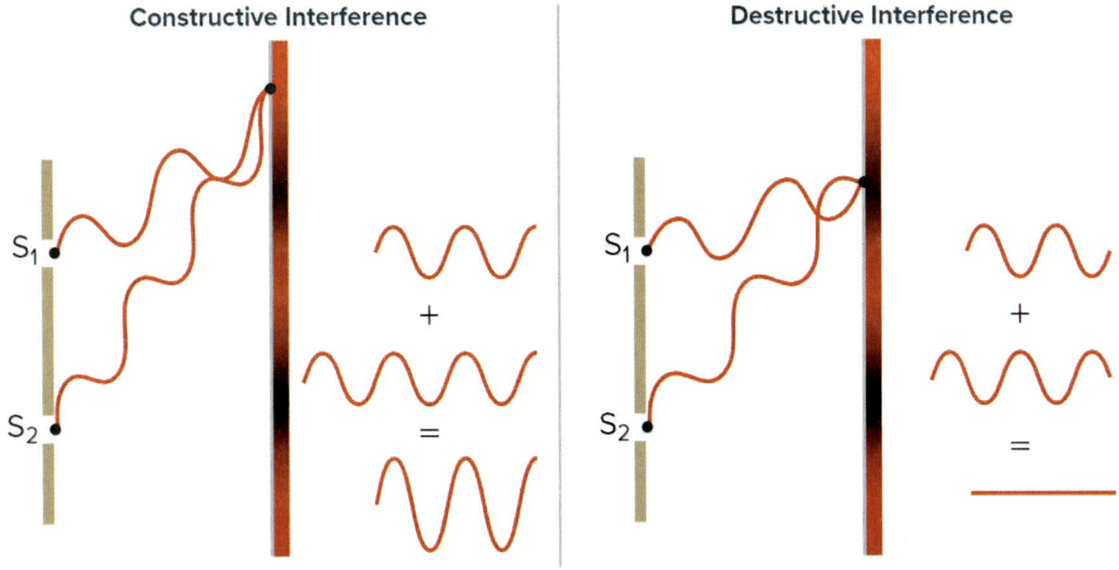

At the points where the waves experience constructive interference, bright bands are seen.

At the points where the waves experience destructive interference, dark bands are seen.

Figure 5 Coherent waves can undergo constructive or destructive interference.

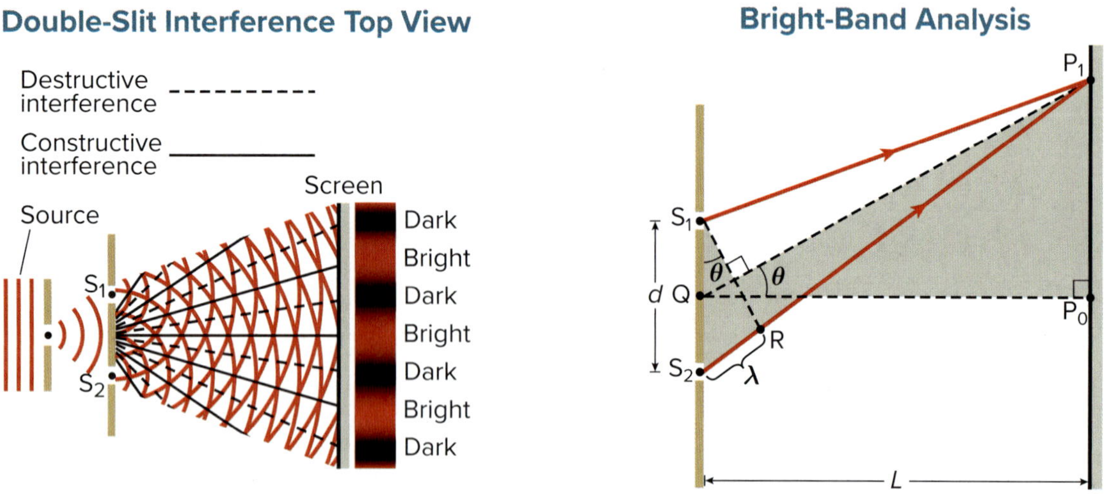

Figure 6 Double-slit interference can be used to determine the wavelength of light. Because *L* is much larger than *d* and the angle *θ* is small, the equation for the wavelength is simplified.

Double-Slit Interference

A top view of the double-slit investigation is shown on the left in **Figure 6.** The wavefronts interfere constructively and destructively to form a pattern of light and dark bands. The right side shows that light that reaches point P_0 travels the same distance from each slit. Because the waves are in phase, they interfere constructively on the screen to create the central bright band at P_0. There is also constructive interference at the first bright band (P_1) on either side of the central band because line segment P_1S_2 is one wavelength (λ) longer than the line segment P_1S_1. Thus, the waves arrive at P_1 in phase.

There are two triangles shaded on the right side. The larger triangle is a right triangle, so $\tan \theta = x/L$. In the smaller triangle ΔRS_1S_2, the side $\overline{S_2R}$ is the length difference of the two light paths, which is one wavelength. There are now two simplifications for wavelength calculations.

1. If L is much larger than d, then line segments S_1P_1 and S_2P_1 are nearly parallel to each other and to line segment QP_1, and ΔRS_1S_2 is very nearly a right triangle. Thus, $\sin \theta \approx \lambda/d$.
2. If the angle θ is small, then $\sin \theta$ is very nearly equal to $\tan \theta$.

With the above simplifications, the relationships $\tan \theta = x/L$, $\sin \theta \approx \lambda/d$, and $\sin \theta \approx \tan \theta$ combine to form the equation $x/L = \lambda/d$. Solving for λ gives the following.

Wavelength from Double-Slit Investigation

The wavelength of light, as measured by a double slit, is equal to the distance on the screen from the central bright band to the first bright band (*x*), multiplied by the distance between the slits (*d*), divided by the distance from the slits to the screen (*L*).

$$\lambda = \frac{xd}{L}$$

Constructive interference occurs at locations x_m on either side of the central bright band, which are determined by the equation $m\lambda = x_m d/L$, where $m = 0$, 1, 2, etc. The central bright band occurs at $m = 0$. The band located where $m = 1$ is often called the first-order band, $m = 2$ the second-order band, and so on.

STEM CAREER Connection

Optician

Opticians understand how thin films can be applied to eyeglasses to reduce glare. With their understanding of light and lenses, they check patients' needs and prescriptions for appropriateness. They can work in a variety of offices, such as with an optometrist, a store that sells eyeglasses and contact lenses, or within a department store to help fit people with visual aids, such as eyeglasses and contact lenses.

EXAMPLE Problem 1

WAVELENGTH OF LIGHT A double-slit investigation is performed to measure the wavelength of red light. The slits are 0.0190 mm apart. A screen is placed 0.600 m away, and the first-order bright band is 21.1 mm from the central bright band. What is the wavelength of the red light?

1 ANALYZE AND SKETCH THE PROBLEM

- Sketch the investigation, showing the slits and the screen.
- Draw the interference pattern with bands in appropriate locations.

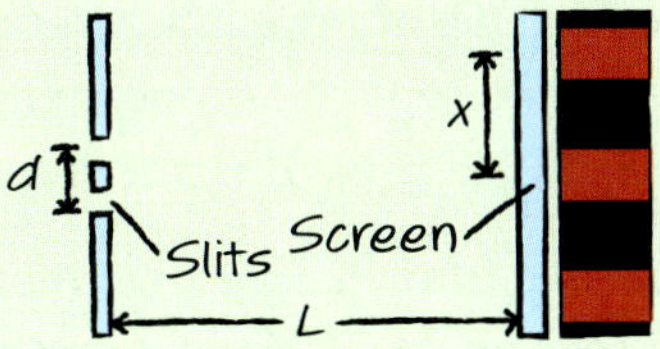

Known	Unknown
$d = 1.90\times10^{-5}$ m	$\lambda = ?$
$x = 2.11\times10^{-2}$ m	
$L = 0.600$ m	

2 SOLVE FOR THE UNKNOWN

$$\lambda = \frac{xd}{L}$$

$$= \frac{(2.11\times10^{-2}\text{ m})(1.90\times10^{-5}\text{ m})}{(0.600\text{ m})}$$ Substitute $x = 2.11\times10^{-2}$ m, $d = 1.90\times10^{-5}$ m, $L = 0.600$ m.

$$= 6.68\times10^{-7}\text{ m} = 668\text{ nm}$$

3 EVALUATE THE ANSWER

- **Are the units correct?** The answer is in units of length, which is correct for wavelength.
- **Is the magnitude realistic?** The wavelength range of red light is about 600 nm to 700 nm. Thus, the answer is reasonable for red light.

PRACTICE Problems

ADDITIONAL PRACTICE

1. Violet light falls on two slits separated by 1.90×10^{-5} m. A first-order bright band appears 13.2 mm from the central bright band on a screen 0.600 m from the slits. What is λ?
2. Yellow-orange light of wavelength 596 nm from a sodium lamp is aimed at two slits that are separated by 1.90×10^{-5} m. What is the distance from the central band to the first-order yellow band if the screen is 0.600 m from the slits?
3. In a double-slit investigation, physics students use a laser with $\lambda = 632.8$ nm. A student places the screen 1.000 m from the slits and finds the first-order bright band 65.5 mm from the central line. What is the slit separation?
4. **CHALLENGE** Yellow-orange light with a wavelength of 596 nm passes through two slits that are separated by 2.25×10^{-5} m and makes an interference pattern on a screen. If the distance from the central line to the first-order yellow band is 2.00×10^{-2} m, how far is the screen from the slits?

Young's findings faced opposition from most physicists because they were supporters of Newton's particle model of light. However, in a competition Jean Fresnel proposed a mathematical solution for the wave nature of light. One of the judges, Siméon Denis Poisson, showed that if Fresnel was correct, the shadow of a circular object illuminated with coherent light would have a bright spot at its center. Jean Arago did the investigation and saw the spot, convincing many of the wave nature of light.

Get It?

Describe how the pattern on the hummingbird at the beginning of this module formed.

Thin-Film Interference

Have you ever seen a spectrum of colors produced by a soap bubble or by the oily film on a water puddle in a parking lot, as in **Figure 7?** These colors are not the result of separation of white light by a prism or of absorption by a pigment. The colors are a result of the constructive and destructive interference when light waves reflect from separate surfaces of a thin film, a phenomenon called **thin-film interference.**

In a vertical soap film, like the one in **Figure 8,** the film's weight makes it thicker at the bottom than at the top. The thickness increases gradually from top to bottom. When a light wave strikes the front surface of the film, it is partially reflected, as shown by ray 1 in **Figure 8,** and partially transmitted. The reflected and transmitted waves have the same frequency as the original.

The transmitted wave travels through the film to the back surface where, again, part is reflected, as shown by ray 2. The splitting continues as the light makes many passes through the film. Because matched sets of waves came from the same source, they are coherent.

Color reinforcement How is the reflection of one color enhanced? This happens when the two reflected waves are in phase for a given wavelength. If the thickness of the soap film in **Figure 8** is one-fourth the wavelength of the light in the film ($\lambda/4$), then the round-trip path length in the film is $\lambda/2$. In this case, you might expect that ray 2 would return to the front surface one-half wavelength out of phase with ray 1 and that the two waves would cancel each other based on the superposition principle.

Figure 7 The swirl of colors seen in an oily film is the result of thin-film interference.

But when a transverse wave is reflected from a medium in which its speed is slower, the wave is inverted. With light, this happens at the boundary of a medium with a larger index of refraction. As a result, ray 1 is inverted on reflection, whereas ray 2 is reflected from a medium with a smaller index of refraction (air) and is not inverted. Thus, ray 1 and ray 2 are in phase.

If the film thickness (d) satisfies the requirement $d = \lambda/4$, then the color of light with that wavelength will be most strongly reflected. Note that because the wavelength of light in the film is shorter than the wavelength in air, $d = \lambda_{film}/4$, or, in terms of the wavelength in air, $d = \lambda_{vacuum}/4n_{film}$. The two waves reinforce each other as they leave the film. Light with other wavelengths undergoes destructive interference.

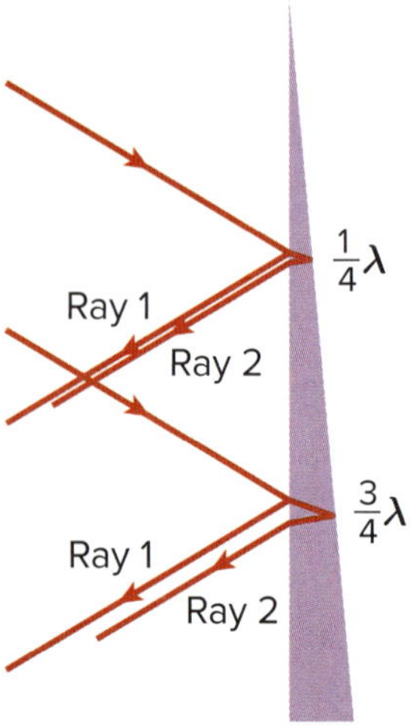

Figure 8 At soap film thicknesses of $\lambda/4$, $3\lambda/4$, $5\lambda/4$, etc., light with wavelength λ is in phase. Bands of that color of light are visible at those thicknesses.

As you know, different colors of light have different wavelengths. For a film of varying thickness, the wavelength requirement will be met at different thicknesses for different colors. The result is a rainbow of color. Where the film is too thin to produce constructive interference for any wavelength of visible light, the film appears to be black. Notice in **Figure 8** that the pattern of colors that appear on the film repeats. When the thickness of the film is $3\lambda/4$, the round-trip distance is $3\lambda/2$, and constructive interference occurs for light with a wavelength λ again. Any thickness equal to $1\lambda/4$, $3\lambda/4$, $5\lambda/4$, and so on, satisfies the conditions for constructive interference for a given wavelength.

Refractive indexes determine wave inversion The example of a film of soapy water in air involves constructive interference with one of two waves inverted upon reflection. In the example of a bubble solution or thin film of oil on a puddle of water, as the thickness of the film or the angle the light makes with the film changes, the wavelength undergoing constructive interference changes. This creates a shifting color on the surface of the film when it is under white light.

In other examples of thin-film interference, neither wave or both waves might be inverted. Whether a wave is inverted depends on the indices of refraction of the mediums involved. If both waves are traveling from a lower to a higher index of refraction, they will both be inverted. In this case, the film thicknesses for constructive interference are $1\lambda/2$, λ, $3\lambda/2$, 2λ, $5\lambda/2$, and so on. You can develop a solution for any problem involving thin-film interference by using the following strategy.

Real-World Physics

A THIN FILM can be placed on the lenses of eyeglasses to keep them from reflecting wavelengths of light that are highly visible to the human eye. This reduces the glare of reflected light and permits more light to be transmitted.

PROBLEM-SOLVING STRATEGY

Thin-Film Interference

When solving thin-film interference problems, construct an equation that is specific to the problem by using the following strategy.

1. Make a sketch of the thin film and the two coherent waves. For simplicity, draw the waves as rays.
2. Read the problem. Is the reflected light of this wavelength brightened or dimmed? When it is bright, the two reflected waves undergo constructive interference. When the reflected light is dimmed, the waves undergo destructive interference.
3. Are either or both waves inverted on reflection? If the index of refraction changes from a lower to a higher value, then the wave is inverted. If it changes from a higher to a lower value, there is no inversion.
4. Find the extra distance the second wave must travel through the thin film to create the needed interference.
   a. If you need constructive interference and one wave is inverted OR you need destructive interference and either both waves or none are inverted, then the difference in distance is an odd number of half wavelengths: $(m + 1/2)\lambda_{film}$, where $m = 0, 1, 2$, etc.
   b. If you need constructive interference and either both waves or none are inverted OR you need destructive interference and one wave is inverted, then the difference is an integer number of wavelengths: $m\lambda_{film}$, where $m = 1, 2, 3$, etc.
5. Set the extra distance traveled by the second ray to twice the film thickness, $2d$.
6. Recall from studying refraction that $\lambda_{film} = \lambda_{vacuum}/n_{film}$.

Reflection from a Thin Film

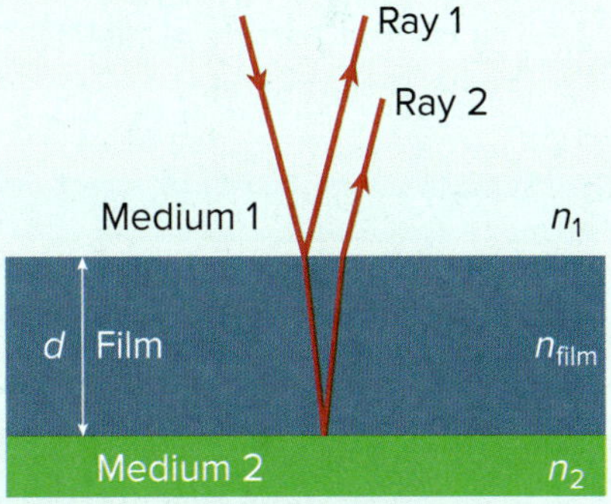

Flickr Open/Nenov/Getty Images

EXAMPLE Problem 2

OIL AND WATER You observe colored rings on a puddle and conclude that there must be an oil slick on the water. You look directly down at the puddle and see a yellow-green ($\lambda = 555$ nm) region. If the index of refraction of oil is 1.45 and that of water is 1.33, what is the minimum thickness of oil that could cause this color?

1 ANALYZE AND SKETCH THE PROBLEM

- Sketch the thin film and layers above and below it.
- Draw rays showing reflection off the top of the film as well as the bottom.

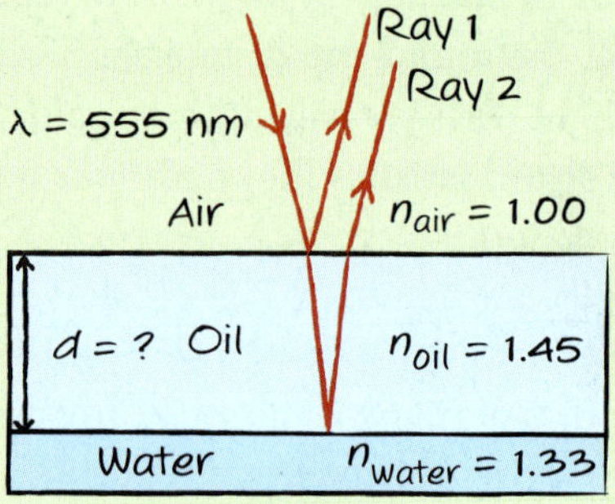

Known	Unknown
$n_{water} = 1.33$	$d = ?$
$n_{oil} = 1.45$	
$\lambda = 555$ nm	

2 SOLVE FOR THE UNKNOWN

Because $n_{oil} > n_{air}$, the wave is inverted on the first reflection. Because $n_{water} < n_{oil}$, there is no inversion on the second reflection. Thus, there is one wave inversion. The wavelength in oil is less than it is in air.

Follow the problem-solving strategy to construct the equation.

$$2d = \left(m + \frac{1}{2}\right)\left(\frac{\lambda}{n_{oil}}\right)$$

Because you want the minimum thickness, $m = 0$.

$$d = \frac{\lambda}{4n_{oil}}$$

$$= \frac{555 \text{ nm}}{(4)(1.45)}$$ Substitute $\lambda = 555$ nm, $n_{oil} = 1.45$.

$$= 95.7 \text{ nm}$$

3 EVALUATE THE ANSWER

- **Are the units correct?** The answer is in nanometers, which is correct for thickness.
- **Is the magnitude realistic?** The minimum thickness is smaller than one wavelength, which is what it should be.

PRACTICE Problems

ADDITIONAL PRACTICE

5. In the situation in Example Problem 2, what would be the thinnest film that would create a reflected red ($\lambda = 635$ nm) band?
6. A glass lens has a nonreflective coating of magnesium fluoride placed on it. How thick should the nonreflective layer be to keep yellow-green light with a wavelength of 555 nm from being reflected? See **Figure 9.**
7. You can observe thin-film interference by dipping a bubble wand into some bubble solution and holding the wand in the air. What is the thickness of the thinnest soap film at which you would see a black stripe if the light illuminating the film has a wavelength of 521 nm? Use $n = 1.33$ for the bubble solution.
8. What is the thinnest soap film ($n = 1.33$) for which light of wavelength 521 nm will constructively interfere with itself?
9. **CHALLENGE** A silicon solar cell has a nonreflective coating placed on it. If a film of silicon monoxide, $n = 1.45$, is placed on the silicon, $n = 3.5$, how thick should the layer be to keep yellow-green light ($\lambda = 555$ nm) from being reflected?

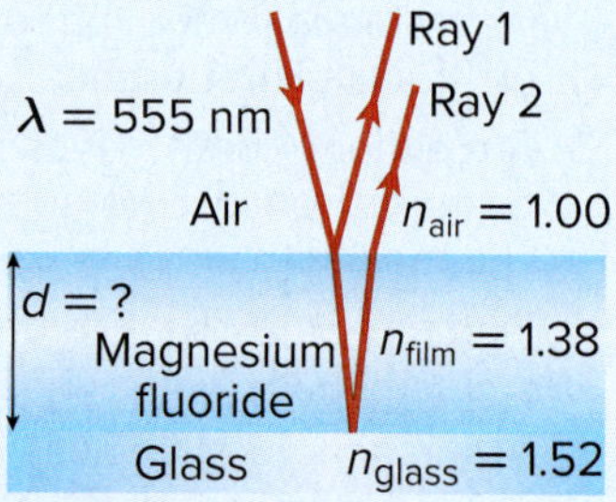

Figure 9

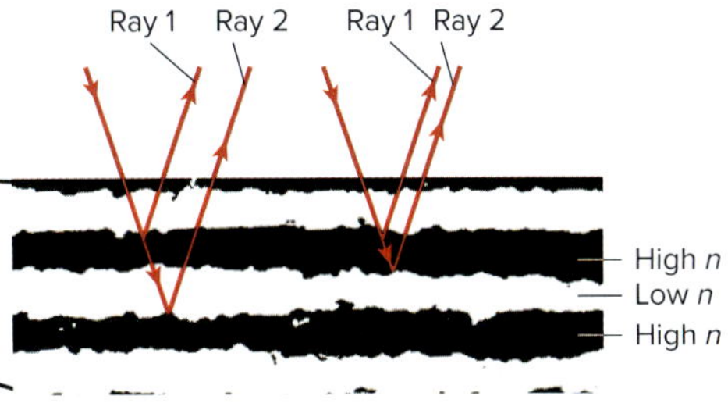

Interference can occur from individual structures and from multiple structures.

Figure 10 The tiger beetle owes its iridescent green color to thin-film interference.

Interference in nature Light interference also occurs naturally in the outer layer of the shells of many beetles, as shown in **Figure 10.** The shimmering green of the tiger beetle is the result of reflection from thin, parallel layers of chitin and sometimes other materials that differ in index of refraction. A diagram showing how these multilayer reflectors work is shown in **Figure 10.** The layers of the exoskeleton reflect light such that the result is constructive interference of green light. A shimmering appearance results.

Check Your Progress

10. **Interference** Two narrow slits are cut close to each other in a large piece of cardboard and illuminated by monochromatic red light. A sheet of white paper placed far from the slits shows a pattern of bright and dark bands on the paper. Describe how a wave behaves when it meets a slit and why some regions are bright and others are dark. Sketch the pattern.
11. **Interference Patterns** Sketch what happens to the pattern in the previous problem when the red light is replaced by blue light.
12. **Interference** Lucien holds a bubble wand with a soap film ($n = 1.33$) in it vertically.
    a. What is the second thinnest width of the soap film at which he could see a bright stripe if the light illuminating the film has a wavelength of 575 nm?
    b. What other widths produce a bright stripe at 575 nm?
13. **Double-Slit Interference** Light of wavelength 542 nm falls on a double slit. Use the values from **Figure 11** to determine how far apart the slits are.

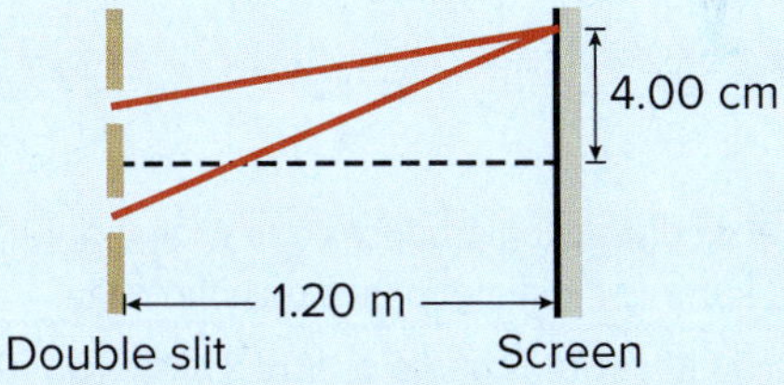

Figure 11

14. **Critical Thinking** The equation for wavelength from a double-slit investigation uses the simplification that θ is small so that $\sin \theta \approx \tan \theta$. Up to what angle is this a good approximation when your data has two significant figures? Would the maximum angle for a valid approximation increase or decrease as you increase the precision of your angle measurement?

LEARNSMART® Go online to follow your personalized learning path to review, practice, and reinforce your understanding.

(l)tacojim/iStock/Getty Images; (r)ilbusca/iStock/Getty Images

LESSON 2

DIFFRACTION

FOCUS QUESTION

How do DVDs produce a rainbow effect?

Single-Slit Diffraction

When studying light, you learned that wavefronts of light diffract when they pass around an edge. Diffraction can be explained by using Huygens' principle that a wavefront is made up of many small point-source wavelets. When light passes through a slit that has two closely spaced edges, a pattern is produced on a screen. This pattern, called a **diffraction pattern,** results from constructive and destructive interference of Huygens' wavelets.

When coherent, green light passes through a single, small opening that is between about 10 and 100 light wavelengths, the light is diffracted by both edges, and a series of bright and dark bands appears on a distant screen, as shown in **Figure 12.**

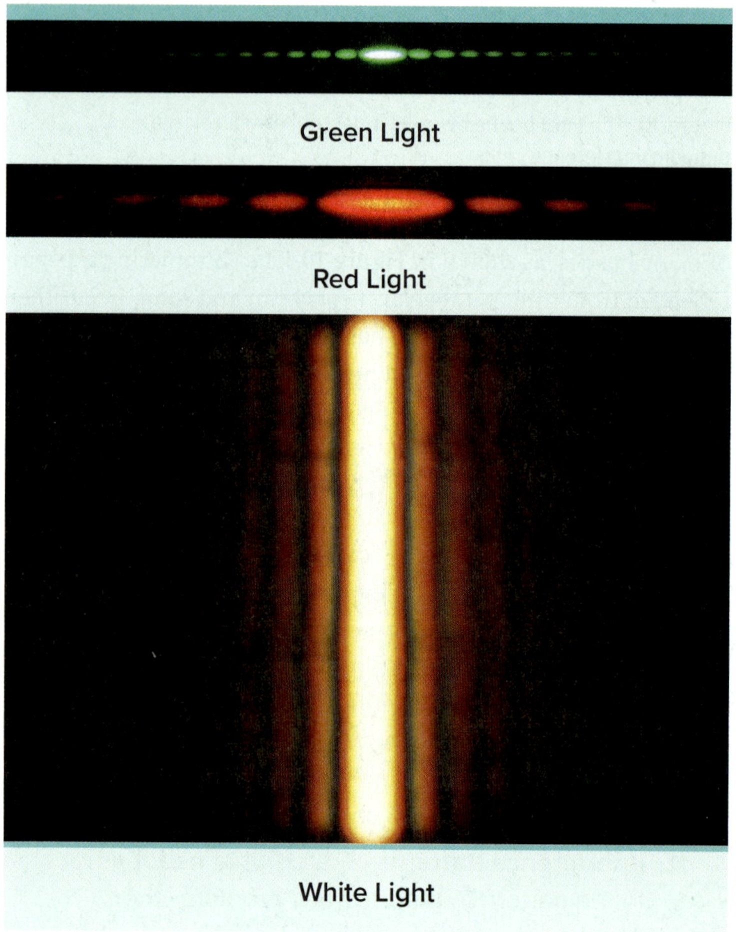

Figure 12 Single-slit diffraction produces one wide, bright central band and narrower, dimmer bands on either side.

Compare *the colors and widths of the central bands produced by the different colors of light.*

3D THINKING **DCI** Disciplinary Core Ideas **CCC** Crosscutting Concepts **SEP** Science & Engineering Practices

COLLECT EVIDENCE

Use your Science Journal to record the evidence you collect as you complete the readings and activities in this lesson.

INVESTIGATE

GO ONLINE to find these activities and more resources.

Quick Investigation: Diffraction Gratings
Carry out an investigation to discover the patterns produced when light is transmitted through a diffraction grating.

Revisit the Encounter the Phenomenon Question
What information from this lesson can help you answer the Unit and Module questions?

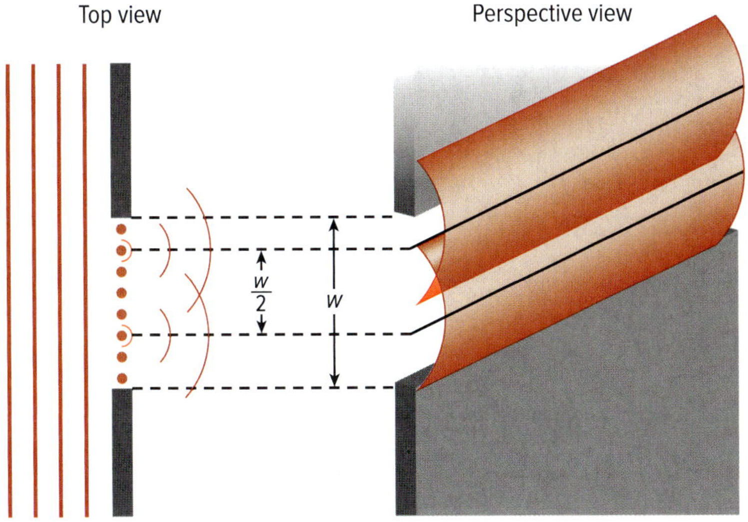

Figure 13 To illustrate diffraction patterns with Huygens' wavelets, a pair of points is chosen such that the separation between the points is $\frac{w}{2}$.

Instead of the nearly equally spaced bands produced by two coherent sources in Young's double-slit investigation, this pattern has a wide, bright central band with dimmer, narrower bands on either side. When using red light instead of green, the width of the bright central band increases. With white light, the pattern is a combination of patterns of all the colors of the spectrum (**Figure 12**).

Huygens' wavelets To see how Huygens' wavelets produce the diffraction pattern, imagine a slit of width w as being divided into an even number of Huygens' points, as shown in **Figure 13.** Each point acts as a source of Huygens' wavelets. Divide the slit into two equal halves, and choose one source from each part so that the pair is separated by a distance $w/2$. This pair of sources produces coherent, cylindrical waves that will interfere.

For any Huygens' wavelet produced in the top half, there will be another Huygens' wavelet in the bottom half, a distance $w/2$ away, that it will interfere with destructively to create a dark band on the screen. All similar pairings of Huygens' wavelets interfere destructively at dark bands. Conversely, a bright band on the screen is where pairings of Huygens' wavelets interfere constructively. In the dim regions between bright and dark bands, partial destructive interference occurs.

Diffraction pattern When the single slit is illuminated, a central bright band appears at location P_0 on the screen, as shown in **Figure 14** on the next page. The first dark band is at position P_1. At this location, the path lengths r_1 and r_2 of the two Huygens' wavelets differ by $\lambda/2$, producing destructive interference.

This model is mathematically similar to that of double-slit interference. A comparison of a single-slit diffraction pattern with a double-slit interference pattern using slits of the same width reveals that the entire single-slit diffraction pattern is covered by the narrow dark and light interference bands. Double-slit patterns result from the interference of the light from each single slit.

We will now develop an equation for the diffraction pattern produced by a single slit. We use the same simplifications that were used for double-slit interference, assuming that the distance to the screen is much larger than w. As described above, the separation distance between the sources of the two interfering waves is $w/2$. To find the distance measured on the screen to the first dark band (x_1), note that the path length difference is $\lambda/2$ because at the dark band there is destructive interference. As a result, $x_1/L = \lambda/w$.

Get It?

Identify the type of interference of Huygens' wavelets that creates a dark band on the screen.

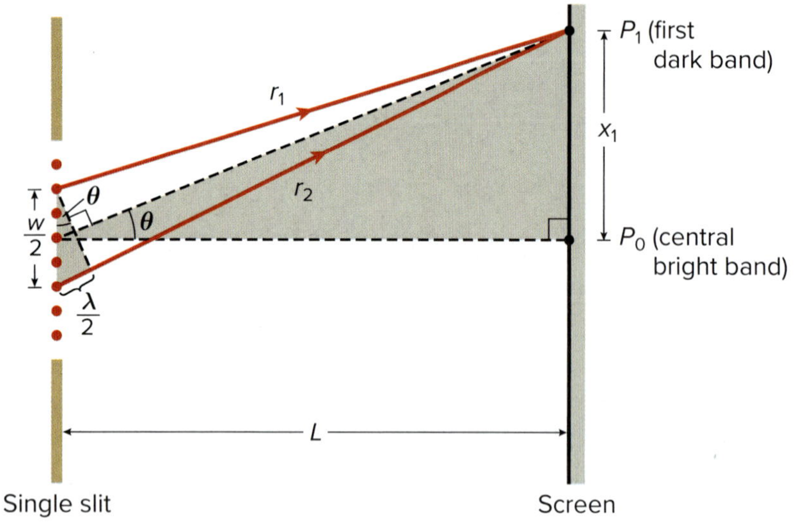

Figure 14 The width of the bright band in single-slit diffraction is related to the wavelength of light, the distance from the slit to the screen, and the width of the slit.

Instead of measuring the distance to the first dark band from the center of the central bright band (x_1), it is better to determine the width of the central bright band ($2x_1$), as in the equation below.

Width of Bright Band in Single-Slit Diffraction

The width of the central bright band ($2x_1$) is equal to the product of twice the wavelength times the distance to the screen (L), divided by the width of the slit (w).

$$2x_1 = \frac{2\lambda L}{w}$$

Single-slit diffraction patterns make the wave nature of light noticeable when the slits are 10 to 100 times the wavelength of the light. Larger openings cast sharp shadows, as Isaac Newton first observed. While the single-slit pattern depends on the wavelength of light, it is not very useful for measuring wavelength. A large number of slits put together provides a more useful tool for measuring wavelength.

Get It?

Describe the assumption that is made concerning w and L for the single-slit diffraction equation.

PHYSICS Challenge

You have several unknown substances and wish to use a single-slit diffraction apparatus to determine what each one is. You decide to place a sample of an unknown substance in the region between the slit and the screen and use the data you obtain to determine the identity of each substance by calculating its index of refraction.

1. Come up with a general formula for the index of refraction of an unknown substance in terms of the wavelength of the light (λ_{vacuum}), the width of the slit (w), the distance from the slit to the screen (L), and the distance between the central bright band and the first dark band (x_1).
2. If the source you used had a wavelength of 634 nm, the slit width was 0.10 mm, the distance from the slit to the screen was 1.15 m, and you immersed the apparatus in water ($n_{substance} = 1.33$), then what would you expect the width of the center band to be?

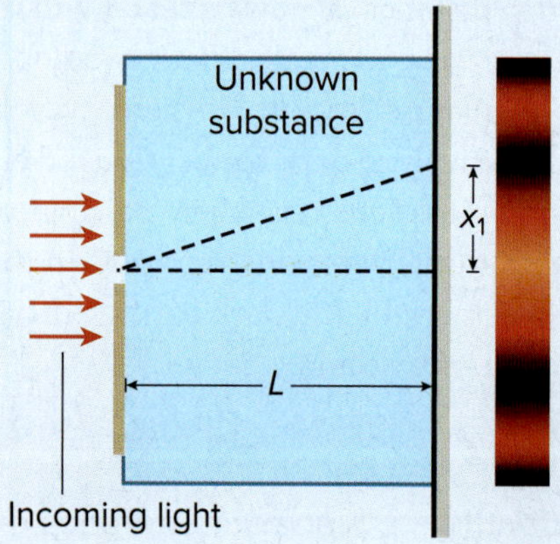

PRACTICE Problems

ADDITIONAL PRACTICE

15. Monochromatic green light of wavelength 546 nm falls on a single slit with a width of 0.095 mm. The slit is located 75 cm from a screen. How wide will the central bright band be?

16. Yellow light with a wavelength of 589 nm passes through a slit of width 0.110 mm and makes a pattern on a screen. If the width of the central bright band is 2.60×10^{-2} m, how far is it from the slit to the screen?

17. Light from a He-Ne laser ($\lambda = 632.8$ nm) falls on a slit of unknown width. A pattern is formed on a screen 1.15 m away, on which the central bright band is 15.0 mm wide. How wide is the slit?

18. Yellow light falls on a single slit 0.0295 mm wide. On a screen that is 60.0 cm away, the central bright band is 24.0 mm wide. What is the wavelength of the light?

19. CHALLENGE White light falls on a single slit that is 0.050 mm wide. A screen is placed 1.00 m away. A student first puts a blue-violet filter ($\lambda = 441$ nm) over the slit, then a red filter ($\lambda = 622$ nm). The student measures the width of the central bright band.

a. Which filter produced the wider band?

b. Calculate the width of the central bright band for both filters.

Diffraction Gratings

Diffraction gratings, such as the one shown in **Figure 15,** are often used to make precise measurements of wavelength. A **diffraction grating** is a device that is made up of many small slits that diffract light and form a pattern that is an overlap of single-slit diffraction patterns. This pattern is similar to that of a two-slit interference pattern, but with much narrower and brighter bands. Diffraction gratings can have as many as 10,000 slits per centimeter, which means the spacing between the slits can be as small as 10^{-6} m. A diffraction grating is a useful tool for the study of light and objects that emit or absorb light.

One type of diffraction grating is called a transmission grating. A transmission grating can be made by scratching very fine lines with a diamond point on glass that transmits light. The spaces between the scratched lines act like slits.

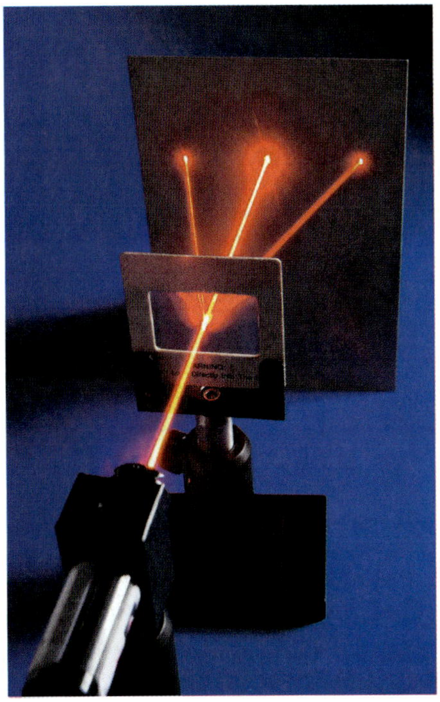

Diffraction gratings can be used to enhance the appearance of diamonds. The gratings are etched into certain surfaces of the diamond to improve the dispersion of light and make the gems appear more brilliant.

Holographic diffraction gratings produce the brightest spectra. They are made by using a laser and mirrors to create a diffraction pattern consisting of parallel bright and dark lines. The pattern is projected on a piece of metal that is coated with a light-sensitive material. The light from the laser produces a chemical reaction that hardens the material.

Figure 15 Diffraction gratings are used in various devices and instruments. The effects they produce also make them appealing for use in jewelry.

The metal is then placed in acid, which attacks the metal where unprotected by the hardened material, resulting in a series of hills and valleys in the metal identical to the original diffraction pattern. The metal itself can be used as a reflection grating. In some cases, a plastic film is placed on the heated metal, producing the hills and valleys in the plastic. Because of the sinusoidal shape of the hills and valleys, the diffraction patterns are very bright.

Figure 16 A CD is a reflection grating, producing a light spectrum. A magnified view of the surface of a CD shows the arrangement of pits and lands.

Reflection gratings You might have seen that light that reflects off CDs, DVDs, or Blu-ray Discs (or optical discs) creates a spectrum diffraction pattern, as in **Figure 16.** A type of diffraction grating made by inscribing fine lines on metallic or reflective glass surfaces is called a reflection grating. Optical discs are examples of reflection gratings. If you were to shine monochromatic light on an optical disc, the reflected light would produce a reflection pattern on a screen. Transmission and reflection gratings produce similar patterns, which can be analyzed in the same manner.

The surfaces of optical discs are covered with lines of microscopic indentations called pits separated by flat areas called lands, as shown in **Figure 16,** arranged in a spiral. The turns of the spiral act as a diffraction grating, separating colors by interference. The fact that optical discs are diffraction gratings is not important to their function, but the way they interact with different wavelengths of light is important.

Storing and reading information A laser is used to "read" the pattern of pits and lands on the optical disc. This is similar to the way a visually impaired person reads braille. The light from the laser is reflected from the surface of the disc into a light detector. The laser is focused so that when it reflects from the lands, its bright spot falls on the detector. When it reflects from the pits, it is spread out and dimmer.

The spot size is limited by diffraction, so if shorter laser wavelength is used, the spot size can be reduced and the pits can be closer together, allowing more information to be stored. As laser technology has advanced, shorter wavelength lasers have been used, permitting more information to be put on the disc. Music CDs use infrared light with a wavelength of 780 nm. A CD can hold about 700 megabytes of information. DVDs use red lasers (650 nm), permitting more than 4 gigabytes to be recorded. Blu-ray Discs use violet lasers (405 nm). The light appears blue, giving the disc its name. Single-layer Blu-ray Discs can hold up to 25 gigabytes of information.

Get It?

Explain how the amount of information an optical disc can store and the wavelength of light used to read it are related.

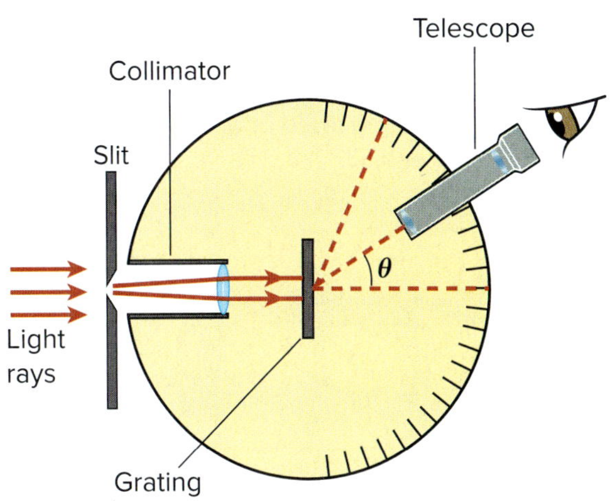

Figure 17 A grating spectroscope is used to accurately measure the wavelength of light.

State *the simplification used for double-slit wavelength calculations that does not apply for gratings.*

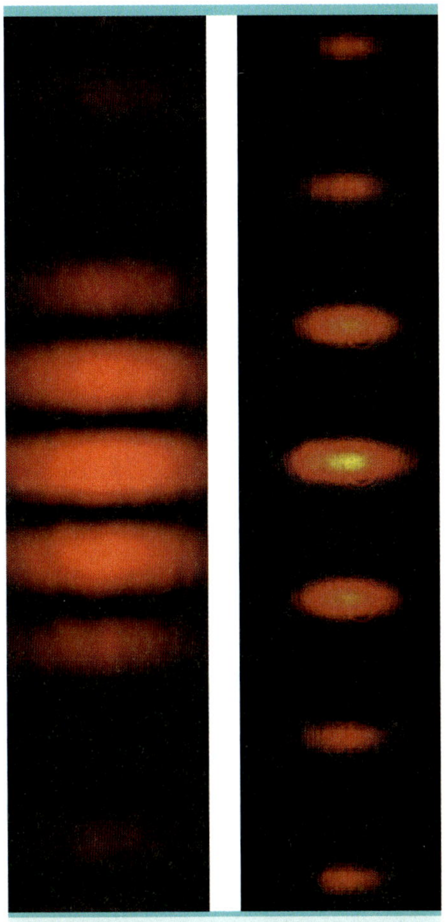

Figure 18 Compare the two diffraction patterns for red light. The diffraction grating pattern provides a more precise measurement.

GIPhotoStock/Science Source

Measuring wavelength

An instrument used to measure wavelengths of light using a diffraction grating is called a grating spectroscope, like the one shown in **Figure 17.** The source to be analyzed emits light that is directed through a slit and a collimator and then to a diffraction grating. The grating produces a diffraction pattern that is viewed through a telescope.

If the source of light is a single color, the diffraction pattern produced by a grating has narrow, equally spaced, bright lines, as shown in **Figure 18.** The larger the number of slits per unit length of the grating, the narrower the lines in the diffraction pattern. The narrower the lines, the more precisely the distance between the bright lines can be measured.

Earlier in this module, you read that the diffraction pattern produced by a double slit could be used to calculate wavelength. An equation for the diffraction grating can be developed in the same way as for the double slit. However, with a diffraction grating, θ could be large, so the small angle simplification does not apply. Wavelength can be found by measuring the angle (θ) between the central bright line and the first-order bright line.

Wavelength from a Diffraction Grating

The wavelength of light is equal to the slit separation (d) distance times the sine of the angle (θ) at which the first-order bright line occurs.

$$\lambda = d \sin \theta$$

Constructive interference from a diffraction grating occurs at angles on either side of the central bright line given by the equation $m\lambda = d \sin \theta$, where $m = 0, 1, 2$, etc. The central bright line occurs at $m = 0$. Spectroscopists often use the $m = 2$ or 3 lines because measurements of the spacing between bright lines can be made more precisely.

Notice that the diffraction grating pattern contains more dark space than the double-slit pattern in **Figure 18.** This is because there is more destructive interference in a diffraction grating than in a double-slit. This results in narrower lines, which also improves the precision of the measurements.

Diffraction gratings are incorporated into spectroscopes used to analyze gemstones. Experienced gemologists recognize the patterns of bands produced by white light passing through different stones. For example, three bright bands of green, yellow, and orange are a strong indication that cobalt is present. This likely means that a blue stone is not an expensive gem such as sapphire or topaz, but rather a cheap piece of glass that has been tinted blue.

EXAMPLE Problem 3

USING A DVD AS A DIFFRACTION GRATING A student noticed the beautiful spectrum reflected off a DVD. She directed a beam from her teacher's green laser pointer at the DVD and found three bright spots reflected on the wall. The label on the pointer indicated that the wavelength was 532 nm. The student found that the spacing between the spots was 1.29 m on the wall, which was 1.25 m away. What is the spacing between the rows on the DVD?

1 ANALYZE AND SKETCH THE PROBLEM

- Sketch the investigation, showing the DVD as a grating and the spots on the wall.
- Identify and label the knowns.

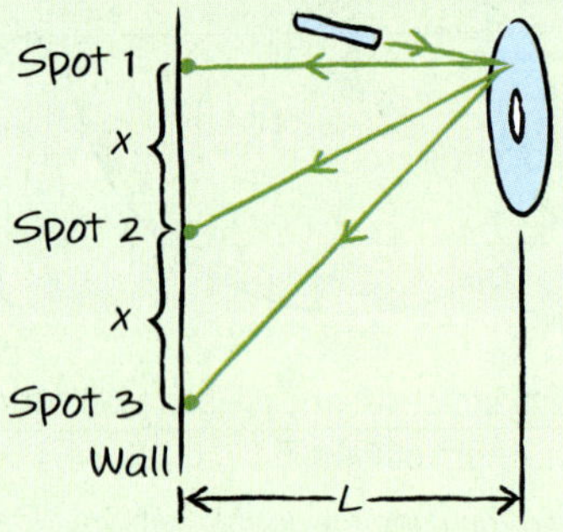

Known	Unknown
$x = 1.29$ m	$d = ?$
$L = 1.25$ m	
$\lambda = 532$ nm	

2 SOLVE FOR THE UNKNOWN

Find the angle between the central bright spot and the one next to it using $\tan \theta = \frac{x}{L}$.

$$\theta = \tan^{-1}\left(\frac{x}{L}\right)$$

$$= \tan^{-1}\left(\frac{1.29 \text{ m}}{1.25 \text{ m}}\right)$$ Substitute $x = 1.29$ m, $L = 1.25$ m.

$$= 45.9°$$

Use the diffraction grating wavelength and solve for d.

$$\lambda = d \sin \theta$$

$$d = \frac{\lambda}{\sin \theta}$$

$$= \frac{532\times10^{-9} \text{ m}}{\sin 45.9°}$$ Substitute $\lambda = 532\times10^{-9}$ m, $\theta = 45.9°$.

$$= 7.41\times10^{-7} \text{ m} = 741 \text{ nm}$$

3 EVALUATE THE ANSWER

- **Are the units correct?** The answer is in meters, which is correct for separation.
- **Is the magnitude realistic?** With x and L almost the same size, d is close to λ.

PRACTICE Problems

ADDITIONAL PRACTICE

20. White light shines through a grating onto a screen. Describe the pattern that is produced.

21. If blue light of wavelength 434 nm shines on a diffraction grating and the spacing of the resulting lines on a screen that is 1.05 m away is 0.55 m, what is the spacing between the slits in the grating?

22. A diffraction grating with slits separated by 8.60×10^{-7} m is illuminated by violet light with a wavelength of 421 nm. If the screen is 80.0 cm from the grating, what is the separation of the lines in the diffraction pattern?

23. For the DVD in Example Problem 3, suppose now that blue light shines on it. If the spots produced on a wall are separated by 58.0 cm, what is the wavelength of the light? The wall is 0.65 m away.

24. CHALLENGE Light of wavelength 632 nm passes through a diffraction grating and creates a pattern on a screen that is 0.55 m away. If the first bright band is 5.6 cm from the central bright band, how many slits per centimeter does the grating have?

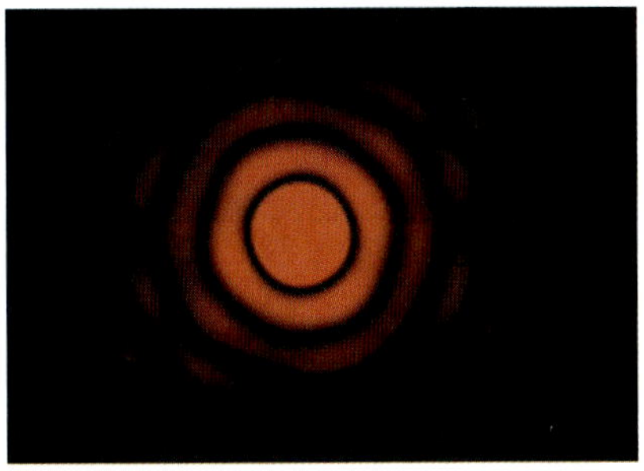

Figure 19 An aperture diffracts light to create a diffraction pattern with a central bright spot with dark and bright rings around it.

Resolving Power of Lenses

The circular lens of a telescope, a microscope, and even your eye acts as a hole, called an aperture, through which light passes. An aperture diffracts light, just as a single slit does. Alternating bright and dark rings occur with a circular aperture, as shown in **Figure 19.** The equation for determining the size of an aperture is similar to that for a single slit. An aperture, however, has a circular edge rather than the two edges of a slit, so slit width (w) is replaced by aperture diameter (D), and a geometric factor of 1.22 is needed, resulting in $x_1 = \frac{1.22\lambda L}{D}$.

ASTRONOMY Connection When light from a distant star is viewed through the aperture of a telescope, the image is spread out due to diffraction. If two stars are close together, their images may blur together. In 1879, Lord Rayleigh, a British physicist, mathematician, and Nobel Prize winner, established a criterion for determining whether there are one or two stars in such an image. The **Rayleigh criterion** states that if the center of the bright spot of one source's image falls on the first dark ring of the second, the two images are at the limit of resolution. If the images of two stars are at the limit of resolution, a viewer can tell that there are two stars rather than only one.

If two objects are at the limit of resolution, how can you find the distance between the objects (x_{obj})? From the Rayleigh criterion, the distance between the centers of the bright spots of the two images is x_1. **Figure 20** shows that similar triangles can be used to find that $\frac{x_{obj}}{L_{obj}} = \frac{x_1}{L}$. You can combine this equation with the equation for aperture size $\left(\frac{x_1}{L} = \frac{1.22\lambda}{D}\right)$ and solve for the distance between the objects (x_{obj}).

Rayleigh Criterion

The separation distance between objects (x_{obj}) that are at the limit of resolution is equal to 1.22 times the wavelength of light, times the distance from the circular aperture to the objects (L_{obj}), divided by the diameter of the circular aperture (D).

$$x_{obj} = \frac{1.22\lambda L_{obj}}{D}$$

Figure 20 The separation distance of objects can be calculated with similar-triangle geometry. The blue and red colors are for illustration only. (Illustration is not to scale.)

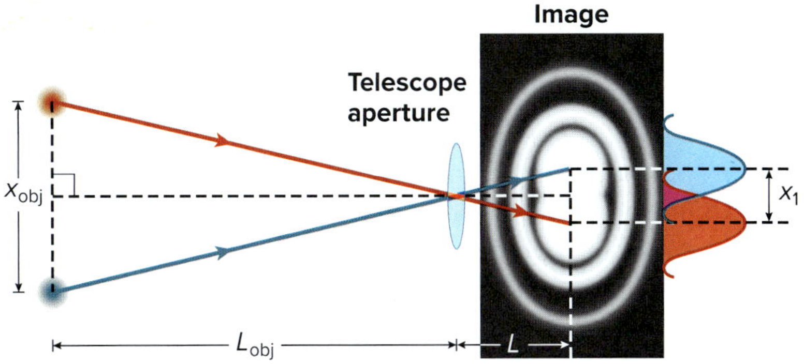

Diffraction in the eye In bright light, the eye's pupil is about 3 mm in diameter. The eye is most sensitive to yellow-green light where $\lambda = 550$ nm. So the Rayleigh criterion applied to the eye gives $x_{obj} = (2\times10^{-4})\,L_{obj}$. The distance between the pupil and the retina is about 2 cm, so by using $x_1 = 1.22\lambda L/D$, the centers of the bright spots of two barely resolved point sources would be separated by about 4 μm on the retina. The spacing between the cones, which are the light detectors in the retina, in the most sensitive part of the retina, the fovea, is about 2 μm. Thus, in the ideal case, three adjacent cones would record light, dark, and light, as illustrated in **Figure 21.** The distance between the centers of the bright spots from two point sources must be at least the distance between two light-recording cones to be resolved. It seems that the eye is ideally constructed.

Applying the Rayleigh criterion to find the ability of the eye to separate two distant sources shows that the eye could separate two automobile headlights (1.5 m apart) at a distance of 7 km. In practice, however, the eye is not limited by diffraction. Imperfections in the lens and the liquid that fills the eye reduce the eye's resolution to about five times that set by the Rayleigh criterion. The vision-processing centers in the human brain also limit detection of small point objects.

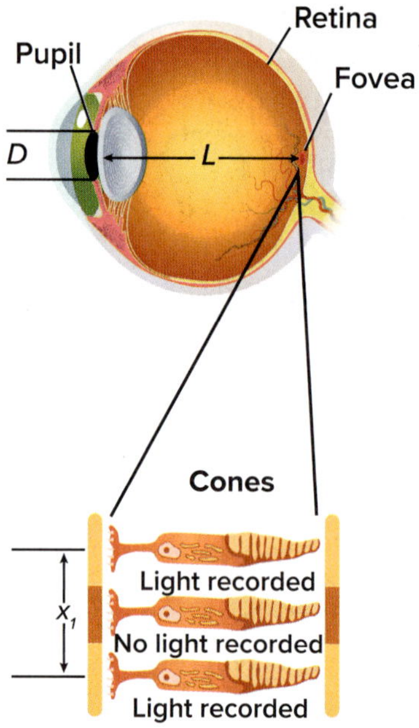

Figure 21 The eye's pupil is an aperture that diffracts light.

Check Your Progress

25. **Slit Diffraction** Several equally spaced narrow slits on a cardboard are illuminated by monochromatic red light. A pattern of bright and dark bands is visible on a sheet of white paper. Sketch the pattern on the paper.

26. **Rayleigh Criterion** The star Sirius is a system of two stars that orbit each other 8.44 light-years from Earth. If the *Hubble Space Telescope* (diameter 2.4 m) is pointed at the Sirius system, what is the minimum separation there would need to be between the stars in order for the telescope to be able to resolve them? Assume that the light coming from the stars has a wavelength of 550 nm.

27. **Line Spacing** You shine a red laser light through one diffraction grating and form a pattern of red dots on a screen. Then you substitute a second diffraction grating for the first one, forming a different pattern. The dots produced by the first grating are spread out more than those produced by the second. Which grating has more lines per millimeter?

28. **First-Order Dark Bands** Monochromatic green light with a wavelength of 546 nm falls on a single slit of width and location from the screen shown in **Figure 22.** What is the separation of the first-order dark bands?

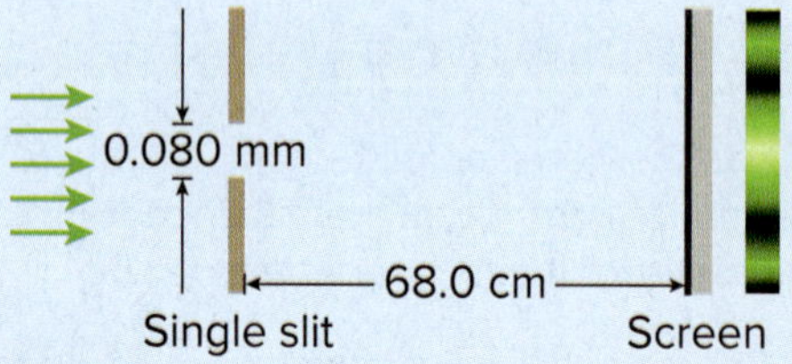

Figure 22

29. **Critical Thinking** Your colleague shows you a spectrometer, but she does not tell you whether it has been constructed with a prism or a diffraction grating. You decide to look at a spectrum produced by white light passing through the spectrometer. Describe how you could determine which device produced the spectrum.

LEARNSMART® Go online to follow your personalized learning path to review, practice, and reinforce your understanding.

Beckoning Bees with Blue Halos

Scientific studies show that several species of bees prefer flower petals whose color is in the blue spectrum. Their eyes' enhanced photoreceptors allow them to "see" in this color range. Many plants lack the ability to produce blue flowers, but they have a structural adaptation that can attract bee pollinators by bending and scattering light to create a "blue halo."

Some flowers, such as *Ursinia anethoides*, have diverse nanostructures that enable them to diffract light. This light interference creates a blue halo, which is visible to bees.

All in the Optics

Flower petals have tiny ridges. When light reflects off the ridges on some flower petals, the light is scattered in such a way that it casts a bluish glow over the blooms. This is referred to as the "blue halo" effect.

Scientists studied the nanostructures responsible for this effect using a variety of microscopy techniques, including scanning electron microscopy. Computer models showed variations in the height and spacing of ridges on petals.

What physicists call "disorder" in the height and spacing of ridges diffracts light, resulting in the iridescent blue and ultraviolet wavelength spectrum, or blue halo effect. The more variability, the stronger the blue halo.

Scientists have also found evidence to support the fact that many different species of flowers independently evolved the structures to produce the blue halo. This is an example of natural selection resulting in convergent evolution—the plants with disordered nanostructures are best adapted to survive because of increases in optical visibility to attract bee pollinators that see blue wavelengths.

Irresistible Iridescence

To test the role of the blue halo in attracting bees, scientists created artificial nanostructures that matched those found in some species of flowers. They attached the nanostructures to the petals of artificial flowers.

In the study, bees could tell the difference between otherwise identical looking flowers based on the blue halo effect. The optical effect of the disordered nanoparticle structures on petals improved the efficiency of foraging bees to detect flowers. They were able to locate flowers with the blue halo faster. This, in turn, increases pollination of plants, which is an evolutionary advantage for survival.

COMMUNICATE SCIENTIFIC INFORMATION

Draw the nanostructures of ridges on flower petals, and write a brief explanation of how they diffract light.

CALIFORNIA CONNECTION

Research a flower species native to California that has the blue halo adaptation. Make a poster that shows the flower and describes the adaptation.

MODULE 17
STUDY GUIDE

GO ONLINE to study with your Science Notebook.

Lesson 1 INTERFERENCE

- incoherent light
- coherent light
- interference fringes
- monochromatic light
- thin-film interference

- The superposition of light waves from coherent light sources can produce an interference pattern. Light passing through two closely spaced, narrow slits produces a pattern of dark and light bands on a screen called interference fringes.
- Interference patterns can be used to measure the wavelength of light.

$$\lambda = \frac{xd}{L}$$

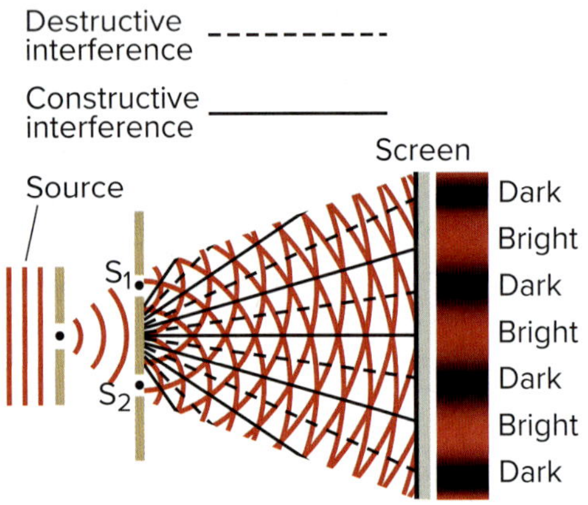

- Interference patterns can result from multiple passes of light through a thin film. Thin-film interference can be modeled by rays reflecting from multiple surfaces of a thin film. The indices of refraction of the mediums the light travels in and the thickness of the film determine how different wavelengths of light will interfere.

Lesson 2 DIFFRACTION

- diffraction pattern
- diffraction grating
- Rayleigh criterion

- Light passing through a narrow slit is diffracted, which means spread out from a straight-line path, producing a diffraction pattern on a screen. The width of the bright central band of a single-slit diffraction pattern is related to the wavelength of light used.
- Diffraction gratings consist of large numbers of slits that are very close together and produce narrow spectral lines that result from interference of light diffracted by all the slits.
- Diffraction gratings can be used to measure the wavelength of light precisely or to separate light composed of different wavelengths.

$$\lambda = d \sin \theta$$

- Diffraction limits the ability of an aperture to distinguish two closely spaced objects, because the resulting image contains a diffuse central bright spot. If two bright spots are closer than the limit of resolution, they will overlap and the objects cannot be distinguished.
- The Rayleigh criterion states that if the center of the bright spot of one source's image falls on the first dark ring of the second, the two images are at the limit of resolution. If the images of two stars are at the limit of resolution, a viewer can tell that there are two stars rather than only one.

THREE-DIMENSIONAL THINKING

Module Wrap-Up

REVISIT THE PHENOMENON

What makes this hummingbird's feathers appear shiny and shimmery?

CER Claim, Evidence, Reasoning

Explain Your Reasoning Revisit the claim you made when you encountered the phenomenon. Summarize the evidence you gathered from your investigations and research and finalize your Summary Table. Does your evidence support your claim? If not, revise your claim. Explain why your evidence supports your claim.

STEM UNIT PROJECT

Now that you've completed the module, revisit your STEM unit project. You will apply your evidence from this module and complete your project.

GO FURTHER

SEP Data Analysis Lab

How can you determine the wavelength of a laser?

Your project team has two laser pointers, one red and one green. Mark and Carlos disagree about which laser has the longer wavelength. Mark insists that red light has a longer wavelength, while Carlos is certain the green has a longer wavelength. You have a DVD and shine each laser on the grooves, reflecting onto a nearby wall.

CER Analyze and Interpret Data

1. Describe what data you can collect about the lasers by observing the reflections.
2. **Claim** Is Mark or Carlos correct?
3. **Evidence and Reasoning** How would you explain the results to Mark and Carlos to settle their disagreement?

Anton Gvozdikov/Shutterstock

UNIT 5

ELECTRICITY AND MAGNETISM

ENCOUNTER THE PHENOMENON

What role do electricity and magnetism play in the technology we use every day?

SEP Ask Questions

What questions do you have about the phenomenon? Write your questions on sticky notes and add them to the driving question board for this unit.

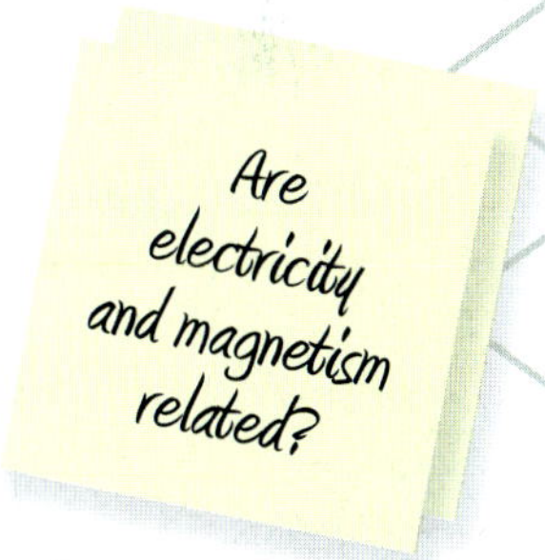

Look for Evidence

As you go through this unit, use the information and your experiences to help you answer the phenomenon question as well as your own questions. For each activity, record your observations in a Summary Table, add an explanation, and identify how it connects to the unit and module phenomenon questions.

Solve a Problem

STEM UNIT PROJECT

From Raw Resource to Usable Energy Natural energy resources, such as fossil fuels or solar energy, are not directly usable. Their energy must be converted to a more usable form—usually electrical energy. Investigate how energy can be transformed from one form to another. Use the results of these investigations and the evidence you collected during the unit to design and test an energy conversion device.

GO ONLINE In addition to reading the information in your Student Edition, you can find the STEM Unit Project and other useful resources online.

Massimiliano Trevisan/Alamy Stock Photo

MODULE 18

ELECTROSTATICS

ENCOUNTER THE PHENOMENON

Why does the rod cause the water to bend?

GO ONLINE to watch a video about early investigations into electric forces.

SEP Ask Questions

Do you have other questions about the phenomenon? If so, add them to the driving question board.

CER Claim, Evidence, Reasoning

Make Your Claim Use your CER chart to make a claim about why the rod causes the water to bend. Explain your reasoning.

Collect Evidence Use the lessons in this module to collect evidence to support your claim. Record your evidence as you move through the module.

Explain Your Reasoning You will revisit your claim and explain your reasoning at the end of the module.

GO ONLINE to access your CER chart and explore resources that can help you collect evidence.

LESSON 1: Explore & Explain: A Microscopic View of Charge

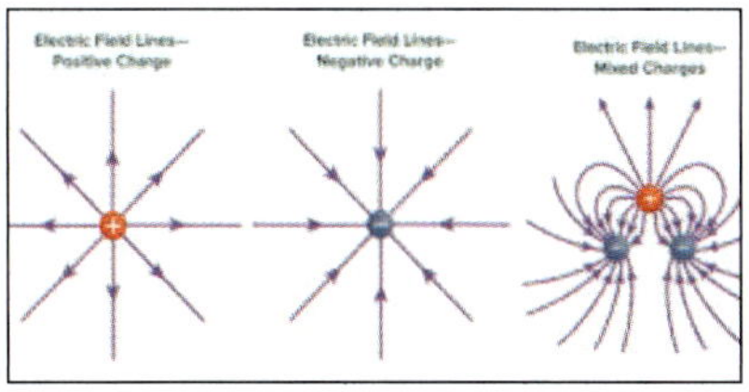

LESSON 3: Explore & Explain: Modeling the Electric Field

Smithsonian

Additional Resources

(t)Video Supplied by BBC Worldwide Learning, (b)Richard Hutchings/Digital Light Source

LESSON 1

ELECTRIC CHARGE

FOCUS QUESTION

Why does rubbing your shoes on the carpet lead to a shock when you touch a metal door knob?

Evidence of Charge

You might have rubbed your shoes on a carpet and then produced a spark when you touched someone. Electric effects produced in this way are called static electricity. In this module, you will investigate **electrostatics,** the study of electric charges that can be collected and held in one place, or static electricity. The effects of static electricity are observable over a vast range of scales, from huge displays of lightning to interactions between electrons on the atomic scale. In later modules, you will study electric current, which is the net movement of electric charge.

Electrostatic force Have you ever noticed that on a dry day your hair is attracted to your comb after you comb it? Perhaps you have noticed that socks sometimes stick together when you take them out of a clothes dryer. You might also recognize what happens after you pull off a wool hat, as shown in **Figure 1.** There must be a strong force pulling upward on the hair because it overcomes the gravitational force pulling downward on the hair. The hat produces an electrostatic force on the hair that is greater than the gravitational force from Earth.

The electrostatic force and the gravitational force are both forces, but they act differently. For example, hair is attracted to a hat only after the hat has been rubbed against the hair; if you wait a while, the attractive property of the hat diminishes. The gravitational force, on the other hand, does not require rubbing and does not decrease. **Figure 2** on the next page shows one way to investigate the electrostatic force and electric charge.

Figure 1 Pulling off a wool hat, especially on a day when the air is dry, can cause your hair to stand up.

Infer *Why would hair be more likely to stand up on a dry day than on a wet day?*

3D THINKING DCI Disciplinary Core Ideas CCC Crosscutting Concepts SEP Science & Engineering Practices

COLLECT EVIDENCE

Use your Science Journal to record the evidence you collect as you complete the readings and activities in this lesson.

INVESTIGATE

GO ONLINE to find these activities and more resources.

Forensics Lab: Static Charge
Carry out an investigation to determine how static electric charges affect each other and neutral objects.

Revisit the Encounter the Phenomenon Question
What information from this lesson can help you answer the Unit and Module questions?

Figure 2 You can use two strips of tape to demonstrate the force between charged objects.

Investigating Charge You can investigate electric charge using strips of transparent tape. Fold over about 5 mm of the end of a 10-cm strip of tape to use as a handle. Stick the tape strip to a dry, smooth surface, such as a desk, as shown here. Attach a second, similar piece of tape next to the first.

These tape strips become charged when you pull them off the desk.

Like Charges Quickly pull both strips off the desk and bring them near each other. What happens? Pulling the strips of tape off the desk gives them the ability to repel each other. The strips repel each other because they have accumulated electric charge. The strips were prepared in the same way, so they have the same charge. Two objects with like charges always repel each other.

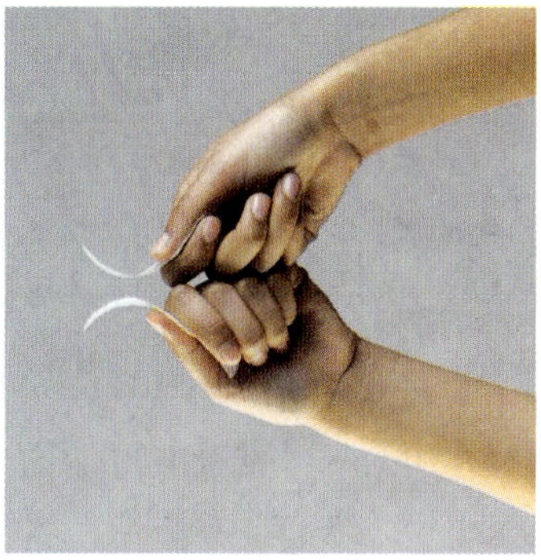

Tape strips repel each other when each has the same charge.

Unlike charges Now, stick one strip of tape on the desk and place the second strip on top of the first. Use the handle of the bottom strip to slowly pull both strips off the table. Rub them with the fingers of your other hand until they are no longer attracted to you. This removes their charge. Now, with one hand holding the handle of one strip and the other hand holding the handle of the second, quickly pull the strips apart. The strips are once more charged—they are attracted to your hands. Bring the strips close together. They now attract each other. Because they attract, you know they do not have the same charge. This suggests there is a second type of charge and unlike charges attract.

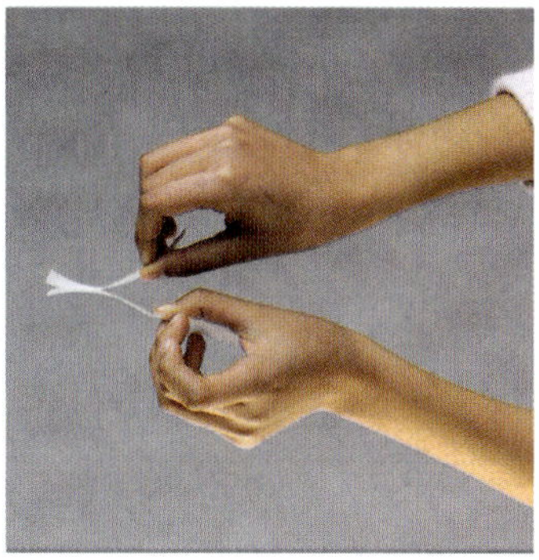

Tape strips attract each other when each has a different charge.

Interactions between charged objects Can you use your tape to explore the charge of other objects? Rub a plastic comb on your clothing, and place the comb on a desk. Then, prepare your strips of tape as you did before. Label the handle end of the bottom strip B and the handle end of the top strip T. Pull the pair off the desk together. Discharge them, and then pull them apart. Now, holding your two strips of tape by the sticky ends, one in each hand, bring first one and then the other close to the comb. You will find that one strip will be attracted to the comb, and the other will be repelled by it.

Repeat this investigation with materials such as glass, silk, or wool.

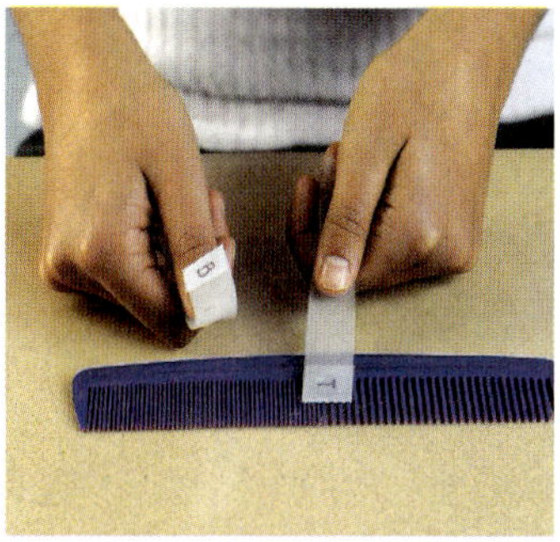

A charged comb attracts one strip and repels the other when each has a different charge.

Types of charge From your investigations with two strips of tape, you have discovered that there are only two types of charge. Benjamin Franklin called them positive charges and negative charges. Matter normally contains both types of charges. Materials have varying degrees of ability to acquire charge. Hard rubber and plastic have a tendency to become negatively charged. Glass and wool have a tendency to become positively charged. Rubbing materials together separates the two types of charge. To explore this further, consider a microscopic view.

Get It?
Predict the type of charge on the pen shown at the beginning of the module.

A Microscopic View of Charge

Electric charges exist within atoms. In 1897 J.J. Thomson discovered that all materials contain low-mass, negatively charged particles. These particles are called electrons. Between 1909 and 1911, Ernest Rutherford, who had earlier worked as Thomson's assistant, discovered that the atom has a massive, positively charged nucleus surrounded by a cloud of orbiting electrons. From these experiments and many others, scientists know that atoms are normally neutral. For a **neutral** object, the amount of negative charge exactly balances the amount of positive charge.

Transfer of electrons With the addition of energy, the outer electrons of an atom can be removed from the atom. An atom that is missing electrons has a net positive charge, and matter that has electron-deficient atoms is positively charged. The electrons can remain unattached or can attach to other atoms, resulting in atoms with net negative charge. From a microscopic viewpoint, acquiring charge is a process of transferring electrons, as modeled in **Figure 3.**

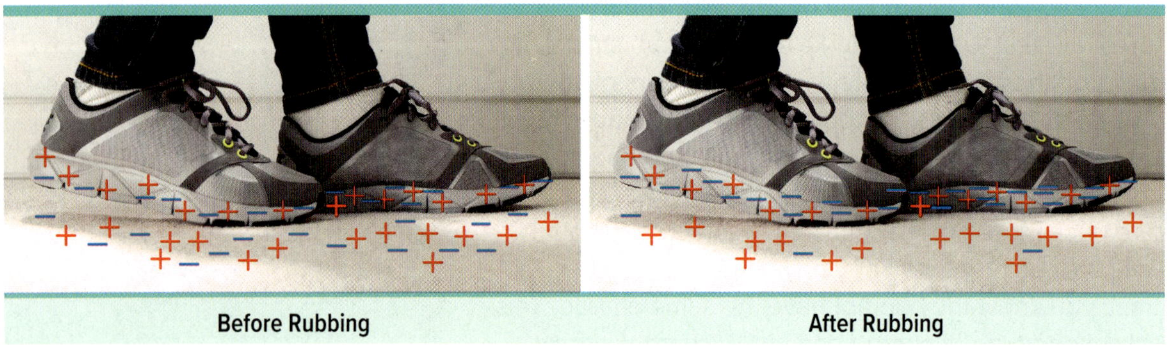

Figure 3 Electrons can be transferred from the wool rug to the rubber shoe.

COLOR CONVENTION

Positive charge	+	red
Negative charge	−	blue

SCIENCE USAGE v. COMMON USAGE

Neutral

Science usage: having exact balance between positive and negative charges
Neutral objects do not experience an electric attraction.

Common usage: not engaged or supporting either side
Historically, Switzerland has been a neutral country.

Separation of charge If you rub two neutral objects together, each object can become charged. For instance, when you rub rubber shoes on a wool rug, the energy from the rubbing removes outer electrons from atoms in the wool, and they transfer to the rubber shoe, as shown in **Figure 3.** The extra electrons on the shoe result in a net negative charge on the shoe. The electrons missing from the wool rug result in a net positive charge on the rug. The combined total charge of the two objects remains the same. Charge is conserved, which is one way of saying that individual charges are never created or destroyed. A net positive or negative charge means that electrons have been transferred.

Effects of Separation of Charge Processes inside a thundercloud can cause the cloud bottom to become negatively charged and the cloud top to become positively charged. Electric charge can be transferred from a road to the car traveling on it. The rubbing of your MP3 player in your pocket can sometimes cause your earbuds to become charged and shock you. In all of these cases, charges are not created, but separated.

Conductors and Insulators

Hold a plastic rod or comb at its midpoint and rub one end on your clothing. You will find that only the rubbed end becomes charged. In other words, the electrons that transferred to the plastic object stayed where they were; they did not move. A material through which a charge will not move easily is called an electric **insulator.** The strips of tape that you charged earlier in this chapter acted as insulators because the charge that accumulated on them did not move; it stayed localized. Glass, dry wood, most plastics, cloth, and dry air are all good insulators.

Suppose you support a metal rod on an insulator so that it is completely surrounded by the insulator. When you touch a charged comb to one end of the metal rod, you will find that the charge spreads very quickly over the entire rod. A material that allows charges to move about easily is called an electric **conductor.** Electrons move and thus conduct electric charge through the metal. **Figure 4** contrasts how excess charges behave on a conductor and on an insulator.

Metals From past experience, we know that most metals are good conductors. This is because at least one electron of each atom of a metal can be removed easily. These electrons no longer remain with any particular atom, but move through the metal as a whole. Copper and aluminum are both excellent conductors and are used commercially to carry electric charge. Some nonmetal materials also are good conductors. These include plasma, which consists of negative electrons and positive ions, and graphite.

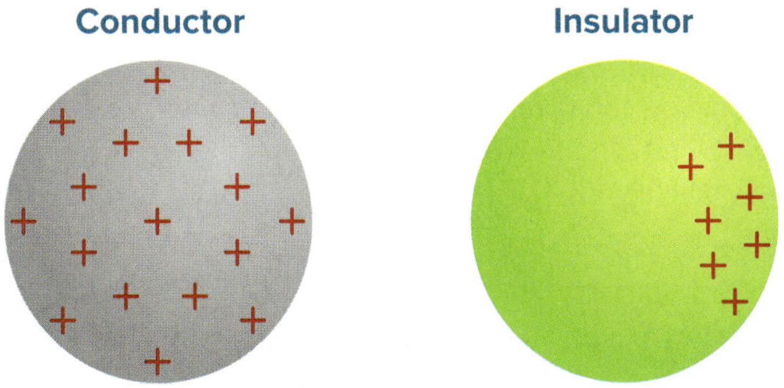

Figure 4 Charge spreads out evenly on a conductor. An insulator holds charges where they are placed.

Consider *why electrical wires are coated with rubber.*

Air as a conductor You read that air is an insulator. Under certain conditions, charges can move through air as if it were a conductor. The spark that jumps between your finger and a doorknob after you have rubbed your feet on a carpet discharges you. In other words, you have become neutral because the excess charges have left your body. Similarly, lightning that you see in a thunderstorm discharges a thundercloud. In both of these cases, air becomes a conductor for a brief moment.

When does air become a conductor? To answer this question, recall that electrons are free to move in conductors. For a spark or lightning to occur, freely moving charged particles must be formed in air, which is normally neutral. In the case of lightning, excess charges in the cloud and on the ground exert enough force to temporarily remove electrons from the molecules in the air. The electrons and positively charged ions form a plasma, which is a conductor.

Real-World Physics

Conductor or Insulator?

It might be tempting to classify an element as solely a conductor or solely an insulator, but the classification can change depending on various factors, including the form the element takes. For example, carbon in the form of diamond is an insulator, but carbon in the form of graphite conducts charge. This is because of differences in the bonding of the carbon atoms. The carbon bonds in graphite allow some electron movement. As a result, graphite is a much better conductor than diamond, even though both are made of carbon atoms.

Check Your Progress

1. **Charged Objects** In the investigations with tape described in this section, how could you find out which strip of tape, B or T, is positively charged?
2. **Charged Objects** After you rub a comb on a wool sweater, you can use the comb to pick up small pieces of paper. Why does the comb lose this ability after a few minutes?
3. **Types of Charge** A pith ball is a small sphere made of a light material, such as plastic foam, that is often coated with a layer of graphite or aluminum paint. How could you determine whether a pith ball suspended from an insulating thread is neutral, charged positively, or charged negatively?
4. **Charge Separation** You can give a rubber rod a negative charge by rubbing the rod with wool. What happens to the charge of the wool? Why?
5. **Net Charge** An apple contains approximately 10^{26} charged particles. Why don't two apples repel each other when they are brought together?
6. **Charging a Conductor** Suppose you hang a long metal rod from silk threads so that the rod is electrically isolated. You then touch a charged glass rod to one end of the metal rod. Describe the charges on the metal rod.
7. **Charging by Friction** You can charge a rubber rod negatively by rubbing it with wool. What happens when you rub a copper rod with wool?
8. **Critical Thinking** Some scientists once proposed that electric charge is a type of fluid that flows from objects with an excess of the fluid to objects with a deficit. How is the current two-charge model more accurate than the single-fluid model?

LEARNSMART® Go online to follow your personalized learning path to review, practice, and reinforce your understanding.

LESSON 2

ELECTROSTATIC FORCE

FOCUS QUESTION

How can objects be electrically charged?

Forces on Charged Objects

Electrostatic forces have been recognized for thousands of years, and over the years many scientists have tried to measure them. Daniel Bernoulli, best known for his work with fluids, made some crude measurements in the 1750s. In the 1770s, Henry Cavendish demonstrated a mathematical relationship between forces on charged particles, but he did not publish his work. The discovery of this relationship, described later in this lesson, is instead credited to the French physicist Charles Coulomb.

Demonstrating forces You can demonstrate electrostatic forces by suspending a negatively charged, hard rubber rod so it turns easily, as illustrated in **Figure 5.** If you bring another negatively charged rod near the suspended rod, the suspended rod will turn away. The negative charges on the rods will repel each other even if the rods do not touch. The electrostatic force acts at a distance. If a negatively charged rod is brought near a positively charged rod, the two will attract each other. These actions of charged rods can be summarized as follows:

- There are two kinds of electric charge: positive and negative.
- Like charges repel; unlike charges attract.
- Charges exert forces on other charges at a distance.
- The force is stronger when the charges are closer together.

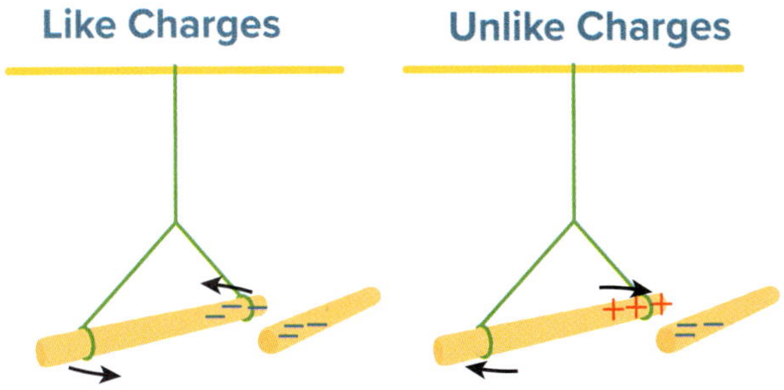

Figure 5 Two charged rods brought near each other will either attract each other or repel each other, depending on their charges.

3D THINKING DCI Disciplinary Core Ideas CCC Crosscutting Concepts SEP Science & Engineering Practices

COLLECT EVIDENCE

Use your Science Journal to record the evidence you collect as you complete the readings and activities in this lesson.

INVESTIGATE

GO ONLINE to find these activities and more resources.

Applying Practices: Gravitational and Electrostatic Forces

HS-PS2-4. Use mathematical representations of Newton's Law of Gravitation and Coulomb's Law to describe and predict the gravitational and electrostatic forces between objects.

PhysicsLAB: Charge It Up

Carry out an investigation into the effects of various methods of electrically charging objects.

Determining charge Neither a strip of tape nor a rubber rod that is hanging in the air is a very sensitive or convenient way to determine charge. Instead, you can use a device called an **electroscope.** An electroscope consists of a metal knob connected by a metal stem to two thin, lightweight pieces of metal foil, called leaves, that are enclosed to eliminate air currents. **Figure 6** shows a neutral, uncharged electroscope. Note that the leaves hang loosely.

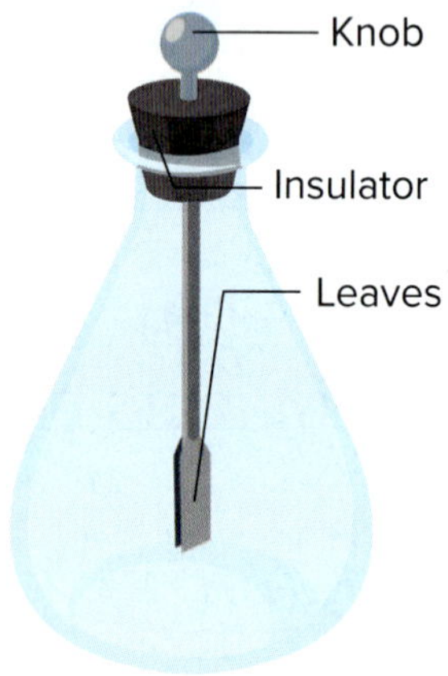

Figure 6 The metal foil leaves on a neutral electroscope hang loosely together.

Charging by conduction When you touch a negatively charged rod to the knob of an electroscope, electrons are transferred to the knob and spread over all the metal surfaces. As shown in the left image of **Figure 7,** the two leaves become negatively charged and repel each other, so they move apart. The leaves also will move apart if you positively charge the electroscope. In either case, you have given the electroscope a net charge. Scientists call charging a neutral object by touching that object with a charged object **charging by conduction.**

If the leaves of an electroscope behave the same no matter the charge, how can you find out whether you have charged an electroscope positively or negatively? You can determine the charge of an electroscope by observing the leaves when you bring a rod of known charge close to but not touching the knob. The leaves will spread farther apart if the rod and the electroscope have the same charge, as shown in the middle image of **Figure 7.** The leaves will fall slightly if the electroscope's charge is opposite that of the rod, as in the far-right image of **Figure 7.**

Separation of charge on neutral objects In your investigations with tape, you found that when you held charged tape, the tape was attracted to your fingers. The tape was attracted to your fingers whether the tape was positively charged or negatively charged. Your fingers, however, remained neutral; they had equal amounts of positive and negative charge. Why were the tape strips attracted to them? The tape was attracted to your fingers because the charges in your fingers had separated.

Figure 7 Charging by conduction occurs when charge is transferred between two objects that are touching.
Explain *what would happen to the electroscope leaves if the positively charged rod actually touched the knob in the image on the right.*

Charging by Conduction

To charge an electroscope by conduction, let a charged metal rod touch the electroscope's knob.

Bringing a negatively charged rod near a negatively charged electroscope causes the leaves to spread apart farther.

Bringing a positively charged rod near a negatively charged electroscope causes the leaves to fall closer together.

Figure 8 A buildup of negative charge in the bottoms of clouds causes lightning.

Suppose you place a neutral metal rod, or any uncharged conductor, near a positively charged object. The object attracts the electrons in the rod. You know that electrons in conductors move easily. Therefore, some of the rod's electrons move toward the object. The end closest to the object becomes negatively charged, and the other end positively charged. Charges have separated, but the rod remains neutral. The closer charges are to each other, the stronger is the electrostatic force between them; therefore, the separation of charges in the rod results in an attractive force between the rod and the charged object.

Lightning The negative charges at the bottoms of thunderclouds can cause the separation of charges on Earth. Negative charges in the ground below a cloud are repelled from Earth's surface. The forces between the charges in the cloud and those on Earth's surface can break molecules in the air into positively and negatively charged particles. The charged particles establish a conducting path from the ground to the cloud. A bolt of lightning occurs when electrons travel along the conducting path and discharge the cloud and Earth, as shown in **Figure 8.**

Charging by induction Suppose that two identical insulated metal spheres are touching, as in **Figure 9.** If you bring a negatively charged rod close to one sphere, as in the middle image of **Figure 9,** electrons will move from this sphere onto the sphere farther from the rod. This will negatively charge the farther sphere and positively charge the closer sphere. If you separate the spheres while the rod is nearby, each sphere will have an equal but opposite charge, as shown on the right in **Figure 9.** This process of charging a neutral object by bringing a charged object near it is called **charging by induction.**

Figure 9 The result of charging by induction is spheres with charges that are equal in magnitude but opposite in sign. No charges were added to the spheres from the rod. (Note: Only the net charge is shown in the figure.)

Charging by Induction

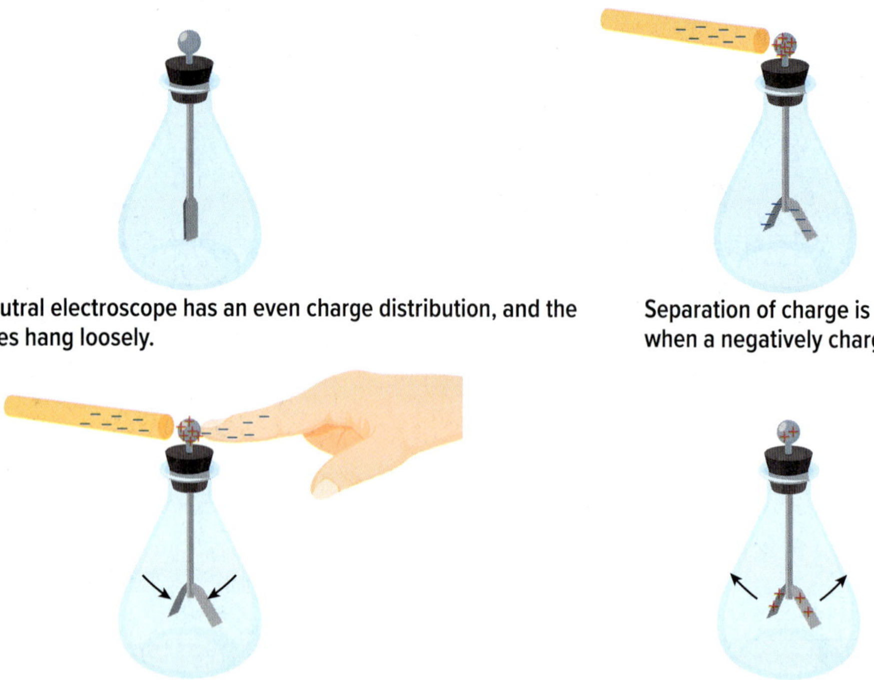

A neutral electroscope has an even charge distribution, and the leaves hang loosely.

Separation of charge is induced in the electroscope when a negatively charged rod is brought near it.

Touching the electroscope allows the charged rod to push electrons out into the hand instead of down into the leaves.

When the ground is removed from the electroscope before the rod is removed, an excess of positive charge is left on the electroscope.

Figure 10 Grounding an electroscope charges the electroscope by induction.

Grounding **Figure 10** shows how to charge an electroscope by induction through **grounding,** which is the process of removing excess charge by connecting an object to Earth. Because Earth is very large, it can absorb great amounts of charge without becoming noticeably charged itself.

If you bring a negatively charged rod close to the knob of the electroscope in **Figure 10,** as in the top-right image, the rod repels electrons onto the leaves. If you ground the knob on the side opposite the charged rod by touching it, electrons will be pushed from the electroscope through your hand and into the ground until the leaves are neutral. If you then remove your hand, leaving the charging rod in place, the electroscope will have a deficit of electrons and a positive charge.

You also can use the ground as a source of electrons. If you bring a positive rod near the knob of a grounded electroscope, electrons will be attracted from the ground and the electroscope will obtain a negative charge. When you employ this process, the charge induced on the electroscope is opposite that of the object used to charge it. Because the charged rod never touches the electroscope, no electrons are transferred between the rod and the electroscope, and the rod can be used many times to charge objects by induction.

CCC CROSSCUTTING CONCEPTS

Cause and Effect Investigate the cause-and-effect relationship that explains how grounding is an important part of safety in the home in the event of a power surge. Given that the power surge is the cause, what effect would the surge have with and without grounding? Make a table to compare the cause-and-effect relationships.

Coulomb's Law

In your investigations with tape, you demonstrated that a force acts between two or more charged objects. You probably found that the strength of the force depended on distance. The force also gets stronger as the strength of the charge on the object increases.

How can you vary the quantity of charge in a controlled way? In the late 1770s, Charles Coulomb devised an apparatus that enabled him to determine the amount of charge between two objects. The type of apparatus Coulomb used is shown in **Figure 11.** An insulating rod with a small conducting sphere at one end (A) was suspended from the top by a thin wire that was free to rotate. An equal-sized conducting sphere (B) at the end of another rod was placed in contact with sphere A. When Coulomb touched the rod attached to B with a charged object, the charge spread evenly over the two spheres. Because the two spheres were the same size, they received an equal amount of charge. The symbol for charge is q. We can represent the amount of charge on the spheres as q_A for sphere A and q_B for sphere B; in this case, $q_A = q_B$.

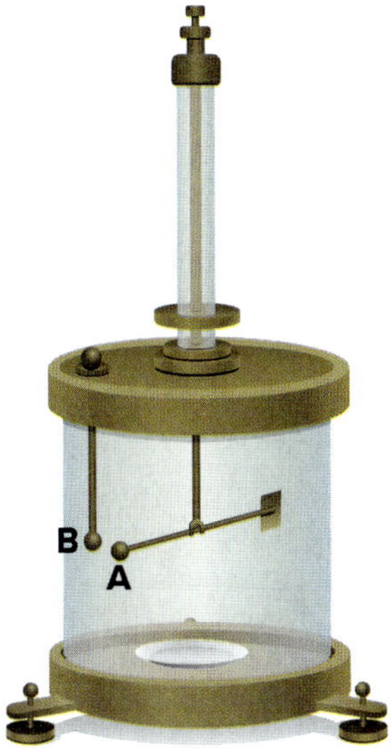

Figure 11 Coulomb used an apparatus similar to this one to measure the electrostatic force between two charged spheres. He observed differences in the amount of deflection of the spheres while varying the distance between them.

Force depends on distance Coulomb found that the force between two charged spheres depends on the distance between them. First, he measured the amount of force needed to twist the suspending wire through a given angle. He then placed equal charges on spheres A and B and varied the distance (r) between them. The force moved A, which twisted the suspending wire. By measuring the deflection of A, Coulomb could calculate the force of repulsion. He showed that the magnitude of the force (F) varies inversely with the square of the distance between the centers of the spheres.

$$F \propto \frac{1}{r^2}$$

Force depends on charge To investigate the way in which the force depends on the amount of charge, Coulomb had to change the charges on the spheres in a measured way. He first charged spheres A and B equally, as before. Then he selected an uncharged sphere (C) of the same size as spheres A and B. When C was placed in contact with B, the spheres shared the charge that had been on B alone. Because the two were the same size, B then had only half of its original charge. Therefore, the charge on B was only one-half the charge on A. After Coulomb adjusted the position of B so that the distance (r) between A and B was the same as before, he found that the force between A and B was half of its former value. That is, he found that the force varies directly with the charge of the bodies.

$$F \propto q_A q_B$$

After many similar measurements, Coulomb summarized the results in a law now known as **Coulomb's law:** the magnitude of the force between point charge q_A and point charge q_B, separated by a distance r, is proportional to the product of the magnitudes of the charges and inversely proportional to the square of the distance between them. Coulomb's law provides a mathematical model to describe and predict the effects of electrostatic forces between distant objects.

$$F \propto \frac{q_A q_B}{r^2}$$

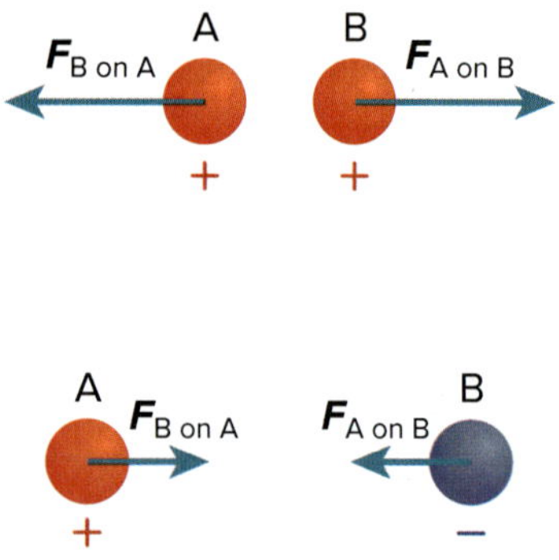

Figure 12 The direction of the force between charged particles depends on the charges. Like charges repel, and unlike charges attract. (Forces are not to scale.)

The unit of charge: the coulomb

The amount of charge an object has is difficult to measure directly. Coulomb's experiments, however, showed that the quantity of charge is related to force. Thus, Coulomb could define a standard quantity of charge in terms of the amount of force that it produces. The SI standard unit of charge is called the **coulomb** (C).

One coulomb is the charge of 6.24×10^{18} electrons or protons. (A proton is an atom's positive charge.) The magnitude of the charge of a single electron or proton is called the **elementary charge.** The elementary charge has a value of 1.602×10^{-19} C. All objects contain charge. In neutral objects the positive and negative charges are balanced, and the net charge is therefore zero. If the negative charge of a book is -1×10^{6} C, and the positive charge of the book is 1×10^{6} C, then the net charge is zero.

In Coulomb's law equation, the magnitude of the force on charge q_A from charge q_B a distance r away from each other can be calculated by incorporating a proportionality constant (K).

Coulomb's Law

The force between two charges is equal to a constant times the product of the two charges, divided by the square of the distance between them.

$$F = K\frac{q_A q_B}{r^2}$$

When the charges are measured in coulombs, the distance in meters, and the force in newtons, the constant is 9.0×10^{9} N·m²/C².

Vector quantity

The Coulomb's law equation gives the magnitude of the force that q_A exerts on q_B and also the magnitude of the force that q_B exerts on q_A. These magnitudes are equal. You can observe this example of Newton's third law of motion when you bring two strips of tape with like charges together. Each exerts a force on the other. If you bring a charged comb near either strip, the strip, with its small mass, moves readily. The acceleration of the comb is much less because of its much greater mass. The electrostatic force, like all forces, is a vector quantity. Force vectors have both a magnitude and a direction. The Coulomb's law equation gives only the magnitude of the force. To determine direction, you can draw a diagram using force vectors, such as that in **Figure 12.** The direction of force between two objects depends on the objects' charges.

PROBLEM-SOLVING STRATEGY

Electrostatic Force Problems

Use these steps to find the magnitude and the direction of the force between charges.

1. Sketch the system, showing all distances and angles to scale.
2. Diagram the vectors of the system.
3. Use Coulomb's law to find the magnitude of the force.
4. Use your diagram, along with trigonometric relations, to find the direction of the force.
5. Perform all algebraic operations on both the numbers and the units. Make sure the units match the variables in the question.
6. Consider the magnitude of your answer. Is the magnitude of the net force relatively close to those of the component forces?

EXAMPLE Problem 1

COULOMB'S LAW IN TWO DIMENSIONS Sphere A, with a charge of +6.0 μC, is located near another charged sphere, B. Sphere B has a charge of −3.0 μC and is located 4.0 cm to the right of A.

a. What is the force of sphere B on sphere A?

b. A third sphere, C, with a +1.5-μC charge, is added. If it is located 3.0 cm directly beneath A, what is the new net force on sphere A?

1 ANALYZE AND SKETCH THE PROBLEM

- Establish coordinate axes, and sketch the spheres.
- Show and label the distances between the spheres.
- Diagram and label the force vectors.

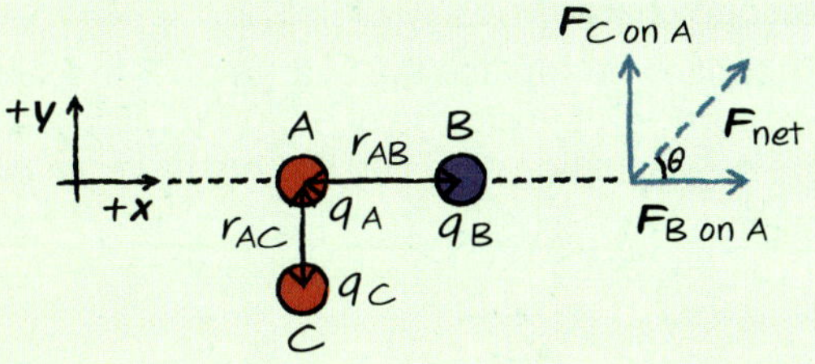

Known		**Unknown**
$q_A = +6.0\ \mu C$	$r_{AB} = 4.0$ cm	$F_{B\ on\ A} = ?$
$q_B = -3.0\ \mu C$	$r_{AC} = 3.0$ cm	$F_{C\ on\ A} = ?$
$q_C = +1.5\ \mu C$	$K = 9.0\times10^9\ N{\cdot}m^2/C^2$	$F_{net} = ?$

2 SOLVE FOR THE FORCES ON SPHERE A

a. Find the force of sphere B on sphere A.

$$F_{B\ on\ A} = K\frac{q_A q_B}{r_{AB}^{\ 2}}$$

$$= (9.0\times10^9\ N{\cdot}m^2/C^2)\frac{(6.0\times10^{-6}\ C)(3.0\times10^{-6}\ C)}{(4.0\times10^{-2}\ m)^2}$$ Substitute $q_A = 6.0\ \mu C$, $q_B = 3.0\ \mu C$, $r_{AB} = 4.0$ cm.

$$= 1.0\times10^2\ N$$

Because spheres A and B have unlike charges, the force of B on A is to the right.

b. Find the net force on sphere A.

$$F_{C\ on\ A} = K\frac{q_A q_C}{r_{AC}^{\ 2}}$$

$$= (9.0\times10^9\ N{\cdot}m^2/C^2)\frac{(6.0\times10^{-6}\ C)(1.5\times10^{-6}\ C)}{(3.0\times10^{-2}\ m)^2}$$ Substitute $q_A = 6.0\ \mu C$, $q_C = 1.5\ \mu C$, $r_{AC} = 3.0$ cm.

$$= 9.0\times10^1\ N$$

Spheres A and C have like charges, which repel. The force of C on A is upward.

Find the vector sum of $F_{B\ on\ A}$ and $F_{C\ on\ A}$ to find F_{net} on sphere A.

$$F_{net} = \sqrt{F_{B\ on\ A}^{\ 2} + F_{C\ on\ A}^{\ 2}}$$

$$= \sqrt{(1.0\times10^2\ N)^2 + (9.0\times10^1\ N)^2}$$ Substitute $F_{B\ on\ A} = 1.0\times10^2$ N, $F_{C\ on\ A} = 9.0\times10^1$ N.

$$= 130\ N$$

$$\tan\theta = \frac{F_{C\ on\ A}}{F_{B\ on\ A}}$$

$$\theta = \tan^{-1}\left(\frac{F_{C\ on\ A}}{F_{B\ on\ A}}\right)$$

$$= \tan^{-1}\left(\frac{9.0\times10^1\ N}{1.0\times10^2\ N}\right)$$ Substitute $F_{C\ on\ a} = 9.0\times10^1$ N, $F_{B\ on\ A} = 1.0\times10^2$ N.

$$= 42°$$

$F_{net} = 1.3\times10^2$ N, 42° above the *x*-axis

3 EVALUATE THE ANSWER

- **Are the units correct?** $(N{\cdot}m^2/C^2)(C)(C)/m^2 = N$. The units work out to be newtons.
- **Does the direction make sense?** Like charges repel; unlike charges attract.
- **Is the magnitude realistic?** The net force is relatively close to the component forces. 130 N is close enough to 90 N and 100 N to be reasonable.

PRACTICE Problems

ADDITIONAL PRACTICE

9. A negative charge of -2.0×10^{-4} C and a positive charge of 8.0×10^{-4} C are separated by 0.30 m. What is the force between the two charges?

10. A negative charge of -6.0×10^{-6} C exerts an attractive force of 65 N on a second charge that is 0.050 m away. What is the magnitude of the second charge?

11. Suppose you replace the charge on B in Example Problem 1 with a charge of $+3.00\ \mu$C. Diagram the new situation, and find the net force on A.

12. Describe how the electrostatic force between two charges changes when the distance between those two charges is tripled.

13. Sphere A is located at the origin and has a charge of $+2.0\times10^{-6}$ C. Sphere B is located at +0.60 m on the *x*-axis and has a charge of -3.6×10^{-6} C. Sphere C is located at +0.80 m on the *x*-axis and has a charge of $+4.0\times10^{-6}$ C. Determine the net force on sphere A.

14. CHALLENGE Determine the net force on sphere B in the previous problem.

Using Coulomb's law As you use the Coulomb's law equation, keep in mind that Coulomb's law is valid only for point charges or for charges that are spread uniformly over a sphere. A charged sphere may be treated as if all the charge were located at its center if the charge is spread evenly across its entire surface or throughout its volume.

If the sphere is a conductor and another charge is brought near it, the charges on the sphere will be attracted or repelled and will move, making the sphere no longer uniformly charged. The charge no longer will act as if it were at the sphere's center. Therefore, it is important to consider how large and how far apart two charged spheres are before applying Coulomb's law.

The problems in this textbook assume that charged spheres are small enough and far enough apart to be considered point charges, unless otherwise noted. When you consider shapes such as long wires or flat plates, you must modify Coulomb's law to account for the non-point charge distributions. To do this, you must model the charge distribution as an additive collection of point charges.

Applications of Electrostatic Forces

There are many commercial and industrial applications that take advantage of electrostatic forces. For example, tiny paint droplets can be charged by induction. When sprayed on automobiles and other objects, the droplets repel each other and paint spreads uniformly. Photocopy machines take advantage of static electricity to place black toner on a page so that a precise reproduction of the original document is achieved. Laser printers use static electricity in a similar way. Toner particles are attracted to charged characters on a drum.

PHYSICS Challenge

CHARGES ON SPHERES Two spheres of equal mass (*m*) and equal positive charge (*q*) are a distance *r* apart, as shown in the figure on the right

1. Derive an expression for the quantity of charge (*q*) that must be on each sphere so that the spheres are in equilibrium; that is, so that the attractive and repulsive forces between them are balanced.
2. How would doubling the distance between the spheres affect the expression for the value of *q* you determined in the previous problem? Explain.
3. Assuming that the mass of each sphere is 1.50 kg, determine the charge on each sphere needed to maintain the equilibrium.

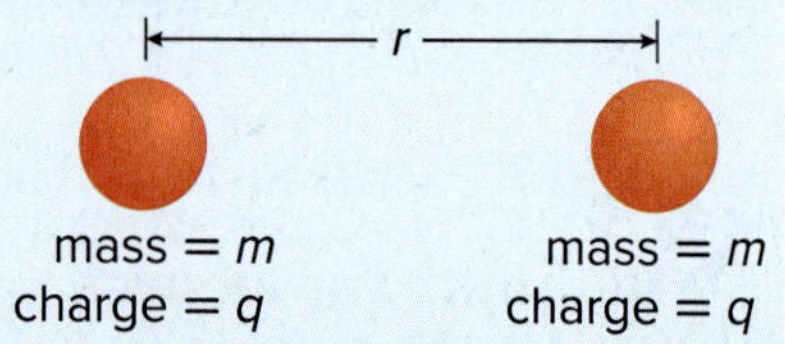

Emissions Without Electrostatic Precipitators | Emissions with Electrostatic Precipitators

Figure 13 Fly ash from coal combustion is an industrial pollutant. Electrostatic precipitators collect fly ash so it is not released into the atmosphere.

Electrostatic forces also are used to collect emissions in smokestacks. Emissions are charged when they are released from combustion processes. They are then attracted to collectors that have been given an opposite charge. A liquid spray removes the collected emissions safely. Electrostatic precipitators make a huge difference in the amount of pollution released into the air, as shown in **Figure 13.**

Check Your Progress

15. **Charge and Distance** What is Coulomb's law and how does it allow you to describe and predict the effects of electrostatic forces between distant objects?
16. **Force and Charge** How are electrostatic force and charge related? Describe the force when the charges are like charges and the force when the charges are opposite.
17. **Force and Distance** How are electrostatic force and distance related? How would the force change if the distance between two charges were tripled?
18. **Charging by Induction** In an electroscope being charged by induction, what happens when the charging rod is moved away before the ground is removed from the knob?
19. **Electroscopes** Why do the leaves of a charged electroscope rise to a certain angle and no farther?
20. **Attraction of Neutral Objects** What properties explain how both positively charged objects and negatively charged objects can attract neutral objects?
21. **Charging an Electroscope** How can you charge an electroscope positively using a positively charged rod? Using a negatively charged rod?
22. **Electrostatic Forces** Two charged spheres are held a distance *r* apart, as shown in **Figure 14.** Compare the force of sphere A on sphere B with the force of sphere B on sphere A.

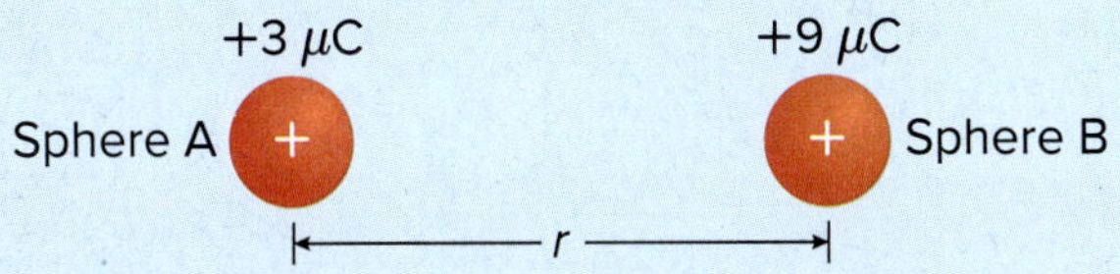

Figure 14

23. **Critical Thinking** Suppose you are testing Coulomb's law using a small, positively charged plastic sphere and a large, positively charged metal sphere. According to Coulomb's law, the force depends on $1/r^2$, where *r* is the distance between the centers of the spheres. As you bring the spheres close together, the force is smaller than expected from Coulomb's law. Explain.

LEARNSMART® Go online to follow your personalized learning path to review, practice, and reinforce your understanding.

LESSON 3

MEASURING ELECTRIC FIELDS

FOCUS QUESTION

How can we explain how electric forces work at a distance?

Defining the Electric Field

In science-fiction movies or books, characters use force fields to create invisible barriers to stop invading armies or to hold prisoners in a cell. Force fields really exist, but they are not quite like the ones in science fiction. Recall that according to Coulomb's law, the force between two point charges varies inversely as the distance between those charges changes. How can one object exert a force on another object across empty space? Michael Faraday suggested the idea of a force field. The electric field is one example of a force field. An **electric field** is a property of the space around a charged object that exerts forces on other charged objects. Electric charges cause electric fields, and electric fields affect charges. An electric field permeates space and exerts forces at a distance by transferring energy through space. You cannot see an electric field, but there are ways to detect that an electric field is present. For example, the streams in the plasma sphere shown in **Figure 15** show the effects of the electric field around the plasma sphere's central electrode.

Figure 15 A plasma sphere's central electrode is a charged object whose net charge alternates rapidly between positive and negative. The glass globe contains charged particles (ions). The electric field from the electrode accelerates the ions, resulting in streams of light that illustrate the ion paths and thus the electric field inside the plasma sphere.

Identify *the cause of the electric field inside the plasma sphere.*

Sinisa Botas/Shutterstock

3D THINKING DCI Disciplinary Core Ideas CCC Crosscutting Concepts SEP Science & Engineering Practices

COLLECT EVIDENCE

Use your Science Journal to record the evidence you collect as you complete the readings and activities in this lesson.

INVESTIGATE

GO ONLINE to find these activities and more resources.

CCC **Identify Crosscutting Concepts**

Create a table of the crosscutting concepts and fill in examples you find as you read.

Review the News

Obtain information from a current news story about electric charge, electric forces, or electric fields. Evaluate your source and communicate your findings to your class.

The strength of an electric field You might wonder how you can detect an electric field if you cannot see it. You can envision the force from an electric field by modeling its effects on a small, charged object—a test charge (q')—at some location. If there is an electrostatic force on the object, then there is an electric field at that point. **Figure 16** illustrates a charged object with a net charge of q. Suppose you place the positive test charge at point (A) and measure a force (F_A). According to Coulomb's law, the force is directly proportional to the strength of the test charge (q'). That is, if the charge on the test charge is doubled, so is the force $\left(\frac{2F}{2q'} = \frac{F}{q'}\right)$. Therefore, the ratio of the force to the strength of the test charge is a constant. When you divide force (F) by the strength of the test charge (q'), you obtain a vector quantity $\left(\frac{F}{q'}\right)$. The electric field at point A, the location of q', is represented by the following equation.

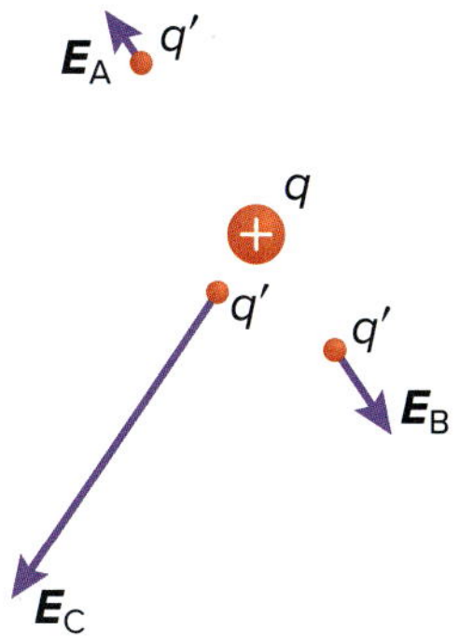

Figure 16 An electric field surrounds particle q. The forces exerted on a positively charged test particle (q') at locations A, B, and C are represented by force vectors. The vectors represent the magnitude and direction of the force from the electric field on the test particle at that point.

Electric Field Strength

The strength of an electric field is equal to the force on a positive test charge divided by the strength of the test charge.

$$E = \frac{F_{\text{on } q'}}{q'}$$

The direction of an electric field is the direction of the force on a positive test charge. The magnitude of the electric field is measured in newtons per coulomb, N/C.

COLOR CONVENTION

Electric field line (E)	purple
Positive charge	red
Negative charge	blue

Get It?

Explain how the strength of an electric field can be determined.

Field vectors You can make a model of an electric field by using arrows to represent the field vectors at various locations, as shown in **Figure 16.** The length of the arrow represents the field strength. The direction of the arrow represents the field direction. To find the field from two charges, add the fields from the individual charges through vector addition. Some typical electric field strengths are shown in **Table 1.**

You also can use a test charge to map the electric field resulting from any collection of test charges. A test charge should be small enough so that its effect on the charge you are testing (q) is negligible. Remember that, according to Newton's third law, the test charge exerts forces back on the charges that produce the electric field. It is important that these forces do not redistribute the charges that you are trying to measure, thereby affecting the electric field you are mapping.

Table 1 Typical Electric Field Strengths (Approximate)

Field	Value (N/C)
Near a charged, hard-rubber rod	1×10^3
Needed to create a spark in air	3×10^6
At an electron's orbit in a hydrogen atom	5×10^{11}

Coulomb's law and electric fields When two charges interacting through a field change relative position, the energy stored in the field is changed. If, and only if, the charge q is a point charge or a uniformly charged sphere, you can calculate its electric field from Coulomb's law. Use Coulomb's law in the electric field equation above to find the magnitude of the force exerted on the test charge ($F_{\text{on } q'}$):

$$E = \frac{F_{\text{on } q'}}{q'} = F_{\text{on } q'} \times \frac{1}{q'} = \frac{Kqq'}{r^2} \times \frac{1}{q'} = \frac{Kq}{r^2}$$

EXAMPLE Problem 2

ELECTRIC FIELD STRENGTH Suppose that you are measuring an electric field using a positive test charge of 3.0×10^{-6} C. This test charge experiences a force of 0.12 N at an angle of 15° north of east. What are the magnitude and direction of the electric field strength at the location of the test charge?

1 ANALYZE AND SKETCH THE PROBLEM

- Draw and label the test charge, q'.
- Show and label the coordinate system centered on the test charge.
- Diagram and label the force vector at 15° north of east.

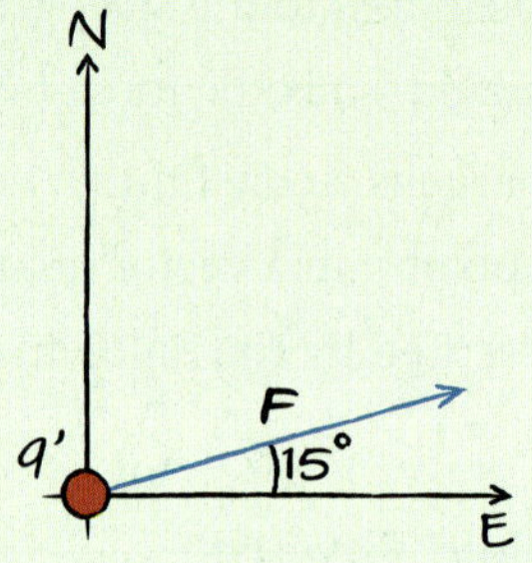

Known	Unknown
$q' = 3.0\times10^{-6}$ C	$E = ?$
$\mathbf{F} = 0.12$ N at 15° N of E	

2 SOLVE FOR THE UNKNOWN

$$E = \frac{F}{q'}$$

$$= \frac{0.12\ \text{N}}{3.0\times10^{-6}\ \text{C}}$$ Substitute $F = 0.12$ N, $q' = 3.0\times10^{-6}$ C.

$$= 4.0\times10^{4}\ \text{N/C}$$

The force on the test charge and the electric field are in the same direction.

$$E = 4.0\times10^{4}\ \text{N/C at 15° N of E}$$

3 EVALUATE THE ANSWER

- **Are the units correct?** Electric field strength is correctly measured in N/C.
- **Does the direction make sense?** The field direction is in the direction of the force because the test charge is positive.
- **Is the magnitude realistic?** This field strength is similar to those in **Table 1.**

PRACTICE Problems

ADDITIONAL PRACTICE

24. A positive test charge of 5.0×10^{-6} C is in an electric field that exerts a force of 2.0×10^{-4} N on it. What is the magnitude of the electric field at the location of the test charge?

25. A negative charge of 2.0×10^{-8} C experiences a force of 0.060 N to the right in an electric field. What are the field's magnitude and direction at that location?

26. A positive charge of 3.0×10^{-7} C is located in a field of 27 N/C directed toward the south. What is the force acting on the charge?

27. Complete **Table 2.**

Table 2 Sample Data

Test Charge Strength (C)	Force Exerted on Test Charge (N)	Electric Field Intensity (N/C)
1.0×10^{-6}	0.30	
2.0×10^{-6}		3.3×10^{5}
	0.45	1.5×10^{5}

28. A negative test charge is placed in an electric field as shown in **Figure 17.** It experiences the force shown. What is the magnitude of the electric field at the location of the charge?

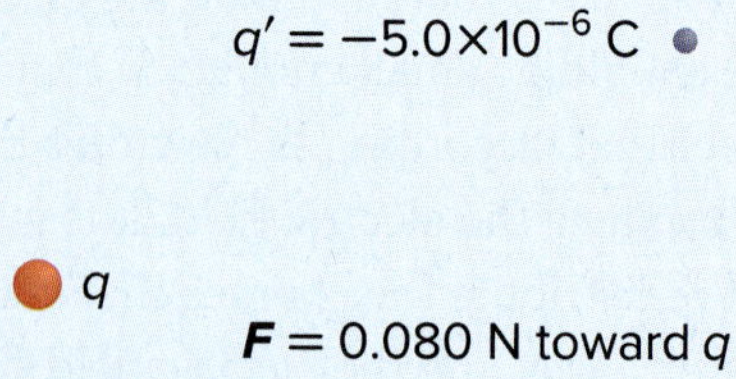

Figure 17

29. CHALLENGE You are probing the electric field of a charge of unknown magnitude and sign. You first map the field with a 1.0×10^{-6}-C test charge, then map it with a 2.0×10^{-6}-C test charge.

a. Would you measure the same forces at the same place with the two test charges? Explain.

b. Would you find the same field strengths? Explain.

EXAMPLE Problem 3

ELECTRIC FIELD STRENGTH AND COULOMB'S LAW What is the magnitude of the electric field at a point that is 0.30 m to the right of a small sphere with a net charge of -4.0×10^{-6} C?

1 ANALYZE AND SKETCH THE PROBLEM

- Draw and label the sphere with its net charge (q) as well as the test charge (q'), 0.30 m away.
- Show and label the distance between the charges.
- Diagram and label the force vector acting on q'.

Recall that a test charge is usually positive.

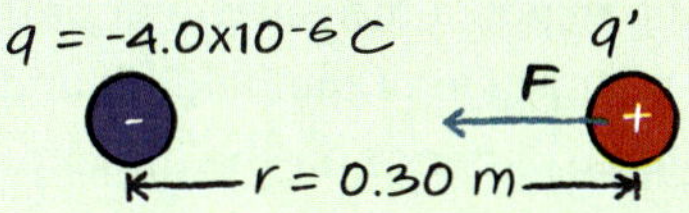

Known	**Unknown**
$q = -4.0\times10^{-6}$ C	$\mathbf{E} = ?$
$r = 0.30$ m	

2 SOLVE FOR THE UNKNOWN

The force and the magnitude of the test charge are unknown, so use Coulomb's law in combination with the electric field strength equation.

$$E = \frac{Kq}{r^2}$$

$$= \frac{(9.0\times10^{9}\ \text{N·m}^2/\text{C}^2)(-4.0\times10^{-6}\ \text{C})}{(0.30\ \text{m})^2}$$

Substitue $K = 9.0\times10^{9}$ N·m²/C², $q = -4.0\times10^{-6}$ C, $r = 0.30$ m.

$$= -4.0\times10^{5}\ \text{N/C}$$

$\mathbf{E} = 4.0\times10^{5}$ N/C toward the sphere, or to the left

3 EVALUATE THE ANSWER

- **Are the units correct?** $\frac{(\text{N·m}^2/\text{C}^2)(\text{C})}{\text{m}^2} = \text{N/C}$. The units work out to be N/C, which is correct for electric field strength.
- **Does the direction make sense?** The negative sign indicates that the negative charge attracts the positive test charge.
- **Is the magnitude realistic?** This field strength is similar to those in **Table 1.**

PRACTICE Problems

ADDITIONAL PRACTICE

30. What is the magnitude of the electric field at a position that is 1.2 m from a 4.2×10^{-6} C point charge?

31. What is the magnitude of the electric field at a distance twice as far from the point charge in the previous problem?

32. What is the electric field at a position that is 1.6 m east of a point charge of $+7.2\times10^{-6}$ C?

33. The electric field that is 0.25 m from a small sphere is 450 N/C toward the sphere. What is the net charge on the sphere?

34. How far from a point charge of $+2.4\times10^{-6}$ C must you place a test charge in order to measure a field magnitude of 360 N/C?

35. Explain why the strength of the electric field exerted on charge q' by the charged body q is independent of the charge on q'. *Hint: Use mathematics to prove your point.*

36. What is the magnitude of the electric field exerted on the test charge shown in **Figure 18**?

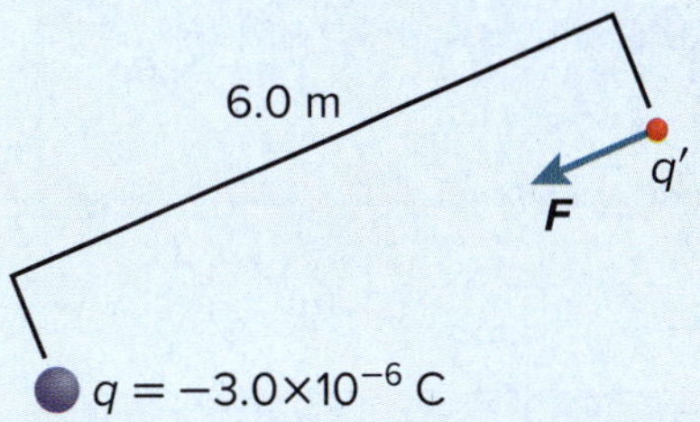

Figure 18

37. CHALLENGE You place a small sphere with a net charge of 5.0×10^{-6} C at one corner of a square that measures 5.0 m on each side. What is the magnitude of the electric field at the opposite corner of the square?

Modeling the Electric Field

An electric field is present even if there is no test charge to measure that field because a test charge is just an imaginary model. Any charge placed in an electric field experiences a force on it resulting from the electric field at that location. So far, you have measured an electric field at a single location. Now, imagine moving the test charge to another location. Measure the force on it again and calculate the electric field again. Repeat this process again and again until you assign to every location in a region a measurement of the electric field. The magnitude of the force depends on the magnitude of the field (E) and the magnitude of the charge (q). Thus, $F = E_q$. The direction of the force depends on the direction of the field and the sign of the charge.

Electric field lines You have already seen how the streams in a plasma sphere represent an electric field. You also can represent electric fields with electric field lines, such as those shown in **Figure 19.** An **electric field line** indicates the direction of the force due to the electric field on a positive test charge. The spacing between the lines indicates the electric field's strength. The field is stronger where the lines are spaced more closely. The field is weaker where the lines are spaced farther apart.

Because electric field lines indicate the force on a positive test charge, they are directed toward negative charges and away from positive charges. The direction of the electric field (E) at any point is the tangent drawn to a field line at that point, as shown in **Figure 20** on the next page.

Figure 20 shows two ways to represent various electric fields—with electric field lines and with grass seed suspended in oil. An electric field affects electric charges whether or not you represent that field with field lines and whether or not you try to visualize that field with grass seed suspended in oil. Grass seed can be useful for visualizing electric fields in three dimensions. Both grass seed and electric field lines are only ways of *representing* electric fields, however. Neither is the same as the electric field itself.

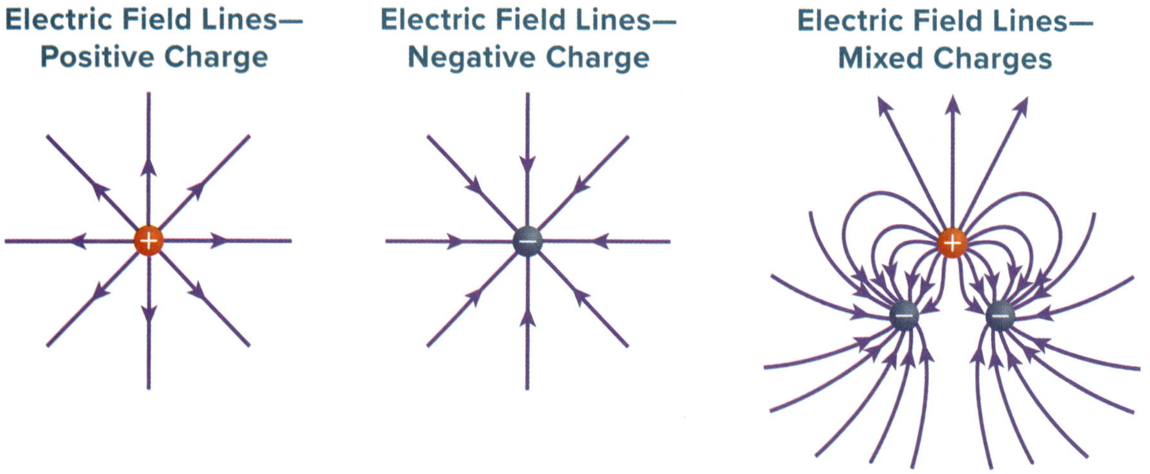

Figure 19 Two-dimensional models of electric fields are shown, but remember that electric fields exist in three dimensions.

CCC CROSS-CUTTING CONCEPTS

Cause and Effect Describe how modeling small-scale, simple electric fields caused by single point charges can help you suggest and predict more complex fields caused by multiple charges. Cite evidence from **Figure 19** and **Figure 20** to support your description.

CCC CROSS-CUTTING CONCEPTS

Patterns Describe how the patterns made by field lines in models such as those in **Figure 19** and **Figure 20** can provide evidence to help you explain phenomena at different scales.

Electrostatic Forces

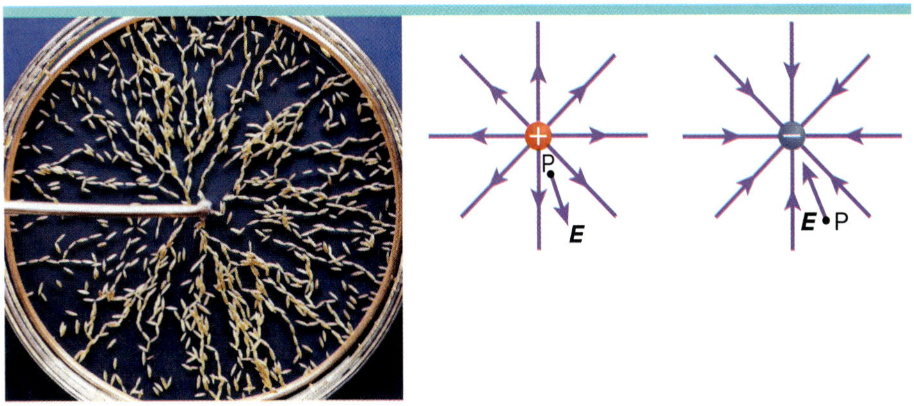

Single Charge For a single charge, the electric field lines radiate from that charge. The field lines radiate out from a positive charge, and they radiate inward toward a negative charge. The direction of the electric field ***(E)*** at any point is the tangent to the field line.

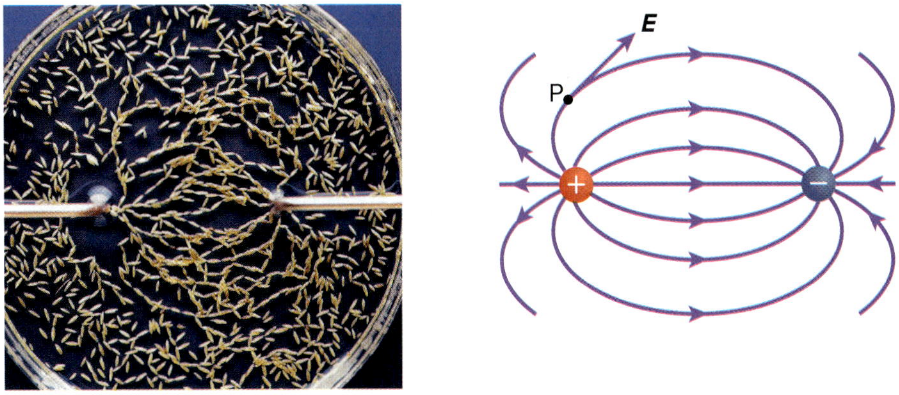

Two Equal and Unlike Charges The electric field lines of two equal and unlike charges form continuous lines from the positive charge to the negative charge. The vector at point P shows the direction of the electric field ***(E)*** at point P. Where the lines are closest, a charged particle would experience the most force.

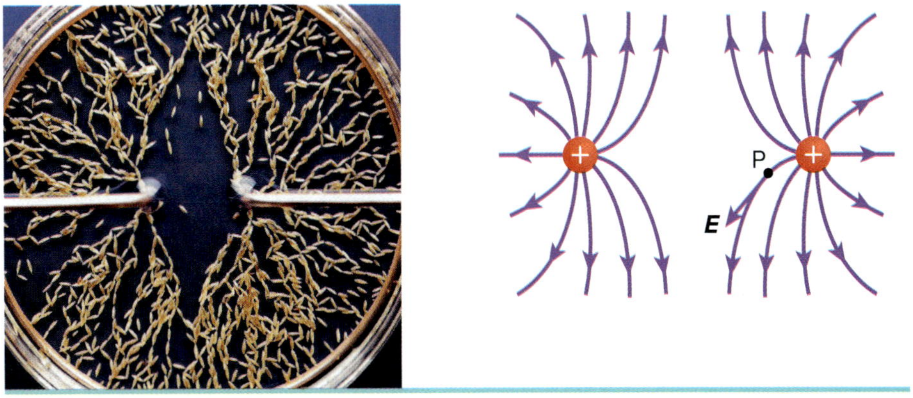

Two Like Charges The electric field lines around two like charges never connect with each other. The vector at point P shows the direction of the electric field ***(E)*** at point P. A charge would experience no electrostatic force in the center, where there are no electric field lines.

Figure 20 Models of electric fields around electric charges are shown. Electrostatic forces cause a separation of charge in each long, thin grass seed. The seeds then turn so that they line up along the direction of the electric field. The lines and vectors drawn in the diagrams represent other ways to model electric fields. These models help show how the relative position energy of two charges can be thought of as being stored in the electric fields.

Describe *how the models show where charged particles would experience more force or less force.*

Figure 21 Robert Van de Graaff devised his generator in the 1930s to aid in the study of particle acceleration and the study of the structure of the atom. When a person touches the generator, the charges on the person's hairs repel each other, causing the hairs to align with the electric field of the metal dome.

Van de Graaff Generators

A Van de Graaff generator, shown in **Figure 21,** can build up a large amount of net charge on its metal dome. In a Van de Graaff generator, a belt is driven by two rollers or pulleys. When someone turns on the machine, the bottom roller rotates and moves the belt. The roller builds up a strong negative net charge, and the belt builds a weaker positive net charge. The strong negative electric field on the roller repels electrons in the tips of a brush assembly near position A. The tips of the brushes become positively charged because the electrons move away from them. At the same time, the strong negative electric field of the roller strips electrons from nearby molecules in the air. The stripped electrons are repelled by the roller and are attracted to the positively charged tips of the brush. The positive ions from the air start to move to the negatively charged roller. The belt is between the positive ions and the negatively charged roller, however. The positive ions collect on the belt and are transported to the metal dome.

When a person touches the metal dome, the positively charged dome attracts the electrons in the person's body. The person's hair becomes positively charged, and the individual hairs repel each other. The hairs align with the electric field around the person's head.

Check Your Progress

38. **Measuring Electric Fields** Suppose you are asked to measure the electric field at a point in space. How do you detect the field at a point? How do you determine the magnitude of the field? How do you choose the magnitude of the test charge?
39. **Field Strength and Direction** A positive test charge of magnitude 2.40×10^{-8} C experiences a force of 1.50×10^{-3} N toward the east. What is the electric field at the position of the test charge?
40. **Field Lines** How can you tell which charges are positive and which are negative by examining the electric field lines?
41. **Field Energy** When two charges interacting through a field change relative position, what happens to the energy in the field?
42. **Critical Thinking** Suppose the top charge in the field on the far right in **Figure 19** is a test charge measuring the field resulting from the two negative charges. Is it small enough to produce an accurate measurement? Explain.

LEARNSMART® Go online to follow your personalized learning path to review, practice, and reinforce your understanding.

LESSON 4

APPLICATIONS OF ELECTRIC FIELDS

FOCUS QUESTION

What is a capacitor and how does it work?

Energy and Electric Potential

As you have read, the concept of energy is extremely useful in mechanics. The law of conservation of energy allows you to predict the motion of objects without knowing the forces in detail. Consider the change in gravitational potential energy of a ball when you lift it, as shown in **Figure 22.** Both the gravitational force ($\boldsymbol{F}_g$) and gravitational field $\left(g = \frac{\boldsymbol{F}_g}{m}\right)$ point toward Earth. If you lift the ball against the gravitational force, you do work on the ball. The gravitational potential energy in the ball-Earth system increases.

The same is true in electrical interactions. The work performed moving a charged particle in an electric field can result in the particle gaining electric potential energy, kinetic energy, or both.

The situation is similar with two unlike charges. The charges attract each other, so you must do work to pull one charge away from the other. When you do this work, you increase the electric potential energy of the two-charge system. The larger the charge or the distance moved, the greater the increase in the electric potential energy (ΔPE). In this module, only charges at rest and changes in electric potential energy are considered and discussed.

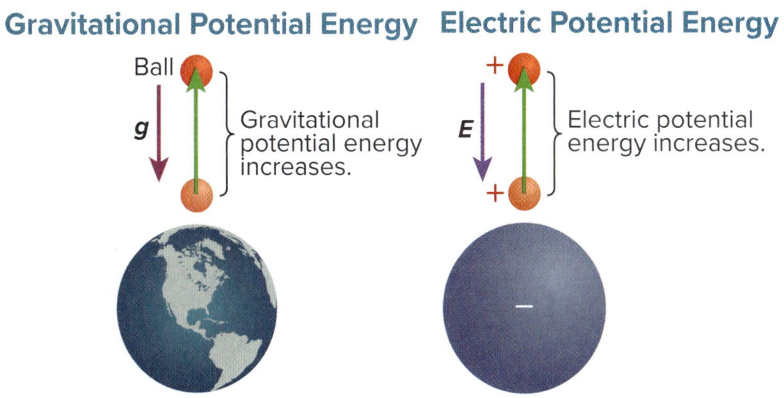

Figure 22 When you lift the ball, you increase the gravitational potential energy of the ball-Earth system. When you separate two unlike charges, you increase the electric potential energy of the two-charge system.

Infer *how you would increase the potential energy of a system that contains two like charges.*

3D THINKING DCI Disciplinary Core Ideas CCC Crosscutting Concepts SEP Science & Engineering Practices

COLLECT EVIDENCE

Use your Science Journal to record the evidence you collect as you complete the readings and activities in this lesson.

INVESTIGATE

GO ONLINE to find these activities and more resources.

Quick Investigation: Electric Fields

Carry out an investigation into the effects of an electric field on a carged object.

PhysicsLAB: Storing Charge

Carry out an investigation to demonstrate that energy can be stored in electric fields.

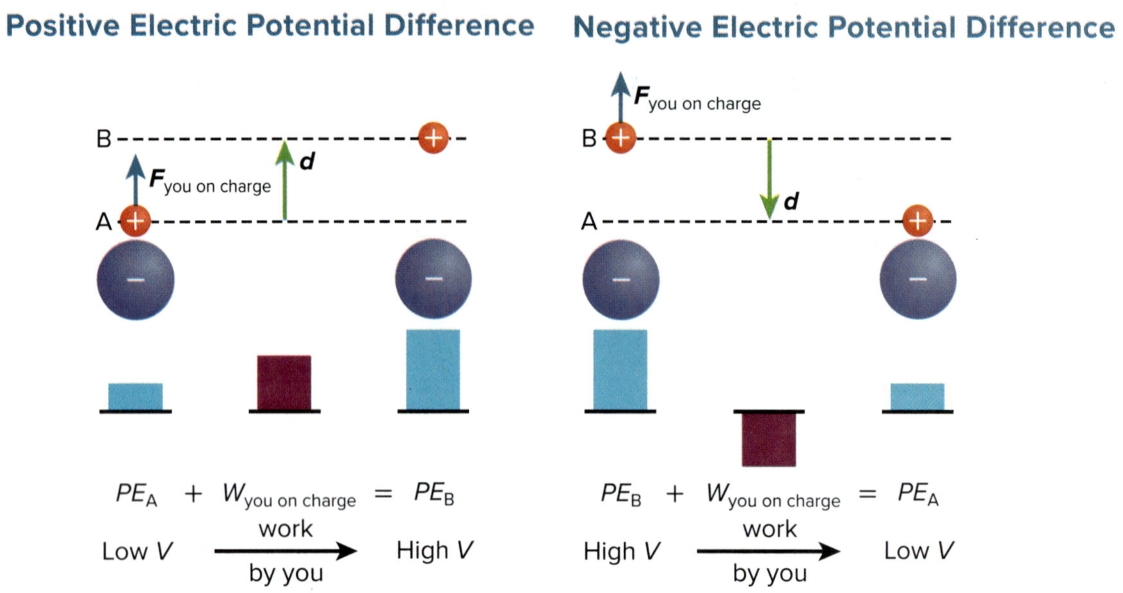

Figure 23 If you move unlike charges apart, then the electric potential difference is positive. If you move unlike charges closer together, then the electric potential difference is negative.

Explain *how work and electric potential are related in regard to the test charge.*

Electric potential difference

You know from your study of mechanics that work done on an object is found using the relationship $W = Fd$. This same relationship is used to find work done to move a charge, $W_{\text{on } q'} = Fd$.

There are differences, however. The work done on a charge is expressed as work done per unit charge and it is called the electric potential difference. The **electric potential difference** (ΔV), which often is called potential difference, is the work ($W_{\text{on } q'}$) needed to move a positive test charge from one point to another, divided by the magnitude of the test charge. You also can think of electric potential difference as the change in electric potential energy (ΔPE) per unit charge.

Electric Potential Difference

Electric potential difference is the ratio of the work needed to move a charge to the magnitude of that charge.

$$\Delta V \equiv \frac{W_{\text{on } q'}}{q'}$$

Electric potential difference is measured in joules per coulomb (J/C). One joule per coulomb is called a **volt** (V).

Positive electric potential difference

The negative charge, in the left panel in **Figure 23,** produces an electric field toward itself. Suppose you place a small positive test charge in the field at position A. That test charge experiences a force in the direction of the field. If you now move the test charge away from the negative charge to position B, you will have to exert a force (***F***) on that test charge. Because the force you exert is in the same direction as the displacement, the work you do on the test charge is positive, and the change in potential energy is positive, as represented by the bar diagram in the figure. Therefore, there is a positive electric potential difference from point A to point B. The potential difference between two points does not depend on the magnitude of the test charge. It depends only on the field and the displacement.

Negative electric potential difference

Suppose you now move the test charge back to position A from position B, as in the right panel in **Figure 23.** The work you do on the test charge and the change in electric potential energy are now negative, also shown in the bar diagram in the figure. The electric potential difference is also negative. In fact, it is equal and opposite to the potential difference for the move from position A to position B. The electric potential difference does not depend on the path used to go from one position to another. It depends only on the starting and final positions on that path.

Zero electric potential difference Is there always an electric potential difference between two positions? Suppose you move the test charge in a circle around the negative charge. The force the electric field exerts on the test charge is always perpendicular to the direction in which you moved it, so you do no work on the test charge. Therefore, the electric potential difference is zero in this case. Whenever the electric potential difference between two or more positions is zero, those positions are said to be at **equipotential.**

Electric potential for like charges You have seen that electric potential difference increases as you pull a positive test charge away from a negative charge. What happens when you move a positive test charge away from another positive charge? There is a repulsive force between these two charges. Potential energy decreases as you move the two charges farther apart. Therefore, the electric potential around the positive charge decreases as you move away from that positive charge, as shown in **Figure 24.**

Reference point in a system Recall that the gravitational potential energy of a system can be defined as zero at any reference point. In the same way, the electric potential of any point can be defined as zero. No matter what reference point you choose, the value of the electric potential difference from point A to point B always will be the same.

Voltage versus volts Only differences in gravitational potential energy can be measured. The same is true of electric potential; thus, only differences in electric potential are important. The electric potential difference from point A to point B is defined as $\Delta V = V_B - V_A$. Electric potential differences are measured with a voltmeter. Sometimes, the electric potential difference is simply called the voltage. Do not confuse electric potential difference (ΔV) with the unit for volts (V).

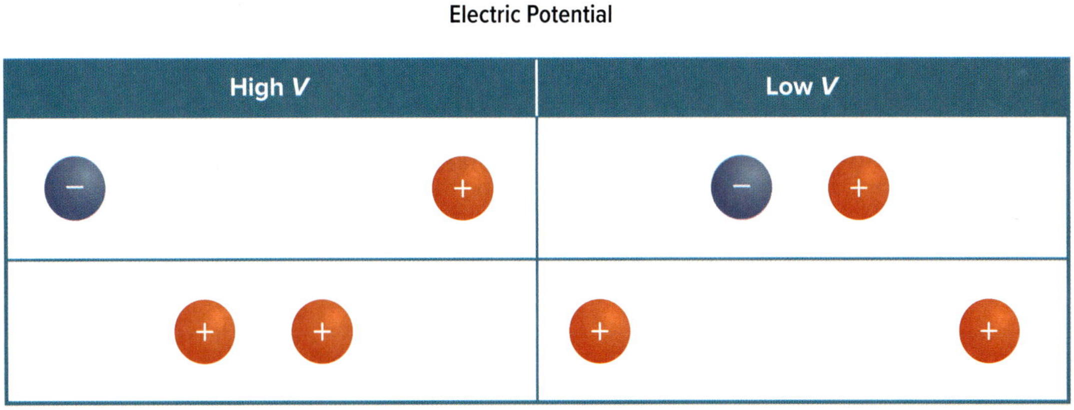

Figure 24 When you push two unlike charges farther apart, the electric potential increases. When you move two like charges farther apart, the electric potential decreases.

STEM CAREER Connection

Power Plant Worker

A power plant worker is responsible for controlling, operating, or maintaining equipment that generates electricity. People doing this job must have a good understanding of how to work safely in situations where they may be exposed to high voltages.

Figure 25 A model of an electric field between two oppositely charged parallel plates is shown. Grass seed is used to model the electric field lines.

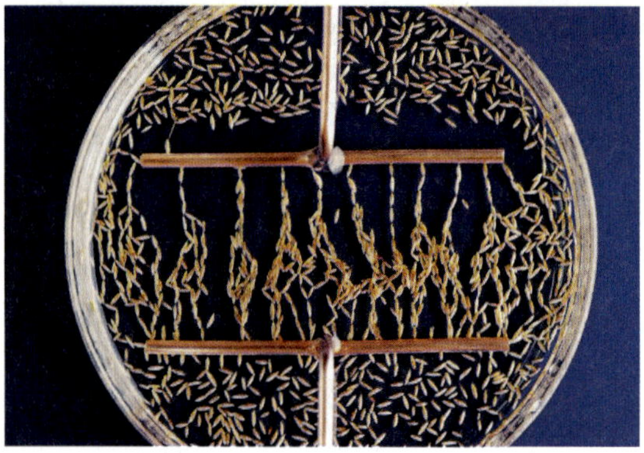

Electric Potential in a Uniform Field

You can produce a uniform electric field by placing two large, flat conducting plates parallel to each other. One plate is charged positively, and the other plate is charged negatively. The magnitude and the direction of the electric field are the same at all points between the plates, except at the edges of the plates, and the electric field points from the positive plate to the negative plate. The pattern formed by the grass seeds pictured in **Figure 25** represents the electric field between parallel plates.

Imagine that you move a positive test charge (q') a distance (d) in the direction opposite the electric field direction. Because the field is uniform, the force from that field on the charge is constant. As a result, you can use the relationship $W_{\text{on } q'} = Fd$ to find the work done on the charge. Then, the electric potential difference, the work done per unit charge, is $\Delta V = \frac{Fd}{q'} = \left(\frac{F}{q'}\right)d$. Remember that electric field intensity is force per unit charge $\left(E = \frac{F}{q'}\right)$. Therefore, you can represent the electric potential difference (ΔV) between two points a distance (d) apart in a uniform field (E) with the following equation.

Electric Potential Difference in a Uniform Field

The potential difference between two locations in a uniform electric field equals the product of electric field intensity and the distance between the locations parallel to the direction of the field.

$$\Delta V = Ed$$

The electric potential is higher near the positively charged plate and lower near the negatively charged plate.

PRACTICE Problems

ADDITIONAL PRACTICE

43. The electric field intensity between two large, charged parallel metal plates is 6000 N/C. The plates are 0.05 m apart. What is the electric potential difference between them?

44. A voltmeter reads 400 V across two charged, parallel plates that are 0.020 m apart. What is the magnitude of the electric field between them?

45. What electric potential difference is between two metal plates that are 0.200 m apart if the electric field between those plates is 2.50×10^3 N/C?

46. When you apply a potential difference of 125 V between two parallel plates, the field between them is 4.25×10^3 N/C. How far apart are the plates?

47. **CHALLENGE** You apply a potential difference of 275 V between two parallel plates that are 0.35 cm apart. How large is the electric field between the plates?

EXAMPLE Problem 4

WORK REQUIRED TO MOVE A PROTON BETWEEN CHARGED PARALLEL PLATES Two charged parallel plates are 1.5 cm apart. The magnitude of the electric field between the plates is 1800 N/C.

a. What is the electric potential difference between the plates?

b. How much work is required to move a proton from the negative plate to the positive plate?

1 ANALYZE AND SKETCH THE PROBLEM

- Draw the plates separated by 1.5 cm.
- Label one plate with positive charges and the other with negative charges.
- Draw uniformly spaced electric field lines from the positive plate to the negative plate.
- Indicate the electric field strength between the plates.
- Place a proton in the electric field.

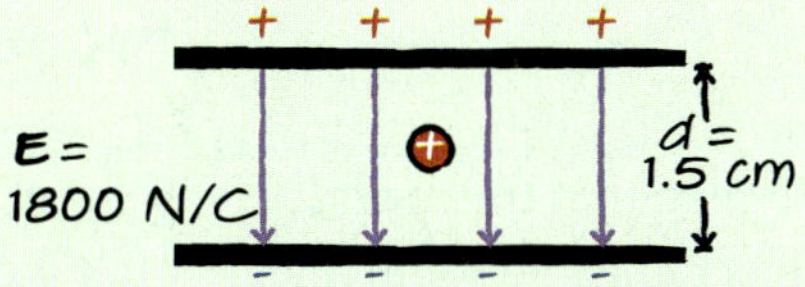

Known	Unknown
$E = 1800$ N/C	$\Delta V = ?$
$d = 1.5$ cm	$W = ?$
$q = 1.602\times10^{-19}$ C	

2 SOLVE FOR THE UNKNOWN

a. $\Delta V = Ed$

$= (1800 \text{ N/C})(0.015 \text{ m})$ — Substitute $E = 1800$ N/C, $d = 0.015$ m.

$= 27 \text{ V}$

b. $\Delta V = \frac{W}{q}$

$W = q\Delta V$

$= (1.602\times10^{-19} \text{ C})(27 \text{ V})$ — Substitute $q = 1.602\times10^{-19}$ C, $\Delta V = 27$ V.

$= 4.3\times10^{-18} \text{ J}$

3 EVALUATE THE ANSWER

- **Are the units correct?** (N/C)(m) = N·m/C = J/C = V. The units work out to be volts. C·V = C(J/C) = J, the unit for work.
- **Does the sign make sense?** You must do positive work to move a positive charge toward a positive plate.
- **Is the magnitude realistic?** When you move such a small charge through a potential difference of less than 100 volts, the work done is small.

PRACTICE Problems

ADDITIONAL PRACTICE

48. What work is done on a 3.0-C charge when you move that charge through a 1.5-V electric potential difference?

49. What is the magnitude of the electric field between the two plates shown in **Figure 26**?

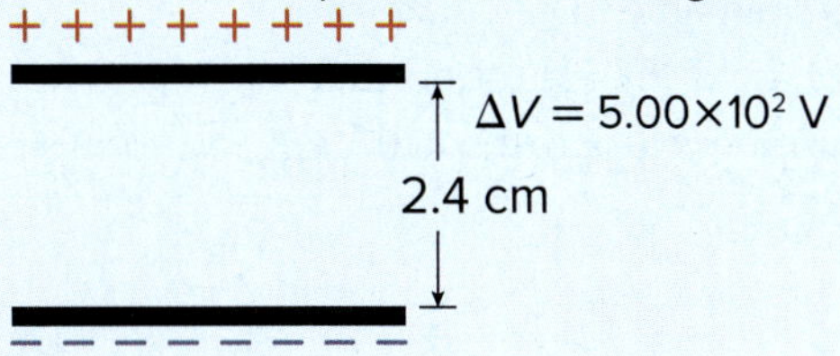

Figure 26

50. An electron in an old television picture tube passes through a potential difference of 18,000 V. How much work is done on the electron as it passes through that potential difference?

51. The electric field in a particle accelerator has a magnitude of 4.5×10^5 N/C. How much work is done to move a proton 25 cm through that field?

52. **CHALLENGE** A 12-V car battery has 1.44×10^6 C of usable charge on one plate when it is fully energized. How much work can this battery do before it needs to be energized again?

Figure 27 The American physicist Robert A. Millikan used an apparatus similar to this one to determine the charge of a single electron in 1909.

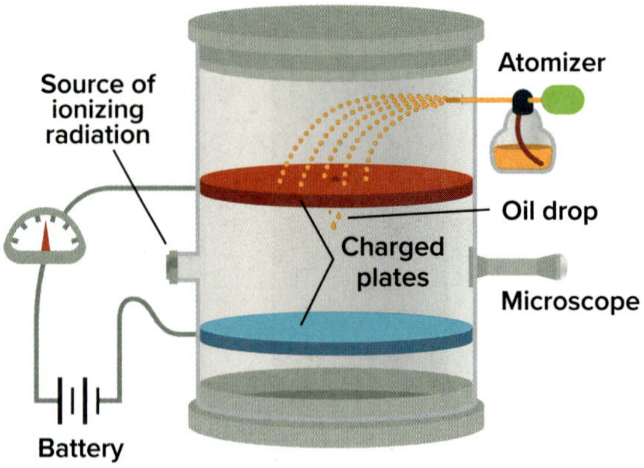

Millikan's Oil-Drop Experiment

You have read that the magnitude of the elementary charge is e = 1.602×10^{-19} C. The net charge on an object must be some integer multiple of this value. In other words, the net charge on an object can be 2e, −5e, 7e, or even −3157e. The net charge cannot be a fractional charge, such as 0.5e, −23.7e, 6.2e, or −31,524.6e. How do we know?

In 1909, Robert Millikan performed an experiment to test whether charge exists in discrete amounts. Millikan was able to measure the magnitude of the elementary charge with his experiment. A diagram of Millikan's apparatus for this experiment is shown in **Figure 27.** Notice that Millikan used two parallel plates to produce a uniform electric field in his apparatus. How did Millikan perform his experiment?

First, Millikan sprayed fine oil drops from an atomizer into the air. These drops were charged by friction as they were sprayed from the atomizer. Earth's gravitational force pulled the drops downward, and a few of those drops entered the hole in the top plate of Millikan's apparatus. Millikan then adjusted the electric field between the plates until he had suspended a negatively charged drop. At this point, the downward force from Earth's gravitational field and the upward force from the electric field were equal in magnitude.

Get It?

Explain why the following net charges are impossible: 0.66e, 1.554e, and 3.504e.

Calculating charge Millikan calculated the magnitude of the electric field (E) from the electric potential difference between the plates $\left(E = \frac{\Delta V}{d}\right)$. Millikan had to make a second measurement to find the weight (mg) of the tiny drop. To find the weight of a drop, Millikan first suspended it. Then, he turned off the electric field and measured the rate of the fall of the drop.

Because of friction with air molecules, the oil drop quickly reached terminal velocity, which was related to the weight of the drop by a complex equation. Millikan was then able to calculate the charge (q) on the drop from the weight of the drop (mg) and the electric field (**E**). He found that the net charge on an oil drop was always some integer multiple of a number close to 1.6×10^{-19} C. In later experiments, others have refined this result to 1.60217662×10^{-19} C.

EXAMPLE Problem 5

FINDING THE CHARGE ON AN OIL DROP In a Millikan oil-drop experiment, a particular oil drop weighs 2.4×10^{-14} N. The parallel plates are separated by a distance of 1.2 cm. When the potential difference between the plates is 450 V, the drop is suspended.

a. What is the net charge on the oil drop?

b. If the upper plate is positive, how many excess electrons are on the oil drop?

1 ANALYZE AND SKETCH THE PROBLEM

- Draw the plates with the oil drop suspended between them.
- Draw and label vectors representing the forces.
- Indicate the potential difference and the distance between the plates.

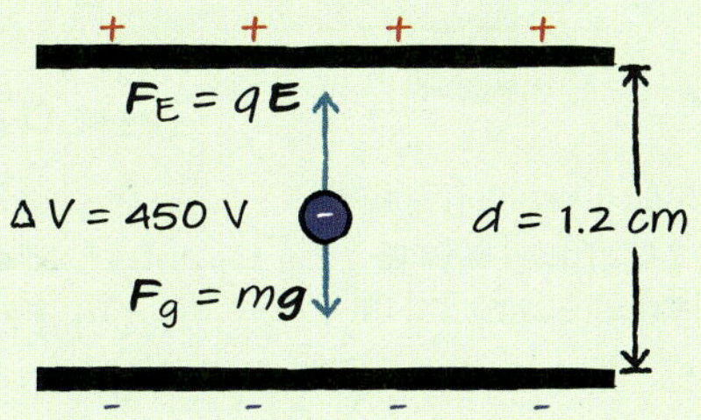

Known	Unknown
$\Delta V = 450$ V	net charge on drop, $q = ?$
$F_g = 2.4\times10^{-14}$ N	number of electrons, $n = ?$
$d = 1.2$ cm	

2 SOLVE FOR THE UNKNOWN

To be suspended, the electric force and gravitational force must be balanced.

$$F_e = F_g$$

$$qE = F_g \quad \text{Substitute } F_e = qE.$$

$$q\frac{(\Delta V)}{d} = F_g \quad \text{Substitute } E = \Delta V/d.$$

Solve for q.

$$q = \frac{F_g d}{\Delta V}$$

$$= \frac{(2.4\times10^{-14}\text{ N})(0.012\text{ m})}{450\text{ V}} \quad \text{Substitute } F_g = 2.4\times10^{-14}\text{ N}, d = 0.012\text{ m}, \Delta V = 450\text{ V}.$$

$$= 6.4\times10^{-19}\text{ C}$$

Solve for the number of electrons on the drop.

$$n = \frac{q}{e}$$

$$= \frac{6.4\times10^{-19}\text{ C}}{1.602\times10^{-19}\text{ C}} \quad \text{Substitute } q = 6.4\times10^{-19}\text{ C}, e = 1.602\times10^{-19}\text{ C}.$$

$$= 4.0$$

3 EVALUATE THE ANSWER

- **Are the units correct?** N·m/V = J/(J/C) = C, the unit for charge.
- **Is the magnitude realistic?** This is a small oil drop. It makes sense that the drop would have a small charge and contain only a few electrons.

PRACTICE Problems

ADDITIONAL PRACTICE

53. A drop is falling in a Millikan oil-drop apparatus with no electric field. What forces are acting on the oil drop, regardless of its acceleration? If the drop is falling at a constant velocity, describe the forces acting on it.

54. An oil drop weighs 1.9×10^{-15} N. You suspend it in an electric field of 6.0×10^{3} N/C. What is the net charge on the drop? How many excess electrons does it carry?

55. An oil drop carries one excess electron and weighs 6.4×10^{-15} N. What electric field strength do you need to suspend the drop so it is motionless?

56. CHALLENGE You suspend a positively charged oil drop that weighs 1.2×10^{-14} N between two parallel plates that are 0.64 cm apart. The potential difference between the plates is 240 V. What is the net charge on the drop? How many electrons is the drop missing?

Conducting Sphere

A

On a conducting sphere, the charge is evenly distributed around the surface.

Hollow Sphere

B

The charges on a hollow sphere are entirely on the outer surface.

Irregular Surface

C

On an irregular conducting surface, the charges are closest together at sharp points.

Figure 28 Charges on a conducting sphere spread far apart to minimize their potential energy.

Electric Fields Near Conductors

Recall that many of the electrons in a conductor are free to move. Consider the charges on the conducting sphere in **Figure 28.** Because these electrons have like charges, they repel each other. They spread far apart in a way that minimizes their potential energy. The result is these charges come to rest on the surface of the conductor. It does not matter if the conducting sphere is solid or hollow. The excess charges move to the outer surface of the conductor.

Closed metal containers What happens if a closed metal container, such as a box, is charged? You can use a voltmeter to measure the electric potential difference between any two points inside the container. You will find that this potential difference is zero no matter which two points you choose inside the container. What are the consequences of this measurement for the electric field inside of the closed, metal container?

Recall that potential difference is equal to the product of the electric field and distance, or $\Delta V = Ed$. Because the potential difference between any two points inside the container is zero, the equation implies that the field is zero everywhere inside a closed, charged metal container.

Cars are a good example of this scenario. A car is a closed metal box that protects passengers from electric fields generated by lightning. On the outside of the conductor, the electric field often is not zero. The field is always perpendicular to the surface of the conductor. This makes the surface an equipotential; the potential difference between any two locations on the surface is zero.

Irregular surfaces The electric field at the surface does depend on the shape of the conductor, as well as on the electric potential difference between it and other objects. Free charges are closer together at the sharp points of a conductor, as indicated in **Figure 28.** Therefore, the field lines are closer together, and the field is stronger. This field can become so strong that when electrons are knocked off of atoms, the electrons and resulting ions are accelerated by the field, causing them to strike other atoms, resulting in more ionization of atoms. This chain reaction produces the pink glow seen inside a gas discharge sphere.

Lightning rods If an electric field is strong enough, when the particles hit other molecules they will produce a stream of ions and electrons that form a plasma, which is a conductor. The result is a spark or, in extreme cases, lightning. In order to protect buildings from lightning, builders install lightning rods. The electric field is strong near the pointed end of a lightning rod. As a result, charges in the clouds spark to the rod, rather than to another point on the building. From the rod, a conductor takes the charges to the ground. A lightning rod safely diverts lightning into the ground and away from the building.

Get It?
Describe how a lightning rod works.

Capacitors

When you lift a book, you increase the gravitational potential energy of the book-Earth system. You can interpret this as storing energy in a gravitational field. In a similar way, you can store energy in an electric field. In 1746, Dutch physician and physicist Pieter Van Musschenbroek invented a device to store electrical energy. In honor of the city in which he worked, he called it a Leyden jar. The modern, much smaller device for storing electrical energy is called a **capacitor.** Manufacturers make capacitors from two conducting plates separated by a thin insulator. They often roll these layers into a cylinder. Capacitors are key components in computers and other electronic devices, as shown in **Figure 29.**

Figure 29 A computer circuit board uses many capacitors in its operation. The gray cylinders, the disk-shaped objects, and the tiny colored tear-drop shaped objects are all capacitors.

Capacitance

Suppose you connect the positive terminal of a 1-V power source to one of the conducting plates of a capacitor and connect the negative terminal of that power source to the other plate of that capacitor. What happens? The power source would produce a potential difference (ΔV) of 1 V between the two plates. This would result in a net positive charge ($+q$) on one plate and a net negative charge of equal magnitude ($-q$) on the other plate.

Now suppose that you increase the voltage of the power source. What happens to the amount of net charge on each plate? **Figure 30** shows the results of an experiment in which the experimenter increases the potential difference across a capacitor's plates from 0 V to 12 V. Notice that the graph of q v. ΔV is a straight line. The slope of the line in a net charge versus potential difference graph is a constant and is called the **capacitance** (C) of the capacitor. You can measure the capacitance of a capacitor in farads (F), where 1 F = 1 C/V.

Capacitance

Capacitance is the ratio of the magnitude of the net charge on one plate of the capacitor to the potential difference across the plates.

$$C = \frac{q}{\Delta V}$$

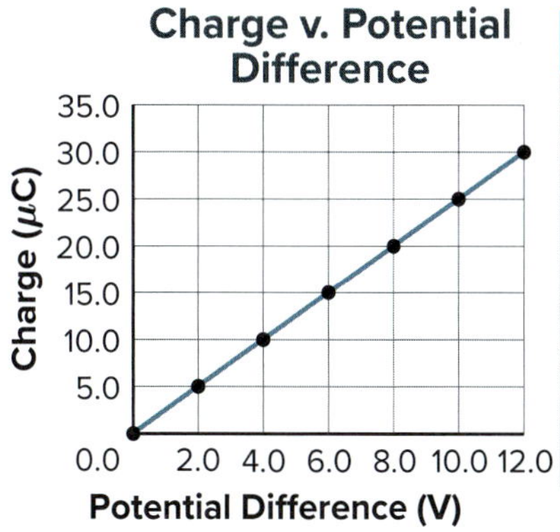

Data Table

Potential Difference (V)	Charge on a Plate (μC)
0.0	0.0
2.0	5.2
4.0	9.7
6.0	15.0
8.0	20.3
10.0	24.7
12.0	30.1

Figure 30 In an experiment, the potential difference results on a charge on each plate. A graph of the data shows the relationship is linear. The slope of this line is the capacitance of the capacitor.

The farad as a unit of measure One farad (F), named after Michael Faraday, is one coulomb per volt (C/V). Just as 1 C is a large amount of net charge, 1 F is also a fairly large capacitance. Most capacitors used in modern electronics have capacitances between 10 picofarads (10×10^{-12} F) and 500 microfarads (500×10^{-6} F). However, the memory capacitors that are used to prevent loss of memory in some computers can have capacitance from 0.5 F to 1.0 F. Note that if the electric potential difference between the plates increases, the net amount of charge on each plate also increases.

EXAMPLE Problem 6

FINDING CAPACITANCE A sphere was connected to the + pole of a 40-V battery while the − pole was connected to Earth. After a period of time, the sphere was charged to 2.4×10^{-6} C. What is the capacitance of the sphere-Earth system?

1 ANALYZE AND SKETCH THE PROBLEM

Draw a sphere above Earth, and label the net charge and potential difference.

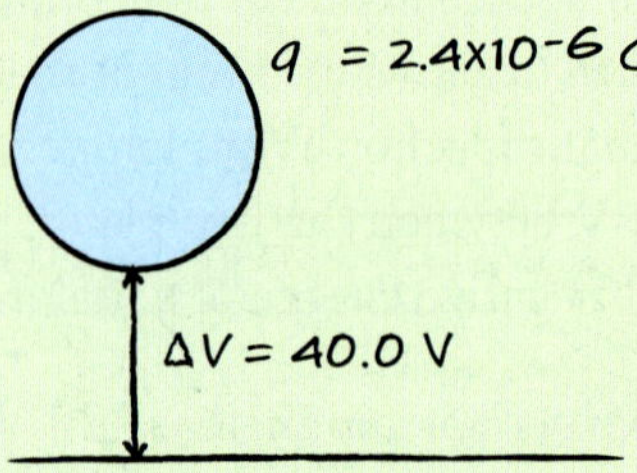

Known	Unknown
$\Delta V = 40.0$ V	$C = ?$
$q = 2.4\times10^{-6}$ C	

2 SOLVE FOR THE UNKNOWN

$$C = \frac{q}{\Delta V}$$

$$= \frac{2.4\times10^{-6}\text{ C}}{40.0\text{ V}}$$ Substitute $\Delta V = 40.0$ V, $q = 2.4\times10^{-6}$ C.

$$= 6.0\times10^{-8}\text{ F}$$

$$= 0.060\ \mu\text{F}$$

3 EVALUATE THE ANSWER

- **Are the units correct?** C/V = F. The units are farads.
- **Is the magnitude realistic?** The amount of net charge stored on the sphere equals the capacitance multiplied by 40.0 V.

PRACTICE Problems

ADDITIONAL PRACTICE

57. A 27-μF capacitor has an electric potential difference of 45 V across it. What is the amount the net charge on the positively charged plate of the capacitor?

58. Suppose you connect both a 3.3-μF and a 6.8-μF capacitor across a 24-V electric potential difference. Which capacitor has the greater net charge on its positively charged plate, and what is its magnitude?

59. You later find that the magnitude of net charge on each of the plates for each of the capacitors in the previous problem is 3.5×10^{-4} C. Which capacitor has the larger potential difference across it? What is that potential difference?

60. Suppose that you apply an electric potential difference of 6.0 V across a 2.2-μF capacitor. What does the magnitude of the net charge on one plate need to be to increase the electric potential difference to 15.0 V?

61. A sphere is charged by a 12 V battery and suspended above Earth as shown in **Figure 31.** What is the net charge on the sphere?

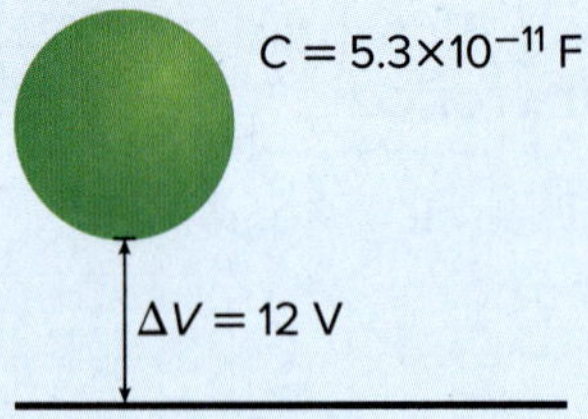

Figure 31

62. CHALLENGE You increase the potential difference across a capacitor from 12.0 V to 14.5 V. As a result, the magnitude of the net charge on each plate increases by 2.5×10^{-5} C. What is the capacitance of the capacitor?

PHYSICS Challenge

The plates of a capacitor attract each other because they have equal and opposite net charge. A capacitor consisting of two parallel plates that are separated by a distance (d) has capacitance (C).

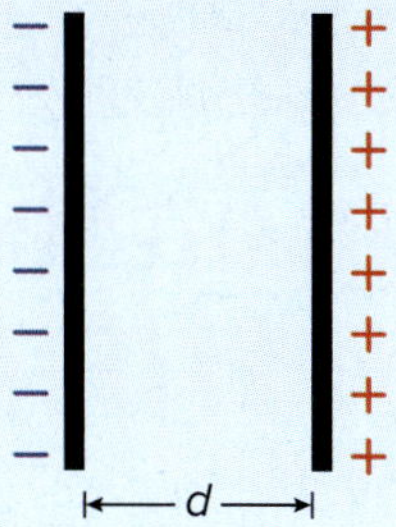

1. Derive an expression for the force between the two plates when the magnitude of net charge on one of the capacitor's plates is q.
2. The electrostatic force between the plates of a 22-μF capacitor is 2.0 N. What is the magnitude of the net charge on one plate if the separation between those plates is 1.5 mm?

Varieties of capacitors Capacitors have many shapes and sizes. Some are large enough to fill whole rooms and can store enough electrical energy to create artificial lightning or power giant lasers that release thousands of joules of energy in a few billionths of a second. Capacitors are also used in computers, televisions, and digital cameras. These capacitors store much less energy than those used to make artificial lightning, but they can still be dangerous. Charge can remain for hours after the device is turned off. This is why many electronic devices carry warnings not to open the case.

Check Your Progress

63. **Electric Potential** Suppose a friend asks you to explain how electric potential relates to potential energy. Write a brief explanation that you could use to explain this concept to a friend who does not understand the relationship between the two concepts.
64. **Potential Difference** Contrast electric potential energy and electric potential difference.
65. **Electric Field and Potential Difference** Show that a volt per meter is the same as a newton per coulomb.
66. **Millikan Experiment** When the net charge on an oil drop suspended in a Millikan apparatus is changed, the drop begins to fall. How should you adjust the potential difference between the conducting plates to bring the drop back into balance?
67. **Charge and Potential Difference** In the previous problem, if changing the potential difference between the plates has no effect on the falling drop, what does this tell you about the new net charge on the drop?
68. **Capacitance** What is the magnitude of net charge on each conductor plate of a 0.47-μF capacitor when a potential difference of 12 V is applied across that capacitor?
69. **Charge Sharing** If you touch a large, positively charged, conducting sphere with a small, negatively charged, conducting sphere, what can be said about the
    a. potentials of the two spheres;
    b. charges on the two spheres?
70. **Critical Thinking** Explain how the charge in **Figure 32** continues to build up on the metal dome of a Van de Graaff generator. In particular, why isn't charge repelled back onto the belt at point B?

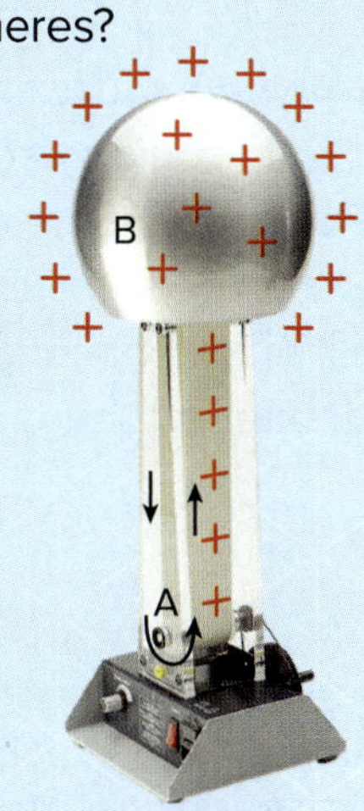

Figure 32

LEARNSMART® Go online to follow your personalized learning path to review, practice, and reinforce your understanding.

Cloud on the Horizon

Since the dawn of human history, people have looked into the night sky and wondered what lies beyond our world. As the millennia passed and our scientific knowledge grew, we learned more about our solar system, but our knowledge about space outside our Sun's reach was still limited. Now technology, both older and newer, is bringing us news. To scientists' surprise, the area directly outside our solar system is magnetic.

NASA's Interstellar Boundary Explorer (IBEX) has collected data about the Local Interstellar Cloud's magnetic field.

The Local Interstellar Cloud

Solar winds blowing outward from the Sun expand the Sun's magnetic field into a bubble called the heliosphere. Our solar system is inside the heliosphere. Since 1977, the two Voyager probes launched by the National Aeronautics and Space Administration (NASA) have been making their way from Earth to the edges of the solar system. As they travel, they return data to Earth. Recent data shows that the area outside the heliosphere is a mass of extremely hot hydrogen and helium atoms that scientists call the Local Interstellar Cloud, or "Local Fluff." Our solar system is moving through this cloud.

Scientists have been puzzled about how the Local Interstellar Cloud could exist, because it is surrounded by a high-temperature, high-pressure bubble of gas formed by a cluster of exploding supernovas. This bubble of gas should have crushed the Local Interstellar Cloud. How does it remain intact? Data from the Voyager probes shows that the Local Fluff is able to resist pressure from the hot gas bubble due to its extremely strong magnetic field.

In 2008, NASA launched an Earth-orbiting satellite called the Interstellar Boundary Explorer (IBEX), which collects data about particles streaming into the solar system from space. The data provides information about the boundaries between the solar system and the surrounding galaxy. Data from IBEX has shown the direction and strength of the Local Fluff's magnetic field, which pushes against and shapes the heliosphere. IBEX data also has shown that particles are entering the heliosphere from a different direction than in decades past.

USE A MODEL TO ILLUSTRATE

Work with a partner to create a 2D or 3D model of the Local Interstellar Cloud, including our solar system within the heliosphere. Display your model in the classroom.

CALIFORNIA CONNECTION

Use an online search engine to find research that a state college or university is doing on the Local Interstellar Cloud. Choose one study that interests you and write a short summary of its findings.

MODULE 18
STUDY GUIDE

GO ONLINE to study with your Science Notebook.

Lesson 1 ELECTRIC CHARGE

- There are two kinds of electric charge, positive and negative. The interactions of these charges explain the attraction and repulsion you observed using strips of tape.
- Electric charge is not created or destroyed; it is conserved. Charging is the separation, not creation, of electric charges.
- Charges added to one part of an insulator remain on that part. Charges added to a conductor spread throughout the conductor.

- electrostatics
- neutral
- insulator
- conductor

Lesson 2 ELECTROSTATIC FORCE

- Electrostatic force acts over a distance and decreases as the distance between charges increases.
- Charging by conduction charges a neutral object by touching that object with a charged object. Charging by induction charges a neutral object by bringing a charged object near the neutral object.
- Coulomb's law states that the force between two charged particles varies directly with the product of their charges and inversely with the square of the distance between them. It is used to find the magnitude of the force two charged particles exert on each other.

$$F = K\frac{q_A q_B}{r_{AB}^{2}}$$

- electroscope
- charging by conduction
- charging by induction
- grounding
- Coulomb's law
- coulomb
- elementary charge

Lesson 3 MEASURING ELECTRIC FIELDS

- An electric field exists around any charged object. The field produces forces on other charged objects.
- Quantities relating to charge, electric fields, and forces are related and can be calculated using these formulas:

$$E = \frac{F}{q'} \qquad E = \frac{Kqq'}{r^2q'} = \frac{Kq}{r^2}$$

- Electric field lines provide a pictorial model of the electric field. They are directed away from positive charges and toward negative charges. They never cross, and their density is related to the strength of the field. The direction of the electric field is the direction of the force on a tiny, positive test charge.

- electric field
- electric field line

Lesson 4 APPLICATIONS OF ELECTRIC FIELDS

- Electric potential difference is the change in potential energy per unit charge in an electric field.
- Potential difference is related to the work required to move a charge and is represented by the following equation:

$$\Delta V = \frac{W}{q'}$$

- Capacitors are used to store electrical energy. A capacitor consists of two conducting plates separated by an insulator.

- electric potential difference
- volt
- equipotential
- capacitor
- capacitance

THREE-DIMENSIONAL THINKING

Module Wrap-Up

REVISIT THE PHENOMENON

Why does the rod cause the water to bend?

CER Claim, Evidence, Reasoning

Explain your Reasoning Revisit the claim you made when you encountered the phenomenon. Summarize the evidence you gathered from your investigations and research and finalize your Summary Table. Does your evidence support your claim? If not, revise your claim. Explain why your evidence supports your claim.

STEM UNIT PROJECT

Now that you've completed the module, revisit your STEM unit project. You will summarize your evidence and apply it to the project.

GO FURTHER

SEP Data Analysis Lab

How do electric fields compare?

Point charges placed at various locations in an electric field are summarized below.

Point Charge	Charge (nC)	Electric Field (N/C)
A	3	70
B	5	600
C	3	20
D	6	35
E	8	10

CER Analyze and Interpret Data

1. **Claim** Rank the point charges according to the magnitude of the electric force experienced, from greatest to least. Specifically indicate any ties.
2. **Evidence and Reasoning** Explain how you decided your rankings.

MODULE 19

ELECTRIC CURRENT AND CIRCUITS

NASA's Earth Observatory/NOAA/DoD

MODULE 19

ELECTRIC CURRENT AND CIRCUITS

ENCOUNTER THE PHENOMENON

How does energy get from the power plant to the lights in your home?

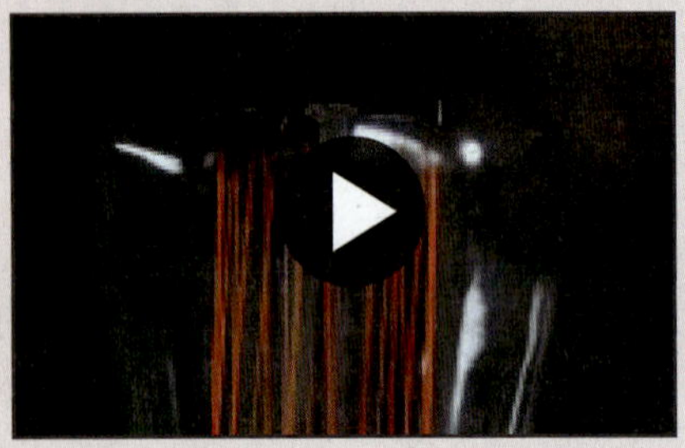

GO ONLINE to watch a video about electric current.

SEP Ask Questions

Do you have other questions about the phenomenon? If so, add them to the driving question board.

CER Claim, Evidence, Reasoning

Make Your Claim Use your CER chart to make a claim about how energy gets from the power plant to the lights in your home. Explain your reasoning.

Collect Evidence Use the lessons in this module to collect evidence to support your claim. Record your evidence as you move through the module.

Explain Your Reasoning You will revisit your claim and explain your reasoning at the end of the module.

GO ONLINE to access your CER chart and explore resources that can help you collect evidence.

LESSON 2: Explore & Explain: Electrical Energy, Resistance, and Power

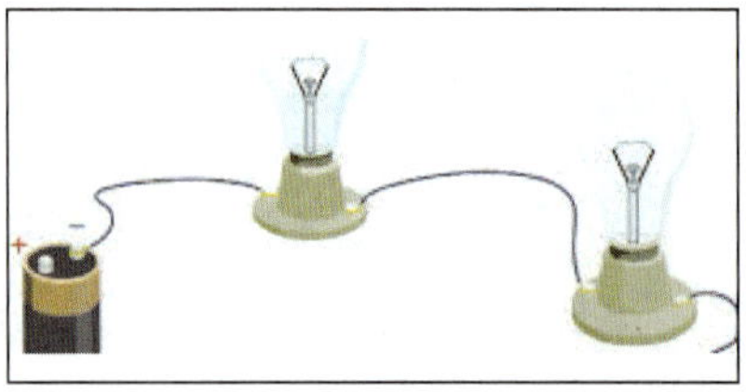

LESSON 3: Explore & Explain: Series Circuits

Smithsonian

Additional Resources

LESSON 1
CURRENT AND CIRCUITS

FOCUS QUESTION

What is an electric circuit?

Producing Electric Current

Many appliances use electricity, such as hair dryers, MP3 players, stoves, and refrigerators. In this chapter, you will learn more about such devices and electrical energy. Recall that when two conducting spheres touch, charges flow from the sphere at a higher potential to the one at a lower potential. The flow continues until there is no potential difference between the two spheres. A flow of charged particles is an **electric current.**

The diagrams in **Figure 1** show charges flowing through a conductive material. In the diagram on the left, two conductors, A and B, are connected by conductor C. Positive charges flow from the higher potential of B to A through C. The electric current stops when the potential difference between A, B, and C is zero. **Conventional current** is the direction in which a positive test charge moves. However, it usually is the negative charges (electrons) that flow. The flow of electrons and the direction of conventional current are in opposite directions.

Figure 1 Positive charges flow from the higher potential at B through the conductive wire C to A, which has a lower potential than B. When the potential difference between B and A is zero, the flow stops. The flow continues in the diagram on the right because a charge pump maintains the potential difference between conductors A and B.

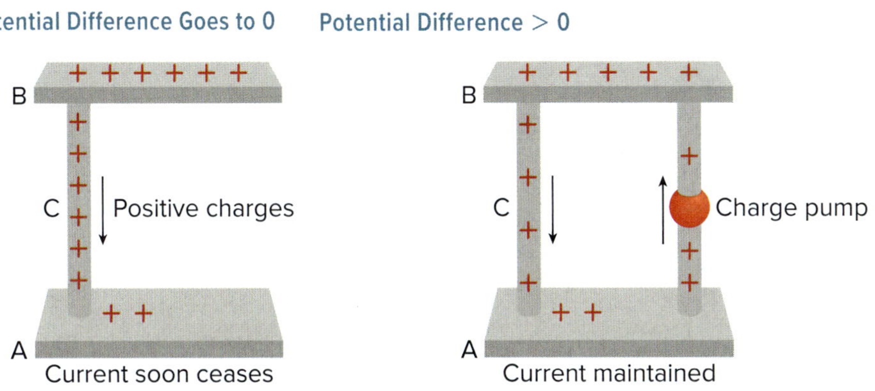

3D THINKING | DCI Disciplinary Core Ideas | CCC Crosscutting Concepts | SEP Science & Engineering Practices

COLLECT EVIDENCE

Use your Science Journal to record the evidence you collect as you complete the readings and activities in this lesson.

INVESTIGATE

GO ONLINE to find these activities and more resources.

Applying Practices: Develop and Use Models for Energy

HS-PS3-2. Develop and use models to illustrate that energy at the macroscopic scale can be accounted for as either motions of particles or energy stored in fields.

Consider **Figure 1** again. You could maintain the electric potential difference between B and A by pumping charged particles from A back to B, as shown on the right in **Figure 1.** Because the pump increases the electric potential energy of the charges, it requires an external energy source. One familiar source, a voltaic or galvanic cell (a common dry cell), transforms chemical energy to electrical energy. Several galvanic cells connected together are called a **battery.** Electrical energy may be stored in a battery.

Electric Circuits

The charges in the diagram on the right in **Figure 1** move around a closed loop, cycling from the pump to B, through C, to A, and back to the pump. Any closed loop or conducting path allowing electric charges to flow is called an **electric circuit.** A circuit includes a charge pump, which increases the electric potential energy of the charges flowing from A to B, and a device that reduces the electric potential energy of the charges flowing from B to A. Thus, these different manifestations of electrical energy can be described as a combination of energy associated with the motion of charged particles in an electric current and energy associated with the configuration, or relative position, of the particles. When the charges flow through the device that reduces the electric potential energy ($q\Delta V$), the energy is transformed into some other form of energy. For example, electrical energy transmitted by electric currents is transformed into kinetic energy by a motor, to light energy by a lamp, and to thermal energy by a heater.

Comparing water flow and current A charge pump creates the flow of charged particles that make up a current. Consider a generator driven by a waterwheel, such as the one shown in **Figure 2.** The water falls and rotates the waterwheel and the generator. Thus, the kinetic energy of the water is transformed to electrical energy by the generator. The generator, like the charge pump, increases the electric potential difference (ΔV) between its two wire connections, which gives energy ($q\Delta V$) to each charge (q) flowing through the generator. This energy comes from the transformation of kinetic energy of the water.

Energy Transformations

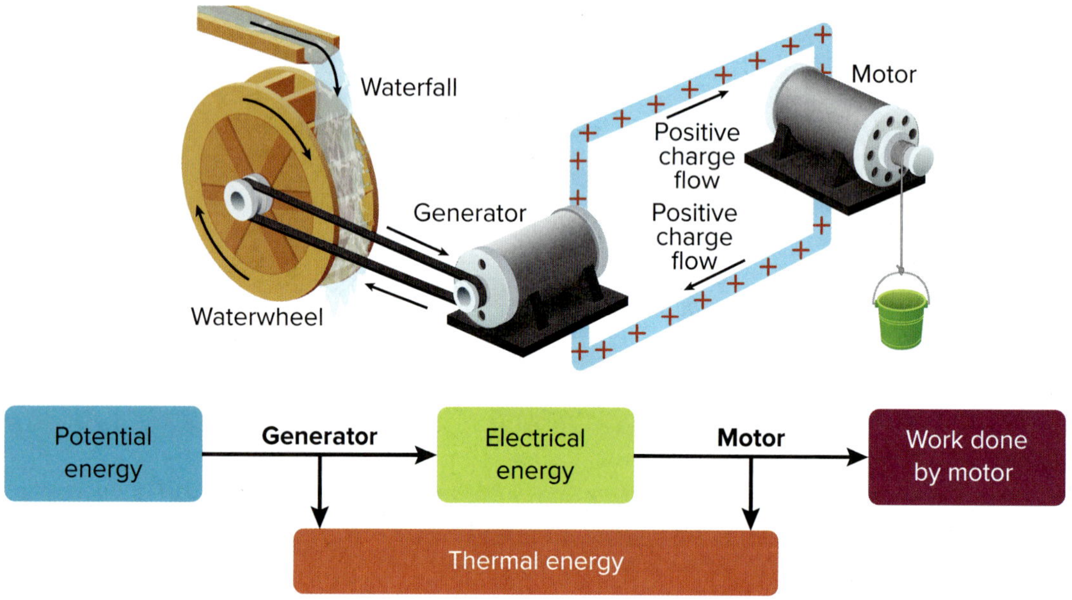

Figure 2 The gravitational potential energy of the water is transformed into kinetic energy, then to electrical energy and thermal energy. Energy transformations are not 100 percent efficient. Thermal energy is produced by the splashing water, friction, and electric resistance.

Not all the water's kinetic energy, however, is transformed to electrical energy, as shown in the flowchart in **Figure 2** on the previous page. If the generator attached to the waterwheel is connected to a motor, the charges in the wire flow into the motor. The flow of charges continues through the circuit back to the generator. The motor transforms electrical energy to kinetic energy.

Conservation of charge Charges cannot be created or destroyed, but they can be separated. Thus, the total amount of charge—the number of negative electrons and positive ions—in the circuit does not change. If one coulomb flows through the generator in 1 s, then one coulomb also will flow through the motor in 1 s. Thus, charge is a conserved quantity. Energy also is conserved. The change in electrical energy (ΔE) equals $q\Delta V$. Because q is conserved, the net change in potential energy of the charges going completely around the circuit must be zero. The increase in potential difference produced by the generator equals the decrease in potential difference across the motor.

If the potential difference between two wires is 120 V, the waterwheel and the generator must do 120 J of work on each coulomb of charge that is delivered. Every coulomb of charge moving through the motor delivers 120 J of energy to the motor.

Get It?

Draw a diagram of an electric circuit, and explain how energy is conserved in it.

Rates of Charge Flow and Energy Transfer

Power, which is measured in watts (W), is the rate at which energy is transferred or transformed, as noted in **Figure 3.** If a generator transfers 1 J of kinetic energy to electrical energy each second, it is transferring energy at the rate of 1 J/s, or 1 W. The energy carried by an electric current depends on the charge transferred (q) and the potential difference across which it moves (ΔV). Thus, $E = q\Delta V$. Recall that the unit for the quantity of electric charge is the coulomb. The rate of flow of electric charge $\frac{q}{t}$ is called electric current and is measured in coulombs per second. Electric current is represented by I, so $I = \frac{q}{t}$. A flow of electric charge equal to one coulomb per second (1 C/s) is called an **ampere** (A).

Figure 3 Some appliances, such as hair dryers, are rated in watts. The rating, 1875 W, indicates the rate at which energy changes. In the hair dryer, electrical energy is transformed into thermal energy and kinetic energy.

The energy carried by an electric current is related to the voltage, $E = q\Delta V$. Since current $\left(I = \frac{q}{t}\right)$ is the rate of charge flow, multiplying voltage and current yields the power $\left(P = \frac{E}{t}\right)$ of an electrical device. To derive a useful form of the equation for the power delivered to an electrical device, you can use $P = \frac{E}{t}$ and substitute $E = q\Delta V$ and $q = It$.

POWER

Power is equal to the current times the potential difference.

$$P = I\Delta V$$

If the current through the motor in **Figure 2** is 3.0 A and the potential difference is 120 V, the power in the motor is calculated using the expression $P = I\Delta V = (3.0\ \text{C/s})(120\ \text{J/C}) = 360\ \text{J/s}$, which is 360 W.

EXAMPLE Problem 1

ELECTRIC POWER AND ENERGY A 6.0-V battery delivers a 0.50-A current to an electric motor connected across its terminals.

a. What power is delivered to the motor?

b. If the motor runs for 5.0 min, how much electrical energy is delivered?

1 ANALYZE AND SKETCH THE PROBLEM

- Draw a circuit showing the positive terminal of a battery connected to a motor and the return wire from the motor connected to the negative terminal of the battery.
- Show the direction of conventional current.

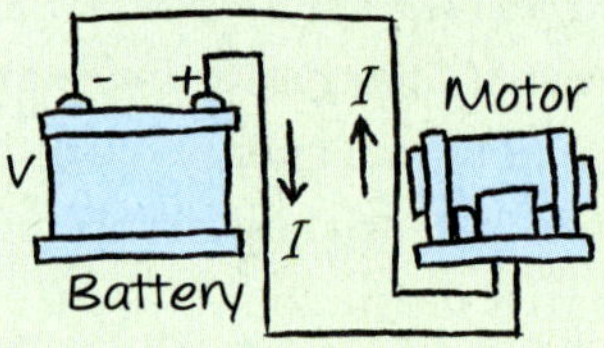

Known	Unknown
$\Delta V = 6.0$ V	$P = ?$
$I = 0.50$ A	$E = ?$
$t = 5.0$ min	

2 SOLVE FOR THE UNKNOWN

a. Use $P = I\Delta V$ to find the power.

$$P = I\Delta V$$

$$P = (0.50\text{ A})(6.0\text{ V}) \quad \text{Substitute } I = 0.50\text{ A}, \Delta V = 6.0\text{ V.}$$

$$= 3.0\text{ W}$$

b. You learned that $P = \frac{E}{t}$. Solve for E to find the energy.

$$E = Pt$$

$$= (3.0\text{ W})(5.0\text{ min}) \quad \text{Substitute } P = 3.0\text{ W}, t = 5.0\text{ min.}$$

$$= (3.0\text{ J/s})(5.0\text{ min})\left(\frac{60\text{ s}}{1\text{ min}}\right)$$

$$= 9.0\times10^{2}\text{ J}$$

3 EVALUATE THE ANSWER

- **Are the units correct?** Power is measured in watts, and energy is measured in joules.
- **Is the magnitude realistic?** With relatively low voltage and current, a few watts of power is reasonable.

PRACTICE Problems

ADDITIONAL PRACTICE

1. A car battery causes a current through a lamp and produces 12 V across it as shown in **Figure 4.** What is the power used by the lamp?

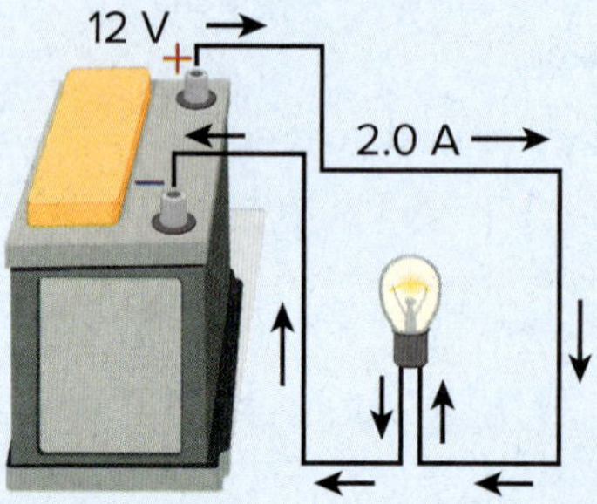

Figure 4

2. What is the current through a 75-W lightbulb that is connected to a 125-V outlet?

3. The current through a lightbulb connected across the terminals of a 125-V outlet is 0.50 A. At what rate does the bulb transform electrical energy to light? (Assume 100 percent efficiency.)

4. The current through the starter motor of a car is 210 A. If the battery maintains 12 V across the motor, how much electrical energy is delivered to the starter in 10.0 s?

5. A 75-V generator supplies 3.0 kW of power. How much current can the generator deliver?

6. A flashlight bulb is rated at 0.90 W. If the lightbulb produces a potential drop of 3.0 V, how much current goes through it?

7. **CHALLENGE** A circuit is changed so the potential difference across a motor doubles and the current through the lightbulb triples. How does this change the motor's power?

Diagramming Circuits

A simple circuit can be described in words. It also can be depicted by photographs or artists' drawings of the parts. You might have noticed, however, that circuits are often diagrammed with simple symbols. Such a diagram is called a circuit schematic. An artist's drawing and a schematic of the same circuit are shown in **Figure 5** at the bottom of this page.

Some of the symbols used in circuit schematics are shown in **Figure 6** on the next page. Notice in both the drawing and the schematic that the electric charge is shown flowing out of the positive terminal of the battery. To draw schematic diagrams, use the problem-solving strategy on the next page.

Get It?

Explain how a schematic of an electric circuit compares to an artist's drawing of the same circuit.

An ammeter measures current, and a voltmeter measures potential differences. Often these devices are combined in a single device called a multimeter. Each instrument has two terminals, usually labeled + and −. A voltmeter measures the potential difference between any two points in a circuit. When connecting the voltmeter in a circuit, always connect the + terminal to the end of the circuit component that is closer to the positive terminal of the battery, and connect the − terminal to the other side of the component. The ammeter should be connected so that the current goes *through* the ammeter.

Get It?

Identify what an ammeter and a voltmeter measure.

Real-World Physics

RESISTANCE The resistance of an operating 100-W lightbulb is about 140 Ω. When the lightbulb is turned off and at room temperature, its resistance is only about 10 Ω. This is because of the great difference between room temperature and the lightbulb's operating temperature, and the temperature dependence of the resistance of the tungsten filament.

Figure 5 Simple circuit diagrams can be represented pictorially and schematically.

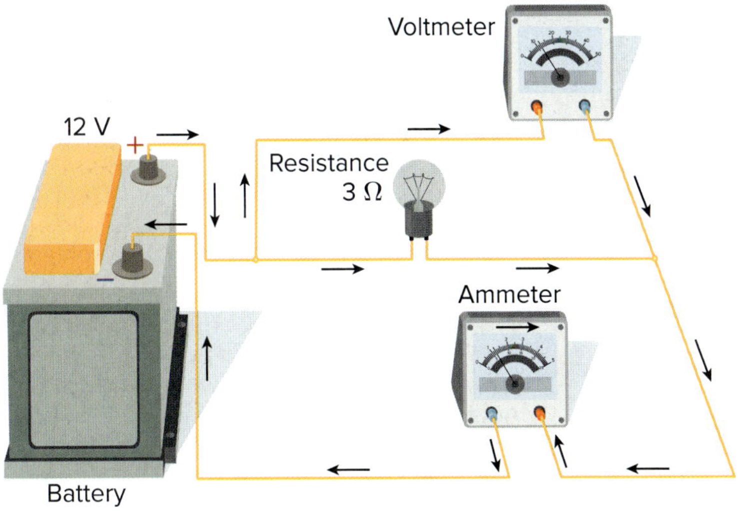

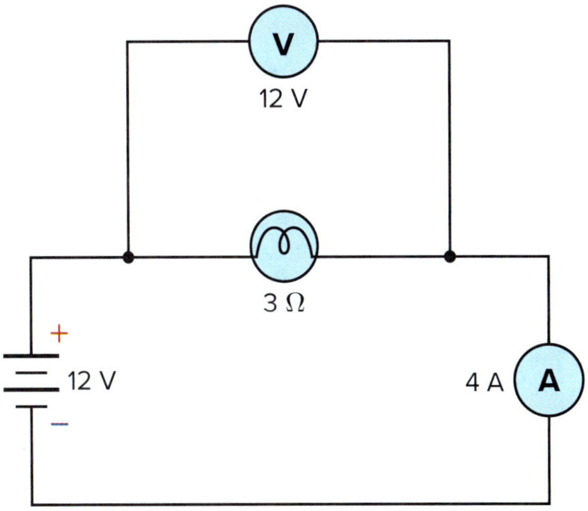

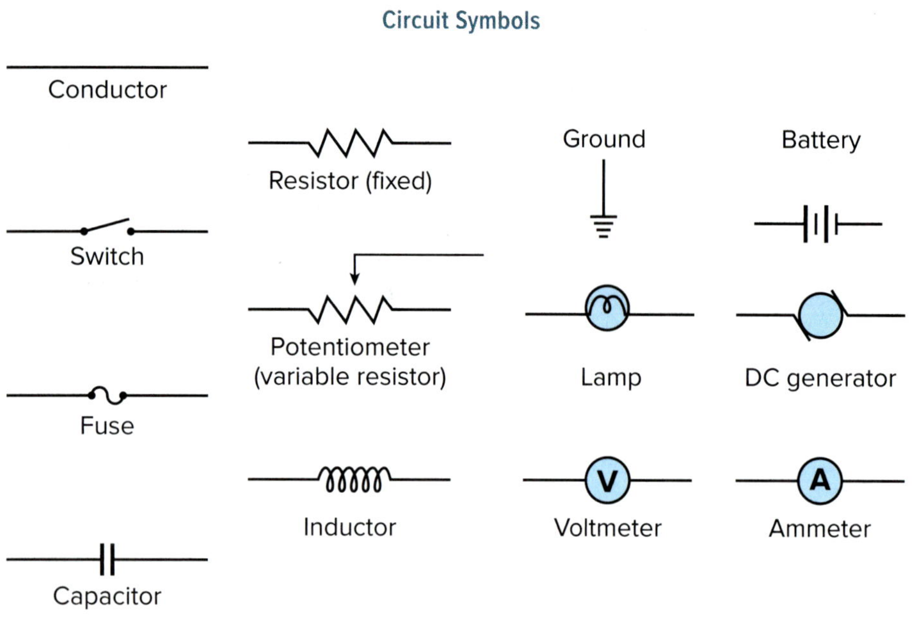

Figure 6 Electric circuit diagrams are commonly drawn using these symbols.

PROBLEM-SOLVING STRATEGIES

DRAWING SCHEMATIC DIAGRAMS

Follow these steps when drawing schematic diagrams.

1. Draw the symbol for the battery or other source of electrical energy, such as a generator, on the left side of the page. Put the positive terminal on top.
2. Draw a wire coming out of the positive terminal. When you reach a resistor or other device, draw the symbol for it.
3. If you reach a point where there are two current paths, such as at a voltmeter, draw a 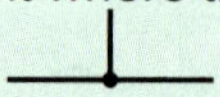 in the diagram. Follow one path until the two current paths join again. Then draw the second path.
4. Follow the current path until you reach the negative terminal of the battery.
5. Check your work to make sure that you have included all parts and that there are complete paths for the current to follow.

PRACTICE Problems

ADDITIONAL PRACTICE

8. Draw a circuit diagram to include a 60.0-V battery, an ammeter, and a resistance of 12.5 Ω in series. Draw arrows on your diagram to indicate the direction of the current.
9. Draw a circuit diagram showing a 4.5-V battery, a resistor, and an ammeter that reads 85 mA. Show the direction of the current using conventional rules, and indicate the positive terminal of the battery.
10. Add a voltmeter to measure the potential difference across the resistors in the previous two problems. Label the voltmeters.
11. Draw a circuit using a battery, a lamp, a potentiometer to adjust the lamp's brightness, and an on-off switch.
12. **CHALLENGE** Repeat the previous problem, adding an ammeter and a voltmeter across the lamp.

Resistance and Ohm's Law

Georg Ohm (1787–1854) studied the relationship between current and potential difference. Ohm's law states that current through a wire is directly proportional to the potential difference between its ends. Suppose a battery causes a potential difference between its two terminals. If they are connected with a copper rod, a large current is created through the rod and the battery. On the other hand, connecting the battery terminals with a glass rod allows almost no current. The measure of how strongly an object or a material impedes current produced by a potential difference is **resistance.** **Table 1** lists some of the factors that impact resistance. Resistance is measured by placing a potential difference across a conductor and dividing the voltage by the current. The resistance (R) is defined as the ratio of electric potential difference (ΔV) to the current (I).

RESISTANCE

Resistance is defined as potential difference divided by current.

$$R \equiv \frac{\Delta V}{I}$$

Table 1 Changing Resistance

Factor	How Resistance Changes	Example	
Length	Resistance increases as length increases.	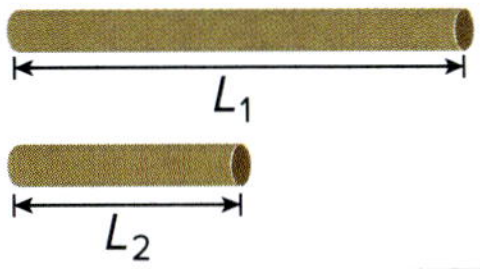	$R_{L1} > R_{L2}$
Cross-sectional area	Resistance increases as the cross-sectional area decreases.	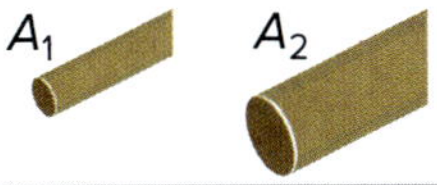	$R_{A1} > R_{A2}$
Temperature	Resistance usually increases as temperature increases.	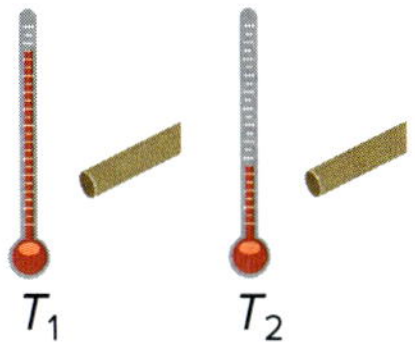	$R_{T1} > R_{T2}$
Material	Keeping length, cross-sectional area, and temperature constant, resistance varies with the material used.	silver, copper, gold, aluminum, iron, platinum R increases →	

SCIENCE USAGE v. COMMON USAGE

Resistance

Science usage: the measure of how much an object or a material impedes the flow of electric current induced by a voltage

The resistance of silver is lower than the resistance of copper.

Common usage: the act of opposing

Customer resistance to the menu changes made the chefs reconsider the changes.

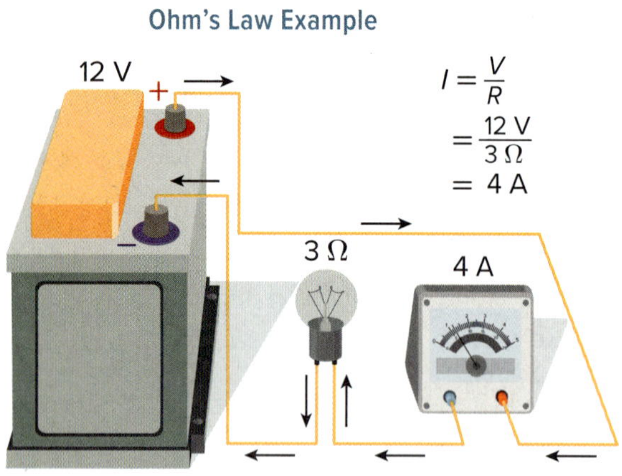

Figure 7 The ammeter and the equation give the same value for the current through the circuit. For a simple circuit with a 3-Ω resistance and a 12-V battery, there is a 4-A current throughout that circuit. The resistance of the ammeter is insignificant.

The resistance of the conductor (R) is measured in ohms. One ohm (1 Ω) is the resistance permitting an electric charge of 1 A to flow when a potential difference of 1 V is applied across the resistance. A simple circuit relating resistance, current, and potential difference is shown in **Figure 7.** The circuit is completed by a connection to an ammeter, which is a device that measures current.

Get It?
Define *resistance* in your own words.

The ohm The unit for resistance is named for German scientist Georg Ohm, who found that the ratio of potential difference to current is constant for a given conductor. The resistance for most conductors does not vary as the magnitude or the direction of the potential difference applied to it changes. A device having constant resistance independent of the potential difference obeys Ohm's law.

Get It?
Explain which devices obey Ohm's law.

Resistivity Most metallic conductors obey Ohm's law, at least over a limited range of voltages. Many important devices, however, such as a phone or calculator, contain transistors and diodes, which do not obey Ohm's law.

Wires used to connect electrical devices have low resistance. A 1-m length of a typical wire used in physics labs has a resistance of about 0.03 Ω. Wires used in home wiring offer as little as 0.004 Ω of resistance for each meter of length.

Because wires have so little resistance, there is almost no potential drop across them. To produce greater potential drops, a large resistance concentrated into a small volume is necessary. A **resistor** is a device designed to have a specific resistance. Resistors may be made of carbon, semiconductors, or wires that are long and thin.

Controlling current There are two ways to control the current in a circuit. Because $I = \frac{\Delta V}{R}$, I can be changed by varying ΔV, R, or both. **Figure 8** shows a simple circuit. When ΔV is 6 V and R is 30 Ω, the current is 0.2 A. According to Ohm's law, the greater the voltage placed across a resistor, the larger the current passing through it. If the potential difference across a resistor is halved, the current through that resistor is also halved. In the middle circuit in **Figure 8,** the voltage applied across the resistor is reduced from 6 V to 3 V to reduce the current to 0.1 A. A second way to reduce the current to 0.1 A is to replace the 30-Ω resistor with a 60-Ω resistor, as shown in the bottom circuit in **Figure 8.**

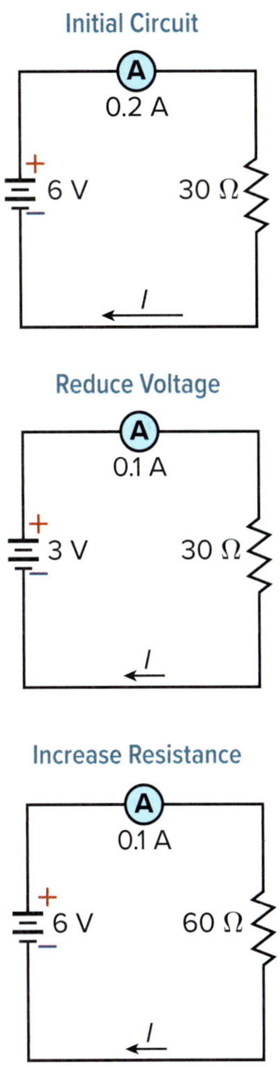

Figure 8 There are two ways to decrease the current through a simple circuit. The voltage can be reduced, or the resistance can be increased.

Get It?

List two ways to control the current flowing through a simple circuit.

Variable resistors Resistors often are used to control the current in circuits or parts of circuits. Sometimes, a smooth, continuous variation of the current is desired. For example, the speed control on some electric motors allows continuous, rather than step-by-step, changes in the rotation of the motor. To achieve this kind of control, a variable resistor, called a potentiometer, is used. A circuit containing a potentiometer is shown in **Figure 9.**

Some variable resistors consist of a coil of resistance wire and a sliding contact point. Moving the contact point to various positions along the coil varies the amount of wire in the circuit. As more wire is placed in the circuit, the resistance of the circuit increases; thus, the current changes in accordance with the equation $I = \frac{\Delta V}{R}$. In this way, the speed of a motor can be adjusted from fast, with little wire in the circuit, to slow, with a lot of wire in the circuit.

HEALTH Connection The human body acts as a variable resistor. Dry skin's resistance is high enough to keep currents that are produced by small and moderate voltages low. If skin becomes wet, however, its resistance is lowered, and an electric current can rise to dangerous levels. A current as low as 1 mA can be felt as a mild shock, while currents of 15 mA can cause loss of muscle control, and currents of 100 mA can cause death. For safety reasons you should be careful with any electric current, even from a lantern battery.

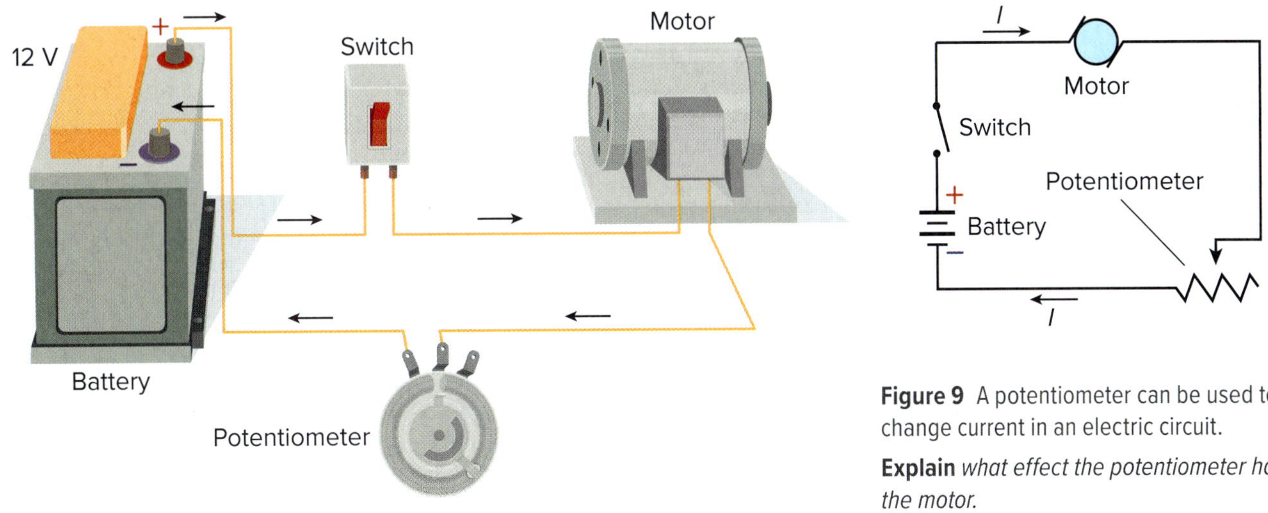

Figure 9 A potentiometer can be used to change current in an electric circuit.

Explain *what effect the potentiometer has on the motor.*

EXAMPLE Problem 2

CURRENT THROUGH A RESISTOR A 30.0-V battery is connected to a 10.0-Ω resistor. What is the current in the circuit?

1 ANALYZE AND SKETCH THE PROBLEM

- Draw a circuit containing a battery, an ammeter, and a resistor.
- Show the direction of the conventional current.

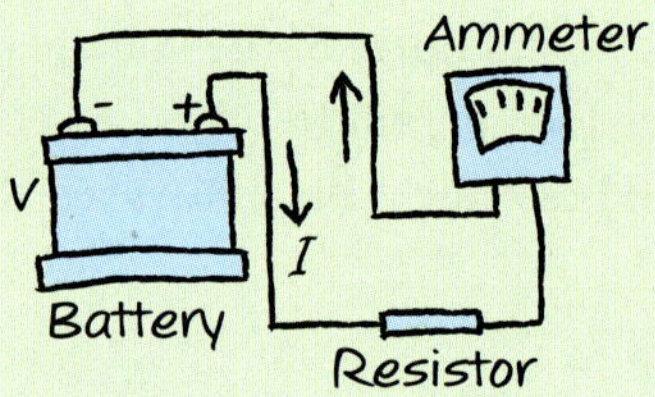

Known	Unknown
$\Delta V = 30.0\ V$	$I = ?$
$R = 10.0\ \Omega$	

2 SOLVE FOR THE UNKNOWN

Use $I = \frac{\Delta V}{R}$ to determine the current.

$$I = \frac{\Delta V}{R}$$

$$= \frac{30.0\ V}{10.0\ \Omega} \quad \text{Substitute } \Delta V = 30.0\ V,\ R = 10.0\ \Omega.$$

$$= 3.00\ A$$

3 EVALUATE THE ANSWER

- **Are the units correct?** Current is measured in amperes.
- **Is the magnitude realistic?** There is a fairly large potential difference and a small resistance, so a current of 3.00 A is reasonable.

PRACTICE Problems

ADDITIONAL PRACTICE

For all problems, assume the battery voltage and the lamp resistances are constant.

13. An automobile panel lamp with a resistance of 33 Ω is placed across the battery shown in **Figure 10.** What is the current through the circuit?

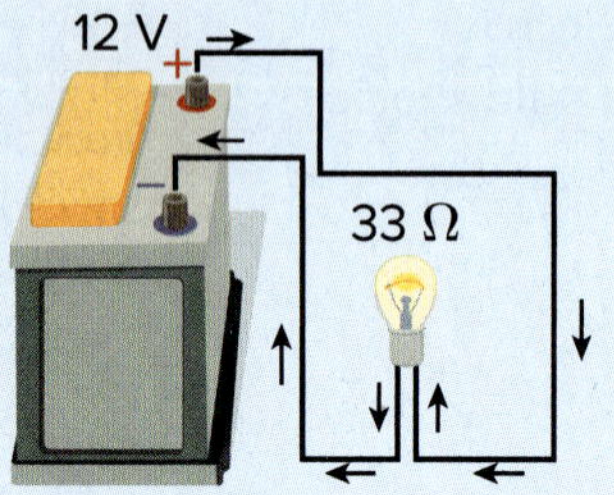

Figure 10

14. A sensor uses 2.0×10^{-4} A of current when it is operated by the battery shown in **Figure 11**. What is the resistance of the sensor circuit?

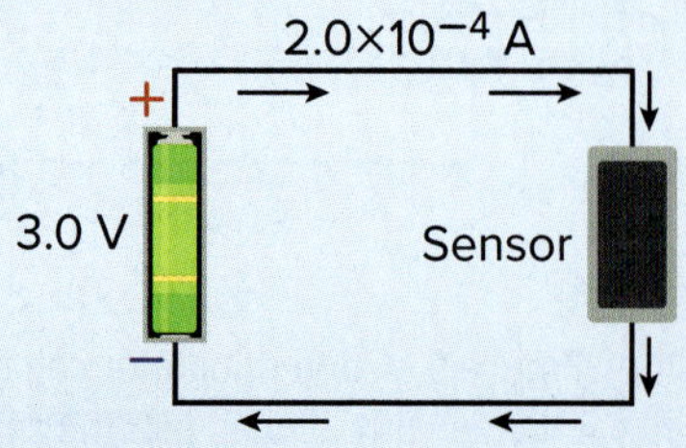

Figure 11

15. A motor with the operating resistance of 32 Ω is connected to a voltage source as shown in **Figure 12.** What is the voltage of the source?

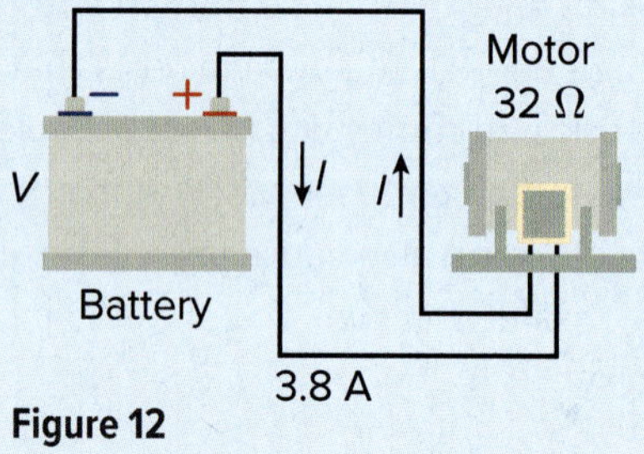

Figure 12

16. A lamp draws a current of 0.50 A when it is connected to a 120-V source.

a. What is the resistance of the lamp?

b. What is the power consumption of the lamp?

17. A 75-W lamp is connected to 125 V.

a. What is the current through the lamp?

b. What is the resistance of the lamp?

18. CHALLENGE A resistor is added to the lamp in the previous problem to reduce the current to half its original value.

a. What is the potential difference across the lamp?

b. How much resistance was added to the circuit?

c. At what rate does the lamp transform electrical energy into radiant and thermal energy?

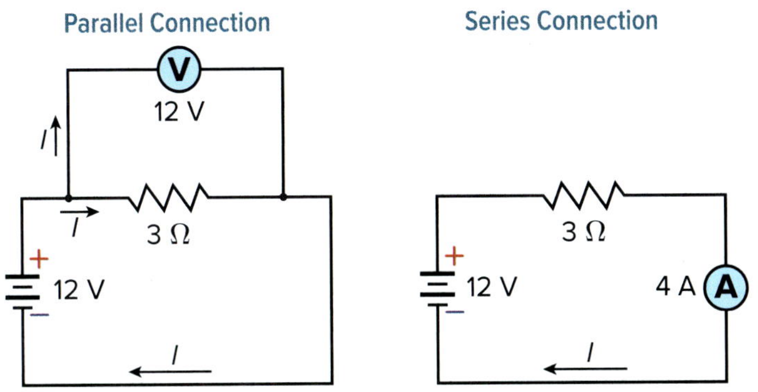

Figure 13 These schematics show a parallel circuit and a series circuit.

Parallel and Series Connections

When a voltmeter is connected across a circuit component, it is called a parallel connection because the component and the voltmeter are aligned parallel to each other in the circuit, as diagrammed in **Figure 13.** Any time the current has two or more paths to follow in a circuit, the connection is a **parallel connection.** The potential difference across the voltmeter is equal to that across the component. Always associate the words *voltage across* with a parallel connection.

An ammeter measures the current through a circuit component. The same current going through the component must go through the ammeter, so there can be only one current path. A connection with only one current path in a circuit is a **series connection,** which also is shown in **Figure 13.** To add an ammeter to a circuit, the wire connected to the circuit component must be removed and connected to the ammeter instead. Then, another wire is connected from the second terminal of the ammeter to the circuit component. Always associate the words *current through* with a series connection.

Check Your Progress

19. **Electrical Energy** Identify two forms in which electrical energy exists.
20. **Schematic** Draw a schematic diagram of a circuit that contains a battery and a lightbulb. Make sure the lightbulb will light in this circuit.
21. **Resistance** Joe states that because $R = \frac{\Delta V}{I}$, if he increases the voltage, the resistance will increase. Is Joe correct? Explain.
22. **Power** A circuit has 12 Ω of resistance and is connected to a 12-V battery. Determine the change in power if the resistance decreases to 9.0 Ω.
23. **Resistance** You want to measure the resistance of a long piece of wire. Show how you would construct a circuit with a battery, a voltmeter, an ammeter, and the wire to be tested to make the measurement. Specify what you would measure and how you would compute the resistance.
24. **Energy** A circuit transforms 2.2×10^3 J of energy when it is operated for 3.0 min. Determine the amount of energy it will transform when it is operated for 1 h.
25. **Critical Thinking** We sometimes say that power is "dissipated" in a resistor. To dissipate is to spread out or disperse. In what sense is something being dispersed when charge flows through a resistor?

LEARNSMART® Go online to follow your personalized learning path to review, practice, and reinforce your understanding.

LESSON 2

USING ELECTRICAL ENERGY

FOCUS QUESTION

How do we use electrical energy?

Electrical Energy, Resistance, and Power

Energy that is supplied to a circuit can be transformed in many useful ways. For example, a lamp changes electrical energy into radiant energy. Unfortunately, not all the energy delivered to a lamp ends up in a useful form. Lightbulbs, especially incandescent lightbulbs, become hot. Some of the electrical energy is transformed into thermal energy.

Get It?

Identify how electric circuits are being used to transform energy based on the evidence in the image of Earth at the beginning of this module.

Heating a resistor Current moving through a resistor increases its thermal energy because flowing electrons bump into the atoms in the resistor. Thus these collisions increase the motion of the atoms and the temperature of the resistor. A space heater, a hot plate, and the heating element in a hair dryer all are designed to transform electrical energy into thermal energy. These and other household appliances, such as those pictured in **Figure 14,** act like resistors when they are in a circuit.

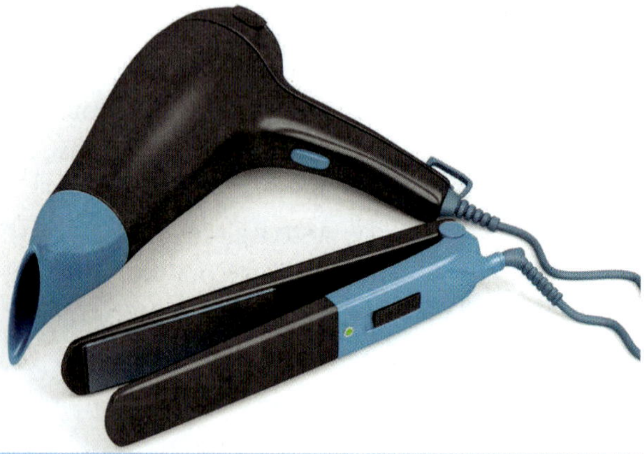

Figure 14 These appliances transform electrical energy into thermal energy.

Explain *why the appliances heat up when they are turned on.*

Panther Media GmbH/Alamy Stock Photo

3D THINKING DCI Disciplinary Core Ideas CCC Crosscutting Concepts SEP Science & Engineering Practices

COLLECT EVIDENCE

Use your Science Journal to record the evidence you collect as you complete the readings and activities in this lesson.

INVESTIGATE

GO ONLINE to find these activities and more resources.

Applying Practices: Modeling Changes in Energy

HS-PS3-1. Create a computational model to calculate the change in the energy of one component in a system when the change in energy of the other component(s) and energy flows in and out of the system are known.

When charge (q) moves through a resistor, its potential is reduced by an amount (ΔV). As you have learned, the energy decrease is represented by $q\Delta V$. In practical use, the rate at which energy is changed—the power $\left(P = \frac{E}{t}\right)$—is more important. Earlier, you learned that current is the rate at which charge flows $\left(I = \frac{q}{t}\right)$ and that power transformed in a resistor is represented by $P = I\Delta V$. For a resistor, $\Delta V = IR$. If you know I and R, you can substitute $\Delta V = IR$ into the equation for electrical power to obtain the equation below. Thus, the rate at which energy is transformed in a resistor is proportional both to the square of the current passing through it and to the resistance.

Power

Power is equal to current squared times resistance.

$$P = I^2R$$

Power and thermal energy

If you know ΔV and R, but not I, you can substitute $I = \frac{\Delta V}{R}$ into $P = I\Delta V$ to obtain the following equation.

Power

Power is equal to the potential difference squared divided by the resistance.

$$P = \frac{(\Delta V)^2}{R}$$

Power is the rate at which energy is transformed from one form to another. Energy is changed from electrical to thermal energy, and the temperature of the resistor rises. The resistors in an immersion heater, for example, convert electrical energy to thermal energy fast enough to bring water to a boil in a few minutes. If power continues to be transformed at a uniform rate, then after time t, the energy transformed to thermal energy will be $E = Pt$. Because $P = I^2R$ and $P = \frac{(\Delta V)^2}{R}$, the total energy to be converted to thermal energy can be written in the following ways.

$$E = Pt \qquad E = I^2Rt \qquad E = \left(\frac{(\Delta V)^2}{R}\right)t$$

Superconductors

A **superconductor** is a material with zero resistance. There is no restriction of current in superconductors, so there is no potential difference (ΔV) across a superconducting wire. Because the rate of energy transformation in a conductor is given by the product $I\Delta V$, a superconductor can conduct electricity without thermal energy transformations. At present, almost all superconductors must be kept at temperatures below 100 K. The practical uses of superconductors today include MRI magnets. Someday superconducting cables may efficiently carry electrical power to cities from distant power plants.

Get It?

State the amount of resistance in a superconductor.

CCC CROSSCUTTING CONCEPTS

Cause and Effect Investigate the expected effects of using superconducting cables to carry electrical power. Develop a brochure to highlight the changes that might result from the change in materials.

EXAMPLE Problem 3

ELECTRIC HEAT A heater has a resistance of 10.0 Ω. It operates on 120.0 V.

a. What is the power of the heater?

b. What thermal energy is supplied by the heater in 10.0 s?

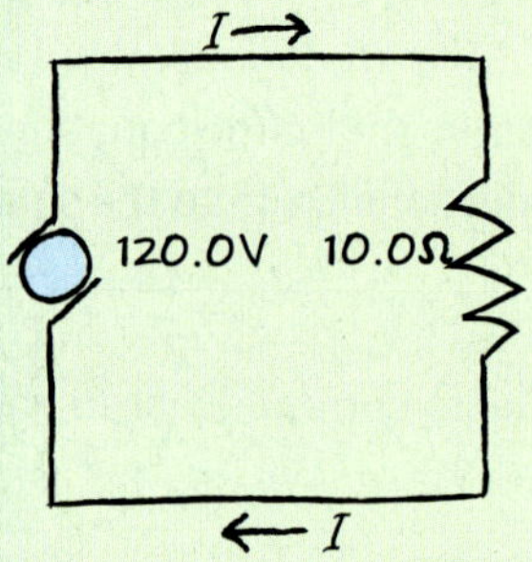

1 ANALYZE AND SKETCH THE PROBLEM

- Sketch the situation.
- Label the known circuit components, which are a 120.0-V potential difference source and a 10.0-Ω resistor.

Known	Unknown
$R = 10.0\ \Omega$	$P = ?$
$\Delta V = 120.0\ \text{V}$	$E = ?$
$t = 10.0\ \text{s}$	

2 SOLVE FOR THE UNKNOWN

a. Because R and ΔV are known, use $P = \frac{(\Delta V)^2}{R}$.

$$P = \frac{(120.0\ \text{V})^2}{10.0\ \Omega}$$ Substitute $\Delta V = 120.0$ V, $R = 10.0\ \Omega$.

$$= 1440\ \text{W} = 1.44\ \text{kW}$$

b. Solve for the energy.

$$E = Pt$$

$$= (1.44\ \text{kW})(10.0\ \text{s})$$ Substitute $P = 1.44$ kW, $t = 10.0$ s.

$$= 14.4\ \text{kJ}$$

3 EVALUATE THE ANSWER

- **Are the units correct?** Power is measured in watts, and energy is measured in joules.
- **Are the magnitudes realistic?** For power, $10^2 \times 10^2 \times 10^{-1} = 10^3$, so kilowatts is reasonable. For energy, $10^3 \times 10^1 = 10^4$, so an order of magnitude of 10,000 joules is reasonable.

PRACTICE Problems

ADDITIONAL PRACTICE

26. A 15-Ω electric heater operates on a 120-V outlet.

a. What is the current through the heater?

b. How much energy is used by the heater in 30.0 s?

c. How much thermal energy is liberated in this time?

27. A 39-Ω resistor is connected across a 45-V battery.

a. What is the current in the circuit?

b. How much energy is used by the resistor in 5.0 min?

28. A 100.0-W lightbulb is 22 percent efficient. This means that 22 percent of the electrical energy is transformed to radiant energy.

a. How many joules does the lightbulb transform into radiant energy each minute it is in operation?

b. How many joules of thermal energy does the lightbulb output each minute?

29. The resistance of an electric stove element at operating temperature is 11 Ω.

a. If 220 V are applied across it, what is the current through the stove element?

b. How much energy does the element transform to thermal energy in 30.0 s?

c. The element is used to heat a kettle containing 1.20 kg of water. Assume 65 percent of the thermal energy is absorbed by the water. What is the water's increase in temperature during the 30.0 s?

30. CHALLENGE A 120-V water heater takes 2.2 h to heat a given volume of water to a certain temperature. How long would a 240-V unit operating with the same current take to accomplish the same task?

Providing Electrical Energy

Figure 15 This hydroelectric dam is one of eleven dams on the Columbia River.

How can the energy produced at hydroelectric facilities, such as the one shown in **Figure 15,** be transmitted over long distances with as little transformation to thermal energy as possible? Electrical energy is transformed at a rate represented by $P = I^2R$. Electrical engineers call the resulting unwanted thermal energy the Joule heating loss, or I^2R loss. All wires have some resistance that depends on its length. The thick wire used to carry electric current into a home has a resistance of 0.20 Ω for a length of 1 km of wire. Thus, the resistance of the wires needed to carry a current from a power plant to a house 3.5 km away and back is: $R = 2(3.5\ \text{km})(0.20\ \Omega/\text{km}) = 1.4\ \Omega$. An electric stove might cause a 41-A current through the wires. The rate of energy transformation (power) in the wires is represented by the following relationship.

$$P = I^2R = (41\ \text{A})^2(1.4\ \Omega) = 2400\ \text{W}$$

Electric transmission lines All this electrical energy is transformed to thermal energy and is wasted. Reducing either the current (I) or the resistance (R) would minimize the waste. Cables of high conductivity and large diameter (and therefore low resistance) are expensive and heavy. It is more important to minimize the current because the transformation is proportional to the square of the current. The electrical energy per second (power) is $P = IV$. Increasing the voltage reduces the current without changing the power. Long-distance transmission lines always operate at voltages much higher than household voltages, with some at more than 500,000 V. The resulting lower current reduces the I^2R loss by keeping the I^2 factor low. The output voltage from the generating plant is reduced at electric substations to 2400 V, and again to 240 V or 120 V before being used in homes.

The kilowatt-hour Power is the rate at which energy is delivered. When consumers pay their home electric bills, an example of which is shown in **Figure 16,** they pay for electrical energy, not power. The amount of electrical energy used by a device is its rate of energy consumption, in joules per second or watts (W), multiplied by the number of seconds the device is operated. Joules per second times seconds $\left(\frac{\text{J}}{\text{s}} \times \text{s}\right)$ equals the total amount of joules of energy. The joule, also defined as a watt-second, is too small an amount of energy for commercial sales use. Instead, electric companies measure energy sales in a larger unit called a **kilowatt-hour** (kWh). A kilowatt-hour is equal to 1000 watts delivered continuously for 3600 s (1 h), or 3.6×10^6 J. Not many household devices other than hot-water heaters, stoves, clothes dryers, microwave ovens, heaters, and hair dryers require more than 1000 W of power.

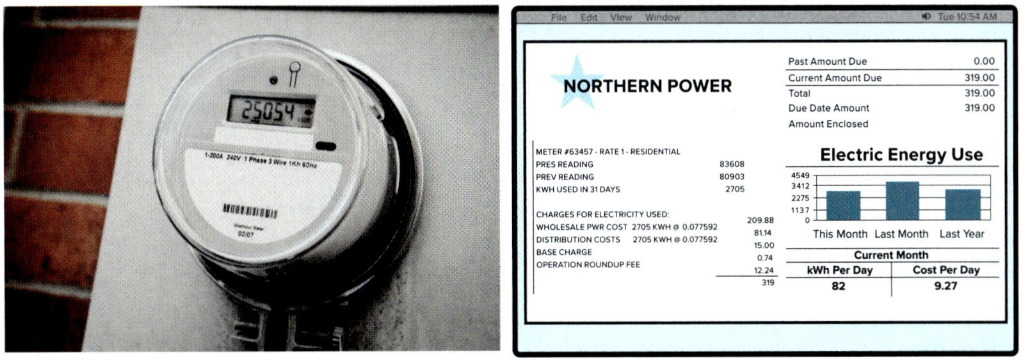

Figure 16 Smart meters measure the amount of electrical energy used by a consumer. Electric bills often detail the amount of electrical energy used.

PRACTICE Problems

ADDITIONAL PRACTICE

31. An electric space heater draws 15.0 A from a 120-V source. It is operated, on the average, for 5.0 h each day.

a. How much power does the heater use?

b. How much energy in kWh does it consume in 30 days?

c. At $0.12 per kWh, how much does it cost to operate the heater for 30 days?

32. A digital clock has a resistance of 12,000 Ω and is plugged into a 115-V outlet.

a. How much current does it draw?

b. How much power does it use?

c. If the owner of the clock pays $0.12 per kWh, how much does it cost to operate the clock for 30 days?

33. An automotive battery can deliver 55 A at 12 V for 1.0 h and requires 1.3 times as much energy for recharge due to its less-than-perfect efficiency. How long will it take to charge the battery using a current of 7.5 A? Assume the charging voltage is the same as the discharging voltage.

34. CHALLENGE Rework the previous problem by assuming the battery requires the application of 14 V when it is recharging.

PHYSICS Challenge

Use the diagram to the right to help you answer the questions below.

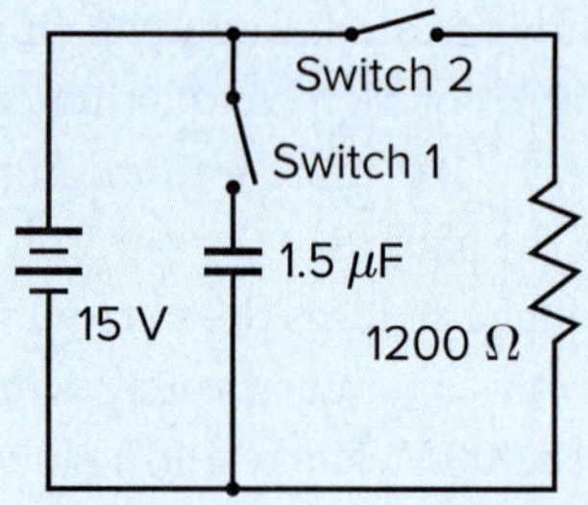

1. Initially, the capacitor is uncharged. Switch 1 is closed, and switch 2 remains open. What is the potential difference across the capacitor?

2. Switch 1 is now opened, and switch 2 remains open. What is the potential difference across the capacitor? Why?

3. Next, switch 2 is closed, while switch 1 remains open. What is the potential difference across the capacitor and the current through the resistor immediately after switch 2 is closed?

4. As time goes on, what happens to the voltage measured across the capacitor and the current through the resistor?

Check Your Progress

35. **Energy Forms** A car engine drives a generator, which transfers electrical energy to the car's battery. The headlights use the energy stored in the car battery to produce light. List the forms of energy in these three operations.

36. **Resistance** A hair dryer operating from 120 V has two settings, hot and warm. In which setting is the resistance likely to be smaller? Why?

37. **Efficiency** Evaluate the impact of research to improve power transmission lines on society and the environment.

38. **Voltage** Why would an electric range and an electric hot-water heater be connected to a 240-V circuit rather than a 120-V circuit?

39. **Energy Cost** A consumer uses 3098 kWh in 29 days. She is charged $0.077592 per kWh for the electricity plus $0.029998 per kWh for its distribution. What is the bill for the 29 days?

40. **Resistance and Power** A toaster is connected to the circuit shown in **Figure 17.**

a. What is the resistance of the toaster?

b. At what rate does the toaster transform energy?

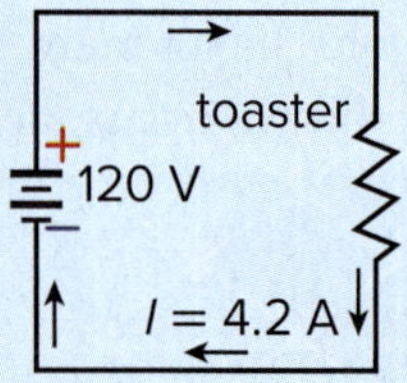

Figure 17

41. **Critical Thinking** When demand for electric power is high, power companies sometimes reduce the voltage, thereby producing a "brown-out." What is being saved?

LEARNSMART® Go online to follow your personalized learning path to review, practice, and reinforce your understanding.

LESSON 3

SIMPLE CIRCUITS

FOCUS QUESTION

How can you build a simple circuit to fit the needs of a given situation?

A River Model

You can use a mountain river to model an electric circuit. From its source high in the mountains, the river flows downhill to the plains below. No matter which path the river takes, its change in elevation, from the mountaintop to the plain, is the same. Some rivers flow downhill in a single stream. Other rivers might split into two or more smaller streams as they flow over a waterfall or through a series of rapids. In this case, part of the river follows one path, while other parts of the river follow different paths. No matter how many paths the river takes, however, the total amount of water flowing down the mountain remains unchanged. In other words, the number of paths water takes does not affect the total amount of water flowing downhill.

How does the river shown in **Figure 18** model an electric circuit? The distance the river drops is similar to the potential difference in a circuit. The rate of water flow is similar to current in a circuit. Large rocks and other obstacles produce resistance to water flow and are similar to resistors in a circuit. Which part of the water cycle is similar to a battery or a generator in an electric circuit? The energy source needed to raise water to the top of the mountain is the Sun. Solar energy evaporates water from lakes and seas, leading to the formation of clouds that release rain or snow that falls on the mountaintops. Continue to think about the mountain river model as you read more about electric circuits.

Figure 18 The flow of a mountain stream is similar to electric current because, regardless of the path, the amount of water and the amount of electric charges remain the same.

Ken Eis/WeatherVideoHD.TV

3D THINKING DCI Disciplinary Core Ideas CCC Crosscutting Concepts SEP Science & Engineering Practices

COLLECT EVIDENCE

Use your Science Journal to record the evidence you collect as you complete the readings and activities in this lesson.

INVESTIGATE

GO ONLINE to find these activities and more resources.

Forensics Lab: Circuit Configurations

Conduct an investigation into how the structure of an electric circuit effects its function.

Review the News

Obtain information from a current news story about circuits or electrical energy in everyday life. Evaluate your source and communicate your findings to your class.

Series Circuits

Suppose three students connect two identical lamps to a battery, as illustrated in **Figure 19.** Before they connect the battery, their teacher asks them to predict the brightnesses of the two lamps. The first student predicts that only the lamp close to the positive (+) terminal of the battery will light because all the current will be used up as light energy. The second student predicts that only part of the current will be used up, and the second lamp will glow, but less brightly than the first. The third student predicts that the lamps will be of equal brightness because current is a flow of charge, and the charge leaving the first lamp has nowhere else to go in the circuit except through the second lamp. The third student reasons that because the current will be the same in each lamp, the brightness also will be the same.

Figure 19 Students constructed this circuit and used it to verify their predictions of the brightness in each lightbulb when the final connection was made.

Current in a series circuit

If you consider the mountain river model for this circuit, you will realize that the third student is correct. Recall that charge cannot be created or destroyed. Because the charge in the circuit has only one path to follow and cannot be destroyed, the same amount of charge that enters the first lamp must also leave that lamp. The current is the same everywhere in the circuit. If you connect three ammeters in the circuit, as shown in **Figure 20,** they all will show the same current. A circuit such as this, in which there is only one path for the current, is called a **series circuit.**

> **Get It?**
> **Define** the term *series circuit.*

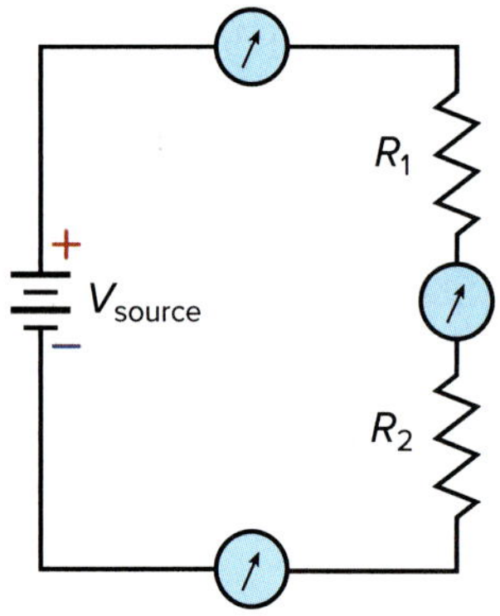

Figure 20 Ammeters can be used in a series circuit to show that the current is the same throughout the circuit.

If current does not decrease when it passes through a lamp, where does the energy to light the lamps come from? The first two students above based their predictions on ideas about energy. The energy to light the lamps must come from somewhere. To these students, it made sense that current would decrease due to a conversion of electrical energy into light energy.

Recall that power, the rate at which electrical energy is converted, can be represented as $P = I\Delta V$. This means that conversions of electrical energy into other forms depend on both current and potential difference. Thus, because current does not change when the lamp converts electrical energy to light energy, there must be a potential difference across the lamp. The resistance of the lamp is defined as $R = \frac{\Delta V}{I}$. Thus, the potential difference across the lamp is $\Delta V = IR$.

STEM CAREER Connection

Electrical Engineer

Are you fascinated by electric motors and the devices that use them? Are you interested in how electrical energy is generated? Electrical engineers help design and develop electrical devices. They are responsible for testing the devices and making sure they function safely and efficiently.

Equivalent resistance of a series circuit From the river model, you know that the height from sea level to the top of the mountain equals the height that the water drops from the top of the mountain to sea level. In an electric circuit, the potential difference across the generator or other energy source (ΔV_{source}) equals the sum of the potential differences across lamps 1 and 2 and is represented by the following equation:

$$\Delta V_{source} = \Delta V_1 + \Delta V_2.$$

To find the potential difference across a resistor, multiply the current in the circuit by the resistance of the individual resistor. Because the current through the lamps is the same, $\Delta V_1 = IR_1$ and $\Delta V_2 = IR_2$. Therefore, $\Delta V_{source} = IR_1 + IR_2$, or $\Delta V_{source} = I(R_1 + R_2)$. The current through the circuit is represented by the following equation:

$$I = \frac{\Delta V_{source}}{(R_1 + R_2)}.$$

The same idea can be extended to any number of resistances in series, not just two. If you replaced all the resistors in a circuit with a single resistor that resulted in the same current, that resistor's value would be the **equivalent resistance** of the circuit. For resistors in series, the same current would exist in the circuit with a single resistor (R) that has a resistance equal to the sum of the individual resistances.

Equivalent Resistance for Resistors in Series

The equivalent resistance of resistors in series equals the sum of the individual resistances of the resistors.

$$R = R_1 + R_2 + \ldots$$

Notice that the equivalent resistance is greater than that of any individual resistor. Therefore, if the battery voltage does not change, adding more devices in series always decreases the current. To find the current through a series circuit, first calculate the equivalent resistance and then use the following equation.

Current

The current through a series circuit is equal to the potential difference across the power source divided by the equivalent resistance.

$$I = \frac{\Delta V_{source}}{R}$$

PRACTICE Problems

ADDITIONAL PRACTICE

42. Three 22-Ω resistors are connected in series across a 125-V generator. What is the equivalent resistance of the circuit? What is the current in the circuit?

43. A 12-Ω, a 15-Ω, and a 5-Ω resistor are connected in a series circuit with a 75-V battery. What is the equivalent resistance of the circuit? What is the current in the circuit?

44. A string of lights has ten identical bulbs with equal resistances connected in series. When the string of lights is connected to a 117-V outlet, the current through the bulbs is 0.06 A. What is the resistance of each bulb?

45. A 9-V battery is in a circuit with three resistors connected in series.

a. If the resistance of one of the resistors increases, how will the equivalent resistance change?

b. What will happen to the current?

c. Will there be any change in the battery voltage?

46. CHALLENGE Calculate the potential differences across three resistors, 12-Ω, 15-Ω, and 5-Ω, that are connected in series with a 75-V battery. Verify that the sum of their potential differences equals the potential difference across the battery.

Voltage dividers

Suppose you have a 9-V battery, but you need a potential difference of 5 V for an electric circuit. How can you produce a potential difference of 5 V if your battery produces a potential difference of 9 V? You could use a type of series circuit called a **voltage divider.** A voltage divider produces a source of potential difference that is less than the potential difference across the battery.

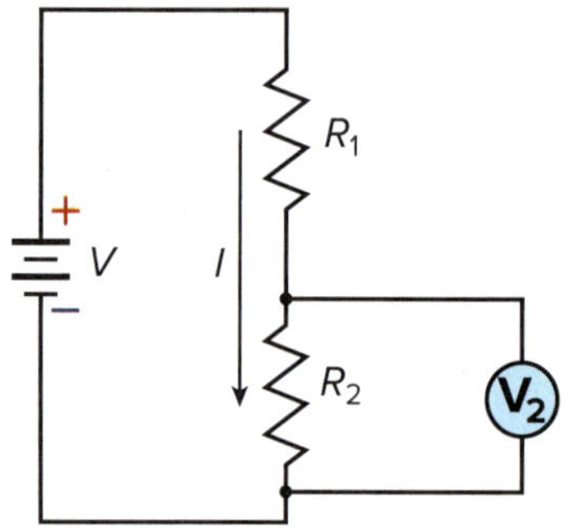

Figure 21 This voltage-divider circuit demonstrates how a voltage of desired magnitude can be achieved by choosing the right combination of resistors.

Explain *why the current direction arrow in the diagram is pointing in that direction.*

Consider the circuit shown in **Figure 21.** Two resistors (R_1 and R_2) are connected in series across a battery with potential difference V. The equivalent resistance of the circuit is

$$R = R_1 + R_2.$$

The current is represented by the following equation:

$$\begin{aligned} I &= \frac{\Delta V}{R} \\ &= \frac{\Delta V}{R_1 + R_2} \end{aligned}$$

The desired voltage (5 V) is the voltage drop (ΔV_2) across resistor R_2: $\Delta V_2 = IR_2$. Substitute the earlier equation into this equation as shown:

$$\begin{aligned} \Delta V_2 &= IR_2 \\ &= \left(\frac{\Delta V}{R_1 + R_2}\right)R_2 \\ &= \frac{\Delta V R_2}{R_1 + R_2} \end{aligned}$$

By choosing the right resistors, you can produce a potential difference of 5 V across a portion of an electric circuit even if you only have a 9-V battery available.

Photoresistors

Voltage dividers often are used with sensors, such as photoresistors. A photoresistor's resistance depends on the amount of light that strikes it. A typical photoresistor has a resistance of 400 Ω when light is striking it compared with a resistance of 400,000 Ω when the photoresistor is in the dark. The voltage output of a voltage divider that uses a photoresistor depends upon the amount of light striking that photoresistor at the time.

This circuit can be used as a light meter, such as the one shown in **Figure 22.** In this device, an electronic circuit detects the potential difference and converts it to a measurement of illuminance that can be read on the digital display.

Figure 22 Light meters used in photography use a voltage divider. The amount of light striking the photoresistor sensor determines the voltage output of the voltage divider.

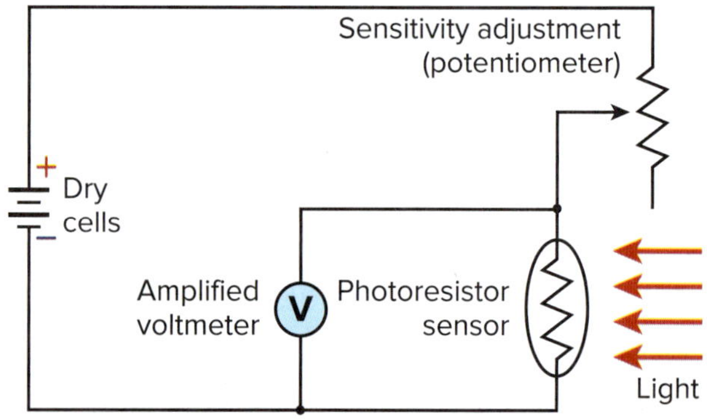

EXAMPLE Problem 4

POTENTIAL DIFFERENCE IN A SERIES CIRCUIT Two resistors, 47 Ω and 82 Ω, are connected in series across a 45-V battery.

a. What is the current in the circuit?

b. What is the potential difference across each resistor?

c. If you replace the 47-Ω resistor with a 39-Ω resistor, will the current increase, decrease, or remain the same?

d. What is the new potential difference across the 82-Ω resistor?

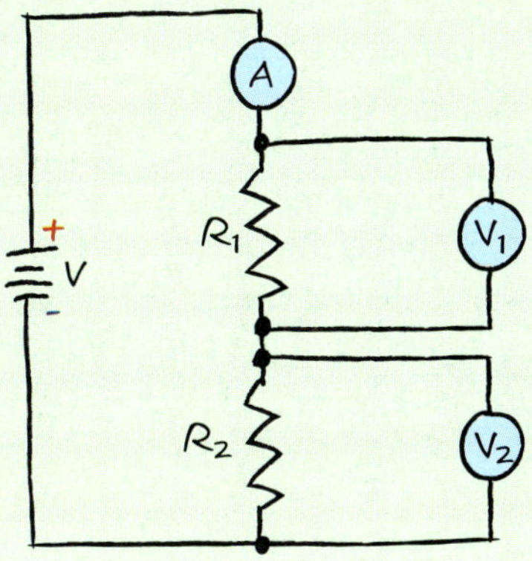

1 ANALYZE AND SKETCH THE PROBLEM

Draw a schematic of the circuit.

Known	Unknown
$\Delta V_{source} = 45$ V	$I = ?$
$R_1 = 47\ \Omega$	$\Delta V_1 = ?$
$R_2 = 82\ \Omega$	$\Delta V_2 = ?$

2 SOLVE FOR THE UNKNOWN

a. To determine the current, first find the equivalent resistance.

$$I = \frac{\Delta V_{source}}{R} \text{ and } R = R_1 + R_2$$

$$= \frac{\Delta V_{source}}{R_1 + R_2}$$ Substitute $R = R_1 + R_2$.

$$= \frac{45\text{ V}}{47\ \Omega + 82\ \Omega}$$ Substitute $\Delta V_{source} = 45$ V, $R_1 = 47\ \Omega$, $R_2 = 82\ \Omega$.

$$= 0.35\text{ A}$$

b. Use $\Delta V = IR$ for each resistor.

$$\Delta V_1 = IR_1$$

$$= (0.35\text{ A})(47\ \Omega)$$ Substitute $I = 0.35$ A, $R_1 = 47\ \Omega$.

$$= 16\text{ V}$$

$$\Delta V_2 = IR_2$$

$$= (0.35\text{ A})(82\ \Omega)$$ Substitute $I = 0.35$ A, $R_2 = 82\ \Omega$.

$$= 29\text{ V}$$

c. Calculate current, this time using 39 Ω as R_1.

$$I = \frac{\Delta V_{source}}{R_1 + R_2}$$

$$= \frac{45\text{ V}}{39\ \Omega + 82\ \Omega}$$ Substitute $\Delta V_{source} = 45$ V, $R_1 = 39\ \Omega$, $R_2 = 82\ \Omega$.

$$= 0.37\text{ A}$$

The current will increase.

d. Determine the new voltage drop in R_2.

$$\Delta V_2 = IR_2$$

$$= (0.37\text{ A})(82\ \Omega)$$ Substitute $I = 0.37$ A, $R_2 = 82\ \Omega$.

$$= 3.0\times10^1\text{ V}$$

3 EVALUATE THE ANSWER

- **Are the units correct?** Current is measured in $\text{A} = \frac{\text{V}}{\Omega}$; potential difference is measured in $\text{V} = \text{A}\cdot\Omega$.
- **Is the magnitude realistic?** The potential difference across any one resistor must be less than the potential difference across the battery. Both the value of ΔV_2 and the value of ΔV_1 are less than ΔV_{source}, which is 45 V.

PRACTICE Problems

ADDITIONAL PRACTICE

47. The circuit shown in Example Problem 4 is producing these symptoms: the ammeter reads 0 A, ΔV_1 reads 0 V, and ΔV_2 reads 45 V. What has happened?

48. Suppose the circuit shown in Example Problem 4 has these values: $R_1 = 255\ \Omega$, $R_2 = 290\ \Omega$, and $\Delta V_1 = 17$ V. No other information is available.

a. What is the current in the circuit?

b. What is the potential difference across the battery?

c. What is the total power used in the circuit, and what is the power used in each resistor?

d. Does the sum of the power used in each resistor in the circuit equal the total power used in the circuit? Explain.

49. Holiday lights often are connected in series and use special lamps that short out when the voltage across a lamp increases to the line voltage. Explain why. Also explain why these light sets might blow their fuses after many bulbs have failed.

50. The circuit in Example Problem 4 has unequal resistors. Explain why the resistor with the lower resistance will operate at a lower temperature.

51. CHALLENGE A series circuit is made up of a 12-V battery and three resistors. The potential difference across one resistor is 1.2 V, and the potential difference across another resistor is 3.3 V. What is the voltage across the third resistor?

EXAMPLE Problem 5

VOLTAGE DIVIDER A 9.0-V battery and two resistors, 390 Ω and 470 Ω, are connected as a voltage divider. What is the potential difference across the 470-Ω resistor?

1 ANALYZE AND SKETCH THE PROBLEM

Draw the battery and resistors in a series circuit.

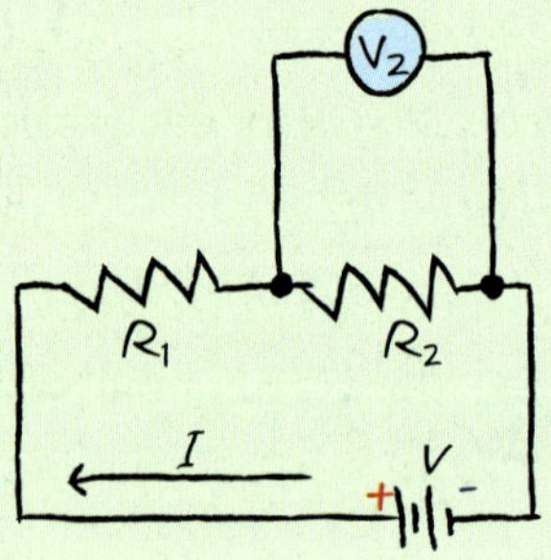

Known

$\Delta V_{source} = 9.0$ V

$R_1 = 390\ \Omega$

$R_2 = 470\ \Omega$

Unknown

$\Delta V_2 = ?$

2 SOLVE FOR THE UNKNOWN

$$R = R_1 + R_2$$

$$I = \frac{\Delta V_{source}}{R}$$

$$= \frac{\Delta V_{source}}{R_1 + R_2}$$ Substitute $R = R_1 + R_2$.

$$\Delta V_2 = IR_2$$

$$= \frac{\Delta V_{source} R_2}{R_1 + R_2}$$ Substitute $I = \frac{\Delta V_{source}}{R_1 + R_2}$.

$$= \frac{(9.0\ \text{V})(470\ \Omega)}{390\ \Omega + 470\ \Omega}$$ Substitute $\Delta V_{source} = 9.0$ V, $R_1 = 390\ \Omega$, $R_2 = 470\ \Omega$.

$$= 4.9\ \text{V}$$

3 EVALUATE THE ANSWER

- **Are the units correct?** Potential difference is measured in $\text{V} = \frac{\text{V}\cdot\Omega}{\Omega}$. The ohms cancel, leaving volts.
- **Is the magnitude realistic?** The potential difference across the resistor is less than the potential difference across the battery. Because 470 Ω is more than half of the equivalent resistance, the potential difference across that resistor is more than half of the potential difference across the battery.

PRACTICE Problems

ADDITIONAL PRACTICE

52. A 22-Ω resistor and a 33-Ω resistor are connected in series and are connected to a 120-V power source.

a. What is the equivalent resistance of the circuit?

b. What is the current in the circuit?

c. What is the potential difference across each resistor?

53. Three resistors of 3.3 kΩ, 4.7 kΩ, and 3.9 kΩ are connected in series across a 12-V battery.

a. What is the equivalent resistance?

b. What is the current through the resistors?

c. Find the total potential difference across the three resistors.

54. CHALLENGE Select a resistor to be used as part of a voltage divider along with a 1.2-kΩ resistor. The potential difference across the 1.2-kΩ resistor is to be 2.2 V when the supply is 12 V.

Parallel Circuits

Look at the circuit shown in **Figure 23.** How many current paths are there? The current from the generator can go through any of the three resistors. A circuit in which there are several current paths is called a **parallel circuit.** The three resistors are connected in parallel; both ends of the three paths are connected together. In the mountain river model, such a circuit is illustrated by three paths for the water over a waterfall. Some paths might have a large flow of water, while others might have a small flow. The sum of the water in the flows, however, is equal to the total flow of water over the falls. In addition, regardless of which channel the water flows through, the drop in height is the same. Similarly, in a parallel electric circuit, the total current is the sum of the currents through each path, and the potential difference across each path is the same.

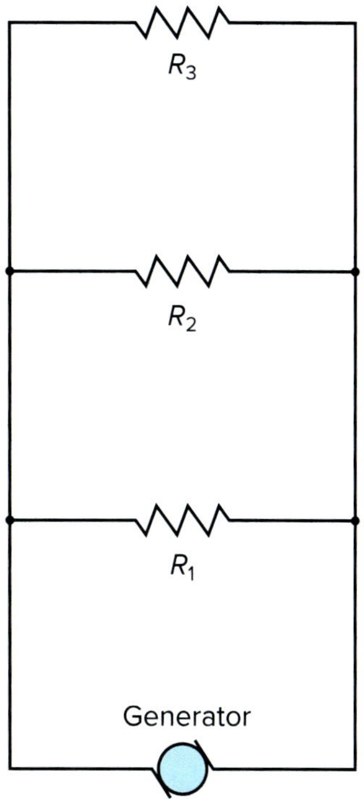

Figure 23 The parallel paths for current in this diagram are analogous to the multiple paths that a river might take down a mountain.

Current in a parallel circuit

What is the current through each resistor in a parallel circuit? The answer to this question depends on the individual resistances. For example, in **Figure 24,** the potential difference across each resistor is 120 V. The current through a resistor is given by $I = \frac{\Delta V}{R}$, so you can calculate the current through the 24-Ω resistor as $I = \frac{120\ \text{V}}{24\ \Omega} = 5.0\ \text{A}$. You can calculate the current through the other two resistors in a similar way. The total current through the generator is the sum of the currents through the three paths. In this case, the current through the generator is 38 A.

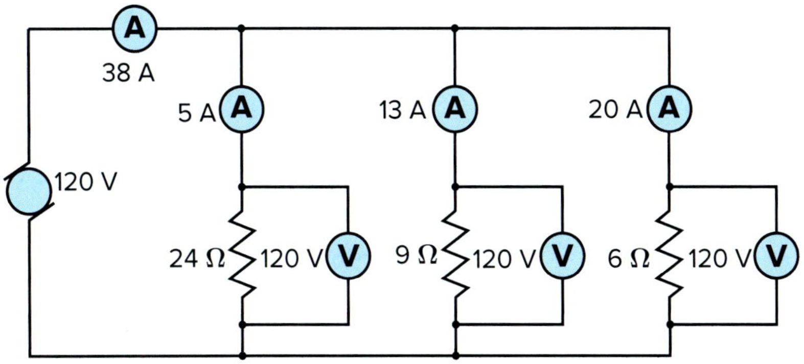

Figure 24 To find the total current in a parallel circuit, find the sum of the currents in each individual path.

State *what the symbols* V *and* A *mean in the circuit.*

Changing the circuit What would happen if you removed the 6-Ω resistor from the circuit in **Figure 24**? Would the current through the 24-Ω resistor change? The current through the 24-Ω resistor depends on the resistance of and the potential difference across that resistor; because neither has changed, the current also has not changed. The same is true for the current through the other resistors. The branches of a parallel circuit are independent of each other. Removing a resistor from the circuit would decrease the total current through the power source, however. For example, if you remove the 6-Ω resistor, the current through the power source would decrease to 18 A.

Get It?

Identify the two factors on which the current through a resistor depends.

Equivalent resistance of a parallel circuit

How can you find the equivalent resistance of a parallel circuit? In **Figure 24,** the total current through the power source is 38 A. The value of a single resistor that results in a 38-A current when a 120-V potential difference is placed across it can be calculated by using the following equation:

$$R = \frac{\Delta V}{I}$$

$$= \frac{120\text{ V}}{38\text{ A}}$$

$$= 3.2\ \Omega$$

Notice that this resistance is smaller than that of any of the three resistors in parallel. Placing two or more resistors in parallel always decreases the equivalent resistance of a circuit. The resistance decreases because each new resistor provides an additional path for current, thereby increasing the total current while the potential difference remains unchanged.

To calculate the equivalent resistance of a parallel circuit, first note that the total current is the sum of the currents through the branches. If I_1, I_2, and I_3 are the currents through the branches and I is the total current, then

$$I = I_1 + I_2 + I_3.$$

The potential difference across each resistor is the same, so the current through each resistor, for example, R_1, can be found from $I_1 = \frac{\Delta V}{R_1}$. Therefore, the equation for the sum of the currents can be written as:

$$\frac{\Delta V}{R} = \frac{\Delta V}{R_1} + \frac{\Delta V}{R_2} + \frac{\Delta V}{R_3}$$

This equation can be extended for any number of resistors in parallel. Dividing both sides of the equation by V provides an equation for the equivalent resistance of a parallel circuit.

Equivalent Resistance for Resistors in Parallel

The reciprocal of the equivalent resistance is equal to the sum of the reciprocals of the individual resistances.

$$\frac{1}{R} = \frac{1}{R_1} + \frac{1}{R_2} + \frac{1}{R_3} + \ldots$$

EXAMPLE Problem 6

EQUIVALENT RESISTANCE AND CURRENT IN A PARALLEL CIRCUIT Three resistors, 60.0 Ω, 30.0 Ω, and 20.0 Ω, are connected in parallel across a 90.0-V battery.

a. Find the current through each branch of the circuit.

b. Find the equivalent resistance of the circuit.

c. Find the current through the battery.

1 ANALYZE AND SKETCH THE PROBLEM

- Draw a schematic of the circuit.
- Include ammeters to show where you would measure each of the currents.

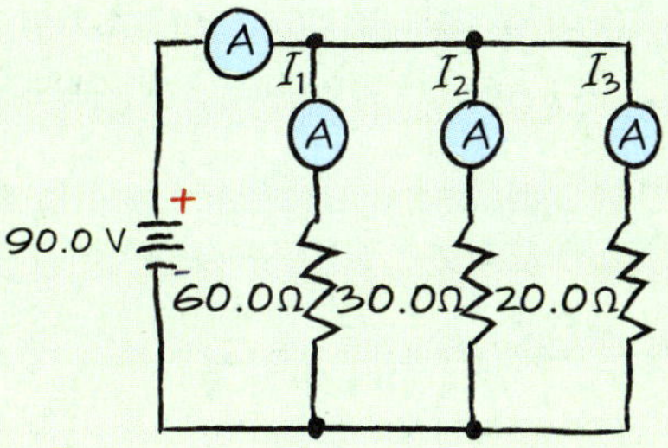

Known	Unknown	
$R_1 = 60.0\ \Omega$	$I_1 = ?$	$I = ?$
$R_2 = 30.0\ \Omega$	$I_2 = ?$	$R = ?$
$R_3 = 20.0\ \Omega$	$I_3 = ?$	
$\Delta V = 90.0\ \text{V}$		

2 SOLVE FOR THE UNKNOWN

a. Because the voltage across each resistor is the same, use $I = \frac{\Delta V}{R}$ for each branch.

$$I_1 = \frac{\Delta V}{R_1}$$

$$= \frac{90.0\ \text{V}}{60.0\ \Omega} \quad \text{Substitute } \Delta V = 90.0\ \text{V},\ R_1 = 60.0\ \Omega.$$

$$= 1.50\ \text{A}$$

$$I_2 = \frac{\Delta V}{R_2}$$

$$= \frac{90.0\ \text{V}}{30.0\ \Omega} \quad \text{Substitute } \Delta V = 90.0\ \text{V},\ R_2 = 30.0\ \Omega.$$

$$= 3.00\ \text{A}$$

$$I_3 = \frac{\Delta V}{R_3}$$

$$= \frac{90.0\ \text{V}}{20.0\ \Omega} \quad \text{Substitute } \Delta V = 90.0\ \text{V},\ R_3 = 20.0\ \Omega.$$

$$= 4.50\ \text{A}$$

b. Use the equivalent resistance equation for parallel circuits.

$$\frac{1}{R} = \frac{1}{R_1} + \frac{1}{R_2} + \frac{1}{R_3}$$

$$= \frac{1}{60.0\ \Omega} + \frac{1}{30.0\ \Omega} + \frac{1}{20.0\ \Omega} \quad \text{Substitute } R_1 = 60.0\ \Omega,\ R_2 = 30.0\ \Omega,\ R_3 = 20.0\ \Omega.$$

$$= 0.100\ \Omega^{-1}$$

$$R = 10.0\ \Omega$$

c. Use $I = \frac{\Delta V}{R}$ to find the total current.

$$I = \frac{\Delta V}{R}$$

$$= \frac{90.0\ \text{V}}{10.0\ \Omega} \quad \text{Substitute } \Delta V = 90.0\ \text{V},\ R = 10.0\ \Omega.$$

$$= 9.00\ \text{A}$$

3 EVALUATE THE ANSWER

- **Are the units correct?** Current is measured in amps; resistance is measured in ohms.
- **Is the magnitude realistic?** The equivalent resistance is less than the resistance of any single resistor. The currents through the resistors are inversely proportional to the resistance. The current for the circuit (I) equals the sum of the currents found for each resistor ($I_1 + I_2 + I_3$).

PRACTICE Problems

ADDITIONAL PRACTICE

55. You connect three 15.0-Ω resistors in parallel across a 30.0-V battery.

a. What is the equivalent resistance of the parallel circuit?

b. What is the current through the entire circuit?

c. What is the current through each branch of the circuit?

56. Suppose you replace one of the 15.0-Ω resistors in the previous problem with a 10.0-Ω resistor.

a. How does the equivalent resistance change?

b. How does the current through the entire circuit change?

c. How does the current through one of the 15.0-Ω resistors change?

57. You connect a 120.0-Ω resistor, a 60.0-Ω resistor, and a 40.0-Ω resistor in parallel across a 12.0-V battery.

a. What is the equivalent resistance of the parallel circuit?

b. What is the current through the entire circuit?

c. What is the current through each branch of the circuit?

58. CHALLENGE You are trying to reduce the resistance in a branch of a circuit from 150 Ω to 93 Ω. You add a resistor to this branch of the circuit to make this change. What value of resistance should you use, and how should you connect this resistor?

Kirchhoff's Rules

Gustav Robert Kirchhoff was a German physicist who, in 1845, when he was only 21 years old, formulated two rules that govern electric circuits–the loop rule and the junction rule. You can use these two rules to analyze complex electric circuits. Both rules are based on fundamental scientific laws.

The loop rule The loop rule is based on the law of conservation of energy. Imagine you are standing on the side of a hill. You hike around the side of this hill in a loop, as shown in the top of **Figure 25.** How does your elevation before your hike compare to your elevation after your hike? Because you begin and end your hike at the same place, your elevation after your hike equals your elevation before your hike. The sum of increases in height equals the sum of decreases in height during your hike.

A similar situation occurs as an electric charge moves around any loop in an electric circuit. Instead of increases and decreases in height, however, electric charges move through increases and decreases in electric potential. The sum of increases in electric potential around a loop in an electric circuit equals the sum of decreases in electric potential around that loop.

Get It?

Explain how Kirchhoff's loop rule and the conservation of energy are related.

For an application of the loop rule, look at the bottom of **Figure 25.** Picture an electric charge traveling clockwise around the red loop. Electric potential increases by 9 V as this charge travels through the battery, and electric potential drops by 5 V as this charge travels through resistor 1. What will be the change in potential as the charge travels through resistor 2? Because the increases in electric potential around a loop must equal the decreases in electric potential around that loop, the drop in electric potential across resistor 2 must be 9 V − 5 V = 4 V.

Notice that resistor 3 does not affect our answer. Why is this? Resistor 3 is not a part of the loop that includes the battery, resistor 1, and resistor 2.

Get It?

State the loop rule in your own words.

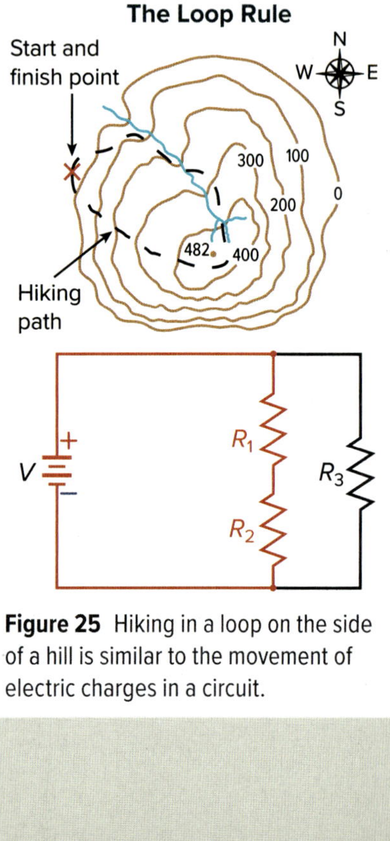

Figure 25 Hiking in a loop on the side of a hill is similar to the movement of electric charges in a circuit.

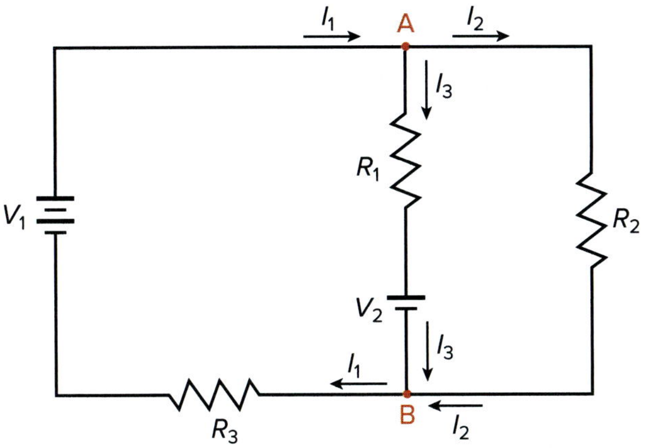

Figure 26 An application of Kirchhoff's junction rule is shown. At junction A the total current entering (I_1) equals the total current leaving ($I_2 + I_3$). At junction B, the total current entering is $I_2 + I_3$, and the total current leaving is I_1.

The junction rule The loop rule describes electric potential differences and is based on the law of conservation of energy. The junction rule describes currents and is based on the law of conservation of charge. Recall that the law of conservation of charge states that charge can neither be created nor destroyed. So, the total current into a section of an electric circuit must equal the total current out of that same section. The circuit shown in **Figure 26** has two junctions—points where three or more wires join—at points A and B. According to the law of conservation of charge, the sum of currents entering a junction is equal to the sum of currents leaving that junction. Otherwise, charge would build up at the junction. This is Kirchhoff's junction rule. In **Figure 26,** $I_1 = I_2 + I_3$ at junction A, and $I_2 + I_3 = I_1$ at junction B. For example, if $I_2 = 0.3$ A and $I_3 = 0.7$ A, then $I_1 = 1.0$ A.

Check Your Progress

59. **Circuit Comparison** Compare and contrast the voltages and the currents in series and parallel circuits.

60. **Total Current** A parallel circuit has four branch currents: 120 mA, 250 mA, 380 mA, and 2.1 A. How much current passes through the power source?

61. **Total Current** A series circuit has four resistors. The current through one resistor is 810 mA. What current passes through the power source?

62. **Circuits** You connect a switch in series with a 75-W bulb to a 120-V power source.
    a. What is the potential difference across the switch when it is closed (turned on)?
    b. What is the potential difference across the switch when it is opened (turned off)?

63. **Loop Rule** Compare Kirchhoff's loop rule to walking around in a loop on the side of a hill.

64. **Junction Rule** Explain how Kirchhoff's junction rule relates to the law of conservation of charge.

65. **Critical Thinking** The circuit in **Figure 27** has four identical resistors. Suppose that a wire is added to connect points A and B. Answer the following questions, and explain your reasoning.
    a. What is the current through the wire?
    b. What happens to the current through each resistor?
    c. What happens to the current through the battery?
    d. What happens to the potential difference across each resistor?

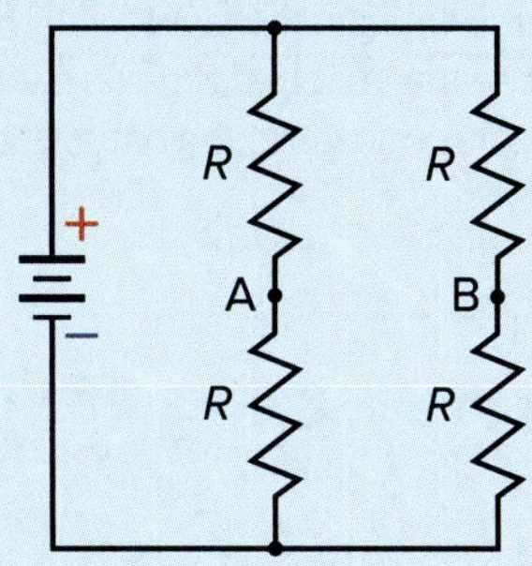

Figure 27

LEARNSMART® Go online to follow your personalized learning path to review, practice, and reinforce your understanding.

LESSON 4

APPLICATIONS OF CIRCUITS

FOCUS QUESTION

How are circuits used in your home and school?

Safety Devices

In an electric circuit, circuit breakers and fuses prevent circuit overloads that can occur when too many appliances are turned on at the same time or when a short circuit occurs in one appliance. A **short circuit** occurs when a circuit with very low resistance is formed. When appliances are connected in parallel, each additional appliance placed in operation reduces the equivalent resistance in the circuit and increases the current through the wires. This additional current might produce enough thermal energy to melt the wiring's insulation, cause a short circuit, or even begin a fire. A **circuit breaker,** shown in **Figure 28,** is an automatic switch that acts as a safety device by stopping the current if the current gets too large and exceeds a threshold value.

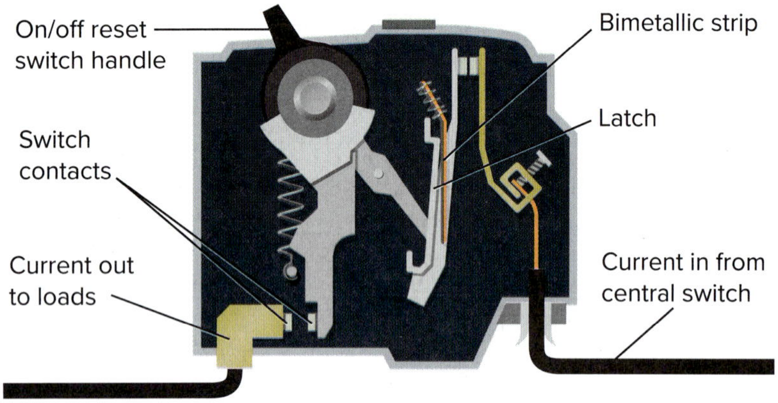

Figure 28 When there is too much current through the bimetallic strip in a circuit breaker, the heat that is generated causes the strip to bend and release the latch. The handle moves to the *off* position, causing the switch to open and the current to stop.

3D THINKING DCI Disciplinary Core Ideas CCC Crosscutting Concepts SEP Science & Engineering Practices

COLLECT EVIDENCE

Use your Science Journal to record the evidence you collect as you complete the readings and activities in this lesson.

INVESTIGATE

GO ONLINE to find these activities and more resources.

Virtual Investigation: Electric Circuits

Use a computer model to investigate the relationships among volatage, resistance, and current in an electric circuit.

Revisit the Encounter the Phenomenon Question

What information from this lesson can help you answer the Unit and Module questions?

A **fuse** is a short piece of metal that acts as a safety device by melting and stopping the current when too large a current passes through it. Engineers design fuses to melt before other elements in a circuit are damaged.

Usually, current follows a single path from the power source through an electrical appliance and back to the source. An appliance malfunction or an accidental drop of the appliance into water can result in additional current pathways. A **ground-fault interrupter** (GFI) is a device that contains an electronic circuit that detects small current differences between the two wires in the cord connected to an appliance. An extra current path, such as one through water, could cause this difference. The GFI stops the current when it detects such differences. This often protects a person from electrocution.

Get It?

Compare how a fuse works with how a ground-fault interrupter works.

Household circuits Some of the most common uses of fuses, circuit breakers, and GFIs are in household circuits. How are these household circuits designed? Are the lights and outlets in a household circuit wired in series or in parallel? Think about what would happen if all the lights and outlets in a household circuit were wired in series. You would only be able to turn everything on and off at once. This is not what we observe in real houses.

The left panel of **Figure 29** shows a simplified household circuit. Notice that this is a parallel circuit with a 15-A fuse connected in series with the 120-V power source. Each resistor represents a light or an electrical appliance. The current in any one branch of the circuit does not depend on the currents in the other branches. This explains how you can turn off the bathroom light without turning off the television. Everything turns off if the fuse wire melts, however.

Let's consider a specific example. Suppose that the three resistors in **Figure 29** represent a 240-W television, a 720-W straightening iron, and a 1440-W hair dryer. We want to know if the fuse will melt, or "blow," if all three appliances are turned on at the same time.

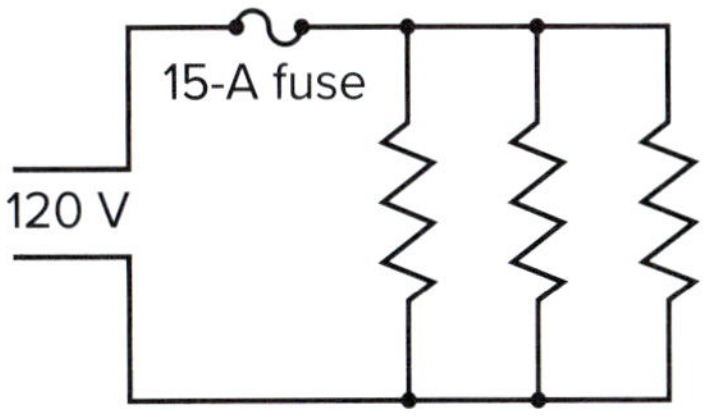

Richard Hutchings/Digital Light Source

Figure 29 Homes are wired using parallel circuits so that more than one appliance can be used simultaneously.
Identify *the current rating of the fuse in the circuit shown.*

You can use $I = \frac{P}{\Delta V}$ to find the current through each appliance.

For the television: $I_{TV} = \frac{240\text{ W}}{120\text{ V}} = 2.0\text{ A}.$

For the straightening iron: $I_{iron} = \frac{720\text{ W}}{120\text{ V}} = 6.0\text{ A}.$

For the hair dryer: $I_{dryer} = \frac{1440\text{ W}}{120\text{ V}} = 12\text{ A}.$

From Kirchhoff 's junction rule, we can see that the current through the fuse equals the sum of the currents through the television, the straightening iron, and the hair dryer.

$$I_{fuse} = I_{TV} + I_{iron} + I_{dryer} = 2.0\text{ A} + 6.0\text{ A} + 12\text{ A} = 20\text{ A}$$

Is this enough current to blow the fuse? The fuse in **Figure 29** is rated at 15 A, so the 20-A current from the television, the straightening iron, and the hair dryer would blow the fuse. This would stop current throughout the circuit.

PHYSICS Challenge

When the galvanometer, a device used to measure very small currents, in this circuit measures zero, the circuit is said to be balanced.

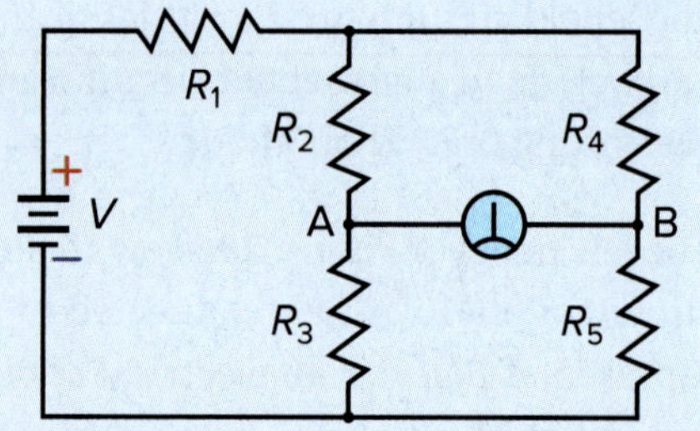

1. Your lab partner states that the only way to balance this circuit is to make all the resistors equal. Will this balance the circuit? Is there more than one way to balance this circuit? Explain.
2. Derive a general equation for a balanced circuit using the given labels. *Hint: Treat the circuit as two voltage dividers.*
3. Which of the resistors can be replaced with a variable resistor and then used to balance the circuit?
4. Which of the resistors can be replaced with a variable resistor and then used as a sensitivity control? Why would this be necessary? How would it be used in practice?

Combined Series-Parallel Circuits

Have you ever noticed the light dim when you turned on a hair dryer? The light and the hair dryer are connected in parallel across 120 V. Because of the parallel connection, the current through the light should not have changed when you turned on the hair dryer. Yet the light did dim, so the current must have changed. The dimming occurred because the household wiring has a small resistance. As shown in **Figure 30,** this resistance was in series with the parallel circuit. A circuit that includes series and parallel branches is a **combination series-parallel circuit.** Strategies for analyzing such circuits are listed on the next page.

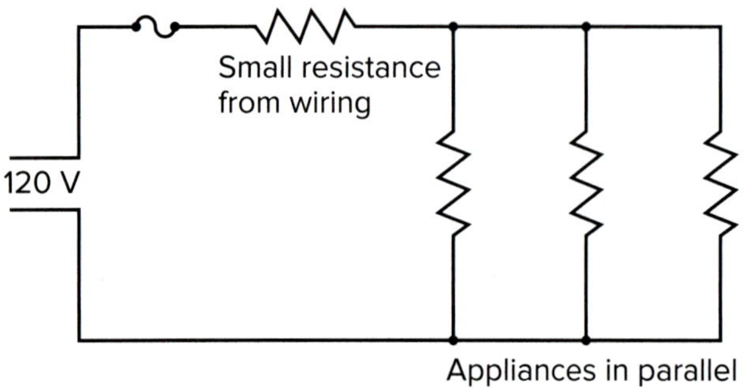

Figure 30 Circuits that contain series and parallel branches are called combination series-parallel circuits.

Explain *why the wiring has resistance.*

PROBLEM-SOLVING STRATEGY

SERIES-PARALLEL CIRCUITS

When analyzing a combination series-parallel circuit, use the following steps to break down the problem.

1. Draw a schematic diagram of the circuit.
2. Find any parallel resistors. Resistors in parallel have separate current paths. They must have the same potential differences across them. Calculate the single equivalent resistance of a resistor that can replace them. Draw a new schematic using that resistor.
3. Are any resistors (including the equivalent resistor) now in series? Resistors in series have one and only one current path through them. Calculate a single new equivalent resistance that can replace them. Draw a new schematic diagram using that resistor.
4. Repeat steps 2 and 3 until you can reduce the circuit to a single resistor. Find the total circuit current. Then go backward through the circuits to find the currents through and the potential differences across individual resistors.

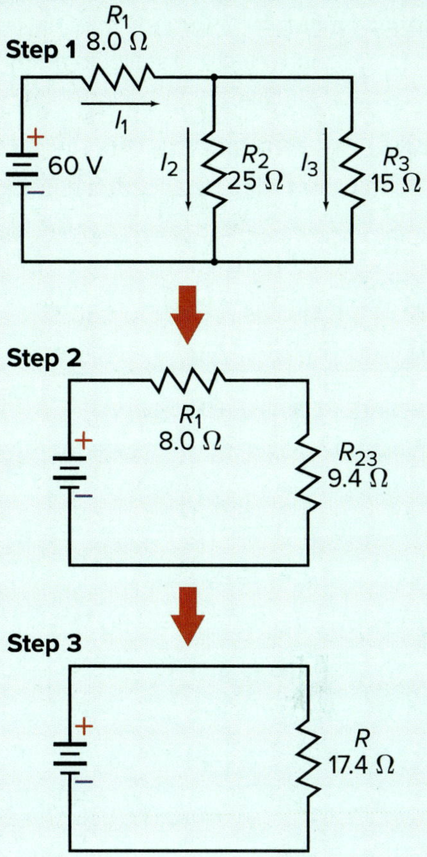

Ammeters

Recall that an ammeter is a device that is used to measure the current in any branch or part of a circuit. If, for example, you wanted to measure the current through a resistor, you would place an ammeter in series with the resistor. This would require opening the current path and inserting an ammeter. Ideally, the use of an ammeter should not change the current in the circuit. Because the current would decrease if the ammeter increased the resistance in the circuit, the resistance of an ammeter is designed to be as low as possible. **Figure 31** shows an ammeter as a high-resistance meter placed in parallel with a 0.01-Ω resistor. Because the resistance of the ammeter is much less than that of the resistors, the current decrease is negligible.

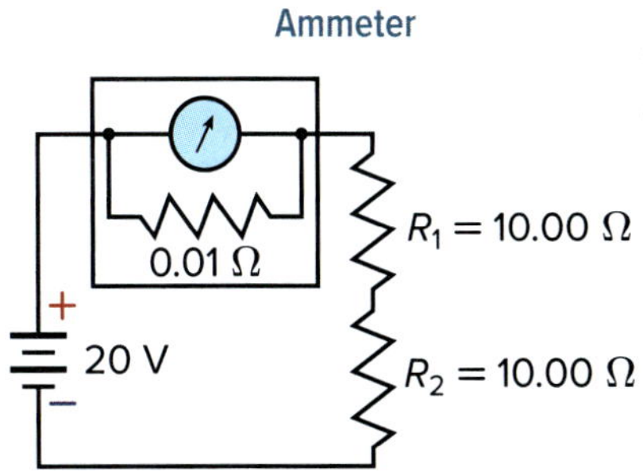

$$R_{net} = 0.01\ \Omega + 10.00\ \Omega + 10.00\ \Omega$$
$$= 20.01\ \Omega$$

Figure 31 An ammeter is connected in series with two resistors. The small resistance of the ammeter slightly alters the current in the circuit.

EXAMPLE Problem 7

SERIES-PARALLEL CIRCUIT A hair dryer with a resistance of 12.0 Ω and a lamp with a resistance of 125 Ω are connected in parallel to a 125-V source through a 1.50-Ω resistor in series. Find the current through the lamp when the hair dryer is on.

1 ANALYZE AND SKETCH THE PROBLEM

- Draw the series-parallel circuit including the hair dryer and the lamp.
- Replace R_1 and R_2 with a single equivalent resistance, R_p.

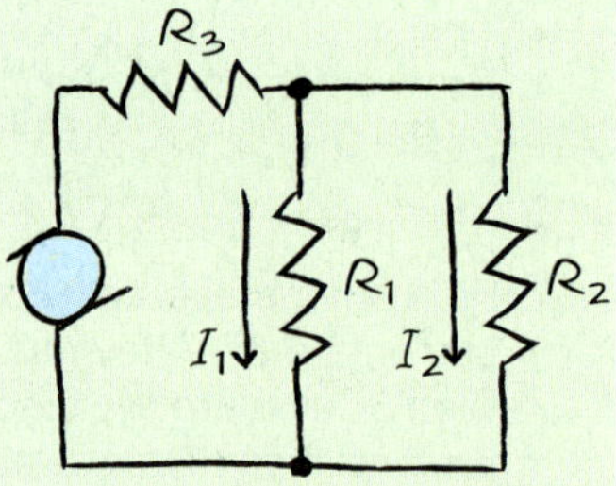

Known		Unknown	
$R_1 = 125\ \Omega$	$R_3 = 1.50\ \Omega$	$I = ?$	$I_1 = ?$
$R_2 = 12.0\ \Omega$	$\Delta V_{source} = 125\ V$	$R = ?$	$R_p = ?$

2 SOLVE FOR THE UNKNOWN

Find the equivalent resistance for the parallel circuit, then find the equivalent resistance for the entire circuit, and then calculate the current.

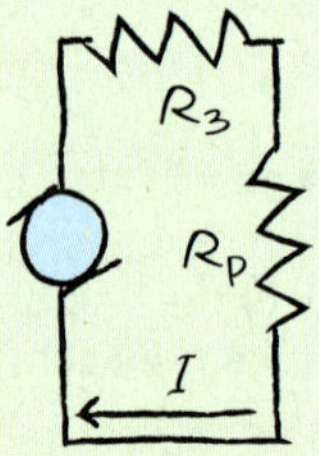

$$\frac{1}{R_p} = \frac{1}{R_1} + \frac{1}{R_2} = \frac{1}{125\ \Omega} + \frac{1}{12.0\ \Omega}$$ Substitute $R_1 = 125\ \Omega$, $R_2 = 12.0\ \Omega$.

$$R_p = 10.9\ \Omega$$

$$R = R_3 + R_p = 1.50\ \Omega + 10.9\ \Omega$$ Substitute $R_3 = 1.50\ \Omega$, $R_p = 10.9\ \Omega$.

$$= 12.4\ \Omega$$

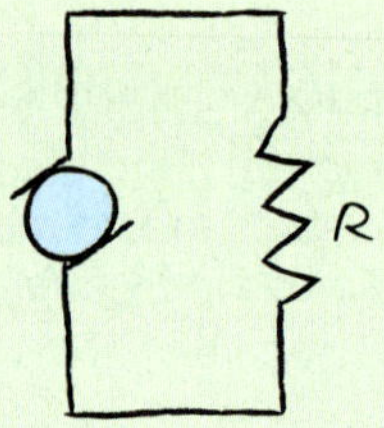

$$I = \frac{\Delta V_{source}}{R} = \frac{125\ V}{12.4\ \Omega}$$ Substitute $\Delta V_{source} = 125\ V$, $R = 12.4\ \Omega$.

$$= 10.1\ A$$

$$\Delta V_3 = IR_3 = (10.1\ A)(1.50\ \Omega) = 15.2\ V$$ Substitute $I = 10.1\ A$, $R_3 = 1.50\ \Omega$.

$$\Delta V_1 = \Delta V_{source} - \Delta V_3 = 125\ V - 15.2\ V$$ Substitute $\Delta V_{source} = 125\ V$, $V_3 = 15.2\ V$.

$$= 1.10\times10^2\ V$$

$$I_1 = \frac{\Delta V_1}{R_1} = \frac{1.10\times10^2\ V}{125\ \Omega}$$ Substitute $\Delta V_1 = 1.10\times10^2\ V$, $R_1 = 125\ \Omega$.

$$= 0.880\ A$$

3 EVALUATE THE ANSWER

- **Are the units correct?** Current is measured in amps, and potential differences are measured in volts.
- **Is the magnitude realistic?** The resistance in ohms is greater than the potential difference in volts, so the current should be less than 1 A.

PRACTICE Problems

ADDITIONAL PRACTICE

66. A series-parallel circuit, similar to the one in Example Problem 7, has three resistors: one uses 2.0 W, the second 3.0 W, and the third 1.5 W. How much current does the circuit require from a 12-V battery?

67. If the 13 lights shown in **Figure 32** are identical, which of them will burn brightest?

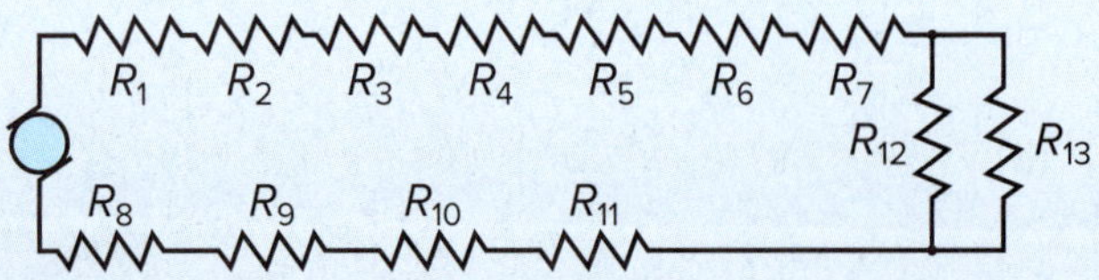

Figure 32

68. CHALLENGE A series-parallel circuit has three appliances on it. A blender and a stand mixer are in parallel, and a toaster is connected in series as shown in **Figure 33.** Find the current through the blender.

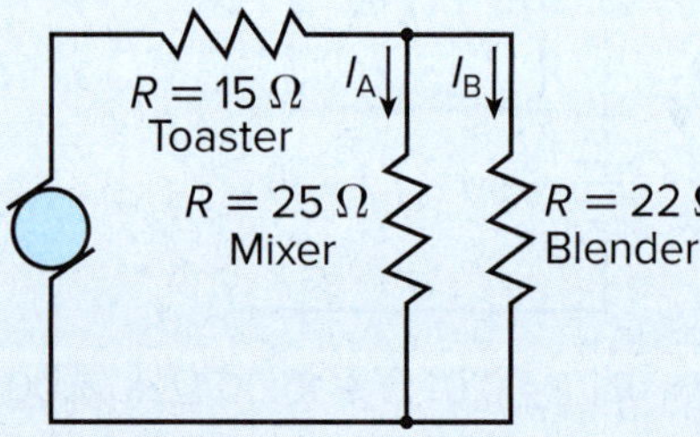

Figure 33

Voltmeters

Another instrument, called a voltmeter, is used to measure the potential difference across a portion of a circuit. To measure the potential difference across a resistor, a voltmeter is connected in parallel with the resistor. Voltmeters are designed to have a very high resistance so as to cause the smallest possible change in currents and voltages in the circuit. Consider the circuit shown in **Figure 34.** A voltmeter is shown as a meter in series with a 10-kΩ resistor. When the voltmeter is connected in parallel with R_2, the equivalent resistance of the combination is smaller than R_2 alone. Thus, the total resistance of the circuit decreases, and the current increases. The value of R_1 has not changed, but the current through it has increased, thereby increasing the potential difference across it. The battery, however, holds the potential difference across R_1 and R_2 constant. Thus, the potential difference across R_2 must decrease. The result of connecting a voltmeter across a resistor is to lower the potential difference across it. If the resistance of the voltmeter is higher, then the change in potential difference across the resistor is lower. Practical meters have resistances of 10 MΩ.

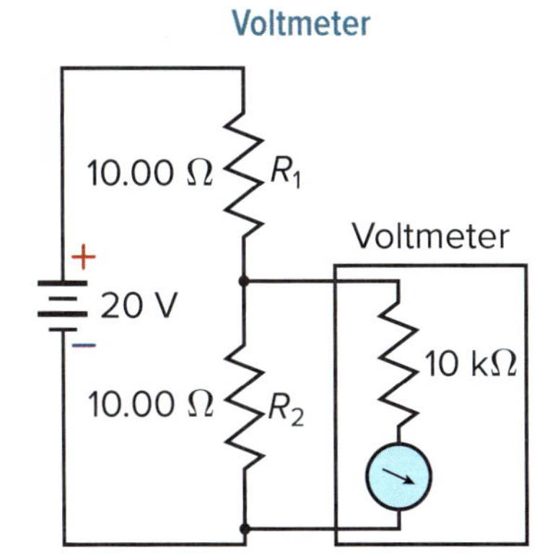

$$R_{net} = 10.00\ \Omega + \frac{1}{\frac{1}{10.00\ \Omega} + \frac{1}{10{,}000\ \Omega}}$$
$$= 19.99\ \Omega$$

Figure 34 A voltmeter is connected in parallel with a resistor. The high resistance of the voltmeter results in a negligible change in the circuit current and voltage.

Check Your Progress

69. **Circuits** Explain in your own words what a combination series-parallel circuit is.

*Refer to **Figure 35** for questions 70–74 and 76.*

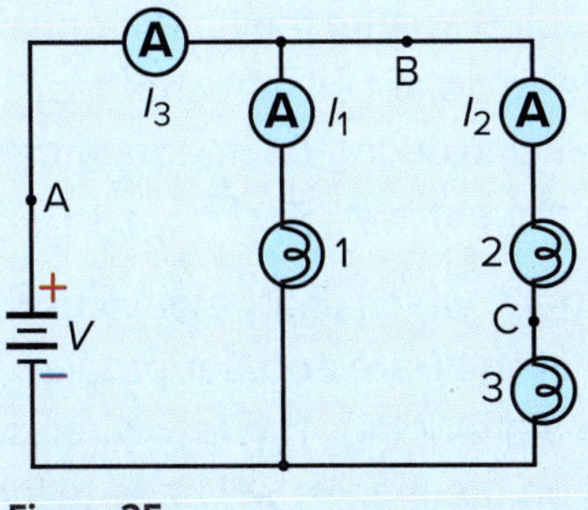

Figure 35

70. **Brightness** How do the brightness of the bulbs compare?
71. **Current** If I_3 is 1.7 A and I_1 is 1.1 A, what is the current through bulb 2?
72. **Circuits in Series** The wire at point C is broken, and a small resistor is inserted in series with bulbs 2 and 3. What happens to the brightness of the two bulbs? Explain.
73. **Battery Voltage** A voltmeter connected across bulb 2 measures 3.8 V, and a voltmeter connected across bulb 3 measures 4.2 V. What is the potential difference across the battery?
74. **Circuits** Using information from the previous problem, determine whether bulbs 2 and 3 are identical.
75. **Circuit Protection** Describe three common safety devices associated with household wiring.
76. **Critical Thinking** How could you rearrange the circuit to make the three bulbs in **Figure 35** burn with equal intensity? Is there more than one way to do this?

LEARNSMART® Go online to follow your personalized learning path to review, practice, and reinforce your understanding.

ENGINEERING & TECHNOLOGY

Leading the Charge

Have you ever been away from home when you discover that the battery is running low on your smartphone or fitness monitor? Finding a place to charge your device can be an impossible task when you are in the middle of a busy day. With ever increasing needs during this fast-paced technological age, there is impetus to invent new, more accessible ways to charge devices "on the go." Engineers are currently developing flexible, wearable supercapacitors and batteries that can tackle this everyday issue.

One day people may charge their smartphones with graphene supercapacitors printed directly on their clothes.

Transportable Power Sources

People carry many electronic devices with them, including health monitors, smartwatches, and medical sensors. All of these devices must be recharged on a regular basis. Thus far, engineers have found it difficult to design a way to power these mobile devices that does not involve plugging them into a wall socket, car, or computer. The transportable power source would have to be tough, flexible, quick and easy to charge, and capable of holding a charge for a long time.

Some engineers are experimenting with flexible graphene supercapacitors, which store energy electrostatically on their surfaces instead of employing chemical reactions, as batteries do. These supercapacitors can be printed on clothes, which people can wear and wash without damaging them. The supercapacitors can charge in seconds, and even after millions of recharges, they will not lose their capacity to store energy.

However, engineers have several problems to solve before flexible graphene supercapacitors become a viable option. Making high-quality graphene is expensive and labor-intensive. In addition, graphene does not conduct electricity as well as metals.

Other engineers are designing a flexible battery. The inflexible metal components of the battery are wound into a spiral so that they can bend. The battery includes a solar cell that charges it. Engineers are still working on integrating the solar cell with the battery.

Wireless chargers, which charge devices that are located in the vicinity, are another product that engineers are developing. The few wireless chargers now on the market will have to increase in speed and decrease in price before they become a popular option.

COMMUNICATE TECHNICAL INFORMATION

Design a pamphlet comparing the pros and cons of flexible graphene supercapacitors, flexible batteries, and wireless chargers.

CALIFORNIA CONNECTION

Enter the keywords *flexible batteries California* or *flexible supercapacitors California* into a search engine. Use the information you find as an aid to write your pamphlet.

MODULE 19
STUDY GUIDE

GO ONLINE to study with your Science Notebook.

Lesson 1 CURRENT AND CIRCUITS

- Electric current is a flow of charged particles. By convention, current direction is the direction in which a positive test charge moves.
- A circuit transforms electrical energy to thermal energy, radiant energy, or some other form of energy.
- Ohm's law states that the ratio of potential difference to current is a constant for a given conductor. Any resistance that does not change with potential difference or the direction of charge flow obeys Ohm's law.
- The following equations show how power, current, potential difference, and resistance are mathematically related.

$$P = I\Delta V \qquad R = \frac{\Delta V}{I}$$

- electric current
- conventional current
- battery
- electric circuit
- ampere
- resistance
- resistor
- parallel connection
- series connection

Lesson 2 USING ELECTRICAL ENERGY

- Electrical energy is transformed into thermal energy whenever moving charges transfer energy to other particles. If energy is transformed at a uniform rate, the total energy transformed can be represented by the following equations.

$$E = Pt \qquad E = I^2Rt \qquad E = \left(\frac{(\Delta V)^2}{R}\right)t$$

- Transmitting electrical energy at higher voltages reduces current without reducing power, which minimizes Joule heating loss.

- superconductor
- kilowatt-hour

Lesson 3 SIMPLE CIRCUITS

- The current is the same everywhere in a simple series circuit. If any branch of a parallel circuit is opened, there is no current in that branch. The current in the other branches is unchanged.
- The current in a series circuit is equal to the potential difference divided by the equivalent resistance. The sum of the voltage drops across resistors that are in series is equal to the potential difference applied across the combination. The equivalent resistance of a series circuit is the sum of the resistances of its parts.
- In a parallel circuit, the total current is equal to the sum of the currents in the branches. The voltage drops across all branches of a parallel circuit are the same. The reciprocal of the equivalent resistance of parallel resistors is equal to the sum of the reciprocals of the individual resistances.

- series circuit
- equivalent resistance
- voltage divider
- parallel circuit

Lesson 4 APPLICATIONS OF CIRCUITS

- A fuse, a circuit breaker, and a ground-fault interrupter create an open circuit and stop current when currents are dangerously high.
- To analyze complex circuits, reduce any parallel branch to a single equivalent resistance. Then, replace resistors in series by a single resistance.
- A voltmeter measures the potential difference (voltage) across any part of a circuit. It has a high resistance and is connected in parallel. An ammeter is used to measure the current in a branch or part of a circuit. It has a low resistance and is connected in series.

- short circuit
- circuit breaker
- fuse
- ground-fault interrupter
- combination series-parallel circuit

THREE-DIMENSIONAL THINKING

Module Wrap-Up

REVISIT THE PHENOMENON

How does energy get from the power plant to the lights in your home?

CER Claim, Evidence, Reasoning

Explain Your Reasoning Revisit the claim you made when you encountered the phenomenon. Summarize the evidence you gathered from your investigations and research and finalize your Summary Table. Does your evidence support your claim? If not, revise your claim. Explain why your evidence supports your claim.

STEM UNIT PROJECT

Now that you've completed the module, revisit your STEM unit project. You will summarize your evidence and apply it to the project.

GO FURTHER

SEP Data Analysis Lab

Does Ohm's law apply to nichrome wires?

Some students connected a length of nichrome wire to a variable power supply to produce between 0.00 V and 10.00 V across the wire. They then measured the current through the wire for several voltages. The students recorded the data for the voltages used and the currents measured, as shown in the table.

ΔV (V)	I (A)	ΔV (V)	I (A)
2.00	0.0140	−2.00	−0.0140
4.00	0.0270	−4.00	−0.0280
6.00	0.0400	−6.00	−0.0390
8.00	0.0520	−8.00	−0.0510
10.00	0.0630	−10.00	−0.0620

CER Analyze and Interpret Data

1. For each measurement, calculate the resistance.
2. Graph I v. ΔV.
3. **Claim** Does the nichrome wire obey Ohm's law?
4. **Evidence and Reasoning** Defend your claim using your calculations and graph.

NASA's Earth Observatory/NOAA/DoD

MODULE 20
MAGNETISM

Stockbyte/Getty Images

MODULE 20
MAGNETIC FIELDS

ENCOUNTER THE PHENOMENON

What makes this electromagnet stronger than a typical refrigerator magnet?

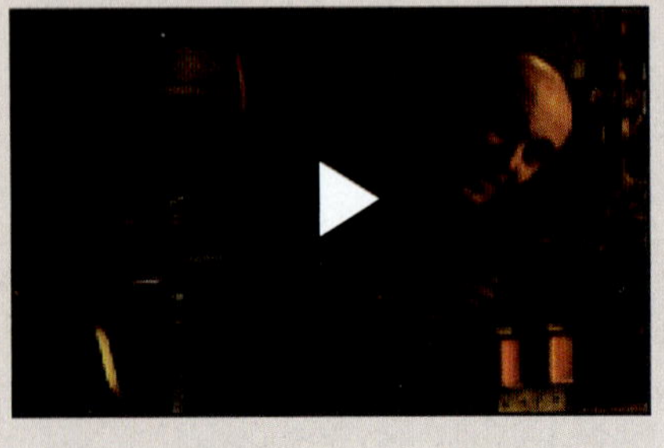

GO ONLINE to watch a video about the discovery of the connection between electricity and magnetism.

SEP Ask Questions

Do you have other questions about the phenomenon? If so, add them to the driving question board.

CER Claim, Evidence, Reasoning

Make Your Claim Use your CER chart to make a claim about why the electromagnet is stronger than a refrigerator magnet. Explain your reasoning.

Collect Evidence Use the lessons in this module to collect evidence to support your claim. Record your evidence as you move through the module.

Explain Your Reasoning You will revisit your claim and explain your reasoning at the end of the module.

GO ONLINE to access your CER chart and explore resources that can help you collect evidence.

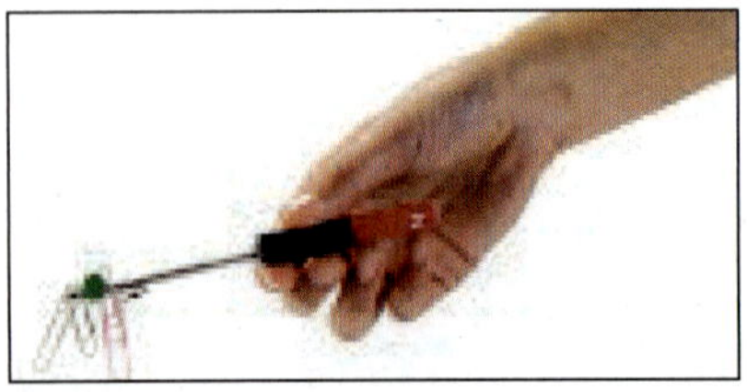

LESSON 1: Explore & Explain: Magnetic Domains

LESSON 2: Explore & Explain: Forces on Current-Carrying Wires

Smithsonian

Additional Resources

LESSON 1

UNDERSTANDING MAGNETISM

FOCUS QUESTION

Why are some materials magnetic and others are not?

Properties of Magnets

Magnets have been known and used for more than 2000 years. Ancient sailors used naturally magnetic rocks, called lodestones, as compasses. Ancient physicians thought lodestones could cure disease. Today, many objects we use every day rely on magnets to work. Electric motors, earbuds, loudspeakers, and computer hard drives all depend on the interaction between magnetic fields and electric currents.

Poles of magnets As a child, you likely did investigations with simple magnets, such as those shown in **Figure 1.** You probably noticed that the two ends of a magnet behave in different ways. That is because magnets are **polarized:** they have two opposite ends, called poles.

Richard Hutchings/Digital Light Source

Figure 1 All magnets, no matter their size or shape, have two poles.

3D THINKING DCI Disciplinary Core Ideas CCC Crosscutting Concepts SEP Science & Engineering Practices

COLLECT EVIDENCE

Use your Science Journal to record the evidence you collect as you complete the readings and activities in this lesson.

INVESTIGATE

GO ONLINE to find these activities and more resources.

Applying Practices: Modeling Magnetic Fields

HS-PS3-5. Develop and use a model of two objects interacting through electric or magnetic fields to illustrate the forces between objects and the changes in energy of the objects due to the interaction.

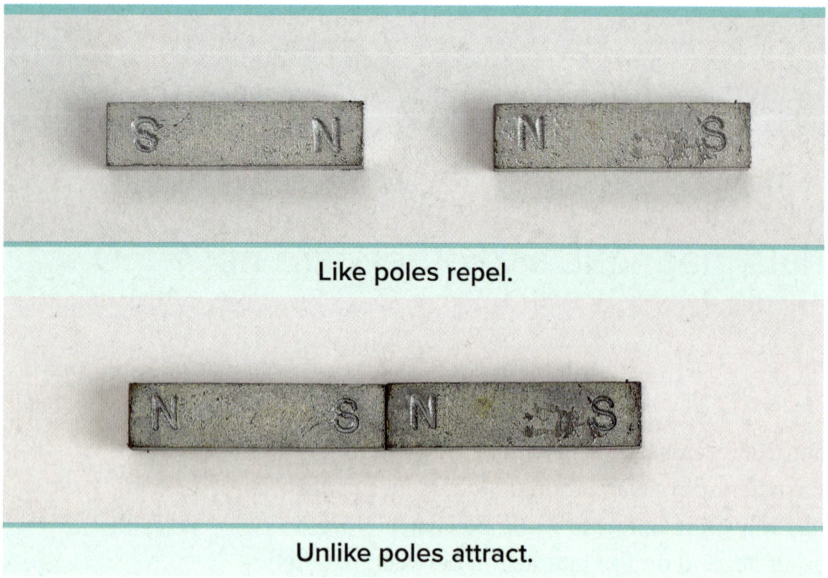

Figure 2 Like poles of two magnets repel each other (top), while unlike poles attract each other (bottom).

Think about a bar magnet suspended on a string. In what direction do you think it will point when it comes to rest? A magnet that is free to rotate always comes to rest pointing in the north-south direction. The pole pointing north is called the north-seeking pole or, more simply, the north pole. The opposite pole is the south pole. A compass is just a small magnet, mounted so that it is free to rotate.

Earth as a magnet The needle of a compass points in a north-south direction because Earth itself is a giant magnet. A compass's north pole points to Earth's geographic North Pole. As you will read, however, a magnet's north pole is always attracted to a magnetic south pole. Therefore, what we call the North Pole is actually near Earth's magnetic south pole, and the South Pole is near Earth's magnetic north pole.

Get It?

Describe examples of how humans have used Earth's magnetic poles through the centuries.

Opposite poles What's inside a magnet that makes it polarized? You know that when you bring a metal rod near an electric charge, one end of the rod becomes negatively charged and the other end becomes positively charged, polarizing the rod. You might think a magnet, like the ones shown in **Figure 2,** is similar, with one half of the magnet positive and the other half negative, but this is not the case. No matter how you cut or break a magnet, a magnet always has two poles. There have been many searches for objects, called monopoles, with only a north pole or only a south pole, but no monopole has ever been found.

SCIENCE USAGE v. COMMON USAGE

Polarized

Science usage: having two opposite ends
All magnets are polarized.

Common usage: broken into opposing factions or groups
Members of Congress were polarized on the issue of Social Security reform.

Poles repel or attract You have likely noticed that forces between two magnets differ depending on how you orient the magnets. When you place the north pole of one magnet next to the north pole of another magnet, the magnets repel each other, as they do in the top of **Figure 2.** The same is true when you bring two south poles together. If, however, you brought the north pole of one magnet next to the south pole of another magnet, the poles would attract each other, as they do in the bottom of **Figure 2.** Like poles repel; unlike poles attract.

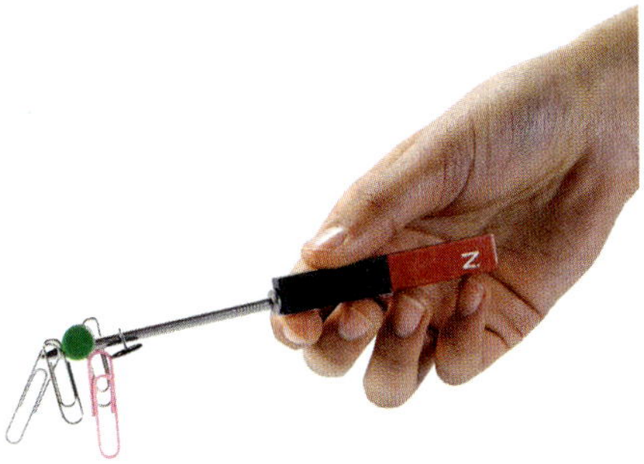

Figure 3 A common nail attached to a magnet becomes a temporary magnet by induction.

Identify *the north and south poles of the nail.*

Temporary magnets

Magnets also attract nails, paper clips, tacks, and other metal objects. These objects have no poles, and both the north and south poles of a magnet attract them. When a magnet touches one of these objects, such as the nail in **Figure 3,** the magnet polarizes the object, making it a temporary magnet. This process is called magnetization by induction.

Magnets only attract some metals. Brass, copper, and aluminum are common metals that are not attracted to magnets. Iron, nickel, and cobalt are strongly attracted. Materials containing these elements, called ferromagnetic materials, can become temporary magnets. A steel nail can become a temporary magnet because it is made of iron with tiny amounts of carbon and other materials. When you remove a nail from a magnet, the nail gradually loses most of its magnetism.

Magnetic domains

What gives a permanent or temporary magnet its magnetic properties? Each atom in a ferromagnetic material acts like a tiny magnet; each has two poles. Each is part of a **domain,** which is a group of neighboring atoms whose poles are aligned. Look at the arrows in **Figure 4.** Each arrow represents a domain. Although domains can contain as many as 10^{20} individual atoms, they are tiny—usually from 10 to 1000 microns across. Even a small sample of a ferromagnetic material contains a huge number of domains.

In a ferromagnetic material that is not magnetized, each domain points in a random direction, as shown in the top panel of **Figure 4.** But if the ferromagnetic material is next to a strong magnet, most of the object's domains preferentially align to point in the same direction as the poles of the external magnet, as shown in the bottom panel of **Figure 4.** When its domains are aligned in the same direction, the material becomes a temporary magnet. When an external magnet is removed from a temporary magnet, the domains of the temporary magnet return to a random arrangement, and the material loses its magnetization. How long it takes for a temporary magnet to lose its magnetization depends on the interactions between the atoms, which depend on the microscopic structure of the material.

Nonmagnetized Material

Magnetized Material

Figure 4 Domains in a nonmagnetized ferromagnetic material point in random directions (top). When a strong magnet is placed near a ferromagnetic material, the domains in that ferromagnetic material align with those of the external magnet (bottom).

Creating permanent magnets

The only naturally occurring magnet is the mineral magnetite. The lodestones that ancient sailors used were nothing more than pieces of magnetite. If magnetite is the only naturally occurring magnet, how, then, are commercial permanent magnets made?

Get It?

Cite Evidence Infer what the magnetic domains look like in the magnet shown on the first page of this module.

PRACTICE Problems

ADDITIONAL PRACTICE

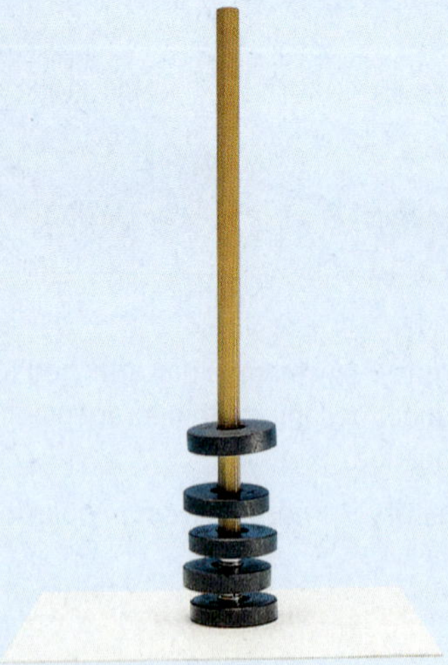

Figure 5

1. If you hold a bar magnet in each hand and bring your hands close together, will the force be attractive or repulsive if the magnets are held in the following ways?
   a. The two north poles are brought close together.
   b. A north pole and a south pole are brought together.
2. **Figure 5** (at left) shows five disk magnets floating above one another. The north pole of the top-most disk faces up. Which poles are on the top side of each of the other magnets?
3. The ends of a compass needle are marked N and S. How would you explain to someone why the pole marked N points north? A complete answer should involve Earth's magnetic poles.
4. **CHALLENGE** When students use magnets and compasses, they often touch the magnets to the compasses. Then they find that the compasses point south. Explain why this might occur.

When an object containing certain ferromagnetic materials is heated in the presence of a strong magnet, thermal energy frees the atoms in each of the object's domains. The domains can rotate and align with the magnet's poles. The object is then cooled while it is still in the presence of the strong magnet. After cooling, the object's atoms are less free to rotate. Therefore, when the strong magnet is removed from the object, the object remains magnetized. A permanent magnet has been created. If this permanent magnet is later reheated or dropped, however, the atoms can jostle out of alignment, reordering the domains and removing the magnetic properties.

History of Earth's magnetism Magnetic domains in rocks containing iron record the history of Earth's magnetism. Rocks on the seafloor form when molten rock (magma) pours out of cracks in the bottom of the oceans. As the magma cools into rock, the domains in the iron-containing rocks align in the direction of Earth's magnetic field. These rocks become weak permanent magnets. As more magma pours out of the cracks, the older rocks are pushed away from the cracks. As a result, rocks farther from the cracks are older than those near the cracks.

Scientists who first examined seafloor rocks were surprised to find that the alignment of domains varied in the iron of rocks of different ages. They concluded that Earth's magnetic north and south poles have exchanged places many times during Earth's history.

Magnetic Fields Around Magnets

When you investigate simple magnets, you notice that the forces between magnets are present not only when magnets touch each other, but also when magnets are held apart. Just as the existence of long-range electric and gravitational forces can be interpreted as being the result of electric and gravitational fields, long-range magnetic forces can be interpreted as being the result of magnetic fields. These fields transfer energy through space.

Magnetic fields are fields that exist in space where magnets would experience a force. They are vector quantities because they have magnitude and direction. The needle of a compass in Earth's magnetic field aligns in the direction of Earth's field. When in a stronger magnetic field, the needle realigns in the direction of the stronger field.

Doug Martin

Figure 6 Iron filings on paper illustrate the magnetic field of a bar magnet two-dimensionally (left) and iron filings on an upright magnet illustrate the magnetic field three-dimensionally (right).

Visualizing magnetic fields What do magnetic fields look like? Like electric and gravitational fields, magnetic fields are invisible. But we can visualize them in a few different ways. One way is to place iron filings around a magnet. Each long, thin iron filing around a magnet becomes a temporary magnet by induction. Just like thousands of tiny compass needles, each iron filing rotates until it is parallel to the magnetic field. You can see the results both two-dimensionally and three-dimensionally in **Figure 6.**

Magnetic field lines Scientists visualize magnetic fields using magnetic field lines, such as those shown in **Figure 7.** Like electric field lines, magnetic field lines are not real. They are used to show the direction as well as the strength of a magnetic field. The number of magnetic field lines passing through a surface perpendicular to the lines is the **magnetic flux.** The flux per unit area is proportional to the strength of the magnetic field. Magnetic flux is most concentrated at magnetic poles, where magnetic field strength is the highest.

The direction of a magnetic field line is defined as the direction in which the north pole of a compass points when placed in a magnetic field. Therefore, field lines emerge from a magnet's north pole and enter at its south pole, as in **Figure 7.** The field lines form closed loops, continuing through a magnet from its south pole to its north pole.

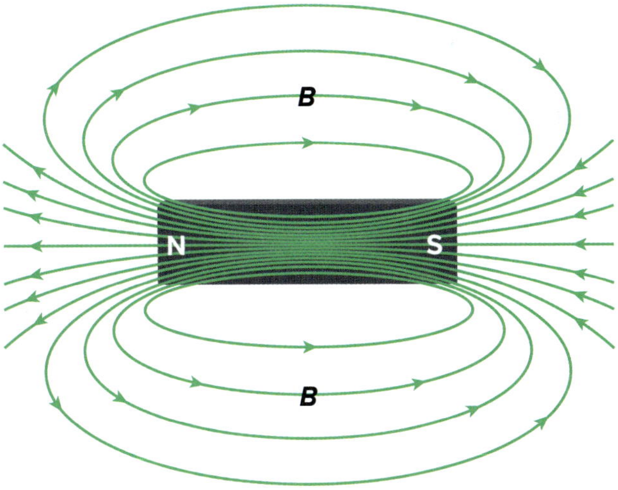

Figure 7 Magnetic field lines can be visualized as lines leaving the north pole of a magnet, entering the south pole, and passing through the magnet, forming closed loops. Magnetic fields are traditionally represented by the letter *B*.

COLOR CONVENTION
Magnetic field line (*B*) ⟷ teal

Forces on objects in magnetic fields

You read earlier that a magnet can polarize ferromagnetic materials. How can this be explained in terms of magnetic fields and forces?

Forces on permanent magnets Magnetic fields exert forces on magnets. When like poles of two magnets are close together, the field produced by the north pole of one magnet pushes the north pole of the second magnet away in the direction of the field lines, as shown by the iron filings in the top panel of **Figure 8.** Now look at the bottom panel of **Figure 8.** The field from the north pole of one magnet now acts on the south pole of the second magnet, attracting it in a direction opposite the field lines. The magnetic field is continuous, forming arcs from one magnet to the other.

Forces on temporary magnets Magnetic fields exert forces not only on other magnets. They also exert forces on ferromagnetic materials. When an object containing a ferromagnetic material is placed in the field of a permanent magnet, field lines leave the magnet's north end and enter the end of the object that is closest to the magnet. The field lines pass through the object and loop back to the magnet's south pole. The domains in the object align their poles along the field lines, making the end of the object closest to the magnet's north pole the object's south pole. The object's new south pole is then attracted to the magnet's north pole, and the object's new north pole is repelled.

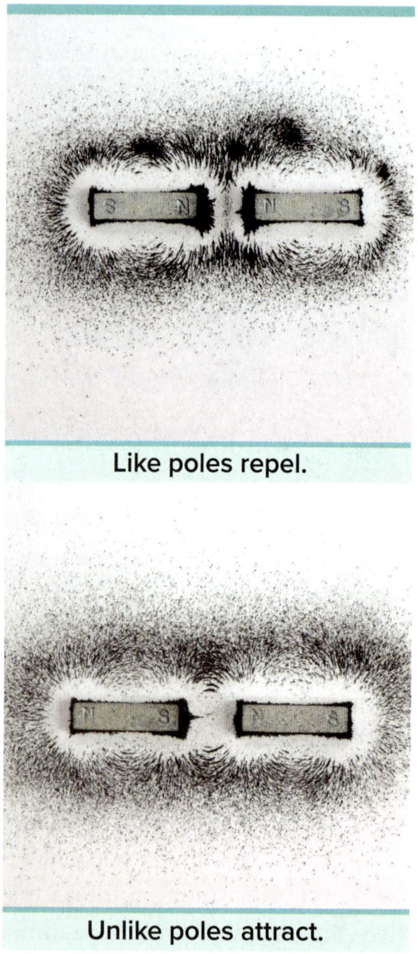

Figure 8 Iron filings can be used to visualize the magnetic field around two like poles (top) and around two unlike poles (bottom). The iron filings help us understand how like poles repel and unlike poles attract.

Electromagnetism

In 1820, while doing a lecture demonstration, Danish physicist Hans Christian Oersted laid a wire across the top of a compass and connected the ends to a battery to complete an electric circuit. The compass was oriented so its needle was parallel to the wire, as shown in the left side of **Figure 9.**

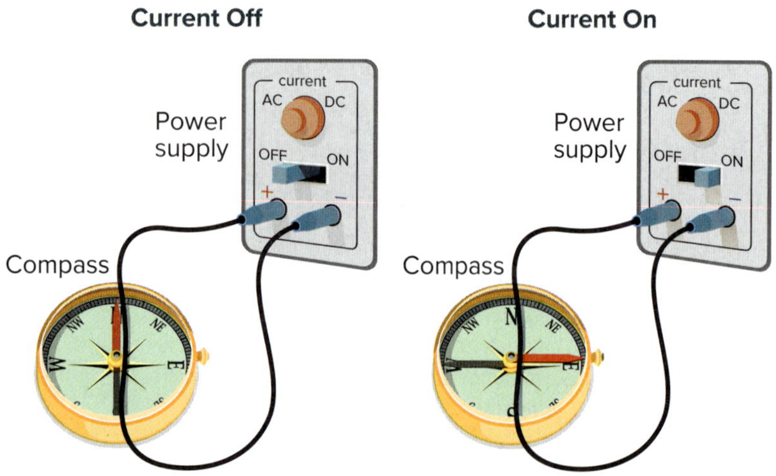

Figure 9 The needle of a compass under a wire and originally parallel to the wire when current is off (left) moves so it is perpendicular to the wire when current is on (right).

Richard Hutchings/Digital Light Source

When Oersted turned the current on, he was amazed to see that the needle moved so it was perpendicular to the wire, as it is in the right side of **Figure 9.** When Oersted placed the compass on top of the wire, the needle again became perpendicular to the wire, but it pointed in the other direction. The same thing happened when he reversed the current's direction: the compass needle reversed direction. When he turned off the current, the needle returned to its original position.

Oersted's conclusion—that a current produces a magnetic field—was the first hint that a connection exists between magnetism and electric currents. As you will read, the relationship between magnetism and electric current underlies the design and operation of many modern devices.

Get It?

Summarize Oersted's conclusion in your own words.

Magnetic fields from current-carrying wires

The magnetic field around a current-carrying wire is always perpendicular to that wire. Just as field lines around permanent magnets form closed loops, the field lines around current-carrying wires also form closed loops. The circular pattern of iron filings shown in the top panel of **Figure 10** represents these loops. The strength of the magnetic field around a long, straight wire is proportional to the current in that wire. Magnetic field strength also varies inversely with distance from the wire.

Direction of the magnetic field

How can you find the direction of the magnetic field around a current-carrying wire? Scientists use right-hand rules to describe how the directions of electric and magnetic properties relate. In this case, imagine holding a length of wire with your right hand, as shown in **Figure 10.** If your thumb points in the direction of the conventional (positive) current, as it does in the bottom panel of **Figure 10,** the fingers of your hand encircling the wire will point in the direction of the magnetic field.

Electromagnets

You just read that a current in a wire produces a magnetic field encircling that wire. What do you think happens to the magnetic field around a wire formed into a loop? An electric current in a single loop of wire forms a magnetic field all around the loop, as shown in the left panel of **Figure 11** on the next page. By applying a right-hand rule to any part of the loop in **Figure 11,** you can see that the direction of the magnetic field inside the loop is always the same.

Magnetic Field Around a Wire

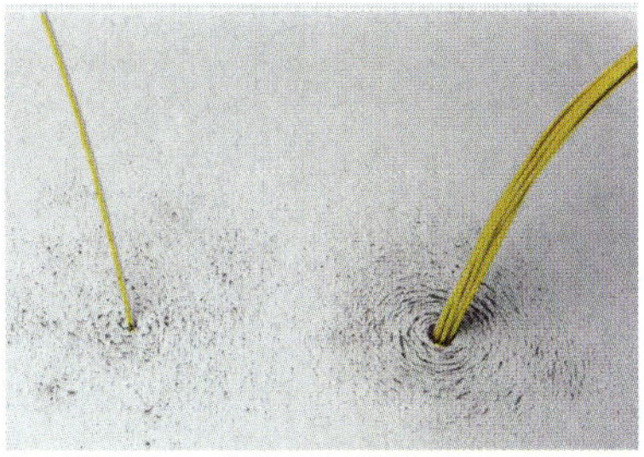

Right-Hand Rule

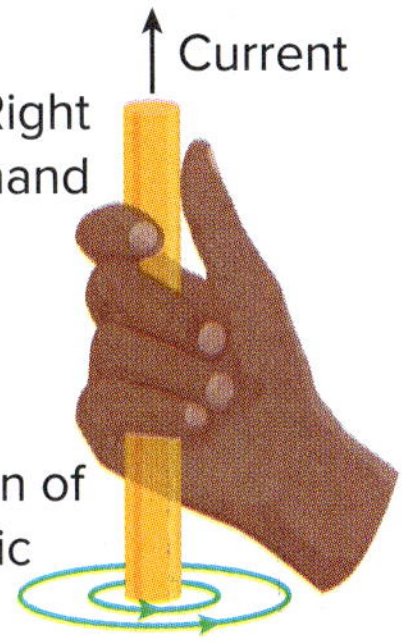

Figure 10 The circular patterns formed by iron filings around a current-carrying wire (top) represent the magnetic field around the wire. You can determine the direction of the magnetic field around the wire using a right-hand rule (bottom).

Analyze *What happens to the magnetic field around a wire when current changes direction?*

CCC CROSSCUTTING CONCEPTS

Cause and Effect Study **Figure 9** on the previous page. What empirical evidence did Oersted observe that led him to the conclusion that electric current produces a magnetic field?

Magnetic Field Around a Loop

Inside the loop, the field is toward you. Outside the loop, it is away from you.

Magnetic Field in a Solenoid

The magnetic fields of the loops inside a solenoid are all in the same direction.

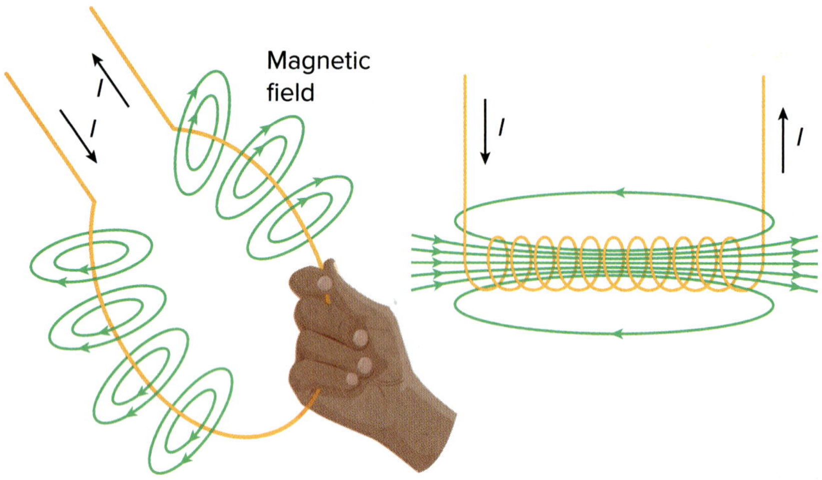

Figure 11 You can model the direction of the magnetic field around a loop of current-carrying wire and around a solenoid.

Assess *Is the magnetic field greater inside or outside the solenoid?*

Now think about a wire with many loops. A wire connected to a circuit and coiled into many spiral loops is a **solenoid.** When current is turned on in a solenoid, each loop produces its own magnetic field. The fields are all in the same direction, as shown in the right panel of **Figure 11,** so the fields add together. When there is an electric current in a solenoid, the solenoid has a magnetic field similar to the field of a permanent magnet. This kind of magnet is an electromagnet. An **electromagnet** is a magnet whose magnetic field is produced by electric current.

Loops and field strength Solenoids can be exceptionally strong electromagnets, producing magnetic fields much stronger than those around permanent magnets. The strength of the magnetic field in a solenoid is proportional to the current in the solenoid's loops. It is also proportional to the number and spacing of loops. The more loops there are in a solenoid and the closer they are spaced, the greater the solenoid's magnetic field strength. The magnetic field strength of a solenoid also can be increased by placing an iron-containing rod inside it. An iron rod strengthens the solenoid's magnetism because the solenoid's field produces a temporary magnetic field in the iron, just as a permanent magnet produces a temporary magnet in a ferromagnetic object.

Right-hand rule for a solenoid You can use a right-hand rule to determine the direction of the magnetic field around a solenoid when current is on. Imagine holding a solenoid with your right hand. If you curl your fingers around the solenoid in the direction of the conventional (positive) current, as in **Figure 12,** your thumb will point toward the solenoid's north pole.

Right-Hand Rule

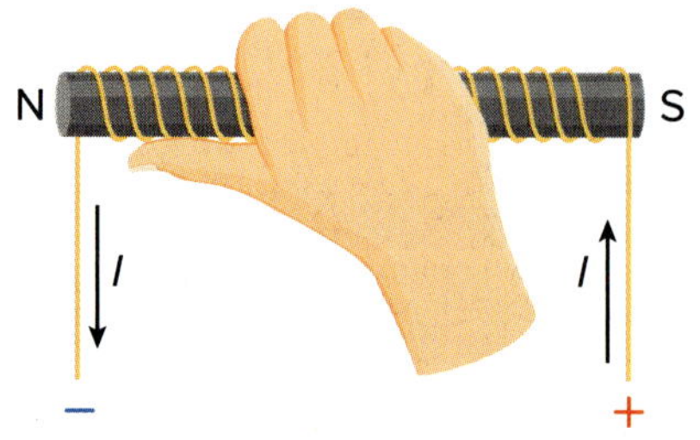

Figure 12 Imagine you are holding the solenoid with your right hand. Your thumb will point toward the solenoid's north pole when you curl your fingers in the direction of the conventional current.

PRACTICE Problems

ADDITIONAL PRACTICE

5. How does the strength of a magnetic field that is 1 cm from a current-carrying wire compare with each of the following?
   a. the strength of the field 2 cm from the wire
   b. the strength of the field 3 cm from the wire
6. A long, straight current-carrying wire lies in a north-south direction.
   a. The north pole of a compass needle placed above this wire points toward the east. In what direction is the current?
   b. If a compass were placed underneath this wire, in which direction would the compass needle point?
7. A student makes a magnet by winding wire around a nail and connecting it to a battery, as shown in **Figure 13.** Which end of the nail—the pointed end or the head—is the north pole?
8. You have a battery, a spool of wire, a glass rod, an iron rod, and an aluminum rod. Which rod could you use to make an electromagnet that can pick up steel objects? Explain.
9. **CHALLENGE** The electromagnet in the previous problem works well, but you would like to make the strength of the electromagnet adjustable by using a potentiometer as a variable resistor. Is this possible? Explain.

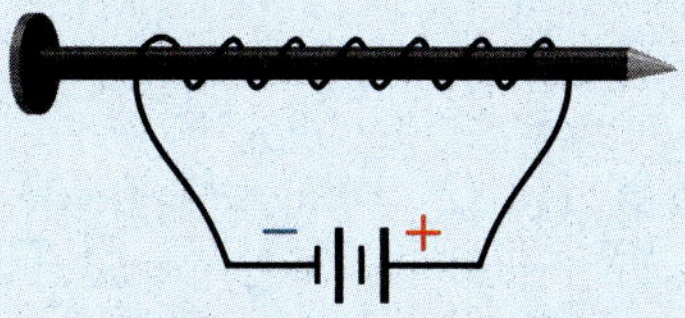

Figure 13

Check Your Progress

10. **Electromagnets** Explain how to construct an electromagnet.
11. **Magnetic Fields** What two things about a magnetic field can magnetic field lines represent?
12. **Magnetic Forces** Considering magnetic forces, how are forces at distance explained?
13. **Magnetic Fields** Where on a bar magnet is the magnetic field the strongest?
14. **Magnetic Fields** Two current-carrying wires are close to and parallel to each other and have currents with the same magnitude. If the two currents were in the same direction, how would the magnetic fields of the wires be affected? How would the fields be affected if the two currents were in opposite directions?
15. **Direction of the Field** Describe how to use a right-hand rule to determine the direction of a magnetic field around a straight, current-carrying wire.
16. **Electromagnets** A glass sheet with iron filings sprinkled on it is placed over an active electromagnet. The iron filings produce a pattern. If this scenario were repeated with the direction of current reversed, what observable differences would result? Explain.
17. **Magnetic Domains** Explain what happens to the domains of a temporary magnet when the temporary magnet is removed from a magnetic field.
18. **Critical Thinking** Imagine a toy containing two parallel, horizontal metal rods, one above the other. The top rod is free to move up and down.
    a. The top rod floats above the lower rod. When the top rod's direction is reversed, however, it falls down onto the lower rod. Explain how the rods could behave in this way.
    b. Assume the toy's top rod was lost and another rod replaced it. The new rod falls on top of the bottom rod no matter its orientation. What type of material is in the replacement rod?

LEARNSMART® Go online to follow your personalized learning path to review, practice, and reinforce your understanding.

LESSON 2
APPLYING MAGNETIC FORCES

FOCUS QUESTION

What role do magnetic forces play in everyday life?

Forces on Current-Carrying Wires

When you put a magnet in a magnetic field, the magnet can move. What happens when you put a current-carrying wire in a magnetic field? Michael Faraday, who performed many electricity and magnetism experiments during the nineteenth century, discovered that a magnetic field produces a force on a current-carrying wire. The force on the wire is always at right angles to both the direction of the magnetic field and the direction of current, as shown in the left part of **Figure 14.** When current changes direction, so does the force.

Direction of force You can use a right-hand rule to determine the direction of force on a current-carrying wire in a magnetic field. Point the fingers of your right hand in the direction of the magnetic field. Point your thumb in the direction of the wire's conventional (positive) current. The palm of your hand will face in the direction of the force acting on the wire, as shown in the right part of **Figure 14.**

Arrows in three dimensions The relationship among magnetic field, electric current, and force is three-dimensional. How do you accurately represent directional arrows in three dimensions on a two-dimensional piece of paper?

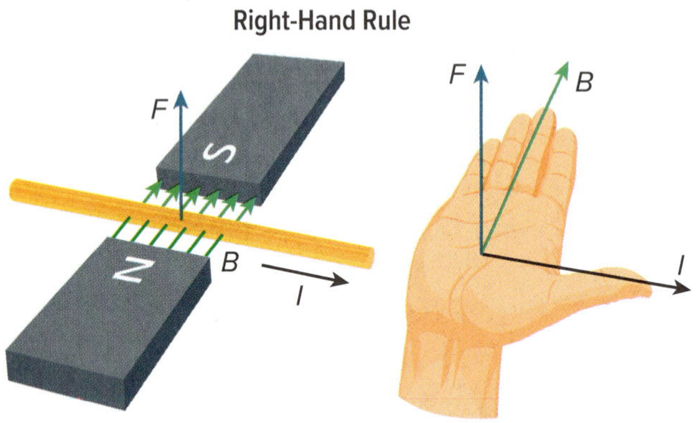

Figure 14 You can use a right-hand rule to determine the direction of force when the current (*I*) and the magnetic field (*B*) are known.

Predict *what would happen to the force if the current changed direction.*

3D THINKING DCI Disciplinary Core Ideas CCC Crosscutting Concepts SEP Science & Engineering Practices

COLLECT EVIDENCE

Use your Science Journal to record the evidence you collect as you complete the readings and activities in this lesson.

INVESTIGATE

GO ONLINE to find these activities and more resources.

Virtual Investigation: Charge in a Magnetic Field
Carry out an investigation to determine the effect a magnetic field has on a moving, electrically charged particle.

Revisit the Encounter the Phenomenon Question
What information from this lesson can help you answer the Unit and Module questions?

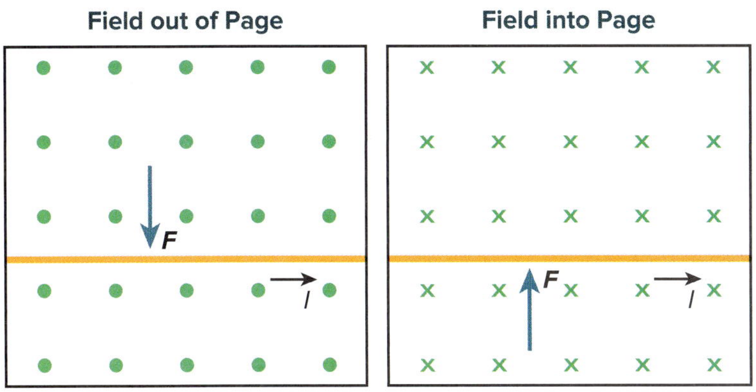

Figure 15 Dots represent a magnetic field coming out of the page, toward you (left). Crosses represent a magnetic field going into the page, away from you (right). Note that the force on each wire is perpendicular to both the magnetic field and the current.

ARROW CONVENTION

- ● ***B*** **out of the page**
- x ***B*** **into the page**

Imagine an archer shooting an arrow toward you. The arrow looks like a dot. Now imagine the same arrow going away from you. The arrow looks like a cross. You can use dots to represent magnetic fields that go into a piece of paper, and crosses to represent fields that go out of the paper, as shown in **Figure 15.**

Magnitude of force You read that you use a right-hand rule to find the direction of the force from a magnetic field on a current-carrying wire. How do you find the magnitude of this force? Experiments show that the magnitude of the force (F) on a current-carrying wire is proportional to the wire's current (I), the wire's length (L), the strength of the magnetic field (B), and the sine of the angle between the current and the magnetic field ($\sin \theta$). Recall that you measure force in newtons (N) and current in amperes (A). You measure the strength of a magnetic field (B) in teslas (T). One T equals 1 N/(A·m).

Force on a Current-Carrying Wire in a Magnetic Field

The magnitude of the force on a current-carrying wire in a magnetic field is equal to the product of the current, the length of the wire, the field strength, and the sine of the angle between the current and the magnetic field.

$$F = ILB\ (\sin \theta)$$

Note that $\sin 0° = 0$, and $\sin 90° = 1$. This means that when the current and the magnetic field are parallel to each other, the force on a current-carrying wire is zero. The force on the wire is greatest when the current and the magnetic field are perpendicular to each other.

Figure 16 An earbud works because an electric current in a wire is affected by a magnetic field.

Richard Hutchings/Digital Light Source

Earbuds You might wonder how people apply the relationship among magnetic fields, electric currents, and force in today's technology. You are probably familiar with one example—earbuds. If you look inside an earbud, such as the one in **Figure 16,** you will find a tiny coil of wire attached to a thin plastic membrane. Beneath the membrane is a permanent magnet. The magnetic field from the permanent magnet is oriented radially so it is perpendicular to both the coil of wire and the direction of motion of the coil.

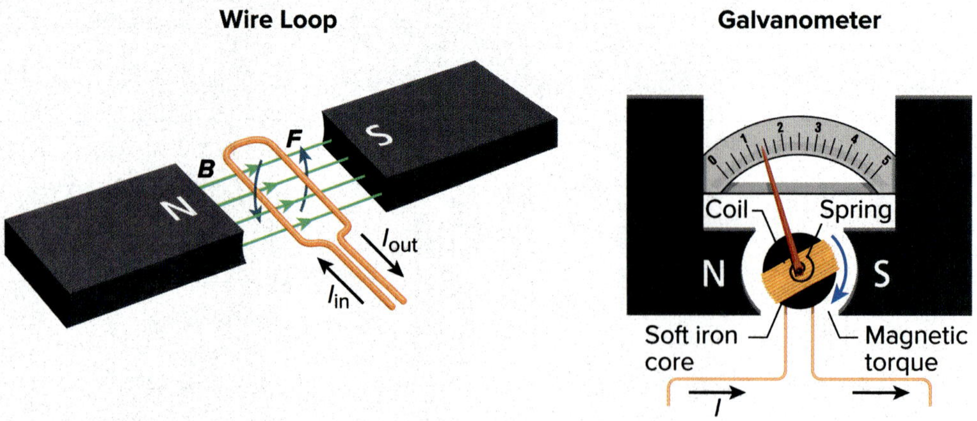

Figure 17 When a wire loop is placed in a magnetic field and current is turned on, the loop rotates (left). The coil in a galvanometer (right) rotates in proportion to the magnitude of the current.

Explain *what causes the wire loop to rotate.*

A music player sends current through an earbud's wires. The current enters the coil, changing direction between 40 and 40,000 times each second, depending on the pitches of the tones it represents. The force from the magnetic field on the coil pushes the coil in and out, depending on the direction of current. This causes the membrane to vibrate, thereby producing sound waves. Each time the current changes direction twice, the membrane vibrates back and forth once.

Most loudspeakers and headphones work in a similar way. A magnetic field exerts a force on a coil of wire mounted on a paper or plastic cone. As the wire moves, it pushes the coil into and out of the field. This motion causes the cone to vibrate and produce sound waves.

Galvanometers You can use the force that a magnetic field exerts on a loop of current-carrying wire to measure currents. How does this work? Current in the wire loop shown in the left of **Figure 17** passes in one end of the loop and out the other. As it does, the force on the loop pushes one side of the loop down and the other side up. The resulting torque causes the loop to rotate. The magnitude of the torque on the loop is proportional to the magnitude of the current. This is the principle used in a galvanometer.

A **galvanometer** is a device used to measure very small currents. The torque of a small coil spring in this device opposes the torque from the current in the wire loop. The amount of rotation is proportional to the current. The meter is calibrated by determining how much the coil turns when a known current is sent through it, as shown in the right of **Figure 17.** The galvanometer can then be used to measure unknown currents.

STEM CAREER Connection

Magnetic Resonance Imaging (MRI) Technologist

MRI technologists use an MRI machine, which is essentially a large tube-shaped magnet, and radio waves to create detailed cross-sectional images of the organs and tissues in the body. These images are tools used to diagnose and treat diseases and other medical conditions. If you like helping people, then a career as an MRI technologist might be for you.

EXAMPLE Problem 1

CALCULATE THE STRENGTH OF A MAGNETIC FIELD A straight wire carrying a 5.0-A current is in a uniform magnetic field oriented at right angles to the wire. When 0.10 m of the wire is in the field, the force on the wire is 0.20 N. What is the strength of the magnetic field (B)?

1 ANALYZE AND SKETCH THE PROBLEM

- Sketch the wire and show the direction of the current with an arrow, the magnetic field as B, and the force on the wire as F.
- Determine the direction of the force using the right-hand rule for the force on a current-carrying wire in a magnetic field. The field, the wire, and the force are all at right angles.

Known	Unknown
$I = 5.0$ A	$B = ?$
$L = 0.10$ m	
$F = 0.20$ N	

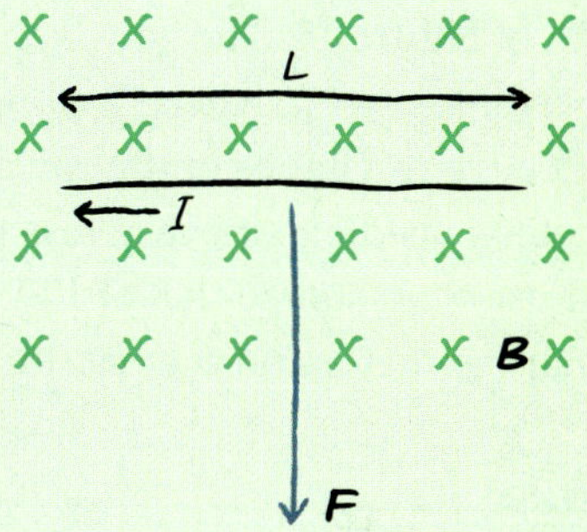

2 SOLVE FOR THE UNKNOWN

B is uniform, and because B and I are perpendicular to each other, $F = ILB$.

$$F = ILB$$

Solve for B.

$$B = \frac{F}{IL}$$

$$= \frac{0.20\text{ N}}{(5.0\text{ A})(0.10\text{ m})}$$ Substitute $F = 0.20$ N, $I = 5.0$ A, $L = 0.10$ m.

$$= 0.40\text{ N/A·m} = 0.40\text{ T}$$

B is 0.40 T from left to right and perpendicular to I and F.

3 EVALUATE THE ANSWER

- **Are the units correct?** The answer is in teslas, the correct unit for a magnetic field.
- **Is the magnitude realistic?** A magnetic field with a strength under 1 T is realistic.

PRACTICE Problems

ADDITIONAL PRACTICE

19. Explain the method you could use to determine the direction of force on a current-carrying wire at right angles to a magnetic field. Identify what must be known to use this method.

20. A wire that is 0.50 m long and carrying a current of 8.0 A is at right angles to a 0.40-T magnetic field. How strong is the force that acts on the wire?

21. A wire that is 75 cm long and carrying a current of 6.0 A is at right angles to a uniform magnetic field. The magnitude of the force acting on the wire is 0.60 N. What is the strength of the magnetic field?

22. A 40.0-cm-long copper wire carries a current of 6.0 A and weighs 0.35 N. A certain magnetic field is strong enough to balance the force of gravity on the wire. What is the strength of the magnetic field?

23. How much current would be required to produce a force of 0.38 N on a 10.0-cm length of wire at right angles to a 0.49-T field?

24. CHALLENGE You are making your own loudspeaker. You make a 1-cm-diameter coil with 20 loops of thin wire. You use hot glue to fasten the coil to an aluminum pie plate. The ends of the wire are connected to a plug that goes into the earphone jack on an MP3 music player. You have a bar magnet to produce a magnetic field. How would you orient the magnetic field to make the plate vibrate and produce sound?

Ammeter A galvanometer can produce full-scale deflections with currents as small as 50 μA (50×10^{-6} A). The resistance of the wire loop in a sensitive galvanometer is about 1000 Ω. To measure larger currents, a galvanometer can be converted to an ammeter. To do this, you would place a resistor with small resistance in parallel with the meter, as shown in the top of **Figure 18.** Because current is inversely proportional to resistance, most of the current (I_s) passes through this resistor, called the shunt because it shunts, or bypasses, much of the current around the galvanometer. Only a few microamps (I_m) pass through. The shunt's resistance is chosen to produce the desired meter sensitivity.

Voltmeter You can also connect a galvanometer as a voltmeter. To make a voltmeter, you would place a resistor, called the multiplier, in series with the meter, as in the bottom of **Figure 18.** The galvanometer measures the current through the multiplier. The current is $I = \frac{V}{R}$, where V is the potential difference across the voltmeter, and R is the effective resistance of the galvanometer and the multiplier. Suppose you want the voltmeter's needle to move across the entire scale when 10 V is placed across it. You would use a resistor so that at 10 V the meter is deflected full-scale by a 50 μA current through the meter and the resistor.

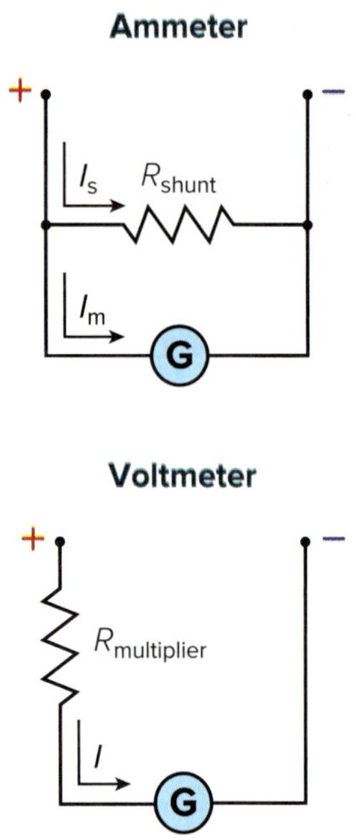

Figure 18 A galvanometer can be connected for use as an ammeter (top) and as a voltmeter (bottom).

Get It?

Compare How would you use a resistor to convert a galvanometer first to an ammeter and then to a voltmeter?

Electric motors The simple loop of wire in a galvanometer cannot rotate more than 180°. The force from the magnetic field pushes one side of the loop up and the other side down until the loop reaches the vertical position. The loop will not continue to turn because the forces are still up and down, parallel to the loop. If you could make the loop rotate continuously, you would have an electric motor.

An **electric motor** is an apparatus that converts electrical energy into mechanical energy. Electric motors rely on a multilooped wire coil called an **armature,** which is mounted on an axle that rotates in a magnetic field. Current enters the armature through a split-ring commutator, which reverses the direction of current as the armature turns, as shown in the simple electric motor in **Figure 19** on the next page.

Although only one loop is shown in **Figure 19,** the armature in most electric motors has many loops. The total force acting on the armature is proportional to $nILB$, where n is the total number of turns on the armature (each completing 360°), I is the current, L is the length of wire in each turn that moves perpendicular to the magnetic field, and B is the strength of the magnetic field. The magnetic field is produced either by a permanent magnet or by an electromagnet (called a field coil). The torque on the armature is controlled by varying the current through the motor. The larger the torque, the faster the armature turns.

Get It?

Describe an object that you use at home that uses an electric motor.

Electric Motor

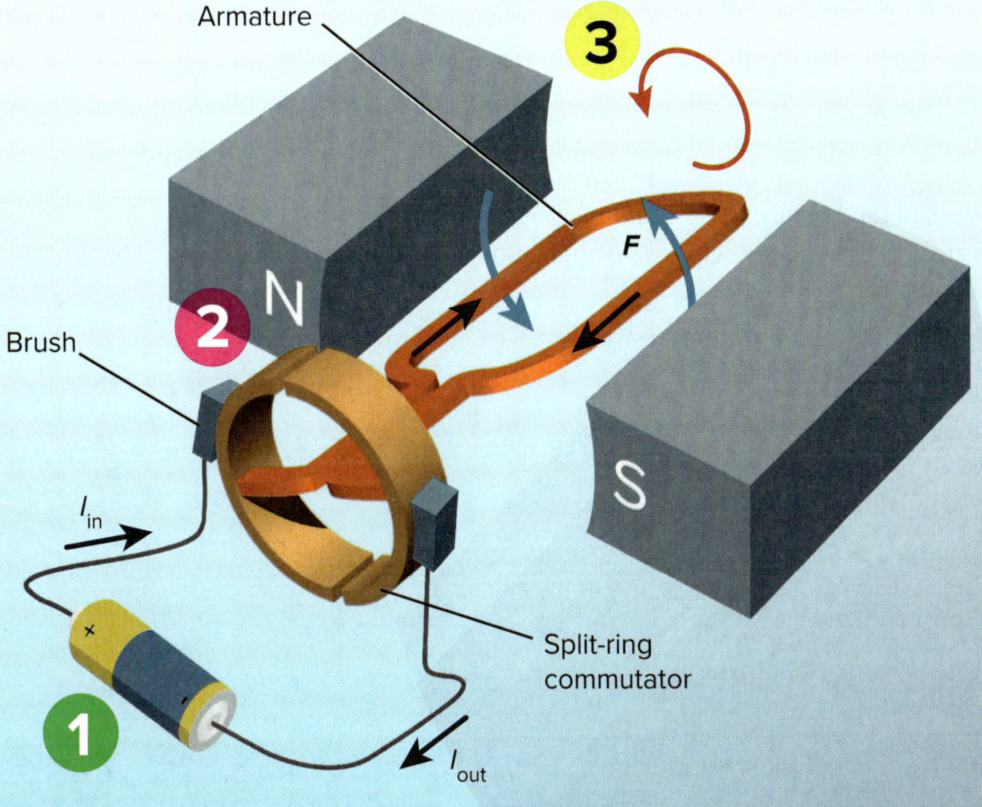

1 Electric Connection

Current from a battery passes through graphite brushes fixed in position but pressed against a split-ring commutator. The commutator is attached to a wire loop—the armature—along an axle in a magnetic field.

2 Current Reversal

Current goes through the brush to the split-ring commutator, which passes it on to the armature. The magnet either repels or attracts the armature, depending on the direction of current. When the armature reaches the vertical position, each half of the commutator changes brushes. This reverses current direction so that the armature can rotate an additional 180°.

3 Continuous Rotation

Current reverses with each half-turn of the armature; this results in continuous rotation.

mike davies/Alamy Stock Photo

Figure 19 In an electric motor, such as the one powering this drone, an armature in a magnetic field rotates 360°. A split-ring commutator changes the direction of the current every 180°, allowing this rotation.

Forces on Single Charged Particles

You use $F = ILB(\sin \theta)$ to determine the force on a current-carrying wire in a magnetic field. A current is simply a stream of charged particles. How do you determine the force on a single charged particle?

Equation of force The magnetic force on a single charged particle depends on the velocity of the particle, the strength of the magnetic field, and the angle between the directions of the velocity and the field. Consider a single electron moving in a wire of length L that is perpendicular to a magnetic field (B). Current (I) is equal to the charge per unit time entering the wire, $I = \frac{q}{t}$. In this case, q is the charge of the electron and t is the time it takes for the electron to move the distance L.

To find the time required for a particle with speed v to travel distance L, you would use the equation of motion, $x = vt$, or, in this case, $t = \frac{L}{v}$.

As a result, you can replace the equation for the current, $I = \frac{q}{t}$, by $I = \frac{qv}{L}$.

Force of a Magnetic Field on a Moving Charged Particle

The amount of force from a magnetic field on a particle equals the product of the particle's charge, its speed, the magnetic field strength, and the sine of the angle between the particle's velocity and the magnetic field.

$$F = qvB\ (\sin \theta)$$

Recall that charge is measured in coulombs (C), velocity in meters per second (m/s), and magnetic field strength in teslas (T). For a particle moving at right angles to a magnetic field, $\sin \theta = 1$, so $F = qvB$.

The direction of the force on a charged particle is perpendicular to that particle's velocity and to the magnetic field. To find the direction of force, you can use the same right-hand rule you use for finding the direction of the force on a current-carrying wire, where the moving charge is the current. If the moving particle is an electron (with a negative charge), the direction of force is reversed.

Get It?

Describe a real-life situation in which you might want to calculate the force of a magnetic field on a moving charged particle.

CCC CROSSCUTTING CONCEPTS

Cause and Effect Using the equation for the force of a magnetic field on a moving charged particle, demonstrate what the force on a charged particle moving parallel to a magnetic field would be.

EXAMPLE Problem 2

FORCE ON A CHARGED PARTICLE IN A MAGNETIC FIELD A beam of electrons travels at 3.0×10^6 m/s through a uniform magnetic field of 4.0×10^{-2} T at right angles to the field. How strong is the force acting on each electron?

1 ANALYZE AND SKETCH THE PROBLEM

Draw the beam of electrons and its direction of motion (v). Indicate the magnetic field (B) and the force on the electron beam (F). Note that the direction of force is opposite that given by the right-hand rule because of the electron's negative charge.

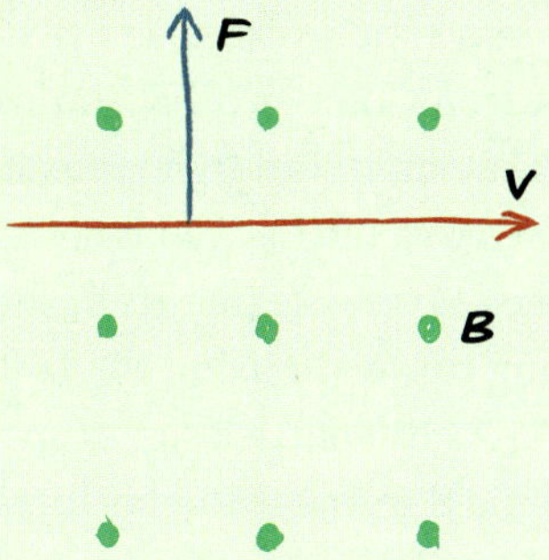

Known	**Unknown**
$v = 3.0 \times 10^6$ m/s	$F = ?$
$B = 4.0 \times 10^{-2}$ T	
$q = -1.602 \times 10^{-19}$ C	

2 SOLVE FOR THE UNKNOWN

$$F = qvB$$

$$= (-1.602 \times 10^{-19}\ \text{C})(3.0 \times 10^6\ \text{m/s})(4.0 \times 10^{-2}\ \text{T})$$

Substitute $q = -1.602 \times 10^{-19}$ C, $v = 3.0 \times 10^6$ m/s, $B = 4.0 \times 10^{-2}$ T.

$$= -1.9 \times 10^{-14}\ \text{N}$$

3 EVALUATE THE ANSWER

- **Are the units correct?** T = N/(A·m) and A = C/s, so T = N·s/(C·m). Thus, (T·C·m)/s = N, the unit for force.
- **Does the direction make sense?** Use the right-hand rule to verify the direction of the force, recalling that the force on the electron is opposite the force given by the right-hand rule due to the electron's negative charge.
- **Is the magnitude realistic?** Forces on electrons and protons are always small fractions of a newton.

PRACTICE Problems

ADDITIONAL PRACTICE

25. In what direction is the force on an electron if that electron is moving east through a magnetic field that points north?

26. What are the magnitude and direction of the force acting on the proton shown in **Figure 20**?

27. A stream of doubly ionized particles (missing two electrons and thus carrying a net positive charge of two elementary charges) moves at a velocity of 3.0×10^4 m/s perpendicular to a magnetic field of 9.0×10^{-2} T. How large is the force acting on each ion?

28. Triply ionized particles in a beam carry a net positive charge of three elementary charge units. The beam enters a magnetic field of 4.0×10^{-2} T. The particles have a speed of 9.0×10^6 m/s and move at right angles to the field. How large is the force acting on each particle?

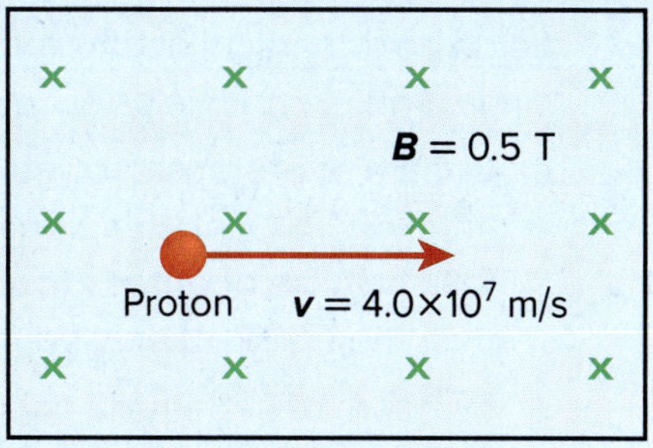

Figure 20

29. A singly ionized particle experiences a force of 4.1×10^{-13} N when it travels at a right angle through a 0.61-T magnetic field. What is the particle's velocity?

30. CHALLENGE Doubly ionized helium atoms (alpha particles) are traveling at right angles to a magnetic field at a speed of 4.0×10^4 m/s. The force on each particle is 6.4×10^{-16} N. What is the magnetic field strength?

Synchrotrons Because the direction of force is always perpendicular to a charged particle's velocity in a magnetic field, magnets can be used to direct a charged particle's path. For example, accelerating particles in a synchrotron, such as the Large Hadron Collider (LHC), move in a circle as they maintain their velocity at right angles to a uniform magnetic field. You can see several segments of the 27-km-long tunnel housing the LHC in **Figure 21.** As the particles gain speed, the magnetic field in the tunnel is increased to keep the radius of the circle constant. Additional magnets provide horizontal and vertical forces to focus the beam. Additional segments along the LHC tunnel add fixed amounts of energy that accelerate the particles.

Figure 21 Powerful electromagnets inside the LHC tunnel provide the uniform magnetic field that makes charged particles move in a circular path.

Check Your Progress

31. **Motors** Explain how electric motors use magnets to convert electrical energy to mechanical energy.
32. **Magnetic Forces** Imagine that a current-carrying wire is perpendicular to Earth's magnetic field and runs east-west. If the current is east, in which direction is the force on the wire?
33. **Synchrotrons** In a synchrotron, magnetic fields bend particle beams into segments of a circle, and electric fields accelerate the beams.
    a. A beam of protons circulates in a clockwise direction. In what direction must the magnetic field be oriented? In what direction must the electric fields be oriented?
    b. If a beam of negatively charged antiprotons is to circulate in a counterclockwise direction, must the direction of the magnetic field be changed? Must the direction of the electric fields be changed?
34. **Galvanometers** Compare the diagram of a galvanometer in the image on the right in **Figure 17** with the electric motor in **Figure 19.** How is the galvanometer similar to an electric motor? How is it different?
35. **Motors** When the plane of an armature in a motor is perpendicular to the magnetic field, the forces do not exert a torque on the coil. Does this mean that the coil does not rotate? Explain.
36. **Resistance** A galvanometer requires 180 μA for full-scale deflection. When it is used as a voltmeter, what total resistance of the meter and the multiplier resistor is needed for a 5.0-V full-scale deflection?
37. **Critical Thinking** Two current-carrying wires move toward each other when they are placed parallel to each other. Compare the directions of the two currents. Explain your reasoning.

LEARNSMART® Go online to follow your personalized learning path to review, practice, and reinforce your understanding.

ENGINEERING & TECHNOLOGY

Accelerating a Solution

For nearly a century, viewers with scrutinizing eyes have detected hints of another painting concealed beneath French impressionist Edward Degas' painting *Portrait of a Woman.* Now, with the help of a particle accelerator, scientists have solved the mystery. Hidden beneath the contemplative woman in the painting is another picture of the same woman, painted seven years earlier.

Scientists in the Netherlands are using special techniques to uncover an earlier painting that was later painted over by Vincent van Gogh.

Portrait of a Woman

Degas painted *Portrait of a Woman* in the late 1870s. The model is tentatively identified as French model Emma Dobigny. As time wore away some of the oil paint from Degas' canvas, the faint outline of another face became visible beneath the painting. Recently, Australian scientists used a type of particle accelerator called a synchrotron (similar to the Large Hadron Collider) to reveal the original painting of the model Dobigny.

Synchrotron Technology

How did scientists use the synchrotron to "look" beneath the top layers of paint without ruining the completed painting? They employed a technique called high-resolution scanning X-ray fluorescence (XRF), which uses a beam of radiation to excite atoms, causing them to emit X-ray photons. Each element's photons have a unique wavelength, enabling scientists to create elemental and scatter maps.

The scans of Degas' painting revealed the presence of eleven element maps, including calcium, chromium, manganese, iron, cobalt, nickel, copper, zinc, arsenic, barium, and mercury. Scientists reconstructed colors by inferring the pigments of paint based on the elemental maps. When two or more elements were located in the same area, scientists were able to better identify the color used by Degas.

The elemental maps from XRF are composed of millions of pixels. These high definition imaging scans are used to make a "false" color representation of the hidden picture. Instead of seeing what the naked eye sees, false color representation most often incorporates the visible light spectrum and the electromagnetic spectrum. When these elemental map images were layered on top of each other, they formed an image behind the painting of Dobigny's older picture.

ASK QUESTIONS TO CLARIFY

Write three questions that you have about XRF and how it was used to reveal Degas' hidden painting. Use print or online sources to find answers to your questions. Share your questions and answers with your class.

CALIFORNIA CONNECTION

Research a flower species native to California that has the blue halo adaptation. Make a poster that shows the flower and describes the adaptation.

MODULE 20

STUDY GUIDE

GO ONLINE study with your Science Notebook.

Lesson 1 UNDERSTANDING MAGNETISM

- polarized
- domain
- magnetic field
- magnetic flux
- solenoid
- electromagnet

- All magnets have north poles and south poles and are surrounded by magnetic fields.
- Ferromagnetic materials become magnetic when their domains are in alignment with each other.
- Magnetic fields are vector quantities because they have direction and magnitude. They exist in any region in space where a magnet would experience a force. Magnetic fields can be represented by field lines, which exit from a north pole and enter at a south pole, forming closed loops.

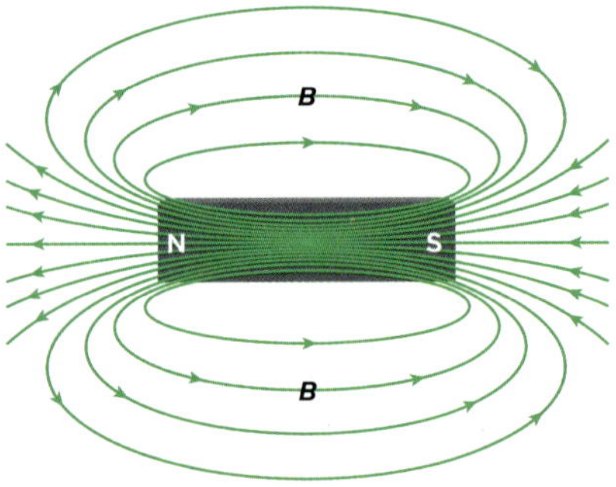

- A magnetic field exists around any current-carrying wire. The magnetic field around a coil of wire is proportional to the number of loops in the coil because the individual fields of the loops are in the same direction.

Lesson 2 APPLYING MAGNETIC FORCES

- galvanometer
- electric motor
- armature

- When a current-carrying wire is placed in a magnetic field, a force is exerted on the wire that is perpendicular to both the field and the wire.
- The force on a current-carrying wire in a magnetic field is proportional to the current times the length of the wire times the field strength times the sine of the angle between the current and the magnetic field.

$$F = ILB(\sin \theta)$$

- An electric motor consists of a coil of wire in a magnetic field. When there is current in the coil, the coil rotates as a result of the force on the wire in the magnetic field. Complete 360° rotation is achieved by using a split-ring commutator to switch the direction of the current in the coil as the coil rotates.
- The force that a magnetic field exerts on a moving charged particle depends on the charge of the particle, the velocity of the particle, the strength of the magnetic field, and the angle between the directions of the velocity and the field. The direction of the force is perpendicular to both the field and the particle's velocity.

$$F = qvB(\sin \theta)$$

THREE-DIMENSIONAL THINKING
Module Wrap-Up

REVISIT THE PHENOMENON

What makes this electromagnet stronger than a typical refrigerator magnet?

CER Claim, Evidence, Reasoning

Explain Your Reasoning Revisit the claim you made when you encountered the phenomenon. Summarize the evidence you gathered from your investigations and research and finalize your Summary Table. Does your evidence support your claim? If not, revise your claim. Explain why your evidence supports your claim.

STEM UNIT PROJECT

Now that you've completed the module, revisit your STEM unit project. You will summarize your evidence and apply it to the project.

GO FURTHER

SEP Data Analysis Lab

Do magnets obey an inverse square law?

The repulsive force between two magnets was measured and found to depend on distance, as given in the table.

Separation (cm)	Force (N)	Separation (cm)	Force (N)
1.0	3.93	2.2	0.011
1.2	0.40	2.4	0.0076
1.4	0.13	2.6	0.0053
1.6	0.057	2.8	0.0038
1.8	0.030	3.0	0.0028
2.0	0.018		

CER Analyze and Interpret Data

1. Plot the force as a function of distance.
2. **Claim** Does this force follow an inverse square law?
3. **Evidence and Reasoning** Defend your claim using your graph.

Ken Phung/Shutterstock

MODULE 21
ELECTROMAGNETISM

ENCOUNTER THE PHENOMENON

What causes the Northern Lights?

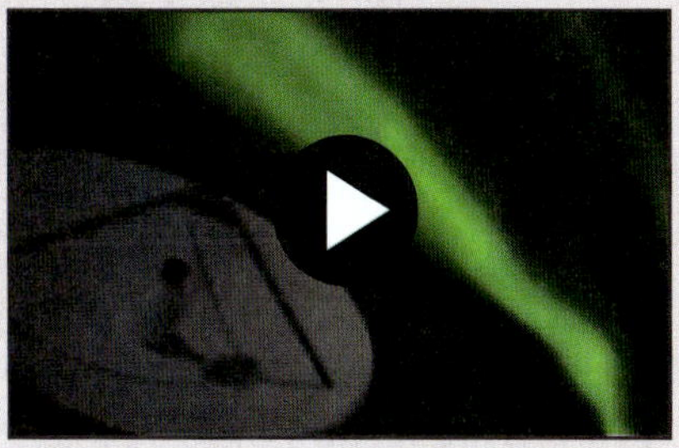

GO ONLINE to watch a video about the Northern Lights.

SEP Ask Questions

Do you have other questions about the phenomenon? If so, add them to the driving question board.

CER Claim, Evidence, Reasoning

Make Your Claim Use your CER chart to make a claim about what causes the Northern Lights. Explain your reasoning.

Collect Evidence Use the lessons in this module to collect evidence to support your claim. Record your evidence as you move through the module.

Explain Your Reasoning You will revisit your claim and explain your reasoning at the end of the module.

GO ONLINE to access your CER chart and explore resources that can help you collect evidence.

LESSON 2: Explore & Explain: Transformers

LESSON 4: Explore & Explain: Producing Electromagnetic Waves

Additional Resources

LESSON 1

INDUCING CURRENTS

FOCUS QUESTION

How can a magnet generate a current?

Changing Magnetic Fields

Following Oersted's discovery that a current produces a magnetic field, Michael Faraday became convinced the reverse was possible: that a magnetic field could produce a current. In 1822 he wrote this goal in his notebook: "Convert magnetism into electricity." After ten years of unsuccessful experiments, he succeeded. He induced a current in a circuit by moving a wire through a magnetic field. In the same year in America, Joseph Henry, a high-school teacher who later became the first secretary of the Smithsonian Institution, made the same discovery. **Figure 1** illustrates a modern version of one of Faraday's and Henry's experiments.

Relative motion An electric current can be generated in a wire in a circuit when at least part of the wire moves through, and cuts, magnetic field lines. Field lines can be cut when a segment of wire moves through a stationary magnetic field, as the wire does in **Figure 1.** Field lines also can be cut when a magnetic field moves past a stationary wire or when the strength of a magnetic field changes around a wire. **Electromagnetic induction** is the process of generating current through a wire in a circuit in a changing magnetic field.

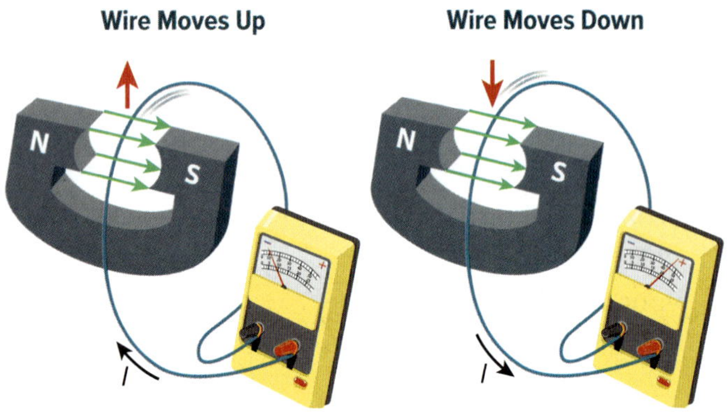

Figure 1 The direction of current induced in a wire moving through a magnetic field depends on the direction the wire moves. When motion stops, so does the current.

3D THINKING DCI Disciplinary Core Ideas CCC Crosscutting Concepts SEP Science & Engineering Practices

COLLECT EVIDENCE

Use your Science Journal to record the evidence you collect as you complete the readings and activities in this lesson.

INVESTIGATE

GO ONLINE to find these activities and more resources.

Applying Practices: Investigate Electromagnetism

HS-PS2-5. Plan and conduct an investigation to provide evidence that an electric current can produce a magnetic field and that a changing magnetic field can produce an electric current.

CCC **Identify Crosscutting Concepts**

Create a table of the crosscutting concepts and fill in examples you find as you read.

Electromotive force How is current produced by electromagnetic induction? You know that a current requires a source of electrical energy, such as a battery. A battery maintains an electric potential difference in a circuit; charges flow from higher to lower potential. The potential difference across the battery's terminals is the electromotive force, or *EMF*. An *EMF* is a difference in potential, not actually a force. Like many historical terms, it was coined before related principles were well understood.

Induced *EMF* You don't need a chemical reaction in a battery to create an *EMF*. When a wire moves perpendicular to a magnetic field, there is a force on the charges in the wire. The force causes negative charges to move to one end of the wire, leaving positive charges at the other end. This separation of charge produces an electric field and therefore a potential difference across the length of the wire. This potential difference is called the **induced electromotive force,** or induced *EMF*.

The magnitude of an induced *EMF* in a wire in a magnetic field depends on the strength of the magnetic field (B), the length of the wire within the field (L), and the component of the velocity of the length of wire that is perpendicular to the field ($v(\sin\theta)$).

Induced Electromotive Force in a Wire

EMF is equal to the strength of the magnetic field times the length of the wire times the component of the velocity of the wire in the field that is perpendicular to the field.

$$EMF = BLv(\sin\theta)$$

If a wire moves perpendicular to a magnetic field, the above equation reduces to $EMF = BLv$, because $\sin 90° = 1$. Note that no *EMF* is induced in a length of wire that moves parallel to a magnetic field because $\sin 0° = 0$.

As in a battery, induced *EMF* is measured in volts (V). Why? Recall that magnetic fields (B) are measured in teslas (T). You can write $B = \frac{F}{IL}$, so the units for B are also N/(A·m). Velocity is measured in meters per second. Using dimensional analysis, you could write the following:

$$\left(\frac{\text{N}}{\text{A·m}}\right)(\text{m})\left(\frac{\text{m}}{\text{s}}\right) = \frac{(\text{N·m})}{(\text{A·s})} = \frac{\text{J}}{\text{C}} = \text{V}$$

Therefore, *EMF* is measured in volts.

Magnitude of current Recall that resistance is equal to potential difference divided by current. Thus, if an *EMF* generates a current in a wire that is part of a circuit with an effective resistance R, and you know the *EMF*, you can determine the current's magnitude by the following equation:

$$I = \frac{EMF}{R}$$

Direction of current How do you determine the direction of an induced current in a wire? As a wire moves through a magnetic field, a force is exerted on the charges in the wire, and they move in the direction of that force, as shown in the top image of **Figure 2.** To find this direction, use the right-hand rule illustrated in the bottom of **Figure 2.** Point your thumb in the direction the wire moves and your fingers in the direction of the magnetic field. Your palm points in the direction of the force on the positive charges and, thus, in the direction of current.

Direction of Current

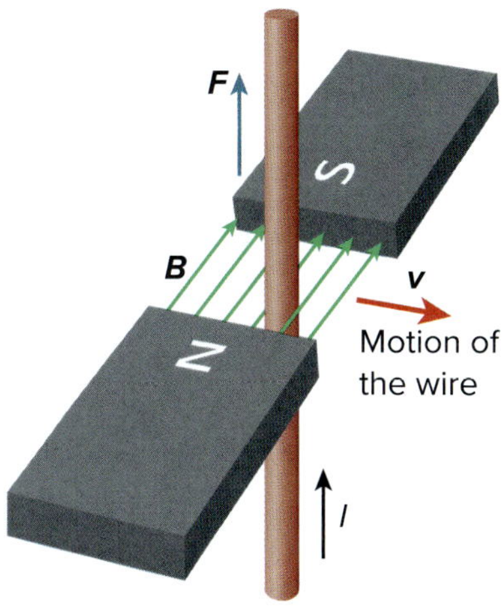

Right-Hand Rule

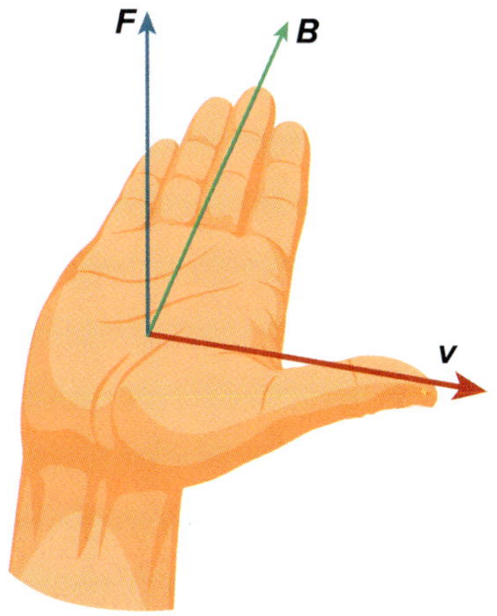

Figure 2 The direction of current in a wire moving through a magnetic field can be found using a right-hand rule. Note that the direction of current is the same as the direction of the force on the charges in the wire.

EXAMPLE Problem 1

INDUCED *EMF* A straight wire is part of a circuit that has a resistance (R) of 0.50 Ω. The wire is 0.20 m long and moves at a constant speed of 7.0 m/s perpendicular to a magnetic field of strength 8.0×10^{-2} T.

a. What *EMF* is induced in the wire?

b. What is the current through the wire?

c. If a different metal were used for the wire, increasing the circuit's resistance to 0.78 Ω, what would the new current be?

1 ANALYZE AND SKETCH THE PROBLEM

- Draw a straight wire of length L. Connect an ammeter to close the circuit so that there will be a current.
- Show a direction for the magnetic field (B) that is perpendicular to the wire length.
- Show a direction for the velocity (v) that is perpendicular to both the length of the wire and the magnetic field.
- Use the right-hand rule to determine the direction of the current, and show it on the diagram.

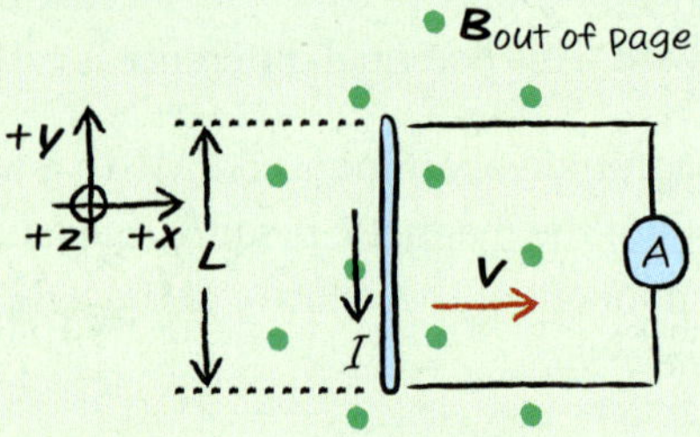

Known	Unknown
$v = 7.0$ m/s	$EMF = ?$
$L = 0.20$ m	$I = ?$
$B = 8.0\times10^{-2}$ T	
$R_1 = 0.50\ \Omega$	
$R_2 = 0.78\ \Omega$	

2 SOLVE FOR THE UNKNOWN

a. $EMF = BLv$

$= (8.0\times10^{-2}\ \text{T})(0.20\ \text{m})(7.0\ \text{m/s})$ — Substitute $B = 8.0\times10^{-2}$ T, $L = 0.20$ m, $v = 7.0$ m/s.

$= 0.11\ \text{T}\cdot\text{m}^2/\text{s}$

$= 0.11$ V

b. $I = \frac{EMF}{R}$

$= \frac{0.11\ \text{V}}{0.50\ \Omega}$ — Substitute $EMF = 0.11$ V, $R_1 = 0.50\ \Omega$.

$= 0.22$ A

c. $I = \frac{EMF}{R}$

$= \frac{0.11\ \text{V}}{0.78\ \Omega}$ — Substitute $EMF = 0.11$ V, $R_2 = 0.78\ \Omega$.

$= 0.14$ A

The current is counterclockwise.

3 EVALUATE THE ANSWER

- **Are the units correct?** Volt is the correct unit for *EMF*. Current is measured in amperes.
- **Does the direction make sense?** The direction obeys a right-hand rule: v is the direction of the thumb, B is the direction the fingers point, and F is the direction the palm faces. Current is in the same direction as the force.
- **Is the magnitude realistic?** The answers are near 10^{-1} A, typical for circuits with small *EMF*s and resistances.

PRACTICE Problems

ADDITIONAL PRACTICE

1. You move a straight wire that is 0.5 m long at a speed of 20 m/s vertically through a 0.4-T magnetic field pointed in the horizontal direction.
   a. What *EMF* is induced in the wire?
   b. The wire is part of a circuit with a total resistance of 6.0 Ω. What is the current?
2. A straight wire that is 25 m long is mounted on an airplane flying at 125 m/s. The wire moves in a perpendicular direction through Earth's magnetic field ($B = 5.0\times10^{-5}$ T). What *EMF* is induced in the wire?
3. A straight wire segment in a circuit is 30.0 m long and moves at 2.0 m/s perpendicular to a magnetic field.
   a. A 6.0-V *EMF* is induced. What is the magnetic field?
   b. The total resistance of the circuit is 5.0 Ω. What is the current?
4. **CHALLENGE** A horseshoe magnet is mounted so that the magnetic field lines are vertical. You pass a straight wire between the poles and pull it toward you. The current through the wire is from right to left. Which is the magnet's north pole? Explain.

Induced *EMF* in microphones In a previous chapter, you learned how earbuds use magnetic fields to work. Microphones do, too. They are constructed like earbuds, but they work in reverse; they convert sound to electrical energy by electromagnetic induction.

A microphone has a thin aluminum diaphragm attached to a coil in a magnetic field, as shown in **Figure 3.** Sound waves cause the diaphragm to vibrate. This moves the attached coil. The motion of the coil in the magnetic field induces an *EMF* across the coil's ends that generates a current. The induced *EMF* and current vary as the frequency of the sound varies. In this way, the sound waves are converted to electrical signals. The potential differences generated are small, typically 10^{-3} V, but they can be increased, or amplified, by electronic devices.

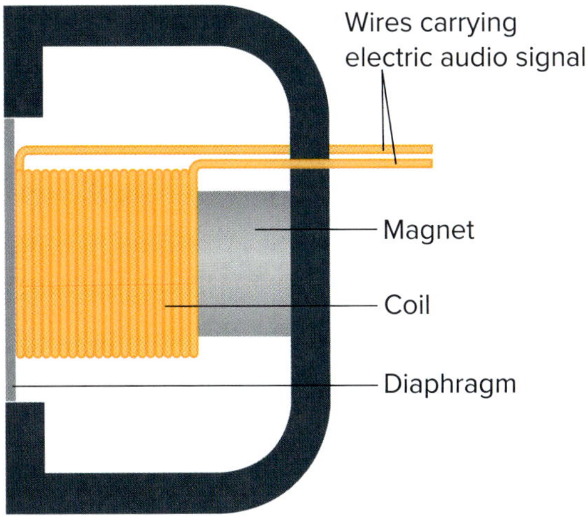

Figure 3 The microphone converts sound waves from the singer into electrical signals. The microphone's aluminum diaphragm is attached to a coil in a magnetic field. When sound waves vibrate the diaphragm, the coil moves. The current that is generated is proportional to the sound wave.

CCC CROSSCUTTING CONCEPTS

Cause and Effect Examine **Figure 3.** Describe how knowing about the effect caused by the movement of a wire in an electric field, as well as the effect caused by sound on a diaphragm, lets you predict how sound affects a microphone. Cite evidence to support your explanation about how the two mechanisms in the system work together.

STEM CAREER Connection

Audio Engineer

There's more to recording music than just turning on the microphone. Audio engineers help shape how a recording sounds by choosing a microphone, setting levels, and manipulating the sound produced. If you have a great ear and love music, audio engineer might be the career for you.

Electric Generators

Microphones are just one of many technologies that rely on electromagnetic induction. Electromagnetic induction is also the principle governing the operation of electric generators.

An **electric generator** converts mechanical energy to electrical energy. It consists of a number of wire loops in a magnetic field. The wire is wound around an iron core, which concentrates the magnetic flux through the wire. The iron and wire make up the generator's armature, also called a rotor.

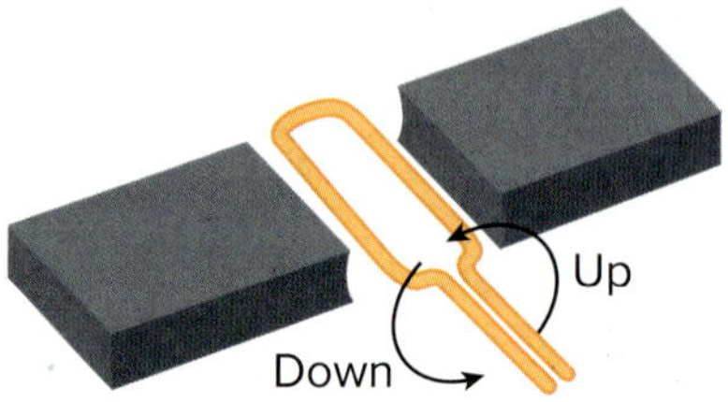

Figure 4 An electric current is induced in a rotating wire loop in a circuit that is in a magnetic field. Direction and strength of the induced current change as the loop rotates.

Determine *What is the direction of the current when the loop is horizontal to the field, as shown?*

Current from a generator Like an armature in a motor, the armature in a generator can rotate freely in a magnetic field. As the armature is turned, its wire loops cut through magnetic field lines. This induces an *EMF* in each loop. Recall that *EMF* depends on wire length, magnetic field strength, and speed. Therefore, increasing the number of loops increases wire length and thus increases the total induced *EMF*. The larger the *EMF*, the stronger the resulting current.

Orientation of loop To understand the steps by which current is produced from the induced *EMF* in a generator, consider the single loop of wire in **Figure 4.** The loop is in a magnetic field. It is also in a closed circuit, so the induced *EMF* produces a current in the loop. As the loop rotates, the strength and the direction of the current change as the orientation of the loop's motion relative to the magnetic field changes.

Recall that only the component of the wire's velocity perpendicular to a magnetic field can induce an *EMF*. Therefore, the current is strongest when the motion of the loop is perpendicular to the field, that is, when the loop is horizontal, as shown in **Figure 4** and in the far left of **Figure 5.**

Current reversals As the loop in **Figure 5** rotates to the vertical position, it moves through the magnetic field lines at an ever-decreasing angle. Thus, it cuts through fewer magnetic field lines per unit of time, and the current decreases. When the loop is in the vertical position, as it is in the middle image of **Figure 5,** current is zero because the wire moves parallel to the magnetic field.

As the loop continues to rotate, the current changes sinusoidally from zero to some maximum value and back to zero during each half-turn. Current reverses each time the loop turns through 180°. A graph of current versus time is shown in the far right of **Figure 5.**

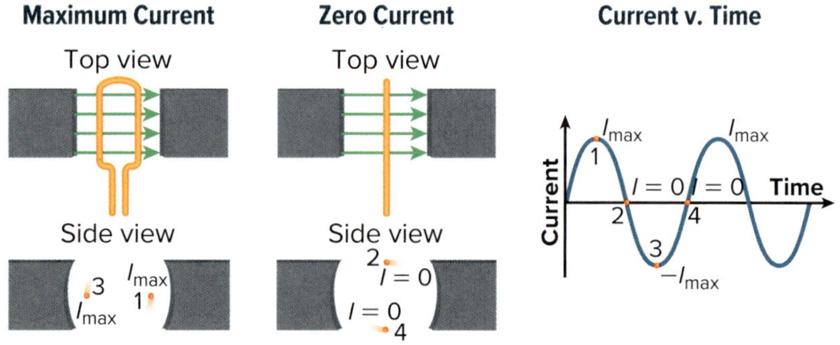

Figure 5 Maximum current is generated when the rotating loop is horizontal to the magnetic field (left). When the loop is vertical to the field (middle), the current is zero. The graph shows that current varies regularly with time as the loop rotates. The numbers on the graph correspond to positions of the loop in the side views. A graph of the variation of *EMF* with time (not shown) would be similar.

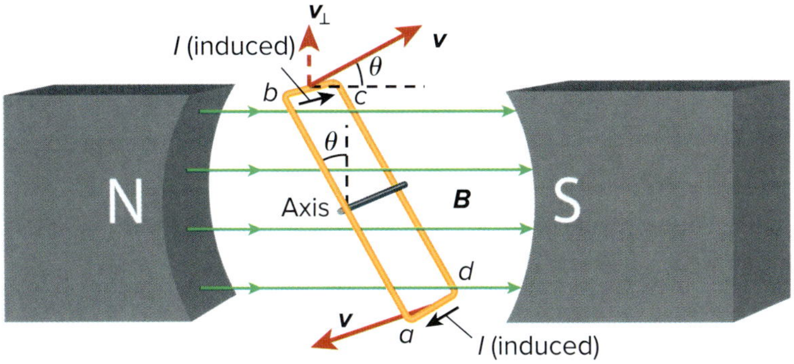

Figure 6 Current is induced in the shorter segments, *bc* and *ad*. No current is induced in the longer segments, *ab* and *cd*. If you pointed your thumb in the direction a long segment is moving and your fingers in the direction of the magnetic field, your thumb would point to the side of the wire. Therefore, no current is induced.

Where is current induced? Not all parts of a loop contribute to an induced *EMF*. Look at **Figure 6.** Only two sides of the wire loop in this figure contribute to an induced *EMF*. When you apply a right-hand rule to segment *ab*, you see that the direction of the induced current would be toward the side of the wire—not along the wire. The same is true of segment *cd*. Thus, no current is induced along either length. In segment *bc*, however, the direction of the induced current is from *b* to *c*, and in segment *ad*, the direction of the current is from *d* to *a*.

Because the conducting loop in **Figure 6** is rotating, the relative angle between a point on the loop and the magnetic field is constantly changing. The *EMF* can be calculated by the equation given earlier, $EMF = BLv(\sin \theta)$, where L is now the combined length of the segments $ad + bc$. The maximum potential difference is induced when a conductor is moving perpendicular to the magnetic field, that is, when $\theta = 90°$.

Get It?

Apply At what position is the wire loop in **Figure 6** when potential difference is greatest?

Sources of mechanical energy Electric generators, such as those in wind turbines, have many wire loops in their armatures, but the total *EMF* induced in them produces current in the same way that an *EMF* produces current in a single loop. Energy from an external source—in this case, the wind—is converted to rotational kinetic energy spinning the turbines. The turbines are attached mechanically to an armature in a magnetic field. As the armature rotates, an *EMF* is induced in each loop. When the generator is connected to an external circuit, the *EMF*s produce current.

Generators and electric motors are almost identical in construction, but they convert energy in opposite directions. A generator converts mechanical energy to electrical energy, while a motor converts electrical energy to mechanical energy. Generators in electrical-generating plants are traditionally powered by steam from water heated by burning fossil fuels or by fissioning uranium. Wind and moving water also can provide the mechanical energy. For example, the power generators in **Figure 7** convert energy from falling water at the Hoover Dam to electrical energy.

Figure 7 These generators at the Hoover Dam, on the Nevada-Arizona border, convert mechanical energy from moving water into electrical energy.

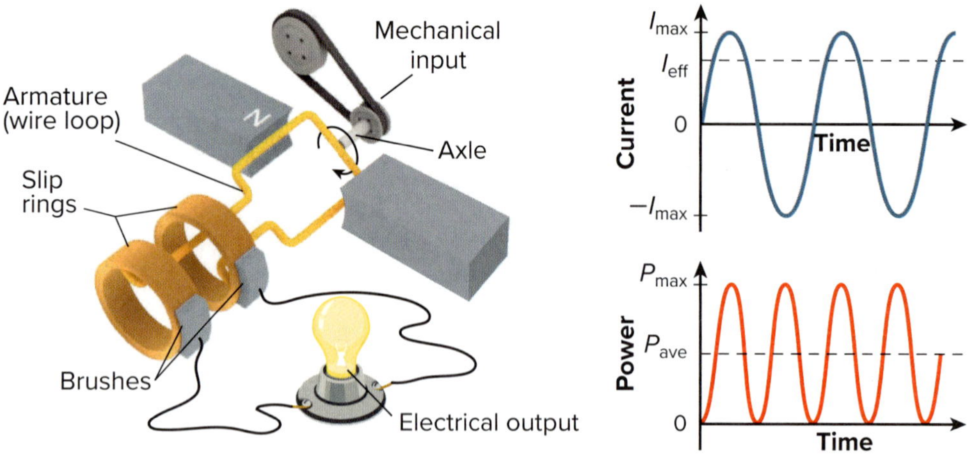

Figure 8 This AC generator is similar in construction to an electric motor, except it connects to a circuit using a brush-slip-ring device instead of a commutator. An outside source rotates the armature. As the armature rotates, the direction of the current alternates in time (top right). The power delivered by the generator is always positive (bottom right).

Identify *In what position is the armature when current is zero?*

Alternating currents and generators There are two main types of generators. In direct current (DC) generators, charges move in a single direction, as they do in batteries. Current is in one direction because the wires of the armature connect to a circuit by means of a commutator.

In contrast, alternating current (AC) generators use a slip-ring device to connect wires to a circuit. In this device, one ring is connected to one end of the armature, and the other ring is connected to the other end of the armature, as shown in the left of **Figure 8.** As the armature rotates through 180°, the induced EMF reverses direction. This means the current also reverses direction. The current alternates from positive to negative. In a consistently turned generator, the current alternates at a fixed rate and varies sinusoidally, as shown in the top graph of **Figure 8.**

The frequency of alternating current is different in different countries. In the United States, all electric utilities use a 60-Hz frequency in which current goes from one direction to the other and back to the first 60 times per second. This means that an armature makes 60 complete rotations in one second. Many other countries use a 50-Hz frequency.

Average power The power produced by an AC generator is the product of the current and the potential difference. Because both current and potential difference vary, the power in the circuit varies. The graph at the bottom right of **Figure 8** shows the power versus time produced by an AC generator. Note that power is always positive because I and V are either both positive or both negative. Average power, P_{AC}, is half the maximum power; thus, $P_{AC} = \frac{1}{2}P_{AC\,max}$.

Effective current It is common to describe alternating current and potential difference in terms of effective current and effective potential difference, rather than referring to their maximum values. Recall that $P = I^2R$. Thus, effective current (I_{eff}) can be expressed in terms of the average AC power, as $P_{AC} = I_{eff}^{\,2}R$. To determine I_{eff} in terms of maximum current (I_{max}), start with the power relationship, $P_{AC} = \frac{1}{2}P_{AC\,max}$, and substitute in I^2R. Then solve for I_{eff}.

Effective Current

Effective current is equal to $\frac{\sqrt{2}}{2}$ times the maximum current.

$$I_{eff} = \left(\frac{\sqrt{2}}{2}\right)I_{max} = 0.707I_{max}$$

Effective potential difference Similarly, because $P = \frac{V^2}{R}$, the effective potential difference is given by the following equation.

Effective Potential Difference

Effective potential difference is equal to $\frac{\sqrt{2}}{2}$ times the maximum potential difference.

$$V_{eff} = \left(\frac{\sqrt{2}}{2}\right)V_{max} = 0.707V_{max}$$

Effective potential difference is commonly referred to as RMS (root mean square) potential difference. In the United States, the RMS potential difference generally available at wall outlets is described as 120 V. In many other countries, the RMS potential difference is 240 V.

War of currents The first electrical distribution system in the United States was developed by Thomas Edison in the late 1800s. Edison's system used direct current. Soon, however, a competing system that used alternating current was developed by Nikola Tesla. Tesla partnered with George Westinghouse to promote AC current, and a "war of the currents" ensued. Eventually, AC won. In the next lesson, you will read why.

PRACTICE Problems

ADDITIONAL PRACTICE

5. A generator develops a maximum potential difference of 170 V.
   a. What is the effective potential difference?
   b. A lamp is placed across the generator with an I_{max} of 0.70 A. What is I_{eff} through the lamp?
   c. What is the resistance of the lamp when it is on?

6. The RMS potential difference of an AC household outlet is 117 V. What is the maximum potential difference across a lamp connected to the outlet?

7. If the average power used over time by an electric light is 75 W, what is the peak power?

8. CHALLENGE An AC generator delivers a peak potential difference of 425 V.
   a. What is the V_{eff} in a circuit connected to the generator?
   b. The resistance is 5.0×10^2 Ω. What is the effective current?

Check Your Progress

9. **Electric Generator** Explain how an electric generator works.

10. **Generator** Could you make a generator by mounting permanent magnets on a rotating shaft and keeping the coil stationary? Explain.

11. **Bike Generator** A small generator on your bike lights the bike's headlight. What is the source of the energy for the bulb when you ride along a flat road?

12. **Microphone** Consider the microphone shown in **Figure 3.** What happens when the diaphragm is pushed in?

13. **Frequency** What changes to an electric generator are required to increase frequency?

14. **Output Potential Difference** Explain why the output potential difference of an electric generator increases when the magnetic field is made stronger. What is another way to increase the output potential difference?

15. **Critical Thinking** A student asks, "Why does AC dissipate power? The energy going into a lamp when the current is positive is removed when the current is negative. The net current is zero." Explain why this reasoning is wrong.

LEARNSMART® Go online to follow your personalized learning path to review, practice, and reinforce your understanding.

LESSON 2

APPLICATIONS OF INDUCED CURRENTS

FOCUS QUESTION

How do transformers help with the distribution of electricity in the power grid?

Lenz's Law

Imagine the hand pulling the wires through the magnetic field in **Figure 9** is yours. As you pull, the movement of the wire induces an *EMF.* The wire slides on the wire loop, making a complete circuit, so the *EMF* produces a current. Because the magnetic field (*B*) is out of the page and the velocity is to the right, the right-hand rule shows that the direction of current in the wire is downward and clockwise in the circuit.

Opposing motion The downward-moving charges in the wire moving through the magnetic field that is out of the page experience a force. The right-hand rule shows that the direction of this force is opposite the movement of the wire. These charges exert a force on the wire that opposes the motion of the wire and makes it harder for you to pull it. The physicist H.F.E. Lenz first demonstrated this force in 1834.

Opposing change While Lenz stated his law in terms of force, there is a second way of interpreting it. The current in the loop produces its own magnetic field. The right-hand rule shows that the direction of this induced field inside the loop is into the paper. That is, it is in the direction that opposes the change in the field that caused it. Thus, **Lenz's law** says that the magnetic field produced by an induced current is in the direction that is opposite to the change in the original field.

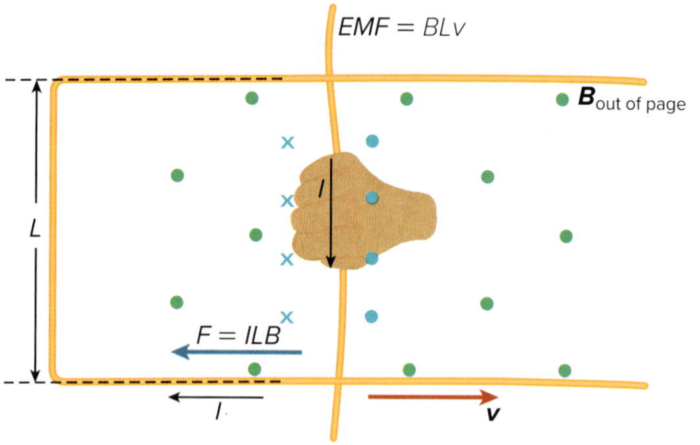

Figure 9 A wire pulled through a magnetic field generates an *EMF*. The *EMF* produces a current in the circuit (*I*). Motion of the charges in the wire produces a force (*F*). The current also produces a magnetic field (blue dots and crosses) that, within the loop, is in the direction opposite that of the change in the field which caused it.

3D THINKING DCI Disciplinary Core Ideas CCC Crosscutting Concepts SEP Science & Engineering Practices

COLLECT EVIDENCE

Use your Science Journal to record the evidence you collect as you complete the readings and activities in this lesson.

INVESTIGATE

GO ONLINE to find these activities and more resources.

PhysicsLAB: Induction and Transformers

Construct an explanation of how the input current affects the output current for a transformer.

Review the News

Obtain information from a current news story about applications of electromagnetism or electromagnetic waves. Evaluate your source and communicate your findings to your class.

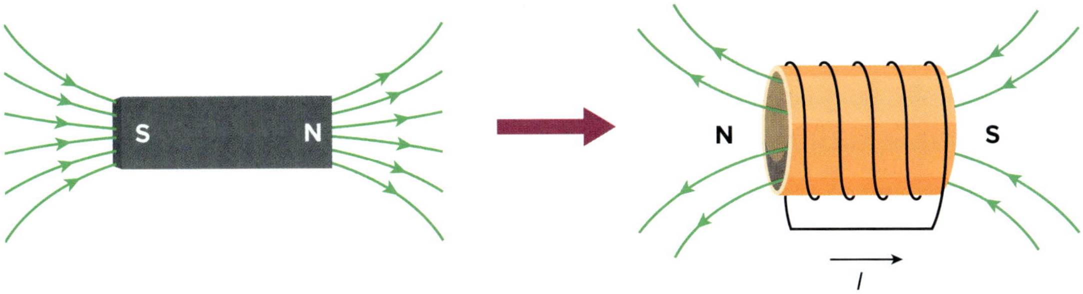

Figure 10 The magnet approaching the coil induces a current in the coil. The right-hand rule shows that the current is in the counterclockwise direction. The current produces a magnetic field in the direction opposite the change in the field that produced it.

Apply *How would the force on the coil change if the magnet were oriented so that its south pole faced the coil?*

Lenz's law explains the direction of the current in the coil of wire in **Figure 10.** The approaching magnet causes an increase in the magnetic field inside the coil. The current produced by the induced *EMF* must be in the direction shown in **Figure 10** to produce a magnetic field that opposes the increase in field. The induced field exerts a repelling force on the approaching magnet. By Newton's third law, there must be an equal and opposite force on the coil. This force pushes the coil away from the magnet. If you moved the magnet away from the coil, the field within the coil would decrease. By Lenz's law, the induced field would oppose this change by being in the same direction as the field of the magnet and adding to the field. In this case, the current producing the induced field in the coil would be in the clockwise direction.

In either case—moving the magnet toward or away from the coil—an induced magnetic field opposes the change in the field that created it. A decreasing magnetic field induces a field to oppose the decrease; an increasing magnetic field induces a field to oppose the increase.

Generators Lenz's law affects the operation of generators and motors. When a generator is not in a circuit, you can turn its armature, but the induced *EMF* generates no current. Therefore, there is no magnetic field. No force acts on the armature, and the armature is easy to turn.

However, when you turn the armature of a generator in a circuit, the induced *EMF* generates current, and the resulting magnetic field exerts a force on the armature. The force on the armature is opposite the external force that turns the armature, and the armature is difficult to turn. A generator supplying a large current produces a large amount of electric energy. The opposing force on the armature means that a proportionately greater amount of mechanical energy must be supplied to produce it. Note this is consistent with the law of conservation of energy.

Motors When you first turn on a motor, such as when you start a vacuum cleaner, current is large. As soon as the motor begins to turn, however, the motion of the wires across the magnetic field induces an *EMF* that opposes the current. This results in reduced current through the motor.

More load, more energy transfer When a mechanical load is placed on a motor, as when work is being done to lift a weight, the rotation of the motor slows. The slowing of the motor decreases the induced *EMF* opposing the current. Therefore, net current through the motor increases—consistent with the law of conservation of energy. When current increases, so does the rate at which electrical energy is being sent to the motor. This energy is delivered in mechanical form to the load. If the mechanical load is sufficient to stop the motor, the current can be so high that wires overheat.

Why lights dim when a motor starts When a motor starts, it draws a large amount of current. Because the wires between a motor and the motor's source of electrical energy have some resistance, the large current causes an increase in the potential difference across the wires. If another device, such as a lightbulb, is in a circuit parallel with the motor, there is a decrease in potential difference across this device. This is why the lightbulbs dim in your home when an air conditioner, table saw, or other motor-driven appliance starts operating.

Why a switch sparks When the current through a motor is interrupted, such as when a switch in the circuit is turned off or when a motor's plug is pulled from an outlet, the magnetic field in the motor drops to zero. Whenever there is a change in a magnetic field around a wire, an *EMF* is induced across the wire that opposes the change that created it. The induced *EMF* can be large enough to cause a spark to jump across the switch or between the plug and the outlet.

Braking effect The effects of Lenz's law can be beneficial. You might have used a laboratory balance to measure masses. A balance takes advantage of induced currents to stop its oscillations when an object is placed on its pan. A piece of metal attached to the balance arm in **Figure 11** rests between the poles of a horseshoe magnet. When an object is placed in the pan, the balance arm swings, and the metal moves through the magnetic field. The induced *EMF* creates currents in the metal, called eddy currents, that slow the metal down.

Eddy currents are currents generated in any piece of metal moving through a magnetic field; the magnetic field they produce opposes the motion that caused the currents. The force on the metal opposes the motion of the metal in either direction, but it does not act when the metal is still. Thus, it does not change measurements of the mass of an object on the pan.

When eddy currents have a braking effect on metal, the effect is called eddy-current damping. Eddy-current damping is commonly used to slow the movement of metal parts. The brakes of some trains and roller coasters are designed to take advantage of eddy-current damping. Often, the damping effect is unwelcome. To reduce eddy-current circulation in the metal parts in motors, the motor cores are constructed from thin metal layers that have insulation added between the layers.

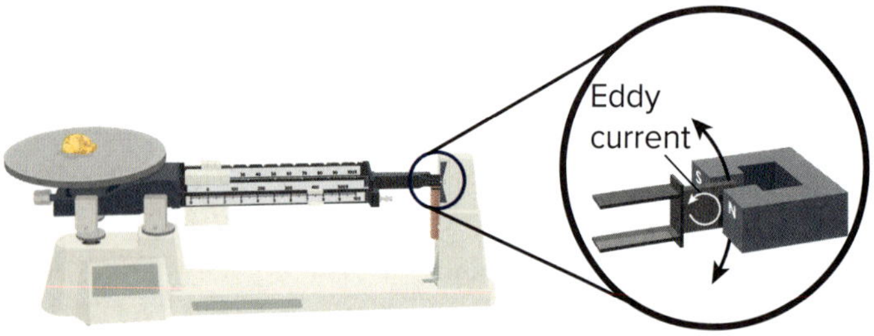

Figure 11 As the metal plate on the end of the balance beam moves through the magnetic field, an eddy current is generated in the metal. The eddy current produces a magnetic field that opposes the motion that caused it. This dampens the beam's oscillations.

CCC CROSSCUTTING CONCEPTS

Cause and Effect Analyze the effects that running a motor can have on other electrical devices in a household and on the device that uses the motor itself. Describe related safety precautions that should be taken when using a device with a motor. Cite evidence to support the cause-and-effect relationships underlying your recommendations.

Levitation Lenz's law dictates that the current in a wire in a changing magnetic field is such that the magnetic field it generates opposes the change of the external field. This effect can cause objects to levitate. How does this work?

The AC current in the coil of wire in **Figure 12** generates a changing magnetic field that induces an *EMF* in both of the aluminum rings above it. In the uncut ring on top, the *EMF* produces a current that generates a magnetic field opposing the change in the generating field. Because like poles face each other, the coil and uncut ring repel, and the ring rises.

The lower ring in **Figure 12** does not rise. Because it has been sawed through, it is an incomplete circuit and the *EMF* does not generate a current. Therefore, there is no force on the ring and no levitation.

Levitation caused by repulsion underlies the operation of very fast trains called magnetic levitation (Maglev) trains. As these trains move, they make no physical contact with the tracks.

Figure 12 An EMF is induced in both aluminum rings, but current is produced only in the continuous ring because only this ring completes a circuit. Thus, only this ring experiences a force, and only this ring levitates. If the rings were made from a nonconducting material, such as nylon or wood, an EMF would not be induced in either one.

Self-inductance An *EMF* can be induced in a wire when the magnetic field in the region of the wire changes. The field can be external, or it can be generated by the current in the wire itself, as in **Figure 13.**

Imagine that the coil of wire in **Figure 13** is suddenly connected to a battery. This causes a change in the potential difference across the coil, producing a changing current through the coil. The changing current generates a changing magnetic field in the coil. This changing field induces an *EMF* in the direction that opposes the change. The induced *EMF* reduces the potential difference across the coil. The result is a net decrease in current.

Therefore, initially the current through the coil is small but increasing. However, as the rate of change of current decreases, so does the opposing *EMF*. When the current reaches a constant final value, the current change is zero, so the induced *EMF* is also zero.

If the potential difference supplied by a battery is decreased, the current is also decreased. So, the induced *EMF* changes to be in a direction to maintain the magnetic field. Therefore, the induced current is in the same direction as the current from the battery. If the coil were suddenly disconnected from the battery, the induced *EMF* could be large enough to produce a spark. The property of a wire, either straight or in a coil, to create an induced *EMF* that opposes the change in the potential difference across the wire is called **self-inductance.**

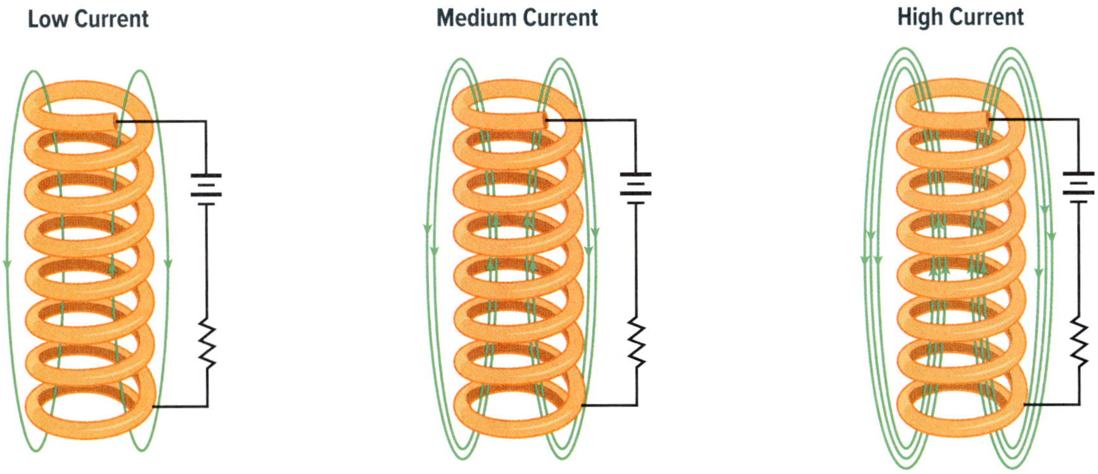

Figure 13 As the current in the coil increases, the magnetic field generated by the current also increases. The increase in the magnetic field induces an EMF that opposes the direction of current, and more energy is needed to increase the current further.

Transformers

You have probably seen metal cylinders attached to utility poles, such as the cylinders on the pole in **Figure 14.** Inside each of these cylinders is a transformer. **Transformers** are devices that increase or decrease potential differences with relatively little waste of energy. Only alternating current (AC) can be sent through a transformer. Direct current (DC) cannot pass through a transformer.

Figure 14 Inside these cylinders are transformers. Transformers reduce high voltages to usable levels before the electrical energy enters your home.

How transformers work

You read that self-inductance produces an *EMF* when current changes in a single coil. A transformer has two coils, electrically insulated from each other but wound around the same iron core, as you can see in **Figure 15.** When one coil—the primary coil—is connected to an AC source, the changing current creates a changing magnetic field that is carried through the core to the other coil—the secondary coil. In the secondary coil, the changing magnetic field induces a varying *EMF* and current. An *EMF* and current in one coil due to changing current in another coil is called **mutual inductance.**

The *EMF* induced in the secondary coil, called the secondary potential difference, is proportional to the potential difference provided to the primary coil. The secondary potential difference also depends on what is called the turns ratio. The turns ratio is the number of turns of wire in the secondary coil divided by the number of turns in the primary coil, as shown on the right in the following expressions.

$$\frac{\text{primary potential difference}}{\text{secondary potential difference}} = \frac{\text{number of turns in primary coil}}{\text{number of turns in secondary coil}}$$

$$\frac{V_p}{V_s} = \frac{N_p}{N_s}$$

If the secondary potential difference is larger than the primary potential difference, as it is in the left part of **Figure 15,** the transformer is called a **step-up transformer.** If the secondary potential difference is smaller than the primary potential difference, as in the right part of **Figure 15,** the transformer is called a **step-down transformer.**

Get It?

Compare How do step-up transformers differ from step-down transformers?

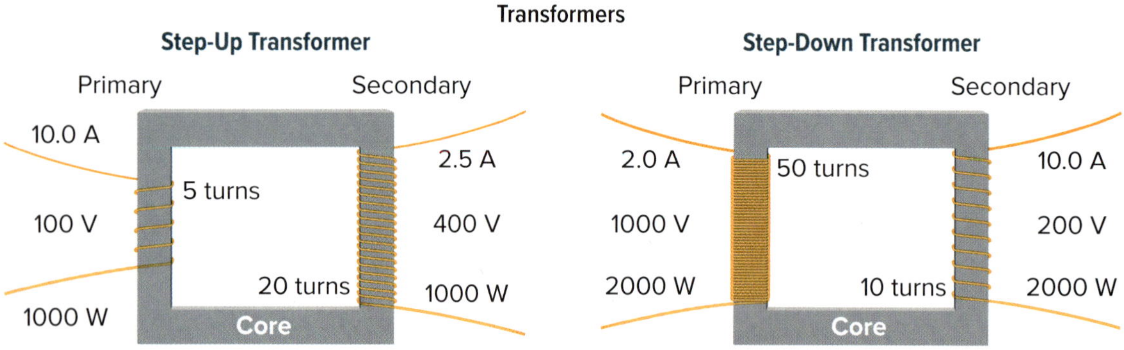

Figure 15 The ratio of primary potential difference to secondary potential difference in a transformer depends on the ratio of the number of turns on the primary coil to the number of turns on the secondary coil. Secondary potential difference can be greater than the primary (left) or less than the primary (right).

montreehanlue/123RF

The ideal transformer In an ideal transformer, the electrical power delivered to the secondary circuit equals the power supplied to the primary circuit. An ideal transformer dissipates no power itself. Because it is 100 percent efficient, it can be represented by the following equations:

$$P_p = P_s$$

$$V_pI_p = V_sI_s$$

Rearranging the equation to find the ratio $\frac{V_p}{V_s}$ shows that the current in the primary circuit depends on how much current is required by the secondary circuit. This relationship can be combined with the relationship shown earlier (between potential difference and the number of turns) to result in the following equation.

CONNECTING MATH to Physics

Inequalities Study the following expressions to help you understand the relationships among potential difference (*V*), current (*I*), and the number of coils in transformers (*N*) in primary and secondary circuits.

Step-Up Transformer	Step-Down Transformer
$V_p < V_s$	$V_p > V_s$
$I_p > I_s$	$I_p < I_s$
$N_p < N_s$	$N_p > N_s$

Transformer Equation

The ratio of the current in the secondary coil to the current in the primary coil is equal to the ratio of the potential difference in the primary coil to the potential difference in the secondary coil, which is also equal to the ratio of the number of turns on the primary coil to the number of turns on the secondary coil.

$$\frac{I_s}{I_p} = \frac{V_p}{V_s} = \frac{N_p}{N_s}$$

Changes in potential difference You read that a step-up transformer increases potential difference in the secondary coil. Because transformers cannot increase power output, there must be a corresponding decrease in current that can be supplied to the secondary circuit. Similarly, in a step-down transformer, the current is greater in the secondary circuit than it is in the primary circuit. A decrease in potential difference corresponds to an increase in current, as you can see in the Connecting Math to Physics. Even in real transformers, which are not 100 percent efficient, the current supplied by a step-down transformer is larger than the current in the primary circuit, and a step-up transformer supplies less current than it draws.

Isolation transformers Not all transformers are step-up or step-down. In some transformers, the primary and secondary coils have the same number of turns, so the input and output potential differences are identical. These transformers, called isolation transformers, are often used for safety reasons. They isolate the DC portion of current from sensitive electronic devices where it otherwise might cause interference or electric shock. These sensitive electronic devices include computers, recording instrumentation, and medical tools, such as those used for ultrasound and other diagnostic imaging. Isolation transformers also can be used to reduce electrical noise.

Real transformers Ideal transformers do not convert any electrical energy to thermal energy. In real transformers, however, some of the electrical energy they deliver to the primary circuit is converted into thermal energy, heating the transformer and the air around it. The power that can be delivered by the secondary circuit is thus reduced. The efficiency of a transformer is given by the ratio of the output power to the input power. Typical transformers are 95 to 98 percent efficient. Though the amount of energy lost to thermal energy might seem small, large transformers handle so much power that they can become very hot.

EXAMPLE Problem 2

STEP-UP TRANSFORMERS A step-up transformer has a primary coil consisting of 200 turns and a secondary coil consisting of 3000 turns. The primary coil is supplied with an effective AC potential difference of 90.0 V.

a. What is the potential difference in the secondary circuit?

b. The current in the secondary circuit is 2.0 A. What is the current in the primary circuit?

1 ANALYZE AND SKETCH THE PROBLEM

- Draw an iron core that has turns of wire on either side.
- Label the variables *I*, *V*, and *N*.

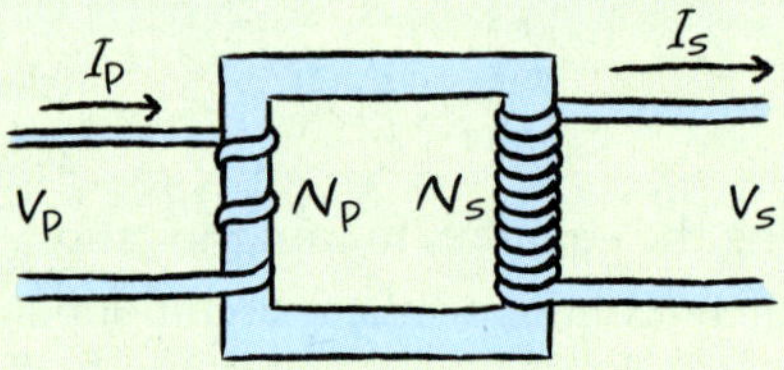

Known		Unknown
$N_p = 200$	$V_p = 90.0$ V	$V_s = ?$
$N_s = 3000$	$I_s = 2.0$ A	$I_p = ?$

2 SOLVE FOR THE UNKNOWN

a. Solve for V_s.

$$\frac{V_s}{V_p} = \frac{N_s}{N_p}$$

$$V_s = \frac{N_s V_p}{N_p}$$

$$= \frac{(3000)(90.0\text{ V})}{200}$$ Substitute $N_s = 3000$, $V_p = 90.0$ V, $N_p = 200$.

$$= 1350\text{ V}$$

b. The power in the primary and secondary circuits is equal assuming 100 percent efficiency.

$$P_p = P_s$$

$$V_p I_p = V_s I_s$$ Substitute $P_p = V_p I_p$, $P_s = V_s I_s$.

Solve for I_p.

$$I_p = \frac{V_s I_s}{V_p}$$

$$= \frac{(1350\text{ V})(2.0\text{ A})}{90.0\text{ V}}$$ Substitute $V_s = 1350$ V, $I_s = 2.0$ A, $V_p = 90.0$ V.

$$= 3.0\times10^1\text{ A}$$

3 EVALUATE THE ANSWER

- **Are the units correct?** Potential difference is in volts, and current is in amps.
- **Is the magnitude realistic?** A large step-up ratio of turns results in a large secondary potential difference yet a smaller secondary current. The answers agree.

PRACTICE Problems

ADDITIONAL PRACTICE

For the following problems, effective currents and potential differences are indicated.

16. A step-down transformer has 7500 turns on its primary coil and 125 turns on its secondary coil. The potential difference across the primary circuit is 7.2 kV. What is the potential difference across the secondary circuit? If the current in the secondary circuit is 36 A, what is the current in the primary circuit?

17. CHALLENGE A step-up transformer that is 95 percent efficient has 300 turns on its primary coil and 90,000 turns on its secondary coil. The potential difference of the generator to which the primary circuit is attached is 60.0 V. What is the potential difference across the secondary circuit? The current in the secondary circuit is 0.50 A. What current is in the primary circuit?

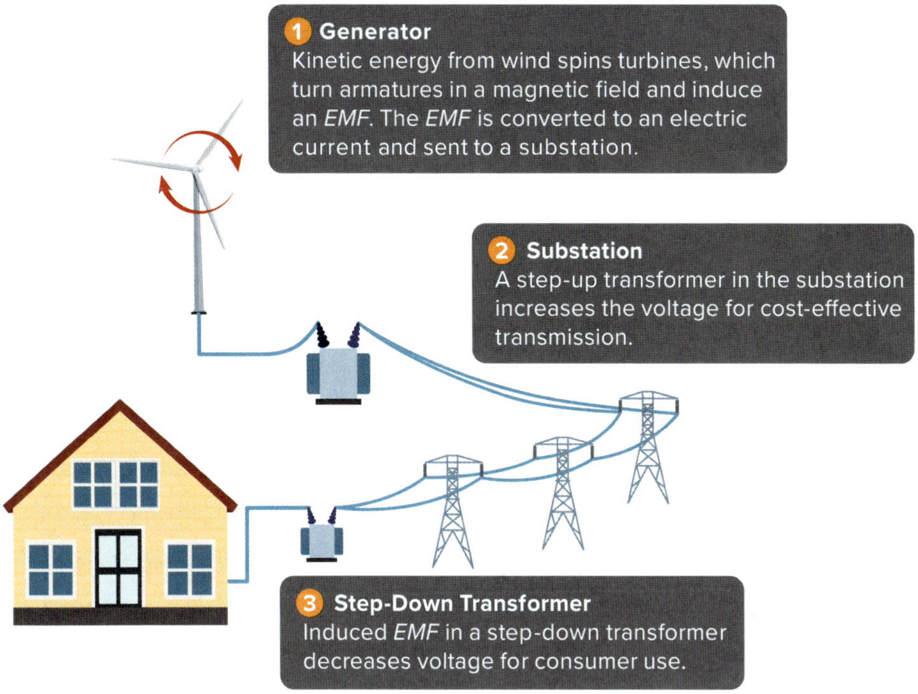

Figure 16 Step-up transformers increase potential differences (voltages) in overhead power lines. Step-down transformers decrease potential differences for consumer use.

Everyday uses of transformers Long-distance transmission of electrical energy is economical only if very high potential differences are used. High potential differences reduce the current required in the transmission lines, keeping the wasteful energy transformations low. As shown in **Figure 16,** step-up transformers are used at power sources, where they can develop potential differences up to 480,000 V. When the energy reaches homes, step-down transformers reduce the potential difference to 120 V. Game systems, printers, laptop computers, and rechargeable toys have transformers inside their casings or in blocks attached to their cords. These small transformers further reduce potential differences from wall outlets to the 3 V–26 V range.

Check Your Progress

18. **Lenz's Law** You hang a coil of wire with its ends joined so that it can swing easily. If you now plunge a magnet into the coil, the coil will start to swing. Which way will it swing relative to the magnet and why?
19. **Motors** If you unplugged a running vacuum cleaner from a wall outlet, you would be much more likely to see a spark than you would if you unplugged a lighted lamp from the wall. Why?
20. **Transformers and Current** Explain why a transformer may be operated only on AC.
21. **Transformers** Frequently, transformer coils that have only a few turns are made of very thick (low-resistance) wire, while those with many turns are made of thin wire. Why?
22. **Step-Up Transformers** Refer to the step-up transformer shown in **Figure 15.** Explain what would happen to the primary current if the secondary coil were short-circuited.
23. **Critical Thinking** Would permanent magnets make good transformer cores? Explain.

LESSON 3

ELECTRIC AND MAGNETIC FORCES ON PARTICLES

FOCUS QUESTION

How does a mass spectrometer work?

Thomson's Experiments

Electromagnetic waves bring you messages on your cell phone, carry music over your radio, and even enable you to see. Before you can understand how electric and magnetic fields combine to make electromagnetic waves, it is necessary to understand the electron.

Discovery of the electron In the nineteenth century, evidence that atoms contained negatively charged particles had been growing. In 1897, while doing experiments with a cathode-ray tube, J.J. Thomson was able to extract the negatively charged particles (later named *electrons*) from atoms of various materials. How did he do this?

Thomson evacuated nearly all the air from a cathode-ray tube, similar to the tube in **Figure 17.** A battery connected to the tube produced a large potential difference between the cathode and the anodes. At the end of the tube opposite the cathode, Thomson observed a glowing spot, made by an invisible beam—the cathode ray—that was accelerated from the cathode toward the anodes by the potential difference. As it traveled, the cathode ray passed through slits in the anodes. Thomson realized that the ray consisted of negatively charged particles from the gases in the tiny amount of air remaining in the tube.

Figure 17 In this cathode-ray tube, similar to Thomson's tube, a potential difference between the anodes and the cathode accelerates electrons toward the anodes.

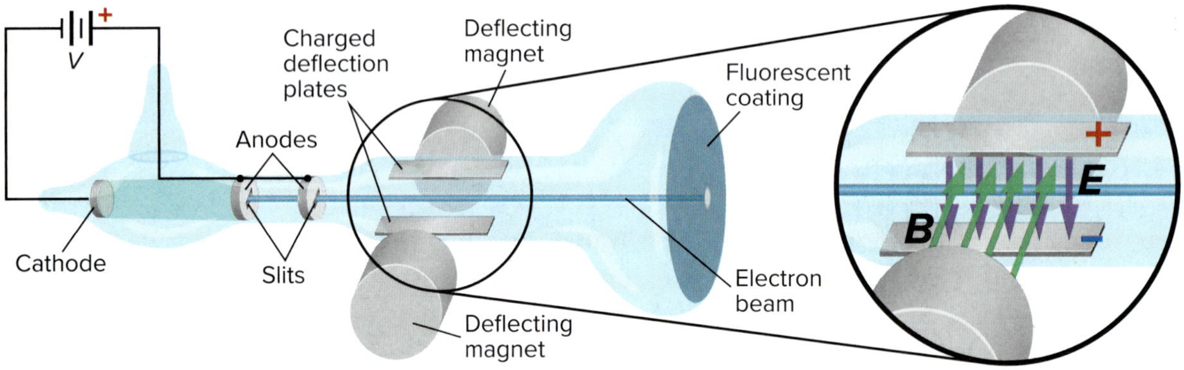

3D THINKING DCI Disciplinary Core Ideas CCC Crosscutting Concepts SEP Science & Engineering Practices

COLLECT EVIDENCE

Use your Science Journal to record the evidence you collect as you complete the readings and activities in this lesson.

INVESTIGATE

GO ONLINE to find these activities and more resources.

PhysicsLAB: Mass of an Electron

Carry out an investigation into how a magnetic field affects a moving electron and calculate the mass of an electron.

Revisit the Encounter the Phenomenon Question

What information from this lesson can help you answer the Unit and Module questions?

Although Thomson knew neither the charge (q) nor the mass (m) of the particles that made up the cathode ray, he was able to determine the particles' charge-to-mass ratio. This ratio, he found, was very high, presumably because the particles had very small mass. Thomson further found that the charge-to-mass ratio remained the same no matter what kind of gas he placed in the tube. He concluded that the particles were negative constituents of all atoms—electrons.

Charge-to-mass ratio A closer look at Thomson's experiment reveals how he was able to determine the electron's charge-to-mass ratio. Inside the cathode-ray tube, as in the tube in **Figure 17,** oppositely charged metal plates produced an electric field. Electromagnets outside the tube produced a magnetic field. By manipulating either field, Thomson found he could change the path of the electron beam.

Thomson oriented the electric field perpendicular to the electron beam. This field (of strength E) produced a force (equal to qE) that deflected the beam upward, toward the positive plate. Thomson oriented the magnetic field at right angles to both the beam and the electric field. Recall that the force exerted by a magnetic field is perpendicular both to the field and to the direction of motion of electrons. Thus, the magnetic field in the tube (of strength B) produced a force (equal to Bqv, where v is the electron velocity) that deflected the electron beam downward.

Get It?

Apply How could you change the magnetic field so the electrons deflected upward?

Thomson adjusted the strengths of the electric and magnetic fields until the electron beam followed a straight, undeflected path. When this occurred, the forces due to the two fields were equal in magnitude and opposite in direction. Mathematically, this can be represented as follows:

$$Bqv = Eq$$

Solving this equation for v yields the following expression:

$$v = \frac{Eq}{Bq} = \frac{E}{B}$$

The above equation shows that the forces are balanced only for electrons that have a specific velocity (v). If the electric field is turned off, only the force due to the magnetic field remains. You know that the direction of motion of an electron is perpendicular to the magnetic force. Therefore, an electron in a uniform magnetic field undergoes centripetal acceleration, having a circular path of radius r. Using Newton's second law of motion for circular motion, the following equation describes the electron's path:

$$Bqv = \frac{mv^2}{r}$$

Solving for $\frac{q}{m}$ results in the following equation.

Charge-to-mass Ratio of an Electron in a Magnetic Field

In a magnetic field, the ratio of an electron's charge to its mass is equal to the ratio of the electron's velocity divided by the product of the magnetic field strength and the radius of the electron's circular path.

$$\frac{q}{m} = \frac{v}{Br}$$

Mass of an electron To find the value of the charge-to-mass ratio of an electron, Thomson first calculated the straight trajectory velocity (v) of the electron beam's path using measured values of E and B. Next, he turned off the electric field and measured the distance between the fluorescent spot formed by the undeflected beam and the spot formed when the magnetic field alone acted on the beam. Using this distance, he calculated the radius (r) of the electron beam's circular path and obtained q/m as -1.759×10^{11} C/kg.

Calculating the charge-to-mass ratio was the first step in calculating the electron's mass. When Thomson was doing his experiments, the average charge of an electron was known from electrolysis experiments involving many electrons as $q = -1.602 \times 10^{-19}$ C. Thomson assumed this value could be used for individual electrons. Later, in 1909, Robert Millikan confirmed this value, finding that all electrons have the same charge (e). Therefore, the mass of the electron (m_e) could be calculated.

$$m_e = \frac{e}{q/m} = \frac{-1.602\times10^{-19}\text{ C}}{-1.759\times10^{11}\text{ C/kg}} = 9.107\times10^{-31}\text{ kg}$$

To three significant figures, $m_e = 9.11\times10^{-31}$ kg.

Experiments with positive ions Thomson also used his cathode-ray apparatus to study positive ions. An ion is a charged atom or molecule. Positively charged particles in an electric or magnetic field undergo a deflection opposite to that of electrons, as shown in **Figure 18.**

Mass of a proton To make positive ions, Thomson added a small amount of hydrogen gas to his tube and reversed the direction of the electric field between the cathode and the anodes. The high electric field pulled electrons off the hydrogen atoms, turning the atoms into positive ions. The electric field then accelerated the ions through a tiny slit into the deflection region of the tube. The resulting ion beam passed through electric and magnetic fields on its way toward the fluorescent screen at the end of the tube.

Thomson determined the charge-to-mass ratio of a positive ion of hydrogen (later shown to be a proton) in the same manner as he determined the charge-to-mass ratio of an electron. From this ratio, the mass of a single proton was found to be 1.67×10^{-27} kg. Thomson went on to use his cathode-ray tube to determine the masses of positive ions produced when one or more electrons were stripped from heavier gases, such as helium, neon, and argon.

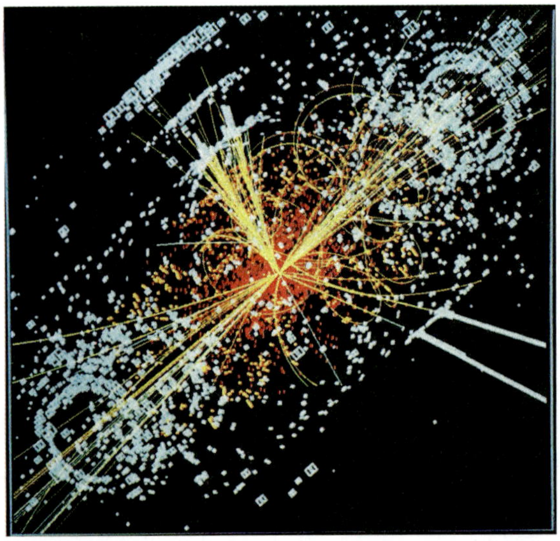

Figure 18 The tracks of negatively and positively charged particles moving through a magnetic field curve in opposite directions. This computer-generated image simulates a collision of two protons at the Large Hadron Collider. The yellow lines represent possible paths of the resulting particles, and the blue rectangles represent the energy deposited by these particles.

EXAMPLE Problem 3

PATH RADIUS An electron (with mass of 9.11×10^{-31} kg) moves through a cathode-ray tube at 2.0×10^{7} m/s perpendicular to a magnetic field of 3.5×10^{-3} T. The electric field is turned off. What is the radius of the electron's circular path?

1 ANALYZE AND SKETCH THE PROBLEM

- Draw the path of the electron and label the velocity (v).
- Sketch the magnetic field perpendicular to the velocity.
- Diagram the force acting on the electron. Add the unknown quantity r of the electron's path.

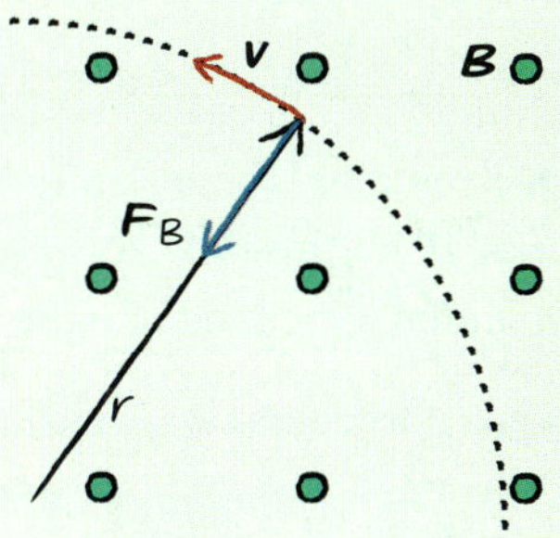

Known	Unknown
$v = 2.0\times10^{7}$ m/s	$r = ?$
$B = 3.5\times10^{-3}$ T	
$m = m_e = 9.11\times10^{-31}$ kg	
$q = e = -1.602\times10^{-19}$ C	

2 SOLVE FOR THE UNKNOWN

Use Newton's second law of motion to describe an electron in a cathode-ray tube subjected to a magnetic field.

$$Bqv = \frac{mv^2}{r}$$

$$r = \frac{mv}{Bq}$$

$$= \frac{(9.11\times10^{-31}\text{ kg})(2.0\times10^{7}\text{ m/s})}{(3.5\times10^{-3}\text{ T})(-1.602\times10^{-19}\text{ C})}$$

Substitute $m = 9.11\times10^{-31}$ kg, $v = 2.0\times10^{7}$ m/s, $B = 3.5\times10^{-3}$ T, $q = -1.602\times10^{-19}$ C.

$$= -3.2\times10^{-2}\text{ m}$$

3 EVALUATE THE ANSWER

- **Are the units correct?** The radius of the circular path is a length measurement, given in units of meters.
- **Are the values reasonable?** Radii of a few centimeters are typical of this kind of apparatus and can easily be measured. The negative sign means that the circle curves in the direction opposite that of a positive ion.

PRACTICE Problems

ADDITIONAL PRACTICE

For the following questions, assume all charged particles move perpendicular to a uniform magnetic field.

24. A proton moves at a speed of 7.5×10^{4} m/s as it passes through a magnetic field of 0.080 T. Find the radius of the circular path. Note that the charge carried by the proton is equal to that of the electron but is positive.

25. Electrons move through a magnetic field of 3.0×10^{-3} T balanced by an electric field of 2.4×10^{4} N/C.

a. What is the speed of the electrons?

b. If the electric field were produced by a pair of plates 0.50 cm apart, what would be the potential difference between the two plates?

c. If the electric field were removed, what would be the radius of the circular path that the electrons followed?

26. Protons passing without deflection through a magnetic field of 0.060 T are balanced by an electric field of 9.0×10^{3} V/m. What is the speed of the moving protons?

27. CHALLENGE What trajectory would a positive ion follow moving in a magnetic field that increases linearly with time?

The Mass Spectrometer

An interesting thing happened when Thomson put neon gas into his cathode-ray tube. He observed two glowing dots on the screen instead of just one. Calculating a separate value of q/m for each dot, he concluded that two different atoms of neon could have identical chemical properties but different masses. Each form of the same atom that has the same chemical properties but a different mass is an **isotope.**

Thomson's separation of the isotopes of neon by their mass was the first example of mass spectrometry. A **mass spectrometer** is an instrument that measures the charge-to-mass ratio of positive ions within a material. From this ratio, it is possible to determine the masses of the atomic isotopes that make up the material. One kind of mass spectrometer is illustrated in **Figure 19.**

Material introduced into a mass spectrometer must either be naturally a gas or be heated to the gaseous state. The gas is inserted into a component called the ion source, which you can see on the bottom of **Figure 19.** In the ion source, an energetic beam of electrons collides with the gas's atoms and rips one or more electrons off, producing positive ions. An electric field—the result of a potential difference between two electrodes in the ion source—then accelerates the ions, which enter a magnetic field in the vacuum chamber. The uniform magnetic field causes the ions to move in a circular path before they collide with an electronic detector.

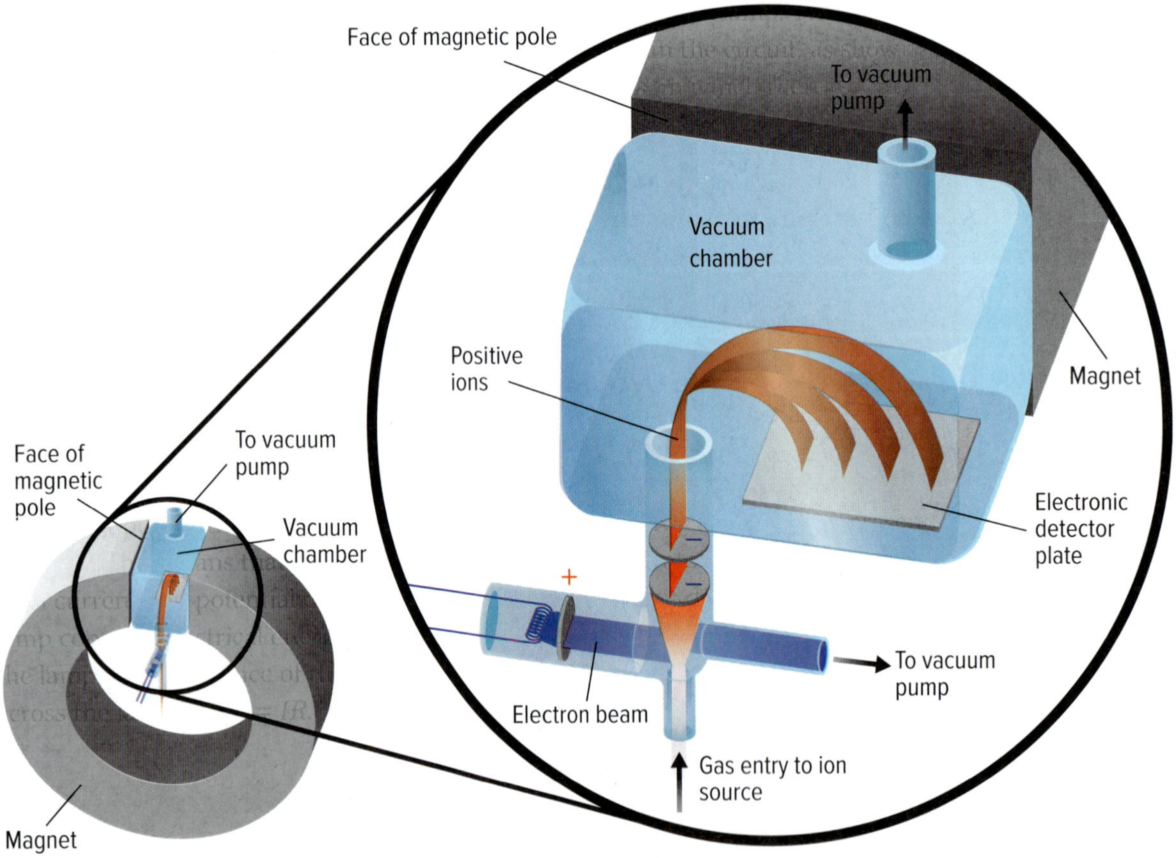

Figure 19 Inside this mass spectrometer (blowout on right), a magnet causes the positive ions in a vacuum chamber to be deflected according to their mass. The mass of each particle is recorded in the vacuum chamber on a detector plate.

Identify *the north pole of the magnet.*

Calculating charge-to-mass ratios Once the ions in a mass spectrometer move into a magnetic field, they follow a circular path. The radius of the path depends on the ion's mass. The lighter the ion, the more it bends and the smaller is the radius of the circle. The radius of the path for each ion can be used to determine the ion's charge-to-mass ratio. The path radius (r) for an ion can be calculated from Newton's second law of motion:

$$Bqv = \frac{mv^2}{r}$$

Solving for r yields the following equation:

$$r = \frac{mv}{qB}$$

The velocity of an accelerated ion can be calculated from the equation for the kinetic energy of ions accelerated from rest through a known potential difference (V_{accel}):

$$KE = \frac{1}{2}mv^2 = qV_{accel}$$

$$v = \sqrt{\frac{2qV_{accel}}{m}}$$

Substituting this expression for v in the equation $r = \frac{mv}{qB}$ gives the radius of the ion's circular path:

$$r = \frac{mv}{qB}$$

$$= \frac{m}{qB}\sqrt{\frac{2qV_{accel}}{m}}$$

$$= \frac{1}{B}\sqrt{\frac{2V_{accel}m}{q}}$$

Simplifying this equation by multiplying both sides by B yields the following:

$$Br = \sqrt{\frac{2mV_{accel}}{q}}$$

This equation can be rearranged as shown below.

Charge-to-mass Ratio of an Ion in a Mass Spectrometer

The ratio of an ion's charge to its mass is equal to twice the accelerating potential difference divided by the product of the square of the magnetic field strength and the square of the radius of the ion's circular path.

$$\frac{q}{m} = \frac{2V_{accel}}{B^2r^2}$$

To calculate the mass of an ion, you would divide the charge on the ion by the measured charge-to-mass ratio. The charge of a proton is the same as the charge of an electron except it is positive: 1.602×10^{-19} C. Note that if more than one electron is removed from the atoms of a gas in the ion source, the charge must be multiplied by the number of electrons removed. This number—the number of electrons removed—can be controlled by the mass spectrometer's operator.

Get It?

Calculate What is the charge on an ion that has three electrons removed?

EXAMPLE Problem 4

MASS OF A NEON ATOM The operator of a mass spectrometer produces a beam of neon ions from which two electrons have been removed. The ions are thus doubly ionized. The ions first are accelerated by a potential difference of 34 V. When they pass through a magnetic field of 0.050 T, the radius of their path is 53 mm. Determine the mass of the neon atom to the closest whole number of proton masses.

1 ANALYZE AND SKETCH THE PROBLEM

- Draw the circular path of the ions. Label the radius.
- Draw and label the potential difference between the electrodes.

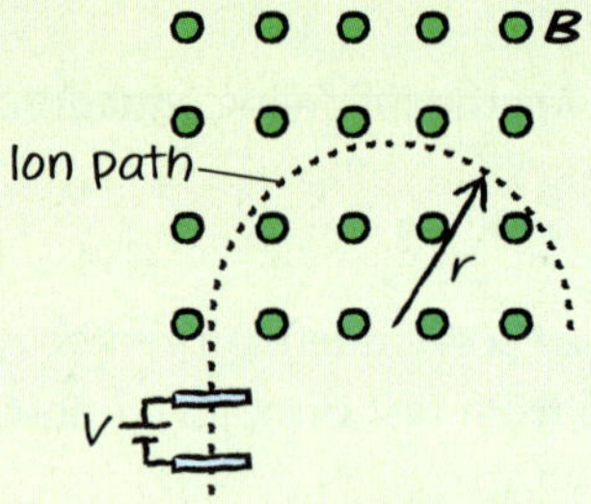

Known

$V_{accel} = 34$ V

$B = 0.050$ T

$r = 0.053$ m

$m_{proton} = 1.67\times10^{-27}$ kg

$q = 2(1.602\times10^{-19}$ C)

$= 3.204\times10^{-19}$ C

Unknown

$m_{neon} = ?$

$N_{proton} = ?$

2 SOLVE FOR THE UNKNOWN

Use the equation for the charge-to-mass ratio of an ion in a mass spectrometer.

$$\frac{q}{m_{neon}} = \frac{2V_{accel}}{B^2r^2}$$

$$m_{neon} = \frac{qB^2r^2}{2V_{accel}}$$

$$= \frac{(3.204\times10^{-19}\text{ C})(0.050\text{ T})^2(0.053\text{ m})^2}{2(34\text{ V})}$$

Substitute $q = 3.204\times10^{-19}$ C, $B = 0.050$ T, $r = 0.053$ m, and $V_{accel} = 34$ V.

$$= 3.3\times10^{-26}\text{ kg}$$

Divide the mass of neon by the mass of a proton to find the number of proton masses.

$$N_{proton} = \frac{m_{neon}}{m_{proton}}$$

$$= \frac{3.3\times10^{-26}\text{ kg}}{1.67\times10^{-27}\text{ kg/proton}}$$

$$\cong 20\text{ protons}$$

3 EVALUATE THE ANSWER

- **Are the units correct?** Mass should be measured in grams or kilograms. The number of protons is not represented by any unit.
- **Is the magnitude realistic?** Neon has two isotopes, with masses of approximately 20 and 22 proton masses.

PRACTICE Problems

ADDITIONAL PRACTICE

28. A beam of singly ionized (1+) oxygen atoms is sent through a mass spectrometer. The values are $B = 7.2\times10^{-2}$ T, $q = 1.602\times10^{-19}$ C, $r = 0.085$ m, and $V_{accel} = 110$ V. Find the mass of an oxygen atom.

29. A mass spectrometer analyzes and gives data for a beam of doubly ionized (2+) argon atoms. The values are $q = 2(1.602\times10^{-19}$ C), $B = 5.0\times10^{-2}$ T, $r = 0.106$ m, and $V_{accel} = 66.0$ V. Find the mass of an argon atom.

30. A beam of singly ionized (1+) lithium atoms ($m \approx 7m_{proton}$) is accelerated by a potential difference of 320 V. The beam passes through a magnetic field of 1.5×10^{-2} T. What is the radius of curvature of the beam in the magnetic field?

31. **CHALLENGE** Regardless of the energy of the electrons used to produce ions, J.J. Thomson never could remove more than one electron from a hydrogen atom. What could he have concluded about the positive charge of a hydrogen atom?

Isotopic analysis The ions in the mass spectrometer in **Figure 19** strike the detector in different places. Where an ion strikes depends on its mass. The more massive the ion, the larger is the diameter of its curved path. This diameter can be easily measured because it is the distance between the location where the ion strikes and the slit in the electrode. The radius of the path (r) is half this measured distance. The approximate spacing between ion strikes on a detector for an ionized chromium sample is shown in **Figure 20.** The four distinct marks indicate that a naturally occurring sample of chromium is composed of four isotopes. The abundance of each isotope corresponds to the width of the mark it makes on the detector.

All the chromium ions in **Figure 20** are singly charged because only one electron was removed from the original chromium atoms in the ion source. Charge depends on the number of electrons removed from the original (neutral) atoms. After the first electron is removed, producing a singly ionized (1+) atom, more energy is required to remove the second electron and produce a doubly ionized (2+) atom. The operator can produce doubly-ionized atoms by increasing the electric field, which gives the electron beam more kinetic energy.

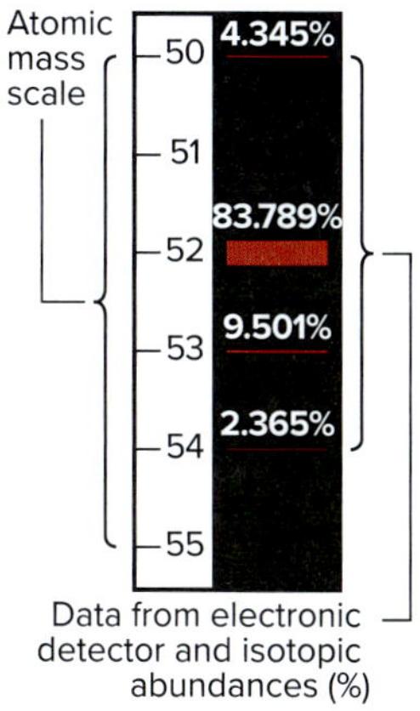

Figure 20 The weighted average of the masses of an element's isotopes provides the mass of the element shown on the periodic table.

Calculate *the average mass of the chromium isotopes.*

Mass spectrometry applications Mass spectrometers are so sensitive that they can separate ions with mass differences as small as one ten-thousandth of one percent. Because of this precision, scientists can use mass spectrometers to identify the presence of a single molecule in a 10 billion–molecule sample.

Mass spectrometers have a wide range of applications. Mass spectrometers are used to analyze the atmospheres and soils of Mars and other objects in the solar system. They also are commonly used in the geological, pharmaceutical, biological, and even the forensic sciences. For example, airports use mass spectrometers to detect traces of molecules found in explosives that might be in the luggage or on the hands or shoes of passengers.

Check Your Progress

32. **Mass Spectrometer** The radius of the circular path of an ion in a mass spectrometer is given by $r = \frac{1}{B}\sqrt{\frac{2V_{accel}m}{q}}$. Use this equation to explain how a mass spectrometer is able to separate ions of different masses.
33. **Cathode-Ray Tube** Describe how the cathode-ray tube used by J.J. Thomson forms an electron beam.
34. **Magnetic Field** A mass spectrometer can analyze molecules having masses of hundreds of proton masses. If the singly charged ions of these molecules were produced using the same accelerating potential difference as used for smaller ions, how would the mass spectrometer's magnetic field have to be changed for the ions to hit the detector?
35. **Path Radius** A proton moves at a speed of 8.4×10^4 m/s as it passes through a magnetic field of 12.0 mT. Find the radius of the circular path.
36. **Mass** A beam of doubly ionized (2+) oxygen atoms is accelerated by a potential difference of 232 V. The oxygen then enters a magnetic field of 75 mT and follows a curved path with a radius of 8.3 cm. What is the mass of the oxygen atom?
37. **Critical Thinking** In Example Problem 4 (on the previous page), the mass of a neon isotope is determined. Another neon isotope has a mass of 22 proton masses. How far apart on a detector would these two isotopes land?

LEARNSMART® Go online to follow your personalized learning path to review, practice, and reinforce your understanding.

LESSON 4

ELECTRIC AND MAGNETIC FIELDS IN SPACE

FOCUS QUESTION

What do electric and magnetic fields have to do with light?

What Are Electromagnetic Waves?

You read in Lesson 3 that electromagnetic waves are a combination of electric and magnetic fields. Recall that Hans Christian Oersted was the first to make the connection between electricity and magnetism when he found, in 1820, that moving charges produce magnetic fields. Ten years later, Michael Faraday and Joseph Henry independently discovered the converse—that a wire's magnetic field can induce an electric field.

Moving through space Electric fields can be induced even without a wire. As shown on the left in **Figure 21,** a changing magnetic field alone produces an electric field. The reverse is also true. A changing electric field produces a magnetic field, as shown in the right of **Figure 21.** Scottish physicist James Maxwell quantified the relationship between electric and magnetic fields in 1873, when he summarized the work of Coulomb, Oersted, and Faraday into four equations. These equations predicted that electromagnetic waves could move through space. In 1887, this was confirmed by German physicist Heinrich Hertz. Hertz's name was later given to the unit that measures a wave's oscillation frequency. Recall that 1 Hz equals one cycle per second.

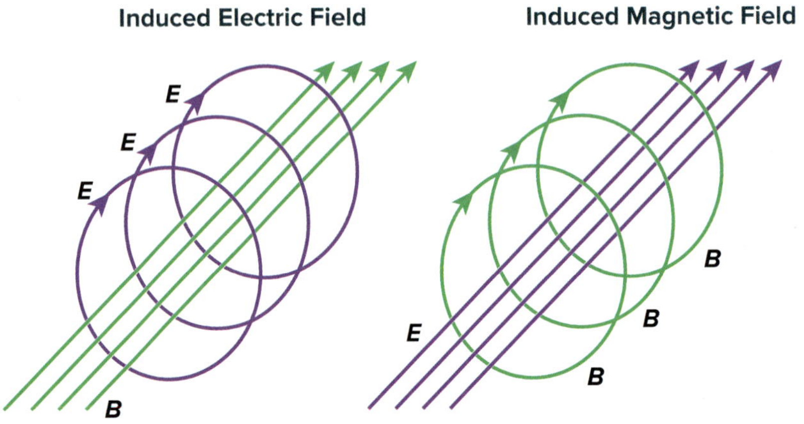

Figure 21 An electric field can be induced by a changing magnetic field (left). Notice that the field lines of the electric field are closed loops. Unlike in an electrostatic field, there are no charges on which the lines begin or end. A magnetic field also can be induced by a changing electric field (right).

3D THINKING   DCI Disciplinary Core Ideas   CCC Crosscutting Concepts   SEP Science & Engineering Practices

COLLECT EVIDENCE

Use your Science Journal to record the evidence you collect as you complete the readings and activities in this lesson.

INVESTIGATE

GO ONLINE to find these activities and more resources.

Applying Practices: Catching Waves

HS-PS4-5. Communicate technical information about how some technological devices use the principles of wave behavior and wave interactions with matter to transmit and capture information and energy.

Today, it is known that **electromagnetic waves** are coupled, oscillating electric and magnetic fields that propagate through space and matter.

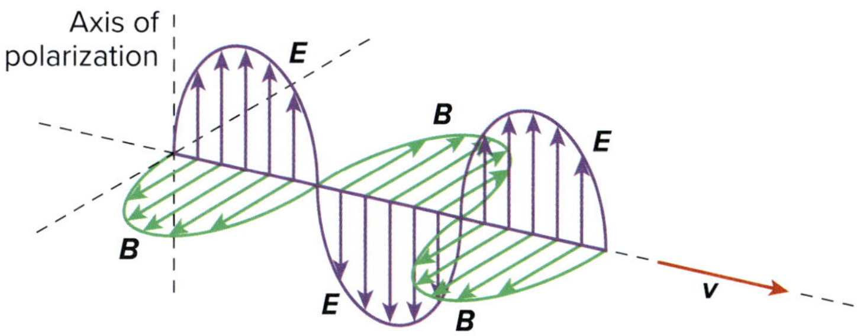

Figure 22 The electric and magnetic fields that make up an electromagnetic wave oscillate at right angles to each other and to the direction of the wave's velocity (v).

Electromagnetic wave properties

How are the magnetic and electric fields related in an electromagnetic wave? Accelerating electrons produce the electric field of a wave, and the varying electric field produces the wave's magnetic field. The orientations of these two fields are shown in **Figure 22.** As the wave moves, its electric field oscillates up and down while its magnetic field oscillates at right angles to the electric field. Both the electric field and the magnetic field are at right angles to the wave propagation direction.

Travel in a vacuum All electromagnetic waves move in the same way. Like waves on a rope, electromagnetic waves are transverse waves and can travel through a medium. Unlike other types of waves, electromagnetic waves also can travel in a vacuum. Recall that the speed of a given type of wave depends upon the medium. The speed at which electromagnetic waves travel in a vacuum is 299,792,458 m/s, approximated as 3.00×10^8 m/s and denoted as c, the speed of light. The waves travel only slightly slower in air. Recall that the wavelength, frequency, and speed of a wave are related by the following equation:

$$\lambda = \frac{v}{f}$$

In the equation above, the wavelength (λ) is measured in meters, the speed (v) is measured in meters per second, and the frequency (f) is measured in hertz. For an electromagnetic wave traveling in a vacuum, the speed (v) is equal to the speed of light (c). Thus, for an electromagnetic wave, the equation becomes the following:

$$\lambda = \frac{c}{f}$$

Note that in the wavelength-frequency equation, the product of frequency and wavelength is constant—equal to c—for any electromagnetic wave. Thus, as wavelength increases, frequency decreases, and vice versa.

Get It?

Calculate If the green light of the aurora borealis shown at the beginning of the module has a wavelength of 525 nm, what is the wavelength of an electromagnetic wave with a frequency that is 1.5 times that of the green light?

PRACTICE Problems

ADDITIONAL PRACTICE

38. What is the wavelength of green light that has a frequency of 5.70×10^{14} Hz?

39. An electromagnetic wave has a frequency of 8.2×10^{14} Hz. What is the wavelength of the wave?

40. What is the frequency of an electromagnetic wave that has a wavelength of 2.2×10^{-2} m?

41. CHALLENGE If an electromagnetic wave is propagating to the right and the electric field is in and out of the page, in what direction is the magnetic field?

Types of electromagnetic waves The range of frequencies that make up the continuum of electromagnetic waves, the **electromagnetic spectrum,** is shown in **Figure 23**. Note that light—the only part of the spectrum your eyes can detect—makes up only a tiny fraction of the electromagnetic spectrum. Electromagnetic waves carry energy in their changing electric and magnetic fields. Energy that is carried, or radiated, by an electromagnetic wave is called **electromagnetic radiation.** Electromagnetic radiation can be thought of as a phenomenon in which energy stored in fields moves across space. The energy carried by an electromagnetic wave is proportional to the square of the amplitude of the electric field and the area the wave crosses. Multiple technologies exist that are based on the understanding of electromagnetic waves and their interactions with matter. As shown in **Figure 24** on the next page, they are part of everyday experiences in the modern world and in scientific research.

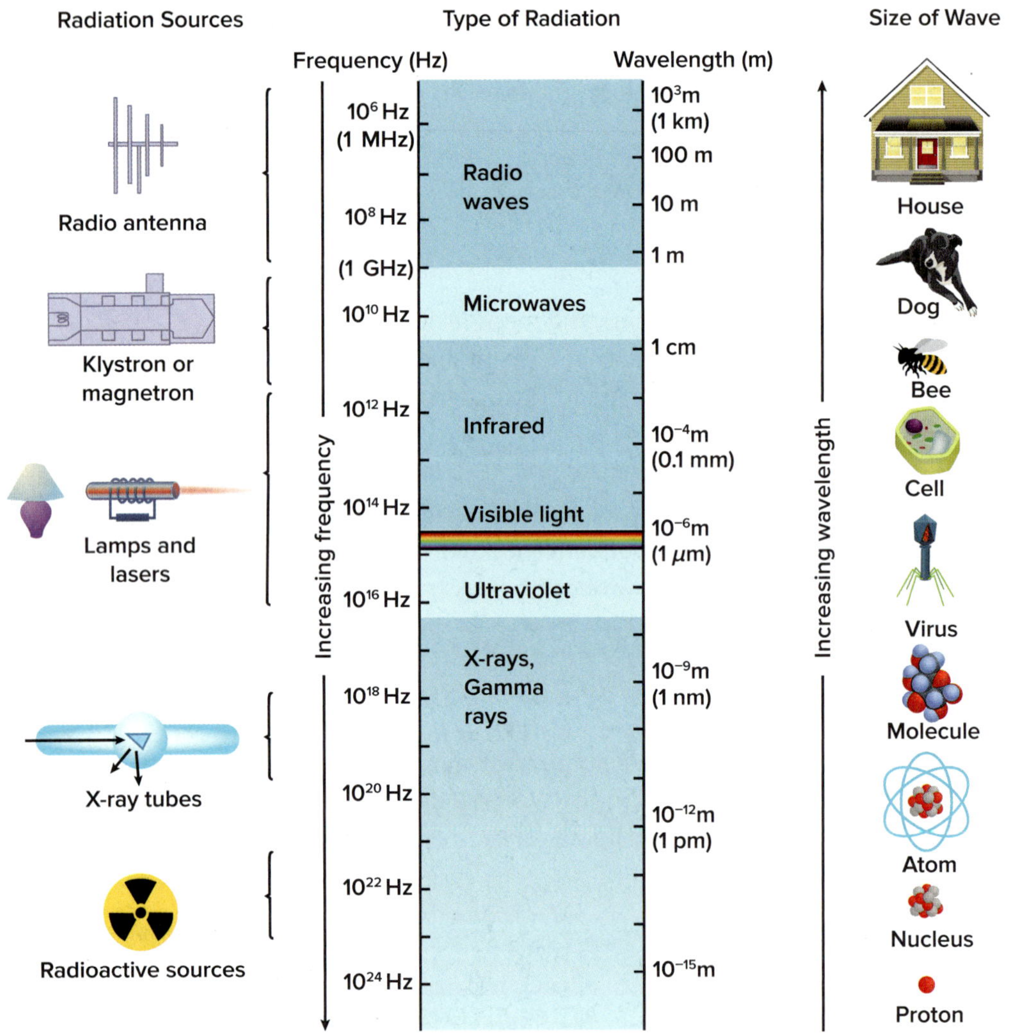

Figure 23 The electromagnetic spectrum contains waves that range from long radio wavelengths larger than houses to short gamma wavelengths smaller than atoms. The left column shows examples of radiation sources. Note that wave frequency extends beyond the frequencies shown in this image.

Observe *Which kind of electromagnetic wave is the size of a proton?*

Figure 24 Electromagnetic radiaion has many applications but can also be harmful. When light or longer wavelength (lower-frequency) electromagnetic radiation is absorbed in matter, it is generally converted into thermal energy (heat). Shorter wavelength (higher-frequency) electromagnetic radiation can ionize atoms and damage living cells.

Uses of lower-frequency waves

Radio waves The lowest-frequency waves, radio waves, are used mainly to broadcast information. Radio waves with long wavelengths can be transmitted long distances because they reflect off ions in the atmosphere. Shorter radio waves used for FM radio and television travel in straight paths and so must be relayed from station to station across Earth's curved surface.

Microwaves Cellular phones and the Global Positioning System transmit information using very-short-wavelength radio waves called microwaves. You also use microwaves to cook food. The water and fat in food absorb microwaves, and the waves' energy turns to thermal energy to cook the food.

Infrared waves Cameras with infrared sensors can produce images, and infrared night-vision goggles and cameras allow people to see in the dark. Because hot objects emit far infrared waves (with long wavelengths), infrared detectors can measure the temperature of buildings and other objects. Near infrared radiation (with shorter wavelengths and higher frequencies) can carry signals on optical fiber systems or through the air, programmed from remote-control devices.

UV radiation With higher frequencies than infrared waves, ultraviolet (UV) radiation can ionize molecules and atoms and cause chemical reactions, resulting in harmful effects such as sunburn. UV radiation also is used in industry to cure polymers and sterilize instruments. In the semiconductor industry, UV radiation is used to etch patterns on silicon wafers in integrated circuits. UV radiation can also be used to reveal forensic evidence.

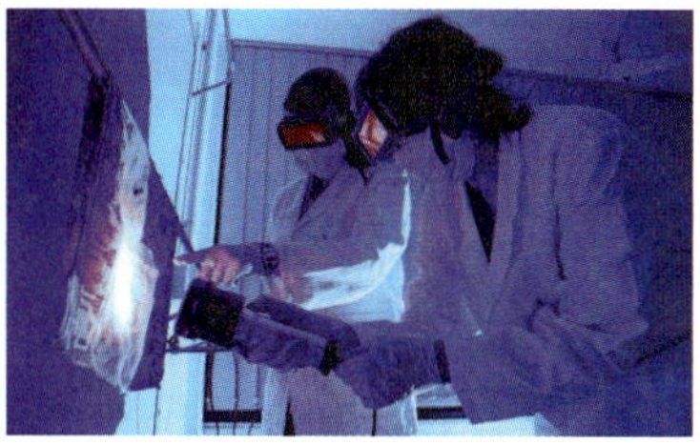

Uses of higher-frequency waves

X-rays X-rays are produced when high-energy electrons rip tightly bound electrons off atoms. When the electrons in the atoms rearrange themselves, they emit X-rays. German physicist William Roentgen discovered X-rays in 1895 using a vacuum glass tube. You are familiar with X-ray pictures of bones and teeth. X-rays are also widely used to kill cancerous cells.

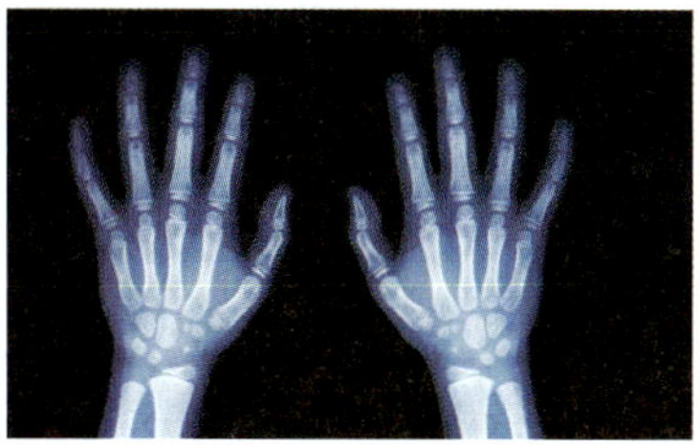

Gamma rays High-frequency gamma rays come from the radioactive nuclei of atoms. Gamma rays can be used to detect dangerous substances in shipping containers. In medicine, they are used to treat cancer by destroying cells.

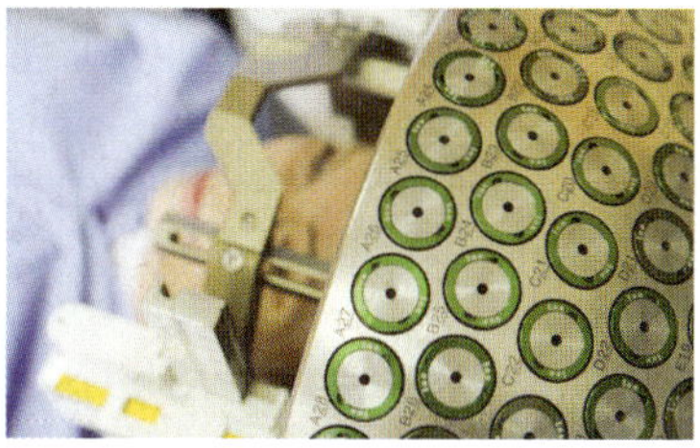

(t to b)FabrikaSimf/Shutterstock; grzymkiewicz/iStockphoto/Getty Images; Ivan Smuk/Alamy Stock Photo; Jochen Tack/Alamy Stock Photo; Puwadol Jaturawutthichai/Shutterstock; MBI/Alamy Stock Photo

Propagating Electromagnetic Waves

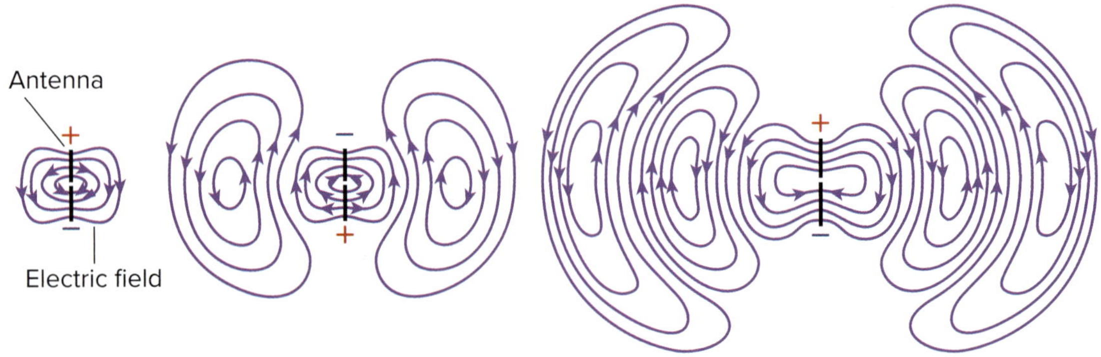

Figure 25 An AC current source connected to an antenna produces an oscillating potential difference across the antenna. This oscillating potential difference accelerates electrons, which generate an oscillating electric field. The oscillating electric field produces a changing magnetic field (not shown), and this magnetic field, in turn, generates a changing electric field. As the process continues, the electromagnetic wave propagates away from the antenna.

Transmitting Electromagnetic Waves

You read that radio waves and microwaves can carry information. How are these waves—and the information they carry—broadcast?

Propagation through space Radio waves and microwaves are broadcast into space by transmitters connected to antennas. A **transmitter** is a device that converts voice, music, pictures, or data to electronic signals, amplifies the signals, and then sends the signals to an antenna. The **antenna** creates the electromagnetic waves that propagate through the air. How does the antenna do this?

The transmitter produces an oscillating potential difference across the metallic antenna that accelerates electrons in the metal. The acceleration generates an oscillating electric field that propagates away from the antenna. You can see how an electric field forms from an antenna in **Figure 25.**

Not shown in **Figure 25** is the changing magnetic field generated by the changing electric field. The magnetic field also propagates away from the antenna at the same speed as the electric field but at right angles to it and to the direction of propagation. Note that an electromagnetic wave produced by an antenna is polarized; that is, its electric field is parallel to the antenna's conductor.

Propagation through matter Electromagnetic waves can travel through matter other than air. Sunlight shining through a glass of water is an example where light waves travel through three forms of matter: air, glass, and water. These materials are dielectrics. A **dielectric** is a poor conductor of electric current whose electric charges partially align with an electric field. The velocity of an electromagnetic wave in a dielectric is always less than the wave's speed in a vacuum; you can calculate the velocity of a wave through any dielectric using the following equation:

$$v = \frac{c}{\sqrt{k}}$$

In this equation, the wave velocity (v) is measured in meters per second; the speed of light (c) is 3.00×10^{8} m/s; and the relative dielectric constant (k) is a dimensionless quantity. In a vacuum, $k = 1.00000$, and the wave velocity is equal to c. In air, $k = 1.00054$, and electromagnetic waves move just slightly slower than c. The dielectric coefficient is the square of the index of refraction, $k = n^2$, so that $\sqrt{k} = n$.

PRACTICE Problems

ADDITIONAL PRACTICE

42. What is the speed of an electromagnetic wave traveling through air? Use c = 299,792,458 m/s in your calculation.

43. Water has a dielectric constant of 1.77. What is the speed of light in water?

44. The speed of light traveling through a material is 2.43×10^{8} m/s. What is the dielectric constant of the material?

45. CHALLENGE A radio signal is transmitted from Earth's surface to the Moon's surface, 376,290 km away. What is the shortest time a reply can be expected?

Producing Electromagnetic Waves

Suppose you've just tuned to your favorite radio station. How does that station generate the radio waves you hear as music?

Carrier waves The Federal Communications Commission assigns every commercial radio station in the United States a specific wavelength in the radio part of the electromagnetic spectrum called the station's **carrier wave.** A station broadcasts music or other information by varying its carrier wave, modulating either its frequency or its amplitude. This is done by the station's transmitter. A transmitter contains three parts. The oscillator generates the carrier wave. The modulator uses music, pictures, words or other data to change the carrier wave's amplitude or frequency. And the amplifier increases the potential difference of the resulting signal.

Setting the oscillation frequency To produce carrier waves up to 400 MHz, an oscillator uses a coil and a capacitor connected in parallel. The oscillator circuit creates a potential difference across the capacitor that produces an electric field and stores charges in the capacitor. When the potential difference is removed, the capacitor discharges and the stored electrons flow through the coil. The current creates a changing magnetic field that induces an *EMF* across the circuit. The *EMF* recharges the capacitor in the reverse direction; the process then repeats in reverse. One complete oscillation cycle is shown in **Figure 26.**

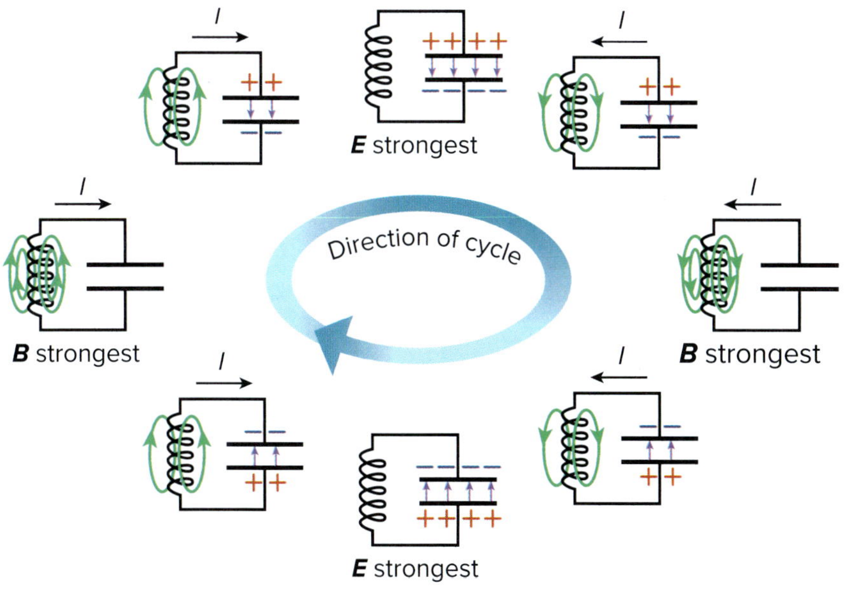

Figure 26 In a complete oscillation cycle of a coil-and-capacitor circuit, the magnetic field is strongest when the electric field is weakest. The number of oscillations per second equals the frequency of the waves produced. If the coil's ability to store magnetic energy or the capacitance changed, so would the frequency of the oscillations

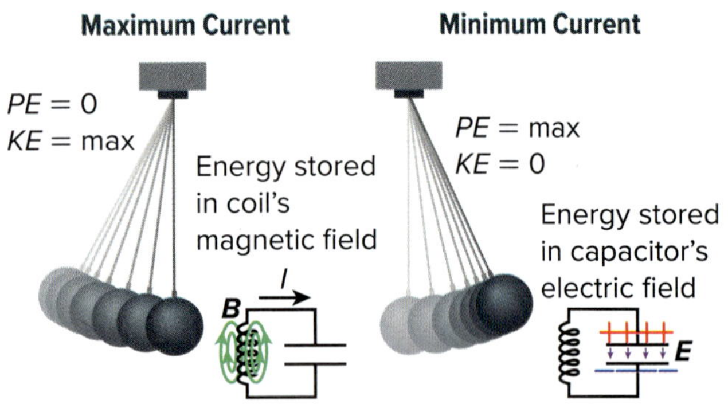

Figure 27 The electrons in a coil-and capacitor circuit are like a pendulum's bob. As the bob swings, its displacement changes. The point where the pendulum's motion comes to a stop is analogous to zero current in the circuit.

Determine *What is the charge on the capacitor when the current is maximum?*

Swinging pendulum analogy You can compare the discharging and recharging process in a coil-and-capacitor circuit to the cyclic oscillations of a swinging pendulum, as shown in **Figure 27.** The pendulum in **Figure 27** has kinetic energy (*KE*) as a result of its motion. It has potential energy (*PE*) due to its displacement. The sum of the *PE* and *KE*—the total energy—is constant throughout the pendulum's motion. The coil-and-capacitor circuit is similar. There is energy both in the circuit's magnetic field, produced by the coil, and in the circuit's electric field, produced by the capacitor. When the current is largest, the energy is all stored in the magnetic field. When the current is zero, the energy is all stored in the electric field. The total energy of the circuit (the sum of the magnetic and electric field energies) is constant.

Steady oscillations Just as friction causes a pendulum to stop swinging if it is left alone, the oscillations in a coil and capacitor die out over time due to the energy being propagated in the electromagnetic wave and due to resistance in the circuit. By adding energy to either system, the oscillations can be made to continue. Gentle pushes applied to a pendulum at appropriate times will keep the pendulum swinging. The largest amplitude swings occur when the frequency of pushes matches the frequency of the swinging motion and is in phase. This is the condition of resonance, which you read about in an earlier chapter. Just as gentle pushes keep a pendulum swinging, potential differences applied to a coil-and-capacitor circuit at the right frequency can keep the oscillations steady. One way of doing this is to add a second coil to the circuit, forming a transformer.

How does a transformer keep the circuit's oscillations going? As shown in **Figure 28,** the transmitter's oscillator increases the AC potential difference induced in the secondary coil of the transformer. The current is then added back to the coil and capacitor. This enables the circuit to maintain its oscillations.

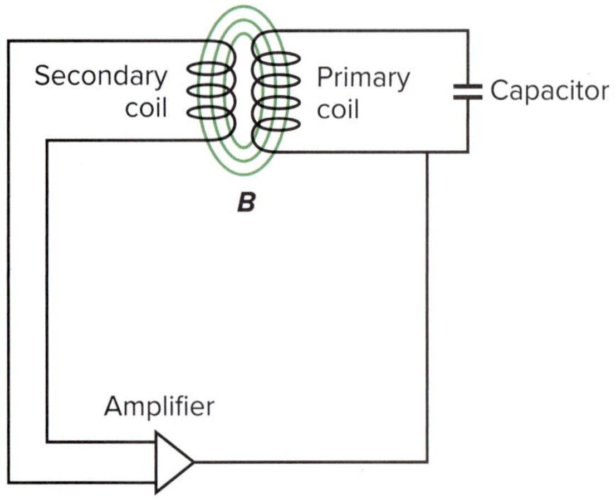

Figure 28 The amplified oscillating current from the secondary coil of the transformer is at the same frequency as the coil-and-capacitor circuit. Therefore, the amplified current keeps the oscillations going.

CCC CROSSCUTTING CONCEPTS

Cause and Effect How could you gather evidence to support the cause-and-effect relationship between the current and the electric and magnetic field energies in a coil-and-capacitor circuit? Describe what you would need to measure.

CCC CROSSCUTTING CONCEPTS

Stability and Change Why do the oscillations in a coil and capacitor die out over time? Cite evidence to explain how adding a transformer maintains stability in the system.

Frequencies set by a resonant cavity The frequency produced by a coil-and-capacitor circuit can be increased by decreasing the coil's ability to store magnetic energy and the capacitor's capacitance. Above frequencies of about 400 MHz, however, individual coils and capacitors are no longer useful.

Microwaves, with frequencies from 0.4 GHz to 100 GHz, are produced using a resonant cavity, a rectangular metal box that acts both as a coil and as a capacitor. The size of the box determines the frequency of oscillation. Note that in a microwave oven, the size of the oven itself has no effect on wave frequency; only the size of the oven's internal resonant cavity affects frequency.

To produce waves with frequencies higher than 100 GHz, a resonant cavity would have to be the size of a molecule. Infrared waves, for example, are generated by vibrating nuclei within molecules. Higher-frequency light, ultraviolet waves, and X-rays are generated by electrons within atoms. The highest-frequency waves, gamma rays, are generated by accelerated charges in atomic nuclei.

Get It?

Assess Why can't resonant cavities be used to produce infrared waves?

Frequencies set by piezoelectricity There are other ways of generating oscillating potential differences for transmitters. For example, quartz crystals deform when a potential difference is applied across them, a property known as **piezoelectricity.** The application of an AC potential difference to a cut section of quartz crystal results in sustained oscillations. Just as a piece of metal vibrates at a specific frequency when it is bent and released, so does a quartz crystal. The thinner the crystal, the higher is the vibration frequency.

A crystal's piezoelectric property also generates an EMF when the crystal is deformed. This EMF is produced at the vibrating frequency of the crystal, so it can be amplified and returned to the crystal to keep it vibrating. Because of their nearly constant vibration frequencies, quartz crystals are commonly used to generate electromagnetic waves in cell phones, television, wireless phones, wireless WiFi routers, and computers.

Receiving Electromagnetic Waves

Antennas propagate electromagnetic waves into space. Antennas also capture electromagnetic waves, converting the waves' oscillating electric fields back to potential differences. As shown in **Figure 29,** a wave's electric field accelerates electrons in the metal of an antenna. The acceleration is largest when the antenna is positioned in the same direction as the wave polarization; that is, when it is parallel to the direction of the wave's electric field. A potential difference across the antenna's terminals oscillates at the frequency of the wave.

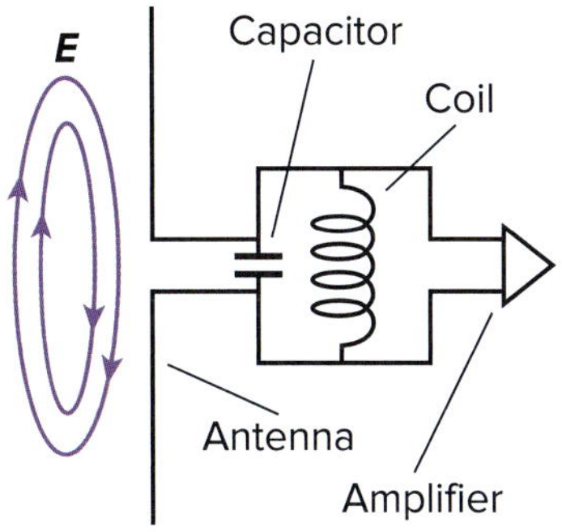

Figure 29 The changing electric fields from a radio station signal cause electrons in an antenna to accelerate. The information carried by the signal can then be decoded and amplified and used to drive a loudspeaker.

Get It?

Infer why the direction in which an antenna points affects its function.

Wire antennas When an antenna is one-half the length of the wave it is designed to detect, the potential difference across its terminals is largest and the antenna is most efficient. Therefore, an antenna designed to receive radio waves is longer than one designed to receive microwaves.

Though antennas that are one-half wavelength long are most efficient, antennas that are one-quarter wavelength long are often used when the connection between the antenna and receiver is at the end rather than the middle of the antenna. Antennas can be made shorter by constructing them from a helical coil or by adding a dielectric material, such as a ceramic, with a dielectric coefficient higher than air.

Cell phones have as many as seven antennas. These phones typically communicate at frequencies near 850, 1700, 1900, and 2100 MHz. Further, they receive GPS signals at 1.575 GHz. They send and receive remote earpiece and WiFi signals at 2.4 GHz. Cell phone antennas built on ceramic dielectric blocks using printed conductors are typically only a few millimeters long. Laptop computers also have several antennas to accommodate WiFi and remote devices.

Dish antennas All electromagnetic waves, not just light waves, undergo reflection, refraction, and diffraction. Dish antennas, such as the one in **Figure 30,** reflect short-wavelength radio signals, just as parabolic mirrors reflect light waves. A parabolic dish antenna reflects and focuses signals off its surface and into a horn. The horn, supported by a tripod structure over the main dish, contains a short dipole antenna. Like a telescope that shows only a narrow portion of sky, a dish antenna is sensitive only to signals coming from specific directions.

Converting waves to information After an antenna converts the electric fields of electromagnetic waves to potential differences, it sends the potential differences to a **receiver,** which converts them to usable information—sound, pictures, or data. Waves of different frequencies strike an antenna and enter the receiver simultaneously. To select waves of a particular frequency only, a receiver uses a tuner. A tuner has a coil-and-capacitor circuit or a resonant cavity. When you turn a radio dial, you select a radio station by adjusting the capacitance until the oscillation frequency of the circuit equals the frequency of the desired wave. Then, only waves of the desired frequency can produce oscillating potential differences of significant amplitude in the receiver.

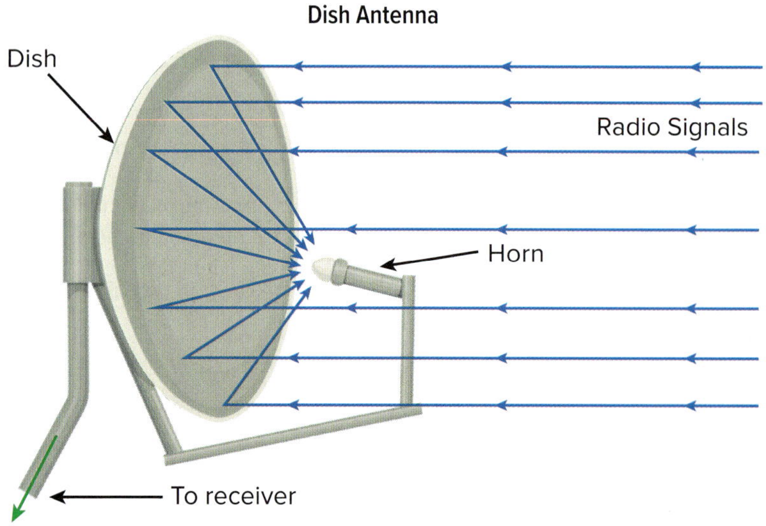

Figure 30 A received signal is reflected off the surface of a dish and focused into the horn, which contains the antenna. The large surface area of the dish collects more electromagnetic wave energy than a wire antenna does, making it well suited for receiving weak radio signals.

FM and AM signals Recall that an electromagnetic wave can carry information only if some property of the wave is changed, or modulated, by the information. A radio station's carrier wave is varied either in amplitude (AM, or Amplitude Modulation) or in frequency (FM, or Frequency Modulation). Commercial AM stations broadcast in the 530–1700 kHz band, while commercial FM stations broadcast between 88 and 108 MHz. FM signals have less noise because most noise sources, such as lightning, create waves of varying amplitude to which FM receivers are insensitive. Many radio services, including commercial, emergency, and two-way walkie talkies, use other frequencies and either FM or AM modulation.

Digital signals Information can be digitized. For example, a picture can be stored as the values of an array of pixels. In this form, it can be stored reliably in computer memory and sent over long distances. Cell phones, televisions, and computers convert digital signals to sound, pictures, and data. Digital signals can be sent and received as a series of wave pulses that are usually encoded and interpreted using a binary system of 1s and 0s that vary in duration, not amplitude or frequency. Digital signals can send more information in the same amount of time as AM or FM and are less affected by noise.

Real-World Physics

DIGITAL TV On June 12, 2009, all full-power U.S. television stations ceased broadcasting analog signals. Analog TV signals, like FM and AM radio signals, are broadcast by modulating carrier waves. Television stations now broadcast only digital signals, which are encoded into 0s and 1s—the digital code used in computers. Digital TV bandwidth can be compressed to provide four, five, or more channels in the same carrier frequency used by a single channel in analog TV. Digital TV also has higher-fidelity sound and carries about five times more picture information.

Check Your Progress

46. **Electromagnetic Waves** Explain how electromagnetic waves propagate through space.
47. **Interaction with Matter** How do different wavelengths of electromagnetic radiation interact with matter? How does this affect how we use and interact with the radiation?
48. **Frequency** An electromagnetic wave has a wavelength of 1.5×10^{-5} m. What is its frequency?
49. **Radio Signals** Radio antennas normally have metal rod elements that are oriented horizontally. From this information, what can you deduce about the directions of the electric fields in radio signals?
50. **Digital Signals** What are the advantages of storing and transmitting sound, pictures, and data as digital signals?
51. **Antenna Design** Would an FM antenna designed to be most sensitive to stations near 88 MHz be shorter or longer than one designed to receive stations near 108 MHz? Explain your reasoning.
52. **Dielectric Constant** The speed of light traveling through an unknown material is 1.98×10^{8} m/s. Given that the speed of light in a vacuum is 3.00×10^{8} m/s, what is the dielectric constant of the unknown material?
53. **Critical Thinking** Most of the UV radiation from the Sun is blocked by the ozone layer in Earth's atmosphere. Scientists have found that the ozone layer over Antarctica and the southern hemisphere has thinned. Use what you have learned about electromagnetic waves to explain why some scientists are very concerned about the thinning ozone layer.

LEARNSMART® Go online to follow your personalized learning path to review, practice, and reinforce your understanding.

Answering the Call

After a storm or natural disaster, cell phone service is often unavailable because cell phone towers have been damaged or destroyed. In the past, residents might be without cell service—including Internet connectivity—for weeks or months. However, the impact to storm-ravaged communities has recently been reduced by new and innovative ways of providing communication services.

Operators move balloons into air currents that carry the balloons to the desired location.

Sending Up a Trial Balloon

In 2013, a technology company launched a project to deliver Internet service to over half of the world's population living in remote areas via solar-powered balloons. When fully operational, the project's balloons will carry transceivers that receive wireless high-speed Internet signals from telecommunications stations on the ground. The balloons will float in the stratosphere high above Earth, transmitting the signals to each other and then down to the phones and other devices of users on the ground.

After the massive spring 2017 floods and mudslides in Peru, the technology company was able to use its balloons to provide cell phone service to the beleaguered country. Because the company had already been working with local wireless carriers to test the project, service was quickly restored.

Hurricane Maria Strikes

When Hurricane Maria, a Category 4 storm, hit U.S. territory Puerto Rico in September 2017, more than 90 percent of the island lost cell phone service. The technology company quickly began working with local wireless carriers to integrate the balloons with their networks. The balloons took the place of destroyed cell phone towers, bringing cell service back to island residents, who were soon able to access voice and data services.

Another company also came to Puerto Rico's aid. They sent drones to deliver cell phone service to San Juan, Puerto Rico's capital. The drones are connected to the ground by a tether that provides power and a secure data connection. The drones also have antennas that pick up satellite signals. From their position in the air, the drones supply Internet connectivity to a small area on the ground.

CONSTRUCT AN ARGUMENT FROM EVIDENCE

Use print or online sources to research restoring cell phone service via balloon or via drone. Using evidence from your sources, write a short paper that argues why the balloon or the drone method is superior.

CALIFORNIA CONNECTION

Research one of California's many observatories. Develop a presentation that both summarizes how the observatory's telescope works and discusses what types of objects it makes visible.

Andrej Sokolow/dpa picture alliance/Alamy Stock Photo

More Than a Light Show

On March 13, 1989, two strange events occurred. First, the Northern Lights were visible as far south as the southern United States. Second, the whole of Quebec, a province in Canada, experienced a power outage. Surprisingly, these two events were related.

1. Particles from the Sun The Sun is constantly emitting a stream of charged particles known as the solar wind, which can interact with magnetic fields. Occasionally, the Sun will emit a large amount of plasma, called a coronal mass ejection (CME).

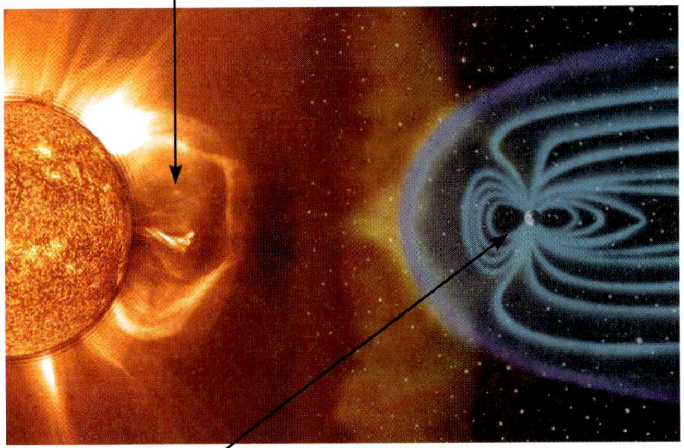

2. The magnetosphere is a region around Earth where Earth's magnetic field deflects most of these charged particles. Some energy, however, enters the magnetosphere and causes electrons there to accelerate towards Earth. When a CME reaches the magnetosphere, its magnetic field interacts with Earth's. This disturbance is called a *geomagnetic storm.*

3. The aurora occurs when those electrons enter Earth's atmosphere and excite oxygen and nitrogen molecules. These molecules release this energy as visible light. During geomagnetic storms, the aurora is visible much father from the poles than normal.

4. Geomagnetically-induced currents During a geomagnetic storm, the changing magnetic field induces a current in any nearby conductors. Long, metal structures—such as power lines, pipes, and train tracks—are particularly susceptible. The induced current in power lines can be large enough to trip safety devices such as surge protectors and circuit breakers, causing power outages.

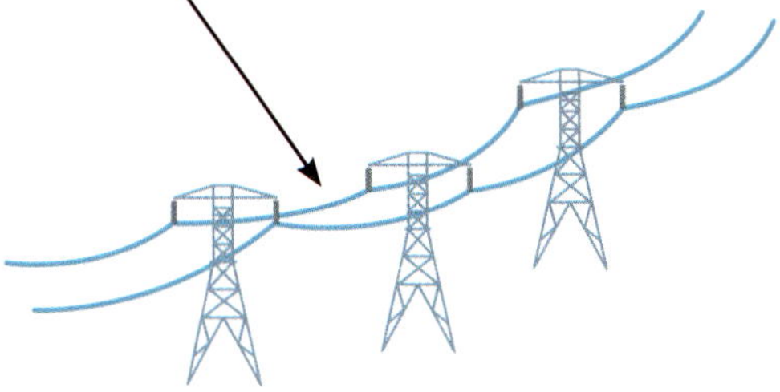

OBTAIN AND COMMUNICATE INFORMATION

Determine what research is currently being done on the aurora, geomagnetic storms, and geomagnetically-induced currents. Summarize your findings in a short news item for a science magazine.

CALIFORNIA CONNECTION

In your news item, include any possible effects of geomagnetic storms on California.

MODULE 21
STUDY GUIDE

GO ONLINE study with your Science Notebook.

Lesson 1 INDUCING CURRENTS

- electromagnetic induction
- induced electromotive force
- electric generator

- An *EMF* is induced in a wire when the magnetic field at the location of the wire changes, as when the wire moves through the magnetic field.
- The *EMF* induced in a length of wire moving through a uniform magnetic field is the product of the magnetic field (B), the length of the wire (L), and the component of the wire's velocity that is perpendicular to the field ($v(\sin\theta)$).

$$EMF = BLv(\sin\theta)$$

- As the armature in a generator is turned by mechanical force, the induced *EMF* creates a current that changes direction each time the armature turns 180°.
- In an AC circuit, the effective current and potential difference are related to their maximum values in the following way:

$$I_{eff} = 0.707 I_{max}$$
$$V_{eff} = 0.707 V_{max}$$

Lesson 2 APPLICATIONS OF INDUCED CURRENTS

- Lenz's law
- eddy current
- self-inductance
- transformer
- mutual inductance
- step-up transformer
- step-down transformer

- Lenz's law states that a current created by an induced *EMF* is in a direction that produces a magnetic field that opposes the change in the magnetic field that generated the current.
- The turns ratio in a transformer affects potential difference and current as expressed by the following equation:

$$\frac{I_s}{I_p} = \frac{V_p}{V_s} = \frac{N_p}{N_s}$$

Lesson 3 ELECTRIC AND MAGNETIC FORCES ON PARTICLES

- isotope
- mass spectrometer

- The charge-to-mass ratio of an electron or positive ion is found by first using crossed electric and magnetic fields to determine the particle's velocity, then using a magnetic field to deflect the particle.
- A mass spectrometer uses both electric and magnetic fields to separate and measure the masses of ionized atoms and molecules.

Lesson 4 ELECTRIC AND MAGNETIC FIELDS IN SPACE

- electromagnetic wave
- electromagnetic spectrum
- electromagnetic radiation
- transmitter
- antenna
- dielectric
- carrier wave
- piezoelectricity
- receiver

- The electric and magnetic fields that make up an electromagnetic wave oscillate at right angles to each other and to the direction of the wave's velocity (v).
- Wavelength is equal to wave speed divided by wave frequency. For an electromagnetic wave traveling in a vacuum, v is equal to the speed of light (c). For an electromagnetic wave traveling through a dielectric, v is equal to c divided by the square root of the relative dielectric constant (k).
- Electromagnetic waves can carry information if their amplitude or frequency is varied by the data, voice, or video information that is to be transmitted. Waves also can be encoded with digitized information.

THREE-DIMENSIONAL THINKING

Module Wrap-Up

REVISIT THE PHENOMENON

What causes the Northern Lights?

CER Claim, Evidence, Reasoning

Explain Your Reasoning Revisit the claim you made when you encountered the phenomenon. Summarize the evidence you gathered from your investigations and research and finalize your Summary Table. Does your evidence support your claim? If not, revise your claim. Explain why your evidence supports your claim.

STEM UNIT PROJECT

Now that you've completed the module, revisit your STEM unit project. You will apply your evidence from this module and complete your project.

GO FURTHER

SEP Data Analysis Lab

Can you evaluate scenarios in science fiction novels?

You and your friends are reading a collection of science fiction books. After you read a book, you meet to discuss whether the science presented in the book is possible.

a. In one novel, alien beings have eyes that are sensitive to microwaves.

b. In H. G. Wells's novel *The Invisible Man*, a man drinks a potion and becomes invisible, although he retains all his other faculties including his ability to see.

c. A scientist in a third novel discovers an "anti-Lenz's law" that states an induced current results in a force that increases the change in a magnetic field. Thus, when more energy was demanded, the force needed to turn a generator was reduced.

CER Analyze and Interpret Data

1. **Claim** Answer the following questions.
   a. In the first novel, how should the aliens' eyes compare in size to human eyes?
   b. Is Wells correct that the invisible man is able to see?
   c. Is the "anti-Lenz's law" possible?
2. **Evidence** What scientific information or principles support each of your claims?
3. **Reasoning** Explain your reasoning for each of the claims that you have made.

Ken Phung/Shutterstock

NASA/JPL-Caltech/STScI/CXC/UofA/ESA/AURA/JHU

UNIT 6

SUBATOMIC PHYSICS

ENCOUNTER THE PHENOMENON

What is the universe made of?

SEP Ask Questions

What questions do you have about the phenomenon? Write your questions on sticky notes and add them to the driving question board for this unit.

What nuclear reactions release energy?

Look for Evidence

As you go through this unit, use the information and your experiences to help you answer the phenomenon question as well as your own questions. For each activity, record your observations in a Summary Table, add an explanation, and identify how it connects to the unit and module phenomenon questions.

Solve a Problem

STEM UNIT PROJECT

The History of Everything People were not around to observe the origin of the universe or the formation of the Earth, but scientists have can use evidence to understand what happened. Investigate the evidence scientists have found for the origin of the universe and the formation of Earth. Use the results of these investigations and your knowledge from this unit to explain the evidence for the current understanding of the history of the universe and Earth.

GO ONLINE In addition to reading the information in your Student Edition, you can find the STEM Unit Project and other useful resources online.

Franz Himpsel/Univ. of Wisconsin/Science Source

MODULE 22

QUANTUM THEORY AND THE ATOM

ENCOUNTER THE PHENOMENON

How could the manipulation of individual atoms be useful in future computing?

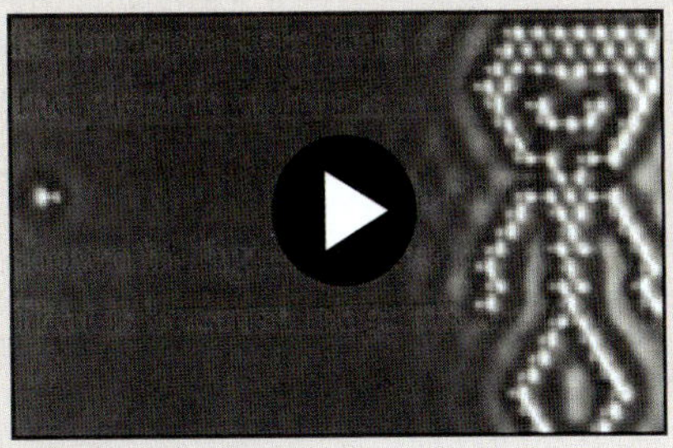

GO ONLINE to watch a video about atomic manipulation and its role in computing.

SEP Ask Questions

Do you have other questions about the phenomenon? If so, add them to the driving question board.

CER Claim, Evidence, Reasoning

Make Your Claim Use your CER chart to make a claim about how manipulation of individual atoms could be useful in future computing. Explain your reasoning.

Collect Evidence Use the lessons in this module to collect evidence to support your claim. Record your evidence as you move through the module.

Explain Your Reasoning You will revisit your claim and explain your reasoning at the end of the module.

GO ONLINE to access your CER chart and explore resources that can help you collect evidence.

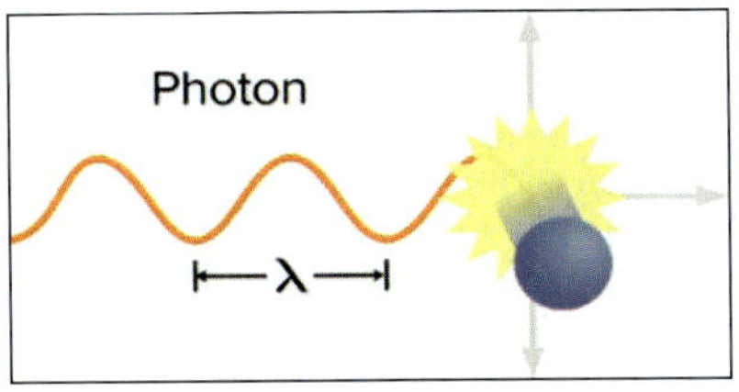

LESSON 2: Explore & Explain: Heisenberg Uncertainty Principle

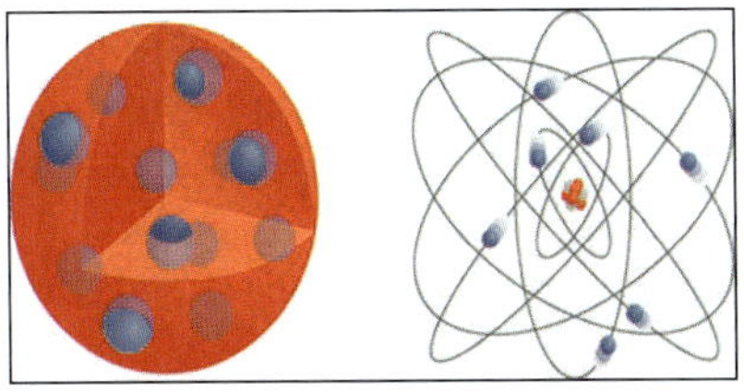

LESSON 3: Explore & Explain: The Nuclear Model

Smithsonian

Additional Resources

Video Supplied by BBC Worldwide Learning

LESSON 1

A PARTICLE MODEL OF WAVES

FOCUS QUESTION

Can light behave like a particle?

A New Model Based on Packets of Energy

James Maxwell's electromagnetic wave theory, that electromagnetic radiation can be modeled as a wave of changing electric and magnetic fields, is useful for explaining many features of electromagnetic radiation, including interference, diffraction, and polarization. But it could not explain others, including the spectrum of electromagnetic waves emitted by all objects and the emission of electrons from metals exposed to ultraviolet radiation.

Electromagnetic radiation from objects The lightbulb shown in **Figure 1** is connected to a dimmer control that varies the potential difference to the bulb. As the potential difference increases, the temperature of the bulb's glowing filament increases. As a result, the color changes from deep red to orange to yellow and finally to white as the hotter filament emits higher-frequency colors blue and violet.

Figure 1 The color of the glow of the lightbulb depends on the temperature of the filament.

Richard Hutchings/Digital Light Source

3D THINKING DCI Disciplinary Core Ideas CCC Crosscutting Concepts SEP Science & Engineering Practices

COLLECT EVIDENCE

Use your Science Journal to record the evidence you collect as you complete the readings and activities in this lesson.

INVESTIGATE

GO ONLINE to find these activities and more resources.

Applying Practices: Is light a wave or a particle?

HS-PS4-3. Evaluate the claims evidence, and reasoning behind the idea that electromagnetic radiation can be described either by a wave model or a particle model, and that for some situations one model is more useful than the other.

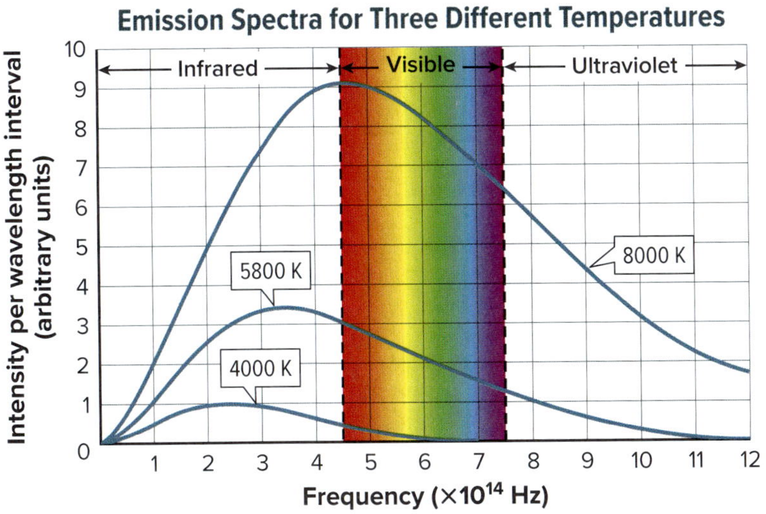

Figure 2 The frequency at which an object's emission spectrum peaks depends on that object's temperature. As the temperature of the object increases, the peak frequency also increases.

As predicted by electromagnetic wave theory, the vibrating charged particles within each bulb filament emit light and infrared radiation. While all objects, from very cold to extremely hot objects, emit a range of electromagnetic waves, the filament glows brightly in the visible light range because it is hot. (This is called incandescence; hence, the name incandescent lightbulb.) The colors you see depend on the relative intensities of the emitted electromagnetic waves of various frequencies and the sensitivity of your eyes to those waves.

Emission spectra What would you expect to see if you viewed the glowing filament through a diffraction grating? When viewed in this way, all the colors of the rainbow would be visible. **Figure 2** shows the visible colors that correspond to the frequencies around 6×10^{14} Hz. The bulb also emits other radiation that you would not see. A graph of the intensity of the radiation emitted from an object over a range of frequencies is known as an **emission spectrum.** Emission spectra of objects at temperatures of 4000 K, 5800 K, and 8000 K are shown in **Figure 2.** Note that at each temperature there is a frequency at which the maximum amount of energy is emitted. If you compare the curves, you will see that as the temperature increases, the frequency at which the maximum amount of energy is emitted also increases.

Get It?

Explain the changes observed in the spectrum of the glowing lightbulb at the beginning of this module as the dimmer control was turned up.

ASTRONOMY Connection The total power emitted by an object also increases with temperature. The power (energy emitted per second) of an electromagnetic wave is proportional to the hot object's kelvin temperature raised to the fourth power (T^4). Thus, hotter objects radiate considerably more energy per second than do cooler objects. Probably the most common example of a hot object radiating a great amount of power is the Sun. The Sun's surface temperature is about 5800 K, and it radiates 4×10^{26} W of power—an enormous quantity. On average, each square meter of Earth's surface receives about 1000 J of energy per second (1000 W) from the Sun. This is enough energy to power ten 100-W lightbulbs.

Get It?

Explain how the power emitted by an object changes if the object's kelvin temperature is doubled.

Explaining emission spectra Between 1887 and 1900, many physicists used electromagnetic wave theory to try to predict emission spectra, but they all failed. **Figure 3** shows the difference between the predicted spectrum and the observed spectrum. In 1900, German physicist Max Planck found that he could calculate the correct spectrum if an atom could only emit or absorb particular amounts of energy. Planck hypothesized an atom's energy changes in a solid were proportional to a frequency of vibration times an integer.

Energy of Vibration

The energy emitted or absorbed by a vibrating atom is equal to the product of an integer, Planck's constant, and the frequency of vibration.

$$E = n\mathrm{h}f$$

In the equation, f is the frequency of vibration of the atom, h is a constant called Planck's constant, which has a value of 6.626×10^{-34} J/Hz, and n is an integer such as 0, 1, 2, 3 . . .

$$n = 0: E = (0)\mathrm{h}f = 0$$

$$n = 1: E = (1)\mathrm{h}f = \mathrm{h}f$$

$$n = 2: E = (2)\mathrm{h}f = 2\mathrm{h}f$$

The energy of the radiation (E) can have values of hf, 2hf, 3hf, and so on, but it can never have fractional values such as $\frac{2}{3}$ hf or $\frac{3}{4}$ hf. In other words, energy is **quantized**—it exists only in bundles of specific amounts. For calculations, the constant h is usually rounded to 6.63×10^{-34} J/Hz.

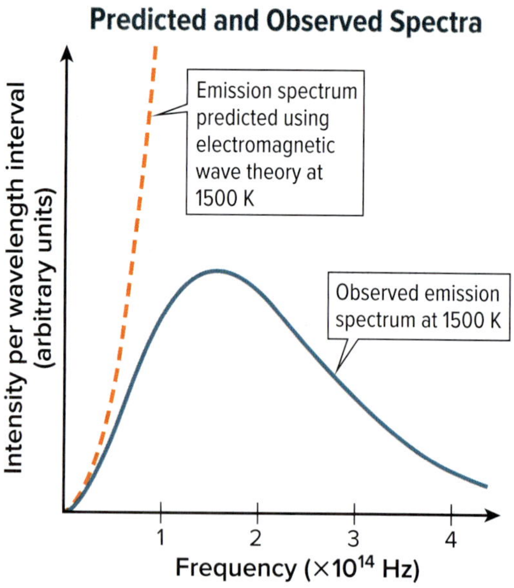

Figure 3 The red, dashed line shows the emission spectrum predicted using the electromagnetic wave theory. The theory predicts any object will radiate an infinite amount of energy. This does not match the observed emission spectrum, shown in blue.

Identify *the region of the graph where electromagnetic wave theory is a good model for the observed emission spectrum.*

Changes in vibrations Electromagnetic wave theory would allow atoms to continuously emit radiation at all times. Instead, Planck proposed that atoms emit radiation only at the particular moments when their vibrational energy changes. For example, if the energy of an atom changes from 3hf to 2hf, according to wave theory, the atom emits radiation only while the energy changes. The energy radiated is equal to the change in energy of the atom, in this case hf. Similarly if the atom absorbed energy hf, its energy could change from 2hf to 3hf.

Planck found that because the constant h has an extremely small value, the energy-changing steps are very small. You would not notice them in the motions of objects in everyday life. Still, the idea of quantized energy troubled physicists, even Planck himself. It hinted that the classical physics of Newton and Maxwell might apply only under certain conditions. Planck's troubling idea was the first step on the way to the development of solar power cells, modern electronics, and computers.

CCC CROSSCUTTING CONCEPTS

Systems and System Models The wave model is useful for explaining many features of electromagnetic radiation, and the particle model explains other features. Create a two-sided poster. On one side identify and explain one feature using the wave model. On the other side identify and explain one feature using the particle model.

The Photoelectric Effect

Physicists of the early 1900s were also challenged by another experimental result that could not be explained by electromagnetic wave theory. When ultraviolet light shines on a negatively charged zinc plate, the plate can produce an electrical discharge—a stream of electrons jumping from the plate. Even a low-intensity ultraviolet light can produce this effect. When visible light shines on the same charged plate, the plate does not discharge. This is true even when the visible light has a high intensity. This result is contrary to electromagnetic wave theory, which predicts the plate would emit electrons when any frequency of electromagnetic radiation shines on the plate. Higher-frequency ultraviolet radiation and lower frequency visible light are forms of electromagnetic radiation, so why would one cause the zinc plate to discharge but not the other? The emission of electrons when electromagnetic radiation falls on an object is called the **photoelectric effect.**

The photocell The photoelectric effect can be studied in a photocell, shown in **Figure 4.** The cell contains two metal electrodes sealed in a tube from which the air has been removed. The evacuated tube prevents the metal surfaces from oxidizing and keeps emitted electrons from being slowed or stopped by gas particles. The larger electrode, called the cathode, usually is coated with cesium or another alkali metal. The smaller electrode, called the anode, is made of a thin wire so that it blocks only a very small amount of radiation. The tube allows ultraviolet light to pass through it. A potential difference placed across the electrodes attracts electrons ejected from the cathode to the anode.

Photovoltaic materials emit electrons when they absorb light of a high-enough frequency. When no light falls on the cathode, which is the negative electrode, there is no current in the circuit. However, when light with certain frequencies falls on the cathode, a current is produced, which is measured by the ammeter, as shown in **Figure 4.** Electrons in the cathode absorb the energy from the light. The electrons escape the potential energy holding them in the cathode and travel to the anode, which is the positive electrode. This flow of photoelectrons forms the current.

Figure 4 Charges flow in a photocell only if the radiation striking the cathode has enough energy.

Compare *the ultraviolet radiation's wavelength to visible light's wavelength. What does the comparison tell you about the relative energies of the two types of radiation?*

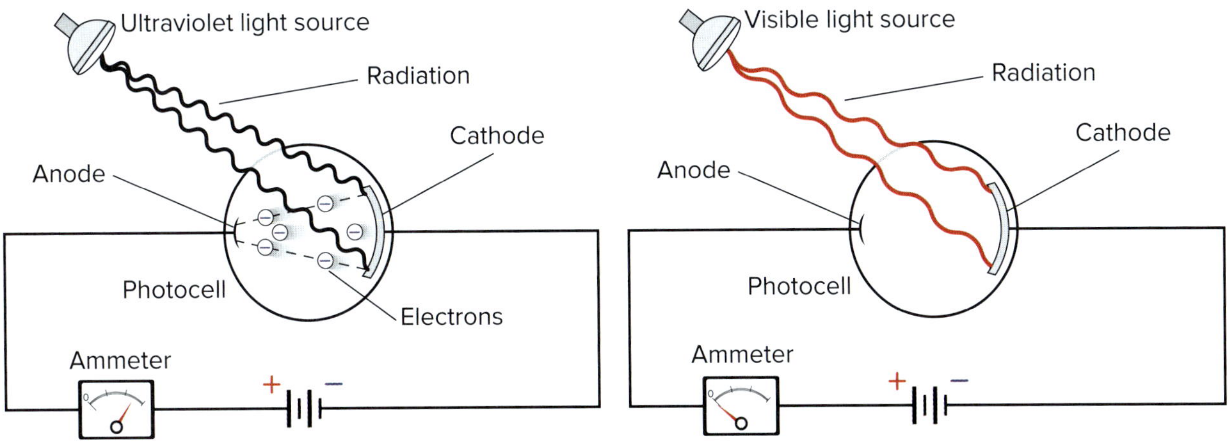

The photoelectric effect is demonstrated by the photocell illuminated by ultraviolet light.

The same photocell, illuminated by visible light, does not display the photoelectric effect.

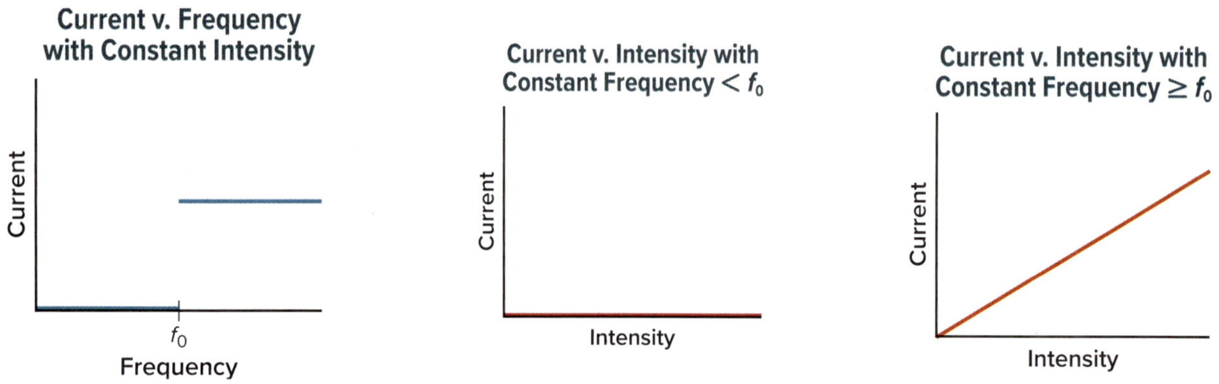

Figure 5 The emission of photoelectrons depends on the incident light's frequency. The threshold frequency must be exceeded in order to produce photoelectrons.

Threshold frequency

Not all radiation falling on a photocell causes a current. Electrons eject from the cathode only if the frequency of the incident radiation is greater than a certain minimum value, called the **threshold frequency** (f_0). The graph on the left in **Figure 5** shows there is a current only when the threshold frequency is reached. The value of the threshold frequency varies widely and depends on the type of metal used for the cathode. For example, all frequencies of visible light except low-frequency red eject electrons from cesium. No frequency of visible light ejects electrons from zinc. Higher-frequency ultraviolet radiation ejects electrons from zinc.

The center graph in **Figure 5** shows that even intense radiation with a frequency below f_0 will not cause the emission of electrons. Conversely, the graph on the right shows that even very low-intensity radiation with a frequency at or above f_0 causes the immediate emission of electrons from the cathode. When the incoming radiation, also called the incident radiation, has a frequency equal to or greater than the threshold frequency, increasing the intensity of the radiation increases the current through the photocell.

The electromagnetic wave theory would say that the electric field in the light wave accelerates and ejects the electrons from the metal. The strength and energy of the electric field are related to the intensity of the light (not to its frequency). Thus, it follows that electrons in the metal could absorb energy from a low-intensity, low-frequency light source for a long time and eventually gain enough energy to be ejected. As you just read, however, this is not the case. Observations show that electrons are ejected immediately when even low-intensity radiation at or above the threshold frequency shines on the metal.

Photons and quantized energy

In 1905 Albert Einstein published a revolutionary theory that explained the photoelectric effect. According to Einstein, visible light and other forms of electromagnetic radiation consist of discrete, quantized bundles of energy, each of which was later called a **photon.** The energy of a photon depends on its frequency.

Energy of a Photon

The energy of a photon is equal to the product of Planck's constant and the frequency of the photon.

$$E = hf$$

Real-World Physics

SOLAR PANELS A solar panel consists of a grid of solar cells made of semiconducting material. Solar cells are human-made devices that capture the sun's energy and produce electrical energy. Photons from the Sun that have a certain threshold frequency can cause electrons to break free from atoms of the semiconducting material, resulting in an electric current within the solar panel.

Electron volts In the equation for the energy of a photon, *f* is frequency, which has units of hertz (Hz), and h is Planck's constant, which has units of joules per hertz (J/Hz). Recall that the unit Hz is defined as $\frac{1}{s}$ or s^{-1}. Therefore, the $\frac{J}{Hz}$ unit of Planck's constant is also equivalent to J·s.

Get It?

Demonstrate how the unit $\frac{J}{Hz}$ is equivalent to J·s.

Because the joule is too large a unit of energy to use with atomic-sized systems, the more convenient energy unit of the electron volt (eV) is often used. One electron volt is the energy of an electron accelerated across a potential difference of 1 V.

$$1\text{ eV} = (1.602\times10^{-19}\text{ C})(1\text{ V})$$

$$= 1.602\times10^{-19}\text{ C·V}$$

$$= 1.602\times10^{-19}\text{ J}$$

Using the definition of an electron volt allows the photon energy equation to be rewritten in a simplified form, as shown below.

Energy of a Photon

The energy of a photon is equal to the constant 1240 eV·nm divided by the wavelength of the photon.

$$E = \frac{hc}{\lambda} = \frac{1240\text{ eV·nm}}{\lambda}$$

PROBLEM-SOLVING STRATEGIES

Units of hc and Photon Energy

Converting the quantity hc to the unit eV·nm results in a simplified equation that you can use to solve problems involving photon wavelength.

1. The energy of a photon of frequency *f* is given by this equation:
$$E = hf$$
2. Recall that $f = \frac{c}{\lambda}$, where c is the speed of light in a vacuum. The equation above can therefore be written
$$E = \frac{hc}{\lambda}$$
3. When using the equation $E = \frac{hc}{\lambda}$, if the value of hc in eV·nm is divided by λ in nm, you will obtain the energy in eV. Thus, it is useful to know the value of hc in eV·nm. Because h and c are constants, the value of hc is also a constant.
4. The conversion of hc to the unit eV·nm is as follows:
$$hc = (6.626\times10^{-34}\text{ J/Hz})\,(2.998\times10^{8}\text{ m/s})\left(\frac{1\text{ eV}}{1.602\times10^{-19}\text{ J}}\right)\left(\frac{10^{9}\text{ nm}}{1\text{ m}}\right)$$
$$= 1240\text{ eV·nm}$$
5. Substituting hc = 1240 eV·nm into the equation for the energy of a photon yields the following, where λ is in nm and *E* is in eV:
$$E = \frac{hc}{\lambda} = \frac{1240\text{ eV·nm}}{\lambda}$$
6. Use the above equation to solve photon energy problems when energy in eV is desired.

PRACTICE Problems

ADDITIONAL PRACTICE

Use E = 1240 eV·nm/λ to solve the following problems.

1. What is a photon's energy if the photon's wavelength is 515 nm?
2. A photon's energy is 2.03 eV. What is the photon's wavelength?
3. Rank the following photons from least to greatest energy.
   a. 4.0 eV
   b. 320 nm
   c. 811 nm
   d. 2.1 eV
4. **CHALLENGE** The diagram in **Figure 6** shows the visible light spectrum. What is the range of energies associated with photons in the visible light spectrum?

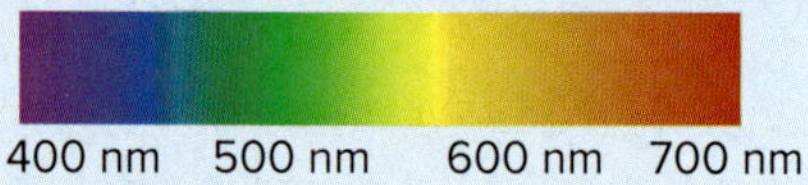

Figure 6

Planck and Einstein

It is important to note that Einstein's photon theory goes further than Planck's hypothesis of radiation from hot objects. While Planck proposed that atoms vibrating at frequency f emit electromagnetic radiation with energy equal to nhf, he did not suggest that light and other forms of electromagnetic radiation act like particles. Einstein's photon theory reinterpreted and extended Planck's hypothesis of radiation emitted by objects.

Einstein's photon theory explains the existence of a threshold frequency and the prompt emission of electrons during the photoelectric effect. A photon with a minimum frequency and energy (hf_0) can eject an electron from metal. If the photon frequency is less than f_0, the photon will not have the energy needed to pry the electron from the metal. Because one photon interacts with one electron, an electron cannot simply store energy from low frequency photons until it collects enough energy to break free. On the other hand, radiation with a frequency greater than f_0 provides more than enough energy to eject an electron. In fact, the excess energy, $hf - hf_0$, becomes the ejected electron's kinetic energy.

Kinetic Energy of an Electron Ejected Due to the Photoelectric Effect

The kinetic emergy of an ejected electron is equal to the difference between the incident photon energy (hf) and the energy of a photon with the threshold frequency (hf_0).

$$KE = hf - hf_0$$

Note that hf_0 is the minimum energy needed to free the most loosely held electron in an atom. Not all electrons in a metal have the same energy; some will require more than this minimum energy in order to escape. As a result, the ejected electrons have differing kinetic energies. Thus, it is important to realize that the phrase "kinetic energy of the ejected electrons" refers to the maximum kinetic energy that an ejected electron could have. Some of the ejected electrons will have less.

Get It?

Explain how an electron emitted by the photoelectric effect can have kinetic energy less than the threshold energy.

Testing the photon theory One way to test the photon theory is to measure the electrons' kinetic energy and compare the results to the value predicted by the theory. The device in **Figure 7** can indirectly measure the electrons' kinetic energy. In the device, the potential difference across the tube is adjustable. When the potential difference is adjusted to make the anode negative, the ejected electrons experience a force opposing their motion due to the electric field between the cathode and anode. As a result, the electrons' kinetic energy decreases as they approach the anode. Only electrons ejected with enough kinetic energy reach the anode.

Increasing the opposing potential difference makes the anode more negative, which means the electron's initial kinetic energy must be greater to reach the anode. At a certain potential difference, called the stopping potential, the current stops because no electrons ejected from the cathode have enough kinetic energy to reach the anode.

At the stopping potential, the electrons' kinetic energy at the cathode equals the work done by the electric field to stop them. In equation form this is

$$KE = -e\Delta V_0$$

In this equation, ΔV_0 is the magnitude of the stopping potential difference in volts (J/C), and e is the charge of the electron (-1.602×10^{-19} C). Note that the negative sign in the equation along with the negative value of e yield a positive value for *KE*. Calculations of electrons' kinetic energy based on this experiment support Einstein's photon theory.

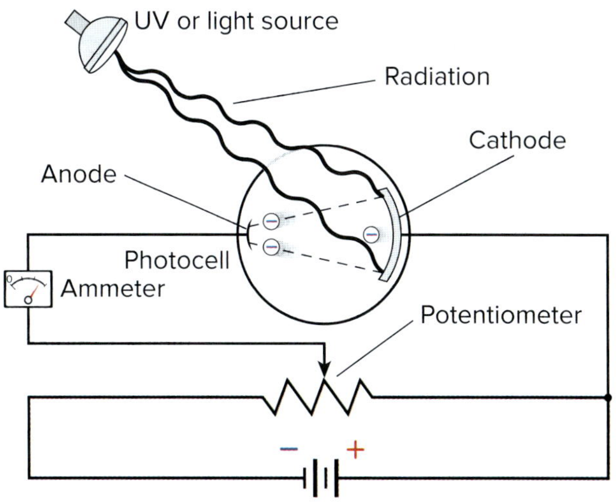

Figure 7 The potentiometer controls the potential difference inside the photocell. By carefully adjusting the potentiometer, you can determine the potential that results in zero current. At the zero current threshold, all electrons fail to reach the anode. Instead, they fall back on the cathode due to the force caused by the electric field between the cathode and the anode.

Application of the photoelectric effect If you have ever used a digital camera, you have used a device that takes advantage of the photoelectric effect. The digital camera shown in **Figure 8** uses a grid of tiny photoelectric detectors—several million of them packed in a few square centimeters.

In a typical design, light enters the camera through the lens and passes through a grid of filters placed over the photoelectric detectors. The filters allow only certain wavelengths of light, corresponding to specific colors of light, to pass through to an individual detector. Photons that strike a detector result in photoelectrons being stored in the detector. The amount of charge stored at the detector's location indicates the brightness of light. The information from the entire grid of detectors is then compiled to recreate the image. Because the photoelectrons generate a voltage within the detectors, the process is a variation of the photoelectric effect known as the photovoltaic effect.

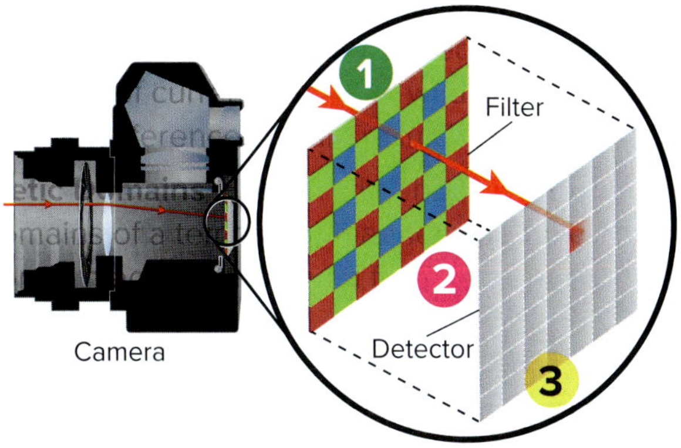

1 The filter grid limits the wavelengths of light reaching a detector. There are more green filters in the grid because the human eye is more sensitive to these wavelengths.

2 Photons that pass through the filter strike the photoelectric detector. This results in a charge being stored in the detector.

3 Information from all the detectors is compiled by the camera's built-in circuitry. Data from adjacent detectors helps the computer determine the best color for each pixel in the picture.

Figure 8 A digital camera's image detector relies on the photoelectric effect to generate a voltage. The picture is digitized and stored as the values of an array of pixels.

EXAMPLE Problem 1

PHOTOELECTRON KINETIC ENERGY The stopping potential difference of a certain photocell is 4.0 V. How much kinetic energy does the incident light give the electrons? Give your answer in both joules and electron volts.

1 ANALYZE AND SKETCH THE PROBLEM

Draw the cathode and the anode, the incident radiation, and the direction of the ejected electron. Note that the stopping potential prevents electrons from flowing across the photocell.

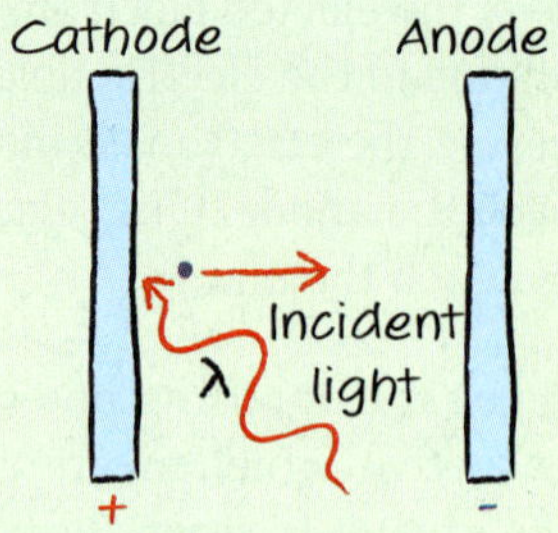

Known	**Unknown**
$\Delta V = 4.0\ \text{V}$	KE (in J and eV) = ?
$e = -1.602\times10^{-19}\ \text{C}$	

2 SOLVE FOR THE UNKNOWN

The electric field does work on the electrons, so the final kinetic energy of an electron equals the initial kinetic energy plus the work done on the electron.

$$KE_{\text{initial}} + W_{\text{electric field on electron}} = KE_{\text{final}}$$

$$KE_{\text{initial}} + W_{\text{electric field on electron}} = 0\ \text{J}$$

The applied potential difference is the stopping potential, so $KE_{\text{final}} = 0$ J.

Solve for KE_{initial}.

$$KE_{\text{initial}} = -W_{\text{electric field on electron}}$$

$$= -e\Delta V_0$$

Use the definition of electric potential difference $\left(\Delta V = \frac{W_{\text{on e}}}{e}\right)$ to find $W_{\text{on e}} = e\Delta V$.

$$= -(-1.602\times10^{-19}\ \text{C})(4.0\ \text{V})$$

Substitute $e = -1.602\times10^{-19}$ C, $\Delta V_0 = 4.0$ V.

$$= +6.4\times10^{-19}\ \text{J}$$

Convert KE from joules to electron volts.

$$KE = (6.4\times10^{-19}\ \text{J})\frac{1\text{eV}}{1.602\times10^{-19}\ \text{J}}$$

$$= 4.0\ \text{eV}$$

3 EVALUATE THE ANSWER

- **Are the units correct?** Joules and electron volts are both units of energy.
- **Does the sign make sense?** Kinetic energy is always positive.
- **Is the magnitude realistic?** One electron volt is the energy of an electron accelerated across a potential difference of 1 V. The electron is accelerated by a potential difference of 4 V, so 4 eV is reasonable.

PRACTICE Problems

ADDITIONAL PRACTICE

5. An electron has an energy of 2.3 eV. What is the kinetic energy of the electron in joules?
6. What is the velocity of the electron in the previous problem?
7. What is the kinetic energy in eV of an electron with a velocity of 6.2×10^6 m/s?
8. The stopping potential for a photoelectric cell is 5.7 V. Calculate the maximum kinetic energy of the emitted photoelectrons in eV.
9. The stopping potential for a photoelectric cell is 5.1 V. How much kinetic energy does the incident light give the electrons in joules?
10. The maximum kinetic energy of emitted photoelectrons in a photoelectric cell is 7.5×10^{-19} J. What is the stopping potential?
11. **CHALLENGE** The stopping potential required to prevent current through a photocell is 3.2 V. Calculate the maximum kinetic energy in joules of the photoelectrons as they are emitted.

Table 1 Threshold Frequency, Threshold Wavelength, and Work Function for Selected Metals

Metal	Threshold Frequency ($\times 10^{14}$ Hz)	Threshold Wavelength (nm)	Work Function (eV)
Cesium	4.70	637	1.95
Magnesium	8.84	339	3.66
Silver	11.1	270	4.6
Sodium	5.70	526	2.36

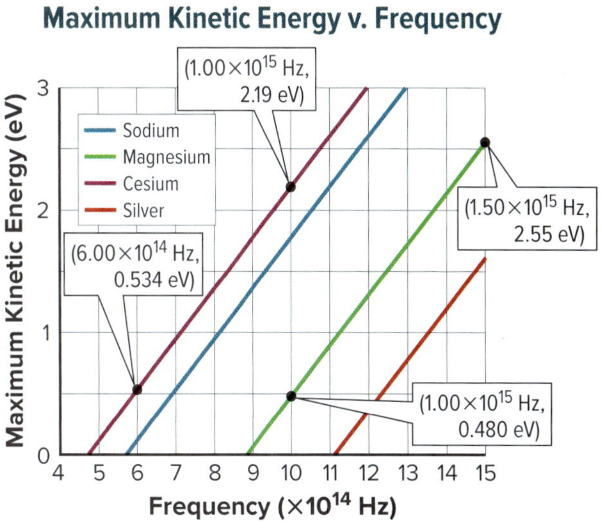

Figure 9 The metal's threshold frequency equals the line's *x*-intercept. Each line's slope equals Planck's constant.

Measuring h

A graph of the maximum kinetic energies of the electrons ejected from a metal versus the frequencies of the incident photons is a straight line, as shown in **Figure 9.** All metals have similar graphs with the same slope. This slope equals the rise/run ratio of the line, which is equal to Planck's constant (h).

The graphs of various metals differ only in the threshold frequency needed to free the electrons. In **Figure 9,** the threshold frequency (f_0) is the point at which $KE = 0$. Looking at the graph, we see that f_0 is located at the intersection of the line with the x-axis. For example, the x-intercept, and therefore the threshold frequency, for cesium is 4.70×10^{14} Hz. A photon with that frequency has just enough energy to eject an electron from the metal. This minimum energy, called the **work function** of a metal, is the energy needed to free the most weakly bound electron from the metal. The magnitude of the work function is equal to hf_0. When a photon of frequency f_0 is incident on a metal, the energy of the photon is sufficient to release an electron but not sufficient to provide the electron with any kinetic energy.

CONNECTING MATH to Physics

Slope of a Line The lines in **Figure 9** are parallel. That means they all have the same slope. Below are the calculations for cesium and magnesium.

Math	Physics	
	Cesium	Magnesium
(x_1, y_1), (x_2, y_2)	(6.00×10^{14} Hz, 0.534 eV), (1.00×10^{15} Hz, 2.19 eV)	(1.00×10^{15} Hz, 0.480 eV), (1.50×10^{15} Hz, 2.55 eV)
$m = \frac{y_2 - y_1}{x_2 - x_1}$	$m = \frac{2.19\text{ eV} - 0.534\text{ eV}}{1.00\times10^{15}\text{ Hz} - 6.00\times10^{14}\text{ Hz}}$	$m = \frac{2.55\text{ eV} - 0.480\text{ eV}}{1.50\times10^{15}\text{ Hz} - 1.00\times10^{15}\text{ Hz}}$
	$m = 4.14\times10^{-15}$ eV/Hz	$m = 4.14\times10^{-15}$ eV/Hz

All lines in **Figure 9** have a slope of 4.14×10^{-15} eV/Hz. When converted to J/Hz, this value becomes the familiar value of Planck's constant:

$$4.14\times10^{-15}\ \frac{\text{eV}}{\text{Hz}}\left(\frac{1.602\times10^{-19}\text{ J}}{1\text{ eV}}\right) = 6.63\times10^{-34}\text{ J/Hz}$$

EXAMPLE Problem 2

WORK FUNCTION AND ENERGY A particular photocell uses a sodium cathode. Sodium has a threshold wavelength of 526 nm.

a. Find the work function of sodium in eV.

b. If ultraviolet radiation with a wavelength of 348 nm falls on sodium, will the cathode discharge electrons? If so, what is the maximum energy of the ejected electrons in eV?

1 ANALYZE AND SKETCH THE PROBLEM

Draw the cathode and anode, the incident radiation, and the direction of the ejected electron.

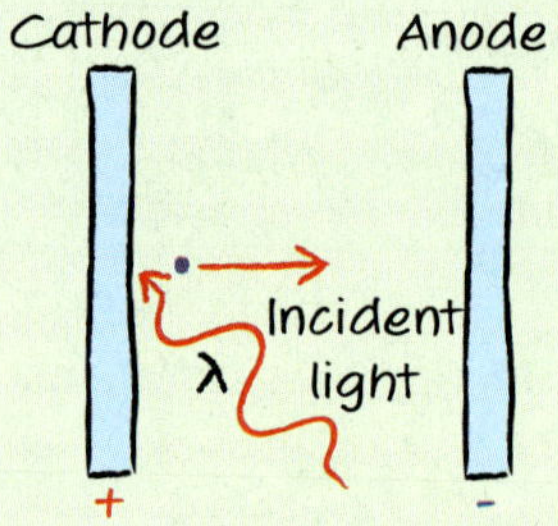

Known		Unknown
$\lambda_0 = 526$ nm	$\lambda = 348$ nm	$W = ?$ $KE = ?$
hc = 1240 eV·nm		

2 SOLVE FOR THE UNKNOWN

a. Find the work function using Planck's constant and the threshold wavelength.

$$W = hf_0 = \frac{hc}{\lambda_0}$$

$$= \frac{1240\ \text{eV·nm}}{526\ \text{nm}}$$ Substitute hc = 1240 eV·nm, $\lambda_0 = 526$ nm.

$$= 2.36\ \text{eV}$$

b. Use the relationship between a photon's energy and wavelength to find the photon's energy.

$$E = \frac{1240\ \text{eV·nm}}{\lambda}$$

$$= \frac{1240\ \text{eV·nm}}{348\ \text{nm}}$$ Substitute $\lambda = 348$ nm.

$$= 3.56\ \text{eV}$$

To calculate the kinetic energy of the ejected electron, subtract the work function from the energy of the incident radiation.

$$KE = hf - hf_0 = \frac{hc}{\lambda} - \frac{hc}{\lambda_0}$$

$$= E - W$$ Substitute $\frac{hc}{\lambda} = E$, $\frac{hc}{\lambda_0} = W$.

$$= 3.56\ \text{eV} - 2.36\ \text{eV}$$ Substitute $E = 3.56$ eV, $W = 2.36$ eV.

$$= 1.20\ \text{eV}$$

3 EVALUATE THE ANSWER

- **Are the units correct?** Performing dimensional analysis on the units verifies that electron volts is the proper unit for kinetic energy.
- **Does the sign make sense?** Kinetic energy is always positive.
- **Are the magnitudes realistic?** Work functions given in Table 1 are a few electron volts, so the magnitude is realistic.

PRACTICE Problems

ADDITIONAL PRACTICE

12. The threshold wavelength of zinc is 310 nm. Find the threshold frequency, in Hz, and the work function, in eV, of zinc.

13. The work function for cesium is 1.95 eV. What is the maximum kinetic energy, in eV, of photoelectrons ejected when 425-nm violet light falls on the cesium?

14. When a metal is illuminated with 193-nm ultraviolet radiation, electrons with kinetic energies of 3.5 eV are emitted. What is the work function of the metal?

15. CHALLENGE A researcher illuminates a sample of metal and finds the longest wavelength to eject electrons is 273 nm. Find the work function. Use **Table 1** to identify the most likely metal.

The Compton Effect

The photoelectric effect demonstrates that a photon has kinetic energy even though it has no mass. In 1916, Einstein predicted that the photon should also have another particle property—momentum. He argued that the momentum of a photon should be equal to $\frac{E}{c}$. Because $E = hf$ and $\frac{f}{c} = \frac{1}{\lambda}$, the photon's momentum is given by the following equation.

Photon Momentum

The momentum of a photon is equal to Planck's constant divided by the photon's wavelength.

$$p = \frac{hf}{c} = \frac{h}{\lambda}$$

Experiments done in 1922 by American physicist Arthur Holly Compton tested Einstein's hypothesis. Compton's results further supported the particle model of light. Compton directed X-rays of a known wavelength at a graphite target, as shown on the left in **Figure 10,** and measured the wavelengths of the X-rays scattered by the target. He observed that some of the X-rays were scattered without change in wavelength, whereas others had a longer wavelength than that of the original radiation. These results are shown on the right panel in **Figure 10.** Note that the peak wavelength for the unscattered X-rays corresponds to the wavelength of the original incident X-rays, whereas the peak wavelength for the scattered X-rays is greater than that of the original incident X-rays.

Recall that the equation for the energy of a photon, $E = hf$, also can be written as $E = \frac{hc}{\lambda}$. This second equation shows that the energy of a photon is inversely proportional to its wavelength. The increase in wavelength that Compton observed meant that the X-ray photons had lost both energy and momentum. The shift in the energy of scattered photons is called the **Compton effect.** This shift in energy, indicated by a shift in wavelength, is very small—only about 10^{-3} nm. It is a measurable effect only when electromagnetic waves have wavelengths less than 10^{-2} nm.

Figure 10 The shift in the peak wavelength of scattered X-rays indicates that the scattered photons lost energy and momentum.

Explain *why an increase in wavelength indicates a decrease in energy.*

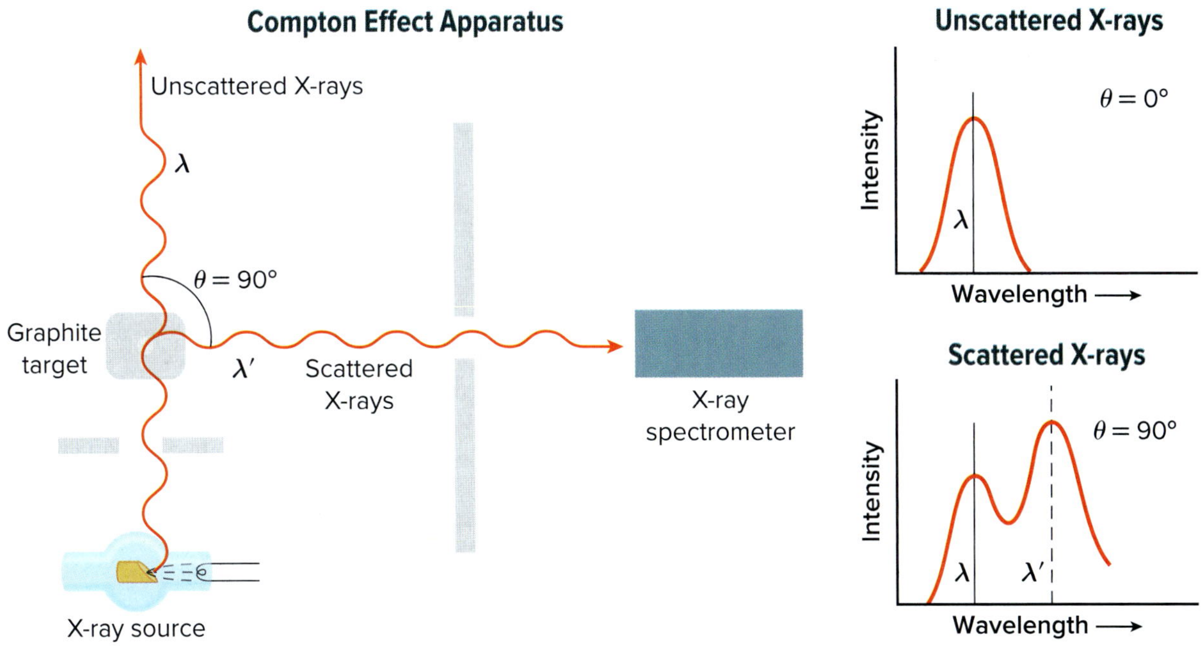

Photons and conservation of energy and momentum In later experiments, Compton observed that electrons were ejected from the graphite block during the experiment. He suggested that the X-ray photons collided with electrons in the graphite target and transferred energy and momentum to them. Compton thought these photon-electron collisions were similar to the elastic collisions experienced by billiard balls, as shown in **Figure 11.** He tested this idea by measuring the energies of the electrons ejected from the graphite block. Compton found that the energy and momentum gained by the ejected electrons equaled the energy and momentum lost by the photons. Thus, photons obey the laws of conservation of energy and momentum when they collide with other particles.

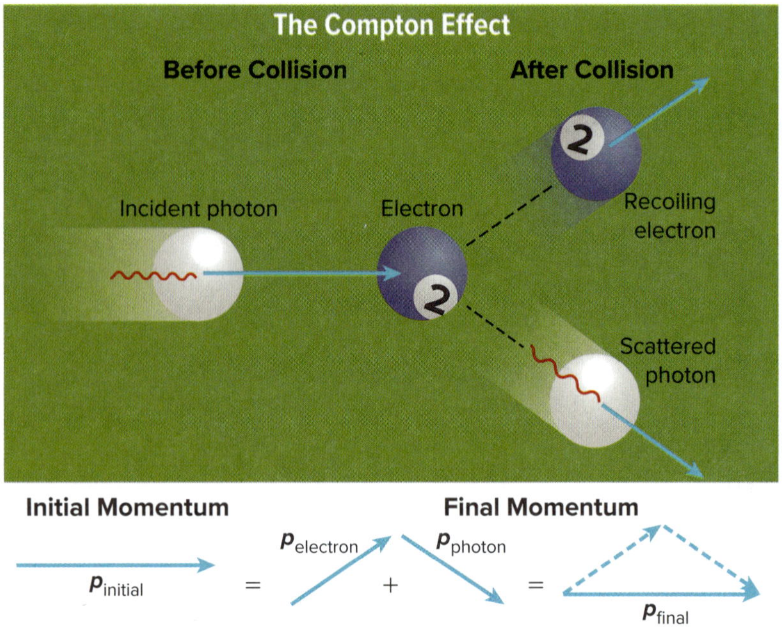

Figure 11 The laws of conservation of energy and momentum apply to collisions between billiard balls and between photons and particles.

Check Your Progress

16. **Photovoltaic Materials** Why is high-intensity, low-frequency light unable to eject electrons from a metal, whereas low-intensity, high-frequency light can?
17. **Frequency and Energy of Hot-Object Radiation** As the temperature of an object is increased, how does the frequency of peak intensity change? How does the total amount of radiated energy from the object change?
18. **Photoelectric and Compton Effects** Distinguish the photoelectric effect from the Compton effect.
19. **Solar Cells** What are solar cells? Describe how they function.
20. **Photoelectric Effect** Green light ($\lambda = 532$ nm) strikes an unknown metal, causing electrons to be ejected. The ejected electrons can be stopped by a potential of 1.44 V. What is the work function, in eV, of the metal?
21. **Energy of a Photon** What is the energy, in eV, of the photons of a 650-nm laser pointer?
22. **Compton Effect** An X-ray strikes a bone, collides with an electron, and scatters. How does the wavelength of the scattered X-ray compare to the wavelength of the incident X-ray?
23. **Photoelectric Effect** An X-ray is absorbed in a bone and releases an electron. The X-ray has a wavelength of about 0.02 nm. Find the electrons' energy, in eV. Assume the work function of bone is negligible compared to the X-ray's energy.
24. **Critical Thinking** Imagine that the collision of two billiard balls models the interaction of a photon and an electron during the Compton effect. Suppose the electron is replaced by a much more massive proton. Would this proton gain as much energy from the collision as the electron does? Would the photon lose as much energy as it does when it collides with the electron?

LEARNSMART® Go online to follow your personalized learning path to review, practice, and reinforce your understanding.

LESSON 2
MATTER WAVES

FOCUS QUESTION

How can particles be described as waves?

de Broglie Waves

If an electromagnetic wave can behave like a collection of particles, could a particle exhibit interference and diffraction as a wave does? In 1923 French physicist Louis de Broglie proposed that a material particle could be described by a wavelength. This proposal was a brave extension of Einstein's photon theory, and Einstein was an early supporter of de Broglie's waves. Recall that the momentum of an object is equal to its mass times its velocity, $p = mv$. If the massive object is also a wave, then its momentum should be like that of a photon, $p = \frac{h}{\lambda}$. De Broglie proposed that these two expressions for momentum could be equated:

$$p = mv = \frac{h}{\lambda}$$

The wavelength λ calculated from the above relationship represents that of the moving particle and is known as the **de Broglie wavelength.** The following equation solves directly for the de Broglie wavelength.

According to de Broglie, any small particle such as an electron or a proton should show wavelike properties. In 1927, independent experiments confirmed electron diffraction, which is evidence of electrons' wavelike properties. Objects you use every day do not exhibit wavelike properties because their wavelengths are too small to observe. For example, the wavelength of a 0.145-kg baseball hit at 38 m/s is:

$$\lambda = \frac{h}{mv} = \frac{6.63\times10^{-34}\ \text{J}\cdot\text{s}}{(0.145\ \text{kg})(38\ \text{m/s})} = 1.2\times10^{-34}\ \text{m}$$

De Broglie Wavelength

The de Broglie wavelength of a moving particle is equal to Planck's constant divided by the particle's momentum.

$$\lambda = \frac{h}{p} = \frac{h}{mv}$$

3D THINKING DCI Disciplinary Core Ideas CCC Crosscutting Concepts SEP Science & Engineering Practices

COLLECT EVIDENCE

Use your Science Journal to record the evidence you collect as you complete the readings and activities in this lesson.

INVESTIGATE

GO ONLINE to find these activities and more resources.

CCC **Identify Crosscutting Concepts**

Create a table of the crosscutting concepts and fill in examples you find as you read.

Review the News

Obtain information from a current news story about technology that makes use of quantum or atomic physics. Evaluate your source and communicate your findings to your class.

EXAMPLE Problem 3

DE BROGLIE WAVELENGTH An electron is accelerated by a potential difference of 75 V. What is its de Broglie wavelength?

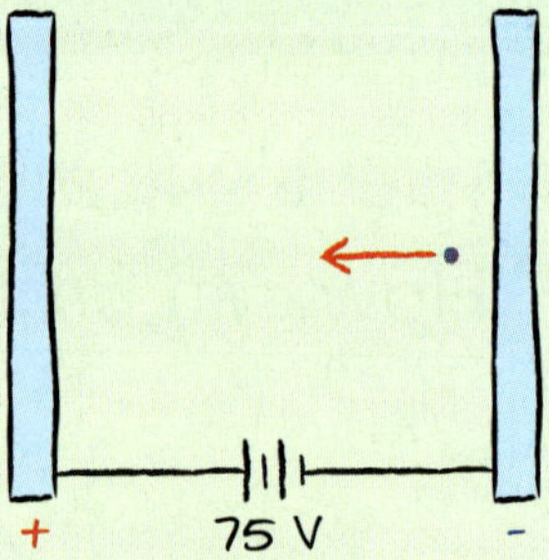

1 ANALYZE AND SKETCH THE PROBLEM

Include the positive and the negative plates in your drawing.

Known		**Unknown**
$\Delta V = 75\ V$	$m_e = 9.11\times10^{-31}\ kg$	$\lambda = ?$
$e = -1.602\times10^{-19}\ C$	$h = 6.63\times10^{-34}\ J{\cdot}s$	

2 SOLVE FOR THE UNKNOWN

Write relationships for the kinetic energy of the electron based on potential difference and motion, and use them to calculate the electron's velocity.

$$KE = -e\Delta V, \text{ and } KE = \frac{1}{2}mv^2$$

$$\frac{1}{2}mv^2 = -e\Delta V$$ Equate the expressions for *KE*.

Solve for *v*.

$$v = \sqrt{\frac{-2e\Delta V}{m}}$$

$$= \sqrt{\frac{-2(-1.602\times10^{-19}\ C)(75\ V)}{9.11\times10^{-31}\ kg}}$$ Substitute $e = -1.602\times10^{-19}$ C, $\Delta V = 75$ V, $m_e = 9.11\times10^{-31}$ kg.

$$= 5.1\times10^{6}\ m/s$$

Use the relationship between mass, velocity, and momentum.

$$p = mv$$

$$= (9.11\times10^{-31}\ kg)(5.1\times10^{6}\ m/s)$$ Substitute $m_e = 9.11\times10^{-31}$ kg, $v = 5.1\times10^{6}$ m/s.

$$= 4.6\times10^{-24}\ kg{\cdot}m/s$$

Use the relationship between momentum and the de Broglie wavelength.

$$\lambda = \frac{h}{p}$$

$$= \frac{6.63\times10^{-34}\ J{\cdot}s}{4.6\times10^{-24}\ kg{\cdot}m/s}$$ Substitute $h = 6.63\times10^{-34}$ J·s, $p = 4.6\times10^{-24}$ kg·m/s.

$$= 1.4\times10^{-10}\ m, \text{ which equals } 0.14\ nm$$

3 EVALUATE THE ANSWER

- **Are the units correct?** Dimensional analysis on the units verifies m/s for *v* and nm for λ.
- **Does the sign make sense?** Positive values are expected for both *v* and λ.
- **Are the magnitudes realistic?** The wavelength is close to 0.1 nm, which is in the X-ray region of the electromagnetic spectrum.

PRACTICE Problems

ADDITIONAL PRACTICE

25. What is the de Broglie wavelength and speed of an electron accelerated by a potential difference of 250 V?

26. A 7.0-kg bowling ball rolls with a velocity of 8.5 m/s.

a. What is the de Broglie wavelength of the bowling ball?

b. Why does the bowling ball exhibit no observable wave behavior?

27. What potential difference is needed to accelerate an electron so it has a 0.125-nm wavelength?

28. **CHALLENGE** The electron in Example Problem 3 has a de Broglie wavelength of 0.14 nm. What is the kinetic energy, in eV, of a proton with the same wavelength?

Location and Momentum

De Broglie's theory of particles' wavelike properties led to a fundamental change in our understanding of atomic-sized objects. In everyday experience, it is logical to think that you can measure any property to any desired level of precision. For example, imagine measuring a ball's position during free fall. You might first try using a meterstick and a stopwatch to make the measurements. If you desire more precise measurements, you might use a video camera and a calculator or a computer. You can imagine using more and more sophisticated equipment to achieve ever greater precision. In fact, classical physics sets no limit on a measurement's precision. De Broglie's revolutionary theory of matter waves forced physicists to redefine the limits of measurement.

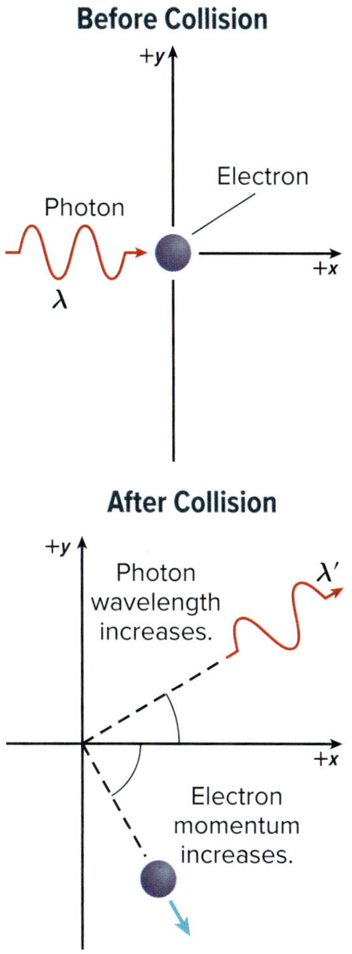

Figure 12 A collision between a photon and the electron is necessary to measure the electron's position. This collision scatters the electron and the photon, changing their momenta.

Compare *the photon wavelength before and after the collision. What does this change in wavelength indicate about the photon's energy?*

Heisenberg uncertainty principle

Consider measuring the location of an atomic or subatomic scale particle by shining light on it and then collecting the reflected light in a measuring device. Because of diffraction, the reflected light you use to detect the particle spreads out and makes it impossible to locate the particle exactly. If more precise measurements are needed, you can use shorter-wavelength radiation, which decreases diffraction and decreases the uncertainty in the particle's location.

As a result of the Compton effect, however, when short-wavelength, high-energy radiation strikes a particle, the particle's momentum is changed, as shown in **Figure 12.** Therefore, the act of measuring the location of a particle has the effect of changing the particle's momentum. The more precise the determination of a particle's location, the greater the uncertainty is in its momentum. Similarly, if the momentum of the particle is precisely known, the position of the particle becomes less certain.

This situation is summarized by the **Heisenberg uncertainty principle,** which states there is a limit to how precisely a particle's position and momentum can simultaneously be measured. This principle, named for German physicist Werner Heisenberg, is the result of light and matter's dual wavelike and particlelike properties. Even though the effects of the limit are only noticeable when measuring atomic-scale particles, Heisenberg's work led to a fundamental change in our understanding of the world. While the classical theories of Newton and Maxwell are successful models of everyday objects, quantum theory and its models of the dual nature of light and matter are required to accurately describe objects on the atomic scale.

Check Your Progress

29. **Waves** Using De Broglie's theory, you can calculate the wavelength of any moving particle. Explain why the wave nature of everyday objects is not obvious.
30. **Wavelength** What is the de Broglie wavelength of an alpha particle (helium nucleus) with mass 6.6×10^{-27} kg and speed 120 m/s?
31. **de Broglie Wavelength** What is the de Broglie wavelength of an electron accelerated through a potential difference of 125 V?
32. **Critical Thinking** How does the Heisenberg uncertainty principle explain double-slit interference patterns even if photons or atoms pass through the slits one at a time?

LEARNSMART® Go online to follow your personalized learning path to review, practice, and reinforce your understanding.

LESSON 3

BOHR'S MODEL OF THE ATOM

FOCUS QUESTIONS

How were atoms and their structure discovered?

The Nuclear Model

By the end of the 1800s, most scientists agreed that atoms exist. J.J. Thomson's discovery of the electron provided evidence that the atom must be made of even smaller, subatomic particles. But atoms were known to be much more massive than the total mass of the electrons, so scientists wondered where the mass was located. Moreover, atoms were known to be electrically neutral. How were the negatively charged electrons arranged? What made the atoms neutral?

Testing Thomson's model J.J. Thomson proposed that the electrons were distributed throughout a positively charged substance like raisins in a muffin. Ernest Rutherford, along with laboratory collaborators Hans Geiger and Ernest Marsden, however, performed a series of experiments that showed the atom had a very different structure. **Figure 13** shows the equipment Rutherford used to investigate the distribution of mass and charge in the atom.

Rutherford's Gold Foil Experiment

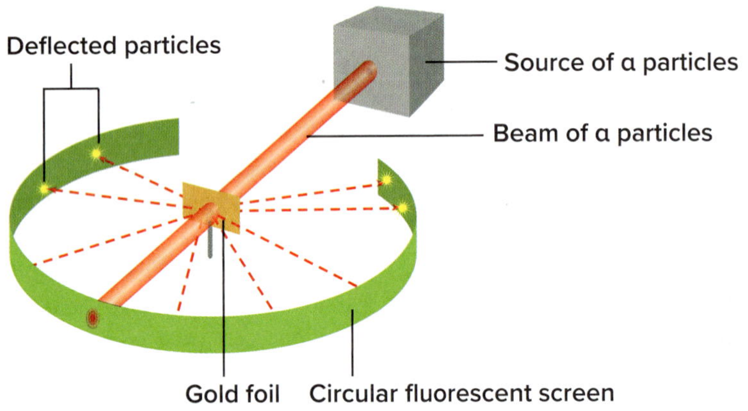

Figure 13 Bombarding metal foil with alpha particles convinced Rutherford's team that most of the atom's mass was concentrated in a small volume in the atom's center.

3D THINKING DCI Disciplinary Core Ideas CCC Crosscutting Concepts SEP Science & Engineering Practices

COLLECT EVIDENCE

Use your Science Journal to record the evidence you collect as you complete the readings and activities in this lesson.

INVESTIGATE

GO ONLINE to find these activities and more resources.

PhysicsLAB: Emission Spectra
Analyze data from the emission specta produced by various gases to identify their elemental composition.

PhysicsLAB: Finding the Size of an Atom
Use a large-scale model to determine the size of an atom.

Rutherford's investigation Rutherford made use of substances that emitted penetrating rays. Some of these emissions, later named **alpha particles,** were known to be massive, positively charged particles that move at high speeds. Alpha particles are represented by the symbol α. The α-particles in Rutherford's experiments could be detected by the small flashes of light produced when the particles collided with a zinc-sulfide-coated screen.

As shown in **Figure 13,** Rutherford directed a beam of α-particles at an extremely thin sheet of gold foil. Rutherford was aware of Thomson's model of the atom, and he expected only minor deflections of the α-particles as they passed through the thin gold foil. He hypothesized that the paths of the massive, high-speed α-particles would be only slightly altered as they passed through the evenly distributed positive charge making up each gold atom. The test results amazed him. While most of the α-particles passed through the gold foil either undeflected or only slightly deflected, a few of the particles were scattered through very large angles. Some even rebounded through angles larger than 90°. **Figure 14** shows a diagram of these results. Rutherford compared his amazement to that of firing a 15-inch cannon shell at tissue paper and then having the shell bounce back and hit him.

Get It?
Explain why Rutherford was so surprised by the results of the gold foil experiment.

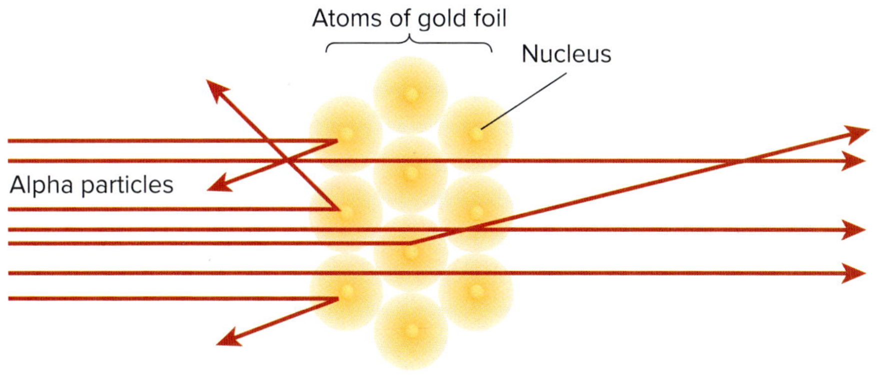

Figure 14 In Rutherford's investigation, most of the alpha particles directed at the thin sheet of gold foil passed through it without deflection. One in 20,000, however, was deflected at a large angle.

Rutherford's revised model of the atom

Using Coulomb's law for the force between charged objects and Newton's laws of motion, Rutherford concluded that the results could be explained if all of an atom's positive charge were concentrated in a tiny, massive core, which is now called the **nucleus.** Therefore, Rutherford's model of the atom is called a nuclear model. Researchers have since determined that all the positive charge and more than 99.9 percent of an atom's mass are contained in its nucleus. The electrons do not contribute a significant amount of mass to the atom and are distributed through a large volume that stretches far away from the nucleus. The volume occupied by the electrons defines the overall size, or diameter, of the atom, which is about 50,000 times larger than the diameter of the nucleus. An atom is mostly empty space.

Emission spectra How are the electrons arranged in the atom? One of the clues scientists used to answer this question came from the study of the light emitted by atoms. As shown in **Figure 15,** atoms of a gaseous sample can be made to emit light in a gas-discharge tube. You probably are familiar with the colorful neon signs used by some businesses. These signs work on the same principles as gas-discharge tubes. A gas-discharge tube consists of a low-pressure gas contained within a glass tube that has electrodes attached to each end. The gas glows when a large potential difference is applied across the tube. What interested scientists the most about this phenomenon was the fact that each different gas glowed with a unique color. The characteristic glows emitted by two gases are shown in **Figure 15.**

When the light emitted by the gas is passed through a diffraction grating or a prism, an emission spectrum is obtained that is a fingerprint of each element in the gas. An emission spectrum can be studied in greater detail using an instrument called a spectroscope. As shown in **Figure 15,** light in a diffraction spectroscope passes through a slit and is diffracted through a diffraction grating. The spectroscope forms an image of the slit at a different position for each wavelength.

As you may have read in an earlier lesson, the spectrum of an incandescent solid, such as a lightbulb filament, is a continuous band of colors from red through violet. The spectrum of a gas, however, is a series of distinct lines of different colors. The bright-line emission spectra for mercury and neon are shown in the bottom portion of **Figure 15.** Each colored line corresponds to a particular wavelength of light emitted by the atoms of that gas.

Figure 15 An element's emission spectrum is a unique pattern that can be used to identify an unknown sample.

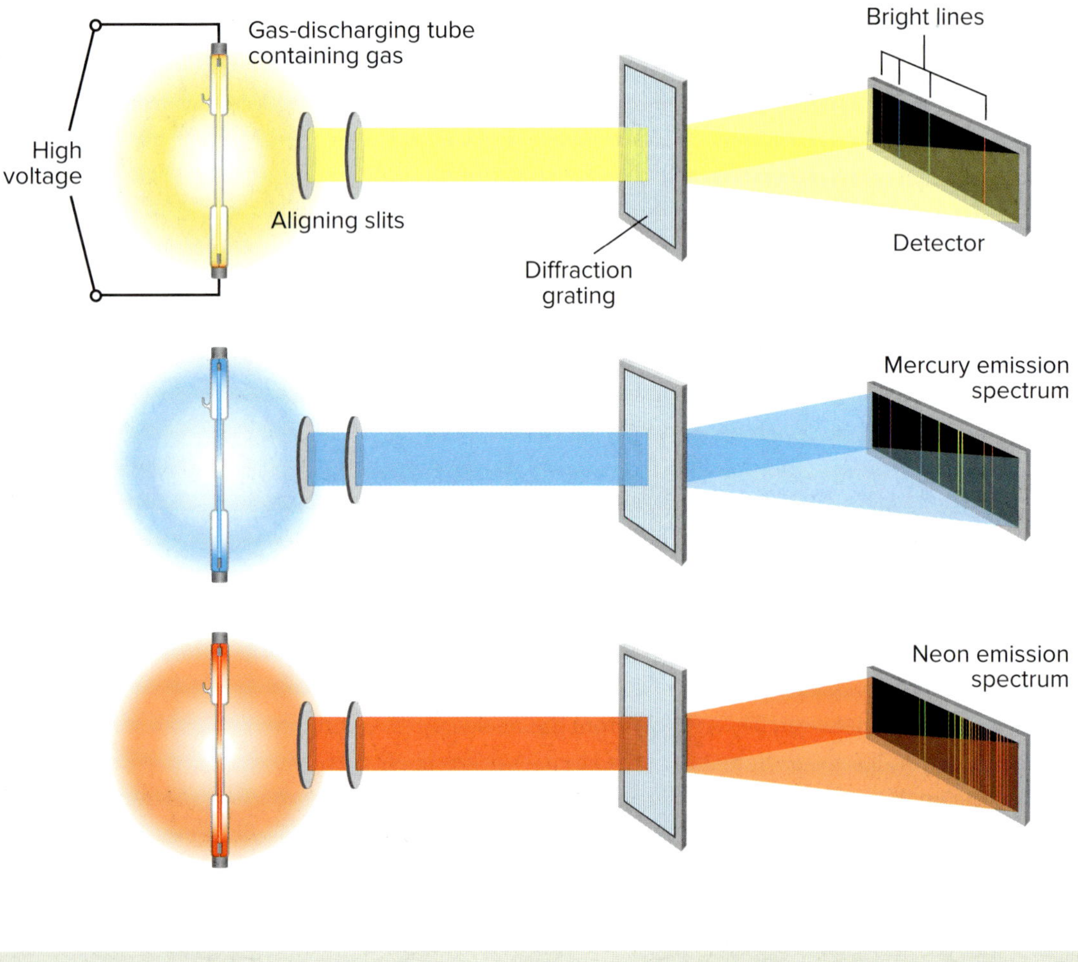

Identifying unknown samples An emission spectrum is a useful analytic tool that can be used to identify an unknown sample of gas. When an unknown gas is placed in a gas-discharge tube, it can be made to emit light. The emitted light includes combinations of wavelengths that are uniquely characteristic of the atoms of that gas. Thus, the unknown gas can be identified by comparing its wavelengths with the wavelengths measured in the spectra of known gas samples.

When the emission spectrum of a combination of elements is photographed, an analysis of the lines on the photograph can indicate the identities and the relative concentrations of the elements present. The elements that are present in the greatest concentrations produce emission lines of greater relative intensities. Through a comparison of the line intensities, the percentage composition of the gaseous material can be determined. This type of analysis is used daily to solve problems in chemistry, engineering, medicine, and environmental science.

Get It?

Explain how an emission spectrum can help to identify the composition and concentrations of elements in a sample.

Absorption spectra **Figure 16** illustrates another technique used to identify unknown samples. Passing white light through a gas sample and then through a prism results in the normally continuous spectrum of white light, but with dark lines in it corresponding to wavelengths of light absorbed by the gas. The set of wavelengths absorbed by a gas is the **absorption spectrum** of the gas. For a gas, the bright lines of the emission spectrum and the dark lines of the absorption spectrum occur at the same wavelengths, as shown in the bottom portion of **Figure 16.** Thus, cool, gaseous elements absorb the same wavelengths that they emit when excited. The composition of a gas can be determined from the wavelengths of the dark lines of the gas's absorption spectrum.

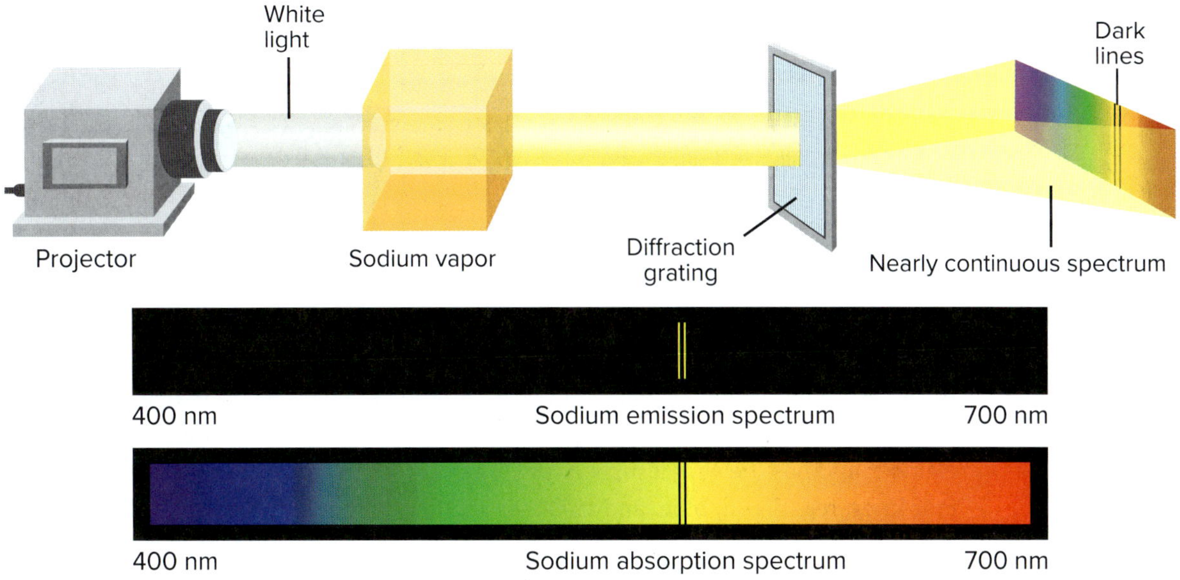

Figure 16 Passing white light through a sample of gas and a prism produces a nearly continuous spectrum. The dark lines missing from the spectrum are the wavelengths absorbed by the gas. Much like emission spectra, absorption spectra can be used to identify unknown samples.

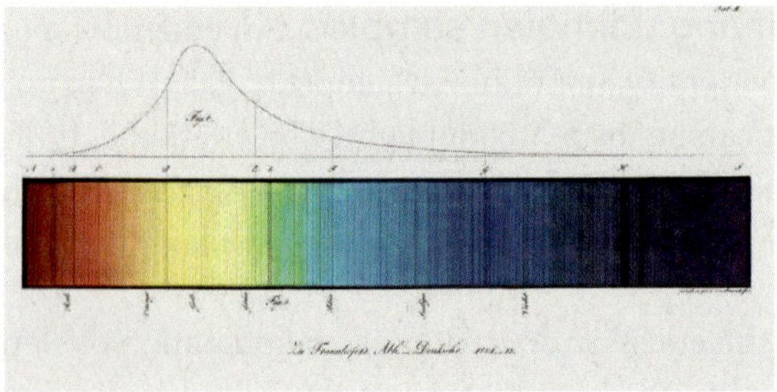

Figure 17 The Sun's absorption spectrum shows characteristic Fraunhofer lines for the elements in the Sun's atmosphere. There are many lines, some of which are faint and others that are very dark, depending on abundance of the elements in the Sun.

ASTRONOMY Connection In 1814 Josef von Fraunhofer observed the presence of several dark lines in the spectrum of sunlight. These dark lines, now called Fraunhofer lines, are seen in **Figure 17.** He reasoned that as sunlight passed through the gaseous atmosphere surrounding the Sun, the gases absorbed certain characteristic wavelengths. These absorbed wavelengths produced the dark lines in the observed spectrum. The composition of the Sun's atmosphere was determined by comparing the missing lines in the observed spectrum with the known emission spectra of various elements. The Sun was found to be mostly hydrogen and helium. The compositions of many other stars have been determined using this technique.

Get It?

Describe how scientists are able to determine the composition of stars.

Spectroscopy As a result of the elements' characteristic spectra, scientists are able to analyze, identify, and quantify unknown materials by observing the spectra that they emit or absorb. The emission and absorption spectra of elements are important in industry as well as in scientific research. For example, steel mills reprocess large quantities of scrap iron of varying compositions. The exact composition of a sample of scrap iron can be determined in minutes by spectrographic analysis. The composition of the steel can then be adjusted to suit commercial specifications.

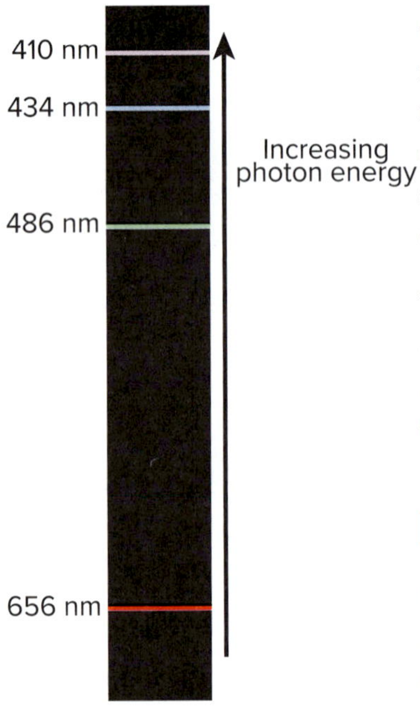

Figure 18 Hydrogen's emission spectrum has four lines in the visible range.

Compare *the difference between adjacent wavelengths of light as the energy increases.*

The study of spectra is a branch of science known as spectroscopy. Spectroscopists are employed in research and industrial settings. Spectroscopy is an effective tool for analyzing materials on Earth, and it is the only currently available tool for studying the composition of stars over the vast expanse of space.

Revising Rutherford's Nuclear Model

In the nineteenth century, many physicists studied atomic spectra to determine the structure of the atom. Hydrogen was studied extensively because it is the lightest element and has the simplest spectrum. The visible spectrum of hydrogen consists of four lines: red, green, blue, and violet, as shown in **Figure 18.** Any theory that explained the structure of the atom would have to account for these wavelengths, as well as support the nuclear model of the atom. The nuclear model as proposed by Rutherford was not without its problems, however. Rutherford had suggested that electrons orbit the nucleus much like the planets orbit the Sun. There was, however, a serious flaw in this planetary model.

Problems with the planetary model An electron in an orbit constantly accelerates toward the nucleus. From previous study, you know that accelerating electrons radiate energy by emitting electromagnetic waves. At the rate that an orbiting electron would lose energy, it should spiral into the nucleus within a fraction of a second. But we know that atoms are stable, so the planetary model does not agree with the laws of electromagnetism. In addition, the planetary model predicted that the accelerating electrons would radiate energy at all wavelengths. However, as you just read, the light emitted by atoms is radiated only at specific wavelengths.

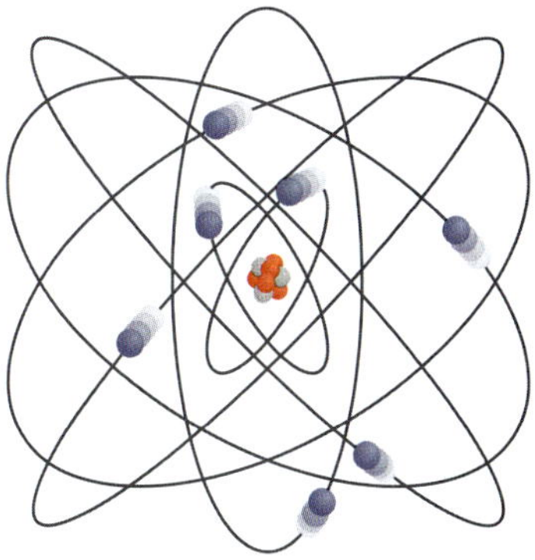

Figure 19 The planetary model of the atom stated electrons move in fixed orbits around the nucleus.

Danish physicist Niels Bohr went to England in 1911 and joined Rutherford's group to work on determining the structure of the atom. He tried to unite the nuclear model of the atom with Planck's quantized energy levels and Einstein's theory of light. This was a courageous idea, because in 1911, neither of these revolutionary ideas was widely understood or accepted.

Quantized Energy

Bohr began with the planetary arrangement of electrons, as diagrammed in **Figure 19,** but then he proposed that the laws of electromagnetism do not apply inside the atom. He postulated that an electron in a stable orbit in an atom does not radiate energy, even though it is accelerating. Bohr referred to this stable condition as a stationary state. He went on to assume that each of the stationary states has its own specific amount of energy. In other words, Bohr considered the energy levels in an atom to be quantized.

As shown in **Figure 20,** the quantization of energies of electrons in atoms can be likened to a flight of stairs with decreasing-height steps. To go up the stairs you must move from one step to the next–it is impossible to stop at a midpoint between steps. Instead of steps, the electrons have quantized amounts of energy, each of which is called an **energy level.** Just as you cannot occupy a position between steps, an electron's energy cannot have a value between allowed energy levels. An atom with the smallest allowable amount of energy is said to be in the **ground state.** When such an atom absorbs energy, one or more of its electrons move to higher energy levels. Any energy level above the ground state is called an **excited state.**

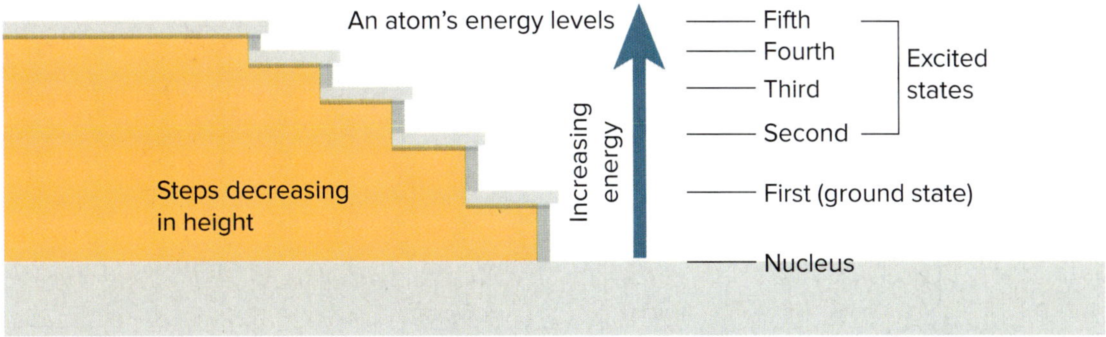

Figure 20 The allowed energy levels in an atom can be compared to a set of stairs. Note how the difference in energy between adjacent energy levels decreases as the energy level increases.

Energy of an atom What determines the amount of energy an atom has? An atom's energy equals the sum of the kinetic energy of the electrons and the potential energy from the attractive force between the electrons and the nucleus. The energy of an atom with electrons in a nearby orbit is less than that of an atom with electrons in a faraway orbit because work must be done to move the electrons away from the nucleus. Thus, atoms in excited, higher-energy states have electrons in larger, more distant, orbits. Because energy is quantized and energy is related to the size of the orbit, the size of the orbit also is quantized. This model of an atom, that of a central nucleus with orbiting electrons having specific quantized energy levels, is the Bohr model of the atom.

If Bohr was correct in hypothesizing that stable atoms do not radiate energy, then what is responsible for an atom's characteristic emission spectrum? To answer this question, Bohr suggested that electromagnetic energy is emitted when the atom changes from one state to another. Incorporating Einstein's photoelectric theory, Bohr knew that the energy of every photon is given by the equation $E_{photon} = hf$. He then postulated that when an atom absorbs a photon, the atom's energy increases by an amount equal to that of the photon. When this excited atom makes a transition to a lower energy level, it emits a photon.

When the atom makes the transition from its initial energy level (E_i) to its final energy level (E_f) the change in energy (ΔE_{atom}) is given by the following equation:

$$\Delta E_{atom} = E_f - E_i.$$

As shown in **Figure 21,** the amount by which the energy of the atom decreases equals the energy of the emitted photon.

$$E_{photon} = -\Delta E_{atom} = E_i - E_f$$

The following equations summarize the relationships between the change in an atom's energy levels and the emitted photon's energy.

Energy of an Emitted Photon

The energy of an emitted photon equals the product of Planck's constant and the emitted photon's frequency. The energy of an emitted photon also is equal to the decrease in the atom's energy.

$$E_{photon} = hf = -\Delta E_{atom}$$

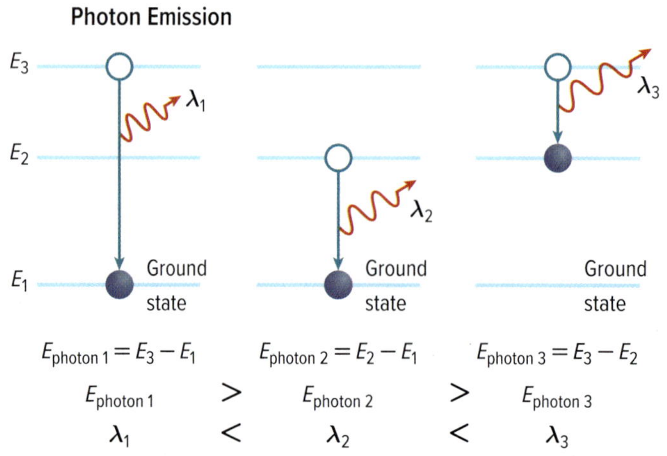

Figure 21 An emitted photon's energy is equal to the amount by which the energy of the atom decreases.

Identify *the transition that produces the photon with the greatest energy.*

Predictions of the Bohr Model

A scientific theory must do more than present postulates; it must allow predictions to be made that can be checked against experimental data. A good theory also can be applied to many different problems, and it ultimately provides a simple, unified explanation of some part of the physical world. Bohr used his theory to calculate the wavelengths of light emitted by a hydrogen atom. The calculations were in excellent agreement with the values measured by other scientists. As a result of this agreement between Bohr's predictions and the measured values, Bohr's model was widely accepted.

Unfortunately, the model only worked for the element hydrogen; it could not predict the spectrum of helium, the next-simplest element. In addition, there was not a good explanation as to why the laws of electromagnetism should work everywhere but inside the atom.

Development of Bohr's model Bohr developed his model by applying Newton's second law of motion ($F_{net} = ma$) to the electron. The net force is described by Coulomb's law for the interaction between an electron of charge −e that is a distance r from a proton of charge +e. That force is given by the following equation:

$$F = \frac{-\mathrm{Ke}^2}{r^2}$$

The acceleration of the electron in a circular orbit about a much more massive proton is given by $a = -\frac{v^2}{r}$, where the negative sign shows that the direction is inward. Thus, Bohr obtained the following relationship:

$$\frac{\mathrm{Ke}^2}{r^2} = \frac{\mathrm{m_e}v^2}{r}$$

Solving this equation for r gives the following relationship:

$$r = \frac{\mathrm{Ke}^2}{\mathrm{m_e}v^2}$$

In the equation, K, the Coulomb's law constant, is $9.0\times10^9\ \mathrm{N{\cdot}m^2/C^2}$.

Next, Bohr considered the angular momentum of the orbiting electron, which is equal to the product of an electron's momentum and the radius of its circular orbit. The angular momentum of the electron is thus mvr. Bohr postulated that angular momentum also is quantized; that is, an electron's angular momentum can have only certain values. Using n to represent an integer, Bohr proposed that the electron's angular momentum equaled an integer multiple of $\frac{\mathrm{h}}{2\pi}$, where h is Planck's constant.

$$\mathrm{m_e}vr = \frac{n\mathrm{h}}{2\pi}$$

Rearranging the above equation to solve for v yields the following:

$$v = \frac{n\mathrm{h}}{2\pi \mathrm{m_e} r}$$

Using this relationship for velocity and the equation for a hydrogen atom's radius given above, Bohr found that the orbital radii of the electrons in a hydrogen atom are given by the following equation.

Electron Orbital Radius in Hydrogen

$$r_n = \frac{\mathrm{h}^2 n^2}{4\pi^2 \mathrm{K m_e e^2}}$$

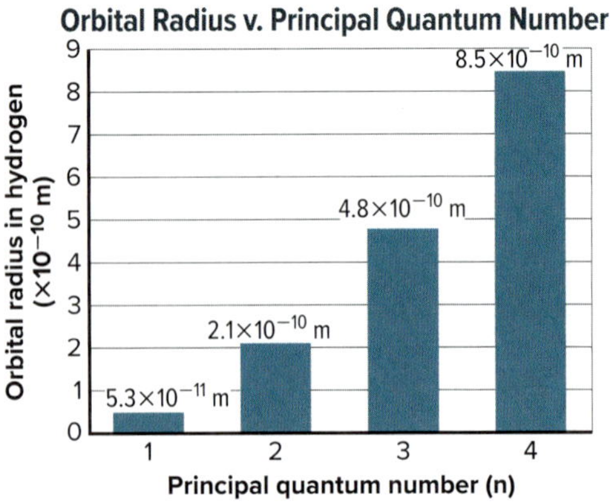

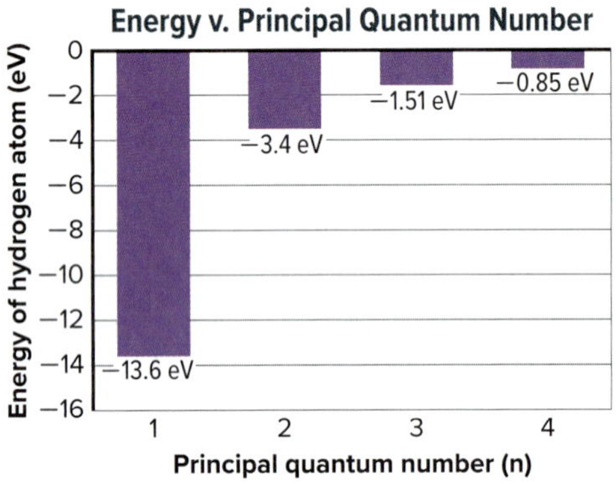

Figure 22 The radius of the hydrogen atom increases as n^2. The energy of the hydrogen atom increases from a minimum of −13.6 eV when $n = 1$ (the ground state) to values closer to 0 eV as n increases.

Calculating orbital radius You can calculate the radius of the innermost orbit of a hydrogen atom, also known as the Bohr radius, by substituting known values and $n = 1$ into the equation for electron orbital radius on the previous page.

$$r_1 = \frac{(6.63\times10^{-34}\ \text{J}\cdot\text{s})^2\ (1)^2}{4\pi^2(9.0\times10^{9}\ \text{N}\cdot\text{m}^2/\text{C}^2)(9.11\times10^{-31}\ \text{kg})(1.6\times10^{-19}\ \text{C})^2}$$

$$= 5.3\times10^{-11}\ \text{J}^2\cdot\text{s}^2/(\text{N}\cdot\text{m}^2\cdot\text{kg})$$

$$= 5.3\times10^{-11}\ \text{m, or } 0.053\ \text{nm}$$

By performing a little more algebra, you can show that the total energy of the atom, which is the sum of the kinetic energy of the electron and the potential energy, and is given by $\frac{-Ke^2}{2r}$, is represented by the following equation:

$$E_n = \frac{-2\pi^2 K^2 m_e e^4}{h^2} \times \frac{1}{n^2}$$

By substituting numerical values for the constants, you can calculate the total energy of the atom in joules, which yields the following equation:

$$E_n = -2.17\times10^{-18}\ \text{J} \times \frac{1}{n^2}$$

Converting the relationship to units of electron volts (1 eV = 1.6×10^{-19} J) yields the following equation.

Energy of a Hydrogen Atom

The total energy of an atom with principal quantum number n is equal to the product of −13.6 eV and the inverse of n^2.

$$E_n = -13.6\ \text{eV} \times \frac{1}{n^2}$$

You previously read that the electron's orbital radius is quantized. The equation above shows that the energy of the atom is quantized as well. The integer (n) that appears in the equations for hydrogen's energy and electron orbital radius is called the **principal quantum number.** The principal quantum number determines the quantized values of r and E.

Figure 22 summarizes these relationships for the first four values of n. The graph at the top of the figure shows that the electron's orbital radius (r) increases as the square of n. The graph at the bottom of the figure shows that the energy (E) of the hydrogen atom depends on $\frac{1}{n^2}$. As the magnitude of the atom's energy decreases, the electron's orbital radius increases.

Get It?

Describe how the orbital radius and the energy depend on the principal quantum number.

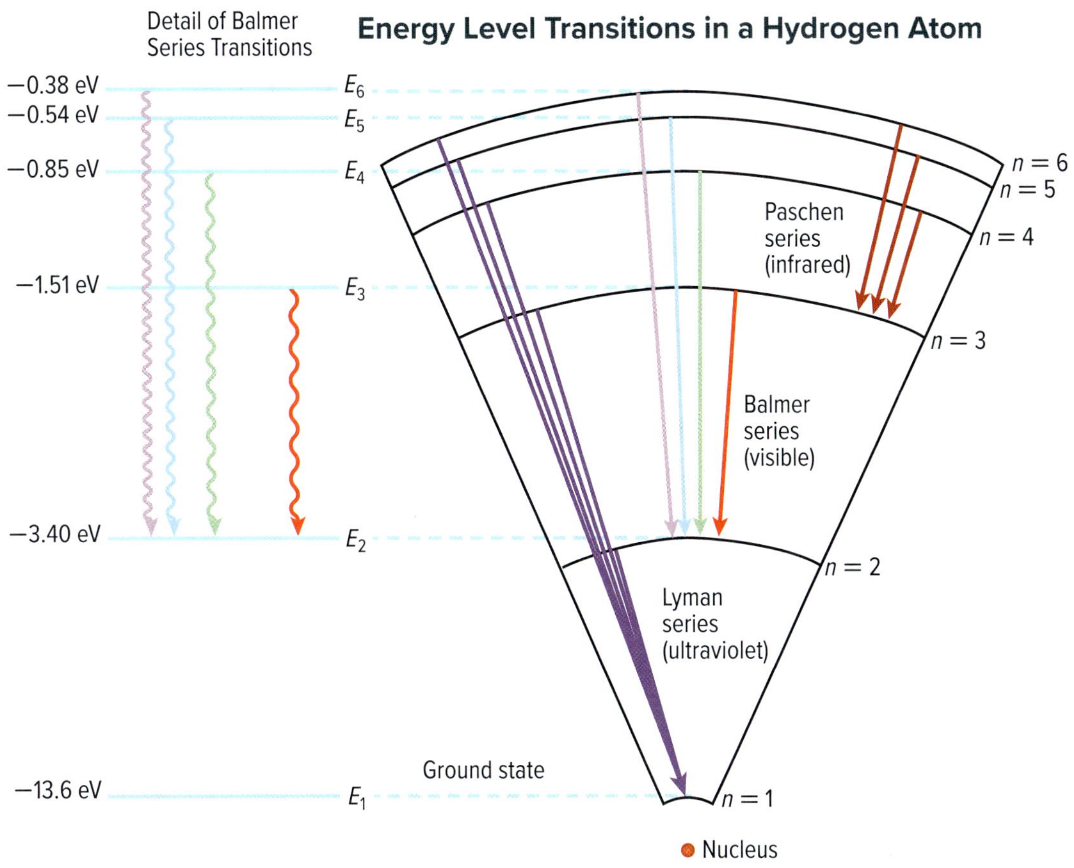

Figure 23 A hydrogen atom emits visible light when electrons transition from higher energy levels to the $n = 2$ energy level. Transitions to other energy levels result in emission of ultraviolet and infrared electromagnetic energy.

Energy and electron transitions You may be wondering why the energy of an atom in the Bohr model has a negative value. Recall from your study of energy that only energy differences affect the motion of objects. The zero energy level can be chosen at will. In this case, zero energy is defined as the energy of the atom when the electron is infinitely far from the nucleus and has no kinetic energy. This condition exists when the atom has been ionized; that is, when the electron has been removed from the atom. Because work has to be done to ionize an atom, the total energy of an atom that has an orbiting electron is less than zero. When an atom makes a transition from a lower to a higher energy level, the total energy becomes less negative, but the overall energy change is positive.

Some of hydrogen's energy levels and the possible energy level transitions that it can undergo are shown in **Figure 23.** Note that an excited hydrogen atom can emit electromagnetic energy in the infrared, visible, or ultraviolet range, depending on the transition that occurs. Ultraviolet light is emitted when the atom drops into its ground state from any excited state. The four visible lines in the hydrogen spectrum are produced when the atom drops from the $n = 3$ or higher energy state into the $n = 2$ energy state.

Get It?

Describe the electron transitions that result in the visible lines in the hydrogen emission spectrum.

EXAMPLE Problem 4

ENERGY LEVELS In order for an electron in a hydrogen atom to move from the lowest energy level ($n = 1$) to the second energy level ($n = 2$), the atom must absorb energy. Determine the energy of the first and second energy levels and the energy absorbed by the atom.

1 ANALYZE AND SKETCH THE PROBLEM

- Diagram the energy levels E_1 and E_2.
- Indicate the direction of increasing energy on the diagram.

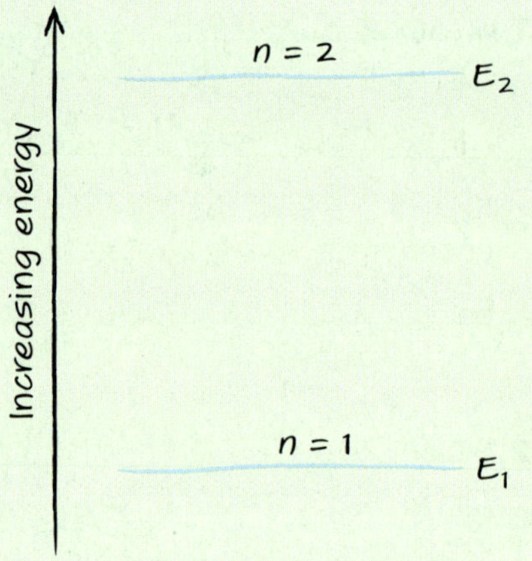

Known	**Unknown**
quantum number of innermost energy level, $n = 1$	energy of level $E_1 = ?$
quantum number of second energy level, $n = 2$	energy of level $E_2 = ?$
	energy difference, $\Delta E = ?$

2 SOLVE FOR THE UNKNOWN

Use the principal quantum number to calculate the energy of each level.

$$E_n = -13.6 \text{ eV} \times \frac{1}{n^2}$$

$$E_1 = -13.6 \text{ eV} \times \frac{1}{(1)^2} \quad \text{Substitute } n = 1.$$

$$= -13.6 \text{ eV}$$

$$E_2 = -13.6 \text{ eV} \times \frac{1}{(2)^2} \quad \text{Substitute } n = 2.$$

$$= -3.40 \text{ eV}$$

The energy absorbed by the atom, ΔE, is equal to the energy difference between the final energy level of the atom, E_f, and the initial energy level of the atom, E_i.

$$\Delta E = E_f - E_i$$

$$= E_2 - E_1 \quad \text{Substitute } E_f = E_2, E_i = E_1.$$

$$= -3.40 \text{ eV} - (-13.6 \text{ eV}) \quad \text{Substitute } E_2 = -3.40 \text{ eV}, E_1 = -13.6 \text{ eV}.$$

$$= 10.2 \text{ eV} \quad \text{Energy is absorbed.}$$

3 EVALUATE THE ANSWER

- **Are the units correct?** Orbital energy values should be measured in electron volts.
- **Is the sign correct?** The energy difference is positive when electrons move from lower energy levels to higher energy levels.
- **Is the magnitude realistic?** The energy needed to move an electron from the first energy level to the second energy level should be approximately 10 eV, which it is.

PRACTICE Problems

ADDITIONAL PRACTICE

33. Calculate the energies of the second, third, and fourth energy levels in the hydrogen atom.

34. Calculate the energy difference between E_3 and E_2 in the hydrogen atom.

35. Calculate the energy difference between E_4 and E_2 in the hydrogen atom.

36. CHALLENGE The text shows the solution of the equation $r_n = \frac{h^2 n^2}{4\pi^2 K m_e e^2}$ for $n = 1$, the innermost orbital radius of the hydrogen atom. Note that with the exception of n^2, all factors in the equation are constants. The value of r_1 is 5.3×10^{-11} m, or 0.053 nm. Use this information to calculate the radii of the second, third, and fourth energy levels in the hydrogen atom.

EXAMPLE Problem 5

FREQUENCY AND WAVELENGTH OF EMITTED PHOTONS An excited electron in a hydrogen atom drops from the second energy level ($n = 2$) to the first energy level ($n = 1$). Calculate the energy and the wavelength of the emitted photon. Use E_1 and E_2 values from Example Problem 4.

1 ANALYZE AND SKETCH THE PROBLEM

- Diagram the energy levels E_1 and E_2.
- Indicate the direction of increasing energy, and show the emission of a photon on the diagram.

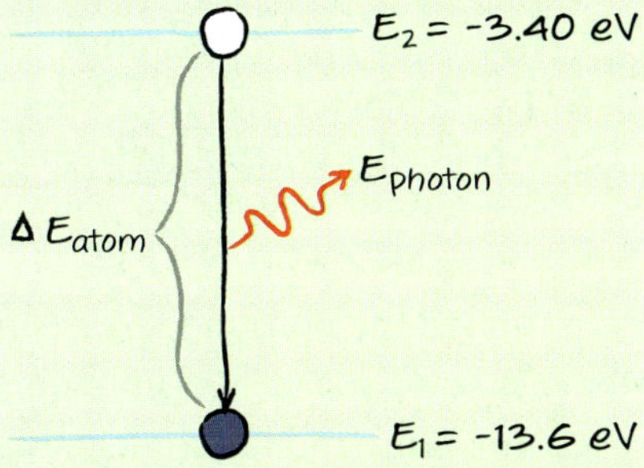

Known	Unknown
energy of level $E_1 = -13.6$ eV	frequency, $f = ?$
energy of level $E_2 = -3.40$ eV	wavelength, $\lambda = ?$
	energy difference, $\Delta E = ?$

2 SOLVE FOR THE UNKNOWN

The energy of the emitted photon is equal to $-\Delta E$, the energy decrease from the initial energy level (E_i) to the final energy level (E_f).

$-\Delta E = E_i - E_f$

$= E_2 - E_1$ — Substitute $E_f = E_1$, $E_i = E_2$.

$= -3.40 \text{ eV} - (-13.6 \text{ eV})$ — Substitute $E_1 = -13.6$ eV, $E_2 = -3.40$ eV.

$= 10.2$ eV Energy is emitted.

To determine the wavelength of the photon, use the relationships between energy, frequency, and wavelength.

$\Delta E = hf$, so $f = \frac{\Delta E}{h}$ — Solve the photon energy equation for frequency.

$c = \lambda f$, so $\lambda = \frac{c}{f}$ — Solve the wavelength-frequency equation for wavelength.

$\lambda = \frac{c}{(\Delta E/h)}$ — Substitute $f = \frac{\Delta E}{h}$.

$= \frac{hc}{\Delta E}$

$= \frac{1240 \text{ eV·nm}}{10.2 \text{ eV}}$ — Substitute $hc = 1240$ eV·nm, $\Delta E = 10.2$ eV.

$= 122$ nm

3 EVALUATE THE ANSWER

- **Are the units correct?** Energy is measured in electron volts. The prefix nano– modifies the base SI unit, the meter, which is the correct unit for the wavelength.
- **Are the signs correct?** Energy is released when the atom emits a photon during the transition from the second energy level to the first; thus, the energy difference is negative.
- **Is the magnitude realistic?** Energy released in this transition produces light in the ultraviolet region below 400 nm.

PRACTICE Problems

ADDITIONAL PRACTICE

37. Find the wavelength of the light emitted in Practice Problems 34 and 35. Which lines in **Figure 18** correspond to each transition?

38. For a particular transition, the energy of a mercury atom drops from 8.82 eV to 6.67 eV.

a. What is the energy of the photon emitted by the mercury atom?

b. What is the wavelength of the photon emitted by the mercury atom?

39. CHALLENGE The ground state of a helium ion is −54.4 eV. A transition to the ground state emits a 304-nm photon. What was the energy of the excited state?

PHYSICS Challenge

Although the Bohr atomic model accurately explained the behavior of a hydrogen atom, it was unable to explain the behavior of any other atom. Verify the limitations of the Bohr model by analyzing an electron transition in a neon atom. Unlike a hydrogen atom, a neon atom has ten electrons. One of these electrons makes a transition between the $n = 5$ and the $n = 3$ energy states, emitting a photon in the process.

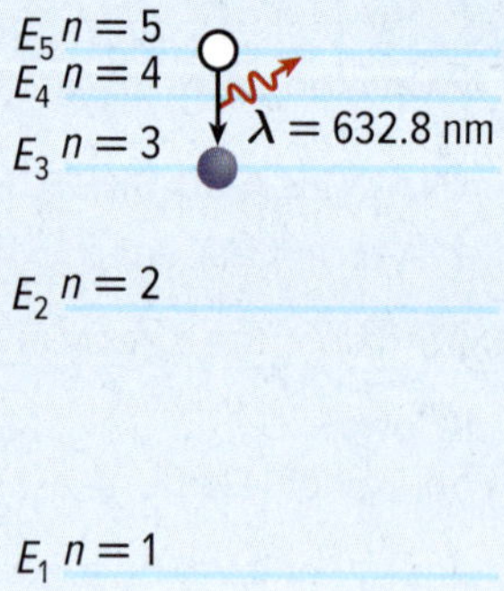

1. Assuming that the neon atom's electron can be treated as an electron in a hydrogen atom, what photon energy does the Bohr model predict?
2. Assuming that the neon atom's electron can be treated as an electron in a hydrogen atom, what photon wavelength does the Bohr model predict?
3. The actual wavelength of the photon emitted during the transition is 632.8 nm. What is the percent discrepancy of the Bohr model's prediction and the measured photon wavelength?

Even though the Bohr model does not correctly predict electron energies for multi-electron atoms, it was a major step in understanding atomic structure. Besides calculating the emission spectrum, Bohr and his students were able to calculate the ionization energy of a hydrogen atom. The ionization energy of an atom is the energy needed to completely free an electron from the atom. The calculated value closely agreed with experimental data. In addition to ionization energies, the Bohr model also provided an explanation of some chemical properties of the elements. Bohr's model led to the idea that each element has a unique electron arrangement, which is the foundation of much of our knowledge of chemical reactions and bonding. Niels Bohr received a Nobel prize in 1922 for his contributions to the understanding of the atom.

Check Your Progress

40. **Photon Absorption** Explain how energy is conserved when an atom absorbs a photon of light.
41. **Rutherford's Nuclear Model** Summarize the structure of the atom according to Rutherford's nuclear model.
42. **Absorption Spectrum** The absorption spectrum for a sample of hydrogen gas is shown in **Figure 24.** Explain how the absorption spectrum of a gas can be determined. Describe the reasons for the spectrum's appearance.

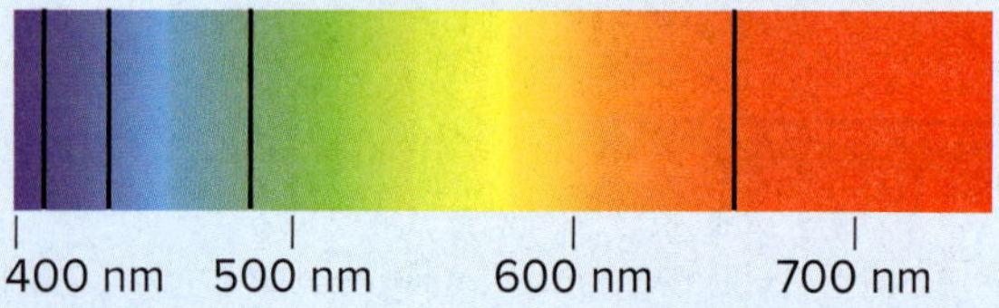

Figure 24

43. **Spectra** How do the emission spectra of incandescent solids and atomic gases differ? In what ways are they similar?
44. **Orbit Radius** A helium ion behaves like a hydrogen atom. The radius of the ion's lowest energy level is 0.0265 nm. According to Bohr's model, what is the radius of the second energy level?
45. **Bohr Model** Hydrogen has been detected transitioning from the 101st to the 100th energy levels. What is the radiation's wavelength? Where in the electromagnetic spectrum is this emission?
46. **Critical Thinking** The nucleus of a hydrogen atom has a radius of about 1.5×10^{-15} m. If you were to build a model of the hydrogen atom using a softball ($r = 5$ cm) to represent the nucleus, where would you locate an electron in the $n = 1$ Bohr orbit? Would it be in your classroom?

LEARNSMART® Go online to follow your personalized learning path to review, practice, and reinforce your understanding.

LESSON 4

THE QUANTUM MODEL OF THE ATOM

FOCUS QUESTION

How do lasers work?

From Orbits to an Electron Cloud

Bohr's model of quantized electron orbits could not correctly predict electron energies for atoms with more than one electron. Recall that de Broglie proposed that particles have wave properties. The de Broglie wavelength of a particle with momentum mv is $\lambda = \frac{h}{mv,}$ and the angular momentum is $mvr = \frac{hr}{\lambda}$. Thus, Bohr's required quantized angular-momentum condition, $mvr = \frac{nh}{2\pi}$, can be written $\frac{hr}{\lambda} = \frac{nh}{2\pi}$ or $2\pi r = n\lambda$.

Figure 25 illustrates this relationship that the circumference of the Bohr orbit ($2\pi r$) must be equal to a whole number multiple (n) of de Broglie wavelengths (λ).

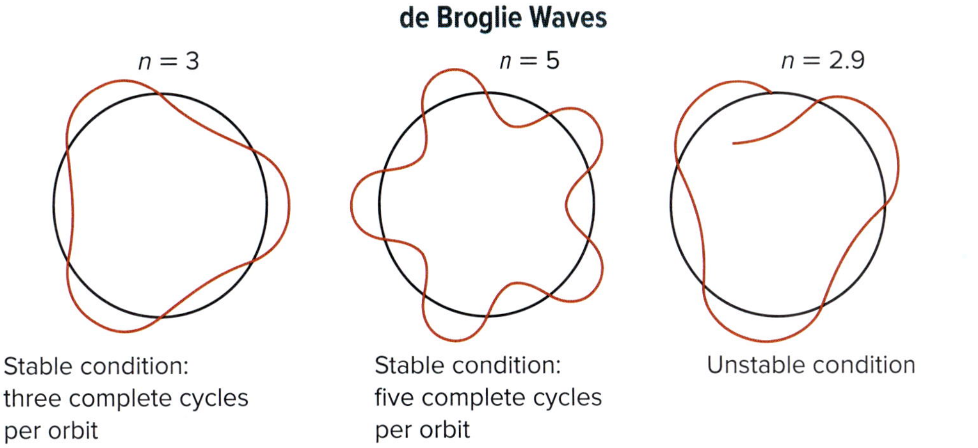

Figure 25 In the Bohr model of the atom, stable electron orbits around the nucleus must have a circumference equal to a whole-number multiple (n) of the de Broglie wavelengths.

3D THINKING DCI Disciplinary Core Ideas CCC Crosscutting Concepts SEP Science & Engineering Practices

COLLECT EVIDENCE

Use your Science Journal to record the evidence you collect as you complete the readings and activities in this lesson.

INVESTIGATE

GO ONLINE to find these activities and more resources.

Applying Practices: Communicate Information About Multiple Technologies

HS-PS4-5. Communicate technical information about how some technological devices use the principles of wave behavior and wave interactions with matter to transmit and capture information and energy.

In 1926, Austrian physicist Erwin Schrödinger extended de Broglie's wave model to create a quantum theory of particles, such as electrons, based on waves. However, the electron itself is not a wave. The wave describes the probability of finding an electron in a particular location or having a particular momentum or energy.

Uncertainty principle applied to atomic theory

Heisenberg proposed that it is impossible to measure exactly both the position and momentum of a particle at the same time. You can only know them approximately. Building on Heisenberg's ideas, the modern **quantum model** of the atom predicts only the probability that an electron is in a specific region. The quantum model predicts that the most probable distance between the electron and the nucleus for a hydrogen atom equals the radius predicted by the Bohr model.

The probability of the electron being at any specific radius can be calculated, and a three-dimensional plot will then show regions of equal probability. The region in which there is a high probability of finding the electron is called the **electron cloud.** **Figure 26** shows a slice through the electron cloud for the two lowest energy states of a hydrogen atom.

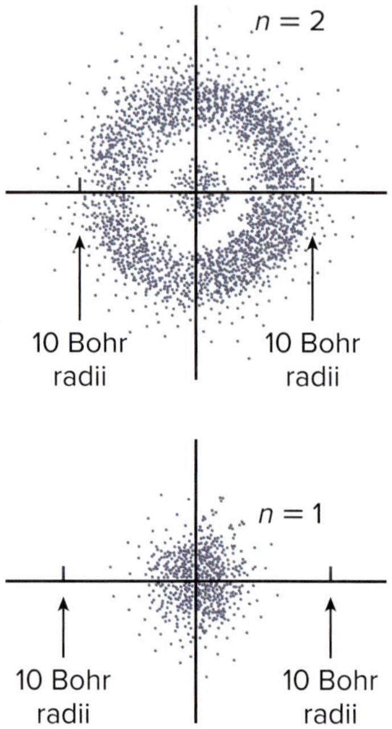

Figure 26 The density of dots in these plots corresponds to the probability of finding the electron in a hydrogen atom in the first and second energy levels.

Get It?
Explain the meaning of the electron cloud.

Even though the quantum model of the atom is difficult to visualize, it was an early success in the development of **quantum mechanics,** the study of the wave properties of matter. Quantum mechanics is a useful tool for physicists, chemists, biologists, and engineers. It has been used to develop new chemicals, new computer chips, new solar power cells, and new medicines. Quantum mechanics also helped analyze the details of atoms' emission and absorption of light, leading to the development of a new type of light source.

Lasers

Most of the light you see is a jumble of uncoordinated waves having many different wavelengths. It is like music from a choir in which every member is singing a different song. As you know, light emitted by an incandescent source such as a filament lightbulb has a continuum of wavelengths, whereas light produced by an atomic gas consists of only a few distinct wavelengths. Light from each of these sources travels in all directions. The waves are not necessarily all at the same point in their cycle at the same time. Such out-of-step light waves are called incoherent light. Waves that are in phase are said to be coherent. Light waves that are coherent are referred to as coherent light. Both types of light waves are shown in **Figure 27.**

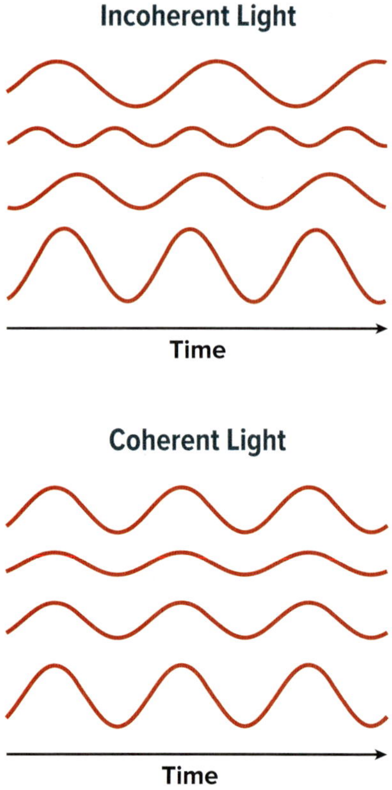

Figure 27 Coherent light is in phase when each wave is at the same point in its cycle at the same time.

Light is emitted by atoms that move from a higher energy state to a lower energy state. So far, you have learned two ways in which atoms can be excited to the higher energy state: thermal excitation and electron collision. In a third process, a photon with exactly the right energy can collide with an atom and excite it. In a laser this process can produce intense, coherent light.

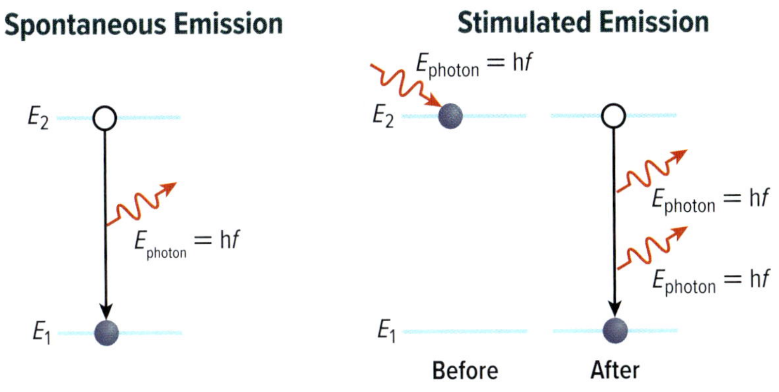

Figure 28 Spontaneous emission occurs when an electron in an atom drops from an excited state to the ground state, emitting a photon. During stimulated emission, an excited atom is struck by a photon with energy $E_2 - E_1$. The atom drops to the ground state and emits two photons. Both of these photons have the same energy: $E_{photon} = E_2 - E_1$.

Compare *the wavelengths of the incident and emitted photons.*

Spontaneous and stimulated emission

What happens when an atom is in an excited state? After a very short time, it normally returns to the ground state, giving off a photon of the same energy that it absorbed, as shown in the left panel of **Figure 28.** This process is called spontaneous emission.

In 1917, Einstein considered what happens to an atom already in an excited state that is struck by a photon with an energy equal to the energy difference between the excited state and the ground state. He showed that the atom, by a process known as **stimulated emission,** returns to the ground state and emits a photon with an energy equal to the energy difference between the two states. The photon that caused, or stimulated, the emission is not affected. The two photons leaving the atom not only will have the same frequency, but they will be in step, or coherent, as shown in the right panel of **Figure 28.** Either of the two photons can now strike other excited atoms, thereby producing additional photons that are in step with the original photons. This process can continue and produce an avalanche of photons, all of the same wavelength and all in phase.

Get It?

Compare the two photons emitted by stimulated emission.

For this process to occur, certain conditions must be met. First, there must be other atoms in the excited state. Second, the atoms must remain in the excited state long enough to be struck by a photon. Third, the photons must be contained so that they are able to strike other excited atoms. In 1960, a device called a **laser** was invented that fulfilled all the conditions needed to produce coherent light. The word laser is an acronym that stands for light amplification by stimulated emission of radiation. An atom that emits light when it is stimulated in a laser is said to lase.

Atom excitation

The atoms in a laser can be excited, or pumped. An intense flash of light with a wavelength shorter than that of the laser can be used to pump the atoms. The shorter-wavelength, higher-energy photons produced by the flash collide with and excite the lasing atoms. When one of the excited atoms decays to a lower energy state by emitting a photon, the avalanche of photons begins. This results in the emission of a brief flash, or pulse, of laser light. Alternatively, the lasing atoms can be excited by collisions with other atoms. In the helium-neon lasers often seen in science classrooms, an electric discharge excites the helium atoms. These excited helium atoms collide with the neon atoms, pumping them to an excited state and causing them to lase. The laser light resulting from this process is a continuous beam rather than pulsed flashes.

Lasing The lasing atoms and their emitted photons are contained within a chamber that has parallel mirrors at each end, one of which allows only about 1 percent of the light hitting it to pass through. Photons emitted toward the mirrors are reflected back into the gas where they strike more atoms that release more photons. Because all the stimulated photons are emitted in step with the photons that struck the atoms, laser light is coherent. The light may also be of one wavelength, or monochromatic. The parallel mirrors produce a highly directional, intense beam with a small diameter. The beam diverges very little and can be focused to a fine point. Finally, stimulated emission can be switched on and off to produce brief, high-power light pulses. These characteristics of laser light make it particularly well-suited to many different applications.

Laser Applications

If you have used a CD, DVD, or Blu-ray player, then you have used a laser. These lasers, as well as the ones used in laser pointers, are made of semiconducting solids. The laser in a CD player is made of layers of gallium arsenide (GaAs) and gallium aluminum arsenide (GaAlAs). The lasing layer is only 200-nm thick, and each side of the crystal is only 1–2 mm long. The atoms in the semiconducting solid are pumped by an electric current, and the resulting photons are amplified as they bounce between the polished ends of the crystal. Shorter wavelength laser beams, such as those used in the DVD player shown in **Figure 29,** can focus on smaller spots. This allows DVDs to store more data than CDs. Blu-ray players use even shorter wavelength lasers than DVD players so that they can store five times more data than DVDs. Because laser beams are narrow and highly directional, surveyors use them in applications such as checking the straightness of long tunnels and pipes. Lasers reflecting off mirrors left on the Moon are used to measure the distance from Earth to the Moon and to measure the movement of Earth's tectonic plates.

Figure 29 The laser used in this DVD player emits light with a wavelength of about 650 nm.

Get It?

Identify the characteristics of laser light that make it well-suited to use in surveying and in measuring large distances.

Because lasers can emit light bursts as brief as 10^{-15} s, they can be used to investigate very fast changes in the structures of atoms or molecules. Other scientists are trying to produce energy from high temperature plasmas using very large lasers that produce petawatt (10^{15} watts) pulses. **Table 2** on the next page summarizes several applications of lasers.

STEM CAREER Connection

Registered Nurse

Do you like helping people? Would you like to work with lasers? The connection between lasers and nursing might not be readily apparent, but lasers have many uses in improving the health and well-being of patients. Under a physician's supervision, registered nurses are able to use lasers to remove spider veins, unwanted hair, or an unwanted tattoo!

Richard Hutchings/Digital Light Source

Table 2 Laser Applications

	**Fiber-Optic Communications** Laser light is commonly used in fiber-optic communications. This technology is based on an understanding of waves and their interactions with matter. A fiber optic cable uses total internal reflection to transmit light over long distances with little loss of signal energy. The laser is rapidly switched on and off, transmitting information as a series of pulses through the fiber. All over the world, optical fibers have replaced copper wires for the transmission of telephone calls, computer data, and television signals.
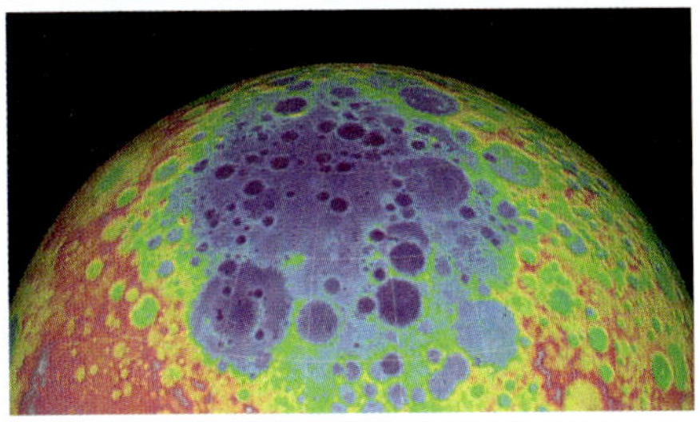	**Lunar Research** The Lunar Reconnaissance Orbiter pointed a laser at the surface of the Moon and recorded the time required for the laser's light to bounce off the Moon's surface and return to the satellite. The data collected during this mission produced topographical maps accurate to within 10 cm. In addition to providing clues about the Moon's history, these maps can be used to select safe landing sites for future robotic and manned missions to the Moon.

Laser Printers

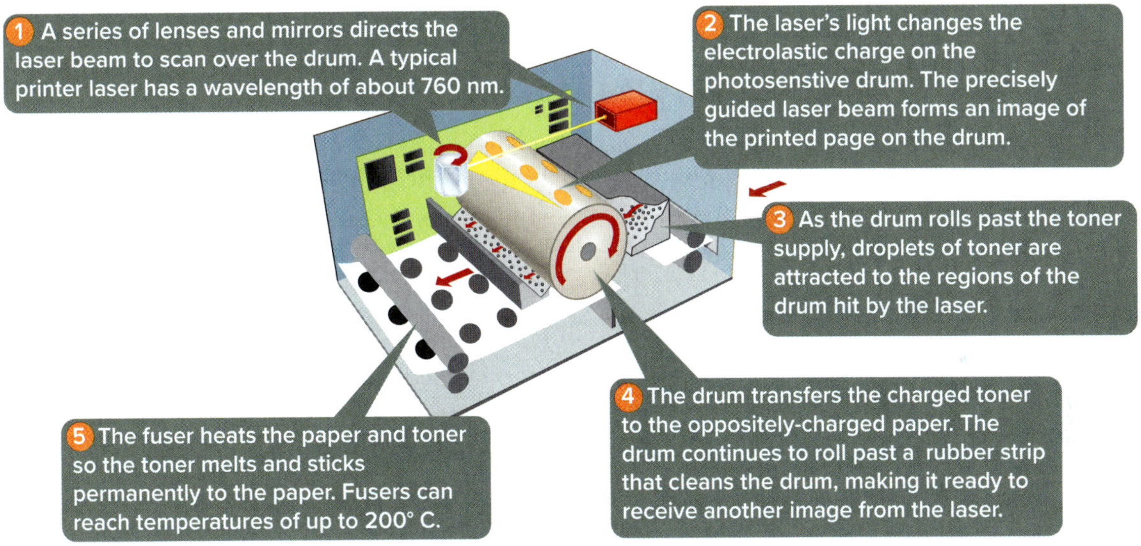

Check Your Progress

47. **Bohr Model Limitations** Although it was able to accurately predict the behavior of hydrogen, in what ways did Bohr's atomic model have serious shortcomings?
48. **Pumping Atoms** Explain whether green light could be used to pump a red laser. Why could red light not be used to pump a green laser?
49. **Laser Light** What are four characteristics of laser light that make it useful?
50. **Critical Thinking** Suppose an electron cloud were reduced to almost the size of the nucleus. Use the Heisenberg uncertainty principle to explain why this process would require adding a tremendous amount of energy.

LEARNSMART® Go online to follow your personalized learning path to review, practice, and reinforce your understanding.

Using Lasers to Detect Tooth Decay

Most American adults have had dental caries, also known as cavities. Dental disease is also common in children. Cavities are generally treated with dental fillings. The treatment can be painful and expensive, as well as frightening for many people. Some people avoid visiting their dentist for these reasons. However, untreated dental disease is also painful, and the pain gets progressively worse without treatment.

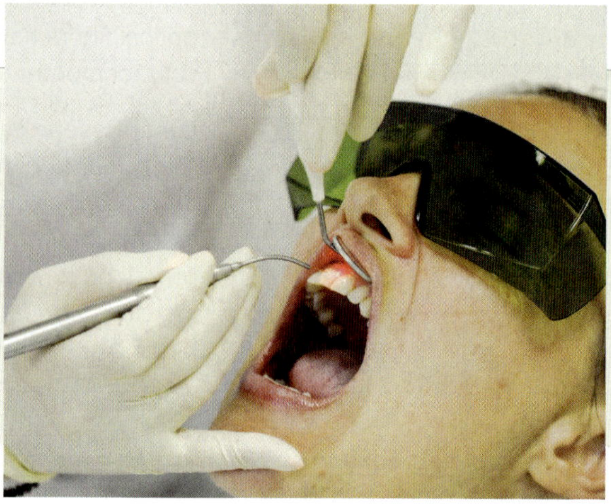

Laser technology can be used to detect cavities in teeth.

zlikovec/iStock/Getty Images

Dentists generally diagnose cavities using X-rays and by visual inspections. However, these methods generally detect only larger cavities, which have been progressing for some time, on easy-to-see surfaces of teeth. They often miss cavities on other surfaces and tooth decay that is hidden below the enamel, which is the hard layer on the outside of teeth.

Now, newer technology that uses thermal radiation and light waves is making detection of these cavities easier. The new technology uses a laser and a sensor. The new diagnostic tool measures the way the light from the laser reflects off a tooth. The tool sends heat-emitting infrared light to the tooth being examined. An infrared sensor in the tool measures the heat and light waves that are reflected, and it can identify the location of cavities.

Reversing Decay

Scientists are also studying lasers that could be used to reverse tooth decay. Experiments on rats show that lasers can be used to stimulate the formation of new dentin. Dentin is found in the core of teeth and is prone to decay. Other studies using cultured cells show that lasers can be used to stimulate a reaction that activates a protein that then activates stem cells to form dentin.

USE A MODEL TO DESCRIBE

Find out more about how laser technology is used to detect dental cavities. Summarize your findings in a visual model.

CALIFORNIA CONNECTION

Photonics involves the transmission and detection of photons (light). Biophotonics involves the study and manipulation of biological systems using photonics. Dental laser technology is an example of a biophotonic application. There are many research groups in California developing biophotonic applications. Research one of these applications. Share your findings with your classmates.

MODULE 22

STUDY GUIDE

GO ONLINE to study with your Science Notebook.

Lesson 1 A PARTICLE MODEL OF WAVES

- The spectrum emitted by an object depends on the object's temperature and covers a broad range of wavelengths. Planck explained an object's spectrum by supposing that a particle can absorb or emit only certain energies that are integer multiples of a constant, now called Planck's constant.
- The photoelectric effect is the emission of electrons by certain metals when they are exposed to electromagnetic radiation. Einstein explained the photoelectric effect by postulating that light exists in bundles of energy called photons.
- The Compton effect demonstrates that photons have momentum, as predicted by Einstein. Even though photons, which travel at the speed of light, have zero mass, they do have energy and momentum. A photon's momentum, frequency, and wavelength are related by the following equation:

$$p = \frac{hf}{c} = \frac{h}{\lambda}$$

- emission spectrum
- quantized
- photoelectric effect
- threshold frequency
- photon
- work function
- Compton effect

Lesson 2 MATTER WAVES

- The wave nature of material particles was suggested by de Broglie and verified experimentally by the diffraction of electrons through crystals.
- The particle and the wave aspects are complementary parts of the complete nature of both matter and light. The Heisenberg uncertainty principle states that it is not possible to simultaneously measure the precise position and momentum of any particle of light or matter.

- de Broglie wavelength
- Heisenberg uncertainty principle

Lesson 3 BOHR'S MODEL OF THE ATOM

- Ernest Rutherford showed that an atom is mostly empty space with a tiny, massive, positively charged nucleus at its center.
- Atoms make transitions between allowable energy levels, absorbing or emitting energy in the form of photons (electromagnetic waves). The photon's energy is equal to the decrease from the initial to the final state of the atom.
- According to the Bohr model, the radius of an electron's orbit can have only certain (quantized) values.
- Niels Bohr's model of the atom correctly showed that the energy of an atom can have only certain values; thus, it is quantized. He showed that the energy of a hydrogen atom in energy level n is equal to the product of −13.6 eV and the inverse of n^2.

- alpha particles
- nucleus
- absorption spectrum
- energy level
- ground state
- excited state
- principal quantum number

Lesson 4 THE QUANTUM MODEL OF THE ATOM

- In the quantum model of the atom, the atom's energy has only specific, quantized values. Only the probability of finding the electron in a specific region can be determined.
- Lasers produce light that is directional, powerful, monochromatic, and coherent. Each of these properties has useful applications.

- quantum model
- electron cloud
- quantum mechanics
- stimulated emission
- laser

THREE-DIMENSIONAL THINKING

Module Wrap-Up

REVISIT THE PHENOMENON

How could the manipulation of individual atoms be useful in future computing?

CER Claim, Evidence, Reasoning

Explain your Reasoning Revisit the claim you made when you encountered the phenomenon. Summarize the evidence you gathered from your investigations and research and finalize your Summary Table. Does your evidence support your claim? If not, revise your claim. Explain why your evidence supports your claim.

STEM UNIT PROJECT

Now that you've completed the module, revisit your STEM unit project. You will summarize your evidence and apply it to the project.

GO FURTHER

SEP Data Analysis Lab

Can you analyze mercury's spectrum?

The result of projecting the spectrum of a high-pressure mercury vapor lamp onto a wall in a dark room is shown in the figure.

Data and Observations

Energy Levels for Mecury	
n	Energy (eV)
1	−10.44
2	−5.77
3	−5.55
4	−4.98
5	−3.73
6	−2.71
7	−2.51
8	−1.90
9	−1.82
10	−1.80
11	−1.61
12	−1.60
13	−1.59
Ionization	0.00

CER Analyze and Interpret Data

1. **Calculate** the differences in energy levels for each of the three visible lines and identify the transitions that took place.
2. **Claim** After the emission of the visible photons, the mercury atom continues to emit photons until it reaches the ground state. Determine whether any of these photons would be visible. Explain.
3. **Evidence and Reasoning** Use the figure and any calculations you've made as evidence and explain your reasoning to justify your claim.

MODULE 23
SOLID-STATE ELECTRONICS

Andrew Brookes/Corbis

MODULE 23

SOLID STATE ELECTRONICS

ENCOUNTER THE PHENOMENON

What innovations allowed computers to shrink from filling a room to fitting in your pocket?

GO ONLINE to watch a video about how an important innovation that helped make computers smaller.

SEP Ask Questions

Do you have other questions about the phenomenon? If so, add them to the driving question board.

CER Claim, Evidence, Reasoning

Make Your Claim Use your CER chart to make a claim about the innovations that led to smaller computers. Explain your reasoning.

Collect Evidence Use the lessons in this module to collect evidence to support your claim. Record your evidence as you move through the module.

Explain Your Reasoning You will revisit your claim and explain your reasoning at the end of the module.

GO ONLINE to access your CER chart and explore resources that can help you collect evidence.

LESSON 1: Explore & Explain: Conductors and Insulators

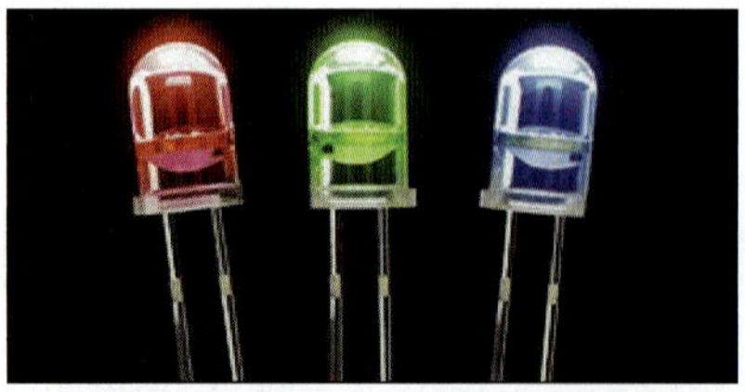

LESSON 2: Explore & Explain: Diodes

Smithsonian

Additional Resources

LESSON 1
CONDUCTION IN SOLIDS

FOCUS QUESTIONS

What is the difference between conductors, insulators, and semiconductors?

Band Theory of Solids

Electric charges can move easily in conductors but not in insulators. A third type of material, called a **semiconductor,** acts as a conductor in certain conditions and an insulator in others. At the atomic level, the difference in the way semiconductors are able to transmit charges becomes apparent.

Crystalline solids consist of atoms bound together in regular arrangements, and an atom consists of a dense, positively charged nucleus surrounded by a cloud of negatively charged electrons. Because the electrons can only occupy certain energy levels and have only certain energies, energy changes are quantized; that is, the energy changes occur in specific, or discrete, amounts.

Energy levels Under most conditions, an atom's electrons occupy the lowest possible energy level. This condition is called the ground state. Consider constructing a solid by bringing atoms together, one by one. You would start with an atom in the ground state. At large interatomic spacings (>0.8 nm), with no near neighbors, there are two discrete energy levels for the atom. As the solid crystal forms by moving atoms closer to the atom, the electric fields of these other neighboring atoms affect the energy levels of its electrons, as shown in **Figure 1.**

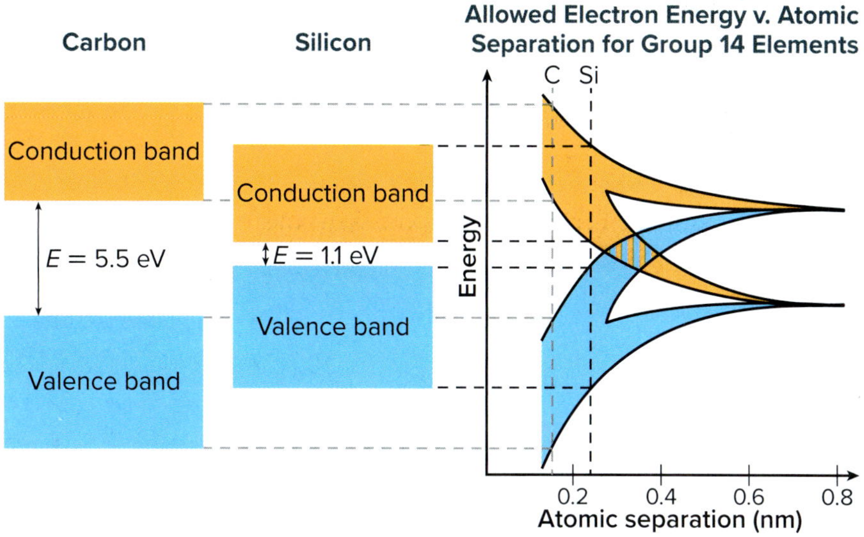

Figure 1 The electric fields of neighboring atoms split the energy levels of an atom. Thus, there is an energy gap between the valence and conduction bands.

Estimate *the energy gap of a material with an atomic separation of 0.2 nm.*

3D THINKING DCI Disciplinary Core Ideas | CCC Crosscutting Concepts | SEP Science & Engineering Practices

COLLECT EVIDENCE

Use your Science Journal to record the evidence you collect as you complete the readings and activities in this lesson.

INVESTIGATE

GO ONLINE to find these activities and more resources.

CCC **Identify Crosscutting Concepts**
Create a table of the crosscutting concepts and fill in examples you find as you read.

Review the News
Obtain information from a current news story about semiconductors or electronic components. Evaluate your source and communicate your findings to your class.

In the solid crystal, the electric fields of the neighboring atoms split the atom's ground-state energy levels into multiple levels. There are so many levels and they are so close together that they no longer appear as distinct levels. Instead, they appear as energy bands, shown in **Figure 1.**

The lower energy regions are called valence bands and are occupied by the bonding electrons in the crystal. The higher energy regions, called conduction bands, are available for electrons to move between atoms.

Get It?

Explain What is the difference between the valence band and the conducting band in a solid?

Energy gaps Notice in **Figure 1** that the atomic separations for crystalline silicon and crystalline carbon (diamond) translate to valence bands and conduction bands that are separated by energy gaps. These gaps have no energy levels available for electrons and are called forbidden energy regions.

This description of valence and conduction bands, separated by forbidden energy gaps, is known as the **band theory** of solids. Band theory can be used to better understand electric conduction.

Conductivity Consider a sample of silicon. At absolute zero, the valence band of silicon would be completely full and the conduction band would be completely empty. But at room temperature, some of the valence electrons have enough thermal energy to jump the 1.1-eV gap to the conduction band and serve as charge carriers. As the temperature increases and more electrons gain enough energy to jump the gap, the conductivity of silicon increases.

Conductivity also depends on the size of the energy gap. The smaller the energy gap, the less energy required to move electrons from the valence band to the conduction band. Thus, the smaller the band gap, the better conductor a material is, as shown in **Figure 2.**

For example, carbon in graphite form is a much better conductor than diamond because graphite has a smaller energy gap. Similarly, crystalline silicon has a smaller energy gap than diamond. It is a better conductor because less energy is required to move electrons to the conduction band.

Germanium has an even smaller energy gap of 0.7 eV. This means that germanium is a better conductor than silicon at any given temperature. However, it also means that germanium is too sensitive to temperature for many electronic applications. Relatively small changes in temperature cause large changes in the conductivity of germanium, making it more difficult to control and keep stable.

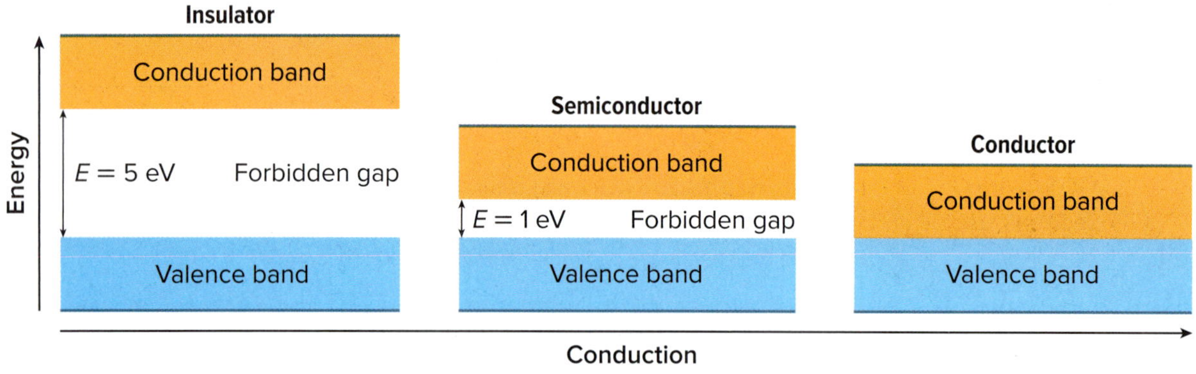

Figure 2 The conductivity of a material increases as the energy gap decreases.
Explain *why a small forbidden gap is more desirable than a large forbidden gap for a conductor.*

Get It?

Explain How would you describe the conductivity of the material at the beginning of this module?

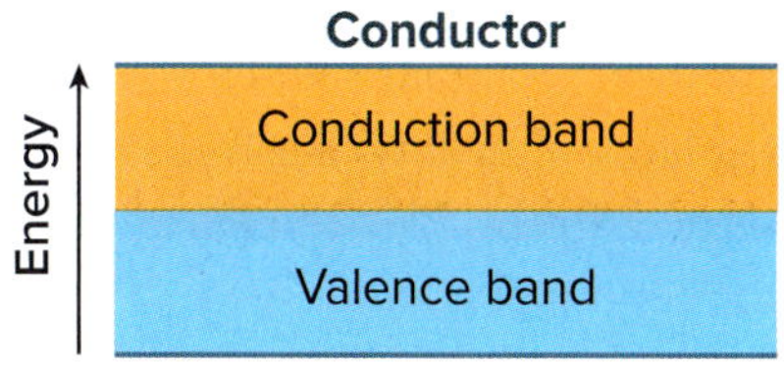

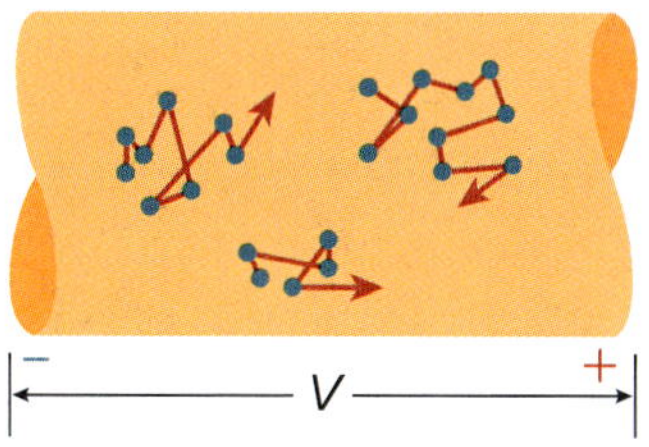

Figure 3 In a conductor, such as a wire, some electrons are in the conduction band. These electrons are free to move between atoms. The electrons move randomly, but if a field is applied across the wire, the electrons drift toward one end.

Conductors and Insulators

When a potential difference is placed across a material, the resulting electric field exerts a force on the electrons and does work on them. The electrons accelerate and gain energy. If there are bands within the material that are only partially filled, then there are energy levels available that are only slightly higher than the electrons' ground-state levels. As a result, the electrons that gain energy from the field can move from one atom to the next.

Such movement of electrons from one atom to the next is an electric current, and the entire process is known as electric conduction. The more easily the charges move, the more conductive the material is. Since conductivity is the reciprocal of resistivity, as a material's conductivity is reduced, its resistivity increases.

Conductors

Materials with partially filled bands, such as copper, conduct electricity easily. Lead has an interatomic spacing of 0.27 nm. The spacing in the graph in **Figure 1** would translate to a conduction band that overlaps the valence band—lead is a good conductor. Materials with overlapping, partially filled bands are conductors, as shown in **Figure 3.**

Free electrons in conductors move about rapidly in a random way and change directions when they collide with the cores of atoms. However, if an electric field is put across a length of wire, there will be a net force pushing the electrons in one direction.

Their motion is not greatly affected, but they have a slow overall movement dictated by the electric field, as shown in **Figure 3.** Electrons continue to move within the wire with speeds of 10^6 m/s in random directions. Although the random motion of the electrons is very fast, they might have drift at speeds of just 10^{-3} m/s or slower toward the positive end of the wire. This model of electron motion in conductors is called the electron-gas model. If the temperature of the conductor is increased, the speed of the electrons increases. The electrons collide more frequently with atomic cores, and the conductivity is reduced.

Insulators

In an insulating material, the valence band is filled and the conduction band is empty. As shown in **Figure 4,** an electron must gain a large amount of energy to go to the next energy level. In an insulator, the lowest energy level in the conduction band is 5 to 10 eV above the highest energy level in the valence band, as shown in **Figure 4.** There is at least a 5-eV gap of energies that no electrons can occupy.

Get It?

Explain why insulators do not conduct electric charges.

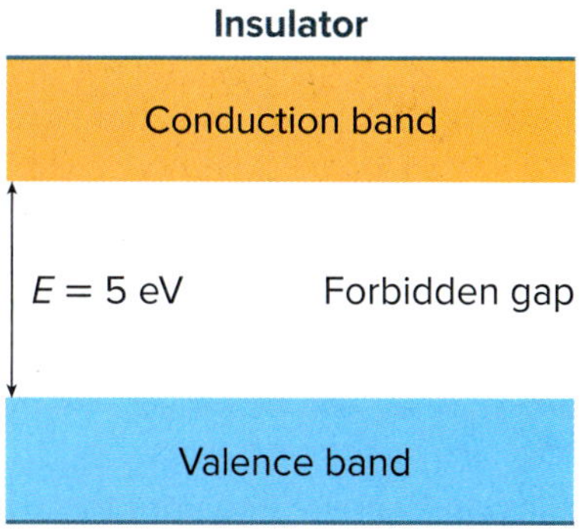

Figure 4 The energy gap in an insulator is large enough that few, if any, electrons have enough thermal energy to reach the conduction band.

Electrons in an insulator have some kinetic energy as a result of their thermal energy. However, the average kinetic energy of these electrons at room temperature is not sufficient for them to jump the forbidden gap. If a small electric field is placed across an insulator, almost no electrons gain enough energy to reach the conduction band. Therefore, there is no current. Electrons in an insulator must be given a large amount of energy to be pulled into the conduction band. As a result, the electrons in an insulator tend to remain in place, and the material does not conduct electricity. Insulators are often used to coat conducting wire to prevent circuits from shorting out.

EXAMPLE Problem 1

THE FREE-ELECTRON DENSITY OF A CONDUCTOR How many free electrons exist in a cubic centimeter of copper? Each atom contributes one electron. The density, atomic mass, and number of atoms per mole of copper can be found in the Reference Tables.

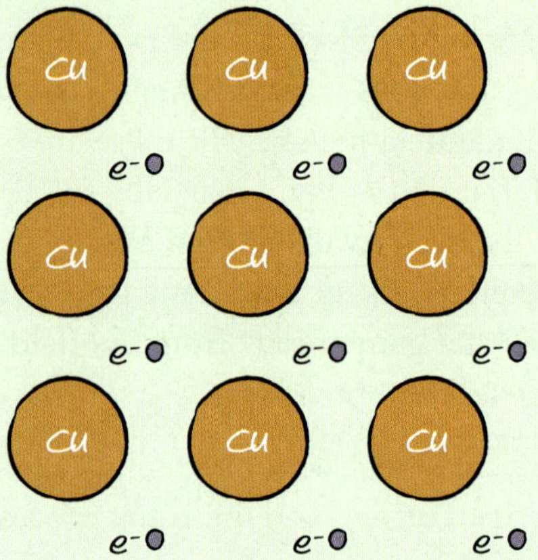

1 ANALYZE THE PROBLEM

Identify the knowns using the Reference Tables.

Known

For copper: 1 free e^- per atom

$\rho = 8.92\ \text{g/cm}^3$

$M = 63.55\ \text{g/mol}$

$N_A = 6.02\times10^{23}$ atoms/mol

Unknown

free $e^-/\text{cm}^3 = ?$

2 SOLVE FOR THE UNKNOWN

$$\frac{\text{free } e^-}{\text{cm}^3} = \frac{(\text{free } e^-)}{\text{atom}}(N_A)\left(\frac{1}{M}\right)(\rho)$$

$$= \left(\frac{1 \text{ free } e^-}{1 \text{ atom}}\right)\left(\frac{6.02\times10^{23} \text{ atoms}}{1 \text{ mol}}\right)\left(\frac{1 \text{ mol}}{63.55 \text{ g}}\right)\left(\frac{8.92 \text{ g}}{1 \text{ cm}^3}\right)$$

Substitute free e^-/atom = 1 free e^-/atom, $N_A = 6.02\times10^{23}$ atoms/mol, $M = 63.55$ g/mol, $\rho = 8.92$ g/cm^3.

$$= 8.45\times10^{22} \text{ free } e^-/\text{cm}^3 \text{ in copper}$$

3 EVALUATE THE ANSWER

- **Are the units correct?** Dimensional analysis on the units confirms the number of free electrons per cubic centimeter.
- **Is the magnitude realistic?** One would expect a large number of electrons in a cubic centimeter.

PRACTICE Problems

ADDITIONAL PRACTICE

1. Zinc, with a density of 7.14 g/cm^3 and an atomic mass of 65.38 g/mol, has two free electrons per atom. How many free electrons are there in each cubic centimeter of zinc?
2. Silver has one free electron per atom. Use the Reference Tables and determine the number of free electrons in 1 cm^3 of silver.
3. Gold has one free electron per atom. Use the Reference Tables and determine the number of free electrons in 1 cm^3 of gold.
4. Aluminum has three free electrons per atom. Use the Reference Tables and determine the number of free electrons in 1 cm^3 of aluminum.
5. **CHALLENGE** The tip of the Washington Monument was made of aluminum because it was a rare and costly metal in the 1800s. Use problem 4 and determine the number of free electrons in the tip of the Washington Monument. The amount of aluminum is 2835 g.

Semiconductors

So far, we have categorized materials as either conductors or insulators. But there are materials called semiconductors that do not strictly fall into either category. As discussed earlier, electrons move more freely in semiconductors than in insulators but not as easily as in conductors. The energy gap between the valence and conduction bands of a semiconductor is approximately 1 eV, as shown in **Figure 5.** A semiconductor's structure explains its conductive characteristics. Atoms of the most common semiconductors, silicon (Si) and germanium (Ge), have four valence electrons. These four electrons are involved in binding atoms together into a solid crystal. The valence electrons form a filled band, as in an insulator, but the forbidden gap is much smaller.

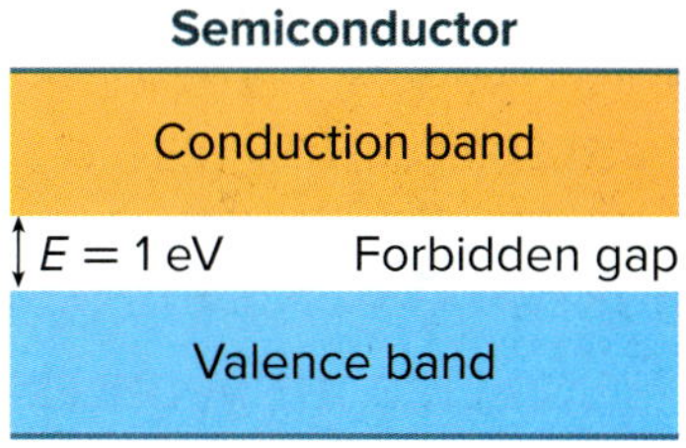

Figure 5 In a semiconductor, the energy gap is small enough that some electrons have enough thermal energy to enter the conduction band.

Holes and free electrons

Not much energy is needed to pull an electron from a silicon atom into the conduction band. The gap is so small that some electrons reach the conduction band as a result of their thermal kinetic energy alone. That is, the random motion of atoms and electrons gives some electrons enough kinetic energy to break free of their home atoms and move around the silicon crystal. Pure semiconductors that conduct as a result of thermally freed electrons are called **intrinsic semiconductors.** There are very few electrons available to carry charge, so intrinsic semiconductors have very low conductivity.

If an electric field is applied to a semiconductor, electrons in the conduction band move through the solid in response. In metals, an increase in temperature reduces the conductivity, but for semiconductors, a higher temperature means that electrons more easily reach the conduction band and the conductivity increases.

Get It?

Contrast How is the effect of temperature on conductivity different for metals than for semiconductors?

An atom from which an electron has broken free has a net positive charge and contains a hole. As shown in **Figure 6,** a hole is an empty energy level in the valence band. An electron from the conduction band can fill this hole and become bound to the atom. When a hole and a free electron recombine, their opposite charges neutralize each other. The electron, however, left behind a hole at its previous location, which can be filled by a different electron. Thus, negatively charged free electrons move in one direction and positively charged holes move in the opposite direction.

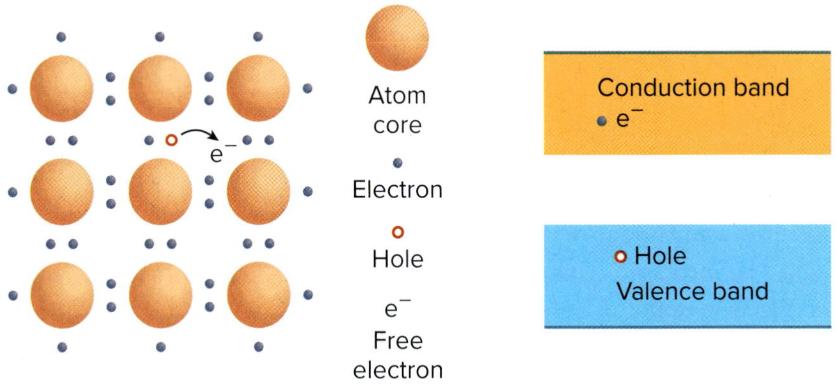

Figure 6 In a semiconductor, some electrons have enough thermal energy to break free and move about the crystal. There are electrons in the conduction band and holes in the valence band.

EXAMPLE Problem 2

FREE-ELECTRON DENSITY IN AN INTRINSIC SEMICONDUCTOR Because of the thermal kinetic energy of solid silicon at room temperature, there are 1.45×10^{10} free electrons/cm^3. What is the number of free electrons per atom of silicon at room temperature?

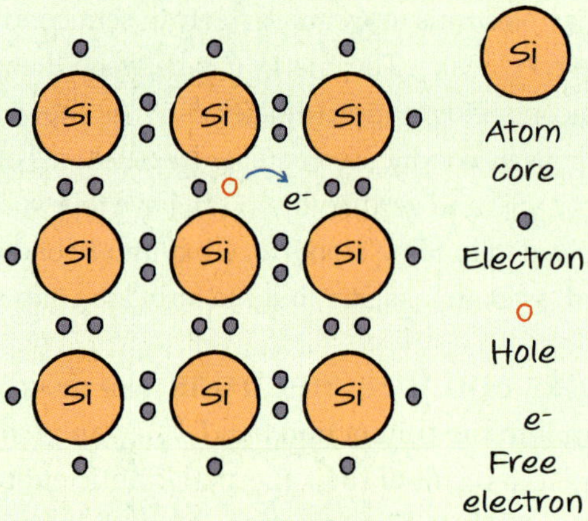

1 ANALYZE THE PROBLEM

Identify the knowns and unknowns.

Known

$\rho = 2.33$ g/cm^3

$M = 28.09$ g/mol

$N_A = 6.02\times10^{23}$ atoms/mol

For Si: 1.45×10^{10} free e$^-$/cm^3

Unknown

free e$^-$/atom of Si = ?

2 SOLVE FOR THE UNKNOWN

$$\frac{\text{free e}^-}{\text{atom}} = \left(\frac{1}{N_A}\right)(M)\left(\frac{1}{\rho}\right)(1.45\times10^{10}\text{ free e}^-/\text{cm}^3\text{ for Si})$$

$$= \left(\frac{1\text{ mol}}{6.02\times10^{23}\text{ atoms}}\right)\left(\frac{28.09\text{ g}}{1\text{ mol}}\right)\left(\frac{1\text{ cm}^3}{2.33\text{ g}}\right)\left(\frac{1.45\times10^{10}\text{ free e}^-}{\text{cm}^3}\right)$$

Substitute $N_A = 6.02\times10^{23}$ atoms/mol, $M = 28.09$ g/mol, $\rho = 2.33$ g/cm^3, free e$^-$/cm^3 Si $= 1.45\times10^{10}$ free e$^-$/cm^3.

$$= 2.90\times10^{-13}\text{ free e}^-/\text{atom of Si}$$

3 EVALUATE THE ANSWER

- **Are the units correct?** Using dimensional analysis confirms the correct units.
- **Is the magnitude realistic?** In an intrinsic semiconductor, such as silicon at room temperature, very few atoms have free electrons.

PRACTICE Problems

ADDITIONAL PRACTICE

6. In pure germanium, which has a density of 5.32 g/cm^3 and an atomic mass of 72.6 g/mol, there are 2.25×10^{13} free electrons/cm^3 at room temperature. How many free electrons are there per atom?
7. At 200.0 K, germanium has 1.16×10^{10} free electrons/cm^3. How many free electrons/atom are there at this temperature?
8. At 100.0 K, germanium has 3.47 free electrons/cm^3. How many free electrons are there per atom at this temperature?
9. A sample of germanium has 4.81×10^{11} free electrons/cm^3. How many free electrons per atom does the sample have?
10. At 100.0 K, silicon has 9.23×10^{-10} free electrons/cm^3. How many free electrons are there per atom at this temperature? What does this temperature represent on the Celsius scale?
11. At 200.0 K, silicon has 1.89×10^{5} free electrons/cm^3. How many free electrons are there per atom at this temperature? What does this temperature represent on the Celsius scale?
12. **CHALLENGE** The density of free electrons in a sample of silicon doubles when the temperature is raised 8.00°C. What is the density of free electrons in a silicon sample at 264 K? *(Hint: Use your answer from the previous question.)*

Doped Semiconductors

To make practical devices, the conductivity of intrinsic semiconductors must be increased greatly. That is, the number of holes and free electrons must be increased. This is done by adding atoms of other elements to the intrinsic semiconductor. Electron donor or acceptor atoms that can be added in low concentrations to intrinsic semiconductors are called **dopants.** The doped semiconductors are known as **extrinsic semiconductors.** There are two types of dopants used to increase the conductivity of semiconductors, *n*-type and *p*-type.

n-type semiconductors In some cases, an electron donor with five valence electrons, such as arsenic (As), is used as a dopant for silicon. The left side of **Figure 7** shows a silicon crystal where a dopant arsenic atom has replaced one of the silicon atoms. Four of the five arsenic valence electrons bind to neighboring silicon. The fifth electron is called the donor electron. The energy of this donor electron is close to the conduction band, as shown in **Figure 7.** Thus, thermal energy can easily move the electron from the dopant atom into the conduction band. Conduction in this kind of semiconductor is increased by the availability of these extra donor electrons to the conduction band. This type of semiconductor is known as an *n*-type semiconductor, since negative begins with *n*.

p-type semiconductors Alternatively, an electron acceptor with three valence electrons, such as gallium (Ga), can dope silicon. When a gallium atom replaces a silicon atom, one binding electron is missing. This creates a hole in the silicon crystal, as shown on the right side of **Figure 7.** Electrons in the conduction band can easily drop into these holes, creating new holes. Conduction in such semiconductors is enhanced by the availability of the extra holes, as shown in **Figure 7.** These kinds of semiconductors are called *p*-type, with the *p* standing for positive. Both *p*-type and *n*-type semiconductors are electrically neutral.

Adding dopant atoms of either type does not add any net charge to a semiconductor. Both types of semiconductor use electrons and holes in conduction. Adding only a few dopant atoms per million silicon atoms increases the semiconductor's conductivity by a factor of 1000 or more.

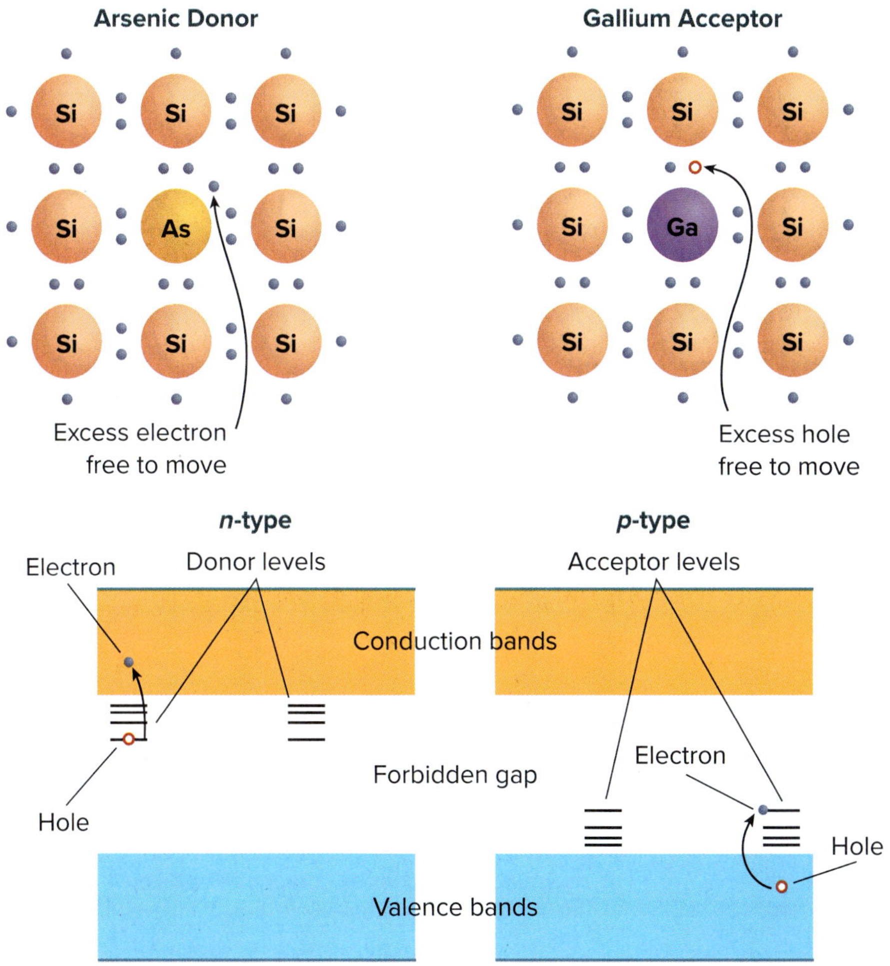

Figure 7 Arsenic, with five valence electrons, can replace a silicon atom to create an *n*-type semiconductor. Such a donor atom provides an unbound electron. Donor energy levels place electrons in the conduction band. Gallium, with three valence electrons, can replace a silicon atom to create a *p*-type semiconductor. Such an acceptor atom creates extra holes.

Infer *Antimony has five valence electrons. Would it be used to make an* n-type *or a* p-type *semiconductor?*

EXAMPLE Problem 3

THE CONDUCTIVITY OF DOPED SILICON A scientist dopes silicon with arsenic so that one in every one million silicon atoms is replaced by an arsenic atom. Each arsenic atom donates one electron to the conduction band.

a. What is the density of free electrons?

b. By what ratio is this density greater than that of intrinsic silicon with 1.45×10^{10} free e^-/cm^3?

c. Is conduction mainly by the electrons of the silicon or the arsenic?

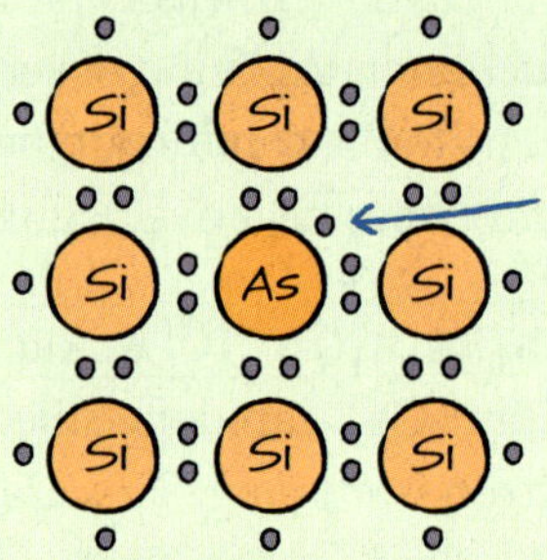

1 ANALYZE THE PROBLEM

Identify the knowns and unknowns.

Known	Unknown
1 As atom/10^6 Si atoms	free e^-/cm^3 donated by As = ?
1 free e^-/As atom	ratio of As-donated free e^- to intrinsic free e^- = ?
4.99×10^{22} Si atoms/cm^3	
1.45×10^{10} free e^-/cm^3 in intrinsic Si	

2 SOLVE FOR THE UNKNOWN

a. $$\left(\frac{\text{free } e^-}{\text{cm}^3}\text{ from As}\right) = \left(\frac{\text{free } e^-}{\text{As atom}}\right)\left(\frac{\text{As atoms}}{\text{Si atoms}}\right)\left(\frac{\text{Si atoms}}{\text{cm}^3}\right)$$

$$\left(\frac{\text{free } e^-}{\text{cm}^3}\right) = \left(\frac{1 \text{ free } e^-}{1 \text{ As atom}}\right)\left(\frac{1 \text{ As atom}}{1\times10^6 \text{ Si atoms}}\right)\left(\frac{4.99\times10^{22} \text{ Si atoms}}{\text{cm}^3}\right)$$

Substitute free e^-/As atom = 1 free e^-/1 As atom, As atoms/Si atoms = 1 As atom/1×10^6 Si atoms, Si atoms/cm^3 = 4.99×10^{22} Si atoms/cm^3.

$= 4.99\times10^{16}$ free e^-/cm^3 from As donor in doped Si

b. $$\text{Ratio} = \left(\frac{\text{free } e^-/\text{cm}^3 \text{ in doped Si}}{\text{free } e^-/\text{cm}^3 \text{ in intrinsic Si}}\right)$$

$$= \left(\frac{4.99\times10^{16} \text{ free } e^-/\text{cm}^3 \text{ in doped Si}}{1.45\times10^{10} \text{ free } e^-/\text{cm}^3 \text{ in intrinsic Si}}\right)$$

Substitute 4.99×10^{16} free e^-/cm^3 in doped Si, 1.45×10^{10} free e^-/cm^3 in intrinsic Si.

$= 3.44\times10^6$ As-donated electrons per intrinsic Si electron

c. Because there are over 3 million arsenic-donated electrons for every intrinsic electron, conduction is mainly by the arsenic-donated electrons.

3 EVALUATE THE ANSWER

- **Are the units correct?** Using dimensional analysis confirms the correct units.
- **Is the magnitude realistic?** The ratio is large enough that intrinsic electrons make almost no contribution to conductivity.

PRACTICE Problems

ADDITIONAL PRACTICE

13. If you wanted to have 1×10^4 as many electrons from arsenic doping as thermally free electrons in silicon at room temperature, how many arsenic atoms should there be per silicon atom?

14. Germanium has 5.19×10^{-9} free electrons per atom. If you wanted to have 5×10^3 as many electrons from arsenic doping as thermally free electrons in a germanium semiconductor, how many arsenic atoms should there be per germanium atom?

15. Silicon at 400.0 K has 4.54×10^{12} thermally liberated carriers/cm^3. If it is doped with 1 arsenic atom per one million silicon atoms, what is the ratio of doped carriers to thermal carriers?

16. CHALLENGE Based on the previous problem, draw a conclusion about the behavior of germanium devices as compared to silicon devices at temperatures in excess of the boiling point of water.

Making and using semiconductors Silicon is doped by putting a pure silicon crystal in a vacuum with a sample of the dopant material. The dopant is heated until it is vaporized, and the atoms condense on the cold silicon. The dopant diffuses into the silicon as it warms. Then, a thin layer of aluminum or gold is evaporated onto the doped crystal. A wire is welded to this metal layer, allowing the user to apply a potential difference across the doped silicon.

Figure 8 Light meters are made with semiconductors that are sensitive to specific wavelengths of light. Photographers use light meters to assess light levels before taking a picture.

Thermistors The electric conductivity of intrinsic and extrinsic semiconductors is sensitive to both temperature and light. In metals, conductivity is reduced when the temperature rises. An increase in the temperature of a semiconductor, however, allows more electrons to reach the conduction band. Thus, conductivity increases and resistance decreases.

One semiconductor device, the thermistor, is designed so that its resistance depends very strongly on temperature. The thermistor can be used as a sensitive thermometer and to compensate for temperature variations of other components in an electric circuit. Thermistors also can be used to detect radio waves and other forms of radiation.

Get It?

Describe the relationship between temperature and conductivity in a thermistor.

Light meters Other useful applications of semiconductors depend on their light sensitivity. When light falls on a semiconductor, the light can excite electrons from the valence band to the conduction band in the same way that other energy sources excite atoms. Thus, the resistance decreases as the light intensity increases.

Extrinsic semiconductors can be tailored to respond to specific wavelengths of light. These include the infrared and visible regions of the spectrum. Materials such as silicon and cadmium sulfide serve as light-dependent resistors in light meters used by lighting engineers to design the illumination of stores, offices, and homes. They are also used by photographers to adjust their cameras to capture the best images, as shown in **Figure 8.**

Check Your Progress

17. **Semiconductor?** Magnesium oxide has a forbidden gap of 8 eV. Is this material a conductor, an insulator, or a semiconductor?
18. **Carrier Mobility** Are electrons most likely to remain with the same atom in a conductor, a semiconductor, or an insulator?
19. **Insulator or Conductor?** Silicon dioxide is widely used in the manufacture of solid-state devices. Its energy-band diagram shows a gap of 9 eV between the valence band and the conduction band. Is it more useful as an insulator or a conductor?
20. **Semiconductors** You are designing an integrated circuit with a crystal of silicon. You want to have a region with good insulating properties. Should you dope this region or leave it as an intrinsic semiconductor? Explain.
21. **Critical Thinking** Silicon produces a doubling of thermally liberated carriers for every 8°C increase in temperature, and germanium produces a doubling of thermally liberated carriers for every 13°C increase. Germanium seems superior for high-temperature applications, but the opposite is true. Explain.

LEARNSMART® Go online to follow your personalized learning path to review, practice, and reinforce your understanding.

LESSON 2

ELECTRONIC COMPONENTS

FOCUS QUESTIONS

How are semiconductors manipulated to make electronic components?

The Birth of Solid-State Electronics

All electronic devices owe their origins to the vacuum tubes of the early 1900s. In vacuum tubes, electron beams flow through space to amplify and control faint electric signals. Vacuum tubes are big, have a high power input, and generate considerable thermal energy. They have heated filaments, which require the replacement of the tubes after one to five years.

In the late 1940s, solid-state devices made of semiconducting materials that could do the jobs of vacuum tubes were invented. These devices amplify and control very weak electric signals through the movement of electrons within a tiny crystalline space. Very few electrons flow in semiconductor devices, and these devices have no filaments. Thus, they operate with a low power input. They are very small, generate little heat, and are inexpensive to manufacture. Their estimated life is 20 years or more.

A decade later, these devices were combined with other electrical components to make the first integrated circuit, shown in **Figure 9.** Today's electronic instruments, such as smart phones, televisions, videodisc players, and laptops, rely on semiconductor devices that are combined on chips of silicon a few millimeters wide. In these devices, current and potential difference vary in ways more complex than Ohm's law. Semiconductor devices can convert current from AC to DC and amplify potential differences.

Figure 9 The first integrated circuit was built in 1958. The strip of germanium in the center of the circuit is about 1 cm long.

Photo courtesy of Texas Instruments

3D THINKING DCI Disciplinary Core Ideas CCC Crosscutting Concepts SEP Science & Engineering Practices

COLLECT EVIDENCE

Use your Science Journal to record the evidence you collect as you complete the readings and activities in this lesson.

INVESTIGATE

GO ONLINE to find these activities and more resources.

PhysicsLAB: Diode Current and Potential Difference

Analyze data to determine how the current through a diode or LED depends on the potential difference applied.

Revisit the Encounter the Phenomenon Question

What information from this lesson can help you answer the Unit and Module questions?

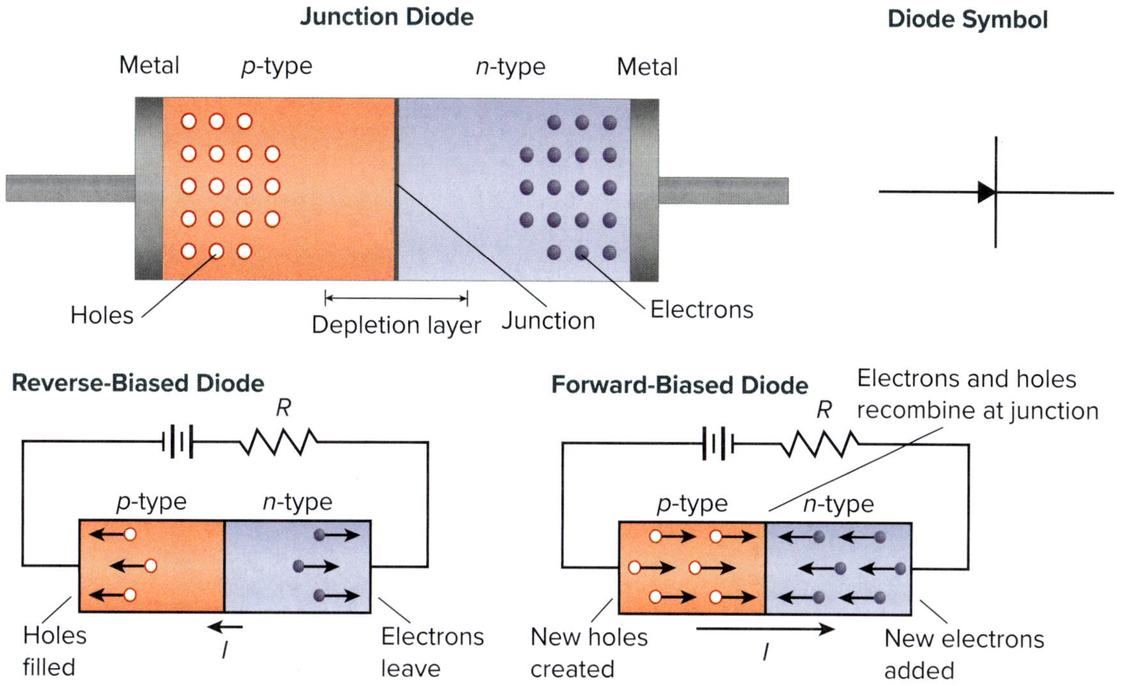

Figure 10 A *pn*-junction diode contains a *p*-type layer and an *n*-type layer. A reverse-biased diode acts as a high-value resistor, and a forward-biased diode acts as a low-value resistor.

Diodes

The simplest semiconductor device is the diode. A **diode** consists of a sandwich of *p*-type and *n*-type semiconductors, as shown in **Figure 10.** To make a diode, a single sample of silicon is treated first with a *p*-dopant and then with an *n*-dopant. The boundary between the *p*-side and the *n*-side is called the junction, so this diode is a *pn*-junction diode. Metal contacts on each end allow wires to be attached.

The free electrons on the *n*-side of the junction are attracted to the positive holes on the *p*-side. The electrons move easily into the *p*-side and recombine with the holes. Similarly, holes from the *p*-side move into the *n*-side and recombine with electrons. As a result of this flow, the *n*-side has a net positive charge, and the *p*-side has a net negative charge.

These charges produce forces in the opposite direction that stop further movement of charge carriers. The region around the junction is left with neither holes nor free electrons and is called the **depletion layer.** It has no charge carriers, so it is a poor conductor of electricity. Thus, a junction diode consists of two relatively good conductors that surround a poor conductor.

Forward and reverse bias

When a diode is connected into a circuit in the way shown on the bottom left side of **Figure 10,** both the free electrons in the *n*-type semiconductor and the holes in the *p*-type semiconductor are attracted toward the battery. The width of the depletion layer is increased, and no charge carriers meet. Almost no current passes through the diode. It acts like a very high-value resistor; it is almost an insulator. A diode oriented in this manner is called a reverse-biased diode.

If the battery is connected in the opposite direction, as shown on the bottom right side of **Figure 10,** charge carriers move toward the junction. If the battery's potential difference is large enough—0.6 V for a silicon diode—electrons reach the *p*-end and fill the holes. The depletion layer is eliminated, and a current passes through the diode. The battery continues to supply electrons for the *n*-end. It removes electrons from the *p*-end, which is the same as supplying holes. A diode in this kind of circuit is a forward-biased diode.

The arrow in the diode symbol indicates the direction that the diode is facing. That is, the arrow indicates the direction that conventional current will be allowed through the diode.

Current and potential difference The graph in **Figure 11** shows the current through a silicon diode as a function of potential difference across it. If the applied potential difference is negative, the diode is reverse-biased and acts like a very high-value resistor. Only a tiny current passes, about 10^{-11} A for a silicon diode.

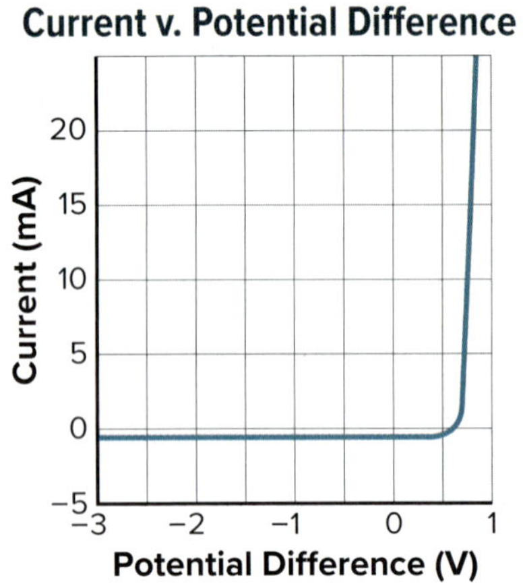

Figure 11 For potential differences below about 0.6 V, the diode is reverse-biased, and there is no current. Above the threshold, the diode is forward-biased, and there is a current.

If the potential difference is positive, the diode is forward-biased and acts like a low-value resistor. Increasing the battery's potential difference increases the current in the diode. The diode does not, however, obey Ohm's law. The potential difference across a typical diode is 0.7 V.

One major use of a diode is to convert alternating current (AC) to direct current (DC). That is, it acts like a one-way door. When a diode is used in such a way, it is called a rectifier.

EXAMPLE Problem 4

A DIODE IN A SIMPLE CIRCUIT A 0.70-V silicon diode, with $\frac{I}{\Delta V}$ characteristics like those shown in **Figure 11,** is connected to a power supply through a 470-Ω resistor. The power supply forward-biases the diode, and its potential difference is adjusted until the diode current is 12 mA. What potential difference does the power supply produce?

1 ANALYZE AND SKETCH THE PROBLEM

Draw a circuit diagram connecting a diode, a 470-Ω resistor, and a power supply. Indicate the direction of current.

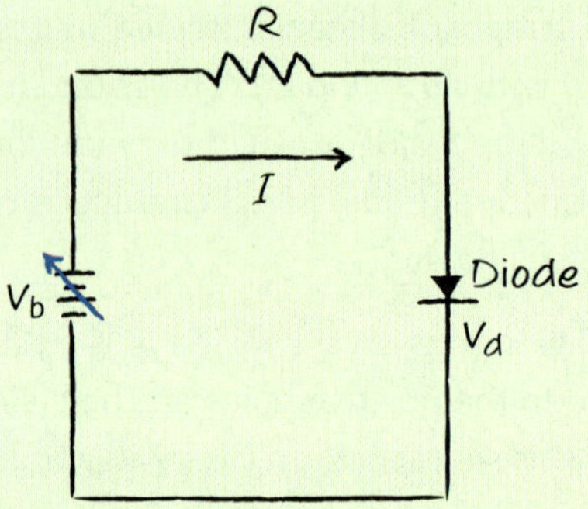

Known	Unknown
$I = 0.012$ A	$\Delta V_b = ?$
$\Delta V_d = 0.70$ V	
$R = 470\ \Omega$	

2 SOLVE FOR THE UNKNOWN

The potential difference across the resistor is known from $\Delta V = IR$. By Kirchhoff's Law, the power supply's potential difference is the sum of the potential differences across the resistor and the diode.

$$\Delta V_b = IR + \Delta V_d$$

$$= (0.012\ \text{A})(470\ \Omega) + 0.70\ \text{V} = 6.3\ \text{V}$$

Substitute $I = 0.012$ A, $R = 470\ \Omega$, $\Delta V_d = 0.70$ V.

3 EVALUATE THE ANSWER

- **Are the units correct?** The power supply's potential difference is in volts.
- **Is the magnitude realistic?** The magnitude is in accord with the current and the resistance.

PRACTICE Problems

ADDITIONAL PRACTICE

Unless otherwise noted, ΔV_{diode} = 0.70 V.

22. What battery potential difference would be needed to produce a current of 2.5 mA in the diode in Example Problem 4?

23. A 0.40-V germanium diode is wired in series with a 470-Ω resistor. What potential difference would a battery need to have in order to produce a diode current of 12 mA?

24. What battery potential difference is needed to produce a current of 2.5 mA if two identical diodes are wired in series with a 470-Ω resistor?

25. Describe how the diodes in the previous problem should be connected.

26. **CHALLENGE** Describe what would happen in the previous two problems if the diodes were connected in series but with improper polarity.

PHYSICS Challenge

Diode Characteristics We often use approximations in diode circuits because diode resistance is not constant. For diode circuits, the first approximation ignores the forward potential difference across the diode. The second approximation takes into account a typical value for the potential difference across the diode. A third approximation uses additional information about the diode, often in the form of a graph, as shown on the right. This is the characteristic current-potential difference curve for the diode. The straight line shows current-potential difference conditions for all possible diode potential differences for a 180-Ω resistor, a 1.8-V battery, and a diode, from a zero diode potential difference and 10.0 mA at one end to a 1.8-V potential difference, 0.0 mA at the other end. Use the diode circuit in Example Problem 4 with ΔV_b = 1.8 V, but with R = 180 Ω.

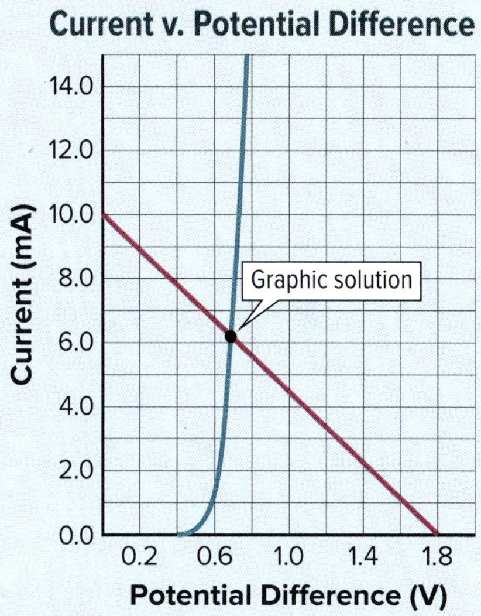

1. Determine the diode current using the first approximation.
2. Determine the diode current using the second approximation and assuming a 0.70-V diode drop.
3. Determine the diode current using the third approximation by using the accompanying diode graph.
4. Estimate the error for all three approximations, ignoring the battery and the resistor. Discuss the impact of greater battery potential differences on the errors.

Light-emitting diodes

Diodes that are made from certain combinations of elements, such as gallium and aluminum with arsenic and phosphorus, emit light when they are forward-biased. When electrons reach the holes in the junction, they recombine and release excess energy as radiant energy. These diodes are called light-emitting diodes (LEDs). Basic LEDs are used as indicator lights on computers, to make display signs, in traffic lights, and in some lamps. LEDs are bright, do not generate much heat, and are extremely efficient.

Figure 12 The diode laser in a bar-code scanner emits light and detects light reflected from the bar code.

Diodes can detect light as well as emit it. Light falling on the junction of a reverse-biased *pn*-junction diode produces electrons and holes, resulting in a current that depends on the light intensity.

Diode laser

Some LEDs are configured to emit a beam of coherent, monochromatic laser light. Such diode lasers are inexpensive, compact, powerful light sources. They are used in optical disk players, laser pointers, and bar-code scanners, like the one shown in **Figure 12.** A typical diode laser emits light at about 800 nm, which is the near infrared. The beam is output from a small spot on a gallium-aluminum-arsenic chip. When powered by 80 mA, the diode has a forward potential difference of about 2 V.

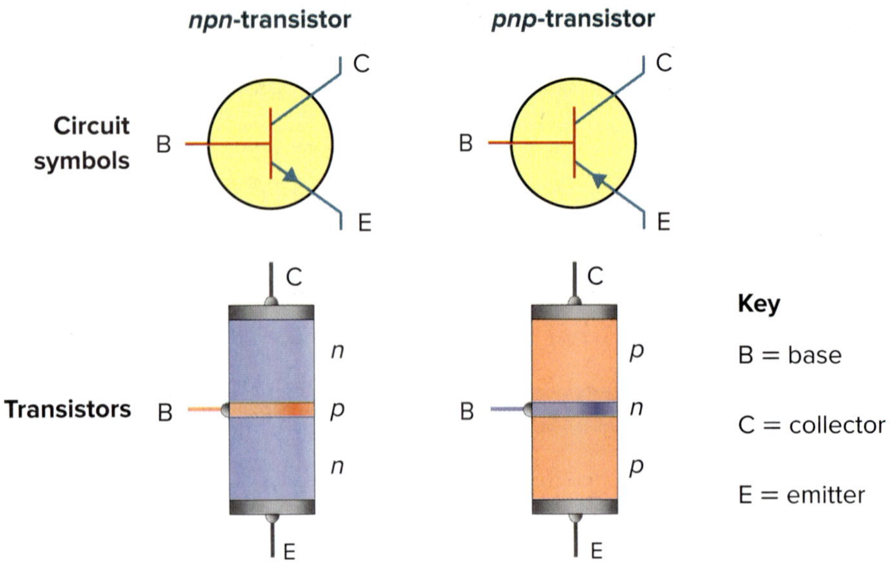

Figure 13 Transistors are made of three layers of semiconductor materials. Both *npn*-transistors and *pnp*-transistors have a collector and an emitter sandwiched around the base layer.

State *what the arrow on the emitter means.*

Transistors

A **transistor** is a device made of doped semiconductor material that can be used to amplify currents. One type is the *npn*-transistor, which consists of layers of *n*-type semiconductor on either side of a thin *p*-type layer, as shown in **Figure 13.** The *pnp*-transistor is similar. The central layer is called the base, and the regions on either side are the emitter and the collector. The schematic symbols for the two transistor types are also shown **Figure 13.** The arrow on the emitter shows the direction of conventional current.

Amplification An *npn*-transistor can be used to amplify current, as illustrated in **Figure 14.** You can think of the two *pn*-junctions in the transistor as two back-to-back diodes. If only the right battery (V_C) is connected, it keeps the collector more positive than the emitter. The base-collector diode is reverse-biased, with a wide depletion layer, so there is no current from the collector to the base. When the battery on the left (V_B) is connected, the base is more positive than the emitter. That makes the base-emitter diode forward-biased, allowing current (I_B) from the base to the emitter.

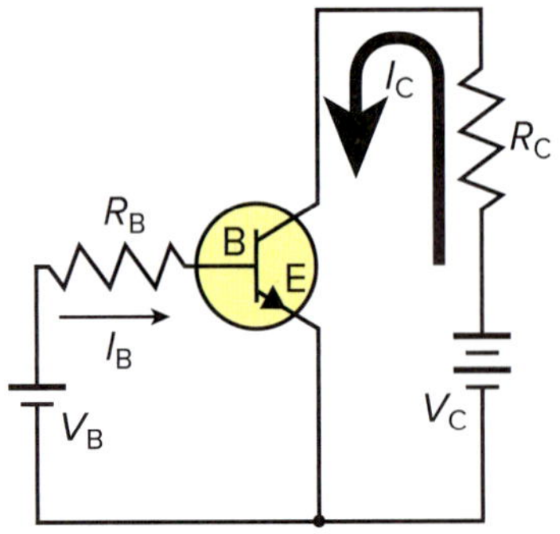

Figure 14 Transistors can be used to amplify currents. This circuit uses an *npn*-transistor.

Draw *a circuit using a pnp-transistor.*

The very thin base region is part of both diodes in the transistor. The charges injected by I_B reduce the reverse bias of the base-collector diode, permitting charge to flow from the collector to the emitter. Thus a small change in I_B produces a large change in I_C.

The collector current causes a potential difference across resistor R_C. If V_B increases, the base current (I_B) increases. The collector current (I_C) will also increase, but many times more (perhaps 100 times or so). Small changes in the base potential (V_B) produce large changes in the collector current. The transistor amplifies small potential differences into much larger ones.

The ratio of the collector-emitter current to the base current $\left(\frac{I_C}{I_B}\right)$ is called the current gain. The current gain is a useful indicator of a transistor's performance. The current gain ranges from 50 to 300 for general-purpose transistors. A *pnp*-transistor works the same way, except that the potentials of both batteries are reversed.

Transistors as switches In computers, small currents in base-emitter circuits can turn on or turn off large currents in collector-emitter circuits. In addition, several transistors can be connected together to perform logic operations or to add numbers together. In these cases, they act as fast switches rather than as amplifiers.

Integrated Circuits

An integrated circuit, called a **microchip,** consists of thousands of transistors, diodes, resistors, and conductors. Each component of a microchip is less than a micrometer across. All these components can be made by doping silicon with donor or acceptor atoms. A microchip begins as an extremely pure single crystal of silicon that is 10 to 30 cm in diameter. The crystal, which may be 1 to 2 m long, is sliced by a diamond-coated saw into wafers less than 1 mm thick, as shown in the left panel of **Figure 15.** The circuit is then built layer by layer on the surface of this wafer.

By a photographic process, most of the wafer's surface is covered by a protective layer, with a pattern of selected areas left uncovered so that they can be doped appropriately. The wafer is then placed in a vacuum chamber. Vapors of a dopant such as arsenic enter the machine, doping the wafer in the unprotected regions. By controlling the amount of exposure, the engineer can control the conductivity of the exposed regions of the chip.

This process creates resistors, as well as one of the two layers of a diode or one of the three layers of a transistor. The protective layer is removed, and another one with a different pattern of exposed areas is applied. Then the wafer is exposed to another dopant, often gallium, producing *pn*-junctions. If a third layer is added, *npn*-transistors can be formed. The wafer also might be exposed to oxygen to produce areas of silicon dioxide insulation. A layer exposed to aluminum vapors can produce a pattern of thin conducting pathways among the resistors, diodes, and transistors. The image on the right side of **Figure 15** shows a wafer after it has completed this process.

Get It?

Explain Why is aluminum, instead of gallium or arsenic, used to make conducting pathways on the silicon wafer?

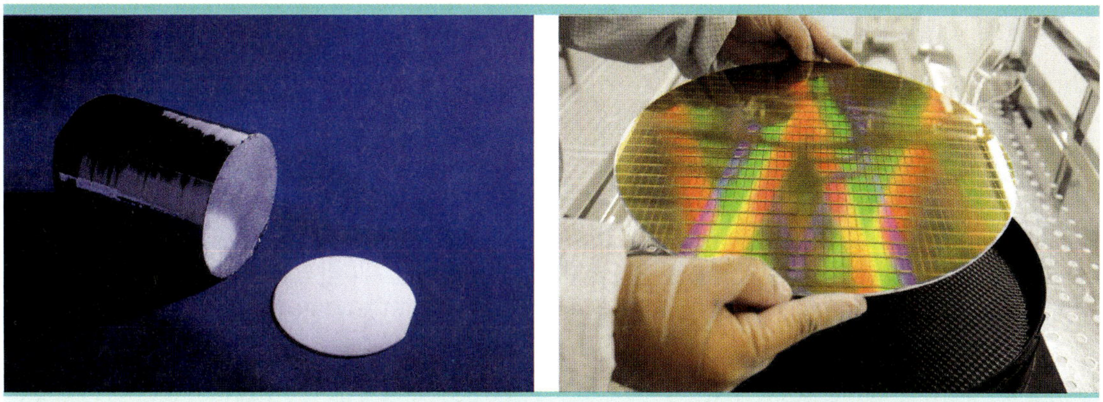

Slice of Silicon Wafer **Processed Silicon Wafer**

Figure 15 Large silicon crystals are sliced into wafers, which are used as the base for microchips.

STEM CAREER Connection

Materials Engineer
Materials engineers must understand the properties and structures of various materials, such as metals, plastics, and nanomaterials, to make any new materials that may be required for a product, such as new computer chips, aerospace materials, and biomedical devices. Many materials engineers work in universities or research and development labs.

CCC CROSSCUTTING CONCEPTS

Structure and Function Construct a circuit similar to the one shown in **Figure 10,** with a battery, a resistor, and a small LED lightbulb. Connect your circuit. Switch the polarity of the battery and try your circuit again. What evidence do you have that one connection is forward-biased? Summarize your findings in terms of semiconductors, electrons, and holes.

Computing and microchips Thousands of identical circuits, usually called chips, are produced at one time on a single wafer. The chips are then tested, sliced apart, and mounted in a carrier. Wires are attached to the contacts, and the final assembly is then sealed into a protective plastic case. Electronic signals need only travel tiny distances, so the microchips are tiny, allowing the placement of complicated circuits in a small space, as shown in **Figure 16.** Such miniaturization has increased the processing speed, the functionality, and the reliability of computers. Chips are used in appliances and automobiles as well as in computers.

Figure 16 The central processing unit of a computer is a collection of microchips.

Kitch Bain/age fotostock

Producing semiconductor electronics requires that physicists, chemists, and engineers work together. Physicists contribute their understanding of the motion of electrons and holes in semiconductors. Physicists and chemists together add precisely controlled amounts of dopants to extremely pure silicon to alter its conductivity. Engineers develop the means of mass-producing chips containing thousands of miniaturized diodes and transistors. Together, their efforts have brought our world into this electronic age.

Check Your Progress

27. **Electronic Components** You have a low-current AC signal. Describe how you could use a diode and a transistor to obtain a larger DC signal.
28. **Diode Resistance** Compare the resistance provided by a *pn*-junction diode when it is forward-biased and when it is reverse-biased.
29. **Diode Polarity** In a light-emitting diode, which battery terminal should be connected to the *p*-end to make the diode light?
30. **Current Gain** The base current in a transistor circuit measures 55 μA, and the collector current measures 6.6 mA. What is the base-to-collector current gain?
31. **Transistor Circuit** The emitter current in a transistor circuit is always equal to the sum of the base current and the collector current: $I_E = I_B + I_C$. If the current gain from the base to the collector is 95, what is the ratio of emitter current to base current?
32. **Diode Circuit** A 0.70-V silicon diode is connected to a power supply through a 110-Ω resistor. The power supply forward-biases the diode, and the adjusted current is 12 mA. What is the potential difference?
33. **Diode Potential Difference** A diode is characterized in **Figure 10.** It is forward-biased by a battery and a series resistor so that there is more than 10 mA of current and the potential difference is always about 0.70 V. The battery potential difference is then increased by 1 V.
    a. By how much does the potential difference across the diode increase? Across the resistor?
    b. By how much does the current through the resistor increase?
34. **Critical Thinking** Could you replace an *npn*-transistor with two separate diodes that are connected by their *p*-terminals? Explain.

LEARNSMART® Go online to follow your personalized learning path to review, practice, and reinforce your understanding.

The Tiniest Transistors

Every year, computers become faster and more powerful. This steady improvement is due to improvements in integrated circuits—microchips. When microchips become more densely packed with transistors and other components, they also become faster and more powerful. Early microchips had several thousand transistors. Today, a microchip of the same size—about the area of a fingernail—has several hundred million transistors.

Scientists can manipulate individual atoms using scanning tunneling microscopes like this one.

How small can a transistor be?

In 1965, scientist and businessman Gordon Moore, prominent in the semiconductor industry, noticed a pattern in the number of transistors and other components on an integrated circuit—the number of components per square inch on an integrated circuit had doubled about every year since integrated circuits were first invented. Moore predicted that the trend would continue, although he later revised the rate at which it would occur to every two years.

This trend became known as Moore's law. Despite its name, Moore's law is not a physical law like Newton's laws of motion, but an observation about how integrated circuit technology had progressed and a prediction about how it would progress in the future. Spurred on by Moore's law, scientists and engineers continue to develop new technologies to enable transistors to become smaller and smaller so that more can be packed onto a microchip.

One team succeeded in making a transistor made of a single atom. Using a scanning tunneling microscope (STM), the scientists were able to manipulate phosphorus atoms to form a pattern on an etched silicon wafer.

They covered the phosphorus atoms with a non-reactive layer of hydrogen. The STM was then used to remove the hydrogen atoms selectively. A chemical reaction was carried out, which incorporated the exposed phosphorus atoms into the silicon surface. Another silicon layer was added to protect the device, which could then be connected to an electric current using the markers that had been etched onto the wafer. The individual phosphorus atoms act as transistors, switching and amplifying electrical current.

Although successful, the transistors are not yet practical for large-scale production and use in commercial devices. They are labor-intensive to make and only function at very low temperatures. For a transistor to take its place in the computers of tomorrow, it must be practical and possible.

COMMUNICATE SCIENTIFIC AND TECHNICAL INFORMATION

Research the history of transistor technology. Make an illustrated timeline to summarize your findings.

CALIFORNIA CONNECTION

Gordon Moore was born, raised, and educated in California. Find out more about Moore's contributions to the semiconductor industry. Write a brief biography describing his education and career.

MODULE 23
STUDY GUIDE

GO ONLINE to study with your Science Notebook.

Lesson 1 CONDUCTION IN SOLIDS

- The band theory of solids states that, in solids, the electric fields of neighboring atoms spread the allowed energy levels for an atom's outer electrons into broad bands. The valence and conduction bands are separated by forbidden energy gaps. The smaller the forbidden gap, the better the material's conduction.
- In conductors, electrons can move through the solid because the conduction band is partially filled. In insulators, more energy is needed to move electrons into the conduction band than is generally available. In semiconductors, the energy gap is small enough that some electrons can reach the conduction band.
- Conductivity of semiconductors increases with increasing temperature or illumination. Conduction in semiconductors is also enhanced by doping pure crystals with small amounts of other kinds of atoms, called dopants.
- The *n*-type semiconductors are doped with electron donor atoms. They conduct by the response of these donor electrons to applied potential differences. The *p*-type semiconductors are doped with electron acceptor atoms. They conduct by making holes available to electrons in the conduction band.

- semiconductor
- band theory
- intrinsic semiconductors
- dopants
- extrinsic semiconductors

Lesson 2 ELECTRONIC COMPONENTS

- A *pn*-junction diode consists of a layer of a *p*-type semiconductor joined with a layer of an *n*-type semiconductor.
- Diodes conduct charges in one direction only. If a potential difference is applied across the diode in the direction of polarity, the diode is forward-biased and conducts charge. If a potential difference is applied opposite its polarity, the diode is reverse-biased and does not conduct charge.
- Diodes can be used to convert AC to DC. Some diodes emit light when a potential difference is applied; these diodes can also detect light.
- A transistor is a sandwich of three layers of semiconductor material, configured as either *npn* or *pnp* layers. The center base layer is very thin compared to the other layers, which are called the emitter and the collector.
- A transistor can amplify a weak signal into a much stronger one. The ratio of the collector-emitter current to the base current is known as the current gain and is a useful measure of transistor amplification.
- A microchip is an integrated circuit. Microchips consist of thousands of transistors, diodes, resistors, and conductors, all of which can be made by doping silicon with donor or acceptor atoms.

- diode
- depletion layer
- transistor
- microchip

THREE-DIMENSIONAL THINKING

Module Wrap-Up

REVISIT THE PHENOMENON

What innovations allowed computers to shrink from filling a room to fitting in your pocket?

CER Claim, Evidence, Reasoning

Explain Your Reasoning Revisit the claim you made when you encountered the phenomenon. Summarize the evidence you gathered from your investigations and research and finalize your Summary Table. Does your evidence support your claim? If not, revise your claim. Explain why your evidence supports your claim.

STEM UNIT PROJECT

Now that you've completed the module, revisit your STEM unit project. You will summarize your evidence and apply it to the project.

GO FURTHER

SEP Data Analysis Lab

Can you analyze a transistor?

You have set up a transistor circuit as shown below. When you close the switch, there is a 0.70-V drop across the base-emitter junction and a current gain from base to collector of 220.

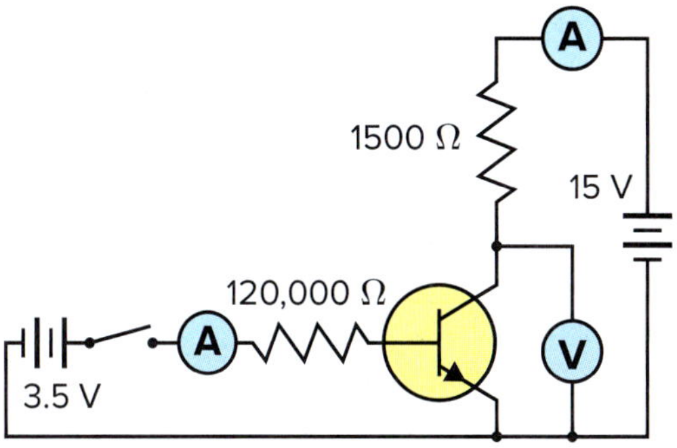

CER Analyze and Interpret Data

1. **Claim** How would you expect each of the following to change when you close the switch?
   a. the base current
   b. the collector current
   c. voltmeter reading
2. **Evidence and Reasoning** Explain your reasoning and any calculations you have made to justify your claim.

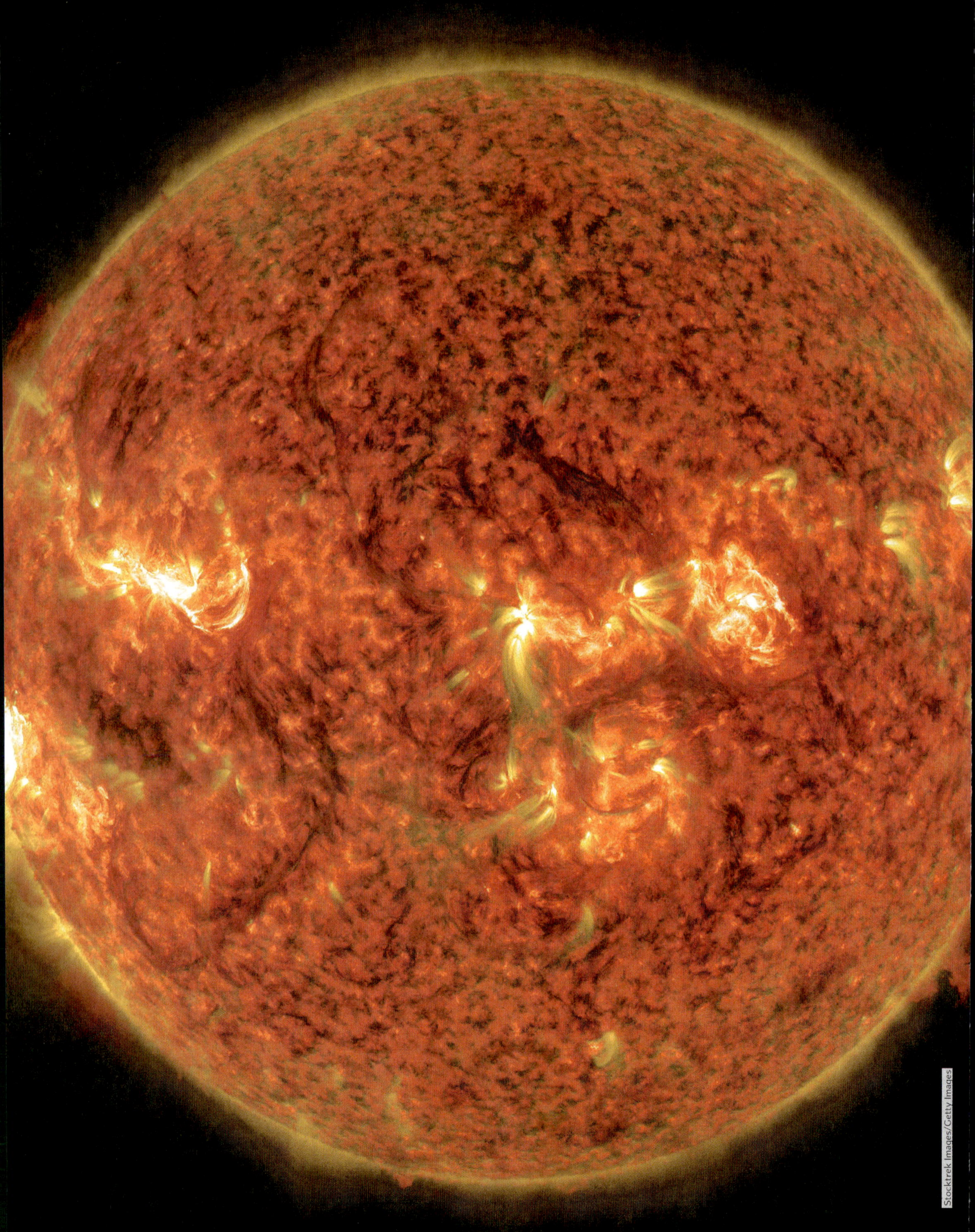
Stocktrek Images/Getty Images

MODULE 24
NUCLEAR AND PARTICLE PHYSICS

ENCOUNTER THE PHENOMENON

How does the Sun produce energy and how can we replicate it?

GO ONLINE to watch a video about how scientists study nuclear fusion.

SEP Ask Questions

Do you have other questions about the phenomenon? If so, add them to the driving question board.

CER Claim, Evidence, Reasoning

Make Your Claim Use your CER chart to make a claim about how the Sun produces energy and how we can replicate it. Explain your reasoning.

Collect Evidence Use the lessons in this module to collect evidence to support your claim. Record your evidence as you move through the module.

Explain Your Reasoning You will revisit your claim and explain your reasoning at the end of the module.

GO ONLINE to access your CER chart and explore resources that can help you collect evidence.

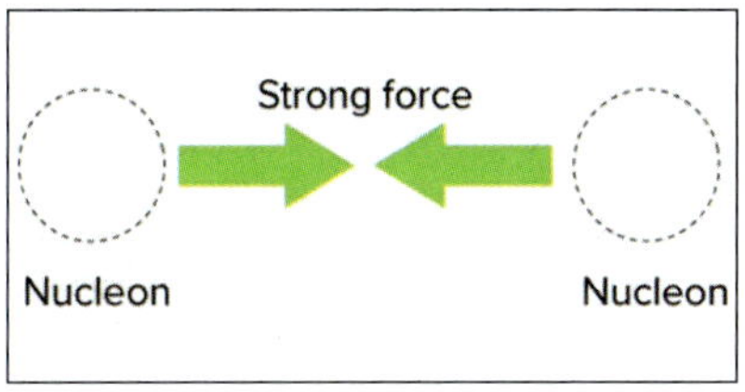

LESSON 1: Explore & Explain: The Strong Nuclear Force

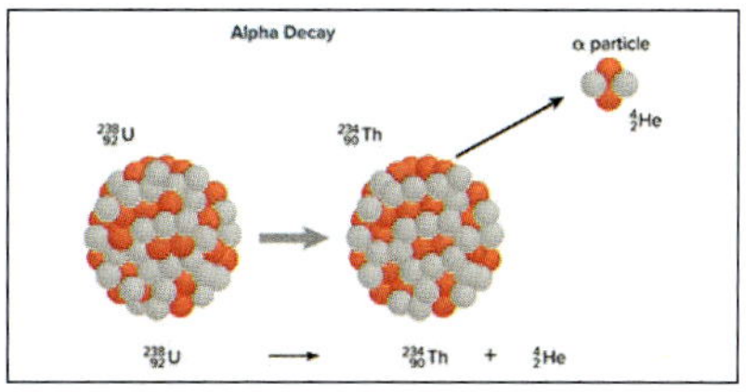

LESSON 2: Explore & Explain: Nuclear Reactions and Equations

Smithsonian

Additional Resources

Video Supplied by BBC Worldwide Learning

LESSON 1
THE NUCLEUS

FOCUS QUESTIONS

What is inside an atom?

Description of the Nucleus

Recall from your study of atoms that Ernest Rutherford and his team measured the deflection of alpha particles as they hit gold foil. From this investigation, Rutherford concluded that the atom has a small, dense, positively charged nucleus surrounded by nearly massless electrons, as shown in **Figure 1.** However, the investigation gave no information about the structure of the nucleus. Henry Mosely, a member of Rutherford's team, found the magnitude of the nuclear charge during X-ray scattering experiments. The results showed that the positively charged protons accounted for roughly half the mass of the nucleus. What accounted for the other half?

The Nucleus

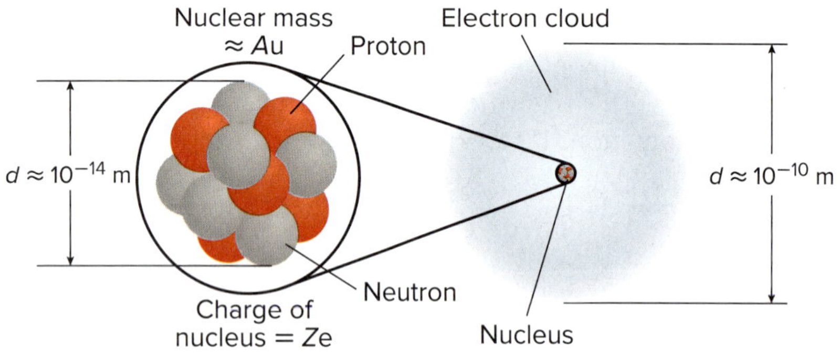

Figure 1 The atom consists of a very dense nucleus surrounded by an electron cloud. The nucleus is made of protons and neutrons.

3D THINKING | DCI Disciplinary Core Ideas | CCC Crosscutting Concepts | SEP Science & Engineering Practices

COLLECT EVIDENCE

Use your Science Journal to record the evidence you collect as you complete the readings and activities in this lesson.

INVESTIGATE

GO ONLINE to find these activities and more resources.

Launch Lab: A Nuclear Model
Use a model to investigate the effects of the forces between protons and neutrons in the nucleus.

CCC **Identify Crosscutting Concepts**
Create a table of the crosscutting concepts and fill in examples you find as you read.

Protons and neutrons One hypothesis was that the nucleus contained protons and enough electrons to reduce the charge to the observed value. A second hypothesis was that the nucleus contained uncharged particles as well as protons. In 1932, English physicist James Chadwick discovered the neutron, a neutral particle with a mass similar to that of a proton. This particle accounted for the missing mass of the nucleus and its charge.

To describe an atom's nucleus, you need to know how many protons and neutrons it has. Protons and neutrons are collectively referred to as **nucleons.** The number of protons is the **atomic number** (Z). To find the charge of the nucleus, multiply this number by the elementary charge (e). For example, a nucleus with five protons has a charge of +5e. The number of nucleons in an atom is the **mass number** (A). Protons and neutrons have almost the same mass—approximately 1.66×10^{-27} kg, or 1 **atomic mass unit** (u). This is more than the mass of 1800 electrons. To find the approximate nuclear mass, multiply the mass number by u.

$$\text{approximate nuclear mass} \cong A \times \text{u}$$

Size of the nucleus Rutherford's investigations produced the first measurements of the size of the nucleus. He found that nuclei have diameters of about 10^{-14} m. A typical atom might have a radius 10,000 times larger than the diameter of its nucleus. Although the nucleus contains nearly all the mass of an atom, proportionally the nucleus occupies less space in the atom than the Sun does in our solar system. The nucleus is incredibly dense—about 1.4×10^{18} kg/m^3. If a nucleus could be 1 cm^3, it would have a mass of about 1 billion tons. This is roughly 90,000 times more massive than the largest ocean liners.

Isotopes

Look at the periodic table in the back of your book, and examine the atomic masses of the elements. The first four elements have atomic masses near whole numbers of atomic mass units. Because the nucleus contains nearly all the mass and each nucleon has a mass of approximately 1 u, this seems to make sense. Boron, however, has a mass of 10.8 u. How can this be?

Nuclei of isotopes The puzzle of atomic masses that are not whole numbers can be solved by applying electromagnetism. Recall that experiments with mass spectrometers, such as the one in **Figure 2,** show that an element can have atoms with different masses. For example, in an analysis of a pure helium sample, two spots appeared on the detector of the spectrometer. The two spots were produced by helium atoms of different masses. One type of helium atom had a mass of 3 u, while the second type had a mass of 4 u. All neutral helium atoms have 2 protons and 2 electrons. But one kind of helium atom has 1 neutron in its nucleus, while the other has 2 neutrons. The two kinds of atoms are called isotopes of helium.

Figure 2 Mass spectrometers are used to measure the mass differences of isotopes.

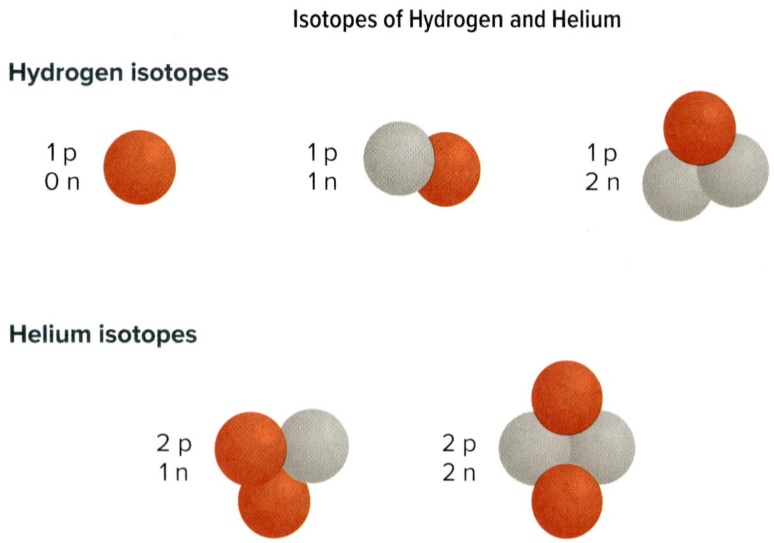

Figure 3 Isotopes have the same number of protons but different numbers of neutrons.
Identify *the mass number and the electric charge of each nucleus.*

All nuclei of an element have the same number of protons but have different numbers of neutrons, as shown by the hydrogen and helium nuclei in **Figure 3.** All isotopes of a neutral element have the same number of electrons as protons and behave chemically in the same way.

Get It?

Compare and contrast the isotopes carbon-12 and carbon-14.

A special notation is used to describe an isotope.

$$^{A}_{Z}X \qquad \begin{aligned} X &= \text{element symbol} \\ A &= \text{mass number} \\ Z &= \text{atomic number} \end{aligned}$$

For example, hydrogen-2 (deuterium) is $^{2}_{1}\mathrm{H}$, and the three naturally occurring isotopes of neon are $^{20}_{10}\mathrm{Ne}$, $^{21}_{20}\mathrm{Ne}$, and $^{22}_{10}\mathrm{Ne}$.

Average mass The measured mass of chlorine is 35.453 u. This figure is now understood to be the average mass of the naturally occurring isotopes of helium. Thus, while the mass of an individual atom is close to a whole number of mass units, the atomic mass determined from an average sample of atoms does not have to be. The atomic mass of chlorine, 35.453 u, is the weighted average of its two isotopes, chlorine-35 and chlorine-37. Most elements have several isotopic forms that occur naturally. The mass of the isotope carbon-12 is now used to define the mass unit. One u is defined to be $\frac{1}{12}$ the mass of the carbon-12 isotope.

Get It?

Explain what a weighted average is and how it is used to determine an element's atomic mass.

PRACTICE Problems

ADDITIONAL PRACTICE

1. Three isotopes of uranium have mass numbers of 234, 235, and 238. The atomic number of uranium is 92. How many neutrons are in the nucleus of each isotope?
2. How many neutrons are in an atom of the mercury isotope $^{200}_{80}Hg$?
3. An isotope of oxygen has a mass number of 15. How many neutrons are in a nucleus of this isotope?
4. **CHALLENGE** Write the symbols for the three isotopes of hydrogen that have zero, one, and two neutrons in the nucleus.

The Strong Nuclear Force

The negatively charged electrons around the positively charged nucleus of an atom are held in place by the attractive electromagnetic force. However, the nucleus consists of positively charged protons and neutral neutrons. The protons repel each other. Why don't the protons cause the nucleus to fly apart? We must conclude that an even stronger attractive force exists within the nucleus to hold the protrons and neutrons together.

When any pair of nucleons gets within about a proton's radius (1.4×10^{-15} m) of each other, they experience an attraction that is more than 100 times stronger than the electromagnetic force. This **strong nuclear force,** also called the strong force, is an attractive force between nucleons that are close together, as they are in a nucleus. The strong force has the same strength between all nucleon pairs (proton-proton, neutron-neutron, or proton-neutron), as shown in **Figure 4.**

Mass defect and binding energy

An initially surprising observation is that the mass of an assembled nucleus is less than the sum of the masses of the protons and neutrons that compose it. For example, the helium nucleus $^{4}_{2}He$ consists of two protons and two neutrons. The mass of a proton is 1.007276 u. The mass of a neutron is 1.008665 u. You might expect the mass of the nucleus to be 4.031882 u, but this is not the case. Careful measurement shows the mass of a helium nucleus is only 4.002603 u. The actual mass of the helium nucleus is less than the mass of its constituent parts by 0.029279 u. This difference between the sum of the masses of the individual nucleons and the actual mass is called the **mass defect.** Why is there a mass defect?

Get It?

Explain what a *mass defect* is.

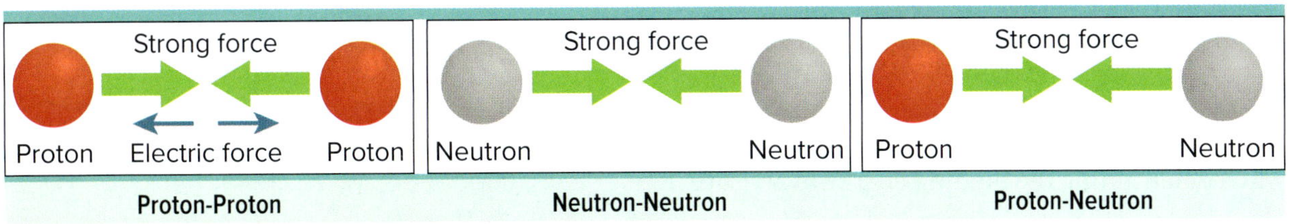

Figure 4 The strong force is the same between all pairs of nucleons.

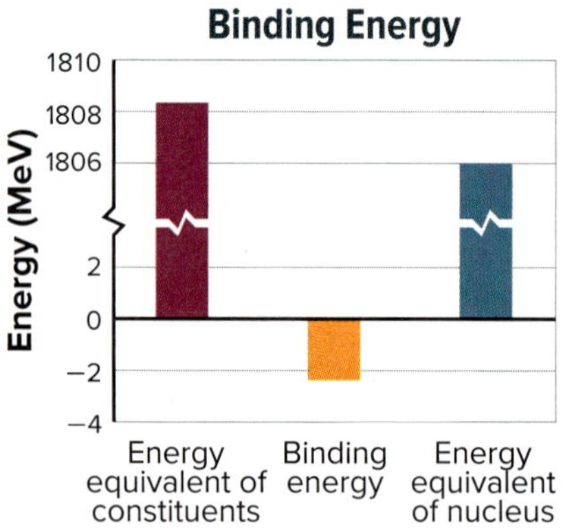

Figure 5 The energy equivalent of the mass of the individual nucleons is equal to the energy equivalent of the mass of the bound nucleus plus the binding energy.

The answer comes from Einstein's work. He showed that mass and energy are the same. To pull a nucleon out of a nucleus, you have to do work to overcome the strong nuclear force. Doing work adds energy to the system. Thus, the system of separated protons and neutrons has more energy than the assembled nucleus. This energy difference between separated nucleons and the assembled nucleus is the **binding energy** of the nucleus. Because the assembled nucleus has less energy, all binding energies are negative, as shown in **Figure 5.** The binding energy can be expressed in the form of an equivalent amount of mass, according to Einstein's famous equation.

Get It?

Explain what *binding energy* is.

Energy Equivalent of Mass

The energy equivalent to a mass is equal to the mass times the square of the speed of light in a vacuum.

$$E = mc^2$$

Determining the mass defect Mass spectrometers usually measure the masses of isotopes; that is, they measure the mass of the nucleus and the electrons. When calculating the mass defect of a nucleus, one must make sure that the mass of the atom's electrons is accounted for properly. Thus, the mass of hydrogen (one proton and one electron) usually is given in mass-defect problems. Furthermore, masses are usually measured in atomic mass units. It is useful, then, to use Einstein's equation to determine the energy equivalent of 1 u (1.6605×10^{-27} kg). In this calculation, each quantity will be expressed with five significant digits.

$$E = mc^2$$

$$= (1.6605\times10^{-27}\text{ kg})(2.9979\times10^{8}\text{ m/s})^2$$

$$= 1.4924\times10^{-10}\text{ J}$$

At this scale, the most convenient unit of energy to use is the electron volt ($1\text{ eV} = 1.60218\times10^{-19}$ J).

$$E = (1.4924\times10^{-10}\text{ J})\left(\frac{1\text{eV}}{1.60218\times10^{-19}\text{ J}}\right)$$

$$= 9.3149\times10^{8}\text{ eV} = 931.49\text{ MeV}$$

Hence, each atomic mass unit of mass is equivalent to 931.49 MeV of energy. This ratio simplifies calculations when determining the binding energy of an isotope.

Get It?

Describe the meaning of the equation $E = 931.49$ MeV/u, and explain why it is important.

EXAMPLE Problem 1

MASS DEFECT AND BINDING ENERGY Find the mass defect and binding energy of tritium (^3_1H). The mass of the tritium isotope is 3.016049 u, the mass of a hydrogen atom is 1.007825 u, and the mass of a neutron is 1.008665 u.

1 ANALYZE THE PROBLEM

Known

mass of 1 hydrogen atom = 1.007825 u

mass of 1 neutron = 1.008665 u

mass of 1 tritium atom = 3.016049 u

energy equivalent of 1 u = 931.49 MeV

Unknown

total mass of tritium's parts = ?

mass defect = ?

binding energy of tritium = ?

2 SOLVE FOR THE UNKNOWN

Determine the total mass of tritium's parts.

mass of 1 hydrogen atom:	1.007825 u	Add the masses of tritium's parts: one hydrogen atom (one proton and one electron) and two neutrons.
mass of 2 neutrons:	+ 2.017330 u	
total mass of nucleons and electron:	3.025155 u	

Determine the mass defect.

mass of tritium:	3.016049 u	The mass defect is equal to the mass of the assembled tritium minus the mass of the sum of its parts.
mass of tritium's pieces:	− 3.025155 u	
mass defect:	− 0.009106 u	

Determine the binding energy of tritium.

E = (mass defect in u)(energy equivalent of 1 u) — The binding energy is the energy equivalent of the mass defect.

$E = (-0.009106 \text{ u})(931.49 \text{ MeV/u})$
$= -8.4821 \text{ MeV}$

Substitute mass defect = −0.009106 u and binding energy per u = 931.49 MeV.

3 EVALUATE THE ANSWER

- **Are the units correct?** Mass is measured in u, and energy is measured in MeV.
- **Does the sign make sense?** Binding energy should be negative.
- **Is the magnitude realistic?** According to **Figure 6 on the next page**, binding energies per nucleon in this range are between −2 MeV and −3 MeV, so the answer for three nucleons is reasonable.

PRACTICE Problems

ADDITIONAL PRACTICE

Use these values to solve the following problems:

mass of hydrogen = 1.007825 u
mass of neutron = 1.008665 u
1 u = 931.49 MeV

5. The carbon isotope $^{12}_6C$ has a mass of 12.000000 u.
   a. Calculate its mass defect.
   b. Calculate its binding energy.

6. Deuterium (^2_1H) has a mass of 2.014102 u.
   a. Calculate its mass defect.
   b. Calculate its binding energy.

7. The nitrogen isotope $^{15}_7N$ has a mass defect of −0.113986 u.
   a. Calculate the mass of this isotope.
   b. Calculate the binding energy of the nucleus.

8. CHALLENGE The oxygen isotope $^{16}_8O$ has a mass of 15.994915 u. A second oxygen isotope is $^{18}_8O$, which has a mass of 17.999160 u.
   a. What is the mass defect of each isotope?
   b. What is the binding energy of each nucleus?
   c. Why do you think the binding energies are different?

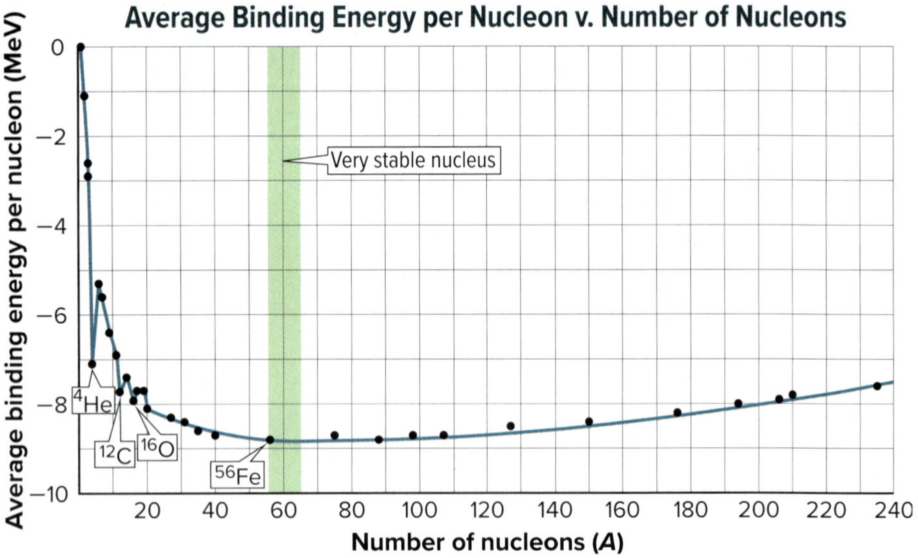

Figure 6 The binding energy per nucleon changes as the number of nucleons (A) varies. Iron-56 has the most tightly bound nucleus and is the most stable isotope known.

Infer *Which nucleus,* $^{131}_{52}I$ or $^{236}_{94}Pu$, *would you expect to have a larger magnitude average binding energy per nucleon?*

Binding energy and nuclear mass **Figure 6** shows that the binding energy per nucleon depends on the mass of the nucleus. Except for a few nuclei, the binding energy per nucleon becomes more negative as the number of nucleons (A) increases to 56, which is that of iron (Fe). A more negative average binding energy indicates tightly bound nucleons and a more stable nucleus. The most stable nucleus is $^{56}_{26}Fe$ because the magnitude of its binding energy is greatest. Smaller nuclei become more stable as the number of nucleons approaches 56. Nuclei with more than 56 nucleons are less strongly bound and therefore less stable. Large, unstable nuclei will readily decay into smaller, more stable nuclei. All nuclei with atomic numbers greater than 92 decay readily. Under certain conditions, such as very high temperature, nuclei smaller than $^{56}_{26}Fe$ can fuse into larger, more stable nuclei.

Check Your Progress

9. **Carbon** The radioactive carbon isotope $^{14}_{6}C$ has a mass of 14.003074 u. What is the mass defect of this isotope? What is the binding energy of its nucleus?
10. **Nuclei** Consider these two pairs of nuclei: $^{12}_{6}C$ and $^{13}_{6}C$ and $^{11}_{5}B$ and $^{11}_{6}C$. In which way are the two pairs alike? In which way are they different?
11. **Binding Energy** When a tritium nucleus ($^{3}_{1}H$) decays, it emits an electron. The tritium isotope becomes the helium isotope $^{3}_{2}He$. Which of the two nuclei would you expect to have a more negative binding energy? Explain.
12. **Mass Defect** Which of the two nuclei in the previous problem has the larger mass defect?
13. **Strong Nuclear Force** The range of the strong nuclear force is so short that only nucleons that are adjacent to each other are affected by the force. Use this fact to explain why in large nuclei the repulsive electromagnetic force can overcome the strong attractive force and make the nucleus unstable.
14. **Critical Thinking** In stars, small nuclei fuse to form nuclei with smaller binding energy per nucleon, such as helium and carbon. Older stars also produce oxygen ($Z = 8$) and silicon ($Z = 14$). What is the atomic number of the heaviest nucleus that could be formed in this way? Explain.

LEARNSMART® Go online to follow your personalized learning path to review, practice, and reinforce your understanding.

LESSON 2

NUCLEAR DECAY AND REACTIONS

FOCUS QUESTIONS

How can nuclear reactions be useful?

Radioactive Decay

In 1896, Henri Becquerel was working with compounds containing the element uranium. He was surprised to find that nearby photographic plates that had been covered to keep out light became fogged, or partially exposed. This fogging suggested that some kind of ray from the uranium had passed through the plate coverings. **Figure 7** shows a photographic plate that Becquerel exposed to this radiation. Several materials other than uranium or its compounds also were found to emit these penetrating rays.

In 1909, after years of experiments, Ernest Rutherford was able to identify the radiation that Becquerel had found. This radiation was composed of helium-4 nuclei. The larger uranium nuclei were decaying and emitting these helium nuclei. Nuclei that emit particles and energy are said to be **radioactive.** A material that emits particles is also said to decay. Nuclei decay in order to become more stable. Radioactive decay is a natural process.

In few areas of physics has basic knowledge led to applications as quickly as in the field of nuclear physics. The medical use of the radioactive element radium began within 20 years of its discovery. Proton accelerators were used for medical applications less than one year after being invented. In the case of nuclear fission (splitting nuclei), the military application was under development before the basic physics was even known. Peaceful applications, such as power plants, followed in less than 10 years.

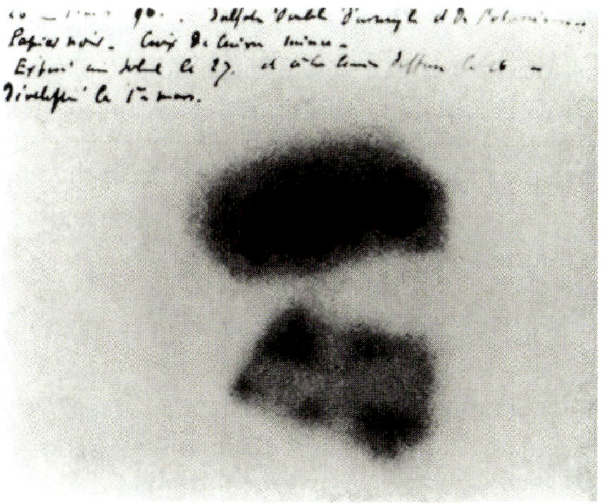

Figure 7 Henri Becquerel found that photographic plates could be exposed by nearby uranium compounds. The dark shadows are the exposed portions of the plates.

3D THINKING DCI Disciplinary Core Ideas CCC Crosscutting Concepts SEP Science & Engineering Practices

COLLECT EVIDENCE

Use your Science Journal to record the evidence you collect as you complete the readings and activities in this lesson.

INVESTIGATE

GO ONLINE to find these activities and more resources.

Applying Practices: Modeling Fission, Fusion, and Radioactive Decay

HS-PS1-8. Develop models to illustrate the changes in the composition of the nucleus of the atom and the energy released during the processes of fission, fusion, and radioactive decay.

Revisit the Encounter the Phenomenon Question

What information from this lesson can help you answer the Unit and Module questions?

Types of Radiation

Rutherford's team found that uranium compounds produce three kinds of radiation. He classified the radiation by how easy it was to stop and named them alpha (α), beta (β), and gamma (γ) radiation. A thick sheet of paper can stop alpha radiation, and 6 mm of aluminum can stop most beta radiation. Gamma rays are stopped by several centimeters of lead. We now know that each type of radiation involves a different particle.

Alpha decay Alpha particles are helium nuclei ($^{4}_{2}He$). The mass number of an α particle is 4, and the atomic number is 2. The emission of an α particle from a nucleus is a process called **alpha decay.** Alpha decay occurs in nuclei with at least 106 nucleons. When a nucleus emits an α particle, its mass number (A) is reduced by 4, and its atomic number (Z) is reduced by 2. The element changes, or transmutes, into a different element. For example, $^{238}_{92}U$ transmutes into $^{234}_{90}Th$ by alpha decay.

Beta decay Beta particles are electrons or positrons (positively charged electrons) emitted by the nucleus. The nucleus does not contain electrons or positrons, so where do these particles come from? **Beta decay** occurs when a neutron changes to a proton or a proton changes to a neutron within the nucleus. In all reactions, charge is conserved. That is, the total charge before the decay equals the total charge after the decay. In beta decay, when a neutron (charge 0) changes to a proton (charge +1e), an electron (charge −1e) also appears. A nucleus with N neutrons and Z protons ends up as a nucleus of $N - 1$ neutrons and $Z + 1$ protons. Another neutral, nearly massless particle called an antineutrino is also emitted. This type of beta decay is called beta minus decay (β^- decay) because an electron is emitted.

When a proton changes to a neutron in the nucleus, a positron and a neutrino are emitted. This type of beta decay is called positron emission or beta plus decay (β^+ decay) because a positron is emitted.

Gamma decay Gamma rays are high-energy photons emitted by the nucleus during gamma decay. **Gamma decay** occurs when there is a redistribution of energy within the nucleus and a gamma ray is emitted. Neither the mass number nor the atomic number changes in gamma decay. Gamma radiation often accompanies alpha and beta decay. The three main types of radiation are summarized in **Table 1.**

Radioactive elements often go through a series of successive decays until they form a stable nucleus. For example, $^{238}_{92}U$ undergoes 14 separate decays before the stable lead isotope $^{206}_{82}Pb$ is produced.

Table 1 Three Types of Radiation

Alpha Decay	Beta Decay	Gamma Decay
α particle ($^{4}_{2}He$)	α particle ($^{0}_{-1}e$)	γ particle (photon)
charge +2	charge −1	no charge
least penetration	medium penetration	highest penetration
transmutes nucleus: $A \rightarrow A-4$ $Z \rightarrow Z-2$ $N \rightarrow N-2$	transmutes nucleus: $A \rightarrow A$ $Z \rightarrow Z+1$ $N \rightarrow N-1$	changes only energy: $A \rightarrow A$ $Z \rightarrow Z$ $N \rightarrow N$

Nuclear Reactions and Equations

Whenever the energy or the number of neutrons or protons in a nucleus changes, the resulting event is called a **nuclear reaction.** Just as in chemical reactions, some nuclear reactions release energy, while others occur only when energy is added to a nucleus. One form of nuclear reaction is the emission of particles by radioactive nuclei. The reaction releases excess energy in the form of the kinetic energy of the emitted particles. Two such reactions are shown in **Figure 8.**

Nuclear reactions can be described by words, diagrams, and equations. The symbols in nuclear equations simplify the calculation of atomic number and mass number in nuclear reactions. For example, the alpha decay in **Figure 8** can be expressed by the following equation:

$$^{238}_{92}\mathrm{U} \rightarrow {}^{234}_{90}\mathrm{Th} + {}^{4}_{2}\mathrm{He}$$

This equation states that uranium-238 decays into thorium-234 by emitting an alpha particle.

The total number of nuclear particles (protons and neutrons) stays the same during the reaction, so the sum of the superscripts on each side of the equation must be equal: $238 = 234 + 4$. The total charge also is conserved during the reaction, so the sum of the subscripts on each side must be equal: $92 = 90 + 2$.

During beta decay, an electron ($^{0}_{-1}\mathrm{e}$) and an antineutrino ($^{0}_{0}\bar{v}$) are produced. **Figure 8** shows the transmutation of a thorium-234 nucleus into a protactinium-234 nucleus by the emission of a β particle. The reaction is described by the following equation. Check that charge is conserved in this reaction.

$$^{234}_{90}\mathrm{Th} \rightarrow {}^{234}_{91}\mathrm{Pa} + {}^{0}_{-1}\mathrm{e} + {}^{0}_{0}\bar{v}$$

Get It?

Cite Evidence In what ways can the nucleus of an atom change?

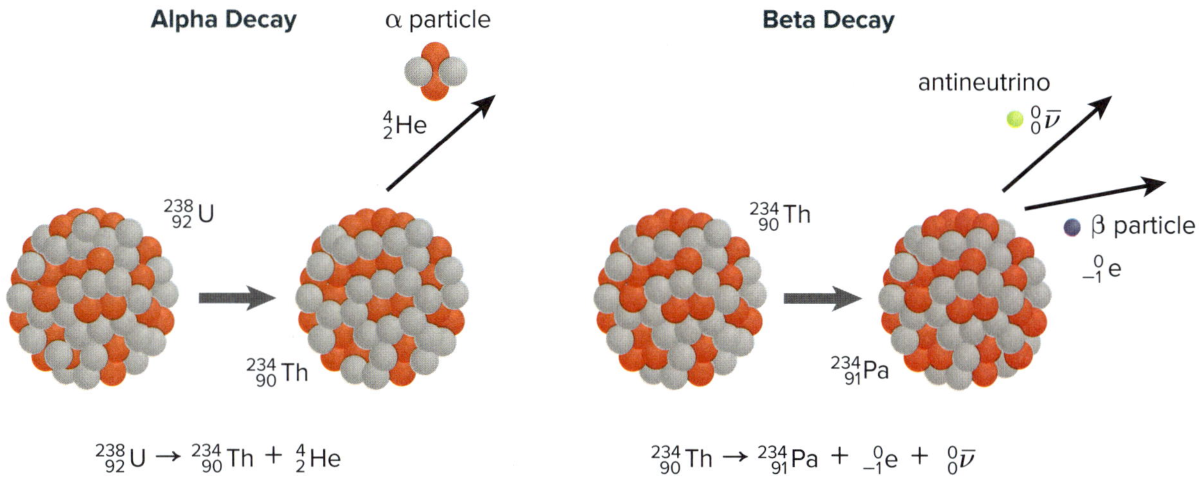

Figure 8 When uranium-238 undergoes alpha decay, an alpha particle is emitted, and the resulting nucleus is thorium-234. Thorium-234 can undergo beta decay to produce protactinium-234 and a beta particle.

Note that the superscript on the left-hand side of the equation equals the sum of the superscripts on the right-hand side. Equality also exists between the subscripts on both sides of the equation. In alpha and beta decay, one nucleus, shown on the left of the equation, decays to another nucleus and one or more radioactive particles, shown on the right of the equation. Another example of transmutation occurs when a particle collides with the nucleus, often resulting in the emission of other particles. For example, ${}^{12}_{6}C + {}^{1}_{1}H \rightarrow {}^{13}_{7}N$. Such reactions are illustrated in Example Problem 3 and in the discussion of fission later in this lesson.

EXAMPLE Problem 2

ALPHA AND BETA DECAY Write the nuclear equation for each radioactive process.

a. The radioactive radium isotope ${}^{226}_{88}Ra$ emits an α particle and becomes the radon isotope ${}^{222}_{86}Rn$.

b. The radioactive lead isotope ${}^{209}_{82}Pb$ decays into the bismuth isotope ${}^{209}_{83}Bi$ by the emission of a β particle and an antineutrino.

1 ANALYZE THE PROBLEM

Known	Unknown
a. Initial: ${}^{226}_{88}Ra$	What is the nuclear equation?
Final: α particle (${}^{4}_{2}He$), ${}^{222}_{86}Rn$	
b. Initial: ${}^{209}_{82}Pb$	What is the nuclear equation?
Final: ${}^{209}_{83}Bi$, β particle (${}^{0}_{-1}e$), antineutrino (${}^{0}_{0}\bar{\nu}$)	

2 SOLVE FOR THE UNKNOWN

a. ${}^{226}_{88}Ra \rightarrow {}^{4}_{2}He + {}^{222}_{86}Rn$ — Substitute ${}^{4}_{2}He$ for α particle.

b. ${}^{209}_{82}Pb \rightarrow {}^{209}_{83}Bi + {}^{0}_{-1}e + {}^{0}_{0}\bar{\nu}$ — Substitute ${}^{0}_{-1}e$ for β particle and ${}^{0}_{0}\bar{\nu}$ for antineutrino.

3 EVALUATE THE ANSWER

- **Is the number of nucleons conserved?**

  a. 226 = 4 + 222, so mass number is conserved.

  b. 209 = 209 + 0 + 0, so mass number is conserved.

- **Is charge conserved?**

  a. 88 = 2 + 86, so charge is conserved.

  b. 82 = 83 − 1 + 0, so charge is conserved.

PRACTICE Problems

ADDITIONAL PRACTICE

15. Write the nuclear equation for the transmutation of the radioactive uranium isotope ${}^{234}_{92}U$ into the thorium isotope ${}^{230}_{90}Th$ by the emission of an α particle.

16. Write the nuclear equation for the transmutation of the radioactive thorium isotope ${}^{230}_{90}Th$ into the radioactive radium isotope ${}^{226}_{88}Ra$ by the emission of an α particle.

17. Write the nuclear equation for the transmutation of americium-241 to neptunium-237 by the emission of an α particle.

18. Write the nuclear equation for the transmutation of the radioactive radium isotope ${}^{226}_{88}Ra$ into the radon isotope ${}^{222}_{86}Rn$ by α decay.

19. The radioactive lead isotope ${}^{214}_{82}Pb$ can change to the radioactive bismuth isotope ${}^{214}_{83}Bi$ by the emission of a β particle and an antineutrino. Write the nuclear equation.

20. CHALLENGE The radioactive carbon isotope ${}^{14}_{6}C$ undergoes β decay. Determine the product of the decay, and write the nuclear equation.

EXAMPLE Problem 3

SOLVE NUCLEAR EQUATIONS When nitrogen gas is bombarded with α particles, high-energy protons are emitted. What new isotope is created? Use the periodic table at the back of the book.

1 ANALYZE THE PROBLEM

Known

nitrogen = $^{14}_{7}N$

$\alpha = ^{4}_{2}He$

proton = $^{1}_{1}H$

Unknown

What isotope belongs on the right side of the equation?

2 SOLVE FOR THE UNKNOWN

Write the equation for the nuclear reaction.

$$^{4}_{2}He + ^{14}_{7}N \rightarrow ^{1}_{1}H + ^{A}_{Z}X$$

Solve for Z and for A.

$$Z = 2 + 7 - 1 = 8$$

$$A = 4 + 14 - 1 = 17$$

The element with $Z = 8$ is oxygen. The isotope must be $^{17}_{8}O$.

3 EVALUATE THE ANSWER

- **Does the equation balance?** The number of nucleons is conserved: 4 + 14 = 1 + 17. The charge is conserved: 2 + 7 = 1 + 8.

PRACTICE Problems

ADDITIONAL PRACTICE

21. Use the periodic table at the back of your book to complete the following:

a. $^{33}_{15}P \rightarrow ? + ^{0}_{-1}e + ^{0}_{0}\bar{\nu}$

b. $^{55}_{24}Cr \rightarrow ? + ^{0}_{-1}e + ^{0}_{0}\bar{\nu}$

22. Write the nuclear equation for the transmutation of the seaborgium isotope $^{263}_{106}Sg$ into the rutherfordium isotope $^{259}_{104}Rf$ by the emission of an alpha particle.

23. A proton collides with the nitrogen isotope $^{15}_{7}N$, forming a new isotope and an alpha particle. What is the isotope? Write the nuclear equation.

24. CHALLENGE Write the nuclear equations for the beta decay of the following isotopes:

a. $^{210}_{82}Pb$ **c.** $^{234}_{90}Th$

b. $^{210}_{83}Bi$ **d.** $^{239}_{93}Np$

Half-Life

Radioactive isotopes vary greatly in stability. Some exist for only a fraction of a second before decaying, while others may exist for billions of years. Scientists describe this property in terms of half-life. The **half-life** is the time required for half of the atoms in any given quantity of a radioactive isotope to decay. After each half-life, approximately half the nuclei remain undecayed, as shown in **Figure 9.** The half-life is an average found by observing a large number of decays. Each isotope has its own half-life.

For example, the half-life of the radium isotope $^{226}_{88}Ra$ is 1600 years. That is, in 1600 years half a given quantity of $^{226}_{88}Ra$ decays into another element—radon. In another 1600 years, half the remaining radium will decay. In other words, after 3200 years, one-fourth of the original amount will remain.

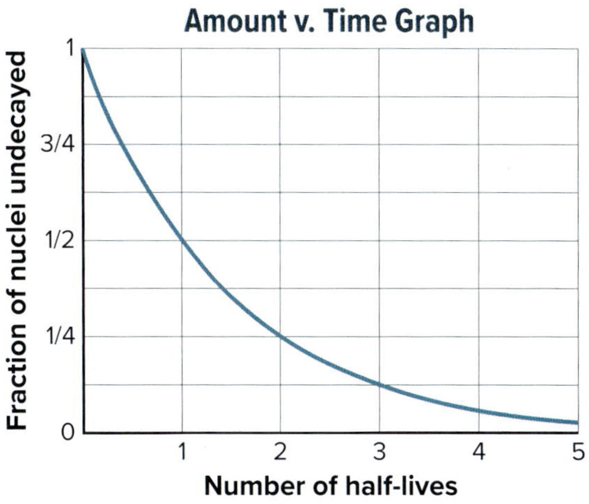

Figure 9 Half of the radioactive nuclei decay each half-life.

Table 2 **Half-Lives of Selected Isotopes**

Element	Isotope	Half-Life	Radiation Produced
Hydrogen	${}^{3}_{1}\text{H}$	12.3 years	β
Carbon	${}^{14}_{6}\text{C}$	5730 years	β
Cobalt	${}^{60}_{27}\text{Co}$	30 years	β, γ
Iodine	${}^{131}_{53}\text{I}$	8.07 days	β, γ
Lead	${}^{212}_{82}\text{Pb}$	10.6 hours	β
Polonium	${}^{194}_{84}\text{Po}$	0.7 seconds	α
Polonium	${}^{210}_{84}\text{Po}$	138 days	α, γ
Uranium	${}^{235}_{92}\text{U}$	7.1×10^{8} years	α, γ
Uranium	${}^{238}_{92}\text{U}$	4.51×10^{9} years	α, γ
Plutonium	${}^{236}_{94}\text{Pu}$	2.85 years	α
Plutonium	${}^{242}_{94}\text{Pu}$	3.79×10^{8} years	α, γ

The half-lives of selected isotopes are shown in **Table 2.** If you know the original amount of a substance and its half-life, you can calculate the amount remaining after a given number of half-lives.

Half-Life

The amount of a radioactive isotope remaining in a sample equals the original amount times $\frac{1}{2}$ to the *t*, where *t* is the number of half-lives that have passed.

$$\text{remaining} = \text{original}\left(\frac{1}{2}\right)^{t}$$

Half-lives of radioactive isotopes are used to date objects. The age of a sample of organic material can be found by measuring the amount of carbon-14 remaining. The age of Earth was calculated based on the decay of uranium into lead.

The number of decays per second of a radioactive substance is called its **activity.** Activity is proportional to the number of radioactive atoms present. Therefore, the activity of a sample also is reduced by one-half in one half-life. Consider ${}^{131}_{53}\text{I}$, with a half-life of 8.07 days. If the activity of a certain sample of iodine-131 is 8×10^{5} decays/s, then 8.07 days later, its activity will be 4×10^{5} decays/s. After another 8.07 days, its activity will be 2×10^{5} decays/s. A sample's activity is also related to its half-life. The shorter the half-life, the higher the activity. Consequently, if you know the activity of a substance and the amount of that substance, you can determine its half-life. The SI unit for decays per second is a Becquerel (Bq).

Get It?

Compare and contrast the half-life and the activity of a radioactive substance.

PRACTICE Problems

ADDITIONAL PRACTICE

Use **Figure 9** *and* **Table 2** *to solve the following problems.*

25. A 1.0-g sample of tritium ($^{3}_{1}H$) is produced. After 24.6 years, what mass of tritium will remain?

26. The isotope $^{238}_{93}Np$ has a half-life of 2.0 days. If 4.0 g of neptunium is produced on Monday, what will be the mass of neptunium remaining on Tuesday of the next week?

27. A sample of $^{210}_{84}Po$ is purchased on September 1. Its activity is 2×10^{6} Bq. The sample is used in an experiment on June 1. What activity can be expected?

28. CHALLENGE Tritium ($^{3}_{1}H$) once was used in watches to produce a fluorescent glow with a brightness proportional to tritium's activity. Compare the watch's brightness after 6 years to its original brightness.

Artificial Radioactivity

Radioactive isotopes can be formed from stable isotopes by bombardment with protons, neutrons, electrons, photons, or other particles. The resulting unstable nuclei emit radiation until they are converted into stable isotopes. The radioactive nuclei might emit alpha, beta, and gamma radiation, as well as neutrinos, antineutrinos, and positrons. A positron $\left(^{0}_{+1}e\right)$ is a positively charged particle with the mass of an electron.

Artificially produced radioactive isotopes often are used in medicine. In some medical applications, patients are given radioactive isotopes of elements that are absorbed by specific parts of the body. A physician uses a radiation counter to monitor the activity in the region in question. A radioactive isotope also can be attached to a molecule that will be absorbed in the area of interest, as is done in positron emission tomography, or PET scan. A PET scan of a brain is shown in **Figure 10.**

Radiation is also used to destroy cancer cells. These cells are more sensitive to the damaging effects of radiation because they divide more often than normal cells. Gamma rays from the isotope $^{60}_{27}Co$ are used to treat cancer patients. Another method is to inject iodine to target thyroid cancer. In a third method, a beam of particles is produced in a particle accelerator and transmitted into tissue in such a way that they decay in the cancerous tissue and destroy the cells.

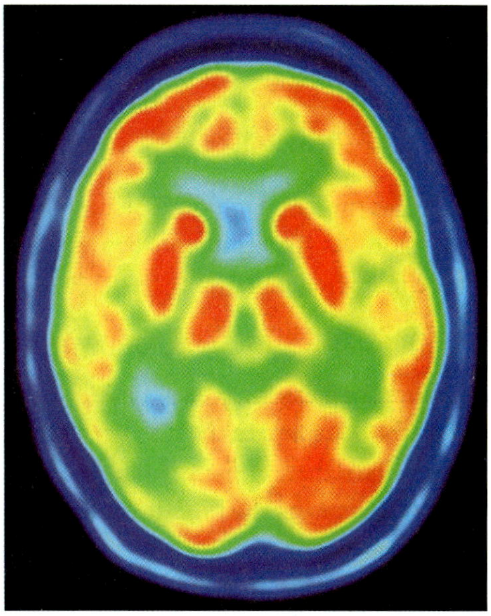

Figure 10 To produce a PET scan, physicians inject a solution into the patient. In the solution, a radioactive isotope is attached to a molecule that will accumulate in the target tissues. When the isotope decays, it produces positrons. These positrons annihilate nearby electrons, producing gamma rays. The PET scanner detects the gamma rays, and the data is used to make a three-dimensional map of the isotope distribution.

Get It?

Summarize three ways to use radiation to destroy cancer cells.

STEM CAREER Connection

Nuclear Engineer

Do nuclear processes interest you? Consider a career as a nuclear engineer. Nuclear engineers research and develop processes, instruments, systems, and procedures using nuclear energy and radiation. Nuclear engineers work in industrial and medical facilities.

CCC CROSSCUTTING CONCEPTS

Energy and Matter In a nuclear process, if atoms are not conserved, what are conserved?

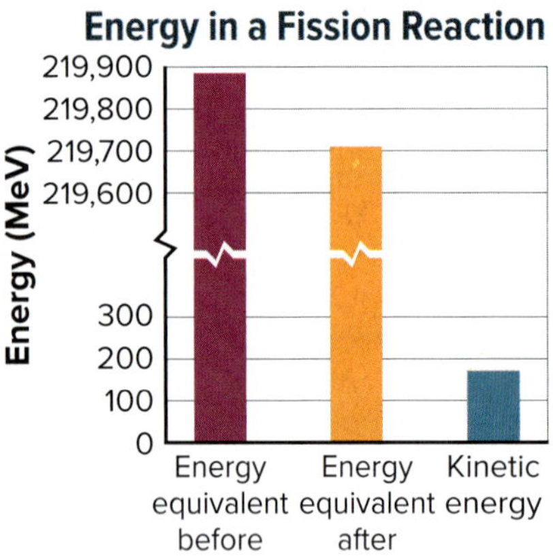

Figure 11 In a fission reaction, the mass of the reactants is equal to the sum of the mass of the products and the mass equivalent of the kinetic energy of the products.

Nuclear Fission

The possibility of obtaining useful forms of energy from nuclear reactions was first discussed in the 1930s. The most promising results came from bombarding substances with neutrons. In Italy in 1934, Enrico Fermi and Emilio Segrè produced many new radioactive isotopes by bombarding uranium with neutrons.

In 1939, German chemists Otto Hahn and Fritz Strassmann showed that the resulting atoms acted chemically like barium. One week later Lise Meitner and Otto Frisch proposed that the neutrons had caused a division of the uranium into two smaller nuclei, resulting in a large release of energy. The possibility that these reactions could be not only a source of energy but also the basis of explosive weapons was immediately realized.

The division of a nucleus into two or more fragments is called **fission.** For example, the uranium isotope $^{235}_{92}\text{U}$ undergoes fission when it is bombarded with neutrons. The elements barium and krypton are typical results of this fission, as described by the following equation:

$$^{1}_{0}\text{n} + {}^{235}_{92}\text{U} \rightarrow {}^{92}_{36}\text{Kr} + {}^{141}_{56}\text{Ba} + 3({}^{1}_{0}\text{n}) + 173\ \text{MeV}$$

The energy released by each fission can be found by calculating the masses of the atoms on each side of the equation. In the reaction above, the total mass on the right side of the equation is 0.186 u smaller than that on the left. The energy equivalent of this mass is 2.78×10^{-11} J, or 173 MeV. This energy appears as the kinetic energy of the products of the fission, as shown in **Figure 11.**

Get It?

Confirm that the fission reaction above releases 173 MeV.

Chain reactions Note that each fission releases three neutrons. These neutrons can collide with additional $^{235}_{92}\text{U}$ nuclei, resulting in more fissions. This continual process of repeated fission reactions caused by the release of neutrons from previous reactions is called a **chain reaction.** The process is illustrated in **Figure 12** on the next page.

Get It?

Explain why the overall release of energy by fission is huge.

Controlling chain reactions To create a controlled and useful chain reaction, the neutrons need to interact with the fissionable uranium at the correct rate. The reaction rate depends on the speed of the neutrons and on the amount of fissionable uranium. Most neutrons released by the fission of $^{235}_{92}\text{U}$ atoms have high speeds and are called fast neutrons. Naturally occurring uranium consists of less than 1 percent $^{235}_{92}\text{U}$ and more than 99 percent $^{238}_{92}\text{U}$. When a $^{238}_{92}\text{U}$ nucleus absorbs a fast neutron, it does not undergo fission but becomes a new isotope, $^{239}_{92}\text{U}$. The $^{238}_{92}\text{U}$ absorbs most of the neutrons. Thus, most neutrons from the original fission do not cause another fission.

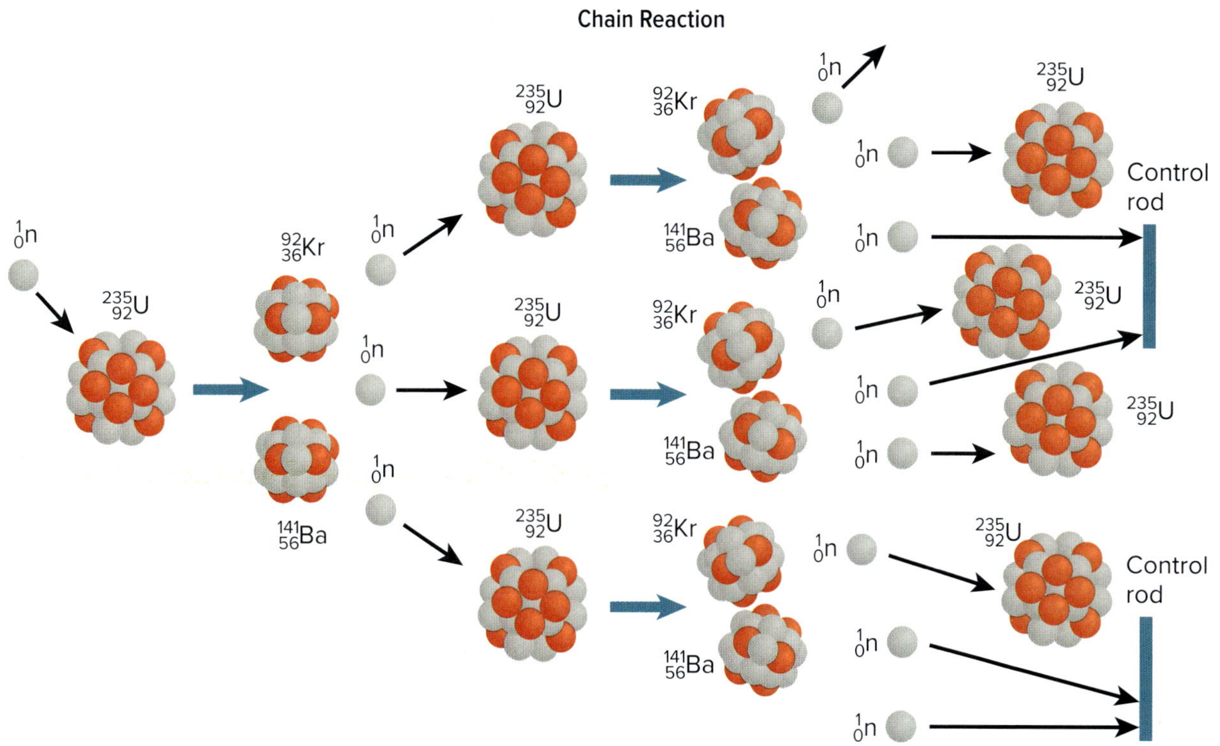

Figure 12 The neutrons produced by the fission of a uranium-235 nucleus can cause more fissions to occur. This can lead to a chain reaction. Controlled chain reactions are required to operate nuclear power plants.

Infer *why rods that absorb neutrons might be added to a reactor.*

To compensate for this, the uranium is broken into small pieces and placed in a moderator, a material that can slow down, or moderate, the fast neutrons. The neutrons collide with relatively light atoms, transfer momentum and energy, and slow down. The moderator slows many fast neutrons to speeds at which they are absorbed more easily by $^{235}_{92}U$ than by $^{238}_{92}U$. This larger number of slow neutrons greatly increases the chances that a neutron released by the fission of a $^{235}_{92}U$ nucleus will cause another $^{235}_{92}U$ nucleus to fission. If there is enough $^{235}_{92}U$ in the sample, a chain reaction can occur. To increase the amount of fissionable uranium, samples are enriched with more $^{235}_{92}U$.

Nuclear reactors

The type of nuclear reactor used in the United States is the pressurized water reactor. It contains about 200 metric tons of uranium sealed in hundreds of metal rods. The rods are immersed in water, as shown in **Figure 13.** The water surrounding the rods not only serves as the moderator, but it also absorbs thermal energy from the fission of uranium.

Figure 13 High-speed particles from fuel rods enter the water faster than the speed of light in water. Photons are emitted, causing the water to glow. This is called the Cerenkov effect and is not the result of radioactivity.

Rods of cadmium metal are placed between the uranium rods. Cadmium absorbs neutrons easily and also acts as a moderator. The cadmium rods are moved in and out of the reactor to control the rate of the chain reaction. Thus, the rods are called control rods. When the control rods are inserted completely into the reactor, they absorb enough of the neutrons released by the fission reactions to prevent any further chain reaction. As the control rods are removed from the reactor, the rate of energy release increases, with more free neutrons available to continue the chain reaction.

Energy released by the fission heats the water surrounding the uranium rods. The water itself doesn't boil because it is under high pressure, which increases its boiling point. As shown in **Figure 14,** this water is pumped to a heat exchanger, where it causes other water to boil, producing steam that turns turbines. The turbines are connected to generators that produce electrical energy.

Fission of $^{235}_{92}U$ nuclei produces krypton, barium, and other atoms in the fuel rods. Most of these atoms are radioactive. About once a year, some of the uranium fuel rods must be replaced. The old rods no longer have enough uranium to be used in the reactor, but they are still extremely radioactive. Because the radiation is so dangerous to people and the environment, the rods must be stored in a location that can be secured. Better methods of permanently storing these radioactive waste products are currently being developed.

Nuclear Fusion

In nuclear **fusion,** nuclei with small masses combine to form a nucleus with a larger mass. In the process, energy is released. You learned earlier in this module that the larger nucleus is more tightly bound, so its mass is less than the sum of the masses of the smaller nuclei. This loss of mass corresponds to a release of energy.

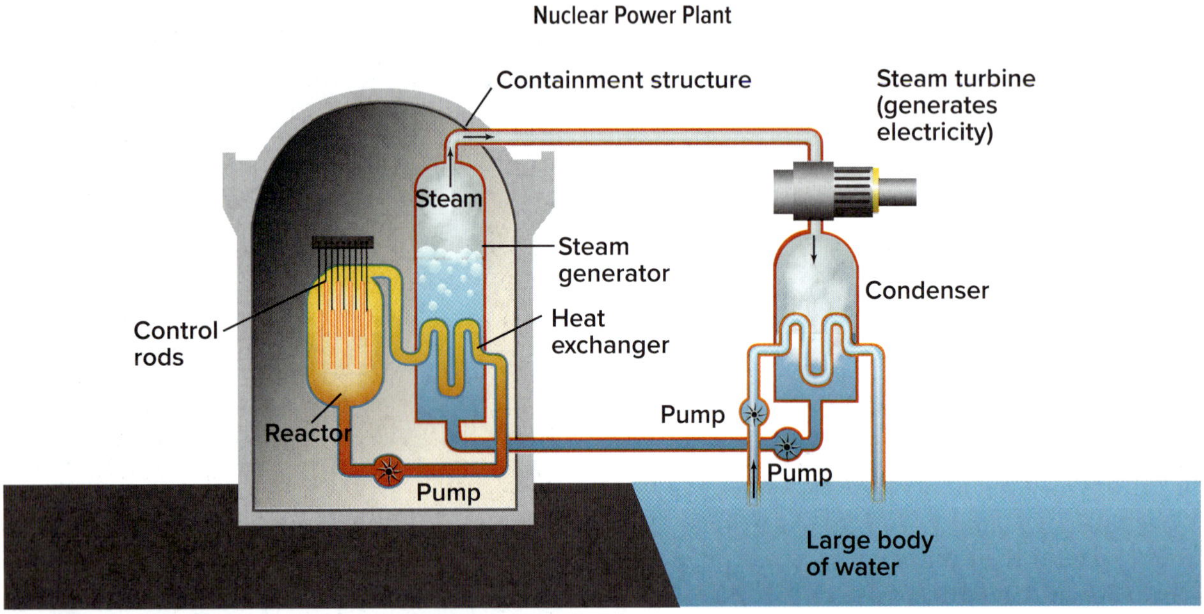

Figure 14 The reactions that occur in nuclear power plants release thermal energy. This energy is ultimately used to turn a turbine, which converts mechanical energy into electrical energy.

An example of fusion is a process that occurs in the Sun. This process, called the proton-proton chain, is shown in **Figure 15.** Four hydrogen nuclei (protons) fuse in several steps to form one helium nucleus. The proton-proton chain can be represented in equation form:

$$^{1}_{1}\text{H} + {^{1}_{1}\text{H}} \rightarrow {^{2}_{1}\text{H}} + {^{0}_{+1}\text{e}} + {^{0}_{0}v} \text{ and } {^{1}_{1}\text{H}} + {^{1}_{1}\text{H}} \rightarrow {^{2}_{1}\text{H}} + {^{0}_{+1}\text{e}} + {^{0}_{0}v}$$

$$^{1}_{1}\text{H} + {^{2}_{1}\text{H}} \rightarrow {^{3}_{2}\text{He}} + \gamma \text{ and } {^{1}_{1}\text{H}} + {^{2}_{1}\text{H}} \rightarrow {^{3}_{2}\text{He}} + \gamma$$

$$^{3}_{2}\text{He} + {^{3}_{2}\text{He}} \rightarrow {^{4}_{2}\text{He}} + 2({^{1}_{1}\text{H}})$$

The net result (subtracting out the two protons produced in the final step) is that four protons produce one $^{4}_{2}$He, two positrons, and two neutrinos.

The mass of the four protons is greater than the mass of the helium nucleus that is produced. The mass difference is equivalent to the sum of the kinetic energy of the resultant particles and the radiant energy of the photons. The energy released by the fusion that produces one helium-4 nucleus is 25 MeV. In comparison, the energy released when one dynamite molecule reacts chemically is about 20 eV, which is about 1 million times smaller.

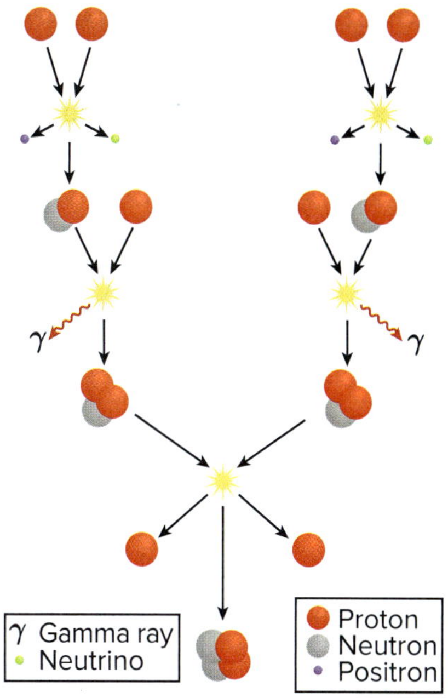

Figure 15 In the Sun, a series of fusion reactions occurs to convert hydrogen into helium.

Confirm *that charge is conserved in this reaction.*

The repulsive force between the charged nuclei requires the fusing nuclei to have high energies. Thus, fusion reactions take place only when the nuclei have large amounts of thermal energy. The proton-proton chain requires a temperature of about 2×10^{7} K, such as that found in the center of the Sun. Stars that are hot enough can also fuse together nuclei to form larger elements such as carbon, nitrogen, and oxygen. The hottest stars have enough thermal energy to fuse nuclei as large as silicon into iron.

Fusion reactions also occur in a hydrogen bomb, which is also called a thermonuclear bomb. In this device, the high temperature necessary to produce the fusion reaction is produced by exploding a uranium fission bomb, which is often called a nuclear bomb.

Check Your Progress

29. **Beta Decay** How can an electron be expelled from a nucleus during beta decay if the nucleus has no electrons?

30. **Nuclear Reactions** The polonium isotope $^{210}_{84}$Po undergoes alpha decay. Write the equation for the reaction.

31. **Half-Life** Use **Figure 9** and **Table 2** to estimate in how many days a sample of $^{131}_{53}$I would have three-eighths its original activity.

32. **Nuclear Reactor** Lead often is used as a radiation shield. Why is it not a good choice for a moderator in a nuclear reactor?

33. **Fusion** Why doesn't helium in a balloon undergo fusion?

34. **Energy** Calculate the energy released in the first fusion reaction in the Sun, $^{1}_{1}\text{H} + {^{1}_{1}\text{H}} \rightarrow {^{2}_{1}\text{H}} + {^{0}_{+1}\text{e}} + {^{0}_{0}v}$.

35. **Critical Thinking** Isotopes that experience alpha decay, called alpha emitters, are used in smoke detectors. An emitter is mounted on one plate of a capacitor, and the α particles strike the other plate. As a result, there is a potential difference across the plates. Explain and predict which plate has the more positive potential.

LEARNSMART® Go online to follow your personalized learning path to review, practice, and reinforce your understanding.

LESSON 3
THE BUILDING BLOCKS OF MATTER

FOCUS QUESTIONS

How do scientists discover particles they can't see?

Particle Accelerators

The first physicists who studied the nucleus with high-speed particles had to use particles from radioactive sources. Other investigations used cosmic rays produced by not-yet-understood processes in stars. In the early 1930s, scientists developed laboratory devices that could accelerate protons and α particles to energies high enough to penetrate the nucleus. Today, the linear accelerator, like the one shown in **Figure 16,** and the synchrotron are regularly used to accelerate particles to high energies and help scientists make further discoveries about the structure of matter.

Figure 16 A linear accelerator accelerates charged particles.

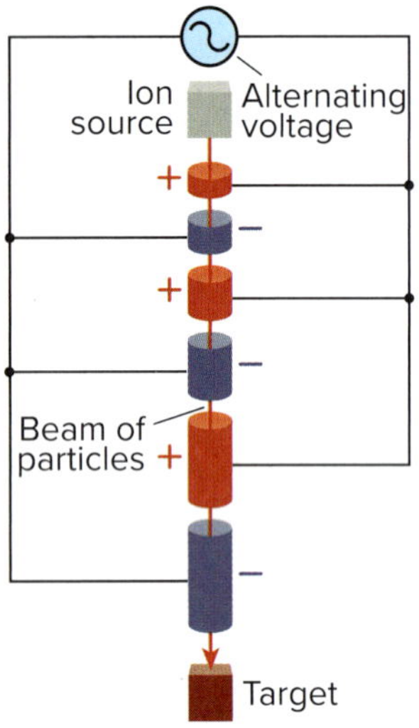

To accelerate charged particles, the charge on the tubes is reversed as the particles move.
(Diagram is not to scale.)

Stanford University's linear accelerator is 3.3 km long. Electrons reach energies of 50 GeV.

3D THINKING **DCI** Disciplinary Core Ideas **CCC** Crosscutting Concepts **SEP** Science & Engineering Practices

COLLECT EVIDENCE

Use your Science Journal to record the evidence you collect as you complete the readings and activities in this lesson.

INVESTIGATE

GO ONLINE to find these activities and more resources.

Quick Investigation: What's in the background?
Carry out an investigation to measure the rate of background radiation and to determine its type.

Review the News
Obtain information from a current news story about current research in nuclear energy, fusion, or particle physics. Evaluate your source and communicate your findings to your class.

Linear accelerators A linear accelerator can be used to accelerate protons, electrons, or their antimatter counterparts (described later in this lesson). It consists of a series of hollow tubes within a long, evacuated chamber. The tubes are connected to a high-frequency alternating potential difference source, as shown in **Figure 16.** Protons are produced in an ion source similar to those in cathode-ray tubes. When the first tube has a negative potential, protons accelerate into it. They move at a constant velocity because there is no electric field within the tube.

The length of the tube and the frequency of the potential difference are adjusted so that the potential reverses at the same time the protons reach the far end of the tube. Thus, the second tube has a negative potential in relation to that of the first. The resulting electric field in the gap between the tubes accelerates the protons into the second tube.

This process continues as the protons receive an acceleration between each pair of tubes. Because the particles move faster as they emerge from each successive tube and all the potential differences change at once, they move farther in the same time. Therefore, each tube must be longer than the one before. The protons ride along the crest of an electric field wave, much as a surfboard moves on the ocean. The proton's energy increases by 1×10^5 eV with each acceleration. By the end, the proton's energy might be billions of electron volts. A similar method is used to accelerate electrons. Note that only charged particles can be accelerated in this way.

Because of their higher mass, protons cannot be accelerated in linear accelerators to the extreme energies needed for modern experiments. Instead, linear proton accelerators are used as the initial process to inject protons into circular accelerators where higher energies can be achieved.

Synchrotrons Using a magnetic field to direct particles in a circle allows an accelerator to be more compact. A circular accelerator, called a synchrotron, is shown in **Figure 17.** Bending magnets are separated by straight regions where a high-frequency potential difference accelerates the particles. The strength of the magnetic field and the length of the path are chosen so that the particles reach the alternating electric field region precisely when the field's polarity will accelerate them.

Figure 17 In a synchrotron, the path and acceleration of the particles are controlled using magnets. The Large Hadron Collider has a circumference of 27 km.

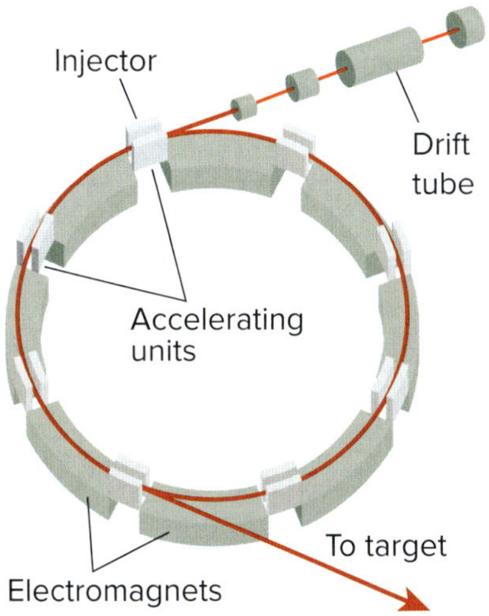

One of the largest synchrotrons is the Large Hadron Collider (LHC) at the European Organization for Nuclear Research (CERN) laboratory near Geneva, Switzerland, shown in **Figure 17** on the previous page. The LHC is so large because it takes a long distance to accelerate particles to the desired energies. The LHC has two proton beams that travel in opposite directions and collide in several interaction regions, and the results are studied. The LHC is designed to accelerate larger nuclei as well.

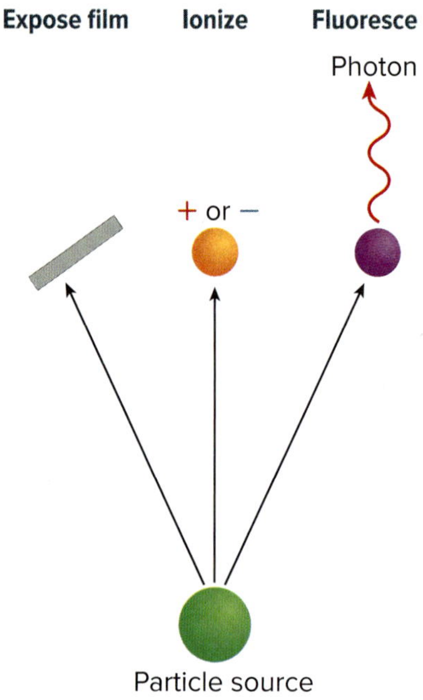

Figure 18 Subatomic particles cannot be seen or felt by humans, but they can be detected when they interact with some types of matter. Particles can expose film, ionize matter, or cause matter to emit photons.

Particle Detectors

The high-speed particles that move through accelerators either collide or crash into targets, producing a spray of subatomic particles. Scientists try to detect these tiny particles to identify and learn about them. To be detected, particles must interact with matter in such a way that people can observe those interactions. Your hand will stop an α particle, yet you will have no idea that the particle struck you. As you read this sentence, billions of neutrinos—many from the Sun—pass undetected through your body. Over the past century, scientists have devised tools to detect and distinguish the products of nuclear reactions.

Recall that radiation from uranium can expose photographic film. Because α particles, β particles, and gamma rays expose photographic film, it is a useful particle detector. Other particle detectors make use of the fact that a high-speed particle can collide with an atom, ejecting one of its electrons. That is, the high-speed particles ionize matter through collisions. In addition, some substances emit photons when exposed to certain types of radiation. These substances also can be used to detect radiation. These three means of detecting radiation are shown in **Figure 18.**

Get It?

Identify three means of detecting subatomic particles.

Geiger-Müller tubes A Geiger-Müller tube contains a copper cylinder with a negative charge. Down the center of this cylinder runs a rigid wire with a positive charge. The potential difference across the wire and the cylinder is kept just below the point at which a spontaneous discharge, or spark, occurs.

When a charged particle or gamma ray enters the tube, it ionizes a gas atom between the copper cylinder and the wire. The resulting positive ion is accelerated toward the copper cylinder by the potential difference, and the electron is accelerated toward the positive wire. As these charged particles move toward the electrodes, they create an avalanche of charged particles, and a pulse of current goes through the tube. Geiger counters are useful detectors because they are small enough to transport.

Condensation trails A device once used to detect particles was the Wilson cloud chamber. The chamber contained an area supersaturated with water vapor or ethanol vapor. Charged particles traveled through the chamber, leaving a trail of ions in their paths. Vapor condensed into small droplets on the ions forming visible trails. In a similar detector, called the bubble chamber, charged particles pass through a liquid held just above the boiling point. The trails of ions cause small vapor bubbles to form, marking the particles' trajectories, as shown in **Figure 19.**

Neutral particles do not leave trails because they do not produce discharges. The laws of conservation of energy and momentum in collisions can be used to infer if any neutral particles were produced.

Figure 19 In a bubble chamber, charged particles leave tracks. This false-color photograph shows the paths of several particles.

Explain *why the trails bend in different directions.*

Modern detectors Recent technology has produced sophisticated detectors to record the paths of particles and analyze properties such as momentum, mass, and energy. A detector called a wire chamber is like a giant Geiger-Müller tube. Huge plates are separated by a small gap filled with a low-pressure gas. A discharge is produced in the path of a particle passing through the chamber. A computer records the discharge position. Solid-state detectors also use the ionization of particles to track their positions. Other detectors emit light when particles pass through them. These are often used to measure a particle's energy or momentum.

The entire array of detectors used in high-energy accelerator experiments can be huge. For example, the detector known as ATLAS at the Large Hadron Collider (LHC) is 25 m high, as shown in **Figure 20.** ATLAS is actually numerous detectors acting together. It is designed to monitor a billion particle collisions each second and select about 100 of these events that may be of interest to scientists. The recorded data can be used to create a computer picture of the collision events, as shown in **Figure 20.**

Figure 20 ATLAS (**A** **T**oroidal **L**HC **A**pparatu**S**) monitors tracks from billions of collisions and produces images of some of the most interesting events. This proton-proton collision was recorded by ATLAS in June 2012, during the search for the Higgs boson. The red tracks show the production of two muon pairs.

Get It?

Explain how the ATLAS detector differs from the Wilson cloud chamber.

Finding New Particles

In 1930 the model of the atom was fairly simple: protons and neutrons surrounded by electrons. More detailed studies of radioactive decay disturbed this simple picture. The α particles and gamma rays emitted by radioactive nuclei have single energies that depend on the decaying nucleus. But β particles are emitted with a wide range of energies that are not always equal to the energy change of the nucleus.

This wide energy range of electrons emitted during beta decay suggested to Niels Bohr that another energy-carrying particle might be involved in nuclear reactions. Wolfgang Pauli in 1931 and Enrico Fermi in 1934 suggested that an unseen neutral particle was emitted with the β particle. Fermi named the particle the neutrino ("little neutral one" in Italian). The particle, which is actually an antineutrino, was not directly observed until 1956.

Get It?
Explain why *neutrino* is an appropriate name.

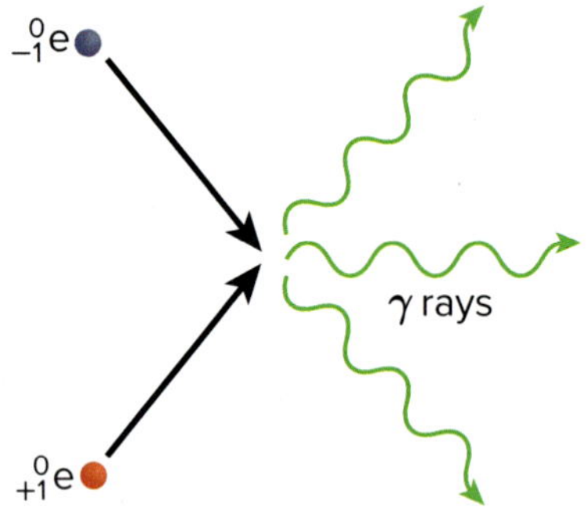

Figure 21 When a particle and its antiparticle collide, they annihilate each other. Here, a positron and an electron collide to produce gamma rays.

More and more particles Other studies discovered more particles. The muon, which seemed to be a heavy electron, was discovered in 1937. In 1947 another particle, the pion, was discovered. Experiments with particle accelerators revealed even more particles, with a wide range of masses. Some were much more massive than the proton. Some had a charge, and others were neutral. Some particles had lifetimes as short as 1×10^{-23} s, while others had no detectable decays. Keeping track of the different particles could be difficult. At one point Fermi was asked to identify a particle track and replied, "If I could remember the names of all these particles, I'd be a botanist!"

Antimatter In the late 1920s, Paul Dirac predicted the existence of an antiparticle associated with each kind of particle. The antiparticle has the same mass and charge magnitude as the particle, but the sign of the charge is opposite. Collectively, these antiparticles are known as antimatter. In some cases, antimatter particles are indicated by a bar over the symbol. In 1932, the electron's positive counterpart, called a positron, was discovered. When a positron and an electron collide, the two can annihilate each other to produce gamma rays, as shown in **Figure 21.** Collisions between other particles and their antiparticles have similar results.

The Standard Model

By the late 1960s, it became clear that protons, neutrons, and pions are not fundamental, or elementary, particles. Physicists divide elementary particles into three families, called quarks, leptons, and bosons, as shown in **Figure 22** on the next page. The theory explaining the interactions of the elementary particles is called the **Standard Model.**

Quarks are elementary particles that have mass and charge, experience the strong force, and can combine into larger particles, called hadrons. There are six types, called flavors, of quarks. Different combinations of quarks make different hadrons. For example, protons and neutrons are made of three quarks and are called baryons. A meson, such as the pion, is made of a quark and an antiquark. **Leptons,** such as electrons and neutrinos, are also elementary particles that have mass and charge but belong to a different family than quarks. There are six flavors of leptons. Unlike quarks, leptons are not affected by the strong force.

Get It?
Distinguish between quarks and leptons.

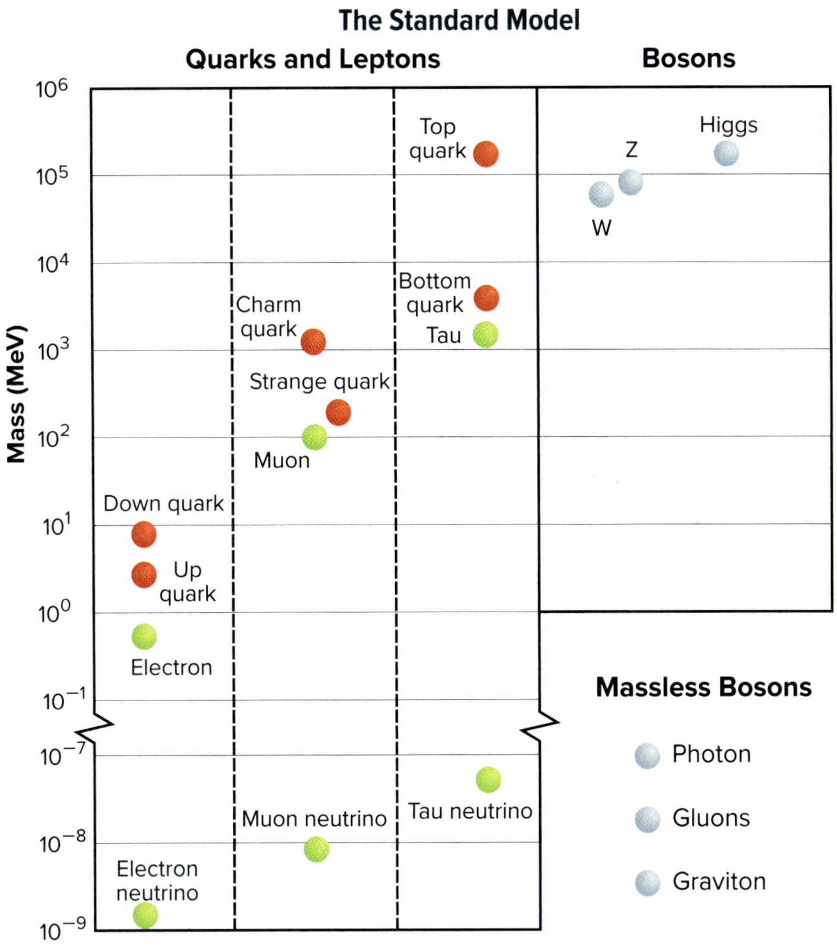

Figure 22 The Standard Model includes three families of particles: quarks, leptons, and bosons. The six known quarks and the six known leptons are divided into three generations, shown as columns in the figure. The everyday world is made from particles in the left-hand column (u, d, e). Particles in the middle column (c, s, μ) are found in cosmic rays and are routinely produced in particle accelerators. Particles in the right-hand column (b, t, τ) are believed to have existed briefly during the earliest moments of the Big Bang and are created in high-energy collisions. The gauge bosons carry the fundamental forces. The masses of the particles are determined by their interactions with the Higgs field, which was confirmed with the observation of the Higgs boson. Masses are stated as energy equivalents, given by Einstein's formula $E = mc^2$.

Bosons are particles that, unlike quarks and leptons, do not obey the Pauli exclusion principle, meaning more than one boson can occupy a given quantum state at the same location at the same time. There are two subsets of elementary bosons: gauge bosons (which are force carriers) and scalar bosons. In 1935, Japanese physicist Hideki Yukawa hypothesized the existence of a particle that could carry the strong nuclear force through space. This idea developed into the model of force carriers, which are particles that transmit forces between objects interacting at a distance. Each force has a corresponding force carrier. Photons carry the electromagnetic force. Gluons carry the strong nuclear force that binds quarks in hadrons. The weak force carriers (W^+, W^-, and Z^0) are involved in β decay. Graviton is the name given to the yet-undetected carrier of gravitational force. Experiments are currently underway to find this elusive particle. The sole elementary scalar boson is the Higgs boson, which results from the quantum excitation of the Higgs field. The masses of all particles are determined by their interactions with this Higgs field.

Protons and Neutrons

Nucleons—the proton and the neutron—are both made of three quarks, as shown in **Figure 23.** The proton has two up quarks (u) with a charge $+\frac{2}{3}$e and one down quark (d) with a charge $-\frac{1}{3}$e . A proton is described as p = uud. The charge on the proton is the sum of the charges of the three quarks, $\left(\frac{2}{3} + \frac{2}{3} + -\frac{1}{3}\right)e = +e$. The neutron is made up of one up quark and two down quarks: n = udd. The charge of the neutron is zero, $\left(\frac{2}{3} + -\frac{1}{3} + -\frac{1}{3}\right)e = 0$.

Groups of Quarks

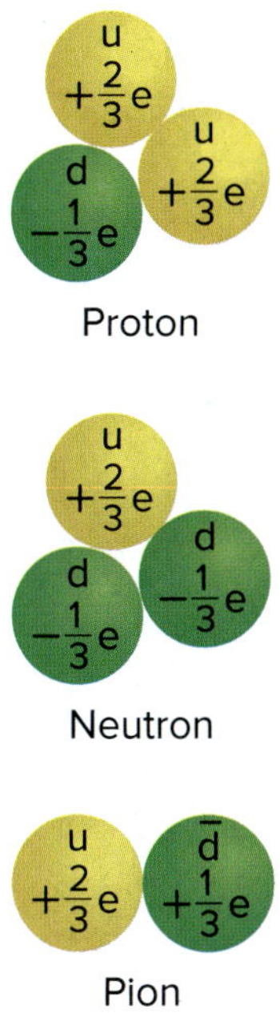

Figure 23 Quarks make up larger particles such as protons, neutrons, and pions. Notice that quarks have fractional charges, but the particles they make have whole-number charges.

Individual free quarks cannot be observed because the strong force that holds them together becomes larger as the quarks are pulled farther apart. In this case, the strong force acts like the force of a spring. It is unlike the electric force, which weakens as charged particles move apart.

Particle Annihilation and Production

Each quark and each lepton has a corresponding antiparticle. A particle and its antiparticle have identical masses. A charged particle and its antiparticle have opposite charges. For example, an up quark (u) has a charge of $+\frac{2}{3}$, while an anti-up quark ($\overline{u}$) has a charge of $-\frac{2}{3}$.

Because a baryon is a group of three quarks, the antiparticle of a baryon is a group of three antiquarks. For instance, a proton (uud) has a charge of +1. The antiproton ($\overline{uud}$) has a charge of $-\frac{2}{3}-\frac{2}{3}+\frac{1}{3}=-1$. Likewise, $\overline{u}\,\overline{u}\,\overline{d}$ is an antineutron, which has a charge 0. Because a meson is a quark-antiquark pair, its antiparticle is also a quark-antiquark pair.

Particle annihilation When a particle and its corresponding antiparticle annihilate each other, the resulting energy can be calculated using $E = mc^2$. An electron and a positron, for example, can collide and produce gamma rays: $e^- + e^+ \rightarrow \gamma + \gamma$. Each has a mass of 9.11×10^{-31} kg. The energy equivalent of the electron-positron pair can be calculated as follows:

$$E = 2(9.11\times10^{-31}\text{ kg})(3.00\times10^{8}\text{ m/s})^2$$

$$= (1.64\times10^{-13}\text{ J})(1\text{eV}/1.60\times10^{-19}\text{ J})$$

$$= 1.02\times10^{6}\text{ eV or } 1.02\text{ MeV}$$

Thus, when an electron and a positron at rest annihilate each other, the sum of the energies of the gamma rays emitted is 1.02 MeV.

Figure 24 Particles are always produced in pairs. If a particle were produced without its antiparticle, charge conservation would be violated. One possibility is for a gamma ray to decay into an electron-positron pair. Two such decays are shown here.

LBNL/Science Source

Particle production The inverse of annihilation also can occur. Energy can be converted directly into matter. The conversion of energy into a matter-antimatter pair of particles is called **pair production.** If a gamma ray with at least 1.02 MeV of energy passes close by a nucleus, for example, a positron and electron pair can be produced: $\gamma \rightarrow e^- + e^+$. This process is shown in a bubble chamber in **Figure 24.** A magnetic field around the bubble chamber caused the oppositely charged particles' tracks to curve in opposite directions. Uncharged gamma rays, entering the detector from the top of the picture, produced no tracks. If the gamma ray's energy is larger than 1.02 MeV, the excess energy goes into the kinetic energy of the positron and electron. The positron soon collides with another electron and they are both annihilated, resulting in two or three gamma rays with a total energy of no less than 1.02 MeV.

Antiprotons and antineutrons also can be created. Recall that protons and neutrons have approximately 1800 times as much mass as electrons. Thus, the energy needed to create proton-antiproton or neutron-antineutron pairs is larger. The first proton-antiproton pair was produced and observed at Berkeley, California, in 1955.

Particle conservation The total number of quarks and the total number of leptons in the universe is constant. That is, quarks and leptons are created or destroyed only in particle-antiparticle pairs. On the other hand, force carriers such as gravitons, photons, gluons, and weak bosons can be created or destroyed if there is enough energy.

Individual reactions, such as $\gamma \rightarrow e^-$ and $\gamma \rightarrow e^+$, however, cannot occur because such events would violate the law of conservation of charge. Reactions such as $\gamma \rightarrow e^- + p$ also do not occur; the pair must be a particle and its corresponding antiparticle.

Beta Decay and the Weak Interaction

In a stable nucleus, neutrons do not decay, but free neutrons and neutrons in an unstable nucleus can undergo beta decay. Recall that during beta decay a neutron is transformed into a proton by emitting a β particle and an antineutrino $\left({}^{0}_{0}\bar{v}\right)$. The antineutrino has a very small mass and is uncharged, but like the photon, it carries momentum and energy. The neutron decay equation is written as follows:

$$ {}^{1}_{0}\text{n} \rightarrow {}^{1}_{1}\text{p} + {}^{0}_{-1}\text{e} + {}^{0}_{0}\bar{v} $$

Protons can decay in a similar way. The decay of a free proton has never been observed, but a proton in the nucleus can decay to a neutron with the emission of a positron $\left({}^{0}_{+1}\text{e}\right)$ and a neutrino $\left({}^{0}_{0}v\right)$. Recall that this is called positron emission.

$$ {}^{1}_{1}\text{p} \rightarrow {}^{1}_{0}\text{n} + {}^{0}_{+1}\text{e} + {}^{0}_{0}v $$

These decays cannot be explained by the strong force. The existence of beta decay indicates that there must be another interaction occurring in the nucleus. The interaction that acts in the nucleus during beta decay is called the **weak nuclear force.** As you might guess, this force is much weaker than the strong nuclear force.

PRACTICE Problems

ADDITIONAL PRACTICE

36. For each particle below, determine its pair production partner.
- **a.** positron
- **b.** neutron
- **c.** antiproton
- **d.** muon

37. The mass of a proton is 1.67×10^{-27} kg.
- **a.** Find the energy equivalent of the proton's mass in joules.
- **b.** Convert this value to electron volts.
- **c.** Find the smallest total γ-ray energy that could result in a proton-antiproton pair.

38. A positron and an electron can annihilate and form three gamma rays. Two gamma rays are detected. One has an energy of 225 keV, the other 357 keV. What is the energy of the third gamma ray?

39. The mass of a neutron is 1.008665 u.
- **a.** Find the energy equivalent of the neutron's mass in megaelectron volts.
- **b.** Find the smallest total γ-ray energy that could result in the production of a neutron-antineutron pair.

40. CHALLENGE The mass of a muon is 0.1135 u. It decays into an electron and two neutrinos. What is the energy released in this decay?

PHYSICS Challenge

Nuclear Decay The uranium isotope $^{238}_{92}U$ decays by α emission and two successive β emissions back into uranium again.

1. Show the three nuclear decay equations.

2. Predict the atomic mass number of the uranium formed.

Quark model of beta decay A proton (uud) and a neutron (udd) differ by only one quark. Beta decay occurs in two steps, as shown in **Figure 25.** First, one d quark in the nucleus changes to a u quark with the emission of a W^- boson: $d \rightarrow u + W^-$. The W^- boson is one of the three carriers of the weak nuclear force. In the second step, the W^- boson decays into an electron and an antineutrino. Similarly, a proton in a nucleus can decay into a neutron and a W^+ boson. The W^+ boson then decays into a positron and a neutrino.

The emission of the third weak-force carrier, the Z^0 boson, is not accompanied by a change from one quark to another. The Z^0 boson produces an interaction between the nucleons and the electrons in atoms. This interaction is similar to, but much weaker than, the electromagnetic force holding the atom together and was first detected in 1979. The W^+, W^-, and Z^0 bosons first were observed directly in 1983.

Neutrinos and antineutrinos have long been thought to be massless, but recent experiments that detect neutrinos from the Sun show that neutrinos do have mass. These masses are much smaller than those of any other known particles and have not yet been measured accurately.

Testing the Standard Model

You can see in **Figure 22** that the quarks and the leptons are separated into three columns. The everyday world is made from particles in the left-hand columns: protons, neutrons, and electrons. Particles in the middle group are found in cosmic rays and are routinely produced in particle accelerators. Particles in the right-hand family are believed to have existed briefly during the earliest moments of the Big Bang and are created in high-energy collisions.

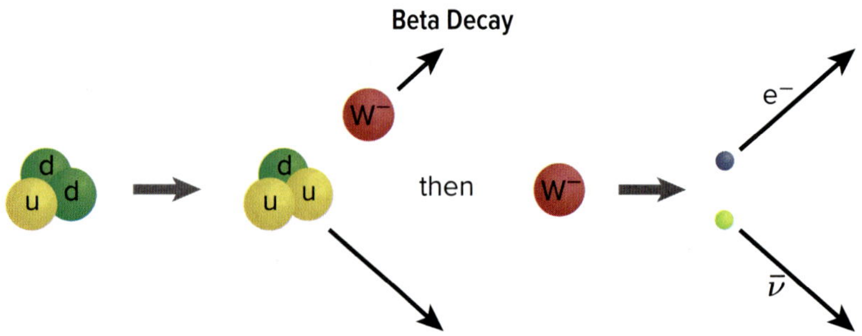

Figure 25 According to the quark model, beta decay of a neutron to a proton occurs in two steps.

Infer *Given that charge is conserved during beta decay, what charge must a W⁻ boson have?*

What determines the masses of the quarks and the leptons? In 2012, physicists observed the Higgs boson at the Large Hadron Collider, which confirmed the existence of the long-hypothesized Higgs field. Interactions with this Higgs field determine a particle's mass. The Standard Model explains the interactions of elementary particles, governed by the four fundamental forces; with the confirmation of the Higgs field, the Standard Model now has an explanation of what causes particles to have mass.

Why are there four forces?

The differences among the four fundamental interactions are evident: the forces act on different quantities such as charge or mass, they have different dependencies on distance, and the force carriers have different properties, as shown in **Figure 26.**

There are, however, some similarities among the interactions. For instance, the force between charged particles, the electromagnetic force, is carried by photons in much the same way as weak bosons carry the weak force. The electromagnetic force acts over a long range because the photon has zero mass, while the weak force acts over short distances because the W and Z bosons are relatively massive. The mathematical structures of the theories of the weak interaction and electromagnetic interaction, however, are similar.

Evidence from the cosmos

Look at the supernova in **Figure 27** on the next page. Astrophysical evidence indicates that during massive stellar explosions, the electromagnetic force and the weak force are identical. According to present theories of the origin of the universe, the two forces were identical during the early moments of the cosmos as well. For this reason, the electromagnetic and weak forces are said to be unified into a single force, called the electroweak force.

In the same way that scientists in the 1970s developed a theory of the unified electromagnetic and weak forces, physicists presently are developing theories that include the strong force as well. Work is still incomplete. Hypotheses are being revised and improved, and experiments to test these hypotheses are being planned. A fully unified theory that includes gravitation will require even more effort.

The Four Fundamental Forces

Gravitational Force	Electromagnetic Force	Weak Force	Strong Force
• affects particles with mass • force carrier: gravitons (not yet found) • infinite range • relative strength: 10^{-36} • always attractive • holds together large-scale structures such as planets and galaxies	• affects electrically charged particles • force carrier: photons • infinite range • relative strength: 1 • can be attractive or repulsive • holds atoms and molecules together	• affects quarks and leptons • force carriers: W^+, W^-, and Z^0 bosons • range: $\approx 10^{-18}$ m • relative strength: 10^{-7} • responsible for beta decay	• affects quarks and hadrons • force carrier: gluons • range: $\approx 10^{-15}$ m • relative strength: 20 • binds quarks into hadrons and holds the nucleus together
	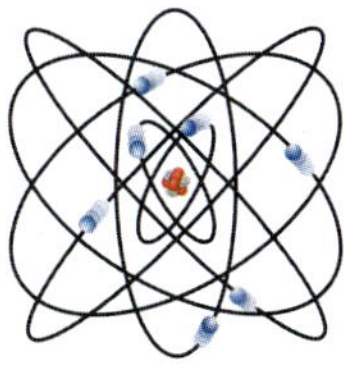	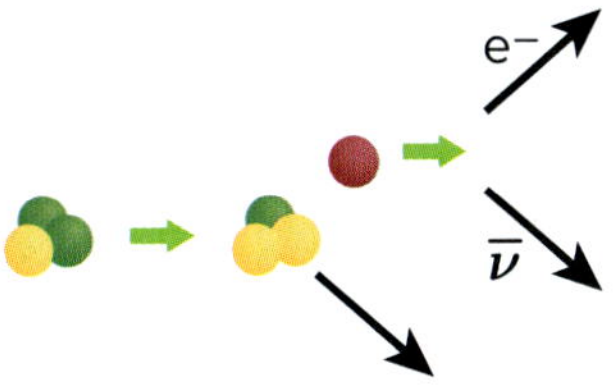	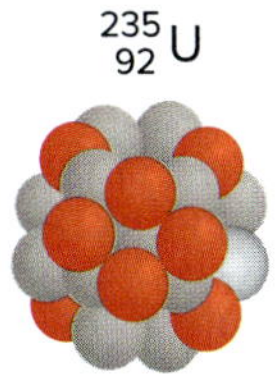

Figure 26 The four fundamental forces are the gravitational force, the electromagnetic force, the weak nuclear force, and the strong nuclear force.

Figure 27 The electromagnetic and weak forces are indistinguishable in a supernova, such as 1987A shown below. When the supernova explodes, it gives off a burst of light and neutrinos. The light and the neutrinos reach Earth at the same time, showing that neutrinos can travel near the speed of light and are produced in supernovas.

Dark matter Results of studies of galaxies are even more perplexing. These results suggest that matter described by the Standard Model makes up only a small fraction of the mass in the universe. A much larger amount is formed of dark matter, so called because it doesn't interact with photons or ordinary matter except through the gravitational force. In addition, there appears to be dark energy, which is thought to be responsible for accelerating the universe's expansion.

Thus the studies of the tiniest particles that make up nuclei are directly connected to investigations of the largest systems, the galaxies that make up the universe. Elementary particle physicists and cosmologists used to be at opposite ends of the length scale; now they are working together as they try to answer the same question together: "What IS the universe made of?"

Check Your Progress

41. **Models** Research the limitations of the Standard Model and possible alternative models.
42. **Nucleus Bombardment** Why would a proton require more energy than a neutron when used to bombard a nucleus?
43. **Particle Accelerator** Protons in the Fermi Laboratory accelerator move counter-clockwise when viewed from above. In what direction is the magnetic field of the bending magnets?
44. **Pair Production** **Figure 24** shows the production of two electron-positron pairs. Why does the bottom set of tracks curve less than the top pair of tracks?
45. **Critical Thinking** Consider the following equations:

    $u \rightarrow d + W^+$ and $W^+ \rightarrow e^+ + \nu$

    How could they be used to explain the radioactive decay of a nucleon that results in the emission of a positron and a neutrino? Write the equation in terms of nucleons, rather than quarks.

Fusion—A Future Energy Source?

Nuclear fusion has the potential to supply Earth's growing population with clean, sustainable, and abundant energy. However, there are many challenges that need to be overcome before fusion can be used to meet society's energy needs.

This photo shows the inside of a fusion reactor.

What is fusion?

Fusion occurs when nuclei with small masses combine, resulting in the formation of a larger nucleus and the release of a vast amount of energy. Temperatures within stars, such as the Sun, are high enough for fusion to occur.

Fusion is considered an ideal energy source. The fuel used for fusion—hydrogen—is available on Earth in vast amounts. Fusion is a "clean" energy source that does not produce by-products that pollute the environment. In addition, fusion is safer than fission, which is what powers currently-used nuclear power plants. Finally, the amount of energy released by fusion reactions is enormous.

Challenges

With all of the advantages of fusion as an energy source, why is it not being used to meet society's growing energy needs? The answer lies with the technological challenges of starting and maintaining a sustainable, controlled fusion reaction.

Fusion requires the particles to have a very high thermal energy in order to overcome repulsive forces. With today's technology, achieving this temperature typically requires an energy input greater than any energy produced.

To initiate a fusion reaction, the hydrogen fuel must be heated, which requires an investment of energy. Temperatures greater than 150 million degrees C are needed to start the reaction, and the super-heated material, called plasma, must be contained. Current technology uses electromagnets to keep the superheated plasma from touching (and melting) surfaces in the reactor. Once fusion begins, it occurs in a chain-reaction, with energy from one reaction fueling more reactions. At that point, the challenge is sustaining and controlling the reaction.

Economic, political, and time constraints are also challenges. Building fusion reactors requires an investment of time, labor, space, and political cooperation between various factions.

ENGAGE IN ARGUMENT FROM EVIDENCE

Develop a written argument about the importance of fusion as a future energy source and the value of continued research in this field. Use evidence to support your argument. Then, engage in a class discussion in which different arguments are shared and discussed.

CALIFORNIA CONNECTION

Research the Department of Energy's Fusion Energy Sciences program to find an affiliated facility in California. Find out more about the fusion-related work being done at this facility. Discuss what you have learned with a small group.

STUDY GUIDE

GO ONLINE to study with your Science Notebook.

Lesson 1 THE NUCLEUS

- The number of protons in a nucleus is given by the atomic number (Z). The sum of the numbers of protons and neutrons in a nucleus is equal to the mass number (A).
- Atoms with the same number of protons but different numbers of neutrons are called isotopes.
- The strong nuclear force binds the nucleus together.
- The mass defect of a nucleus is the difference between the sum of the masses of its parts and the mass of the assembled nucleus. The energy difference between separated nucleons and the assembled nucleus is the binding energy.
- The mass defect is the energy equivalent of the binding energy of the nucleus.

$$E = mc^2$$

- nucleon
- atomic number
- mass number
- atomic mass unit
- strong nuclear force
- mass defect
- binding energy

Lesson 2 NUCLEAR DECAY AND REACTIONS

- An unstable nucleus decays and can transmute into another element. Radioactive decay produces three kinds of particles. Alpha (α) particles are helium nuclei, beta (β) particles are high-speed electrons, and gamma (γ) rays are high-energy photons. In nuclear reactions, the sums of the mass numbers (A) and the nuclear charges (Z) are not changed.
- The half-life of a radioactive isotope is the time required for half of the nuclei to decay. After t half-lives:

$$\text{remaining} = \text{original}\left(\frac{1}{2}\right)^t$$

  The number of decays of a radioactive sample per second is the activity.
- In nuclear fission, a nucleus is split into two smaller nuclei with a release of neutrons and energy. Nuclear reactors use the energy released in fission to generate electrical energy.
- In nuclear fusion, nuclei with small masses combine to form a nucleus with a larger mass. The loss of mass that occurs in this process corresponds to a release of energy.

- radioactive
- alpha decay
- beta decay
- gamma decay
- nuclear reaction
- half-life
- activity
- fission
- chain reaction
- fusion

Lesson 3 THE BUILDING BLOCKS OF MATTER

- Particles are produced by radioactive decay, during high-energy collisions in accelerators, and in stars. They can be detected if they expose film, ionize matter, or cause matter to emit photons.
- The Standard Model includes quarks, leptons, and force carriers. Ordinary matter is made of quarks and leptons. Matter interacts with other matter through particles called force carriers.
- The weak nuclear force is a short-range force that allows quarks and leptons to change flavor. It is responsible for beta decay in which a neutron is transformed into a proton or a proton is transformed into a neutron.

- Standard Model
- quark
- lepton
- force carrier
- pair production
- weak nuclear force

THREE-DIMENSIONAL THINKING

Module Wrap-Up

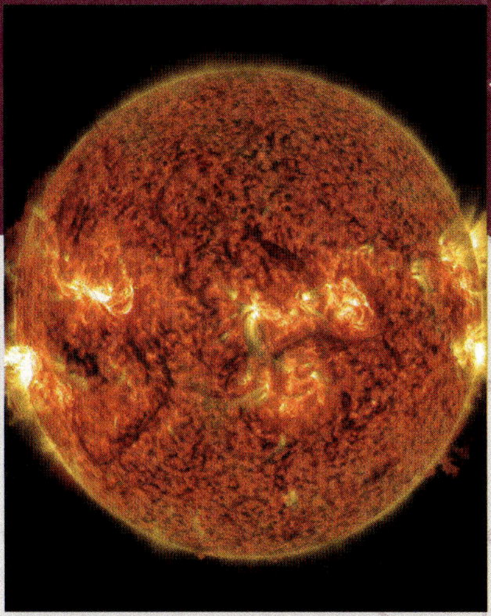

REVISIT THE PHENOMENON

How does the Sun produce energy and how can we replicate it?

CER Claim, Evidence, Reasoning

Explain Your Reasoning Revisit the claim you made when you encountered the phenomenon. Summarize the evidence you gathered from your investigations and research and finalize your Summary Table. Does your evidence support your claim? If not, revise your claim. Explain why your evidence supports your claim.

STEM UNIT PROJECT

Now that you've completed the module, revisit your STEM unit project. You will apply your evidence from this module and complete your project.

GO FURTHER

SEP Data Analysis Lab

Can you evaluate claims from a Web site?

A Web site that appears to be dedicated to reporting scientific breakthroughs makes the following claims.

a. Scientists have been able to cause iron nuclei to undergo fission.

b. Carbon dating reveals that papyrus documents are over 2,000 years old.

c. Scientists have used a linear accelerator to accelerate neutrons to 0.9c.

CER Analyze and Interpret Data

1. **Claim** For each claim the Web site makes, determine if the claim is likely to be true.
2. **Evidence** What scientific information and principles support your claims?
3. **Reasoning** Explain the reasoning you used to justify your claims.

STUDENT RESOURCES

I. Symbols

Δ	change in quantity	$\equiv$	is defined as
$\pm$	plus or minus a quantity	$a \times b$ ab $a(b)$	a multiplied by b
$\propto$	is proportional to		
$=$	is equal to	$a \div b$ a/b $\frac{a}{b}$	a divided by b
$\approx$	is approximately equal to		
$\leq$	is less than or equal to		
$\geq$	is greater than or equal to	$\sqrt{a}$	square root of a
$<<$	is much less than	$\lvert a \rvert$	absolute value of a
$>>$	is much greater than	$\log_b x$	log to the base b of x

II. Measurements and Significant Figures

Connecting Math to Physics Math is the language of physics. Physicists use mathematical equations to describe relationships among the measurements that they make. Each measurement is associated with a symbol that is used in physics equations. These symbols are called variables.

Significant Figures

All measured quantities are approximate and have significant figures. The number of significant figures indicates the precision of the measurement. Precision is a measure of exactness. The number of significant figures in a measurement depends on the smallest unit on the measuring tool. The digit farthest to the right in a measurement is estimated.

Example: In the figure below, what is the estimated digit for each of the measuring sticks used to measure the length of the rod?

By using the lower measuring tool, the length is between 9 and 10 cm. The measurement would be estimated to the nearest tenth of a centimeter. If the length was exactly on the 9-cm or 10-cm mark, record it as 9.0 cm or 10.0 cm.

By using the upper measuring tool, the length is between 9.5 and 9.6 cm. The measurement would be estimated to the nearest hundredth of a centimeter. If the length was exactly on the 9.5-cm or 9.6-cm mark, record it as 9.50 cm or 9.60 cm.

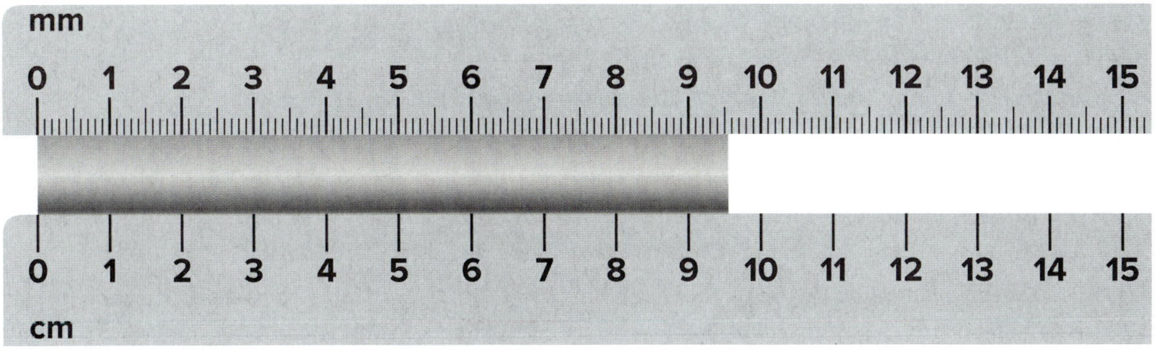

All nonzero digits in a measurement are significant figures. Some zeros are significant, and some are not. All digits between and including the first nonzero digit from the left through the significant figure on the right are significant. Use the following rules when determining the number of significant figures.

1. Nonzero digits are significant.
2. Final zeros after a decimal point are significant.
3. Zeros between two significant figures are significant.
4. Zeros used only as placeholders are not significant.

Example: State the number of significant figures in each measurement.

5.0 g has two significant figures. **Using rules 1 and 2**

14.90 g has four significant figures. **Using rules 1 and 2**

0.0 has one significant figure. **Using rules 2 and 4**

300.00 mm has five significant figures. **Using rules 1, 2, and 3**

5.06 s has three significant figures. **Using rules 1 and 3**

304 s has three significant figures. **Using rules 1 and 3**

0.0060 mm has two significant figures (6 and the last 0). **Using rules 1, 2, and 4**

140 mm has two significant figures (just 1 and 4). **Using rules 1 and 4**

PRACTICE Problems 1 ADDITIONAL PRACTICE

State the number of significant figures in each measurement.

a. 1405 m

b. 2.50 km

c. 0.0034 m

d. 12.007 kg

e. 5.8×10^6 kg

f. 3.03×10^{25} mL

There are two cases in which numbers are considered exact and, thus, have an infinite number of significant figures.

1. Counting numbers have an infinite number of significant figures.
2. Conversion factors have an infinite number of significant figures.

Examples:

The factor "2" in 2*mg* has an infinite number of significant figures.

The number 2 is a counting number. It is an exact integer.

The number "4" in 4 electrons has an infinite number of significant figures.

Because you cannot have a partial electron, the number 4, a counting number, is considered to have an infinite number of significant figures.

60 s/1 min has an infinite number of significant figures.

There are exactly 60 seconds in 1 minute, thus there is an infinite number of significant figures in the conversion factor.

Rounding

You can round a number to a specific place value (such as hundreds or tenths) or to a specific number of significant figures. To do this, determine the place to which you are rounding, and then use the following rules.

1. When the leftmost digit to be dropped is less than 5, that digit and any digits that follow are dropped. Then the last digit in the rounded number remains unchanged.
2. When the leftmost digit to be dropped is greater than 5, that digit and any digits that follow are dropped, and the last digit in the rounded number is increased by one.
3. When the leftmost digit to be dropped is 5 followed by a nonzero number, that digit and any digits that follow are dropped. The last digit in the rounded number increases by one.
4. If the digit to the right of the last significant digit is equal to 5 and 5 is followed by a zero or no other digits, look at the last significant digit. If it is odd, increase it by one; if it is even, do not round up.

Examples: Round the following numbers to the stated number of significant figures.

8.7645 rounded to 3 significant figures is 8.76.	Using rule 1
8.7676 rounded to 3 significant figures is 8.77.	Using rule 2
8.7519 rounded to 2 significant figures is 8.8.	Using rule 3
92.350 rounded to 3 significant figures is 92.4.	Using rule 4
92.25 rounded to 3 significant figures is 92.2.	Using rule 4

PRACTICE Problems 2 ADDITIONAL PRACTICE

Round each number to the number of significant figures shown in parentheses.

a. 1405 m (2)
b. 2.50 km (2)
c. 0.0034 m (1)
d. 12.007 kg (3)

Operations with Significant Figures

If using a calculator, do all of the operations with as much precision as the calculator allows, and then round the result to the correct number of significant figures. The number of significant figures in the result depends on the measurements and on the operation.

Addition and subtraction Round the result to the least precise value among the measurements—the smallest number of digits to the right of the decimal points.

Example: Add 1.456 m, 4.1 m, and 20.3 m.

The least precise values are 4.1 m and 20.3 m because they have only one digit to the right of the decimal points.

$$\begin{array}{rl} 1.456 & \text{m} \\ 4.1 & \text{m} \\ +\ 20.3 & \text{m} \\ \hline 25.856 & \text{m} \end{array}$$

Add the numbers.

25.9 m — Round the result to place value of the least precise value.

Multiplication and division Look at the number of significant figures in each measurement. Perform the calculation. Round the result so that it has the same number of significant figures as the measurement with the least number of significant figures.

Example: Multiply 20.1 m by 3.6 m.

$$(20.1 \text{ m})(3.6 \text{ m}) = 72.36 \text{ m}^2$$

The least precise value is 3.6 m with two significant figures. The product can only have as many digits as the least precise of the multiplied numbers.

72 m — **Round the result to two significant figures.**

PRACTICE Problems 3 — ADDITIONAL PRACTICE

Simplify the following expressions using the correct number of significant figures.

a. 5.012 km + 3.4 km + 2.33 km

b. 45 g − 8.3 g

c. 3.40 cm × 7.125 cm

d. 54 m ÷ 6.5 s

Combination When doing a calculation that requires a combination of addition/subtraction and multiplication/division, use the multiplication/division rule.

Examples:

$$x = 19 \text{ m} + (25.0 \text{ m/s})(2.50 \text{ s}) + \frac{1}{2}(-10.0 \text{ m/s}^2)(2.50 \text{ s})^2$$

$$= 5.0\times10^1 \text{ m}$$

19 m only has two significant figures, so the answer should only have two significant figures.

$$\text{slope} = \frac{70.0 \text{ m} - 10.0 \text{ m}}{29 \text{ s} - 11 \text{ s}}$$

$$= 3.3 \text{ m/s}$$

29 s and 11 s only have two significant figures each, so the answer should only have two significant figures.

Multistep calculations Do not round to significant figures in the middle of a multistep calculation. Instead, round to a reasonable number of decimal places that will not cause you to lose significance in your answer. When you get to your final step where you are solving for the answer asked for in the question, you should then round to the correct number of significant figures.

Example:

$$F = \sqrt{(24 \text{ N})^2 + (36 \text{ N})^2}$$

$$= \sqrt{576 \text{ N}^2 + 1296 \text{ N}^2}$$ **Do not round to 580 N^2 and 1300 N^2.**

$$= \sqrt{1872 \text{ N}^2}$$ **Do not round to 1800 N^2.**

$$= 43 \text{ N}$$ **Final answer, so it should be rounded to two significant figures.**

If you had rounded in each step, you would have obtained an answer of 44 N. This might seem like a small discrepancy, but in more complex calculations, the effects can be large.

For example, if you calculate 5 × 5 × 5 × 5 × 5 without rounding your answer will be 3125, which rounds to 3000. If you round at each step, however, the answer you find is 2500, which rounds to 2000.

III. Fractions, Ratios, Rates, and Proportions

Fractions

A fraction names a part of a whole or a part of a group. It also can express a ratio (see page 710). A fraction consists of a numerator, a division bar, and a denominator.

$$\frac{\text{numerator}}{\text{denominator}} = \frac{\text{number of parts chosen}}{\text{total number of parts}}$$

Simplification Sometimes, it is easier to simplify an expression before substituting the known values of the variables. Variables often cancel out of the expression.

Example: Simplify $\frac{pn}{pw}$.

$\frac{pn}{pw} = \left(\frac{p}{p}\right)\left(\frac{n}{w}\right)$ — Factor out the p in the numerator and the denominator, and break the fraction into the product of two fractions.

$= (1)\left(\frac{n}{w}\right)$ — Substitute (p/p) = 1.

$= \frac{n}{w}$

Multiplication and division To multiply fractions, multiply the numerators and multiply the denominators.

Example: Multiply the fractions $\frac{s}{a}$ and $\frac{t}{b}$.

$\left(\frac{s}{a}\right)\left(\frac{t}{b}\right) = \frac{st}{ab}$ — Multiply the numerators and the denominators.

To divide fractions, multiply the first fraction by the reciprocal of the second fraction. To find the reciprocal of a fraction, invert it—switch the numerator and the denominator.

Example: Divide the fraction $\frac{s}{a}$ by $\frac{t}{b}$.

$\frac{s}{a} \div \frac{t}{b} = \left(\frac{s}{a}\right)\left(\frac{b}{t}\right)$ — Multiply the first fraction by the reciprocal of the second fraction.

$= \frac{sb}{at}$ — Multiply the numerators and the denominators.

Addition and subtraction To add or subtract two fractions, first write them as fractions with a common denominator. To do this, multiply each fraction by the denominator of the other in the form of a fraction equal to one. Then add or subtract the numerators.

Example: Add the fractions $\frac{1}{a}$ and $\frac{2}{b}$.

$\frac{1}{a} + \frac{2}{b} = \left(\frac{1}{a}\right)\left(\frac{b}{b}\right) + \left(\frac{2}{b}\right)\left(\frac{a}{a}\right)$ — Multiply each fraction by a fraction equal to 1.

$= \frac{b}{ab} + \frac{2a}{ab}$ — Multiply the numerators and the denominators.

$= \frac{b + 2a}{ab}$ — Write a single fraction with the common denominator.

PRACTICE Problems 4 ADDITIONAL PRACTICE

Perform the indicated operation. Write the answer in simplest form.

a. $\frac{1}{x} + \frac{y}{3}$

b. $\frac{a}{2b} - \frac{3}{b}$

c. $\left(\frac{3}{x}\right)\left(\frac{1}{y}\right)$

d. $\frac{2a}{5} \div \frac{1}{2}$

Ratios

A ratio is a comparison between two numbers by division. Ratios can be written in several different ways. The ratio of 2 and 3 can be written in four different ways:

2 to 3   2 out of 3   2:3 $\frac{2}{3}$

Rates

A rate is a ratio that compares two quantities with different measurement units. A unit rate is a rate that has been simplified so that the denominator is 1.

Example: Write 98 km in 2.0 hours as a unit rate.

98 km in 2.0 hours is a ratio of $\frac{98\text{ km}}{2.0\text{ hours}}$

$$\frac{98\text{ km}}{2.0\text{ hours}} = \left(\frac{98}{2.0}\right)\left(\frac{\text{km}}{\text{hour}}\right)$$ **Split the fraction into the product of a number fraction and a unit fraction.**

$$= (49)\left(\frac{\text{km}}{\text{hour}}\right)$$ **Simplify the number fraction.**

$$= 49\text{ km per hour or }49\text{ km/h}$$

Example: Write 16 Swedish crowns in 2 U.S. dollars as a unit rate.

16 Swedish crowns in \$2.00 American currency is a ratio of $\frac{16\text{ Swedish crowns}}{2\text{ U.S. dollars}}$

$$\frac{16\text{ Swedish crowns}}{2\text{ U.S. dollars}} = \left(\frac{16}{2}\right)\left(\frac{\text{Swedish crowns}}{\text{U.S. dollars}}\right)$$

$$= (8)\left(\frac{\text{Swedish crowns}}{\text{U.S. dollars}}\right)$$

= 8 Swedish crowns per U.S. dollar or 8 Swedish crowns/U.S. dollar

Proportions

A proportion is an equation that states that two ratios are equal:

$\frac{a}{b} = \frac{c}{d}$, where b and d are not zero.

Proportions used to solve ratio problems often include three numbers and one variable. You can solve the proportion to find the value of the variable. To solve a proportion, use cross multiplication.

Example: Solve the proportion $\frac{a}{b} = \frac{c}{d}$ for b.

Cross multiply: $\frac{a}{b} = \frac{c}{d}$

$bc = ad$ **Write the equation resulting from cross multiplying.**

$b = \frac{ad}{c}$ **Solve for *b*.**

PRACTICE Problems 5   **ADDITIONAL PRACTICE**

Solve the following proportions.

a. $\frac{4}{x} + \frac{2}{3}$

b. $\frac{13}{15} = \frac{n}{75}$

c. $\frac{36}{12} = \frac{s}{16}$

d. $\frac{2.5}{5.0} = \frac{7.5}{w}$

IV. Exponents, Powers, Roots, and Absolute Value

Exponents

An exponent is a number that tells how many times a number (a) is used as a factor. An exponent is written as a superscript. In the term a^n, a is the base and n is the exponent. a^n is called the nth power of a or a raised to the nth power.

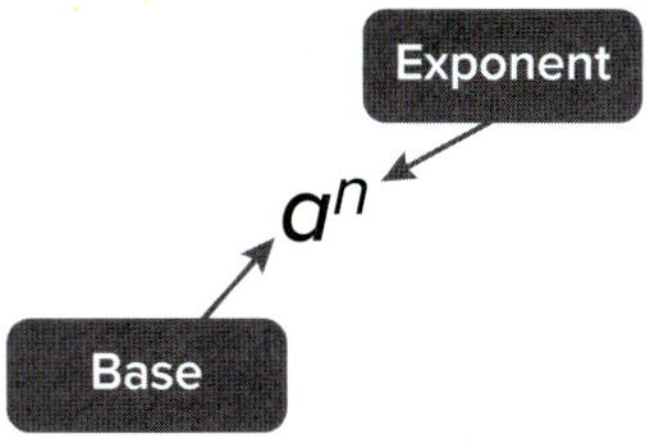

Connecting Math to Physics A subscript is not an exponent. In physics, a subscript is used to further describe the variable. For example, V_0 can be used to represent the velocity at time 0. A subscript is a part of the variable.

Positive exponent For any nonzero number (a) and any integer (n)

$$a^n = (a)(a)(a)...(a) \quad n \text{ times}$$

Examples: Simplify the following exponent terms.

$10^4 = (10)(10)(10)(10) = 10{,}000$

$2^3 = (2)(2)(2) = 8$

Zero exponent For any nonzero number (a)

$$a^0 = 1$$

Examples: Simplify the following zero exponent terms.

$2^0 = 1 \qquad 13^0 = 1$

Negative exponent For any nonzero number (a) and any integer (n)

$$a^{-n} = \frac{1}{a^n}$$

Examples: Write the following negative exponent terms as fractions.

$2^{-1} = \frac{1}{2^1} = \frac{1}{2} \qquad 2^{-2} = \frac{1}{2^2} = \frac{1}{4}$

Square and Cube Roots

A square root of a number is one of its two equal factors. A radical sign ($\sqrt{\ }$) indicates a square root. A square root can be shown as exponent $\frac{1}{2}$, as in $\sqrt{b} = b^{\frac{1}{2}}$. You can use a calculator to find square roots.

Examples: Simplify the following square root terms.

$\sqrt{a^2} = \sqrt{(a)(a)} = a$

$\sqrt{9} = \sqrt{(3)(3)} = 3$

$\sqrt{64} = \sqrt{(8.0)(8.0)} = 8.0$ — Keep two significant figures.

$\sqrt{38.44} = 6.200$ — Place two zeros to the right of the calculator answer to keep four significant figures.

$\sqrt{39} = 6.244997 \ldots = 6.2$ — Round the answer to keep two significant figures.

A cube root of a number is one of its three equal factors. A radical sign with the number 3 ($\sqrt[3]{\ }$) indicates a cube root. A cube root also can be shown as exponent $\frac{1}{3}$, as in $\sqrt[3]{b} = b^{\frac{1}{3}}$.

Example: Simplify the following cube root terms.

$$\sqrt[3]{125} = \sqrt[3]{(5.00)(5.00)(5.00)} = 5.00$$

$$\sqrt[3]{39.304} = 3.4000$$

PRACTICE Problems ADDITIONAL PRACTICE

6. Find each root. Round the answer to the nearest hundredth.

a. $\sqrt{22}$ c. $\sqrt{676}$

b. $\sqrt[3]{729}$ d. $\sqrt[3]{46.656}$

7. Simplify by writing without a radical sign.

a. $\sqrt{16a^2b^4}$ b. $\sqrt{9t^6}$

8. Write using exponents.

a. $\sqrt{n^3}$ b. $\frac{1}{\sqrt{a}}$

Operations with Exponents

In the following operations with exponents, *a*, *b*, *m*, and *n* can be numbers or variables.

Product of powers To multiply terms with the same base, add the exponents, as in $(a^m)(a^n) = a^{m+n}$.

Quotient of powers To divide terms with the same base, subtract the bottom exponent from the top exponent, as in $a^m/a^n = a^{m-n}$.

Power of a power To calculate the power of a power, use the same base and multiply the exponents, as in $(a^m)^n = a^{mn}$.

***n*th root of a power** To calculate the root of a power, use the same base and divide the power exponent by the root exponent, as in $\sqrt[n]{a^m} = a^{\frac{m}{n}}$.

Power of a product To calculate the power of a product of *a* and *b*, raise both to the power and find their product, as in $(ab)^n = a^nb^n$.

PRACTICE Problems ADDITIONAL PRACTICE

9. Write an equivalent form using the properties of exponents.

a. $\frac{x^2t}{x^3}$ b. $\sqrt{t^3}$ c. $(d^2n)^2$ d. $x^2\sqrt{x}$

10. Simplify $\frac{m}{q}\sqrt{\frac{2qv}{m}}$.

Absolute Value

The absolute value of a number (n) is its magnitude, regardless of its sign. The absolute value of n is written as $|n|$. Because magnitudes cannot be less than zero, absolute values always are greater than or equal to zero.

Examples:

$|3| = 3$

$|-3| = 3$

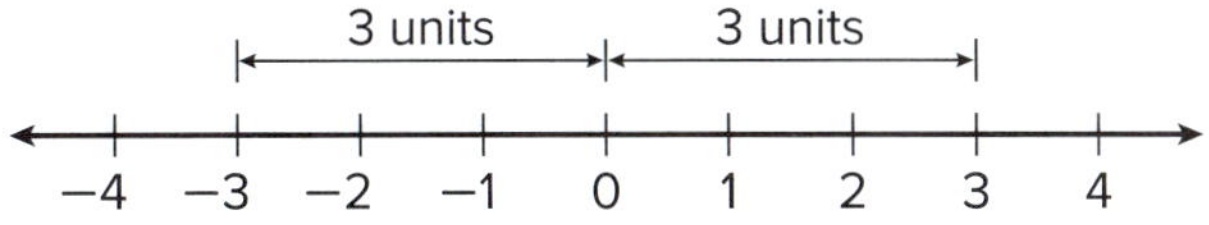

V. Scientific Notation

A number of the form $a \times 10^n$ is written in scientific notation, where $1 \leq a \leq 10$, and n is an integer. The base (10) is raised to a power (n).

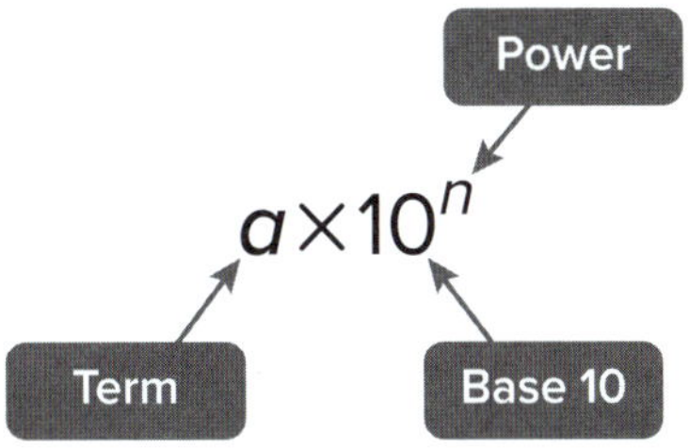

Connecting Math to Physics Physicists commonly use scientific notation to express measurements that are greater than 10 or less than 1. For example, the mass of a proton is written as 6.73×10^{-28} kg. The density of water is written as 1.000×10^3 kg/m^3. This shows, using significant digit rules, that this measurement is exactly 1000 to four significant figures. However, writing the density of water as 1000 kg/m^3 would imply that it has only one significant digit, which is incorrect. Scientific notation helps physicists keep accurate track of significant figures.

Large Numbers—Using Positive Exponents

Multiplying by a power of 10 is like moving the decimal point that same number of places to the right (if the power is positive) or to the left (if the power is negative). To express a large number in scientific notation, first determine the value for a, $1 \leq a \leq 10$. Count the number of decimal places from the decimal point in a to the decimal point in the number. Use that count as the power of 10. A calculator shows scientific notation with e for exponent, as in 2.4e+11 = 2.4×10^{11}. Some calculators use an *E* to show the exponent, or there is often a place on the display where the calculator can show smaller-sized digits representing the exponent.

Example: Write 7,530,000 in scientific notation.

The value for a is 7.53. (The decimal point is to the right of the first nonzero digit.) So the form will be 7.53×10^n.

$7{,}530{,}000 = 7.53 \times 10^6$ There are six decimal places, so the power is 6.

To write the standard form of a number expressed in scientific notation, write the value of a, and place extra zeros to the right of the number. Use the power and move the decimal point in a that many places to the right.

Example: Write the following number in standard form.

$2.389 \times 10^5 = 2.38900 \times 10^5 = 238{,}900$

Small Numbers—Using Negative Exponents

To express a small number in scientific notation, first determine the value for a, $1 \leq a \leq 10$. Then count the number of decimal places from the decimal point in a to the decimal point in the number. Use that number as the power of 10. Multiplying by a number with a negative power is the same as dividing by that number with the corresponding positive power.

Example: Write 0.000000285 in scientific notation.

The value for a is 2.85. (The decimal point is to the right of the first nonzero digit.) So the form will be 2.85×10^n.

$0.000000285 = 2.85 \times 10^{-7}$ — There are seven decimal places, so the power is −7.

To express a small number in standard form, write the value for a and place extra zeros to the left of a. Use the power and move the decimal point in a that many places to the left.

Example:

$1.6 \times 10^{-4} = 00001.6 \times 10^{-4} = 0.00016$

PRACTICE Problems ADDITIONAL PRACTICE

11. Express each number in scientific notation.

a. 456,000,000 **b.** 0.000020

12. Express each number in standard notation.

a. 3.03×10^{-7} **b.** 9.7×10^{10}

Operations with Scientific Notation

Calculating with numbers written in scientific notation uses the properties of exponents.

Multiplication Multiply the terms, and add the powers of 10.

Example: Simplify.

$(4.0 \times 10^{-8})(1.2 \times 10^5) = (4.0 \times 1.2)(10^{-8} \times 10^5)$ — Group terms and bases of 10.

$= (4.8)(10^{-8+5})$ — Multiply terms.

$= (4.8)(10^{-3})$ — Add powers of 10.

$= 4.8 \times 10^{-3}$ — Recombine in scientific notation.

Division Divide the base numbers and subtract the exponents of 10.

Example: Simplify.

$\frac{9.60 \times 10^7}{1.60 \times 10^3} = \left(\frac{9.60}{1.60}\right) \times \left(\frac{10^7}{10^3}\right)$ — Group terms and bases of 10.

$= 6.00 \times 10^{7-3}$ — Divide terms, and subtract powers of 10.

$= 6.00 \times 10^4$

Addition and subtraction To add and subtract numbers in scientific notation, the powers of 10 must be the same. Thus, you may need to rewrite one of the numbers with a different power of 10. With equal powers of 10, use the distributive property.

Example: Simplify.

$(3.2\times10^5) + (4.8\times10^5) = (3.2 + 4.8)\times10^5$ — Group terms.

$= 8.0\times10^5$ — Add terms.

Example: Simplify.

$(3.2\times10^5) + (4.8\times10^4) = (3.2\times10^5) + (0.48\times10^5)$ — Rewrite 4.8×10^4 as 0.48×10^5.

$= (3.2 + 0.48)\times10^5$ — Group terms.

$= 3.68\times10^5$ — Add terms.

$= 3.7\times10^5$ — Round using the addition significant figures rule.

PRACTICE Problems 13 ADDITIONAL PRACTICE

Evaluate each expression; express the result in scientific notation.

a. $(5.2\times10^{-4})(4.0\times10^8)$ **b.** $(2.4\times10^3) + (8.0\times10^4)$

VI. Equations

Order of Operations

Scientists and mathematicians have agreed on a set of steps or rules, called the order of operations, so that everyone interprets mathematical symbols in the same way. Follow these steps in order when you evaluate an expression or use a formula.

1. Simplify the expressions inside grouping symbols, such as parentheses ( ), brackets [ ], braces { }, and fraction bars.
2. Evaluate all powers and roots.
3. Do all multiplications and/or divisions from left to right.
4. Do all additions and/or subtractions from left to right.

A useful way to remember these is the phrase "*P*lease *e*xcuse *m*y *d*ear *A*unt *S*ally." The first letter of each word represents an operation: *p*arentheses, *e*xponents, *m*ultiplication, *d*ivision, *a*ddition, *s*ubtraction.

Example: Simplify the following expression.

$4 + 3(4 - 1) - 2^3 = 4 + 3(3) - 2^3$ — Order of operations step 1.

$= 4 + 3(3) - 8$ — Order of operations step 2.

$= 4 + 9 - 8$ — Order of operations step 3.

$= 5$ — Order of operations step 4.

Connecting Math to Physics The previous example was shown step-by-step to demonstrate the order of operations. When solving a physics problem, do not round to the correct number of significant figures until after the final calculation.

In calculations involving an expression in a numerator and an expression in a denominator, the numerator and the denominator are separate groups and you should calculate them before dividing the numerator by the denominator. The multiplication/division rule for significant figures is used to determine the final number of significant figures.

Solving Equations

To solve an equation means to find the value of the variable that makes the equation a true statement. To solve equations, apply the distributive property and the properties of equality. Any properties of equalities that you apply on one side of an equation, you also must apply on the other side.

Distributive property For any numbers a, b, and c,

$$a(b + c) = ab + ac \qquad a(b - c) = ab - ac$$

Example: Use the distributive property to expand the following expression.

$3(x + 2) = 3x + (3)(2) = 3x + 6$

Addition and subtraction properties of equality If two quantities are equal, and the same number is added to (or subtracted from) each, then the resulting quantities also are equal.

$$\text{If } a = b, \text{ then}$$

$$a + c = b + c \quad \text{and} \quad a - c = b - c$$

Example: Solve $x - 3 = 7$ using the addition property.

$$x - 3 = 7$$
$$x - 3 + 3 = 7 + 3$$
$$x = 10$$

Example: Solve $t + 2 = -5$ using the subtraction property.

$$t + 2 = -5$$
$$t + 2 - 2 = -5 - 2$$
$$t = -7$$

Multiplication and division properties of equality If two equal quantities each are multiplied by (or divided by) the same number, then the resulting quantities also are equal.

$$\text{If } a = b, \text{ then}$$
$$ac = bc \text{ and}$$
$$\frac{a}{c} = \frac{b}{c}, \text{ for } c \neq 0$$

Example: Solve $\frac{1}{4}a = 3$ using the multiplication property.

$$\frac{1}{4}a = 3$$
$$\left(\frac{a}{4}\right)(4) = 3(4)$$
$$a = 12$$

Example: Solve $6n = 18$ using the division property.

$$6n = 18$$
$$\frac{6n}{6} = \frac{18}{6}$$
$$n = 3$$

Example: Solve $2t + 8 = 5t - 4$ for t.

$$2t + 8 = 5t - 4$$
$$8 + 4 = 5t - 2t$$
$$12 = 3t$$
$$4 = t$$

Isolating a Variable

Suppose an equation has more than one variable. To isolate a variable—that is, to solve the equation for a variable—write an equivalent equation so that one side contains only that variable with a coefficient of 1.

Connecting Math to Physics Isolate the variable P(pressure) in the ideal gas law equation.

$$PV = n\text{R}T$$

$$\frac{PV}{V} = \frac{n\text{R}T}{V} \quad \text{Divide both sides by } V.$$

$$P\left(\frac{V}{V}\right) = \frac{n\text{R}T}{V} \quad \text{Group } \frac{V}{V}.$$

$$P = \frac{n\text{R}T}{V} \quad \text{Substitute } \frac{V}{V} = 1.$$

PRACTICE Problems 14 ADDITIONAL PRACTICE

Solve for x.

a. $2 + 3x = 17$

b. $x - 4 = 2 - 3x$

c. $t - 1 = \frac{x + 4}{3}$

d. $a = \frac{b + x}{c}$

e. $\frac{2x + 3}{x} = 6$

f. $ax + bx + c = d$

Square Root Property

If a and n are real numbers, $n > 0$, and $a^2 = n$, then $a = \pm\sqrt{n}$.

Connecting Math to Physics Solve for v in Newton's second law for a satellite orbiting Earth.

$$\frac{mv^2}{r} = \frac{\text{G}m_\text{E}m}{r^2}$$

$$\frac{rmv^2}{r} = \frac{r\text{G}m_\text{E}m}{r^2} \quad \text{Multiply both sides by } r.$$

$$mv^2 = \frac{\text{G}m_\text{E}m}{r} \quad \text{Substitute } \frac{r}{r} = 1.$$

$$\frac{mv^2}{m} = \frac{\text{G}m_\text{E}m}{rm} \quad \text{Divide both sides by } m.$$

$$v^2 = \frac{\text{G}m_\text{E}}{r} \quad \text{Substitute } \frac{m}{m} = 1.$$

$$\sqrt{v^2} = \pm\sqrt{\frac{\text{G}m_\text{E}}{r}} \quad \text{Take the square root.}$$

$$v = \frac{\text{G}m_\text{E}}{r} \quad \text{Use the positive value for speed.}$$

When using the square root property, it is important to consider what you are solving for. Because we solved for speed in the above example, it did not make sense to use the negative value of the square root. Also, you need to consider if the negative or positive value gives you a realistic solution. For example, when using the square root property to solve for time, a negative value might give you a time before the situation even started.

Quadratic Equations

A quadratic equation has the form $ax^2 + bx + c = 0$, where $a \neq 0$. A quadratic equation has one variable with a power (exponent) of 2. It also might include that same variable to the first power. You can estimate the solutions by graphing on a graphing calculator. If $b = 0$, then there is no x-term in the quadratic equation. In this case, you can solve the equation by isolating the squared variable and finding the square root of each side of the equation using the square root property.

Quadratic Formula

You can find the solutions of any quadratic equation by using the quadratic formula. The solutions of $ax2 + bx + c = 0$, where $a \neq 0$, are given by

$$x = \frac{-b \pm \sqrt{b2 - 4ac}}{2a}$$

As with the square root property, it is important to consider whether the solutions to the quadratic formula give you a realistic answer to the problem you are solving. Usually, you can throw out one of the solutions because it is unrealistic. Projectile motion often requires the use of the quadratic formula when solving equations, so keep the realism of the solution in mind when solving.

PRACTICE Problems 15 ADDITIONAL PRACTICE

Solve for x.

a. $4x^2 - 19 = 17$

b. $12 - 3x^2 = -9$

c. $x^2 - 2x - 24 = 0$

d. $24x^2 - 14x - 6 = 0$

Dimensional Calculations

When doing calculations, you must include the units of each measurement that is written in the calculation. All operations that are performed on the number also are performed on its units.

Connecting Math to Physics The acceleration due to gravity (a) is given by the equation $a = \frac{2\Delta x}{\Delta t^2}$. A free-falling object near the Moon drops 20.5 m in 5.00 s. Find the acceleration (a). Acceleration is measured in meters per second squared.

$$a = \frac{2\Delta x}{\Delta t^2}$$

$$a = \frac{2(20.5\text{ m})}{(5.00\text{ s})^2}$$

The number 2 is an exact number, so it does not affect the determination of significant figures.

$$a = \frac{1.64\text{ m}}{\text{s}^2} \text{ or } 1.64\text{ m/s}^2$$

Calculate and round to three significant figures.

Unit conversion Use a conversion factor to convert from one measurement unit to another of the same type, such as from minutes to seconds. This is equivalent to multiplying by one.

Connecting Math to Physics Find Δx when $v_i = 67$ m/s and $\Delta t = 5.0$ min. Use the equation $\Delta x = v_i \Delta t$.

$$\frac{60 \text{ seconds}}{1 \text{ minute}} = 1$$

$$\Delta x = v_i \Delta t$$

$$\Delta x = \frac{67 \text{ m}}{\text{s}}\left(\frac{5.0 \text{ min}}{1}\right)\left(\frac{60 \text{ s}}{1 \text{ min}}\right)$$ Multiply by the conversion factor $\frac{60 \text{ s}}{1 \text{ min}} = 1$

$$\Delta x = 20{,}100 \text{ m} = 2.0\times10^4 \text{ m}$$ Calculate and round to two significant figures. The numbers 60 s and 1 min are exact numbers, so they do not affect the determination of significant figures.

PRACTICE Problems ADDITIONAL PRACTICE

16. Simplify $\Delta t = \frac{4.0 \times 10^2 \text{ m}}{16 \text{ m/s}}$.
17. Find the velocity of a dropped brick after 5.0 s using $v = a\Delta t$ and $a = -9.8 \text{ m/s}^2$.
18. Calculate the product: $\left(\frac{32 \text{ cm}}{1 \text{ s}}\right)\left(\frac{60 \text{ s}}{1 \text{ min}}\right)\left(\frac{60 \text{ min}}{1 \text{ h}}\right)\left(\frac{1 \text{ m}}{100 \text{ cm}}\right)$.
19. An Olympian ran 100.0 m in 9.87 s. What was the speed in kilometers per hour?

Dimensional Analysis

Dimensional analysis is a method of doing algebra with the units. It often is used to check the validity of the units of a final result and the equation being used, without completely redoing the calculation.

Physics Example Verify that the final answer of $x_f = x_i + v_i t + \frac{1}{2}at^2$ will have the units meters (m).

x_i is measured in meters (m).

t is measured in seconds (s).

v_i is measured in meters per second (m/s).

a is measured in meters per second squared (m/s²).

$$x_f = \text{m} + \left(\frac{\text{m}}{\text{s}}\right)(\text{s}) + \frac{1}{2}\left(\frac{\text{m}}{\text{s}^2}\right)(\text{s})^2$$ Substitute the units for each variable.

$$= \text{m} + \text{m}\left(\frac{\text{s}}{\text{s}}\right) + \frac{1}{2}(\text{m})\left(\frac{\text{s}^2}{\text{s}^2}\right)$$ Simplify the fractions using the distributive property.

$$= \text{m} + (\text{m})(1) + \frac{1}{2}(\text{m})(1)$$ Substitute $\text{s/s} = 1$, $\text{s}^2/\text{s}^2 = 1$.

$$= \text{m} + \text{m} + \frac{1}{2}\text{ m}$$ Everything simplifies to m; thus x_f is in m.

The factor of $\frac{1}{2}$ in the above does not apply to the units. It applies only to any number values that would be inserted for the variables in the equation. It is easiest to remove number factors such as the $\frac{1}{2}$ when first setting up the dimensional analysis.

VII. Graphs of Relations

The Coordinate Plane

You can locate points on a plane in reference to two perpendicular number lines, called axes. The horizontal number line is called the x-axis and represents the independent variable. The vertical number line is called the y-axis and represents the dependent variable. A point is represented by two coordinates (x, y), which also is called an ordered pair. The value of the independent variable (x) always is listed first in the ordered pair. The ordered pair (0, 0) represents the origin.

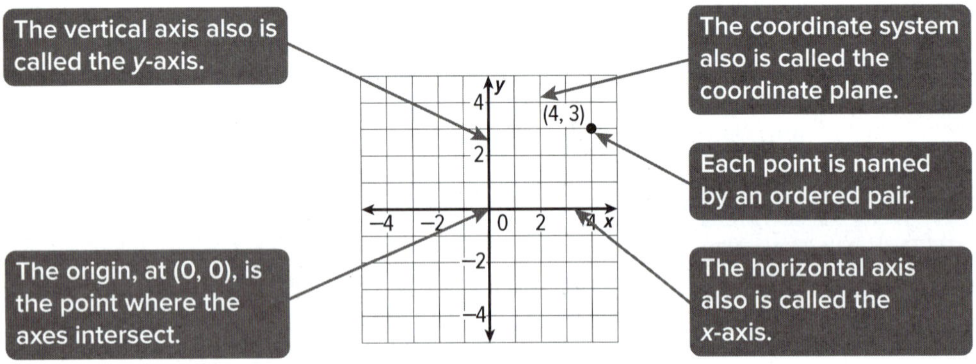

Graphing Data to Determine a Relationship

Use the following steps to graph data.

1. Identify the independent and the dependent variables.
2. Draw two perpendicular axes. Label each axis using the variable names.
3. Determine the range of data for each variable. Use the ranges to decide on a convenient scale for each axis. Mark and number the scales.
4. Plot each data point.
5. If the points seem to lie approximately in a line, draw a best-fit line. The line should be as close to as many points as possible. If the points do not lie in a line, draw a smooth curve through as many points as possible. If there does not appear to be a trend, do not draw any line or curve.
6. Write a title that clearly describes what the graph represents.

Applied Force (N)	Distance Stretched (cm)
10	0.8
20	2.1
30	2.8
40	4.2
50	4.7

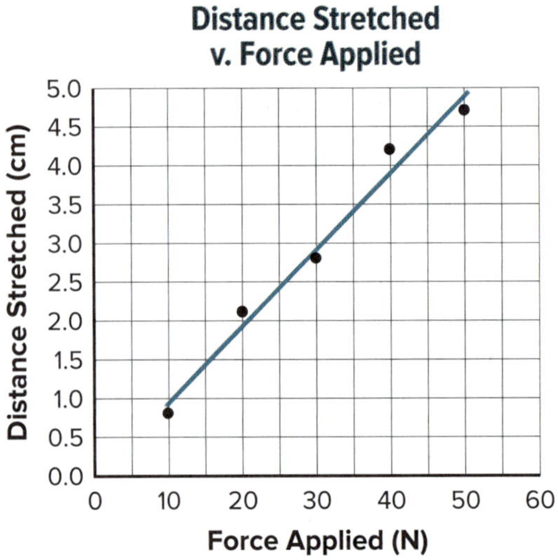

Interpolating and Extrapolating

Interpolation is a process used to estimate a value for a relation that lies between two known values. Extrapolation is a process used to estimate a value for a relation that lies beyond the known values. The equation of the best-fit line helps you interpolate and extrapolate.

Example: Using the data and the graph, estimate how far the spring will stretch if a force of 25 N is applied. Use interpolation.

- Draw a best-fit line.
- Draw a line segment from 25 N on the *x*-axis to the best-fit line.
- Draw a line segment from that intersection point to the *y*-axis.
- Read the scale on the *y*-axis. A force of 25 N will stretch the spring 2.4 cm.

Example: Use extrapolation to estimate how far the spring will stretch if a 60-N force is applied.

- Draw a line segment from 60 on the *x*-axis to the best-fit line. Extend the best-fit line if necessary.
- Read the corresponding value on the *y*-axis. Extend the axis scale if necessary.
- A force of 60 N will stretch the spring 5.8 cm.

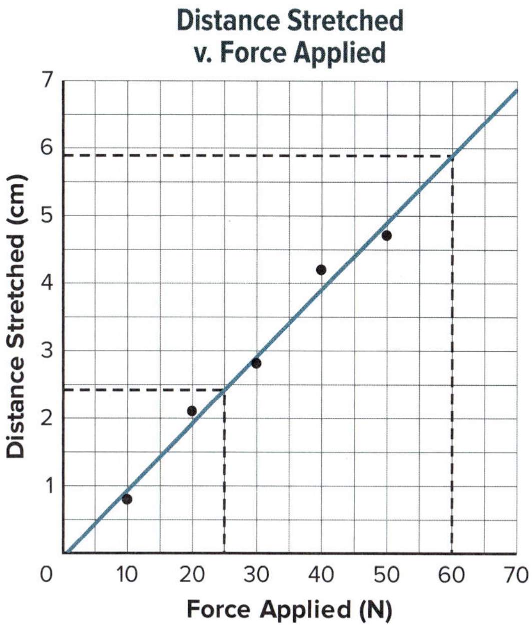

Interpreting Line Graphs

A line graph shows the linear relationship between two variables. Two types of line graphs that describe motion are used frequently in physics.

Connecting Math to Physics

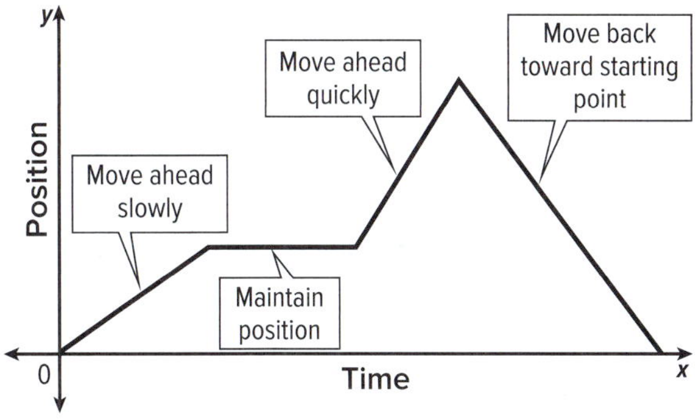

This line graph shows a changing relationship between the two graphed variables.

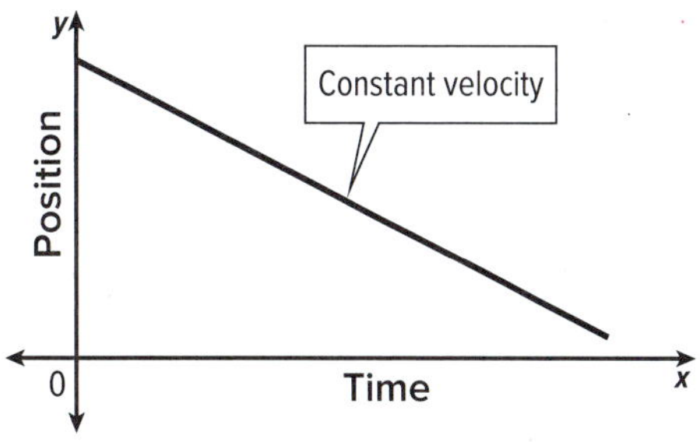

The line graph above shows a constant relationship between the two graphed variables.

Linear Equations

A linear equation can be written as a relation (or a function), $y = mx + b$, where m and b are real numbers, m represents the slope of the line, and b represents the intercept, the point at which the line crosses the y-axis.

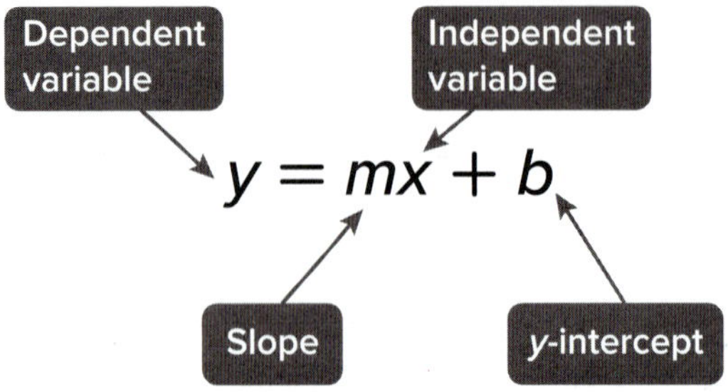

The graph of a linear equation is a line. The line represents all of the solutions of the linear equation. To graph a linear equation, choose three values for the independent variable. (Only two points are needed, but the third point serves as a check.) Calculate the corresponding values for the dependent variable. Plot each ordered pair (x, y) as a point. Draw a line through the points.

Example: Graph $y = -\frac{1}{2}x + 3$

Calculate three ordered pairs to obtain points to plot.

Ordered Pairs	
x	y
0	3
2	2
6	0

Ordered Pairs Plot

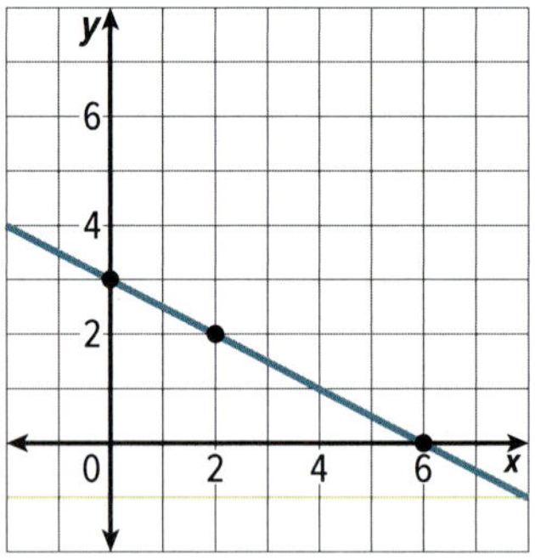

Slope

The slope of a line is the ratio of the change in y-coordinates to the change in x-coordinates. It is the ratio of the vertical change (rise) to the horizontal change (run). This number tells you how steep the line is. It can be a positive number or a negative number.

To find the slope of a line, select two points, (x_1, y_1) and (x_2, y_2). Calculate the run, which is the difference (change) between the two x-coordinates, $x_2 - x_1 = \Delta x$. Calculate the rise, which is the difference (change) between the two y-coordinates, $y_2 - y_1 = \Delta y$. Form the ratio.

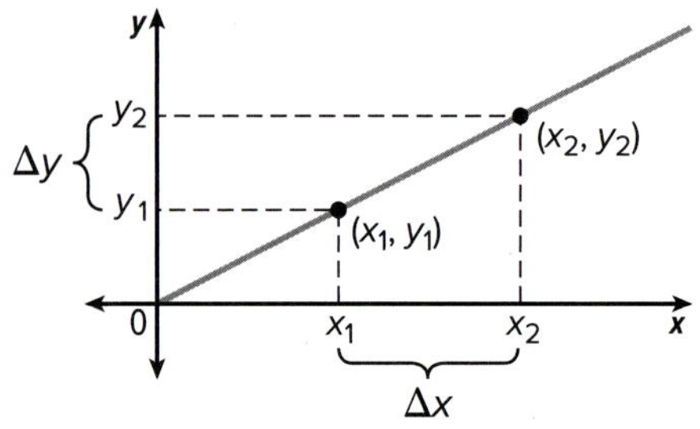

$$\text{Slope } m = \frac{rise}{run} = \frac{y_2 - y_1}{x_2 - x_1} = \frac{\Delta y}{\Delta x}$$

where $x_1 \neq x_2$

Direct Variation

If there is some nonzero constant (m) such that $y = mx$, then y varies directly with x. That means as the independent variable x doubles, the dependent variable y doubles. The variables x and y also are said to be proportional. This is a linear equation of the form $y = mx + b$ in which the value of b is zero. The graph passes through the origin, (0, 0).

Connecting Math to Physics In the force equation for an ideal spring, $F = -kx$, where F is the force on the spring, $-k$ is the spring constant, and x is the spring's displacement, the force on the spring varies directly with (is proportional to) the spring's displacement. That is, the force on the spring increases as the spring's displacement increases.

Inverse Variation

If there is some nonzero constant (m) such that $y = \frac{m}{x}$, then y varies inversely with x.

That means as the independent variable x increases, the dependent variable y decreases. The variables x and y also are said to be inversely proportional. This is not a linear equation because it contains the product of two variables. The graph of an inverse relationship is a hyperbola. This relationship can be written as

$$y = \frac{m}{x}$$
$$y = m\frac{1}{x}$$
$$xy = m$$

Example: Graph the equation $y = \frac{90}{x}$.

Ordered Pairs	
x	y
−10	−9
−6	−15
−3	−30
−2	−45
2	45
3	30
6	15
10	9

Inverse Variation

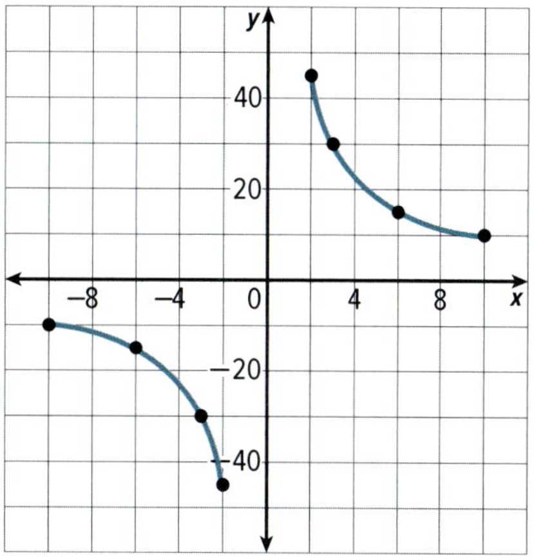

Connecting Math to Physics In the equation for the speed of a wave $\left(\lambda = \frac{v}{f}\right)$ where λ is wavelength, f is frequency, and v is wave speed, wavelength varies inversely with (is inversely proportional to) frequency. That is, as the frequency of a wave increases, the wavelength decreases. v is constant.

Quadratic Graphs

A quadratic relationship is a relationship of the form

$$y = ax^2 + bx + c, \text{ where } a \neq 0.$$

A quadratic relationship includes the square of the independent variable (x). The graph of a quadratic relationship is a parabola. Whether the parabola opens upward or downward depends on whether the value of the coefficient of the squared term (a) is positive or negative.

Example: Graph the equation $y = -x^2 + 4x - 1$.

Ordered Pairs	
x	y
−1	−6
0	−1
1	2
2	3
3	2
4	−1
5	−6

Quadratic Equation Graph

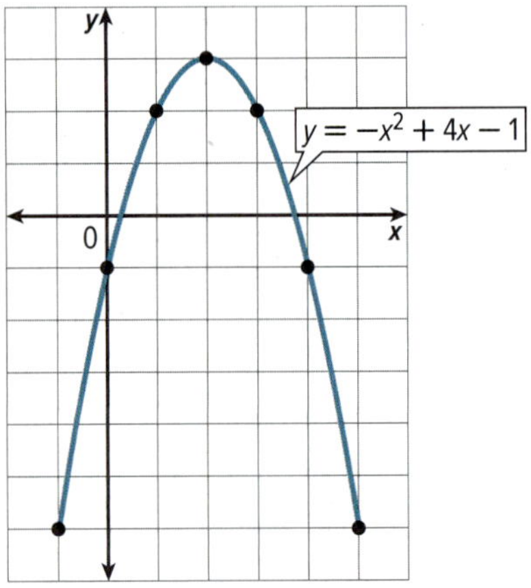

Connecting Math to Physics A position-time graph in the shape of a quadratic relation means that the object is moving at a constant acceleration.

$$x_f = 2\text{ m} + (1\text{ m/s})t + \frac{1}{2}(2\text{ m/s}^2)t^2$$

Ordered Pairs	
Time (s)	Position (m)
0	2
1	4
2	8
3	14
4	22

Position v. Time

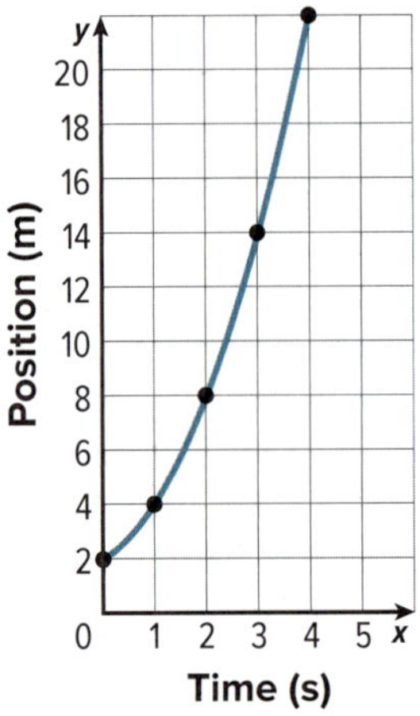

VIII. Geometry and Trigonometry

Perimeter and Area

	Perimeter, Circumference Linear units	Area Squared units
Square side a	$P = 4a$	$A = a^2$
Rectangle length l width w	$P = 2l + 2w$	$A = lw$
Triangle base b height h	$P = \text{side 1} + \text{side 2} + \text{side 3}$	$A = \frac{1}{2}bh$
Circle radius r	$C = 2\pi r$	$A = \pi r^2$

Surface Area and Volume

	Surface Area Squared units	Volume Cubic units
Cube side a	$SA = 6a^2$	$V = a^3$
Cylinder radius r height h	$SA = 2\pi rh + 2\pi r^2$	$V = \pi r^2 h$
Sphere radius r	$SA = 4\pi r^2$	$V = \frac{4}{3}\pi r^3$

Connecting Math to Physics Look for geometric shapes in your physics problems. They could be in the form of objects or spaces. For example, vectors can sometimes form two-dimensional shapes.

Area Under a Graph

To calculate the approximate area under a graph, cut the area into smaller pieces and find the area of each piece using the formulas shown above. To approximate the area under a line, cut the area into a rectangle and a triangle, as shown below on the left.

To approximate the area under a curve, draw several rectangles from the *x*-axis to the curve, as shown below on the right. Using more rectangles with a smaller base will provide a closer approximation of the area.

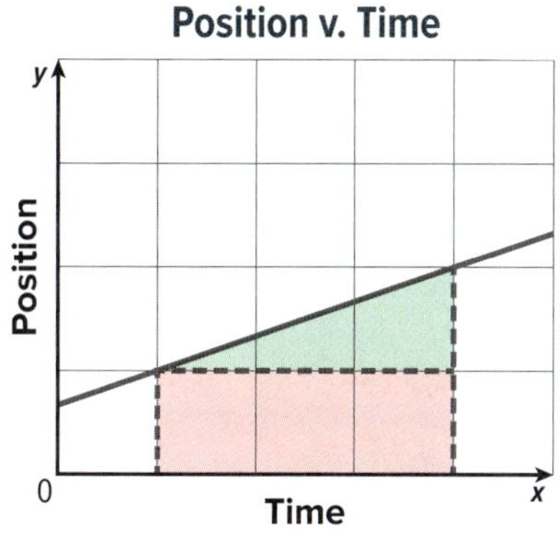

Total area = Area of the rectangle + Area of the triangle

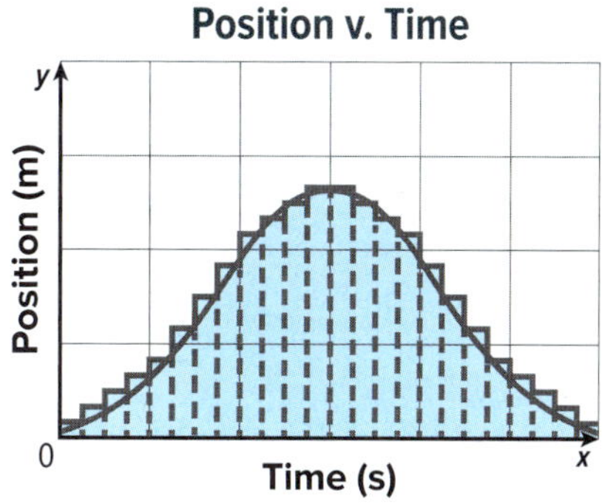

Total area = Area 1 + Area 2 + Area 3 + ...

Right Triangles

The Pythagorean theorem states that if a and b are the measures of the legs of a right triangle and c is the measure of the hypotenuse, then

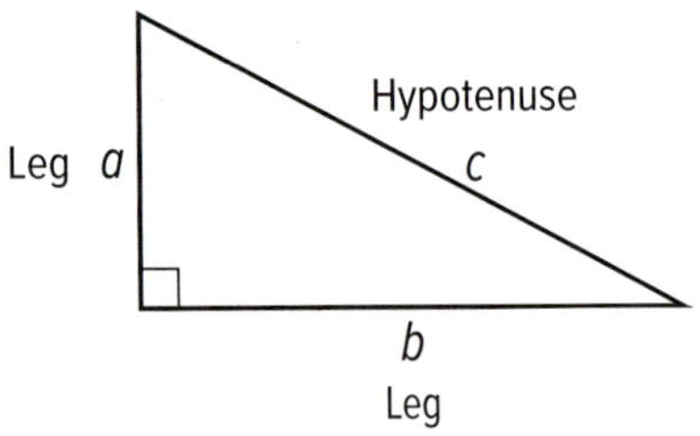

$$c^2 = a^2 + b^2.$$

To determine the length of the hypotenuse, use the square root property. Because distance is positive, the negative value does not have meaning.

$$c = \sqrt{a^2 + b^2}$$

Example: In the triangle, $a = 4$ cm and $b = 3$ cm. Find c.

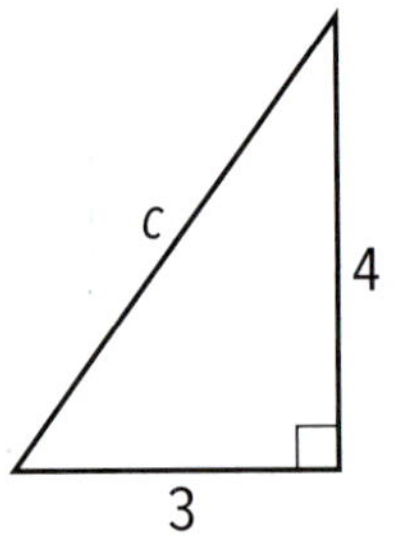

$$\begin{aligned} c &= \sqrt{a^2 + b^2} \\ &= \sqrt{(4\text{ cm})^2 + (3\text{ cm})^2} \\ &= \sqrt{16\text{ cm}^2 + 9\text{ cm}^2} \\ &= \sqrt{25\text{ cm}^2} \\ &= 5\text{ cm} \end{aligned}$$

45°-45°-90° triangles The length of the hypotenuse is $\sqrt{2}$ multiplied by the length of a leg.

45°—45°—90° Right Triangle

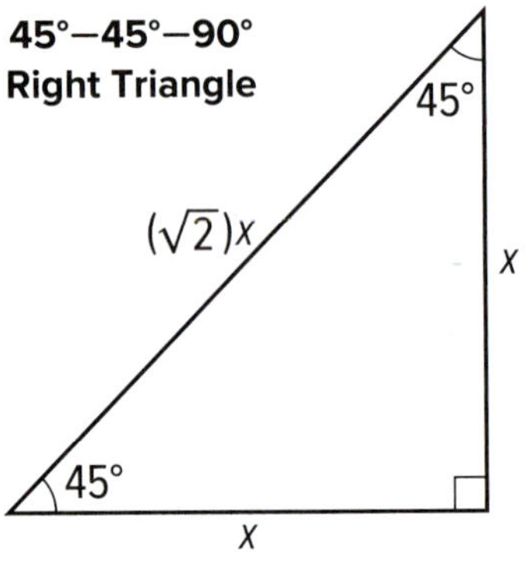

30°-60°-90° triangles The length of the hypotenuse is twice the length of the shorter leg. The length of the longer leg is $\sqrt{3}$ times the length of the shorter leg.

30°—60°—90° Right Triangle

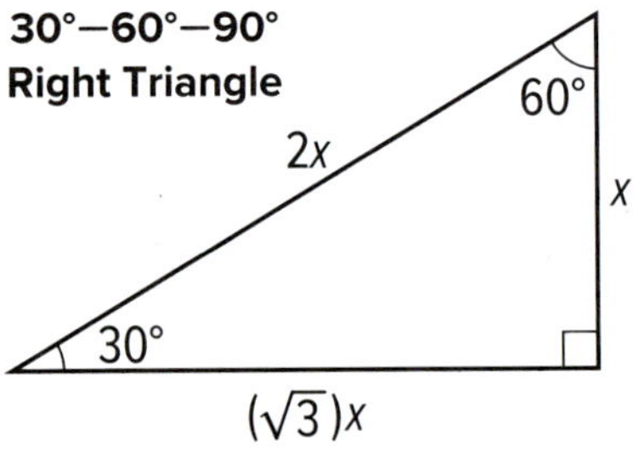

Trigonometric Ratios

A trigonometric ratio is a ratio of the lengths of sides of a right triangle. The most common trigonometric ratios are sine, cosine, and tangent. To memorize these ratios, learn the acronym SOH-CAH-TOA. SOH stands for Sine equals Opposite over Hypotenuse. CAH stands for Cosine equals Adjacent over Hypotenuse. TOA stands for Tangent equals Opposite over Adjacent.

Words	Memory Aid	Symbols
The sine is the ratio of the length of the side opposite to the angle over the length of the hypotenuse.	SOH $\sin\theta = \frac{\text{opposite}}{\text{hypotenuse}}$	$\sin\theta = \frac{a}{c}$
The cosine is the ratio of the length of the side adjacent to the angle over the length of the hypotenuse.	CAH $\cos\theta = \frac{\text{adjacent}}{\text{hypotenuse}}$	$\cos\theta = \frac{b}{c}$
The tangent is the ratio of the length of the side opposite to the angle over the length of the side adjacent to the angle.	TOA $\tan\theta = \frac{\text{opposite}}{\text{adjacent}}$	$\tan\theta = \frac{a}{b}$

Example: In right triangle *ABC*, if $a = 3$ cm, $b = 4$ cm, and $c = 5$ cm, find $\sin\theta$ and $\cos\theta$.

$$\sin\theta = \frac{3\text{ cm}}{5\text{ cm}} = 0.6$$

$$\cos\theta = \frac{4\text{ cm}}{5\text{ cm}} = 0.8$$

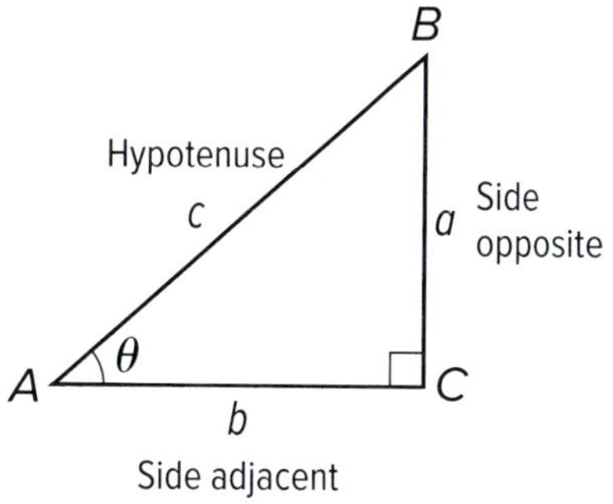

Example: In right triangle *ABC*, if $\theta = 30.0°$ and $c = 20.0$ cm, find a and b.

$$a = (20.0\text{ cm})(\sin 30.0°) = 10.0\text{ cm}$$

$$b = (20.0\text{ cm})(\cos 30.0°) = 17.3\text{ cm}$$

Law of Cosines and Law of Sines

The laws of cosines and sines let you calculate the sides and angles of any triangle.

Law of cosines The law of cosines is the same as the Pythagorean theorem, except for the last term. θ is the angle opposite side c. If the angle θ is 90°, the $\cos\theta = 0$ and the last term equals zero. If θ is greater than 90°, its cosine is a negative number.

$$c^2 = a^2 + b^2 - 2ab\cos\theta$$

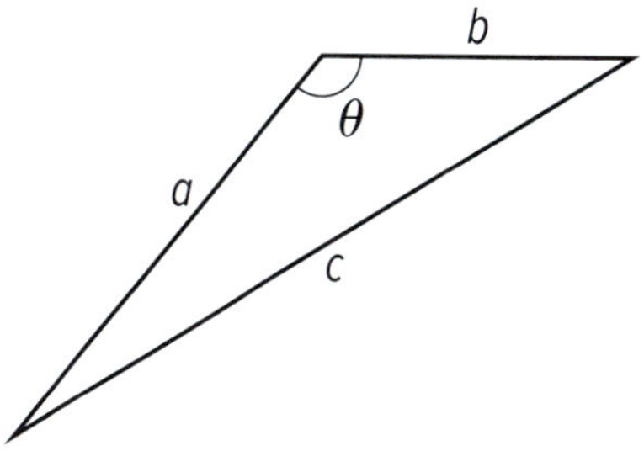

Example: Find the length of the third side of a triangle with $a = 10.0$ cm, $b = 12.0$ cm, $\theta = 110.0°$.

$$c^2 = a^2 + b^2 - 2ab\cos\theta$$

$$c = \sqrt{a^2 + b^2 - 2ab\cos\theta}$$

$$= \sqrt{(10.0\text{ cm})^2 + (12.0\text{ cm})^2 - 2(10.0\text{ cm})(12.0\text{ cm})(\cos 110.0°)}$$

$$= \sqrt{1.00\times10^2\text{ cm}^2 + 144\text{ cm}^2 - (240.0\text{ cm}^2)(\cos 110.0°)}$$

$$= 18.1\text{ cm}$$

Law of sines The law of sines is an equation of three ratios, where a, b, and c are the sides opposite angles A, B, and C, respectively. Use the law of sines when you know the measures of two angles and any side of a triangle.

$$\frac{\sin A}{a} = \frac{\sin B}{b} = \frac{\sin C}{c}$$

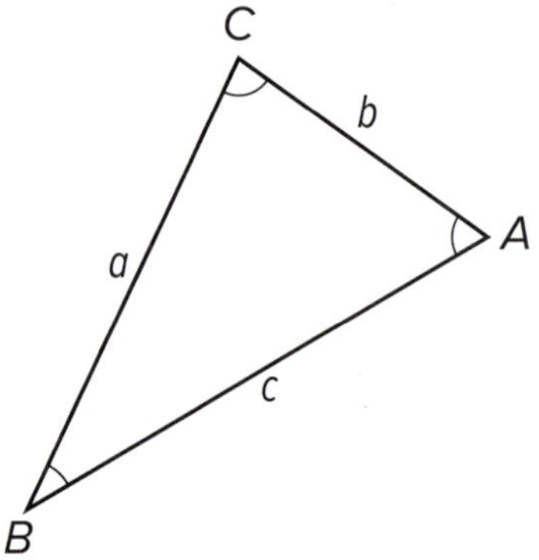

Example: In a triangle, $C = 60.0°$, $a = 4.0$ cm, $c = 54.6$ cm. Find the measure of angle A.

$$\frac{\sin A}{a} = \frac{\sin C}{c}$$

$$\sin A = \frac{a \sin C}{c}$$

$$\sin A = \frac{(4.0\ \text{cm})(\sin 60.0°)}{4.6\ \text{cm}}$$

$$A = 49°$$

Inverses of Sine, Cosine, and Tangent

The inverses of sine, cosine, and tangent allow you to do the reverse of the sine, cosine, and tangent functions and find the angle. The trigonometric functions and their inverses are as follows:

Trigonometric Function	Inverse
$y = \sin x$	$x = \sin^{-1} y$ or $x = \text{arc} \sin y$
$y = \cos x$	$x = \cos^{-1} y$ or $x = \text{arc} \cos y$
$y = \tan x$	$x = \tan^{-1} y$ or $x = \text{arc} \tan y$

Example: Solve $0.62 = \sin \theta$ for θ.

$$\theta = \sin^{-1} 0.62 = 38°$$

PRACTICE Problems 20 ADDITIONAL PRACTICE

Solve each of the following for θ.

a. $0.62 = \cos \theta$

b. $0.13 = \cos \theta$

c. $0.53 = \tan \theta$

d. $0.84 = \sin \theta$

Graphs of Trigonometric Functions

The sine function ($y = \sin x$) and the cosine function ($y = \cos x$) are periodic functions. The period for each function is 2π. x can be any real number. y is a real numbers between -1 and 1, inclusive.

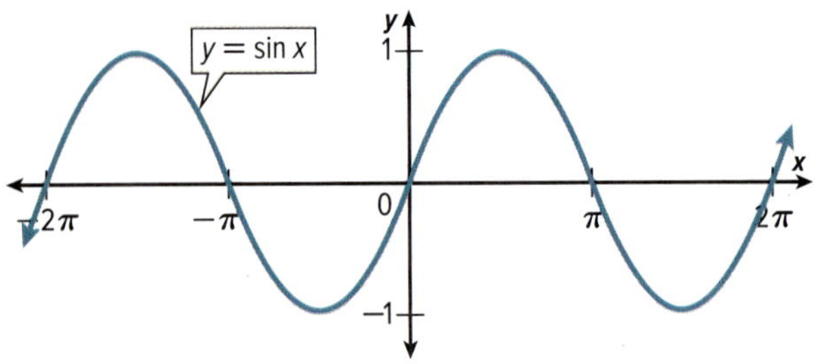

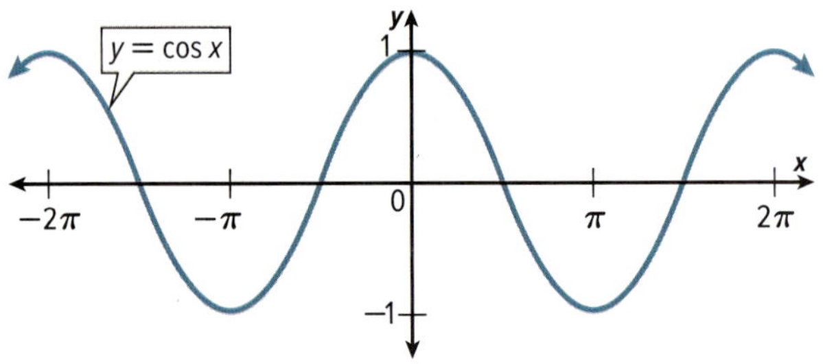

IX. Logarithms

Logarithms with Base b

Let b and x be positive integers with $b \neq 1$. The logarithm of x with base b, written $\log_b x$, equals y, where y is the exponent that makes the equation $b^y = x$ true. The log to the base b of x is the number to which you can raise b to get x.

$$\log_b x = y \text{ if and only if } b^y = x$$

Memory aid: "the log is the exponent"

Examples: Calculate the following logarithms.

$\log_2 \frac{1}{16} = -4$ Because $2^{-4} = \frac{1}{16}$

$\log_{10} 1000 = 3$ Because $10^3 = 1000$

When you want to find the log of a number, you also can use a computer.

Connecting Math to Physics Physicists use logarithms to work with measurements that extend over several orders of magnitude, or powers of 10. Geophysicists use the Richter scale, a logarithmic scale that allows them to rate earthquakes from 0 to 7, or larger, though the power of earthquakes differ by 7 or more powers of 10.

Common Logarithms

Base 10 logarithms are called common logarithms. They often are written without the subscript 10.

$$\log_{10} x = \log x \qquad x > 0$$

Antilogarithms or Inverse Logarithms

An antilogarithm is the inverse of a logarithm. An antilogarithm is the number that has a given logarithm.

Examples: Solve $\log x = 4$ for x.

$\log x = 4$

$x = 10^4$ 10^4 is the antilogarithm of 4.

Connecting Math to Physics An equation for loudness (L) in decibels is $L = 10 \log_{10} R$, where R is the relative intensity of the sound. Calculate R for a fireworks display with a loudness of 130 decibels.

$130 = 10 \log_{10} R$ Divide by 10.

$13 = \log_{10} R$ Use the logarithm rule.

$R = 10^{13}$

When you know the logarithm of a number and want to know the number itself, use a calculator to find the antilogarithm.

PRACTICE Problems ADDITIONAL PRACTICE

21. Write the following in exponential form: $\log_3 81 = 4$

22. Write the following in logarithmic form: $10^{-3} = 0.001$

23. Find x. $\log x = 3.125$

REFERENCE TABLES

Color Conventions

Displacement vector (***x***)		Negative charge	−
Velocity vector (***v***)		Positive charge	+
Acceleration vector (***a***)		Current direction	
Force vector (***F***)		Electron	
Momentum vector (***p***)		Proton	
Light ray		Neutron	
Object		Coordinate axes	
Image			
Electric field line (***E***)			
Magnetic field line (***B***)			

Electric Circuit Symbols

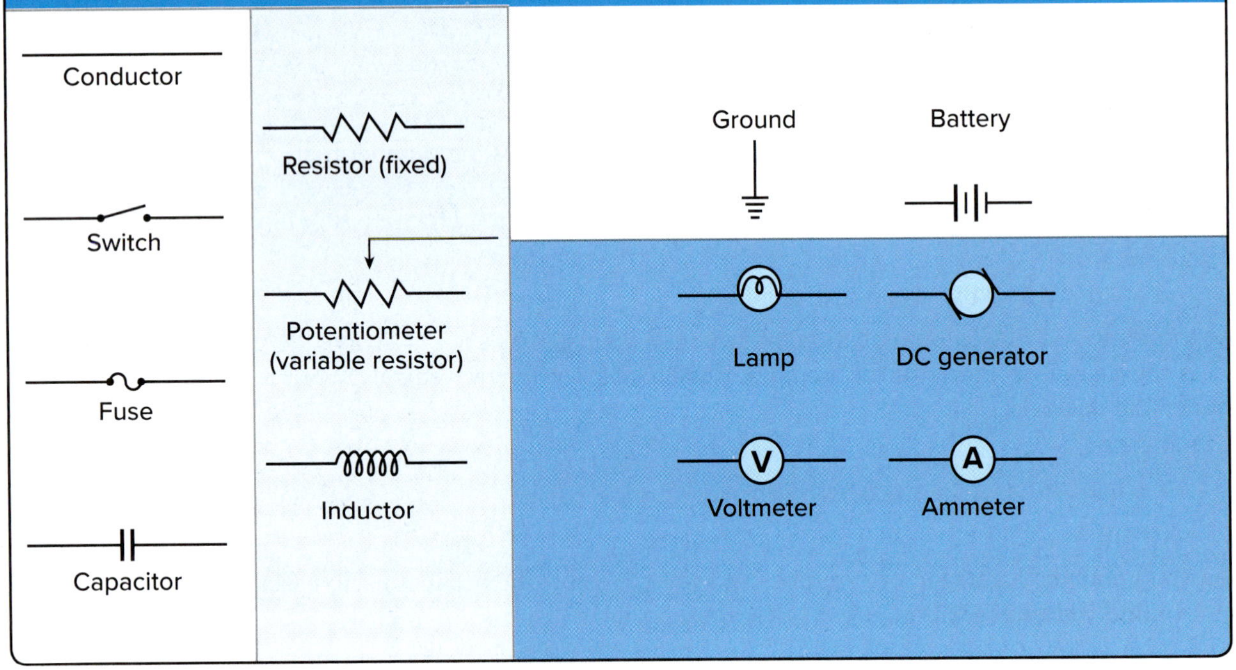

SI Base Units

Quantity	Unit	Unit Abbreviation
Length	meter	m
Mass	kilogram	kg
Time	second	s
Temperature	kelvin	K
Amount of a substance	mole	mol
Electric current	ampere	A
Luminous intensity	candela	cd

SI Derived Units

Quantity	Unit	Unit Symbol	Unit Expressed in Base Units	Unit Expressed in Other SI Units
Acceleration	meters per second squared	m/s^2	m/s^2	
Area	meters squared	m^2	m^2	
Capacitance	farad	F	$A^2{\cdot}s^4/(kg{\cdot}m^2)$	
Density	kilograms per meter cubed	kg/m^3	kg/m^3	
Electric charge	coulomb	C	A·s	
Electric field	newtons per coulomb	N/C	$kg{\cdot}m/(A{\cdot}s^3)$	V/m
Electric resistance	ohm	Ω	$kg{\cdot}m^2/(A^2{\cdot}s^3)$	V/A
Energy, work	joule	J	$kg{\cdot}m^2/s^2$	N·m
Force	newton	N	$kg{\cdot}m/s^2$	
Frequency	hertz	Hz	s^{-1}	
Illuminance	lux	lx	cd/m^2	
Magnetic field	tesla	T	$kg/(A{\cdot}s^2)$	N·s/(C·m)
Potential difference, *EMF*	volt	V	$kg{\cdot}m^2/(A{\cdot}s^3)$	W/A or J/C
Power	watt	W	$kg{\cdot}m^2/s^3$	J/s
Pressure	pascal	Pa	$(kg/m)s^2$	N/m^2
Velocity	meters per second	m/s	m/s	
Volume	meters cubed	m^3	m^3	

Useful Conversions

1 in = 2.54 cm	1 kg = 6.02×10^{26} u	1 lb = 4.45 N	1 cal = 4.184 J
1 mi = 1.61 km	1 oz = 28.4 g	1 atm = 14.7 lb/in^2	1 eV = 1.60×10^{-19} J
1 mi^2 = 640 acres	1 kg = 2.21 lb	1 atm = 1.01×10^5 N/m^2	1 kWh = 3.60 MJ
1 gal = 3.79 L	1 m^3 = 264 gal	1 atm = 101 kPa	1 mol = 6.02×10^{23} particles

Physical Constants

Quantity	Symbol	Value	Approximate Value
Atomic mass unit	u	$1.660538782\times10^{-27}$ kg	1.66×10^{-27} kg
Avogadro's number	N_A	6.02214179×10^{23} mol^{-1}	6.022×10^{23} mol^{-1}
Boltzmann's constant	k	1.3806504×10^{-23} $Pa{\cdot}m^3/K$	1.38×10^{-23} $Pa{\cdot}m^3/K$
Coulomb's constant	K	8.987551788×10^{9} $N{\cdot}m^2/C^2$	9.0×10^{9} $N{\cdot}m^2/C^2$
Elementary charge	e	1.60217653×10^{-19} C	1.602×10^{-19} C
Gas constant	R	8.314472 $Pa{\cdot}m^3/mol{\cdot}K$	8.31 $Pa{\cdot}m^3/mol{\cdot}K$
Gravitational constant	G	6.67428×10^{-11} $N{\cdot}m^2/kg^2$	6.67×10^{-11} $N{\cdot}m^2/kg^2$
Mass of an electron	m_e	9.10938215×10^{-31} kg	9.11×10^{-31} kg
Mass of a proton	m_p	$1.672621637\times10^{-27}$ kg	1.67×10^{-27} kg
Mass of a neutron	m_n	$1.674927211\times10^{-27}$ kg	1.67×10^{-27} kg
Planck's constant	h	6.62606896×10^{-34} $J{\cdot}s$	6.63×10^{-34} $J{\cdot}s$
Speed of light in a vacuum	c	2.99792458×10^{8} m/s	3.00×10^{8} m/s

SI Prefix

Prefix	Symbol	Scientific Notation
femto	f	10^{-15}
pico	p	10^{-12}
nano	n	10^{-9}
micro	μ	10^{-6}
milli	m	10^{-3}
centi	c	10^{-2}
deci	d	10^{-1}
deka	da	10^{1}
hector	h	10^{2}
kilo	k	10^{3}
mega	M	10^{6}
giga	G	10^{9}
tera	T	10^{12}
peta	P	10^{15}

Moments of Inertia for Various Objects

Object	Location of Axis	Diagram	Moment of Inertia
Thin hoop of radius *r*	through central diameter	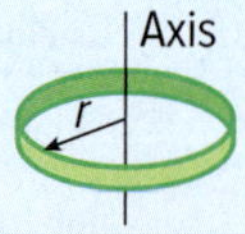	mr^2
Solid, uniform cylinder of radius *r*	through center	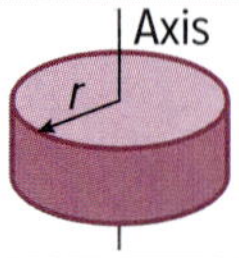	$\frac{1}{2}mr^2$
Uniform sphere of radius *r*	through center	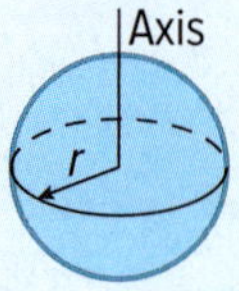	$\frac{2}{5}mr^2$
Long, uniform rod of length *l*	through center	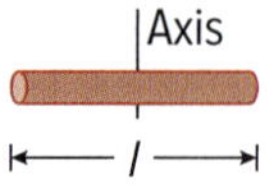	$\frac{1}{12}ml^2$
Long, uniform rod of length *l*	through end	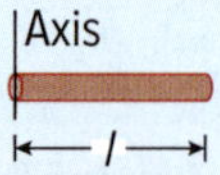	$\frac{1}{3}ml^2$
Thin, rectangular plate of length *l* and width *w*	through center	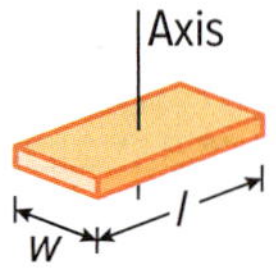	$\frac{1}{12}m(l^2 + w^2)$

Densities of Some Common Substances	
Substance	Density(g/cm^3)
Aluminum	2.70
Cadmium	8.65
Copper	8.92
Germanium	5.32
Gold	19.32
Hydrogen	8.99×10^{-5}
Indium	7.31
Iron	7.87
Lead	11.34
Mercury	13.534
Oxygen	1.429×10^{-3}
Silicon	2.33
Silver	10.5
Water (4°C)	1.000
Zinc	7.14

Melting and Boiling Points		
Substance	Melting Point (°C)	Boiling Point (°C)
Aluminum	660.32	2519
Copper	1084.62	2562
Germanium	938.25	2833
Gold	1064.18	2856
Indium	156.60	2072
Iron	1538	2861
Lead	327.5	1749
Silicon	1414	3265
Silver	961.78	2162
Water	0.000	100.000
Zinc	419.53	907

Reference Tables

Specific Heats			
Material	Specific Heat, C[J/(kg·K)]	Material	Specific Heat, C[J/(kg·K)]
Aluminum	897	Lead	130
Brass	376	Methanol	2450
Carbon	710	Silver	235
Copper	385	Water	4180
Glass	840	Water vapor	2020
Ice	2060	Zinc	388
Iron	450		

Heats of Fusion and Vaporization		
Material	Heat of Fusion, H_f (J/kg)	Heat of Vaporization, H_v (J/kg)
Copper	2.05×10^5	5.07×10^6
Gold	6.30×10^4	1.64×10^6
Iron	2.66×10^5	6.29×10^6
Lead	2.04×10^4	8.64×10^5
Mercury	1.15×10^4	2.72×10^5
Methanol	1.09×10^5	8.78×10^5
Silver	1.04×10^5	2.36×10^6
Water (solid)	3.34×10^5	2.26×10^6

REFERENCE TABLES

Coefficients of Thermal Expansion at 20°C		
Material	**Coefficient of Linear Expansion α (°C^{-1})**	**Coefficient of Volume Expansion β (°C^{-1})**
Solids		
Aluminum	23×10^{-6}	69×10^{-6}
Brass	19×10^{-6}	57×10^{-6}
Concrete	12×10^{-6}	36×10^{-6}
Copper	17×10^{-6}	51×10^{-6}
Glass (soft)	9×10^{-6}	27×10^{-6}
Glass (ovenproof)	3×10^{-6}	9×10^{-6}
Iron, steel	12×10^{-6}	35×10^{-6}
Platinum	9×10^{-6}	27×10^{-6}
Liquids		
Gasoline		950×10^{-6}
Mercury		180×10^{-6}
Methanol		1200×10^{-6}
Water		210×10^{-6}
Gases		
Air (and most other gases)		3400×10^{-6}

Speed of Sound in Various Mediums	
Medium (0°C)	**Speed (m/s)**
Air (0°C)	331
Air (20°C)	343
Helium (0°C)	972
Hydrogen (27°C)	1310
Water (25°C)	1497
Seawater (25°C)	1533
Rubber	1600
Copper (25°C)	3560
Iron (25°C)	5130
Ovenproof glass	5640
Diamond	12,000

Wavelengths of Visible Light	
Color	**Wavelength, λ (nm)**
Violet	380–430
Indigo	430–450
Blue	450–500
Cyan	500–520
Green	520–565
Yellow	565–590
Orange	590–625
Red	625–740

Dielectric Constants, *k* (20°C)	
Vacuum	1.0000
Air (1 atm)	1.00059
Neon (1 atm)	1.00013
Glass	4–7
Quartz	4.3
Fused quartz	3.75
Water	80

Solar System Data								
	Mercury	Venus	Earth	Mars	Jupiter	Saturn	Uranus	Neptune
Mass (kg $\times 10^{24}$)	0.330	4.87	5.97	0.642	1899	569	86.8	102
Average radius (m $\times 10^{6}$)	2.44	6.05	6.38	3.40	71.5	60.3	25.6	24.8
Density (kg/m^3)	5427	5243	5515	3933	1326	687	1270	1638
Albedo	0.068	0.90	0.306	0.250	0.343	0.342	0.300	0.290
Average distance from the Sun (m $\times 10^{9}$)	57.91	108.2	149.6	227.9	778.4	1433.5	2872.5	4498.2
Orbital period (Earth days)	88.0	224.7	365.2	687.0	4332	10,759	30,685	60,189
Orbital inclination (degrees)	7.0	3.4	0.0	1.9	1.3	2.5	0.8	1.8
Orbital eccentricity	0.205	0.007	0.017	0.094	0.049	0.057	0.046	0.011
Rotational period (h)	1407.6	5832.5^R	23.9	24.6	9.9	10.7	17.2^R	16.1
Axial tilt (degrees)	0.03	177.4	23.4	25.2	3.1	26.7	97.8	28.3
Average surface temperature (K)	440	737	288	210	163	133	78	73
Gravitational field strength near the surface (N/kg)	3.7	8.9	9.8	3.7	20.9	10.4	8.4	10.7

R indicates retrograde motion.

The Moon	
Mass	0.073×10^{24} kg
Equatorial radius	1738 km
Mean density	3340 kg/m^3
Albedo	0.11
Average distance from Earth	384×10^{3} km
Orbital period	27.3 Earth days
Synodic period (lunar)	29.53 Earth days
Orbital inclination	5.1°
Orbital eccentricity	0.055
Rotational period	655.7 h
Gravitational field strength near the surface	1.6 N/kg

The Sun	
Mass	1.99×10^{30} kg
Equatorial radius	6.96×10^{8} m
Mean density	1408 kg/m^3
Absolute magnitude	+4.83
Luminosity	3.846×10^{26} J/s
Spectral type	G2 V
Rotational period (equatorial)	609.12 h
Mean energy production	0.1937×10^{-3} J/kg
Average surface temperature	5778 K

Reference Tables

PERIODIC TABLE OF THE ELEMENTS

Key: Atomic number: 1; Symbol: H; Element: Hydrogen; Atomic mass: 1.008

Metal · Metalloid · Nonmetal · Synthetic

	1	2	3	4	5	6	7	8	9	10	11	12	13	14	15	16	17	18
1	1 H Hydrogen 1.008																	2 He Helium 4.003
2	3 Li Lithium 6.941	4 Be Beryllium 9.012											5 B Boron 10.811	6 C Carbon 12.011	7 N Nitrogen 14.007	8 O Oxygen 15.999	9 F Fluorine 18.998	10 Ne Neon 20.180
3	11 Na Sodium 22.990	12 Mg Magnesium 24.305											13 Al Aluminum 26.982	14 Si Silicon 28.086	15 P Phosphorus 30.974	16 S Sulfur 32.066	17 Cl Chlorine 35.453	18 Ar Argon 39.948
4	19 K Potassium 39.098	20 Ca Calcium 40.078	21 Sc Scandium 44.956	22 Ti Titanium 47.867	23 V Vanadium 50.942	24 Cr Chromium 51.996	25 Mn Manganese 54.938	26 Fe Iron 55.847	27 Co Cobalt 58.933	28 Ni Nickel 58.693	29 Cu Copper 63.546	30 Zn Zinc 65.39	31 Ga Gallium 69.723	32 Ge Germanium 72.61	33 As Arsenic 74.922	34 Se Selenium 78.971	35 Br Bromine 79.904	36 Kr Krypton 83.80
5	37 Rb Rubidium 85.468	38 Sr Strontium 87.62	39 Y Yttrium 88.906	40 Zr Zirconium 91.224	41 Nb Niobium 92.906	42 Mo Molybdenum 95.95	43 Tc Technetium (98)	44 Ru Ruthenium 101.07	45 Rh Rhodium 102.906	46 Pd Palladium 106.42	47 Ag Silver 107.868	48 Cd Cadmium 112.411	49 In Indium 114.82	50 Sn Tin 118.710	51 Sb Antimony 121.757	52 Te Tellurium 127.60	53 I Iodine 126.904	54 Xe Xenon 131.290
6	55 Cs Cesium 132.905	56 Ba Barium 137.327	57 La Lanthanum 138.905	72 Hf Hafnium 178.49	73 Ta Tantalum 180.948	74 W Tungsten 183.84	75 Re Rhenium 186.207	76 Os Osmium 190.23	77 Ir Iridium 192.217	78 Pt Platinum 195.08	79 Au Gold 196.967	80 Hg Mercury 200.59	81 Tl Thallium 204.383	82 Pb Lead 207.2	83 Bi Bismuth 208.980	84 Po Polonium 208.982	85 At Astatine 209.987	86 Rn Radon 222.018
7	87 Fr Francium (223)	88 Ra Radium (226)	89 Ac Actinium (227)	104 Rf Rutherfordium * (267)	105 Db Dubnium * (270)	106 Sg Seaborgium * (269)	107 Bh Bohrium * (270)	108 Hs Hassium * (277)	109 Mt Meitnerium * (278)	110 Ds Darmstadtium * (281)	111 Rg Roentgenium * (281)	112 Cn Copernicium * (285)	113 Nh Nihonium * (286)	114 Fl Flerovium * (289)	115 Mc Moscovium * (289)	116 Lv Livermorium * (293)	117 Ts Tennessine * (294)	118 Og Oganesson * (294)

The number in parentheses is the mass number of the longest-lived isotope for that element.

* Properties are largely predicted.

Lanthanide series	58 Ce Cerium 140.115	59 Pr Praseodymium 140.908	60 Nd Neodymium 144.242	61 Pm Promethium (145)	62 Sm Samarium 150.36	63 Eu Europium 151.965	64 Gd Gadolinium 157.25	65 Tb Terbium 158.925	66 Dy Dysprosium 162.50	67 Ho Holmium 164.930	68 Er Erbium 167.259	69 Tm Thulium 168.934	70 Yb Ytterbium 173.04	71 Lu Lutetium 174.967
Actinide series	90 Th Thorium 232.038	91 Pa Protactinium 231.036	92 U Uranium 238.029	93 Np Neptunium (237)	94 Pu Plutonium (244)	95 Am Americium (243)	96 Cm Curium (247)	97 Bk Berkelium (247)	98 Cf Californium (251)	99 Es Einsteinium * (252)	100 Fm Fermium * (257)	101 Md Mendelevium * (258)	102 No Nobelium * (259)	103 Lr Lawrencium * (262)

The Elements

Element	Symbol	Atomic Number	Atomic Mass	Element	Symbol	Atomic Number	Atomic Mass
Actinium	Ac	89	(227)	Mendelevium	Md	101	(258)
Aluminum	Al	13	26.982	Mercury	Hg	80	200.59
Americium	Am	95	(243)	Molybdenum	Mo	42	95.96
Antimony	Sb	51	121.760	Neodymium	Nd	60	144.24
Argon	Ar	18	39.948	Neon	Ne	10	20.180
Arsenic	As	33	74.922	Neptunium	Np	93	(237)
Astatine	At	85	(210)	Nickel	Ni	28	58.693
Barium	Ba	56	137.327	Niobium	Nb	41	92.906
Berkelium	Bk	97	(247)	Nitrogen	N	7	14.007
Beryllium	Be	4	9.012	Nobelium	No	102	(259)
Bismuth	Bi	83	208.980	Osmium	Os	76	190.23
Bohrium	Bh	107	(272)	Oxygen	O	8	15.999
Boron	B	5	10.811	Palladium	Pd	46	106.42
Bromine	Br	35	79.904	Phosphorus	P	15	30.974
Cadmium	Cd	48	112.411	Platinum	Pt	78	195.078
Calcium	Ca	20	40.078	Plutonium	Pu	94	(244)
Californium	Cf	98	(251)	Polonium	Po	84	(209)
Carbon	C	6	12.011	Potassium	K	19	39.098
Cerium	Ce	58	140.116	Praseodymium	Pr	59	140.908
Cesium	Cs	55	132.905	Promethium	Pm	61	(145)
Chlorine	Cl	17	35.453	Protactinium	Pa	91	231.036
Chromium	Cr	24	51.996	Radium	Ra	88	(226)
Cobalt	Co	27	58.933	Radon	Rn	86	(222)
Copernicium	Cn	112	(285)	Rhenium	Re	75	186.207
Copper	Cu	29	63.546	Rhodium	Rh	45	102.906
Curium	Cm	96	(247)	Roentgenium	Rg	111	(280)
Darmstadtium	Ds	110	(281)	Rubidium	Rb	37	85.468
Dubnium	Db	105	(262)	Ruthenium	Ru	44	101.07
Dysprosium	Dy	66	162.500	Rutherfordium	Rf	104	(265)
Einsteinium	Es	99	(252)	Samarium	Sm	62	150.36
Erbium	Er	68	167.259	Scandium	Sc	21	44.956
Europium	Eu	63	151.964	Seaborgium	Sg	106	(271)
Fermium	Fm	100	(257)	Selenium	Se	34	78.96
Fluorine	F	9	18.998	Silicon	Si	14	28.086
Francium	Fr	87	(223)	Silver	Ag	47	107.868
Gadolinium	Gd	64	157.25	Sodium	Na	11	22.990
Gallium	Ga	31	69.723	Strontium	Sr	38	87.62
Germanium	Ge	32	72.63	Sulfur	S	16	32.065
Gold	Au	79	196.967	Tantalum	Ta	73	180.948
Hafnium	Hf	72	178.49	Technetium	Tc	43	(98)
Hassium	Hs	108	(270)	Tellurium	Te	52	127.60
Helium	He	2	4.003	Terbium	Tb	65	158.925
Holmium	Ho	67	164.930	Thallium	Tl	81	204.383
Hydrogen	H	1	1.008	Thorium	Th	90	232.038
Indium	In	49	114.81	Thulium	Tm	69	168.934
Iodine	I	53	126.904	Tin	Sn	50	118.710
Iridium	Ir	77	192.217	Titanium	Ti	22	47.867
Iron	Fe	26	55.847	Tungsten	W	74	183.84
Krypton	Kr	36	83.798	Uranium	U	92	238.029
Lanthanum	La	57	138.906	Vanadium	V	23	50.942
Lawrencium	Lr	103	(262)	Xenon	Xe	54	131.293
Lead	Pb	82	207.2	Ytterbium	Yb	70	173.04
Lithium	Li	3	6.941	Yttrium	Y	39	88.906
Lutetium	Lu	71	174.967	Zinc	Zn	30	65.38
Magnesium	Mg	12	24.305	Zirconium	Zr	40	91.224
Manganese	Mn	25	54.938	Nihonium	Nh	113	(284)
Meitnerium	Mt	109	(276)	Flerovium	Fl	114	(289)
Moscovium	MC	115	(288)	Oganesson	Og	118	(294)
Livermorium	Lv	116	(291)	Tennessine	Ts	117	(294)

Reference Tables

Safety Symbols	Hazard	Examples	Precaution	Remedy
Disposal	Special disposal procedures need to be followed.	Certain chemicals, living organisms	Do not dispose of these materials in the sink or trash can.	Dispose of wastes as directed by your teacher.
Biological	Organisms or other biological materials that might be harmful to humans	Bacteria, fungi, blood, unpreserved tissues, plant materials	Avoid skin contact with these materials. Wear mask and gloves.	Notify your teacher if you suspect contact with material. Wash hands thoroughly.
Extreme Temperature	Objects that can burn skin by being too cold or too hot	Boiling liquids, hot plates, dry ice, liquid nitrogen	Use proper protection when handling.	Go to your teacher for first aid.
Sharp Object	Use of tools or glassware that can easily puncture or slice skin	Razor blades, pins, scalpels, pointed tools, dissecting probes, broken glass	Practice common-sense behavior and follow guidelines for use of the tool.	Go to your teacher for first aid.
Fume	Possible danger to respiratory tract from fumes	Ammonia, acetone, nail polish remover, heated sulfur, moth balls	Be sure there is good ventilation. Never smell fumes directly. Wear a mask.	Leave foul area and notify your teacher immediately.
Electrical	Possible danger from electrical shock or burn	Improper grounding, liquid spills, short circuits, exposed wires	Double-check setup with teacher. Check condition of wires and apparatus.	Do not attempt to fix electrical problems. Notify your teacher immediately.
Irritant	Substances that can irritate the skin or mucous membranes of the respiratory tract	Pollen, moth balls, steel wool, fiberglass, potassium permanganate	Wear dust mask and gloves. Practice extra care when handling these materials.	Go to your teacher for first aid.
Chemical	Chemicals that can react with and destroy tissue and other materials	Bleaches such as hydrogen peroxide; acids such as sulfuric acid, hydrochloric acid; bases such as ammonia, sodium hydroxide	Wear goggles, gloves, and an apron	Immediately flush the affected area with water and notify your teacher.
Toxic	Substance may be poisonous if touched, inhaled, or swallowed.	Mercury, many metal compounds, iodine, poinsettia plant parts	Follow your teacher's instructions.	Always wash hands thoroughly after use. Go to your teacher for first aid.
Flammable	Flammable chemicals may be ignited by open flame, spark, or exposed heat.	Alcohol, kerosene, potassium permanganate	Avoid open flames and heat when using flammable chemicals.	Notify your teacher immediately. Use fire safety equipment if applicable.
Open Flame	Open flame in use, may cause fire.	Hair, clothing, paper, synthetic materials	Tie back hair and loose clothing. Follow teacher's instruction on lighting and extinguishing flames.	Notify your teacher immediately. Use fire safety equipment if applicable.

Eye Safety

Proper eye protection should be worn at all times by anyone performing or observing science activities.

Clothing Protection

This symbol appears when substances could stain or burn clothing.

Radioactivity

This symbol appears when radioactive materials are used.

Handwashing

After the lab, wash hands with soap and water before removing goggles.

GLOSSARY/GLOSARIO

Multilingual Glossary The multilingual science glossary includes Arabic, Bengali, Chinese, English, Haitian Creole, Hmong, Korean, Portuguese, Russian, Spanish, Tagalog, Urdu, and Vietnamese.

Pronunciation Key

Use the following key to help you sound out words in the glossary.

a	back (BAK)	**ew**	food (FEWD)
ay	day (DAY)	**yoo**	pure (PYOOR)
ah	father (FAH thur)	**yew**	few (FYEW)
ow	flower (FLOW ur)	**uh**	comma (CAHM uh)
ar	car (CAR)	**u** (+con)	rub (RUB)
e	less (LES)	**sh**	shelf (SHELF)
ee	leaf (LEEF)	**ch**	nature (NAY chur)
ih	trip (TRIHP)	**g**	gift (GIHFT)
i (i + con + e)	idea, life (i DEE uh, life)	**j**	gem (JEM)
oh	go (GOH)	**ing**	sing (SING)
aw	soft (SAWFT)	**zh**	vision (VIHZH un)
or	orbit (OR but)	**k**	cake (KAYK)
oy	coin (COYN)	**s**	seed, cent (SEED, SENT)
oo	foot (FOOT)	**z**	zone, raise (ZOHN, RAYZ)

Como usar el glosario en español:
1. Busca el termino en ingles que desees encontrar.
2. El termino en español, junto con la definicion, se encuentran en la columna de la derecha.

ENGLISH — A — ESPAÑOL

absorption spectrum: The characteristic set of wavelengths absorbed by a gas, which can be used to identify that gas.

spectrum espectro de absorción: El conjunto característico de longitudes de onda absorbidas por un gas, que puede ser usado para identificarlo.

acceleration: The rate at which the velocity of an object changes.

aceleración: La tasa que expresa la variación de la velocidad de un objeto.

accuracy: A characteristic of a measured value that describes how well the results of a measurement agree with the "real" value, which is the accepted value, as measured by competent experimenters.

exactitud: La característica de un valor medido que describe cuanto concuerdan los resultados de una medición con el valor "real", aceptado, como ha sido medido por experimentadores competentes.

achromatic (a kroh MA tik) lens: A combination of two or more lenses with different indices of refraction (such as a concave lens with a convex lens) that is used to minimize chromatic aberration.

lente acromático: Una combinación de dos o más lentes con índices de refracción diferentes (una lente cóncava y una lente convexa) que es utilizada para minimizar una aberración cromática.

activity: The number of decays per second of a radioactive substance.

actividad: El número de desintegraciones por segundo emitidas por una sustancia radiactiva.

adhesive forces: The forces of attraction that particles of different substances exert on one another; responsible for capillary action.

fuerzas de adhesión: Fuerzas de atracción que se producen cuando entran en contacto partículas de diferentes sustancias, responsables de la acción capilar.

alpha decay: The radioactive decay process in which an alpha (α) particle is emitted from a nucleus.

desintegración alfa: El proceso de desintegración radiactiva donde un núcleo emite una partícula alfa (α).

GLOSSARY/GLOSARIO

alpha particle: Massive, positively charged atomic particle that moves at high speed; represented by the symbol α.

amorphous (uh MOR fus) solid: A substance having a definite shape and volume but lacking a regular crystal structure.

ampere (AM pihr): A flow of electric charge, or electric current, equal to one coulomb per second (1 C/s).

amplitude: In any periodic motion, the maximum distance an object moves from the equilibrium position.

angular acceleration: The change in angular velocity divided by the time needed to make the change; measured in rad/s^2.

angular displacement: The change in the angle as an object rotates.

angular impulse–angular momentum theorem: States that the angular impulse on an object is equal to the change in the object's angular momentum.

angular momentum: The product of a rotating object's moment of inertia and its angular velocity; measured in kg•m^2/s.

angular velocity: The angular displacement of an object divided by the time needed to make the angular displacement.

antenna: A device that propagates electromagnetic waves through the air.

antinode (AN ti nohd): The point with the largest displacement when two wave pulses meet.

apparent weight: The support force acting on an object.

Archimedes' (ahr kuh MEE deez) principle: States that an object immersed in a fluid has an upward force on it that equals the weight of the fluid displaced by the object.

armature (AR muh chur): The wire coil of an electric motor, made up of many loops mounted on an axle or shaft that rotates in a magnetic field; torque on an armature, and the motor's resultant speed, is controlled by varying the current through the motor.

atomic mass unit: A unit of mass, u, where u is equal to 1.66×10^{-27} kg; the approximate mass of a proton or neutron.

atomic number: The number of protons in an atom's nucleus.

average acceleration: The change in an object's velocity during a measurable time interval divided by that specific time interval; measured in m/s^2.

average speed: The distance traveled divided by the time taken to travel that distance; for uniform motion, it is the absolute value of the slope of the object's position-time graph.

average velocity: The ratio of an object's change in position to the time interval during which the change occurred; for uniform motion, it is the slope of the object's position-time graph.

partículas alfa: Partículas atómicas masivas, de carga positiva, que se mueven a alta velocidad; representadas por el símbolo α.

sólido amorfo: Una sustancia que tiene forma y volumen definidos, pero que carece de una estructura cristalina regular.

amperios: Un flujo de carga o de corriente eléctrica, igual a un coulomb de carga por segundo (1 C/s).

amplitud: En un movimiento periódico, es la distancia máxima que se desplaza un objeto desde su posición de equilibrio.

aceleración angular: Cambio que experimenta la velocidad angular necesaria para hacer un cambio por unidad de tiempo; se mide en rad/s^2.

desplazamiento angular: El cambio de ángulo hecho por un objeto en rotación.

teorema del momento angular–impulso angular: Establece que el impulso angular sobre un objeto es igual a la variación del momento angular del objeto.

momento angular: Es el producto del momento de inercia de un objeto en rotación por su velocidad angular, se mide en kg·m^2/s.

velocidad angular: El desplazamiento angular de un objeto dividido entre el tiempo necesario para realizar tal desplazamiento.

antena: Una estructura que propaga ondas electromagnéticas a través del aire.

antinodo: El punto de máximo de desplazamiento en el encuentro de dos pulsos de onda.

peso aparente: La fuerza de apoyo que actúa sobre un objeto.

principio de Arquímedes: Afirma que un cuerpo sumergido en un fluido recibirá una fuerza vertical ascendente igual al peso del fluido desplazado por dicho cuerpo.

armadura: La bobina de alambre de un motor eléctrico, compuesto de muchos circuitos montados sobre un eje, que rota en un campo magnético; el esfuerzo de torsión en una armadura, y la velocidad resultante del motor, es controlada mediante la variación de corriente a través del motor.

unidad de masa atómica: Unidad de masa, u, donde u es igual a 1.66×10^{-27} kg, es la masa aproximada de un protón o un neutrón.

número atómico: El número de protones en el núcleo de un átomo.

aceleración promedio: El cambio en la velocidad de un objeto durante un intervalo de tiempo mensurable dividido entre el mismo intervalo de tiempo; se mide en m/s^2.

rapidez promedio: La distancia recorrida por un objeto dividida entre el tiempo que se toma para recorrer esa distancia; para un movimiento uniforme, es el valor absoluto de la pendiente trazada en la gráfica de posición-tiempo de un objeto.

velocidad media: La relación del cambio de posición de un objeto por un intervalo de tiempo durante el cual se produce el cambio; para un cuerpo en movimiento uniforme, es la pendiente de la recta en la gráfica de posición-tiempo de un objeto.

B

band theory: The theory that electric conduction in solids can be better understood in terms of valance and conduction bands separated by forbidden energy gaps.

teoría de bandas: La teoría en la que la conducción eléctrica en los sólidos puede ser mejor entendida en términos de bandas de valencias y de conducción que están separadas entre sí por espacios sin energía.

battery: A device made up of several galvanic cells connected together that converts chemical energy to electrical energy.

batería: Dispositivo compuesto por varias células galvánicas conectadas entre sí, que convierte la energía química en energía eléctrica.

beat: The oscillation of wave amplitude that results from the superposition of two sound waves with almost identical frequencies.

batimiento: La oscilación de amplitud que resulta de la superposición de dos ondas sonoras con frecuencias ligeramente distintas.

Bernoulli's (bur NEW leez) principle: States that as the velocity of a fluid increases, the pressure exerted by that fluid decreases.

principio de Bernoulli: Afirma que a medida que incrementa la velocidad de un fluido, disminuye la presión que ejerce.

beta decay: The radioactive decay process in which a neutron is changed to a proton or a proton is changed to a neutron within the nucleus; results in the emission of a beta (β) particle and a neutrino or antineutrino.

desintegración beta: El proceso de desintegración radiactiva en la que un neutrón se transforma en un protón o un protón se transforma en un neutrón en el núcleo' resulta en la emisión de una partícula beta (β) y neutrino o un antineutrino.

binding energy: The energy difference between the assembled nucleus and the individual nucleons; it is the energy equivalent of the mass defect and is always negative.

energía de enlace nuclear: La diferencia de energía entre el núcleo ensamblado y los nucleones individuales; es el equivalente energético del defecto de masa y tiene siempre carga negativa.

boson: A particle, such as a photon, that does not obey the Pauli exclusion principle, meaning more than one boson can occupy a given quantum state at the same location at the same time. The two subsets of elementary bosons are gauge bosons (which are force carriers) and scalar bosons.

bosón: Una partícula, como un fotón, que no obedece el principio de exclusión de Pauli, así que más de un bosón puede ocupar un idéntico estado cuántico al mismo lugar al mismo tiempo. Los dos tipos de bosones elementales son bosones de gauge (cuales portan fuerzas) y bosones escalares.

buoyant force: The upward force exerted on an object immersed in a fluid, due to an increase in pressure with increasing depth.

empuje hidrostático: La fuerza vertical ascendente ejercida sobre un objeto que está sumergido en un fluido, debido a un aumento de la presión con un aumento de la profundidad.

C

capacitance (ku PA suh tuns): The ratio of the magnitude of charge on one capacitor plate to the electric potential difference between the plates; the slope of the line of a net charge versus potential difference graph.

capacitancia: La razón entre la magnitud de una carga en un condensador de placas y la diferencia en el potencial eléctrico entre las placas; la pendiente de la recta en el gráfico de una carga neta en relación con la diferencia potencial.

capacitor: An electrical device used to store electrical energy; made up of two conductors separated by an insulator.

condensador: Un dispositivo eléctrico utilizado para almacenar energía eléctrica; se compone de dos conductores separados por un aislante eléctrico.

carrier wave: A specific part of the radio portion of the electromagnetic spectrum assigned to a radio or television station by the Federal Communications Commission. A station broadcasts by varying its carrier wave.

onda portadora: Un parte especifica de la porción del radio del espectro electromagnético que es asignado a una estación de radio o televisión por la Comisión Federal de Comunicaciones. Una estación transmite variando su onda portadora.

center of mass: The point on the object that moves in the same way a point particle would move.

centro de masa: El punto en un objeto que se mueve en la misma forma en la que el punto de una partícula se movería.

centrifugal (sen TRIH fyew gul) "force": The apparent force that seems to pull on a moving object but does not exert a physical outward push on it and is observed only in rotating frames of reference.

"fuerza" centrífuga: La fuerza aparenta que parece empujar a un objeto que se aleja del eje de rotación, pero no ejerce un empuje físico real hacia el exterior, y es observable solamente en sistemas de referencia en rotación.

centripetal (sen TRIH put ul) acceleration: The center-seeking acceleration of an object moving in a circle at a constant speed.

aceleración centrípeta: La aceleración de un objeto que se mueve circularmente a una rapidez constante en búsqueda del centro del eje.

Glossary
Glosario

GLOSSARY/GLOSARIO

centripetal force: The net force exerted toward the center of the circle that causes an object to have a centripetal acceleration.

chain reaction: Continual process of repeated fission reactions caused by the release of neutrons from previous fission reactions.

charging by conduction: The process of charging a neutral object by touching it with a charged object.

charging by induction: The process of charging an object without touching it, which can be accomplished by bringing a charged object close to a neutral object, causing a separation of charges, then separating the object to be charged, trapping opposite but equal charges.

chromatic aberration: A lens defect in which light passing through a lens is dispersed, causing an object viewed through a lens to appear ringed with color.

circuit breaker: An automatic switch that opens when the current through an electric circuit exceeds a threshold value.

closed-pipe resonator: A resonating tube with one end closed to air; its resonant frequencies are odd-numbered multiples of the fundamental.

closed system: A system that does not gain or lose mass.

coefficient of kinetic friction: The slope of a line on a kinetic friction force v. normal force graph, μ_k. Relates frictional force to normal force and depends on the two surfaces in contact.

coefficient of linear expansion: The change in length divided by the original length and the change in temperature.

coefficient of static friction: A dimensionless constant depending on the two surfaces in contact. It is used to calculate the maximum static frictional force that needs to be overcome before motion begins.

coefficient of volume expansion: The change in volume divided by the original volume and the change in temperature; is about three times the coefficient of linear expansion because solids expand in three directions.

coherent light: Light made up of waves of the same wavelength that are in phase with each other.

cohesive forces: The forces of attraction that like particles exert on one another; responsible for surface tension and viscosity.

combination series-parallel circuit: A complex electric circuit that includes both series and parallel branches.

combined gas law: States that, for a fixed amount of an ideal gas, the pressure times the volume, divided by the Kelvin temperature equals a constant; reduces to Boyle's law if temperature is constant and to Charles's law if pressure is constant.

complementary colors: Two colors of light that, when combined, produce white light.

components: Projections of a vector parallel to the x-axis and another parallel to the y-axis.

fuerza centrípeta: La fuerza neta ejercida hacia el centro de un círculo y que produce la aceleración centrípeta.

reacción en cadena: Secuencia de reacciones repetidas de fisión causada por la liberación de neutrones a partir de reacciones de fisión anteriores.

carga por conducción: El proceso de cargar un objeto neutro al tocarlo con un objeto cargado.

carga por inducción: Es el proceso de cargar un objeto sin tocarlo, la carga se logra por acercar un objeto cargado al objeto neutro provocando una separación de cargas, posteriormente se separa el objeto que debía cargarse y este captura cargas iguales pero opuestas.

aberración cromática: Un defecto de una lente en el que la luz que la atraviesa se dispersa, causando que el objeto visto parezca rodeado de color.

disyuntor: Un interruptor automático que se abre si la corriente a través de un circuito eléctrico excede el valor de umbral.

resonador de tubo cerrado: Un tubo resonador con una salida de aire cerrada; sus frecuencias de resonancia son de múltiplos impares de la frecuencia fundamental.

sistema cerrado: Un sistema que no gana ni pierde masa.

coeficiente de fricción cinética: La pendiente de la recta en un grafico de fuerza de fricción cinética-fuerza normal, μ_k. Relaciona la fuerza de fricción con la fuerza normal y es dependiente de las dos superficies en contacto.

coeficiente de expansión lineal: El cambio en la longitud dividida entre la longitud original y el cambio de temperatura.

coeficiente de fricción estática: Una constante sin dimensión que depende de las dos superficies en contacto. Se utiliza para calcular la fuerza de fricción estática máxima que hay que superar para que comience el movimiento.

coeficiente de dilatación: El cambio del volumen dividido entre el volumen original y el cambio de temperatura; es aproximadamente tres veces el coeficiente de expansión lineal, debido a que los sólidos se amplían en tres direcciones.

luz coherente: Es la luz compuesta por ondas de misma longitud y que guardan una relación de fase constante.

fuerzas de cohesión: Las fuerzas de atracción ejercida entre partículas similares; son responsables por la tensión de superficies y la viscosidad.

circuito mixto en serie y paralelo: Un circuito eléctrico complejo que incluye series y ramas paralelas.

ley general de los gases (ley combinada): Establece que para una cantidad fija de un gas ideal, la presión multiplicada por el volumen, dividida entre la temperatura Kelvin es igual a una constante; se reduce a la ley de Boyle cuando la temperatura es constante o a la ley de Charles cuando la presión es constante.

colores complementarios: Dos colores de la luz, que cuando se combinan, producen luz blanca.

componentes: Las proyecciones de un vector paralela al eje-x y al eje-y.

compound machine: A machine consisting of two or more simple machines that are connected so that the resistance force of one machine becomes the effort force of the second machine.

Compton effect: The shift in the energy of scattered photons.

concave lens: A diverging lens, thinner at its middle than at its edges, that spreads out light rays passing through it when surrounded by material with a lower index of refraction; produces a smaller, virtual, upright image.

concave mirror: A mirror that reflects light from its inwardly curving surface, the edges of which curve toward the observer; can produce either an upright, virtual image or an inverted, real image.

conductor: A material, such as copper, through which a charge will move easily.

consonance: A pleasant set of pitches.

convection: A type of thermal energy transfer that occurs due to the motion of fluid in liquid or gas that is caused by differences in temperature.

conventional current: the direction in which a positive test charge moves.

convex lens: A converging lens, thicker at its center than at its edges, that refracts parallel light rays so the rays meet at a point when surrounded by a medium with a lower index of refraction; can produce a smaller or larger, inverted, real image or a larger, upright, virtual image.

convex mirror: A mirror with the edges curved away from the observer that reflects light from its outwardly curving surface; produces an upright, reduced, virtual image.

coordinate system: A system used to describe motion that gives the zero point location of the variable being studied and the direction in which the values of the variable increase.

Coriolis (kor ee OH lus) "force": The apparent force that seems to deflect a moving object from its path and is observed only in rotating frames of reference.

coulomb (KEW lahm): The SI standard unit of charge; one coulomb (C) is the magnitude of the charge of 6.24×10^{18} electrons or protons.

Coulomb's law: States that the force between two point charges varies directly with the product of their charge and inversely with the square of the distance between them.

crest: The high point of a transverse wave.

critical angle: The angle of incidence in which a refracted light ray lies along the boundary between two mediums.

crystal lattice: A fixed, regular pattern formed when the temperature of a liquid is lowered, the average kinetic energy of its particles decreases and, for many solids, the particles become frozen but do not stop moving and instead, vibrate around their fixed positions.

máquina compuesta: Es una maquina formada por dos o más máquinas simples que están conectadas de manera que la fuerza de resistencia de una máquina se convierte en la fuerza de apoyo de la segunda máquina.

efecto Compton: El cambio en la energía de fotones dispersos.

lente cóncava: Una lente divergente, más delgada en su parte media que en sus bordes, extiende los rayos de luz que le atraviesan cuando está rodeada por material de menor índice de refracción; produce una imagen más pequeña, virtual y en posición vertical.

espejo cóncavo: Un espejo con un superficie curvada que refleja la luz de su superficie hacia el interior, sus bordes se curvan al observador; puede producir ya sea una imagen virtual en posición vertical o una imagen real invertida.

conductor: Un material a través del cual una carga se mueve con facilidad, como el cobre.

consonancia: Un conjunto de tonos armónicos.

convección: Un tipo de transferencia de energía térmica que se produce debido al movimiento de fluidos en estado líquido o gaseoso y que es causado por diferencias en la temperatura.

corriente convencional: La dirección en la cual se mueve una prueba de carga positiva.

lente convexa: Una lente convergente, más gruesa en su centro que en sus bordes, que refracta los rayos de luz paralelamente hacia un punto central cuando está en un medio con índice menor de refracción; puede producir imágenes reales, invertidas, grandes o pequeñas; o imágenes virtuales, verticales y grandes.

espejo convexo: Un espejo con bordes curvos alejados del observador que refleja la luz desde su superficie curvada exterior; produce una imagen virtual reducida y en posición vertical.

sistema de coordenadas: Un sistema utilizado para describir desde el lugar del punto cero el movimiento de una determinada variable y la dirección en la cual los valores de esa variable incrementan.

"fuerza" de Coriolis: La supuesta fuerza que parece desviar a un objeto en movimiento de su trayectoria y es observable únicamente en marcos de referencia en rotación.

coulomb: El estándar de la unidad de SI de carga; un coulomb (C) es la magnitud de la carga de $6{,}24\times10^{18}$ electrones o protones.

ley de Coulomb: Establece que la fuerza entre dos cargas de puntos es directamente proporcional al producto de sus cargas e inversamente proporcional al cuadrado de la distancia entre ellos.

cresta: El punto alto de una onda transversal.

angulo crítico: El ángulo de incidencia a partir del cual un rayo de luz refractado pasa a lo largo de la frontera entre dos medios.

red cristalina: Un patrón fijo y regular que se forma cuando se baja la temperatura de un líquido, disminuye el promedio de la energía cinética de sus partículas y, para muchos sólidos, las partículas se congelan pero no dejan de moverse, en cambio, vibran alrededor de sus posiciones fijas.

GLOSSARY/GLOSARIO

D

de Broglie (duh BROY lee) wavelength: The wavelength associated with a moving particle.

decibel: The most common unit of measurement for sound level; also can describe the power and intensity of sound waves. Abbreviated dB.

dependent variable: The factor in an investigation that depends on the independent variable.

depletion layer: The region around a *pn*-junction diode where there are no charge carriers and electricity is poorly conducted.

dielectric: A poor conductor of electric current whose electric charges partially align with an electric field.

diffraction: The bending of light around a barrier.

diffraction grating: A device consisting of many small, closely-spaced slits that diffract light and form a diffraction pattern that is an overlap of single-slit diffraction patterns; can be used to precisely measure light wavelength or to separate light of different wavelengths.

diffraction pattern: A pattern of constructive and destructive interference of Huygens' wavelets, often shown on a screen.

diffuse reflection: A scattered, fuzzy reflection produced by a rough surface.

dimensional analysis: A method of treating units as algebraic quantities that can be cancelled; can be used to check that an answer will be in the correct units.

diode: The simplest semiconductor device; conducts charges in one direction only and consists of a sandwich of *p*-type and *n*-type semiconductors.

dispersion: The separation of white light into a spectrum of colors by such means as a glass prism or water droplets in the atmosphere.

displacement: A change in position having both magnitude and direction; is equal to the final position minus the initial position.

dissonance (DIH suh nunts): An unpleasant, jarring set of pitches.

distance: The entire length of an object's path, even if the object moves in many directions.

domain: A very small group, usually 10–1000 μm on a side, that is formed when the magnetic fields of the electrons in a group of neighboring atoms are aligned in the same direction.

dopant (DOH punt): Material with electron donor or acceptor atoms that can be added in low concentration to intrinsic semiconductors, increasing their conductivity by making either extra electrons or holes available.

longitud de onda de De Broglie: La longitud de onda asociada con el movimiento de una partícula.

decibelios: La unidad de medida del nivel de sonido más utilizado, también describe el poder y la intensidad de las ondas sonoras.

variable dependiente: El factor de una investigación que depende de la variable independiente.

agotamiento de la capa: La región alrededor de un diodo de unión *pn*, donde no hay portadores de carga y la electricidad es débilmente conducida.

dieléctrica: Un mal conductor de corriente eléctrica cuyas cargas eléctricas se alinean parcialmente con un campo eléctrico.

difracción: La curvatura de la luz alrededor de una barrera.

red de difracción: Un dispositivo con varias rendijas cercanas, que difracta la luz, y forma un patrón regular de difracción que es la superposición de los patrones de difracción de cada rendija; se utiliza para medir con precisión la longitud de onda de luz o para separar la luz de diferentes longitudes de onda.

patrón de difracción: Un patrón de interferencia constructiva y destructiva de ondas primarias de Huygens en una pantalla.

reflexión difusa: Una reflexión esparcida y borrosa producida por una superficie rugosa.

análisis dimensional: Un método que trata las unidades como cantidades algebraicas, las cuales pueden ser canceladas; puede ser utilizado para comprobar que la respuesta obtenida estará con las unidades correctas.

diodo: El dispositivo semiconductor más simple, que permite el paso de cargas en una única dirección y consiste en un sándwich de semiconductores tipo-*p* y tipo-*n*.

dispersión: La separación de la luz blanca en un espectro de colores, haciendo uso de medios como un prisma de vidrio o de gotas de agua en la atmósfera.

desplazamiento: Un variación de magnitud y dirección en la posición de un objeto; es igual a la posición final menos la posición inicial.

disonancia: Una serie de tonos desagradable y chocante.

distancia: La longitud total de la trayectoria de un objeto, aun si el objeto se moviera en muchas direcciones.

dominio: Un grupo muy pequeño, generalmente 10-1000 μm de un lado, que se forma cuando los campos magnéticos de los electrones en un grupo de átomos vecinos se alinean en una misma dirección.

dopantes: Material con átomos que son receptores o electrones donantes que se agregan en concentraciones bajas a semiconductores intrínsecos para aumentar su conductividad creando ya sea electrones adicionales o habilitando huecos.

Doppler effect: The change in the frequency of sound or light caused by the movement of either the source, the detector, or both the detector and the source.

drag force: The force exerted by a fluid on an object that opposes the object's motion through the fluid; depends on the object's motion and properties and the fluid's properties.

efecto Doppler: El cambio en la frecuencia del sonido o de la lux causado ya sea por el movimiento de la fuente, por el detector, o bien por ambos el detector y la fuente.

fuerza de arrastre: La fuerza ejercida por un fluido sobre un objeto para oponerse al movimiento del objeto a través del líquido; depende del movimiento y las propiedades del objeto y de las propiedades del fluido.

E

eddy current: A current generated in a piece of metal that is moving through a changing magnetic field, producing a magnetic field that opposes the motion that caused the current.

efficiency: The ratio of output work to input work.

effort force: The force a user exerts on a machine.

elastic collision: A type of collision in which the kinetic energy before and after the collision remains the same.

elastic potential energy: Stored energy due to an object's change in shape.

electric circuit: A closed loop or pathway that allows electric charges to flow.

electric current: A flow of charged particles.

electric field: A property of the space around a charged object that exerts forces on other charged objects.

electric field line: A line that indicates the direction of the force due to the electric field on a positive test charge; the spacing between the lines indicates the field's strength; a closer line spacing indicates a stronger field; the lines never cross and they are directed toward negative charges and away from positive charges.

electric generator: A device that converts mechanical energy to electrical energy and consists of a number of wire loops placed in a strong magnetic field.

electric motor: An apparatus that converts electrical energy into mechanical energy.

electric potential difference: The work needed to move a positive test charge from one point to another, divided by the magnitude of the test charge; also called potential difference.

electromagnet: A type of magnet whose magnetic field is produced by electric current.

electromagnetic induction: The process of generating a current through a circuit due a changing magnetic field or to the relative motion between a wire and a magnetic field.

electromagnetic radiation: Energy that is carried, or radiated, in the form of electromagnetic waves.

corriente de Eddy (o de Foucault): La corriente generada por un pieza de metal que se mueve a lo largo de un campo magnético variable, produciendo un campo magnético que se opone al movimiento que formó la corriente.

eficiencia: La relación entre la energía de salida y energía de entrada.

fuerza aplicada: La fuerza que ejerce un usuario en una máquina simple.

choque elástico: Un tipo de colisión en la cual la energía cinética antes y después de la colisión se mantiene igual.

energía potencial elástica: La energía almacenada de un objeto debido a un cambio de forma.

circuito eléctrico: Un circuito o pasaje cerrado que permite que las cargas eléctricas fluyan.

corriente eléctrica: Un flujo de partículas cargadas.

campo eléctrico: Una propiedad del espacio alrededor de un objeto cargado que ejerce fuerzas hacia otros objetos cargados.

línea de campo eléctrico: una línea que indica la dirección de la fuerza debida al campo eléctrico de una prueba de carga positiva; el espacio entre las líneas indica la fuerza del campo; un espaciamiento angosto indica un campo más fuerte; las líneas nunca se cruzan, y se dirigen hacia las cargas negativas y evadiendo las cargas positivas.

generador eléctrico: Un dispositivo que convierte la energía mecánica en energía eléctrica y consiste en una serie de circuitos de alambres colocados en un campo magnético fuerte.

motor eléctrico: Un aparato que convierte la energía eléctrica en energía mecánica.

diferencia de potencial eléctrico: El trabajo realizado para mover una carga de prueba positiva de un punto a otro, dividido por la magnitud de la carga de prueba; también conocido como la diferencia potencial.

electroimán: Un imán cuyo campo magnético es producido por una corriente eléctrica.

inducción electromagnética: El proceso de generar corriente a través de un circuito debido a un campo magnético variable o al movimiento relativo entre el alambre y el campo magnético.

radiación electromagnética: La energía portada o irradiada, en forma de ondas electromagnéticas.

electromagnetic spectrum: The entire range of frequencies and wavelengths that make up the continuum of electromagnetic waves, including radio waves, microwaves, visible light, and *X*-rays.

electromagnetic wave: Coupled, oscillating electric and magnetic field that travels through space and matter.

electron cloud: The region in which there is a high probability of finding an electron.

electroscope: A device that is used to detect electric charges and consists of a metal knob connected by a metal stem to two thin metal leaves.

electrostatics: The study of electric charges that can be collected and held in one place.

elementary charge: The magnitude of the charge of an electron; approximately 1.602×10^{-19} C.

emission spectrum: A plot of the intensity of radiation emitted from an object over a range of frequencies.

energy: The ability of a system to produce a change in itself or in the world around it; represented by the symbol *E*.

energy level: The quantized amount of energy allowed for electrons in an atom.

entropy (EN truh pee): A measure of the energy dispersal in a system.

equilibrant: A force that places an object in equilibrium; the same magnitude as the resultant but opposite in direction.

equilibrium: The condition in which the net force on an object is zero.

equipotential (ee kwuh puh TEN shul): The electric potential difference of zero between two or more positions in an electric field.

equivalent resistance: The value of a single resistor that, when it replaces all the resistors in the circuit, results in the same current.

excited state: Any energy level of an atom that is higher than its ground state.

extrinsic (ek STRIHN zik) semiconductor: Semiconductor with greatly enhanced conductivity resulting from the addition of dopants.

espectro electromagnético: La gama entera de frecuencias y longitudes de onda que componen el continuum de ondas electromagnéticas; incluye a ondas de radio, microondas, luz visible y rayos X.

onda electromagnética: Un campo eléctrico y magnético conectado y variable que viaja a través del espacio y la materia.

nube de electrones: La región en la cual existe una alta probabilidad de encontrar un electrón.

electroscopio: Instrumento que permite determinar las cargas eléctricas, consiste en una perilla metálica conectada por una varilla metálica a dos laminillas de metal.

electrostática: El estudio de cargas eléctricas que pueden ser recogidas y almacenadas en un lugar.

carga elemental: La magnitud de la carga de un electrón, aproximadamente 1.602×10^{-19} C.

espectro de emisión: La intensidad de la radiación emitida por un objeto sobre un conjunto de frecuencias.

energía: La capacidad de un sistema para producir un cambio en sí mismo o en el mundo que le rodea; representado por el símbolo *E*.

nivel energético: La cantidad de energía cuántica permitida para los electrones de un átomo.

entropía: Es la medición de la energía dispersa en un sistema.

equilibrante: La fuerza que coloca un objeto en equilibrio; tiene la misma magnitud que la fuerza resultante, pero en direcciones opuestas.

equilibrio: La condición en la cual la fuerza neta sobre un objeto es cero.

equipotencial: La diferencia del potencial eléctrico de cero entre dos o más posiciones en un campo eléctrico.

resistencia equivalente: El valor de una sola resistencia que, cuando se reemplaza a todas las resistencias en el circuito, resulta en la misma corriente.

estado excitado: Cualquier nivel de energía de un átomo por encima de su estado fundamental.

semiconductor extrínseco: Semiconductor con gran realce de conductividad producto de la adición de dopantes.

F

farsightedness: A vision defect in which a person cannot see close objects clearly because images are focused behind the retina; also called hyperopia; can be corrected with a convex lens.

first law of thermodynamics: States that the change in thermal energy of a system is equal to the heat that is added to the system, minus the work done by the system.

fission: The process in which a nucleus is divided into two or more fragments, and neutrons and energy are released.

hipermetropía: Un defecto ocular en la que una persona no puede ver claramente los objetos cercanos porque las imágenes son enfocadas en un punto detrás de la retina; se corrige con una lente convexa.

primera ley de la termodinámica: Afirma que la variación de la energía térmica de un sistema es igual al calor agregado al sistema, menos el trabajo realizado por el sistema.

fisión: El proceso en el cual el núcleo se divide en dos o más fragmentos, liberándose neutrones y energía.

fluid: A material, such as a liquid or gas, that can flow and has no definite shape of its own.

focal length: The distance between the focal point and the mirror or lens.

focal point: The point where incident light rays that are parallel to the principal axis converge after reflecting from a mirror or refracting through a lens.

force: A push or a pull exerted on an object; has both direction and magnitude and may be a contact or a field force.

free-body diagram: A physical model that represents the forces acting on a system.

free fall: The motion of an object body when air resistance is negligible and the motion can be considered due to the force of gravity alone.

free-fall acceleration: The acceleration of an object due only to the effect of gravity.

frequency: The number of complete oscillations a point on a wave makes each second; measured in hertz (Hz).

fundamental: For a musical instrument, the lowest frequency of sound that will resonate.

fuse: A short piece of metal that acts as a safety device by melting and stopping the current when too large a current passes through it.

fusion: The process in which nuclei with small masses combine to form a nucleus with a larger mass and energy is released.

fluido: Un material (como un líquido o un gas) que puede fluir y no tiene forma propia definida.

distancia focal: La distancia entre el punto focal y el espejo o la lente.

punto focal: El punto donde los rayos de luz incidente que son paralelos al eje principal convergen después de reflejarse en un espejo o refractarse a tráves de una lente.

fuerza: Un empuje o jalón ejercido sobre un objeto; posee tanto dirección como magnitud y puede ser de contacto o de un campo de fuerza.

diagrama de cuerpo libre: un modelo físico que representa las fuerzas que actúan en un sistema.

caída libre: El movimiento de un objeto para el cual la resistencia del aire es insignificante y su movimiento puede ser considerado únicamente debido a la fuerza de gravedad.

aceleración de caída libre: La aceleración de un objeto debido únicamente al efecto de la gravedad.

frecuencia: El número de oscilaciones completas que un punto en una onda realiza por segundo; se mide en hercios (Hz).

fundamentales: En un instrumento musical, es la menor frecuencia de resonancia del sonido.

fusible: Un trozo de metal que actúa como un dispositivo de seguridad al derretirse y detener el fluyo de la corriente cuando un mayor flujo de corriente pasa.

fusión: El proceso en el que los núcleos de masa pequeña se combinan para formar un núcleo de mayor masa y para liberar energía.

G

galvanometer (gal vuh NAH muh tur): A device that is used to measure very small currents; can be used as a voltmeter or an ammeter.

gamma decay: The radioactive decay process in which there is a redistribution of energy within the nucleus but no change in atomic mass or charge, and a gamma ray is emitted.

gravitational field: A vector quantity that relates the mass of an object to the gravitational force it experiences at a given location; represented by the symbol g.

gravitational force: The attractive force between two objects that is directly proportional to the mass of the objects.

gravitational mass: Mass as used in the law of universal gravitation; the quantity that measures an object's response to gravitational force.

gravitational potential energy: The stored energy in a system resulting from the gravitational force between objects; represented by the symbol *GPE*.

galvanómetro: Un dispositivo que se utiliza para medir corrientes muy pequeñas; puede ser utilizado como un voltímetro o un amperímetro.

desintegración gamma: El proceso de desintegración radiactiva en el cual hay una redistribución de energía dentro del núcleo, sin ocurrir ningún cambio de masa atómica o de carga eléctrica, y emite un rayo gamma.

campo gravitatorio: Magnitud vectorial que relaciona la masa de un objeto con la fuerza gravitatoria experimentada en un lugar determinado; representada por el símbolo g.

fuerza de la gravedad: La fuerza de atracción entre dos objetos, es directamente proporcional a la masa de los objetos.

masa gravitacional: Masa como se define en la ley de la gravedad; cantidad que mide la respuesta de atracción de un objeto ante la fuerza gravitacional.

energía potencial gravitatoria: La energía almacenada en un sistema resultado de la fuerza gravitatoria entre objetos; se representa por el símbolo *EPG*.

ground-fault interrupter (GFI): A device that contains an electronic circuit that detects small current differences caused by an extra current path; it opens the circuit, prevents electrocution, and often is required as a safety measure for bathroom, kitchen, and exterior outlets.

interruptor de circuito por falla a tierra (ICFT): Un dispositivo que contiene un circuito electrónico capaz de detectar diferencias pequeñas de corrientes causadas por la adición de un trayecto de corriente; abre el circuito, previene la electrocución y usualmente es usado como medida de seguridad en los baños, cocinas y tomas de corriente exteriores.

grounding: The process of dispersing excess charge by touching an object to Earth or other large conductive body.

conexión a tierra: El proceso de dispersar carga excesiva por tocar un objeto a la tierra o otro cuerpo conductor y grande.

ground state: State of an atom with the smallest allowable amount of energy.

estado fundamental: Estado en el que un átomo posee la cantidad más pequeña de energía permitida.

H

half-life: The time required for half the atoms in a given quantity of a radioactive isotope to decay.

vida media: El tiempo requerido para que la mitad de los átomos se desintegren en una determinada cantidad de un isótopo radiactivo.

harmonics: Higher frequencies, which are whole-numbered multiples of the fundamental frequency; give certain musical instruments their own unique timbre.

armónicos: Las frecuencias más altas, que son múltiplos de números enteros de la frecuencia fundamental; dota a ciertos instrumentos musicales con su propio y único timbre.

heat: A transfer of thermal energy, which occurs spontaneously from a warmer object to a cooler object; represented by Q.

calor: Transferencia de energía térmica que ocurre espontáneamente de un objeto más caliente a un objeto más frío; representado por Q.

heat engine: A device that continuously converts thermal energy to mechanical energy; requires a high-temperature thermal energy source, a low-temperature receptacle (a sink), and a way to convert the thermal energy into work.

motor térmico: Dispositivo que continuamente convierte energía térmica en energía mecánica; requiere de una fuente de energía térmica a temperatura alta, de un recipiente de temperatura baja (sumidero), y de un mecanismo para convertir la energía térmica en trabajo.

heat of fusion: The amount of thermal energy required to change 1 kg of a substance from a solid state to a liquid state at its melting point.

calor de fusión: Es la cantidad de energía térmica necesaria para hacer que un kilogramo de una sustancia pase del estado sólido a líquido en su punto de fusión.

heat of vaporization: The amount of thermal energy required to change 1 kg of a substance from a liquid state to a gaseous state at its boiling point.

calor de vaporización: La cantidad de energía térmica necesaria para hacer que 1 kg de una sustancia pase de estado liquido a gaseoso en su punto de ebullición.

Heisenberg uncertainty principle: States there is a limit to how precisely a particle's position and momentum can simultaneously be measured.

principio de incertidumbre de Heisenberg: Establece que hay un límite en la precisión en cual se puede medir simultáneamente la posición y el momento de una partícula.

Hooke's law: States that the force acting on a spring is directly proportional to the amount the spring is stretched.

ley de Hooke: Establece que la cantidad de estiramiento de un resorte es directamente proporcional a la fuerza que actúa sobre el resorte.

hypothesis: A possible explanation for a problem using what you know and what you observe.

hipótesis: Una posible explicación de un problema haciendo uso de lo que se sabe y lo que se observa.

I

ideal gas law: For an ideal gas, the pressure times the volume is equal to the number of moles, times the constant (R) and the Kelvin temperature; predicts the behavior of gases remarkably well unless under high-pressure or low-temperature conditions.

ley de los gases ideales: Para un gas ideal, es la ecuación de presión por volumen igual al número de moles, por la constante (R) y la temperatura Kelvin; predice aceptablemente la conducta de los gases a menos que estos se encuentren bajo condiciones de presión alta o temperatura baja.

ideal mechanical advantage: For an ideal machine, is equal to the displacement of the effort force divided by displacement of the resistance force.

ventaja mecánica ideal: Para una maquina ideal, equivale al desplazamiento de la fuerza de esfuerzo dividido entre la fuerza de resistencia.

illuminance (ih LEW muh nunts): The rate at which light strikes a surface, or falls on a unit area; measured in lumens per square meter (lm/m^2) or lux (lx).

image: An optical reproduction of an object formed by the combination of light rays reflected or refracted by an optical device.

impulse: The product of the average net force on an object and the time interval over which the force acts.

impulse-momentum theorem: States that the impulse on an object equals the object's final momentum minus the object's initial momentum.

incident wave: A wave pulse that strikes a boundary between two mediums.

incoherent light: Light whose waves are not in phase.

independent variable: The factor that is changed or manipulated during an investigation.

index of refraction: Determines the angle of refraction of light as it crosses the boundary between mediums; for a given medium, it is the ratio of the speed of light in a vacuum to the speed of light in that medium; represented by the symbol n.

induced electromotive force: The potential difference across a wire that results from the production of an electric field.

inelastic collision: A type of collision in which the kinetic energy after the collision is less than the kinetic energy before the collision.

inertia (ihn UR shuh): The tendency of an object to resist changes in velocity.

inertial mass: A measure of an object's resistance to any type of force.

instantaneous acceleration: The change in an object's velocity at a specific instant.

instantaneous position: The position of an object at any particular instant.

instantaneous velocity: A measure of motion that tells the speed and direction of an object at a specific instant.

insulator: A material, such as glass, through which a charge will not move easily.

interaction pair: A pair of forces that are equal in strength but opposite in direction and act on different objects.

interference: Results from the superposition of two or more waves; can be constructive or destructive.

interference fringes: A pattern of light and dark bands on a screen, resulting from the constructive and destructive interference of light waves passing through two narrow, closely spaced slits in a barrier.

intrinsic semiconductor: Pure semiconductor that conducts charge as a result of thermally freed electrons.

inverse relationship: A hyperbolic relationship that exists when one variable depends on the inverse of the other variable.

iluminancia: La cantidad en la cual la luz incide sobre una superficie por unidad de área; se mide en lúmenes por metro cuadrado (lm/m^2) o lux (lx).

imagen: Una reproducción óptica de un objeto, formado por la combinación de rayos de luz reflejados o refraccionados por un dispositivo óptico.

impulso: Resultado de la fuerza neta promedio en un objeto y el intervalo de tiempo durante el cual actúa la fuerza.

teorema del impulso y momento: Establece que el impulso de un objeto es igual a la resta del impulso final del objeto menos el impulso inicial del mismo.

onda incidente: Un pulso de onda que incide un límite entre dos medios.

luz incoherente: Luz cuyas ondas no están en fase.

variable independiente: El factor que cambia o se manipula durante una investigación.

índice de refracción: Determina el ángulo de refracción de la luz a su paso frontera través del límite entre dos medios, para un medio específico, es el cociente de la velocidad de la luz en el vacío entre la velocidad de la luz en ese medio; representado por el símbolo n.

fuerza electromotiva inducida: La diferencia potencial a lo largo de un alambre que resulta de la producción de un campo eléctrico.

colisión inelástica: Un tipo de choque en el que la energía cinética después del choque es menor que la energía cinética antes del choque.

inercia: La propiedad de un objeto a resistir los cambios en la velocidad.

masa inercial: Es la medida de la resistencia de un objeto a cualquier tipo de fuerza.

aceleración instantánea: El cambio en la velocidad de un objeto en un instante específico.

posición instantánea: La posición de un objeto en cualquier instante en particular.

velocidad instantánea: Una medida del movimiento que indica la velocidad y dirección de un objeto en un instante específico.

aislante: Un material, como el vidrio, a través del cual una carga no se mueva con facilidad.

par de interacción: Un par de fuerzas de igual intensidad, sentidos opuestos, y que actúan sobre objetos diferentes.

interferencia: El resultado de la superposición de dos o más ondas; puede ser constructiva o destructiva.

franjas de interferencia: Un patrón de bandas oscuras e iluminadas sobre una pantalla, resultado de la interferencia constructiva y destructiva de ondas de luz que pasan a través de dos rendijas estrechas y muy próximas entre sí en una barrera.

semiconductor intrínsico: Semiconductor puro que conduce carga como resultado de la liberación térmica de electrones.

relación inversa: Una relación hiperbólica que se da cuando una variable depende de la inversa de la otra variable.

isolated system: A closed system on which the net external force is zero.

isotope: Each of the differing forms of the same atom that have different masses but have the same chemical properties; atoms with the same number of protons, but different numbers of neutrons.

sistema aislado: Un sistema cerrado en el que la fuerza externa neta es cero.

isótopo: Cada una de las diferentes formas de un mismo átomo con masas diferentes, pero con las mismas propiedades químicas; atomos con el mismo numero de protones, pero distinctos números de neutrones.

J

joule (JEWL): The SI unit of work and energy (J); 1 J of work is done when a force of 1 N acts on an object over a displacement of 1 m.

julio: La unidad de trabajo y energía del SI (J); 1 J de trabajo equivale al trabajo producido por una fuerza de 1 N cuyo punto de aplicación se desplaza 1 m en la dirección de la fuerza.

K

Kepler's first law: States that the planets move in elliptical paths with the Sun at one focus.

Kepler's second law: States that an imaginary line from the Sun to a planet sweeps out equal areas in equal time intervals.

Kepler's third law: States that the square of the ratio of the periods of any two planets is equal to the cube of the ratio of their average distances from the Sun.

kilowatt-hour: An energy unit used by electric companies to measure energy sales; 1 kWh is equal to 1000 watts (W) delivered continuously for 3600 s (1 h).

kinetic energy: The energy of an system that is associated with its motion.

kinetic friction: The force exerted on one surface by a second surface when the two surfaces rub against each other because one or both of the surfaces are moving.

primera ley de Kepler: Establece que los planetas se mueven en trayectorias elípticas, con el Sol situado en un foco.

segunda ley de Kepler: Establece que una línea imaginaria desde el Sol a un planeta barre áreas iguales en intervalos iguales de tiempo.

tercera ley de Kepler: Establece que el cuadrado de la relación de los períodos de dos planetas cualesquiera es igual al cubo de la proporción de sus distancias medias desde el sol.

kilovatio-hora: Una unidad de energía utilizada por las empresas eléctricas para medir las ventas de energía, 1 kWh es igual a 1000 vatios (W) emitido continuamente durante 3600 s (1 h).

energía cinética: La energía de un sistema asociada con su movimiento.

fricción cinética: La fuerza ejercida sobre una superficie por una segunda superficie, cuando las dos superficies se rozan entre sí porque uno o ambas de las superficies están en movimiento.

L

laser: A device that produces powerful, coherent, directional, monochromatic light that can be used to excite other atoms; the acronym stands for *l*ight *a*mplification by *s*timulated *e*mission of *r*adiation.

law of conservation of angular momentum: States that if there are no net external torques on a closed system, then its angular momentum is conserved.

law of conservation of energy: States that in a closed, isolated system, energy is not created or destroyed, but rather, is conserved.

law of conservation of momentum: States that the momentum of any closed, isolated system does not change.

law of reflection: States that the angle of incidence is equal to the angle of reflection.

láser: Un dispositivo que produce luz monocromática poderosa, coherente y direccional que puede ser utilizada para excitar a otros átomos, su acrónimo indica las letras iniciales de su nombre en Ingles que significa *luz* amplificada por emisión de radiación estimulada

ley de la conservación del momento angular: Establece que, si no hay esfuerzo de torsión neta externa sobre un sistema cerrado, entonces su momento angular se conserva.

ley de la conservación de la energía: Establece que, en un sistema cerrado y aislado, la energía no se crea ni se destruye, sino que se conserva.

ley de conservación del momento: Establece que el momento de un sistema cerrado y aislado no cambia.

ley de la reflexión: Establece que el ángulo de incidencia es igual al ángulo de reflexión.

law of universal gravitation: States that gravitational force between any two objects is directly proportional to the product of their masses and inversely proportional to the square of the distance between their centers.

ley de la gravitación universal: Establece que la fuerza gravitacional entre dos objetos es directamente proporcional al producto de sus masas e inversamente proporcional al cuadrado de la distancia entre sus centros.

lens: A piece of transparent material, such as glass or plastic, that is used to focus light and form an image.

lente: Una pieza de material transparente, como el vidrio o plástico, que se utiliza para enfocar la luz y formar una imagen.

Lenz's law: States that an induced current always is produced in a direction such that the magnetic field resulting from the induced current opposes the change in the magnetic field that is causing the induced current.

ley de Lenz: Establece que una corriente inducida siempre se produce en una dirección tal que se opone a la variación en el campo magnético que está causando la corriente inducida.

lepton: An elementary particle, such as an electron or a neutrino, that has mass and charge but is not affected by the strong force.

lepton: Una partícula elemental, como un electrón o neutrino, que tiene masa y carga pero no experimenta la fuerza nuclear fuerte.

lever arm: The perpendicular distance from the axis of rotation to the point where force is exerted.

brazo de palanca: La distancia perpendicular desde el eje de rotación hasta el punto donde la fuerza se ejerce.

linear relationship: A relationship in which the dependent variable varies linearly with the independent variable.

relación lineal: Una relación en la que la variable dependiente varía linealmente con la variable independiente.

line of best fit: A line drawn on a graph as close to all the data points as possible; used to describe data and predict where new data will appear on the graph.

línea de mejor ajuste: Una línea trazada sobre un gráfico lo más cercanamente posible a todos los puntos de datos, se utiliza para describir los datos y predecir dónde aparecerán nuevos datos en el gráfico.

longitudinal wave: A mechanical wave in which the disturbance is in the same direction, or parallel to, the direction of wave travel.

onda longitudinal: Una onda mecánica en la que la alteración está en la misma dirección, o en una dirección paralela a la dirección de la propagación de la onda.

loudness: Sound intensity as sensed by the ear and interpreted by the brain; depends mainly on the pressure wave's amplitude.

sonoridad: La intensidad del sonido detectada por el oído e interpretada por el cerebro; depende principalmente de la presión de la amplitud de la onda.

luminous flux: The rate at which light energy is emitted from a luminous source; measured in lumens (lm).

flujo luminoso: La velocidad a la cual se emite la energía luminosa desde una fuente luminosa; se mide en lúmenes (lm).

luminous source: An object, such as the Sun or a lightbulb, that emits light.

fuente luminosa: Un objeto, como el Sol o un bombillo incandescente, que emite luz.

M

machine: A device that makes work easier (but does not change the amount of work) by changing the magnitude or the direction of the force exerted to do work.

máquina: Un dispositivo que facilita el trabajo (pero no cambia la cantidad), modificando la magnitud o la dirección de la fuerza ejercida para hacer el trabajo.

magnetic field: The area around a magnet, or around any current-carrying wire or coil of wire, where a magnetic force exists.

campo magnético: El área alrededor de un imán, o cercano a cualquier cable portador de corriente o rollo de alambre, donde existe una fuerza magnética.

magnetic flux: The number of magnetic field lines that pass through a perpendicular surface.

flujo magnético: El número de líneas de un campo magnético que pasan a través de una superficie perpendicular.

magnification: The amount that an image is enlarged or reduced in size, relative to the object.

magnificación: La cantidad que una imagen es ampliada o reducida en tamaño, en relación con el objeto.

magnitude: A measure of size.

magnitud: Una medida de tamaño.

Malus's law: States that the intensity of light coming out of a second polarizing filter equals the intensity of polarized light coming out of a first polarizing filter times the cosine, squared, of the angle between the polarizing axes of the two filters.

Ley de Malus: Establece que la intensidad de la luz proveniente de un segundo filtro polarizador equivale a la intensidad de la luz polarizada que proviene del primer filtro polarizador por el coseno al cuadro del ángulo entre los ejes de polarización de los dos filtros.

mass defect: The difference between the sum of the masses of individual nucleons and the mass of the assembled nucleus.

defecto de masa: La diferencia entre la suma de las masas de los nucleones individuales y la masa del núcleo montado.

mass number: The number of nucleons in an atom's nucleus.

número másico: El número de nucleones en el núcleo de un átomo.

mass spectrometer: Device that uses both electric and magnetic fields to measure the charge-to-mass ratio of positive ions within a material; can be used to determine the masses of ions.

espectrómetro de masas: Instrumento que utiliza ambos los campos eléctricos y magnéticos para medir la razón entre carga y masa de los iones en una material; puede ser usado para determinar las masas de los iones.

measurement: A comparison between an unknown quantity and a standard.

medición: Una comparación entre una cantidad desconocida y una establecida.

mechanical advantage: The ratio of resistance force to effort force.

ventaja mecánica: La razón entre la fuerza de resistencia y la fuerza de esfuerzo.

mechanical energy: The sum of kinetic energy and potential energy of the objects in a system.

energía mecánica: La suma de energía cinética y energía potenciale de los objetos en un sistema.

microchip: An integrated circuit consisting of thousands of diodes, transistors, resistors, and conductors.

microchip: Un circuito integrado que consta de miles de diodos, transistores, resistencias y conductores.

model: A representation of an idea, event, structure, or object to help people better understand it.

modelo: Representación de una idea, evento, estructura, u objeto para ayudar a la gente a entenderlo mejor.

moment of inertia: The resistance to rotation.

momento de inercia: La resistencia a la rotación.

momentum: The product of the object's mass and the object's velocity; measured in kg·m/s.

momento: El producto de la masa del objeto por la velocidad del objeto; se mide en kg·m/s.

monochromatic light: Light having only one wavelength.

luz monocromática: La luz que tiene sólo una onda longitudinal.

motion diagram: A series of images showing the positions of a moving object taken at regular (equal) time intervals.

diagrama de movimiento: Una serie de imágenes que muestran las posiciones de un objeto en movimiento tomadas en periodos iguales de tiempo.

mutual inductance: Effect in which a changing current in a coil creates a changing magnetic field that induces a varying *EMF* in a second coil.

inductancia mutua: Efecto en el que una corriente variable en una bobina crea un campo magnético cambiante, que induce una *FEM* variable en una segunda bobina.

N

nearsightedness: A vision defect in which a person cannot see distant objects clearly because images are focused in front of the retina; also called myopia; can be corrected with a concave lens.

miopía: Un defecto ocular en el que una persona no puede ver claramente los objetos distantes, porque las imágenes se enfocan en un punto delante de la retina, se puede corregir con una lente cóncava.

net force: The vector sum of all the forces on an object.

fuerza neta: La suma vectorial de todas las fuerzas sobre un objeto.

neutral: An object whose positive charges are exactly balanced by its negative charges.

neutro: Un objeto cuyas cargas positivas están balanceadas con exactitud por sus cargas negativas.

Newton's first law: States that an object at rest will remain at rest and a moving object will continue moving in a straight line with constant speed, if and only if the net force acting on that object is zero.

primera ley de Newton: Establece que un objeto en reposo permanecerá en reposo, y un objeto en movimiento continuará moviéndose a una velocidad constante en línea recta, si y sólo si la fuerza neta que actúa sobre ese objeto es igual a cero.

Newton's second law: States that the acceleration of an object is proportional to the net force and inversely proportional to the mass of the object being accelerated.

segunda ley de Newton: Establece que la aceleración de un objeto es directamente proporcional a la fuerza neta en el e inversamente proporcional a la masa del objeto que se acelera.

Newton's second law for rotational motion: States that the angular acceleration of an object is directly proportional to the net torque on it and inversely proportional to its moment of inertia.

segunda ley de Newton para el movimiento de rotación: Establece que la aceleración angular de un objeto es directamente proporcional a la torsión neta e inversamente proporcional a sul momento de inercia.

Newton's third law: States that all forces come in pairs and that the two forces in a pair act on different objects, are equal in strength, and are opposite in direction.

tercera ley de Newton: Establece que todas las fuerzas vienen en pares y que las dos fuerzas en un par actúan sobre objetos diferentes, son iguales en intensidad y son opuestas en sentido.

node: The stationary point where two equal wave pulses meet and are in the same location, having a displacement of zero.

normal: The line in a ray diagram that shows the orientation of the barrier or mirror and is drawn at a right angle, or perpendicular, to the barrier or mirror.

normal force: The perpendicular contact force exerted by a surface on an object.

nuclear reaction: Occurs when the energy or number of neutrons or protons in a nucleus changes.

nucleon: A particle found in the nucleus of an atom; that is, a proton or a neutron.

nucleus: The tiny, massive, positively charged central core of an atom.

nodo: El punto estacionario en el que dos ondas con igual pulso se encuentran y están en la misma ubicación, teniendo un desplazamiento de cero.

normal: La línea en un diagrama de rayos que muestra la orientación de la barrera o el espejo y esta dibujada en un ángulo recto, o perpendicular, a la barrera o al espejo.

fuerza normal: La fuerza de contacto perpendicular ejercida por una superficie sobre un objeto.

reacción nuclear: Se produce cuando cambia la energía o el número de neutrones o protones en un núcleo.

nucleon: Una partícula encontrada en el núcleo de un átomo; es decir, un protones o un neutron.

núcleo: Un centro pequeño y masiva con carga positiva de un átomo.

Glossary
Glosario

O

object: A luminous or illuminated source of light rays.

opaque (oh PAYK): A property of a medium that allows that medium to absorb light and reflect some light rather than transmitting it, preventing objects from being seen through it.

open-pipe resonator: A resonating tube with both ends open that also will resonate with a sound source; its resonant frequencies are whole-number multiples of the fundamental.

origin: The point at which both variables in a coordinate system have the value zero.

objeto: Una fuente luminosa o iluminada por los rayos de luz.

opaco: Una propiedad de un medio que permite al medio absorber la luz y reflejar algo de la luz sin transmitirla, previene que los objetos sean vistos a través de él.

resonador de tubo abierto: Un tubo de resonancia con ambos extremos abiertos, que también resonara con una fuente de sonido; sus frecuencias de resonancia son múltiplos de números enteros de la fundamental.

origen: El punto en un sistema de coordenadas en que ambas variables tienen el valor cero.

P

pair production: The conversion of energy into a matter-antimatter pair of particles.

parallel circuit: A type of electric circuit in which there are several current paths; its total current is equal to the sum of the currents in the individual branches, and if any branch is opened, the current in the other branches remains unchanged.

parallel connection: A type of connection in which there are two or more current paths to follow.

particle model: A simplified version of a motion diagram in which the moving object is replaced by a series of single points.

pascal: The SI unit of pressure; equal to 1 N/m^2.

Pascal's principle: States that any change in pressure applied at any point on a confined fluid is transferred undiminished throughout the fluid.

period: In any periodic motion, the amount of time required for an object to repeat one complete cycle of motion.

periodic motion: Any motion that repeats in a regular cycle.

producción de pares: La conversión de energía en un par de materia-antimateria de partículas.

circuito paralelo: Un tipo de circuito eléctrico en el que hay varios trayectos de corrientes, y su corriente total es igual a la suma de las corrientes en las ramas individuales, y si cualquiera de las ramas se abre, la corriente en las otras ramas se mantiene sin cambios.

conexión en paralelo: Un tipo de conexión en la que hay dos o más trayectos de corrientes para seguir.

modelo de partículas: Una versión simplificada de un diagrama de movimiento en el que se sustituye el movimiento del objeto por una serie de puntos individuales.

pascal: Unidad de presión en el SI; igual a 1 N/m^2

principio de Pascal: Establece que la presión ejercida en cualquier lugar en un fluido encerrado se transfiere por igual en todas las direcciones a través del fluido.

período: En cualquier movimiento periódico, es la cantidad de tiempo necesaria para que un objeto repita un ciclo completo de movimiento.

movimiento periódico: Movimiento que se repite en intervalos regulares de tiempo.

periodic wave: A wave whose disturbances occur at a constant rate.

photoelectric effect: The emission of electrons by certain metals that occurs when they are exposed to electromagnetic radiation.

photon: A discrete, quantized bundle of radiation that travels at the speed of light, has zero mass, and has energy and momentum.

physics: The branch of science that studies matter and energy and their relationships.

piezoelectricity: The property of a crystal that causes it to bend or deform, producing electric vibrations, when a voltage is applied across it.

pitch: The highness or lowness of a sound, which depends on the frequency of vibration; can be given a name on the musical scale.

plane mirror: A flat, smooth surface from which light is reflected by specular reflection, producing a virtual image that is the same size as the object, has the same orientation, and is the same distance from the mirror as the object.

plasma: A gaslike, fluid state of matter made up of negatively charged electrons and positively charged ions that can conduct electric charge; makes up most of the matter in the universe, such as stars.

polarization: Production of light with a specific pattern of oscillation.

polarized: The characteristic of magnets that they have two opposite ends called poles.

position: The distance and direction from the origin to an object.

position-time graph: A graph that can be used to determine an object's velocity and position, as well as where and when two objects meet, made by plotting the time data on a horizontal axis and the position data on a vertical axis.

potential energy: Stored energy due to the interactions between objects in a system.

power: The rate at which energy is transformed.

precision: A characteristic of a measured value describing the degree of exactness of a measurement.

pressure: The perpendicular component of a force on a surface divided by the surface's area.

primary color: Red, green, and blue, which can be combined to form white light and mixed in pairs to produce the secondary colors yellow, cyan, and magenta.

primary pigment: Cyan, magenta, and yellow, each of which absorbs one primary color from white light and reflects two primary colors; can be mixed in pairs to produce the secondary pigments red, green, and blue.

principal axis: A straight line perpendicular to the surface of a mirror or lens that divides the mirror or lens in half.

onda periódica: Onda cuyas alteraciones ocurren a un ritmo constante.

efecto fotoeléctrico: Consiste en la emisión de electrones por ciertos metales que ocurre cuando estan expuestos a la radiación electromagnética.

fotón: Paquete cuántico, discreto, nulo en masa, que porta radiación y viaja a la velocidad de la luz, tiene energía e impulso.

física: La rama de la ciencia que estudia la materia y la energía así como sus relaciones.

piezoelectricidad: La propiedad del cristal de doblarse o deformarse, que al aplicársele un voltaje produce vibraciones eléctricas.

tono: Lo más alto o más bajo de un sonido, dependiendo de la frecuencia de vibración; puede dársele un nombre en la escala musical.

espejo plano: Superficie plana y pulida que refleja luz por reflexión especular, produciendo imágenes virtuales del mismo tamaño del objeto, tiene la misma orientación, y tiene la misma distancia del espejo que del objeto.

plasma: Un gas constituido, un estado fluido de la material hecho de carga negativa de electrones y de iones cargados positivamente que pueden conducir carga eléctrica; produce la mayoría de la material del universo, tales como las estrella

polarización: La producción de la luz con un patrón especifico de oscilación .

polarizado: La característica de los imanes de tener dos extremos llamados polos.

posición: La distancia y dirección de un objeto con referencia a su origen.

grafico de posición-tiempo: Una gráfica que puede utilizarse para determinar la velocidad y posición de un objeto, así como dónde y cuándo dos objetos se encuentran, hecho por trazar los datos del tiempo en el eje horizontal y los datos de posición en el eje vertical.

potencial energético: Energía almacenada debido a las interacciones entre los objetos en un sistema.

poder: Tasa a la cual la energía se transforma.

precisión: Una característica de la medida de un valor que describe el grado de exactitud de la medición.

presión: El componente perpendicular de una fuerza sobre una superficie, dividida entre el área de la superficie.

colores primarios: rojo, verde y azul, pueden combinarse para formar luz blanca y mezclarse en parejas para producir los colores secundarios: amarillo, cian y magenta.

pigmento primario: cian, magenta y amarillo, cada uno de ellos absorbe un color primario de la luz blanca y refleja dos colores primarios; se pueden mezclar en parejas para producir los pigmentos secundarios: rojo, verde y azul.

eje principal: Una línea recta perpendicular a la superficie de un espejo o una lente que divide el espejo o la lente por la mitad.

principal quantum number: The integer (n) that determines the quantized values of r and E for an electron's orbital radius in hydrogen and the energy of a hydrogen atom.

principle of superposition: States that the displacement of a medium caused by two or more waves is the algebraic sum of the displacements of the individual waves.

projectile: An object shot through the air, such as a football, that has independent vertical and horizontal motions and, after receiving an initial thrust, travels through the air only under the force of gravity.

número cuántico principal: El número entero (n) que determina los valores cuánticos de r y E para el radio del orbita de un electrón en el hidrógeno y la energía de un átomo de hidrógeno.

principio de superposición: Establece que el desplazamiento de un medio causado por dos o más ondas es la suma algebraica de los desplazamientos de las ondas individuales.

proyectil: Objeto lanzado en el aire, como un balón de fútbol, que tiene movimientos verticales y horizontales independientes, que después de recibir un impulso inicial, viaja por el aire únicamente por la fuerza de la gravedad.

Q

quadratic relationship: A parabolic relationship that results when one variable depends on the square of another variable.

quantized: The property of energy that it exists only in bundles of specific amounts.

quantum mechanics: The study of the wave properties of matter.

quantum model: A model of the atom that predicts only the probability that an electron is in a specific region.

quark: An elementary particle that has mass and charge, experiences the strong force, and can combine with other quarks to form larger particles such as protons, neutrons, and pions.

relación cuadrática: Una relación parabólica que se produce cuando una variable depende del cuadrado de otra variable.

cuantizada: La propiedad de la energía que existe en paquetes con cantidades específicas.

mecánica cuántica: El estudio de las propiedades ondulatorias de la materia.

modelo cuántico: Un modelo del atomo que solamente predice la probabilidad de que un electrón se encuentra en una región especifica.

quark: Partícula elemental que tiene masa y carga, experimenta la fuerza nuclear fuerte y puede combinarse con otros quarks para formar una partícula mas grande como un protón, neutrón o pión.

R

radian: $\frac{1}{2}\pi$ of a revolution; abbreviated rad.

radiation: The thermal transfer of energy by electromagnetic waves.

radioactive: Exhibiting spontaneous emission of particles and energy due to nuclear decay.

ray: A line drawn at a right angle to a wavefront; represents the direction of wave travel.

Rayleigh criterion: States that if the central bright spot of one image falls on the first dark ring of the second image, the images are at the limit of resolution.

ray model of light: A model that represents light as a ray that travels in a straight path, whose direction can be changed only by encountering a boundary.

real image: An optical image that is formed by the converging of light rays.

receiver: A device that converts oscillating potential differences in an antenna to sound, pictures, or data.

reference frame: A coordinate system from which motion is viewed.

reference level: The position where gravitational potential energy is defined as zero.

reflected wave: A returning wave that results from some of the energy of the incident wave's pulse being reflected backward.

radián: $\frac{1}{2}\pi$ de una revolución; se abrevia *rad.*

radiación: La transferencia térmica de energía por ondas electromagnéticas.

radiactivo: Que exhibe la emisión espontánea de partículas y energía por la desintegración nuclear.

rayo: Una línea trazada en ángulo recto al frente de una onda; representa el sentido del recorrido de la onda.

criterio de Rayleigh: Establece que, si el punto central brillante de una imagen cae en el primer anillo de oscuridad de una segunda, las imágenes estarán en el límite de la resolución.

modelo de rayos de luz: Un modelo que representa la luz como un rayo que recorre una trayectoria recta, cuya dirección solo puede cambiar al encontrar una barrera.

imagen real: Una imagen óptica que se forma por la convergencia de los rayos de luz.

receptor: Un dispositivo que convierte diferencias potenciales oscilantes en una antena al sonido, imágenes, o data.

marco de referencia: Un sistema de coordenadas a través del cual se observa el movimiento.

nivel de referencia: La posición en la el potencial de energía gravitatoria es igual a cero.

onda reflejada: Onda rebotada que proviene del reflejo del pulso de la energía en retroceso de la onda incidente.

refraction: The change in direction of waves at the boundary between two different mediums.

resistance: The measure of how strongly an object or material impedes the flow of electric charge produced by a potential difference; equal to the potential difference divided by the current.

resistance force: The force exerted by a machine.

resistor: A device with a specific resistance; may be made of long, thin wires, graphite, or semiconductors and often is used to control the current in circuits or parts of circuits.

resonance: A special form of periodic motion that occurs when forces are applied to an oscillating or vibrating object at regular time intervals equal to the period of oscillation and the amplitude of the vibration increases.

resultant: A vector that represents the sum of two other vectors; it always points from the first vector's tail to the last vector's tip.

rotational kinetic energy: Kinetic energy of a system due to its rotational motion, proportional to the system's moment of inertia and the square of its angular velocity.

refracción: El cambio de dirección que experimenta una onda al pasar de un medio a otro.

resistencia: La medida de la intensidad de oposición de un objeto o material ante un flujo de carga eléctrica producida por una diferencia de potencial; es igual a la diferencia de potencial dividida por la corriente.

fuerza de resistencia: La fuerza ejercida por una máquina.

resistor: Un dispositivo con una resistencia específica, puede ser hecho de alambres largos y delgados, de grafito, o de semiconductores y, a menudo se utiliza para controlar la corriente en circuitos o partes de los circuitos.

resonancia: Una forma especial de movimiento periódico que ocurre cuando se aplican fuerzas a un objeto oscilante o vibrante en intervalos regulares e iguales al período de oscilación, aumentando la amplitud de la vibración.

resultante: Un vector que representa la suma de dos vectores; siempre apunta desde la cola del primer vector a la cabeza del último vector.

energía cinética de rotación: La energía cinética de un sistema debido a su movimiento de rotación, es proporcional al momento de inercia del sistema y el cuadrado de su velocidad angular.

S

scalar: A quantity, such as temperature or distance, that is just a number without any direction.

scientific law: A statement about what happens in nature and seems to be true all the time.

scientific methods: The patterns of investigation procedures.

scientific theory: An explanation of things or events based on knowledge gained from many observations and investigations.

secondary color: Yellow, cyan, and magenta, each of which is produced by combining two primary colors.

secondary pigment: Red, green, and blue, each of which absorbs two primary colors from white light and reflects one primary color; can be produced by mixing pairs of cyan, magenta, and yellow pigments.

second law of thermodynamics: States that whenever there is an opportunity for energy dispersal, the energy always spreads out; states that natural processes go in a direction that maintains or increases the total entropy of the universe.

self-inductance: The property of a wire, either straight or in a coil, to create an induced *EMF* that opposes the change in the potential difference across the wire.

semiconductor: Material that behaves as a conductor under certain conditions and as an insulator in others; can be used to make solid-state electronic components.

escalar: Una cantidad que es meramente un numero sin dirección, como la de la temperatura o la de la distancia.

ley científica: Declaración sobre lo que sucede en la naturaleza y parece ser verdadera todo el tiempo.

métodos científicos: Los patrones de los procedimientos de investigación.

teoría científica: Una explicación de las cosas o eventos basándose en conocimientos adquiridos por varias observaciones e investigaciones.

color secundario: Amarillo, cian y magenta, cada uno de los cuales es producido por la combinación de dos colores primarios.

pigmento secundario: Rojo, verde y azul, cada uno de los cuales absorbe dos colores primarios de la luz blanca y refleja un color primario, puede ser producido por la mezcla de pares de pigmentos de cian, magenta y amarillo.

segunda ley de la termodinámica: Establece que cuando hay una oportunidad para la dispersión de energía, la energía siempre se esparcirá; establece que los procesos naturales van en una dirección que mantienen o aumentan la entropía total del universo.

autoinducción: La propiedad de un alambre, recto o en forma de una bobina, de crear una fuerza electromotiva inducida opuesta al cambio en la diferencia potencial a lo largo del alambre.

semiconductor: Un material que en determinadas condiciones se comporta como conductor y en otras como aislante; puede ser usado en la fabrica de componentes electrónicos de estado sólido.

series circuit: A type of electric circuit in which there is only one current path and all current travels through each device; the current is the same everywhere and is equal to the potential difference divided by the equivalent resistance.

series connection: A type of connection in which there is only a single current path.

short circuit: An unintended, very low-resistance circuit that results in a very large current, overloading wires and potentially breaking electrical components and starting fires.

significant figures: All the valid digits in a measurement, the number of which indicates the measurement's precision.

simple harmonic motion: A motion that occurs when the restoring force on an object is directly proportional to the object's displacement from the equilibrium position.

simple pendulum: A device that can demonstrate simple harmonic motion when its bob (a massive ball or weight), suspended by a string or light rod, is pulled to one side and released, causing it to swing back and forth.

solenoid (SOH luh noyd): A long coil of wire with many spiral loops that is attached to a circuit; fields from each loop add to the fields of the other loops, creating a greater total field strength.

sound level: A logarithmic scale that measures sound intensities; depends on the ratio of the intensity of a particular sound wave to the intensity of the most faintly heard sound; unit of measurement is the decibel (dB).

sound wave: A pressure variation transmitted through matter as a longitudinal wave; it reflects and interferes and has frequency, wavelength, speed, and amplitude.

specific heat: The amount of energy that must be added to a material to raise the temperature of a unit mass by one temperature unit; measured in J/kg·K.

specular reflection: A reflection produced by a smooth surface in which parallel light rays are reflected in parallel.

spherical aberration: The image defect of a spherical mirror or lens that does not allow parallel light rays far from the principal axis to converge at the focal point and produces an image that is fuzzy, not sharp.

Standard Model: The theory explaining the interactions of elementary particles, grouped into three families—quarks, leptons, and bosons.

standing wave: A wave that appears to be standing still, produced by the interference of two traveling waves moving in opposite directions.

static friction: The force exerted on one surface by a second surface when there is no motion between the two surfaces.

circuito en serie: Un tipo de circuito eléctrico en el que hay solamenta una trayectoria por el corriente y toda la corriente viaja a través de cada dispositivo; la corriente es la misma en todas partes y es igual a la diferencia de potencial entre el equivalente de resistencia.

conexión en serie: Tipo de conexión en la que sólo existe un trayecto de corriente.

cortocircuito: Un circuito formado sin intención de muy baja resistencia que resulta en una corriente muy grande, sobrecargando los cables y potencialmente rompiendo los componentes eléctricos e iniciando un incendio.

cifras significativas: Todos los dígitos válidos en la medición, indican la precisión de la medición.

movimiento armónico simple: Movimiento periódico que resulta cuando la fuerza restauradora actúa sobre un objeto de manera directamente proporcional a su desplazamiento respecto a su posición de equilibrio.

péndulo simple: Una herramienta que demuestra el movimiento armónico simple, cuando una masa puntual (una bola de masa o peso), suspendida de un hilo o una cuerda sin peso, oscila de un lado a otro.

solenoide: Una bobina larga de cables enrollados en espiral que esta conectagdo a un circuito; los campos de un espiral se añaden a los campos de otras espirales, creando mayor fortaleza al campo total

nivel de sonido: Una escala logarítmica que mide las intensidades de sonido; depende de la relación de la intensidad de una partícula de onda sonora con laintensidad del sonido más débilmente escuchado, su unidad de medida son los decibelios (dB).

onda sonora: Una variación de la presión transmitida a través de la materia como una onda longitudinal reflejando e interfiriendo; tiene frecuencia, longitud de onda, velocidad y amplitud.

calor específico: La cantidad de energía que debe añadirse a un material para elevar la temperatura de una unidad de masa por una unidad de temperatura; se mide en J/kg·K.

reflexión especular: Una reflexión producida por una superficie lisa en la que los rayos paralelos de luz se reflejan paralelamente.

aberración esférica: Defecto en la imagen de un espejo o lente esférico que no permite que los rayos paralelos de luz alejados del eje principal converjan en el punto focal, produciendo una imagen borrosa y poco nítida.

Modelo Estándar: La teoría que explica las interacciones de las partículas elementarías, agrupadas en tres familias—quarks, leptones, y bosones.

onda estacionaria: Una onda que parece permanecer confinada en un espacio, resulta de la interferencia de dos ondas que se mueven en direcciones opuestas.

Fricción estática: La fuerza ejercida por una superficie sobre otra superficie y en donde no existe movimiento entre las dos superficies.

step-down transformer: A type of transformer in which the voltage coming out of the transformer is smaller than the voltage put into the transformer.

step-up transformer: A type of transformer in which the voltage coming out of the transformer is larger than the voltage put into the transformer.

stimulated emission: The process that occurs when an excited atom is struck by a photon having energy equal to the energy difference between the excited state and the ground state—the atom drops to the ground state and emits a photon with energy equal to the energy difference between the two states.

streamlines: Lines representing the flow of fluids around objects.

strong nuclear force: An attractive force between nucleons that binds the nucleus together; is of the same strength between all nucleon pairs.

superconductor: A material with zero resistance that can conduct electricity without thermal energy transformations.

surface wave: Disturbance in which the medium's particles follow a circular path that is at times parallel to the direction of wave travel and at other times perpendicular to the direction of wave travel.

system: Object or objects of interest that can interact with each other and the external world.

transformador de reducción de voltaje: Un tipo de transformador en el que el voltaje de salida es menor que el voltaje de entrada en el transformador.

transformador de elevación de voltaje: Un tipo de transformador en el que el voltaje saliente del transformador es mayor que el voltaje de entrada del transformador.

emisión estimulada: Proceso que ocurre cuando un átomo excitado recibe un estimulo de un fotón que tiene energía igual a la diferencia de energía entre los dos estados el excitado y el estado fundamental—el átomo baja al estado fundamental y emite un fotón con energía igual a la diferencia de energía entre los dos estados.

líneas de corriente: Líneas que representan el flujo de fluidos alrededor de los objetos.

fuerza nuclear fuerte: Una fuerza de atracción entre nucleones y que se enlaza el nucleo; es de la misma fuerza entre todos los pares de nucleones.

superconductor: Un material con resistencia igual a cero que puede conducir electricidad sin transformaciones de energía térmica.

onda de superficie: Alteración en la que las partículas de un medio siguen un trayecto circular que es a veces paralelo al recorrido de la dirección de la onda y en otras ocasiones perpendicular al recorrido de la dirección de la onda.

sistema: Objeto u objetos de interés que pueden interactuar uno con el otro y con el mundo exterior.

T

tension: The specific name for the force exerted by a rope or a string.

terminal velocity: The constant velocity of an object that is reached when the drag force equals the force of gravity.

thermal conduction: The transfer of thermal energy when particles collide.

thermal energy: The sum of the kinetic energies and potential energies of the particles in a system.

thermal equilibrium: The state in which the rates of thermal energy transfer between two objects are equal and the objects are at the same temperature.

thermal expansion: A property of all forms of matter that causes the matter to expand, becoming less dense, when heated.

thin-film interference: A phenomenon in which light waves reflect from separate boundaries of a thin film and experience constructive and destructive interference.

thin lens equation: States that the inverse of the focal length of a spherical lens equals the sum of the inverses of the image position and the object position.

threshold frequency: The certain minimum value at or above which the frequency of incident radiation causes the ejection of electrons from a metal.

time interval: The difference between two times.

tensión: El nombre específico para la fuerza ejercida por una cuerda o una cadena.

velocidad terminal: La velocidad constante que un objeto alcanza cuando la fuerza de arrastre iguala la fuerza de gravedad.

conducción térmica: La transferencia de energía térmica por la colisión de partículas.

energía térmica: La suma de las energías cinéticas y potenciales de las partículas en un sistema.

equilibrio térmico: El estado en cual las tazas de la transferencia de energía térmica entre dos objetos son igual y los objetos se encuentran a la misma temperatura.

expansión térmica: Una propiedad de todas las formas de la materia que la capacita para extenderse, y volverse menos densa, al calentarse.

interferencia en película delgada: Un fenómeno en el que las ondas de luz se reflejan en fronteras distinctas de una película delgada y producen interferencia constructiva y destructiva.

ecuación de la lente delgada: Establece que la inversa de la distancia focal de una lente esférica es igual a la suma de las inversas de la posición de la imagen y la posición del objeto.

umbral de frecuencia: El valor mínimo determinado o superior a él, al cual la frecuencia de la radiación incidente provoca la expulsión de electrones de un metal.

intervalo de tiempo: La diferencia entre dos tiempos.

torque: The combination of force and lever arm that can cause an object to rotate; the magnitude is equal to the force times the perpendicular lever arm.

par de torsión: La combinación de la fuerza y el brazo de palanca que puede causar la rotación de un objeto; la magnitud es igual a la fuerza multiplicada por el brazo de palanca perpendicular.

total internal reflection: An optical phenomenon that occurs when light strikes a boundary between two mediums at an angle of incidence that is greater than the critical angle and all light reflects back into the region of the higher index of refraction.

reflexión interna total: Un fenómeno óptico que ocurre cuando la luz golpea a un límite entre dos medios en un ángulo mayor que el ángulo critico, y toda la luz se refleja de nuevo hacia la región con el mayor índice de refracción.

trajectory: The path of a projectile through space.

trayectoria: El recorrido de un proyectil a través del espacio.

transformer: A device that can decrease or increase the voltages in AC circuits with relatively little energy loss.

transformador: Un dispositivo que puede aumentar o disminuir el voltaje en circuitos CA con una pérdida de energía relativamente pequeña.

transistor: A simple device made of doped semiconducting material that can act as an amplifier, converting a weak signal to a much stronger one.

transmisor: Un dispositivo simple hecho de material semiconductor dopado que puede actuar como amplificador, convirtiendo una señal débil en una señal mucho más fuerte.

translational kinetic energy: The energy of a system due to the system's change in position.

energía cinética de traslación: La energía de un sistema causada por un cambio de posición del sistema.

translucent (trans LEW sunt): A property of a medium that allows that medium to transmit light and reflect a fraction of the light, preventing objects from being seen clearly through it.

translúcida: Una propiedad de un medio que permite transmitir la luz y reflejar una fracción de la luz, sin permitir que los objetos se vean claramente a través de él.

transmitter: a circuit that converts voice, music, pictures, or data to electronic signals, amplifies the signals, and then sends them to an antenna

transmitir: Un circuito que convierte voz, música, fotografías, o data en señales electrónicas, los amplifican, y los manda a una antena.

transparent: A property of a medium that allows that medium to transmit light and reflect a fraction of the light, allowing objects to be seen clearly through it.

transparente: Una propiedad de un medio que permite que el medio transmita la luz y reflejar una fracción de la luz, dejando que los objetos sean claramente visibles a través de él.

transverse wave: A wave that vibrates perpendicular to the direction of the wave's travel.

onda transversal: Una onda que vibra perpendicularmente a la dirección del movimiento de la onda.

trough (TROF): The low point of a transverse wave.

depresión: El punto bajo de una onda transversal.

U

uniform circular motion: The movement of an object at a constant speed around a circle with a fixed radius.

movimiento circular uniforme: El movimiento de un objeto a una velocidad constante alrededor de un círculo de radio fijo.

V

vector: A quantity, such as position, that has both magnitude and direction.

vector: Una cantidad, como la posición, que tiene tanto magnitud y dirección.

vector resolution: The process of breaking a vector into its components.

resolución vectorial: El proceso de romper un vector en sus componentes.

velocity-time graph: A graph that has velocity plotted on the vertical axis and time plotted on the horizontal axis; its slope is the acceleration of the object whose motion is described by the graph.

gráfica velocidad-tiempo: Un gráfico en la que la velocidad es trazada en el eje vertical y el tiempo en el eje horizontal; su pendiente describe la aceleración del objeto cuyo movimiento está descrito en el gráfico.

virtual image: The image formed of diverging light rays.

imagen virtual: La imagen formada por rayos de luces divergentes.

volt: The unit of electric potential difference; equal to one joule per coulomb, 1 J/C.

voltios: La unidad de la diferencia de potencial eléctrico; igual a un julio por culombio, 1 J/C.

voltage divider: A series circuit that is used to produce a potential difference source of desired magnitude from a battery with a higher potential difference; often is used with sensors such as photoresistors.

divisor de tensión: Un circuito en serie que se utiliza para producir un fuente de diferencia de una magnitud deseada por potencial de una batería con una diferencia potencial mayor; a menudo se usa con sensores como fotoresistores.

W

watt (W): Unit of power; 1 J of energy transferred in 1 s.

wave: A disturbance that carries energy through matter or space; transfers energy without transferring matter.

wavefront: A line representing the crest of a wave in two dimensions that can show the wavelength, but not the amplitude, of the wave when drawn to scale.

wavelength: The shortest distance between points on a wave where the wave pattern repeats itself, such as from crest to crest or from trough to trough.

wave pulse: A single disturbance or bump that travels through a medium.

weak nuclear force: The interaction that acts in the nucleus during beta (β) decay; much weaker than the strong nuclear force.

weight: The gravitational force experienced by an object.

weightlessness: An object's apparent weight of zero that results when there are no contact forces supporting the object.

work: The transfer of energy that occurs when a force is applied through a distance; equal to the product of the system's displacement and the force applied to the system in the direction of displacement.

work-energy theorem: States that when work is done on a system, a change in energy occurs.

work function: The energy required to free the most weakly bound electron from a metal; measured by the threshold frequency in the photoelectric effect.

watt (W): Unidad de potencia; 1 J de energía transferida en 1 s.

onda: Una perturbación que transporta energía a través de la materia o del espacio; transfiere energía sin transferir materia.

frente de onda: Una línea que representa la cresta de una onda en dos dimensiones mostrando la longitud de onda, pero no la amplitud, cuando es dibujada a escala.

longitud de onda: La distancia más corta entre dos puntos de una onda donde el patrón de la onda se repite, como el de cresta a cresta o de depresión a depresión.

pulso de onda: Una sola o pulso único que viaja a través de un medio.

fuerza nuclear débil: La interacción que durante la desintegración beta (β), actúa en el núcleo, es mucho más débil que la fuerza nuclear intensa.

peso: La fuerza gravitatoria que experimenta un objeto.

ingravidez: El estado de un objeto en el que su peso aparente es igual a cero que resulta cuando no hay fuerzas de contacto que apoya al objeto.

trabajo: La transferencia de energía que se produce cuando se aplica una fuerza por una distancia; es igual al producto del desplazamiento del sistema y de la fuerza aplicada al sistema en la dirección al desplazamiento.

teorema trabajo-energía: Establece que, cuando se realiza trabajo sobre un sistema, se genera un cambio de energía.

función de trabajo: La energía mínima necesaria para liberar el enlace más débil entre los electrones en un metal; se mide por el umbral de frecuencia en el efecto fotoeléctrico.

INDEX

Index Key

Italic page numbers = illustration/photo **Bold page numbers = vocabulary term**

INDEX

INDEX

F

G

INDEX

L

M

Index

INDEX

Q

R

S

INDEX

Index

INDEX

PERIODIC TABLE OF THE ELEMENTS

Atomic number: 1 | Symbol: H | Element: Hydrogen | Atomic mass: 1.008

- Metal
- Metalloid
- Nonmetal
- Synthetic

	1	2	3	4	5	6	7	8	9	10	11	12	13	14	15	16	17	18
1	1 H Hydrogen 1.008																	2 He Helium 4.003
2	3 Li Lithium 6.941	4 Be Beryllium 9.012											5 B Boron 10.811	6 C Carbon 12.011	7 N Nitrogen 14.007	8 O Oxygen 15.999	9 F Fluorine 18.998	10 Ne Neon 20.180
3	11 Na Sodium 22.990	12 Mg Magnesium 24.305											13 Al Aluminum 26.982	14 Si Silicon 28.086	15 P Phosphorus 30.974	16 S Sulfur 32.066	17 Cl Chlorine 35.453	18 Ar Argon 39.948
4	19 K Potassium 39.098	20 Ca Calcium 40.078	21 Sc Scandium 44.956	22 Ti Titanium 47.867	23 V Vanadium 50.942	24 Cr Chromium 51.996	25 Mn Manganese 54.938	26 Fe Iron 55.847	27 Co Cobalt 58.933	28 Ni Nickel 58.693	29 Cu Copper 63.546	30 Zn Zinc 65.39	31 Ga Gallium 69.723	32 Ge Germanium 72.61	33 As Arsenic 74.922	34 Se Selenium 78.971	35 Br Bromine 79.904	36 Kr Krypton 83.80
5	37 Rb Rubidium 85.468	38 Sr Strontium 87.62	39 Y Yttrium 88.906	40 Zr Zirconium 91.224	41 Nb Niobium 92.906	42 Mo Molybdenum 95.95	43 Tc Technetium (98)	44 Ru Ruthenium 101.07	45 Rh Rhodium 102.906	46 Pd Palladium 106.42	47 Ag Silver 107.868	48 Cd Cadmium 112.411	49 In Indium 114.82	50 Sn Tin 118.710	51 Sb Antimony 121.757	52 Te Tellurium 127.60	53 I Iodine 126.904	54 Xe Xenon 131.290
6	55 Cs Cesium 132.905	56 Ba Barium 137.327	57 La Lanthanum 138.905	72 Hf Hafnium 178.49	73 Ta Tantalum 180.948	74 W Tungsten 183.84	75 Re Rhenium 186.207	76 Os Osmium 190.23	77 Ir Iridium 192.217	78 Pt Platinum 195.08	79 Au Gold 196.967	80 Hg Mercury 200.59	81 Tl Thallium 204.383	82 Pb Lead 207.2	83 Bi Bismuth 208.980	84 Po Polonium 208.982	85 At Astatine 209.987	86 Rn Radon 222.018
7	87 Fr Francium (223)	88 Ra Radium (226)	89 Ac Actinium (227)	104 Rf Rutherfordium * (267)	105 Db Dubnium * (270)	106 Sg Seaborgium * (269)	107 Bh Bohrium * (270)	108 Hs Hassium * (277)	109 Mt Meitnerium * (278)	110 Ds Darmstadtium * (281)	111 Rg Roentgenium * (281)	112 Cn Copernicium * (285)	113 Nh Nihonium * (286)	114 Fl Flerovium * (289)	115 Mc Moscovium * (289)	116 Lv Livermorium * (293)	117 Ts Tennessine * (294)	118 Og Oganesson * (294)

The number in parentheses is the mass number of the longest-lived isotope for that element.

*Properties are largely predicted.

Lanthanide series	58 Ce Cerium 140.115	59 Pr Praseodymium 140.908	60 Nd Neodymium 144.242	61 Pm Promethium (145)	62 Sm Samarium 150.36	63 Eu Europium 151.965	64 Gd Gadolinium 157.25	65 Tb Terbium 158.925	66 Dy Dysprosium 162.50	67 Ho Holmium 164.930	68 Er Erbium 167.259	69 Tm Thulium 168.934	70 Yb Ytterbium 173.04	71 Lu Lutetium 174.967
Actinide series	90 Th Thorium 232.038	91 Pa Protactinium 231.036	92 U Uranium 238.029	93 Np Neptunium (237)	94 Pu Plutonium (244)	95 Am Americium (243)	96 Cm Curium (247)	97 Bk Berkelium (247)	98 Cf Californium (251)	99 Es Einsteinium * (252)	100 Fm Fermium * (257)	101 Md Mendelevium * (258)	102 No Nobelium * (259)	103 Lr Lawrencium * (262)

Physical Constants

Quantity	Symbol	Value	Approximate Value
Atomic mass unit	u	$1.660538782 \times 10^{-27}$ kg	1.66×10^{-27} kg
Avogadro's number	N_A	$6.02214179 \times 10^{23}$ mol^{-1}	6.022×10^{23} mol^{-1}
Boltzmann's constant	k	$1.3806504 \times 10^{-23}$ Pa·m^3/K	1.38×10^{-23} Pa·m^3/K
Coulomb's constant	K	$8.987551788 \times 10^{9}$ N·m^2/C^2	9.0×10^{9} N·m^2/C^2
Elementary charge	e	$1.60217653 \times 10^{-19}$ C	1.602×10^{-19} C
Gas constant	R	8.314472 Pa·m^3/mol·K	8.31 Pa·m^3/mol·K
Gravitational constant	G	6.67428×10^{-11} N·m^2/kg^2	6.67×10^{-11} N·m^2/kg^2
Mass of an electron	m_e	$9.10938215 \times 10^{-31}$ kg	9.11×10^{-31} kg
Mass of a proton	m_p	$1.672621637 \times 10^{-27}$ kg	1.67×10^{-27} kg
Mass of a neutron	m_n	$1.674927211 \times 10^{-27}$ kg	1.67×10^{-27} kg
Planck's constant	h	$6.62606896 \times 10^{-34}$ J·s	6.63×10^{-34} J·s
Speed of light in a vacuum	c	2.99792458×10^{8} m/s	3.00×10^{8} m/s

SI Derived Units

Quantity	Unit	Unit Symbol	Unit Expressed in Base Units	Unit Expressed in Other SI Units
Acceleration	meters per second squared	m/s^2	m/s^2	
Area	meters squared	m^2	m^2	
Capacitance	farad	F	A^2·s^4/(kg·m^2)	
Density	kilograms per meter cubed	kg/m^3	kg/m^3	
Electric charge	coulomb	C	A·s	
Electric field	newtons per coulomb	N/C	kg·m/(A·s^3)	V/m
Electric resistance	ohm	Ω	kg·m^2/(A^2·s^3)	V/A
EMF	volt	V	kg·m^2/(A^2·s^3)	
Energy, work	joule	J	kg·m^2/s^2	N·m
Force	newton	N	kg·m/s^2	
Frequency	hertz	Hz	s^{-1}	
Illuminance	lux	lx	cd/m^2	
Magnetic field	tesla	T	kg/(A·s^2)	N·s/(C·m)
Potential difference	volt	V	kg·m^2/(A·s^3)	W/A or J/C
Power	watt	W	kg·m^2/s^3	J/s
Pressure	pascal	Pa	kg/m·s^2	N/m^2
Velocity	meters per second	m/s	m/s	
Volume	meters cubed	m^3	m^3	